Essentials and Applications of Food Engineering

Essentials and Applications of Food Engineering

C. Anandharamakrishnan and S. Padma Ishwarya

CRC Press
Taylor & Francis Group
Boca Raton London New York

CRC Press is an imprint of the
Taylor & Francis Group, an **informa** business

CRC Press
Taylor & Francis Group
6000 Broken Sound Parkway NW, Suite 300
Boca Raton, FL 33487-2742

Printed on acid-free paper

International Standard Book Number-13: 978-1-138-36655-8 (Hardback)

Library of Congress Cataloging-in-Publication Data

Names: Anandharamakrishnan, C., author. | Ishwarya, S. Padma, 1988- author.
Title: Essentials & applications of food engineering / C.
Anandharamakrishnan, S. Padma Ishwarya.
Other titles: Essentials and applications of food engineering
Description: Boca Raton : CRC Press, Taylor & Francis Group, 2019. | Includes
bibliographical references and index.
Identifiers: LCCN 2018055496 | ISBN 9781138366558 (hardback : alk. paper)
Subjects: LCSH: Food industry and trade—Study and teaching.
Classification: LCC TP370 .A5295 | DDC 664—dc23
LC record available at https://lccn.loc.gov/2018055496

Visit the Taylor & Francis Web site at
http://www.taylorandfrancis.com

and the CRC Press Web site at
http://www.crcpress.com

Cover Credit: Image Created by Fanjianhua—Freepik.com

Contents

Preface

The perspective of *food engineering* is beyond just a classroom topic. Unlike the other branches of engineering, food engineering is not a rigid combination of scientific principles and mathematics. It is about viewing the food we eat daily through the lens of physics to improve the methods of food production, packaging, and storage. While most would see bread as a soft-to-eat, flavorful, and baked food made of flour, water, sugar, fat, and yeast, a food engineer perceives it as a *solid foam* comprising microscopic gas bubbles entrapped in the three-dimensional starch-gluten network, achieved through a series of stages involving aeration of the polymeric dough matrix, a pseudoplastic material. Though food engineering seems to be an intricate theme, it could be transformed into an interesting subject if its essentials and applications are clearly appreciated, and that is the sole objective of this book, *Essentials and Applications of Food Engineering*. This book intends to present even the most complex food engineering concepts in a very simplified manner to the undergraduate and postgraduate students of the Food Engineering, Food Technology, and Food Science majors.

The principles and concepts of food engineering find application in every activity of the food industry. Food engineering is a truly multidisciplinary field, encompassing the design and development of machinery for food production; food product development; and the testing and evaluation of the manufacturing process and final product for quality, energy efficiency, cost-effectiveness, and safety. Accordingly, a food engineer plays a versatile role in any food industry. He or she may be responsible for manufacturing food products; monitoring and testing the quality of the manufactured product; simulating and testing the operation of a machinery or food system; or proposing specifications for foods, machinery, or packaging. At times, a food engineer is also expected to be a troubleshooter! Problem-solving and logical reasoning skills are vital to investigate the root cause for the failure of a component or product and find appropriate solutions for the same. Process costing is also among the other duties of a food engineer.

The topics included in this textbook have been thoughtfully chosen to render the students familiar with all the aforementioned elements of food engineering. We have conceptualized the contents of this textbook based on our strong belief in *learning with a purpose*. Thus, every engineering concept dealt with in this book has been related to its practical utility in food processing operations. Descriptive illustrations, case studies pertinent to the food industry, and numerical calculations have been included in each chapter to provide a comprehensive understanding of the food engineering concepts. The information presented in this textbook would facilitate young food engineering professionals to communicate their ideas to peers and superiors in their future career. Our aspiration through this textbook is that even professionals from an academic background other than food technology/food science, but still involved in the manufacturing of food and ingredients, should gain knowledge of the food engineering concepts.

This textbook includes 16 chapters on various fundamental and novel concepts in the subject of food engineering. The first four chapters encompass the basic discussions on units and dimensions, material balance, energy balance, and fluid flow. The following two chapters of the book present the governing principles and applications of psychrometry and reaction kinetics in food processing. The next two chapters deal with the theory of heat and mass transfer so that readers can appreciate the practical applications of these transport processes in the high- and low-temperature food preservation operations such as pasteurization, sterilization, evaporation, drying, refrigeration, freezing, and separation processes discussed in Chapters nine to twelve. The final three chapters of the book focus on the nonthermal and nanotechnology-based novel food processing techniques, 3D food printing, food packaging, and the applications of computational modeling in food processing.

The aim of this textbook is to improve the students' ability to understand and approach the food engineering concepts cognitively and confidently. We dedicate this book to the enthusiastic young minds and wish that they excel as proficient professionals in the food processing sector!

C. Anandharamakrishnan
S. Padma Ishwarya

Acknowledgments

We are grateful to our teachers, mentors, and students, who always keep inspiring us in the journey of learning. The constant encouragement from our teachers and the inquisitiveness of our students steered us towards conceiving and executing this textbook.

We thank Mr. Stephen Zollo, Senior Editor, CRC Press/Taylor & Francis Group, for his fervor and effective coordination of this project right from the proposal stage to a fruitful completion.

C. Anandharamakrishnan
S. Padma Ishwarya

Authors

C. Anandharamakrishnan is a prominent scientist in the field of food engineering with two decades of experience in research and development. Presently, he is the director of the Indian Institute of Food Processing Technology (IIFPT), Thanjavur, India. Prior to assuming office as director, IIFPT, he was a Principal Scientist at the Food Engineering Department of the CSIR—Central Food Technological Research Institute (CFTRI), Mysore, India. Anandharamakrishnan has made seminal contributions to the field of food engineering, particularly nanoencapsulation and microencapsulation of bioactive food ingredients and computational modeling of food processing operations and biological systems. He holds the acclaim of developing Asia's first "engineered human stomach and small intestinal dynamic model system" with its demonstrated capability to replace the practice of animal studies.

Anandharamakrishnan completed his doctoral degree in chemical engineering from Loughborough University, United Kingdom. For his doctoral thesis, he worked on the topic, "*Experimental and Computational Fluid Dynamics Studies on Spray-Freeze-Drying and Spray Drying of Proteins.*" His research endeavors are well documented in the form of highly cited publications in reputed international journals and nine international and Indian patents. Anandharamakrishnan has also authored 3 books and 20 book chapters published by coveted publishers. He has also edited two books. In recognition of his scientific contributions, Anandharamakrishnan has been conferred with the Fellow of The Royal Society of Chemistry (FRSC) and The Royal Society of Biology (FRSB) in the United Kingdom and the Institute of Engineers (FIE), India. He is also the recipient of various prestigious awards.

S. Padma Ishwarya is currently a Postdoctoral Fellow at the Agro Processing and Technology Division, CSIR - National Institute for Interdisciplinary Science and Technology (NIIST), Thiruvananthapuram, India. She completed her PhD in Engineering and Master of Science in Food Technology from the CSIR—Central Food Technological Research Institute (CFTRI), India. For her doctoral thesis, she worked on the "*Development of a Combined Experimental and Computational Modeling Approach to Investigate the Influence of Bran Addition on the Volume and Structural Development during the Breadmaking Process.*" Before joining her doctoral program, she worked as Quality Assurance Officer at the coffee and beverages division of Nestlé India Limited. Padma's research interests span across the food engineering disciplines including food structure, food foams, and soluble coffee processing. She has published various research and review articles in peer-reviewed international journals and coauthored a book.

1

Units and Dimensions

"If you cannot measure it, you cannot control it."

– *Lord Kelvin*

A food process engineer's jargon on the shop floor is often comprised of measurements to describe a specific resource (machinery), process, or product. *Dimension* or *Measurement* is an indispensable tool to provide understandable instructions to operators on the formulation of the product, process temperature, and time. Any measurement is expressed as a combination of its numerical value and unit (Figure 1.1).

For instance, a tomato sauce manufacturing process can be explained as the blending of 625 *kilogram* of tomato paste, 300 *kilogram* of water, 70 *kilogram* of oil, 15 *kilogram* of flavors, and 3 *kilogram* of spices, followed by cooking at 78.9–80 *°Celsius* for 240 *seconds*. Here, *kilogram*, *°Celsius*, and *seconds* are the respective units of the dimensions, mass, temperature, and time. The numerical value for each dimension is obtained as a result of a comparison between the measured and standard quantities. For example, the weight of an ingredient is expressed as a ratio with the standard weight of 1 *kilogram*. Consequently, the number is expressed in terms of multiples of the standard quantity, i.e., 625 × 1 *kilogram* = 625 *kilogram*. The operator will be clueless if the bill of material sheet is devoid of these units. Therefore, the *units* facilitate effective communication of the measured quantities. Any measurement is meaningful and complete only when expressed along with its unit.

Thus, *units and dimensions* are fundamental to the understanding of food engineering principles. The objective of this chapter is to introduce the standard system and classification of units and dimensions used in the engineering calculations. The chapter begins with a discussion on the different unit systems and the method of interconversion. The subsequent parts of this chapter present the importance of dimensionless groups, dimensional equations, and dimensional analysis. Before studying the concept of units and dimensions in detail, it is important to familiarize with certain terminologies that are most relevant and frequently encountered.

1.1 The Glossary of Units and Dimensions

- **Dimension:** the qualitative measure of a physical variable, without a numerical value. In other words, the dimension is a property or characteristic of an entity that can be quantified using direct measurement or calculation by multiplying or dividing the other dimensions. While the former is called a fundamental or primary dimension, the latter is termed a derived or secondary dimension. Dimension can also be defined as the powers to which the fundamental units are raised to obtain one unit of that quantity.

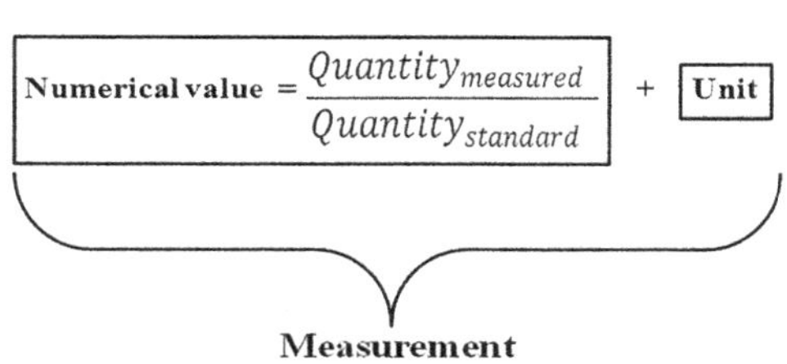

FIGURE 1.1 Constituents of the record of a measurement.

- **Unit:** the quantitative magnitude, size, or amount of a dimension to be measured with an assigned numerical value of unity.
- **Dimensional formula of a quantity:** the expression which signifies the exponent or power to which a fundamental unit is to be raised to obtain the unit of a derived quantity.
- **Dimensional system:** an accepted standard of fundamental dimensions and derived dimensions with corresponding units and scales.
- **Dimensional equation:** an equation that contains numerical value along with its unit is known as a dimensional equation.
- **Conversion factor:** a ratio which defines the equivalence between two units of the same measurement.

$$\text{Unit}_1(\text{Conversion factor}) = \text{Unit}_2 \tag{1.1}$$

where, Conversion factor $= \dfrac{\text{Unit}_2}{\text{Unit}_1}$

- **Dimensional constants:** the physical quantities with dimensions and a fixed value.

  For example: Gravitational constant ($G = 6.67408 \times 10^{-11}\,\text{m}^3\text{kg}^{-1}\text{s}^{-2}$), universal gas constant ($R = 8.314\ \text{Jmol}^{-1}\ \text{K}^{-1}$), and velocity of light in vacuum ($c = 299792458$ m/s).
- **Dimensionless quantities:** a dimensionless quantity can be defined as a number or expression which does not possess a unit but has a fixed value. It is a combination of dimensions such that it eventually cancels out to unity. For example: π or *pi* (= 3.14159) and *e* (Euler's number, a mathematical constant equal to 2.71828).
- **Dimensionless variables:** physical quantities which do not have dimensions as well as a fixed value.

  For example: Specific gravity, refractive index.
- **Dimensional formula:** a formula in which the given physical quantity is expressed in terms of the fundamental dimensions raised to a suitable exponent.
- **Law of homogeneity of dimensions:** the law of dimensional homogeneity states that *the terms on the left- and right-hand sides of the equation should have the same dimensions.* A dimensionally homogeneous equation is also said to be dimensionally consistent.
- **Precision of measurement:** the degree of consistency between the results of independent measurement obtained under standard conditions (Figure 1.2a). Food and beverage industries strive towards strict controls concerning both the process and product. For example, in the production environment, temperature and other factors such as humidity and pressure have to be maintained within stringent limits for which the measurement and control systems need to be very precise.
- **Accuracy of measurement:** the degree of consistency between a measurement result and the accepted standard value (Figure 1.2b). With respect to dairy products for which quality is central, the accuracy of temperature sensors is vital. In general, temperature measurement in food processing demands accuracy in the order of 0.1°C.
- **Error:** the uncertainty or difference between the measured value of a physical quantity and its accurate value.

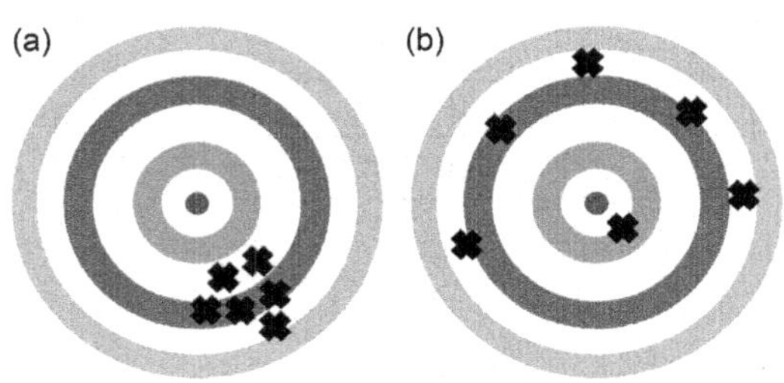

FIGURE 1.2 Schematic representation of (a) precision and (b) accuracy.

- **Significant figures:** the meaningful digits of a number that precisely describe its value. The identification of significant figures is subjected to the following five rules:
  - **Rule 1:** All nonzero digits are significant.
  - **Rule 2:** Zeros between nonzero digits are significant.
  - **Rule 3:** Leading zeros are never significant.
  - **Rule 4:** In a number with a decimal point, trailing zeros (those to the right of the last nonzero digit) are significant.
  - **Rule 5:** In a number without a decimal point, trailing zeros may or may not be significant, which is defined on a case-by-case basis using symbols to clarify the significance of trailing zeros.
    A. An overbar is placed over the last significant figure which indicates that the trailing zeros following this are insignificant. For example, $12\bar{0}0$ has three significant figures showing that the number is precise to the nearest ten.
    B. The last significant figure of a number is underlined. For example, $1\underline{0}00$ has two significant figures.
    C. A decimal point is placed after the number. For example "500." specifies that there are three significant figures (Myers et al., 2000).
- **Rounding off numbers:** a method for the termination of numbers after a predetermined number of digits. This approach accounts for the size of the digits being removed. It is different from the other method of termination, namely, chopping, which does not take into account the size of the digits being removed. The rules for the rounding off numbers (NIST, 2016) are stated as follows (Table 1.1):
  a. When the first digit to be dropped is <5, the last digit retained should not be changed.

     For example, if the quantity 384.3 is to be rounded off to three significant digits, the number 3 to the right of the decimal point must be dropped since it is less than 5, and the last digit to be retained (the number 4) will remain unchanged. Thus, the rounded number would be 384. Similarly, to round off for 48.4 and 6.34 to two significant figures, the final digit is dropped and the last digit retained remains unchanged such that the rounded off numbers are 48 and 6.3, respectively.
  b. When the first digit to be dropped is >5, or it is a 5 followed by at least one digit other than zero, the last digit to be retained should be increased by one unit.

     For example, while rounding off to three significant figures, 384.7 becomes 385; 684.51 becomes 685; 10.86 becomes 10.9; and 11.88 becomes 11.9.
  c. When the first digit to be dropped is exactly 5, followed only by zeros, the final digit to be retained should be rounded up if it is an odd number (1, 3, 5, 7, or 9), but no change should be made if it is an even number (2, 4, 6, or 8).

     For example, 384.50 becomes 384; 485.50 becomes 486 (three significant figures); 58.50 becomes 58; 6.450 becomes 6.4; and 8.550 becomes 8.6 (two significant figures) (NIST, 2016).

TABLE 1.1

Rounding Rules

When the First Digit Discarded Is	The Last Digit Retained Is	Examples
Less than five	Not changed	1.34 rounded to 2 digits is 1.3 1.329 rounded to 2 digits is 1.3
More than five, or five followed by at least one digit other than zero	Increased by one	7.46 rounded to 2 digits is 7.5 7.451 rounded to 2 digits is 7.5
Five followed by zeros	Not changed if even and increased by one if odd	8.450 rounded to 2 digits is 8.4 8.750 rounded to 2 digits is 8.8

Example 1.1

Find out the number of significant digits in the following numeric values:

a. 345
b. 123.45
c. 123.105
d. 0.00012
e. 12.3400
f. 0.00123400

Solution

a. The number of significant digits in 345 is 3 (3, 4, and 5; according to ***Rule 1*** of significant digits).
b. The number of significant digits in 123.45 is 5 (1, 2, 3, 4, and 5; according to ***Rule 1*** of significant digits).
c. The number of significant digits in 123.105 is 6 (1, 2, 3, 1, 0, and 5; according to ***Rule 2*** of significant digits).
d. The number of significant digits in 0.00012 is 2 (1 and 2; according to ***Rule 3*** of significant digits).
e. The number of significant digits in 12.3400 is 6 (1, 2, 3, 4, 0, and 0; according to ***Rule 4*** of significant digits).
f. The number of significant digits in 0.00123400 is 6 (1, 2, 3, 4, 0, and 0; according to ***Rule 4*** of significant digits, i.e., the zeroes to the left of a nonzero digit are not significant).

1.2 Classification of Dimensions

As discussed previously, dimensions can be classified into two types. There are seven primary dimensions which are independent of each other and cannot be resolved further. A combination of primary dimensions results in a secondary dimension (Figure 1.3). The definitions of the different primary and secondary dimensions are given in Section 1.2.1.

1.2.1 Definitions and Applications of Fundamental Dimensions in Food Processing

- **Mass:** the measure of the amount of matter in an object, wherein, the matter is anything that takes up space. The concept of mass is essential for describing the change in material quantities during the food processing operation and controlling the product output. The abovementioned analysis is known as "material balance" based on the law of conservation of mass, which will be dealt elaborately in Chapter 2. Although used interchangeably, mass and weight are entirely

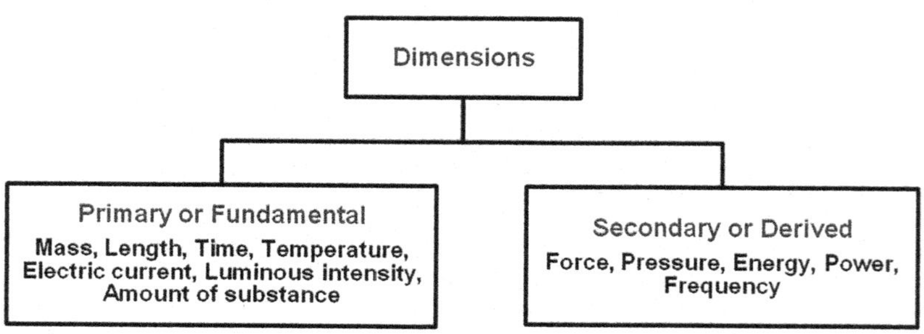

FIGURE 1.3 Classification of dimensions.

different quantities. Defined as the force acting on a body owing to gravity, the weight of an object varies with changes in the gravitational force. However, mass remains constant.

- **Length:** is concerned with the quantitative measurement of the linear extent of the size or position of an object (Jong and Rogers, 1991). The size or length of a commodity such as a fruit or vegetable is an important quality parameter during the sorting step carried out before its processing.
- **Time:** deals with measuring the duration of a phenomenon (Jong and Rogers, 1991). Time is a critical factor in deciding the throughput of a process. Reducing the processing time increases the productivity.
- **Temperature:** the measure of hotness or coldness of a solid body or fluid. Between two bodies in contact, heat always flows from the body at a higher temperature to the one at a lower temperature until the thermal equilibrium is achieved. When the temperatures of the two bodies become equal, they are said to be in thermal equilibrium. In the absence of heat flow, the bodies are considered to be at thermal equilibrium. Temperature is central to food processing as it influences the quality, flavor, freshness, and more importantly, the safety of foods. The use of high temperatures for food preservation is attributed to its destructive effects on the microorganisms (Jay, 1992). Contrarily, the use of low temperatures is based on the fact that food-borne microorganisms are inactivated at temperatures above freezing, and their growth is stopped at sub-freezing temperatures (Jay, 1998).
- **Electric current:** the rate at which charge flows through the cross section of a wire. Electric or *Ohmic* heating of foods is a more recent and innovative food preservation technique, in which the resistance to the flow of current through a food product is converted to heat energy.
- **Luminous intensity:** the rate of flow of visible energy from the source per unit solid angle in a given direction. The luminous intensity of the lighting in food processing plants should be adequate to ensure hygienic processing, proper cleaning, and disinfection (Melero et al., 2014) and to avoid insects in the shop floor (Rahman, 2007).
- **Amount of substance:** the quantity that measures the size of a group of elementary units such as atoms, molecules, electrons, and particles.

The fundamental dimensions and their respective symbol of representation are given in Table 1.2.

1.2.2 Definitions and Applications of Derived Dimensions in Food Processing

- **Area and volume:** Area (*A*) can be defined as a quantity that describes the amount of material with a thickness that constitutes a two-dimensional (2D) shape, in the plane (Weisstein, 2012). It is the 2D counterpart of a 1D magnitude such as length. Surface area is a morphological property of foods which is equivalent to the area on the 2D surface of a three-dimensional (3D) object. Volume (*V*) is the quantity of a 3D space that is defined by a length, a width, and a height and enclosed by a closed surface. Measurement of the volume is vital for aerated food products such as bread and ice cream, which are sold by volume.

TABLE 1.2
Fundamental Dimensions and Their Symbol of Representation

Dimension	Symbol
Mass	M
Length	L
Time	T
Temperature	θ
Electric current	I
Luminous intensity	C
Amount of substance	n

- **Volumetric flow rate and mass flow rate:** Volumetric flow rate (Q) is the volume of fluid which passes per unit time. Similarly, the mass flow rate ($\dot{m}$) is the mass of a substance which passes per unit time. The volumetric flow rate is an important parameter in calculating the efficiency of pumps that are used in food processing facilities to transport liquid foods such as milk and juices. The capacity of food processing equipment which handles fluids is often expressed in terms of volumetric flow rate (Lewis, 1996). Nevertheless, in process industries, measurement of mass flow rate is advantageous over the measurement of volumetric flow rate since mass is not affected by changes in temperature and pressure (Figure 1.4).
- **Velocity:** The velocity (v) of a body is defined as the rate of change of its position with respect to time. The application of velocity is appreciated in the fluidized bed drying or freezing, where hot or cold air is passed from the bottom of a bed of food product to be dried or frozen. The air velocity is set at a value which maintains the product suspended in the drying or freezing medium, without carrying away the material. Also, the application of high fluid velocities has been realized in the cleaning and sanitation of food processing equipment and pipelines during cleaning-in-place. The calculation of the average velocity (v) of a fluid through a pipe or channel Eq. (1.2) is also vital in the transport of liquid foods from one point to another in an industry. The velocity of a fluid (v) flowing through a pipe of a cross-sectional area A is given by

$$v = \frac{Q}{A} \tag{1.2}$$

where Q is the volumetric flow rate of the fluid.

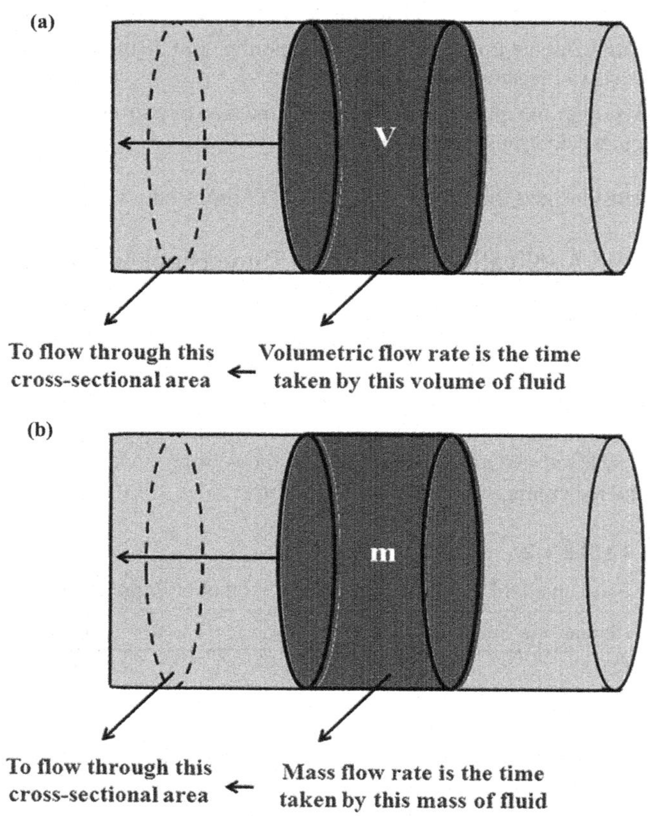

FIGURE 1.4 Schematic representation of (a) volumetric flow rate and (b) mass flow rate.

- **Acceleration (*a*):** It is the rate of change of velocity of an object with respect to time, which is the net result of all forces acting on it, as governed by Newton's second law.
- **Force:** It is the action that changes or tries to change the state of motion or shape of a body. According to Newton's second law, *the force acting on an object is equal to the mass of that object times its acceleration*, which is mathematically expressed as $F = m \times a$. Size reduction is a necessary unit operation in the food processing industry which employs different types of forces (Figure 1.5) such as compression (e.g., crushing of fruits to obtain pulp), impact (e.g., hammer milling of fibrous foods), or attrition (e.g., grinding of spices).
- **Pressure:** It is defined as the force exerted by a substance on the internal surface of the container per unit area. Pressure is an intensive property as it does not depend on the size of the system. Thus, the pressure exerted by a substance will be the same irrespective of the shape of the container and depends only on its depth in the container. Pressure can be measured as gauge pressure or absolute pressure. While gauge pressure is measured relative to local atmospheric pressure, absolute pressure is measured relative to perfect vacuum. Therefore, in gauge pressure, zero corresponds to one atmospheric pressure, and in absolute pressure, zero corresponds to perfect vacuum. Figure 1.6 clearly shows the difference between the absolute and gauge pressure scales. Vacuum is negative on a gauge pressure scale and positive on an absolute pressure scale.

  High-pressure technique for food preservation employs elevated pressure as the main lethal agent for pathogen reduction without compromising the nutritional and organoleptic properties of the food product (Martinez-Monteagudo and Balasubramaniam, 2016).
- **Energy and power:** Energy is defined as the potential to provide useful work or heat. Power is the amount of energy consumed per unit time. Food processing operations such as drying are energy intensive and hence demand proper utilization to prevent unnecessary expenditure of energy. High quality of electrical power is critical for the food and beverage industry. For instance, disturbances in the supply of power to machinery in a dairy plant can result in unexpected downtime in processing and wastage of valuable product. This is because the production of milk is time-critical. If the power supply is interrupted, the machinery may also be blocked, thus, demanding to clear off the product or packaging (Bennett, 2018).
- **Frequency:** It is the number of wavelengths of a wave that passes through a fixed point per unit time. Frequency gains importance in the ultrasound processing of foods owing to its inverse relationship with energy. Ultrasonic frequencies in the range of 20–100 kHz are commonly used for ultrasonic processing such as cleaning, homogenization, and cell disruption (Yamamoto et al., 2015).

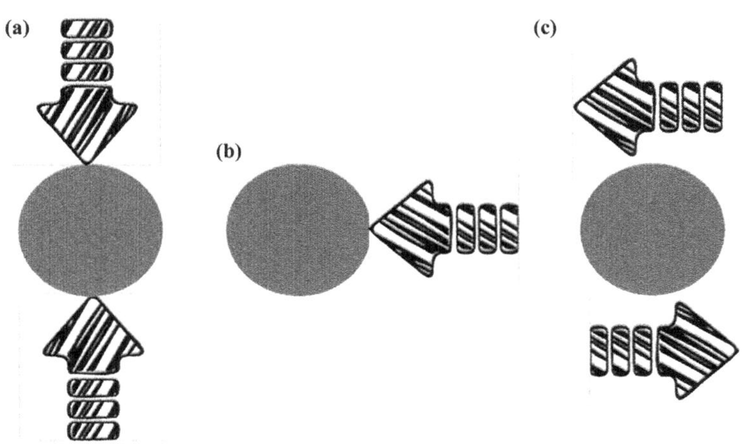

FIGURE 1.5 Types of forces: (a) compression, (b) impact, and (c) attrition.

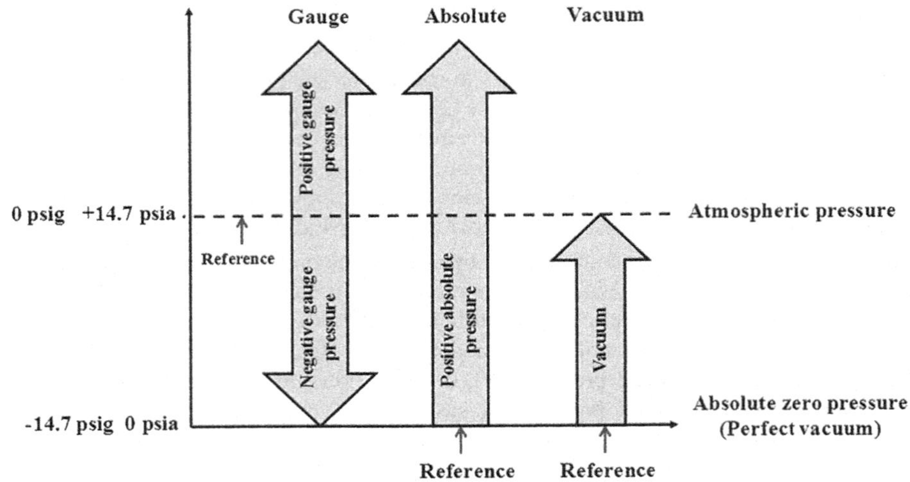

FIGURE 1.6 Schematic representation of absolute and gauge pressure.

- **Density:** It is the measure of the mass of a substance per unit volume. Density is a mechanical property of foods which is temperature-dependent since volume expands with an increase in temperature. It is used to characterize a food product and used in process calculations. The major application of density in the beverage (soft drinks) industry is to provide the Brix value which quantifies the sugar content. On the other hand, in the fragrance and flavors industry, density, or more specifically, the specific gravity (density of a substance in relation to that of water) is a primary quality indicator of the incoming raw materials (Wei et al., 2007; Singh et al., 1997; Mirica et al., 2010). Measurement of density facilitates the detection of adulteration which even the human nose cannot detect. For example, Brazilian Bois de rose oil has a specific gravity of 0.8680–0.8910. Pure synthetic linalool has a specific gravity of 0.8580–0.8620. The main constituent of unadulterated Bois de rose oil is linalool, and if it is adulterated with pure synthetic linalool, its specific gravity is expected to drop below 0.8680. Thus, the adulteration can be easily identified which may be very difficult to identify by odor (Dixit, 2006). Density is also a key determining factor of the packaging cost of a food product as a dry product with high density requires lesser storage space and hence a smaller unit of packaging (bag or can) compared to a moist product with low bulk density. Expressions for the density (ρ) of major constituents in food as a function of temperature (T) are as follows:

  Water: $\rho_W = 997.18 + 0.0031439T - 0.0037574T^2$

  Ice: $\rho_{ic} = 916.89 - 0.13071T$

  Protein: $\rho_P = 1329.9 - 0.5184T$

  Fat: $\rho_F = 925.59 - 0.41757T$

  Carbohydrate: $\rho_C = 1599.1 - 0.31046T$

  Fiber: $\rho_{Fi} = 1311.5 - 0.36589T$

  Ash: $\rho_A = 2423.8 - 0.28063T$

- **Viscosity:** It is the property of a fluid which indicates its flowability. Viscosity is a measure of the fluid's resistance to a shearing action. It can be used as a quality indicator in such liquid foods where a thicker product is considered to be of superior quality than a thinner product (e.g., ketchup, sugar syrup). It also characterizes the texture of food. The applications of viscosity in the food industry would be discussed elaborately in Chapter 4.

The derived dimensions and their symbols of representation are presented in Table 1.3.

TABLE 1.3

Derived Dimensions and Their Symbol of Representation

Quantity	Dimension
Area	L^2
Volume	L^3
Velocity	LT^{-1}
Acceleration	LT^{-2}
Volumetric flow rate	L^3T^{-1}
Mass flow rate	MT^{-1}
Density	ML^{-3}
Dynamic viscosity	$ML^{-1}T^{-1}$
Kinematic viscosity	L^2T^{-1}
Force	MLT^{-2}
Frequency	T^{-1}
Energy, heat, work	ML^2T^{-2}
Power, radiant flux	ML^2T^{-3}
Pressure, stress	$ML^{-1}T^{-2}$
Surface tension	MT^{-2}
Heat flux density, irradiance	MT^{-3}
Heat capacity, entropy	$ML^2T^{-2}\Theta^{-1}$
Specific heat capacity	$L^2T^{-2}\Theta^{-1}$
Thermal conductivity	$MLT^{-3}\Theta^{-1}$
Energy density	$ML^{-1}T^{-2}$
Electric field strength	$MLT^{-3}I^{-1}$
Surface charge density	ITL^{-2}
Permittivity	$M^{-1}L^{-3}T^4I^2$
Permeability	$MLT^{-2}I^{-2}$
Exposure (x- and γ-rays)	ITM^{-1}
Absorbed dose rate	L^2T^{-3}
Radiant intensity	ML^2T^{-3}
Radiance	MT^{-3}

1.3 Classification of Unit Systems

Worldwide, different unit systems are available for the measurement of physical quantities. Considering length and time as fundamental, the unit systems are classified into three types (Figure 1.7) based on the third primary dimension included in it.

1.3.1 Absolute Unit System

The unit system with mass as the third fundamental dimension is known as the absolute unit system. The CGS (centimeter–gram–second), MKS (meter–kilogram–second), and FPS (foot–pound–second) systems fall under this category. Thus, the fundamental dimensions of all these systems are length, mass, and time. The units of length, mass, and time in the CGS, MKS, and FPS systems are presented in Table 1.4. In the absolute unit systems, force and energy are the derived dimensions, the units of which are given in Table 1.5. In the CGS and MKS systems, the unit of temperature is degree Celsius (°C). In the FPS system, the temperature is measured in units of degree Fahrenheit (°F).

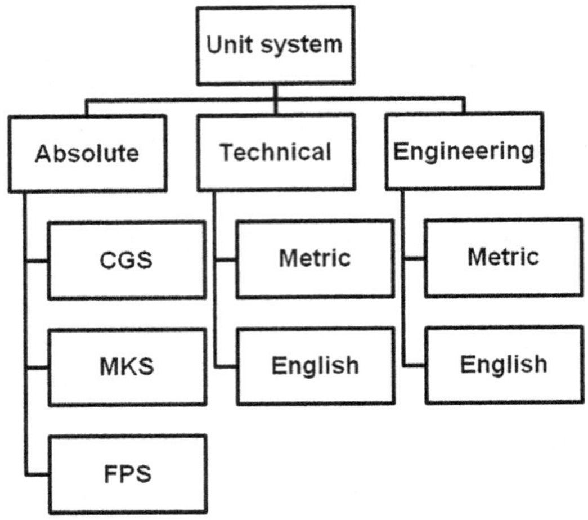

FIGURE 1.7 Classification of unit systems.

TABLE 1.4
Absolute Unit Systems: Fundamental Dimensions and Units

	Fundamental Dimensions and Units		
System	**Dimension**	**Unit**	**Symbol**
CGS	Length (L)	centimeter	cm
	Mass (M)	gram	g
	Time (T)	second	s
MKS	Length (L)	meter	m
	Mass (M)	kilogram	kg
	Time (T)	second	s
FPS	Length (L)	foot	ft
	Mass (M)	pound	lb
	Time (T)	second	s

TABLE 1.5
Absolute Unit Systems: Derived Dimensions and Units

	Derived Dimensions and Units	
System	**Force**	**Energy**
CGS	dyne	erg
MKS	newton (N)	joule (J)
FPS	poundal	(pound) (foot)

1.3.2 Technical Unit Systems

The unit system with force as the third fundamental dimension is known as technical unit systems. The commonly used technical unit systems are the metric and the English systems. Here, mass is a derived dimension, which is expressed in units of TMU (technical mass unit) in the metric system and slug in the English system. The units of length, force, and time in the technical unit system are given in Table 1.6. In the technical unit system, the unit of temperature is degree Celsius (°C) in the metric system and degree Fahrenheit (°F) in the English system.

TABLE 1.6

Technical Unit Systems

	Fundamental Dimensions and Units		
System	**Dimension**	**Unit**	**Symbol**
Metric	Length (L)	meter	m
	Force (F)	kilogram force	kgf
	Time (T)	second	s
English	Length (L)	foot	ft
	Force (F)	pound force	lb_f
	Time (T)	second	s

1.3.3 Engineering Unit Systems

This system is unique in terms of including four fundamental dimensions in contrast to three in the absolute and technical unit systems. Engineering unit systems consider length, time, mass, and force as the primary dimensions (Table 1.7). In the engineering unit system, the unit of temperature is degree Celsius (°C) in the metric system and degree Fahrenheit (°F) in the English system.

1.3.4 International Unit System (SI)

To establish universality and uniformity in the usage of units, the "Systѐme International d'Unités" (SI) or the *International System of Units* emerged in 1960. This unit system was introduced at the 11th General Conference on Weights and Measures (CGPM). Since then, the SI system is the internationally agreed system of units and recognized as the *modern metric system.* The SI system was designed to cater to the needs of both the industrial and the scientific fraternities. Also, it is said to be a *coherent system* of units in which the units of derived dimensions are obtained as multiples or submultiples of two or more fundamental units. There are seven fundamental or base units (Table 1.8) and several derived units (Tables 1.9–1.11). In addition to the fundamental and derived units, there are two supplementary units.

1.3.4.1 Definitions of Fundamental Units

- **meter (m):** It is the unit of length and is defined as the *length of the path traveled by light in vacuum during a time interval of 1/299 792 458 of a second* (CGPM, 1983).
- **kilogram (kg):** It is the unit of mass which is equal to the *mass of the international prototype. The prototype is a cylinder of 39 mm in diameter and 39 mm in height made of an alloy, constituted by 90% platinum and 10% iridium, with a density of 21500 kg/m³ (approximate).*

TABLE 1.7

Engineering Unit Systems

	Fundamental Dimensions and Units		
System	**Dimension**	**Unit**	**Symbol**
Metric	Length (L)	meter	m
	Mass (M)	kilogram	kg
	Force (F)	kilogram force	kgf
	Time (T)	second	s
English	Length (L)	foot	ft
	Mass (M)	pound mass	lb_m
	Force (F)	pound force	lb_f
	Time (T)	second	s

TABLE 1.8

Fundamental Dimensions and Their SI Units

Dimension	Unit	Abbreviation of the Unit
Mass (M)	kilogram	kg
Length (L)	meter	m
Time (T)	second	s
Temperature (θ)	kelvin	K
Electric current (I)	ampere	A
Luminous intensity (C)	candela	cd
Amount of substance (n)	mole	mol

TABLE 1.9

Derived SI Units with Common Symbols

Quantity	Unit
Area (A)	m^2
Volume (V)	m^3
Velocity (u)	$m\ s^{-1}$
Acceleration (a)	$m\ s^{-2}$
Flow rate (Q)	$m^3 s^{-1}$
Mass flow rate ($\dot{m}$)	$kg\ s^{-1}$
Density (ρ)	$kg\ m^{-3}$
Viscosity (μ)	$kg\ m^{-1} s^{-1}$
Kinematic viscosity (η)	$m^2 s^{-1}$

TABLE 1.10

Derived SI Units with Special Names and Symbols

Quantity	SI Unit	Special Name for the Unit	Symbols of the Unit
Force	$kg\ m\ s^{-2}$	newton	N
Frequency	s^{-1}	hertz	Hz
Energy, quantity of heat, work	N m	joule	J
Power, radiant flux	$J\ s^{-1}$	watt	W
Pressure, stress	$N\ m^{-2}$	pascal	Pa

This prototype is maintained at the Bureau International des Poids et Mesures at Sevres, France, under the conditions specified by the first CGPM, in 1889. Accordingly, the prototype is stored at atmospheric pressure in a specially designed triple bell jar.

- **second (s):** It is the unit of time and defined more precisely as the *duration of 9,192,631,770 periods of radiation corresponding to the transition between the two hyperfine levels of the cesium 133 atom in its ground state at a temperature of 0 K* (CGPM, 1967/1968).
- **ampere (A):** It is the unit of constant current which, *if maintained in two straight parallel conductors of infinite length, of a negligible circular cross-section, and placed 1 meter apart in vacuum, would produce a force equal to 2 × 10^{-7} newton per meter of length, between these conductors.*
- **kelvin (K):** It is the unit of thermodynamic temperature and defined as *1/273.16 of the thermodynamic temperature of the triple point of water.* Notable is that the thermodynamic temperature of water at its triple point is exactly 273.16 K.

TABLE 1.11

Example of Other Derived SI Units with Special Names and Symbols

Quantity	SI Unit	Name of the Unit	Abbreviation of the Unit
Dynamic viscosity	kg $m^{-1}s^{-1}$	pascal second	Pa s
Moment of force	kg m^2s^{-2}	newton meter	N m
Surface tension	kg s^{-2}	newton per meter	N/m
Heat flux density, irradiance	kg s^{-3}	watt per square meter	W/m^2
Heat capacity, entropy	kg m^2s^{-2} K^{-1}	joule per kelvin	J/K
Specific heat capacity	m^2s^{-2} K^{-1}	joule per kilogram kelvin	J/(kg K)
Thermal conductivity	kg m s^{-3} K^{-1}	watt per meter kelvin	W/(m K)
Energy density	kg $m^{-1}s^{-2}$	joule per cubic meter	J/m^3
Electric field strength	kg m s^{-3} A^{-1}	volt per meter	V/m
Surface charge density	A s m^{-2}	coulomb per square meter	C/m^2
Permittivity	$kg^{-1}m^{-3}s^4$ A^2	farad per meter	F/m
Permeability	kg m s^{-2} A^{-2}	henry per meter	H/m
Exposure (x- and γ-rays)	A s kg^{-1}	coulomb per kilogram	C/kg
Absorbed dose rate	m^2s^{-3}	gray per second	Gy/s
Radiant intensity	kg m^2s^{-3}	watt per steradian	W/sr
Radiance	kg s^{-3}	watt per square meter steradian	W/(sr m^2)

- **mole (mol):** It is the unit of amount of substance in a system which *contains as many elementary entities as there are atoms in 0.012 kilogram of carbon 12.* The preceding definition is based on an agreement between the International Union of Pure and Applied Physics and the International Union of Pure and Applied Chemistry that designated isotope of carbon with mass number 12 as the base for defining mole.
- **candela (cd):** It is the unit of *luminous intensity, in a given direction, of a source that emits monochromatic radiation of frequency 540 × 10^{12} hertz and that has a radiant intensity in that direction of 1/683 watt per steradian.*

1.3.4.2 Definitions of Supplementary Units

Apart from the seven base units, radian and steradian are known as supplementary units. These are not derived from the base units.

- **radian (rad):** It is the unit of plane angle and is defined as the *plane angle between two radii of a circle that cut off on the circumference of an arc equal in length to the radius* (Figure 1.8).
- **steradian (sr):** It is the unit of solid angle and is defined as the *solid angle that having its vertex in the centre of a sphere, cuts off an area of the surface of the sphere equal to that of a square with sides of length equal to the radius of the sphere* (Figure 1.9).

Although the radian and steradian are units of measurements, they are dimensionless quantities. As understood from the preceding definitions, both radian and steradian are essentially the ratios of the length of the enclosed arc to the length of the circle's radius and the area subtended to the square of its distance from the vertex, respectively.

1.3.4.3 Definitions of Derived Units

- **newton (N):** It is defined as a force that gives an acceleration of 1 m/s^2 to 1 kg mass.
- **joule (J):** It is the quantity of work done when 1 m of displacement occurs from the point of application due to 1 N force in the direction of the force.

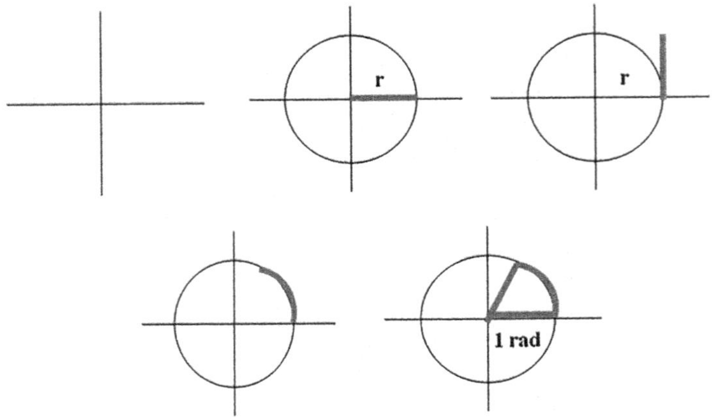

FIGURE 1.8 Schematic representation of radian.

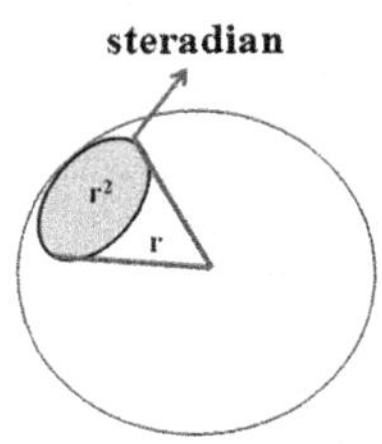

FIGURE 1.9 Schematic representation of steradian.

- **coulomb (C):** It is defined as the quantity of electricity transported by 1 A of current in 1 s.
- **watt (W):** It is defined as the power that produces energy at the rate of 1 J/s.
- **volt (V):** It is described as the electric potential difference between the two points of conducting wire that carries a 1 A constant current and 1 W power dissipated between these two points.
- **ohm (Ω):** It is defined as the electric resistance between two points of a conductor when 1 V potential difference applied between these points which produce 1 A current in the conductor, while the conductor is not the source of any electromotive force.
- **farad (F):** It is defined as the capacitance of a capacitor that is charged by 1 C of electricity with 1 V potential difference between the plates.
- **lumen (lm):** It is defined as the luminous flux emitted by a point source having 1 cd intensity in the 1 steradian solid angle.
- **henry (H):** It is defined as the inductance of a closed circuit in which 1 V of electromotive force is produced when the electric current varies at the rate of 1 A/s in the circuit.
- **weber (Wb):** It is the magnetic flux that, by linking a circuit of one turn, produces in it, 1 V of electromotive force, as it is reduced to zero at a uniform rate in 1 s.

1.3.4.4 Prefixes for SI Units

In some instances, units can be too small or big which necessitates the use of prefixes to express them in multiples or submultiples of the fundamental units. For example, the particle size of a food powder such as soluble coffee is expressed in terms of the micrometer (μm), which is 10^{-6} m, and the frequency of

TABLE 1.12

Prefixes for SI Units

Prefix	Multiplication Factor	Symbol
yotta	10^{24}	Y
zetta	10^{21}	Z
exa	10^{18}	E
peta	10^{15}	P
tera	10^{12}	T
giga	10^{9}	G
mega	10^{6}	M
kilo	10^{3}	K
hecto	10^{2}	h
deca	10	da
deci	10^{-1}	d
centi	10^{-2}	c
milli	10^{-3}	m
micro	10^{-6}	μ
nano	10^{-9}	n
pico	10^{-12}	p
femto	10^{-15}	f
atto	10^{-18}	a
zepto	10^{-21}	z
yocto	10^{-24}	y

microwave radiation is defined in megahertz, where, 1 megahertz = 10^6 Hz. The list of commonly used prefixes for SI units is given in Table 1.12. The advantage of using an appropriate unit prefix in combination with a numerical value is that it avoids the ambiguity over the significant figures. For example, the number of significant figures in a mass specified as 2300 g is unclear, which is resolved when it is expressed as 2.3 kg.

1.3.4.5 Guidelines to Write Units

- Units should always be mentioned next to the associated dimensional quantity.
- Full names of the units, even when they are named after a scientist, should not be written with a capital letter.
  - For example: newton, watt, ampere, meter and not as Newton, Watt, Ampere, Meter
- The unit should be written either in full or using the agreed symbols only.
- Units do not take the plural form.
  - For example: 52 kg but not 52 kgs and 3.5 N but not 3.5 Ns
- No full stop or punctuation mark should be used within or at the end of symbols for units.
  - For example: 2.5 W/m K but not 2.5 W/m.K.

1.4 Conversion of Units

The use of the SI system is widely prevalent in the present days. However, few countries in the world are yet to adopt the SI unit system completely. Therefore, knowledge of the method of interconversion between different unit systems is vital to disseminate the information on measurements without ambiguities across the geographical boundaries of the world. This is because both the scientific and the international business are transacted in SI units. Unit conversion is a multistep process that requires the following information:

1. The starting unit or the unit to be converted.
2. The desired unit or the unit to be obtained.
3. The conversion factor.

Table 1.13 lists the conversion factors for various units. The first column, "To convert from," lists the units commonly used to express the quantities; the second column, "To be converted to," gives SI units or other preferred units; and the third column, "Multiply by," gives the conversion factor by which the numerical value in "To convert from" units must be multiplied to obtain the numerical value in "To be converted to" units.

TABLE 1.13

Conversion Factors

To Convert from	To Be Converted to	Multiply by
Mass		
lb	kg	0.45359
kg	lb	2.2
ton	kg	907.18474
Length		
ft	m	0.3048
in	m	0.0254
mile	m	1609.344
Time		
min	s	60
h	s	3600
day	s	86400
Area		
in^2	mm^2	645.2
in^2	m^2	6.4516×10^{-4}
ft^2	m^2	0.0929
Diffusivity		
ft^2/h	m^2/s	2.581×10^{-5}
Volume		
ft^3	m^3	0.02832
L	cm^3	1000
m^3	L	1000
U.S. gallon	m^3	3.785×10^{-3}
Density		
lb/ft^3	kg/m^3	16.0185
kg/m^3	lb/ft^3	0.062428
Energy, heat, and power		
Btu	J	1055
	kJ	1.055
	cal	252.16

(Continued)

TABLE 1.13 (*Continued*)
Conversion Factors

To Convert from	To Be Converted to	Multiply by
kcal	kJ	4.185
	J	4185
erg	J	10^{-7}
hp	kW	0.7457
	ft lb_f/s	550
W	cal/min	14.34
hp	Btu/s	0.7068
Btu/h	W	0.29307
Btu/min	W	17.58
hp h	kW h	0.7457
	Btu	2544.5
kWh	kJ	3600
kJ/h	kW	2.778×10^{-4}
J/s	W	1
Thermal magnitude (enthalpy, heat flow, heat flux, specific heat)		
Btu/(h ft^2)	W/m^2	3.1546
Btu/(h ft °F)	W/(m K)	1.731
Btu/(h ft^2 °F)	W/(m^2 K)	5.678
Btu/lb	kJ/kg	2.3258
Btu/(lb °F)	kJ/kg K	4.187
Force		
pdl (poundal)	N	0.138
lb_f	N	4.4482
lb_f	pdl	32.2
N	kg m/s^2	1
dyne	g cm/s^2	1
	kg m/s^2 (or) N	10^{-5}
Pressure		
psia	kPa	6.895
	N/m^2	6.895×10^{-3}
bar	Pa	10^5
	kPa	100
Pa	N/m^2	1
mm Hg (tor)	N/m^2 (or) Pa	133.3224
	kgf/cm^2	13.59
Atm	psi	14.22
Atm	Pa	1.01325×10^5
Psi	kgf/m^2	703
lbf/ft^2	$dyne/cm^2$	4.788×10^2
	N/m^2	47.88
Temperature		
Temperature (T) in (°C)	T in (°F)	T(°C) × 1.8+32
T in (°F)	T in (°C)	[T(°F) − 32]/1.8
T in (°C)	T in (K)	T (°C) + 273.15
Viscosity		
N s/m^2	Pa s (or) 1 kg/(m s)	1

(*Continued*)

TABLE 1.13 (*Continued*)
Conversion Factors

To Convert from	To Be Converted to	Multiply by
centipoise (cP)	Pa s	10^{-3}
	poise	10^{-2}
poise	Pa s	0.1
lb/ft h	mPa s	0.414
Mass flow		
lb/h	g/s	0.126
ton/h	kg/s	0.282
lb/ft^{2}h	g/s m^2	1.356

After the abovementioned information is known, the steps involved in the unit conversion task are schematically represented in Figure 1.10.

While the first three steps represented in Figure 1.10 are simple, the fourth and final step, in turn, has a procedure which is described as follows.

1.4.1 Procedure for the Determination of Significant Digits and Rounding Off

The number of significant digits and the rounding off criteria depend on the end application in which the converted measurement (number + unit) will be used. The application has been categorized into two types: (i) to develop a technical document or a specification and (ii) to be used for packaged goods in the commercial marketplace (NIST, 2006).

1.4.1.1 Rounding Procedure for Technical Documents or Specifications

In this case, the number of significant digits is established such that the accuracy is neither compromised nor overstated. The following procedure may be used to achieve the accuracy:

i. If the first significant digit of the converted value is greater than or equal to the first significant digit of the original value, then the converted value is rounded to the same number of significant digits as there are in the original value.

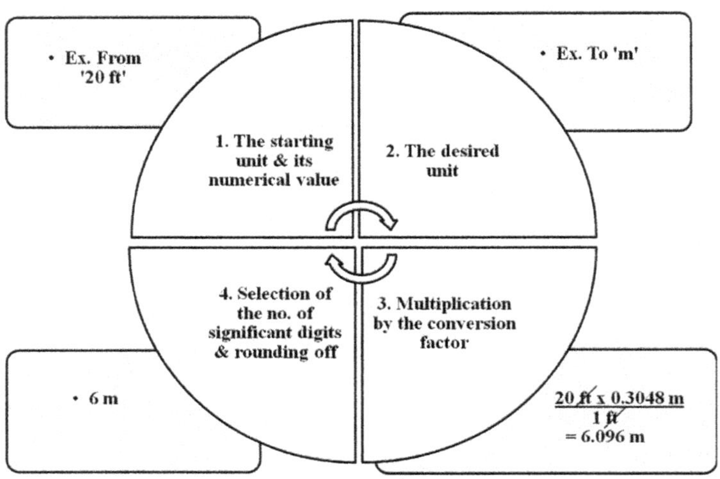

FIGURE 1.10 Steps involved in unit conversion.

ii. If the first significant digit of the converted value is smaller than the first significant digit of the original value, then the converted value is rounded to one more significant digit than the original value.

1.4.1.2 Rounding Practices Used for Packaged Goods in the Commercial Marketplace

It is mandatory for the manufacturers of commercially packaged goods to accurately declare the net quantity of contents of their packages as directed by the labeling regulations of the country. In general, the regulations direct the manufacturers or packers to round the converted values down to avoid overstating the net quantity of contents declared on the label. For example, if a package is labeled 453.59 g, the packer means that the package declaration is accurate within ±0.005 g. With respect to the measurement of liquid volume, a label declaration of 473 mL implies that the package declaration is accurate to within ±0.5 mL (NIST, 2017).

1.4.1.3 Rounding of Temperature Values

As explained previously, the number of significant digits for a temperature value will depend upon the intended accuracy of the original temperature. Temperature is usually expressed in °F as whole numbers and converted to the nearest 0.5°C. This is because the magnitude of °C is approximately twice the size of °F, and rounding to the nearest Celsius would reduce the precision of the original measurement (NIST, 2006).

Example 1.2

The length and breadth of a metal sheet are 5.123 ft and 5.004 ft, respectively. Convert the obtained area value from ft^2 to m^2 and round the converted value to an appropriate number of significant digits.

Solution

Area of metal sheet = length × breadth = 5.123 ft × 5.004 ft
Area = 25.635 ft^2
1 ft = 0.3048 m
1 ft^2 = 0.0929 m^2
25.635 ft^2 = 2.3814915 m^2 = 2.3815 m^2

This calculation would fall under the category of technical documents or specifications. Thus, as described in the Section 1.4.1.1, in this example, the first significant digit of the converted value is equal to the first significant digit of the original value. Therefore, the converted value is rounded to the same number of significant digits as there are in the original value, which is 5. Further, in the converted value, the digit after the fifth digit is greater than 5, due to which the last significant digit is increased by 1.

Answer: Area of the metal sheet is 2.3815 m^2

Example 1.3

Convert the following and round the value to an appropriate number of significant digits:

1. 4535 g to lb
2. 1.23×10^{-13} kg to μg
3. 25.4 in to m
4. 154236 s to min
5. 1345.2 mm^2 to in^2
6. 1.291×10^{-5} m^2/s to ft^2/h
7. 635 mL to m^3
8. 14.0285 kg/m^3 to lb/ft^3
9. 212 °F to °C

10. 2000 Btu to kJ
11. 5000 cP to Pa s
12. 0.5 Btu/(h ft °F) to W/(m K)
13. 50 psia to atm
14. 2 lb/h to kg/s

Solution

1. From Table 1.13, 1 lb = 0.45359 kg = 453.59 g (Since 1 kg = 1000 g)
   ∴ 1 g = 1/453.59 = 0.00220 lb
   And, 4535 g = 9.997964 lb
   Since the first nonzero or significant digit of the converted value (9) is greater than that of the original value (4), it is maintained to have the same number of significant digits as that of the latter. Further, in the converted value, the digit after the fifth digit is greater than 5, due to which the last significant digit is increased by 1.

Answer: 9.998 lb

2. From Table 1.11, 1 g = 10^{-3} kg
   And, 1 g = 10^{6} µg
   ∴ 1 kg = 10^{9} µg
   ∴ 1.23×10^{-13} kg = $1.23 \times 10^{-13} \times 10^{9} = 1.23 \times 10^{-4}$ µg

Answer: 1.23×10^{-4} µg

3. From Table 1.13, 1 in = 2.54 cm = 2.54×10^{-2} m
   ∴ 25.4 in = $25.4 \times 2.54 \times 10^{-2} = 64.516 \times 10^{-2}$ m rounded off to 64.5×10^{-2} m
   The value can also be expressed as 0.645 m

Answer: 0.645 m

4. From Table 1.13, 1 s = 1/60 min
   ∴ 154236 s = 154236/60 = 2570.6 min
   After rounding off to six significant digits, the value is 2570.60 min

Answer: 2570.60 min

5. From Table 1.13, 1 in = 2.54 cm = 25.4 mm
   ∴ 1 mm = 0.0394 in
   1 mm^2 = 0.00155 in^2
   1345.2 mm^2 = 1345.2×0.00155 = 2.08506 in^2
   The same value remains after rounding as the first significant digit of the converted value (2) is greater than that of the original value (1), and hence, the same number of significant digits (5) is retained.

Answer: 2.0851 in^2

6. From Table 1.13, 1 ft^2/h = 2.581×10^{-5} m^2/s
   ∴ 1 m^2/s = $1/2.581 \times 10^{-5}$ ft^2/h = 0.3874×10^{5} ft^2/h
   And, 1.291×10^{-5} m^2/s = $1.291 \times 10^{-5} \times 0.3874 \times 10^{5}$ = 0.5001334 ft^2/h
   After rounding, the value is 0.5001 ft^2/h

Answer: 0.5001 ft^2/h

7. From Table 1.13, 1 m^3 = 1000 L
   ∴ 1 L = 1000 mL = 10^{-3} m^3
   635 mL = 635×10^{-6} m^3

Answer: 635×10^{-6} m^3

8. From Table 1.13, 1 kg/m^3 = 0.062428 lb/ft^3
   ∴ 14.0285 kg/m^3 = 14.0285×0.062428 = 0.875771198 lb/ft^3
   After rounding, the value is 0.875771 lb/ft^3

Answer: 0.875771 lb/ft^3

9. From Table 1.13, T (°C) = [T (°F) − 32]/1.8

   In this case, T (°C) = (212 − 32)/1.8 = 100°C

   After rounding to four significant figures, the corrected value is 100.0°C

   Answer: 100.0°C

10. From Table 1.13, 1 Btu = 1.055 kJ

    ∴ 2000 Btu = 2000 × 1.055 = 2110 kJ

    After rounding to one significant figure, the corrected value is 2000 kJ

    Answer: 2000 kJ

11. From Table 1.13, 1 cP = 10^{-3} Pa s

    ∴ 5000 cP = 5000 × 10^{-3} = 5 Pa s

    Answer: 5 Pa s

12. From Table 1.13, 1 Btu/h ft °F = 1.731 W/m K

    ∴ 0.5 Btu/h ft °F = 0.5 × 1.731 = 0.8655 W/m K

    After rounding to one significant digit, the corrected value is 0.9 W/m K

    Answer: 0.9 W/m K

13. From Table 1.13, 1 psia = 6.895 kPa and 1 atm = 101.325 kPa

    ∴ 1 kPa = 0.00986923 atm

    6.895 kPa = 0.06805 atm = 1 psia

    (or) 1 psia = 0.06805 atm

    ∴ 50 psia = 50 × 0.06805 = 3.4025 atm

    After rounding to two significant digits (one more significant digit than the original value) the corrected value is 3.4 atm

    Answer: 3.4 atm

14. From Table 1.13, 1 lb/h = 0.126 g/s = 0.126 × 10^{-3} kg/s

    ∴ 2 lb/h = 2 × 0.126 × 10^{-3} = 0.252 × 10^{-3} kg/s

    After rounding to one significant digit as in the original value, the corrected value is 0.3 × 10^{-3} kg/s

    Answer: 0.3 × 10^{-3} kg/s

1.5 Dimensional Analysis

Dimensional analysis is an analytical approach based on the law of dimensional homogeneity in which a relationship in the form of an empirical equation is established between the variables that are known to be involved in a physical phenomenon. However, the relationship thus obtained should be further verified with experimentation to finalize the equation including all the relevant variables. The dimensional analysis serves various applications in engineering problems: (i) reducing the number of variables in the experimental data, (ii) decreasing the number of experimental trials without compromising on the necessary information to be obtained from the experimentation, and (iii) scale-up and scale-down of processes. The first step in the dimensional analysis is to assemble the variables into dimensionless groups, to identify the relationship between different variables.

1.5.1 Dimensionless Groups

A dimensionless quantity or a dimensionless group can be a simple ratio of two dimensionally equal quantities, or it can be composed of the products of dimensionally equal quantities in the numerator and the denominator (Kuneš, 2012). Table 1.14 lists the dimensionless quantities commonly encountered in engineering problems. The significance and applications of these dimensionless groups would be appreciated by the readers in the forthcoming chapters.

TABLE 1.14

Dimensionless Numbers

Dimensionless Number	Expression	Explanation of the Variables
Reynolds number (N_{Re})	$\frac{Lv\rho}{\mu}$	L: characteristic length; v: velocity of the fluid; ρ: density of the fluid; μ: dynamic viscosity of the fluid
Nusselt number (N_{Nu})	$\frac{hL}{k}$	h: convective heat transfer coefficient; L: characteristic length; k: thermal conductivity
Prandtl number (N_{Pr})	$\frac{C_P\mu}{k}$	C_P: specific heat; μ: viscosity; k: thermal conductivity
Biot number (N_{Bi})	$\frac{hL}{k}$	h: convective heat transfer coefficient; L: characteristic length; k: thermal conductivity
Fourier number (N_{Fo})	$\frac{\alpha t}{L^2}$	α: thermal diffusivity; t: characteristic time; L: characteristic length
Grashof number (N_{Gr})	$\frac{g\beta d^3\Delta T\rho^2}{\eta^2}$	g: acceleration due to gravity; β: coefficient of thermal expansion; d: characteristic length; ΔT: difference between surface and bulk temperature; ρ: density; η: kinematic viscosity
Peclet number (N_{Pe})	$\frac{\rho v d C_P}{k}$	ρ: density; v: velocity; d: characteristic length; C_P: specific heat; k: thermal conductivity
Power number (N_{Po})	$\frac{P}{D^5N^3\rho}$	P: power consumed by the propeller; D: diameter of the propeller; N: rotational frequency of the propeller; ρ: fluid density
Schmidt number (N_{Sc})	$\frac{\mu}{\rho D}$	μ: dynamic viscosity of the fluid; ρ: density of the fluid; D: mass diffusivity
Sherwood number (N_{Sh})	$\frac{k_gL}{D}$	k_g: convective mass transfer film coefficient; L: characteristic length; D: mass diffusivity
Stanton number (N_{St})	$\frac{h}{C_P\rho v}$	h: convective heat transfer coefficient; C_P: specific heat of the fluid; ρ: density of the fluid; v: velocity of the fluid
Weber number (N_{We})	$\frac{\rho Lv^2}{\sigma}$	ρ: density of the fluid; L: characteristic length; v: velocity; σ: surface tension

Due to practical limitations, experimental trials are often conducted using a laboratory or a pilot-scale model rather than the actual prototype. Usually, the model is smaller than the actual prototype, the condition of which is referred to as the "scale-down." During the "scale-up" of a pilot-scale model to industrial scale, it is necessary to satisfy three conditions in order to extrapolate the experimental results obtained on the model to the actual prototype. These conditions are collectively known as the *criteria of similitude*, as explained subsequently:

i. ***Geometric similarity*:** Two objects are said to be geometrically similar if the ratio of any length in the model to the corresponding length in the actual prototype is the same everywhere. Here, length is said to be the similarity parameter, and the ratio of lengths is known as the *scale factor*. For example, when scaling down spray dryer equipment, the diameter of drying chamber, the height of cylindrical top section, and the height of bottom cone can be reduced by half (scale factor: ½; Figure 1.11).
ii. ***Kinematic similarity (similarity of motion)*:** Two particles in motion are said to be kinematically similar if the ratios of corresponding velocities (similarity parameter) in the model and actual prototype are equal at all geometrically similar points. With the same example of spray dryer as explained previously, in order to achieve kinematic similarity, velocity of the atomized feed droplets and spray kinetics should be maintained the same between the model and actual prototype (which in turn depends on the atomization gas, feed flow rate, nozzle type and size, and properties of feed solution).

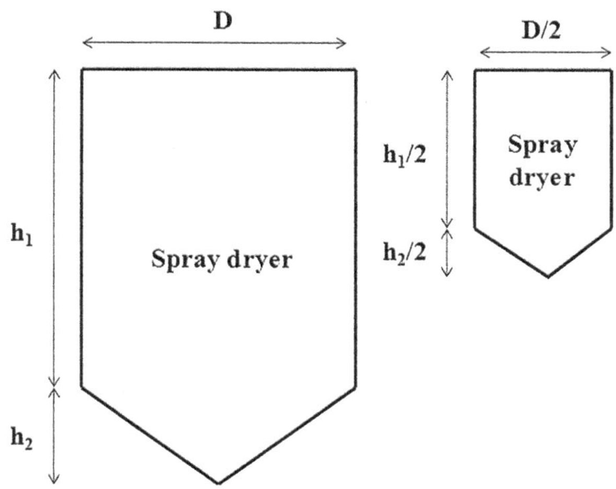

FIGURE 1.11 Example of geometric similarity.

iii. ***Dynamic similarity (similarity of forces)*:** Two systems are said to possess dynamic similarity if they are kinematically similar in addition to having the same ratios of homologous forces. In this criterion of similitude, force is the similarity parameter. Dynamic similarity can also be attained by ensuring that the relevant dimensionless numbers are the same for both the model and the actual prototype.

1.5.2 Dimensional Consistency

Dimensions are vital, as the correctness of any equation is judged by its dimensional consistency as defined in Section 1.1. The concept of dimensional consistency can be understood by considering the well-known equation of the universal gas law, $PV = nRT$, where,

P: pressure (Pa)
V: volume (m^3)
n: number of moles (mol)
R: universal gas constant (J/mol K)
T: temperature (K)

The dimensions of the terms on the right-hand side of the equation are derived as follows:

$$[n] = n \quad (1.3)$$

$$[R] = \mathrm{ML^2T^{-2}n^{-1}}\theta^{-1} \quad (1.4)$$

$$[T] = \theta \quad (1.5)$$

∴ The dimensions on the right-hand side of the equation are as follows:

$$[nRT] = \mathrm{ML^2T^{-2}} \quad (1.6)$$

The dimensions of the terms on the left-hand side of the equation are derived as follows:

$$[P] = \mathrm{ML^{-1}T^{-2}} \quad (1.7)$$

$$[V] = \mathrm{L^3} \quad (1.8)$$

∴ The dimensions on the left-hand side of the equation are as follows:

$$[PV] = \mathrm{ML^2T^{-2}} \tag{1.9}$$

Thus, the dimensional consistency of the universal gas law equation is established. The two widely used methods of dimensional analysis are the Rayleigh's theorem and Buckingham Π (Pi) theorem.

1.5.3 Rayleigh's Theorem of Dimensional Analysis

Rayleigh's method is a simple method for dimensional analysis which can be simplified to provide dimensionless groups that control a phenomenon. The steps involved are listed in the flowchart given by Figure 1.12.

In Figure 1.12, X is one of the variables of major interest to the phenomenon, which is expressed analytically as an exponent function of the remaining variables X_1, X_2, X_3, and X_4. K is the dimensionless constant and a, b, c, and d are the arbitrary exponents. In the next step, the variables are substituted by their corresponding dimensions. Then, by applying the law of dimensional homogeneity (equating the exponents of the magnitudes on both the sides), x number of simultaneous equations are obtained, where x is the number of dimensions involved. In other words, if x is the number of dimensions, the number of unknowns in the set of simultaneous equations would be $(x - 1)$. The simultaneous equations are solved to obtain the relationship between the variables. The preceding steps are explained with the example of the velocity (c) of a wave that passes through a point, which depends on the wavelength (λ) and frequency (v) of the wave. The possible relationship can be obtained by the dimensional analysis as follows:

$$c = \frac{\mathrm{L}}{\mathrm{T}};\ v = \frac{1}{\mathrm{T}};\ \lambda = \mathrm{L} \tag{1.10}$$

$$c = Kv^a\lambda^b \tag{1.11}$$

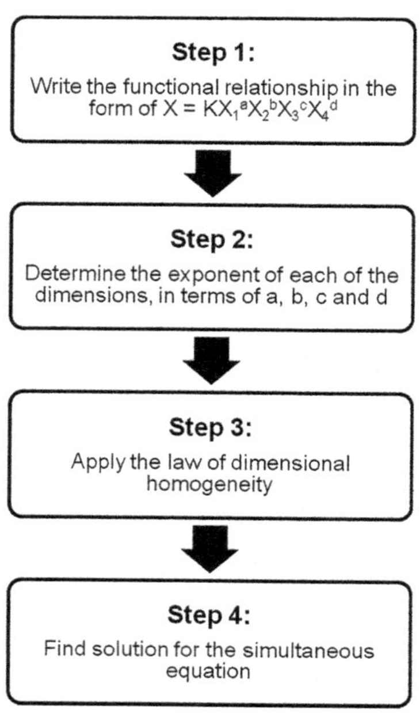

FIGURE 1.12 Steps of Rayleigh's method for dimensional analysis.

$$\frac{L}{T} = \frac{1^a}{T} L^b \tag{1.12}$$

$$L: 1 = b$$

$$\therefore b = 1$$

$$T: -1 = -a$$

$$\therefore a = 1$$

The possible equation is

$$c = Kv\lambda \tag{1.13}$$

The application of Rayleigh's method is limited to those phenomena which involve a lesser number of independent variables. Hence, the Buckingham Pi theorem explained in the subsequent section is practiced in the case of a greater number of variables.

1.5.4 Buckingham's Π (Pi) Theorem of Dimensional Analysis

The factor Π refers to every term that is dimensionless. The Buckingham's Π (Pi) theorem is based on three fundamental principles of dimensional analysis proposed by Bridgman, given as follows.

1. All the physical magnitudes may be expressed as power functions of a reduced number of fundamental magnitudes.
2. The equations that relate physical magnitudes are dimensionally homogeneous.
3. If an equation is dimensionally homogeneous, it may be reduced to a relation among many dimensionless rates or groups. These include all the physical variables that influence the phenomenon, the dimensional constants that may correspond to the selected unit system, and the universal constants related to the phenomenon treated.

According to this theorem, *the number of dimensionless groups (Π) that can be generated from a functional relationship is equal to the difference between the number of relevant variables (n) and the number of independent dimensions (m).*

$$\Pi = n - m \tag{1.14}$$

For each of the abovementioned $n - m$ variables, a nondimensional Π is constructed of the form

$$\Pi = (\text{variable})(\text{scale 1})^a (\text{scale 2})^b (\text{scale 3})^c \tag{1.15}$$

where, a, b, c... are chosen such that each Π is nondimensional. Therefore, the most crucial step involved in this approach is the appropriate selection of relevant variables.

The steps involved in obtaining a nondimensional expression using the Buckingham Pi theorem are explained with the following example of convective heat transfer to a fluid flowing through a cylindrical pipe.

Step 1: Identify the relevant variables.

$$h = f(L, v, \rho, \mu, C_P, k)$$

where h: convective heat transfer coefficient, L: characteristic dimension, v: velocity, ρ: density, C_P: specific heat capacity, and k: thermal conductivity.

Step 2: Write down the dimensions of the abovementioned relevant variables.

Variable	Unit	Dimensions
h	W/m² K	$MT^{-3}\Theta^{-1}$
L	m	L
v	m/s	LT^{-1}
ρ	kg/m³	ML^{-3}
μ	kg/m s	$ML^{-1}T^{-1}$
C_P	J/kg K	$L^2T^{-2}\Theta^{-1}$
k	W/m K	$MLT^{-3}\Theta^{-1}$

Step 3: Establish the number of independent dimensions and dimensionless groups.

No. of relevant variables (n) = 7

No. of independent dimensions (m) = 4 (M, L, T, Θ)

No. of dimensionless groups = $n - m = 7 - 4 = 3$

Step 4: Choose the core variables (or) the dimensionally independent scaling variables (Π_1, Π_2, and Π_3).

No. of core variables = No. of dimensions (m) = 4

The following rules should be considered while choosing the core variables:

i. A dependent variable should not be chosen as a core variable. In this case, h is not eligible to be chosen as a core variable, and hence, the selection should be made between L, v, ρ, μ, C_P, and k.

ii. The core variable should be chosen in such a manner that no two variables have the same dimensions.

iii. The chosen variables should involve all the dimensions.

iv. The core variables chosen should be identifiable as geometric, kinematic/time-dependent, or dynamic/mass dependent.

Therefore, according to the abovementioned rules, in this case, core variables = {L, v, μ, k}.

Step 5: Create the Π's by nondimensionalizing the remaining variables.

$$\Pi_1 = L^{\alpha 1}\, v^{\alpha 2}\, \mu^{\alpha 3}\, k^{\alpha 4}\, h^1$$

$$\Pi_2 = L^{\alpha 5}\, v^{\alpha 6}\, \mu^{\alpha 7}\, k^{\alpha 8}\, \rho^1$$

$$\Pi_3 = L^{\alpha 9} v^{\alpha 10} \mu^{\alpha 11} k^{\alpha 12} C_P^1$$

Considering the first dimensionless group,

$$\Pi_1 = L^{\alpha 1}(LT^{-1})^{\alpha 2}(ML^{-1}T^{-1})^{\alpha 3}(MLT^{-3}\theta^{-1})^{\alpha 4}(MT^{-3}\theta^{-1})^1$$

For dimensional homogeneity,

$$M^0 = M^{\alpha 3}.\, M^{\alpha 4}.\, M^1$$

Equating the exponents,

$$\alpha 3 + \alpha 4 + 1 = 0$$

Repeating the similar procedure for other dimensions, i.e., L, T, and Θ, we get

$$-\alpha 2 - \alpha 3 - 3.\,\alpha 4 - 3 = 0;$$

$-\alpha 4 - 1 = 0$ (therefore, $\alpha 4 = -1$) and,

$\alpha 3 + \alpha 4 + 1 = 0$ (therefore, $\alpha 3 = 0$, on substituting the value of $\alpha 4 = -1$)

From the abovementioned expressions

$\alpha 1 = 1$

$\alpha 2 = 0$

$\alpha 3 = 0$

$\alpha 4 = -1$

Substituting the values of exponents in the expression for $\Pi_{1,}$ we obtain

$$\Pi_1 = L^1 v^0 \mu^0 k^{-1} h^1$$

$\Pi_1 = hL/k$ = Nusselt number (dimensionless number for convective heat transfer; N_{Nu})

Repeating the previous steps for Π_2 and Π_3,

$\Pi_2 = Lv\rho/\mu$ = Reynolds number (dimensionless number for fluid mechanics; N_{Re})

$\Pi_3 = C_P\mu/k$ = Prandtl number (dimensionless number for heat transfer between a moving fluid and a solid body; N_{Pr})

Step 6: Set out the nondimensional relationship.

$$\Pi_1 = f\left(\Pi_2, \Pi_3\right)$$

or

$$\frac{hL}{k} = f\left(\frac{Lv\rho}{\mu}, \frac{C_P\mu}{k}\right)$$

or

$$Nu = f\left(N_{\mathrm{Re}}, N_{\mathrm{Pr}}\right)$$

1.5.5 Limitations of the Dimensional Analysis

- Dimensionless quantities cannot be determined by this method.
- Constant of proportionality cannot be determined by this method. They can be found either by experiment or by theory.
- This method is not applicable to trigonometric, logarithmic, and exponential functions.
- In the case of physical quantities, which are dependent upon more than three physical quantities, this method will be complicated.
- In some cases, the constant of proportionality also possesses dimensions. In such cases, this system cannot be used.
- If one side of the equation contains addition or subtraction of physical quantities, this method cannot be used to derive the expression.

The concept of units and dimensions would be the basis of any food process engineering calculation. This chapter provided an insight into the different aspects of units and dimensions which would aid the readers in appreciating the concepts to be explained in the forthcoming chapters.

1.6 Problems to Practice

1.6.1 Multiple Choice Questions

1. Force per unit area is
   a. pressure
   b. strain
   c. stress
   d. acceleration

Answer: a

2. The dimension of specific volume is
   a. $L^3 M^{-1}$
   b. L^{3M}
   c. L^3
   d. M^{-1}

Answer: a

3. SI unit of heat transfer coefficient is
   a. m K/W
   b. W/m^2 K
   c. W m K
   d. W K/m

Answer: b

4. $MLT^{-3}\Theta^{-1}$ is the dimension of
   a. heat transfer coefficient
   b. thermal diffusivity
   c. thermal conductivity
   d. specific heat capacity

Answer: c

5. The dimension of universal gas constant is
   a. $ML^2T^2\Theta$
   b. $MLT^{-1}\Theta$
   c. ML^2T^{-2}
   d. $ML^2T^{-2}\Theta^{-1}$

Answer: d

6. The number of significant figures in 4500 is
   a. five
   b. seven
   c. two
   d. ten

Answer: c

7. N/m^2 is equal to
   a. henry
   b. tesla
   c. pascal
   d. dyne

Answer: c

8. Which is not a unit of time?
   a. year
   b. leap year
   c. second
   d. light year

Answer: d

9. The length and breadth of a metal sheet are 4.125 m and 4.004 m, respectively. The area of this sheet rounded to four significant figure is
   a. 16.5165
   b. 16.52
   c. 16.516
   d. 16.5

Answer: b

10. A force *F* is given by $F = at + bt^2$, where *t* is time. What are the dimensions of *a* and *b* such that the equation is dimensionally consistent?
    a. MLT^{-3} and MLT^{-4}
    b. MLT^{-2} and MLT
    c. MLT^{-2} and ML^2T^3
    d. ML^2T and ML^2T^{-1}

Answer: a

11. The dimension of Nusselt number is
    a. MT^{-2}
    b. MLT^{-2}
    c. ML^3T
    d. None

Answer: d

12. One atmospheric pressure is equal to
    a. 1 kg/m^2
    b. 1 kg/cm^2
    c. 1 g/m^3
    d. 1 g/cm^3

Answer: b

13. CGS unit of viscosity is
    a. poise
    b. pascal second
    c. pascal meter
    d. pascal

Answer: a

14. The SI unit of heat flow rate is
    a. joule
    b. ampere
    c. watt
    d. coulomb

Answer: c

15. candela is the unit of
    a. magnetic flux
    b. intensity of electric field
    c. luminous intensity
    d. charge

Answer: c

16. The unit system with force as the third fundamental dimension is
    a. absolute unit system
    b. technical unit system
    c. international unit system
    d. none of the above

Answer: b

17. One nanometer is equal to
    a. 10^{-6}m
    b. 10^{-8}m
    c. 10^{-9}m
    d. 10^{-5}m

Answer: c

18. The number of dimensionless groups (Π) that can be generated from six variables and three independent dimensions is
    a. two
    b. one
    c. four
    d. three

Answer: d

19. Which one of the following is not a dimension?
    a. Length
    b. Mass
    c. Time
    d. Celsius

Answer: d

20. The number of significant figures in 0.07500 is
    a. five
    b. four
    c. two
    d. three

Answer: b

1.6.2 Numerical Problems

1. Determine the dimensions of the following quantities:
   (i) Enthalpy, (ii) acceleration due to gravity, (iii) Nusselt number, (iv) density, (v) frequency, (vi) pressure, (vii) viscosity, (viii) momentum, (ix) power, and (x) energy

Solution

i. Enthalpy

$$\text{SI unit of enthalpy} = \text{Joule}$$

$$\text{Joule} = \text{N} \times \text{m}$$

$$\text{N} = \frac{\text{kg m}}{\text{s}^2}; \therefore \text{J} = \frac{\text{kg m}}{\text{s}^2} \times \text{m} = \frac{\text{kg m}^2}{\text{s}^2}$$

$$\therefore \text{Dimension of Enthalpy: } \frac{\text{ML}^2}{\text{T}^2} \text{ or ML}^2\text{T}^{-2}$$

ii. Acceleration due to gravity

$$\text{SI unit of accleration due to gravity} = \frac{\text{m}}{\text{s}^2}$$

$$\therefore \text{Dimension of accleration due to gravity} = \text{LT}^{-2}$$

iii. Nusselt number

$$N_{Nu} = \frac{hD}{k}$$

$$\text{SI unit of } h = \frac{\text{W}}{\text{m}^2\text{K}}$$

$$\text{SI unit of } D = \text{m}$$

$$\text{SI unit of } k = \frac{\text{W}}{\text{m K}}$$

$$\therefore \text{Unit of } N_{Nu} = \frac{\left(\frac{\text{W m}}{\text{m}^2\text{K}}\right)}{\left(\frac{\text{W}}{\text{m K}}\right)} = \frac{\left(\frac{\text{W}}{\text{m K}}\right)}{\left(\frac{\text{W}}{\text{m K}}\right)} = \text{No unit}$$

$\therefore$ Nusselt number is dimensionless.

iv. Density

$$\text{SI unit of density} = \frac{\text{kg}}{\text{m}^3}$$

$$\therefore \text{Dimension of density} = \text{ML}^{-3}$$

v. Frequency

$$\text{SI unit of frequency} = \frac{1}{\text{s}}$$

$$\therefore \text{Dimension of frequency} = \text{T}^{-1}$$

vi. Pressure

$$\text{SI unit of pressure} = \frac{\text{N}}{\text{m}^2}$$

$$\text{N} = \frac{\text{kg m}}{\text{s}^2}$$

$$\therefore \frac{\text{N}}{\text{m}^2} = \frac{\left(\frac{\text{kg} \cancel{\text{m}}}{\text{s}^2}\right)}{\text{m}^{\cancel{2}}} = \left(\frac{\text{kg}}{\text{m s}^2}\right)$$

$$\therefore \text{Dimension of pressure} = \left(\frac{\text{M}}{\text{LT}^2}\right) = \text{ML}^{-1}\text{T}^{-2}$$

vii. Viscosity

$$\text{SI unit of viscosity} = \frac{\text{kg}}{\text{m s}}$$

$$\therefore \text{Dimension of viscosity} = \frac{\text{M}}{\text{L T}} = \text{ML}^{-1}\text{T}^{-1}$$

viii. Momentum

$$\text{SI unit of momentum} = \frac{\text{kg m}}{\text{s}}$$

$$\therefore \text{Dimension of momentum} = \frac{\text{ML}}{\text{T}} = \text{MLT}^{-1}$$

ix. Power

$$\text{SI unit of power} = \text{Watt or W}$$

$$\text{W} = \frac{\text{J}}{\text{s}} = \frac{\text{N m}}{\text{s}} = \frac{\left(\frac{\text{kg m}}{\text{s}^2}\right) \times \text{m}}{\text{s}} = \frac{\text{kg m}^2}{\text{s}^3}$$

$$\therefore \text{Dimension of power} = \frac{\text{ML}^2}{\text{T}^3} = \text{ML}^2\text{T}^{-3}$$

x. Energy

$$\text{SI unit of energy} = \text{joule or J}$$

$$\text{J} = \text{N} \times \text{m} = \left(\frac{\text{kg m}}{\text{s}^2}\right) \times \text{m} = \frac{\text{kg m}^2}{\text{s}^2}$$

$$\therefore \text{Dimension of energy} = \frac{\text{ML}^2}{\text{T}^2} = \text{ML}^2\text{T}^{-2}$$

2. Convert the following:
   i. 1.5 Btu/lb_m °F to J/kg °C
   ii. 0.0175 ft^2/h to m^2/s
   iii. 1013.25 Pa to psia
   iv. 1.5 J/kg K to cal/g °C
   v. 3500 W to Btu/h

Solution

i. 1.5 Btu/lb_m °F to J/kg °C

$$1 \text{ Btu} = 1055 \text{ J}$$

$$1 \text{ lb}_\text{m} = 0.45359 \text{ kg}$$

$$1°\text{F} = (1.8 \times \text{T°C}) + 32$$

$$\therefore 1.5 \text{ Btu/lb}_\text{m} \text{°F} = \left(\frac{1.5 \times 1055}{0.45359}\right) \times 1.8 = 6280.2 \text{ J/kg °C}$$

After rounding to two significant figures, the corrected value is 6300 J/kg °C

ii. 0.0175 ft^2/h to m^2/s

$$1 \text{ ft} = 0.3048 \text{ m}$$

$$1 \text{ ft}^2 = 0.0929 \text{ m}^2$$

$$1 \text{ h} = 3600 \text{ s}$$

$$\therefore 0.0175 \text{ ft}^2/\text{h} = \frac{0.0175 \times 0.0929}{3600} = 4.52 \times 10^{-7} \text{ m}^2/\text{s} \quad \text{(after rounding to three significant figures).}$$

iii. 1013.25 Pa to psia

$$1 \text{ psia} = 6.895 \text{ kPa}$$

$$\therefore 6.895 \text{ kPa} = 1 \text{ psia}$$

$$6895 \text{ Pa} = 1 \text{ psia}$$

$$\therefore 1013.25 \text{ Pa} = \frac{1013.25}{6895} = 0.146954 \text{ psia} \quad \text{(after rounding to six significant figures).}$$

iv. 1.5 J/kg K to cal/g °C

$$1 \text{ cal} = 4.185 \text{ J}$$

$$\therefore 1\ \text{J} = \frac{1}{4.185} = 0.2389\ \text{cal}$$

$$1\ \text{kg} = 1000\ \text{g}$$

$$1.5\ \text{J/kg K} = \frac{1.5 \times 0.2389}{1000} = 0.00036\ \text{cal/g}\ ^\circ\text{C} \quad \text{(after rounding to two significant figures).}$$

v. 3500 W to Btu/h

$$1\ \text{Btu/h} = 0.29307\ \text{W}$$

$$\therefore 1\ \text{W} = \frac{1}{0.29307} = 3.4\ \text{Btu/h}$$

$$\therefore 3500\ \text{W} = 11900\ \text{Btu/h} \approx 12000\ \text{Btu/h}$$

3. A cylindrical tank is to be designed to hold 5000 L of soluble coffee extract. The tank has a length of two times its diameter. Determine the size of the cylindrical tank in meter and ft. Express the volume of the tank in cubic meter.

Given

i. Capacity of the tank = 5000 L
ii. Length of tank = two times its diameter, i.e., $L = 2 \times D$

To find

i. Size of the cylindrical tank in meter (m) and feet (ft)
ii. Volume of the tank in cubic meter

Solution

$$1\ \text{m}^3 = 1000\ \text{L}$$

$$\therefore 5000\ \text{L} = 5\ \text{m}^3$$

$$\text{Volume of cylindrical tank, } V = \pi r^2 L$$

$$r = \frac{D}{2}; \therefore L = 2D,\ D = \frac{L}{2}; \therefore r = \frac{L/2}{2} = \frac{L}{4}$$

$$\therefore V = \pi \times \left(\frac{L}{4}\right)^2 \times L = \frac{\pi L^3}{16}$$

$$L^3 = \frac{16V}{\pi} = \frac{16 \times 5}{\pi} = 25.465\ \text{m}^3$$

$$\therefore L = 2.942\ \text{m}$$

$$1\ \text{ft} = 0.3048\ \text{m}$$

$$\therefore 2.942\ \text{m} = 9.652\ \text{ft}$$

Answer: (i) Size of the cylindrical tank in meter = 2.942 m
(ii) Size of the cylindrical tank in feet = 9.652 ft
(iii) Volume of the cylindrical tank in cubic meter = 5 m^3

4. Calculate the quantity of liquid nitrogen that can be filled in a tank of dimensions, 28 in × 14 in × 10 in, when it is full. The density of liquid nitrogen is 0.807 g/mL. Express the density in kg/L.

Given

i. Dimension of tank = 28 in × 14 in × 10 in
ii. Density of liquid nitrogen = 0.807 g/mL

To find

i. Quantity of liquid nitrogen that can be filled in the tank in full capacity
ii. Density in kg/L

Solution

$$\text{Volume of the tank, } V = 28 \times 14 \times 10 = 3920 \text{ in.}^3$$

$$1 \text{ in} = 0.0254 \text{ m}$$

$$\therefore 1 \text{ in}^3 = 0.0000164 \text{ m}^3$$

$$3920 \text{ in}^3 = 0.0642 \text{ m}^3$$

$$1 \text{ mL} = 10^{-3} \text{ L}$$

$$1 \text{ m}^3 = 1000 \text{ L} = 10^6 \text{ mL}$$

$$\therefore\ 1 \text{ mL} = 10^{-6} \text{ m}^3$$

$$0.807 \text{ g/mL} = \frac{0.807 \times 10^{-3}}{10^{-3}} = 0.807 \text{ kg/L}$$

$$\text{Density} = \frac{\text{mass}}{\text{volume}}$$

$$\therefore \text{Mass} = \text{density} \times \text{volume} = \frac{0.807 \times 10^{-3}}{10^{-6}} \times 0.0642 = 51.8094 \text{ kg}$$

Answer: (i) Quantity of liquid nitrogen = 51.81 kg
(ii) Density in kg/L = 0.807 kg/L

5. Determine the number of significant digits in the following numeric values:

i. 0.00500
ii. 0.03040
iii. 239.59
iv. 505
v. 0.0052
vi. 0.05050
vii. 75381

viii. 549.04188000

ix. $45\bar{0}0$

x. 400.

Solution

i. 0.00500

Number of significant figures = three (According to ***Rule 4***, a number with a decimal point, trailing zeros, i.e., those to the right of the last nonzero digit are significant.)

ii. 0.03040

Number of significant figures = four (According to ***Rule 4***, a number with a decimal point, trailing zeros, i.e., those to the right of the last nonzero digit are significant.)

iii. 239.59

Number of significant figures = five

iv. 505

Number of significant figures = three (According to ***Rule 2***, zeros between nonzero digits are significant.)

v. 0.0052

Number of significant figures = two (According to ***Rule 4***, a number with a decimal point, trailing zeros, i.e., those to the right of the last nonzero digit are significant.)

vi. 0.05050

Number of significant figures = four (According to ***Rule 4***, a number with a decimal point, trailing zeros, i.e., those to the right of the last nonzero digit are significant.)

vii. 75381

Number of significant figures = five

viii. 549.04188000

Number of significant figures = 11 (According to ***Rule 4***, a number with a decimal point, trailing zeros, i.e., those to the right of the last nonzero digit are significant.)

ix. $45\bar{0}0$

Number of significant figures = three (According to ***Rule 5A***, an overbar is placed over the last significant figure which indicates that the trailing zeros following this are insignificant.)

x. 400.

Number of significant figures = three (According to ***Rule 5C***, a decimal point is placed after the number which specifies the number of significant figures.)

6. The thermal conductivity of copper is 386 W/m K. Convert this value in the CGS and British units of measurement.

Given

i. Thermal conductivity of copper = 386 W/m K

To find

ii. Conversion of thermal conductivity (k) value in CGS and British units of measurement.

Solution

iii. CGS

The unit of thermal conductivity in CGS unit system is cal/s cm °C

$$k = 386\ \frac{\text{W}}{\text{m K}} = 386\ \frac{\text{J}}{\text{s m K}}$$

From Table 1.12, 1 kcal = 4185 J; therefore, 1 J = 2.389 × 10^{-4} kcal = 0.239 cal

$$k = 386 \frac{\text{J}}{\text{s m K}} = \frac{386 \times 0.239}{10^2} = 0.922 \frac{\text{cal}}{\text{s cm °C}}$$

iv. British units

The unit of thermal conductivity in British unit system is BTU/h ft °F

From Table 1.12, 1 BTU = 1055 J and 1 ft = 0.3048 m; therefore, 1 J = 9.479 x 10-4 BTU and 1 m = 3.281 ft

$$k = 386 \frac{\text{J}}{\text{s m K}} = \frac{386 \times 9.479 \times 10^{-4}}{(1/3600) \times 3.281 \times (9/5)} = 223.035 \frac{\text{BTU}}{\text{h } ft°F}$$

After rounding to 4 significant figures, the corrected value is 223.0 BTU/h ft °F

Answer: (i) Value of thermal conductivity in CGS unit system: 0.922 cal/s cm °C

(ii) Value of thermal conductivity in British unit system: 223.0 BTU/h ft °F

7. Show the following expressions are dimensionless:

i. $N_{Re} = \frac{Dv\rho}{\mu}$

ii. $N_{Nu} = \frac{hD}{k}$

iii. $N_{Pr} = \frac{C_P\mu}{k}$

iv. $N_{Bi} = \frac{hL}{k}$

v. $N_{Fo} = \frac{\alpha t}{L^2}$

where, D = diameter, v = velocity, ρ = density, μ = viscosity, h = heat transfer coefficient, k = thermal conductivity, L = thickness, α = thermal diffusivity, and t = time.

Solution

i. $N_{Re} = \frac{Dv\rho}{\mu}$

$$\text{SI unit of } D = \text{m}$$

$$\text{SI unit of } v = \frac{\text{m}}{\text{s}}$$

$$\text{SI unit of } \rho = \frac{\text{kg}}{\text{m}^3}$$

$$\text{SI unit of } \mu = \frac{\text{kg}}{\text{m s}}$$

$$\therefore \text{Unit of } N_{Re} = \frac{\text{m} \times (\text{m/s}) \times (\text{kg/m}^3)}{(\text{kg/m s})} = \frac{\text{m}^3}{\text{m}^3} = \text{No units}$$

ii. $N_{Nu} = \frac{hD}{k}$

$$\text{SI unit of } h = \frac{\text{W}}{\text{m}^2\text{K}}$$

$$\text{SI unit of } D = \text{m}$$

$$\text{SI unit of } k = \frac{\text{W}}{\text{m K}}$$

$$\therefore \text{Unit of } N_{Nu} = \frac{\left(\frac{\text{W}}{\text{m}^2\text{ K}}\right)\times \text{m}}{\left(\frac{\text{W}}{\text{m K}}\right)} = \frac{\left(\frac{\text{W}}{\text{m K}}\right)}{\left(\frac{\text{W}}{\text{m K}}\right)} = \text{No units}$$

iii. $N_{Pr} = \frac{C_P \mu}{k}$

$$\text{SI unit of } C_P = \frac{\text{J}}{\text{kg K}}$$

$$\text{SI unit of } \mu = \frac{\text{kg}}{\text{m s}}$$

$$\text{SI unit of } k = \frac{\text{W}}{\text{m K}} = \frac{\text{J}}{\text{s m K}}$$

$$\therefore \text{Unit of } N_{Nu} = \frac{\left(\frac{\text{J}}{kg\text{ K}}\right)\left(\frac{\text{kg}}{\text{m s}}\right)}{\left(\frac{\text{J}}{\text{s m K}}\right)} = \frac{\left(\frac{\text{J}}{\text{s m K}}\right)}{\left(\frac{\text{J}}{\text{s m K}}\right)} = \text{No units}$$

iv. $N_{Bi} = \frac{hL}{k}$

$$\text{SI unit of } h = \frac{\text{W}}{\text{m}^2\text{K}}$$

$$\text{SI unit of } L = \text{m}$$

$$\text{SI unit of } k = \frac{\text{W}}{\text{mK}}$$

$$\therefore \text{Unit of } N_{Bi} = \frac{\left(\frac{\text{W}}{\text{m}^2\text{ K}}\right)\times \text{m}}{\left(\frac{\text{W}}{\text{m K}}\right)} = \frac{\left(\frac{\text{W}}{\text{m K}}\right)}{\left(\frac{\text{W}}{\text{m K}}\right)} = \text{No units}$$

v. $N_{Fo} = \dfrac{\alpha t}{L^2}$

$$\text{SI unit of } \alpha = \frac{\text{m}^2}{\text{s}}$$

$$\text{SI unit of } t = \text{s}$$

$$\text{SI unit of } L^2 = \text{m}^2$$

$$\therefore \text{Unit of } N_{Fo} = \frac{\left(\dfrac{\text{m}^2}{\text{s}}\right) \times \text{s}}{\text{m}^2} = \frac{\text{m}^2}{\text{m}^2} = \text{No units}$$

8. Round off the first five numerical values in problem number 5 to three significant digits and the remaining values to two significant digits.

Solution

i. $0.00500 \rightarrow 0.0050\underline{0}$
ii. $0.03040 \rightarrow 0.030\underline{4}$
iii. $239.59 \rightarrow 240$
iv. $505 \rightarrow 50\underline{5}$
v. $0.0052 \rightarrow 0.0052\underline{0}$
vi. $0.05050 \rightarrow 0.05\underline{0}$
vii. $75381 \rightarrow 7\underline{5}000$
viii. $549.04188000 \rightarrow 5\underline{5}0$
ix. $45\bar{0}0 \rightarrow 4\bar{5}00$
x. $400. \rightarrow 4\underline{0}0$

9. Complete the following table by rounding the original numerical value to (a) three significant figures, (b) two significant figures, and (c) one significant figure

Original Numerical Value	Three Significant Figures	Two Significant Figures	One Significant Figure
3.857			
54.63			
21.29			
632.51			
98.98			

Solution

Original Numerical Value	Three Significant Figures	Two Significant Figures	One Significant Figure
3.857	$3.8\underline{6}$	$3.\underline{9}$	$\underline{4}$
54.63	$54.\underline{6}$	$5\underline{5}$	$\underline{5}0$
21.29	$21.\underline{3}$	$2\underline{1}$	$\underline{2}0$
632.51	$63\underline{3}$	$6\underline{3}0$	$\underline{6}00$
98.98	$99.\underline{0}$	$9\underline{9}$	$\underline{1}00$

10. Check the dimensional consistency of the Einstein's equation for mass–energy conservation: $E = mc^2$.

Solution

$$E = mc^2$$

In the abovementioned equation

$$E = \text{energy};\ m = \text{mass};\ \ c = \text{velocity of light}$$

Dimensions of the left-hand side of the equation

$$\text{SI unit of } E = \text{J} = \text{N m} = \left(\frac{\text{kg m}}{\text{s}^2}\right) \times \text{m} = \frac{\text{kg m}^2}{\text{s}^2}$$

$$\text{Dimension of } E = \frac{\text{ML}^2}{\text{T}^2} = \text{ML}^2\text{T}^{-2}$$

Dimensions of the right-hand side of the equation

$$\text{SI unit of } m = \text{kg}$$

$$\text{SI unit of } c = \frac{\text{m}}{\text{s}}$$

$$\therefore \text{Unit of } c^2 = \frac{\text{m}^2}{\text{s}^2}$$

$$\text{Unit of } mc^2 = \frac{\text{kg m}^2}{\text{s}^2}$$

$$\text{Dimension of } mc^2 = \frac{\text{ML}^2}{\text{T}^2} = \text{ML}^2\text{T}^{-2}$$

Therefore, the dimensions of left- and right-hand sides of the equation are same. Hence, the equation is said to be dimensionally consistent.

BIBLIOGRAPHY

Bennett, B. 2018. The importance of power protection in the food and beverage industry. www.azom.com/article.aspx?ArticleID=15174 (accessed July 23, 2018).

Comptes Rendus de la 1e CGPM. 1889. 1890, 34.

Comptes Rendus de la 13e CGPM. 1967/1968. 1969, 103.

Comptes Rendus de la 17e CGPM. 1983. 1984, 98.

Dixit, S. 2006. Specifications in the flavour and fragrance industry. *The FAFAI Journal*, April–June, 1–5.

Jay, J. M. 1992. *Modern Food Microbiology*. Dordrecht: Springer.

Jay, J. M. 1998. *Modern Food Microbiology*. Boston, MA: Springer.

Jong, I. C. and Rogers, B. G. 1991. *Engineering Mechanics: Statics and Dynamics*. Philadelphia, PA: Saunders College Publishing.

Kuneš, J. 2012. *Dimensionless Physical Quantities in Science and Engineering*. London: Elsevier Inc.

Lewis, M. J. 1996. *Physical Properties of Foods and Food Processing Systems*. Cambridge: Woodhead Publishing, pp. 17–50.

Martinez-Monteagudo, S. I. and Balasubramaniam, V.M. 2016. Fundamentals and applications of high-pressure processing technology. In *High Pressure Processing of Food-Principles, Technology and Application*, eds. V. M. Balasubramaniam, G. V. Barbosa-Canovas, and H. Lelieveld, 3–17. New York: Springer LLC.

Melero, B., Diez, A. M. and Rovira, J. 2014. Basic sanitation. In *Handbook of Fermented Meat and Poultry*, eds. F. Toldrá, Y. H. Hui, I. Astiasaran, J. Sebranek, and R. Talon, 443–450. Chichester: John Wiley & Sons Ltd.

Mirica, K. A., Phillips, S. T., Mace, Ch. R. and Whitesides, G. M. 2010. Magnetic levitation in the analysis of foods and water. *Journal of Agriculture and Food Chemistry* 58: 6565–6569.

Myers, R. T., Oldham, K. B., Tocci, S. 2000. Chemistry. Austin, TX: Holt Rinehart, Winston.

NIST. 2006. The International System of Units (SI) – Conversion Factors for General Use. National Institute of Standards and Technology Special Publications 1038 Natl. Inst. Stand. Technol. Spec. Pub. 1038. Available through NIST Weights and Measures Division STOP 2600, Gaithersburg, MD.

NIST Handbook 130. 2016. Uniform Regulations. www.nist.gov/sites/default/files/documents/pml/wmd/pubs/2016/02/25/iva-pkglblgreg-16-h130-final.pdf (accessed July 23, 2018).

NIST Handbook 130. 2017. Uniform Laws and Regulations in the areas of legal metrology and engine fuel quality. As adopted by the 101st National Conference on Weights and Measures 2016. https://dx.doi.org/10.6028/NIST.HB.130-2017 (accessed July 23, 2018).

Rahman, M. S. 2007. Hygienic design and sanitation. In *Handbook of Food Preservation*, ed. M. S. Rahman, 957–968. Boca Raton, FL: CRC Press.

Singh, P. C., Singh, R. K., Smith, R. S. and Nelson, P. E. 1997. Evaluation of in-line sensors for selected properties measurements in continuous food processing. *Food Control* 8: 45–50.

Wei, G., Liu, J., Sun, J. and Shida, K. 2007. Estimating viscosity and density of ternary solution based on least-squares B-spline approximation. Instrumentation and Measurement Technology Conference Proceedings, IMTC 2007, Warsaw, Poland. IEEE, 1–5.

Weisstein, E. W. 2012. "Area" From: MathWorld—A Wolfram Web Resource. http://mathworld.wolfram.com/Area.html (accessed July 23, 2018).

Yamamoto, K., King, P. M., Wu, X., Mason, T. J. and Joyce, E. M. 2015. Effect of ultrasonic frequency and power on the disruption of algal cells. *Ultrasonics Sonochemistry* 24: 165–171.

2

Material Balance

Material balance refers to the method of accounting the mass of raw materials entering a process and the finished product and by-product streams exiting it. During food processing, the mass entering the system may undergo mixing, heating, drying, evaporation, fermentation, or any other unit operation. In a food industry, *material* comprises the ingredients, the final product, and the stock of ingredients or final product stored in the warehouse. The word *balance* means reconciliation, which involves matching the quantity of raw material received with the quantity of material produced. Thus, the concept of material balance is the physical interpretation and application of the law of conservation of mass which states that *matter is neither created nor destroyed.*

The purpose of conducting material or mass balance exercise in a food industry is manifold. The foremost of all is the traceability of a lot of product, ingredient, or packaging material. Traceability is critical in the current scenario to avoid risks in the supply chain and achieve compliance towards food safety. According to ISO 9001:2015, traceability is defined as the *ability to identify and trace the history, distribution, location, and application of products, parts, materials, and services.* A 100% conformity to mass balance holds importance as it precisely traces back to all the batches of a finished product where all of a contaminated batch of raw material was used. Apart from traceability, material balance is useful in calculating the ingredient quantities in a recipe or formulation, the composition of the final product, process yield, and efficiency (Farkas and Farkas, 1997). Furthermore, mass balance provides an estimate of the levels of solid food wastes emanating from a food processing plant (Mardikar and Niranjan, 1995; Niranjan, 1994). Therefore, obtaining a thorough understanding of the mass balance for a process would lead to the reduction of material wastage and consequent improvement in the process yield.

The concept of material balance can be fully appreciated only after understanding the fundamental principles of thermodynamics which is the branch of science that deals with the properties of matter. Terminologies in thermodynamics which are of relevance to this chapter are defined as follows:

2.1 Terminologies and Definitions

1. **System:** It (Figure 2.1) comprises of any matter identified for investigation in the universe.
2. **Surroundings:** Remainder of the universe which encloses the system is known as the surrounding (Figure 2.1).
3. **System boundary:** The boundary which separates the system and surroundings. The boundary may be real or imaginary. The outer jacket of a heating pan and the walls of a drying chamber are the examples of a real boundary. An imaginary boundary is not based on a physical control surface (Figure 2.1).
4. **Closed system:** A closed system (Figure 2.2) is one where no mass moves across the boundaries; however, energy can be transferred across the boundary.
5. **Open system:** It (Figure 2.3) is one where mass and energy can move across the boundaries (in or out) at some point during the process.
6. **Unit operation:** It can be defined as the subunit or stage of a process, which stands out with a specific function and based on a logical–physical principle.
7. **Process:** It is defined as a set of actions which occur in a specific sequence to attain a defined end. A process begins with the raw material and ends with the products and by-products. In

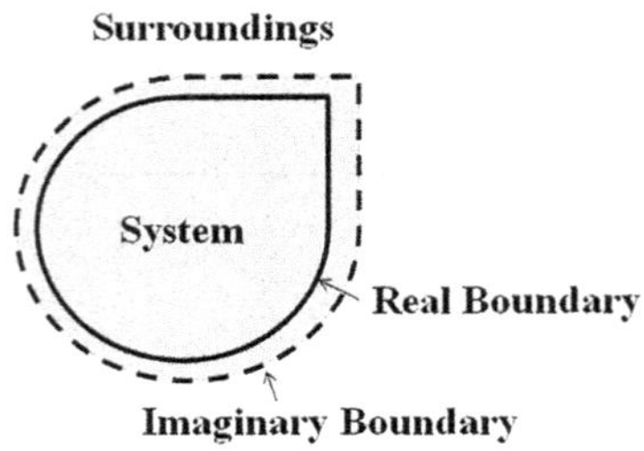

FIGURE 2.1 Schematic representation of a thermodynamic system, surroundings, and boundary.

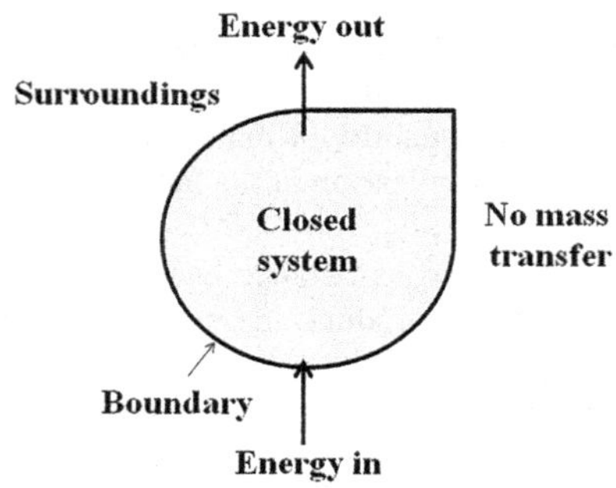

FIGURE 2.2 Schematic representation of a closed system.

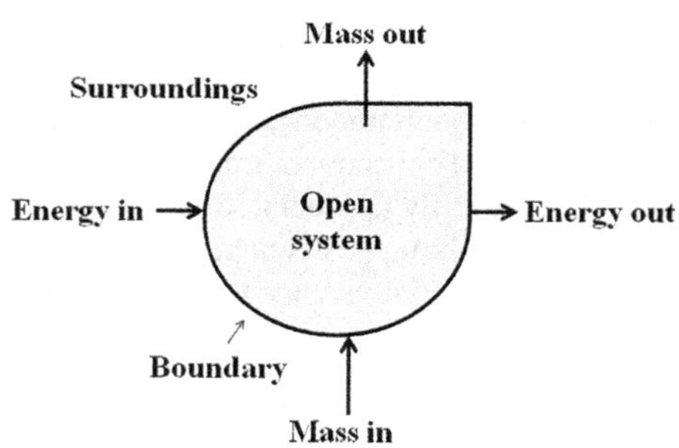

FIGURE 2.3 Schematic representation of an open system.

other words, a process is an engineering system which transforms the raw materials into desired products, through a series of logically related unit operations with specified functions (Evranuz and Kılıç-Akyılmaz, 2012).

8. **Steady state:** If all the properties of a system, such as temperature, pressure, concentration, volume, and mass, do not vary with time at any point of the system, the system or process is said to be at a steady state. Thus, if any variable of a steady-state system is monitored continuously, it can be noted that its value will remain constant with time. However, the properties may vary with position within the system.
9. **Unsteady state:** If all the process variables corresponding to each position within the system vary with time, the process or system is said to be at unsteady state.
10. **State of equilibrium:** A system is said to have attained its state of equilibrium when its properties do not change with time as the opposing forces acting on it are exactly counterbalanced. At equilibrium, there is no net change in the system or the universe, as the driving force for change is absent. Thus, during a processing operation, the state of equilibrium is avoided by disturbing the system so that there is a continuous transformation of raw material to the desired product.

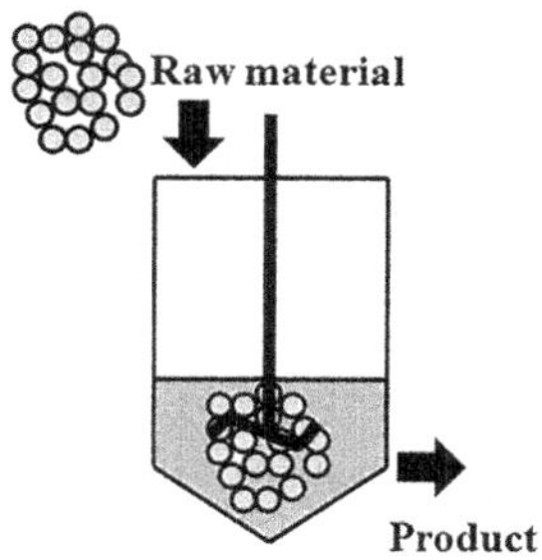

FIGURE 2.4 Schematic representation of a batch process.

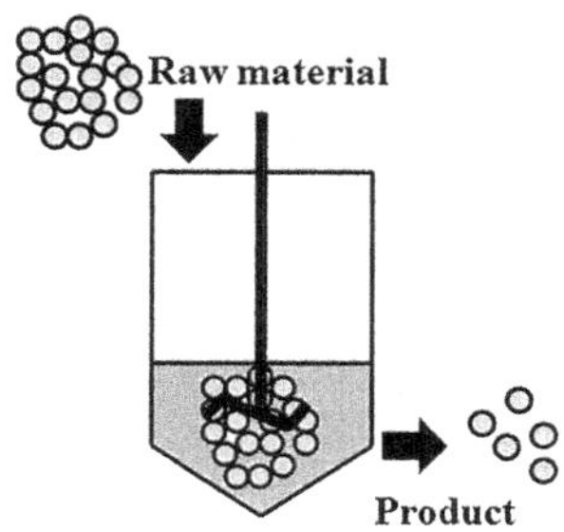

FIGURE 2.5 Schematic representation of a continuous process.

11. **Batch process:** In a batch mode operation, all the materials are added to the system at the beginning of the process; the system is then closed and the products are removed only when the process is complete (Figure 2.4). A batch process operates in a closed system. Thus, there is no exchange of mass between the system and the surroundings.

    For example: Small-scale baking of bread, biscuits, and cookies; freeze-drying or lyophilization; and dry blending of ingredients in a flavored beverage manufacturing unit

12. **Continuous process:** The material flows continuously in and out of the system, throughout the process (Figure 2.5). If the rates of mass input and output are equal, continuous processes can be operated indefinitely. Many of the food processing operations are continuous.

    For example: A continuous centrifuge separating whole milk into skim milk and cream in a dairy industry and spray drying for the production of instant food powders (e.g., instant coffee, milk powder).

13. **Semi-batch or fed-batch process:** This is a combination of batch and continuous processes. A semi-batch process allows either input or output of mass into the system during the processing time but not both.

    For example: Coating, granulation, and fermentation (media is added to increase the yield of production)

2.2 Fundamentals of Material Balance

According to the law of conservation of mass, the equation of material balance can be written as follows:

$$\begin{aligned}&\text{Mass of input material} + \text{mass of material generated} =\\&\text{mass of outflow material} + \text{mass of accumulated material} + \text{mass of material consumed} +\\&\text{mass of material lost}\end{aligned} \quad (2.1)$$

The inflow stream is constituted by the raw material(s) or ingredients and the outflow stream comprises of the processed product(s), waste, losses, and by-products. Accumulation in the system refers to decrease or increase in mass or moles with respect to time. Generation and consumption of mass are not commonly encountered in food processing, except with the involvement of a chemical reaction which results in the formation or consumption of chemical species. Thus, with respect to any food processing operation, Eq. (2.1) can be written as

$$\begin{aligned}\text{Mass of raw materials} = {} & \text{mass of products} + \text{mass of wastes} + \text{mass of} \\ & \text{by-products} + \text{mass of losses} + \text{mass of accumulated} \\ & \text{product}\end{aligned} \tag{2.2}$$

2.3 Classification of Material Balance Equations

Depending on the rate of accumulation, material balance equations can be classified into steady state and unsteady state.

2.3.1 Steady-State Material Balance

As discussed previously, under steady-state conditions, the flow rate and composition of the mass/material remain constant with time. Therefore, the mass of accumulation and losses is zero. Thus, Eq. (2.2) is further reduced to result in Eq. (2.3):

$$\text{Mass of raw materials} = \text{mass of products} + \text{mass of wastes} + \text{mass of by-products} \tag{2.3}$$

Batch and semi-batch processes cannot operate under steady-state conditions as the mass of the system increase or decrease with time, despite the total mass remains constant. However, it is feasible to carry out a continuous process at steady state (Figure 2.6) or near steady-state conditions (Doran, 1995).

2.3.2 Unsteady-State Material Balance

In this case, mass or material accumulates in the system and the concentration or quantity of material varies with the time. Thus, the material balance equation retains the term related to the accumulation of component (as in Eq. (2.2)). In this case, the accumulation term is expressed as a differential term involving the rate of change of the system property with time. Then, the material balance equation is integrated with respect to time to obtain an equation for the value of the dependent variable.

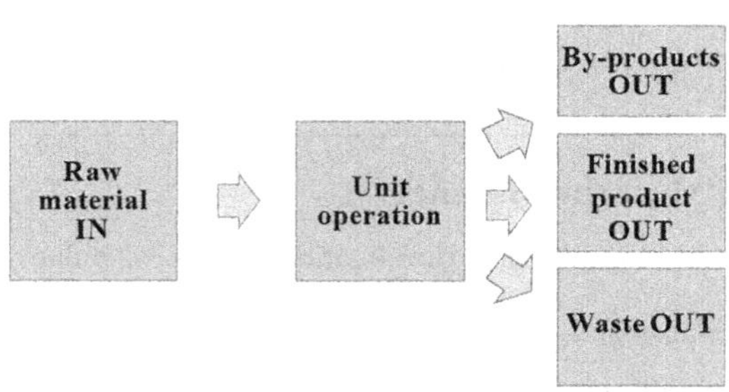

FIGURE 2.6 Schematic representation of a continuous and steady-state process.

2.4 Methodology for Conducting a Material Balance Exercise

The steps involved in carrying out a material balance exercise are depicted in Figure 2.7.

2.4.1 Data Collection

The primary step in formulating a material balance equation is to understand the process. Thus, the first step involves the collation of all known data on the type of unit operation, mass, and composition of inflow and outflow streams of the process as understood from the problem statement.

2.4.2 Construction of Block Diagram

The next step is to identify and set the system boundaries. This can be accomplished by constructing a block diagram (Figure 2.8) to appreciate the various physical principles involved in a process. A block diagram is the simplest form of a process flow sheet. It schematically describes the different steps of the process along with the input and output streams entering into and emerging from these steps, respectively. *Blocks* or *rectangles* are used in the block diagrams to represent the different stages of the process, with all the known data included in the illustration (Figure 2.8). In the block diagram, the unit operation (stage) involved in the process is written within the block or rectangle and the concerned inflow/outflow material to and from the stage is indicated by arrows. At this stage, it is important to select the unit system to be used further in the material balance exercise and express all the quantities involved using the chosen units in a consistent manner. Quantities along with the units should also be indicated in the block diagram.

2.4.3 Selection of Basis and Tie Materials

The next step in solving the material balance problem is the selection of an appropriate *basis for calculation* and identification of the *tie material.* The basis is defined as the reference for a specific calculation. It is a number, relative to which all the other values are calculated. The basis is especially useful when the initial quantity is unknown and the expected outcome is in terms of percentage or mass fraction. In the case of batch processes, the mass of incoming raw material is usually chosen as the basis. Generally, numbers that are convenient to handle during the calculation are preferred as the basis. Usually, with the mass of solids or liquids, 1 kg or 100 kg is used as the basis. Sugar, fat, protein, and salt are the other components involved in material balance problems pertaining to food processing operations.

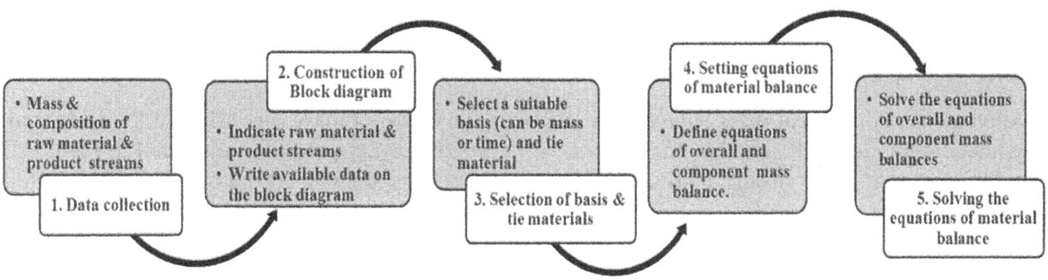

FIGURE 2.7 Steps in conducting a material balance exercise.

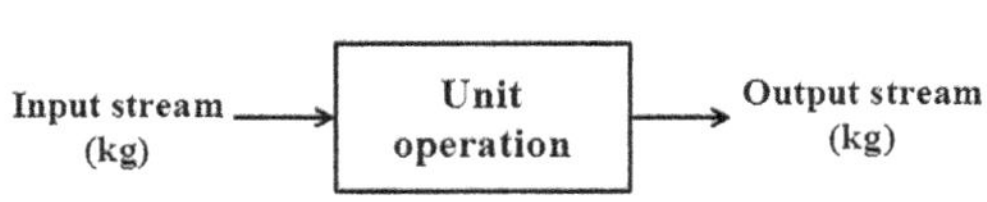

FIGURE 2.8 Block diagram for material balance calculation.

For continuous operations, the mass flow rate of raw material or product (kg/h) or a fixed time frame of operation is normally used as the basis of calculation (1 h or 1 day). Here, the material balance will be arrived based on the input to the system or output of the system during this period. In multistage processes, a block diagram for the process is drawn. System boundaries are moved along the parts of the system to determine the unknown quantities in the material balance. Here, the basis is defined at each stage of the process. It is important to mention the chosen basis while beginning to solve a material balance problem.

Tie material is a component which is used to relate the quantity of one process stream to that of the other. It is the component which does not change from one stream to another. Understanding the tie materials in a process helps in simplifying the problem and determining unknown quantities. For example, in drying, which involves removal of water, the mass of solids remains constant and hence can be considered as the tie material. In certain cases, such as the mixing process the tie material need not be identified.

2.4.4 Setting Up the Equations of Material Balance

To set the simultaneous equations (equations involving two or more unknowns that should have the same values in each equation), the material balance is devised on the entire system, known as the overall mass balance. The material balance devised with respect to individual components is termed as the component mass balance. At this stage, the assumptions, if any, have to be stated. The commonly used assumption is that the system is at steady state, and consequently, the mass flow rates and compositions of the streams do not change with time and the accumulation term is zero.

2.4.4.1 Overall Mass Balance

The overall mass balance considers the totality of the mass of process streams that enter and exit the system (Figure 2.9).

In Figure 2.9, r_1 and r_2 are the inflow streams and p_1 and p_2 are the outflow streams. If these streams are entering and exiting the system at the mass flow rates of $\dot{m}_{r1}$, $\dot{m}_{r2}$, $\dot{m}_{p1}$, and $\dot{m}_{p2}$ (e.g., in the units of kg/s) respectively, according to the law of conservation of mass

$$\dot{m}_{r1} + \dot{m}_{r2} = \dot{m}_{p1} + \dot{m}_{p2} \tag{2.4}$$

In other words, the sum of mass flow rates of the inflow streams is equal to the sum of mass flow rates of the outflow streams. Equation 2.4 represents the overall material balance around the system depicted in Figure 2.9.

2.4.4.2 Component Mass Balance

To establish the component mass balance, the law of conservation of mass is considered for each of the individual components of a stream. For example, if the streams r_1, r_2, p_1, and p_2 indicated in Figure 2.9 contain a common component A of concentration x, then the component material balance on the system with respect to A is written as

$$x_{r1}\dot{m}_{r1} + x_{r2}\dot{m}_{r2} = x_{p1}\dot{m}_{p1} + x_{p2}\dot{m}_{p2} \tag{2.5}$$

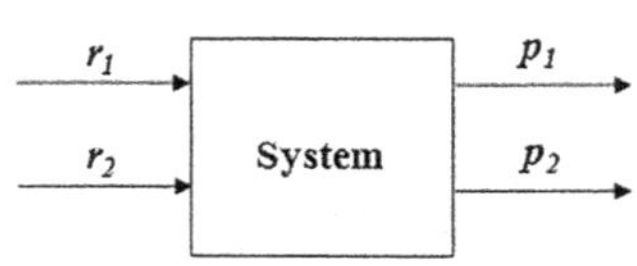

FIGURE 2.9 Schematic of a process with two inflow and two outflow streams.

If n numbers of components are present in the system, then $(n-1)$ numbers of component balance equations can be formulated in addition to the overall mass balance equation. To determine the mass and composition of the inflow and outflow streams, these equations must be solved simultaneously.

The concentration of an individual component in a stream is conventionally expressed in terms of either mass fraction or mole fraction. Mass fraction of a given component is defined as its mass expressed as a fraction of the total mass of the stream containing that component. If a stream contains a mixture of the components A and B of mass m_A and m_B, respectively, then the mass fraction of component A is given by

$$x_A = \frac{m_A}{m_A + m_B} \tag{2.6}$$

And, the mass fraction of component B is given by

$$x_B = \frac{m_B}{m_A + m_B} \tag{2.7}$$

Similarly, mole fraction can be defined as the number of moles of the component expressed as a fraction of the total number of moles in the stream. If the stream contains n_A moles of A and n_B moles of B, mole fractions of A and B are calculated according to Eqs. (2.8) and (2.9), respectively.

$$x_A = \frac{n_A}{n_A + n_B} \tag{2.8}$$

$$x_B = \frac{n_A}{n_A + n_B} \tag{2.9}$$

Conventionally, the notation x is used to denote the mass or mole fraction of a component in liquid state, and the notation y is used for the gaseous state.

It is always desirable to write one component balance equation with respect to the tie material, as it simplifies the solution and facilitates the determination of the unknowns (Toledo, 2007).

2.4.4.3 Recycle and Bypass

A recycle stream can be defined as a portion of the product stream which is separated and then returned to mix again with the fresh feed and reenter the system. For instance, during drying, hot air exiting the dryer is separated, recycled, and allowed to mix with the fresh air entering the dryer. The mixing is carried out using steam in a heat exchanger until the air attains a defined humidity and temperature.

The composition of a recycle stream may be similar or different from that of the product stream. The separation of the recycle stream (R) from the gross product stream (G) is achieved by distillation, filtration, or extraction. The input of the inlet feed (T) to the process is made up by mixing the fresh feed (F) and recycle stream (R) as shown in Figure 2.10.

In processes that involve recycling, material balance calculations can be made around the entire process (or) the mixing unit (or) the process (or) the separation unit. The *recycle ratio* or *reflux ratio* is commonly used in the material balance calculations of processes that involve recycling of streams. It is given by

$$\text{Recycle ratio} = \frac{\text{Amount of recycle}}{\text{Amount of net product}} = \frac{R}{P} \tag{2.10}$$

The quantity of net product (P) depends only on the quantity of the fresh feed. The recycle stream circulates inside the process at a constant flow rate and composition under steady-state conditions. The quality

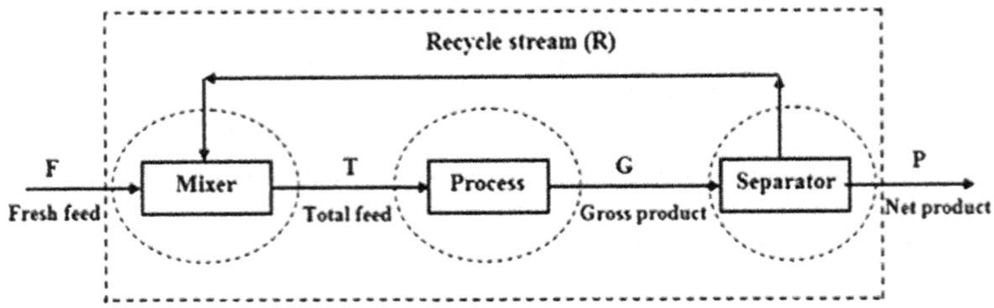

FIGURE 2.10 Block diagram for a process involving recycle of stream.

of the recycle stream depends on the process conditions such as conversion, and it is fixed according to the economic considerations.

Recycling is of significance with respect to the economics of the process. It also increases the rate of conversion of a reactant by enhancing its chances to react and convert to the product. Other reasons for recycling include the following:

- Recovery of valuable materials
- Conservation of energy by recycling high-temperature carrier fluids
- Improvement of temperature control over a process
- To decrease the inlet concentration of a given component by diluting it to a defined level

A bypass stream is a fraction of the fresh feed that skips one or more stages of the process and fed forward into a later stage of the process (Figure 2.11). By varying the proportion of the bypassed feed, the composition and properties of the product can be altered. Bypass is useful for reducing the extent of conversion of input materials or for providing improved control over the temperature of the streams.

Example 2.1

The drying rate of a continuous dryer is 12 kg/min of a food product containing 20% moisture (on a wet basis) to result in a product containing 10% moisture. As the handling capacity of the dryer is limited to a moisture content of less than 15%, a part of the dry product is recycled and mixed with the fresh feed. Calculate the recycle ratio.

Solution

Let *W*, *R*, and *P* be the mass flow rates (kg/min) of evaporated water, recycle, and product, respectively.

From Figure 2.12, an overall material balance over the entire process around boundary 1 is given by

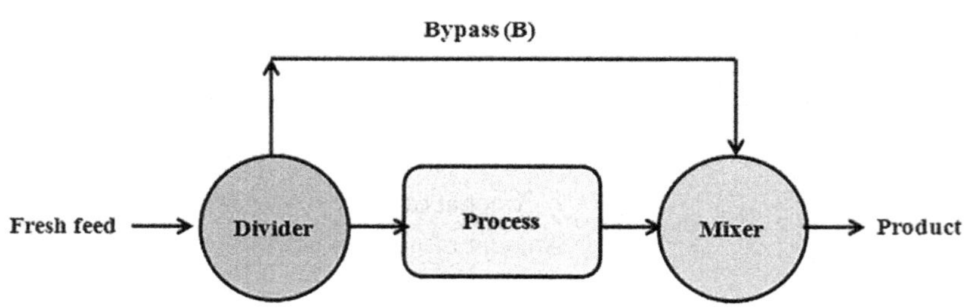

FIGURE 2.11 Block diagram for a process involving bypass stream.

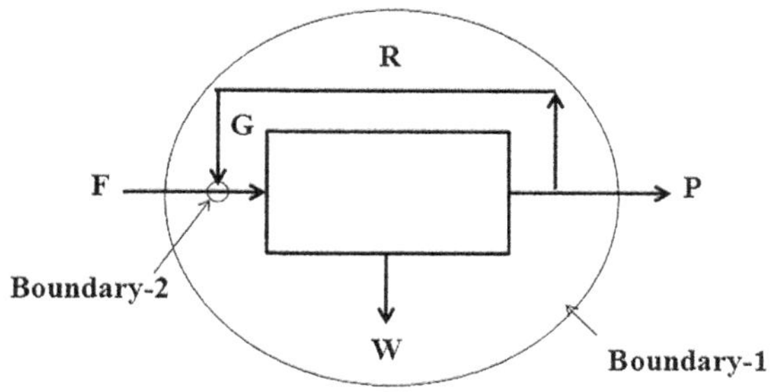

FIGURE 2.12 Material balance around the dryer.

$$F = W + P$$

$$12 = W + P$$

The component material balance with respect to water is

$$0.20 \times 12 = W + 0.1\ P$$

Therefore, $2.4 = W + 0.1\ (10 - W)$

Thus, the evaporation rate is W = 1.55 kg/min.

The recycle stream mixes with the fresh feed to give a combined feed rate to the dryer. Nevertheless, the overall material balance over the whole process does not give any information about the recycle within the boundary 1, and therefore, a second overall material balance around boundary 2 is required. Thus,

$$12 + R = G$$

$$(0.20 \times 12) + (0.1 \times R) = (0.15)G$$

$$2.4 + 0.1\ R = 0.15(10 + R)$$

$$R = 18\ \text{kg/min}$$

Thus, the recycle ratio is equal to 18/12 or 1.5.

2.4.5 Solving the Equations of Material Balance

Solution to the material balance problem is obtained by solving the overall mass balance and component balance equations simultaneously. The equations should be written in an order such that the calculations are simplified. For example, the equation with only one unknown is written first, as it can be solved at once, thereby eliminating an unknown from the subsequent calculations.

2.4.6 Material Balance for a Drying Process

The following section presents a stepwise procedure to establish the material balance for a drying process with an example.

Example 2.2

Determine the weight reduction that would result when potatoes are dried from 85% to 5% moisture.

Solution

- **Unit operation:** Drying
- **System:** Dryer
- **Type of process:** Batch
- **Boundary:** Real boundary formed by the walls of the drying chamber
- **Raw material:** Potatoes having moisture content of 85%
- **Product:** Dried potatoes having moisture content of 5%
- **Waste stream:** Moisture removed from the potatoes during drying.
- In this example, moisture and solid content of potatoes are the two components. The decrease in moisture content will result in a corresponding increase in the solid concentration.
- The solid content is considered as the tie material.

In the block diagram shown in Figure 2.13,

R = Mass of potatoes fed to the dryer (kg)
P = Mass of dried potato chips (kg)
W = Mass of moisture removed during drying (kg)
MC = Moisture content (%)

In the given example, the mass of potatoes (in kg) entering the dryer is assumed and considered as the basis for calculation
Basis: 100 kg of potato inflow into the dryer.
$\therefore R = 100\,\text{kg}$

Overall mass balance equation

$$R = W + P \tag{2.11}$$

i. **Component balance with respect to moisture content**

$$(0.85)R = (1)W + (0.05)P \tag{2.12}$$

ii. **Component balance with respect to solid content (tie material)**

$$(0.15)R = (0)W + (0.95)P \tag{2.13}$$

(or)

$$(0.15)R = 0.95(P) \tag{2.14}$$

$$\therefore P = 0.16(R) \tag{2.15}$$

Thus, substituting Eq. (2.15) in Eq. (2.12),

$$(0.85)R = W + (0.05)(0.16)(R) \tag{2.16}$$

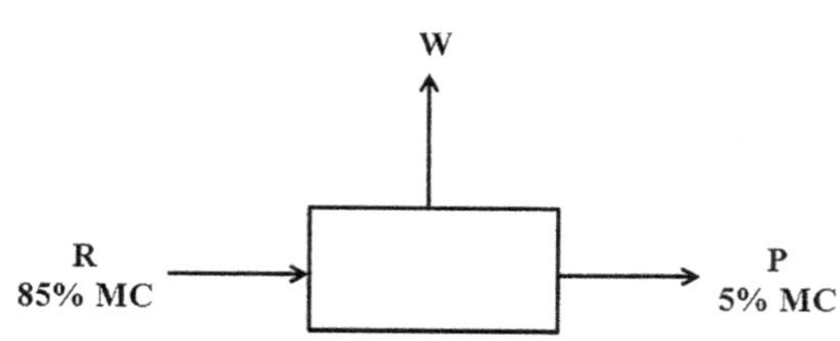

FIGURE 2.13 Block diagram to explain the process flow during drying of potato chips.

Since, $R = 100$,

$$W = 84.2 \text{ kg}$$

From Eq. (2.11),

$$P = 100 - 84.2 = 15.8 \text{ kg}$$

$$\therefore P = 15.8 \text{ kg}$$

Hence, weight reduction on drying of potato is $100 - 15.8 = 84.2\,\text{kg}$.

2.4.7 Material Balance for a Mixing Process

Explained in the following example are the steps involved in estimating the amounts of different ingredients to be mixed to prepare a food product.

Example 2.3

A 100 kg of a frankfurter formulation is to be prepared from lean beef, pork fat, and water. The composition of ingredients and the required composition of the final product are as follows:

- Lean beef—15% fat, 65% water, 20% protein
- Pork fat—85% fat, 10% water, 5% protein
- Frankfurter—20% fat, 15% protein, 65% water

Calculate the amount of lean beef, pork fat, and water required to prepare the frankfurter formulation.

Solution

Step 1: Data collection
From the preceding problem statement, the following information can be inferred:

- **Unit operation:** Mixing
- **Type of process:** Batch
- **System:** Mixer
- **Boundary:** Real boundary formed by the walls of the mixer
- **Raw materials:** Lean beef, pork fat, and water (of known composition as given in the problem statement)
- **Product:** 100 kg of frankfurter (of known composition as given in the problem statement)

Step 2: Construction of block diagram (See Figure 2.14)
Step 3: Selection of basis
Basis: 100 kg of frankfurter with composition 20% fat, 15% protein, 65% water.
Step 4: Setting up the equations of material balance

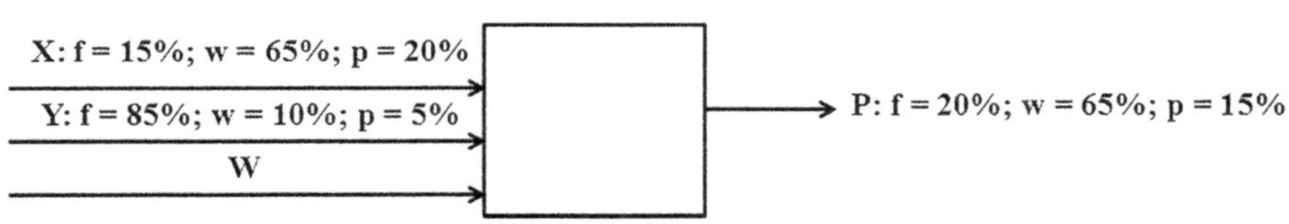

FIGURE 2.14 Block diagram for the mixing process.

i. **Overall mass balance**

In the block diagram shown in Figure 2.14,

P = amount of frankfurter produced (kg)
X = amount of lean beef used to prepare the frankfurter (kg)
Y = amount of pork fat used to prepare the frankfurter (kg)
W = amount of water used to prepare the frankfurter (kg)
f = fat content (%)
w = water content (%)
p = protein content (%)

Therefore, the overall mass balance can be written as

$$X + Y + W = P \tag{2.17}$$

Since, $P = 100\,\text{kg}$ (given)

$$\therefore X + Y + W = 100 \tag{2.18}$$

ii. **Component mass balance**

$P = 100\,\text{kg}$ (given data)

I. **Protein:** $0.20(X) + 0.05(Y) + W(0) = 0.15(100)$

$$0.20(X) + 0.05(Y) = 15 \tag{2.19}$$

II. **Fat:** $0.15(X) + 0.85(Y) + W(0) = 0.20(100)$

$$0.15(X) + 0.85(Y) = 20 \tag{2.20}$$

III. **Water:** $0.65(X) + 0.10(Y) + W(1) = 0.65(100)$

$$0.65(X) + 0.10(Y) + W = 65 \tag{2.21}$$

Step 5: Solving the equations of material balance

First, solving the simultaneous Eqs. (2.19) and (2.20) and changing the sign of Eq. (2.20), in order to eliminate one of the unknown variables, we get

$$(0.20 \times 17)(X) + (0.05 \times 17)(Y) = (15 \times 17)$$
$$(-)\ (0.15 \times 1)(X)\ (-)\ (0.85 \times 1)(Y) = (-)(20 \times 1)$$
$$3.4X \cancel{+0.85Y} = 255$$
$$-0.15X \cancel{-0.85Y} = -20$$
$$3.25X = 235$$

$$\therefore X = 72.31\,\text{kg}$$

Substituting the value of $X = 72.31\,\text{kg}$ in Eqs. (2.19) or (2.20), the value of Y can be obtained. Accordingly,

$$Y = 10.76\,\text{kg}$$

Substituting the values of $X = 72.31\,\text{kg}$ and $Y = 10.76\,\text{kg}$ in the Eq. (2.18), the value of W can be obtained. Accordingly,

$$W = (100) - (72.31 + 10.76)$$

$$\therefore W = 16.93\,\text{kg}$$

Hence, for the preparation of 100 kg frankfurter with 20% fat, 15% protein, and 65% water, 72.31 kg of lean beef 10.76 kg of pork fat and 16.93 kg of water are required.

This problem can also be solved using the determinant method. The matrices for the protein and fat balance equations are

$$\begin{bmatrix} 0.15 & 0.85 \\ 0.2 & 0.05 \end{bmatrix} \begin{bmatrix} X \\ Y \end{bmatrix} = \begin{bmatrix} 20 \\ 15 \end{bmatrix}$$

$$X = \frac{\begin{vmatrix} 20 & 0.85 \\ 15 & 0.05 \end{vmatrix}}{\begin{vmatrix} 0.15 & 0.85 \\ 0.2 & 0.05 \end{vmatrix}} = \frac{(20)(0.05) - (15)(0.85)}{(0.15)(0.05) - (0.20)(0.85)} = 72.31$$

$$Y = \frac{\begin{vmatrix} 0.15 & 20 \\ 0.2 & 15 \end{vmatrix}}{\begin{vmatrix} 0.15 & 0.85 \\ 0.2 & 0.05 \end{vmatrix}} = \frac{(0.15)(15) - (0.2)(20)}{(0.15)(0.05) - (0.20)(0.85)} = 10.76$$

From the overall mass balance: $W = 100 - 72.31 - 10.76 = 16.93\,\text{kg}$

2.4.8 Material Balance for an Evaporation Process

The following discussion is on setting up a material balance around a single-effect evaporator which concentrates a dilute liquid food product. The application of material balance, in this case, is to calculate the amount of water evaporated and the product collected per unit time of the process.

Example 2.4

A single-effect evaporator concentrates 500 kg/h of juice from 10% to 40%. Calculate the amount of water evaporated and the product collected per hour.

Solution

Step 1: Data collection

- **Unit operation:** Evaporation
- **Type of process:** Continuous
- **System:** Single-effect evaporator
- **Boundary:** Real boundary formed by the outer jacket of the evaporator chamber
- **Raw material:** 500 kg juice of concentration 10%
- **Product:** Juice of concentration 40%
- **Waste stream:** Water removed from juice during evaporation

Step 2: Construction of block diagram (Figure 2.15)

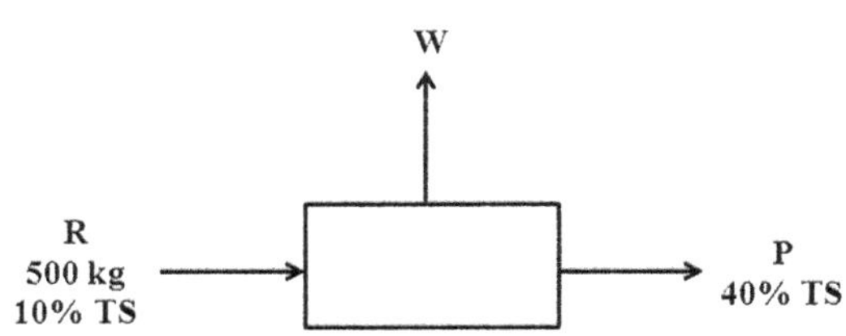

FIGURE 2.15 Block diagram for the evaporation process.

Step 3: Selection of basis and tie materials
Basis: 1 h of operation
In the block diagram,

R = Amount of dilute orange juice fed to the evaporator (kg)
P = Amount of orange juice concentrate (kg)
W = Amount of water removed from the orange juice during evaporation (kg)
TS = Total solid content (%)

Tie material: During evaporation, water is removed from the juice to concentrate it. The mass of solids remain constant and hence can be considered as the tie material.

Step 4: Setting up the equations of material balance

i. **Overall mass balance**

$$R = P + W \tag{2.22}$$

Since, $R = 500\,\text{kg}$ (given)

$$500 = P + W \tag{2.23}$$

ii. **Component mass balance**
In this system, water and solids are the two components of the material involved (inflow of dilute juice and outflow of concentrated juice). Thus, writing the component mass balance equations with respect to water and solid content as follows:
I. **Solids**

$$(0.10)R = (0.40)P + (0)W \tag{2.24}$$

$$(0.10)(500) = (0.40)P$$

$$50 = (0.40)P \tag{2.25}$$

II. **Water**

$$(1-0.10)R = (1-0.40)P + (1)W$$

$$(0.90)R = (0.60)P + W \tag{2.26}$$

$$(0.90)(500) = (0.60)P + W$$

$$450 = (0.60)P + W \tag{2.27}$$

Step 5: Solving the equations of material balance
From Eq. (2.25),
$P = 50/0.40 = 125$
$\therefore P = 125\,\text{kg}$
Substituting the value of $P = 125\,\text{kg}$ in Eq. (2.25),
$450 = (0.60)(125) + W$
$\therefore W = 375\,\text{kg}$
Amount of water evaporated per unit time: 375 kg
Amount of product collected per unit time: 125 kg

In many instances, two or more unit operations may be combined to result in a final product. The principle and steps for the calculation remain the same. The block diagram should represent the different stages. Material balance for such cases is explained with the following examples:

Example 2.5

Onion slices containing 85% moisture is dehydrated in the dryer using hot air. The dry product has 5% moisture content. Find the quantity of fresh onion required to produce 1 ton dry product per day. Preparation losses like peeling and trimming will be 10%.

Solution

Step 1: Data collection

- **Unit operations:** (1) Slicing (peeling and trimming), (2) drying
- **Type of process:** Continuous
- **System:** (1) Slicer, (2) dryer
- **Boundary:** Real boundary formed by the walls of the slicer and drying chamber
- **Raw material:** To the dryer: onion slices containing 85% moisture content
- **Product:** Dry onion containing 5% moisture content at the rate of 1 ton/day
- **Losses:** From slicer: 10% preparation losses (peels and trimmings)
- **Waste stream:** From dryer: moisture removed from the onion slices during drying

Step 2: Construction of block diagram

In the block diagram shown in Figure 2.16,

R = Amount of fresh onion fed to the slicer (kg)
L = Amount of onion peel waste (kg)
X = Amount of onion after slicing (kg)
P = Amount of dried onion slices (kg)
W = Amount of moisture removed during drying (kg)
M = Moisture content (%)

Step 3: Selection of basis and tie materials
Basis: 1 day
Tie material: Solid content of onion slices
Step 4: Setting up the equations of material balance

i. **Overall mass balance**

$$\text{Overall mass balance around the slicer: } R = X + L \tag{2.28}$$

$$R = X + (0.10)R$$

$$\therefore X = 0.9(R) \tag{2.29}$$

$$\text{Overall mass balance around the dryer: } X = P + W \tag{2.30}$$

Since $P = 1$ ton (or) 1000 kg,

$$X = 1000 + W \tag{2.31}$$

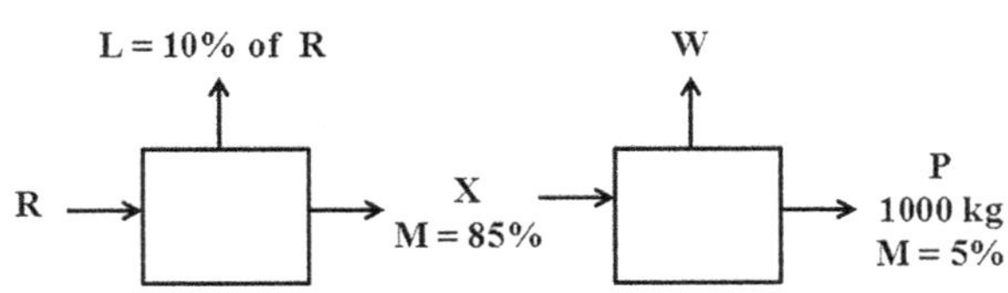

FIGURE 2.16 Block diagram for the onion dehydration process.

ii. **Component mass balance**

In this system, moisture and solid content of onion slices are the two components. The component mass balance equations are considered around the dryer as the mass of abovementioned components are changed only during the drying operation.

I. **Moisture balance**

$$(0.85)X = (0.05)P + (1)W \tag{2.32}$$

II. **Solid balance**

$$(0.15)\,X = (0.95)P + (0)W \tag{2.33}$$

$$(0.15)X = (0.95)(1000)$$

$$(0.15)X = 950 \tag{2.34}$$

Step 5: Solving the equations of material balance

From Eq. (2.34), $X = 950/0.15$

$\therefore X = 6333.33\,\text{kg}$

From Eq. (2.29), $R = X/0.9 = 6333.33/0.9 = 7037.04\,\text{kg}$

$\therefore R = 7037.04\,\text{kg}$

From Eq. (2.28),

$L = 7037.04 - 6333.33 = 703.71\,\text{kg}$

$\therefore L = 703.71\,\text{kg}$

From Eq. (2.31), $W = 6333.33 - 1000$

$\therefore W = 5333.33\,\text{kg}$

Thus, 7037.04 kg of fresh onion was taken for the processing. In the first stage of slicing, 703.71 kg was separated as waste in the form of peels and trimmings. Subsequently, 6333.33 kg of onion slices were dried to yield 1 ton of dried product per day, during which 5333.33 kg of moisture was removed.

Example 2.6

For the production of marmalade, fruit pulp is mixed with sugar and pectin and the mixture is boiled to about 65% solid concentration. Find the amount of fruit pulp, sugar, and pectin that must be used for the production of 100 kg marmalade, if the solid content of fruit pulp is 10%, the ratio of sugar to fruit pulp in the recipe is 56:44, and the ratio of sugar to pectin is 100.

Solution

Step 1: Data collection

- **Unit operations:** (1) Mixing, (2) evaporation
- **Type of process:** Batch
- **System:** (1) Mixer, (2) evaporator
- **Boundary:** Real boundary formed by the walls of mixer and evaporator
- **Raw material:** (1) To mixer: fruit pulp with 10% solids, sugar, and pectin
  (2) To evaporator: mixture of fruit pulp, sugar, and pectin
- **Product:** (1) From mixer: mixture of fruit pulp, sugar, and pectin
  (2) From evaporator: marmalade of 65% solid concentration
- **Waste stream:** From evaporator: moisture removed during the boiling and concentration of the mixture of fruit pulp, sugar, and pectin
- **Ratio of sugar to fruit pulp:** 56:44
- **Ratio of sugar to pectin:** 100

Step 2: Construction of block diagram

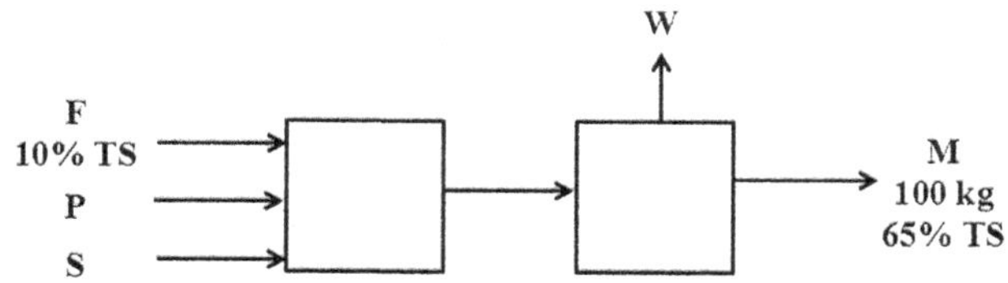

FIGURE 2.17 Block diagram for the manufacturing of marmalade.

Step 3: Selection of basis and tie materials
Basis: 100 kg of marmalade
In the block diagram shown in Figure 2.17,

F = Amount of fruit pulp (kg)
P = Amount of pectin (kg)
S = Amount of sugar (kg)
W = Amount of water removed during marmalade-making (kg)
M = Amount of marmalade produced (kg) = 100 kg
TS = Total solid content (%)

Tie material: Solid content of the (fruit pulp + sugar + pectin) mixture
Step 4: Setting up the equations of material balance

i. **Overall mass balance**

$$F + S + P = M + W \tag{2.35}$$

ii. **Component mass balance**

The solid balance equations are considered around the evaporator as the mass of solids is altered only during the evaporation.
Component mass balance around the evaporator

Solid balance

$$(0.10)F + (1)S + (1)(P) = (0.65)M + (0)W \tag{2.36}$$

$$0.10F + S + P = 65 \tag{2.37}$$

Step 5: Solving the equations of material balance
It is given that the ratio of sugar to fruit pulp in the recipe is 56:44.
Therefore, $\frac{F}{S} = \frac{44}{56}$ and $F = \frac{44}{56}S$

Substituting $F = \frac{44}{56}S$ and $P = 0.01S$ in Eq. (2.37),

$$\left\{(0.10)\left(\frac{44}{56}\right)S\right\} + S + 0.01S = 65$$

$S = 59.71$ kg
And, $F = 46.92$ kg (calculated from the known ratio of fruit pulp to sugar)
Since, the ratio of sugar to pectin is 100
$\therefore\ P = 0.597$ kg
Therefore, for the production of 100 kg of marmalade with 65% solid concentration, 59.71 kg sugar, 46.92 kg of fruit pulp, and 0.597 kg of pectin are required.

2.5 Material Balance for Food Standardization

Standardization can be defined as a *process which is done to ensure that a consistent level of product quality is maintained throughout the period of its availability*. The most common example of food standardization is that of milk practised in the dairy industry. The concentration of fat and protein in milk vary with season and other factors. Hence, standardization of milk is essential to maintain a constant fat content and consistency, irrespective of the period of supply. According to the Food and Drug Administration (CFR, 2017), standardized milk is the lacteal secretion obtained by the complete milking of one or more healthy cows, that in the final package form for beverage use shall contain not less than 8.25% of milk solids not fat and not less than 3.25% milk fat. Milk may be adjusted by separating part of the milk fat therefrom, or by adding thereto cream, concentrated milk, dry whole milk, skim milk, concentrated skim milk, or nonfat dry milk. Thus, the process of milk standardization can be classified into two types, as that involving the partial removal of cream from milk to achieve fat reduction or the mixing of skimmed milk and whole milk to achieve the desired fat content. Thus, material balance plays a major role in the standardization of milk.

The most simple and commonly adopted material balance approach in the food standardization process is the Pearson square. This can also be considered as an easy method to solve simultaneous equations limited to two variables.

The steps involved in using the Pearson square concept are listed as follows:

1. Draw a square.
2. At the center of the square, write the value of the desired percentage of fat, f_x (or any other component) to be present in the standardized milk (or any other food to be standardized).
3. At the upper left-hand corner of the square, write the higher constituent content, f_w (% fat concentration of the most concentrated fat source used; e.g., whole milk or cream).
4. At the bottom left-hand corner of the square, place the lower constituent content, f_s (% fat concentration of the lesser concentrated fat source used; e.g., skim milk).
5. Subtract diagonally the higher value from the smaller value between those at the center and left side corners as shown in Figure 2.18.
6. Write the obtained values on the diagonally opposite right-hand corners of the square.
7. More specifically, write the "parts milk" on the top right-hand side and write "parts skim" on the bottom right-hand side of the Pearson square.
8. The numbers, thus, written on the right corners of the Pearson square are the solutions of material balance equations with two variables. Thus, these values provide the proportions of the ingredients to be mixed to obtain the desired standardized food product.

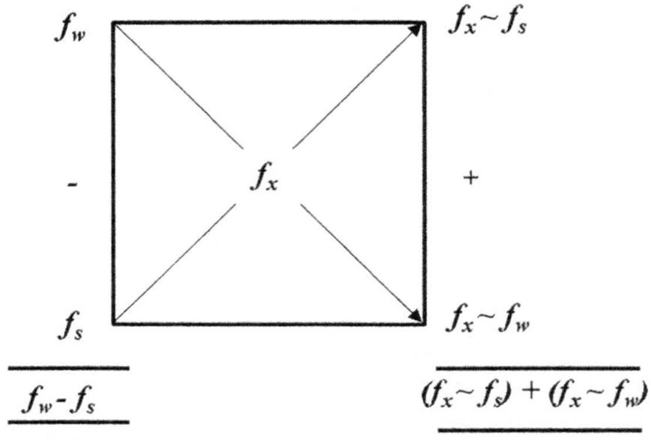

FIGURE 2.18 The concept of Pearson square for mass balance.

It is to be noted that the sum of values on the right-hand side corners of the square equals the difference between the values at the top left and bottom left-hand side corners (Figure 2.18).

The application of Pearson square for mass balance calculation is explained with the following example:

Example 2.7

A dairy industry has whole milk and skim milk with 6.5% and 0.04% fat content, respectively. Calculate the amounts of whole milk and skim milk needed to produce 1000 kg of standardized milk with 4.5% fat.

Solution

I. **Data collection**

- Desired fat percentage in the standardized milk: 4.5%
- Higher constituent content, $f_w = 6.5\%$
- Lower constituent content, $f_s = 0.04\%$
- Amount of standardized milk to be produced = 1000 kg

II. **Construction of Pearson square**

Therefore, from the Pearson square (Figure 2.19), it could be inferred that the

Proportion of whole milk = 4.46/6.46

Amount of whole milk required = $(4.46/6.46) \times 1000 = 690.4$ kg

Proportion of skimmed milk = 2/6.46

Amount of skimmed milk required = $(2/6.46) \times 1000 = 309.6$ kg

Thus, 690.4 kg of whole milk and 309.6 kg of skim milk are required to produce 1000 kg of standardized milk with 4.5% fat content.

Pearson square is not confined to milk standardization but can also be used to calculate the amounts of any two components to be mixed together to result in a desired final concentration. Examples could be that of

i. Mixing of fruit juice and sugar syrup to prepare a fruit squash
ii. Mixing of fruit pulp and sugar in the jam preparation
iii. Mixing rations for animal feeding
iv. In the meat industry to produce meat products such as sausages to a particular fat content
v. Blending of wines and other alcoholic beverages to provide products of a specified alcohol concentration.

However, the Pearson square concept is limited to the blending of only two components. When more than two components are involved, more complex mass balance equations have to be used (Mullan, 2006).

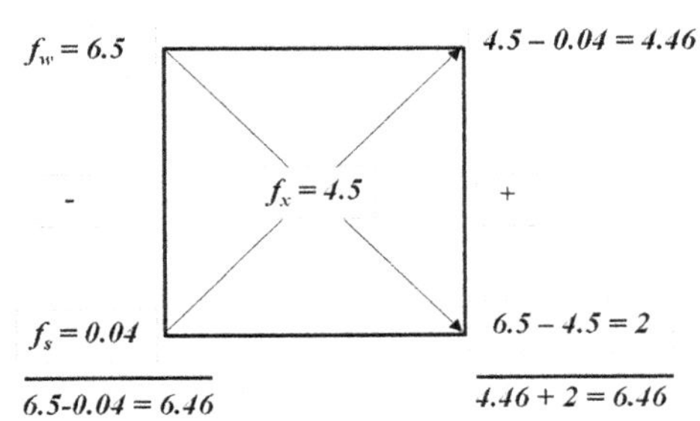

FIGURE 2.19 Pearson square for milk standardization.

2.6 Application of Material Balance in Food Product Traceability

Including a material balance check exercise on at least one key ingredient or packaging material of a product is essential to test the efficacy of a traceability system in a food industry. Here, the purpose of mass balance exercise is to check whether the manufacturer can account for any input material that is involved in the production of a lot or batch of the product. The input material can be a food ingredient, semi-processed product, or packaging material. Mass balance and traceability are extremely crucial in the reported event of a critical food safety (contamination) or regulatory (mislabeling) issue, which demands a complete recall of a particular lot of the product from the market. The application of mass balance in a traceability exercise is explained by the following example. Following are the steps involved in conducting the traceability exercise.

1. To conduct the reconciliation exercise and obtain details on the quantity of the product that has been sold, retained in the inventory, and disposed of for reasons if any.
2. The collection of abovementioned details gives the total quantity of unaccounted material.
3. Determine the percentage effectiveness of the traceability procedure using the following formula:

$$\% \text{ Effectiveness of traceability procedure} = \frac{B+C+D}{A} \times 100$$

Example 2.8

A complete reconciliation/mass balance exercise is to be conducted for the ingredient traceability in a seasoning mix product. The list of details to be collected for the exercise is given in Table 2.1.

Details of reconciliation for the product containing a raw ingredient of the recalled product are listed in Table 2.2.

TABLE 2.1

General Details of the Lot

Name of the Raw Material (Ingredient)	Whey Powder
Name of the recalled product in which the ingredient was used	Seasoning mix
Production date	02/08/2017 (2nd August 2017)
Reason for recall	Report of allergy due to incidental and undeclared dairy ingredient
Lot number of raw material	WP080117
Lot number of product	SM080217
Dispatch invoice number	SM-2023
Date of delivery	05/08/2017 (5th August 2017)
Name of the customer	XYZ

TABLE 2.2

Details of Mass Balance

Total Amount of Product Produced (A)	10000 Packs
Amount still on inventory (B)	3000 packs
Amount delivered to customers (C)	5900 packs
Incidental disposal (D)	50 packs
Total	8950 packs

where,

- A: Total amount of product produced
- B: Amount still on inventory
- C: Amount delivered to customers
- D: Incidental wastage (product dropped on ground and so on)

Thus, total accounted: 8950 packs; total unaccounted: 1050 packs.

$$\% \text{ Effectiveness of traceability} = \frac{(3000 + 5900 + 50) \times 100}{10000}$$

$$= 89.5\%$$

2.7 Problems to Practice

2.7.1 Multiple Choice Questions

1. The concept of material balance is based upon the law of
   a. conservation of mass
   b. conservation of energy
   c. conservation of momentum
   d. conservation of volume

Answer: a

2. In a steady-state process, the following quantity in the material balance equation is zero:
   a. Mass of waste out
   b. Mass of accumulated products
   c. Mass of by-products
   d. Mass of products

Answer: b

3. How many independent mass balance equations are possible, if only one component exists in each stream?

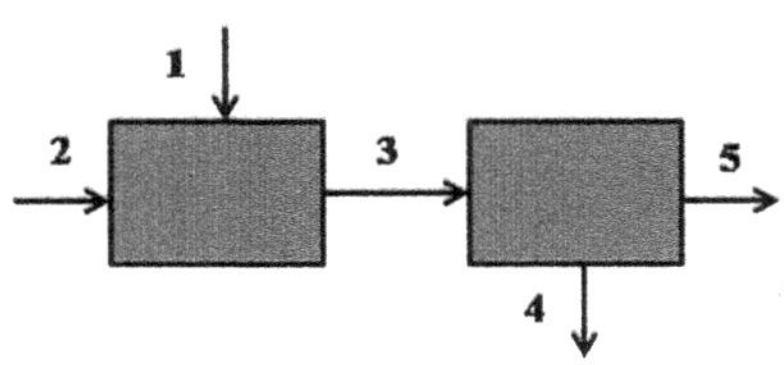

   a. one
   b. two
   c. three
   d. four

Answer: b

4. The reference for material balance equation is
   a. tie
   b. basis
   c. solid content
   d. moisture content

Answer: b

5. The tie material in a process involving the drying of potatoes is
   a. moisture content
   b. solid content
   c. mass of fresh potatoes
   d. mass of dried potatoes

Answer: b

6. The basis for material balance calculation in a continuous process is
   a. mass of raw material
   b. mass flow rate of raw material
   c. time
   d. Both b and c

Answer: d

7. In the Pearson square concept, the value at the upper left-hand corner of the square is
   a. lower constituent content
   b. higher constituent content
   c. desired percentage of component in the standardized product
   d. difference between lower and higher constituent contents

Answer: b

8. In an unsteady-state process, the process variables
   a. vary with time
   b. vary with position
   c. both a and b
   d. remain constant

Answer: c

9. A batch process can operate
   a. in an open system
   b. in a closed system
   c. in both open and closed systems
   d. none of the above

Answer: b

10. A system in which energy can be transferred across its boundaries but not mass is
    a. open system
    b. closed system
    c. both a and b
    d. a or b

Answer: b

11. A component which relates the quantity of one process stream to that of another and does not change from one stream to another is referred to as
    a. basis
    b. inert material
    c. tie material
    d. component

Answer: c

12. Pearson square concept is limited to the blending of
    a. three components
    b. four components
    c. two components
    d. five components

Answer: c

13. The amount of dry sugar that must be added in 10 kg of aqueous sugar solution in order to increase its concentration from 10% to 40% is
    a. 2.5 kg
    b. 10 kg
    c. 5 kg
    d. 40 kg

Answer: c

14. In the general equation of material balance, the sum of the masses of final product, wastes, by-products, and losses is equal to
    a. mass of raw material − mass of accumulated product
    b. mass of raw material + mass of accumulated product
    c. mass of accumulated product
    d. mass of raw material

Answer: a

15. Usually, the basis for material balance calculation involving a batch process is
    a. mass of incoming raw material
    b. time
    c. mass flow rate
    d. volumetric flow rate

Answer: a

2.7.2 Numerical Problems

1. Estimate the mass flow rate (kg/h) of coffee extract with 10% solid content that must be fed to an evaporator to produce 5000 kg/h of coffee concentrate with 50% solid content.

   Given

   i. Unit operation: evaporation
   ii. Solid content of coffee extract = 10%
   iii. Mass flow rate of product = 5000 kg/h
   iv. Solid content of concentrate after evaporation = 50%

To find: Mass flow rate (kg/h) of coffee extract

Solution

Basis: 1 h

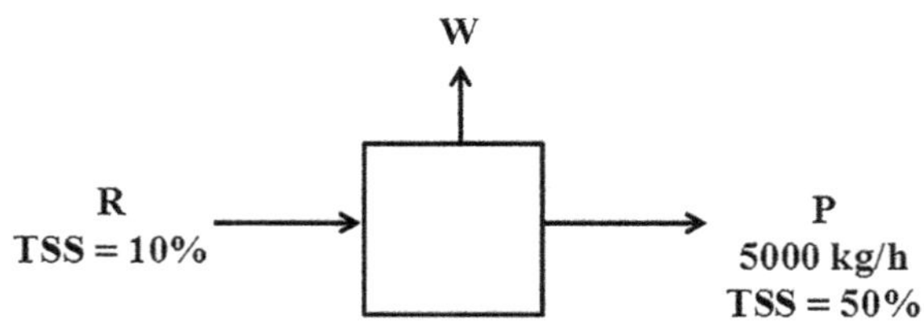

In the preceding block diagram,

R = Amount of coffee extract (kg)

P = Amount of coffee concentrate (kg)

W = Amount of water removed during the evaporation process (kg)

TSS = Total soluble solid content (%)

Overall mass balance equation

$$R = W + P$$

$$R = W + 5000$$

Solid balance

$$R\ (0.1) = W(0) + P(0.5)$$

$$0.1\ R = 0.5 \times 5000 = 2500$$

$$\therefore R = \frac{2500}{0.1} = 25000 \text{ kg / h}$$

Answer: Mass flow rate of coffee extract = 25000 kg/h

2. A 150 kg batch of raw honey has a moisture content of 25%. Calculate the amount of water to be removed to obtain a final moisture content of 20% in the processed honey.

Given

i. Unit operation: evaporation
ii. Batch quantity of raw honey = 150 kg
iii. Moisture content of raw honey = 25%
iv. Final moisture content of processed honey = 20%

To find: Amount of water (W) to be removed from raw honey to obtain a final moisture content of 20% in the processed honey.

Solution

Basis: 150 kg of raw honey

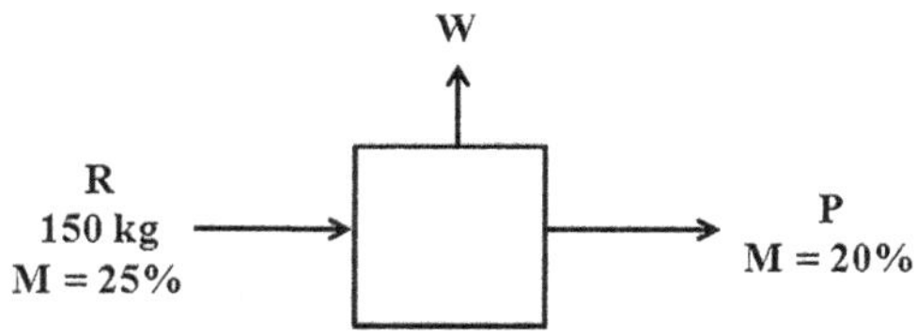

In the preceding block diagram,

R = Amount of raw honey (kg)

P = Amount of processed honey (kg)

W = Amount of water removed from the honey during processing (kg)

M = Moisture content (%)

Overall mass balance

$$R = W + P \tag{1}$$

$$150 = W + P \tag{2}$$

Moisture balance

$$R(0.25) = W(1) + P(0.2)$$

$$37.5 = W + 0.2P \tag{3}$$

Solving Eqs. (2) and (3)

$$W + P = 150$$
$$(-)W + 0.2P = 37.5$$
$$P - 0.2P = 150 - 37.5$$
$$0.8P = 112.5$$
$$P = 140.6$$

From Eq. (2),

$$W = R - P$$

$$W = 150 - 140.6$$

$$W = 9.4 \text{ kg}$$

Answer: Amount of water to be removed from honey = 9.4 kg

3. How many kilograms of a solution containing 6% of glucose can be obtained by diluting 30 kg of 40% glucose solution with water?

Given

i. Unit operation: dilution
ii. Batch quantity of concentrated glucose solution = 30 kg
iii. Initial concentration of glucose solution = 40%
iv. Required final concentration of glucose solution after dilution = 6%

To find: Quantity of diluted solution (D) that can be obtained.

Solution

Basis: 30 kg of glucose solution containing 40% glucose

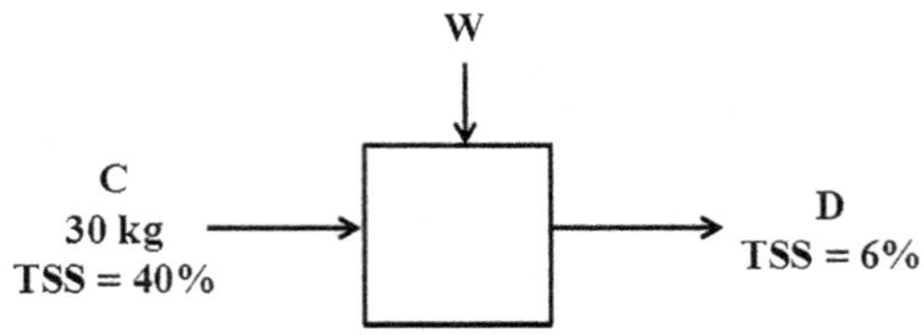

In the preceding block diagram,

C = Amount of concentrated glucose solution (kg)

D = Amount of diluted glucose solution (kg)

W = Amount of water added during dilution (kg)

TSS = Total soluble solid content (%)

Overall mass balance equation

$$C + W = D$$

$$30 + W = D$$

Solid balance equation

$$C\ (0.40) + W(0) = D(0.06)$$

$$(30 \times 0.4) = 0.06\ D$$

$$12 = 0.06\ D$$

$$\therefore D = 200\ \text{kg}$$

From the overall mass balance equation, $W = D - 30 = 200 - 30 = 170$ kg.

Answer: 200 kg of 6% glucose solution can be obtained by diluting 30 kg of 40% glucose solution with 170 kg of water.

4. How much water should be removed from a 5% salt solution in order to form 20% solution?

Given

i. Unit operation: concentration (removal of water)
ii. Initial concentration (TSS) of salt solution = 5%
iii. Final concentration of salt solution = 20%

To find: Amount of water to be removed from the 5% salt solution in order to form the 20% solution.

Solution

Basis: 100 kg of 5% salt solution

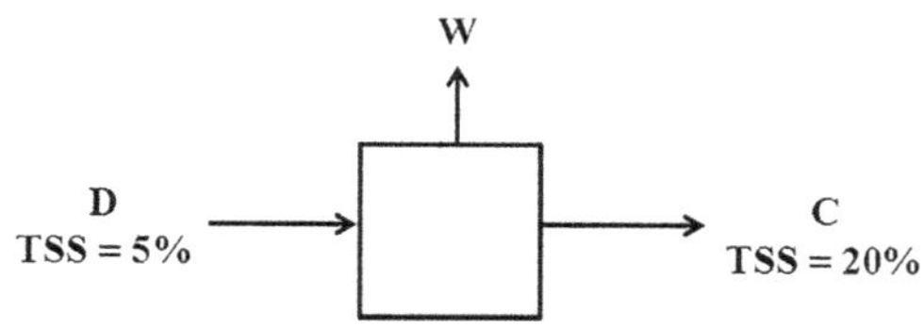

In the preceding block diagram,

D = Amount of dilute salt solution (kg)

C = Amount of concentrated salt solution (kg)

W = Amount of water removed during concentration (kg)

TSS = Total soluble solid content (kg)

Overall mass balance equation

$$D = W + C$$

$$100 = W + C$$

Solid balance equation

$$D\ (0.05) = W(0) + C(0.20)$$

$$5 = 0.2\ C$$

$$\therefore C = 25\ \text{kg}$$

From the overall mass balance equation,

$$W = D - C = 100 - 25 = 75\ \text{kg}$$

Answer: 75 kg of water should be removed from 100 kg of 5% salt solution to form 20% solution.

5. A dairy spray-drying plant produces 300 kg/h of milk powder containing 3% moisture content. The spray dryer is fed with a milk concentrate constituted of 40% solids, resultant from the original feed of 12% solid content. Determine the following:

(a) Mass flow rate of product from evaporator

(b) Amount of moisture removed in the evaporator

(c) Amount of moisture removed in the spray dryer

Given

i. Mass flow rate of milk powder = 300 kg/h

ii. Moisture content of milk powder = 3%

iii. Solid content of milk concentrate fed to the spray dryer = 40%

iv. Initial solid concentration of original feed = 12%

To find

i. Mass flow rate of product from evaporator

ii. Amount of moisture removed in evaporator

iii. Amount of moisture removed in spray dryer

Solution

Basis: 1 h

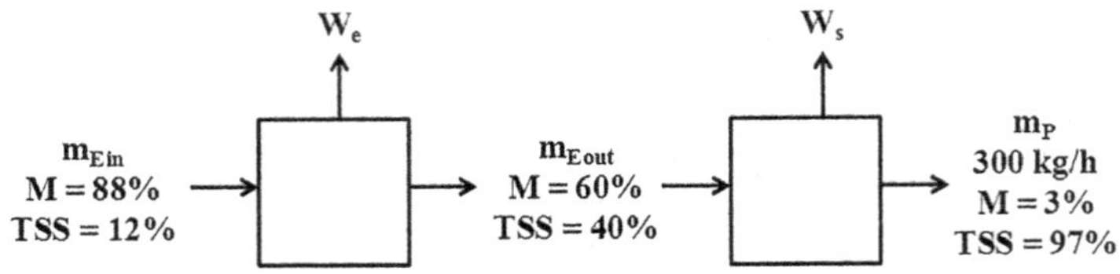

In the preceding block diagram,

$m_{E\text{in}}$ = Amount of liquid milk fed to the evaporator (kg)

$m_{E\text{out}}$ = Amount of concentrated milk exiting the evaporator (kg)

m_p = Amount of milk powder from spray dryer (kg)

W_e = Amount of water removed in evaporator (kg)

W_s = Amount of water removed in spray dryer (kg)

M = Moisture content (%)

TSS = Total soluble solid content (%)

Spray dryer

Overall mass balance

$$m_{E\text{out}} = W_s + m_p$$

$$m_{E\text{out}} = W_s + 300$$

Solid component balance

$$m_{E\text{out}}(0.40) = W_s(0) + 300(0.97)$$

$$m_{E\text{out}}(0.40) = 291$$

$$\therefore m_{E\text{out}} = 727.5 \text{ kg}$$

$$\therefore W_s = 727.5 - 300 = 427.5 \text{ kg}$$

Evaporator

Overall mass balance

$$m_{E\text{in}} = W_e + m_{E\text{out}}$$

$$m_{E\text{in}} = W_e + 727.5$$

Solid component balance

$$m_{E\text{in}}(0.12) = W_e(0) + 727.5(0.40)$$

$$m_{E\text{in}}(0.12) = 291$$

$$\therefore m_{E\text{in}} = 2425 \text{ kg}$$

$$\therefore W_e = 2425 - 727.5 = 1697.5 \text{ kg}$$

Answer (i) Mass flow rate of product from evaporator = 727.5 kg/h
(ii) Moisture removed in evaporator = 1697.5 kg
(iii) Moisture removed in spray dryer = 427.5 kg

6. The evaporation capacity of an evaporator is 60 kg/h. The raw material contains 90% water. Calculate the mass flow rate of juice concentrate from the evaporator which would contain 60% water.

Given

i. Unit operation: Evaporation
ii. Evaporation capacity = 60 kg/h
iii. Water content of raw material = 90%
iv. Water content of final product = 60%

To find: Mass flow rate of juice concentrate from the evaporator

Solution
Basis: 1 h

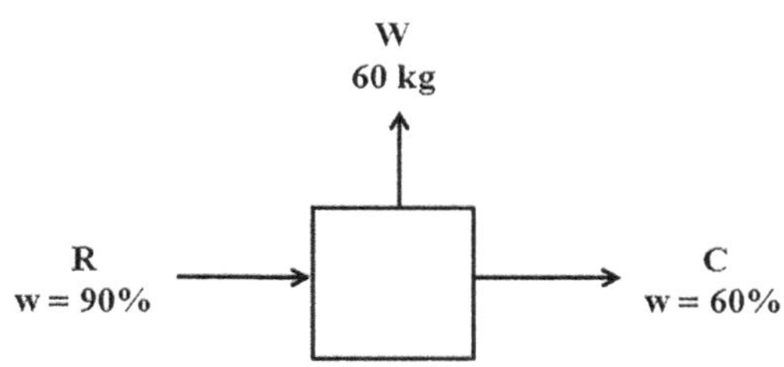

In the preceding block diagram,

R = Amount of dilute juice (raw material) (kg)
C = Amount of juice concentrate (kg)
W = Amount of water removed during evaporation (kg)
w = Water content (%)

Overall mass balance

$$R = W + C \tag{1}$$

$$R = 60 + C \tag{2}$$

Water component balance

$$R(0.90) = 60(1) + C(0.6)$$

$$0.90R = 60 + 0.6C \tag{3}$$

Subtracting Eq. (3) from Eq. (2),

$$
\begin{array}{r}
R - C = 60 \\
(-)0.90R - 0.6C = 60 \\
\hline
0.1R - 0.4C = 0 \\
0.1R = 0.4C \\
R = 4C
\end{array}
$$

From Eq. (2),

$$4C - C = 60$$

$$3C = 60$$

$$\therefore C = 20 \text{ kg}$$

Answer: Mass flow rate of juice concentrate from the evaporator is 20 kg/h.

7. Tomato juice flowing through a pipe at a rate of 100 kg/min is salted by adding saturated salt solution (26% salt) to the pipeline at a constant rate. At what rate would the saturated salt solution be added to provide 2% salt in the product?

Given

i. Unit operation: Mixing
ii. Flow rate of tomato juice = 100 kg/min
iii. Concentration (TSS) of saturated salt solution = 26%
iv. Required final concentration of salt in the product = 2%

To find: Rate of addition of the saturated salt solution

Solution

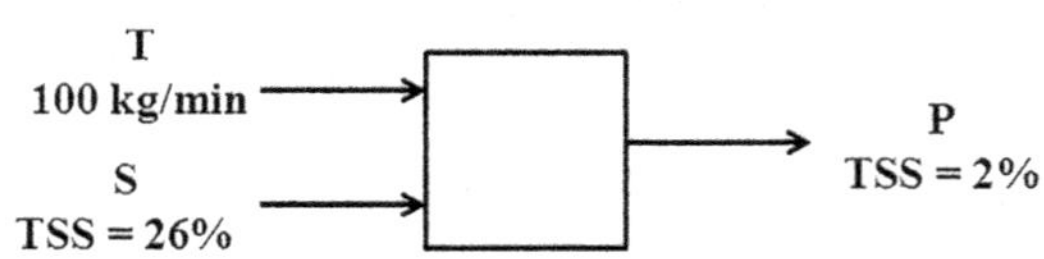

Basis: 1 min

In the preceding block diagram,

T = Amount of tomato juice (kg)
S = Amount of saturated salt solution (kg)
P = Amount of final product (kg)
TSS = Total soluble solid content (%)

Overall mass balance

$$T + S = P$$

$$100 + S = P$$

Salt component balance

$$100(0) + S(0.26) = P(0.02)$$

$$0.26S = 0.02P$$

$$\therefore S = 0.077P$$

Substituting $S = 0.077P$ in overall mass balance equation,

$$100 + 0.077P = P$$

$$100 = 0.923P$$

$$\therefore P = 108.3 \text{ kg}$$

$$\therefore S = P - 100 = 108.3 - 100 = 8.3 \text{ kg}$$

Answer: Rate of addition of the saturated salt solution = 8.3 kg/min

8. Minced meat produced by a meat processing plant is supposed to contain 10% of fat. If the product is to be prepared from beef having 20% of fat and from pork with 5% of fat, determine the proportions in which the constituents should be mixed.

Given

i. Unit operation: Mixing
ii. Required fat content (f) of minced meat = 10%
iii. Fat content of beef = 20%
iv. Fat content of pork = 5%

To find: Proportion in which the constituents should be mixed.

Solution

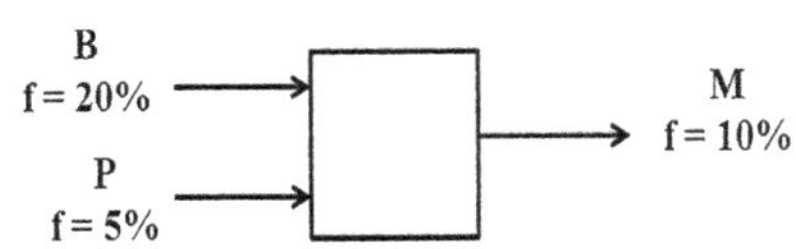

Basis: 100 kg of minced meat

In the preceding block diagram,

M = Amount of minced meat produced (kg)
B = Amount of beef used to prepare the minced meat (kg)
P = Amount of pork used to prepare the minced meat (kg)
f = Fat content (%)

Overall mass balance

$$B + P = M \tag{1}$$

$$B + P = 100 \tag{2}$$

Fat component balance

$$B(0.20) + P(0.05) = M(0.10)$$

$$0.2B + 0.05P = 100(0.10)$$

$$0.2B + 0.05P = 10 \qquad (3)$$

Subtracting Eq. (3) from Eq. (2) multiplied by 0.2,

$$\begin{array}{r} 0.2B + 0.2P = 20 \\ (-)0.2B + 0.05P = 10 \\ \hline 0.15P = 10 \\ P = 66.7 \text{ kg} \end{array}$$

From Eq. (2),

$$B = 100 - P = 100 - 66.7 = 33.3 \text{ kg}$$

Answer: Proportions of beef and pork to be added for the preparation of minced meat are 33.3 kg and 66.7 kg, respectively.

9. Using the concept of Pearson square, calculate the proportions of pineapple juice (10% sugar content) and sugar syrup (70% sugar content) required to manufacture 100 kg of pineapple squash containing 25% sugar.

Given

i. Sugar content of pineapple juice = 10%
ii. Sugar content of sugar syrup = 70%
iii. Quantity of pineapple squash to be produced = 100 kg
iv. Required sugar content in pineapple squash = 25%

To find: Proportions of pineapple juice and sugar syrup required to manufacture 100 kg of pineapple squash.

Solution

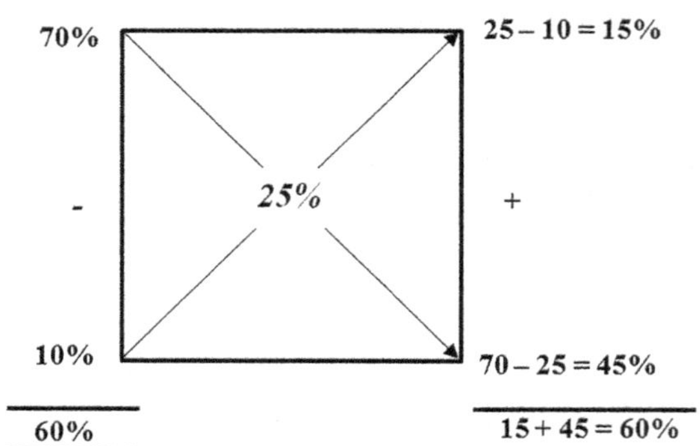

Therefore, from the Pearson square, it could be inferred that,

$$\text{Proportion of sugar syrup} = \left(\frac{15}{60}\right) \times 100 = 0.25 \times 100 = 25 \text{ kg}$$

$$\text{Proportion of pineapple juice} = \left(\frac{45}{60}\right) \times 100 = 0.75 \times 100 = 75 \text{ kg}$$

Answer: 25 kg of sugar syrup and 75 kg of pineapple juice are required to manufacture 100 kg of pineapple squash containing 25% sugar.

10. Calculate the quantity of strawberry pulp, sugar, and pectin required to produce 100 kg of strawberry jam. The strawberry pulp has a soluble solids content of 20%. The pulp to sugar ratio is 45:55 and the sugar-to-pectin ratio is 150. The required soluble solids content in the finished product is 68%. Determine the amount of water removed by evaporation.

Given

i. Process: Production of jam
ii. Batch quantity of strawberry jam = 100 kg
iii. Soluble solids content (TSS) of strawberry pulp = 20%
iv. Pulp to sugar ratio (*P*:*S*) = 45:55
v. Sugar-to-pectin ratio (*S*:*Pe*) = 150
vi. Required soluble solids content (TSS) in the finished product = 68%

To find

i. Quantity of strawberry pulp, sugar, and pectin required to produce 100 kg of strawberry jam
ii. Amount of water removed by evaporation

Solution

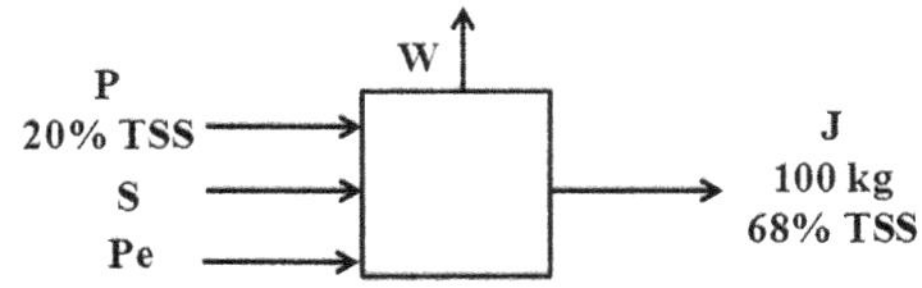

Basis: 100 kg of strawberry jam

In the preceding block diagram,

P = Amount of fruit pulp (kg)
Pe = Amount of pectin (kg)
S = Amount of sugar (kg)
W = Amount of water removed during jam-making (kg)
J = Amount of jam produced (kg)
TSS = Total solid content (%)

Tie material: Solid content of the (fruit pulp + sugar + pectin) mixture

Overall mass balance

$$P + S + Pe = J + W$$

$$P + S + Pe = 100 + W$$

Solid mass balance

$$P(0.2) + S(1) + Pe(1) = J(0.68) + W(0)$$

$$0.2P + S + Pe = 100(0.68) = 68$$

$$0.2P + S + Pe = 68$$

Given that *S*:*Pe* is 150,

$$\frac{S}{Pe} = 150; \; \therefore Pe = \frac{S}{150}$$

$$\therefore 0.2P + S + \frac{S}{150} = 68$$

$$0.2P + \frac{151S}{150} = 68$$

$$0.2P + 1.007S = 68$$

Given that *P*:*S* is 45:55,

$$\frac{P}{S} = \frac{45}{55}; \; \therefore P = 0.818\ S$$

$$\therefore 0.2(0.818\ S) + 1.007\ S = 68$$

$$0.164\ S + 1.007\ S = 68$$

$$1.171\ S = 68$$

$$\therefore S = \frac{68}{1.171} = 58.07 \text{ kg}$$

$$\therefore P = 0.818 \times 58.07 = 47.5 \text{ kg}$$

$$Pe = \frac{S}{150} = \frac{58.07}{150} = 0.387 \text{ kg}$$

From the overall mass balance,

$$W = (P + S + Pe) - J$$

$$W = (47.5 + 58.07 + 0.387) - 100$$

$$\therefore W = 5.957 \text{ kg}$$

Answer (i) 47.5 kg of strawberry pulp, 58.07 kg of sugar, and 0.387 kg of pectin are required to produce 100 kg of strawberry jam
(ii) Amount of water evaporated during jam production = 5.957 kg

BIBLIOGRAPHY

Code of Federal Regulations (CFR). 2017. Part 131—Milk and Cream. Title 21, Volume 2, Revised as of April 1, 2017.

Doran, P. M. 1995. *Bioprocess Engineering Principles*. London: Academic Press.

Evranuz, E. O. and Kılıç-Akyılmaz, M. 2012. Material and energy balances. In *Handbook of Food Process Design*, eds. J. Ahmed and M. S. Rahman, 39–73. West Sussex: John Wiley & Sons Ltd.

Farkas, B. E. and Farkas, D. F. 1997. Material and energy balances. In *Handbook of Food Engineering Practice*, eds. K.J. Valentas, E. Rotstein and R. P. Singh, 253–289. New York: CRC Press.

Mardikar, S. H. and Niranjan, K. 1995. Food processing and the environment. Environmental Management and Health 6: 23–26.

Mullan, W. M. A. 2006. Pearson square or rectangle. www.dairyscience.info/index.php/calculators-models/78-articles/site-calculators-and-models/133-pearson.html (accessed July 31, 2017).

Niranjan, K. 1994. An assessment of the characteristics of food processing wastes. In *Environmentally Responsible Food Processing (Symposium Series No. 300)*, eds. K. Niranjan, M. R. Okos, and M. Rankowitz, 1–7. Washington, DC: American Institute of Chemical Engineers.

Toledo, R. T. 2007. *Fundamentals of Food Process Engineering*. New York: Springer.

3

Energy Balance

Energy is indispensable in food processing as most of the unit operations involve the addition or removal of heat energy to ensure and enhance product quality and shelf life (Rodriguez-Gonzalez et al., 2015). Heat energy is vital due to the requirement for steam at different temperatures and pressures to achieve an acceptable level of food safety (Barron and Burcham, 2001). Numerous definitions exist for the term *energy*. The oldest definitions of energy from an engineer's perspective are (Kent, 1916),

> Energy or stored work is the capacity for performing work.
> Energy is that which is continually passing from one portion of matter to another.

The cost of energy is an integral part of the total cost of processed foods. Also, as energy scarcity prevails worldwide, energy conservation is the need-of-the-hour. Food processing sector is no exception to this as it accounts for a substantial amount of energy consumption (UNIDO and MITI, 1995). In this background, *energy balance* is appreciated as a tool to assist in the measures taken towards energy conservation.

Energy balance can be defined as the *examination of a building or process to identify the energy outputs and equate them with energy inputs, as required by the first law of thermodynamics* (Dalzell, 1994), which states that *energy is neither created nor destroyed, but it is converted from one form to another.* The concept of energy balance is central to the energy audits and energy surveys conducted in industry. An energy audit is the examination of the use of energy in an enterprise and the systems in place to manage the energy use and expression of an opinion on the same. Energy survey is the examination of a building or site to identify the following

- Areas of energy use which give rise to unnecessary energy waste
- Energy using equipment inappropriate to the task
- Equipment that is not functioning as intended (Dalzell, 1994)

Energy audit and energy survey are relevant since some operations demand more energy than the others and hence need to be identified and segregated. The information required for energy audit and energy survey can be obtained during energy balance calculations.

Freezing, canning, and drying are considered to be the most energy-intensive operations, accounting for one-third of the energy consumption in the food sector (Hendrickson, 1996). Computing the energy balance for a manufacturing process can aid in the identification of that bottleneck unit operation to which the application of energy conservation principle would be beneficial in reducing the process cost. Further, knowledge of energy balance can assist in the calculation of the amount of heat to be transferred in a process. It also helps in the appropriate selection of heat transfer equipment with suitable dimensions (Mittal, 2010).

3.1 Forms of Energy

Energy balance is considered to be more complicated than material balance as energy can take different forms, and some of these forms are interconvertible. Primarily, energy can exist in three forms: (i) potential energy, (ii) kinetic energy, and (iii) internal energy.

3.1.1 Potential Energy

Potential energy (E_{PE}) is the energy due to the position of an object placed in a gravitational or electromagnetic field, relative to a specified reference around the object. For an object of mass m (in units of mass; kilogram), which is located at a height h (in units of length; meter), and if the acceleration due to gravity is g (in units of length/time^2; m/s^2), the potential energy (E_{PE}, in units of energy; joule) is given by

$$E_{PE} = mgh \tag{3.1}$$

Example 3.1

Chicory extract is pumped from one storage tank (Tank-1) to the other (Tank-2). Tank-2 is 50 ft above Tank-1. Calculate the increase in potential energy with each kg of chicory extract pumped from Tank-1 to Tank-2.

Solution

$$\text{Increase in potential energy per kg} = \frac{E_{PE}}{m} = gh = 9.806\frac{\text{m}}{\text{s}^2}\times\frac{50\,\text{m}}{3.2808} = 149.445\ \frac{\text{m}^2}{\text{s}^2}$$

$$= 149.445\ \frac{\text{J}}{\text{kg}}\left(\because \text{J} = \text{Nm};\quad \text{N} = \frac{\text{kg m}}{\text{s}^2};\quad \therefore \frac{\text{J}}{\text{kg}} = \frac{\text{m}^2}{\text{s}^2}\right)$$

Answer: 149.445 $\frac{\mathbf{J}}{\mathbf{kg}}$

3.1.2 Kinetic Energy

Kinetic energy (E_{KE}) is the energy due to the motion of a body relative to a reference at rest. In other words, it is the energy acquired by a moving object due to its velocity. For an object of mass m (in units of mass; kg) moving at a velocity u (in units of length/time; m/s), the kinetic energy is given by

$$E_{KE} = \frac{1}{2}mu^2 \tag{3.2}$$

The velocity (u) is always measured relative to a frame of reference which defines the stationary position of the object.

The major application of potential and kinetic energies in the food industry has been realized in the selection and design of conveying systems for the transport of solids and fluids from one location to another. For example, the energy required to lift a wheat flour bag of mass m from the ground to a height h will be equal to the potential energy as calculated from Eq. (3.1). Similarly, when a packaged food product of mass m is dropped onto a horizontal belt conveyor, it will be accelerated to the speed of the conveyor with a kinetic energy of $\frac{1}{2}mu^2$ and then travel along the length of the conveyor.

Example 3.2

Milk is pumped from a storage tank through a pipe of 25 mm inner diameter at the rate of 1 L/s. What is the kinetic energy per unit weight of milk in the pipe?

Solution

$$E_{KE} = \frac{1}{2}mu^2$$

Diameter of the pipe = D = 25 mm = 25×10^{-3} m

Datum: m = 1 kg

Flow rate = Q = 1 L/s

$$\text{Cross-section area of the pipe} = A = \frac{1}{4}\pi D^2 = \frac{1}{4}\pi\left(25\times10^{-3}\right)^2 = 4.909\times10^{-4}\ \text{m}^2$$

$$\text{Average velocity of milk} = v = \frac{Q}{A} = \frac{0.001\dfrac{\text{m}^3}{\text{s}}}{4.909\times10^{-4}\ \text{m}^2} = 2.037\frac{\text{m}}{\text{s}}$$

$$\text{Kinetic energy per kg of milk} = \frac{E_{KE}}{m} = \frac{1}{2}v^2$$

$$= \frac{1}{2}(2.037)^2 = 2.075\frac{\text{J}}{\text{kg}}$$

Answer: $\mathbf{2.075\frac{J}{kg}}$

3.1.3 Internal Energy

Internal energy (U) is the energy due to microscopic interactions involving molecular, atomic, and subatomic motions. Consequently, it cannot be measured as an absolute value. Rather, changes in internal energy can be related to other properties such as temperature and pressure. Internal energy is an extensive property and hence independent of the path of a process.

The abovementioned forms of energy are interconvertible within the system. In turn, the system can also exchange energy with its surroundings in the form of heat (q) and work (W). While heat transfer occurs as a result of temperature difference, transfer of work happens via a moving mechanical part (W_s, Shaft work) or movement of system boundary against pressure ($W_f = \Delta(pV)$; flow work). Heat and work are the energy forms acquired (by transfer) but not possessed by a system. Conversely, kinetic, potential and internal energies are the inherent properties of a system.

According to the law of conservation of energy, an increase or decrease in the total energy of the system (ΔE) must reflect either as gain (or) loss of work (W) (or) heat (q) from/to the surroundings. The preceding statement can be represented as

$$\Delta E = \Delta\left(U + E_{PE} + E_{KE}\right) = q + W \tag{3.3}$$

Equation 3.3 is called the "general energy balance equation."

The major energy forms used in some of the typical food processing operations are given in Table 3.1. However, not all of the abovementioned forms may be involved in a particular process; depending on the process type, the energy form which predominates over the others is considered central to the energy balance calculations. The other insignificant energy forms are neglected with the confidence that it does not introduce errors in the computation. For instance, when wheat flour is tipped from the bags into silos, only the potential and kinetic energy of the wheat flour changes, but other energy forms such as chemical, magnetic, and electrical do not change and, hence, may be neglected in the energy analysis. Irrespective of the form, the SI unit of energy is Joule (as explained in Chapter 1).

TABLE 3.1

Major Energy Forms in Food Processing Operations

Process	Major Energy Form
Pasteurization of milk, drying of food products	Heat energy
Extrusion, pumping of fluid through pipes	Mechanical energy
Wet cleaning operation in any food industry	Chemical energy (provided by the detergent)
Steam generation in heat exchangers to increase the temperature	Electrical energy
Microwave and infrared drying	Electromagnetic energy

3.2 Heat Energy

As mentioned before, heat is the form of energy transferred as a result of the difference in temperature between the system and its surroundings. Heat flows from a region of high temperature to low temperature. In general, when the net heat (q) is transferred from the surrounding into the system, the numerical value of heat transferred is positive ($q > 0$). The abovementioned phenomenon increases the energy of the system. A negative value for q indicates that the net heat is transferred from the system to the surroundings. As a result, the system loses its energy. q is expressed in units of energy, i.e., J, Btu or cal.

Heat energy is relevant concerning food processing as it not only facilitates food preservation but also creates food products with unique textural properties such as that in popcorn, jam, and jellies. Conservation of heat energy is vital in unit operations such as heating and drying. In such operations, the *total heat* is referred to as *enthalpy*. Enthalpy (H) is an extensive property, and thus, its value cannot be measured directly. The value of H is always expressed relative to its increase from a defined reference temperature (T_{ref}), at which the enthalpy is usually zero. Enthalpy is expressed in units of kJ, and its value is calculated as follows.

$$H = C_P(T - T_{\text{ref}}) \tag{3.4}$$

In Eq. (3.4), C_P refers to the specific heat at constant pressure which is explained in Section 3.2.1.

3.2.1 Specific Heat

Specific heat or specific heat capacity (C_P) is a thermal property which is central to energy balance calculations. Heat transfer to a food product during heating or cooling is highly dependent on its C_P. Specific heat can be defined at constant volume or constant pressure. Since most of the food processing applications are carried out at constant pressure, specific heat at constant pressure (C_P) is considered for calculations Eq. (3.5).

$$C_P = \frac{q}{m\Delta T} \tag{3.5}$$

The meaning of Eq. (3.5) is that, to raise the temperature of m kg of product, through a temperature difference of ΔT °C or K, an amount of energy equal to q joules is required. Thus, the specific heat can be expressed in units of J/kg K or kJ/kg K or J/kg °C. Irrespective of the system of units used for the calculation of specific heat, the difference in temperature would be equal, i.e., a difference of 1 K is equal to the difference of 1°C.

Specific heat of any food product can be determined experimentally by differential scanning calorimetry. Besides, it can also be predicted using empirical equations. These empirical equations are obtained based on fitting the experimental data into appropriate numerical or mathematical models. Usually, these models are based on the composition of foods, i.e., carbohydrate, fat, protein, minerals, and moisture content. Some of the commonly used empirical models for prediction of specific heat are described as follows.

3.2.1.1 Siebel's Model

Siebel (1892) proposed a model to calculate the specific heat (J/kg K) based on the moisture content Eqs. (3.6) and (3.7). This model is based on the observation that specific heat of a product varies with its moisture content and it can be determined as the weighted average of the specific heat of water and solids.

$$C_P = 837 + 3348\ X_W\ \left(\text{J/kg K}\right) \tag{3.6}$$

$$C_P = 837 + 1256\ X_W\ \left(\text{J/kg K}\right) \tag{3.7}$$

where X_W is the fraction of moisture content (wet basis) in the food product. The Eqs. (3.6) and (3.7) apply to food materials that are above and below their freezing point, respectively. However, this model does not account for the effect of food components other than water.

3.2.1.2 Charm's Model

The Charm's model (1978) was put forth to address the limitation of Siebel's equation, as mentioned before. This model considers the influence of food constituents such as fat, nonfat solids, and water for the calculation of specific heat, as shown in the following equation.

$$C_P = 2093X_f + 1256X_S + 4187X_W \tag{3.8}$$

where X_f, X_S, and X_W are the mass fractions of fat, nonfat solids, and water, respectively.

3.2.1.3 Heldman and Singh Model

Heldman and Singh (1981) proposed an empirical equation for specific heat, which accounts for all the major food components such as carbohydrate, fat, protein, ash, and moisture content.

$$C_P = 1424(C) + 1549(P) + 1675(F) + 837(A) + 4187(W) \tag{3.9}$$

where C is the mass fraction of carbohydrate, P is the mass fraction of protein, F is the mass fraction of fat, A is the mass fraction of ash, and W is the mass fraction of water, all on a wet basis.

3.2.1.4 Choi and Okos Model

Choi and Okos (1986) proposed an empirical correlation which is computationally tedious but accounts for the effect of temperature change.

$$C_P = P(C_{Pp}) + F(C_{Pf}) + C(C_{Pc}) + Fi(C_{PFi}) + A(C_{PA}) + W(C_{Pw}) \tag{3.10}$$

where the subscript c denotes carbohydrate, p denotes protein, f denotes fat, A denotes ash, Fi denotes fiber, and w denotes water. Specific heat of each of the component mentioned previously is calculated as follows:

$$\text{Protein: } C_{Pp} = 2008.2 + 1208.9 \times 10^{-3}T - 1312.9 \times 10^{-6}T^2 \tag{3.11}$$

$$\text{Fat: } C_{Pf} = 1984.2 + 1473.3 \times 10^{-3}T - 4800.8 \times 10^{-6}T^2 \tag{3.12}$$

$$\text{Carbohydrate: } C_{Pc} = 1548.8 + 1962.5 \times 10^{-3}T - 5939.9 \times 10^{-6}T^2 \tag{3.13}$$

$$\text{Fiber: } C_{PFi} = 1845.9 + 1930.6 \times 10^{-3}T - 4650.9 \times 10^{-6}T^2 \tag{3.14}$$

$$\text{Ash: } C_{PA} = 1092.6 + 1889.6 \times 10^{-3}T - 3681.7 \times 10^{-6}T^2 \tag{3.15}$$

$$\text{Water: } C_{Pw} = 4176.2 - 9.0864 \times 10^{-5}T + 5473.1 \times 10^{-6}T^2 \tag{3.16}$$

3.2.2 Enthalpy

Enthalpy can be obtained from the definition of specific heat at constant pressure (C_P), given by

$$C_P = \left(\frac{\partial H}{\partial T}\right)_P \tag{3.17}$$

where H is the enthalpy and T is the temperature. Similar to the specific heat, enthalpy of foods can also be determined using mathematical models by integrating the equation for specific heat with respect to time. The models for the estimation of enthalpy depend on whether the food is frozen or unfrozen. When the food product is above its freezing point, enthalpy includes only the sensible heat. However, when it is below its freezing point, enthalpy includes both sensible and latent heat (ASHRAE, 2006).

3.2.2.1 Enthalpy Models for Unfrozen Food

For food products at temperatures above their initial freezing point, enthalpy is obtained by integrating the respective expression for C_P above the freezing point, which is

$$C_u = \sum C_i x_i \tag{3.18}$$

where C_u is the specific heat of unfrozen food (kJ/kg K), C_i is the specific heat of the individual food components, and x_i is the mass fraction of the food components. Therefore, the enthalpy (H) of an unfrozen food can be determined by integrating Eq. (3.18) as follows.

$$H = \sum H_i x_i = \sum \int C_i x_i \, dT \tag{3.19}$$

where H_i is the enthalpy of the individual food components and x_i is the mass fraction of the food components.

A simpler model for enthalpy of unfrozen foods was proposed by Chen (1985) by integrating the following equation for specific heat:

$$C_u = 4.19 - 2.30x_s - 0.628x_s^3 \tag{3.20}$$

where x_s is the mass fraction of the solids in the food. On integrating Eq. (3.20), the enthalpy of unfrozen food is given by

$$H = H_f + (t - t_f)(4.19 - 2.30x_s - 0.628x_s^3) \tag{3.21}$$

where H is the enthalpy of food product (kJ/kg), H_f is the enthalpy of food product at initial freezing temperature (kJ/kg), t is the temperature of the food product (°C), t_f is the initial freezing temperature of the food product (°C), and x_s is the mass fraction of solids in the food product.

3.2.2.2 Enthalpy Models for Frozen Food

For frozen food products, the mathematical model for enthalpy is obtained by integrating the following expression for the specific heat of frozen foods (Schwartzberg, 1976).

$$C_a = C_u + (x_b - x_{wo})\Delta C + Ex_s\left(\frac{RT_0^2}{M_w t^2} - 0.8\Delta C\right) \tag{3.22}$$

where C_a is the apparent specific heat, x_b is the mass fraction of bound water, x_{wo} is the mass fraction of water above the initial freezing point, 0.8 = constant, ΔC is the difference between specific heats of water and ice = $C_w - C_{ice}$, E is the ratio of relative molecular masses of water M_w and food solids M_s ($E = M_w/M_s$), R is the universal gas constant (8.314 kJ/kmol K), T_0 is the freezing point of water (273.2 K), M_w is the relative molecular mass, and t is the temperature of the food product. Integrating Eq. (3.22) between a reference temperature (T_r) and the temperature of the food product (T) leads to the following expression for the enthalpy of a food (Schwartzberg, 1976).

$$H = (T - T_r) \times \left\{ C_u + (x_b - x_{wo})\Delta C + Ex_s \left[\frac{RT_0^2}{18(T_0 - T_r)(T_0 - T)} - 0.8\Delta C \right] \right\} \tag{3.23}$$

In general, T_r is considered as 233.2 K (−40°C), at which point the enthalpy is zero.

A much simpler model for the enthalpy of frozen foods was proposed by Chen (1985), which can be obtained by integrating the following equation between T_r and T.

$$C_a = 1.55 + 1.26x_s + \frac{x_s RT_0^2}{M_s t^2} \tag{3.24}$$

where M_s is the relative molecular mass of the soluble solids in the food product. On integrating Eq. (3.24),

$$H = (t - t_r)\left[1.55 + 1.26x_s + \frac{x_s RT_0^2}{M_s t t_r} \right] \tag{3.25}$$

In addition to the abovementioned models, Chang and Tao (1981) developed empirical correlations for the enthalpy of foods, which are expressed as functions of water content, initial and final temperatures, and the type of food product (meat, juice, or fruit/vegetable). The correlations are defined at a reference temperature (T_r) of −45.6°C and have the following form:

$$H = H_f\left[y\bar{T} + (1 - y)\bar{T}^z \right] \tag{3.26}$$

where H_f is the enthalpy of the food product at the initial freezing temperature (kJ/kg), $\bar{T}$ is the reduced temperature given by $(T - T_r)/(T_f - T_r)$, and y and z are the correlation parameters. By performing regression analysis on the experimental data available in the published literature, Chang and Tao (1981) developed the following correlation parameters y and z in Eq. (3.26):

Meat

$$y = 0.316 - 0.247(x_{wo} - 0.73) - 0.688(x_{wo} - 0.73)^2 \tag{3.27}$$

$$z = 22.95 + 54.68(y - 0.28) - 5589.03(y - 0.28)^2 \tag{3.28}$$

Fruit, vegetable, and juice

$$y = 0.362 + 0.0498(x_{wo} - 0.73) - 3.465(x_{wo} - 0.73)^2 \tag{3.29}$$

$$z = 27.2 - 129.04(y - 0.23) - 481.46(y - 0.23)^2 \tag{3.30}$$

Chang and Tao (1981) had also put forth the correlations to determine the initial freezing temperature T_f in Eq. (3.26). The following correlations provide T_f as a function of water content:

Meat

$$T_f = 271.18 + 1.47x_{wo} \tag{3.31}$$

Fruit/vegetable

$$T_f = 287.56 - 49.19x_{wo} + 37.07x_{wo}^2 \quad (3.32)$$

Juice

$$T_f = 120.47 - 327.35x_{wo} - 176.49x_{wo}^2 \quad (3.33)$$

Besides y, z, and T_f, H_f is also required in Eq. (3.26). Chang and Tao (1981) suggested the below correlation for determining the food product's enthalpy at T_f:

$$H_f = 9.79246 + 405.096x_{wo} \quad (3.34)$$

3.2.3 Heat Balance

In practice, energy balance studies focus on the more dominant form of energy, which is heat in most cases of food processing operations such as heating, cooling, drying, cooking, freezing, pasteurization, and sterilization. Thus, heat balance would be useful in explaining the cost and quality aspects of a process. Similar to mass balance, heat balance can be specified around a system (equipment or components of equipment) or process stages, or the entire processing plant (Figure 3.1). Here, the major assumption is that the amount of heat that is converted into other forms of energy is insignificant.

Comparable to the concept of *basis* as explained in mass balance, while writing the enthalpy or heat balance, the quantities are always specified with respect to a *reference level or datum*. A proper selection of datum facilitates the easy calculation of enthalpy and, hence, the energy balances. After defining the datum for calculation, energy balances can be easily worked out by considering the quantities of the raw material, product, waste streams, their specific heats, and changes in their temperature and/or state. The change in temperature and state occurs in conjunction with two corresponding types of heat, namely, sensible heat and latent heat (Figure 3.2).

3.2.3.1 Sensible Heat

Sensible heat (q_S) is the heat which can be sensed as it leads to a change in temperature when it is added or removed from the food product (Figure 3.2). The change in sensible heat can be calculated by multiplying the mass (kg), specific heat (J/kg °C), and the change in temperature (°C) of the food product Eq. (3.35). Thus, the unit of sensible heat is joule (J).

$$\Delta q_S = mC_P\Delta T \quad (3.35)$$

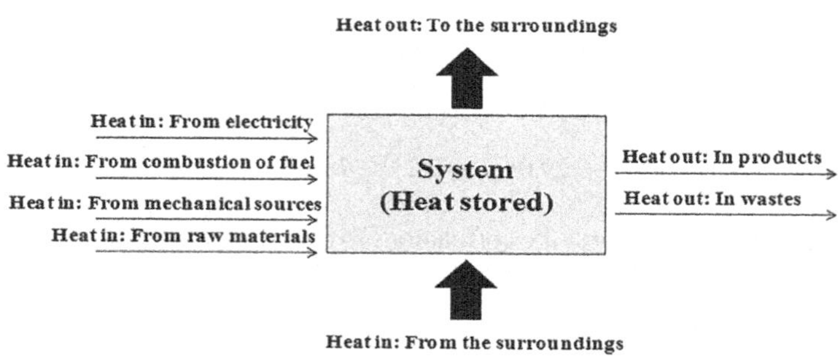

FIGURE 3.1 Schematic of heat balance around a system.

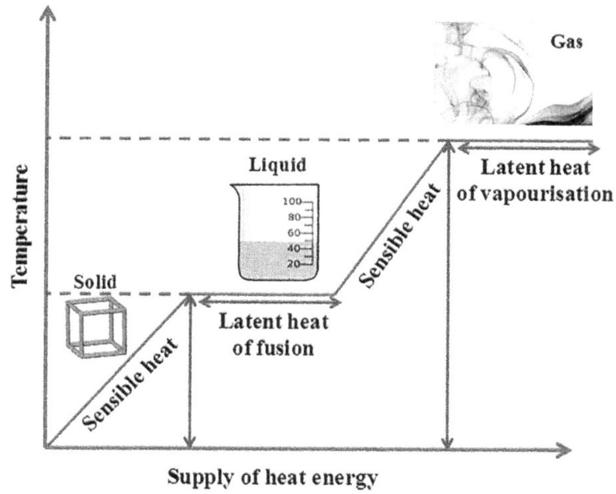

FIGURE 3.2 The concept of sensible and latent heat.

3.2.3.2 Latent Heat

Latent heat is the heat associated with the change in physical state or phase of materials from solid to liquid (melting or fusion), liquid to gas (vaporization), or solid to gas (sublimation) (Earle, 1983). Accordingly, it is expressed in three different forms:

i. *Latent heat of vaporization:* For the transformation of liquid to the gaseous phase.
ii. *Latent heat of fusion:* In case of the conversion of physical state from solid to liquid, wherein the solid melts to form a liquid and thereby gains the energy or heat equivalent to its latent heat of fusion (Figure 3.2).
iii. *Latent heat of sublimation:* The latent heat of sublimation at a particular temperature is the amount of heat required to convert a unit mass of solid into a gas. Accordingly, the latent heat of sublimation of ice at 0°C is equal to 2838 kJ/kg, which is estimated to be the amount of heat required when ice sublimates into vapor at 0°C (Hock and Holmgren, 2005).

Specific latent heat or latent heat (λ) of a food product is a measure of the heat energy (q) released or absorbed during a phase change, per unit mass (m) Eq. (3.36).

$$\lambda = \frac{q}{m} \tag{3.36}$$

Thus, the SI unit of specific latent heat is J/kg.

3.3 The Principle of Energy Balance Calculation

According to the law of conservation of energy, the amount of energy entering a process in any form should be equal to the sum of total amount of energy leaving with the product and waste streams, stored energy, and the energy that is lost to the surroundings Eq. (3.37) (Fellows, 2000).

$$\text{Energy in} = \text{energy out} + \text{stored energy} + \text{energy lost} \tag{3.37}$$

Equation 3.37 is analogous to that of the material balance equation explained in Chapter 2. Similar to the case of material balance, in a steady state system, there is no accumulation of energy; further, the energy losses can be neglected under conditions of adequate insulation. Therefore, under the abovementioned conditions, Eq. (3.37) reduces to

$$\text{Energy in} = \text{Energy out} \tag{3.38}$$

However, in an unsteady-state system, accumulation term should be expressed as differential expression, as its value changes with time.

3.4 The Methodology of Energy Balance Calculation

The energy balance calculation comprises the six steps outlined in Figure 3.3. Each of these steps is explained subsequently by considering the following example of a cooling process.

Example 3.3

A still retort containing 500 cans of tomato concentrate was sterilized at 121°C. After sterilization, the cans need to be cooled down to 37°C before leaving the retort. The specific heats of tomato concentrate and the can material are 3730 J/kg °C and 468 J/kg °C, respectively. Each can weighs 50 g and contains 500 g of tomato concentrate. The retort wall is made of cast iron and weighs 2500 kg. The specific heat of cast iron is 554 J/kg °C. It is assumed that cooling by the surrounding air is negligible. Calculate the amount of cooling water required if it enters at 20°C and exits at 30°C.

Solution

Step 1: Define the unit operation: Unit operation in this example is cooling.
Step 2: Construct a block diagram of the process and identify the sources of heat input and exit to/from the system (Figure 3.4).

In the block diagram,

q_1 = Heat input from cans (J)
q_2 = Heat input from tomato concentrate (J)
q_3 = Heat input from cooling water at 20°C (J)
q_4 = Heat input from retort wall (J)

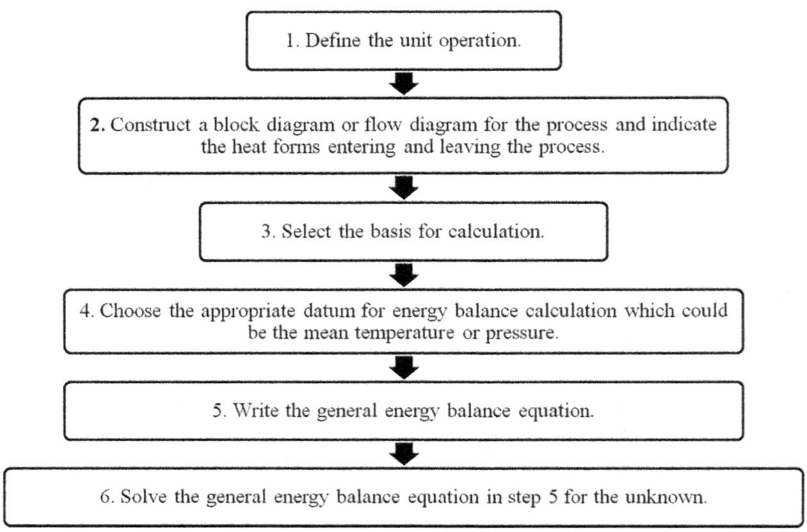

FIGURE 3.3 Steps in energy balance calculation.

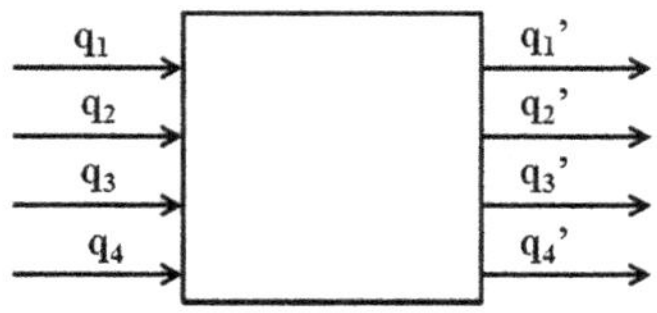

FIGURE 3.4 Block diagram of the cooling process.

q_1' = Heat outflow from cans (J)
q_2' = Heat outflow from tomato concentrate (J)
q_3' = Heat outflow from cooling water at 30°C (J)
q_4' = Heat outflow from retort wall (J)

Step 3: Choose the basis for calculation: 500 metal cans containing the tomato concentrate are selected as the basis.
Step 4: Select the datum for calculation: The reference temperature/datum is chosen as 37°C.
Step 5: Write the general energy balance equation. This equation should balance the total heat (q) exiting the system against all forms of heat entering the system, where

$$q = mC_P\Delta T \tag{3.39}$$

From the block diagram shown in Figure 3.4,

$$q_1 = (\text{weight of cans}) \times (\text{specific heat of can}) \times (\text{temperature above the datum})$$

$$q_1 = 500(\text{can}) \times 50(\text{g/can}) \times 468(\text{J/kg °C}) \times (121 - 37)(°\text{C}) = 982800\ \text{J}$$

$$\therefore q_1 = 982.8\ \text{kJ}$$

$$q_2 = (\text{weight of tomato concentrate}) \times (\text{specific heat of tomato concentrate})$$
$$\times (\text{temperature above the datum})$$

$$q_2 = 500(\text{can}) \times 500(\text{g/can}) \times 3730(\text{J/kg °C}) \times (121 - 37)(°\text{C}) = 7.833 \times 10^7\ \text{J}$$

$$\therefore q_2 = 7.833 \times 10^4\ \text{kJ}$$

$$q_3 = (\text{weight of water}) \times (\text{specific heat of water}) \times (\text{temperature above the datum})$$

$$q_3 = W(\text{kg}) \times 4180(\text{J/kg °C}) \times (20 - 37)(°\text{C}) = -71060 \times W(\text{J}) = -71.060 \times W(\text{kJ})$$

$$q_4 = 2500(\text{kg}) \times 554(\text{J/kg °C}^{-1}) \times (121 - 37)(°\text{C}) = 116340000(\text{J}) = 116340\ \text{kJ}$$

$\therefore$ Total heat entering the system is given by

$$q_{\text{in}} = q_1 + q_2 + q_3 + q_4 = 982.8(\text{kJ}) + (7.833 \times 10^4)(\text{kJ}) + (-71.060)(W)(\text{kJ}) + (116340)(\text{kJ})$$
$$= 195652.8\ \text{kJ} - 71.06\,W(\text{kJ})$$

Similarly, heat exiting the system can be calculated as explained in the following expressions:

$$q_1' = 500(\text{can}) \times 50(\text{g/can}) \times 468(\text{J/kg K}) \times (37 - 37)(\text{K}) = 0$$

$$q_2' = 500(\text{can}) \times 450(\text{g/can}) \times 3730(\text{J/kg K}) \times (37-37)(\text{K}) = 0$$

$$q_3' = W(\text{kg}) \times 4180(\text{J/kg °C}) \times (30-37) \times (\text{K}) = -29260 \times W(\text{J}) = -29.26 \times W(\text{kJ})$$

$$q_4' = 2500(\text{kg}) \times 554(\text{J/kg K}) \times (37-37)(\text{K}) = 0$$

∴ Total heat exiting the system is given by

$$q_{\text{out}} = q_1' + q_2' + q_3' + q_4' = -29.26 \times W(\text{kJ})$$

According to the law of conservation of energy, $q_{\text{in}} = q_{\text{out}}$. ∴ the general energy balance equation for this cooling process can be written as

$$195652.8 - 71.06W = -29.26W$$

Step 6: Solving the general energy balance equation, $W = 4680.7$ kg

Answer: To cool down the tomato concentrate from 121°C to 37°C, 4681 kg of cold water at 20°C is required

3.5 Steam and Its Properties

3.5.1 Steam

Steam is the gaseous form of water, generated when the water undergoes a phase change from the liquid state to the gaseous state. In food processing operations, steam is the commonly used means to transport energy. The source of steam is either solid ice or liquid water. Steam used by the food processors can be classified into two broad categories: *culinary steam* and *plant steam*. The culinary steam is used for direct injection into the product or to clean or sterilize product contact surfaces. Thus, it is also known as *sanitary* or *clean* steam. Any additives in culinary steam must comply with the requirements of the Food and Drug Administration (FDA) and the United States Department of Agriculture (USDA) pertaining to human consumption. The direct ultra-high temperature (UHT) processes adopted in dairy industry use steam injection in which pressurized culinary steam is injected into a milk flow to increase its temperature. Alternatively, for the same purpose, milk can be sprayed into an atmosphere of culinary steam; this technique is termed as *steam infusion*. On the contrary, plant steam is used in most applications that do not involve direct contact with food products or with surfaces that contact food products. It is also termed as *utility steam* or just *steam*. An established application of plant steam is its use as the heat transfer medium in noncontact type heat exchangers such as the plate heat exchanger (Chapter 5). In general, steam exists in the following four states or conditions:

i. *Dry steam:* Steam in which all the water molecules remain in the gaseous state.
ii. *Wet steam:* Steam in which a portion of water molecules give up their latent heat and condense to form tiny water droplets.
iii. *Superheated steam:* Steam which is obtained when the temperature of water vapor is higher than the boiling point temperature of the water. In other words, when the heat content of steam is higher than the saturated vapor at any pressure and temperature, then it is known as superheated steam.
iv. *Supersaturated steam:* The steam having lesser temperature and higher density at a particular saturation pressure is called supersaturated steam. This condition is obtained when the steam is cooled by its own expansion in a nozzle. However, the supersaturated steam is very unstable, and the steam soon resumes the saturated condition.

3.5.2 Formation of Steam

Steam is generated in boilers by the addition of heat to either ice or water, at constant pressure. *T–h* or *Temperature–Enthalpy diagram* (Figure 3.5) is a graphical demonstration of different stages in the transformation of 1 kg of ice into 1 kg of superheated steam.

Different stages in the formation of steam are:

i. At constant pressure, when 1 kg of ice is gradually heated under an absolute pressure, *P* bar and temperature, – *t* °C (below the freezing point), the temperature of ice will steadily increase from – *t* °C to until it reaches the freezing temperature of 0°C (From P to Q; Figure 3.5).

ii. With the addition of more heat, the ice starts melting at a constant temperature by giving off its latent heat till it is completely converted into liquid water (From Q to R; Figure 3.5). The amount of heat lost by the ice during this stage is also known as the *latent heat of fusion of ice* or *latent heat of ice.*

iii. When heated further, water gains the sensible heat and undergoes an increase in temperature to its boiling point (*t* °C), which is known as the *saturation temperature.* The corresponding pressure is termed as the *saturation pressure.* This stage of vaporization corresponds to the atmospheric pressure at 1.01325 bar or 760 mm of Hg at 100°C. The proportion of heat which is added to the liquid water to increase its temperature from 0°C to the boiling point (from R to S; Figure 3.5) is termed as *enthalpy of saturated water* or *total heat of water.* During the vaporization process, a slight increase in the volume of water (saturated water) is observed which is known as the *specific volume of saturated water* (v_f or v_w).

iv. When heated beyond the *saturation temperature* (beyond S; Figure 3.5), the water gradually begins to evaporate and convert to steam at a constant temperature. Until the steam remains in contact with water, it is called *wet steam* or *saturated steam.*

v. When additional heat is applied to the wet steam, the temperature remains constant, but the entire water converts to steam. However, it remains as wet steam. Until this point, the total energy required to generate steam from water at 0°C is called the *enthalpy of wet steam or saturated steam* (h_g). The resultant volume is known as the *specific volume of wet steam* (v_{wet}).

vi. On heating the wet steam further, the fine droplets of water in suspension will start evaporating gradually, and at a particular time point, even the final droplet evaporates. The steam

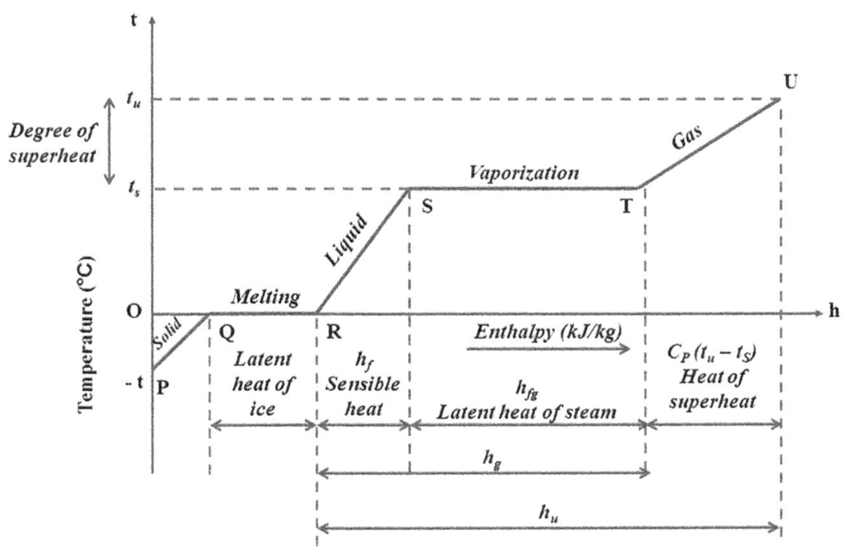

FIGURE 3.5 *T–h* or *Temperature–Enthalpy* diagram for the formation of steam at constant pressure.

corresponding to the abovementioned time point (T; Figure 3.5) is called *dry steam* or *dry saturated steam*. The resulting volume is known as *specific volume of dry steam* (v_g). The amount of energy required to convert the saturated water into dry saturated steam (from S to T; Figure 3.5) is called *latent heat of vaporization of steam* or *latent heat of steam* (h_{fg}). During the process, the saturation temperature remains constant. The total heat supplied from 0°C is called the *enthalpy of dry steam* (h_g). The dry steam does not obey the gas laws (Boyle's Law and Charles' Law).

vii. The process of further heating the wet steam (from T to U; Figure 3.5) is called *superheating*, and the resultant steam is known as the *superheated steam*. Its behavior is close to a perfect gas and obeys the gas laws to a certain extent. The increase in temperature (from t_s to t_u) during this stage is called the *degree of superheat*, and the corresponding amount of heat supplied is referred to as the *heat of superheat*, C_P, where C_P is the mean specific heat of the superheated steam. The total heat supplied from 0°C is called the *enthalpy of superheated steam* (h_u). The resultant volume is known as the *specific volume of superheated steam* (v_u). It is important to note that superheating is always carried out at constant pressure.

3.5.3 Properties of Steam

3.5.3.1 Specific Enthalpy of Steam

It is the total heat absorbed by the steam per unit mass from the freezing point of water (0°C or 273 K) to the saturation temperature (100°C or 373 K) plus the heat absorbed during evaporation. It is expressed in kJ/kg. The specific enthalpy of steam increases with the increase in temperature and pressure.

3.5.3.2 Specific Entropy of Steam

It is a theoretical value of heat energy which cannot be transformed into mechanical work under the given conditions of temperature or pressure. It is also called the degree of disorder of the system. The most common term used is the change of entropy which is mathematically given as

$$\Delta s = \frac{\Delta q}{\Delta T} = \frac{\text{Heat supplied}}{\text{Temperature of the system}} \tag{3.40}$$

Specific entropy is expressed in kJ/kg K. It decreases with an increase in temperature and pressure.

3.5.3.3 Dryness Fraction of Saturated Steam

Dryness fraction (x) is a measure of the quality of wet steam. It is the ratio of the mass of dry steam (m_g) to the mass of total wet steam ($m_g + m_w$), where m_w is the mass of water in suspension.

$$x = \frac{m_g}{m_g + m_w} \tag{3.41}$$

Total enthalpy for the steam with quality below 100% is expressed as

$$h_{\text{total}} = h_f(1 - x) + xh_g \tag{3.42}$$

where h_f is the enthalpy of liquid and h_g is the enthalpy of vapor.

3.5.3.4 Quality of Steam

Steam quality is the representation of dryness fraction in percentage.

$$\text{Quality of steam} = x_s = x \times 100 \tag{3.43}$$

Steam quality is decided based on the percentage of saturated vapor in a vapor–liquid mixture. When steam is said to be of 60% quality, it means that the vapor–liquid mixture contains 60% of steam and 40% of liquid. If further heat is added to this steam having quality less than 100%, the temperature and pressure will remain constant until the all saturated liquid is converted to vapor.

3.5.3.5 Wetness Fraction of Steam

Wetness fraction is yet another measure of the quality of wet steam, calculated as $(1 - x)$.

3.5.3.6 Priming

Priming is the representation of wetness fraction in percentage.

$$\text{Priming} = (1 - x) \times 100 \tag{3.44}$$

Therefore, quality + priming = 100%

3.5.3.7 Density of Steam

The density of steam (ρ, in kg/m^3) is the mass of steam per unit of volume of steam at the given pressure and temperature. It is the reciprocal of the specific volume, given by

$$\rho = \frac{1}{v} \tag{3.45}$$

3.5.3.8 Specific Volume of Steam

The specific volume of steam (v, m^3/kg) is the volume occupied by the steam per unit mass at a given temperature and pressure. It increases with the increase in temperature and decreases with the increase in pressure. The expression for specific volume is given by

$$V = V_l(1 - x_s) + x_s V_v \tag{3.46}$$

where V_l is the volume of liquid, x_s is the quality of steam, and V_v is the volume of the vapor.

Example 3.4

Determine the enthalpy of steam at 110°C with 90% quality.

Solution

Enthalpy of saturated vapor at 110°C = 2691.5 kJ/kg
Enthalpy of liquid at 110°C = 461.30 kJ/kg
Enthalpy of steam at 110°C with 90% quality is given by

$$H = H_l(1 - x_s) + x_s H_v$$

$$= 461.30\,(1 - 0.9) + 0.9\,(2691.5) = 2468.48 \text{ kJ/kg}$$

Example 3.5

Calculate the dryness fraction and quality of steam which has 3 kg of water in suspension with 60 kg of steam.

Solution

Given

Mass of dry steam = m_g = 60 kg

Mass of water in suspension, m_w = 3 kg

$$\therefore \text{Dryness fraction} = x = \frac{m_g}{m_g + m_W} = \frac{60}{60+3} = 0.952$$

$$\text{Quality of steam} = x_s = x \times 100 = 0.952 \times 100 = 95.2\%$$

Answer: Dryness fraction of steam = 0.952; quality of steam = 95.2%

3.5.4 Steam Table

The steam table is a compilation of the different properties of dry saturated steam and superheated steam including specific volume, specific heat or specific enthalpy, and specific entropy. Further, the latent heat of vaporization can be obtained from the steam tables by subtracting the enthalpy of saturated liquid from that of saturated vapor. Similarly, by assuming a linear relationship, other values also can be obtained by interpolation. The steam table is of two types, one that is listed as a function of temperature and the other as a function of pressure. Accordingly, the format and applications of the following tables are explained in the subsequent sections:

1. Saturated steam table (temperature-based)
2. Saturated steam table (pressure-based)
3. Superheated steam table

3.5.4.1 Saturated Steam Table (Temperature-Based)

The temperature-based saturated steam table (Appendix I) comprise the following properties of steam, applicable in the temperature range of 0°C to 374.15°C (critical temperature):

- Absolute pressure (bar)
- Specific volume (m^3/kg)
- Specific enthalpy (kJ/kg)
- Specific entropy (kJ/kg K)

The data listed in a saturated steam table always correspond to steam at the saturation point, also known as the boiling point. The boiling point is the point at which water (liquid) and steam (gas) coexist at the same temperature and pressure. Since water can remain in either liquid or gaseous state at its saturation point, two sets of data are presented in the steam table: (i) data for saturated water (liquid), which is indicated by an *f* in the subscript of all the properties, and (ii) data for saturated steam (gas), marked using a *g* in the subscript of all the properties. The legends used in the saturated steam tables are explained in Table 3.2.

3.5.4.2 Saturated Steam Table (Pressure-Based)

The pressure-based saturated steam table (Appendix II) uses two different types of pressure: absolute pressure and gauge pressure. While absolute pressure is zero referenced against a perfect vacuum, gauge pressure is zero referenced against atmospheric pressure (101.325 kPa or 14.696 psi). To distinguish between absolute pressure and gauge pressure, a suffix *g* is added to the pressure unit of the latter (e.g., kPag or psig).

TABLE 3.2

Explanation of Legends Used in the Saturated Steam Table

Legend	Explanation
P	Pressure of the steam or water
T	Saturation temperature of steam or water (boiling point)
v_f	Specific volume of saturated water (liquid)
v_g	Specific volume of saturated steam (gas)
h_f	Specific enthalpy of saturated water
h_{fg}	Latent heat of evaporation
h_g	Specific enthalpy of saturated steam

The following properties of steam are tabulated in the range of pressure between 0.0061 and 221.2 bar (critical pressure):

- Temperature (t, in °C)
- Specific volume (v, in m^3/kg)
- Specific enthalpy (h, in kJ/kg)
- Specific entropy (s, in kJ/kg K)

Applications of the saturated steam table are as follows:

- To calculate the specific enthalpy and specific entropy of steam
- To determine the temperature of saturated steam from steam pressure
- To determine the pressure of saturated steam from steam temperature

Example 3.6

If the dryness fraction 0.6, calculate the specific enthalpy and specific entropy for

a. 2 kg of saturated steam at 20°C
b. 1 kg of saturated steam at 0.025 bar

Solution

a. Given

- Mass of steam (m) = 2 kg
- Temperature of steam (t) = 20°C
- Dryness fraction = 0.6

From the steam tables, corresponding to $t = 20°C$,

$$h_f = 83.9\,\text{kJ/kg};\quad h_{fg} = 2454.3\,\text{kJ/kg};\quad s_f = 0.296\,\text{kJ/kg K};\quad s_{fg} = 8.372\,\text{kJ/kg K}$$

Specific enthalpy of steam

$$h = m\left[h_f + xh_{fg}\right] = 2\times\left[83.9 + (0.6\times 2454.3)\right] = 3112.96\frac{\text{kJ}}{\text{kg}}$$

Specific entropy of steam

$$s = m\left[s_f + xs_{fg}\right] = 2\times\left[0.296 + (0.6\times 8.372)\right] = 10.6384\frac{\text{kJ}}{\text{kg K}}$$

b. Given

- Mass of steam (m) = 1 kg
- Pressure of steam (p) = 0.025 bar
- Dryness fraction = 0.6

From the steam tables, corresponding to $p = 0.025$ bar,

$$h_f = 88\,\text{kJ/kg};\quad h_{fg} = 2452\,\text{kJ/kg};\quad s_f = 0.310\,\text{kJ/kg K};\quad s_{fg} = 8.336\,\text{kJ/kg K}$$

Specific enthalpy of steam

$$h = m\left[h_f + xh_{fg}\right] = 1\times\left[88 + (0.6\times 2452)\right] = 1559.2\frac{\text{kJ}}{\text{kg}}$$

Specific entropy of steam

$$s = m\left[s_f + xs_{fg}\right] = 1\times\left[0.310 + (0.6\times 8.336)\right] = 5.312\frac{\text{kJ}}{\text{kg K}}$$

Example 3.7

Determine the enthalpy of dry saturated steam at temperature 88°C from the steam table using interpolation.

Solution

From steam tables, enthalpy of saturated steam at 85°C = 2651.9 kJ/kg

Enthalpy of saturated steam at 90°C = 2660.1 kJ/kg

Assuming that the enthalpy and temperature relationship is linear between a temperature of 85°C and 90°C, enthalpy at 88°C can be given by

$$H = \left(\frac{88-85}{90-85}\right)(2660.1 - 2651.9) + 2651.9$$

$$= \left(\frac{3}{5}\right)(8.2) + 2651.9$$

$$= 2656.82$$

Answer: The enthalpy of dry saturated steam at 88°C is 2656.82 kJ/kg

3.5.4.3 Superheated Steam Table

The properties of superheated steam cannot be obtained from the saturated steam table because the temperature of superheated steam, unlike saturated steam, varies considerably even at the same pressure. This signifies the requirement of a superheated steam table (Appendix III, IV, and V). This table provides the values for specific volume, specific enthalpy, and specific entropy of superheated steam from an absolute pressure of 0.02 bar to 221.2 bar (critical pressure), at various temperatures from 100°C to 800°C.

3.5.5 Mollier Diagram

Mollier diagram is the graphical representation of the steam table (Figure 3.6). It is the plot between specific entropy (x-axis; abscissa) and specific enthalpy (y-axis; ordinate). A reasonably horizontal line termed as the *saturation curve* divides the diagram into two parts. The lower part is called the wet steam region and contains the values of wet steam. The upper portion is named as the superheated steam

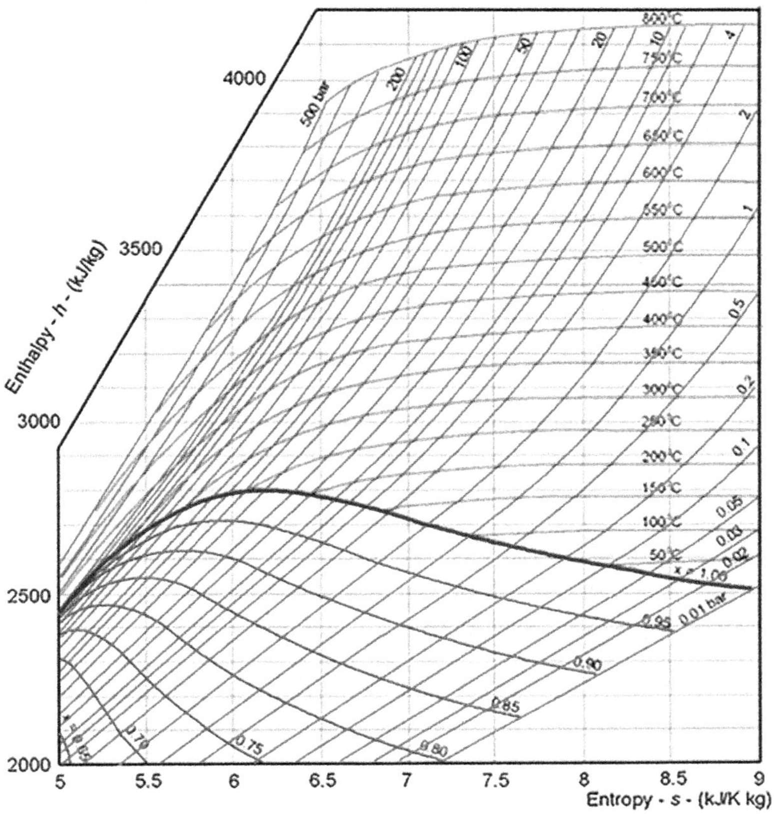

FIGURE 3.6 Mollier diagram. (Reproduced with permission from Demirel, Y. 2016. *Energy. Green Energy and Technology*. Cham: Springer.)

region and includes the values of superheated steam. The following lines constitute the Mollier diagram (Figure 3.6):

- ***Dryness fraction lines:*** These slightly curved lines extending from left to right represent the condition of wet steam. The dryness fraction lines are present in the wet steam region just below the saturation curve corresponding to $x = 1$.
- ***Constant pressure lines:*** These lines start from the wet steam region, move upward, extend beyond the saturation curve to the superheated steam region (Figure 3.6). While the constant pressure lines are straight in the wet steam region, they are curved slightly upward in the superheated region.
- ***Constant temperature lines:*** These slightly curved lines in the horizontal direction are confined to the superheated steam region above the saturation curve (Figure 3.6). The constant temperature lines signify the temperature of steam between different values of enthalpy and entropy.

3.6 Energy Balance Calculations in Food Processing Plants

This section presents the general scheme of energy balance calculation for two important industrial operations: spray drying of milk and pasteurization of fruit juice.

3.6.1 Spray Drying of Milk (Dairy Industry)

Step 1: Define the unit operation

The unit operation is spray drying. The principle and operation of spray drying would be explained in detail in Chapter 10 (Drying). A simple description is given here to acquire a basic understanding of the process. The spray-drying process involves the production of dried milk powder from concentrated milk which in turn is obtained by removing a portion of water from the fresh liquid milk in an evaporator before entering the spray dryer (Figure 3.7). The evaporator uses saturated steam at atmospheric pressure, and the water evaporated is condensed using cooling water that enters the condenser at a lower temperature and leaves it at a higher temperature. An air heater (Figure 3.7) is used to raise the temperature of ambient air to the specified inlet air temperature that enters the spray-drying chamber. The spray-drying chamber is a cylindroconical structure in which the actual drying of concentrated milk takes place. By striking an energy balance around the evaporator, the steam consumption and the amount of requirement of cooling water can be calculated. The process parameters of importance to the energy balance calculation are listed as follows:

1. Mass of steam consumed (kg) = m_s
2. Mass of milk concentrate (kg) = $m_{m.c}$
3. Mass of water evaporated (kg) = m_{wv}
4. Mass of cooling water (kg) = m_{cw}
5. Mass of dry air (kg) = m_a
6. Specific humidity of air (kg water/kg dry air) = w
7. Heating load of the air heater (kJ) = q_{H}
8. Enthalpy or total heat of milk solids (kJ) = H_{ms}
9. Enthalpy or total heat of water (kJ) = H_w
10. Enthalpy or total heat of dry air (kJ) = H_a
11. Enthalpy or total heat of water evaporated (kJ) = H_v
12. Enthalpy or total heat of saturated steam (kJ) = H_s
13. Enthalpy or total heat of condensate (kJ) = H_c

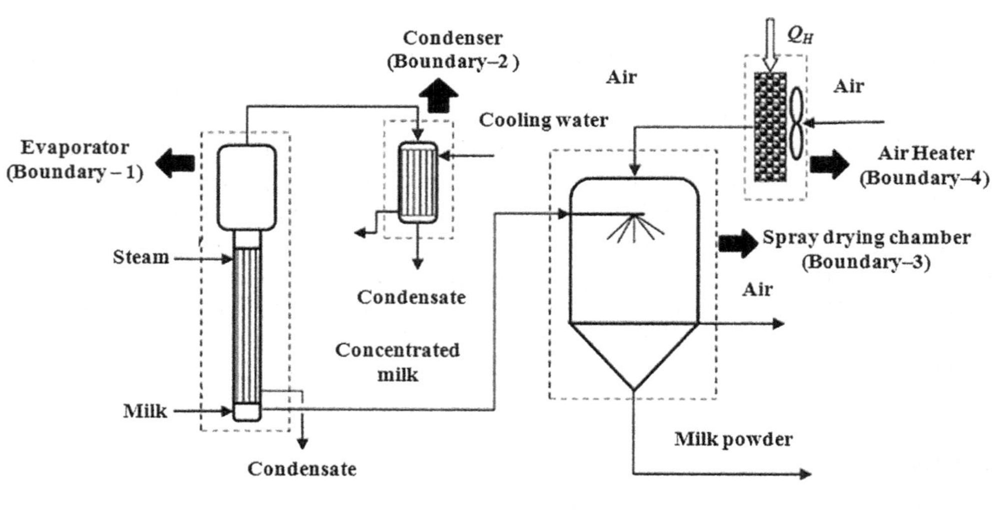

FIGURE 3.7 Flow diagram for the spray-drying process. (Modified from Zogzas, N. 2015. Mass and energy balances. In *Food Engineering Handbook: Food Engineering Fundamentals*, eds. T. Varzakas and C. Tzia, 3–40. Boca Raton, FL: CRC Press.)

Step 2: Construct a block diagram or flow diagram of the process

For the spray-drying operation, the energy balance is applied around the evaporator (boundary—1), spray-drying chamber (boundary—3), and air heater (boundary—4), as depicted in process flow diagram (Figure 3.7). The block diagrams as shown in Figures 3.8–3.10 indicate the input and output of heat to and from the evaporator, spray-drying chamber, and air heater, respectively.

In the block diagram shown in Figure 3.8,

$q_{ms.\text{in}}$ = Heat input from milk solids (J)
$q_{w.\text{in}}$ = Heat input from water (J)
$q_{c.\text{in}}$ = Heat input from steam condensate (J)
$q_{ms.\text{out}}$ = Heat outflow from milk solids (J)
$q_{w.\text{out}}$ = Heat outflow from water (J)
$q_{c.\text{out}}$ = Heat outflow from steam condensate (J)
$q_{v.\text{out}}$ = Heat outflow from water vapor (J)

In the block diagram shown in Figure 3.9,

$q_{ms.\text{in}}$ = Heat input from milk solids (J)
$q_{w.\text{in}}$ = Heat input from water (J)
$q_{a.\text{in}}$ = Heat input from dry air (J)
$q_{v.\text{in}}$ = Heat input from water vapor (J)
$q_{ms.\text{out}}$ = Heat outflow from milk solids (J)
$q_{w.\text{out}}$ = Heat outflow from water (J)
$q_{a.\text{out}}$ = Heat outflow from air (J)
$q_{v.\text{out}}$ = Heat outflow from water vapor (J)

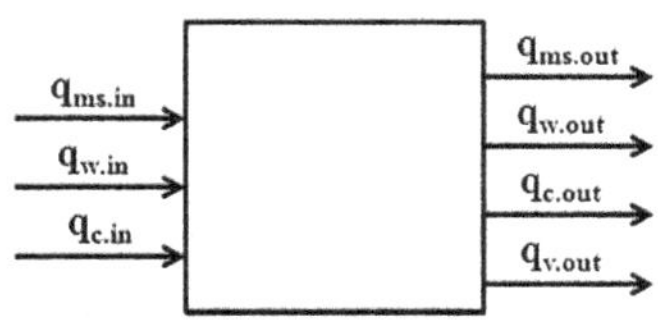

FIGURE 3.8 Block diagram for the evaporation process.

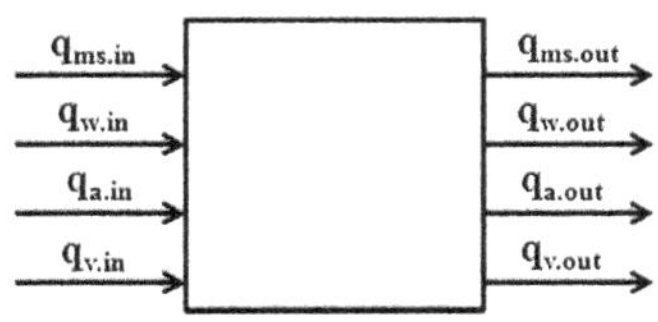

FIGURE 3.9 Block diagram for the spray-drying process.

FIGURE 3.10 Block diagram for the air-heating process.

In the block diagram shown in Figure 3.10,

q_{in} = Heat input from air (J)
q_H = Heat load (J)
q_{out} = Heat outflow from air (J)

Step 3: Selection of basis for calculation
Since spray drying is a continuous process, 1 h of operation is chosen as the basis for calculation.

Step 4: Selection of datum for calculation
A 0°C is chosen as the datum for calculation. Thus, the value for specific enthalpies of milk solids and liquid water is considered as 0 at 0°C.

Step 5: Write the general energy balance equation

$$q_{in} = q_{out} \tag{3.47}$$

i. **Evaporator**

$$q_{ms.in} + q_{w.in} + q_{c.in} = q_{ms.out} + q_{w.out} + q_{c.out} + q_{v.out} \tag{3.48}$$

ii. **Spray-drying chamber**

$$q_{ms.in} + q_{w.in} + q_{a.in} + q_{v.in} = q_{ms.out} + q_{w.out} + q_{a.out} + q_{v.out} \tag{3.49}$$

iii. **Air heater**

$$q_H + q_{a.in} = q_{a.out} \tag{3.50}$$

(or)

$$q_H = q_{a.out} - q_{a.in} \tag{3.51}$$

(or)

$$q_H = m_a \cdot q_{a.out} + w_{out} m_a q_{v.out} - m_a h_{a.in} - w_{in} m_a h_{v.in} \tag{3.52}$$

(or)

$$q_H = m_a \left(h_{a.out} + w_{out} h_{v.out} - h_{a.in} - w_{in} h_{v.in}\right) \tag{3.53}$$

Step 6: Solving the general energy balance equation

i. **Evaporator**

The enthalpies of milk solids and liquid water can be determined by using Eq. (3.48) and the following equations from (3.54) to (3.56):

$$H_{s.in} = m_s \times h_{g.Ts} \tag{3.54}$$

$$H_{c.out} = m_s \times h_{f.Ts} \tag{3.55}$$

$$H_{v.out} = m_{wv} \times h_{g.Tv} \tag{3.56}$$

where $h_{g.Ts}$ and $h_{g.Tv}$ are the specific enthalpies of saturated steam at temperatures T_s and T_v, respectively. $h_{f.Ts}$ is the specific enthalpy of saturated water (condensate) at temperature T_s.

m_s and m_{wv} are the mass of steam consumed (kg) and mass of water evaporated (kg), respectively. Substituting the abovementioned values and solving Eq. (3.48) will provide the amount of steam consumed in the evaporation process.

ii. **Spray-drying chamber**

The enthalpies of milk solids and liquid water can be determined by using Eq. (3.49). The enthalpies of dry air (H_a) and water vapor (H_v) can be calculated using the following equations

$$H_a = m_a \times h_a \tag{3.57}$$

$$H_v = w \times m_a \times h_v \tag{3.58}$$

where w is the specific humidity of air (kg water/kg dry air), h_a and h_v are the specific enthalpies of dry air and water vapor, respectively, given by the following equations (ASHRAE, 2009):

$$h_a \approx 1.006\ T \tag{3.59}$$

$$h_v \approx 2501 + 1.86\ T \tag{3.60}$$

where the number 1.006 in Eq. (3.59) denotes the specific heat capacity of dry air at a constant pressure in kJ/kg °C, which is justified as dry air can be considered ideal at atmospheric pressure. In Eq. (3.60), the specific enthalpy of water vapor mixed with dry air (h_v) is calculated by adding the specific enthalpy of saturated water vapor (~2501 kJ/kg) at 0°C and the product of its specific heat capacity at constant pressure (C_P = 1.86 kJ/kg °C) and temperature (T). Substituting the abovementioned values and solving Eq. (3.49) will provide the amount of steam consumed in the spray-drying process.

iii. **Air heater**

In Eqs. (3.57) and (3.58), h_a and h_v can be calculated using Eqs. (3.59) and (3.60), respectively. w_{in} and w_{out} are the corresponding specific humidity of air (kg water/kg dry air) entering and leaving the air heater. Substituting the abovementioned values of specific enthalpies and specific humidity in Eq. (3.53) gives the heat load (q_H) of the air heater.

3.6.2 Pasteurization of Fruit Juice (Beverage Industry)

Step 1: Define the unit operation

In this case, the process under consideration is pasteurization which involves heating (the unit operation) a defined quantity of fruit juice ($m_{J.\text{in}}$ kg) from its initial temperature (T_i) to the defined temperature of pasteurization (T_f), using steam of quantity (m_s kg) at temperature T_S, to produce $m_{J.\text{out}}$ kg of pasteurized juice.

Step 2: Construct a block diagram or a flow diagram of the process.

Step 3: Selection of basis for calculation

$m_{J.\text{in}}$ kg of incoming fresh juice can be chosen as the basis for calculation in case of a batch process. In case of continuous pasteurization, y hour (say, 1 h), the time of operation can be selected as the basis.

FIGURE 3.11 Block diagram for the pasteurization process.

Step 4: Selection of datum for calculation

The steam temperature (T_S) (usually 100°C) can be selected as the datum for energy balance calculation around the pasteurizer.

Step 5: Write the general energy balance equation

From the block diagram (Figure 3.11), it can be understood that heat is lost by m_s kg of steam to heat the fruit juice from T_i to T_f. According to the law of conservation of energy, heat lost by the steam should be equal to the heat gained by the juice. As the steam loses its latent heat of vaporization (λ_V), it will condense from the vapor phase to the liquid phase, without a change in temperature. Therefore,

$$q_S = m_S \times \lambda_V \tag{3.61}$$

The latent heat of vaporization at atmospheric pressure is 2260 kJ/kg.

$$\therefore q_S = m_S \times 2260 \tag{3.62}$$

In addition, there also occurs a loss in the sensible heat of the condensed steam (water) as it cools down from T_S to T_C, where T_C is the condensate temperature. The loss in the sensible heat of the condensed steam (q_C) can be calculated using Eq. (3.35) by substituting the values for the mass of steam (m_S), the heat capacity of water ($C_{P\text{water}} = 4.18$ kJ/kg °C), and the temperature difference between steam and condensate ($T_S - T_C$). Therefore, the total heat lost by the steam is given by

$$q_{\text{heat lost by steam}} = q_S + q_C \tag{3.63}$$

The sensible heat gained by the pasteurized juice ($q_{\text{gained by the juice}}$) can also be calculated using Eq. (3.35) by substituting the values for the mass of juice ($m_{J.\text{in}}$), the heat capacity of juice ($C_{P\text{juice}}$), and the difference between initial and final temperatures ($T_f - T_i$) of the juice. Hence, the general energy balance equation can be written as

$$q_{\text{heat lost by steam}} = q_{\text{gained by the juice}} \tag{3.64}$$

Step 6: Solving the general energy balance equation

$$(m_S \times \lambda_V) + (m_S \times C_{P_{\text{water}}} \times (T_S - T_C)) = m_{J.\text{in}} \times C_{P_{\text{juice}}} \times (T_f - T_i) \tag{3.65}$$

The solution for the preceding general energy balance equation gives the mass of steam required for the pasteurization process.

3.7 Problems to Practice

3.7.1 Multiple Choice Questions

1. The numerical value of net heat transferred from the system to surroundings is
   a. positive
   b. negative
   c. zero
   d. unity

Answer: b

2. Dryness fraction of saturated steam is always
   a. less than or equal to 1
   b. greater than or equal to 1
   c. greater than 1
   d. equal to 1

Answer: a

3. The proportion of heat which is added to the liquid water to increase its temperature from 0°C to the boiling point is
   a. enthalpy of saturated water
   b. enthalpy of wet steam
   c. latent heat of vaporization of steam
   d. enthalpy of superheated steam

Answer: a

4. The horizontal line that divides the Mollier diagram into two parts is
   a. dryness fraction lines
   b. saturation curve
   c. constant pressure lines
   d. constant temperature lines

Answer: b

5. The quality of saturated steam with dryness fraction 0.7 is
   a. 30%
   b. 100%
   c. 70%
   d. 35%

Answer: c

6. The enthalpy of superheated steam with pressure 150 kPa and temperature of 121.1°C is
   a. 2767.3 kJ/kg
   b. 1767.3 kJ/kg
   c. 3767.3 kJ/kg
   d. 2713.6 kJ/kg

Answer: d

7. Find the enthalpy of saturated steam at 53°C, if the enthalpy at 50°C and 55°C is 2591.3 kJ/kg and 2600.1 kJ/kg, respectively.
   a. 2197.4 kJ/kg
   b. 2397.4 kJ/kg
   c. 2596.6 kJ/kg
   d. 2797.4 kJ/kg

Answer: c

8. Saturated steam at 160°C is flowing in a steel pipe of inner diameter 4.5 cm. If the average velocity of the steam is 5 m/s, the mass flow rate of steam is
   a. 0.0485 kg/s
   b. 0.0259 kg/s
   c. 0.0385 kg/s
   d. 0.2285 kg/s

Answer: b

9. The dryness fraction of 1 kg of steam at 1400 kPa with a total enthalpy content of 2202.09 kJ is
   a. 0.3
   b. 0.2
   c. 0.5
   d. 0.7

Answer: d

10. The specific heat of orange juice concentrate with 35% solids is
    a. 3013.2 J/kg K
    b. 4008.8 J/kg K
    c. 3200 J/kg K
    d. 1008.9 J/kg K

Answer: a

11. The potential energy acquired by a 100 kg bag of wheat flour lifted from ground to a silo at a height of 50 ft is
    a. 14935.2 J
    b. 1493.52 J
    c. 149.352 J
    d. 14.9352 J

Answer: a

12. Internal energy is an
    a. intensive property
    b. extensive property
    c. colligative property
    d. none of the above

Answer: b

13. The predominant form of energy in microwave heating of foods is
    a. electrical energy
    b. chemical energy
    c. mechanical energy
    d. electromagnetic energy

Answer: d

14. The heat which can be sensed as it leads to a change in temperature when it is added or removed from the food product is
    a. latent heat
    b. sensible heat
    c. total heat
    d. none of the above

Answer: b

15. Steam of quality 40% contains
    a. 60% of steam and 40% of liquid
    b. 50% of steam and 50% of liquid
    c. 40% of steam and 60% of liquid
    d. 40% of steam and 40% of liquid

Answer: c

16. The dryness fraction of steam containing 2 kg of water in 50 kg of steam is
    a. 0.96
    b. 0.861
    c. 0.096
    d. 0.038

Answer: a

17. The quality of steam in Problem 16 is
    a. 86.1%
    b. 9.61%
    c. 96%
    d. 3.8%

Answer: c

18. When a system does not accumulate of mass or energy over a specific time, it is said to be
    a. at steady state
    b. at unsteady state
    c. closed
    d. none of the above

Answer: a

19. In the Mollier diagram, the lines which start from the wet steam region, move upward, and extend beyond the saturation curve to the superheated steam region are
    a. dryness fraction lines
    b. constant pressure lines
    c. constant temperature lines
    d. saturation curve

Answer: b

20. The total energy required to generate steam from water at 0°C is called
    a. enthalpy of superheated steam
    b. enthalpy of saturated water
    c. latent heat of vaporization of steam
    d. enthalpy of wet steam or saturated steam

Answer: d

3.7.2 Numerical Problems

1. Estimate the quantity of heat required to increase the temperature of 20 kg of tomato pulp from 30°C to 70°C if the mean heat capacity of tomato pulp is 4 kJ/kg K.

   Given

   i. Unit operation: Heating
   ii. Quantity of tomato pulp (m_t) = 20 kg
   iii. Initial temperature of tomato pulp (T_i) = 30°C = 303 K
   iv. Final temperature of tomato pulp (T_f) = 70°C = 343 K
   v. Mean heat capacity of tomato pulp (C_{Pt}) = 4 kJ/kg K

   To find: Quantity of heat (q) required to increase the temperature of 20 kg of tomato pulp from 30°C to 70°C.

   Solution

$$q = m_t C_{Pt} \left(T_i - T_f\right)$$

$$q = 20 \times 4 \times (343 - 303) = 3200\,\text{kJ}$$

Answer: 3200 kJ of heat (q) is required to increase the temperature of 20 kg of tomato pulp from 30°C to 70°C

2. A heat exchanger is used for the heating of 500 kg/h of milk from 30°C to 70°C using water as the heating medium. The heating medium enters the heat exchanger at 80°C and exits at 60°C. Determine the mass flow rate of the heating medium if the heat is lost to the environment at the rate of 1 kW. The heat capacity of water is 4.2 kJ/kg °C and that of milk is 3.9 kJ/kg °C.

   Given

   i. Unit operation: Heating
   ii. Heating medium: Water
   iii. Mass flow rate of milk (m_m) = 500 kg/h
   iv. Initial temperature of milk (T_i) = 30°C
   v. Final temperature of milk (T_f) = 70°C
   vi. Initial temperature of heating medium $\left(T_{hi}\right)$ = 80°C
   vii. Final temperature of heating medium $\left(T_{hf}\right)$ = 60°C
   viii. Rate of heat loss (q_L) = 1 kW
   ix. Heat capacity of water $\left(c_{Pw}\right)$ = 4.2 kJ/kg °C
   x. Heat capacity of milk $\left(c_{Pm}\right)$ = 3.9 kJ/kg °C

   To find

   i. Mass flow rate of the heating medium $\left(\dot{m}_H\right)$

Solution
Basis: 1 h

In the preceding block diagram,

$$q_1 = \text{Heat input from milk (kJ)}$$

$$q_2 = \text{Heat input from heating medium (kJ)}$$

$$q_1' = \text{Heat out from milk (kJ)}$$

$$q_2' = \text{Heat out from heating medium (kJ)}$$

$$q_L = \text{Heat loss (kJ)}$$

Datum (T_D): 30°C
General energy balance equation: $q_{\text{in}} = q_{\text{out}}$

$$q = (\text{mass flow rate}) \times (\text{specific heat}) \times (\text{temperature above the datum})$$

In this problem,

$$q_1 + q_2 = q_1' + q_2' + q_L$$

$$q_1 = \text{Heat input from milk} = m_m c_{Pm}(T_i - T_D) = 500 \times 3.9 \times (30 - 30) = 0$$

$$q_2 = \text{Heat input from heating medium} = m_H c_{Pw}(T_{hi} - T_D) = (x) \times 4.2 \times (80 - 30) = 210x$$

$$q_1' = \text{Heat out from milk} = m_m c_{Pm}(T_f - T_D) = 500 \times 3.9 \times (70 - 30) = 78000\,\text{kJ}$$

$$q_2' = \text{Heat out from heating medium} = m_H c_{PH}(T_{hf} - T_D) = x \times 4.2 \times (60 - 30) = 126x$$

$$q_L = 1\,\text{kW} = 1\frac{\text{kJ}}{\text{s}} = 1\frac{\text{kJ}}{(1/3600)\,\text{h}} = 3600\frac{\text{kJ}}{\text{h}}$$

$$0 + 210x = 78000 + 126x + 3600$$

$$210x - 126x = 81600$$

$$84x = 81600$$

$$\therefore x = \frac{81600}{84} = 971.4\frac{\text{kg}}{\text{h}}$$

Answer: Mass flow rate of the heating medium = 971.4 kg/h

3. Calculate the amount of saturated steam at 121 kPa pressure required to heat 750 g/h of juice from 8°C to 80°C if the heat capacity of juice is 4 kJ/ kg °C.

Given

i. Unit operation: Heating
ii. Steam pressure (P) = 121 kPa
iii. Mass flow rate of juice $(\dot{m}_J) = 750$ g/h = 0.75 kg/h
iv. Initial temperature of juice (Ti) = 8°C
v. Final temperature of juice (T_f) = 80°C
vi. Heat capacity of juice $\left(c_{P_{\text{juice}}}\right) = 4$ kJ/kg °C

To find

i. Amount of saturated steam required to heat the juice from 8°C to 80°C.

Solution

Basis: 1 h

Datum: 8°C

Energy balance equation

$$\text{Heat given by steam} = \text{Heat received by juice}$$

$$q_S = q_J$$

$$q_J = \dot{m}_J \times c_{P_{\text{juice}}} \times (80 - 8) = 0.75 \times 4 \times 72 = 216\,\text{kJ}$$

$$q_S = \dot{m}_S \lambda_S$$

$$\lambda_S = \text{Latent heat of steam at } 121\,\text{kPa} = 439.3686\,\text{kJ/kg}$$

$$q_S = \dot{m}_S \lambda_S$$

$$q_S = \dot{m}_S\,(439.3686) = 216\,\text{kJ}$$

$$\therefore \dot{m}_S = \frac{216}{439.3686} = 0.492\,\text{kg}$$

Answer: 0.492 kg of saturated steam is required to heat the juice from 8°C to 80°C

4. Calculate the specific heat of a food composition containing 15% protein, 15% fat, and 70% water.

Given

i. Protein content of food product = 15%
ii. Fat content of food product = 15%
iii. Water content of food product = 70%

To find: Specific heat of the food composition.

Solution

Applying the Charm's model and considering that the proteins constitute the nonfat solids,

$$C_P = 2093X_f + 1256X_S + 4187X_W$$

$$C_P = (2093 \times 0.15) + (1256 \times 0.15) + (4187 \times 0.7) = 3433.25\,\text{kJ/kg}\,^\circ\text{C}$$

Answer: Specific heat of the food composition is 3433.25 kJ/kg °C

5. Determine the specific heat capacity at constant pressure of the following compositions by mass: (i) 88% water, 0.5% ash, 10% carbohydrate, 0.7% protein, and 0.8% fat; (ii) 10.0% water, 0.5% ash, 80% carbohydrate, 8% protein, and 1.5% fat.

Given

i. Food composition-1: 88% water, 0.5% ash, 10% carbohydrate, 0.7% protein, and 0.8% fat
ii. Food composition-2: 10.0% water, 0.5% ash, 80% carbohydrate, 8% protein, and 1.5% fat

To find: Specific heat capacity at constant pressure (C_P)

Solution

Applying the Heldman and Singh model for specific heat capacity,

$$C_P = 1424(C) + 1549(P) + 1675(F) + 837(A) + 4187(W)$$

Composition-1

$$C_{P_1} = (1424 \times 0.1) + (1549 \times 0.007) + (1675 \times 0.008) + (837 \times 0.005) + (4187 \times 0.88)$$

$$C_{P_1} = 3855.4 \ \text{kJ/kg}\,^\circ\text{C}$$

$$C_{P_2} = (1424 \times 0.80) + (1549 \times 0.008) + (1675 \times 0.0015) + (837 \times 0.005) + (4187 \times 0.1)$$

$$C_{P_2} = 1576.9\,\text{kJ/kg}\,^\circ\text{C}$$

Answer: Specific heat of the first food composition is 3855.4 kJ/kg °C and that of the second food composition is 1576.9 kJ/kg °C

6. Milk is flowing through a heat exchanger at the rate of 2000 kg/h. The heat exchanger supplies heat at the rate of 111600 kJ/h. The outlet temperature of the product is 95°C. Determine the inlet temperature of the milk (specific heat of milk = 3.8 kJ/kg K).

Given

i. Unit operation: Heating
ii. Mass flow rate of milk $(\dot{m}) = 2000$ kg/h
iii. Heat supplied by the heat exchanger (q) = 111600 kJ/h
iv. Outlet temperature of milk (T_o) = 95°C = 368 K
v. Specific heat of milk (C_{Pm}) = 3.8 kJ/kg K

To find: Inlet temperature of milk (T_i)

Solution

Basis: 1 h

$$q = m\ C_{Pm}(T_o - T_i)$$

$$111600 = 2000 \times 3.8 \times (368 - T_i)$$

$$368 - T_i = 14.684$$

$$\therefore T_i = 353.316 \approx 353\,\text{K} = 80°\text{C}$$

Answer: Inlet temperature of milk (T_i) = 80°C

7. A liquid food product is being cooled from 90°C to 30°C in an indirect heat exchanger using cold water as a cooling medium. The mass flow rate of the product is 1800 kg/h. Determine the water flow rate required to accomplish product cooling if the temperature of water increases from 10°C to 30°C (specific heat of product and water is 3.8 and 4.1 kJ/kg K, respectively).

Given

i. Unit operation: Cooling
ii. Initial temperature of liquid food product (T_i) = 90°C
iii. Final temperature of liquid food product (T_f) = 30°C
iv. Mass flow rate of product $(\dot{m}_p)$ = 1800 kg/h
v. Cooling medium: Water
vi. Initial temperature of water (T_{wi}) = 10°C
vii. Final temperature of water (T_{wo}) = 30°C
viii. Specific heat of liquid food product (C_{Pp}) = 3.8 kJ/kg K
ix. Specific heat of water (C_{Pw}) = 4.1 kJ/kg K

To find: Flow rate of water required to accomplish product cooling $(\dot{m}_w)$

Solution
Basis: 1 h

$$\text{Heat given out by the product} = \text{Heat absorbed by the cooling medium}$$

$$m_p C_{Pp}\left(T_i - T_f\right) = m_w C_{Pw}\left(T_{wo} - T_{wi}\right)$$

$$1800 \times 3.8 \times (90 - 30) = m_w \times 4.1 \times (30 - 10)$$

$$410400 = 82(m_w)$$

$$\therefore m_w = 5004.9 \approx 5005\,\text{kg}$$

Answer: 5005 kg/h of water is required to cool the product from 90°C to 30°C

8. Steam is used for the peeling of potato in a semicontinuous operation. 4 kg of steam is supplied per 100 kg of unpeeled potatoes. The unpeeled potatoes enter the system with a temperature 17°C, and the peeled potatoes leave at 35°C. A waste stream from the system leaves at 60°C. Assuming reference temperature as 0°C, heat content of the steam is 2750 kJ/kg. Determine the quantities of the waste stream and the peeled potatoes from the process (specific heat of unpeeled potatoes, waste stream, and peeled potatoes is 3.7 kJ/kg K, 4.2 kJ/kg K, and 3.5 kJ/kg K, respectively). Consider the steam pressure to be 121 kPa.

Given

i. Process: Peeling of potatoes
ii. Steam supply (m_s) = 4 kg per 100 kg of unpeeled potatoes

iii. Heat content of steam $(\lambda_s) = 2750$ kJ/kg
iv. Temperature of unpeeled potatoes (T_i) = 17°C
v. Temperature of peeled potatoes (T_o) = 35°C
vi. Temperature of waste stream (T_w) = 60°C
vii. Specific heat of unpeeled potatoes $(C_{Pupp}) = 3.7$ kJ/kg K
viii. Specific heat of peeled potatoes $(C_{Ppp}) = 3.5$ kJ/kg K
ix. Specific heat of waste stream $(C_{Pw}) = 4.2$ kJ/kg K

To find: Quantities of the peeled potatoes (m_{pp}) and waste stream (m_w) from the process.

Solution

Basis: 100 kg of unpeeled potatoes

Reference temperature: 0°C

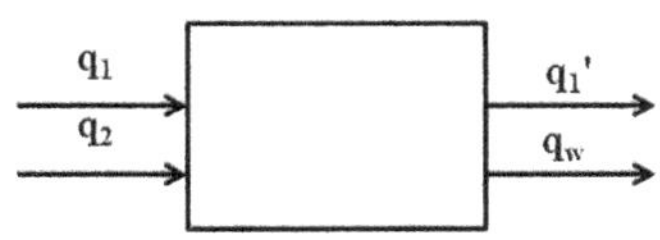

In the preceding block diagram,

q_1 = Heat input from unpeeled potatoes (kJ)
q_2 = Heat input from steam (kJ)
q_1' = Heat output from peeled potatoes (kJ)
q_w = Heat output from waste stream (kJ)

General energy balance equation: $q_{in} = q_{out}$

$$q = (\text{mass flow rate}) \times (\text{specific heat}) \times (\text{temperature above the datum})$$

In this problem,

$$q_1 + q_2 = q_1' + q_w$$

$$q_1 = \text{Heat in from unpeeled potatoes} = 100 \times 3.7 \times (17 - 0) = 6290\,\text{kJ}$$

$$q_2 = \text{Heat in from steam} = m_s\ \lambda_s = 4 \times 2750 = 11000\,\text{kJ}$$

$$q_1' = \text{Heat out from peeled potatoes} = m_{pp} \times 3.5 \times (35 - 0) = 122.5\ (m_{pp})$$

$$q_w = \text{Heat out from waste stream} = m_w \times 4.2 \times (60 - 0) = 252\ (m_w)$$

$$6290 + 11000 = (122.5 \times m_{pp}) + (252 \times m_w)$$

$$122.5\ m_{pp} + 252\ m_w = 17290 \tag{1}$$

Overall mass balance equation

$$m_{upp} = m_{pp} + m_w$$

$$m_{pp} + m_w = 100 \tag{2}$$

Multiplying Eq. (2) by 252 and subtracting Eq. (1) from Eq. (2),

$$\begin{array}{rl} & 252m_{pp} + 252m_w = 25200 \\ (-) & 122.5\ m_{pp} + 252\ m_w = 17290 \\ \hline & 129.5m_{pp} = 7910 \end{array}$$

$$\therefore m_{pp} = 61.1\text{kg}$$

From Eq. (2),

$$m_w = 100 - m_{pp} = 100 - 61.1 = 38.9\,\text{kg}$$

Answer: (i) Quantity of peeled potatoes = 61.1 kg
(ii) Quantity of waste stream = 38.9 kg

9. A closed rigid vessel contains 250 g of liquid and 15 g of water vapor in equilibrium at 1000 kPa. Calculate the volume of the mixture.

Given

i. Proportion of liquid in the liquid–vapor mixture = 250 g = 0.25 kg
ii. Proportion of vapor in the liquid–vapor mixture = 15 g = 0.015 kg
iii. Pressure of the liquid–vapor mixture = 1000 kPa

To find: Volume of the liquid–vapor mixture

Solution

From the steam table,

$$\text{volume of the liquid, } V_l = 0.25\text{kg}(0.001128)\text{m}^3/\text{kg} = 0.000282\ \text{m}^3$$

$$\text{volume of the vapor, } V_v = 0.015\,\text{kg}(0.19380)\text{m}^3/\text{kg} = 0.002907\,\text{m}^3$$

$$\therefore \text{Total volume} = V_l + V_v = 0.000282 + 0.002907 = 0.003189\,\text{m}^3$$

Answer: Volume of the liquid–vapor mixture is 0.003189 m^3

10.

a. Calculate the specific volume of a 50% quality steam at a pressure of 10 kPa.
b. A tank of volume 1.5 m^3 contains saturated steam at 100 kPa. If the saturated steam is cooled to 80°C, calculate the quality of the new vapor–liquid mixture.
c. Calculate the mass of 30 m^3 of water (mixed phase liquid–vapor) at 500 kPa and quality of 60%.
d. 1 kg of steam at 14 bar has a total enthalpy content of 2202090 J.
Calculate the dryness fraction of the steam.

a. Given

i. Quality of steam = 50%
ii. Steam pressure = 10 kPa = 0.1 bar

To find: Specific volume of steam

Solution

$$V = V_l(1 - x_s) + x_s V_v$$

$$V = V_l(1 - 0.5) + 0.5\ V_v = 0.5\ V_l + 0.5\ V_v$$

From the steam table,

V_l at a steam pressure of 0.1 bar is 0.001010 $\frac{\text{m}^3}{\text{kg}}$

V_v at a steam pressure of 0.1 bar is 14.557 $\frac{\text{m}^3}{\text{kg}}$

$$\therefore V = (0.5 \times 0.001010) + (0.5 \times 14.557) = 7.279\,\text{m}^3/\text{kg}$$

Answer: Specific volume of steam = 7.279 m³/kg

b. Given

i. Volume of tank = 1.5 m³
ii. Steam pressure = 100 kPa = 1 bar
iii. Saturated steam is cooled to 80°C

To find: Quality of the new vapor–liquid mixture

Solution

Since the water exists in the state of saturated steam, its specific volume will be v_g, the value of which can be determined from the steam table

$$v = v_g = 1.726\frac{\text{m}^3}{\text{kg}}$$

When cooling the water to 80°C, some of the water will condense giving a quality of less than 1. In any case, the specific volume remains constant. Therefore, under the new conditions,

$$1.726\frac{\text{m}^3}{\text{kg}} = v_f(1 - x) + v_g\ (x)$$

The value of x can be obtained by determining the values of v_f and v_g from the steam table.

$$1.726 = 0.001029(1 - x) + 3.4091\ (x)$$

$$1.726 = 0.001029 - 0.001029x + 3.4091x$$

$$1.726 = 0.001029 + 3.408071x$$

$$x = \frac{1.726 - 0.001029}{3.408071} = 0.5061$$

Answer: The quality of the new vapor–liquid mixture is 50.61%

c. Given

i. Volume of the liquid–vapor mixture = 30 m^3
ii. Pressure of the liquid–vapor mixture = 500 kPa = 5 bar
iii. Quality of the liquid–vapor mixture = 60%

To find: Mass of the liquid–vapor mixture

Solution

The specific volume is given by

$$v = v_f(1-x) + v_g(x)$$

$$v = v_f(1-0.6) + v_g(0.6)$$

From the steam table, at a pressure corresponding to 5 bar,

$$v_f = 0.001093 \ \frac{m^3}{kg}$$

$$v_g = 0.376 \frac{m^3}{kg}$$

$$\therefore v = (0.001093)(0.4) + (0.376)(0.6)$$

$$v = 0.226 \frac{m^3}{kg}$$

$$\text{Mass of the liquid – vapor mixture} = \frac{\text{volume}}{\text{specific volume}} = \frac{30}{0.226} = 132.7 \text{ kg}$$

Answer: Mass of the liquid–vapor mixture is 132.7 kg

d. Given

i. Quantity of steam = 1 kg
ii. Steam pressure = 14 bar
iii. Total enthalpy content = 2202090 J = 2202.09 kJ

To find: Dryness fraction of steam

Solution

From the steam tables, at steam pressure = 14 bar,

h_f = 830.1 kJ/kg
h_g = 2787.8 kJ/kg

$$h_{\text{total}} = h_f(1-x) + xh_g$$

$$2202.09 = (830.1)(1-x) + x(2787.8)$$

$$2202.09 = 830.1 - 830.1x + 2787.8x$$

$$2202.09 = 830.1 + 1957.7x$$

$$x = \frac{2202.09 - 830.1}{1957.7} = 0.7$$

Answer: Dryness fraction of the steam = 0.7

BIBLIOGRAPHY

ASHRAE 2006. Thermal properties of foods. In *ASHRAE Handbook 2006 Refrigeration*, ed. M. S. Owen, 9.1–9.31. American Society of Heating, Refrigerating and Air Conditioning Engineers, Inc: Atlanta, GA.

ASHRAE 2009. *ASHRAE Handbook—Fundamentals for SI Units*. Atlanta, GA: American Society of Heating, Refrigerating and Air Conditioning Engineers, Inc.

Barron, F. and Burcham, J. 2001. Recommended energy studies in the food processing and packaging industry: identifying opportunities for conservation and efficiency. *Journal of Extension* 39. www.joe.org/joe/2001april/tt3.html (accessed June 1, 2018).

Chang, H. D. and Tao, L. C. 1981. Correlations of enthalpies of food systems. *Journal of Food Science* 46: 1493–1497.

Charm, S. E. 1978. *The Fundamentals of Food Engineering*. Westport, CT: AVI Publishing Co.

Choi, Y. and Okos, M. R. 1986. Effects of temperature and composition on the thermal properties of food. In *Food Engineering and Process Applications*, Volume 1: Transport Phenomena, eds. M. Le Maguer and P. Jelen, 93–101. London: Elsevier Applied Science Publishers.

Dalzell, J. M. 1994. Glossary. In *Food Industry and the Environment Practical Issues and Cost Implications*, ed. J. M. Dalzell, xxvii–xxxxi. Dordrecht: Springer Science+Business Media; originally published by Chapman and Hall.

Demirel, Y. 2016. *Energy. Green Energy and Technology*. Cham: Springer.

Earle, R. L. 1983. *Unit Operations of Food Processing*. New York: Pergamon Press.

Fellows, P. 2000. *Food Processing Technology: Principles and Practice*. Cambridge, England: Woodhead Publishing Limited.

Heldman, D. R. and Singh, R. P. 1981. *Food Process Engineering*. Westport, CT: AVI Publ. Co.

Hendrickson, J. 1996. *Energy Use in the U.S. Food System: A Summary of Existing Research and Analysis*. Madison, WI: Center for Integrated Agricultural Systems.

Hock, R. and Holmgren, B. 2005. A distributed surface energy balance model for complex topography and its applications to Storglaciaren Sweden. *Journal of Glaciology* 51: 25–36.

Kent, W. M. 1916. The definition of energy. *Science* 43: 820–821.

Mittal, A. 2010. *Energy-Water Nexus: Many Uncertainties Remain about National and Regional Effects of Increased Biofuel Production on Water Resources*. Washington, DC: Government Accountability Office (GAO). www.gao.gov/products/GAO-10-116 (accessed June 1, 2018).

Rodriguez-Gonzalez, O., Buckow, R., Koutchma, T. and Balasubramaniam, V. M. 2015. Energy requirements for alternative food processing technologies—Principles, assumptions, and evaluation of efficiency. *Comprehensive Reviews in Food Science and Food Safety* 14: 536–554.

Schwartzberg, H. G. 1976. Effective heat capacities for the freezing and thawing of food. *Journal of Food Science* 41: 152–156.

Siebel, E. 1892. Specific heats of various products. *Ice and Refrigeration* 2: 256–257.

UNIDO and MITI 1995. Handy Manual on "Food Processing Industry". Output of a seminar on Energy Conservation in Food Processing Industry. Sponsored by United Nations Industrial Development Organization (UNIDO) and Ministry of International Trade and Industry (MITI), Japan. Organized by The Energy Conservation Center (ECC), Japan.

Zogzas, N. 2015. Mass and energy balances. In *Food Engineering Handbook: Food Engineering Fundamentals*, eds. T. Varzakas and C. Tzia, 3–40. Boca Raton, FL: CRC Press.

4

Fluid Flow

Fluid can be defined as a substance which experiences endless motion and deformation. It flows in response to an externally applied shear stress and continues to do so until the shear stress is removed. Understanding of fluid flow phenomenon is important for food process engineers to design fluid transport systems in industries. Fluids encountered in a process industry can be classified as liquids and gases. In turn, liquids can be categorized as thin liquids and thick liquids. In a food industry, thin liquids include water, juices and milk; thick liquids comprise honey, oil and jams and steam is the predominant form of gas used. While liquids are pumped from one location to another, steam is distributed to the entire processing plant through pipelines. The scope of this chapter is to explain the concepts of fluid flow along with the governing laws and principles.

4.1 Terminologies of Fluid Flow

- **Uniform flow:** Fluid flow is said to be uniform if its velocity is the same in magnitude and direction at every point.
- **Nonuniform flow:** Fluid flow is termed non-uniform, if its velocity is not the same at every point, at a given instance of time.
- **Steady flow:** In a steady flow, the flow of fluid may differ in velocity and pressure at different points but remains constant with time.
- **Unsteady flow:** If the flow conditions of a fluid change with time at any point, it is said to be unsteady. In practice, there are always slight variations in flow conditions. Therefore, if the average flow conditions remain constant, flow can be treated as a steady flow.
- **Compressibility:** Compressibility refers to the change in density of the fluid in response to the application of pressure. In general, liquids are incompressible and gases are compressible.
- **Shear stress:** It is the stress component applied tangentially to the plane on which the force acts. Shear stress is expressed as force per unit area, and it is a vector quantity possessing both magnitude and direction. Shear stress is denoted by σ, and its SI unit is Pascal (Pa). Shear stress acts as the resistance to flow or deformation of a material.
- **Shear rate:** It is the velocity gradient established in a fluid as a result of an applied shear stress. It can also be defined as the relative change in velocity divided by the distance between the fluid layers. Shear rate is denoted by γ and can be calculated from the following equation:

$$\gamma = \frac{du}{dy} \tag{4.1}$$

  where du represents the change in velocity and dy is the distance. The unit of shear rate is s^{-1}.
- **Mass flow rate:** It is defined as the mass of fluid that flows through a conduit per unit time (Eq. (4.2)).

$$\text{Mass flow rate} = \dot{m} = \frac{\text{Mass of fluid}}{\text{time}} \tag{4.2}$$

  The unit of mass flow rate is kg/s.

- **Volumetric flow rate:** It is defined as the volume of fluid that flows through a channel per unit time (Eq. (4.3)).

$$\text{Volumetric flow rate} = V = \frac{\text{Volume of fluid}}{\text{time}} \tag{4.3}$$

The unit of volumetric flow rate is m^3/s.

4.2 Properties of Fluids

4.2.1 Mass Density or Density

Mass density or density of a fluid (ρ) is defined as its mass per unit volume, expressed in units of kg/m^3. The density of a fluid varies with temperature and pressure. Liquids being incompressible show constant density with a change in pressure, whereas the density of gas varies with pressure.

$$\text{Density} = \frac{\text{Mass}}{\text{Volume}} \tag{4.4}$$

Density is an intrinsic property of a fluid. The power required to cause fluid flow is a function of the fluid density in addition to the fluid viscosity and diameter of the conduit. The density of selected liquid foods is compiled in Table 4.1.

4.2.2 Specific Gravity

Specific gravity is the ratio of the density of a given liquid to the density of water at a given temperature. If the specific gravity is known, the density of the liquid can be determined. An instrument namely hydrometer (Figure 4.1) is used to determine the specific gravity of liquids.

4.3 The Concept of Viscosity

Viscosity is defined as the internal resistance offered by a fluid to the externally acting forces of deformation. In accordance with Newton's first law, a fluid continues to flow or remain at rest, until an

TABLE 4.1

Density of Selected Liquid Foods at Atmospheric Pressure

Liquid Food	Density (kg/m^3)
Liquid water (at 20°C)	998
Honey	1464
Apple juice	1044
Carbonated beverage	1020
Grape juice	1054
Orange juice	1038
Cow's milk (whole)	1040
Soybean oil	917
Sunflower oil	960
Coconut oil	924
Red wine	990
Beer	1000

Source: Data from Min et al. (2010), Charrondiere et al. (2012).

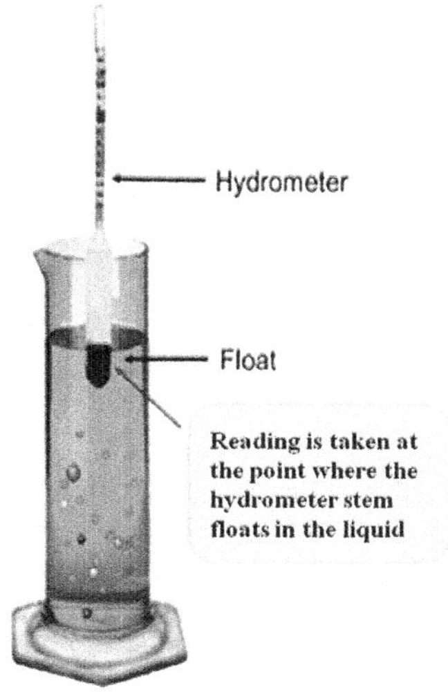

FIGURE 4.1 Schematic diagram of a hydrometer. (Reproduced with permission from Maheshwari, R., Todke, P., Kuche, K., Raval, N. and Tekade, R. K. 2018. Micromeritics in pharmaceutical product development. In *Advances in Pharmaceutical Product Development and Research, Dosage Form Design Considerations*, ed. R. K. Tekade, 599–635. London: Academic Press.)

external force acts on it. The extent of the force required to induce flow of a fluid at a particular velocity depends on its viscosity. The viscosity of a fluid can be expressed in different ways, as explained in the forthcoming sections.

4.3.1 Dynamic Viscosity

According to Newton, direct proportionality exists between the shear stress and the shear rate. Therefore,

$$\sigma \propto \gamma \tag{4.5}$$

where σ is the shear stress (Pa) and γ is the shear rate (s^{-1}).

A constant of proportionality is introduced to remove the proportionality sign. Thus, Eq. (4.5) changes to

$$\sigma = \mu\gamma \tag{4.6}$$

where μ is the dynamic viscosity often called *absolute viscosity* or *viscosity*. Thus, viscosity can be defined as the ratio between shear stress and shear strain Eq. (4.7).

$$\mu = \frac{\sigma}{\gamma} \tag{4.7}$$

γ can also be written as shear deformation rate $\left(\frac{u}{y}\right)$ or shear velocity $\left(\frac{du}{dy}\right)$. SI unit of viscosity is Pascal second (Pa s), and its counterpart in the CGS (centimeter–gram–second) system is poise.

$$1 \text{ Pa s} = 10 \text{ poise} = 1000 \text{ centipoise (cP)} = 1000 \text{ millipascal second (mPa s)} \tag{4.8}$$

TABLE 4.2
Viscosity of Some Common Fluids

Substance	Viscosity (cP or mPa s)
Air	1.86×10^{-4}
Water (0°C)	1.7921
Water (20°C)	1.000
20% Sucrose solution (20°C)	1.967
Glycerol	1759

Table 4.2 lists the viscosity of some common fluid-foods.

Equation 4.7 is known as Newton's law. This law forms the main criterion for the classification of fluids based on the nature of the relationship that exists between the shear stress and shear rate and the value of the coefficient of proportionality, i.e., viscosity (μ).

4.3.1.1 Newtonian and Non-Newtonian Fluids

4.3.1.1.1 Newtonian Fluids

Sir Isaac Newton (1642–1727) stated that fluid flow is directly proportional to the force applied to it. For many fluids, the viscosity is independent of shear rate and, therefore, obeys the linear relationship given by Eq. (4.5), i.e., $\sigma \propto \gamma$. Such fluids are termed as Newtonian fluids. The flow behavior of a fluid is graphically represented by a *rheogram*, which is the plot between shear stress and shear rate (Figure 4.2). On a *rheogram*, a Newtonian fluid is represented by a straight line through the origin (Figure 4.2). Water, milk, honey, clear fruit juices, vegetable oil, and dilute solutions of low molecular weight solutes are some well-known examples of Newtonian liquids. In general, liquid foods with more than 90% water content depict Newtonian flow behavior.

4.3.1.1.2 Non-Newtonian Fluids

On the other hand, fluids with shear-dependent viscosity are known as non-Newtonian fluids. For non-Newtonian fluids, the relationship between shear stress and shear rate is not linear (Figure 4.2). These are in turn classified as time dependent and time independent, as shown in Figure 4.3.

4.3.1.1.2.1 Time-Independent Non-Newtonian Fluids Time-independent non-Newtonian fluids respond instantaneously to the application of shear stress, indicated by their spontaneous flow. Two major classes of time-independent non-Newtonian fluids are the shear-thinning and the shear-thickening fluids. These fluid types are differentiated by the pattern in which their apparent viscosity changes with

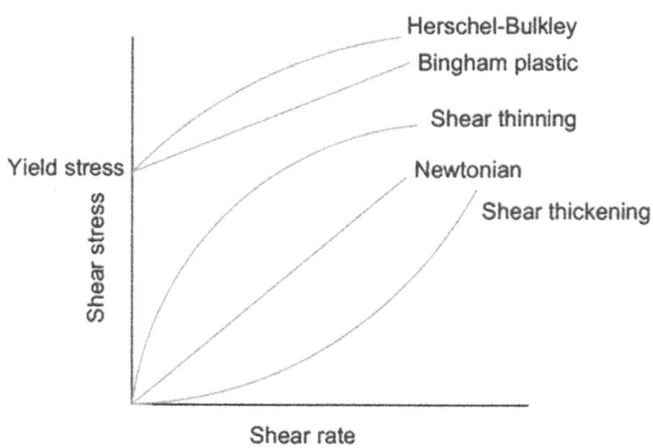

FIGURE 4.2 A *Rheogram* showing the flow behavior of different fluid types.

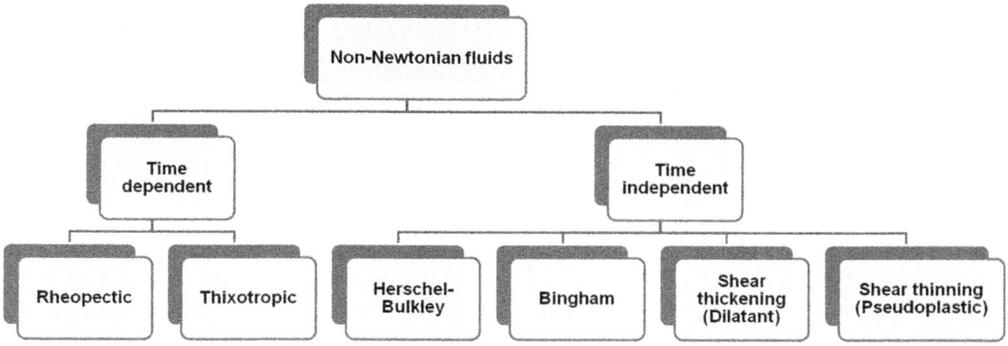

FIGURE 4.3 Classification of non-Newtonian fluids.

the shear rate. Apparent viscosity (μ_a) is obtained by approximating the viscosity of a non-Newtonian fluid to a Newtonian fluid. In this approach, a coefficient is calculated from the empirical data considering that the fluid obeys Newton's law. The calculation of apparent viscosity involves the following steps: A straight line is drawn from the selected point on the curve to the origin at any selected shear rate (Figure 4.4). The apparent viscosity value is obtained from the slope which depends on the selected shear rate. Therefore, the values of apparent viscosity and shear rate are always mentioned together.

i. **Shear thinning or pseudoplastic fluids**

   In shear-thinning fluids, as the shear rate increases, the apparent viscosity decreases. Shear-thinning fluids are also known as pseudoplastic or power law fluids. On a *rheogram*, the line representing the flow behavior of a pseudoplastic fluid starts from the origin and proceeds concave upward. On shaking, the viscosity of the shear-thinning liquid decreases. This is because, on the application of shear, the randomly distributed particles in the fluid align themselves in the direction of flow and the agglomerated particles separate. This increases fluidity, thus decreasing the apparent viscosity. Applesauce, condensed milk, mayonnaise, banana purees, orange juice concentrate, dairy cream, and vegetable soup are examples of shear-thinning liquids.

ii. **Shear thickening or dilatant fluids**

   In the case of shear-thickening fluids, apparent viscosity increases with an increase in shear rate. On applying shear, the dispersed phase swells or changes in shape or the molecules cross-link with each other and trap the molecules of the dispersed medium, thus, increasing the viscosity. At a low shear rate, the fluid keeps the solid particles in suspension. On increasing the shear rate, the solid particles separate out and increase the overall volume of suspension

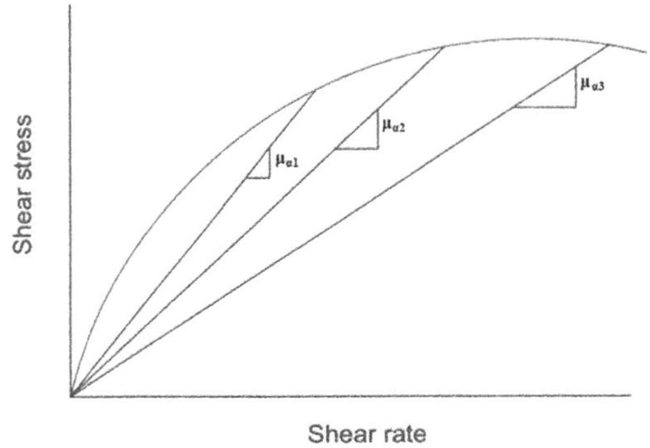

FIGURE 4.4 Determination of apparent viscosity from the plot of shear rate versus shear stress.

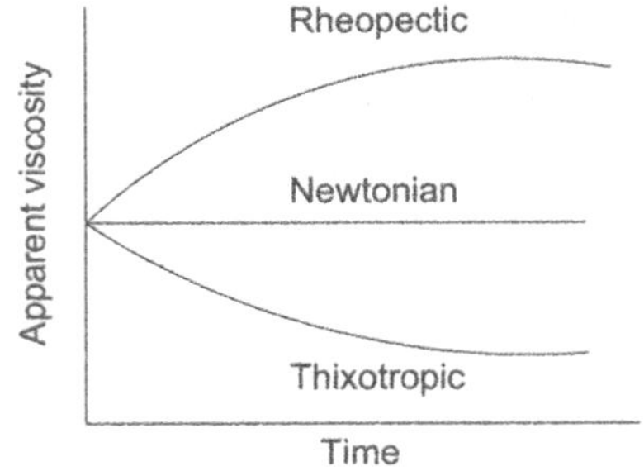

FIGURE 4.5 Time-dependent flow behavior of fluids.

which increases the resistance to flow and thereby raises the fluid viscosity. On a *rheogram*, the flow behavior of a dilatant fluid is represented by a line that begins at the origin and moves concave downward. A well-known example of shear-thickening liquid is a 60% suspension of cornstarch in water.

Other types of time-independent non-Newtonian fluids include the Bingham plastic and Herschel–Bulkley. These fluids are characterized by their requirement of a certain amount of *yield stress* before flowing or responding to the application of shear stress. The Bingham plastic and Herschel–Bulkley fluids exhibit a combination of the Newtonian and non-Newtonian flow behaviors. If a fluid adopts the Newtonian behavior at an applied stress greater than the yield stress, it is called Bingham plastic (Figure 4.2). On the other hand, if the fluid exhibits a shear-thinning behavior beyond its yield stress, then it is said to be of the Herschel–Bulkley type (Figure 4.2). While tomato paste exhibits the behavior of a Bingham plastic fluid, minced fish paste and raisin paste represent the Herschel–Bulkley fluids.

4.3.1.1.2.2 Time-Dependent Non-Newtonian Fluids The two major classes of this category include the thixotropic and rheopectic fluids. In the case of thixotropic fluids, the apparent viscosity decreases with the time of shearing (Figure 4.5). This is due to the decrease in intermolecular interactions between the molecules in the fluid. The starch gel is a common example of thixotropic liquid.

On contrary to the thixotropic type, in the rheopectic fluids, apparent viscosity increases with the time of shearing (Figure 4.5). This type of behavior is rarely observed in food systems.

4.3.2 Kinematic Viscosity

Absolute viscosity of a fluid divided by its density yields the kinematic viscosity Eq. (4.9). It is denoted by ν.

$$\nu = \frac{\mu}{\rho} = \frac{\sigma}{\rho\gamma} \tag{4.9}$$

where μ is the absolute viscosity of fluid (Pa s) and ρ is the fluid density (kg/m^3). The SI unit of kinematic viscosity is meter-square-per-second (m^2/s). Stokes is the absolute unit of kinematic viscosity, and it has the dimensions M^2T^{-1}.

4.4 Empirical Models Governing the Flow Behavior of Non-Newtonian Fluids

4.4.1 Power Law Model

The non-Newtonian fluids are governed by the power law model. This model is based on the empirical relationship given by

$$\sigma = K\gamma^n \tag{4.10}$$

where σ is the shear stress (Pa), γ is the shear rate (s^{-1}), and K and n are the constants which are characteristic of fluids, known as the consistency index (Pa s^n) and flow behavior index, respectively. n is a dimensionless number, which is equal to 1 for a Newtonian fluid. For pseudoplastic fluids, n is < 1, and for a dilatant fluid, n is > 1.

4.4.2 Herschel–Bulkley Model

The fluids that obey this model are characterized by their requirement for the application of stress greater than a critical value known as the yield stress Eq. (4.11).

$$\sigma = \sigma_0 + K\gamma^n \tag{4.11}$$

where σ_0 is the yield stress (Pa). According to this model, the plot between shear stress (σ) and shear rate (γ) is a straight line, with its slope equal to K.

4.4.3 Casson Model

Casson developed this equation for the printing ink (Casson, 1959), but it has been found effective in explaining the flow behavior of chocolates in the confectionery industry.

$$\sqrt{\sigma} = \sqrt{\sigma_0} + K\sqrt{\gamma} \tag{4.12}$$

The International Office of Cocoa and Chocolate has approved this model as the official method for interpretation of the flow behavior of chocolates (Servais et al., 2003). From Eq. (4.12), the plot between the square root of the shear rate and the square root of the shear stress yields a straight line. The flow consistency index is obtained from the slope of the line, and the square of the intercept gives the value of yield stress (σ_0).

The practical application of the Casson model has been realized in the confectionery industry. At forces below the yield stress, the chocolate does not flow. The yield stress is set in the molten and stationary chocolate by the network of particles resting on each another (Figure 4.6a). At this point, apparent viscosity is not mathematically defined as there is no movement. Subsequently, when the force is increased to just above the value of yield stress, the agglomerated particles in the molten mass of chocolate begin to separate and the flow occurs but only gradually since some networks are still present (Figure 4.6b). The slow movement caused by the residual aggregates leads to a high apparent viscosity. When the force is increased further (Figure 4.6c), the agglomerates are disrupted to a greater extent, and the irregularly shaped particles align themselves in the plane of flow. This results in proportionately greater relative movement. The consequent high shear lowers the apparent viscosity (Seguine, 1988).

The plot between the apparent viscosity of chocolate and shear rate (Figure 4.7) shows that the viscosity is high at a lower shear rate but begins to approach a constant value as the shear rate is increased. This value of limiting viscosity is termed as the Casson plastic viscosity. From Figure 4.7, it is evident that the

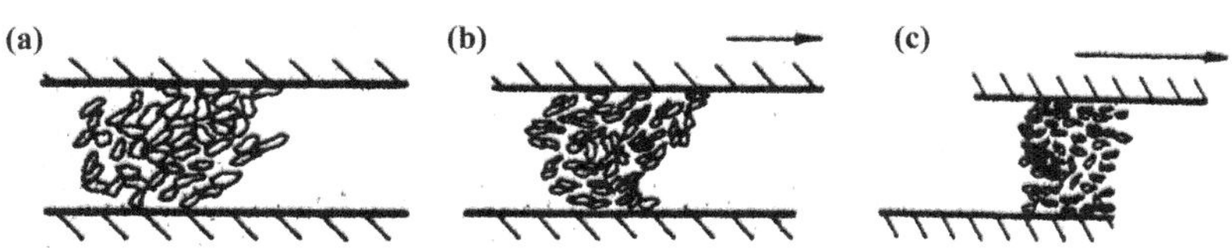

FIGURE 4.6 (a) Chocolate at rest; (b) chocolate under low shear; (c) chocolate under high shear. (Reproduced with permission from Seguine, E. S. 1988. Casson plastic viscosity and yield value. What they are what they mean to the confectioner. *The Manufacturing Confectioner*, November: 57–63.)

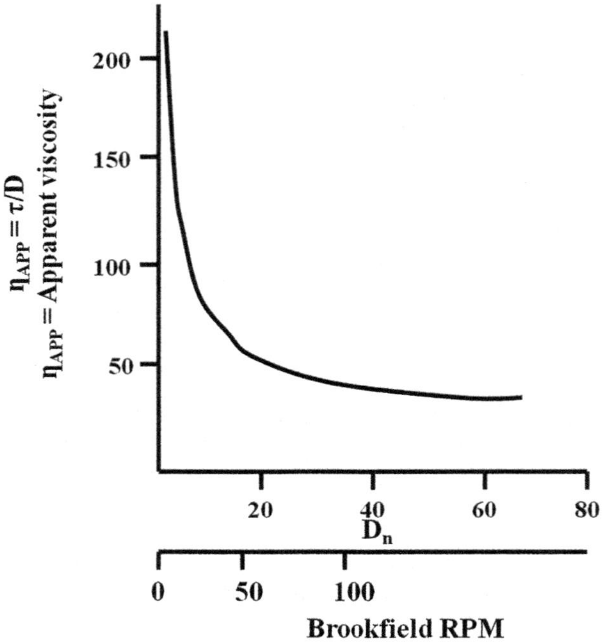

FIGURE 4.7 Chocolate apparent viscosity as a function of shear rate. (Reproduced with permission from Seguine, E. S. 1988. Casson plastic viscosity and yield value. What they are what they mean to the confectioner. *The Manufacturing Confectioner*, November: 57–63.)

power needed to agitate a tank of chocolate when it has not been agitated for some time can be significantly greater than the power required to maintain its motion once it has started. The aforesaid behavior of fluids can be attributed to the need to break up the aggregates from rest. In an industrial operation, the equipment is designed such that the agitator drive assembly is oversized to a large extent in order to handle this startup situation (Seguine, 1988).

4.5 Temperature Dependence of Viscosity

The viscosity of liquids is independent of pressure but is strongly dependent on temperature. In case of gases, viscosity increases with increase in temperature owing to an increase in the frequency of intermolecular collisions at higher temperatures. During processing, food material is subjected to different temperatures, and hence, it is important to determine the effect of temperature on the viscosity of the fluid. The relationship between temperature and viscosity is explained by the Arrhenius relationship given as follows:

$$\mu = \mu_0 e^{E_a/RT} \tag{4.13}$$

where μ_0 is the Arrhenius factor, E_a is the activation energy (J/mol), T is the temperature (K), and R is the universal gas constant (J/mol K). Equation 4.13 can be linearized by applying natural logarithm on both the sides as follows:

$$\ln(\mu) = \ln(\mu_0) + \frac{E_a}{R}\left(\frac{1}{T}\right) \tag{4.14}$$

4.6 Measurement of Viscosity

The instrument used for measuring liquid viscosity is termed as a viscometer. The viscosity of a Newtonian fluid can be easily measured with a relatively simple viscometer. However, measurement of the viscosity of a non-Newtonian fluid involves certain complexities. Viscometer to measure the viscosity of the non-Newtonian fluid requires a mechanism to measure the inducing flow; the applied force and the geometry of the system should be simple so that the force and flow can be easily translated into the shear rate and the shear stress.

4.6.1 Bostwick Consistometer

The measurement of viscosity by a Bostwick consistometer (Figure 4.8) is based on the distance travelled by the product under the influence of its own weight along a sloped surface within a defined period of time. The construction of a Bostwick consistometer includes an inclined plane which is nothing but a rectangular stainless steel trough comprising two compartments. The first compartment is of dimensions 5 cm (length) × 5 cm (width) × 3.8 cm (height). It is separated from the second compartment by a spring-loaded gate. The second compartment which is located adjacent to the first one is a trough of size 24 cm (length) × 5 cm (width) × 2.5 cm (height). The floor of the second compartment has a series of parallel lines drawn across it at 0.5 cm intervals, commencing at the gate and extending until the opposite end. In order to measure the viscosity of a product (sample), the gate is closed and locked by a trigger. Then, the first compartment is filled with the product (e.g., comminuted fruit or vegetable such as applesauce, carrot puree or baby foods) the consistency of which is to be evaluated. Subsequently, the consistometer is leveled with the help of a spirit level (located at the right-hand side end of the instrument as shown in Figure 4.8), and the trigger is pressed to release the gate, which springs up out of the way. As a result, the fluid product is now free to flow under the force of gravity from the first compartment to the second compartment. The Bostwick consistometer reading is given by the distance traversed by the fluid from the gate after 30 s, in cm (Bourne, 1982). Thus, in this test method, the height of the incline signifies the shear stress applied to the sample and the time for flow is related to the flow rate or shear rate of the sample. The Bostwick consistometer is used as an indicator for the quality control purposes. For example, to confirm whether the product complies with the given specification of flow time, the technician simply needs to measure the time for the flow of the sample down the incline and compare it against the specified standard value for the product being tested (Figura and Teixeira, 2007).

Bostwick consistometer has been accepted by the National Canners Association as the official test method for measuring the consistency of catsup. According to the United States Standard for Tomato Catsup, the grades A and B quality should be of good consistency with their flow not more than 9 cm in 30 s at 20°C in a Bostwick consistometer. On the other hand, grade C tomato catsup must have a fairly good consistency and flow not more than 14 cm in 30 s at 20°C in a Bostwick consistometer

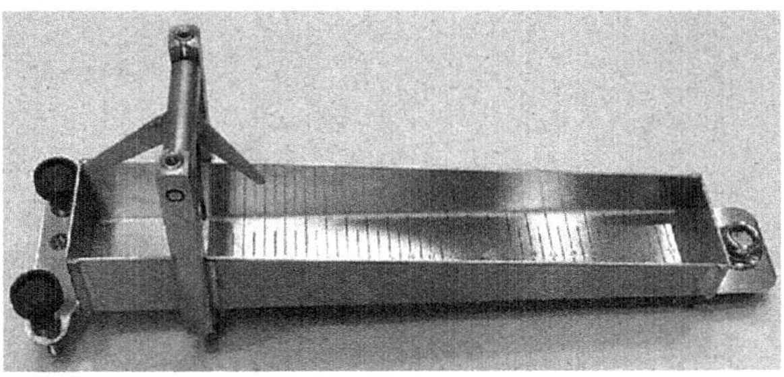

FIGURE 4.8 Bostwick consistometer. (Reproduced with permission from Steele, C. M., James, D. F., Hori, S., Polacco, R. C. and Yee, C. 2014. Oral Perceptual discrimination of viscosity differences for non-Newtonian liquids in the nectar- and honey-thick ranges. *Dysphagia* 29: 355–364.)

(Bourne, 1982). Apart from tomato catsup, Bostwick consistometer can also be used for other concentrated tomato products such as paste, sauce, puree, and for the quality control of pureed foods such as baby foods.

4.6.2 Capillary Tube Viscometer

In a capillary tube viscometer, the pressure drop (ΔP) overcomes the shear force within the fluid and produces a flow of a given rate. For the entire length of the tube (L) and distance from the tube center (h), the shear force operates on all internal liquid surfaces. The design of any capillary tube viscometer can be obtained using the Poiseuille's equation given by

$$\dot{V} = \frac{\pi \Delta P R^4}{8 \mu L} \tag{4.15}$$

where μ is the viscosity of the fluid flowing through a viscometer having tube length L and radius R with a volumetric flow rate $\dot{V}$ and pressure ΔP. Rearranging Eq. (4.15),

$$\mu = \frac{\pi \Delta P R^4}{8 L \dot{V}} \tag{4.16}$$

Equation 4.16 is applicable for the Newtonian fluid. Same viscosity rates are obtained with any combination of pressure drop and flow rate.

Figure 4.9 shows the Cannon–Fenske type viscometer that operates on the principle of gravity-induced flow. It is made up of glass and available in different capillary sizes. The selection of capillary size should be such that it reduces the time of efflux for the viscous fluid. Liquid flows through the glass capillary tube as a result of the pressure created by the gravitational force (Eq. 4.17).

$$\Delta P = \rho L g \tag{4.17}$$

The volumetric flow rate through the capillary tube $(\dot{V})$ is given by V/t where V is the volume of bulb and t is the discharge time. Thus, Eq. (4.16) becomes

$$\mu = \frac{\pi \rho g R^4 t}{8V} \tag{4.18}$$

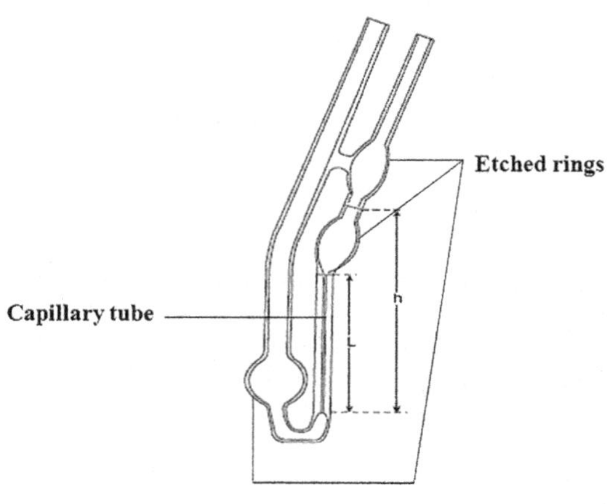

FIGURE 4.9 Capillary tube viscometer.

Equation 4.18 shows that the viscosity of a liquid measured by the glass capillary tube is a function of the liquid volume in the bulb, density of the liquid, and acceleration due to gravity. Viscosity can be obtained by determining the time required to drain the liquid from the bulb.

4.6.3 Rotational Viscometer

The operation of a rotational viscometer is based on the principle that the torque required to turn an object in a fluid is a function of the fluid viscosity. A rotational viscometer estimates viscosity by applying a known force or torque and measuring the rate of rotation of a solid shape placed inside a viscous fluid, at a defined angular velocity. With rotational viscometers, it is possible to perform continuous and multiple measurements on a sample as a function of shear rate and temperature, under steady-state conditions.

4.6.3.1 Coaxial Cylinder Viscometer

Coaxial cylinder viscometer is the most common type of viscometer (Figure 4.10) used for determining the flow characteristics of food liquids in process industries. It can be used for the analysis of both Newtonian and non-Newtonian liquids. In a coaxial cylinder viscometer, the estimation of viscosity is based on the Margules equation (Eq. (4.19)) (Margules, 1881) which applies to the Newtonian fluids.

$$\mu = \left(\frac{M}{4\prod h\Omega}\right)\left(\frac{1}{R_b^2} - \frac{1}{R_c^2}\right) \tag{4.19}$$

where μ is the absolute viscosity, M is the torque on solid material (bob or disk) that rotates inside the test fluid, Ω is the angular velocity of rotating element, h is the length of bob in contact with the fluid, R_b is the radius of the bob, and R_c is the radius of sample cup. For a defined geometry of a given instrument, an instrument constant, K is specified by grouping the constants in Eq. (4.19). Thus, Eq. (4.19) becomes $\mu = \frac{KM}{\Omega}$, where K is the instrument constant given by $(1/4\prod h)(1/R_{b^2} - 1/R_{c^2})$ (Bourne, 2002).

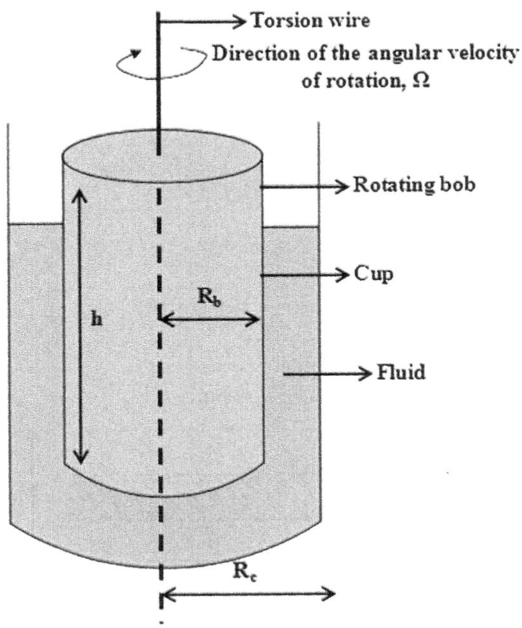

FIGURE 4.10 Coaxial cylinder viscometer.

4.6.3.2 Cone and Plate Viscometer

In this type of viscometer, the fluid is held between a cone and a flat surface, by virtue of its own surface tension. The cone is of small angle (usually, 2°) that is in contact with a flat surface (Figure 4.11). While the cone rotates at an angular velocity, Ω, the flat surface remains stationary. The torque on the cone caused by the drag of the fluid is measured. Eq. (4.20) is useful in determining the viscosity of a Newtonian fluid.

$$\mu = \frac{3\alpha M}{2\prod R_b^3 \Omega} \tag{4.20}$$

where μ is the absolute viscosity, α is the angle of the cone (usually less than 2°), M is the torque, R_b is the radius of the cone, and Ω is the angular velocity of rotation. Similar to the coaxial cylinder viscometer, for a given instrument of fixed geometry, Eq. (4.20) reduces to

$$\mu = \frac{KM}{\Omega} \tag{4.21}$$

where K is the instrument constant, given by, $3\alpha/2\prod R_b^3$ (Bourne, 2002).

4.6.3.3 Parallel Plate Viscometer

Here, the measurement of viscosity is based on the torsional flow of liquid between plates that are parallel to each other (Figure 4.12). The shear rate at the edge of the plate is given by

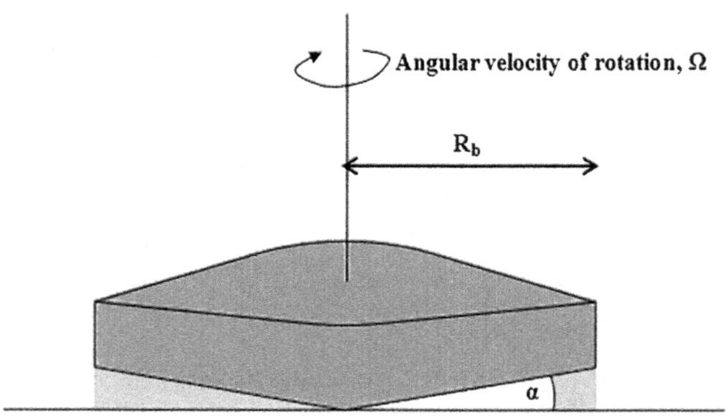

FIGURE 4.11 Cone and plate viscometer.

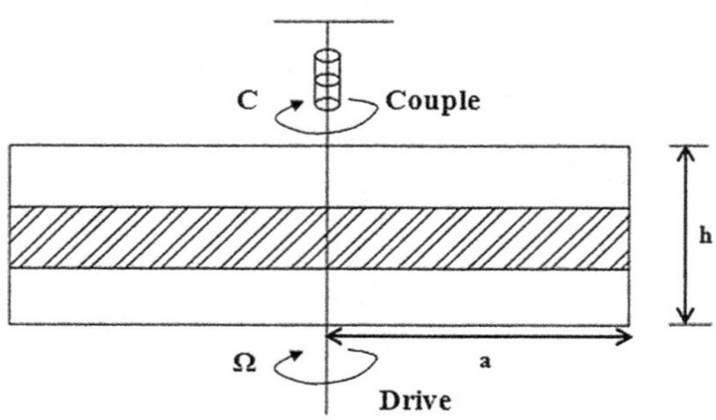

FIGURE 4.12 Parallel plate viscometer.

$$\gamma = \frac{a\Omega}{h} \tag{4.22}$$

where γ is the shear rate, a is the distance between the axis of rotation and edge of the plate, Ω is the angular velocity of rotation, and h is the height of the plate. The liquid viscosity is given by

$$\mu = \frac{3Ch}{2\Pi\alpha^4\Omega}\left[1+\frac{1}{3}\frac{d\ln C}{d\ln\Omega}\right] \tag{4.23}$$

where C is the couple on one of the plates (Barnes et al., 1989). *Couple* refers to a pair of parallel forces equal in magnitude, which acts in opposite directions on an object, but not through the same point so as to produce a rotating or turning effect (Figure 4.12).

4.7 Viscosity as a Process and Quality Control Tool in the Food Industry

Viscosimetry or *measurement of viscosity* is an indispensable tool for process control and quality management in any food industry which manufactures flowable products. Viscosity measurement assists in maximizing the production efficiency and cost-effectiveness. This is because the raw materials, semifinished products, and final products have different viscosities (Kulicke and Clasen, 2004). Viscosity influences the rate of fluid transport through a pipe and the setting time/drying time of liquid food products. Knowledge of product viscosity is required in designing a production process involving pipes and dispensers to achieve optimal flow induced by the right amount of force without overfilling the package. Viscosity is also a measure of food texture, which is an important quality attribute. Measurement and monitoring of product viscosity during and after production is vital to ensure batch-to-batch consistency. Thus, viscosity measurement is also an integral part of the research and development and quality control laboratories of food industries. In general, the procedure for measurement is aligned in accordance with the standard references depending on the product type. Irrespective of the product, the following information is sought before beginning the viscosity measurement (McGregor, 2016):

- The quantity of food material available for testing (sample quantity) and clarification on any limitation in the available quantity.
- Type of spindle required to test the material.
- A suitable range of torque measurement for the instrument.
- Whether or not the temperature measurement and control sample are required for the test.
- The duration for which the spindle must rotate in the material before taking the reading.

The subsequent sections present three case studies on the importance of viscosity as a process and quality control tool in food industries manufacturing beer, chocolate, and tomato products.

4.7.1 Beer

Beer is a fermented alcoholic beverage prepared from barley, flavored with hops (flowers of the hop plant). The manufacturing process comprises milling of the grains, mashing of the milled grist with water, boiling of the wort (aqueous solution of the extract), followed by the fermentation and maturation of the wort to result in beer. The beer exhibits a nearly ideal viscous behavior. Therefore, it is a Newtonian liquid (Steffe, 1996). Knowledge of viscosity is of crucial importance in the beer brewing process. Here, the major objectives of viscosimetry are the optimization of filtration process (in designing filters and defining the working pressures) and the quality control of malt (raw material), wort (intermediate product), and beer (final product). The Newtonian behavior facilitates the viscosity determination using relatively simple principles of measurement (Severa et al., 2009). Usually, glass viscometers are used which measure the rate at which liquid flows through a capillary tube, relative to a standard of water. The viscosity of wort is given by

$$\text{Viscosity of wort} = \left(\frac{\text{Flow time of wort at } 20^\circ\text{C}}{\text{Flow time of water at } 20^\circ\text{C}}\right) \times \text{specific gravity of wort} \times 1.002 \tag{4.24}$$

where 1.002 is the viscosity of water at 20°C. Here, the unit of viscosity is centipoise (cP). From Eq. (4.24), it is evident that, higher the specific gravity of the wort, greater would be its viscosity (Bamforth, 2002).

Wort viscosity is used by the brewer as an indication of potential wort separation and beer filtration difficulties. The high viscosity of beer (>1.7 mPa s) leads to difficulties during filtration and thereby causes a reduction in capacity (Severa et al., 2009).

Viscosity measurements have also been used to detect the specific point at which the transition from yeast to beer occurs during the purge of a fermentation tank. This analysis led to a significant reduction in downtime. Downtime is any unexpected event that halts the production for any amount of time. A detailed case study was done in a Montreal-based brewery (Ostand, 2012), which is presented as follows.

At the end of fermentation, the yeast used in the process settles at the bottom of the tank. Subsequently, the tank is emptied by pumping out the yeast which is then subjected to reactivation. Then, the portion of beer that is left behind in the tank is centrifuged to separate the trace amounts of residual yeast from the beer. The remaining batch is directly forwarded to the filtration unit. Here, the bottleneck is the difficulty in determining the definite time point during the pumping process at which all the yeast has passed, leaving only the beer. The complexity arises as both yeast and beer are brown in color. Further, centrifugation is a time-consuming process (as long as 3 h per batch) with limited efficiency in separating the proteins which tend to clog the filters. Customarily, the decision to stop the pumping process has merely been subjective. The operator, who views the process through a glass section of the pipe, guesses the point of separation by spotting the subtle color change between yeast and beer.

Addition of yeast to the tank was also done on a trial-and-error basis after guessing the quantity of yeast in the tank depending on temperature, yeast type, and pH. Due to the instability in yeast production between batches, the brewer also had to pump different quantities of yeast for each batch, depending on the type of beer. In case of an incorrect judgment, either the beer may be directed to the yeast recycling or the yeast may be forwarded to the filtration operation. While the former leads to the loss of beer (loss estimated at about $100,000 worth of product per year), the latter leads to quick plugging of filters after which the operators have to stop the process to clean the filters at the expense of time and money. The abovementioned events increase the process downtime.

To overcome the limitation mentioned previously, a viscosity measurement system was installed on the main line, just below the main transfer pump. With this system, the fermented product containing the beer and the yeast flows through a continuous pipe containing two low-pressure drop static mixers. A sensor device measures the pressure drop at both static mixers by means of two differential pressure measurements: P_1 and P_2. Pipe flow rate is obtained using a flow meter which is used for the accurate determination of the flow regime to which the viscosity is to be calculated. The viscosity of the product flowing through the pipeline is estimated based on the pressure drop measurements and the flow meter reading (ENICO Technologies Inc., 2010). Since the concentrated yeast is highly viscous than beer, the measured values of viscosity are significantly high and the yeast fraction is directed to the recycling operation. When the measurement drops down, it can be ascertained that the beer is separated from the residual yeast. This beer fraction is forwarded to the filtration unit, thus completely avoiding the centrifugation process.

The advantages of installing the viscosity measurement system are the savings of operation time at about 3 h per batch and two times less filter plugging as the highly concentrated yeast is avoided from entering the filtration unit. The advanced viscosity measurement systems have more precise yeast detection threshold (<1%) and are capable of being used as an indicator to detect a viscosity variation of $\frac{1^{\text{th}}}{10}$ of a cP. This translated to a significant drop in yeast content per batch of beer from 2% to 0.25% and a corresponding saving of $1 for every four barrels (636 L) of beer produced (Ostand, 2012).

4.7.2 Chocolate

The flow behavior of molten chocolate is defined by the Casson plastic viscosity (apparent viscosity) and yield value (yield stress), often mentioned as PV and YV in the confectionery industry. While PV is a measure of how easily a material flows once it has started flowing, YV is the force required to start the material flow. If the force which causes the chocolate coating to flow is reduced below the yield value, the coating will cease to flow. On the other hand, the plastic viscosity is important in controlling the pumping characteristics, filling of rough surfaces, coating, and sensory attributes of the chocolate mass. Chocolates of high viscosity have a pasty mouthfeel and last longer in the mouth (Gonçalves and Lannes, 2010). Furthermore, the yield value and plastic viscosity influence the performance of chocolate during the other stages of production namely transport, dipping, and dosing.

The quality characteristics of chocolate including aesthetic appeal (surface appearance) and mouthfeel are directly related to viscosity. The viscosity of liquid chocolate intended for molded or solid bars is substantially different from that used in bars with multiple ingredients. If the viscosity is too low, the weight of the chocolate over an enrobed candy will be very less. The chocolate coating would be very thin with the center coated insufficiently. Contrarily, if the viscosity is too high, the mold might not be filled with the chocolate within the defined time (usually controlled by adjusting the speed of a conveyor belt which carries the molds), or the chocolate might not flow throughout the entire mold, leaving air bubbles or gaps. In addition, the flavor and taste perceived in the mouth depend upon the order and rate of contact, which are related to the viscosity and rate of melt (Beckett, 2009). Not limited to the enrobing and coating applications, the viscosity of chocolate is also important in designing bulk handling systems, wherein the fully liquid properties of chocolate are considered. The aforesaid reasons justify the use of viscosity as a standard quality control tool in any industry manufacturing chocolate or using chocolate as an ingredient to produce related confectionery products (Oldörp, 2014).

Figure 4.13 shows the typical flow properties of chocolate. The values of actual viscosity and yield value depend on the fat content, grind, and ingredients of the chocolates; their processing methodology; and temperature. The cookie drop chocolates are characterized by high plastic viscosity and yield value, whereas, a very high fat molding chocolate falls at the low end of the range depicted in Figure 4.13.

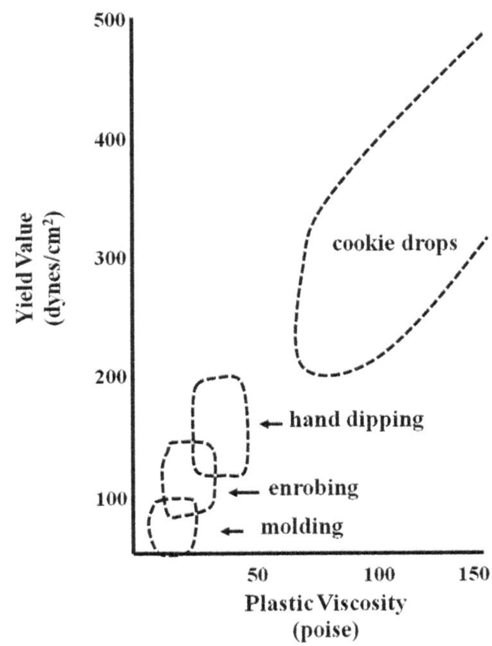

FIGURE 4.13 Flow behavior of chocolates. (Reproduced with permission from Seguine, E. S. 1988. Casson plastic viscosity and yield value. What they are what they mean to the confectioner. *The Manufacturing Confectioner*, November: 57–63.)

In the confectionery industry, the viscosity is measured by the standard methodology approved by the International Office of Cocoa (International Office of Cocoa, 2000). This method is based on the use of a rotational viscometer fitted with a concentric cylinder. The spindle rotates inside a sample of chocolate. If the chocolate has a high viscosity, there will be more resistance to the turning of the spindle. Viscosity tests are normally conducted on molten chocolate heated to 40°C. The stress and viscosity are measured at shear rates between 2 s^{-1} and 50 s^{-1}, preceded by a pre-shear at 5 s^{-1} for 5 min (Afoakwa et al., 2009).

4.7.3 Tomato Products

Tomato is a widely processed vegetable and a key ingredient in products such, as soups, sauces and ketchup (Kaur et al., 2007). Viscosity is one of the important quality attributes which is directly related to the consumer acceptance of tomato products. In addition, the viscosity is used to optimize unit operations such as mixing, pumping, and filling of tomato products (Valencia et al., 2003). Also, the process cost can be reduced and the company's profit can be improved by using viscosity as a control parameter. Because, higher the viscosity of the tomato paste, the lesser the raw material required to attain the desired final product consistency.

Characteristically, tomato products are pseudoplastic fluids. Serum viscosity of tomato products is an index of product consistency, especially in tomato juice and ketchup. Serum separation is a significant quality defect in liquid tomato products, which occurs when the solids begin to settle down and a layer of clear, straw-colored serum forms on top of the product. To prevent the aforementioned phenomenon, the insoluble particles should remain in a stable suspension throughout the serum. In general, higher viscosity lowers serum separation, which is obtained by homogenizing the tomato juice, by forcing it through a narrow orifice at high pressure to shred the suspended solids. The resultant large particle surface area leads to increased product viscosity. The viscosity value for tomato products also depends on the amount of pectin retained and the inactivation of pectolytic enzymes such as pectin methyl esterase and polygalacturonase, during the heat treatment. Heat inactivation of pectolytic enzymes results in increased viscosity of the tomato products (Hayes et al., 1998).

Among the various tomato products, tomato ketchup is a heterogeneous and spiced condiment produced from either cold or hot extracted tomatoes or directly from the concentrates, purees, or tomato paste (Sahin and Ozdemir, 2004). With respect to its flow behavior, ketchup is a non-Newtonian fluid exhibiting a slight thixotropy (Bottiglieri et al., 1991). Ketchup with high solid content acquires its viscosity from a combination of water retention of the fibrous strands in the paste and the gelling effect of pectin inherently available in the tomatoes. In ketchup with lower solid content, wherein starch-based thickeners are added, high-pressure homogenization assists in obtaining the desired viscosity (Silverson®, 2018). Glass capillary viscometer, concentric cylinder or cone, and plate viscometers are commonly used for the viscosity determination of tomato products.

4.8 Governing Laws of Fluid Flow

4.8.1 Principle of Continuity

An ideal fluid does not have a uniform viscosity and velocity throughout the cross section of the flow passage. Therefore, the mass balance for the given flow system can be written as

$$\text{Mass entering per unit time} = \text{Mass leaving per unit time} + \text{Accumulation of mass} \quad (4.25)$$

The mass flow rate of a fluid at two different points A and B is given by

$$\rho_A u_A A_A = \rho_B u_B A_B + \frac{d(\rho_{av} V)}{dt} \quad (4.26)$$

where A is the area of cross section (m^2), ρ is the fluid density (kg/m^3), u is the velocity of the fluid (m/s), ρ_{av} is the average fluid density (kg/m^3), and V is the volumetric flow rate of the fluid between the points A and B (m^3). The differential term on the right-hand side of Eq. (4.26) denotes the accumulation of mass. Equation 4.26 is known as the continuity equation.

At steady state, there is no mass accumulation, and hence, Eq. (4.26) can be written as

$$\rho_A u_A A_A = \rho_B u_B A_B \tag{4.27}$$

Since liquids are incompressible fluids $(\rho_A = \rho_B)$, density remains almost constant. Therefore, Eq. (4.27) can be written as

$$u_A A_A = u_B A_B = V = \text{constant} \tag{4.28}$$

Thus, the principle of continuity in fluid dynamics states that *in any steady state process, the rate at which the mass enters a system is equal to the rate at which mass leaves the system.* This is in accordance with the law of conservation of mass.

4.8.2 Bernoulli's Equation

Bernoulli's equation is named after the Swiss mathematician Daniel Bernoulli. It states that *under a steady state condition, the rate of change of momentum is equal to the resultant force acting on the liquid.* In other words, the mechanical energy and pressure remain constant along a streamline for a stationary, incompressible, and ideal flow if no power is supplied to the flow. In a closed pipe, Bernoulli's equation relates the energy of the fluid at any two different points. It can be used for the flow through a pipe, both in the presence and absence of friction.

Considering a case where the friction of the pipe is neglected, as shown in the Figure 4.14, when an incompressible fluid (density remains constant along the entire pipe length i.e., $\rho_A = \rho_B = \rho$) passes through the pipe without any frictional resistance, the energy balance between any two points, say, A and B is given by

$$\frac{P_A}{g\rho} + \frac{u_A^2}{2g} + Z_A = \frac{P_B}{g\rho} + \frac{u_B^2}{2g} + Z_B \tag{4.29}$$

On multiplying Eq. (4.29) by g on both sides,

$$\frac{P_A}{\rho} + \frac{u_A^2}{2} + gZ_A = \frac{P_B}{\rho} + \frac{u_B^2}{2} + gZ_B \tag{4.30}$$

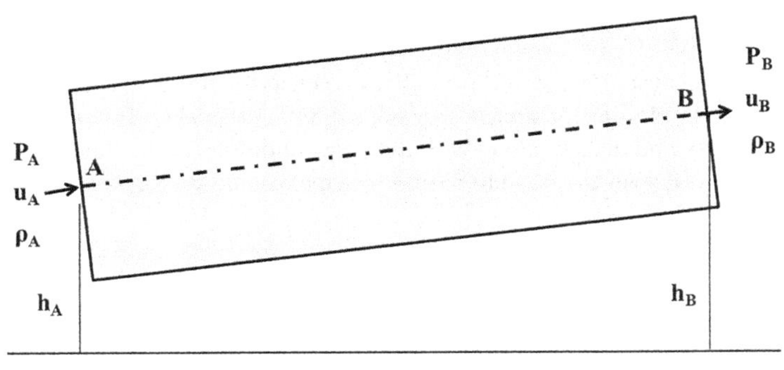

FIGURE 4.14 Flow of fluid through a pipe.

In Eq. (4.30),

$\frac{P}{\rho}$ represents the mechanical work done by the force
$\frac{u_A^2}{2}$ is the kinetic energy at point A
$\frac{u_B^2}{2}$ is the kinetic energy at point B
gZ_A is the potential energy at point A
gZ_B is the potential energy at point B

where P is pressure (Pa), ρ is the density of the fluid (kg/m^3), Z_A and Z_B are the heights of the pipe (m) at the points A and B, respectively, and g is the acceleration due to gravity (m/s^2).

All the terms in Eq. (4.30) have the units of meter. However, Eq. (4.30) does not consider the friction for flow through pipes. Bernoulli's equation is useful in finding the pressure and fluid velocity at the outlet of a pipe if the values of inlet flow parameters are known. From the Bernoulli's equation, it can be inferred that the potential energy, pressure energy, and kinetic energy of the fluid are interconvertible. Moreover, the kinetic energy varies as the velocity changes along the length of the pipe. Therefore, to account such variation in the kinetic energy along the pipe's length, a kinetic energy correction factor is introduced, which is denoted by α. To calculate the kinetic energy correction factor, one should know the local velocity as a function of the location. Generally, α is considered as 2.0 for laminar flow and 1.05 for a highly turbulent flow. On adding the kinetic energy correction factor to the Eq. (4.30),

$$\frac{P_A}{\rho} + \frac{\alpha_A u_A^2}{2} + gZ_A = \frac{P_B}{\rho} + \frac{\alpha_B u_B^2}{2} + gZ_B \quad (4.31)$$

During fluid flow through a pipe, some loss of mechanical energy occurs due to the friction at the wall of the pipe. This energy loss due to friction should be accounted for at the outlet side of the Bernoulli's equation using a frictional loss factor (H_f). Thus, Eq. (4.31) can be written as

$$\frac{P_A}{\rho} + \frac{\alpha_A u_A^2}{2} + gZ_A = \frac{P_B}{\rho} + \frac{\alpha_B u_B^2}{2} + gZ_B \quad (4.32)$$

where H_f represents the energy loss due to friction over the entire length of the pipe. In Eq. (4.32), friction is not interconvertible with the other mechanical energy quantities.

4.9 Fluid Flow Regimes

4.9.1 The Concept of Reynolds Number

Fluid flow is characterized by a dimensionless number known as the Reynolds number. Reynolds number is named after the British physicist Osborne Reynolds (1842–1912). The Reynolds number takes into account the fluid viscosity and density along with the flow conditions of the fluid, i.e., fluid velocity and system geometry by considering the conduit's (tube or pipe) diameter. Reynolds number is defined by the following equation:

$$N_{Re} = \frac{\rho D u}{\mu} \quad (4.33)$$

where ρ is the density of the fluid (kg/m^3), D is the diameter of the pipe (m), u is the velocity of the fluid in the pipe (m/s), and μ is the viscosity of the fluid (Pa s).

Thus, Reynolds number is the ratio of inertial forces to the viscous forces of fluid flow. Inertial force is represented by the fluid velocity whereas viscous force is characterized by the viscosity of the fluid. From Eq. (4.33), it is evident that a fluid of higher viscosity will have a smaller value for Reynolds number than the fluids with lower viscosity. If the fluid velocity is unknown and the mass flow rate $(\dot{m})$ is known, Eq. (4.34) can be used for the determination of Reynolds number.

$$N_{Re} = \frac{4\dot{m}}{\mu \pi D} \tag{4.34}$$

4.9.2 Laminar and Turbulent Flow

Based on the value of Reynolds number, flow through a conduit is classified into three categories

- If $N_{Re} < 2100$, the flow is laminar.
- If $N_{Re} > 4000$, the flow is turbulent.
- When $N_{Re} > 2100$ and < 4000, the flow is transitional, where it can be either laminar or turbulent.

In the laminar flow, the different layers of a fluid move past one another in a parallel manner (Figure 4.15a), i.e., in the absence of transfer of matter between the layers. However, beyond a certain fluid velocity, the molecules or particles move from one layer to another (Figure 4.15b) with the dissipation of a considerable amount of energy. This phenomenon is known as the turbulent flow. An increased energy requirement in a turbulent flow is implied by the greater amount of shear stress applied to initiate the flow compared to that in a laminar flow at the same shear rate.

Example 4.1

At what velocity does the flow of air and water convert from laminar to transitional and transitional to turbulent flow in a pipe of diameter, 6 cm at a temperature of 20°C?

Solution

Given

Pipe diameter = 6 cm = 0.06 m
Temperature = 20°C
Density of water at 20°C = 998.2 kg/m^3
Viscosity of water at 20°C = 993.414 × 10^{-6} Pa s
Density of air at 20°C = 1.164 kg/m^3
Viscosity of air at 20°C = 18.24 × 10^{-6} Pa s

Velocity of the fluid can be obtained by following equation:

$$u = \frac{N_{Re}\mu}{\rho D}$$

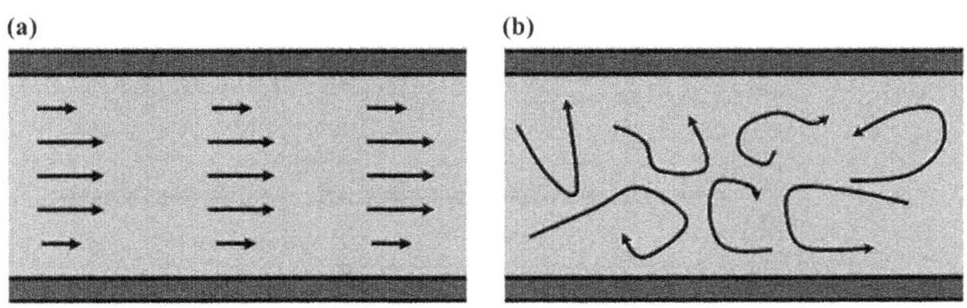

FIGURE 4.15 The flow of fluids through pipes: (a) laminar flow; (b) turbulent flow.

For air,

$$u = \frac{2100 \times 18.24 \times 10^{-6}}{1.164 \times 0.06} = 0.5485 \text{ m/s}$$

For water,

$$u = \frac{2100 \times 993.414 \times 10^{-6}}{998.2 \times 0.06} = 0.0348 \text{ m/s}$$

To change the flow regime from laminar to transitional, the velocities of air and water should be 0.5485 and 0.0348 m/s, respectively.

Similarly, for attaining the turbulent flow characteristics:

For air,

$$u = \frac{4000 \times 18.24 \times 10^{-6}}{1.164 \times 0.06} = 1.0447 \text{ m/s}$$

For water,

$$u = \frac{4000 \times 993.414 \times 10^{-6}}{998.2 \times 0.06} = 0.0663 \text{ m/s}$$

To change the flow regime from transitional to turbulent, the velocities of air and water should be 1.0447 and 0.0663 m/s, respectively.

4.10 Flow of Fluid through Pipes

4.10.1 Entrance Region and Fully Developed Flow

Flow through the pipe is divided as that in the entrance region and fully developed flow (Figure 4.16). The point from which fluid enters the pipe up to a certain length is known as the entrance region. In the entrance region of flow, fluid velocity changes with distance.

Initially, fluid velocity is uniform while entering the pipe. As the fluid moves inside the pipe, due to the frictional resistance offered by the pipe wall, the velocity of the fluid varies along the radial axis. Near the wall, fluid experiences zero velocity. Velocity increases as the fluid moves towards the center in the radial direction.

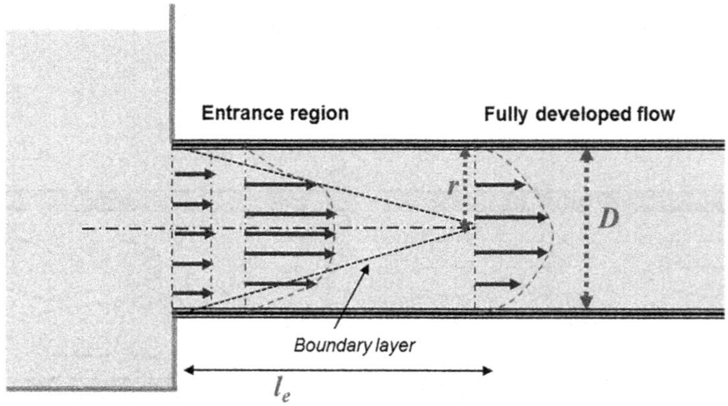

FIGURE 4.16 Velocity profile of fluid flowing through a pipe.

A boundary layer develops which affects the velocity profile. The flow beyond the boundary layer is known as the fully developed flow. Here, the fluid velocity remains constant and does not vary with distance.

The length of entrance region (l_e/D) is a function of Reynolds number.

$$\text{For laminar flow} \quad \frac{l_e}{D} = 0.06\ N_{Re} \tag{4.35}$$

$$\text{For turbulent flow} \quad \frac{l_e}{D} = 4.4\ (N_{Re})^{1/6} \tag{4.36}$$

Example 4.2

If a 10 m pipe of 4 cm diameter delivers water at 50 L/min at 20°C. Find out the length of the entrance region of the pipe.

Solution

Given

Length of pipe = 10 m
Diameter of pipe = 4 cm = 0.04 m
Therefore, radius of pipe = 0.04/2 = 0.02 m
Flow rate = 50 L/min = 8.33×10^{-4} m^3/s

1. The average velocity of fluid

$$\bar{u} = \frac{Q}{A} = \frac{8.33 \times 10^{-4}}{3.14 \times 0.02^2}$$
$$= 0.6635 \text{ m / s}$$

2. The Reynolds number

$$N_{Re} = \frac{\rho D u}{\mu} = \frac{998.2 \times 0.6635 \times 0.04}{993.414 \times 10^{-6}}$$
$$= 26667$$

3. Entrance region length
   Reynolds number is greater than 4000. Hence, the flow is turbulent. Thus, the equation for turbulent flow is chosen for calculating the length of the entrance region.

$$l_e = 4.4 \times D \times (N_{Re})^{1/6}$$

$$= 0.04 \times 4.4 \times (26667)^{1/6} = 0.962 \text{ m}$$

   Therefore, the length of the entrance region is 0.962 m.

Calculation of pressure distribution and velocity profile at the entrance region is very complex. Once the flow reaches the fully developed region, it is much easier to determine the velocity profile. Knowledge of the velocity profile is important in determining the pressure drop, flow rate, and head loss.

4.10.2 Velocity Profile in the Fully Developed Region

Velocity profile in the fully developed region mainly depends on the radial distance from the centerline of pipe and does not change with the distance in the x-direction. To explain the velocity profile, a fully developed, steady flow is considered with no acceleration in a horizontal pipe (assuming the negligible effect of gravitation force on fluid flow) having a constant diameter.

The pressure gradient is the major driving force for the fluid flow through a pipe in the fully developed region. During the flow, the pressure force is balanced by the viscous force. Pressure force provides the force required to cause the fluid flow, whereas viscous force opposes the flow of fluid. To determine the velocity profile in fully developed flow, Newton's second law of motion is used. As shown in Figure 4.17, a fluid element of cylindrical shape has length l, radius r centered on the axis of a horizontal pipe of diameter D at a time t. It is assumed that, due to a steady flow, no acceleration is experienced by the fluid element. Also, it is presumed that the pressure at face A is p_1 and that at face B is p_2 and pressure decreases by Δp *(therefore,* $p_2 = p_1 - \Delta p$*).*

Pressure force acting on cylinder face $A = p_1 \pi r^2$
Pressure force acting on cylinder face $B = (p_1 - \Delta p)\pi r^2$
Viscous force opposing the pressure force on the circumference of cylinder $= \sigma 2\pi r l$

As the acceleration is zero, force in the x-direction is also zero (according to Newton's second law, $F = ma$; as $a = 0$, $F = 0$).

Forces acting on the fluid element must be balanced, hence,

$$p_1 \pi r^2 - (p_1 - \Delta p)\pi r^2 - \sigma 2\pi r l = 0 \tag{4.37}$$

$$\frac{\Delta p}{l} = \frac{2\sigma}{r} \tag{4.38}$$

$$\sigma = \frac{\Delta p r}{2l} \tag{4.39}$$

Shear stress for a Newtonian fluid is given by

$$\sigma = -\mu \frac{du}{dr} \tag{4.40}$$

Velocity of the fluid decreases with increase in radial distance and, therefore, du/dr carries a negative sign in Eq. (4.40).

$$\frac{du}{dr} = -\left(\frac{\Delta p}{2\mu l}\right) r \tag{4.41}$$

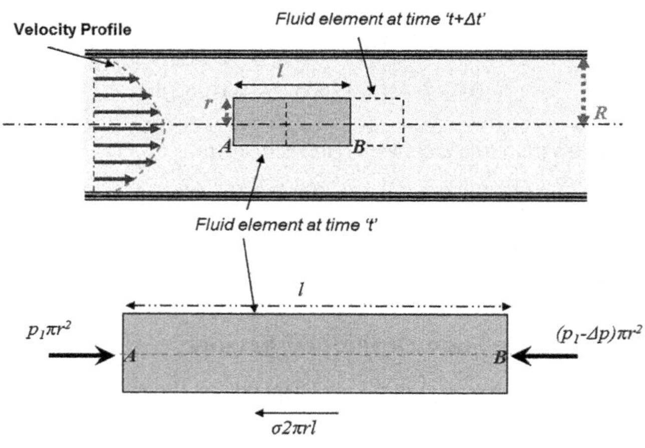

FIGURE 4.17 Movement of a cylindrical fluid element within the pipe.

Integrating to obtain the velocity profile

$$\int du = -\left(\frac{\Delta p}{2\mu l}\right)\int r \cdot dr \tag{4.42}$$

$$u(r) = -\left(\frac{\Delta p}{4\mu l}\right)r^2 + C_1 \tag{4.43}$$

where C_1 is constant. At the wall of the pipe $r = R$ and velocity $u = 0$

$$C_1 = \left(\frac{\Delta p R^2}{4\mu l}\right) \tag{4.44}$$

Therefore, Eq. (4.43) can be written as

$$u(r) = \left(\frac{\Delta p}{4\mu l}\right)\left(R^2 - r^2\right) \tag{4.45}$$

$$u(r) = \frac{\Delta p R^2}{4\mu l}\left[1 - \left(\frac{r}{R}\right)^2\right] \tag{4.46}$$

Equation 4.46 represents a parabolic equation. Therefore, a parabolic velocity profile is obtained for the fully developed flow conditions.

At the centerline of the pipe, the maximum velocity is obtained where $r = 0$ and $u = u_{max}$

$$u_{max} = \frac{\Delta p R^2}{4\mu l} \tag{4.47}$$

On integrating the velocity profile through the pipe, volumetric flow rate for a pipe can be obtained.

$$Q = \int u\, dA = \int_{r=0}^{r=R} u(r) 2\pi r dr = \frac{2\pi \Delta p R^2}{4\mu l}\int_0^R \left[1 - \left(\frac{r}{R}\right)^2\right] r dr \tag{4.48}$$

$$Q = \frac{\pi R^4 \Delta p}{8\mu l} = \frac{\pi D^4 \Delta p}{128\mu l} \tag{4.49}$$

Equation 4.49 is known as the Poiseuille's equation. From Poiseuille's equation, it can be inferred that an increase in the radius of the pipe increases the flow rate (doubling the radius or doubling the diameter increases the flow rate by 16-fold). It is also clear that the fully developed laminar flow in a pipe is directly proportional to the pressure drop and fourth power of diameter and inversely proportional to the viscosity of fluid and pipe length.

Mean velocity $(\bar{u})$ of the fluid can be determined from the flow rate, on dividing it by the cross-sectional area $\left(\pi R^2\right)$,

$$\bar{u} = \frac{Q}{A} = \frac{\Delta p R^2}{8\mu l} \tag{4.50}$$

On dividing Eq. (4.50) by Eq. (4.47),

$$\frac{\bar{u}}{u_{max}} = 0.5 \tag{4.51}$$

Therefore, for a fully developed laminar flow, the average velocity is half of the maximum velocity.

Example 4.3

Fluid with viscosity 2 Pa s is flowing through a pipe with diameter 4 cm and length 5 m. The pressure drop is 500 Pa. Determine the mean velocity and velocity of the fluid at 0.5, 1.0, 1.5, and 2.0 cm radial location in the pipe.

Solution

Given

Length of pipe = 5 m, diameter of pipe = 4 cm = 0.04 m, pressure drop = 500 Pa
Viscosity of fluid = 2 Pa s
Velocity of fluid is determined using the equation

$$u(r) = \left(\frac{\Delta p}{4\mu l}\right)\left(R^2 - r^2\right)$$

Velocity of fluid at $r = 0.5$ cm

$$u(0.5\text{ cm}) = \left(\frac{500}{4\times2\times5}\right)\left(0.02^2 - 0.005^2\right)$$
$$= 4.688\times10^{-3}\text{ m/s}$$

Similarly, Velocity of fluid at $r = 1.0$ cm $u(1.0\text{ cm}) = 3.75\times10^{-3}\text{ m/s}$

Velocity of fluid at $r = 1.5$ cm $u(1.5\text{ cm}) = 2.188\times10^{-3}\text{ m/s}$

Velocity of fluid at $r = 2.0$ cm $u(2.0\text{ cm}) = 0\text{ m/s}$

Mean velocity of the fluid is half of the maximum velocity for laminar flow.
Therefore, mean velocity $\bar{u} = 2.344\times10^{-3}\text{ m/s}$

4.11 Friction Force during Fluid Flow

Viscous and friction force oppose the flow of a fluid through a pipe. The friction force is the resistance offered by the inner walls of the pipe. Frictional energy loss occurs due to friction between the surface and the fluid, which results in mechanical energy loss. Frictional forces depend on the flow rate, Reynolds number, and surface roughness. The friction force is represented by the Friction factor, which is the ratio between the shear stress at the wall and the kinetic energy of the fluid.

$$f = \frac{\sigma_w}{\rho\bar{u}^2/2} \tag{4.52}$$

The shear stress at the wall is given by

$$\sigma_w = \frac{D\Delta p}{4l} \tag{4.53}$$

Substituting Eq. (4.53) in Eq. (4.52),

$$f = \frac{D\Delta p}{2\rho \bar{u}^2 l} \tag{4.54}$$

Pressure drop is given by

$$\Delta p = \frac{32\bar{u}\mu l}{D^2} \tag{4.55}$$

Substituting Eq. (4.55) in Eq. (4.54),

$$f = \frac{D}{2\rho \bar{u}^2 l} \times \frac{32\bar{u}\mu l}{D^2} = \frac{16\mu}{\rho \bar{u} D} = \frac{16}{N_{Re}} \tag{4.56}$$

where f is termed as the Fanning friction factor.

4.12 Flow Measuring Instruments

Measuring and controlling the flow of fluids holds importance during various food processing operations. Different types of the flow measuring instruments are used in industries, based on their applicability. These include the manometer, orifice meter, venturi meter, and rotameter. Orifice and venturi meter are used for the direct quantification of mass flow rate, whereas, the manometer is used to determine the pressure difference and flow.

4.12.1 Manometer

A manometer is generally known as a U-tube manometer, due to the U-shape of the tube (Figure 4.18). Two arms of the manometer are connected to a pipe through which the fluid passes. It consists of a small diameter tube, which is partially filled with the manometer fluid, generally mercury.

In general, the manometer fluid must be different from the fluid for which the pressure and flow rate are to be determined. In addition, the density of the fluid used in the manometer should be higher than that of the fluid in the pipe. Due to the pressure created by the fluid flow, manometer liquid is pushed down in the left-hand arm and it rises in the right-hand arm by an equal distance. After the initial displacement, manometer fluid comes to rest. Considering the pressure at various locations within the U-tube manometer, it is observed that, there is an increase in pressure from the point *A* to *B*. At points *B* and *C*, the pressure is same due to same elevation and the presence of the same fluid between them.

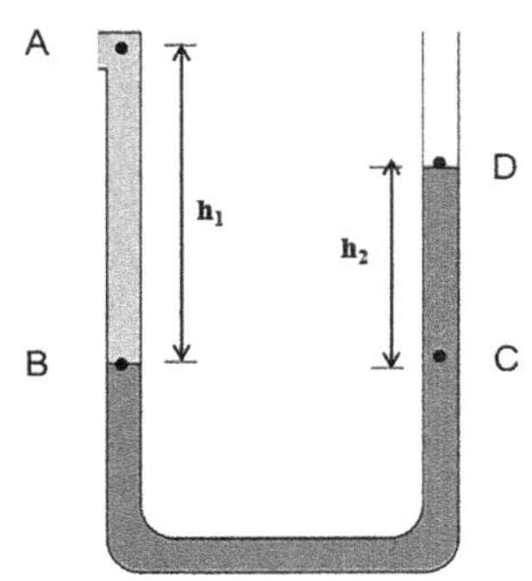

FIGURE 4.18 Schematic diagram of the manometer.

There is a decrease in pressure from points C to D, and point D is exposed to the atmosphere. The pressure at point A can be expressed as

$$P_A = \rho_m g z_m - \rho_l g z_l + P_{atm} \tag{4.57}$$

If the density of the fluid flowing through the pipe is very low compared to the density of manometer fluid, the pressure at point A will be

$$P_A = \rho_m g z_m + P_{atm} \tag{4.58}$$

Hence, once the difference between the heights of the manometer fluid in the two arms is known along with the manometer fluid density, the pressure at any desired location in the pipe can be determined.

4.12.2 Orifice Meter

Orifice meter as shown in Figure 4.19 consists of an orifice plate that is carefully machined and drilled. Orifice plate with a hole is mounted between the two flanges of the pipe.

The diameter of the orifice aperture is lower than the diameter of the pipe that accelerates the flow of fluid and creates the pressure drop. Thus, the velocity head is increased at the cost of the decrease in pressure head. Pressure taps are provided one before the orifice plate and another after the orifice plate. These taps are connected to the manometer. Determination of the pressure between the taps will help to calculate the flow rate and velocity from the Bernoulli's equation.

The velocity of the fluid through the orifice meter can be determined using Eq. (4.59) as follows:

$$u_0 = \frac{C_0}{\sqrt{(1-\beta^4)}}\sqrt{\frac{2(P_A - P_B)}{\rho}} \tag{4.59}$$

where u_0 is the velocity of the fluid through an orifice (m/s), C_0 is the orifice coefficient, β is the ratio of the orifice diameter to pipe diameter, P_A and P_B are the pressure at points A and B, respectively. Orifice coefficient is always determined experimentally as it varies considerably with a change in β and Reynolds number at the orifice.

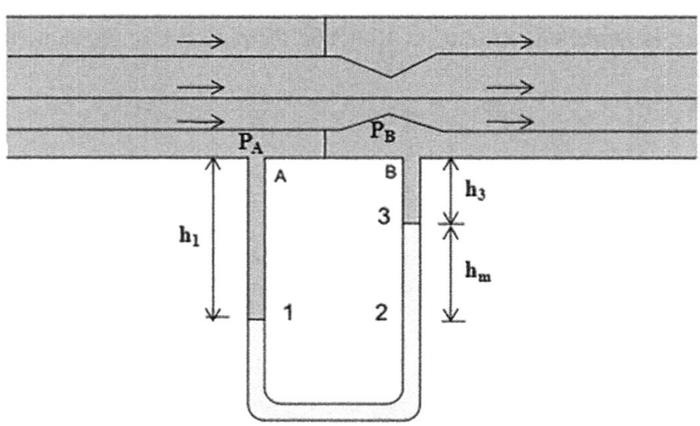

FIGURE 4.19 Schematic diagram of orifice meter.

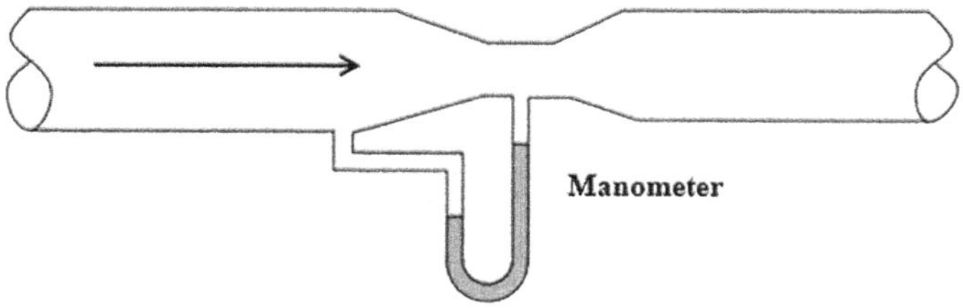

FIGURE 4.20 Schematic diagram of venturi meter.

4.12.3 Venturi Meter

The venturi meter overcomes the major limitation of an orifice meter, wherein the sudden contraction in flow causes friction which leads to energy loss. Venturi meter consists of a short conical inlet section followed by a throat section and finally a long discharge cone (Figure 4.20). Pressure taps are connected at the inlet and throat section to a manometer. Venturi meters are used for gases and liquids. It is mostly used for large flow of water, as venturi requires less power compared to the other flow meters.

The velocity of the fluid at the venturi throat (u_v) is given by,

$$u_v = \frac{C_v}{\sqrt{(1-\beta^4)}}\sqrt{\frac{2(P_A - P_B)}{\rho}} \tag{4.60}$$

where C_v is the venturi coefficient, β is the ratio of the throat diameter of the meter to the pipe diameter, P_A and P_B are the pressures at the inlet cone of the pipe and throat of venturi meter, respectively. Generally, for a well-designed venturi meter, the coefficient of venturi is about 0.98 for pipe diameter of 2–8 in (5.1–20.3 cm). For larger sized pipes, the coefficient is about 0.99.

4.12.4 Rotameter

Rotameter is a fairly common and inexpensive flow meter device. The area through which the fluid flows inside a rotameter varies with the flow rate at a constant pressure drop. Thus, rotameter is referred to as a variable area meter.

Rotameter consists of a transparent glass tube having a large end at the top (see Figure 4.21). When fluid flows upward through the tapered glass tube, it lifts the higher density bobbin upward, relative to the flow rate. This is attributed to the buoyant effect of the fluid. Higher the flow rate, the bobbin will ride higher in the rotameter. The glass tube is marked in divisions and vertical displacement of the bobbin can be read from the calibrated scale by reading at the edge of the bobbin. Bobbins may be constructed of any metal of variable densities, such as lead, aluminum, glass or plastics.

4.13 Pumps

Pumps are intended to achieve the transportation of fluids in a process industry. A pump is the most essential component of a fluid transportation system. Pump increases the mechanical energy of the fluid by increasing its pressure, velocity or raising the height. During pumping, the fluid density remains unchanged. Generally, a pump will have a suction line and delivery line. Fluid is drawn through the suction line and delivered from the pump to the delivery point, via the delivery line.

A graphical representation of the relationship between the volumetric flow rate and the increment in the total head i.e. increase in pressure is called the *characteristic curve* (Figure 4.22). The characteristic curve provides vital information about the pump efficiency and power.

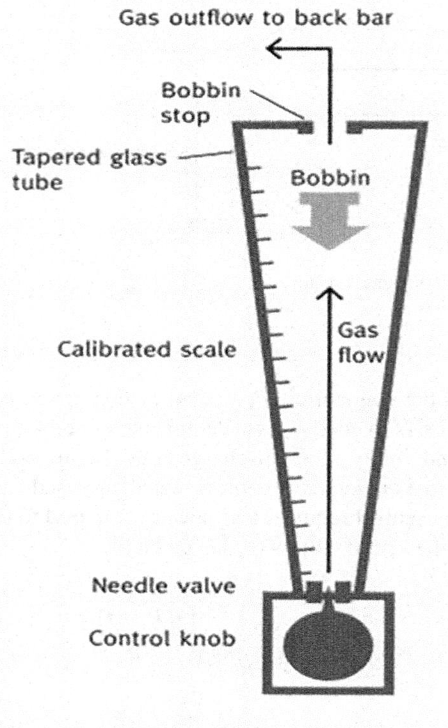

FIGURE 4.21 Schematic diagram of a rotameter. (Reproduced with permission from Jewitt, H. and Thomas, G. 2012. Measurement of flow and volume of gases. *Anaesthesia and Intensive Care Medicine* 13: 106–110.)

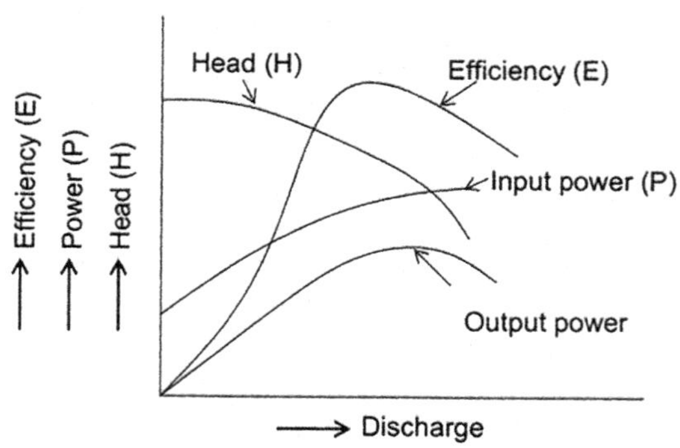

FIGURE 4.22 The characteristic curve of a pump.

Theoretical power requirement (W_{th}) of a pump depends on the volumetric flow rate (Q) and pressure drop (ΔP) given by,

$$W_{th} = Q\Delta P \tag{4.61}$$

However, actual power requirement (W) will be higher than the theoretical power requirement (W_{th}) due to the mechanical efficiency (η_m) of the pump Eq. (4.62).

$$W = \frac{W_{th}}{\eta_m} \tag{4.62}$$

4.13.1 Types of Pumps

Pumps are classified into two major types based on their operational mechanism into two major types, namely, kinetic pumps and positive displacement pumps. In positive displacement pumps, the pressure is applied directly on the fluid by a reciprocating piston. Whereas, the kinetic pumps generate a high rotational velocity which is further converted into pressure energy.

4.13.1.1 Centrifugal Pumps

Centrifugal pump is the most commonly used kinetic pump. Its working is based on the principle that, *centrifugal force increases the fluid pressure.* Centrifugal pumps consist of a motor-driven impeller with curved vanes (Figure 4.23). Fluid enters through the suction line at the centre of the rotating impeller. Rotating impeller imparts high rotational movement to the fluid. Due to the centrifugal force, fluid moves to the periphery and leaves tangentially. The velocity of the fluid is decreased in the volute region (space between the vanes and outer casing) due to an increase in the cross-sectional area. With the volute velocity, head of the fluid is converted to pressure head.

The simple construction of the centrifugal pump allows adaptability of this pump to cleaning-in-place (CIP) operations due to easy maintenance of hygienic flow conditions. However, in the centrifugal pump, the pressure generated is low and hence it is suitable only for the transportation of low viscous fluids such as water, milk or fruit juices. In addition, a centrifugal pump is not self-priming (self-priming pump is one which clears its passages of air by itself and begins pumping); thus, it should not be operated dry or with air. Further, pumping action is not possible until the suction line is again filled with the liquid.

4.13.1.2 Positive Displacement Pumps

In the positive displacement pumps, a confined portion of the fluid in a chamber is mechanically displaced forward. Here, the flow rate can be precisely controlled as the product movement is directly related to the moving parts of the pump. Thus, these pumps can be used for high viscous fluids. The positive displacement pumps are further classified into reciprocating pumps, rotary pumps, and peristaltic pumps.

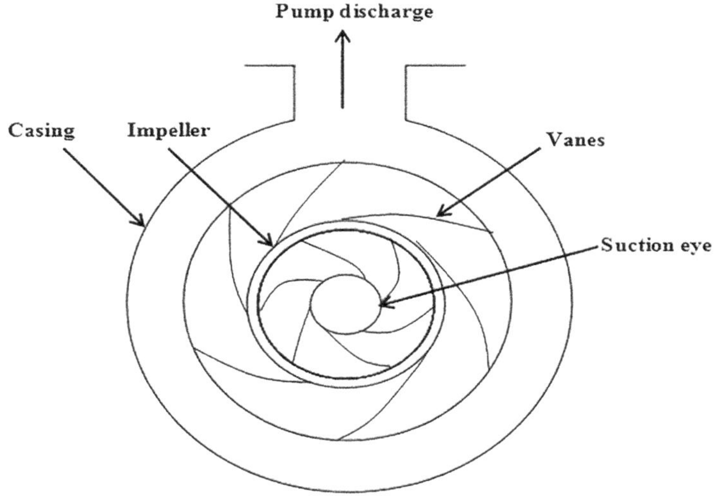

FIGURE 4.23 Schematic diagram of a centrifugal pump.

4.13.1.2.1 *Reciprocating Pumps*

Diaphragm pump and piston pump fall under the category of reciprocating pumps. In the piston pump, on the withdrawal of piston, the liquid is drawn into the cylinder through an inlet check valve and on the return stroke, it is forced through the discharge check valve. It is widely used for low viscosity liquids that require low flow rate at higher pressure. For a commercial piston pump, the maximum discharge pressure obtained is 50 atm.

In the diaphragm pumps, a flexible diaphragm made up of metal, plastic or rubber is used. It can be used for toxic or corrosive fluids. It develops pressure in excess of 100 atm and can handle the flow of small to moderate liquid flow rate. The volume of the chamber can be reduced or expanded by a diaphragm that is sealed to one side of the chamber. The diaphragms are replaceable and inexpensive. Mechanical efficiency of the reciprocating pump varies from 70% to 90% for large pumps and 40% to 50% for small pumps.

4.13.1.2.2 *Rotary Pump*

Rotary pump facilitates the movement of the liquid by exerting a rotary action on a pocket of liquid that is enclosed between the rotating part of the pump and pump housing. The pump steadily discharges a defined volume of liquid from the inlet to the pump outlet. The design of a rotary pump should ensure tight seals with a suitable material, to withstand the rubbing action of the pump. Reversal of flow direction is possible by reversing the direction of rotor rotation. Rotary pumps include sliding vane, lobe type, internal gear, and gear type pumps.

4.13.1.2.3 *Peristaltic Pump*

In a peristaltic pump (Figure 4.24), a lengthy flexible tube is squeezed by a series of moving rollers, which cause the trapped liquid to move along with the tube. Here, the discharge rate is constant and these pumps are mostly suited for small flow rate applications without leakage and air exposure. Peristaltic pumps are more hygienic in operation compared to the other pumps and thus find wide applications in the biotechnology and pharmaceutical industries.

4.13.2 Selection Criteria for Pumps

Pumps are designed based on the amount of fluid to be delivered and the discharge pressure. The viscosity of the fluid determines the head against which the pump has to work and the pressure drop during the transportation of fluid. Fluid temperature is also one of the important factors as it affects the fluid vapor pressure which can create a vacuum lock at high vapor pressure and impede the flow. To manage such conditions, generally, a fluid reservoir is kept elevated above the pump level, as gravity flow will keep the pump inlet flooding. Fluids containing suspended fragile particles require pumps with large cavities such that the particles are not compressed. Easy cleaning of pumps should be possible to prevent any microbial contamination. Pump material that comes in contact with the fluid should be non-corrosive in nature.

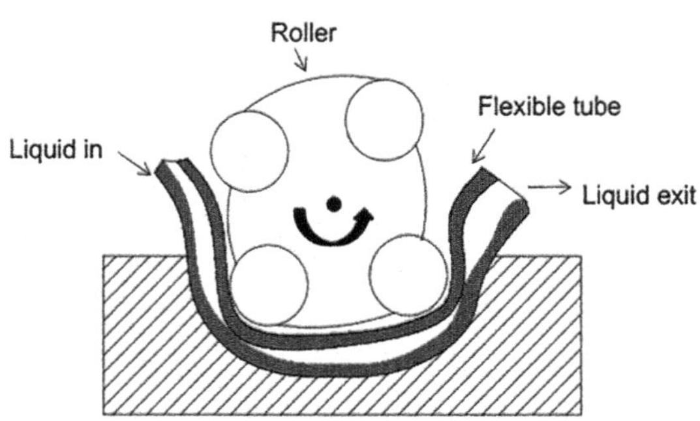

FIGURE 4.24 Schematic diagram of a peristaltic pump.

4.13.3 Energy Requirement of Pumps

The power requirement of the pump can be computed if all the energy heads associated with the pumping liquid from one place to another are known. Energy requirement of the pump should account for the energy losses due to friction, sudden contraction and expansion and the energy loss due to the pipe fittings. Equation 4.63 shows the energy requirement of the pump in terms of pressure head (h_{pump}).

$$h_{pump} = \frac{P_2 - P_1}{\rho g} + \frac{1}{2g}(u^2{}_2 - u^2{}_1) + (z_2 - z_1) + \frac{2fu^2L}{gD} + C_{fe}\frac{u^2}{2g} + C_{fc}\frac{u^2}{2g} + C_{ff}\frac{u^2}{2g} \tag{4.63}$$

$\frac{P_2 - P_1}{\rho g}$ represents the pressure head

$\frac{1}{2g}(u^2{}_2 - u^2{}_1)$ is the velocity head

$(z_2 - z_1)$ is the elevation head

$\frac{2fu^2L}{gD}$ represents the friction losses

$C_{fe}\frac{u^2}{2g}$ is the energy loss due to sudden expansion, where $C_{fe} = \left(1 - \frac{A_1}{A_2}\right)^2$

$C_{fc}\frac{u^2}{2g}$ corresponds to the energy losses due to sudden contraction

where $C_{fc} = 0.4\left(1.25 - \frac{A_2}{A_1}\right)$ when $\frac{A_2}{A_1} < 0.715$, or $C_{fc} = 0.75\left(1 - \frac{A_2}{A_1}\right)$ when $\frac{A_2}{A_1} > 0.715$,

$C_{ff}\frac{u^2}{2g}$ denotes the energy losses due to pipe fittings. Friction losses due to various fittings can be obtained from the material handbooks.

A_1 and A_2 are the cross-sectional area of the pipe (before and after the expansion/contraction); f is the friction factor. If the mass flow rate of the fluid is known, then the power requirement of the pump is given by,

$$\text{Power} = \Phi = \text{Mass flow rate} \times \text{Pump head} \tag{4.64}$$

4.14 Problems to Practice

4.14.1 Multiple Choice Questions

1. A fluid whose apparent viscosity increases with shear rate is termed as
   a. Newtonian
   b. Viscous
   c. Dilatant
   d. Non-viscous

Answer: c

2. A rotameter measures
   a. flow rate of fluid
   b. speed of turbines
   c. viscosity of fluid
   d. density of fluid

Answer: a

3. For the laminar flow, the Reynolds number should be less than,
   a. 2100
   b. 3100
   c. 4100
   d. 5100

Answer: a

4. Newton's law of viscosity relates the
   a. shear stress and viscosity
   b. velocity gradient and pressure intensity
   c. shear stress and rate of angular deformation in a liquid
   d. pressure gradient and rate of angular deformation

Answer: c

5. Reynold's number is
   a. the ratio between inertial force and viscous force
   b. the ratio between viscous force and inertial force
   c. the ratio between inertial force and pressure
   d. the ratio between viscous force and pressure difference

Answer: a

6. With increase in temperature, viscosity of a liquid
   a. increases
   b. decreases
   c. does not change
   d. may increase or decrease depending on the liquid

Answer: b

7. Manometer is a simple device for measuring
   a. force
   b. pressure
   c. density
   d. temperature

Answer: b

8. The viscosity of a gas
   a. decreases with increase in temperature
   b. increases with increase in temperature
   c. does not change with temperature
   d. may increase or decrease

Answer: b

9. An example of Newtonian fluid is
   a. sewage sludge
   b. tomato ketchup
   c. starch slurry
   d. milk

Answer: d

10. The angle between the plate and cone in a cone-and-plate viscometer is usually
    a. more than 5°
    b. 2°
    c. 3°
    d. 1°

Answer: b

11. The class of fluids having a constant viscosity which is temperature-dependent but independent of the applied shear rate is
    a. rheopectic
    b. thixotropic
    c. pseudoplastic
    d. Newtonian

Answer: d

12. The stress–strain relationship for Newtonian fluid is
    a. hyperbolic
    b. parabolic
    c. linear
    d. exponential

Answer: c

13. For non-Newtonian fluids, apparent viscosity is a function of
    a. shear rate
    b. flow rate
    c. viscous rate
    d. specific rate

Answer: a

14. The example of a thixotropic fluid is
    a. milk
    b. yogurt
    c. water
    d. diluted fruit juice

Answer: b

15. For a Newtonian fluid
    a. shear stress is proportional to shear strain
    b. rate of shear stress is proportional to shear strain
    c. rate of shear stress is proportional to rate of shear strain
    d. shear stress is proportional to rate of shear strain

Answer: d

16. For a Newtonian fluid, the slope of the plot between shear stress and shear rate is
    a. tan 45°
    b. tan 60°
    c. tan 30°
    d. tan 90°

Answer: a

17. The ratio between shear stress and shear rate is known as
    a. density
    b. specific gravity
    c. viscosity
    d. yield stress

Answer: c

18. Which of the following is the equation of continuity?
    a. $m = \rho u A = \text{constant}$
    b. $\frac{P}{\rho} + gZ + \frac{u^2}{2} = \text{constant}$
    c. $\frac{m}{\rho} = uA = \text{constant}$
    d. Both a and c

Answer: d

19. The pump in which the kinetic energy imparted to the liquid is converted to pressure energy is
    a. centrifugal
    b. rotary
    c. peristaltic
    d. all the above

Answer: a

20. A fluid exhibiting reversible time dependency is said to be
    a. Newtonian fluid
    b. pseudoplastic fluid
    c. thixotropic fluid
    d. dilatant fluid

Answer: c

4.14.2 Numerical Problems

1. Milk at 20°C is flowing in a steel pipe of inner diameter 4.5 cm. If the average velocity of milk is 5 m/s, calculate the mass flow rate of milk.

   Given
   i. $T = 20°C$
   ii. Inner diameter of steel pipe $(D) = 4.5$ cm $= 0.045$ m
   iii. Velocity of milk flow $(u) = 5$ m/s

   To find: Mass flow rate of milk $(\dot{m})$

   Solution

   Inner cross-sectional area of the pipe: $A = \frac{\pi D^2}{4} = \frac{\pi (0.045)^2}{4} = 0.0016 \text{ m}^2$

   Volumetric flow rate: $Q = A \times u = 0.0016 \times 5 = 0.008 \text{ m}^3/\text{s}$

   Density of milk at 20°C: $\rho = 1030 \text{ kg/m}^3$

   Therefore, specific volume of milk: $v = \frac{1}{\rho} = \frac{1}{1030} = 9.709 \times 10^{-4} \text{ m}^3/\text{kg}$

$$\dot{m} = \frac{Q}{v} = \frac{0.008}{9.709 \times 10^{-4}} = 8.24 \text{ kg/s}$$

Answer: Mass flow rate of milk = 8.24 kg/s

2. Sunflower oil is flowing in tube of inner diameter 0.0525 m and length 10 m. Calculate the velocity of flow if the pressure drop is 1200 Pa. The kinematic viscosity of sunflower oil is 48.46 mm^2/s and its density is 911 kg/m^3.

Given

i. Inner diameter of tube = 0.0525 m
ii. Length of tube = 10 m
iii. ΔP = 1200 Pa
iv. Kinematic viscosity of sunflower oil = 48.46 mm^2/s = 48.46×10^{-6} m^2/s
v. Density of sunflower oil = 911 kg/m^3

To find: Velocity of flow of sunflower oil

Solution

Dynamic viscosity, μ = kinematic viscosity × density = $48.46 \times 10^{-6} \times 911 = 0.044 \dfrac{\text{kg}}{\text{m s}}$

$$\bar{u} = \frac{Q}{A} = \frac{\Delta p R^2}{8 \mu l}$$

$$\bar{u} = \frac{1200 \times (0.0525/2)^2}{8 \times 0.044 \times 10} = 0.235 \text{ m/s}$$

Answer: Velocity of flow of sunflower oil = 0.235 m/s

3. Orange juice at 20°C is flowing in a tube of inner diameter 10 cm, at a volumetric flow rate of 6 m^3/h. Using this data, calculate the Reynolds number for the flow, if the viscosity and density of orange juice are 0.63 kg/m s and 1038 kg/m^3, respectively.

Given

i. Inner diameter of tube (D) = 10 cm = 10×10^{-2} = 0.1 m
ii. Volumetric flow rate (V) = 6 m^3/h
iii. Density of orange juice (ρ) = 1038 kg/m^3
iv. Viscosity of orange juice (μ) = 0.63 kg/m s
v. Temperature of orange juice (T) = 20°C

To find: Reynolds number for the flow of orange juice.

Solution

$$N_{Re} = \frac{Dv\rho}{\mu}$$

$$\text{Area of tube} = \frac{\pi D^2}{4} = \frac{\pi \times (0.1)^2}{4} = 0.0079 \text{ m}^2$$

$$\text{Flow velocity, } v = \frac{\text{Volumetric flow rate}}{\text{Area}} = \frac{6}{0.0079} = 759.494 \frac{\text{m}}{\text{h}} = 0.211 \frac{\text{m}}{\text{s}}$$

$$N_{Re} = \frac{0.1 \times 0.211 \times 1038}{0.63} = 34.765$$

Answer: Reynolds number = 34.765

4. The viscosity of a diluted fruit juice is measured using a capillary tube viscometer. The tube is of diameter 5 cm and length 20 cm. If a pressure of 2 kPa is required to maintain a flow rate of 0.5 kg/s and the density of fruit juice is 1030 kg/m^3, calculate the viscosity of the fruit juice sample.

Given

i. Diameter of tube = 5 cm = 5×10^{-2} m = 0.05 m
ii. Length of tube = 20 cm = 20×10^{-2} m = 0.2 m
iii. Pressure = 2 kPa = 2000 Pa
iv. Flow rate = 0.5 kg/s
v. Density of fruit juice = 1030 kg/m^3

To find: Viscosity of the fruit juice sample

Solution

$$\mu = \frac{\pi \Delta P R^4}{8L\dot{V}}$$

$$\dot{V} = \frac{\text{Mass flow rate}}{\text{Density}} = \frac{0.5}{1030} = 4.854 \times 10^{-4} \text{ m}^3/\text{s}$$

$$\mu = \frac{\pi \times 2000 \times (0.05/2)^4}{8 \times 0.2 \times \left(4.854 \times 10^{-4}\right)} = \frac{0.00245}{0.00078} = 3.141 \text{ Pa s}$$

Answer: Viscosity of fruit juice = 3.141 Pa s

5. Calculate the pipe diameter that would result in a flow velocity of 0.02 m/s for a fluid pumped at the rate of 1.27×10^{-4} m^3/s through a pipe.

Given

i. Flow velocity (u) = 0.02 m/s
ii. Volumetric flow rate (V) = 1.27×10^{-4} m^3/s

To find: Pipe diameter

Solution

$$\text{Area of pipe} = \frac{V}{u} = \frac{1.27 \times 10^{-4}}{0.02} = 63.5 \times 10^{-4} \text{ m}^2$$

$$\therefore \text{Area of pipe} = \frac{\pi D^2}{4} = 63.5 \times 10^{-4}$$

$$D^2 = \frac{63.5 \times 10^{-4} \times 4}{\pi} = 0.008$$

$$\therefore D = 0.09 \text{ m}$$

Answer: Diameter of pipe = 0.09 m

6. Determine the pressure drop for apple juice of density 1040 kg/m^3 and viscosity 0.0021 Pa s, which flows at an average velocity of 0.2 m/s through a pipe of diameter 0.0254 m and length 5 m.

Given

i. Density of apple juice (ρ) = 1040 kg/m^3
ii. Viscosity of apple juice (μ) = 0.0021 Pa s
iii. Flow velocity (u) = 0.2 m/s
iv. Diameter of pipe (D) = 0.0254 m
v. Length of pipe (L) = 5 m

To find: Pressure drop (ΔP)

Solution

$$A = \frac{\pi D^2}{4} = \frac{\pi \times 0.0254^2}{4} = 5.067 \times 10^{-4} \text{ m}^2$$

$$\dot{V} = A \times u = 5.067 \times 10^{-4} \times 0.2 = 1.013 \times 10^{-4} \text{ m}^3/\text{s}$$

$$\mu = \frac{\pi \Delta P R^4}{8L\dot{V}}$$

$$\Delta P = \frac{8\mu L\dot{V}}{\pi R^4} = \frac{8 \times 0.0021 \times 5 \times 1.013 \times 10^{-4}}{\pi \times (0.0254/2)^4} = \frac{8.51 \times 10^{-6}}{8.17 \times 10^{-8}} = 104.2 \text{ Pa}$$

Answer: Pressure drop = 104.2 Pa

7. Calculate the length of the entrance region of a pipe having a diameter of 0.03 m which would give a pressure drop of 70 Pa due to fluid friction for milk flowing at the rate of 0.04 kg/s with a viscosity of 3 cP and density of 1030 kg/m^3.

Given

i. Diameter of pipe (D) = 0.03 m
ii. Pressure drop (ΔP) = 70 Pa
iii. Mass flow rate of milk ($\dot{m}$) = 0.04 kg/s
iv. Viscosity of milk (μ) = 3 cP = 0.003 kg/m s
v. Density of milk (ρ) = 1030 kg/m^3

To find: Length of entrance region of the pipe

Solution

Volumetric flow rate of milk

$$Q = \frac{\dot{m}}{\rho} = \frac{0.04}{1030} = 3.883 \times 10^{-5} \text{ m}^3/\text{s}$$

Average velocity of milk

$$\bar{u} = \frac{Q}{A} = \frac{3.883 \times 10^{-5}}{(\pi \times 0.03^2)/4} = \frac{3.883 \times 10^{-5}}{7.069 \times 10^{-4}} = 0.055 \text{ m/s}$$

Reynolds number

$$N_{Re} = \frac{Du\rho}{\mu} = \frac{0.03 \times 0.055 \times 1030}{0.003} = 566.5$$

Length of the entrance region

Reynolds number is lesser than 4000. Hence, the flow is laminar. Thus, the equation for laminar flow is chosen for calculating the length of the entrance region.

$$l_e = D \times 0.06 \times N_{Re} = 0.03 \times 0.06 \times 566.5 = 1.019 \text{ m}$$

Answer: Length of the entrance region for laminar flow = 1.019 m

8. Determine the pipe diameter which results in a turbulent fluid flow having Reynolds number equal to 5000. The length of the entrance region of the pipe is 0.9 m.

Given

i. $N_{Re} = 5000$
ii. Length of entrance region of the pipe (l_e) = 0.9 m

To find: Diameter of pipe (D)

Solution

The turbulent flow equation for calculating the length of the entrance region is

$$\frac{l_e}{D} = 4.4\,(N_{Re})^{1/6}$$

$$D = \frac{l_e}{4.4 \times (N_{Re})^{1/6}} = \frac{0.9}{4.4 \times (5000)^{1/6}} = 0.049 \approx 0.05 \text{ m}$$

Answer: Diameter of pipe = 0.05 m

9. At what velocity does the flow of air and water convert from laminar to transitional and transitional to turbulent flow in a pipe of diameter, 5 cm at a temperature of 30°C?

Given

i. Pipe diameter = 5 cm = 0.05 m
ii. Temperature = 30°C
iii. Density of water at 30°C = 995.7 kg/m^3
iv. Viscosity of water at 30°C = 0.000797 Pa s
v. Density of air at 30°C = 1.164 kg/m^3
vi. Viscosity of air at 30°C = 1.868×10^{-5} kg/m s

To find: Velocity at which the flow of air and water convert from laminar to transitional and from transitional to turbulent flow.

Solution

Velocity of the fluid can be obtained by following equation:

$$u = \frac{N_{Re}\mu}{\rho D}$$

For air,

$$u = \frac{2100 \times 1.868 \times 10^{-5}}{1.164 \times 0.05} = 0.674 \text{ m/s}$$

For water,

$$u = \frac{2100 \times 0.000797}{995.7 \times 0.05} = 0.034 \text{ m/s}$$

To change the flow regime from laminar to transitional, the velocities of air and water should be 0.674 m/s and 0.034 m/s respectively.

Similarly, for attaining the turbulent flow characteristics:

For air,

$$u = \frac{4000 \times 1.868 \times 10^{-5}}{1.164 \times 0.05} = 1.284 \text{ m/s}$$

For water,

$$u = \frac{4000 \times 0.000797}{995.7 \times 0.05} = 0.064 \text{ m/s}$$

To change the flow regime from transitional to turbulent, the velocities of air and water should be 1.284 m/s and 0.064 m/s respectively.

10. Following is the data of relationship between shear stress and shear rate for a power law fluid. Determine the flow consistency index and flow behavior index.

Shear Stress (Pa)	Shear Rate (1/s)
50	100
61	150
71	200
79	250
112	500

Solution

$$\sigma = K\gamma^n$$

Applying logarithm on both sides,

$$\log\sigma = \log K + n\log\gamma$$

A plot as shown should be constructed between $\log\sigma$ and $\log\gamma$, in which the slope gives flow behavior index (n) and the intercept gives flow consistency index (K).

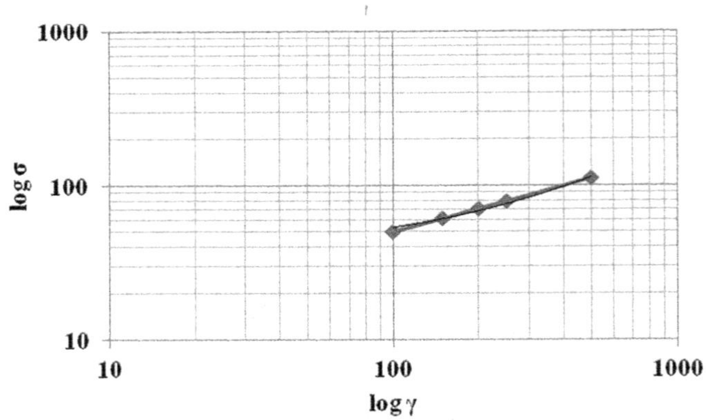

From the plot,

Slope = n = 0.5

Intercept = log K = 0.695

$$\therefore K = 4.95$$

Answer: (i) Flow consistency index (K) = 4.95 Pa s^n
(ii) Flow behavior index (n) = 0.5

BIBLIOGRAPHY

Afoakwa, E. O., Paterson, A., Fowler, M. and Vliera, J. 2009. Comparison of rheological models for determining dark chocolate viscosity. *International Journal of Food Science and Technology* 44: 162–167.

Bamforth, C. 2002. Nutritional aspects of beer: a review. *Nutritional Research* 22: 227–237.

Barnes, H. A., Hutton, J. F. and Walters, K. 1989. *An Introduction to Rheology*. New York: Elsevier Science Polishing Company Inc.

Beckett, S.T. 2009. *The Science of Chocolate*. Cambridge: Royal Society of Chemistry Paperbacks.

Bottiglieri, P., De Sio, F., Fasanaro, G., Mojoli, G., Impembo, M. and Castaldo, D. 1991. Rheological characterization of ketchup. *Journal of Food Quality* 14: 497–512.

Bourne, M.C. 1982. *Food Texture and Viscosity. Concept and Measurement*. New York: Academic Press.

Bourne, M.C. 2002. *Food Texture and Viscosity. Concept and Measurement*. New York: Academic Press.

Casson, N. 1959. A flow equation for pigment oil suspensions of the printing ink type. In *Rheology of Disperse Systems*, ed. C. C. Mill. London: Pergamon Press.

Charrondiere, U. R., Haytowitz, D. and Stadlmayr, B. 2012. FAO/INFOODS density database version. www.fao.org/docrep/017/ap815e/ap815e.pdf (accessed June 5, 2018).

ENICO Technologies Inc. 2010. Viscoline – CVL. Laval, QC. http://krohne.com/fileadmin/content/files-northam/Viscoline_brochure_2010_-_KROHNE.pdf (accessed June 5, 2018).

Figura, L. O. and Teixeira, A. A. 2007. *Food Physics*. Berlin: Springer.

Gonçalves, E. V. and Lannes, S. C. d. 2010. Chocolate rheology. *Food Science and Technology (Campinas)* 30: 845–851.

Hayes, W. A., Smith, P. G. and Morris, A. E. J. 1998. The production of quality and tomato concentrates. *Critical Reviews in Food Science and Nutrition* 38: 537–564.

International Office of Cocoa. 2000. Viscosity of cocoa and chocolate products. Analytical method 46. Available from CAOBISCO, Brussels.

Jewitt, H. and Thomas, G. 2012. Measurement of flow and volume of gases. *Anaesthesia and Intensive Care Medicine* 13: 106–110.

Kaur, C., George, B., Deepa, N., Jaggi, S. and Kapoor, H.C. 2007. Viscosity and quality of tomato juice as affected by processing methods. *Journal of Food Quality* 30: 864–877.

Kulicke, W.M. and Clasen, C. 2004. *Viscosimetry of Polymers and Polyelectrolytes*. Berlin: Springer-Verlag.

Maheshwari, R., Todke, P., Kuche, K., Raval, N. and Tekade, R. K. 2018. Micromeritics in pharmaceutical product development. In *Advances in Pharmaceutical Product Development and Research, Dosage Form Design Considerations*, ed. R. K. Tekade, 599–635. London: Academic Press.

Margules, M. 1881. On the determination of the coefficients of friction and of gliding in the plane motions of a fluid. *Wien. Akad. Sitzber. [Ser. 2]* 83: 588–602 (in German).

McGregor, R. 2016. Fundamentals of viscosity in quality control. www.foodqualityandsafety.com/article/fundamentals-viscosity-quality-control/ (accessed July 10, 2018).

Min, S., Sastry, S. K. and Balasubramaniam, V. M. 2010. Compressibility and density of select liquid and solid foods under pressures up to 700MPa. *Journal of Food Engineering* 96: 568–574.

Oldörp, K. 2014. Flow behaviour of chocolate melts-working according to ICA Standards. Application Notes V-269. Karlsruhe: Thermo Fisher Scientific Inc., Material Characterization.

Ostand, P. 2012. Part II: Obtaining accurate viscosity measurements for food and beverages. www.flowcontrolnetwork.com/part-ii-obtaining-accurate-viscosity-measurements-for-food-beverages/ (accessed July 10, 2018).

Sahin, H. and Ozdemir, F. 2004. Effect of some hydrocolloids on the rheological properties of different formulated ketchups. *Food Hydrocolloids* 18: 1015–1022.

Seguine, E. S. 1988. Casson plastic viscosity and yield value. What they are what they mean to the confectioner. *The Manufacturing Confectioner*, November: 57–63.

Servais, C., Ranc, H. and Roberts, I. D. 2003. Determination of chocolate viscosity. *Journal of Texture Studies* 34: 467–497.

Severa, L., Los, J., Nedomová, Š. and Buchar, J. 2009. On the rheological profile of malt wort during processing of substrate for lager beer. *Journal of Food Physics* 22: 5–16.

Silverson®. 2018. Manufacturing tomato sauces and ketchups. Issue No. 24FA4, Silverson Machines, Inc. East Longmeadow, MA.

Steele, C. M., James, D. F., Hori, S., Polacco, R. C. and Yee, C. 2014. Oral perceptual discrimination of viscosity differences for non-Newtonian liquids in the nectar- and honey-thick ranges. *Dysphagia* 29: 355–364.

Steffe, J. F. 1996. *Rheological Methods in Food Process Engineering*. East Lansing, MI: Freeman Press.

Valencia, C., Sánchez, M. C., Ciruelos, A., Latorre, A., Madiedo, J. M. and Gallegos, C. 2003. Non-linear viscoelasticity modeling of tomato paste products. *Food Research International* 36: 911–919.

5

Heat Transfer

Heat is a form of energy which is utilized in food processing to ensure food safety, promote digestibility, and enhance the palatability of foods. Heating of foods renders the food safe by the destruction of deteriorative microorganisms. It also leads to the development of flavor and color and improved texture of foods. The exchange of thermal or heat energy between two bodies is termed as *heat transfer.* Heat transfer can also be defined as the *discipline of engineering that deals with the generation, use, conversion, and exchange of heat energy between physical systems and determination of the rate of transfer of heat energy.* A striking difference between heat transfer and thermodynamics is that the latter deals with the *amount* of heat transferred and the former focuses on the *rate* of heat transferred.

Heat transfer is an integral component of food preservation methods such as evaporation, drying, refrigeration, freezing, pasteurization, and sterilization. In principle, the abovementioned processes involve either the addition or removal of heat to achieve the desired physical, chemical, and storage characteristics in the food product. Also, heat transfer is an important factor that affects the final product quality. Thus, it is inevitable for food engineers to obtain a fundamental understanding of the physical principles governing heat transfer. Knowledge of heat transfer would help them to determine the optimum process conditions, design heat transfer equipment, and modify an existing process design to achieve better performance. Further, a deeper qualitative and quantitative understanding of the heat transfer mechanism underlying various techniques for food production, processing, preservation, and storage is important from the perspectives of achieving better food quality and quality control, development of new food sources, and for more economical and energy-efficient processing of food from the existing sources (King, 1977). This chapter is intended to provide a comprehensive understanding of the principles governing heat transfer and the three mechanisms by which it occurs.

5.1 Theory of Heat Transfer

The rate of any transfer process is given by the following generalized equation:

$$\text{Rate of transfer} = \frac{\text{Driving force}}{\text{Resistance}} \tag{5.1}$$

Equation 5.1 would be explained with reference to heat transfer in the forthcoming sections.

5.1.1 Driving Force for Heat Transfer

According to the natural laws of physics, the energy that drives a transport phenomenon in a system always flows until the equilibrium is reached. Heat energy that leaves a warmer body would be transferred to a cold body, as long as a temperature gradient exists between the two. In other words, heat always flows from a region of higher temperature to lower temperature. The heat transfer process comes to an end when a state of thermal equilibrium is attained, i.e., when both the bodies are at the same temperature. Thus, during heat transfer, the spatial temperature difference between two objects drives the transfer of thermal energy between them.

5.1.2 Resistance to Heat Transfer

The heat transferred from a warmer body to a colder body travels through a medium, which in general offers resistance to the heat flow. The sources of resistance to heat transfer include the thickness or length crossed by the heat flow, the area of heat transfer, and the thermal conductivity of the food product. Thus, for heat transfer, the Eq. (5.1) can be rewritten as

$$\text{Rate of heat transfer} = \frac{\text{Temperature difference between the hotter and colder bodies exchanging heat}}{\text{Resistance offered by the medium}} \tag{5.2}$$

In general, heat transfer between two bodies is governed by the following rules:

- Heat will always be transferred from a hot body to a cold body through the surrounding medium.
- There must always be a temperature difference between the bodies which exchange heat.
- The heat lost by the hot body is equal to the amount of heat gained by the cold body, neglecting the heat lost to the surroundings.

5.2 Classification of Heat Transfer Processes

Heat transfer can be classified as a steady- or unsteady-state (transient) process, depending on whether the temperature of heat transfer medium and food product changes with respect to time and/or location.

5.2.1 Steady-State Heat Transfer

In the context of any field, the term *steady state* refers to a condition under which the system has finished to evolve and consequently all its physical properties remain constant with time. Nevertheless, the properties may change with the location within the product but not always. With respect to heat transfer, the property of interest is the temperature distribution within the product under study. During steady-state heat transfer, the temperature within the food product does not vary with time, but it may or may not differ with location. Thus, in this case, the rate of heat transfer is constant with time. A common example of steady-state heat transfer is that which prevails in continuously operating ovens.

5.2.2 Unsteady-State Heat Transfer

Unsteady state refers to the initial stages of heat transfer that prevails before the system attains a steady state. On contrary to the steady-state heat transfer, under unsteady state or transient conditions, the temperature distribution within the product varies as a function of both time and location. Common examples of unsteady-state heat transfer are bread baking, pasteurization, and heating or cooling of cans in retorts to sterilize the food product.

In food processing operations, heat transfer often occurs in the unsteady state. However, the system cannot remain under unsteady-state conditions indefinitely. Eventually, the system would attain a steady state once the temperature of the food product turns equivalent to that of the heating medium. Once the steady-state conditions are attained, any additional amount of heat energy being transferred into the system is utilized for phase change rather than to cause an increase in temperature.

During heat transfer analysis, the transitional period of unsteady-state heat transfer is often neglected and only the phase of steady-state heat transfer is considered for the simplicity of mathematical analysis. In other words, a steady-state condition is assumed to obtain relevant information for the design and optimization of processes and equipment. Nevertheless, the period of unsteady-state heat transfer cannot be ignored in such heat transfer operations where the temperature changes rapidly with time throughout the system. The analysis of transient state heat transfer problems involves finding solutions for the relevant

TABLE 5.1
Difference between Steady-State and Unsteady-State Heat Transfer

Steady-State Heat Transfer	Unsteady-State Heat Transfer
The temperature within the food product at any location along the heat flow path does not vary with time.	The temperature within the food product at any location along the heat flow path varies with time.
The rate of heat transfer is constant with time.	The rate of heat transfer is not constant with time.
$\frac{dT}{dt} = 0$	$\frac{\partial T}{\partial t}$ = Not constant
Problems of steady-state heat transfer are solved by integrating the differential equation for different modes of heat transfer.	Problems of unsteady-state heat transfer are deciphered by solving expressions in the form of partial differential equations.

governing equations written as partial differentials in three dimensions. The key differences between steady-state and unsteady-state heat transfer are mentioned in Table 5.1.

5.3 Mechanisms of Heat Transfer

Heat transfer occurs by three modes, namely, ***conduction, convection, and radiation***. These three modes contribute to the overall heat transfer process either individually or in combination with each other in different proportions with one of them dominating the others.

5.3.1 Heat Transfer by Conduction

Conduction is the direct transfer of heat energy from a hotter to a cooler zone within a solid or stationary fluid through vibration and collision of molecules and free electrons (Figure 5.1). The molecules vibrate by virtue of the heat energy absorbed from the heating medium. Consequently, the molecules at higher temperature vibrate faster and transfer part of their kinetic energy by means of collision to the neighboring molecules at a lower temperature. The resultant energy of friction between the molecules is ultimately converted to heat energy. During the conductive heat transfer, only energy is transferred between the molecules and there is no change in their position (Anandharamakrishnan, 2017). Heat transfer by conduction is also termed as *diffusive*, *molecular*, or *microscopic* heat transfer.

A typical example of conductive heat transfer encountered in daily life is the stovetop cooking of food in a metal pan (Figure 5.2). Here, conduction occurs in two steps. Initially, heat is transferred from the molecules of the heating source (flame/burner) to the molecules within the adjacent pan bottom. Subsequent heat transfer occurs from the bottom of the pan to its sides and further to the food contained

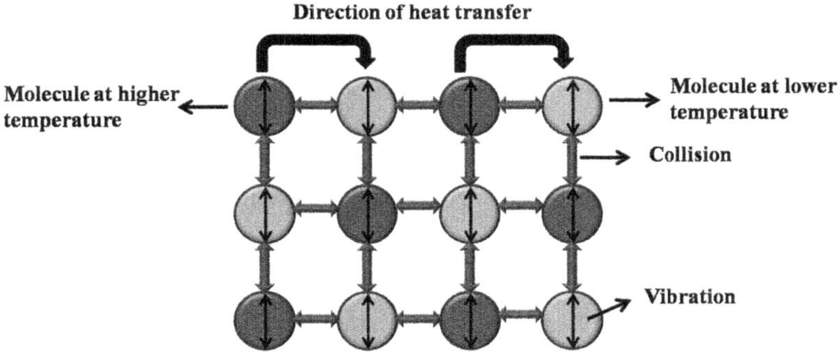

FIGURE 5.1 The principle of heat transfer by conduction. (Reproduced with permission from Anandharamakrishnan, C. 2017. Introduction to drying. In *Handbook of Drying for Dairy Products*, ed. C. Anandharamakrishnan, 1–14. Chichester, West Sussex, United Kingdom: John Wiley and Sons.)

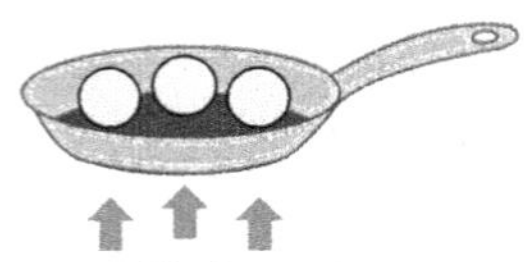

FIGURE 5.2 A typical example of conductive heat transfer in foods.

within the pan. In other words, heat from the burner passes into the metal pan, and in turn, the heat from the metal pan is transferred into the food, thus heating it up.

In general, to facilitate conductive heat transfer, materials with good thermal conductivity are used in the fabrication of the interior surfaces of food processing equipment such as a baking oven, hot plate, and baking tray and tin. Another major prerequisite for conductive heat transfer between two bodies is the direct physical contact between them. Owing to the abovementioned reason, conduction is a relatively slow method of heat transfer. However, direct transfer of heat between the adjacent molecules is advantageous in terms of its outside-in heating effect which results in a completely cooked exterior with a moist and succulent interior. Further, heat transfer by conduction always occurs in the presence of a surrounding medium. As mentioned previously, during conductive heat transfer, molecules in the medium vibrate in their own place and transfer the heat to the neighboring particles. In the absence of a medium, the transfer of heat by exchange of kinetic energy between the molecules would not be plausible. Therefore, heat transfer by conduction cannot occur in a vacuum as there would be no matter to transfer the energy.

5.3.1.1 Fourier's Law of Conductive Heat Transfer

Heat transfer by conduction is governed by the *Fourier's law* which states that "the rate of heat transfer through a material is proportional to the area and the negative gradient of temperature". The mathematical expression of Fourier's law is given by

$$q_{\text{conduction}} \propto A\frac{dT}{dx} \tag{5.3}$$

where q is the rate of heat transfer (W), dT is the temperature gradient between the heating medium (T_1, K) and the surface temperature of the product which is at or slightly above the boiling point of water (T_2, K), A is the surface area available for heat transfer, and x is the characteristic dimension of the product or thickness of the wall across which heat is transferred. On including a constant of proportionality (k), Eq. (5.3) becomes

$$q_{\text{conduction}} = -kA\frac{dT}{dx} \tag{5.4}$$

where k is the thermal conductivity (W/m K). Fourier's law is an empirical relationship which applies to isotropic or homogeneous substances. Figure 5.3 illustrates the principle of Fourier's law of conductive heat transfer.

The expression of Fourier's law signifies that conductive heat transfer is a directional quantity. This supports the second law of thermodynamics, which states that heat will always be transferred from a region of higher temperature to a region of lower temperature. Thus, a minus sign is included in Eq. (5.4) to indicate that heat transfer occurs in the direction of decreasing temperature. Using a minus sign in Eq. (5.4) returns a positive value for heat flow in the direction of decreasing temperature. Equation 5.4 holds for steady-state conductive heat transfer, and its solution can be obtained on integration between the limits T_1 and T_2.

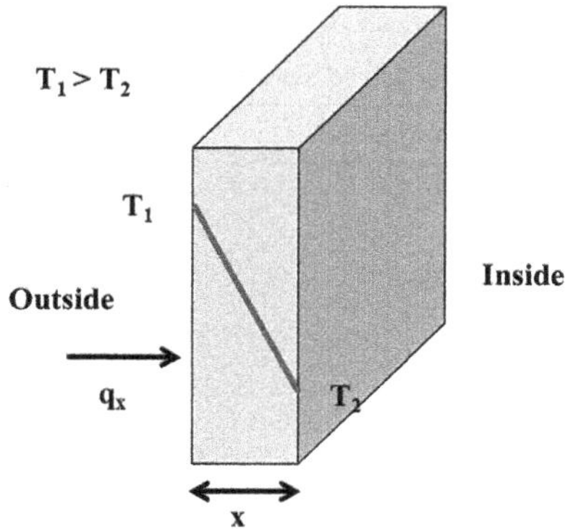

FIGURE 5.3 Conductive heat transfer through an isotropic solid (wall).

5.3.1.2 Unsteady-State Heat Transfer by Conduction

Practically, unsteady-state heat transfer dominates, where the temperature changes with time and location within the product to be heated. Here, partial derivatives are used as the temperature T changes with time (t) and location (x). In this case, Fourier's law is expressed as a partial differential equation, given by

$$\frac{\partial T}{\partial t} = \alpha \left[\frac{\partial^2 T}{\partial x^2} + \frac{\partial^2 T}{\partial y^2} + \frac{\partial^2 T}{\partial z^2} \right] \tag{5.5}$$

where x, y, and z are the rectangular coordinates and α is the thermal diffusivity.

5.3.1.3 Thermal Properties of Foods

Knowledge of the thermal properties of foods is essential to understand the different types of conductive heat transfer. It is also relevant from the perspective of heat transfer calculations performed during the design of heat transfer equipment and estimating the process times for refrigeration, freezing, heating, or drying of foods.

The commonly used thermal properties of foods in heat transfer calculations involving conduction, convection, and radiation modes are as follows:

Thermal conductivity	Thermal diffusivity	Enthalpy	Specific heat
Density	Heat transfer coefficient	Surface conductance	Absorptivity
Emissivity	Transmissivity		

The thermal properties of foods have a strong dependence on the chemical composition, structure, temperature, and state. Also many types of food are available. Therefore, it is nearly impossible to experimentally determine and tabulate the thermal properties of foods for all possible conditions and compositions. Hence, these are predicted using temperature-dependent mathematical models that account for the thermal properties of individual food constituents. The thermal property models for food components are compiled in Table 5.2. Since density, specific heat, and enthalpy have already been discussed in Chapters 1 and 3, other thermal properties of significance to conductive heat transfer such as thermal conductivity and thermal diffusivity would be explained in Sections 5.3.1.3.1 and 5.3.1.3.2, respectively. Thermal properties pertaining to convection and radiation modes of heat transfer would be discussed in the corresponding sections.

TABLE 5.2

Mathematical Models for the Thermal Properties of Food Components (Temperature: −40°C to 150°C)

Thermal Property	Food Component	Thermal Property Model
Thermal conductivity (W/m K)	Protein	$k = 1.7881 \times 10^{-1} + 1.1958 \times 10^{-3}t - 2.7178 \times 10^{-6}t^2$
	Fat	$k = 1.8071 \times 10^{-1} - 2.7604 \times 10^{-4}t - 1.7749 \times 10^{-7}t^2$
	Carbohydrate	$k = 2.0141 \times 10^{-1} + 1.3874 \times 10^{-3}t - 4.3312 \times 10^{-6}t^2$
	Fiber	$k = 1.8331 \times 10^{-1} + 1.2497 \times 10^{-3}t - 3.1683 \times 10^{-6}t^2$
	Ash	$k = 3.2962 \times 10^{-1} + 1.4011 \times 10^{-3}t - 2.9069 \times 10^{-6}t^2$
Thermal diffusivity (m²/s)	Protein	$\alpha = 6.8714 \times 10^{-8} + 4.7578 \times 10^{-10}t - 1.4646 \times 10^{-12}t^2$
	Fat	$\alpha = 9.8777 \times 10^{-8} - 1.2569 \times 10^{-11}t - 3.8286 \times 10^{-14}t^2$
	Carbohydrate	$\alpha = 8.0842 \times 10^{-8} + 5.3052 \times 10^{-10}t - 2.3218 \times 10^{-12}t^2$
	Fiber	$\alpha = 7.3976 \times 10^{-8} + 5.1902 \times 10^{-10}t - 2.2202 \times 10^{-12}t^2$
	Ash	$\alpha = 1.2461 \times 10^{-7} + 3.7321 \times 10^{-10}t - 1.2244 \times 10^{-12}t^2$
Density (kg/m³)	Protein	$\rho = 1.3299 \times 10^{3} - 5.1840 \times 10^{-1}t$
	Fat	$\rho = 9.2559 \times 10^{2} - 4.1757 \times 10^{-1}t$
	Carbohydrate	$\rho = 1.5991 \times 10^{3} - 3.1046 \times 10^{-1}t$
	Fiber	$\rho = 1.3115 \times 10^{3} - 3.6589 \times 10^{-1}t$
	Ash	$\rho = 2.4238 \times 10^{3} - 2.8063 \times 10^{-1}t$
Specific heat (kJ/kg K)	Protein	$C_p = 2.0082 + 1.2089 \times 10^{-3}t - 1.3129 \times 10^{-6}t^2$
	Fat	$C_p = 1.9842 + 1.4733 \times 10^{-3}t - 4.8008 \times 10^{-6}t^2$
	Carbohydrate	$C_p = 1.5488 + 1.9625 \times 10^{-3}t - 5.9399 \times 10^{-6}t^2$
	Fiber	$C_p = 1.8459 + 1.8306 \times 10^{-3}t - 4.6509 \times 10^{-6}t^2$
	Ash	$C_p = 1.0926 + 1.8896 \times 10^{-3}t - 3.6817 \times 10^{-6}t^2$

Source: Choi and Okos (1986).

5.3.1.3.1 Thermal Conductivity

Thermal conductivity (k) is the measure of a material's ability to transfer heat energy by conduction. It can be defined as the amount of heat that is conducted through a material of unit thickness per unit time when a temperature gradient exists across the thickness. Its value is expressed in units of W/m K. In general, solids have a higher thermal conductivity than liquids. Generally, food products with higher moisture content have a thermal conductivity value closer to that of water. Dried porous food materials and those products with air or void such as bread show low thermal conductivity. With respect to food structure, thermal conductivity varies with the orientation of fibers. For instance, in meats, thermal conductivity measured along the fibers is 15%–30% higher than that measured across the fibers (Dickerson, 1968). Similarly, the state of foods also influences the thermal conductivity. The thermal conductivity of ice ($k = 2.24$ W/m K) is about four times higher than that of water ($k = 0.56$ W/m K). As a result, the thermal conductivity of frozen foods will be threefold to fourfold higher than that of unfrozen foods (Marella and Muthukumarappan, 2013).

Choi and Okos (1986) gave the following expression to determine the thermal conductivity of a material, as a function of composition and temperature.

$$k = \sum_{i=1}^{n} k_i V_i \tag{5.6}$$

where n is the number of components in a food material and k_i is the thermal conductivity of ith component (Table 5.2); V_i is the volume fraction of ith component, obtained from the weight fraction (X_i) and density (ρ_i) using the following expression

$$V_i = \frac{X_i/\rho_i}{\sum_{i=1}^{n}(X_i/\rho_i)} \tag{5.7}$$

5.3.1.3.2 Thermal Diffusivity

Thermal diffusivity (α) is a thermal property of relevance to unsteady-state heat transfer where the temperature varies with time and location. It is the measure of a material's ability to respond to changes in its thermal environment. Thermal diffusivity (α) is a combination of three basic thermal properties: thermal conductivity, density, and specific heat (Eq. (5.8)).

$$\alpha = \frac{k}{\rho C_P} \tag{5.8}$$

The unit of thermal diffusivity is m²/s. Thus, from Eq. (5.8), it is apparent that α can be readily computed when density, specific heat, and thermal conductivity data are known. The values of thermal diffusivity for food products range from 1×10^{-7} to 2×10^{-7} m²/s. However, in the absence of experimental data, Riedel's equation (Riedel, 1969) (Eq. (5.9)) can be used for the calculation of α.

$$\alpha = 0.088 \times 10^{-6} + \left(\alpha_W - 0.088 \times 10^{-6}\right) X_W \tag{5.9}$$

where α is the thermal diffusivity, α_W is the thermal diffusivity of water, and X_W is the mass fraction of water in the food product.

Thermal diffusivity is dependent on k and C_P, wherein k of solid water (ice) is around four times higher than that of liquid water and C_P of ice is half of that of water. Consequently, the thermal diffusivity values of frozen foods are approximately nine to ten times higher when compared to their unfrozen counterparts (Desrosier and Desrosier, 1982).

Example 5.1

Estimate the rate of heat transfer through 2.0 m² of the wall of an oven insulated with 3.0 cm of glass wool where the inside surface temperature is 220°C and the outside surface temperature is 30°C. The thermal conductivity of glass wool is approximately 0.038 W/m °C.

Solution

Given

Area of the wall = 2 m²; thickness of the glass wool = 3.0 cm = 0.03 m
Inside temperature of oven = 220°C; outside surface temperature = 30°C
Thermal conductivity of glass wool = 0.038 W/m °C
Heat transfer through oven wall can be determined by the following equation:

$$q = \frac{kA\Delta T}{x}$$

$$q = \frac{0.038 \times 2 \times (220 - 30)}{0.03} = 481.33\ \text{W/m}^2$$

Answer: Heat transfer through the oven wall is 481.33 W/m²

5.3.1.4 Conductive Heat Transfer through a Rectangular Slab

Consider a rectangular slab (Figure 5.4) of a constant cross-sectional area with sides C and D. While the temperature T_2 on the side D is unknown, temperature T_1 on the side C is known. Under steady-state conditions, the temperature on the opposite side D and/or at any location inside the slab can be determined using the following procedure:

On rearranging Eq. (5.4) of Fourier's law, $q_{\text{conduction}} = -kA\dfrac{dT}{dx}$,

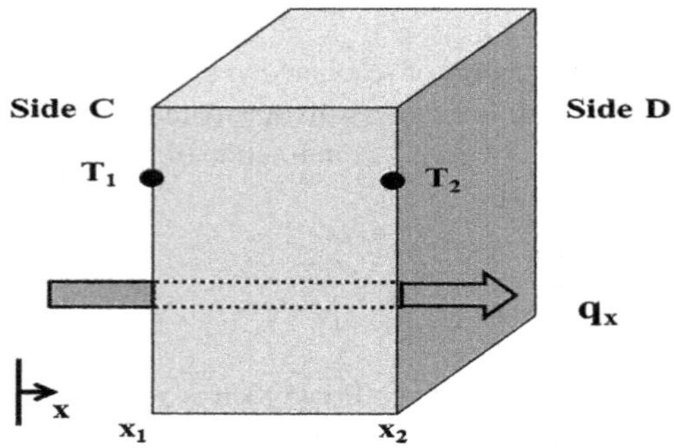

FIGURE 5.4 Schematic of heat transfer through a rectangular slab.

$$\frac{q_{\text{conduction}}}{A} = \frac{-dT}{(dx / k)} \tag{5.10}$$

Applying the boundary conditions,

$$\begin{aligned} x = x_1 \quad & T = T_1 \\ x = x_2 \quad & T = T_2 \end{aligned} \tag{5.11}$$

On integrating Eq. (5.10) between the limits specified in Eq. (5.11),

$$\int_{x_1}^{x_2} \frac{q_{\text{conduction}}}{A} dx = -\int_{T_1}^{T_2} kdT \tag{5.12}$$

As k is assumed to be independent of T and q_x is independent of x, Eq. (5.12) can be rearranged to yield

$$\frac{q_{\text{conduction}}}{A} \int_{x_1}^{x_2} dx = -k \int_{T_1}^{T_2} dT \tag{5.13}$$

On integrating Eq. (5.13),

$$\frac{q_{\text{conduction}}}{A}(x_2 - x_1) = -k(T_2 - T_1) \tag{5.14}$$

or

$$\frac{q_{\text{conduction}}}{A} = -k\frac{(T_2 - T_1)}{(x_2 - x_1)} \tag{5.15}$$

As T_2 is the temperature on face D, on rearranging Eq. (5.15),

$$T_2 = T_1 - \frac{q_{\text{conduction}}}{kA}(x_2 - x_1) \tag{5.16}$$

For estimating the temperature T, within a slab at any location, x, the terms T_2 and x_2 can be replaced with the unknown T and distance variable x, respectively, as given in Eq. (5.17):

$$T = T_1 - \frac{q_{\text{conduction}}}{kA}(x - x_1) \tag{5.17}$$

Example 5.2

One face of the stainless steel plate 2 cm thick is maintained at 220°C, while the other face is at 90°C. Calculate the heat flux through the plate. The thermal conductivity of stainless steel is 17 W/m °C. Assume steady-state conductions.

Solution

Given

Thickness of the plate = 2 cm = 0.02 m
Surface temperature of the plate = 220°C
Opposite side surface temperature of the plate = 90°C
Thermal conductivity of stainless steel = 17 W/m °C

$$\frac{q}{A} = -k\frac{dT}{dx} = -\frac{17 \times (220-90)}{(0-0.02)} = 110500 \text{ W/m}^2$$

Answer: Heat flux through the plate is 110.5 kW/m²

5.3.1.5 The Concept of Thermal Resistance

As aforementioned, rate of heat transfer is directly proportional to the temperature gradient and inversely proportional to thermal resistance which can be obtained by rearranging Eq. (5.15) as follows:

$$q_{\text{conduction}} = \frac{(T_1 - T_2)}{\left(\frac{x_2 - x_1}{kA}\right)} \tag{5.18}$$

where $\frac{(x_2 - x_1)}{kA}$ or $\frac{\Delta x}{kA}$ is the thermal resistance, denoted by R_{thermal}. Thus, thermal resistance decreases with reduced product thickness, higher thermal conductivity of the food product, and larger surface area for heat transfer.

Conductive heat transfer is analogous to the conduction of electricity in many aspects. The counterpart of the heat transfer rate ($q_{\text{conduction}}$) is current and that of the temperature difference (ΔT) is the voltage or potential difference (ΔV). The thermal resistance in conductive heat transfer is similar to the electrical resistance (R). During electric conduction, electric charge is transported from one point to another in a conductor by the motion of the electrons. This is similar to the transfer of heat from one point to another within a solid by the vibration of its molecules owing to their increased energy during thermal conduction. The analogy between heat and electrical conduction is schematically depicted in Figure 5.5.

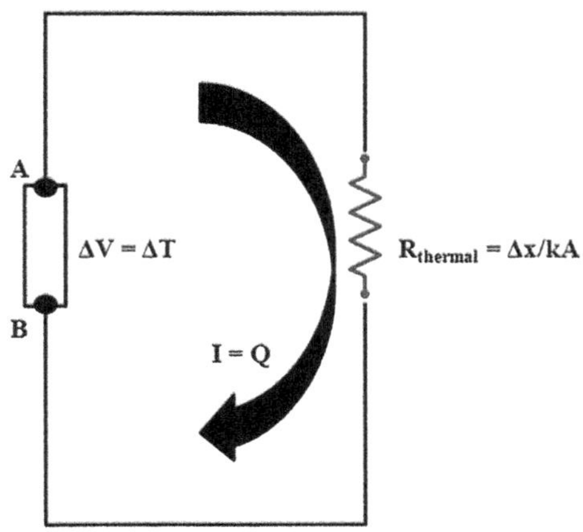

FIGURE 5.5 The analogy between the conductive transfer of heat and electricity.

The concept of analogy between thermal and electrical resistance is useful in solving heat transfer problems that involve composite wall or cylinder, to be discussed in the forthcoming sections.

5.3.1.6 Conductive Heat Transfer through a Composite Wall

Generally, composite walls are used to reduce heat losses in furnaces and refrigerated storage rooms. This is based on the principle of increasing the thermal resistance by increasing the thickness of insulating materials. Glass wool, cotton, and silk wool are the more commonly used insulating materials. Alternatively, materials having low thermal conductivity such as brick and asbestos are also used owing to their ability to withstand high temperature.

In the case of heat transfer through composite walls, total heat loss or the amount of heat transferred through the composite walls can be determined using the following methodology.

A composite wall of the furnace (Figure 5.6) is made up of several layers of insulating materials with different thermal conductivities (k_C, k_D, and k_E) and thicknesses (Δx_C, Δx_D, and Δx_E). The insulating materials are arranged in series in the direction of heat transfer. From Fourier's law, temperature, T, can be determined by the equation

$$\Delta T = -\frac{q_{\text{conduction}}\Delta x}{kA} \tag{5.19}$$

From Figure (5.6),

$$\Delta T = T_1 - T_2 = \Delta T_C + \Delta T_D + \Delta T_E \tag{5.20}$$

In Eq. (5.20), temperature for the materials C, D, and E is given by

$$\Delta T_C = -\frac{q_{\text{conduction}}\Delta x_C}{k_C A} \tag{5.21}$$

$$\Delta T_D = -\frac{q_{\text{conduction}}\Delta x_D}{k_D A} \tag{5.22}$$

$$\Delta T_E = -\frac{q_{\text{conduction}}\Delta x_E}{k_E A} \tag{5.23}$$

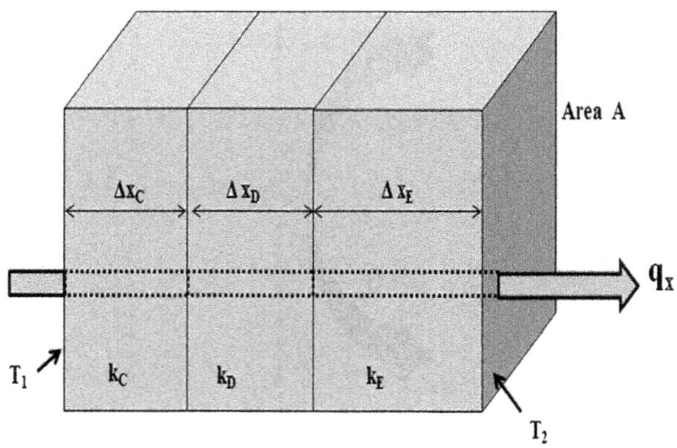

FIGURE 5.6 Conductive heat transfer through a composite rectangular wall.

From Eqs. (5.19–5.23),

$$T_1 - T_2 = -\left(\frac{q_{\text{conduction}} \Delta x_C}{k_C A} + \frac{q_{\text{conduction}} \Delta x_D}{k_D A} + \frac{q_{\text{conduction}} \Delta x_E}{k_E A} \right) \tag{5.24}$$

On rearranging the terms of Eq. (5.24),

$$T_1 - T_2 = -\frac{q_{\text{conduction}}}{A} \left(\frac{\Delta x_C}{k_C} + \frac{\Delta x_D}{k_D} + \frac{\Delta x_E}{k_E} \right) \tag{5.25}$$

Rewriting Eq. (5.25) for thermal resistance as given in the Eq. (5.16)

$$q_{\text{conduction}} = \frac{T_2 - T_1}{\left(\frac{\Delta x_C}{k_C A} + \frac{\Delta x_D}{k_D A} + \frac{\Delta x_E}{k_E A} \right)} \tag{5.26}$$

or

$$q_{\text{conduction}} = \frac{T_2 - T_1}{\left(R_{\text{thermal}C} + R_{\text{thermal}D} + R_{\text{thermal}E} \right)} \tag{5.27}$$

where

$$R_{\text{thermal}C} = \frac{\Delta x_C}{k_C A} \; R_{\text{thermal}D} = \frac{\Delta x_D}{k_D A} \; R_{\text{thermal}E} = \frac{\Delta x_E}{k_E A} \tag{5.28}$$

Overall heat transfer resistance is given by

$$R_{\text{thermal}} = R_{\text{thermal}C} + R_{\text{thermal}D} + R_{\text{thermal}E} \tag{5.29}$$

or

$$\frac{1}{R_{\text{thermal}}} = \frac{1}{R_{\text{thermal}C}} + \frac{1}{R_{\text{thermal}D}} + \frac{1}{R_{\text{thermal}E}} \tag{5.30}$$

Thus, similar to electrical circuits, thermal resistances occurring in series are additive. For example, when heat flows through two components with a resistance of 1 and 2 K/W, respectively, then the overall resistance to heat transfer is 3 K/W.

Example 5.3

A compound wall of a cold storage consists of a glass wool insulation (0.2 m, $R_{\text{glass wool}} = 26$ m °C/W) and plywood on both the sides (0.01 m, $R_{\text{plywood}} = 8.7$ m °C/W). The outside and inside wall temperatures are 13°C and −18°C, respectively. Find the overall thermal resistance of the wall and the heat flow through the wall.

Solution

Given

Plywood wall has a thickness of 0.01 m with R = 8.7 m °C/W
Glass wool thickness 0.2 m with R = 26 m °C/W
Inside and outside temperature of the cold storage is −18°C and 13°C

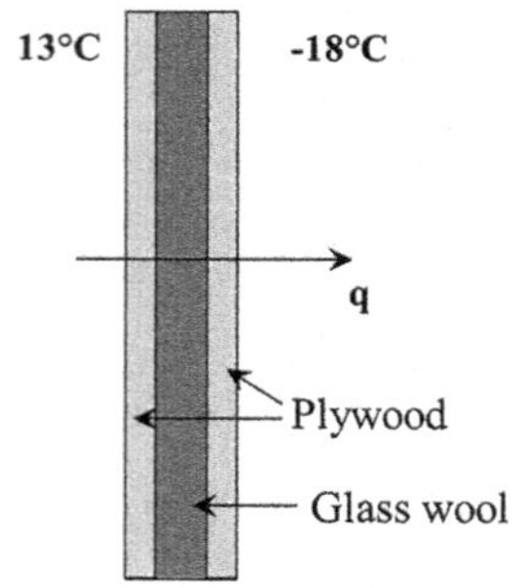

Overall thermal resistance can be calculated using the following equation:

$$R_{\text{thermal}} = \frac{\Delta x_1}{k_1 A}$$

Therefore, $R_{\text{thermal}} = \frac{0.01}{1/8.7} + \frac{0.2}{1/26} + \frac{0.01}{1/8.7} = 5.374$

Heat flow through the wall can be obtained using the following equation:

$$q = \frac{\Delta T}{R_{\text{thermal}}}$$

$$q = \frac{[13-(-18)]}{5.374} = 5.769\ \text{W/m}^2$$

Therefore, overall heat transfer through the compound wall of cold storage is 5.769 W/m^2 with the overall thermal resistance of 5.374 m^2 °C/W.

Example 5.4

A furnace is constructed with 0.2 m of firebrick, 0.1 m of insulating brick, and 0.2 m of building brick. The inside temperature is 1000°C and the outside temperature is 45°C. If the thermal conductivities are 1.4, 0.21, and 0.7 W/m °C, respectively, find the heat loss per unit area and temperature at the junction of the firebrick and the insulating brick.

Solution

Given

Thickness and thermal conductivity data

Firebrick $x_1 = 0.2$ m, $k_1 = 1.4$ W/m °C
Insulating brick $x_2 = 0.1$ m, $k_2 = 0.21$ W/m °C
Building brick $x_3 = 0.2$ m, $k_3 = 0.7$ W/m °C
Total resistance can be obtained from the equation

$$R = \left(\frac{x_1}{k_1} + \frac{x_2}{k_2} + \frac{x_3}{k_3}\right)$$

$$R = \left(\frac{0.2}{1.4} + \frac{0.1}{0.21} + \frac{0.2}{0.7}\right) = 0.905$$

Heat transfer through the walls can be obtained as

$$q = \frac{T_3 - T_1}{R} = \frac{1000-45}{0.905} = 1055.249\ \text{W/m}^2$$

The temperature at the junction of firebrick and the insulating brick can be determined using the following expression:

$$q = \frac{T_3 - T_2}{R} = \frac{1000 - T_2}{0.143}$$

$$\therefore T_2 = T_3 - (q \times R_1) = 1000 - (1055.249 \times 0.143) = 849.25$$

Answer: Temperature at the junction of firebrick and the insulating brick is 849.25°C

5.3.1.7 Conductive Heat Transfer through a Cylinder

In the shop floor of a food industry, food engineers come across cylindrical geometry during the analysis of heat transfer through pipes. The striking difference between heat conduction in a rectangular geometry and that in a cylindrical geometry is the change in area for heat flow from one radial location to another in the latter. While the steady temperature profile of one-dimensional conduction in a rectangular slab is a straight line (in the presence of constant thermal conductivity), it is logarithmic in the radial coordinate of a cylindrical geometry in a similar situation.

The conductive heat transfer through a cylinder can be understood by considering a cylindrical pipe of length L, inner radius r_i, and outer radius r_o (Figure 5.7). The thermal conductivity of the pipe material is assumed to remain constant with temperature.

Fourier's law for heat transfer rate along the radial direction (q_r) through a tubular pipe is given by

$$q_r = -kA\frac{dT}{dr} \tag{5.31}$$

Substituting the circumferential area of pipe in Eq. (5.31)

$$q_r = -k(2\pi rL)\frac{dT}{dr} \tag{5.32}$$

The boundary conditions are given by

$$\begin{aligned} r &= r_i \quad & T &= T_i \\ r &= r_o \quad & T &= T_o \end{aligned} \tag{5.33}$$

Rearranging Eq. (5.32) and applying the boundary conditions

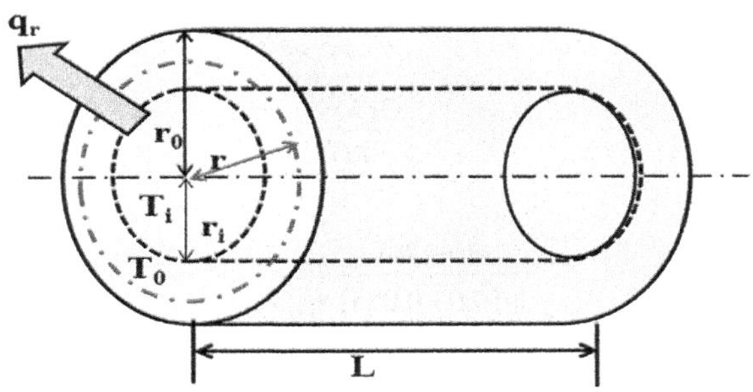

FIGURE 5.7 Conductive heat transfer through a tubular pipe.

$$\frac{q_r}{2\pi L}\int_{r_i}^{r_o}\frac{dr}{r} = -k\int_{T_i}^{T_o} dT \tag{5.34}$$

On integrating Eq. (5.34)

$$\frac{q_r}{2\pi L}\left|\ln r\right|_{r_i}^{r_o} = -k\left|T\right|_{T_i}^{T_o} \tag{5.35}$$

$$q_r = \frac{2\pi Lk(T_i - T_o)}{\ln\left(\frac{r_o}{r_i}\right)} \tag{5.36}$$

or

$$q_r = \frac{(T_i - T_o)}{\left|\frac{\ln(r_o/r_i)}{2\pi Lk}\right|} \tag{5.37}$$

Thus, it is clear that the radial conduction through a cylindrical wall is logarithmic and not linear as observed in the case of a rectangular wall, under the same conditions. The expression for thermal resistance can be derived from Eq. (5.37) as follows:

$$R_t = \frac{\ln r_o/r_i}{2\pi Lk} \tag{5.38}$$

Example 5.5

A steel pipe is 10 m in length, with an inner radius of $r_i = 2.5$ cm, an outer radius of $r_o = 3$ cm, inside wall temperature of 80°C, outside wall temperature of 25°C, and $k = 45$ W/(m K). Find the heat transfer (kW) through the steel pipe due to the temperature difference between the inside and outside pipe walls.

Solution

Given

Pipe length = 10 m, inner radius of pipe = 0.025 m, outer radius of pipe = 0.03 m, inside temperature of pipe wall = 80°C, outside temperature of pipe wall = 25°C, thermal conductivity of pipe wall = 45 W/(m K).

Heat transfer through the pipe wall can be determined by the following equation:

$$q_r = \frac{(T_i - T_o)}{\left|\frac{\ln(r_o/r_i)}{2\pi Lk}\right|}$$

$$q_r = \frac{(80 - 25)}{\left|\frac{\ln(0.03/0.025)}{2\times 3.14\times 10\times 45}\right|} = 852505\text{ W}$$

Heat transfer through the pipe wall is 852.5 kW.

5.3.1.8 Conductive Heat Transfer through a Composite Cylinder

The heat transfer rate through a composite cylinder made up of two layers (see Figure 5.8) can be calculated as follows:

In the earlier section, the rate of heat transfer through a single-walled cylinder was given by

$$q_r = \frac{(T_i - T_o)}{\left|\frac{\ln(r_o/r_i)}{2\pi Lk}\right|} = \frac{(T_i - T_o)}{R_t}$$

In this case, the composite cylinder contains two layers, with thermal resistance R_{tC} and R_{tD}. Therefore, the rate of heat transfer is given by

$$q_r = \frac{(T_1 - T_3)}{\frac{\ln(r_2/r_1)}{2\pi Lk_C} + \frac{\ln(r_3/r_2)}{2\pi Lk_D}} \tag{5.39}$$

or, expressing q_r in terms of thermal resistance

$$q_r = \frac{(T_1 - T_3)}{R_{tC} + R_{tD}} \tag{5.40}$$

If more cylinders are present in the system with surface temperatures between T_1 and T_3, the thermal resistance of these cylinders should be added as denoted in the denominator of Eq. (5.40). To determine the temperature of middle layer C (T_2), the equation for the rate of heat transfer Eq. (5.39) is considered. Under steady-state conditions, q_r has the same value through each layer of the composite cylinder. Therefore, using Eq. (5.39), the unknown temperature of the middle layer can be determined.

$$T_2 = T_1 - q\left(\frac{\ln(r_2/r_1)}{2\pi Lk_C}\right) \tag{5.41}$$

5.3.2 Heat Transfer by Convection

Convection is the predominant mode of heat transfer in fluids. Fluids include liquids and gases which contain loosely packed molecules that can move easily from one point to another. Generally, heat transfer by convection occurs between a moving fluid and a stationary solid surface (Figure 5.9), with a difference in temperature between them. In contrast to conductive heat transfer, during convection, heat energy is

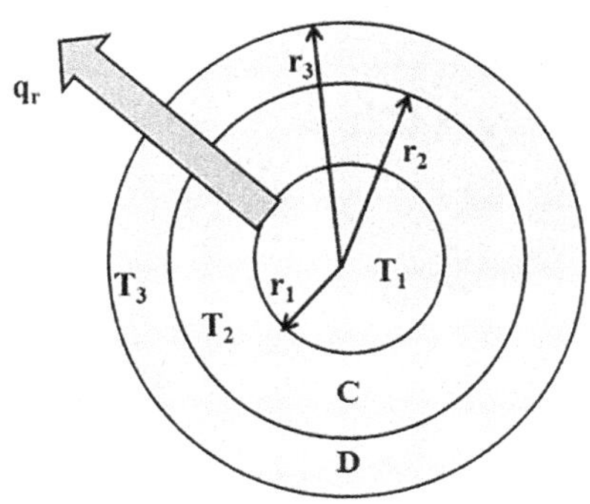

FIGURE 5.8 Conductive heat transfer through a composite cylinder.

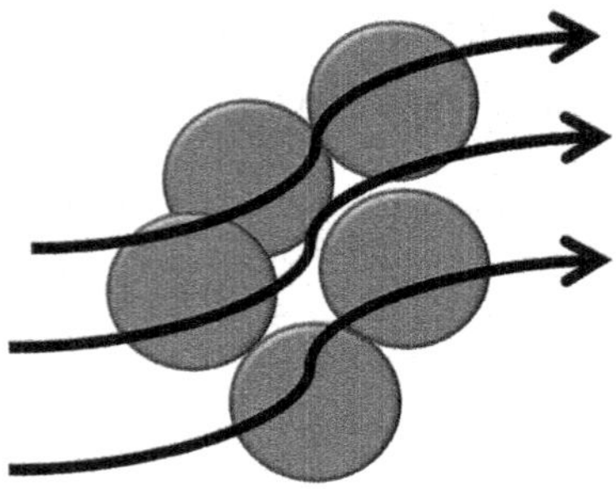

FIGURE 5.9 The principle of convection heating. (Reproduced with permission from Anandharamakrishnan, C. 2017. Introduction to drying. In *Handbook of Drying for Dairy Products*, ed. C. Anandharamakrishnan, 1–14. Chichester, West Sussex, United Kingdom: John Wiley and Sons.)

transferred by the physical movement of fluid layers in the form of eddies, also known as convection currents. While the molecules at lower temperature are displaced downward owing to their higher density, the vice versa happens with those at a higher temperature. In other words, molecules with lower kinetic energy replace the molecules having higher kinetic energy. Thus, molecules undergo a change in position within the fluid as a function of their temperature and density. The magnitude of the fluid motion also plays an important role in convective heat transfer. A day-to-day life example of convective heat transfer is cooling a cup of coffee by blowing air over it (Figure 5.10). Other examples include drying of vegetables using hot air at high velocity to achieve faster drying when compared to using stagnant hot and use of air heating units to warm the air in a room or the production area of a food processing unit.

5.3.2.1 Newton's Law for Convective Heat Transfer

Newton's law of cooling states that *the rate of heat loss from a solid surface to the fluid is directly proportional to the area of solid surface* and *difference in temperature between the surface and the fluid.* It can be expressed as

$$q \propto A(T_S - T_\infty) \tag{5.42}$$

FIGURE 5.10 A daily life example of heat transfer by convection.

where q is the rate of heat transfer, A is the area of the solid surface, and T_S and T_∞ are the temperatures of solid surface and fluid, respectively. When the constant of proportionality is removed, Eq. (5.42) becomes

$$q = h\,A\left(T_S - T_\infty\right) \tag{5.43}$$

where, h, the constant of proportionality is termed as the convective heat transfer coefficient, expressed in units of W/(m^2 K). Equation 5.43 is also known as the *Newton's law of cooling.*

Example 5.6

The rate of heat transfer per unit area from a metal plate is 1700 W/m^2. The surface temperature of the plate is 200°C, and the ambient temperature is 30°C. Find out the convective heat transfer coefficient.

Solution

Given

Plate surface temperature = 200°C
Ambient temperature = 30°C
The rate of heat transfer per unit area = 1700 W/m^2
As the rate of heat transfer per unit area is known, convective heat transfer coefficient can be obtained directly from Newton's law of cooling Eq. (5.43).

$$h = \frac{1700[\mathrm{W/m^2}]}{(200-30)[^\circ\mathrm{C}]}$$

$$h = 10\left[\mathrm{W/(m^2\,^\circ C)}\right]$$

Answer: The convective heat transfer coefficient is found to be 10 W/m^2°C

Example 5.7

The rate of heat flux from a metal plate is 1000 W/m^2. The surface of the plate is at a temperature of 220°C and the ambient temperature is 20°C. Estimate the convective heat transfer coefficient.

Solution

Given

Heat flux = 1000 W/m^2; surface temperature of plate = 220°C
Ambient temperature = 20°C
Convective heat transfer coefficient can be obtained using expression

$$h = \frac{q}{(T_S - T_\infty)} = \frac{1000}{220-20} = 5\,\mathrm{W/m^2\,^\circ C}$$

Answer: Convective heat transfer coefficient is 5 W/m^2°C

5.3.2.2 Types of Convective Heat Transfer

Based on the driving mechanism by which the convection currents are induced in the fluid, convective heat transfer can be classified into two types as follows: natural or free convection and forced convection.

5.3.2.2.1 Natural Convection

In natural convection, the fluid motion is induced by buoyancy forces. Heat is transferred by the movement of bulk fluid, instigated by the temperature-induced density differences in the fluid.

Consider a heated vertical flat plate at a surface temperature of T_S, in contact with a cold fluid at temperature T_b, on one side of the plate. The temperature of the hot plate will drop as a result of heat transfer with the cold fluid and the temperature of the fluid will increase. As a result, the hot plate would be surrounded by a thin layer of warm air, and subsequently, heat will be transferred from this layer to the outer layers of air. Owing to its higher temperature, the density of the thin layer of warm air adjacent to the hot plate is lower. Thus, the heated air rises and the resultant fluid motion is called the natural convection current. Under the influence of a gravitational field, the net force that pushes the low-density air (fluid) upward through the high-density air is called the buoyancy force.

During natural convection, a boundary layer forms close to the hot plate. A boundary layer can be defined as the region wherein the fluid exerts significant viscous effects and undergoes considerable change in velocity. The velocity profile of the fluid in the boundary layer during natural convection is shown in Figure 5.11. Adjacent to the plate surface, the velocity is zero. Hence, under the abovementioned condition of no fluid motion, heat transfer occurs merely by conduction but at a very low rate. Subsequently, the velocity increases to attain its maximum and then reduces to zero at the end of the boundary layer as the fluid is at rest in the bulk. Thus, the fluid flow in the boundary layer is laminar during the initial period of natural convection. During the latter stages, at a certain distance from the leading edge, turbulent eddies appear in the fluid, depending on the fluid properties and the temperature difference between the plate and the bulk fluid. As a result, the fluid enters the transition zone between the laminar and turbulent flow profiles. Further away from the plate, the boundary layer may become entirely turbulent leading to boundary-layer instabilities.

According to the velocity profile described previously, a density gradient would be established as a function of temperature between the fluid near the solid surface and that at a farther distance from the plate. The density difference would generate a positive or negative buoyancy force. While hot surface establishes a positive buoyancy force, the cold surface will lead to negative buoyancy force. Thus, buoyancy force is the driving force which initiates and maintains the natural convective heat transfer process. The variation of fluid density with the temperature at constant pressure can be expressed in terms of the coefficient of thermal expansion (β), given by

$$\beta = -\frac{1}{\rho}\cdot\frac{\partial \rho}{\partial T}\bigg|_{P=\text{Constant}} \tag{5.44}$$

where T is the absolute temperature, ρ is the density, and P is the pressure. As the buoyancy force is proportional to the density gradient, the larger the temperature difference between the fluid and solid

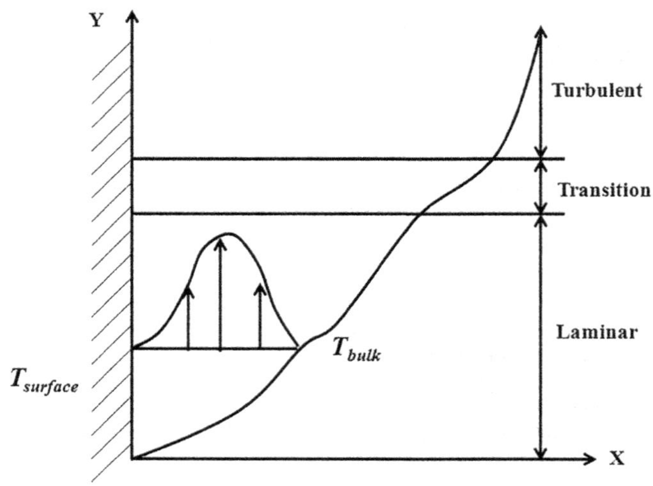

FIGURE 5.11 Velocity profile of the fluid in the boundary layer during natural convection. (T_{surface}, surface temperature; T_{bulk}, bulk fluid temperature).

surface, greater would be the buoyancy force. According to Charles' law, for an ideal gas, the coefficient of thermal expansion is given by

$$\beta_{\text{ideal gas}} = \frac{1}{T} \quad (5.45)$$

Heat transfer by natural convection depends on physical properties of the fluid including density (ρ), viscosity (μ), thermal conductivity (k), specific heat at constant pressure (C_P), and coefficient of thermal expansion (β). In addition, temperature gradient (ΔT) and acceleration due to gravity (g) are also responsible for the generation of natural convection currents.

Natural convection is encountered in the food industry and household situations in several instances, of which, a few are listed below:

- Vertical or horizontal retorts, exposed to ambient air, with or without insulation.
- Food placed in a chiller or freezer in which circulation is not aided by fans.
- Food product placed in ovens without fans.
- Air cooling of cooked/baked product.

5.3.2.2.1.1 Dimensional Groups for Natural Convection

i. Grashof number

Grashof number (N_{Gr}) is a dimensionless group which represents the ratio of the buoyancy force that arises due to a temperature difference in the fluid, to the viscous force acting on the fluid. N_{Gr} is given by

$$N_{Gr} = \frac{d^3 g\beta\Delta T}{\nu^2} \quad (5.46)$$

where d is the characteristic dimension of the product (m), g is the acceleration due to gravity (m/s^2), β is the coefficient of thermal expansion coefficient, ΔT is the temperature difference between the bulk of the fluid and the surface (K), and ν is the kinematic viscosity of the fluid (m^2/s), given by μ/ρ, where μ and ρ are the dynamic viscosity (kg/m s) and density of the fluid (kg/m^3), respectively.

By substituting the value of kinematic viscosity in Eq. (5.46), the physical meaning of the Grashof number can be interpreted as follows:

$$N_{Gr} = \frac{d^3 g\beta\Delta T}{\nu^2} = \frac{\rho^2 d^3 g\beta\Delta T}{\mu^2} = \frac{\left(\frac{\text{Buoyant force}}{\text{Area}}\right)\cdot\left(\frac{\text{Momentum}}{\text{Area}}\right)}{\left(\frac{\text{Viscous force}}{\text{Area}}\right)^2} \quad (5.47)$$

Similar to the Reynolds number in fluid flow, Grashof number can also be used to determine whether the flow is laminar or turbulent. Fluid flow over a plate is considered to be turbulent if the Grashof number is higher than 10^9.

ii. Rayleigh number (N_{Ra}):

Another parameter, the Rayleigh number, used for perfect natural convection is the product of Grashof and Prandtl number, given by

$$N_{Ra} = N_{Gr} \cdot N_{Pr} = \frac{d^3 g\beta C_P \rho^2 \Delta T}{\mu k} \quad (5.48)$$

where k is the thermal conductivity (W/m K), β is the thermal expansion coefficient, d is the characteristic dimension of the product (m), g is the acceleration due to gravity (m/s^2), ρ is the density of fluid (kg/m^3), μ is the viscosity (kg/m.s), and ΔT is the temperature difference between the fluid and the surface (K). However, natural convection is not always limited to laminar flow. In general, it has been observed that the transition of flow regime from laminar to turbulent occurs at the critical Rayleigh number of ~10^9.

5.3.2.2.1.2 Natural Convection Correlations The heat transfer coefficient for natural convection is given by

$$N_{Nu} = f(N_{Gr}, N_{Pr}) \tag{5.49}$$

If the dimensionless groups of convective heat transfer can be related by a simple power function, then the general equation for natural convection can be written as

$$N_{Nu} = KN_{Pr}^{k} N_{Gr}^{m} (L/D)^{n} \tag{5.50}$$

The correlations pertaining to heat transfer by natural convection are based on the shape and characteristic dimension of the food product or equipment geometry, which is typically the length in case of rectangular-shaped products and diameter for spherical/cylindrical products.

i. ***For vertical retorts and oven walls*** (vertical cylinders and planes)

$$N_{Nu} = 0.53(N_{Pr} \times N_{Gr})^{0.25}, \quad \text{for } 10^4 < (N_{Pr} \times N_{Gr}) < 10^9 \tag{5.51}$$

$$N_{Nu} = 0.12(N_{Pr} \times N_{Gr})^{0.33}, \quad \text{for } 10^9 < (N_{Pr} \times N_{Gr}) < 10^{12} \tag{5.52}$$

In the case of air, the preceding equations can be approximated respectively as given in Eqs. (5.53) and (5.54):

$$h_c = 1.3(\Delta T/L)^{0.25} \tag{5.53}$$

$$h_c = 1.8(\Delta T)^{0.25} \tag{5.54}$$

where Eqs. (5.53) and (5.54) are the dimensional equations in standard units: ΔT in °C, L or D in m, and h_c in W/m^2°C. In the Eqs. (5.53) and (5.54), the height of the plane or cylinder is used as the characteristic dimension for the calculation of N_{Nu} and N_{Gr}. The physical constants of the fluid in these natural convection equations are considered at the average temperature between the surface and the bulk fluid.

ii. ***For steam pipe or sausages lying on a rack*** (horizontal cylinders)

$$N_{Nu} = 0.54(N_{Pr} \times N_{Gr})^{0.25}, \quad \text{for laminar flow, } 10^3 < (N_{Pr} \times N_{Gr}) < 10^9 \tag{5.55}$$

Similar to the abovementioned case, in case of air (applicable when in contact with hotter or colder foods), Eq. (5.55) can be simplified as follows:

$$h_c = 1.3(\Delta T/D)^{0.25}, \quad \text{for } 10^4 < (N_{Pr} \times N_{Gr}) < 10^9 \tag{5.56}$$

$$h_c = 1.8(\Delta T)^{0.33}, \quad \text{for } 10^9 < (N_{Pr} \times N_{Gr}) < 10^{12} \tag{5.57}$$

iii. ***For the cooling of slabs of cake*** (horizontal planes)

In this case, the corresponding cylinder equations may be applied, using the length of the plane instead of the diameter of the cylinder to calculate N_{Nu} and N_{Gr}.

5.3.2.2.2 Forced Convection

Forced convection heat transfer occurs under the influence of an external force. In the forced convection mechanism, food products of varied geometrical shapes are heated or cooled by inducing convection currents in the surrounding fluid by mechanical means, using a pump, impeller, or fan. Thus, in addition to the thermal properties of the fluid, the rate of forced convection heat transfer also depends on the type of mechanical force used to generate the convection currents.

In the food industry, forced convection is the mechanism of heat transfer in baking ovens, blast freezers, fluidized-bed freezers, meat chillers, ice-cream hardening rooms, and agitated retorts. Most of the industrial tunnel dryers consist of a fan or blower to induce turbulence in the air. As the fluid is constantly replaced during forced convection, the rates of heat transfer are higher than that obtained with natural convection. Further, the heat transfer rate increases with the increase in the velocity of the fluid. Typical values of heat transfer coefficient for natural and forced convection are listed in Table 5.3. The effect of thermal expansion coefficient and Grashof number on the rate of heat transfer during natural convection is replaced by the influence of circulation velocities and Reynolds number in the case of forced convection.

5.3.2.2.2.1 Dimensionless Groups for Forced Convection

i. **Nusselt number (N_{Nu})**

The Nusselt number for convective heat transfer involves the heat transfer coefficient (h, in W/m^2 K), the thermal conductivity of the fluid (k, in W/m K), and the characteristic dimension of the product (d, in m) Eq. (5.58).

$$N_{Nu} = \frac{hd}{k} \tag{5.58}$$

Nusselt number is also considered as the ratio of the characteristic dimension of the product to the thickness of the laminar layer. Thus, N_{Nu} represents the convection-mediated improvement in the rate of heat transfer through a fluid over and above the conductive heat transfer that occurs across the same fluid layer. For example, if $N_{Nu} = 2$, it can be inferred that the convective heat transfer due to fluid motion is two-fold higher than that obtained with conductive heat transfer wherein the fluid is stationary. Also, from the Nusselt number, the heat transfer coefficient can be estimated from which the rate of convective heat transfer can be determined by applying Newton's law of cooling Eq. (5.43).

ii. **Prandtl number (N_{Pr})**

The Prandtl number includes specific heat (C_P, in J/kg K), viscosity (μ, in kg/m s), and thermal conductivity (k, in W/m K), as derived below.

$$N_{Pr} = \frac{\text{Molecular diffusivity of momentum}}{\text{Molecular diffusivity of heat}} = \frac{\text{kinematic viscosity}}{\text{thermal diffusivity}} = \frac{v}{\alpha}$$

Substituting Eq. (5.8) in the above relationship,

$$N_{Pr} = \frac{C_P\mu}{k} \tag{5.59}$$

TABLE 5.3

Typical Values of Heat Transfer Coefficient

Natural convection	Gases: 2–25 W/m^2 K Liquids: 50–100 W/m^2 K
Forced convection	Gases: 25–250 W/m^2 K Liquids: 50–20000 W/m^2 K

Before proceeding to understand the physical meaning and applications of the dimensionless Prandtl number, it is important to be familiar with the terms *hydrodynamic boundary layer* and *thermal boundary layer.*

Hydrodynamic boundary layer

The hydrodynamic boundary layer is a region of fluid flow near a solid surface, where the flow patterns are directly influenced by the viscous drag from the surface wall.

Thermal boundary layer

Similar to the velocity boundary layer, a thermal boundary layer develops due to the temperature gradient that exists between the fluid stream and surface. The thermal boundary layer is a region of a fluid flow near a solid surface, where the fluid temperatures are directly influenced by the heating or cooling from the surface wall. The concept of thermal boundary layer finds application in determining the temperature of the surface, which is in contact with the fluid. The temperature of the surface which is in contact with a liquid food product should not be very high as the product would attain the same temperature as that of the surface, which is detrimental to the product quality.

Consider a fluid flowing over an isothermal flat plate (at constant temperature), as shown in Figure 5.12. At the leading edge, the temperature profile is uniform as the fluid in contact with the surface quickly reaches the surface temperature. When the temperature gradient is set between the fluid and the surface, a thermal boundary layer develops. Thermal boundary layer grows with the increase in distance from the edge of the surface (Figure 5.12), which influences the rate of heat transfer between the surface of the flat plate and the fluid.

In case of liquids such as water, N_{Pr} >> 1.0 indicates that the hydrodynamic boundary layer is thicker than the thermal boundary layer. On the other hand, when N_{Pr} is equivalent to 1.0, it denotes that both the layers are of the same thickness. This condition is encountered in gases and vapors. The condition wherein N_{Pr} is << 1.0 rarely occurs in the case of molten metal, which indicates that the thermal boundary layer is thicker than the hydrodynamic boundary layer. The Prandtl number may also be perceived as a ratio between the rates at which the viscous forces and the thermal energy penetrate the product. Thus, N_{Pr} is proportional to the rate of growth of the two boundary layers, which is expressed as

$$\frac{\partial}{\partial t} = Pr^{1/3} \tag{5.60}$$

iii. **Peclet number (N_{Pe})**

The Peclet number is the product of Reynolds and Prandtl number. It involves the density of fluid (ρ, in kg/m^3), characteristic dimension of the product (d, in m), and the velocity (v, in m/s), specific heat (C_P, in J/kg K), and thermal conductivity (k, in W/m K) of the fluid.

$$N_{Pe} = N_{Re} \cdot N_{Pr} = \frac{v\rho C_P d}{k} \tag{5.61}$$

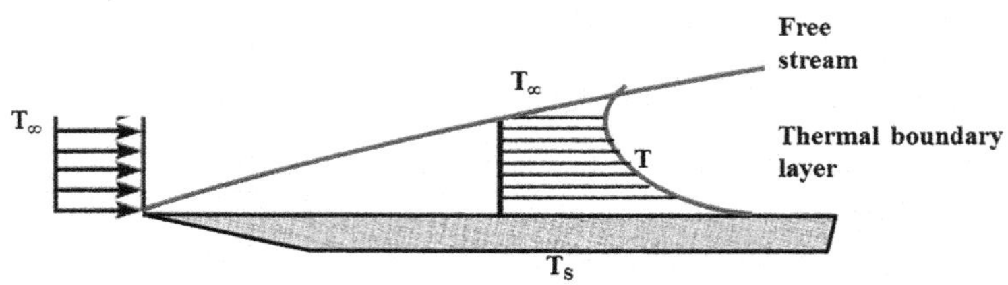

FIGURE 5.12 Thermal boundary layer development on a flat plate.

iv. **Biot number (N_{Bi})**

Biot number is the ratio of the internal conductive thermal resistance of a solid to the boundary-layer convective thermal resistance.

$$N_{Bi} = \frac{hd}{k_s} \tag{5.62}$$

where k_s is the thermal conductivity of the solid material (W/m K), d is the characteristic dimension of the system (m), and h is the heat transfer coefficient (W/m^2 K). The convective resistance offered by the fluid is small if $N_{Bi} > 40$. If N_{Bi} is close to zero, it means that the heat transfer is influenced by the boundary layer resistance offered by the fluid. A striking difference between the Nusselt and the Biot number is that the denominator of N_{Nu} involves thermal conductivity of the fluid at the solid–fluid convective interface and the denominator of N_{Bi} includes the thermal conductivity of the solid at the solid–fluid convective interface.

v. **Graetz number (N_{Gz})**

Graetz number is similar to the Peclet number. It is used to explain the heat transfer between a surface and a fluid in movement along that surface, in laminar flow.

$$N_{Gz} = 0.25\pi[N_{\text{Re}}N_{\text{Pr}}(d/L)] = \frac{\dot{m}C_P}{kL} \tag{5.63}$$

where $\dot{m}$ is the mass flow rate of the fluid (kg/s).

In addition to the aforementioned nondimensional groupings, Reynolds number (the ratio of inertia forces in the fluid to the viscous forces) is also of relevance to the forced convection heat transfer, the expression and physical meaning of which have already been discussed in Chapter 4 on "Fluid Flow."

5.3.2.2.2.2 Forced Convection Correlations As the heat transfer coefficient (h) is a function of the temperature gradient, the physical variables on which it depends is expressed as

$$h = f(\text{fluid properties, velocity field, geometry, temperature}\ldots)$$

Since the function is dependent on several parameters, for the ease of understanding, heat transfer coefficient is usually expressed in terms of correlations involving appropriate dimensionless numbers, among those discussed previously. Specifically, for forced convection, the heat transfer correlation is generally expressed in the following form:

$$N_{Nu} = f(N_{Re}, N_{Pr}) \tag{5.64}$$

On comparing Eqs. (5.49) and (5.64), it is evident that in forced convection, Reynolds number takes over the function of Grashof number in natural convection. For different flow conditions, the Nusselt number bears different empirical correlations with the Reynolds and Prandtl number, which are discussed in the following sections.

i. ***For heating and cooling of fluid foods being pumped through pipes or tubes***

Under the conditions of moderate temperature differences, laminar flow, and reasonably long tubes, it has been observed that

$$N_{Nu} = 4 \tag{5.65}$$

and in the zones of turbulence within the pipe wherein $N_{Re} > 2100$ and $N_{Pr} > 0.5$,

$$N_{Nu} = 0.023 N_{Re}^{0.8} N_{Pr}^{0.4} \tag{5.66}$$

In more viscous liquids such as oils and syrups, the surface heat transfer varies depending upon whether the fluid is heated or cooled. Thus, in this case, the effect of viscosity is considered for the calculation of h. For $N_{Re} > 10000$, h can be calculated using the following equation:

$$N_{Nu} = 0.027 \left(\mu/\mu_s\right)^{0.14} N_{Re}^{0.8} N_{Pr}^{0.33} \tag{5.67}$$

In both the cases discussed previously, the fluid properties are those of the bulk fluid except for μ_s, which is the fluid viscosity at the temperature of the tube surface.

For gases, as the variation of Prandtl number with temperature or between different gases is not significant, it can be considered as 0.75. As a result, for gases, Eq. (5.67) simplifies to

$$N_{Nu} = 0.02 N_{Re}^{0.8} \tag{5.68}$$

Equation 5.68 is obtained from Eq. (5.67) by assuming that the viscosity ratio is negligible, and hence, all the quantities are calculated at the temperature of the bulk gas. If all the other fluid properties such as thermal conductivity and density are also constant, Eq. (5.68) further reduces to

$$h_c = k' v^{0.8} \tag{5.69}$$

ii. ***For heating or cooling over plane surfaces***

The shape of many food products such as cartons of meat, slabs of cheese, and ice cream approximate to plane surfaces. With plane surfaces, since the type of length to be chosen for the calculation of Reynolds number is uncertain, the flow characterization is of concern. However, it has been ascertained that the experimental data correlates reasonably well if the length of the plate measured in the direction of the flow is considered as the dimension (d) during the calculation of Reynolds number. Accordingly, the correlation is given by

$$N_{Nu} = 0.036\, N_{Re}^{0.8} N_{Pr}^{0.33}, \quad \text{for } N_{Re} > 2 \times 10^4 \tag{5.70}$$

The simplified equations for fluid flow over a smooth surface (plate) are as follows:

$$h = 5.7 + 3.9v \quad \text{for } v < 5\,\text{m/s} \tag{5.71}$$

$$h = 7.4 v^{0.8} \quad \text{for } 5 < v < 30\,\text{m/s} \tag{5.72}$$

The values of heat transfer coefficient for rough plates are comparatively higher than that for smooth plates.

iii. ***For heating and cooling outside the tubes***

In the context of food processing, heating or cooling outside the tubes is observed during the chilling of water and sausages and in spaghetti processing. Experimental data for the aforesaid case fits well with the correlation of the following form:

$$N_{Nu} = K N_{Re}^{n} N_{Pr}^{m} \tag{5.73}$$

where K is a constant and the powers n and m vary with Reynolds number. The diameter of the tube over which the fluid flow occurs is used for the calculation of N_{Re}. The same values of N_{Re} cannot be used to represent streamline or turbulent conditions as it is done for the fluids flowing inside the pipes.

For gases and liquids at high or moderate Reynolds numbers, Eq. (5.73) is specified as

$$N_{Nu} = 0.26 N_{Re}^{0.6} N_{Pr}^{0.3} \tag{5.74}$$

For liquids at low Reynolds numbers, i.e., $1 < N_{Re} < 200$, Eq. (5.73) is given by

$$N_{Nu} = 0.86 N_{Re}^{0.43} N_{Pr}^{0.3} \tag{5.75}$$

In the earlier discussed forced convection equations, fluid properties are computed at the mean film temperature, which in turn is the arithmetic mean of the temperatures of tube walls and bulk fluid.

iv. ***Laminar flow in pipes***

For laminar flow, the Nusselt number is given by

$$N_{Nu} = 1.86\left[N_{Re} \cdot N_{Pr}\left(\frac{d}{L}\right)\right]^{0.33}\left(\frac{\mu}{\mu_w}\right)^{0.14} \tag{5.76}$$

where μ_w is the viscosity of fluid (kg/m s) at the surface temperature of the wall, L is the length of pipe (m), and d is the inner diameter of the pipe (m).

By substituting the Graetz number, Eq. (5.76) can be rewritten as

$$N_{Nu} = 2.014\left(N_{Gz}\right)^{0.33}\left(\frac{\mu}{\mu_w}\right)^{0.14} \tag{5.77}$$

v. ***Turbulent flow in pipes***

For turbulent flow in forced convection, Dittus–Boelter equation (Eq. (5.78)) is used to determine the Nusselt number, which is based on the Reynolds and Prandtl number.

$$N_{Nu} = 0.023(N_{Re})^{0.8}(N_{Pr})^{0.33}\left(\frac{\mu}{\mu_w}\right)^{0.14} \tag{5.78}$$

vi. ***Transitional flow in pipes***

For Reynolds number in the range of 2100–10000, Eq. (5.79) is used to determine the Nusselt number

$$N_{Nu} = \frac{(f/8)(N_{Re} - 1000)N_{Pr}}{1 + 12.7(f/8)^{0.5}(N_{Pr}^{0.66} - 1)} \tag{5.79}$$

where f is the friction factor and is obtained for the smooth pipes using the following equation

$$f = \frac{1}{(0.790 \ln N_{Re} - 1.64)^2} \tag{5.80}$$

vii. ***Flow past immersed objects***

In some of the food processing applications, the object or the food product is immersed in the fluid. An example is that of canned pineapple tidbits, where the fruit cubes are immersed in the sugar solution during sterilization of the canned product. In this case, the heat transfer rate depends on the fluid properties, the relative position of the object, the flow rate, and the geometrical shape of the object.

For flow past a single sphere, the Nusselt number for heating or cooling is given by

$$N_{Nu} = 2 + 0.6(N_{Re})^{0.5}(N_{Pr})^{0.33} \quad \text{for} \begin{cases} 1 < N_{Re} < 70000 \\ 0.6 < N_{Pr} < 400 \end{cases} \tag{5.81}$$

Example 5.8

Find the Prandtl number for water at 100°C. The properties of water at 100°C, are $k = 0.682$ W/m K; $C_P = 4.211$ kJ/kg K; $\mu = 2.775 \times 10^{-4}$ Ns/m^2.

Solution

Given

Properties of water at 100°C

$K = 0.682$ W/m K; $C_P = 4.211$ kJ/kg K; $\mu = 2.775 \times 10^{-4}$N s/m^2.

$$N_{Pr} = \frac{C_P \mu}{k} = \frac{4211 \times 2.775 \times 10^{-4}}{0.682} = 1.713$$

Answer: Prandtl number for water is 1.713

5.3.2.3 Estimation of Convective Heat Transfer Coefficient

5.3.2.3.1 Methodology to Calculate Heat Transfer Coefficient Using Empirical Correlations

The convective heat transfer coefficient h is predicted from empirical correlations. It is mostly influenced by the velocity of fluid, physical properties of the fluid, temperature gradient, and geometrical shape of the surface. The steps involved in calculating the heat transfer coefficient are listed as follows:

1. Identify the geometry which is in contact with the fluid, as for whether it is a pipe, sphere, rectangular duct, or a plate. Spot whether the fluid is flowing inside the geometry or over the surface.
2. Estimate the fluid properties at the average fluid temperature. Average fluid temperature (T_∞) can be determined as the mean of average inlet fluid temperature (T_i) and average outlet fluid temperature (T_e)

$$T_\infty = \frac{T_i + T_e}{2} \tag{5.82}$$

3. Calculate the Reynolds number to determine the nature of the flow as whether it is laminar, turbulent, or transitional. This information is required to choose the appropriate empirical correlation to determine the heat transfer coefficient.
4. Selection of a suitable empirical correlation from those to be discussed subsequently is based on the conditions from steps (1) to (3), followed by determination of the Nusselt number to finally obtain the heat transfer coefficient.

5.3.2.3.2 Convective Heat Transfer Coefficient for Natural Convection

As described earlier, natural convection occurs in a fluid due to gravitational and buoyancy forces. Empirical expressions for predicting convective heat transfer coefficient are given using the Rayleigh number (N_{Ra}).

$$N_{Nu} = \frac{hd}{k} = a(N_{Ra})^m \tag{5.83}$$

where a and m are the constants which can be chosen based on the criteria listed in Table 5.4.

TABLE 5.4

Coefficients for the Equation of Natural Convection

Geometry and Characteristic Length	Range of N_{Ra}	a	m	Equation
Vertical plate	$10^4 - 10^9$	0.59	0.25	$N_{Nu} = a(N_{Ra})^m$
L	$10^9 - 10^{13}$	0.1	0.333	
Horizontal plate	$10^4 - 10^7$	0.54	0.25	$N_{Nu} = a(N_{Ra})^m$
A/P	$10^7 - 10^{11}$	0.15	0.333	
A = area; P = perimeter				
Upper hot surface of a plate				
Horizontal plate	$10^5 - 10^{11}$	0.27	0.25	$N_{Nu} = a(N_{Ra})^m$
A/P				
A = area; P = perimeter				
Bottom hot surface of a plate				
Horizontal cylinder	$10^{-5} - 10^{12}$	—	—	$N_{Nu} = \left\{0.6 + \frac{0.387 N_{Ra}^{1/6}}{\left[1+\left(\frac{0.559}{N_{Pr}}\right)^{9/16}\right]^{8/27}}\right\}^2$
D				
Sphere	$N_{Ra} \le 10^{11}$	—	—	$N_{Nu} = 2 + \frac{0.589 N_{Ra}^{1/4}}{\left[1+\left(\frac{0.469}{N_{Pr}}\right)^{9/16}\right]^{4/9}}$
D				

5.3.2.4 *Thermal Resistance in Convective Heat Transfer*

The term thermal resistance for convective heat transfer can be defined in the similar manner as discussed in Section 5.3.1.5 on "conductive heat transfer." Convective heat transfer rate is given by

$$q = hA(T_S - T_\infty) \tag{5.84}$$

On rearranging the terms in Eq. (5.84),

$$q = \frac{(T_S - T_\infty)}{\left(\frac{1}{hA}\right)} \tag{5.85}$$

or

$$q = \frac{(T_S - T_\infty)}{(R_t)_{\text{convection}}} \tag{5.86}$$

where $(R_t)_{\text{convection}}$ is the thermal resistance due to convection, as given by Eq. (5.87)

$$(R_t)_{\text{convection}} = \frac{1}{hA} \tag{5.87}$$

5.3.2.5 *Overall Heat Transfer Coefficient*

In many instances, heat transfer occurs by a combination of conduction and convection. A known example is a pipe carrying a fluid at a higher temperature than the surrounding environment. In such a case, heat must be transferred by convection to the pipe's inner surface from the hot fluid, followed by conduction of heat through the pipe wall material, and finally, the heat transfer will occur from the outer surface of the pipe to the surrounding environment by natural or free convection (Figure 5.13). Hence, in this case, heat transfer occurs in series through three layers. With this approach, the denominator in the equation for heat transfer rate is given by the sum of resistances offered by the simultaneous convective and conductive heat transfer through the three layers as discussed previously.

$$q = \frac{(T_i - T_\infty)}{R_t} \tag{5.88}$$

$$R_t = (R_t)_{\text{inside convection}} + (R_t)_{\text{conduction}} + (R_t)_{\text{outside convection}} \tag{5.89}$$

Thermal resistance for the convection inside with h_i as the inside convective heat transfer coefficient and A_i as the inside surface area of pipe is given by

$$(R_t)_{\text{inside convection}} = \frac{1}{h_i A_i} \tag{5.90}$$

Thermal resistance for conduction through the pipe wall with thermal conductivity k (W/m K), length of pipe L (m), and inner and outer radii of pipe as r_i and r_o (m), respectively, is given by

$$(R_t)_{\text{inside conduction}} = \frac{\ln\left(\frac{r_o}{r_i}\right)}{2\pi kL} \tag{5.91}$$

Thermal resistance for convection outside with h_o as the outside convective heat transfer coefficient and A_o as the outside surface area of pipe is given by

$$(R_t)_{\text{outside convection}} = \frac{1}{h_o A_o} \tag{5.92}$$

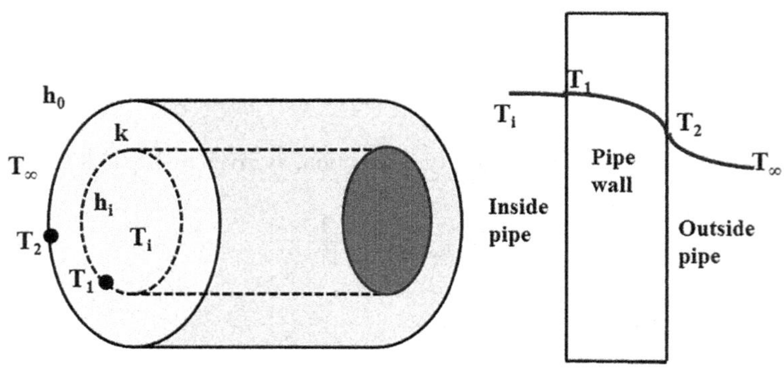

FIGURE 5.13 Combined convective and conductive heat transfer.

Therefore, Eq. (5.88) can be written as

$$q = \frac{(T_i - T_\infty)}{\frac{1}{h_i A_i} + \frac{\ln(r_o/r_i)}{2\pi kL} + \frac{1}{h_o A_o}} \tag{5.93}$$

Equation 5.93 can also be written in terms of the overall heat transfer coefficient based on the inner area of the pipe (U_i) as

$$q = U_i A_i (T_i - T_\infty) \tag{5.94}$$

or

$$q = \frac{(T_i - T_\infty)}{\left(\frac{1}{U_i A_i}\right)} \tag{5.95}$$

Therefore, from Eqs. (5.93) and (5.95)

$$\frac{1}{U_i A_i} = \frac{1}{h_i A_i} + \frac{\ln(r_o/r_i)}{2\pi kL} + \frac{1}{h_o A_o} \tag{5.96}$$

The heat transfer coefficient based on the inside area (U_i) can be calculated from Eq. (5.96).

For a slab or wall made up of two different materials with thicknesses x_1 and x_2, respectively, and if the area A is same across the thickness of the slab, the overall heat transfer is given by

$$\frac{1}{U} = \frac{1}{h_i} + \frac{x_1}{k_1} + \frac{x_2}{k_2} + \frac{1}{h_o} \tag{5.97}$$

The overall heat transfer coefficient is a measure of the efficiency of heat transfer. It gives an insight to the amount of heat that passes through 1 m^2 of the heat transfer surface per 1°C of temperature difference.

Example 5.9

Using the following data, calculate the heat transfer rate in watt for the composite pipe, which is used for pumping heated puree.

- Length of pipe $L = 5$ m, inside fluid temperature = 85°C, outside fluid temperature = 23°C.
- Inside and outside radii of pipe are 2.5 and 2.7 cm, respectively. Radii of the first layer and the second layer are 8 and 9 cm, respectively.
- Surface heat transfer coefficient at the inner pipe wall is 100 W/m^2°C and for the outside layer is 5 W/m^2°C. The thermal conductivity of the first, second and third layers are 45, 0.0389 and 0.0228 W/m°C, respectively.

Solution

Given

Length of pipe $L = 5$ m; inside fluid temperature = 85°C
Outside fluid temperature = 23°C; inner radius of pipe = 2.5 cm
Outer radius of pipe = 2.7 cm; radius of first layer = 8 cm
Radius of second layer = 9 cm. $h_i = 100$ W/m^2°C and $h_o = 5$ W/m^2°C

The composite pipe is represented by the following figure:

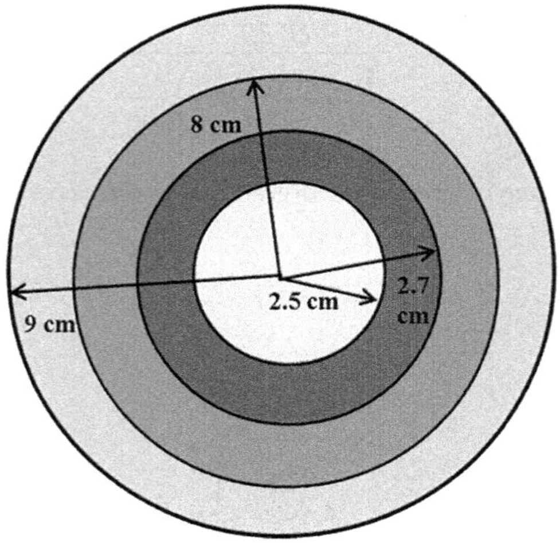

The conductive resistance for the pipe is given by

$$R = \frac{\ln r_o/r_i}{2\pi Lk}$$

$$\therefore R_1 = \frac{\ln(0.027/0.025)}{2\times 3.14\times 5\times 45} = 5.447\times 10^{-5}$$

$$\therefore R_2 = \frac{\ln(0.08/0.027)}{2\times 3.14\times 5\times 0.0389} = 0.889$$

$$\therefore R_1 = \frac{\ln(0.09/0.08)}{2\times 3.14\times 5\times 0.0228} = 0.165$$

Overall conductive resistance is

$$R_{\text{conduction}} = 5.447\times 10^{-5} + 0.889 + 0.165 = 1.054$$

Similarly, convective resistance can be obtained as

$$R_i = \frac{1}{h_i A}$$

$$\therefore R_i = \frac{1}{100\times 2\times 3.14\times 0.025\times 5} = 0.013$$

$$\therefore R_o = \frac{1}{5\times 2\times 3.14\times 0.09\times 5} = 0.071$$

Overall convective resistance is

$$R_{\text{convection}} = 0.013 + 0.071 = 0.084$$

Overall resistance = overall conductive resistance + overall convective resistance

$$R_{\text{overall}} = 0.084 + 1.054 = 1.138$$

Overall resistance is 1.138 m² °C/W.

Heat transfer rate through the composite pipe can be determined using the equation

$$q = \frac{\Delta T}{R_{\text{overall}}} = \frac{85-23}{1.138} = 54.48$$

Answer: Heat transfer rate through the composite pipe is 54.48 W

5.3.2.6 Unsteady-State Heat Transfer during Convection

As explained in the case of conduction, convection also has a phase of transient heat transfer during various processing applications. Therefore, it is necessary to account for the unsteady-state conditions to determine the temperature change during the process.

Considering the convective heat transfer occurring in an object which is immersed in a hot fluid, the heat balance during the unsteady-state heat transfer is given by

$$q = \rho C_P V \frac{dT}{dt} = hA(T_\infty - T) \tag{5.98}$$

where V is the volume of fluid (m³), h is the heat transfer coefficient (W/m² K), A is the surface area of the object (m²), and T_∞ is the fluid temperature (K). On rearranging Eq. (5.98),

$$\frac{dT}{(T_\infty - T)} = \frac{hA}{\rho C_P V} dt \tag{5.99}$$

The boundary conditions for the integration of Eq. (5.99) are as given in Eq. (5.100).

$$\begin{aligned} T &= T_o \quad & t &= 0 \\ T &= T \quad & t &= t \end{aligned} \tag{5.100}$$

$$\int_{T_o}^{T} \frac{dT}{(T_\infty - T)} = \frac{hA}{\rho C_P V} \int_0^t dt \tag{5.101}$$

On integration,

$$\ln \frac{(T_\infty - T)}{(T_\infty - T_o)} = \frac{hA}{\rho C_P V} t \tag{5.102}$$

According to Eq. (5.102), a semilogarithmic plot of $(T - T_\infty)/(T_o - T_\infty)$ against time (t) would be linear and its slope will be equal to $-hA/\rho C_P V$. Equation 5.102 is applicable where there is no internal resistance to heat transfer. This situation is observed in a steam-jacketed kettle where liquid food is well stirred.

5.3.2.7 Heat Exchangers

Heating or cooling of different food products is carried out in an equipment known as *heat exchanger*. A heat exchanger works on the principle of indirect heating, to facilitate heat transfer between fluids. It is symbolically represented by two channels separated by a tubular partition (Figure 5.14). The hot and cold fluids flow through their respective channels, and the heat transfer between them occurs through the partition. The hot fluid enters its respective channel at a temperature of T_{Hi} and is cooled to a temperature of T_{Ho} at the outlet. The cold fluid enters the heat exchanger at a temperature of T_{ci} and is heated by the hot

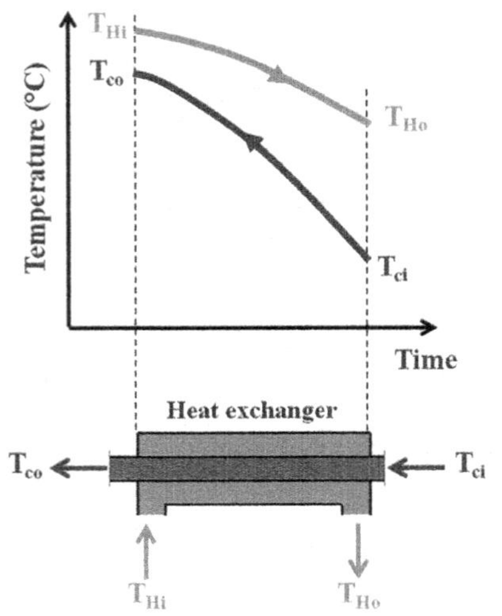

FIGURE 5.14 Symbolic representation of a heat exchanger and the temperature profile of hot and cold fluids within the heat exchanger.

fluid to an exit temperature of T_{co}. The change in temperature of the fluids during their passage through the heat exchanger is symbolized by the curves shown in Figure 5.14.

Different types of heat exchangers are available for varied applications in the food industry, which are described in the forthcoming sections.

5.3.2.7.1 Tubular Heat Exchanger

Tubular or double pipe heat exchanger (Figure 5.15) is the simplest of all the heat exchanger types. It comprises of a concentric tube arrangement, wherein the heating/cooling medium flows in the annular region and the food product to be heated flows in the inner tube. The fluid streams may flow in cocurrent or countercurrent direction with respect to each other. The temperature profile of fluids within the heat exchanger would vary with their flow direction.

5.3.2.7.1.1 Cocurrent Flow The flow pattern in a heat exchanger is said to be cocurrent when the heating/cooling medium and the liquid food product flow in the same direction. The inlet streams in a cocurrent heat exchanger are the liquid food product to be heated and the heating medium. The exit streams comprise the heated food product and the heating medium at a lower temperature which has transferred its heat to the product within the heat exchanger. The cocurrent flow pattern and the temperature profiles along the length of the heat exchanger are shown in Figure 5.16. In this configuration, the temperature difference at the inlet ($\Delta T_1 = T_{Hi} - T_{ci}$) is very high compared to that at the outlet ($\Delta T_2 = T_{Ho} - T_{co}$). Also, the temperature difference gradually decreases along the length of the heat exchanger (Figure 5.16). Consequently, the heat transfer rate is initially high and gradually decreases along the length of the heat exchanger. This may be advantageous or disadvantageous based on the

FIGURE 5.15 Double pipe heat exchanger.

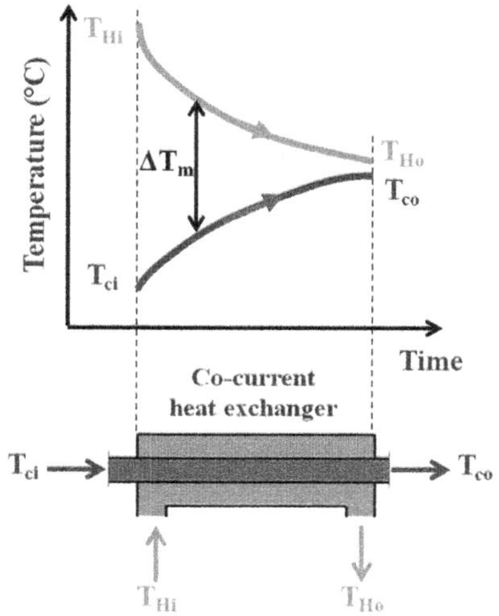

FIGURE 5.16 Cocurrent flow system with the temperature profile.

nature of the product. Viscous products such as corn soup need an initially high heat input which can be effectively provided in a cocurrent heat exchanger.

5.3.2.7.1.2 Countercurrent Flow In the countercurrent flow pattern, the heating/cooling medium and the liquid food product flow in the opposite directions within the heat exchanger (Figure 5.17). The liquid food product to be heated and the heating medium at a lower temperature which has transferred its heat to the product within the heat exchanger are on one side of the heat exchanger. The heated food product and the heating medium are on the other end of the heat exchanger. Thus, in a countercurrent flow, there is no significant difference between ΔT_1 (= $T_{Ho} - T_{ci}$) and ΔT_2 (= $T_{Hi} - T_{co}$) and the fluid is exposed to a constant temperature difference (Figure 5.17). Countercurrent flow has a distinct advantage over the concurrent flow, as the exit temperature of the hot medium may be lower than the exit temperature of the cold liquid product.

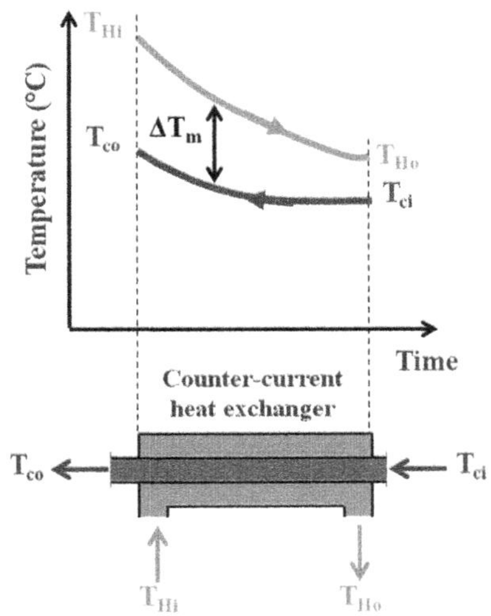

FIGURE 5.17 Countercurrent flow system with the temperature profile.

5.3.2.7.1.3 Log Mean Temperature Difference

i. Log mean temperature difference (LMTD) for cocurrent flow heat exchanger

To predict the performance of or to design a heat exchanger, it is important to correlate the total heat transfer rate to the required heat transfer area, inlet and outlet temperatures of the fluid, and the overall heat transfer coefficient. When the temperature of hot fluid changes from T_{Hi} at the inlet to T_{Ho} at the outlet, the heat transfer rate (q) is given by

$$q = \dot{m}_H C_{PH} \left(T_{Hi} - T_{Ho} \right) \tag{5.103}$$

where $\dot{m}_H$ is the mass flow rate of the fluid (kg/s) and C_{PH} is the specific heat of the fluid (kJ/kg °C). The subscript H refers to the hot fluid. Similarly, for the cold fluid, the heat transfer rate is given by

$$q = \dot{m}_c C_{Pc} \left(T_{co} - T_{ci} \right) \tag{5.104}$$

where the subscript c refers to the cold fluid. Therefore, the energy balance between the hot and the cold fluid streams traversing inside the heat exchanger can be written as

$$q = \dot{m}_H C_{PH} \left(T_{Hi} - T_{Ho} \right) = \dot{m}_c C_{Pc} \left(T_{co} - T_{ci} \right) \tag{5.105}$$

Equation 5.105 is useful in determining the inlet and outlet temperature of the hot or cold fluid. Also, it can be applied to calculate the mass flow rate of the hot or cold fluids, provided the other values are known.

Consider a cocurrent flow heat exchanger, as shown in Figure 5.16. According to Newton's law of cooling, heat transfer rate (q) is given by

$$q = UA\Delta T_m \tag{5.106}$$

where U is the overall heat transfer coefficient, A is the total heat transfer area, and ΔT_m is the mean temperature difference.

Considering a thin slice of the heat exchanger, the rate of heat transfer, dq, from the hot fluid (H) to the cold fluid (c), may be expressed based on Eq. (5.106) as given in Eq. (5.107):

$$dq = UdA\Delta T \tag{5.107}$$

where ΔT is the temperature difference between the hot and the cold fluids $\left(\Delta T = T_H - T_c \right)$. The temperature difference ΔT varies from the inlet to the outlet of the heat exchanger. Initially, the temperature difference between the hot and the cold fluids at the inlet is ΔT_1 and that at the outlet is ΔT_2.

$$\Delta T_1 = T_{Hi} - T_{ci} \tag{5.108}$$

$$\Delta T_2 = T_{Ho} - T_{co} \tag{5.109}$$

As the temperature inside the heat exchanger varies nonlinearly, considering the arithmetic mean for temperature difference is not appropriate. From Eq. (5.107), energy balance for the hot and cold fluids in the considered thin slice of the heat exchanger can be written as

$$\text{For hot fluid}: dq = -\dot{m}_H C_{PH} dT_H \tag{5.110}$$

$$\text{For cold fluid}: dq = \dot{m}_c C_{Pc} dT_c \tag{5.111}$$

Since dT_H is a negative quantity in case of hot fluids, a negative sign has been included in Eq. (5.110) to obtain a positive value for dq. dT_H and dT_c can be obtained from the Eqs. (5.110) and (5.111), respectively, as follows:

$$dT_H = -\frac{dq}{\dot{m}_H C_{PH}} \tag{5.112}$$

$$dT_c = \frac{dq}{\dot{m}_c C_{Pc}} \tag{5.113}$$

Subtracting Eq. (5.113) from Eq. (5.112),

$$dT_H - dT_c = -\frac{dq}{\dot{m}_H C_{PH}} - \frac{dq}{\dot{m}_c C_{Pc}} = -dq\left(\frac{1}{\dot{m}_H C_{PH}} + \frac{1}{\dot{m}_c C_{Pc}}\right) \tag{5.114}$$

Substituting dq from Eq. (5.107),

$$\frac{d\Delta T}{\Delta T} = -U\left(\frac{1}{\dot{m}_H C_{PH}} + \frac{1}{\dot{m}_c C_{Pc}}\right)dA \tag{5.115}$$

Integrating Eq. (5.115) between the specified limits,

$$\int_{\Delta T_1}^{\Delta T_2} \frac{d\Delta T}{\Delta T} = -U\left(\frac{1}{\dot{m}_H C_{PH}} + \frac{1}{\dot{m}_c C_{Pc}}\right)\int_0^A dA \tag{5.116}$$

$$\ln\frac{\Delta T_2}{\Delta T_1} = -UA\left(\frac{1}{\dot{m}_H C_{PH}} + \frac{1}{\dot{m}_c C_{Pc}}\right) \tag{5.117}$$

From Eqs. (5.103) and (5.104),

$$\ln\frac{\Delta T_2}{\Delta T_1} = -UA\left(\frac{T_{Hi} - T_{Ho}}{q} + \frac{T_{co} - T_{ci}}{q}\right) \tag{5.118}$$

Rearranging the terms in Eq. (5.118),

$$\ln\frac{\Delta T_2}{\Delta T_1} = -\frac{UA}{q}(T_{Hi} - T_{ci}) - (T_{Ho} - T_{co}) \tag{5.119}$$

Substituting Eqs. (5.108) and (5.109) in Eq. (5.119),

$$\ln\frac{\Delta T_2}{\Delta T_1} = -\frac{UA}{q}(\Delta T_1 - \Delta T_2) \tag{5.120}$$

Rearranging the terms in Eq. (5.120),

$$q = UA\frac{\Delta T_2 - \Delta T_1}{\ln\dfrac{\Delta T_2}{\Delta T_1}} \tag{5.121}$$

And, introducing the term ΔT_{lm} in Eq. (5.121)

$$q = UA\Delta T_{lm} \tag{5.122}$$

The important application of Eq. (5.122) is in finding the area of the heat exchanger and the overall resistance to heat transfer. In Eq. (5.122), ΔT_{lm} is known as LMTD, given by

$$\Delta T_{lm} = \frac{\Delta T_2 - \Delta T_1}{\ln \frac{\Delta T_2}{\Delta T_1}} = \frac{\Delta T_1 - \Delta T_2}{\ln \frac{\Delta T_1}{\Delta T_2}} \tag{5.123}$$

For a cocurrent flow heat exchanger, LMTD can be expanded further as

$$\Delta T_{lm} = \frac{(T_{Hi} - T_{ci}) - (T_{Ho} - T_{co})}{\ln \frac{(T_{Hi} - T_{ci})}{(T_{Ho} - T_{co})}} \tag{5.124}$$

ii. LMTD for countercurrent flow heat exchanger
For a countercurrent flow heat exchanger,

$$\Delta T_1 = T_{Hi} - T_{co} \tag{5.125}$$

$$\Delta T_2 = T_{Ho} - T_{ci} \tag{5.126}$$

Thus, LMTD for the countercurrent flow heat exchanger is given by

$$\Delta T_{lm} = \frac{(T_{Hi} - T_{co}) - (T_{Ho} - T_{ci})}{\ln \frac{(T_{Hi} - T_{co})}{(T_{Ho} - T_{ci})}} \tag{5.127}$$

At the same inlet and outlet temperatures, LMTD for the countercurrent flow is higher than that for the cocurrent flow. Therefore, a countercurrent flow heat exchanger requires a lesser surface area than its cocurrent flow counterpart, to achieve the same rate of heat transfer.

Example 5.10

Water at 20°C is heated to 80°C by condensing steam in a heat exchanger. Saturated steam is supplied at a rate of 100 kg/h and a temperature of 200°C. It leaves the exchanger at a quality of 0.05. How much water can be heated? Find the LMTD and compute the value of U if the heat exchanger surface area is 0.35 m^2.

Solution

Given

Surface area of the heat exchanger = 0.35 m^2
Final quality of the steam leaving the system = 0.05 = 5%
Steam is supplied at the rate of 100 kg/h at 200°C
From the steam table, enthalpy of steam at 200°C = 2793.2 kJ/kg
Enthalpy of liquid at 200°C = 852.45 kJ/kg
Heat given by the steam = 2793.2 − (2793.2 × 0.05 + 852.452 × 0.95) = 1843.741 kJ/kg
100 kg/h of steam would give 184374.1 kJ/h of heat.

The amount of water it will cool can be obtained by

$$Q = mC_P\Delta T \quad \therefore 184374.1 = m \times 4.21 \times (80 - 20)$$

$$m = 729.905\,\text{kg}$$

Water cooled by the steam per hour in the heat exchanger is 729.905 kg

LMTD for the parallel type heat exchanger

$$\Delta T_{lm} = \frac{(T_{Hi} - T_{ci}) - (T_{Ho} - T_{co})}{\ln \dfrac{(T_{Hi} - T_{ci})}{(T_{Ho} - T_{co})}}$$

$$\Delta T_{lm} = \frac{(200 - 20) - (200 - 80)}{\ln\left(\dfrac{(200 - 20)}{(200 - 80)}\right)} = 147.978 \cong 148°\text{C}$$

Heat transfer rate can be obtained as

$$q = \frac{184374.1 \times 10^3\,[\text{J}]}{3600\,[\text{s}]} = 51215.03\,\text{W}$$

Overall heat transfer coefficient for the heat exchanger can be obtained by the following equation:

$$q = UA\Delta T_{lm}$$

$$U = \frac{q}{A\Delta T_{lm}} = \frac{51215.03}{0.35 \times 148} = 988.71\,\text{W/m}^2\,°\text{C}$$

Overall heat transfer coefficient for the given heat exchanger is 988.71 W/m² °C.

Example 5.11

Compute the LMTD for a one shell pass–four tubes pass baffled heat exchanger, with the hot fluid entering at 200°C and leaving at 100°C. The cold fluid enters at 10°C and leaves at 80°C.

Solution

Given

Hot fluid temperature at inlet = 200°C; hot fluid temperature at outlet = 100°C
Cold fluid temperature at inlet = 10°C; cold fluid temperature at outlet = 80°C
LMTD for the heat exchanger

$$\Delta T_{lm} = \frac{(T_{Hi} - T_{co}) - (T_{Ho} - T_{ci})}{\ln \dfrac{(T_{Hi} - T_{co})}{(T_{Ho} - T_{ci})}}$$

$$\Delta T_{lm} = \frac{(200 - 80) - (100 - 10)}{\ln \dfrac{(200 - 80)}{(100 - 10)}} = 104.281°\text{C}$$

Answer: LMTD for the heat exchanger is 104.281°C

Example 5.12

A heat exchanger is required to cool 20 kg/s of water from 87°C to 67°C by means of 25 kg/s water entering at 27°C. If the overall coefficient of heat transfer is constant at 2 kW/m²°C, calculate the surface area required in (a) countercurrent concentric tubes heat exchanger and (b) a cocurrent flow concentric tubes heat exchanger.

Solution

Given

Hot fluid temperature at inlet = 87°C; hot fluid temperature at outlet = 67°C
Cold fluid temperature at inlet = 27°C
Flow rate of hot fluid = 20 kg/s; flow rate of cold fluid = 25 kg/s
Overall heat transfer coefficient of heat exchanger = 2 kW/m²°C.
Heat supplied by the hot fluid is equal to the heat gain by the cold fluid; therefore,

$$(mC_P\Delta T)\text{hot} = (mC_P\Delta T)\text{cold}$$

As both the fluids are same; therefore, specific heat of the fluid is taken out

$$20(87-67) = 25(T_{co}-27)$$

$$T_{co} = \frac{20\times 20}{25} + 27 = 43°\text{C}$$

Answer: The outlet temperature of the cold fluid is 43°C

a. Countercurrent concentric tube heat exchanger
   LMTD for the heat exchanger

$$\Delta T_{lm} = \frac{(T_{Hi}-T_{co})-(T_{Ho}-T_{ci})}{\ln\frac{(T_{Hi}-T_{co})}{(T_{Ho}-T_{ci})}}$$

$$\Delta T_{lm} = \frac{(87-43)-(67-27)}{\ln\frac{(87-43)}{(67-27)}} = \frac{(44)-(40)}{\ln\frac{(44)}{(40)}} = 41.968 \cong 42°\text{C}$$

Area required for the heat exchanger can be obtained from the equation

$$q = UA\Delta T_{lm} \quad \therefore A = \frac{q}{U\Delta T_{lm}}$$

$$\therefore A = \frac{20\times 4.2\times 20}{2\times 42} = 20\,\text{m}^2$$

Answer: Area required for the countercurrent heat exchanger is 20 m² and LMTD is 42°C

b. Cocurrent concentric tube heat exchanger
   LMTD for the heat exchanger

$$\Delta T_{lm} = \frac{(T_{Hi}-T_{ci})-(T_{Ho}-T_{co})}{\ln\frac{(T_{Hi}-T_{ci})}{(T_{Ho}-T_{co})}}$$

$$\Delta T_{lm} = \frac{(87-27)-(67-42)}{\ln\frac{(87-27)}{(67-42)}} = \frac{(60)-(25)}{\ln\frac{(60)}{(25)}} = 40°\text{C}.$$

Area required for the heat exchanger can be obtained from the equation

$$q = UA\Delta T_{lm} \quad \therefore A = \frac{q}{U\Delta T_{lm}}$$

$$\therefore A = \frac{20 \times 4.2 \times 20}{2 \times 40} = 21\,\text{m}^2$$

Answer: Area required for the countercurrent heat exchanger is 21 m^2 and LMTD is 40°C

5.3.2.7.2 Shell and Tube Heat Exchanger

Shell and tube heat exchangers (SHEs) are widely used in the food industry for the evaporation, sterilization, and condensation of liquid foods. A SHE comprises a bundle of tubes of very small diameter in the range of 1–2 cm, placed inside a cylindrical shell (Figure 5.18). The tube geometry is either of the square or triangular type, of which, the triangular arrangement allows a higher number of tubes than the square one, for the same shell diameter. Stainless steel is the preferred material of construction for the shell and tubes as it provides better hygienic conditions. Usage of high-quality stainless steel for the fabrication of shell is of relevance as they come into contact with the heating or cooling medium. Further, the construction of tubes should comply with the hygienic rules as they come into direct contact with the liquid food product. The length of SHE is in the order of several meters, with a diameter of maximum 1 m. Thus, it provides a large surface area (of up to 100 m^2) within a compact volume. While one fluid flows inside the tube, the other flows inside the shell. According to the standards laid down by the Tubular Exchanger Manufacturers Association (TEMA), the least distance between the tube centers should be 1.25 times the outside tube diameter, for the triangular arrangement.

On either side of the SHE, there are metal plates which separate the heating and the cooling streams. The tubes of SHE are stuck on to these plates, which in turn are supported by the tie rods. The shell side of SHE includes metallic discs, known as "baffles," supported by spacers located on the tie rods. The baffles are usually in the form of the segment of a circular disc having holes to accommodate tubes. The baffles of SHE have two important functions: to enhance the heat transfer process and to support the tubes along the length of the heat exchanger, thus preventing the tubes from bending. The transversely arranged baffles increase the convective heat transfer coefficient by inducing turbulence in the shell-side fluid, which is achieved by changing the direction of fluid movement, either cocurrent or countercurrent with respect to the tube bundles. This arrangement also provides an option for the liquid food product to traverse the tube, one time or multiple times, referred to as "one-pass" or "multi-pass" arrangement, respectively. In the one-pass arrangement, the liquid product enters at one end and exits

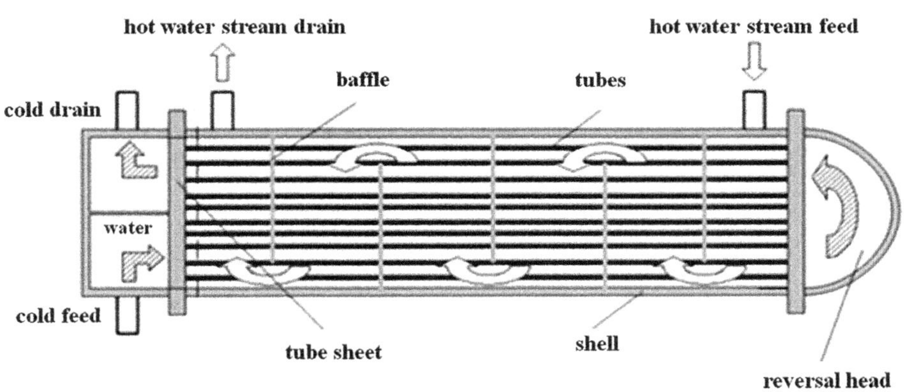

FIGURE 5.18 Shell and tube heat exchanger. (Reproduced with permission from Correa, D. J. and Marchetti, J. L. 1987. Dynamic simulation of shell-and-tube heat exchangers. *Heat Transfer Engineering* 8: 50–59.)

at the opposite end. In the multi-pass arrangement, during each pass, the product travels back and forth through different tubes, before finally exiting the heat exchanger.

Baffle spacing is an important parameter which influences the rate of heat transfer. While larger baffle spacing reduces the shell-side pressure drop in conjunction with a decrease in the turbulence and heat transfer coefficient; smaller baffle spacing increases the turbulence and heat transfer coefficient. Nevertheless, the significant increase in pressure drop outweighs the advantage of increased heat transfer coefficient obtained with the smaller baffle spacing and thus nullifies it. Thus, the baffle spacing is chosen after careful consideration of the allowable shell-side pressure drop and the required heat transfer coefficient. As recommended by the TEMA standards, the minimum space between the baffles should be one-fifth (1/5th) of the shell diameter.

5.3.2.7.3 Plate Heat Exchanger

Plate heat exchanger (PHE) is the predominantly used heat exchanger type in the dairy industry. It consists of a series of corrugated and pressed stainless steel plates, clamped together in a frame (Figure 5.19a). The plates are corrugated in a pattern to enhance heat transfer (Figure 5.19b). Points of support on the corrugations hold the plates apart, which facilitate the formation of thin channels of fluid between them. The spacing between the plates is usually 2–3 mm. The gaskets around the edges of

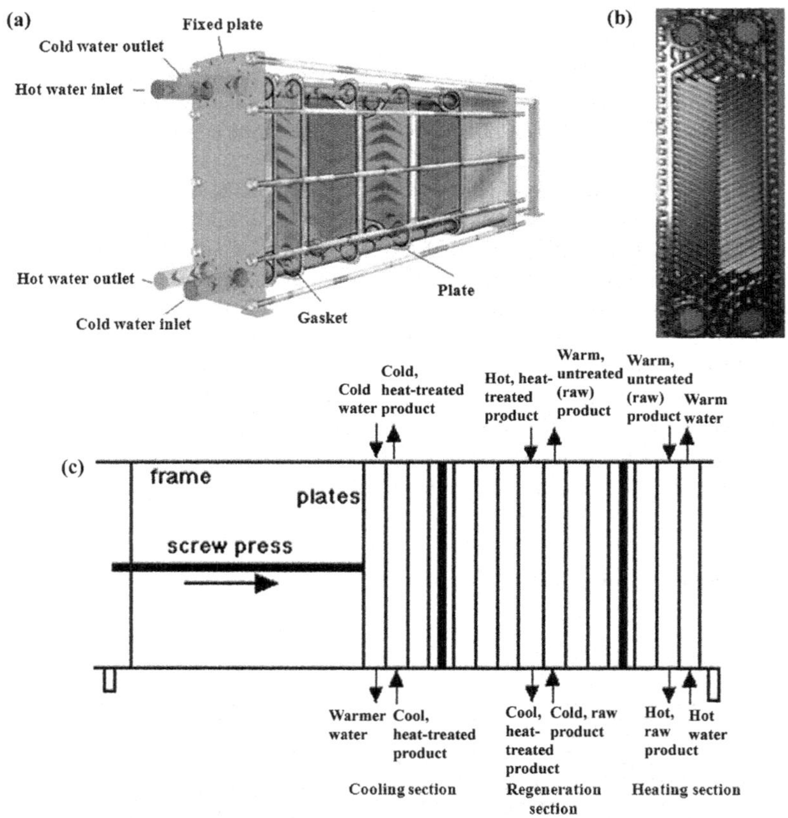

FIGURE 5.19 (a) Schematic of PHE (Modified and reproduced with permission from Khodamorad, S. H., Alinezhad, N., Fatmehsari, D. H. and Ghahtan, K. 2016. Stress corrosion cracking in Type.316 plates of a heat exchanger. *Case Studies in Engineering Failure Analysis* 5–6: 59–66.); (b) Photograph of the original heat exchanger plate (Reproduced with permission from Wajs, J. and Mikielewicz, D. 2016. Influence of metallic porous microlayer on pressure drop and heat transfer of stainless steel plate heat exchanger. *Applied Thermal Engineering* 93: 1337–1346.); (c) Three zones of a PHE. (Reproduced with permission from Goff, H. D. 2018. Dairy science and technology education series. University of Guelph, Canada. www.uoguelph.ca/foodscience/book-page/htst-milk-flow-overview.)

the plates prevent leakage and intermixing of fluids. The frame consists of three sections: the heating, regeneration, and cooling zones, each comprising several packs of plates (Figure 5.19c). The liquid food product will traverse all these three sections of the PHE. In the heating section, the product temperature is raised up to the process temperature using the heating medium, which is usually steam or steam/water mixture. Before being chilled in the cooling section, the heated liquid food is allowed to flow through the regeneration section to preheat the product stream which is just entering the system. The latent heat of outgoing food product at high temperature is utilized to heat the incoming cold product. This concept is known as *regeneration.* In the cooling zone, the product is cooled down to an average temperature of 4°C using cold water.

The heating or cooling medium and the liquid food product enter and leave alternative channels through holes in the corners of the plates. The fluids flow in the form of a thin layer and at high velocity, thus leading to a high heat transfer rate. For efficient heat transfer, the channels between the plates should be as narrow as possible. However, when a large volume of product passes through narrow channels, the flow velocity and pressure drop will increase to a greater extent. The direction of product and heating/cooling medium may be cocurrent (parallel) or countercurrent (opposite direction) with respect to each other.

The foremost advantage of PHEs is the regeneration of energy used in the system. Regeneration improves the thermal efficiency of the PHE and leads to significant energy saving. Also, in a PHE, the rate of heat transfer is high owing to the maximum surface area and turbulence induced in the liquid to be heat treated. Consequently, its overall heat transfer coefficient is two times higher than that of a SHE. This reduces the processing time and minimizes nutrient loss. PHE also provides other advantages such as compactness, sanitary design for food processing applications, the possibility of increasing the capacity by addition of more plates to the frame, simple maintenance, and low cost compared to other noncontact-type heat exchangers. However, the major drawback of a PHE is its inability to handle high viscous liquids.

5.3.2.7.4 Scraped Surface Heat Exchanger

Scraped surface heat exchangers (SSHEs) are utilized in the food, pharmaceutical, and chemical industries for heat transfer, crystallization, and other continuous processes. This type of heat exchanger consists of a central rotating shaft or rotor fitted with scrapping blades, housed inside a double pipe (stator) (Figure 5.20). The fluid is pumped through the inlet bowl using a volumetric pump. The heat transfer surface is continuously scraped off by the scraping blades, which reduces fouling, maintains a high and consistent heat exchange, and eventually leads to improved heat transfer rate. Thus, the surface is continuously exposed to the passage of untreated product. Consequently, high heat transfer coefficients are obtained as the boundary layer is continuously replenished with the fresh product. The scraper blades

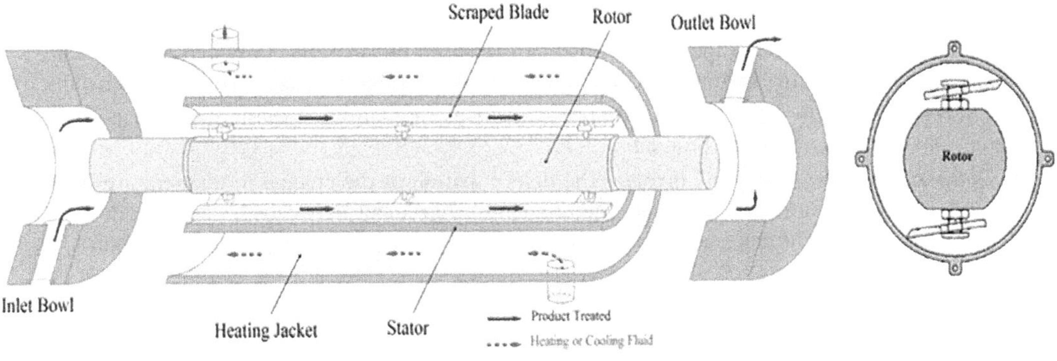

FIGURE 5.20 Schematic representation of an SSHE: Longitudinal and transversal cross section. (Reproduced with permission from Bayareh, M., Pordanjani, A. H., Nadooshan, A. A. and Dehkordi, K. S. 2017. Numerical study of the effects of stator boundary conditions and blade geometry on the efficiency of a scraped surface heat exchanger. *Applied Thermal Engineering* 113: 1426–1436.)

also provide simultaneous mixing and agitation of the product. Since the scraper blades come into direct contact with the food product, stainless steel, nickel, and corrosion-resistant materials are their preferred materials of construction. In the SSHE, the heating or cooling medium flows inside the jacketed outer tubes. Steam, hot water, and brine are commonly used as the heating medium.

The applications of SSHEs have been realized in closed, continuous processes involving heating, cooling, mixing, cooking, and gelling of foods. SSHE is the most appropriate heat exchanger for the processing of viscous, sticky products containing particulate matter, pumpable products, and those requiring crystallization. Heavy salad dressings, margarine, chocolate, peanut butter, fondant, ice cream, shortenings, fruit juices, honey, soups and tomato paste are some of the food products that are processed in an SSHE.

The main advantage of using an SSHE is that the product remains in contact with the heating surface for only a few seconds. Thus, high-temperature gradients can be used without subjecting the product to detrimental reactions. The other advantages of SSHE are listed as follows:

- ***Lesser requirement for floor space:*** SSHEs require lesser floor space owing to their minimal overall length and flexibility which allows them to be mounted either vertically or horizontally, thus rendering them ideal for a wide range of available space. Lower floor space requirement is directly related to more production.
- ***Consistent quality of final products:*** This advantage of SSHE can be attributed to its superior heat transfer performance and versatility with respect to handling a broad range of product viscosities. Certain products require rapid cooling to ensure that the entire batch has reached the desired low temperature, the failure of which leads to batch freezing and inconsistent product quality. In the aforementioned context, SSHE is ideal as the product is pumped into the lower end of the heat exchange cylinder which facilitates uniform cooling.
- ***Easy cleaning and maintenance:*** SSHEs can be easily disassembled for cleaning-in-place, inspection, and routine maintenance, without the need to detach their major modules.
- ***Reduced risk of contamination:*** As the product is subjected to rapid heating/cooling in an SSHE, the problem of bacterial growth due to unsafe temperatures is not encountered.

5.3.2.7.5 Guidelines for the Design and Selection of Heat Exchanger

Size and configuration are the main parameters to be considered during the design of heat exchanger. The term *size* refers to the heat transfer area of the heat exchanger. Calculation of size involves a complex protocol and depends on the following factors:

1. ***Product flow rate:*** The volumetric flow rate of the product depends on the capacity of the food processing plant. Greater the capacity, higher should be the flow rate and larger would be the size of the heat exchanger. For instance, if the plant capacity is doubled, then the original size of the heat exchanger should also be increased by two-fold, provided the flow rates of the heating/cooling media are also doubled, while other factors are maintained constant.
2. ***Thermal properties of the food product:*** It includes the density, specific heat capacity, and thermal conductivity of the product.
3. ***Temperature program:*** This involves the determination of the change in temperature of the food product during heat exchange, the temperature gradient between the food product and heating/cooling medium, and the flow direction of the liquids (cocurrent or countercurrent). While the inlet temperature of the heating/cooling medium is determined based on the processing conditions, the outlet temperature is calculated by conducting an energy balance analysis around the heat exchanger. In advanced heat exchanger types, energy lost to the surroundings is negligible and hence can be neglected. Therefore, according to the law of conservation of energy, the heat energy lost by the hot liquid [denoted by subscript 1 in Eq. (5.128)] is equal to the heat energy gained by the cold liquid [denoted by subscript 2 in Eq. (5.128)], which can be expressed as follows:

$$V_1\rho_1C_{P1}\Delta T_1 = V_2\rho_2C_{P2}\Delta T_2 \tag{5.128}$$

To be more accurate, the LMTD (ΔT_{lm}) is used in the general formula given in Eq. (5.128).

4. ***Overall heat transfer coefficient:*** With respect to the heat exchanger, the overall heat transfer coefficient (U) should be as high as possible. The value of U depends on the following factors:
    a. **Allowed pressure drops for the food product and heating/cooling medium:** Greater the pressure drops for the food product and heating/cooling medium, larger is the amount of heat transferred and smaller would be the size of the heat exchanger. The pressure drop for liquids can be increased by reducing the size or narrowing of the channel through which the food product flows. This lessens the distance across which the heat must be transferred. However, this reduces the cross-section area of flow which results in the increase of flow velocity through the channel and a consequent creation of turbulence in the liquid. In order to force the liquid through narrow channels, a booster pump may be installed, which achieves a higher pressure on the product side and thus prevents intermixing of untreated and treated product.
    b. **Viscosity of the liquids:** While flowing through the heat exchanger, a high-viscosity product develops lesser turbulence than a low-viscosity product. This translates to the requirement of a larger size for the heat exchanger, provided other factors remain constant.
    c. **Shape and thickness of the heat transfer surface:** The heat transfer surface or the partitioning plate that separates the heating medium and the product is often corrugated. When compared to a smooth surface, a corrugated one would create more intense turbulence. In a PHE, the shape of the partition is varied depending on the product to be treated and thermal efficiency requirements. Plates with different types of corrugations lead to different thermal properties and pressure drops. In general, the number of contact points should be less to facilitate handling of liquids with smaller sized particles or fibers. In case of a tubular heat exchanger, the surface of the inner tube is corrugated to enhance the heat transfer. Here, the corrugation also results in a higher pressure drop. The choice between smooth and corrugated tubes is governed by the need to optimize the heat transfer–pressure drop relationship. The thickness of the heat transfer surface is yet another parameter that influences the overall heat transfer coefficient. Lesser the thickness of the partition, higher would be the heat transfer. However, the partition should be strong enough to resist the fluid pressure.
    d. **Material of heat transfer surface:** For food processing applications, stainless steel is the preferred material of construction for the heat transfer surface, owing to its reasonably good heat transfer characteristics.
    e. **Presence of fouling matter:** Fouling can be defined as the deposition of solids from the food material in a thin layer on the partition. Fouling reduces the overall heat transfer coefficient as it increases the thermal resistance by increasing the thickness of the layer through which the heat must be transferred. As a result, a greater driving force, i.e., larger differential temperature between the heating medium and product would be required to transfer the same amount of heat as before.

        Fouling occurs when the heat transfer surface is too hot when compared to the product. Therefore, fouling can be avoided when the temperature difference between the heating medium and the food product is maintained at an optimal low value. Fouling also depends upon the product quality, air content of the product, and pressure conditions in the heating section. Air occluded in the product tends to increase fouling, and therefore, it is important to keep the air content as low as possible.
    f. **Flow rate:** The value of the overall heat transfer coefficient in a heat exchanger also depends on the flow rate and flow characteristics of the fluid. While higher flow rate leads to turbulent flow and increases the value of U, throttling the flow results in less turbulence and reduces the value of U. Hence, variations in the flow rate through a heat exchanger should be preferably avoided.

Considering all the aforementioned factors, the general formula for calculating the size of a heat exchanger is given by

$$A = \frac{V\rho C_P \Delta T}{\Delta T_{lm} U} \tag{5.129}$$

where A is the heat transfer area, V is the product flow rate, ρ is the product density, C_P is the specific heat of the product, ΔT is the temperature change in the product during heating/cooling, ΔT_{lm} is the LMTD, and U is the overall heat transfer coefficient. Nevertheless, the value of heat transfer area obtained from Eq. (5.129) is to be considered only as a theoretical value. Practically, the product nature and process requirements must also be taken into account while deciding the size of a heat exchanger. Requirements for cleaning and processing time are two such key factors which are not included in the general formula but holds importance.

1. ***Cleaning requirements:*** The design of a heat exchanger must permit effective cleaning. When several parallel channels are present in a heat exchanger, the turbulence during cleaning should be high for the effective removal of fouling deposits. Conversely, if only a few parallel channels are present, the resultant high turbulence may increase the pressure drop to a greater extent, which will reduce the flow velocity of the cleaning solution, thus lowering the effectiveness of cleaning. In general, for liquid products containing particles or fibers, backflush is required during cleaning. Backflush refers to the condition where the flow is reversed for a certain period during the cleaning program.
2. ***Processing time:*** When fouling occurs, the processing time will always be limited before the heat exchanger must be stopped for cleaning. The duration of processing depends on the amount of fouling deposited on the heat transfer surface. Apart from fouling, the processing time may also be limited by the growth of microorganisms in the heat exchanger.

5.3.2.7.6 Regeneration

Regeneration is the method of using the heat energy of the hot product exiting the heat exchanger to preheat the incoming cold product. This is done for the purpose of economizing on water and energy. The regeneration efficiency (R, in %) can be calculated using Eq. (5.130):

$$R = \frac{(T_r - T_i) \times 100}{(T_p - T_i)} \tag{5.130}$$

where T_r is the product temperature after regeneration (°C), T_i is the temperature of raw incoming product (°C), and T_p is the processing temperature (°C).

5.3.2.7.7 Design of the Holding Tube

The appropriate tube length to achieve the required holding time for the product at the specified temperature can be calculated using the hourly capacity and the inner diameter of the holding tube. Some molecules in the liquid product will move faster than the average due to the nonuniform velocity profile in the holding tube. To assure that even the fastest molecule is sufficiently heated, an efficiency factor (η) must be used. Although the value of η depends on the design of the holding tube, it is often in the range of 0.8–0.9, if the flow is turbulent. For more viscous fluids, the flow may be laminar, and hence, the efficiency factor would be lower. The formula for calculation of the length of holding tube is given by

$$V_P = \frac{Q \times HT}{3600 \times \eta} \tag{5.131}$$

$$L = \frac{V \times 4}{\pi \times D^2} \tag{5.132}$$

where V_P is the volume of the liquid product (Liter, L or m^3), Q is the flow rate of the product in the heat exchanger (L/h), HT is the holding time (s), η is the efficiency factor, L is the length of holding tube (m), corresponding to Q and HT, and D is the inner diameter of holding tube (m), to be known or modified according to Q and HT (Bylund, 1995).

5.3.3 Heat Transfer by Radiation

Radiation is a distinct mode of heat transfer in which energy emitted by a source in the form of electromagnetic waves is absorbed by the product which subsequently converts the radiant energy into heat energy. Radiation is propagated in the form of electromagnetic waves, and the standard wave properties such as wavelength (λ) and frequency (v) can be correlated by the following equation:

$$\lambda = \frac{c}{v} \tag{5.133}$$

where c is the speed of light in the medium which has a value of 2.998×10^8 m/s, under vacuum.

Heat transfer by radiation is different from conduction and convection in that it does not require any medium for propagation. This is why we can experience the heat energy emitted by the sun which is millions of kilometers away from the earth. Energy from the sun reaches the earth in the form of waves by radiation mode through a vacuum.

Radiation is the mode of heat transfer that governs food processing operations such as baking and drying. Although a vast array of phenomena exists under the radiation heat transfer, only thermal radiations are dealt with in this chapter. Thermal radiations are a part of the electromagnetic spectrum, ranging in wavelength between 0.1 and 100 µm, comprising the infrared, visible, and ultraviolet regions. As the heating and drying techniques based on radiation would be dealt elaborately in Chapter 14, the scope of discussion in this chapter is limited to the fundamental principles and the laws governing radiative heat transfer.

5.3.3.1 Principles of Radiative Heat Transfer

When radiation of a given wavelength is incident on a body, a portion of the incident radiation is reflected back, some are transmitted, and the remaining is absorbed by the body (Figure 5.21). This can be expressed as

$$\alpha + \gamma + \tau = 1 \tag{5.134}$$

where α is the fraction of heat absorbed by the body, termed as absorptivity, γ is the fraction of heat reflected by the body, known as reflectivity, and τ is the fraction of incident energy transmitted from the body, referred to as transmissivity. The portion of radiation energy absorbed contributes to the increase in the temperature of the body.

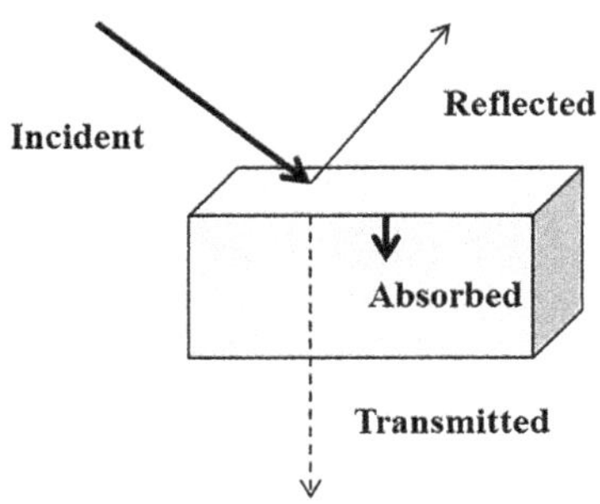

FIGURE 5.21 Radiant energy dissipation by an object.

An ideal blackbody is one which can absorb all the incident radiation ($\alpha = 1.0$). Thus, it is considered as the reference for calculating absorptivity. However, on earth, nothing is a true blackbody. Any object with $\alpha < 1.0$ is called a gray body. In order to describe the radiative heat transfer of such materials, a thermal property was introduced, namely, the *emissivity* (ε). Emissivity can be defined as the *fraction of radiation actually emitted or absorbed by a surface at a given temperature to the fraction that is emitted or absorbed by a black body at the same temperature.* In other words, emissivity is the ratio of the emissive power of a body (W) to that of a blackbody (W_b), given by

$$\varepsilon = \frac{W}{W_b} \tag{5.135}$$

Thus, ε is a measure of a material's ability to absorb and radiate energy. The emissivity of a material depends on its composition and surface characteristics, apart from being wavelength-dependent [$\varepsilon(\lambda)$]. The value of ε ranges between 0 for a perfect reflector and 1 for an ideal blackbody. Most food products have relatively high emissivities in the range of 0.90–0.95.

5.3.3.2 Laws Governing the Radiative Heat Transfer

5.3.3.2.1 Kirchhoff's Law

Kirchhoff's law states that *emissivity of a body will be exactly equal to its absorptivity if the body is in thermal equilibrium with its surroundings.* The statement of Kirchhoff's law can be expressed as

$$\alpha = \varepsilon \tag{5.136}$$

While solving problems of radiative heat transfer, it is assumed that the emissivity and absorptivity of a system are equal even if it is not in thermal equilibrium with the surroundings. This is because absorptivity of most real surfaces is relatively insensitive to temperature and wavelength. The aforementioned assumption justifies the concept of the gray body for which emissivity is considered to be independent of the wavelength.

5.3.3.2.2 Stefan–Boltzmann Law

Stefan–Boltzmann law states that *the energy flux emitted by a black body is proportional to the fourth power of its absolute temperature.*

$$W_b = \frac{q}{A} = \sigma T^4 \tag{5.137}$$

Thus, the rate of heat emission (q) from an object can be expressed as

$$q = A\sigma T^4 \tag{5.138}$$

For a gray body, Eq. (5.137) can be rewritten as

$$W = \frac{q}{A} = \sigma\varepsilon T^4 \tag{5.139}$$

or

$$q = \sigma\varepsilon A T^4 \tag{5.140}$$

where A is the surface area of the object (m^2), σ is the Stefan–Boltzmann constant with a value of 5.669×10^{-8} ($W/m^2\,K^4$), ε is the emissivity, and T is the absolute temperature of the object (K). This law forms the basis of radiative heat transfer.

5.3.3.2.3 Wien's Displacement Law

Wien's displacement law states that *for a blackbody, the curve of spectral energy density (kJ/m³ nm) as a function of wavelength for different temperatures peaks at a wavelength, λ_{max} which is inversely proportional to the temperature* Eq. (5.141).

$$\lambda_{\max} \propto \frac{1}{T} \tag{5.141}$$

or

$$\lambda_{\max} = \frac{b}{T} \tag{5.142}$$

where T is the absolute temperature (K) and b is the constant of proportionality known as the Wien's displacement constant, equal to 2.898×10^{-3} m K.

Equation 5.142 can be rewritten as

$$\lambda_{\max} T = 2.898 \times 10^{-3}\ \text{m K} \tag{5.143}$$

or

$$\lambda_{\max} = \frac{2898}{T}\ \mu\text{m} \tag{5.144}$$

Thus, according to Wien's displacement law, the product of absolute temperature and wavelength at maximum flux intensity is constant. Therefore, the maximum emissive power emitted by the blackbody at the peak wavelength over the spectral distribution of radiation can be calculated using this law.

Example 5.13

A polished metal surface of area 75 m² is at a temperature of 30 °C. The surface shows an emissivity of 0.05. Calculate the rate of heat energy emitted by the metal surface.

Solution

Given

$\varepsilon = 0.05$
$A = 75\ \text{m}^2$
$T = 30°\text{C} = 303\ \text{K}$

Using the equation of Stefan–Boltzmann law (Eq. (5.140)),

$$q = A\varepsilon\sigma T^4$$

$$q = (5.669 \times 10^{-8}\ \text{W/[m}^2\ \text{K}^4] \times 0.05 \times 75\ \text{m}^2 \times (303\ \text{K})^4) = 1792\ \text{W}$$

Answer: The rate at which the heat energy is emitted by the polished metal surface is 1792 W

5.3.3.3 The Concept of View Factor

The radiation heat transfer between surfaces depends on the on the properties of the radiating surface such as the temperature, emissivity, and absorptivity, and orientation of the surfaces relative to each other. *View factor*, also termed as *shape factor*, is a geometrical parameter that accounts for the effect of orientation on the net heat exchange between two radiating surfaces. It can be defined as the fraction of radiant energy leaving a surface which strikes another surface directly (reflected and re-radiated energy is not taken into account). In such cases, where it is not possible for the materials to absorb all the emitted energy, *view factor* is used to parameterize the fraction of radiant energy leaving the first body and reaching the second body. During the calculation of view factor, the space/medium between the radiating surfaces is assumed to be devoid of bodies that absorb, emit, or scatter radiation. The value of view factor ranges between 0 and 1. While a view factor of zero indicates that two surfaces do not see each other directly, the value of one indicates that the first surface completely surrounds the second surface.

Considering two surfaces A and B, the view factor $F_{A\rightarrow B}$ or F_{AB} is the fraction of radiant energy emitted by surface A that impinges on surface B. In this case, the fraction of heat leaving surface A and reaching surface B ($q_{A\rightarrow B}$) is given by

$$q_{A\rightarrow B} = A_A \sigma F_{A\rightarrow B} T_A^4 \tag{5.145}$$

Likewise, the fraction of heat leaving surface B and reaching surface A ($q_{B\rightarrow A}$) is given by

$$q_{B\rightarrow A} = A_B \sigma F_{B\rightarrow A} T_B^4 \tag{5.146}$$

Using the areas of the surfaces A and B (A_A and A_B), which are in thermal equilibrium with each other ($T_A = T_B$), the reciprocity relationship for view factors can be derived, which is given by

$$A_A F_{A\rightarrow B} = A_B F_{B\rightarrow A} \tag{5.147}$$

Thus, if $F_{A\rightarrow B}$ is known, $F_{B\rightarrow A}$ can be easily calculated.

On the contrary, if A and B are gray bodies with emissivities ε_A and ε_B, respectively, wherein the two surfaces see only a fraction of each other, the rate of heat flow from A to B is given by

$$q_{A\rightarrow B} = \left(\left(\frac{1-\varepsilon_A}{\varepsilon_A}\right) + \frac{1}{F_{A\rightarrow B}} + \left(\frac{1-\varepsilon_B}{\varepsilon_B}\right)\frac{A_A}{A_B}\right)^{-1} A_A \sigma \left(T_A^4 - T_B^4\right) \tag{5.148}$$

5.4 Problems to Practice

5.4.1 Multiple Choice Questions

1. SI unit of thermal conductivity is
   a. $J/m^2 s$
   b. J/m K s
   c. W/m K
   d. (b) and (c)

Answer: d

2. When heat transfer occurs in the absence of a medium, it is referred to as
   a. conductive heat transfer
   b. convective heat transfer
   c. radiative heat transfer
   d. both a and b

Answer: c

3. When heat transfer occurs by molecular collision, it is known as
    a. conductive heat transfer
    b. convective heat transfer
    c. radiative heat transfer
    d. both b and c

Answer: a

4. Heat transfer in liquid and gases takes place by
    a. conduction
    b. convection
    c. radiation
    d. conduction and convection

Answer: b

5. The concept of overall heat transfer coefficient is used in heat transfer problems pertaining to
    a. conduction
    b. convection
    c. radiation
    d. conduction and convection

Answer: b

6. The equation for unsteady-state heat conduction through solids is given by
    a. $\delta T/\delta x = \alpha\ (\delta^2 T/\delta t^2)$
    b. $\delta T/\delta x = \alpha\ (\delta^2 T/\delta x^2)$
    c. $\delta T/\delta t = \alpha\ (\delta^2 T/\delta x^2)$
    d. $\delta T/\delta t = \alpha\ (\delta^2 T/\delta t^2)$

Answer: c

7. Natural convection is characterized by
    a. low heat transfer
    b. occurrence due to gravity and natural buoyant forces
    c. dependence on viscosity, density and thermal conductivity
    d. all of the above

Answer: d

8. The reciprocal of heat transfer coefficient is
    a. conductance
    b. resistance
    c. density
    d. temperature difference

Answer: b

9. In order to cool the food product by radiation, a food industry paints its walls white but not black. The abovementioned approach is based on
   a. Kirchhoff's law
   b. Stefan–Boltzmann law
   c. Fourier's law
   d. Newton's law of cooling

Answer: a

10. The heat exchanger which leads to quick exchange of heat is
   a. SHE
   b. PHE
   c. SSHE
   d. tubular heat exchanger

Answer: b

11. For the same inlet and outlet temperatures of hot and cold fluids, the LMTD would be
   a. greater for cocurrent flow heat exchanger than for countercurrent flow heat exchanger.
   b. greater for countercurrent flow heat exchanger than for cocurrent flow heat exchanger.
   c. same for both cocurrent and countercurrent flow heat exchangers.
   d. does not depend on the flow direction

Answer: b

12. Which of the following has the highest thermal conductivity?
   a. Steam
   b. Solid ice
   c. Melting ice
   d. Water

Answer: b

13. Thermal diffusivity is
   a. a dimensionless group
   b. a function of temperature
   c. used as mathematical model
   d. a parameter related to radiative heat transfer

Answer: b

14. A non-dimensional number generally associated with natural convection heat transfer is
   a. Grashof number
   b. Nusselt number
   c. Reynolds number
   d. Prandtl number

Answer: a

15. In a blackbody, all the radiations that are incident on it are
    a. reflected
    b. refracted
    c. transmitted
    d. absorbed

Answer: d

16. The value of the wavelength for maximum emissive power is given by
    a. Wien's law
    b. Planck's law
    c. Stefan's law
    d. Fourier's law

Answer: a

17. In natural convection heat transfer, Nusselt number is function of
    a. Grashof number and Reynolds number
    b. Grashof number and Prandtl number
    c. Prandtl number and Reynolds number
    d. Grashof number, Prandtl number, and Reynolds number

Answer: b

18. According to Wien's law, the wavelength corresponding to maximum energy is proportional to
    a. absolute temperature
    b. emissivity
    c. surface area
    d. frequency

Answer: a

19. In natural convection, transition of flow regime from laminar to turbulent occurs at Rayleigh number
    a. $\sim 10^9$
    b. 10^{12}
    c. 10^{15}
    d. 100

Answer: a

20. If the value of $N_{Bi} < 0.01$, it means that
    a. the convective resistance of the fluid is negligible
    b. the conductive resistance of the solid is negligible
    c. the conductive resistance of the fluid is negligible
    d. the convective resistance of the solid is negligible

Answer: b

21. The ratio of energy transferred by convection to that by conduction is termed as
    a. Stanton number
    b. Graetz number
    c. Biot number
    d. Peclet number

Answer: c

22. In cocurrent flow heat exchangers,
    a. the exit temperature of hot fluid is always equal to the exit temperature of cold fluid
    b. the exit temperature of hot fluid is always less than the exit temperature of cold fluid
    c. the exit temperature of hot fluid is always more than the exit temperature of cold fluid
    d. the exit temperature of hot fluid is always less than the inlet temperature of cold fluid

Answer: c

23. The temperature difference which is appropriate for usage in the design calculations of heat exchangers is
    a. arithmetic mean temperature difference
    b. logarithmic mean temperature difference
    c. both a and b
    d. either a or b

Answer: b

24. The Grashof number in natural convection plays same role as
    a. Prandtl number (N_{Pr}) in forced convection
    b. Reynolds number (N_{Re}) in forced convection
    c. Nusselt number (N_{Nu}) in forced convection
    d. Peclet number (N_{Pe}) in forced convection

Answer: b

25. The unit for the LMTD is
    a. °C
    b. 1/°C
    c. dimensionless
    d. °C/s

Answer: a

5.4.2 Numerical Problems

1. The heat flux through a solid food product of thickness 25 mm is estimated to be 40 W/m². The inner and outer surface temperatures of the product are 50°C and 20°C, respectively. Determine the thermal conductivity of the food product.

   Given

   i. Heat flux through the food product $(q/A) = 40\,\text{W/m}^2$
   ii. Thickness of the food product $(\Delta x) = 25$ mm $= 0.025$ m
   iii. Inner temperature of the food product $(T_i) = 50°\text{C}$
   iv. Outer temperature of the food product $(T_o) = 20°\text{C}$

   To find: Thermal conductivity of the food product (k)

Solution

The mode of heat transfer through the solid food product is conduction. Hence, applying the Fourier's law of conductive heat transfer,

$$q = kA\frac{\Delta T}{\Delta x}$$

$$k = \frac{q\Delta x}{A\Delta T} = \frac{40 \times 0.025}{(50 - 20)} = 0.033\,\text{W/m}\,^\circ\text{C}$$

Answer: Thermal conductivity of the food product (*k*) = 0.033 W/m °C

2. An acrylic glass window has a surface area of 5 m² and thickness of 7.5 mm. The temperatures on the either side of the window are 20°C and 30°C, respectively. Thermal conductivity of acrylic glass is 0.2 W/m K. Assuming steady-state condition, calculate the rate of heat transfer through the acrylic window.

Given

i. Surface area of acrylic glass window (A) = 5 m^2
ii. Thickness of the glass window (Δx) = 7.5 mm = 0.0075 m
iii. Temperatures on either sides of the window = 20°C and 30°C
iv. Thermal conductivity of acrylic glass = 0.2 W/m K

To find: Rate of heat transfer through the acrylic window

Solution

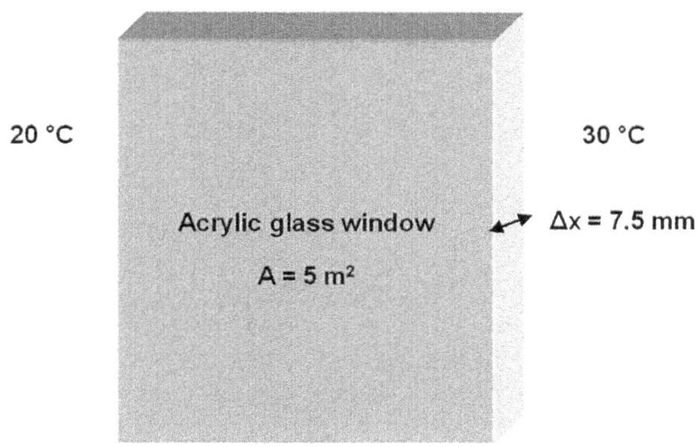

The mode of heat transfer through the window would be conduction. Hence, applying the Fourier's law of conductive heat transfer,

$$q = kA\frac{\Delta T}{\Delta x}$$

$$\therefore q = \frac{0.2 \times 5 \times (30 - 20)}{\left(7.5 \times 10^{-3}\right)} = 1333.3\,\text{W}$$

Answer: Rate of heat transfer through the acrylic window is 1333.3 W

3. A composite wall consists of two layers X and Y of thickness 5 and 7.5 mm, respectively. The corresponding values of thermal conductivity are 0.05 and 20 W/m °C. If the flux of heat transfer is 150 W/m^2 at steady state, calculate the driving force for heat transfer.

Given

a. Composite wall made up of two layers
b. Thickness of the first layer (Δx_1) = 5 mm = 5×10^{-3} m = 0.005 m
c. Thickness of the second layer (Δx_2) = 7.5 mm = 7.5×10^{-3} m = 0.0075 m
d. Thermal conductivity of the first layer (k_1) = 0.05 W/m °C
e. Thermal conductivity of the second layer (k_2) = 20 W/m °C
f. Flux of heat transfer (q/A) = 150 W/m^2

To find: Driving force for heat transfer (ΔT)

Solution

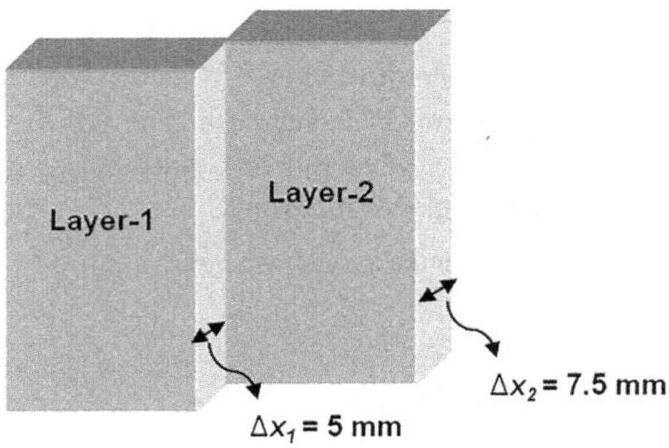

For a composite wall,

$$\frac{q}{A} = \frac{\Delta T}{\left(\frac{\Delta x_1}{k_1} + \frac{\Delta x_2}{k_2}\right)}$$

$$150 = \frac{\Delta T}{\left(\frac{0.005}{0.05} + \frac{0.0075}{20}\right)}$$

$$150 = \frac{\Delta T}{0.100375}$$

$$\therefore \Delta T = 150 \times 0.100375 = 15.05625 \approx 15.1°\text{C}$$

Answer: The driving force for heat transfer through the composite wall is 15.1°C

4. Hot oil flows through a stainless steel pipe of inner diameter 0.05 m, length 10 m, and wall thickness 3 mm. The inside wall temperature is 95°C and the outside surface temperature of pipe is 90°C. Thermal conductivity of stainless steel is 16 W/m °C. Calculate the rate of heat loss from the pipe under steady-state conditions.

Given

a. Inner diameter of the stainless steel pipe (D_i) = 0.05 m
b. Length of the stainless steel pipe (L) = 10 m
c. Thickness of the stainless steel pipe (x) = 3 mm = 3×10^{-3} m = 0.003 m
d. Temperature of inner wall of the pipe (T_i) = 95°C
e. Temperature of outer surface of the pipe (T_o) = 90°C
f. Thermal conductivity of stainless steel (k) = 16 W/m °C

To find: Rate of heat loss from the pipe under steady state conditions

Solution

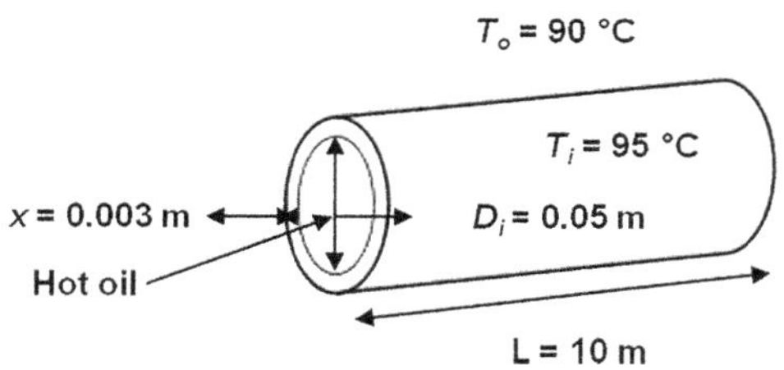

$$q_r = \frac{(T_i - T_o)}{\left|\frac{\ln(r_o/r_i)}{2\pi Lk}\right|}$$

From the preceding figure,

$$r_i = D_i/2 = 0.05/2 = 0.025\,\text{m}$$

$$r_o = r_i + x = 0.025 + 0.003 = 0.028\,\text{m}$$

$$q_r = \frac{(95-90)}{\left|\frac{\ln(0.028/0.025)}{(2\times\pi\times10\times16)}\right|}$$

$$\therefore q_r = \frac{5}{(1.127\times10^{-4})} = 4.435372\times10^4\ \text{W (or) } 44353.72\,\text{J/s}$$

Answer: The rate of heat loss through the stainless steel pipe is 44353.72 J/s

5. A cocurrent flow heat exchanger is used to cool milk (C_P = 3.9 kJ/kg K) which is flowing at a rate of 2 kg/s from its pasteurization temperature to 30°C. Chilled water at 4°C is used as the cooling medium, at a flow rate of 2.5 kg/s. The outer diameter of the outer pipe is 4 cm.

The overall heat transfer coefficient is 450 W/m²°C. Determine the outlet temperature of the cooling medium and the length of the heat exchanger.

Given

i. Process: Cooling of pasteurized milk in a cocurrent flow heat exchanger
ii. Mass flow rate of milk $(\dot{m}_H) = 2$ kg/s
iii. Outlet temperature of milk $(T_{Ho}) = 30°C = 303$ K
iv. Cooling medium: Chilled water
v. Inlet temperature of cooling medium $(T_{ci}) = 4°C = 277$ K
vi. Flow rate of cooling medium $(\dot{m}_c) = 2.5$ kg/s
vii. Overall heat transfer coefficient $(U) = 450$ W/m²°C
viii. Outer diameter of outer pipe $(D_o) = 4$ cm $= 4 \times 10^{-2}$ m $= 0.04$ m
ix. Specific heat capacity of milk at constant pressure $(C_{P_{\text{milk}}}) = 3.9$ kJ/kg K

To find

i. Outlet temperature of the cooling medium (T_{co})
ii. Length of the heat exchanger (L)

Solution

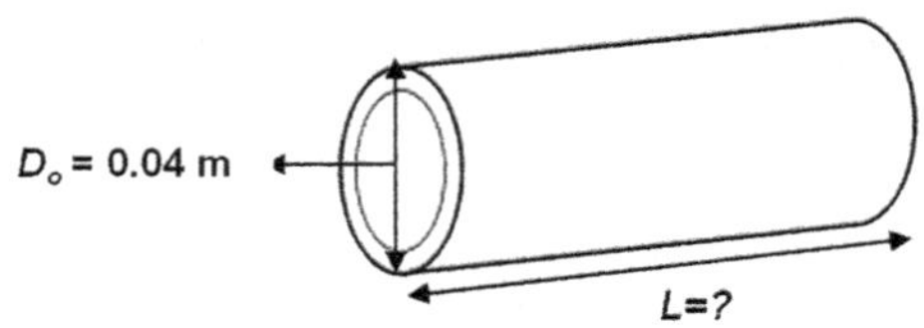

$$q = mC_P\Delta T$$

Considering the pasteurization temperature as 63°C or 336 K (T_{Hi}),

$$q = \dot{m}_H C_{PH}(T_{Hi} - T_{Ho}) = 2 \times 3.9 \times (336 - 303) = 257.4\,\text{kW (or) } 257.4\,\text{kJ/s}$$

Similarly for the cooling medium,

$$q = \dot{m}_c C_{Pc}(T_{co} - T_{ci}) = 2.5 \times 4.2 \times (T_{co} - 277)$$

(Since the cooling medium is water, $C_{Pc} = 4.2$ kJ/kg K)

Applying the principle of energy balance,

$$q = \dot{m}_H C_{PH}(T_{Hi} - T_{Ho}) = \dot{m}_c C_{Pc}(T_{co} - T_{ci})$$

$$257.4 = 2.5 \times 4.2 \times (T_{co} - 277)$$

$$257.4 = 10.5 \times (T_{co} - 277)$$

$$(T_{co} - 277) = \frac{257.4}{10.5} = 24.5$$

$$\therefore T_{co} = 24.5 + 277 = 301.5\,\text{K} = 28.5°\text{C}$$

According to Newton's law of cooling, heat transfer rate (q) is given by

$$q = UA\Delta T_{lm}$$

$$\therefore A = \frac{q}{U\Delta T_{lm}}$$

For a cocurrent flow heat exchanger, the LMTD is given by

$$\Delta T_{lm} = \frac{(T_{Hi} - T_{ci}) - (T_{Ho} - T_{co})}{\ln\frac{(T_{Hi} - T_{ci})}{(T_{Ho} - T_{co})}}$$

$$\Delta T_{lm} = \frac{(63-4)-(30-28.5)}{\ln\frac{(63-4)}{(30-28.5)}} = \frac{57.5}{\ln\left(\frac{59}{1.5}\right)} = \frac{57.5}{3.672} = 15.659°\text{C} = 15.7°\text{C}$$

$$\therefore A = \frac{257.4 \times 1000\,\text{J/s}}{(450\,\text{W/m}^2\,°\text{C})(15.7)°\text{C}} = 36.4\,\text{m}^2$$

$$A = 2\pi rL$$

$$36.4\,\text{m}^2 = \left(2 \times \pi \times \left(\frac{0.04}{2}\right) \times L\right)$$

$$36.4 = (0.126 \times L)$$

$$\therefore L = \frac{36.4}{0.126} = 288.9\,\text{m}$$

Answer: (i) Outlet temperature of cold fluid = 28.5°C
(ii) Length of the heat exchanger = 288.9 m

6. A four-layered composite pipe has an inner diameter of 0.8 cm and an internal surface temperature of 150°C. The first layer from inside is 2.5 cm thick with a thermal conductivity of 20 W/(m °C), the second layer is 3.5 cm thick with a thermal conductivity of 0.02 W/(m °C), the third layer is 1.5 cm thick with a thermal conductivity of 150 W/(m °C), and the fourth layer is 1 cm thick with a thermal conductivity of 0.04 W/(m °C). Outside surface temperature of the pipe is 80°C. Length of the pipe is 10 m. Determine the rate of heat transfer through the pipe under steady-state conditions.

Given

i. Four-layered composite pipe
ii. Inner diameter of the pipe (D_1) = 0.8 cm = 0.8×10^{-2} m = 0.008 m
iii. Length of the pipe = 10 m

iv. Thickness of first layer $(\Delta x_1) = 2.5$ cm $= 2.5 \times 10^{-2}$ m $= 0.025$ m
v. Thickness of second layer $(\Delta x_2) = 3.5$ cm $= 3.5 \times 10^{-2}$ m $= 0.035$ m
vi. Thickness of third layer $(\Delta x_3) = 1.5$ cm $= 1.5 \times 10^{-2}$ m $= 0.015$ m
vii. Thickness of fourth layer $(\Delta x_4) = 1$ cm $= 1 \times 10^{-2}$ m $= 0.01$ m
viii. Thermal conductivity of first layer $(k_1) = 20$ W/(m °C)
ix. Thermal conductivity of second layer $(k_2) = 0.02$ W/(m °C)
x. Thermal conductivity of third layer $(k_3) = 150$ W/(m °C)
xi. Thermal conductivity of fourth layer $(k_4) = 0.04$ W/(m °C)
xii. Internal surface temperature of the pipe $(T_i) = 150$°C
xiii. Outer surface temperature of the pipe $(T_o) = 80$°C

To find: Rate of heat transfer through the pipe under steady-state conditions

Solution

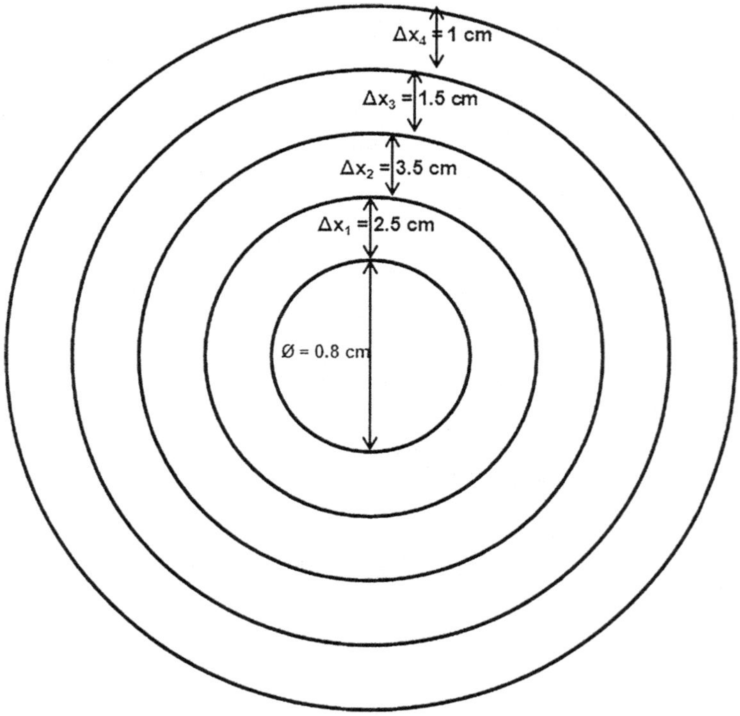

The rate of heat transfer is given by

$$q_r = \frac{(T_i - T_o)}{\left\{ \dfrac{\ln(r_2/r_1)}{2\pi L k_1} + \dfrac{\ln(r_3/r_2)}{2\pi L k_2} + \dfrac{\ln(r_4/r_3)}{2\pi L k_3} + \dfrac{\ln(r_5/r_4)}{2\pi L k_D} \right\}}$$

$$r_1 = \frac{D_1}{2} = \frac{0.008}{2} = 0.004\ \text{m}$$

$$r_2 = r_1 + x_1 = 0.004 + 0.025 = 0.029\,\text{m}$$

$$r_3 = r_2 + x_2 = 0.029 + 0.035 = 0.064\,\text{m}$$

$$r_4 = r_3 + x_3 = 0.064 + 0.015 = 0.079\,\text{m}$$

$$r_5 = r_4 + x_4 = 0.079 + 0.01 = 0.089\,\text{m}$$

$$q_r = \frac{(150-80)}{\left\{\frac{\ln(0.029/0.004)}{(2\pi\times10\times20)} + \frac{\ln(0.064/0.029)}{(2\pi\times10\times0.02)} + \frac{\ln(0.079/0.064)}{(2\pi\times10\times150)} + \frac{\ln(0.089/0.079)}{(2\pi\times10\times0.04)}\right\}}$$

$$\therefore q_r = 103.093\,\text{W}$$

Answer: Rate of heat transfer through the pipe is 103.093 W

7. Grape juice is flowing in a pipe with 0.03 m inner diameter and 0.035 m outer diameter, while steam is condensing on the outside. If the heat transfer coefficient on the juice side is 1000 W/m²°C, on the steam side 2000 W/m²°C, and thermal conductivity of the tube is 20 W/m °C, calculate the overall heat transfer coefficient based on the inside area.

Given

i. Inner diameter of the pipe (D_i) = 0.03 m
ii. Outer diameter of the pipe (D_o) = 0.035 m
iii. Heat transfer coefficient on the juice side (h_i) = 1000 W/m²°C
iv. Heat transfer coefficient on the steam side (h_o) = 2000 W/m²°C
v. Thermal conductivity of the tube (k) = 20 W/m °C

To find: Overall heat transfer coefficient based on the inside area (U_i)

Solution

$$\frac{1}{U_i} = \frac{1}{h_i} + \frac{\ln(r_o/r_i)}{2\pi kL} + \frac{A_i}{h_o A_o}$$

$$r_o = \frac{D_o}{2} = \frac{0.035}{2} = 0.0175\,\text{m}$$

$$r_i = \frac{D_i}{2} = \frac{0.03}{2} = 0.015\,\text{m}$$

Assuming the length of the pipe to be 1 m,

$$\frac{1}{U_i} = \frac{1}{1000} + \frac{\ln(0.0175/0.015)}{(2\pi\times20\times1)} + \frac{(2\pi\times0.015\times1)}{(2000)(2\pi\times0.0175\times1)}$$

$$\frac{1}{U_i} = 0.001 + 0.0012 + \left(4.286\times10^{-4}\right) = 0.0026$$

$$\therefore U_i = 380.4\,\text{W/m}^2\,°\text{C}$$

Answer: Overall heat transfer coefficient based on the inside area, $U_i = 380.4\,\text{W/m}^2\,°\text{C}$

8. A square plate with side 5 cm is at a temperature of 30°C. Calculate the rate of radiative heat transfer if the emissivity of tube surface is 0.6.

Given

i. Side of square plate (a) = 5 cm = 5×10^{-2} m = 0.05 m
ii. Temperature (T) = 30°C
iii. Emissivity of tube surface (ε) = 0.6

To find: Rate of radiative heat transfer

Solution

Based on the Stefan–Boltzmann law of radiative heat transfer,

$$q = \sigma \varepsilon A T^4$$

$$A = a^2 = 0.05^2 = 0.0025\,\text{m}^2$$

$$q = \left(5.669 \times 10^{-8}\right) \times 0.6 \times 0.0025 \times (303)^4$$

$$\therefore q = 0.717\,\text{W}$$

Answer: Rate of radiative heat transfer = 0.717 W or 0.717 J/s

9. A biscuit at a surface temperature of 50°C with 40 cm² surface area and emissivity 0.8 is baked in a deck oven. Calculate the flux of radiative heat transfer to the biscuit if the walls of oven are at 200°C.

Given

i. Surface temperature of biscuit (T_B) = 50°C = 323 K
ii. Surface area of biscuit (A) = 40 cm^2
iii. Emissivity (ε) = 0.8
iv. Oven wall temperature (T_W) = 200°C = 473 K

To find: Flux of radiative heat transfer

Solution

Based on the Stefan–Boltzmann law of radiative heat transfer,

$$q = \sigma \varepsilon A (T_W^4 - T_B^4)$$

$$\text{Flux of radiative heat transfer} = \frac{q}{A} = \sigma \varepsilon (T_W^4 - T_B^4) = \left(5.669 \times 10^{-8}\right) \times 0.8 \times \left(473^4 - 323^4\right)$$

$$\therefore \text{Flux} = 1776.4\,\text{W/m}^2$$

Answer: Flux of radiative heat transfer = 1776.4 W/m²

10. The wall of a hot air oven consists of two metal sheets with an intermediate layer of insulation. The surface temperature of the inner wall is 150°C and that of the outer wall is 70°C. Each of the metal sheets is of thickness 1.5 mm, the thickness of the insulation is 4 cm, and the values of thermal conductivity for metal sheet and insulation are 15 and 0.045 W/m °C, respectively. Calculate the total resistance to heat transfer provided by the wall and the amount of heat lost through the wall per m² of wall area.

Given

i. Surface temperature of the inner wall of hot air oven (T_S) = 150°C
ii. Temperature of the outer wall of hot air oven (T_o) = 70°C
iii. Thickness of the metal sheet (Δx_1, Δx_3) = 1.5 mm = 0.0015 m
iv. Thickness of the insulation layer (Δx_2) = 4 cm = 0.04 m
v. Thermal conductivity of the metal sheet (k_1, k_3) = 15 W/m °C
vi. Thermal conductivity of the insulation (k_2) = 0.045 W/m °C

To find

i. Total resistance to heat transfer provided by the wall (R_t)
ii. Amount of heat lost through the wall per m² of wall area (q_L)

Solution

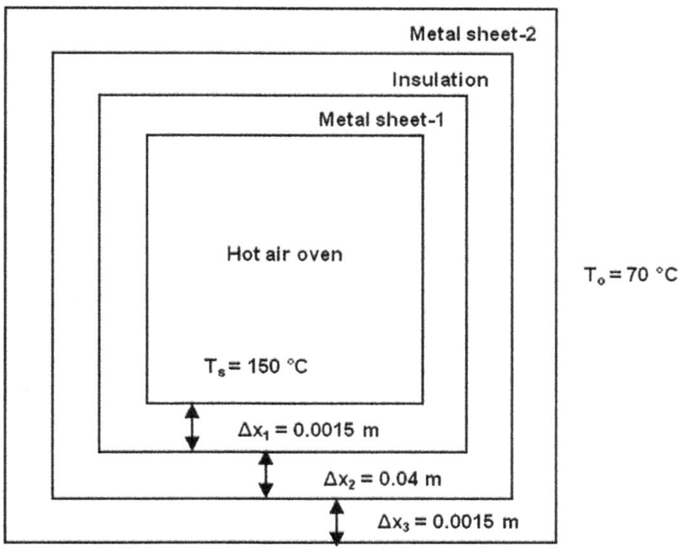

$$q_L = \frac{q}{A} = \frac{(T_s - T_o)}{\left(\frac{\Delta x_1}{k_1}\right)+\left(\frac{\Delta x_2}{k_2}\right)+\left(\frac{\Delta x_3}{k_3}\right)}$$

where $\left(\frac{\Delta x_1}{k_1}\right)+\left(\frac{\Delta x_2}{k_2}\right)+\left(\frac{\Delta x_3}{k_3}\right) = R_t$

$$R_t = \left(\frac{0.0015}{15}\right)+\left(\frac{0.04}{0.045}\right)+\left(\frac{0.0015}{15}\right) = 0.889\,\text{m}^2\,°\text{C/W}$$

$$q_L = \frac{(150-70)}{\left(\frac{0.0015}{15}\right)+\left(\frac{0.04}{0.045}\right)+\left(\frac{0.0015}{15}\right)} = \frac{80}{0.889} = 89.989\ \text{W/m}^2$$

Answer: (i) Total resistance to heat transfer provided by the wall (R_t) = 0.889 m² °C/W
(ii) Amount of heat lost through the wall per m² of wall area (q_L) = 89.989 W/m²

BIBLIOGRAPHY

Anandharamakrishnan, C. 2017. Introduction to drying. In *Handbook of Drying for Dairy Products*, ed. C. Anandharamakrishnan, 1–14. Chichester, West Sussex, UK: John Wiley & Sons.

Bayareh, M., Pordanjani, A. H., Nadooshan, A. A. and Dehkordi, K. S. 2017. Numerical study of the effects of stator boundary conditions and blade geometry on the efficiency of a scraped surface heat exchanger. *Applied Thermal Engineering* 113: 1426–1436.

Bylund, G. 1995. *Dairy Processing Handbook*. Lund, Sweden: Tetra Pak Processing Systems AB.

Choi, Y. and Okos, M. R. 1986. Effects of temperature and composition on the thermal properties of food. In *Food Engineering and Process Applications, Volume 1, "Transport Phenomena"*, eds. M. Le Maguer and P. Jelen, 93–101. London: Elsevier Applied Science Publishers.

Correa, D. J. and Marchetti, J. L. 1987. Dynamic simulation of shell-and-tube heat exchangers. *Heat Transfer Engineering* 8: 50–59.

Desrosier, N. W. and Desrosier, J. W. 1982. *The Technology of Food Preservation*. Westport, CT: AVI Publishing Company, Inc.

Dickerson Jr, R. W. 1968. Thermal properties of food. In *The Freezing Preservation of Food*, eds. D. K. Tressler, W. B. Van Arsdel, and M. R. Copley, 26–51. Westport, CT: AVI Publishing Company.

Goff, H. D. 2018. Dairy science and technology education series. Canada: University of Guelph. www.uoguelph.ca/foodscience/book-page/htst-milk-flow-overview (accessed June 5, 2018).

Khodamorad, S. H., Alinezhad, N., Fatmehsari, D. H. and Ghahtan, K. 2016. Stress corrosion cracking in Type.316 plates of a heat exchanger. *Case Studies in Engineering Failure Analysis* 5–6: 59–66.

King, C. J. 1977. Heat and mass transfer fundamentals applied to food engineering. *Journal of Food Process Engineering* 1: 3–14.

Marella, C. and Muthukumarappan, K. 2013. Food freezing technology. In *Handbook of Farm, Dairy and Food Machinery Engineering*, ed. M. Kurtz, 355–378. London, UK: Academic Press, Elsevier Inc.

Riedel, L. 1969. Temperaturfaehigkeits Messungen an wasserreichen. *Lebensm. Kaeltetechn. Klimatisierung* 21: 315.

Wajs, J. and Mikielewicz, D. 2016. Influence of metallic porous microlayer on pressure drop and heat transfer of stainless steel plate heat exchanger. *Applied Thermal Engineering* 93: 1337–1346.

6

Mass Transfer

Mass transfer phenomena are encountered in everyone's daily routine. Dissolution of sugar in a cup of tea, percolation of water through roast and ground coffee granules to prepare an aromatic brew in a coffee maker, perception of rich aroma from the vapors above a cup of coffee, refrigeration and air conditioning are all the different forms of mass transfer in day-to-day life. However, moving a mass of fluid or solid from one place to another is not mass transfer. Similar to any transport process, the mass transfer also occurs as a result of the balance between two forces: resisting force and driving force. While proximity (distance) between the components acts as the resisting force, concentration gradient acts as the driving force. Concentration gradient (ΔC) is the difference in concentrations ($C_1 - C_2$) of a component with respect to different positions within a system (Figure 6.1). Thus, mass transfer can be defined as the *transport of one component in a mixture from a region of higher concentration to that of a lower concentration.*

Mass transfer always occurs in the direction of reducing concentration gradient (Figure 6.1). In a batch process, the mass transfer continues to occur until equilibrium is attained with the disappearance of the concentration gradient. If the concentration gradient is maintained constant by supplying the component to the region of high concentration and removing it from the region of low concentration, then the mass transfer process is said to be continuous.

6.1 Criteria for the Classification of Mass Transfer Phenomena

6.1.1 Phases Involved in Mass Transfer

> Mass transfer of a component may take place within a single phase or across multiple phases. In most of the food engineering operations, mass transfer involves at least one fluid phase (gas or liquid), and in certain cases, a solid phase is also involved.

Migration of a component in a mixture can occur within the same phase to an interface between two similar phases in contact. Extraction is one such mass transport operation occurring within the same phase, in which a liquid solute is transferred from a first liquid (solution of the solute in solvent-1) to a second liquid (solvent-2 selective to solute of interest). Contrarily, the component in the first phase can penetrate the interface and then transfer into the bulk of a second phase. The above mentioned

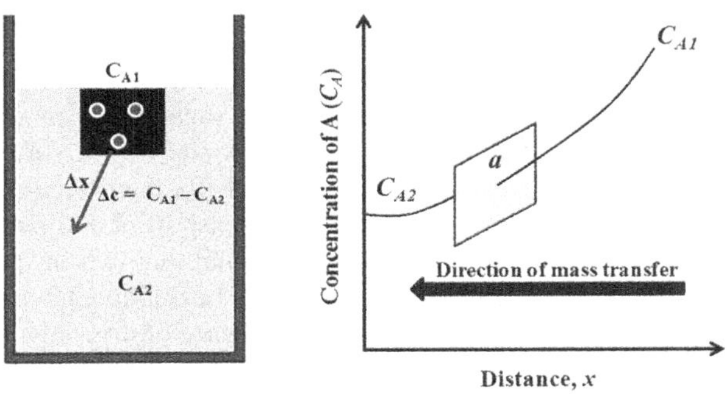

FIGURE 6.1 Schematic representations of the concentration gradient and direction of mass transfer.

TABLE 6.1

Applications of Mass Transfer in Food Processing

Mass Transfer Operation	Phases Involved in the Transport Process	Example
Humidification	Transfer of water from liquid to the gaseous phase.	Humidification of air is carried out in the food industry to ensure quality and freshness in food production and to protect the food from drying, shrinkage, and weight loss.
Crystallization	Transport of liquid solute from liquid to solid phase.	Production of sugar from sugar cane and sugar beet.
Distillation	Separation of two liquids by converting one of the liquids into the vapor phase.	Separation of volatile flavors; industrial production of alcoholic beverages from fruits and grains.
Leaching	Migration of solid solute from solid to liquid phase.	Transfer of soluble coffee solids from roasted coffee grounds by pressurized hot water during the soluble coffee manufacturing process.
Fermentation	Transfer of gaseous component to the liquid phase.	Transfer of oxygen in the gaseous phase to the liquid culture medium.
Membrane processing	Separation of one liquid from the other by selective permeability of membranes.	Removal of water during the concentration of fruit juices.
Absorption	Removal of one or more solutes from the gas phase by contacting with a solvent (liquid phase) which selectively dissolves the components of interest.	Carbonation of beverages.
Adsorption	Adhesion of gas or liquid particles onto a solid or liquid surface termed as the adsorbent.	During processing of vegetable oils, color compounds and minor impurities are removed by selective adsorption of these components on to solid adsorbents such as activated carbon and fuller's earth.

phenomenon happens during drying in which water is removed in gaseous (vapor) form from either a liquid or a solid food into a warm stream of gas (air). Other examples of single- and two-phase mass transfer in food processing are compiled in Table 6.1.

6.1.2 Modes of Mass Transfer

There are two different modes of mass transport: *diffusive* and *convective*, similar to the conductive and convective heat transfer, respectively. Often, these two modes of mass transfer co-occur. However, one can dominate over the other, thus, approximating the solutions to only the dominating mode involved.

6.1.2.1 Diffusive Mass Transfer

When sugar crystals or cubes are placed at the bottom of a tall glass tumbler filled with water, the sugar will slowly disperse and dissolve in the water. Initially, the sugar cubes will remain concentrated in the bottom of the tumbler. Subsequently, the sugar particles would dissolve from the surface of each crystal into the neighboring layer of water (liquid). After a time t, the sugar particles will penetrate upward by a few centimeters. As time proceeds further, the sugar would completely dissolve in water and the solution will appear homogeneous. The process which is responsible for the movement of the sugar particles is termed as *diffusion*. Thus, *diffusive mass transfer* is the transport of a component from a region of higher concentration to lower concentration in a mixture containing two or more molecular species, whose relative concentrations vary with position (Figure 6.2). It begins and proceeds as a natural process leading to diminished concentration gradient without the influence of internal currents of circulation or external force.

Mass transfer by diffusion occurs between two stationary layers in contact by random movement of molecules or species or particles of a component. The random molecular motion is driven by a

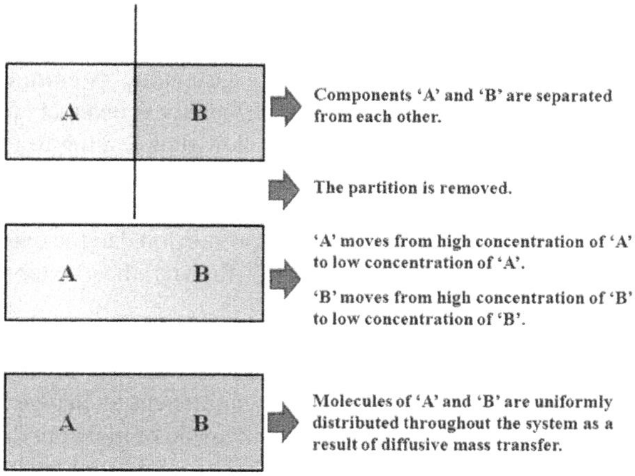

FIGURE 6.2 Diffusive mass transfer.

concentration gradient resulting from thermally induced agitation. Thus, *diffusive mass transfer* is also termed as *molecular diffusion*, *molecular mass transport,* or *mass transport on a microscopic scale.* Molecular diffusion occurs more intensively at high temperatures owing to high mean molecular velocities. Diffusion can be a slow or fast process depending on the medium in which it occurs. While diffusion proceeds at a rate of about 5 cm/min in gases, it progresses at the rate of ~0.05 cm/min in liquids. The rate of diffusion is the slowest in solids at about 0.00001 cm/min (Cussler, 2007).

Consider a binary (two component) system in which components A and B are separated by a membrane (Figure 6.2). On removing the membrane, molecules of components A and B will move across the membrane to attain an equilibrium concentration throughout the system (Figure 6.2). *Diffusive mass transfer* is a slow process leading to complete homogenization of the mixture (Figure 6.2) after a time when the concentrations of the two components become uniform throughout the entire system, and their concentration gradients disappear. Diffusion is the predominant mode of mass transfer in a stationary solid or fluid.

6.1.2.1.1 Fick's Law

The diffusive mass transfer is governed by the *Fick's law of diffusion* which states that *mass flux per unit area of a component is proportional to its concentration gradient in a mixture.* While the Fick's first law explains the steady-state diffusive mass transfer, unsteady-state mass transfer is described by the Fick's second law.

6.1.2.1.1.1 Fick's First Law In a mixture of A and B as shown in Figure 6.2, the transfer rate of component A is determined by the diffusion of A. Thus, Fick's law for the diffusion of component A is given by

$$J_A = -D_{AB}\frac{dC_A}{dx} \tag{6.1}$$

where J_A is the molar flux of A which has units of amount of component diffused per unit area per unit time (kmol/m^{2}s), $\frac{dC_A}{dx}$ is the concentration gradient (kmol/m^3) in the x-direction (direction of flow), and D_{AB} is the constant of proportionality, known as the binary diffusion coefficient or mass diffusivity of A in B (m^2/s). The negative sign is to convert the flow in the direction of reducing concentration gradient into a positive quantity. This is justified as $\frac{dC_A}{dx}$ is a negative quantity since concentration decreases in the flow direction. The mass diffusivity is a measure of the rate at which a component diffuses through an area in a given medium. The value of the diffusion coefficient for a component depends on the concentration and the medium. Further, the diffusion rate (D) depends directly on temperature and molecular spacing. As a result, diffusion in gases proceeds at a faster rate (5 cm/min) compared to liquids (0.05 cm/min) and solids (10^{-5} cm/min). In gases, mass diffusivity is inversely proportional to pressure. At low pressure, the concentration of molecules would be low thus leading to fewer collisions between them. A larger number of collisions per unit time hinder the molecular motion. This reduces the mean free path which is the distance traveled by a molecule in the gaseous phase between two subsequent collisions. On the other hand, diffusivity in liquids is inversely proportional to the concentration and viscosity of the solution. Also, from Eq. (6.1), the higher the mass diffusivity, the greater the mass transfer flux.

6.1.2.1.1.2 Fick's Second Law Fick's first law provides an expression for the flux of mass transfer by diffusion. However, an additional equation is required to predict the unsteady-state diffusive mass transfer, wherein diffusion causes the concentration of a component to change with time. Examples of unsteady-state mass transfer in food applications include diffusion of moisture during continuous drying of foods, moisture uptake by a dry powdered food due to diffusion of water vapor within the food matrix during storage, migration of oils and volatile flavors through packaging material, and diffusion of salt within a food matrix. The equation for unsteady-state mass transfer is given by the Fick's second law, which is derived by considering a control volume (CV) of the system as depicted in Figure 6.3.

In the CV, the time-dependent change in mass m of the diffusing component is given by the law of conservation of mass Eq. (6.2).

$$\frac{\partial m}{\partial t} = \sum m_{\text{in}} - \sum m_{\text{out}} \tag{6.2}$$

The flux of diffusive mass transfer in and out of the CV is given by the Fick's first law, which for the x-direction is given by

$$J_{x.\text{in}} = -D\left.\frac{\partial C}{\partial x}\right|_1 \tag{6.3}$$

$$J_{x.\text{out}} = -D\left.\frac{\partial C}{\partial x}\right|_2 \tag{6.4}$$

where the subscripts 1 and 2 denote the inflow and outflow faces in Figure 6.3. Therefore, to obtain the total mass flux m, the diffusive flux in x-direction (J_x) is multiplied by the surface area of CV, which is $\delta y \delta z$. Thus, the net flux in the x-direction can be written as

$$\partial m\big|_x = -D\delta y\delta z\left[\left.\frac{\partial C}{\partial x}\right|_1 - \left.\frac{\partial C}{\partial x}\right|_2\right] \tag{6.5}$$

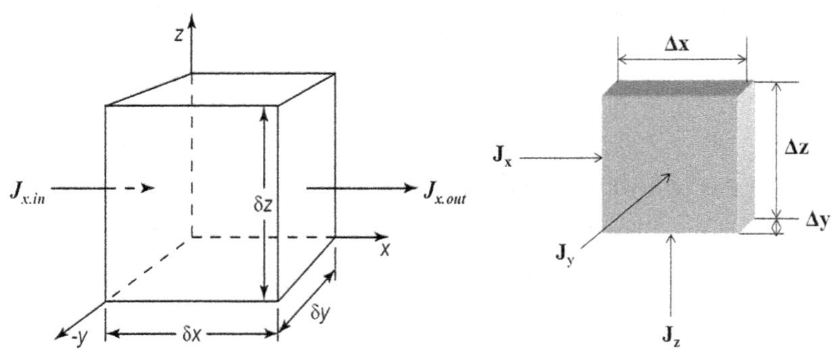

FIGURE 6.3 Schematic of a system to derive the expression for Fick's second law.

Further, the linear Taylor series expansion is used to evaluate $\partial C/\partial x$ at point 2. The general form of Taylor series expansion is given by

$$f(x) = f(x_0) + \left.\frac{\partial f}{\partial x}\right|_{x0} \delta x \tag{6.6}$$

$$\therefore \left.\frac{\partial C}{\partial x}\right|_2 = \left.\frac{\partial C}{\partial x}\right|_1 + \frac{\partial}{\partial x}\left(\left.\frac{\partial C}{\partial x}\right|_1\right)\delta x \tag{6.7}$$

Substituting Eq. (6.7) in Eq. (6.5) and dropping the subscript 1 gives

$$\partial m|_x = -D\delta y\delta z\left[\frac{\partial^2 C}{\partial x^2}\right]\delta x \tag{6.8}$$

Similarly, the net fluxes through the CV in the y- and z- directions are

$$\partial m|_y = -D\delta x\delta z\left[\frac{\partial^2 C}{\partial y^2}\right]\delta y \tag{6.9}$$

$$\partial m|_z = -D\delta x\delta y\left[\frac{\partial^2 C}{\partial z^2}\right]\delta z \tag{6.10}$$

Before substituting the Eqs. (6.8)–(6.10) in Eq. (6.2), the mass m in Eq. (6.2) should be converted to concentration as

$$m = C\delta x\delta y\delta z \tag{6.11}$$

After substituting the concentration C and net fluxes in Eq. (6.2), the three-dimensional diffusion equation can be obtained, which is given by Eq. (6.12).

$$\frac{\partial C}{\partial t} = D\left(\frac{\partial^2 C}{\partial x^2} + \frac{\partial^2 C}{\partial y^2} + \frac{\partial^2 C}{\partial z^2}\right) = D\nabla^2 C \tag{6.12}$$

where C is the mass or molar concentration of the diffusing component within the food (mol/m^3); D is the diffusion coefficient (m^2/s); x, y, and z are the distances in the direction of diffusion (m) and t is time (s). The solution of Eq. (6.12) gives the variation in concentration distribution within the food with respect to time.

6.1.2.1.1.3 Application of Fick's Law in Reducing the Cooking Time of Foods An insight into the practical application of Fick's first law can be obtained by considering the example of preparing a flavorful soup stock. The surface area of meat or vegetable pieces used for making the soup plays an essential role in reducing the cooking time of this diffusive process. Conventionally, preparation of soup stock involves simmering the meat and vegetables for many hours to achieve maximal extraction of flavor. For instance, a popular recipe for the preparation of soup stock prescribes 6–8 h of uncovered simmering for 4 lb of chicken carcass pieces in a large water-filled stockpot along with one-fourth of onion and some halved carrots, celery, and leek (Brown, 2018). However, it was found that an equally flavorful stock can be obtained in 2 h by the diminution of the vegetables to a very fine dice (Myhrvold et al., 2011). Fick's law can effectively explain the reason for the reduction in cooking time after the dicing of vegetables. Reducing the size of vegetable pieces increases the area of the vegetable–water interface. Consequently, a greater number of molecules can diffuse into the stock per unit time as the larger interface and smaller dice effectively increases the concentration gradient (dC/dx; Eq. (6.1)). The flux for mass transfer (J) also

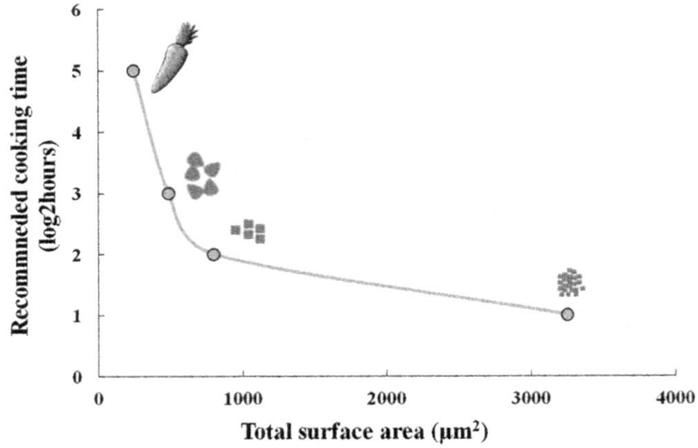

FIGURE 6.4 Schematic showing cooking time as a function of carrot piece size. (Redrawn using data from Zhou, L., Nyberg, K. and Rowat, A. C. 2015. Understanding diffusion theory and Fick's law through food and cooking. *Advances in Physiology Education* 39: 192–197.)

increases to achieve the same level of flavor in the stock within a shorter period. Figure 6.4 depicts over tenfold increase in total surface area by chopping a carrot of a given volume into increasingly smaller pieces and the resultant reduction in cooking time (Zhou et al., 2015).

6.1.2.2 Convective Mass Transfer

In contrast to the diffusive mass transfer, the constituents of a mixture can be transferred between a stationary surface (solid or liquid) and a moving fluid or between two relatively immiscible moving fluids, aided by dynamic characteristics of the flow. This bulk migration of components is said to be *convective mass transfer.*

Based on the dynamic flow characteristics in the moving fluid, the convective mass transfer can be categorized into two types:

1. Natural convection: Natural convection currents develop due to variation in density within the fluid phase resulting from temperature differences or relatively larger concentration differences.
2. Forced convection: Forced convection occurs under the influence of external forces such as the turbulent motion or eddies caused by mechanical agitation or pressure difference created by the use of pumps and compressors.

Facilitating a faster dissolution of sugar in milk by stirring with a spoon is a well-known example of mass transfer by forced convection, which we come across in daily life.

6.1.2.2.1 The Rate Equation for Convective Mass Transfer

The rate of mass transfer is directly proportional to the driving force for transfer and the area available for the transport process to occur. Thus,

$$\text{Transfer rate} \propto \text{transfer area} \times \text{driving force} \tag{6.13}$$

The coefficient of proportionality in Eq. (6.13) is known as the mass transfer coefficient. Hence,

$$\text{Transfer rate} = \text{mass transfer coefficient} \times \text{transfer area} \times \text{driving force} \tag{6.14}$$

Expressing Eq. (6.14) in mathematical form results in the following rate equation for forced or natural convective mass transfer, which is analogous to Newton's law of cooling.

$$J_A = k_m\left(C_{Ai} - C_A\right) \tag{6.15}$$

where J_A is the molar-mass flux of component A (kmol/m^{2}s), k_m is the convective mass transfer coefficient (m/s), which defines the volume of component A transferred across a boundary of unit area per second, C_{Ai} is the concentration of component A at the interface, and C_A is the concentration at a randomly defined location in the moving fluid or bulk phase. Thus, it is evident that the rate of mass transfer depends upon the concentration gradient of the component, physical properties of the phases, interfacial area, and turbulence in the fluid phases. Therefore, the mass transfer rate can be enhanced by increasing the interfacial area or turbulence in the fluid.

6.1.2.2.2 Convective Mass Transfer Coefficient

From the rate equation of convective mass transfer Eq. (6.15), k_m can be defined as the rate of mass transfer per unit concentration difference per unit area. It indicates how fast the mass transfer occurs by convection. If c is the concentration of the component x (kmol/m^3), $\dot{m}$ is the mass flux (mol/s), and A is the area (m^2), k_m can be expressed as

$$k_m = \frac{\dot{m}}{A\left(c_{x1} - c_{x2}\right)} \tag{6.16}$$

k_m varies with the system geometry, fluid properties, and the dynamic characteristics of the flowing fluid. The subsequent sections explain the predominantly used methods for the calculation of the mass transfer coefficient.

6.1.2.2.2.1 Determination of Mass Transfer Coefficient by Dimensional Analysis A prior understanding on the various dimensionless numbers in mass transfer is essential to conduct the dimensional analysis for the calculation of mass transfer coefficient.

6.1.2.2.2.1.1 Dimensionless Numbers in Mass Transfer

i. **Sherwood number (N_{Sh})**

Sherwood number represents the concentration gradient at the surface. It is analogous to the Nusselt number (N_{Nu}) of heat transfer. Sherwood number is the ratio of convective mass transfer to diffusive mass transfer. The convective mass transfer is represented by the mass transfer coefficient (k_m), and molecular diffusion is represented by mass diffusivity (D_{AB}). The expression of the Sherwood number is given by Eq. (6.18).

$$N_{Sh} = \frac{\text{Total mass transferred by convection}}{\text{Total mass transferred by molecular diffusion}} \tag{6.17}$$

$$N_{Sh} = \frac{k_m d}{D_{AB}} \tag{6.18}$$

Where d is a characteristic length (m).

ii. **Schmidt number (N_{Sc})**

Schmidt number is the ratio of kinematic viscosity ($\mu/\rho = \eta$) to diffusivity. This provides the link between momentum and mass transfer, similar to the Prandtl number (N_{Pr}) of heat transfer which provides the link between the heat and momentum transfer. Equation 6.20 gives the expression for the Schmidt number.

$$N_{Sc} = \frac{\text{Molecular diffusion of momentum}}{\text{Molecular diffusion of mass}} \tag{6.19}$$

$$N_{Sc} = \frac{\mu}{\rho D_{AB}} \tag{6.20}$$

iii. **Lewis number (N_{Le})**

Lewis number is the ratio between thermal and mass diffusivities, and it is expressed as,

$$N_{Le} = \frac{\text{Thermal diffusivity}}{\text{Mass diffusivity}} \tag{6.21}$$

$$N_{Le} = \frac{\alpha}{D_{AB}} \tag{6.22}$$

iv. **Biot number (N_{Bim})**

Biot number for mass transfer is given by the expression

$$N_{Bim} = \frac{k_m L}{D_{AB}} \tag{6.23}$$

where L is a characteristic dimension. If $N_{Bim} < 0.1$, internal resistance to mass transfer is negligible, and mass transfer is influenced by the convective mass transfer coefficient at the surface. If $N_{Bim} > 100$, external resistance to mass transfer is negligible and mass transfer turns dependent on diffusivity.

v. **Fourier number (N_{Fom})**

Fourier number for mass transfer is given by the expression

$$N_{Fom} = \frac{D_{AB} t}{L^2} \tag{6.24}$$

Unsteady-state mass transfer relationship between Biot and Fourier numbers is given by

$$\frac{C_A - C_{A\infty}}{C_{A0} - C_{A\infty}} = f(N_{Bim}, N_{Fom}) \tag{6.25}$$

where C_{A0} is the initial concentration of component A, $C_{A\infty}$ is the concentration of component A at the surface, and C_A is the concentration of component A at any time.

6.1.2.2.2.1.2 Dimensional Analysis

Determination of convective heat transfer coefficient by dimensional analysis was explained in Chapter 1. Similarly, mass transfer coefficient can also be determined using a combination of dimensionless numbers explained in the previous section. The stepwise procedure of dimensional analysis for the calculation of the mass transfer coefficient is explained as follows:

Step 1: Identification of relevant variables

- Velocity of the fluid (v)
- Density of the fluid (ρ)
- Fluid viscosity (μ)
- Characteristic dimension (d)
- Mass diffusivity (D_{AB})
- Mass transfer coefficient (k_m)

In the case of natural convection, density difference ($\Delta\rho$) and acceleration due to gravity (g) can also affect the mass transfer.

Step 2: Listing the dimensions of problem variables and parameters

Quantity	Dimensions
v	LT^{-1}
ρ	ML^{-3}
μ	$ML^{-1}T^{-1}$
d	L
D_{AB}	L^2T^{-1}
k_m	LT^{-1}

Step 3: Number of independent dimensions and dimensionless groups

Number of relevant variables $(n) = 6$

Number of independent dimensions $(m) = 3$ (M, L, T)

Number of dimensionless groups $= n - m = 6 - 3 = 3$

Step 4: Choice of core variables

Number of core variables = number of dimensions $(m) = 3$

Applying the rules explained in the Section 1.5.4, the core variables in this case would be as follows:

$$\{\Pi_1 = d; \Pi_2 = v \text{ and } \Pi_3 = \rho\}$$

Step 5: Solving the dimensional equations for the dimensions (L, M, and T) in terms of the core variables

$$\mathrm{L} = d$$

$$\mathrm{T} = d/v$$

$$\mathrm{M} = \rho d^3$$

Step 6: Solving the dimensional equations for the dimensions (L, M, and T) in terms of the core variables

$$\mu = \frac{\mathrm{M}}{\mathrm{LT}} = \frac{\rho d^3}{d\left(\dfrac{d}{v}\right)} = \rho v d \tag{6.26}$$

$$D_{AB} = \frac{\mathrm{L}^2}{\mathrm{T}} = \frac{d^2}{d/v} = vd \tag{6.27}$$

$$k_m = \frac{\mathrm{L}}{\mathrm{T}} = \frac{d}{d/v} = v \tag{6.28}$$

Dividing Eq. (6.26) by Eq. (6.27) and Eq. (6.28) by Eq. (6.27)

$$\frac{\mu}{D_{AB}} = \rho \tag{6.29}$$

$$\frac{k_m}{D_{AB}} = \frac{1}{d} \tag{6.30}$$

The resulting equations are each a dimensional identity, so dividing one side by the other results in one dimensionless group from each equation. Therefore, from Eqs. (6.26), (6.29), and (6.30),

$$N_1 = \frac{\rho v d}{\mu} = N_{Re} \tag{6.31}$$

$$N_2 = \frac{\mu}{\rho D_{AB}} = N_{Sc} \tag{6.32}$$

$$N_3 = \frac{k_m d}{D_{AB}} = N_{Sh} \tag{6.33}$$

These dimensionless groups can be used as the primary variables to define the system behavior instead of the original six variables. The dimensionless groups can also be written in the form of dimensionless equations to represent the system (Eqs. (6.34) and (6.35)):

$$Sh = f\left(N_{Re}, N_{Sc}\right) \tag{6.34}$$

$$Sh = a\left(N_{Re}^{\alpha} N_{Sc}^{\beta}\right) \tag{6.35}$$

Equation 6.35 is termed as the mass transfer correlation and the constants (a, α, and β) in the correlation are determined experimentally.

6.2 Theories of Mass Transfer

6.2.1 Two Film Theory

This theory postulated by Lewis and Whitman (1924) states that, *wherever a liquid and a gas come into contact, there exists on the gas side of the interface a layer of gas in which motion by convection is less compared to that in the main body of the gas, and that similarly on the liquid side of the interface there is a surface layer of liquid which is practically free from mixing by convection.* Thus, the two-film theory proposes the presence of stagnant films of gas and liquid of constant thickness, respectively, on the gas and liquid sides of the interface. The interfacial mass transfer occurs due to diffusion through these films. Therefore, resistance to diffusional mass transfer depends on the film properties, given by Eq. (6.36).

$$k_m = \frac{D_{AB}}{d} \tag{6.36}$$

where d is the thickness of the film. Film thickness which is not possible to be determined experimentally can be calculated from Eq. (6.36), if the values of the mass transfer coefficient and diffusivity are known. This theory postulates that the mass transfer coefficient is proportional to the diffusion coefficient and independent of the fluid velocity. Hence, diffusion can be increased by increasing the mass transfer coefficient. Thus, film theory ignores the effect of fluid velocity on mass transfer.

For a gas dissolving in a liquid, in the first stage, gas diffuses through the film on the gas side to the interface and dissolves (Figure 6.5). During this phase, the gas transfer rate (N_G) over the gas film is given by

$$N_G = k_G A\left(p_b - p_i\right) \tag{6.37}$$

where k_G is the gas film coefficient, A is the interfacial area (m^2), p_b is the partial pressure of the gas in the bulk, and p_i is the partial pressure of gas at the interface.

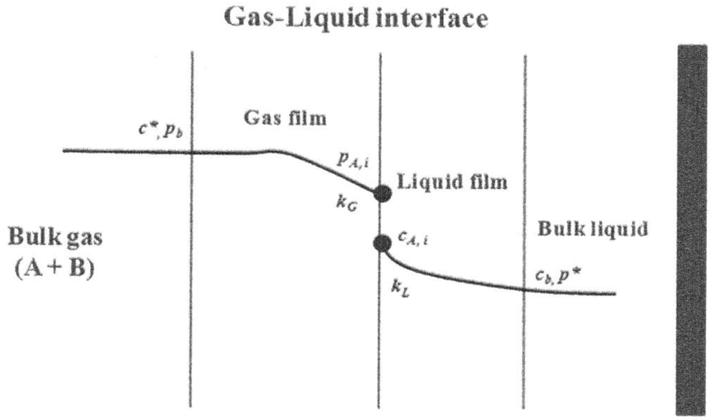

FIGURE 6.5 The concept of two film theory.

Subsequently, the dissolved gas diffuses through the liquid side film into the bulk liquid (Figure 6.5). Therefore, mass transfer rate (N_L) through the liquid film is given by

$$N_L = k_L A(c_i - c_b) \tag{6.38}$$

where k_L is the liquid film coefficient, c_i is the gas concentration at the interface, and c_b is the gas concentration in the bulk liquid.

Under steady-state condition, when $N_L = N_G$,

$$k_L A(c_i - c_b) = k_G A(p_b - p_i) \tag{6.39}$$

Therefore,

$$\frac{k_L}{k_G} = \frac{p_b - p_i}{c_i - c_b} \tag{6.40}$$

If $k_L \gg k_G$, the gas film controls the mass transfer process. In this case, partial pressure difference over the gas phase is higher than the concentration gradient on the liquid film. If $k_L \ll k_G$, liquid film controls the mass transfer process and the concentration gradient over liquid is greater than the partial pressure difference over the gas phase.

The overall mass transfer coefficient can be determined from the film coefficients. Considering a unit surface area ($A = 1\,\text{m}^2$), the overall mass transfer coefficient can be expressed as

$$N = K_G\left(p_b - p^*\right) \tag{6.41}$$

and

$$\frac{1}{K_G} = \frac{\left(p_b - p^*\right)}{N} \tag{6.42}$$

However, $\left(p_b - p^*\right)$ can be written as $p_b - p_i + p_i - p^*$.

Therefore,

$$\frac{1}{K_G} = \frac{(p_b - p_i)}{N} + \frac{\left(p_i - p^*\right)}{N} \tag{6.43}$$

According to Henry's law (refer Section 6.3), $(p_i - p^*) = (c_i - c_b)/H$ then,

$$\frac{1}{K_G} = \frac{1}{k_G} + \frac{(c_i - c_b)}{HN} \tag{6.44}$$

Thus,

$$\frac{1}{K_G} = \frac{1}{k_G} + \frac{1}{Hk_L} \tag{6.45}$$

and

$$\frac{1}{K_L} = \frac{1}{Hk_G} + \frac{1}{k_L} \tag{6.46}$$

6.2.2 Penetration Theory

Penetration theory postulates that, *diffusion is an unsteady state process, and that the molecules of solute are in constant random motion, where clusters of these molecules arrive at the interface, remain there for a fixed period, and some of them penetrate while the rest mixes back into the bulk of the phase.* The process is described by Fick's second law under the initial boundary conditions at $t = 0$ and $C_A = C_{Ai}$, given by Eq. (6.47).

$$\frac{\partial C_A}{\partial t} = D_{AB}\left(\frac{\partial^2 C}{\partial y^2}\right) \tag{6.47}$$

The boundary conditions are given by

$$\left\{t = 0,\ y > 0 : C_A = C_{Ab} \text{ and } t > 0,\ \ y = 0 : C_A = C_{Ai}\right\}$$

where C_{Ab} is the concentration of solute at an infinite distance from the surface and C_{Ai} is the solute concentration at the surface. The solution of the partial differential equation given in Eq. (6.47) for the previously stated boundary conditions is

$$\frac{C_{Ai} - C_A}{C_{Ai} - C_{Ab}} = \text{erf}\left(\frac{y}{2\sqrt{D_{AB}t}}\right) \tag{6.48}$$

where erf(x) is the error function defined as

$$\text{erf}(x) = \frac{2}{\sqrt{\pi}}\int_0^x \exp(-y^2)\,dy \tag{6.49}$$

If the mass transfer occurs by unidirectional diffusion, and if $C_{Ab} \sim 0$ the mass flux of component A, J_A (kg/m^{2}s) can be calculated from the following equation:

$$J_A = \frac{-\rho D_{AB}}{1 - C_{Ab}}\left(\frac{\partial c}{\partial y}\right)_{y=0} \approx -\rho\left(\frac{\partial c}{\partial y}\right)_{y=0} \tag{6.50}$$

On substituting Eq. (6.48) in Eq. (6.50), the rate of mass transfer at the time, t would be,

$$J_A(t) = \rho\sqrt{\frac{D_{AB}}{\pi t}}(C_{Ai} - C_{Ab}) \tag{6.51}$$

Hence, the mass transfer coefficient is given by

$$k_m = \left(\frac{D_{AB}}{\pi t} \right)^{1/2} \tag{6.52}$$

Thus, according to the penetration theory, mass transfer coefficient is proportional to the square root of diffusivity. However, mass transfer coefficient cannot be calculated due to the experimental limitations in determining the exact contact time.

6.2.3 Surface Renewal Theory

In the year 1951, Danckwerts proposed the surface renewal theory to provide a clear insight into the mass transfer. This concept is an extension of the penetration theory. According to the surface renewal theory, mass transfer occurs in the interfacial region and elements in this region are constantly exchanged with the new element from the bulk region. In other words, a portion of the mass transfer surface is constantly renewed with a new surface by the movement of eddies next to the surface. This theory also suggests that the mass transfer coefficient is proportional to the square root of diffusivity. It expresses the mass transfer coefficient as

$$k_m = \sqrt{D_{AB} s} \tag{6.53}$$

where s is the fractional rate of the surface renewal with the unit per second.

6.3 Laws of Mass Transfer

6.3.1 Raoult's Law

Raoult's law states that the partial vapor pressure of each component of an ideal mixture of liquids is equal to the vapor pressure of the pure component multiplied by its mole fraction in the mixture. The following equation explains the Raoult's law

$$p_A = p'_A x_A \tag{6.54}$$

where p_A is the partial pressure of component A, p'_A is the pure component vapor pressure of A, and x_A is the mole fraction of component A in the liquid phase.

6.3.2 Henry's Law

Henry's law provides a quantitative relationship between the pressure and solubility of a gas in a solvent, wherein, the solubility of a gas is defined as the concentration of the dissolved gas in equilibrium with the substance in the gaseous state. Under conditions of equilibrium, the rate at which the solute gas molecules escape the solution and enter the gas phase is equal to the rate at which gas molecules reenter the solution. Henry's law states that *at a constant temperature, the amount of a given gas that dissolves in a specific volume and type of a liquid is directly proportional to the partial pressure of that gas in equilibrium with that liquid.* In simple terms, the amount of gas that can be solubilized in a liquid is directly dependent on the partial pressure exerted by that gas on the liquid at a constant temperature. Henry's law is given by the expression

$$c_A = Hp_A \tag{6.55}$$

where p_A is the partial pressure of component A in the vapor phase, c_A is the concentration of component A in the liquid phase, usually expressed in molarity (M), and H is Henry's law constant or solubility.

To facilitate easy calculation, Henry's law may be written in different forms such as $H = p/c$ (or) $H = c/p$ (or) $H = p/x$, where, x is the mole fraction of the gas in solution. Respectively, the unit of H would be $L_{solution}$ atm/mol_{gas}, $mol_{gas}/L_{solution}$ atm, or atm $mol_{solution}/mol_{gas}$.

The Henry's law constant is unique for each solute-and-solvent pair. It is typically measured in moles per liter atmosphere (mol/L atm). At 25°C, Henry's law constant for carbon dioxide, nitrogen, and oxygen is 3.4×10^{-2}, 6.1×10^{-4}, and 1.3×10^{-3} mol/L atm, respectively (Sander, 2015). Therefore, at a given pressure and temperature, the solubility of CO_2 in water is 56 times more than that of nitrogen (N_2) and 26 times more than that of oxygen (O_2). According to Eq. (6.55), at constant pressure and temperature of 1 atm and 25°C, respectively, the amount of CO_2 that dissolves in 1 L of water is 1.496 g. Alternatively, if the carbonation is done under pressure at 5 atm, 7.480 g of CO_2 can be dissolved in 1 L of water at the same temperature. This shows that a fivefold increase in pressure leads to the dissolution of five times more carbon dioxide in water. An increase in pressure increases the frequency of collisions between the gas particles within the surface of the solution. Consequently, more particles of the solute dissolve in the liquid.

6.3.2.1 Applications of Henry's Law

Henry's law finds practical applications in many food processing operations. The most common application is realized in the production of carbonated soft drinks, which would be elaborated in this section.

6.3.2.1.1 Carbonation of Soft Drinks

Carbonation is the process of dissolving carbon dioxide into a solution of water under pressure. Carbonated soft drinks are packaged in bottles or cans under high pressure in a chamber filled with carbon dioxide gas. In a carbonated beverage industry, Henry's law can be applied to determine the pressure level to attain the required level of carbonation, if the required amount of CO_2 (total number of moles) to be incorporated in the soft drink is known and vice versa. Example 6.1 explains the practical application of Henry's law in beverage processing.

Before the bottle of carbonated drink is opened, nearly all of the gas above the drink is almost pure carbon dioxide at a pressure slightly higher than atmospheric pressure. When the bottle or can is opened to the air, the gas escapes as the partial pressure of carbon dioxide above the drink drops. This decreases the solubility and the concentration of CO_2 in the drink, thus causing the gas bubbles to pop out on opening the bottle. This is the reason behind the fizzy nature of carbonated beverages. However, the can or the bottle containing carbonated beverage does not explode on opening because the surface tension of the liquid prevents the formation and expansion of CO_2 bubbles.

Example 6.1

A carbonated soft drink is set to contain 4.9 g of carbon dioxide dissolved in each liter of the drink.

a. Calculate the pressure under which the carbonation should be done at 25°C.
b. At the same temperature, if the pressure applied in the bottling process is 5 atm, how many grams of carbon dioxide can be dissolved in a 1 L bottle of carbonated water?
c. What is the solubility of CO_2 after the bottle is opened if its partial pressure is 4×10^{-4} atm?

Solution

a. Given

- Concentration of CO_2 in solution = 4.9 g/L = 4.9/44 = 0.1 mol/L (since, 1 mol of CO_2 = 44 g).
- Henry's law constant for CO_2 at 25°C is 3.4×10^{-2} mol/L atm

Thus, from Henry's law expression,

$$p_{CO_2} = \frac{c_{CO_2}}{H_{CO_2}} = \frac{0.1}{3.4 \times 10^{-2}} = 2.9 \text{ atm}$$

∴ Carbonation should be done at a pressure of 2.9 atm to achieve a CO_2 concentration of 4.9 g in every 1 L bottle of the carbonated water.

b. Given: Pressure applied in the bottling process (p) = 5 atm
Thus, from Henry's law expression,

$$c_{CO_2} = H_{CO_2} \times p_{CO_2} = 3.4 \times 10^{-2} \times 5 = 0.17 \text{ mol/L} = 7.5 \text{ g/L}$$

∴ 7.5 g of CO_2 can be dissolved in every 1 L bottle of carbonated water.

c. Given: Partial pressure of CO_2 after opening the bottle = 4×10^{-4} atm
Thus, from Henry's law expression,

$$c_{CO_2} = H_{CO_2} \cdot p_{CO_2} = 3.4 \times 10^{-2} \times 4 \times 10^{-4} = 13.6 \times 10^{-6} \text{ mol/L} = 598.4 \times 10^{-6} \text{ g/L} = 0.598 \text{ mg/L}$$

∴ Solubility of CO_2 after opening the bottle is 0.598 mg/L

6.4 Analogies between Heat, Mass, and Momentum Transfer

Mass transfer is analogous to heat and momentum transfer in many aspects. The concentration gradient of mass transfer is comparable to the temperature gradient and velocity gradient of heat transfer and momentum transfer, respectively. Flux, gradient, transfer coefficient, diffusivity, boundary layer, and resistance are the terms in common to transport processes. The analogies between heat, mass, and momentum transfer compiled in Tables 6.2–6.9 are very useful as by applying them it is possible to determine the mass transfer coefficient from the parameters of the other transport processes.

Based on the dimensional similarity, mass diffusivity (L^2T^{-1}) can be considered analogous to the thermal diffusivity term (k) of the Fourier's law for heat transfer by conduction and the momentum diffusivity.

TABLE 6.2

Mathematical Laws Governing Heat, Mass, and Momentum Transfer

Transfer Process	Governing Equation	Expression
Momentum	Equation of motion (Navier–Stokes equation)	$u_x \frac{\partial u}{\partial x} + u_y \frac{\partial u}{\partial y} = \nu\left(\frac{\partial^2 u}{\partial y^2}\right)$ $\nu = \text{kinematic viscosity} = \frac{\mu}{\rho} = \frac{\text{dynamic viscosity}}{\text{density}}$
Heat	Fourier law of unsteady-state heat transfer	$u_x \frac{\partial T}{\partial x} + u_y \frac{\partial T}{\partial y} = \alpha\left(\frac{\partial^2 T}{\partial y^2}\right)$ $\alpha = \text{Thermal diffusivity} = \frac{k}{\rho C_P} = \frac{\text{Thermal conductivity}}{\text{Density} \times \text{specific heat capacity}}$
Mass	Equation of continuity (under non-stationary conditions in all the three directions)	$u_x \frac{\partial C_A}{\partial x} + u_y \frac{\partial C_A}{\partial y} = D\left(\frac{\partial^2 C_A}{\partial y^2}\right)$

TABLE 6.3

Flux of the Transport Process

Transfer Process	Expression for Flux
Momentum	τ = Shear stress = $-v\dfrac{\partial(\rho u)}{\partial y}$, where u is the local velocity
Heat	$\dfrac{\alpha}{A}=\dfrac{\partial(\rho C_P T)}{\partial y}$, where ρ and C_P are the density and specific heat capacity, respectively, at temperature T.
Mass	$N_A=-D\left(\dfrac{\partial C_A}{\partial y^2}\right)$

TABLE 6.4

Analogous Quantities in Momentum, Heat, and Mass Transfer

Transfer Process	Transfer Properties	Flux	Diffusivity	Gradient
Momentum	ρ, u	τ	v	$\dfrac{\partial u}{\partial y}$
Heat	ρ, C_P, T	$\dfrac{\alpha}{A}$	α	$\dfrac{\partial T}{\partial y}$
Mass	C_A	N_A	D_A	$\dfrac{\partial C_A}{\partial y}$

TABLE 6.5

Prandtl, Reynolds, and Schmidt Number for Heat, Mass, and Momentum Transfer

Transfer Process	Dimensionless Number	Expression
Momentum	Reynolds number	$N_{Re}=\dfrac{D\rho u}{\mu}$
Heat	Prandtl number	$N_{Pr}=\dfrac{C_P\mu}{k}$
Mass	Schmidt number	$N_{Sc}=\dfrac{\mu}{\rho D_{AB}}$

TABLE 6.6

Nusselt and Sherwood Numbers for Heat and Mass Transfer

Transfer Process	Dimensionless Number	Expression
Heat	Nusselt number	$N_{Nu}=\dfrac{hd}{k}$
Mass	Sherwood number	$N_{Sh}=\dfrac{k_m d}{D_{AB}}$

TABLE 6.7

Stanton Number for Heat and Mass Transfer

Transfer Process	Expression for the Stanton Number
Heat	$N_{St\,H} = \dfrac{N_{Nu}}{N_{Re}N_{Pr}} = \dfrac{\frac{hd}{k}}{\frac{du\rho}{\mu}\frac{C_P\mu}{k}} = \dfrac{h}{u\rho C_P}$
Mass	$N_{St\,M} = \dfrac{N_{Sh}}{N_{Re}N_{Sc}} = \dfrac{\frac{k_m d}{D_{AB}}}{\frac{du\rho}{\mu}\frac{\mu}{\rho D_{AB}}} = \dfrac{k_m}{u}$

TABLE 6.8

Analogy for Experimental Determination of Heat and Mass Transfer Coefficients

Transfer Process	Governing Equation	Expression for Transfer Coefficient
Heat	Dittus–Boelter equation for forced convective heat transfer in a pipe under turbulent flow conditions	$N_{Nu} = 0.023(N_{Re})^{0.8}(N_{Pr})^{0.33}$ $\dfrac{hd}{k} = 0.023\left(\dfrac{du\rho}{\mu}\right)^{0.8}\left(\dfrac{C_P\mu}{k}\right)^{0.33}$
Mass	Gilliland equation for the flow inside the pipe for turbulent flow conditions	$N_{Sh} = 0.023(N_{Re})^{0.8}(N_{Sc})^{0.33}$ $\dfrac{k_m d}{D_{AB}} = 0.023\left(\dfrac{du\rho}{\mu}\right)^{0.8}\left(\dfrac{\mu}{\rho D_{AB}}\right)^{0.33}$

TABLE 6.9

Ranz–Marshall Equation: Heat and Mass Transfer Coefficients for Spherical Particles

Transfer Process	Expression for the Ranz–Marshall Equation
Heat	$N_{Nu} = \dfrac{hd}{k} = 2 + 0.6(N_{Re})^{0.5}(N_{Pr})^{0.33}$
Mass	$N_{Sh} = \dfrac{k_m d}{D_{AB}} = 2 + 0.6(N_{Re})^{0.5}(N_{Sc})^{0.33}$

The following example problem explains the application of analogies between the transfer processes in the estimation of the mass transfer coefficient.

Example 6.2

Use the Ranz–Marshall correlation to estimate the film gas mass transfer coefficient for the evaporation of water from atomized spray droplets, 100 μm in diameter, falling in the air at 77°C with a relative velocity between the air and the droplets of 0.4 m/s. (Viscosity of air at 77°C is 2.075×10^{-5} Pa s; density of air at 77°C is 1.009 kg/m^3, and diffusivity of water vapor at 77°C is 3.29×10^{-5} m^2/s).

Solution

Diameter of atomized particle = 100 μm; Air temperature = 77°C; velocity of air = 0.4 m/s; Viscosity of air at 77°C = 2.075×10^{-5} Pa s; Density of air at 77°C = 1.009 kg/m^3; Diffusivity of water vapor at 77°C = 3.29×10^{-5} m^2/s.

1. Reynolds number for the droplet is given by

$$N_{Re} = \frac{\rho u d}{\mu}$$

$$N_{Re} = \frac{1.009 \times 0.4 \times 100 \times 10^{-6}}{2.075 \times 10^{-5}} = 1.945$$

2. Schmidt number is

$$N_{Sc} = \frac{\mu}{\rho D_{AB}} = \frac{2.075 \times 10^{-5}}{1.009 \times 3.29 \times 10^{-5}} = 0.625$$

3. Ranz–Marshall equation for the Sherwood number

$$N_{Sh} = 2 + 0.6 N_{Re}^{0.5} N_{Sc}^{0.33}$$

$$N_{Sh} = 2 + 0.6\left[1.945^{0.5} \times 0.625^{0.33}\right] = 2.717$$

4. Film gas mass transfer coefficient

$$k_m = \frac{N_{Sh} D_{AB}}{d} = \frac{2.717 \times 3.29 \times 10^{-5}}{100 \times 10^{-6}} = 0.894\ \text{m/s}$$

Answer: The film gas mass transfer coefficient for evaporation of water is 0.894 m/s

6.5 Problems to Practice

6.5.1 Multiple Choice Questions

1. Fick's law is given by the expression
   a. $N = -D\, dc/dx$
   b. $N = D\, dc/dx$
   c. $N = dc/dx$

Answer: a

2. The SI unit of mass diffusivity is
   a. s/m^2
   b. m^2/s
   c. $m^2 s$

Answer: b

3. The dimensionless number in mass transfer which relates velocity and temperature profiles is
   a. Fourier number
   b. Prandtl number
   c. Biot number

Answer: b

4. The velocity and concentration profiles are related by
   a. Schmidt number
   b. Reynolds number
   c. Stanton number

Answer: a

5. The mass transfer rate is independent of
   a. turbulence effect
   b. physical properties
   c. chemical properties

Answer: c

6. According to Henry's law, concentration is directly related to
   a. pressure
   b. temperature
   c. viscosity

Answer: a

7. The driving force in the Fick's law of diffusion is
   a. concentration gradient
   b. characteristics of diffusing components
   c. both (a) and (b)

Answer: a

8. Mass transfer law which is similar to Fourier's law of heat transfer is
   a. Raoult's law
   b. Fick's law
   c. Henry's law

Answer: b

9. According to the two film theory, the diffusion coefficient and the mass transfer coefficient are
   a. directly proportional to each other
   b. inversely proportional to each other
   c. not related to each other

Answer: a

10. The dimensionless number pertaining to mass transfer which relates the temperature and concentration profiles are
   a. Prandtl number
   b. Lewis number
   c. Schmidt number

Answer: b

11. Lewis number of a mixture is equal to one when
   a. the mass diffusivity is equal to momentum diffusivity
   b. the mass diffusivity is equal to thermal conductivity
   c. the mass diffusivity is equal to thermal diffusivity

Answer: c

12. Molecular diffusivity of a liquid
   a. increases with temperature
   b. decreases with temperature
   c. is independent of temperature

Answer: a

13. Sherwood number in mass transfer is analogous to the following dimensionless number of heat transfer.
    a. Graetz number
    b. Grashof number
    c. Nusselt number

Answer: c

14. Mass transfer coefficient of liquid is
    a. influenced more by temperature than that for gases
    b. influenced much less by temperature than that for gases
    c. independent of temperature

Answer: a

15. If Biot number is less than 0.1, it indicates that
    a. external resistance to mass transfer is negligible
    b. internal resistance to mass transfer is negligible
    c. internal resistance to mass transfer is high

Answer: b

16. Mass diffusivity is considered analogous to the following heat transfer term:
    a. thermal diffusivity
    b. thermal conductivity
    c. heat transfer coefficient

Answer: a

17. If $k_L \gg k_G$, the mass transfer process is controlled by
    a. liquid film
    b. gas film
    c. partial pressure gradient

Answer: b

18. The dimensionless number which relates momentum and mass transfer is
    a. Schmidt number
    b. Lewis number
    c. Sherwood number

Answer: a

19. The quantitative relationship between pressure and solubility of a gas in a solvent is given by
    a. Raoult's law
    b. Dalton's law
    c. Henry's law

Answer: c

20. The resistance to mass transfer is offered by
    a. concentration gradient
    b. distance between the components
    c. temperature

Answer: b

6.5.2 Numerical Problems

1. If CO_2 gas is bubbled through water at 298 K, how many millimoles of CO_2 gas would dissolve in 1 L of water? Assume that CO_2 exerts a partial pressure of 1.52 atm. Henry's law constant for CO_2 in water at 298 K is 1.67 kbar.

Given

i. Temperature (T) = 298 K
ii. Volume of water (V) = 1 L
iii. Partial pressure of CO_2 (p_{CO_2}) = 1.52 atm = 1.54 bar
iv. Henry's law constant for CO_2 at 298 K (H_{CO_2}) = 1.67 kbar = 1670 bar

To find: Number of millimoles of CO_2 gas that would dissolve in 1 L of water.

Solution

Solubility of a gas is related to its mole fraction in aqueous solution. Mole fraction of the gas in the solution can be calculated by applying Henry's law, given by

$$x(CO_2) = \frac{p_{CO_2}}{H_{CO_2}} = \frac{1.54}{1670} = 9.2 \times 10^{-4}$$

$$1\text{ L of water} = 1000\,\text{mL} = 1000\,\text{g} = \frac{1000}{18} = 55.5\,\text{moles of water}$$

Therefore, if n represents the number of moles of CO_2 in the solution,

$$x(CO_2) = \frac{n\,\text{mol}}{n\,\text{mol} + 55.5\,\text{mol}} = \frac{n}{55.5} = 9.2 \times 10^{-4}$$

n in the denominator is neglected as it is << 55.5.

$$\therefore n = 0.051\,\text{mol} = 51\,\text{mmol}$$

Answer: Number of millimoles of CO_2 gas that would dissolve in 1 L of water at 298 K = 51 mmol

2. Calculate the quantity of CO_2 in 2.5 L of carbonated drink when packed under a pressure of 3 atm at 298 K.

Given

i. $p_{CO_2} = 3\,\text{atm}$
ii. T = 298 K
iii. Volume of carbonated soft drink = 2.5 L

To find

a. Quantity of dissolved CO_2

Solution

According to Henry's law,

$$p_{CO_2} = Hx_{CO_2}$$

where x is the mole fraction of CO_2 in the CO_2 + water mixture.

$$p_{CO_2} = 3\,\text{atm} = 3 \times 1.01325 \times 10^5\,\text{Pa} = 3.03975 \times 10^5\,\text{Pa}$$

Henry's law constant for CO_2 at 298 K = 1.67 kbar = 1670 bar = 1.67×10^8 Pa

$$x_{CO_2} = \frac{3.03975 \times 10^5}{1.67 \times 10^8} = 0.00182$$

$$x_{CO_2} = \frac{n_{CO_2}}{n_{CO_2} + n_{H_2O}} = \frac{n_{CO_2}}{n_{H_2O}}$$

n_{CO_2} in the denominator is neglected as it is negligible in comparison to the number of moles of water.

$$2.5\text{ L of water} = 2500\text{ mL} = 2500\,\text{g} = \frac{2500}{18} = 138.89\,\text{moles of water}$$

$$\therefore n_{CO_2} = x_{CO_2} \times n_{H_2O} = 0.00182 \times 138.89 = 0.25\,\text{moles}$$

$$\therefore \text{Amount of } CO_2 \text{ in 2.5L of carbonated drink} = 0.25 \times 44 = 11\,\text{g}$$

Answer: Quantity of CO_2 dissolved in 2.5 L of carbonated drink = 11 g

3. Compute the convective mass transfer coefficient for a water vapor transport process, if the Sherwood number is given by 2.5, characteristic dimension of the product is 10 cm, and the mass diffusivity for water vapor in air is $1.5 \times 10^{-5}\,m^2/s$.

Given

i. $N_{Sh} = 2.5$
ii. Characteristic dimension of the product (d) = 10 cm = 0.1 m
iii. Mass diffusivity for water vapor (D) = $1.5 \times 10^{-5}\,m^2/s$

To find: Convective mass transfer coefficient

Solution

$$N_{Sh} = \frac{k_m d}{D}$$

$$\therefore k_m = \frac{D \times N_{Sh}}{d} = \frac{1.5 \times 10^{-5} \times 2.5}{0.1} = 3.75 \times 10^{-4}\,\text{m/s}$$

Answer: Convective mass transfer coefficient = 3.75×10^{-4} m/s

4. A spherical ball of glucose with diameter 0.3 cm is placed in a water stream (30°C) flowing at a rate of 0.2 m/s. The diffusivity of glucose in water is $0.69 \times 10^{-9}\,m^2/s$. Determine the mass transfer coefficient if the viscosity and density of water at 30°C are 798×10^{-6} kg/m s and 995.7 kg/m^3, respectively.

Given

i. Diameter of glucose ball (d) = 0.3 cm = 0.3×10^{-2} m = 0.003 m
ii. Temperature of water stream (T) = 30°C
iii. Flow rate of water stream (u) = 0.2 m/s

iv. Diffusivity of glucose in water (D) = $0.69 \times 10^{-9}\, m^2/s$
v. Viscosity of water (μ) at 30°C = 798×10^{-6} kg/m s
vi. Density of water (ρ) at 30°C = 995.7 kg/m^3

To find: Mass transfer coefficient (k_m)

Solution

$$N_{Sc} = \frac{\mu}{\rho D_{AB}}$$

$$N_{Sc} = \frac{798 \times 10^{-6}}{995.7 \times 0.69 \times 10^{-9}} = 1.161 \times 10^3$$

$$N_{Re} = \frac{Du\rho}{\mu} = \frac{0.003 \times 0.2 \times 995.7}{798 \times 10^{-6}} = 0.0749 \times 10^4$$

Ranz–Marshall equation for the Sherwood number is given by

$$N_{Sh} = 2 + 0.6 N_{Re}^{0.5} N_{Sc}^{0.33}$$

$$N_{Sh} = 2 + \left\{0.6 \times \left(0.0749 \times 10^4\right)^{0.5} \times \left(1.161 \times 10^3\right)^{0.33}\right\} = 170.575 = \frac{k_m d}{D}$$

$$\therefore k_m = \frac{170.575 \times 0.69 \times 10^{-9}}{0.3 \times 10^{-2}} = 392.32 \times 10^{-7}\ m/s$$

Answer: Mass transfer coefficient = 3.92×10^{-5} m/s

BIBLIOGRAPHY

Brown, A. 2018. Chicken stock (online). www.foodnetwork.com/recipes/alton-brown/chicken-stock-recipe/index.html (accessed July 11, 2018).

Cussler, E. L. 2007. *Diffusion Mass Transfer in Fluid Systems* (Third edition). Cambridge: Cambridge University Press.

Danckwerts, P. V. 1951. Significance of liquid-film coefficients in gas absorption. *Industrial and Engineering Chemistry* 43: 1460–1467.

Lewis, W. K. and Whitman, W. G. 1924. Principles of gas absorption. *Industrial and Engineering Chemistry* 16: 1215–1220.

Myhrvold, N., Young, C. and Bilet, M. 2011. *Modernist Cuisine: The Art and Science of Cooking*. Bellevue, WA: The Cooking Lab.

Sander, R. 2015. Compilation of Henry's law constants (version 4.0) for water as solvent. *Atmospheric Chemistry and Physics* 15: 4399–4981.

Zhou, L., Nyberg, K. and Rowat, A. C. 2015. Understanding diffusion theory and Fick's law through food and cooking. *Advances in Physiology Education* 39: 192–197.

7

Psychrometry

In the food industry, air is an inevitable transfer medium in the commonly encountered transport phenomena. Air plays a vital role as the heating, cooling, or drying medium in the unit operations such as drying and refrigeration. The air in storage areas and packaging atmosphere is also vital as it influences the shelf life of fresh produce and processed food products. Thus, gaining an understanding of the properties of air holds significance. Ideally, air is never pure. It exists as moist air which is a two-component (gas–vapor) mixture of dry air and water vapor.

Psychrometry is the *branch of science which deals with the thermodynamic properties of air-water vapor mixture.* A better understanding of psychrometry is useful in the designing of dryers, cooling towers, humidifiers, dehumidifiers, and refrigeration systems for the processing and storage of fresh agricultural commodities. More specifically, the application of psychrometry will facilitate the vegetable producers, packing-house operators, and commercial cooler operators to improve the postharvest cooling and storage conditions for fresh vegetables (Talbot and Baird, 1993). Owing to its great practical importance, psychrometry is an essential part of the curriculum of chemical engineering, food engineering, and allied fields (Erdélyi and Rajkó, 2016).

The number of independent variables (properties) required to describe the thermodynamic state of air is given by the Gibbs' phase rule, which states that,

$$F = C - P + 2 \tag{7.1}$$

where F is the number of possible variables (also, known as the degrees of freedom), C is the number of components, and P is the number of phases.

For moist air which is homogeneous in composition, $C = 2$ (dry air and water vapor) and $P = 1$. Thus, from Eq. (7.1), three independent variables are required to describe the state of the air. These variables include the temperature, pressure, and moisture content. However, psychrometry is often concerned with the study of air at atmospheric pressure. Therefore, pressure is eliminated from the list of variables, thus limiting the psychrometric properties to temperature and moisture content (or humidity) of the air. With this background, the psychrometric properties of importance to postharvest storage of agricultural products include dry-bulb temperature (DBT), wet-bulb temperature, dew point temperature, relative humidity, and humidity ratio (Talbot and Baird, 1993) (Figure 7.1).

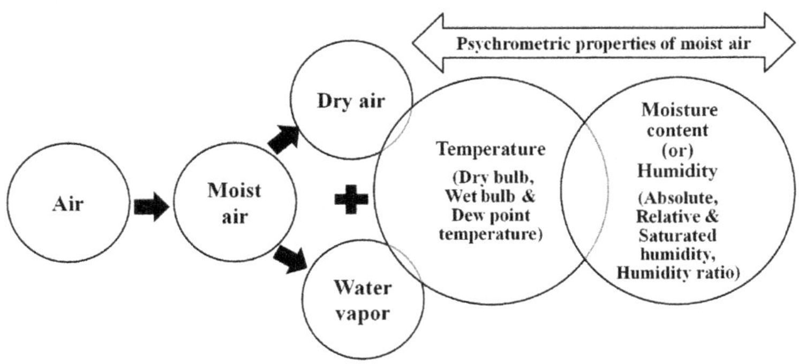

FIGURE 7.1 An overview of the psychrometric properties of air.

A detailed explanation of each of the abovementioned properties along with the methodology for calculating the same from the psychrometric chart would be dealt in this chapter. Further, the various applications of psychrometry in food processing industry will be discussed in the latter part of this chapter.

7.1 The Governing Laws of Psychrometry

The psychrometric relationships are based on three fundamental laws. It is important to familiarize with these laws before proceeding to understand the psychrometric properties.

7.1.1 The Ideal Gas Law (Perfect Gas Equation)

The ideal or universal gas law is a physical law which describes the relationship between the measurable properties of an ideal gas. The gases which obey the ideal gas law are said to be perfect or ideal. The perfect gas equation is obtained by combining the principles of the three gas laws, namely, Boyle's law, Charles's law, and Avogadro's law (Figure 7.2).

Boyle's law states that *at a constant temperature, the volume (V) of a gas varies inversely with its pressure (P)* (Eq. (7.2)).

$$V \propto \frac{1}{P} \text{ (or) } P \times V = \text{Constant} \tag{7.2}$$

Charles's law states that *volume and absolute temperature are directly proportional at constant pressure* (Eq. (7.3)).

$$V \propto T \quad \text{(or)} \quad \frac{V}{T} = \text{Constant} \tag{7.3}$$

Avogadro's law states that *one mole of any gas contains the same number of atoms or molecules and occupies the same volume at a given temperature* (Eq. (7.4)).

$$V \propto n \quad \text{(or)} \quad \frac{V}{n} = \text{Constant} \tag{7.4}$$

Thus, the ideal gas equation (Eq. (7.5)) is obtained by combining the equations of Boyle's law, Charles's law, and Avogadro's law.

$$P \times V = n \times R \times T \tag{7.5}$$

where P is the pressure (Pa), V is the volume (m^3), n is the amount of substance of the gas (mol), T is the absolute temperature (K), and R is the universal gas constant with the value of 8314.41 (m^3 Pa)/(mol K).

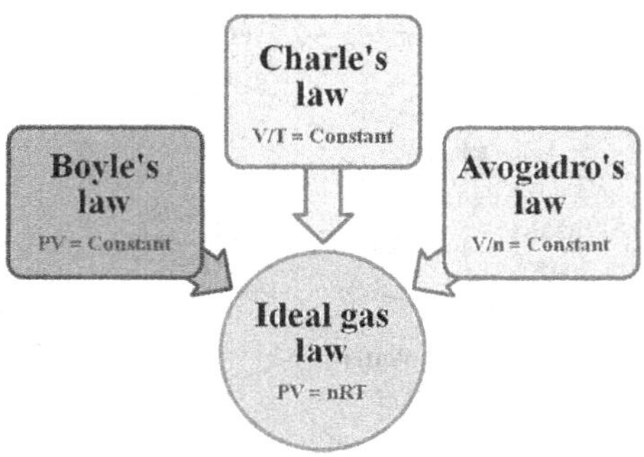

FIGURE 7.2 The ideal gas law.

7.1.2 Gibbs–Dalton Law of Partial Pressures

According to the Gibbs–Dalton law, the total pressure (P) exerted by a mixture of air and water vapor is equal to the sum of the partial pressure exerted by air (p_a) and water vapor (p_w) individually (Eq. (7.6)).

$$P = p_a + p_w \tag{7.6}$$

7.1.3 The First Law of Thermodynamics

The first law of thermodynamics is nothing but the law of conservation of energy which states that, *if any system undergoes a process during which energy is added to or removed from it (in the form of work or heat), none of the added energy is destroyed within the system, and none of the removed energy is created within the system.* The mathematical expression of the first law of thermodynamics is simple, given by

$$\text{Energy in} = \text{Energy out} \tag{7.7}$$

7.2 The Terminologies of Psychrometry

The terminologies of relevance to psychrometry are presented in this section.

- **Moist air:** It is a binary mixture of the dry air and water vapor. The proportion of dry air and water vapor in the moist air depends on the absolute pressure and temperature of the mixture.
- **Dry air:** It can be defined as the purified atmospheric air which is rendered free of moisture by the removal of water vapor. It is a mixture of nitrogen (78%), oxygen (21%), 0.9% of argon, 0.03% of carbon dioxide, 0.002% of neon, 0.0005% of helium, and 0.0006% of other gases in trace amounts. Dry air has a molecular weight of 28.9645 g/mol.
- **Water vapor:** The water vapor in the air can be considered as superheated steam at a low temperature and partial pressure, The amount of water vapor in the air can range between 0 and 20 g water/kg dry air, depending on the temperature and pressure. Water vapor is a significant factor pertaining to the effect of air conditions on the postharvest life of perishable commodities (Talbot and Baird, 1993).
- **Saturated air:** Air is said to be saturated when it has imbibed the maximum amount of water vapor into it at a particular temperature and pressure. When the saturated air is cooled, the water vapor present in it starts condensing which is visualized as moist, fog, or condensation on cold surfaces.
- **Dew point temperature (T_{dew}):** Temperature at which the moist air becomes saturated on cooling at constant pressure. Cooling of moist air to the dew point at constant pressure causes the condensation of water vapor.
- **Dry-bulb temperature (T_{db}):** It is defined as the temperature of dry air measured using a standard thermometer or thermocouple (an unmodified temperature sensor).
- **Wet-bulb temperature (T_{wb}):** It is the temperature measured using the wet sensing element which is cooled by the evaporating water (Figure 7.3). When a thermometer is covered with a wet wick and placed in the air stream, water evaporates from the wick due to the higher vapor pressure of wet wick compared to the surrounding air. Evaporating water cools the bulb to a temperature which is lower than that indicated by a thermometer with a dry sensing element (dry-bulb thermometer). Also, wet-bulb temperature is the lowest temperature to which an air mixture can be cooled solely by the addition of water with absolutely no removal of heat (Talbot and Baird, 1993).
- **Dew point depression:** It is the difference between the DBT and dew point temperature of air $(T_{db} - T_{dew})$.

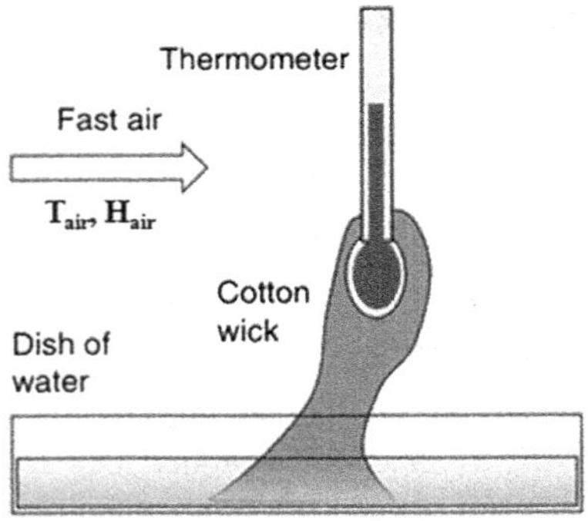

FIGURE 7.3 Wet-bulb temperature. (Modified and reproduced with permission from Driscoll, R. H. 2014. Dehydration. In *Food Processing: Principles and Applications* (Second edition), eds. S. Clark, S. Jung and B. Lamsa, 61–78. Hoboken, NJ: John Wiley and Sons, Ltd.)

- **Wet-bulb depression:** It is the difference between the DBT and wet-bulb temperature of air $(T_{db} - T_{wb})$. Wet-bulb depression is a function of the relative humidity of the air. If the air is more humid, it will restrict the evaporation of water and thus reduce the wet-bulb depression. The mathematical relationship between partial pressure and wet-bulb depression of air–water vapor mixture is given by the following expression:

$$p_W = p_{W0} - \frac{(P - p_{W0})(T_{db} - T_{wb})}{1555.56 - 0.722T_{wb}} \tag{7.8}$$

where p_W is the partial pressure of water vapor at the dew point temperature (kPa), p_{W0} is the saturation pressure of water vapor at the wet-bulb temperature (kPa), and P is the atmospheric pressure (kPa).

- **Absolute humidity:** Mass of water vapor associated with the unit mass of dry air is known as absolute humidity (H). Humidity is also known as the moisture content of the air. Absolute humidity is dimensionless as it is a ratio (kg/kg) between the mass of water vapor and the mass of dry air. Assuming ideal behavior, expressions for the mass of vapor and mass of dry air are given by

$$\text{Mass of vapor} = \frac{p_W M_W}{RT} \tag{7.9}$$

$$\text{Mass of air} = \frac{(P - p_W)M_A}{RT} \tag{7.10}$$

where M_w is the molecular weight of water vapor, p_W is the partial pressure of water vapor, P is the total pressure of the system, T is the temperature, M_A is the mean molecular weight of air, and $(P - p_W)$ is the partial pressure of air. Therefore, from Eqs. (7.9) and (7.10), the humidity of the air–water vapor mixture is

$$H = \frac{p_W M_W}{(P - p_W)M_A} \tag{7.11}$$

$$H = \frac{p_W}{(P - p_W)}\frac{M_W}{M_A} = \frac{18.01534}{28.9645} \times \frac{p_W}{(P - p_W)} = 0.622\frac{p_W}{(P - p_W)} \tag{7.12}$$

- **Saturated humidity:** At any given temperature, the amount of water vapor present in air attains a limit before the vapor condenses to form liquid water. The abovementioned condition is explained by the term saturated humidity (H_0), which is the absolute humidity of the air when it is saturated with the water vapor. Increase in temperature increases the saturated humidity or the water vapor holding capacity of air.

Saturated humidity is given by the expression

$$H_0 = \frac{p_{W0}M_W}{(P - p_{W0})M_A} = 0.622\frac{p_{W0}}{(P - p_{W0})} \tag{7.13}$$

where p_{W0} is the partial pressure of water vapor at saturation.

- **Relative humidity (RH):** It is defined as the ratio of the humidity of air to the saturated humidity at the same temperature, expressed in percentage.

$$\%\text{RH} = \frac{H}{H_0} \times 100 \tag{7.14}$$

Relative humidity can also be defined as the ratio of partial pressure of water vapor in the air to the saturated vapor pressure of water in the air at the same temperature (Eq. (7.15)).

$$\%\text{RH} = \frac{p_W}{p_{W0}} \times 100 \tag{7.15}$$

%RH can also be expressed as mole fraction and ratio of densities as given in Eqs. (7.16) and (7.17), respectively.

$$\%\text{RH} = \frac{x_W}{x_{W0}} \times 100 \tag{7.16}$$

$$\%\text{RH} = \frac{\rho_W}{\rho_{W0}} \times 100 \tag{7.17}$$

where x_W is the mole fraction of water vapor in air, x_{W0} is the mole fraction of saturated water vapor in air at the same pressure and temperature, ρ_W is the density of the water vapor in the air (kg/m^3), ρ_{W0} is the density of the saturated water vapor in air (kg/m^3) at the DBT.

- **Humidity ratio:** The amount of moisture contained in the air per unit mass of dry air is the humidity ratio of moist air. Humidity ratio is often expressed as pounds of moisture per pound of dry air. Unlike the relative humidity, the humidity ratio of moist air is not dependent on temperature. Therefore, it is easier to use in calculations.
- **Degree of saturation:** It is the ratio between the actual humidity ratio and the humidity ratio of saturated air at the same temperature and pressure.
- **Humid heat:** It is defined as the amount of heat required to raise the temperature of 1 kg moist air through 1 K. It can be expressed as the sum of the heat capacity of dry air and the product of heat capacity of water vapor and humidity (Eq. (7.18)).

$$C_s = 1.005 + 1.88H \tag{7.18}$$

where C_s is the humid heat of moist air (kJ/kg dry air K), and H is the humidity (kg water/kg dry air).

- **Specific volume (V'):** Specific volume of a substance is defined as the volume occupied by the unit weight of the specific substance. It is expressed in units of m^3/kg and, hence, can also be termed as the reciprocal of density.
- **Specific heat (C_P):** Specific heat of a substance is the amount of heat required to raise the temperature of a unit mass of the substance by one degree Celsius.
- **Enthalpy (H):** It is a thermodynamic quantity which is equal to the total heat content of a system. The mathematical expression for enthalpy is given by the sum of internal energy of the system (E) and the product of pressure (P) and volume (V).

$$H = E + PV \tag{7.19}$$

Example 7.1

The partial pressure of water vapor in a warehouse maintained at the normal temperature and pressure conditions (NTP) is 1.5 kPa. Determine the relative humidity and percentage saturation, if the vapor pressure of pure water at the specified conditions is 2.339 kPa.

Solution

Given: NTP conditions: Temperature (T): 20°C and pressure (P): 101.325 kPa
$p_w = 1.5\,\text{kPa}$; $p_{wo} = 2.339\,\text{kPa}$

$$\therefore \%\text{RH} = \frac{p_W}{p_{WO}} \times 100 = \frac{1.5}{2.339} \times 100 = 64.129\%$$

$$\%\text{saturation} = \frac{(P - p_{WO})}{(P - p_W)} \frac{p_W}{p_{WO}} \times 100 = \left(\frac{101.325 - 2.339}{101.325 - 1.5}\right) \frac{1.5}{2.339} \times 100$$

$$\%\text{saturation} = 63.591$$

Therefore, percentage saturation is 63.591% and percentage RH is 64.129%.

Example 7.2

A wet-bulb thermometer reads 60°C, and a dry-bulb thermometer reads 72°C. Determine the wet-bulb depression and the partial pressure of water vapor at the dew point temperature under NTP conditions.

Solution

Given:
$T_{wb} = 60°\text{C}$; $T_{db} = 72°\text{C}$; P = 101.325 kPa
∴ Wet-bulb depression = $T_{db} - T_{wb} = 72 - 60 = 12°\text{C}$

$$p_W = p_{WO} - \frac{(P - p_{WO})(T_{db} - T_{wb})}{1555.56 - 0.722 T_{wb}} = 2.339 - \frac{(101.325 - 2.339)(72 - 60)}{1555.56 - (0.722 * 60)} = 1.554\,\text{kPa}$$

Answer: The wet-bulb depression is 12°C, and the partial pressure of water vapor is 1.554 kPa

7.3 Properties of the Constituents of Moist Air

7.3.1 Properties of Dry Air

- **Gas constant:** The molecular weight (M_a) of dry air is 28.9645 g/mol. From the perfect gas equation (Eq. (7.5)), the gas constant (R_a) for dry air can be calculated as

$$R_a = \frac{8314.41}{28.9645} = 287.055\,\text{m}^3\,\text{Pa/g K} \tag{7.20}$$

- **Specific volume (V'_a):** From the equation of ideal gas law (Eq. (7.5)), the specific volume of dry air is given by

$$V'_a = \frac{R_a T_{db}}{p_a} \tag{7.21}$$

where T_{db} is the absolute (dry bulb) temperature (K), p_a is the partial pressure of dry air (Pa), and R_a is the gas constant of dry air ([m^3 Pa]/[kg K]).

- **Specific heat (C_{Pa}):** Specific heat of dry air at the atmospheric pressure (101.325 kPa) varies from 0.997 to 1.022 kJ/kg K in the temperature range of −40°C to 60°C. Thus, for most calculations, an average value of 1.005 kJ/kg K is used.
- **Enthalpy (H_a):** Enthalpy of dry air, in kJ/kg, is determined using the following equation:

$$H_a = 1.005(T_{db} - T_0) \tag{7.22}$$

where T_{db} is the DBT and T_0 is the reference temperature, which is generally considered as 0°C.

7.3.2 Properties of Water Vapor

- **Gas constant:** The molecular weight (M_w) of water is 18.01534 g/mol. From the perfect gas equation (Eq. (7.5)), the gas constant (R_w) for water vapor can be calculated using Eq. (7.23).

$$R_w = \frac{R}{M} = \frac{8314.41}{18.01534} = 461.52\,\mathrm{m^3\,Pa/g\,K} \tag{7.23}$$

- **Specific volume (V'_w):** Similar to that of dry air, specific volume of water vapor can also be calculated using the ideal gas law, under the condition that at a temperature below 66°C, the saturated or superheated water vapor behaves like an ideal gas. Under this condition, the specific volume of water vapor is given by

$$V'_w = \frac{R_w T_{wb}}{p_W} \tag{7.24}$$

where T_{wb} is the absolute (wet bulb) temperature (K), p_W is the partial pressure of water vapor (kPa), and R_w is the gas constant of water vapor ([m^3 Pa]/[kg K]).
- **Specific heat:** Specific heat of saturated and superheated vapor is 1.88 kJ/kg K, and it changes insignificantly in the temperature range of −71°C to 124°C.
- **Enthalpy:** Enthalpy of saturated or superheated water vapor, (H_W) in kJ/kg, is determined using the following expression:

$$H_W = 2501.4 + 1.88(T_{db} - T_0) \tag{7.25}$$

7.3.3 Adiabatic Saturation of Air

Saturation and cooling of air find applications in various industrial operations. During the convective drying of a food product, gas is passed over the food surface at a rate such that the equilibrium temperature is achieved. Considering an adiabatic system (no gain or loss of heat to the surrounding), the sensible heat of the dry air will be reduced by an amount equal to the latent heat of the water evaporated, and it results in the adiabatic saturation of air. Adiabatic saturation can be mathematically expressed as

$$H - H_S = -\frac{C_P}{L}(T - T_S) \tag{7.26}$$

where H_S is the humidity at saturation, T_S is the adiabatic saturation temperature, and L is the latent heat of vaporization of water.

7.4 Psychrometric Chart

A psychrometric chart is the graphical representation of the psychrometric properties of moist air. This chart (Figure 7.4) includes the following:

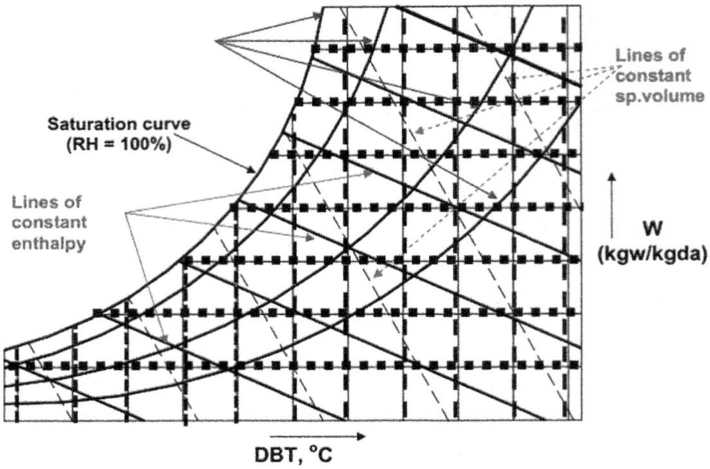

FIGURE 7.4 Explanation of the components of a psychrometric chart. (Modified from NPTEL 2018. Lesson 27 Psychrometry. https://nptel.ac.in/courses/112105129/pdf/R&AC%20Lecture%2027.pdf (accessed October 15, 2018.))

- DBT on the horizontal axis (abscissa; x-axis)
- Humidity ratio on the vertical axis (ordinate; y-axis)
- Saturated air line (100 % humidity) along its upper curved boundary.

Apart from the abovementioned properties, wet-bulb temperature (T_{wb}), percentage relative humidity ($\%RH$), dew point temperature (T_{dew}), enthalpy (H), and specific volume (V') of the air–water mixture can also be obtained from the psychrometric chart.

The psychrometric chart appears complicated with many lines inclined in different directions and angles. However, with a thorough understanding, one can become well versed with its interpretation and application in the appropriate situations. The subsequent sections present different exercises to obtain a systematic understanding of the usage of a psychrometric chart. The first step is to be familiar with the method to read different properties represented by the psychrometric chart.

7.4.1 Components of the Psychrometric Chart

7.4.1.1 Lines of Dry-Bulb Temperature

The base of the psychrometric chart shows the scale of DBT (°C), which increases from left to right. Vertical dashed lines in the chart represent the constant DBT lines (Figure 7.4). Thus, all the points located along these constant DBT lines are to be read as the same temperature.

7.4.1.2 Lines of Constant Humidity

The humidity of moist air (g/kg of dry air) is indicated by the vertical scale located at the right of the psychrometric chart. The horizontal dotted lines originating from the vertical scale correspond to the constant humidity lines (Figure 7.4).

7.4.1.3 Lines of Wet-Bulb Temperature

The upper curved boundary on the left side of the psychrometric chart signifies the scale of wet-bulb temperature (°C). The diagonal lines extending from this curved scale constitute the constant wet-bulb temperature lines which move in the downward direction towards the right side of the chart (Figure 7.4).

7.4.1.4 Lines of Dew Point Temperature

Dew point temperature of the air depends on its humidity. Consequently, the scale and lines of constant dew point temperature coincide with those of the constant humidity (Figure 7.4).

A striking difference between the lines of constant wet-bulb temperature and constant dew point temperature is that the former is represented by diagonal lines extending downward and the horizontal lines symbolize the latter.

7.4.1.5 Lines of Relative Humidity

The curved lines of relative humidity represent different values of relative humidity expressed in units of percentage. The lines of RH begin at the upper curved boundary of the psychrometric chart known as the saturation curve (100% RH) and decrease while moving downward towards the horizontal base of the chart (0% RH). Generally, these lines are showed in intervals of 10% RH (curved lines; Figure 7.4). Along the saturation curve, the values of the DBT, wet-bulb, and dew point temperatures are equal.

7.4.1.6 Lines of Constant Enthalpy

Total enthalpy of air is the sum of its sensible and latent heat. The scale for constant enthalpy is located outside the main body of the psychrometric chart. The lines of constant enthalpy extend from this outer scale and move diagonally downward from left to right across the chart (Figure 7.4). It is necessary to know the wet-bulb temperature to determine the total enthalpy of air. The constant enthalpy line passing through the known value of wet-bulb temperature indicates the enthalpy of air. Lines of constant enthalpy and constant wet-bulb temperature coincide on the chart shown in Figure 7.4. However, the values are read from different scales. More precise psychrometric charts use slightly different lines to represent wet-bulb temperature and enthalpy.

7.4.1.7 Lines of Constant Specific Volume

The lines of constant specific volume (m^3/kg) begin from the saturation curve and descend at an angle with respect to the vertical lines of the DBT (dashed lines; Figure 7.4). The total volume of air needed for a unit operation can be calculated by multiplying the values of specific volume obtained from the psychrometric chart as explained previously and the total amount of air required (Dossat, 1997; Ananthanarayan, 2005).

Figure 7.5 shows the psychrometric chart in SI units at barometric pressure 101.325 kPa. The psychrometric charts are available in different temperature ranges which find applications in unit operations that operate at the appropriate range of temperature. Similarly, psychrometric charts are also available for different pressures, corresponding to varied elevations from the sea level.

7.4.2 Methodology for Using the Psychrometric Chart

Various properties of the air–water vapor mixture are interrelated. If any of the two properties are known, a point can be established on the chart, and the remaining properties can be determined. In this section, the procedure to obtain the details from a psychrometric chart is illustrated. A simplified schematic of the psychrometric chart (not drawn to scale) shown in Figure 7.6 would help in understanding the steps involved in obtaining the different properties of moist air from the chart.

I. Steps to determine relative humidity

1. Locate the DBT (point A) and the specific or absolute humidity (point B) (Figure 7.6).
2. Draw a vertical line from the point A and a horizontal line from B.
3. Locate the point C, where these two lines coincide.
4. Read the relative humidity from the curve passing through the point C.

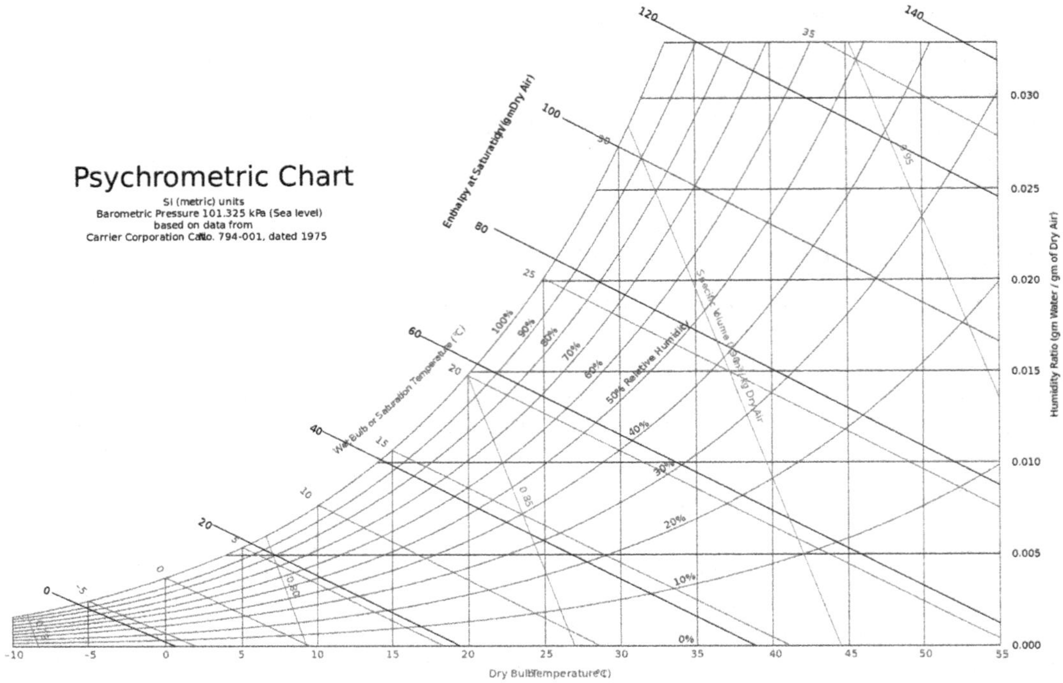

FIGURE 7.5 Psychrometric chart. (Ogawa, A. 2009. CC BY-SA 3.0 (Creative Commons Attribution-Share Alike 3.0 Unported). https://commons.wikimedia.org/wiki/File:PsychrometricChart.SeaLevel.SI.svg (accessed December 7, 2018))

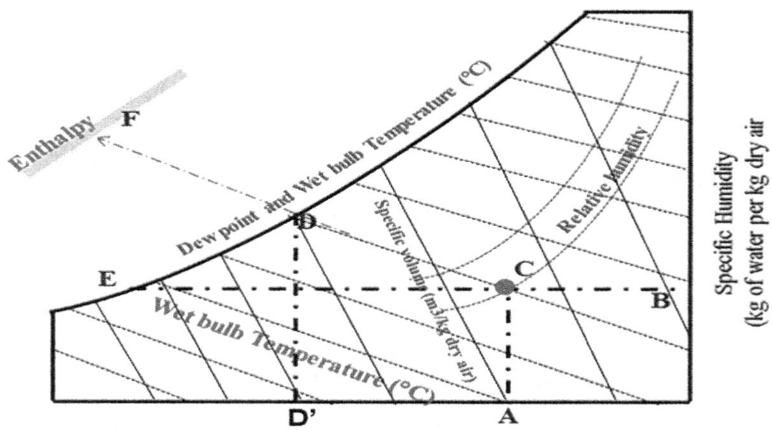

FIGURE 7.6 Illustration to determine the various properties of the air–vapor mixture.

II. Steps to determine wet-bulb temperature

1. Draw a line parallel to the wet-bulb line and passing through the point C (Figure 7.6).
2. Locate the wet-bulb temperature on the line where it coincides with the saturation curve (100% relative humidity) at point D.
3. Draw a vertical line, DD' from the point D which coincides with the DBT axis.
4. Read the corresponding temperature as the wet-bulb temperature at point D'.

III. Steps to determine dew point temperature

1. Draw a horizontal line from point C that intersects the wet bulb and dew point line at point E (Figure 7.6).
2. Read the dew point temperature at point E.

IV. Steps to determine absolute humidity

1. Draw a line parallel to the x-axis that passes through the point D (Figure 7.6).
2. Read the saturation humidity at the wet-bulb temperature on the ordinate (y-axis).

V. Steps to determine the enthalpy

1. Extend the wet-bulb temperature line to the enthalpy scale given on the chart (Figure 7.6).
2. Read the corresponding enthalpy.

VI. Steps to determine specific volume

1. Specific volume is pointed on the specific volume line passing through the point C (Figure 7.6).
2. If it is falling between the two specific volume lines, interpolation can be used to determine the specific volume.

Example 7.3

Determine the properties of moist air which is at a DBT of 30°C with a relative humidity of 60%. Determine absolute humidity, wet-bulb temperature, dew point temperature, and total enthalpy of the air stream.

Solution

From the psychrometric chart,

1. To determine the absolute humidity, draw a vertical line from T_{db} = 30°C until it intersects with the 60% RH curve, which gives
   Absolute humidity = 0.016 kg H_2O/kg of air
2. To determine the wet-bulb temperature, dew point temperature, and total enthalpy of air, draw a vertical line from T_{db} = 30°C and a horizontal line from the absolute humidity = 0.016 kg H_2O/kg. The lines of constant wet-bulb temperature, dew point temperature, and total enthalpy that pass through the point of intersection of the above-mentioned two lines give
   Wet-bulb temperature = 24°C
   Dew point temperature = 21°C
   Total enthalpy = 72.5 kJ/kg of dry air

7.4.3 Applications of Psychrometry

Psychrometry deals with the changes that occur in the properties of the air–water vapor mixture during food processing operations. Figure 7.7a–f shows the psychrometric charts representing different operations explained as follows:

7.4.3.1 Heating

The heating process is represented on a psychrometric chart by a straight line that starts from the given value of the DBT and extends to the right (Figure 7.7a).

Generally, heating of air is achieved without the addition of any moisture. The increase in air temperature due to the addition of sensible heat maintains the absolute humidity constant while decreasing the percentage saturation, i.e., reduction of relative humidity. It is due to the increase in capacity of air to hold moisture due to increase in temperature.

7.4.3.2 Cooling

The cooling process is represented on a psychrometric chart by a straight line that starts from the given value of the DBT and extends to the left (Figure 7.7b)

During cooling operation, heat is removed from the air, and thus, DBT decreases at a constant absolute humidity until the saturation occurs. The temperature at the saturation is the dew point where the formation of water droplets occurs.

7.4.3.3 Mixing

Mixing process is represented on a psychrometric chart by two straight lines indicating the two different air streams facing each other, connected by a dotted line to signify the mixing operation (Figure 7.7c).

Mixing of different air streams having different absolute humidity content or relative humidity results in a mixture having an absolute humidity, which is the average of the humidity of the initial streams before mixing.

7.4.3.4 Drying

The drying process is represented on a psychrometric chart by a straight line starting from a specific DBT at the right and extending upward to the left (Figure 7.7d).

During the drying of moist food sample, hot air is forced through the dryer. Drying is an adiabatic saturation process, wherein the evaporation of water occurs due to the heat provided by the hot air. Relative and absolute humidity of the air increase during the drying process.

7.4.3.5 Heating-Cum-Humidification

On the psychrometric chart, the heating-cum-humidification process is represented by an angular line that starts from a given value of the DBT and extends upward towards the right (Figure 7.7e).

During the process of heating and humidification, there is a simultaneous increase in the DBT and the humidity of the air. In general, this process is carried out by two methods, either by passing the air through a spray of water which is maintained at a temperature higher than the DBT of air or by mixing air and steam. The air experiences an increase in humidity as it absorbs the moisture evaporated from the

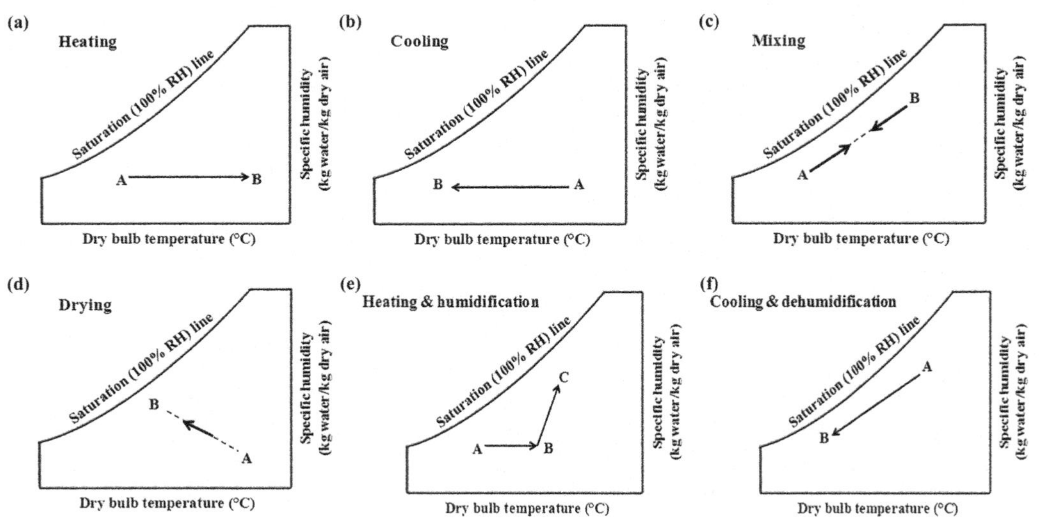

FIGURE 7.7 Representation of (a) heating, (b) cooling, (c) mixing, (d) drying, (e) heating and humidification, and (f) cooling and dehumidification operations, on the psychrometric chart.

spray. In unison, as the temperature of the moisture is greater than the DBT of the air, there is an overall increase in its temperature. Also, the wet-bulb and dew point temperature of air increases owing to the increase in its relative humidity.

7.4.3.6 Cooling-Cum-Dehumidification

The cooling-cum-dehumidification process is represented on the psychrometric chart by a straight angular line. The line starts from the given value of the DBT and extends downwards towards left (Figure 7.7f).

During the process of cooling and dehumidification, air at a particular DBT undergoes sensible cooling below its dew point temperature along with the removal of moisture. This simultaneous process is accomplished by contacting the air with a cooling coil which is maintained at a temperature less than the dew point temperature of the air. As a result, the DBT of air starts reducing. The cooling of air continues until it reaches the value of the dew point temperature. At this point, the water vapor in the air is converted into dew particles leading to dew formation on the surface of the cooling coil. Consequently, moisture content in the air reduces thereby causing a drop in its humidity.

Example 7.4

Moist air enters a heater at 30°C, 100 kPa, and 70% relative humidity. The temperature of the air is increased to 120°C in the heater while the pressure remains constant at 100 kPa. Determine (a) dew point temperature of the incoming mixture, (b) absolute humidity of incoming mixture, (c) absolute humidity of the outgoing mixture and, (d) relative humidity of the existing mixture. Vapor pressure at 30°C and 120°C are 2.339 and 198.541 kPa, respectively.

Solution

Given

i. Vapor pressure at 30°C = 2.339 kPa
ii. Vapor pressure at 120°C = 198.541 kPa

Moist air enters at 30°C with 70% RH and leaves at 120°C without the addition of any moisture. From the psychrometric chart (Figure 7.11),

a. Dew point temperature at 30°C with 70% RH is 24°C.
b. Absolute humidity of moist air at the entry point to the heater is 18 g/kg of dry air = 0.018 kg/kg of dry air.
c. Absolute humidity of the exiting air stream will remain same as that of the incoming air, as moisture content is constant = 0.018 kg/kg of dry air.
d. Relative humidity of the moist air is calculated from the following equation:

$$\% \text{RH} = \frac{p_W}{p_{W0}} \times 100$$

The partial pressure of water vapor at the entry point to the heater is

$$70 = \frac{p_W}{2.339} \times 100$$

$$\therefore\ p_W = 1.637\,\text{kPa}$$

Moisture content remains constant throughout the operation. Therefore, the partial pressure of water vapor in the exiting air stream is 1.637 kPa

$$\therefore \% \text{RH} = \frac{1.637}{198.541} \times 100 = 0.825$$

Therefore, the relative humidity of the exiting air stream is 0.825%.

Example 7.5

Air at 25°C temperature and 50% saturation is heated to 80°C. Find the partial pressure of water vapor and percentage saturation at 80°C. Further, the air passes through the dryer and is discharged at 40°C and 25% saturation. Find the moisture picked up and partial pressure of water vapor in the outgoing air. Vapor pressure at 25°C, 40°C, and 80°C are 3.1599, 7.3749, and 47.3601 kPa, respectively.

Solution

Given

i. Vapor pressure at 25°C, 40°C, and 80°C are 3.1599, 7.3749, and 47.3601 kPa, respectively.
ii. Air enters the heater at 25°C with 50% saturation; air enters the dryer at 80°C; air leaves dryer at 40°C and 25% saturation.

a. **The partial pressure of water vapor at 25°C**

$$p_W = \frac{\%\text{RH} \times p_{W0}}{100} = \frac{50}{100} \times 3.1599 = 1.58\,\text{kPa}$$

As the moisture content of air remains constant during the heating operation, the partial pressure of water vapor at 80°C leaving the heater is also equal to that of the incoming stream as calculated above, i.e., 1.58 kPa.

b. **The partial pressure of water vapor at 40°C**

$$p_W = \frac{\%\text{RH} \times p_{W0}}{100} = \frac{25}{100} \times 7.3749 = 1.844\,\text{kPa}$$

c. **Relative humidity of air exiting the heater at 80°C**

$$\%\text{RH} = \frac{p_W}{p_{W0}} \times 100$$

$$\%\text{RH} = \frac{1.58}{47.3601} \times 100 = 3.336\%$$

d. **Absolute humidity of air exiting the dryer at 40°C and 25% saturation**

$$H = \frac{p_W}{(P - p_W)} \frac{M_W}{M_A} = \frac{1.844}{101.32 - 1.844} \times \frac{18}{29} = 0.0115\,\text{kg/kg of dry air}$$

e. **Absolute humidity of air entering dryer at 80°C and 3.336% saturation**

$$H = \frac{p_W}{(P - p_W)} \frac{M_W}{M_A} = \frac{1.58}{101.32 - 1.58} \times \frac{18}{29} = 0.0098\,\text{kg/kg of dry air}$$

f. **Moisture picked up by the air during drying**
= 0.0115 − 0.0098 = 0.0017 kg water/kg of dry air

Answer: Moisture picked up by the air during drying is 1.7 g/kg of dry air.

7.4.3.7 Estimation of Wet-Bulb and Outlet Particle Temperatures

The outlet particle temperature (T_p) can be calculated from the wet-bulb lines of the psychrometric chart (Figure 7.8), which is explained as follows with the example of whey protein (Anandharamakrishnan, 2008).

$$\text{Mass transfer rate from particle} = -\dot{m} = k_c A_p (Y_s^* - Y_{\text{air}}) \tag{7.27}$$

$$\text{Heat transfer rate from particle} = -h_{fg}\dot{m} = hA_p(T_{\text{air}} - T_p) \tag{7.28}$$

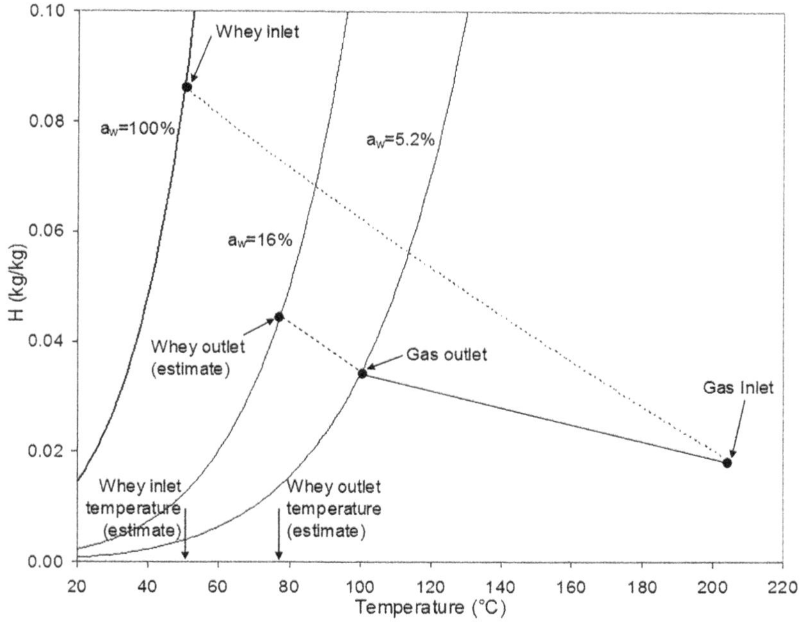

FIGURE 7.8 Psychrometric chart showing typical gas and particle inlet and outlet conditions. (Reproduced with permission from Anandharamakrishnan, C. 2008. Experimental and computational fluid dynamics studies on spray-freeze-drying and spray-drying of proteins. *PhD thesis*. UK: Loughborough University; The wet bulb temperature is constant along the dotted lines. Curved lines of constant water activity, aw, are also shown: 5.2% corresponds to the relative humidity of the outlet gas, and 16% is the equilibrium relative humidity corresponding to the bulk moisture content of the whey product.)

where k_c is the mass transfer coefficient (m/s), h is the heat transfer coefficient (W/m^2 K), and Y_s^* is the water activity at the surface of particles.

Combining the Eqs. (7.27) and (7.28) gives

$$hA_p(T_{\text{air}} - T_p) = k_c A_p h_{fg}(Y_s^* - Y_{\text{air}}) \tag{7.29}$$

Hence, the wet-bulb line is

$$Y_s^* - Y_{\text{air}} = \frac{h}{h_{fg}k_c}(T_{\text{air}} - T_p) \tag{7.30}$$

The outlet particle temperature can be estimated by following the wet-bulb lines from the gas outlet conditions to the relative humidity line corresponding to an estimate of the water activity (equilibrium relative humidity), a_w, on the surface of the whey protein isolate (WPI) product. The latter can be determined from the moisture sorption isotherm of WPI powder (Figure 7.9) at the corresponding temperature, using the measured product moisture content.

7.5 Measurement of Psychrometric Properties

7.5.1 Psychrometer

A psychrometer is an instrument which is capable of measuring the psychrometric state of the air. Generally known as the dry- and wet-bulb instrument, psychrometer can be used for the determination of different psychrometric properties of air such as the relative humidity, absolute humidity, humidity ratio, and enthalpy. Measurement of humidity by psychrometer is based on the principle of *evaporative cooling*

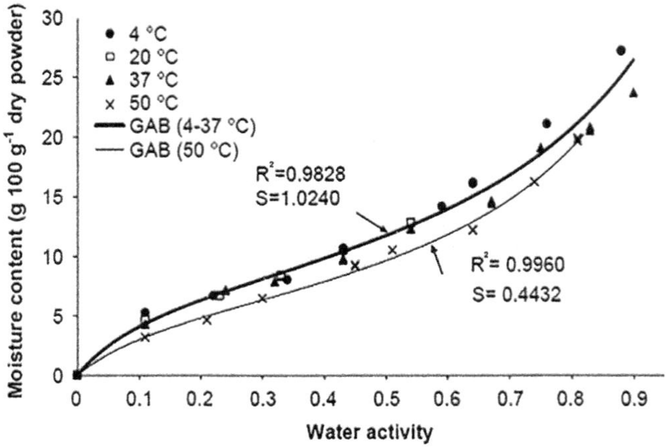

FIGURE 7.9 Moisture sorption isotherm of a WPI powder (Guggenheim–Anderson–De Boer sorption isotherm model data). (Reproduced with permission from Foster, K. D., Bronlund, J. E. and Paterson, A. H. J. 2005. The prediction of moisture sorption isotherms for dairy powders. *International Dairy Journal* 15: 411–418.)

by which the temperature decreases due to evaporation of moisture from a wet surface. A psychrometer includes two thermometers, of which, the bulb of one is covered with a moistened wick made of cotton or linen, and that of the other thermometer is left uncovered, known as the "wet-bulb thermometer" and "dry-bulb thermometer," respectively. The two sensing bulbs are separated and shielded from each other such that the radiation heat transfer between them is insignificant. Evaporation of water from the moistened wick of wet-bulb thermometer causes a reduction in the temperature. The value of humidity can then be determined from the paired temperature values (T_{db} and T_{wb}) using the following empirical relations for the vapor pressure of water in the moist air.

i. Apjohn equation

$$p_W = p_{W0} - \left(\frac{1.8p(T_{db} - T_{wb})}{2700}\right) \quad (7.31)$$

ii. Ferrel equation

$$p_W = p_{W0} - 0.00066p(T_{db} - T_{wb})\left[1 + \frac{1.8(273 + T_{db})}{1571}\right] \quad (7.32)$$

iii. Carrier equation

$$p_W = p_{W0} - \left(\frac{(p - p_{W0})(T_{db} - T_{wb})}{1547 - 1.44(273 + T_{wb})}\right) \quad (7.33)$$

where

T_{db} = DBT (°C)

T_{wb} = wet-bulb temperature (°C)

P_p = atmospheric pressure (Pa)

p_W = vapor pressure of water in the air (Pa)

p_{W0} = saturation vapor pressure of water in air at the wet-bulb temperature (Pa)

After calculating the vapor pressure of water in the moist air, the other properties can be determined from the psychrometric equations specified in Section 7.2.

There are different types of psychrometers, of which, sling psychrometer and aspirated psychrometer are commonly used. The sling psychrometer consists of a dry- and wet-bulb thermometer mounted side

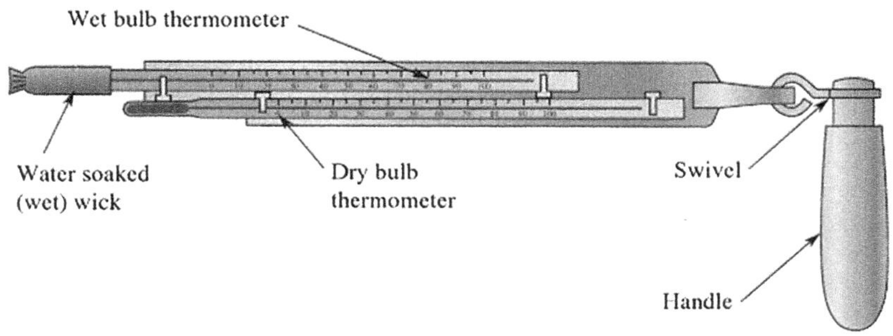

FIGURE 7.10 Sling psychrometer. (Reproduced with permission from Balmer, R. T. 2011. *Modern Engineering Thermodynamics*. Burlington, MA: Academic Press, Elsevier Inc.)

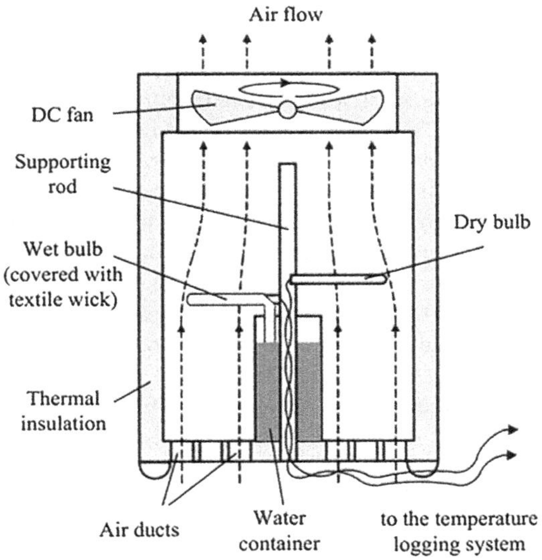

FIGURE 7.11 Aspirated psychrometer. (Reproduced with permission from Sapundjiev, P. V. 2009. Air humidity measurement using the psychrometric method. *Annual Journal of Electronics* 1: 168–171.)

by side and fitted in a frame which is attached to a handle by a swivel connection that permits the device to be rotated through the air (see Figure 7.10). The required air circulation around the sensing bulbs, usually in the range of 3–5 m/s, is obtained by rotating the psychrometer in the air at ~300 RPM for approximately 1 min. The temperature readings are recorded after both the thermometers attain equilibrium.

An aspirated psychrometer operates on the same principle as that of the sling psychrometer, except that its thermometers remain stationary, and a battery-powered fan, blower, or syringe moves the air across the thermometer bulbs (see Figure 7.11). Nevertheless, the aspirated psychrometer is more reliable as it demonstrates various advantages over the sling psychrometer. First, the airspeed over the wet wick can be better controlled by an aspirated psychrometer than by the rotational movement used in sling psychrometer. Second, to record the reading on a sling psychrometer, the rotation of the psychrometer must be stopped which can change the properties of the wet wick.

7.5.2 Optical Dew Point Hygrometer

The psychrometers of present days are capable of directly displaying the values of the psychrometric properties of moist air. These modern psychrometers are known as *optical dew point hygrometers*, which work on the principle of condensation. This instrument directly measures the dew point temperature of moist

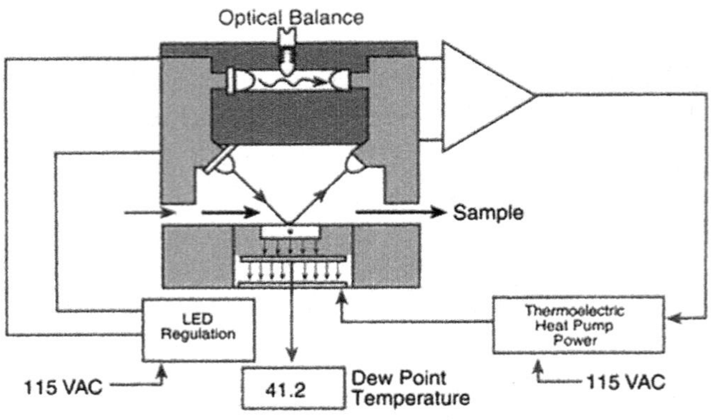

FIGURE 7.12 Schematic of optical dew point hygrometer. (Reproduced with permission from Wiederhold, P. R. 1997. *Water Vapor Measurement: Methods and Instrumentation*. New York: Marcel Dekker Inc.)

air by observing the temperature at which dew or frost starts appearing on an artificially cooled and polished surface (Genskow et al., 2008). The surface is usually a mirror-like metallic plane which is cooled by evaporation of a low-boiling solvent such as ether, by vaporization of a condensed permanent gas such as CO_2 or liquid air, or by a stream of water at controlled temperature. The mirror is illuminated with a light source, and the condensation of moisture on the mirror surface is optically detected by monitoring the reflectivity. As the dew forms on the mirror surface, the light scatters and decreases the signal at the detector (Moreno-Bondi et al., 2004). After the optical detection, the surface temperature is measured using a platinum resistance thermometer, thermocouple, or thermistor fixed on the mirror surface (see Figure 7.12). The measured temperature is reported as the dew point temperature. These hygrometers are often more reliable with excellent accuracy and repeatability (Wiederhold, 1995). However, calibration at a defined frequency of at least once per year is required to maintain the efficiency of dew point hygrometers.

7.5.3 Electric Hygrometer

Determination of relative humidity by electric hygrometers is based on measuring the change in electrical impedance (resistance or capacitance) of a film of hygroscopic materials exposed to the air. The electrical capacitance of certain polymers and salt films change in response to the humidity of the surrounding air. The electric hygrometers use a sensor with hygroscopic material sandwiched between two electrodes (Figure 7.13). The moisture will be adsorbed by or desorbed from the material, depending on the water vapor pressure of the surrounding air (Fontana and Campbell, 2004). An electric hygrometer is usually of "probe type" attached directly or by a cable to an electronic unit that displays the relative humidity reading (Bell, 2011).

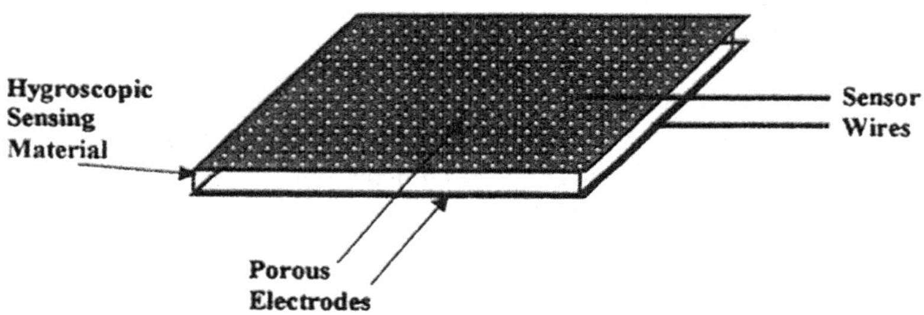

FIGURE 7.13 Schematic of the design of an electric hygrometer. (Reproduced with permission from Fontana, A. J. and Campbell, C. S. 2004. Water activity. In *Handbook of Food Analysis: Physical characterization and nutrient analysis*, ed. L. M. L. Nollet, 39–54. New York: Marcel Dekker Inc.)

7.6 Problems to Practice

7.6.1 Multiple Choice Questions

1. In the psychrometric chart, the DBT at any point on the saturation curve is always
   a. less than the corresponding wet-bulb temperature
   b. more than the corresponding wet-bulb temperature
   c. equal to the corresponding wet-bulb temperature

Answer: c

2. The horizontal line in psychrometric chart joining the change of state of air represents
   a. humidification
   b. sensible cooling or heating
   c. sensible cooling or heating with humidification

Answer: b

3. Humidification is the process of addition moisture in air at
   a. constant wet-bulb temperature
   b. constant DBT
   c. constant latent heat

Answer: b

4. During dehumidification of air,
   a. air is heated above its dew point temperature
   b. air is cooled below its dew point temperature
   c. air is cooled up to its dew point temperature

Answer: b

5. The dew point temperature is less than the wet-bulb temperature for
   a. unsaturated air
   b. saturated air
   c. both saturated and unsaturated air

Answer: a

6. A psychrometer measures
   a. DBT
   b. wet-bulb temperature
   c. both a and b

Answer: c

7. During heating and humidification process,
   a. relative humidity of air increases
   b. relative humidity of air decreases
   c. relative humidity of air remains constant

Answer: a

8. Specific humidity is the mass of water vapor present in
   a. 1 m^3 of dry air
   b. 1 kg of wet air
   c. 1 kg of dry air

Answer: c

9. On a psychrometric chart, the humidification process is represented by
   a. inclined line
   b. horizontal line
   c. vertical line

Answer: c

10. The atmospheric air is to be conditioned from 75% relative humidity and 40°C DBT to 50% relative humidity and 30°C DBT. The aforementioned sequence of events corresponds to the process of
    a. cooling and humidification
    b. cooling and dehumidification
    c. dehumidification

Answer: b

11. On psychrometric chart, wet-bulb temperature lines are
    a. horizontal with uniformly spaced
    b. inclined with uniformly spaced
    c. inclined with nonuniformly spaced

Answer: b

12. When the dew point temperature is equal to the air temperature then the relative humidity is
    a. 0%
    b. 50%
    c. 100%

Answer: c

13. The temperature, at which the air cannot hold all the water vapor mixed in it and some vapor starts condensing is called as
    a. humidification temperature
    b. dehumidification temperature
    c. dew point temperature

Answer: c

14. When the moisture evaporation no longer occurs, then the relative humidity of air is
    a. 0%
    b. 100%
    c. 50%

Answer: b

15. At 100% relative humidity, the wet-bulb temperature is
    a. lower than the dew point temperature
    b. higher than the dew point temperature
    c. equal to the dew point temperature

Answer: c

16. The specific humidity lines in psychrometric chart are also called as
    a. relative humidity lines
    b. moisture content lines
    c. both a and b

Answer: b

17. The ratio of partial pressure of water vapor in a mixture to the saturation pressure of water at the same temperature of the mixture is called as
    a. relative humidity
    b. humidity
    c. specific humidity

Answer: a

18. The temperature of air recorded by a thermometer when the bulb is covered by a cotton wick saturated by water is called as
    a. DBT
    b. wet-bulb temperature
    c. psychrometric temperature

Answer: b

19. Humid air A has the same moisture content as humid air B. However, the DBT of B is higher than that of A, which infers that
    a. A and B have the same relative humidity
    b. A has higher relative humidity
    c. B has higher relative humidity

Answer: b

20. For air at a given temperature, as the relative humidity is increased isothermally,
    a. the wet-bulb temperature and specific enthalpy increases
    b. the wet-bulb temperature and specific enthalpy decreases
    c. the wet-bulb temperature decreases and specific enthalpy increases

Answer: a

7.6.2 Numerical Problems

1. The air in a room has a DBT of 30°C and a wet-bulb temperature of 18°C. At the atmospheric pressure (1 atm), determine the specific humidity and relative humidity of the air using the psychrometric chart.

   Given
   i. DBT = 30°C
   ii. Wet-bulb temperature = 18°C
   iii. Pressure = 1 atm

   To find
   i. Specific humidity of the air
   ii. Relative humidity of the air

Solution

a. **Step 1:** Locate the properties T_{db} = 30°C and T_{wb} = 18°C on the psychrometric chart.
b. **Step 2:** Draw a vertical line from T_{db} = 30°C which crosses the T_{wb} = 18°C diagonal line. The intersection of these two lines denotes the state of the atmospheric air.
c. **Step 3:** Draw a line from the point of intersection towards the right to meet the vertical axis representing the humidity ratio of air, which gives a value of 11 g moisture content/kg dry air.
d. **Step 4:** To determine the relative humidity, follow the curved line that passes through the state point, which gives a value of 40%.

Answer: (i) Specific humidity of air = 11 g moisture content/kg dry air
(ii) Relative humidity of air = 40%

2. If the DBT and relative humidity of air are 40°C and 60%, respectively, find its specific volume, wet-bulb temperature, and moisture content.

Given

i. DBT = 40°C
ii. Relative humidity = 60%

To find

i. Specific volume
ii. Wet-bulb temperature
iii. Moisture content of air

Solution

In the psychrometric chart, a vertical line is drawn upwards from the DBT of 40°C until it intersects with the 60% RH line. With reference to the intersection point or state point, the specific volume is 0.925 m^3/kg, wet-bulb temperature is 31°C, and the moisture content is 25.5 g/kg dry air.

Answer: (i) Specific volume of air = 0.925 m^3/kg
(ii) Wet-bulb temperature of air = 31°C
(iii) Moisture content of air = 25.5 g/kg dry air

3. Calculate the total enthalpy and wet-bulb temperature of the air at 30°C DBT and 50% saturation.

Given

i. DBT = 30°C
ii. Relative humidity = 50%

To find

i. Total enthalpy
ii. Wet-bulb temperature of the air

Solution

In the psychrometric chart, a vertical line is drawn upwards from the DBT of 30°C until it intersects with the 50% RH line. With reference to the intersection point or state point, the total enthalpy of air is 65 kJ/kg dry air and the wet-bulb temperature of the air is 22°C.

Answer: (i) Total enthalpy of air = 65 kJ/kg dry air
(ii) Wet-bulb temperature of air = 22°C

4. Find the DBT and wet-bulb temperature of air which has a relative humidity of 60% and total enthalpy of 61 kJ/kg of dry air.

Given

i. Relative humidity of air = 60%
ii. Total enthalpy of air = 61 kJ/kg dry air

To find

i. DBT of air
ii. Wet-bulb temperature of air

Solution

In the psychrometric chart, an inclined line is drawn downwards from the total enthalpy line of 61 kJ/kg until it intersects with the 60% RH line. From the point of intersection, a vertical line is drawn downward to the horizontal axis representing DBT, which gives a value of $T_{db} = 27°C$. From the point of intersection, an inclined line is drawn upward until it meets the upper curved boundary on the left side of the psychrometric chart which signifies the scale of wet-bulb temperature (°C). Accordingly, the T_{wb} was found to be 21.5°C.

Answer: (i) T_{db} = 27°C; (ii) T_{wb} = 21.5°C

5. Air at a DBT of 30°C and relative humidity of 20% is humidified to attain a relative humidity of 50%. Determine the amount of moisture added in the humidifier per kg of dry air.

Given

i. DBT of air = 30°C
ii. Relative humidity of air = 20%
iii. Operation: Humidification of air
iv. Relative humidity of air after humidification = 50%

To find: Amount of moisture added in the humidifier per kg of dry air.

Solution

From the psychrometric chart, when a horizontal line is drawn from the point of intersection between the DBT line corresponding to 30°C and RH line corresponding to 20% to meet the vertical axis at the right, the amount of moisture in the air before humidification is 5.4 g/kg dry air. Similarly, with reference to the DBT of 30°C and RH of 50%, the amount of moisture in the air after humidification is 13 g/kg dry air.

$\therefore$ Amount of moisture added in the humidifier per kg of dry air = 13 − 5.4 = 7.6

Answer: Amount of moisture added in the humidifier = 7.6 g/kg dry air

6. For the drying of potato slices in a tray dryer, ambient air at 25°C and 70% relative humidity was heated to 70°C. The relative humidity of heated air is 40%. The air exits the dryer at 40°C and 50% saturation. Calculate (a) partial pressure of water vapor of the inlet air to the dryer and (b) partial pressure of water vapor in the outlet air. Vapor pressure of air at 25°C, 40°C, and 70°C is 3.1599, 7.3749, and 38.5575 kPa, respectively.

Given

i. Temperature of ambient air entering the dryer = 25°C
ii. Relative humidity of ambient air entering the dryer = 70%
iii. Temperature of heated air = 70°C
iv. Temperature of air exiting the dryer = 40°C

v. Vapor pressure at 25°C = 3.1599 kPa
vi. Vapor pressure at 40°C = 7.3749 kPa
vii. Vapor pressure at 70°C = 38.5575 kPa

To find

i. Partial pressure of the water vapor in the inlet air to the dryer
ii. Partial pressure of the water vapor in the air exiting the dryer

Solution

In the psychrometric chart, draw a vertical upward from $T_{db} = 70°C$ to intersect the 40% RH line. From the state point or point of intersection between the DBT and RH lines, draw a horizontal line towards the right to meet the vertical axis, which gives the value for absolute humidity of air entering the dryer equal to 88 g/kg dry air. It is known that, humidity is given by

$$H = \frac{p_W}{(P - p_W)}\frac{M_W}{M_A} = \frac{18.01534}{28.9645} \times \frac{p_W}{(P - p_W)} = 0.622\frac{p_W}{(P - p_W)}$$

$$0.088 = 0.622\frac{p_W}{38.5575 - p_W}$$

$$\therefore p_W = 4.765\,\text{kPa}$$

Similarly, for the outlet air from the dryer, from the state point or point of intersection between 50% RH line and 40°C DBT line, draw a horizontal line towards the right to meet the vertical axis, which gives the value for absolute humidity of air exiting the dryer equal to 26 g/kg dry air.

$$0.026 = 0.622\frac{p_W}{7.3749 - p_W}$$

$$\therefore p_W = 0.296\,\text{kPa}$$

Answer: (i) Partial pressure of the water vapor in the inlet air to the dryer is 4.765 kPa
(ii) Partial pressure of the water vapor in the air exiting the dryer is 0.296 kPa

7. Moist air at 30°C, 100 kPa, and 70% RH steadily enters a heater, where its temperature is increased to 120°C. If the pressure remains constant at 100 kPa, find the absolute humidity of the entering mixture and the relative humidity of the outgoing mixture. Vapor pressure of air at 30°C and 120°C are 2.339 and 198.541 kPa, respectively.

Given

i. Temperature of moist air = 30°C
ii. Pressure of moist air = 100 kPa
iii. Relative humidity of moist air = 70%
iv. Temperature of heated air = 120°C
v. Vapor pressure of air at 30°C = 2.339 kPa
vi. Vapor pressure of air at 120°C = 198.541 kPa

To find

i. Absolute humidity of entering mixture
ii. Relative humidity of outgoing mixture

Solution

In the psychrometric chart, draw a vertical upward from $T_{db} = 30°C$ to intersect the 70% RH line. From the state point or point of intersection between the DBT and RH lines, draw a horizontal line towards the right to meet the vertical axis, which gives the value for absolute humidity of air entering the dryer equal to 19 g/kg dry air.

$$\%RH = \frac{p_W}{p_{W0}} \times 100$$

During heating, the absolute humidity of air (H) remains constant. The relative humidity of heated air can be calculated from

$$\%RH = \frac{H}{H_0} \times 100$$

$$H = \frac{p_W}{(P - p_W)} \frac{M_W}{M_A} = \frac{18.01534}{28.9645} \times \frac{p_W}{(P - p_W)} = 0.622 \frac{p_W}{(P - p_W)}$$

$$0.019 = 0.622 \frac{p_w}{100 - p_w}$$

$$\therefore\ p_W = 2.964\,\text{kPa}$$

T_{wb} corresponding to $H = 0.019$ kg/kg dry air is 24°C.

$$p_W = p_{W0} - \frac{(P - p_{W0})(T_{db} - T_{wb})}{1555.56 - 0.722 T_{wb}}$$

$$2.964 = p_{W0} - \frac{(100 - p_{W0})(120 - 24)}{1555.56 - 0.722(24)}$$

$$2.964 = p_{W0} - \left(\frac{(100 - p_{W0})(96)}{1538.232} \right)$$

$$2.964 = \frac{1538.232 p_{W0} - 9600 + 96 p_{W0}}{1538.232}$$

$$4559.502 = 1634.232 p_{W0} - 9600$$

$$\therefore\ p_{W0} = \frac{14159.5}{1634.232} = 8.664$$

Therefore, saturated humidity (H_0) is given by

$$H_0 = \frac{p_{W0} M_W}{(P - p_{W0}) M_A} = 0.622 \frac{p_{W0}}{(P - p_{W0})}$$

$$H_0 = 0.622 \frac{8.664}{(100 - 8.664)} = 0.059$$

$$\%RH = \frac{0.019}{0.059} \times 100 = 32.2$$

Answer: (i) Absolute humidity of entering mixture = 0.019 kg/kg dry air
(ii) Relative humidity of exiting mixture = 32.2%

8. To prevent absorption of moisture owing to hygroscopic nature, the air in the packing room of an instant coffee manufacturing unit must have relative humidity of less than or equal to 20%. This is done by subjecting the air at 30°C and 70% RH to the cooling and dehumidification process, to result in air at 25°C and 20% RH. Represent the abovementioned processes on the psychrometric chart and find the amount of moisture removed from the air during the dehumidification process.

Given

i. Temperature of air before the cooling and dehumidification process = 30°C
ii. Relative humidity of air before cooling and dehumidification process = 70%
iii. Relative humidity of air after cooling and dehumidification process = 20%
iv. Temperature of air after cooling and dehumidification process = 25°C

To find

i. Representation of the cooling and dehumidification and heating processes on the psychrometric chart.
ii. Amount of moisture removed from the air during the dehumidification process.

Solution

i. In the psychrometric chart, draw a vertical upward from T_{db} = 30°C to intersect the 70% RH line. From the state point or point of intersection between the DBT and RH lines, draw a horizontal line towards the right to meet the vertical axis, which gives the value for absolute humidity of air entering the dryer equal to 0.0185 kg moisture/kg dry air. Similarly, for the cooled and dehumidified air draw a vertical upward from T_{db} = 25°C to intersect the 20% RH line. From the state point or point of intersection between the DBT and RH lines, draw a horizontal line towards the right to meet the vertical axis, which gives the value for absolute humidity of air entering the dryer equal to 0.0039 kg moisture/kg dry air. Thus, the cooling and dehumidification process can be represented on the psychrometric chart as indicated in the following figure:

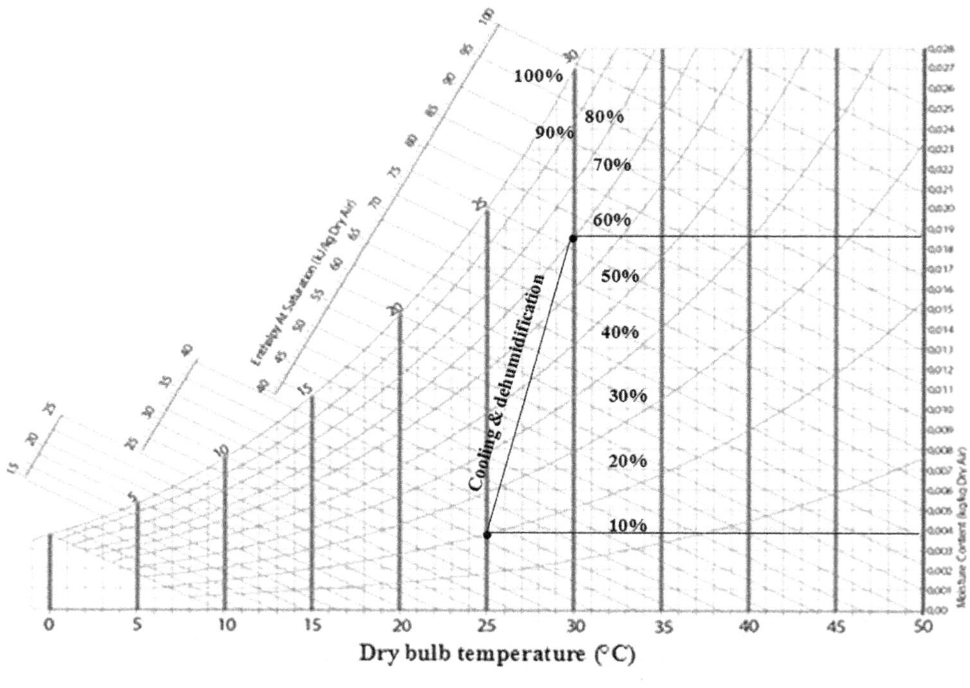

ii. Amount of moisture removed from the air during the dehumidification process is given by, 0.0185 − 0.0039 = 0.0146 kg = 14.6 g

Answer: Amount of moisture removed from the air during the dehumidification process is 14.6 g

BIBLIOGRAPHY

Anandharamakrishnan, C. 2008. Experimental and computational fluid dynamics studies on spray-freeze-drying and spray-drying of proteins. *PhD thesis*. UK: Loughborough University.

Ananthanarayanan, P. N. 2005. *Basic Refrigeration and Air Conditioning* (Third edition). New Delhi: Tata McGraw-Hill Education.

Balmer, R. T. 2011. *Modern Engineering Thermodynamics*. Burlington, MA: Academic Press, Elsevier Inc.

Bell, S. 2011. *A Beginner's Guide to Humidity Measurement*. Teddington, Middlesex, UK: NPL Management Ltd.

Dossat, R. J. 1997. *Principles of Refrigeration* (Fourth edition). Hoboken, NJ: Pearson Education Inc.

Driscoll, R. H. 2014. Dehydration. In *Food Processing: Principles and Applications* (Second edition), eds. S. Clark, S. Jung, and B. Lamsa, 61–78. Hoboken, NJ: John Wiley and Sons, Ltd.

Erdélyi, P. and Rajkó, R. 2016. Using interactive psychrometric charts to visualize and explore psychrometric processes. *Journal of Chemical Education* 93: 391–393.

Fontana, A. J. and Campbell, C. S. 2004. Water activity. In *Handbook of Food Analysis: Physical Characterization and Nutrient Analysis*, ed. L. M. L. Nollet, 39–54. New York: Marcel Dekker Inc.

Foster, K. D., Bronlund, J. E. and Paterson, A. H. J. 2005. The prediction of moisture sorption isotherms for dairy powders. *International Dairy Journal* 15: 411–418.

Genskow, L. R., Beimesch, W. E., Hecht, J. P., Kemp, I., Langrish, T., Schwartzbach, C. and Smith, F. L. 2008. Psychrometry, evaporative cooling, and solids drying (Section 12). In *Perry's Chemical Engineering Handbook* (Eighth edition), eds. R. Perry and D. Green, 12-1–12-109. New York: Mc-Graw Hill Education.

Moreno-Bondi, M. C., Orellana, G. and Bedoya, M. 2004. Fibre-optic sensors for humidity monitoring. In *Optical Sensors: Industrial Environmental and Diagnostic Applications*, eds. R. Narayanaswamy and O. S. Wolfbeis, 251–280. New York: Springer-Verlag Berlin Heidelberg.

NPTEL. 2018. Lesson 27 psychrometry. https://nptel.ac.in/courses/112105129/pdf/R&AC%20Lecture%2027.pdf (accessed October 15, 2018).

Ogawa, A. 2009. https://commons.wikimedia.org/wiki/File:PsychrometricChart.SeaLevel.SI.svg (accessed December 7, 2018).

Sapundjiev, P. V. 2009. Air humidity measurement using the psychrometric method. *Annual Journal of Electronics* 1: 168–171.

Talbot, M. T. and Baird, D. 1993. Psychrometrics and Postharvest Operations. CIR1097. Series of the Agricultural and Biological Engineering Department, Florida Cooperative Extension Service, Institute of Food and Agricultural Sciences, University of Florida. Original publication date May, 1993. Reviewed July, 2002, http://edis.ifas.ufl.edu (accessed August 11, 2017).

Wiederhold, P. 1995. Humidity measurements. In *Handbook of Industrial Drying* (Second edition), Revised and Expanded, Volume 1, ed. A. S. Mujumdar, 1313–1342. New York: Marcel Dekker Inc.

Wiederhold, P. R. 1997. *Water Vapor Measurement: Methods and Instrumentation*. New York: Marcel Dekker Inc.

8

Fundamentals and Applications of Reaction Kinetics

Food processing involves a sequence of changes that converts fresh agricultural commodities, meat, and marine products to edible and safer food products. Processing encompasses development of desirable components (flavors, aroma) and elimination of undesirable traits and constituents (off-odor, microbial load, degrading enzymes, and foreign substances). The changes that occur during food processing are quite complex and interactive. The changes could be chemical, physical, biological, nutritional, or sensorial (Figure 8.1). The abovementioned changes can be collectively termed as *reactions*. Reactions cause changes in composition and other characteristics of the raw material or finished product in response to the processing conditions and environmental stimuli during the preparation and storage phases, respectively. Processing conditions are the external factors which bring about the reactions. Temperature is an important external factor in food processing which brings about reactions such as browning, caramelization, and gelation. Formation of melanoids due to the browning reaction during baking contributes to the brown coloration and pleasant flavor of bread crust (Stauffer, 1990). The reactions that occur during one phase of the processing chain can continue in its next phase (Earle and Earle, 2003). The other factors include light, concentration of oxygen, presence of catalysts, pH, and water activity (Kumar et al., 2016).

Reactions are not limited to occur during food processing but are also responsible for the deterioration of food products during storage. Thus, the reactions can be desirable or undesirable concerning the quality of the food product. The examples of desirable reactions in food processing are hydrogenation of oils to produce solid fats and hydrolysis of collagen during cooking of meat. On the contrary, undesirable reactions that lead to food deterioration include enzymatic and chemical changes; nonenzymatic browning; rancidity due to lipid oxidation and hydrolysis; loss or degradation of nutrients, color, and texture.

A food technologist's key objective is to predict the rate at which the reactions occur in different food products under specified conditions. It is essential to know the reactions and their interrelationships to control them and obtain the desired products (Earle and Earle, 2003). The aforementioned reasons necessitate the knowledge of *kinetics*. It is a branch of science that involves the study of reaction rates and mechanisms, through the development of empirical and mathematical models. In simple terms, kinetics indicates how fast a reaction occurs. Kinetics also provides information on the factors that affect the reaction rate and the reaction mechanisms. The study of reaction kinetics is useful in the prediction of change in quality as a function of time and environmental conditions (Labuza, 1984). Kinetic studies of food systems are predominantly empirical owing to the complexities of measuring the concentrations of multiple reacting components in foods (Cavalieri and Reyes De Corcuera, 2009). Nevertheless, with

Chemical	Physical	Biological	Nutritional	Sensory
• Hydrolysis • Oxidation • Browning • Hydrogenation • Esterification	• Gelation • Hardening • Softening • Colour loss • Emulsification	• Microbial growth and death • Ripening of fruits	• Loss of vitamins • Destruction of anti-nutrients	• Loss of aroma and flavor • Changes in texture and color

FIGURE 8.1 The type of reactions occurring in foods.

an understanding of reaction kinetics, a food engineer can design novel processing techniques to achieve the finest product quality.

This chapter is intended to present the fundamental concepts of reaction kinetics pertaining to the food system with suitable examples. Further, the influence of temperature on reaction rates would be elaborated. The application of reaction kinetics in determining the shelf life of food product would be detailed in the latter part of this chapter with a case study.

8.1 Glossary of Reaction Kinetics

- **Reactors:** Reactors are the physically defined boundaries within which the reactions take place to convert the raw materials into products. A fermentor used in a brewery is a typical example of a reactor.
- **Stoichiometry of reactions:** Stoichiometry accounts for the mass balance of reactants entering a reaction and the resultant products. It facilitates the calculation of relative quantities of reactants and products in a reaction.

  Equation 8.1 is the general form of a stoichiometric equation for a reaction, which denotes that a moles of reactant A react with b moles of reactant B to produce x moles of product X and y moles of product Y. a, b, x, and y are termed as the stoichiometric coefficients of a chemical reaction.

$$aA + bB \xrightarrow{k} xX + yY \tag{8.1}$$

  The reaction given in Eq. (8.2) represents the oxidation of ascorbic acid ($C_6H_8O_6$) by hexacyanoferrate(III) ion to form dehydroascorbic acid ($C_6H_6O_6$). This reaction is an important example of a deteriorative change in food products which are rich in vitamin C.

$$H_2Asc + 2\left[Fe(CN)_6\right]^{3-} \xrightarrow{k} Asc + 2\left[Fe(CN)_6\right]^{4-} + 2H^+ \tag{8.2}$$

  In Eq. (8.2), H_2Asc stands for ascorbic acid, and Asc stands for dehydroascorbic acid. The stoichiometry shows that 1 mole of ascorbic acid is oxidized by 2 moles of hexacyanoferrate(III) ion to form 1 mole of dehydroascorbic acid and 2 moles of hexacyanoferrate(IV) ion (reduced), with the transfer of two electrons.
- **Reaction mechanism:** Reaction mechanism describes the stepwise sequence of elementary reactions that leads to the conversion of reactants into products. Food products often have multiple reactants. The major reacting components in foods include water, carbohydrates, lipids, amino acids, peptides, and proteins. The minor reactants are vitamins, minerals, colorants, and flavors (Fennema, 1996). In the abovementioned example of ascorbic acid oxidation, the reaction occurs through a sequence of four steps, as described in Eqs. (8.3)–(8.6).

  1. First ionization of ascorbic acid

$$H_2Asc \underset{k_{-1}}{\overset{k_1}{\rightleftarrows}} HAsc^- + H^+ \tag{8.3}$$

  2. First oxidation step (or first electron transfer)

$$HAsc^- + \left[Fe(CN)_6\right]^{3-} \xrightarrow{k_2} HAsc^{\cdot} + \left[Fe(CN)_6\right]^{4-} \tag{8.4}$$

  3. Second ionization step

$$HAsc^{\cdot} \xrightarrow{k_3} Asc^{-\cdot} + H^+ \tag{8.5}$$

4. Second oxidation step (or second electron transfer)

$$Asc^{-\cdot} + [Fe(CN)_6]^{3-} \xrightarrow{k_4} Asc + [Fe(CN)_6]^{4-} \tag{8.6}$$

- **Reaction rate:** In general, reaction rates are reported as the change in concentration of a reactant as a function of time, i.e., the rate of consumption of reactants. Reaction rate provides a measure of the reactivity and stability of a given component in a particular system. Alternatively, the rate of a reaction can also be defined as the change in concentration of the products with time, i.e., the rate of formation of products. Therefore, the rate of reaction given in Eq. (8.1) can be calculated as follows:

$$-r_A = -\frac{d[A]}{dt}\left[\frac{\text{mol}}{\text{m}^3 \cdot \text{s}}\right] \tag{8.7}$$

In the preceding equation, r_A denotes the rate of reaction with respect to the consumption of reactant *A* and negative sign indicates the reduction in its concentration with time. The unit of reaction rate is usually expressed in moles/liter second. The IUPAC recommends that the time should always be expressed in units of second (IUPAC, 1997). The relationship between the rates of all the reactants involved and the products formed is given by

$$\frac{-r_A}{a} = \frac{-r_B}{b} = \frac{r_X}{x} = \frac{r_Y}{y} \tag{8.8}$$

where r_B is the rate of reaction with respect to reactant *B*, r_x is the rate of reaction with respect to product *X*, and r_y is the rate of reaction with respect to product *Y*.

- **Rate law and rate constant:** The rate law or rate equation for a reaction is defined as an equation that defines the relationship between reaction rate, the concentration of reactants, and certain constant parameters such as the rate constant and partial order of the reaction. If Eq. (8.1) corresponds to an elementary gas-phase reaction, then the rate law is simple of the form

$$r = k[A]^a[B]^b \tag{8.9}$$

where [*A*] and [*B*] indicate the concentrations of reactants *A* and *B*, respectively. The proportionality constant (*k*) which relates the reaction rate and the concentrations of the reactants (or products) is termed as the *rate constant*. The rate constant is expressed in units of time inverse (s^{-1}, min^{-1}, or $days^{-1}$). However, in general, the rate law is expressed in terms of the concentration of both the reactants and products as given in Eq. (8.10), where [*X*] and [*Y*] indicate the concentrations of the products *X* and *Y*, respectively. Here, the exponents of the rate law (*p, q, r,* and *s*) are not necessarily related to the stoichiometric coefficients (*a, b, x,* and *y*) of the reaction.

$$r_A = k[A]^p[B]^q[X]^r[Y]^s \tag{8.10}$$

- **Activation:** It refers to the action of input of external energy into a chemical system to bring about activation of the system. In other words, it is the addition of a substance to increase the rate of a catalyzed reaction (IUPAC, 1996).
- **Energy of activation (E_a):** Also known as the activation energy, E_a is the quantity which explains the dependence of rate constant on temperature (IUPAC, 1996).
- **Order of reaction:** *Order* of a reaction is defined with respect to a specific substance, i.e., reactant, product, or catalyst, as the index or exponent to which its concentration term is raised to

in the rate equation (IUPAC, 2014). In general, the concentration of one of the reactants would be much higher than that of the other reactants. Hence, the concentration of other reactants does not affect the reaction rate significantly. In such a case, the equation for reaction rate is given by

$$-\frac{d[A]}{dt} = k[A]^n \tag{8.11}$$

The exponent n in Eq. (8.11) is termed the *order* of the reaction. Order of a reaction is defined as the sum of the exponents of the reactant concentration terms in the rate equation. Classification of reactions is done based on their order.

8.2 Classification of Reactors

A simple scheme for the classification of reactors is based on the mode of operation. Accordingly, reactors can be classified into three types:

1. Batch reactors
2. Semi-batch reactors
3. Continuous reactors

8.2.1 Batch Reactors

In a batch reactor, all the reactants are added to the reactor before commencing the reaction. The final product is removed after the completion of the reaction which is carried out under the desired conditions until the target conversion is attained. During the reaction, no material is added or withdrawn. Thus, the key variable in the operation of a batch reactor is the residence time for which the reactants are allowed to remain in the reactor to achieve the desired level of conversion. Typically, batch reactors are stirred tanks (Figure 8.2) attached to impellers, gas bubbles, or a pump. The stirrers facilitate the mixing of reactor contents. Owing to this simplicity of design, batch reactors are easy to be monitored and operated for almost all types of reaction. Temperature inside a batch reactor is regulated using internal cooling surfaces (coils or tubes), jackets, reflux condensers, or heat exchangers. A batch reactor is operated either isothermally or adiabatically.

Batch reactors are appropriate for lower production rates and longer reaction times. The advantage of this reactor type is its capacity to handle reactions with high degree of selectivity (e.g., polymerization reactions) and widely varied products in the same equipment. Fermentors used in the beverage plants (e.g., brewery), effluent treatment plants, and those used in the production of enzyme and yeast are all examples of batch reactors.

8.2.2 Continuous Reactors

In a continuous reactor, addition of reactants and withdrawal of products are carried out continuously at a constant mass flow rate. This is suitable for large production rates and can be further classified into the following two types:

8.2.2.1 Continuous Stirred Tank Reactors

A continuous stirred-tank reactor (CSTR), also called a mixed flow reactor (Figure 8.3), is a vessel to which reactants are added and products are removed while the reaction proceeds. The addition and removal of material to and from the vessel is carried out simultaneously. The contents within the CSTR are vigorously stirred by internal agitation or by recycling the contents internally or externally. Attributed to the stirred state of contents in a CSTR, the outlet concentration is assumed to be the same

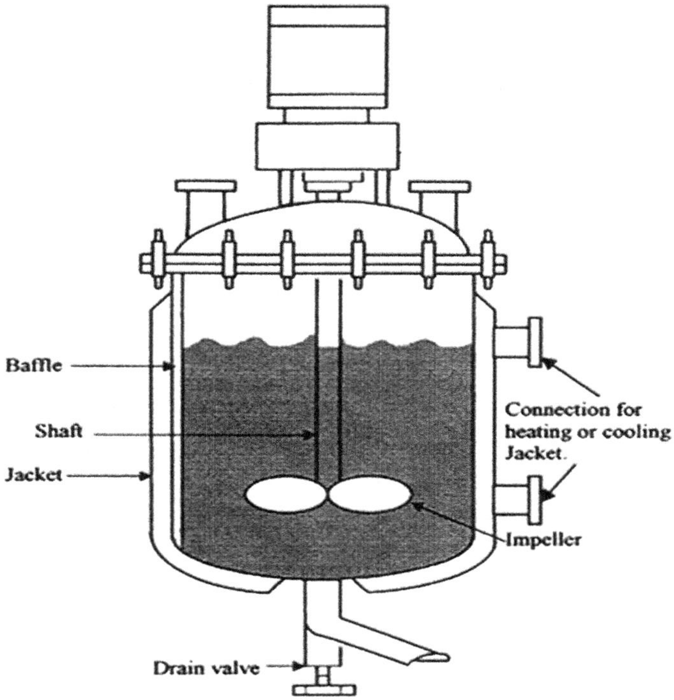

FIGURE 8.2 Schematic diagram of a batch reactor. (Reproduced with permission from Ahmed, E. M. 2015. Hydrogel: Preparation, characterization, and applications: A review. *Journal of Advanced Research* 6: 105–121.)

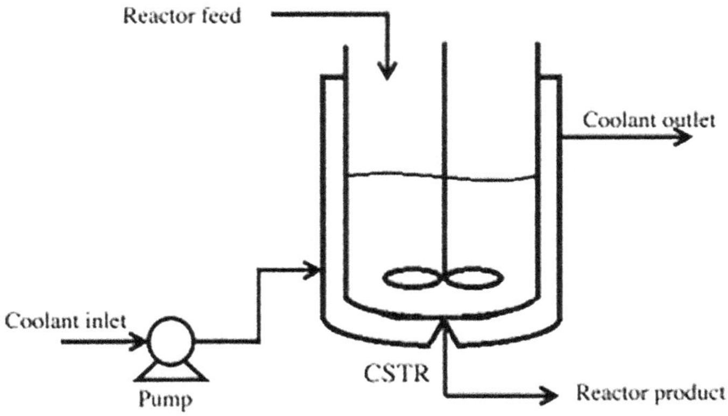

FIGURE 8.3 Schematic diagram of a CSTR. (Reproduced with permission from Zhao, D., Zhu, Q. and Dubbeldam, J. 2015. Terminal sliding mode control for continuous stirred tank reactor. *Chemical Engineering Research and Design* 94: 266–274.)

as the concentration at any point within the reactor. To improve the process economy, CSTRs can be arranged either in series or in parallel.

Residence time of reactants within a CSTR is the average amount of time, a discrete quantity of reactants spends inside the tank. It is given by

$$\text{Residence time} = \frac{\text{volumetric flow rate}}{\text{volume of the tank}}$$

CSTRs are predominantly used in industries handling gas–liquid reactions (e.g., hydrogenation of fats and oils) and homogeneous liquid-phase flow reactions (e.g., vinegar production), which require constant agitation. In waste treatment plants, a number of CSTRs are arranged in series such that the chemical and biological oxygen demands are significantly reduced before the water is discharged into river or urban sewage systems.

The CSTRs are economical in construction, easy to maintain, and exhibit good temperature control. In addition, a CSTR has large heat capacity and easy accessibility to its interior. However, the conversion of reactant to product per unit volume of a CSTR is less compared to other reactors.

In batch reactors and CSTRs, the type of stirrer used has an influence on the motion within the reactor contents and the effectiveness of mass transfer. An overview of the different types of stirrers used in reactors is provided in Table 8.1.

8.2.2.2 Plug Flow Reactors

Plug flow reactors (PFRs) comprise a hollow pipe or tube (Figure 8.4) through which the reactants (nutrients or microorganisms) are pumped continuously. The reactants flow through the reactor in the form of a "plug" with a constant velocity profile across the radius of the pipe. The reaction proceeds until the reactants move inside the PFR. Water at a controlled temperature is circulated through the tank to maintain the reactants at a constant temperature. A PFR may be operated as a closed (as in an oil pipeline) or open system. In an ideal PFR, it is assumed that there is no mixing of the medium along the

TABLE 8.1

An Overview of Stirrer Types in Reactors

Type of Stirrer	Schematic Diagram	Description
Paddle	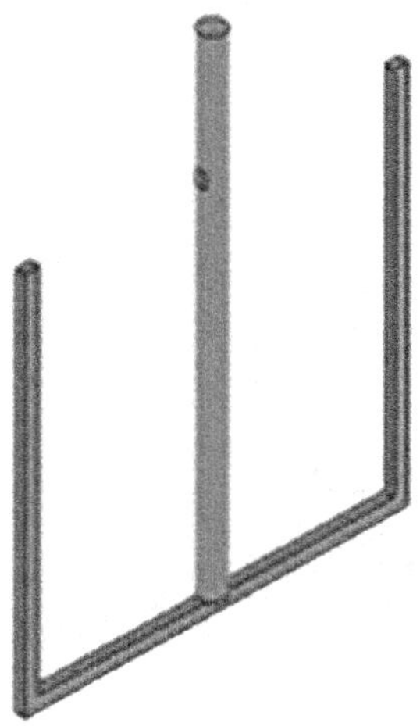	• Paddle agitators are used when a uniform laminar flow of liquids within the tank is desired. • The blades of a paddle agitator can reach up to the walls of the tank. • Paddle agitators are suitable for low speed mixing in the range of 20–200 rpm.
Turbine	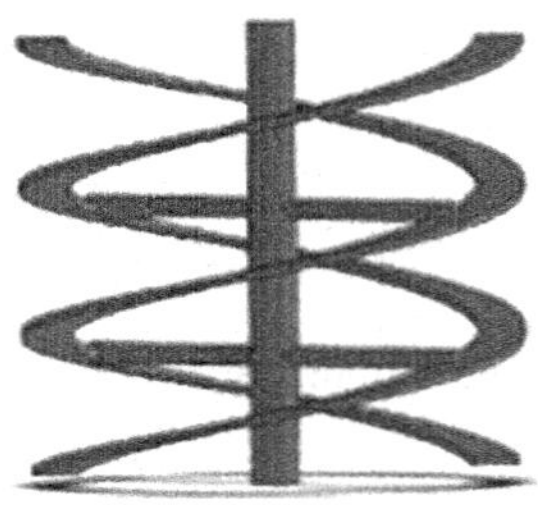	• Turbine agitators cause a turbulent movement of the reactor contents by a combination of centrifugal and rotational motions. • These are high-speed stirrers with its value of diameter ratio D/d ranging between 3 and 5, where, D is the diameter of the tank and d is the diameter of turbine blade. • There are four types of turbine stirrers: straight blade, pitched blade, curved blade, and disk blade. • Turbine agitators are suitable for low viscosity liquids and predominantly used in baffled tanks.

(*Continued*)

TABLE 8.1 ***(Continued)***

An Overview of Stirrer Types in Reactors

Type of Stirrer	Schematic Diagram	Description
Propeller	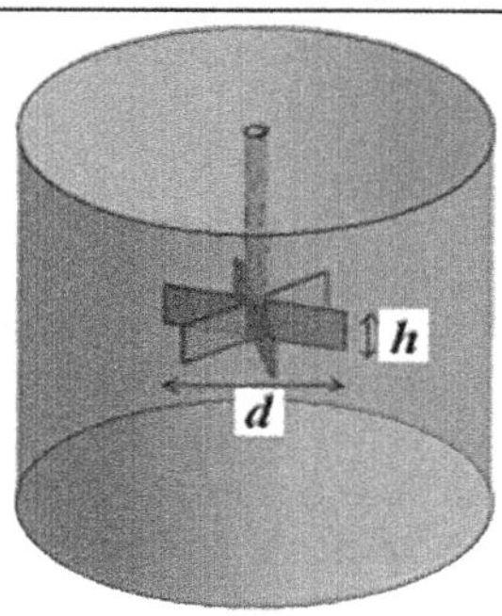 (*d* = diameter of propeller blade; *h* = height of propeller blade)	• Propeller agitators are high-speed stirrers, which accelerate the liquid in the axial direction. • These include blades tapering towards the shaft which minimize the centrifugal force and maximize the axial flow. • Propeller agitators are appropriate for low-viscosity liquids and baffled tanks.
Helical ribbon		• Helical ribbon stirrer is a slow-speed and close-clearance stirrer with its *D*/*d* ratio ≈ 1.05. • It operates in such a manner that the liquid on the wall is transported downward. • Helical ribbon stirrer is highly suitable for high-viscosity liquids.
Anchor	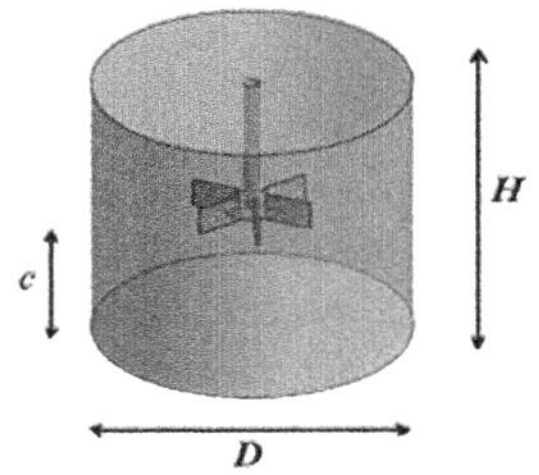 (*D* = diameter of tank; *H* = height of tank; *c* = clearance between the impeller and tank surface)	• Anchor agitator consists of a shaft and an anchor-type propeller. • These stirrers can be mounted within the reactor vessel either centrally or at an angle. • In general, the slow-speed anchor stirrers are used when a close wall clearance (*D*/*d* ≤1.05) is required to increase heat transfer in high-viscosity liquids.

Paddle, Turbine: Reproduced with permission from Ameur, H., Kamla, Y. and Sahel, D. 2017. Data on the agitation of a viscous Newtonian fluid by radial impellers in a cylindrical tank. *Data in Brief* 15: 752–756; **Propeller:** Reproduced with permission from Jirout, T. and Rieger, F. 2011. Impeller design for mixing of suspensions. *Chemical Engineering Research and Design* 89: 1144–1151; **Helical ribbon:** Reproduced with permission from Robinson, M. and Cleary, P. W. 2012. Flow and mixing performance in helical ribbon mixers. *Chemical Engineering Science* 84: 382–398; **Anchor:** Reproduced with permission from Ameur, H. 2015. Energy efficiency of different impellers in stirred tank reactors. *Energy* 93: 1980–1988.

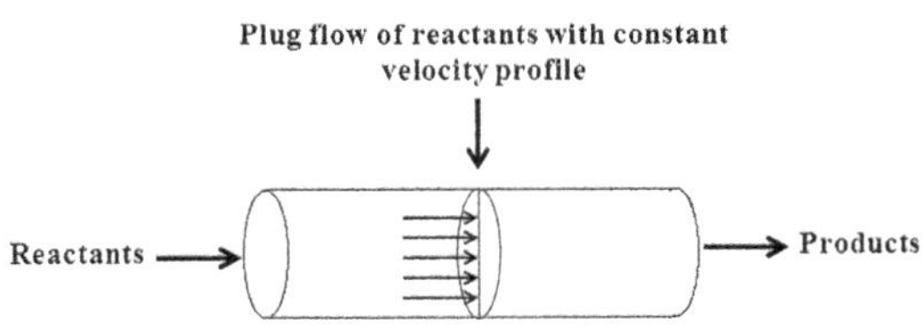

FIGURE 8.4 Schematic diagram of a PFR.

horizontal length (x-axis) of the reactor. Thus, PFR demonstrates a continuous concentration gradient in the direction of flow. However, there may be lateral mixing in the medium at any point along the vertical length (y-axis) of the reactor.

In general, a PFR operates at a higher efficiency than a CSTR of the same volume. The PFR is highly suitable for reactions that require a substantial heat transfer and where high pressures and extremely high or extremely low temperatures occur.

8.2.3 Semi-Batch Reactors

A semi-batch reactor is essentially a batch reactor but has either a continuous input or output stream during its operation. In this reactor, certain reactants are loaded in the beginning, while the others are fed gradually and continuously as the reaction proceeds. When the reactor is full, it is operated in the batch mode to complete the reaction. Thus, a semi-batch reactor is considered to be a combination of the batch and CSTRs.

Semi-batch reactors are chosen under the conditions of large heat effects and limited heat-transfer capability. Consequently, in a semi-batch reactor, exothermic reactions can be slowed down and endothermic reactions could be controlled by limiting the reactant concentration. This type of reactor is suitable for biological reactions which demand limiting of reactant concentration to reduce toxicity. Semi-batch reactors are also appropriate for the control of undesirable by-products and when one of the reactants is used in the gaseous state which has limited solubility and is fed continuously at the dissolution rate. A typical example is chlorination wherein the chlorine is used in the gaseous form. When introduced continuously, the chlorine gas would bubble off. Therefore, a continuous feed of gas is injected into a batch of liquid.

8.3 Classification of Reactions

8.3.1 Zero-Order Reaction

If $n = 0$ in Eq. (8.11), then the reaction is called as a *zero-order* reaction. Chemical reactions in foods that follow zero-order kinetics exhibit a constant rate of change in the concentration of a reactant or product. In other words, with zero-order reactions, the rate of reaction is independent of the reactant concentration and is equal to the rate constant. Zero-order reactions occur when a reactant is present in an excessively high concentration such that the reaction turns independent of the reactant concentration.

Therefore, the rate equation for a zero-order reaction can be written as

$$r = -\frac{d[A]}{dt} = k \tag{8.12}$$

In Eq. (8.12), the SI unit of the rate constant (k) is mol/L s or M/s. When Eq. (8.12) is integrated between the limits $t = 0$ and $t = t$, by assuming k as constant,

$$C_A = C_{A_0} - kt \tag{8.13}$$

where C_{A0} and C_A are the initial concentration and concentration at any time $= t$ of the reactant A, respectively.

In the zero-order reaction, a linear relationship is observed between the reaction time and concentration of the reactant which implies that the concentration of the reactant decreases linearly with time. The plot between concentration and time yields a straight line, and its negative slope gives the rate constant (Figure 8.5). Some of the zero-order reactions related to food processing include nonenzymatic browning, sourness due to inhibited rancidity, and quality losses in frozen food products.

The time required for the concentration of reactant A to be reduced to half its initial value is called as *half-life* or $t_{1/2}$. The half-life for a zero-order reaction is given by

$$t_{1/2} = \frac{C_{A_0}}{2k} \tag{8.14}$$

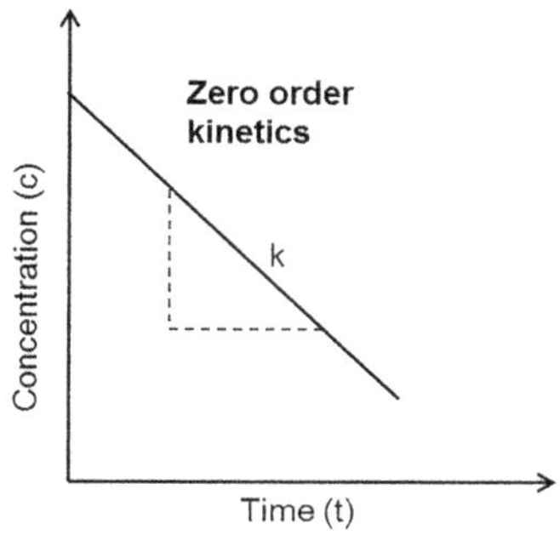

FIGURE 8.5 Graphical representation of zero-order kinetics.

8.3.2 First-Order Reaction

When $n = 1$, Eq. (8.11) becomes

$$-\frac{d[A]}{dt} = kC_A \tag{8.15}$$

For the first-order reaction, the SI unit of k is s^{-1}. The integration of Eq. (8.15) yields

$$C_A = C_{A_0} e^{-kt} \tag{8.16}$$

Thus, in the first-order reactions, the concentration of a reactant would undergo an exponential decline with time (Figure 8.6a).

Applying natural logarithm to both the sides of Eq. (8.16) gives

$$\ln \frac{C_A}{C_{A_0}} = -kt \tag{8.17}$$

The reaction rate constant is given by the slope of the plot between $\ln C_A/C_{A_0}$ and time t (Figure 8.6b).

According to Eq. (8.15), reactant A cannot be completely depleted. Examples of the first-order reaction related to food processing include starch gelatinization, color degradation during processing, and inactivation of enzymes and microorganisms.

The half-life for a first-order reaction is given by

$$t_{1/2} = \frac{\ln 2}{k} = \frac{0.693}{k} \tag{8.18}$$

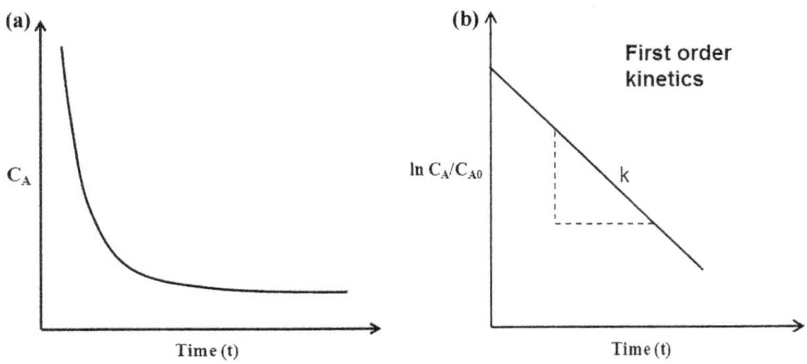

FIGURE 8.6 Graphical representation of first-order kinetics: (a) Change in the concentration of reactant with time; (b) Determination of reaction rate constant.

While determining the half-life of any process, it is important to note that the first half-life of the process reduces the original value to 50%. Further, during the second and third half-life periods, the concentration will decrease to 75% and 87.5% of its initial value, respectively. Therefore, one can get only infinitesimally close to the end but can never reach the true end.

Example 8.1

The half-life of a first-order reaction is 200 s. Find out the percent of the initial concentration of reactant remains after 1000 s?

Solution

Half-life of reaction = 200 s

Half-life is given by the equation

$$t_{1/2} = \frac{\ln 2}{k} = \frac{0.693}{k}$$

$$\therefore k = \frac{0.693}{t_{1/2}} = \frac{0.693}{200} = 3.465 \times 10^{-3}\ \text{s}^{-1}$$

$$\frac{C_A}{C_{A0}} = e^{-kt} = e^{-3.465\times10^{-3}\times1000} = 0.03127 = 3.13\%$$

Answer: 3.13% of initial concentration of the reactant remains after 1000 s

8.3.3 Second-Order Reaction

A second-order reaction is characterized by a hyperbolic relationship between the reactant or the product concentration and the reaction time. The rate constant for this reaction type is obtained by plotting $1/A$ versus time (t).

In the second-order reaction, a decrease in the concentration of reactant is proportional to the square of its concentration Eq. (8.19). Here, the unit of rate constant (k) is m^3/mol s or $M^{-1}s^{-1}$ where M = molarity = mol/m^3 or mol/L.

$$-\frac{d[A]}{dt} = k([A])^2 \tag{8.19}$$

On integrating Eq. (8.19),

$$\frac{1}{C_A} - \frac{1}{C_{A0}} = kt \tag{8.20}$$

Second-order reactions are infrequent with respect to food processing and are seldom reported in the case of Maillard reactions involving specific amino acids.

In the expression for the half-life of a second-order reaction, the rate constant depends on the initial concentration of a reactant.

$$t_{1/2} = \frac{1}{kC_{A0}} \tag{8.21}$$

8.3.4 n^{th} Order Reaction

Determination of the reaction order is based on a trial-and-error methodology, which involves assuming various values for n until the best fit is obtained. The general rate equation for an nth order reaction is given by

$$-\frac{d[A]}{dt} = k([A])^n; \quad n > 1 \tag{8.22}$$

On integrating Eq. (8.22),

$$C_A^{(1-n)} - C_{A_0}^{(1-n)} = -(1-n)kt \tag{8.23}$$

8.4 Temperature Dependence of Reaction Rates

8.4.1 Arrhenius Relationship

In general, reaction rate increases with increase in temperature. According to the collision model theory, for a reaction to occur, the molecules should collide. Molecules move faster at an increased temperature, thus increasing their collision frequency. Increase in the collision frequency will result in a faster rate of the reaction. Further, a greater number of molecules (due to the higher concentration) lead to a higher probability of collision. Therefore, an increase in the reaction temperature and the reactant concentration results in the enhancement of the reaction rate. However, Arrhenius found that the abovementioned theory has certain limitations as all the collisions do not lead to a reaction. Therefore, on careful observation of the reaction-rate data, he stated that molecules must possess a minimum amount of energy called the *activation energy*, to facilitate the reaction. This energy is required to initiate a chemical reaction and is specific for a particular reaction. For example, the activation energy for surface browning of bread is different than that for enzymatic surface browning of apple chunks. Arrhenius found that the reaction rate in most cases is based on three factors: the number of collisions per unit time, collision fractions with the correct orientation, and the fraction of colliding molecules possessing energy greater than or equal to the activation energy. Thus, the Arrhenius equation intends to describe the relationship between the temperature and the reaction rate.

$$k = A_0 e^{-(E_a/RT)} \tag{8.24}$$

where E_a is the activation energy [kJ/kmol], k is the rate constant, and A_0 is the pre-exponential factor. The activation energies for reactions that occur during food processing are compiled in Table 8.2. It is to be noted that the activation energy and pre-exponential factors are specific for each reaction. Arrhenius equation can be represented graphically as shown in Figure 8.7.

Therefore, from the Arrhenius equation, it can be inferred that the first-order rate constant increases exponentially with the rise in temperature. Arrhenius model is widely used in the food processing operations to define changes occurring in the final product quality, such as starch gelatinization and browning of bread surface.

TABLE 8.2

Activation Energies for Food Processing Reactions

Type of Chemical Reaction	Activation Energy (E_a) (kJ/mol)
General chemical reactions	40–100
Hydrolysis reactions	60–120
Lipid oxidation	40–100
Browning or nonenzymatic reactions	100–200
Degradation of vitamin	70–150
Protein denaturation or coagulation	200–500
Enzymatic reactions	100–200
Microbial growth	100–150
Inactivation of vegetative microorganisms	300–500
Inactivation of spores	250–350

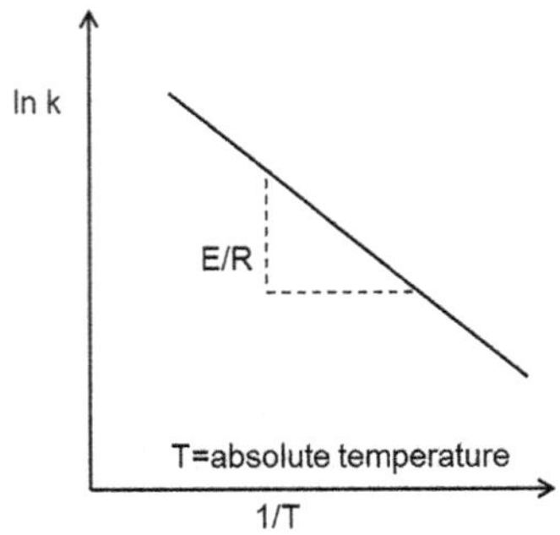

FIGURE 8.7 Schematic representation of the Arrhenius law.

The activation energy for a specific reaction can be determined experimentally by determining the rate constant at two different temperatures. Furthermore, if the rate constant and the activation energy at a particular temperature are known, then the rate constant at any other temperature can be determined by the following expression:

$$\ln\frac{k_2}{k_1}=\frac{E_a}{R}\left(\frac{1}{T_1}-\frac{1}{T_2}\right) \tag{8.25}$$

where k_1 is the rate constant at temperature T_1, and k_2 is the rate constant at temperature T_2. During the accelerated shelf-life study (ASLS) of a food product, the rate constant is determined at a higher temperature than that encountered in the real time. Then, the plot between the natural logarithm of k ($\ln k$) and $1/T$ gives a straight line which can be extrapolated to lower temperatures.

Example 8.2

Determine the temperature at which the rate constant for a reaction will be doubled if activation energy is 38.9 kJ/mol at 25°C.

Solution

Activation energy of the reaction (E_a) = 38.9 kJ/mol

The temperature at which $k_2 = 2k_1$ is to be determined.

According to the equation, $\ln\frac{k_2}{k_1}=\frac{E_a}{R}\left(\frac{1}{T_1}-\frac{1}{T_2}\right)$

$$\ln\frac{k_2}{k_1}=\ln\frac{2k_1}{k_1}=\ln 2=\frac{E_a}{R}\left(\frac{1}{T_1}-\frac{1}{T_2}\right)$$

$$\therefore 0.693=4.678\times10^3\left(\frac{1}{298}-\frac{1}{T_2}\right)$$

$$\therefore T_2=311.76\,\text{K}$$

$$\Delta T=311.76-298=13.76\,\text{K}$$

Answer: The reaction rate will double at 311.76 K or with a rise in temperature by 13.76 K

8.4.2 Q_{10} Value

The temperature dependence of reactions can also be defined by the Q_{10} value. It is defined as the factor by which the reaction rate is changed when the temperature is increased by 10°C. Suppose, if the reaction rate doubles with 10°C change in temperature, then, $Q_{10} = 2$. For nonenzymatic and enzymatic browning reactions and flavor changes in food, the Q_{10} value is close to 2. Q_{10} is defined by Eq. (8.26) as follows:

$$Q_{10} = \frac{k_{T+10}}{k_T} \tag{8.26}$$

If the Arrhenius equation holds true, then

$$Q_{10} = \exp\left(\frac{10E_a}{RT^2}\right) \tag{8.27}$$

However, the limitation of this parameter is that at a higher temperature, the Q_{10} value is largely influenced by the deactivation energy. It does not include any pre-exponential factor such as that present in the Arrhenius equation. In other words, Q_{10} has a strong dependence on temperature and thus needs to be reported along with its applicable temperature range.

8.4.3 *z* Value

The z value expresses the increase in temperature that would produce an increase in reaction rate by a factor of 10. This parameter is widely used to describe the thermal inactivation of microorganisms. The z value is different for each microorganism. The z value can be expressed in relation to the Arrhenius equation and Q_{10} value as given in Eq. (8.28).

$$z = \frac{2.303RT^2}{E_a} = \frac{10}{\log Q_{10}} \tag{8.28}$$

However, similar to the Q_{10} value, the z value is temperature dependent which restricts its applications.

8.5 Applications of Reaction Kinetics

The important applications of reaction kinetics of relevance to food system are (Labuza, 1984) as follows:

- Prediction of quality loss and deterioration of a food product.
- In deciding the *use-by* date for the food which is declared on the package to promote consumer awareness of product handling.
- To evaluate the effect of the addition of new ingredients, change in the packaging material and environmental conditions on the shelf life of a food product.

8.5.1 Determination of *Use-By* Date by Kinetics Study

A *use-by* date is applicable if the nutrient levels decline to unacceptable levels in a food product intended by the manufacturer to form the sole source of nutrition for a person's diet for a specified period (or) if the food product can become microbiologically unsafe before its spoilage which proceeds subsequently (Food Standards Australia New Zealand, 2013). A food scientist requires the following information to determine the *use-by* date of a food product using the principle of reaction kinetics (Labuza, 1984) as follows:

1. The most labile nutrient or ingredient of the specified food product.
2. The number of servings per package of the food product.

3. The minimum value for the nutrient (which constitutes the label claim of the product), to be present in the package at the time of consumption.
4. The percentage daily value (% DV) to be met by each serving of the packaged product.
5. The predominant mode of quality deterioration for the food product under study.
6. The processing conditions which control the initial quality or nutritional value of the food product.
7. The conditions including temperature, relative humidity, and light to which the food product would possibly be exposed during its storage.
8. The nature of packaging material and its characteristics such as permeability to oxygen, water vapor, and light.

8.5.1.1 Determination of Use-By Date for Packaged Guava Fruit Drink

The example of a packaged guava fruit drink product is considered to understand the methodology of determining the *use-by* date. Vitamin C is the most labile nutrient in this product. The data required for the estimation of *use-by* date for the packaged guava fruit drink are as follows:

- Label declaration states that each serving of the packaged guava fruit drink provides 100% of the DV for vitamin C.
- The DV for vitamin C is 60 mg (as stated by the U.S. Food and Drug Administration).
- Number of servings per package = 8.
- Total minimum vitamin C needed in the package at the time of consumption = number of servings per package × DV = 8 × 60 = 480 mg.
- To account for processing losses, vitamin C is supplemented to make up the total content in the product after processing to be 960 mg.

After the collection of the required data, an ASLS should be carried out at different temperatures, higher than the actual storage temperature of the product. This is practiced because the chosen quality parameter depletes slowly under actual storage conditions. The ASLS is conducted by subjecting the food to environmental conditions that speed up the quality deterioration, followed by extrapolation of the results to milder conditions which the product encounters in real time (Mizrahi, 2000). In general, the degradation of ascorbic acid in juices during the refrigerated storage follows the first-order reaction kinetics (Polydera et al., 2005; Burdurlu et al., 2006). Specifically, the degradation of ascorbic acid in guava juices upon refrigerated storage was evaluated using the first-order kinetic models Eq. (8.16). In this case, C_{A_0} in Eq. (8.16) represents the initial concentration of ascorbic acid in the fruit juice and C_A represents the concentration of ascorbic acid at time t, where t is the storage time and k is the degradation rate constant for ascorbic acid (day^{-1}). Half-life ($t_{1/2}$) is the estimated time at which the concentration of ascorbic acid would decrease by 50% from the initial value (i.e., from 960 to 480 mg). The rate constant at each temperature is determined from the Arrhenius plot as shown in Figure 8.8. The kinetics data required for the construction of the Arrhenius plot is given in Table 8.3. Then the k value can be extrapolated to the actual storage temperature. In this example, the temperatures for the ASLS are 5°C, 10°C, and 15°C and the actual storage temperature is 4°C (Sinchaipanit et al., 2015).

After data extrapolation using the Arrhenius plot (Figure 8.8) at 4°C, $k = 0.0486$ days^{-1}. The maximum allowable loss of vitamin C is said to occur when its content in the food product reaches one-half of its initial level, i.e., at the half-life for vitamin C under the specified conditions (Sinchaipanit et al., 2015). Thus, the half-life for a first-order reaction is given by

$$t_{1/2} = \frac{0.693}{0.0486} = 14 \text{ days}$$

So, if the guava drink product is packed on October 8, 2018 then its *use-by* date is October 22, 2018.

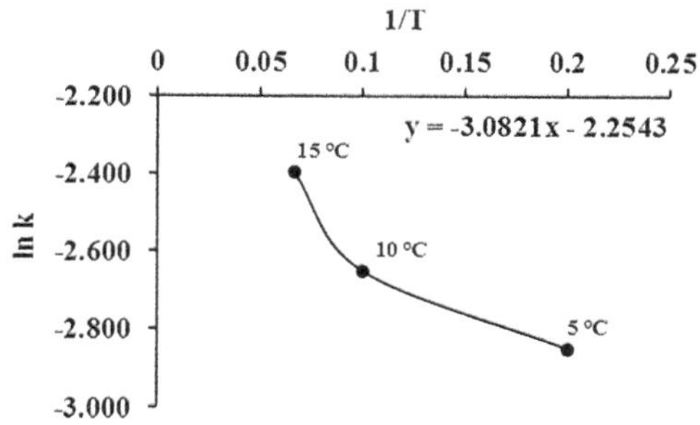

FIGURE 8.8 Arrhenius plot for the degradation of vitamin C (Plot prepared using the data from Sinchaipanit et al., 2015).

TABLE 8.3

Rate Constants for Ascorbic Acid Degradation in Guava Juice

Storage Temperature (°C)	**k (day^{-1})**
5	0.0579
10	0.0707
15	0.0912

Source: Data from Sinchaipanit et al. (2015).

8.6 Problems to Practice

8.6.1 Multiple Choice Questions

1. The SI unit of the rate constant for a first order reaction is,
   a. mol/L s
   b. s^{-1}
   c. mol^{-1}L s^{-1}
   d. mol/L

Answer: b

2. For the reaction to occur, the minimum energy required by the reacting molecules is
   a. potential energy
   b. kinetic energy
   c. thermal energy
   d. activation energy

Answer: d

3. The half-life of a first-order reaction is 5 min. The concentration of A would be reduced to 25% of its initial concentration in
   a. 10 min
   b. 90 min
   c. 33 min
   d. 70 min

Answer: a

4. The value of activation energy can be found from the plot between
   a. ln k versus T
   b. ln k versus 1/T
   c. k versus T
   d. 1/k versus T

Answer: b

5. The half-life time of a reaction is constant if the order of the reaction is
   a. zero
   b. one
   c. two
   d. three

Answer: b

6. For a first-order reaction, $t_{1/2}$ is 693 s. Then, the rate constant is given by
   a. $10^{-1} s^{-1}$
   b. $10^{-3} s^{-1}$
   c. $10^{-2} s^{-1}$
   d. $10^{-4} s^{-1}$

Answer: b

7. The effect of increasing the temperature of a reaction by 10°C on the reaction rate is
   a. reaction rate is doubled
   b. reaction rate is reduced by half
   c. reaction rate is tripled
   d. reaction rate remains constant

Answer: a

8. Reaction rate is equal to the slope of plot between
   a. time and rate
   b. mass and time
   c. volume and time
   d. time and concentration

Answer: d

9. The reaction rate doubles for every 10°C increase in temperature because of
   a. increase in activation energy
   b. decrease in activation energy
   c. increase in the number of molecular collisions
   d. increase in the number of activated molecules

Answer: d

10. The activation energy of a reaction is zero. The rate constant of the reaction
   a. increases with increase in temperature
   b. decreases with increase in temperature
   c. decreases with decrease in temperature
   d. is nearly independent of temperature

Answer: d

11. The rate constant of a reaction depends on
    a. concentration of reactants
    b. concentration of products
    c. time
    d. temperature

Answer: d

12. The SI unit of rate constant in a zero-order reaction is
    a. L mol/s
    b. mol/L s
    c. dimensionless
    d. s^{-1}

Answer: b

13. Order of reaction is decided by
    a. temperature
    b. mechanism of reaction as well as relative concentration of reactants
    c. volume
    d. pressure

Answer: b

14. For the reaction $A + B \rightarrow C$, it is found that doubling the concentration of A increases the rate by four times, and doubling the concentration of B doubles the reaction rate. What is the overall order of the reaction?
    a. 4
    b. 3/2
    c. 3
    d. 1

Answer: c

15. If the reaction rate doubles with 10°C change in temperature, then, Q_{10} is
    a. 2
    b. 3
    c. 1
    d. 4

Answer: a

16. Rate constant for a first-order reaction is $10^{-3}\,s^{-1}$. $t_{0.75}$ is
    a. 288 s
    b. 693 s
    c. 2880 s
    d. 28 s

Answer: a

17. The rate constant of a reaction has the same units as the rate of reaction. The reaction is of
    a. zero order
    b. first order
    c. second order
    d. third order

Answer: a

18. The rate constant of a reaction is $4 \times 10^{-3}\ M^{-1}s^{-1}$. The order of this reaction is
    a. zero
    b. 1
    c. 3
    d. 2

Answer: d

19. With an increase in temperature, the rate of a reaction
    a. increases
    b. decreases
    c. may increase or decrease
    d. remains constant

Answer: a

20. A reaction has the rate law expression of rate $= k\,[A]^2\,[B]^{1/2}$. The order of the reaction is
    a. 1
    b. 2
    c. 5/2
    d. 1/2

Answer: c

8.6.2 Numerical Problems

1. Consider the reaction of the form, $2A + B_2 \rightarrow 2AB$. The chemical kinetics of this reaction was studied at 25°C and the experimental data thus obtained are as follows:

Experiment Number	$[A]$ (mol/L)	$[B_2]$ (mol/L)	Initial rate (mol/L.s)
1	0.010	0.010	1.2×10^{-4}
2	0.010	0.020	2.4×10^{-4}
3	0.020	0.020	9.6×10^{-4}

i. Determine the order of the abovementioned reaction.
ii. Write the rate law for the reaction.
iii. Determine the rate constant's (k) value and unit.

Given

i. $2A + B_2 \rightarrow 2AB$
ii. Temperature of reaction = 25°C
iii. Experimental data on initial reaction rate

To find

i. Order of the reaction

ii. Rate law for the reaction
iii. Value of rate constant (k) with unit

Solution

i. Order of the reaction:
Comparing the given reaction with the general form of a reaction,

$$aA + bB \xrightarrow{k} xX$$

$$a = 2;\ b = 1;\ x = 2$$

$$\therefore \text{Order of the reaction} = 2 + 1 = 3$$

ii. Rate law for the reaction:
In general, rate law is given by, $r = k[A]^a[B]^b$
For the given reaction, $r = k[A]^2[B_2]^1$
$r = k[A]^2[B_2]$

iii. Value of rate constant (k) with unit
From the data for Experiment Number 1,

$$1.2\times 10^{-4} = k(0.010)^2(0.010)^1$$

$$\therefore k = \frac{1.2\times 10^{-4}}{(0.010)^3} = 120\ \text{s}^{-1}$$

Answer: (i) Order of the reaction = 3
(ii) Rate law for the reaction: $r = k[A]^2[B_2]$
(iii) $k = 120\,\text{s}^{-1}$

2. A first-order reaction is 40% complete in 600 s.
   i. Calculate the value of rate constant.
   ii. Determine the half-life of the reaction.
   iii. Determine the time taken for the reaction to attain 90% completion.

Given

i. $t_{40\%} = 600\,\text{s}$

To find

i. Value of rate constant (k)
ii. Half-life time of the reaction ($t_{1/2}$)
iii. Time taken for the reaction to attain 90% completion ($t_{90\%}$)

Solution

For a first-order reaction,

$$k = \frac{2.303}{t}\log\left(\frac{a}{a-x}\right)$$

Here, $x = \left(\frac{40}{100}\right)a = 0.4a$

Given that $t_{40\%} = 600\ \text{s} = 10\,\text{min}$

$$\therefore k = \frac{2.303}{10}\log\left(\frac{a}{a-0.4a}\right)$$

$$k = \frac{2.303}{10}\log\left(\frac{1}{0.6}\right)$$

$$k = 0.0511\ \text{min}^{-1}$$

$$t_{1/2} = \frac{0.693}{k} = \frac{0.693}{0.0511} = 13.564 \approx 13.6\ \text{min}$$

when $x = 0.9a$ and $k = 0.0511\ \text{min}^{-1}$,

$$0.0511 = \frac{2.303}{t_{90\%}}\log\left(\frac{a}{a-0.9a}\right)$$

$$t_{90\%} = \frac{2.303}{0.0511}\log\left(\frac{1}{0.1}\right)$$

$$\therefore t = 45.068\ \text{min} \approx 45\ \text{min and } 4\ \text{s}$$

Answer: (i) Value of rate constant, $k = 0.0511\ \text{min}^{-1}$
(ii) Half-life time of the reaction, $t_{1/2} = 13.6\ \text{min}$
(iii) Time taken for the reaction to attain 90% completion, $t_{90\%} = 45\ \text{min and } 4\ \text{s}$

3. Rate constant for a reaction is 1.5×10^{-4} L/mol s at 20°C and 1.4×10^{-3} L/mol s at 40°C. Evaluate the activation energy for the reaction.

Given

i. $T_1 = 20°\text{C}$
ii. $T_2 = 40°\text{C}$
iii. $k_1 = 1.5 \times 10^{-4}$ L/mol s
iv. $k_2 = 1.4 \times 10^{-3}$ L/mol s

To find: Activation energy for the reaction

Solution

$$\ln\frac{k_2}{k_1} = \frac{E_a}{R}\left(\frac{1}{T_1} - \frac{1}{T_2}\right)$$

$$\ln\frac{\left(1.4\times10^{-3}\right)}{\left(1.5\times10^{-4}\right)} = \frac{E_a}{8.314}\left(\frac{1}{20} - \frac{1}{40}\right)$$

$$2.2336 = E_a(0.003)$$

$$\therefore E_a = 742.8\ \text{J/mol}$$

Answer: Activation energy for the reaction, $E_a = 742.8\ \text{J/mol}$

4. A first-order reaction depicts a half-life of 360 s. Determine the percentage of initial concentration of the reactant that remains after 1800 s?

Given

i. First-order reaction
ii. $t_{1/2} = 360\ \text{s}$

To find: Percentage of initial concentration of the reactant that remains after 1800 s.

Solution

For a first-order reaction,

$$t_{1/2} = \frac{0.693}{k}$$

$$\therefore k = \frac{0.693}{360} = 1.925 \times 10^{-3}\ \text{s}^{-1}$$

$$\ln \frac{C_A}{C_{A0}} = -kt$$

$$\ln \frac{C_A}{C_{A0}} = -\left(1.925 \times 10^{-3} \times 1800\right)$$

$$\ln \frac{C_A}{C_{A0}} = -3.465$$

$$\frac{C_A}{C_{A0}} \times 100 = 3.127\%$$

Answer: 3.127% of the initial concentration of reactant will remain after 1800 s

5. Determine the temperature at which the rate constant doubles for a reaction if the activation energy is 36 kJ/mol at 30°C.

Given

i. $E_a = 36$ kJ/mol = 36000 J/mol
ii. T = 30°C = 303 K

To find: Temperature at which the rate constant doubles for the reaction, i.e., when $k_2 = (2 \times k_1)$.

Solution

$$\ln \frac{k_2}{k_1} = \frac{E_a}{R}\left(\frac{1}{T_1} - \frac{1}{T_2}\right)$$

$$\ln 2 = \frac{36000}{8.314}\left(\frac{1}{303} - \frac{1}{T_2}\right)$$

$$0.693 = 4330.046\left(0.0033 - \frac{1}{T_2}\right)$$

$$1.6 \times 10^{-4} = 0.0033 - \frac{1}{T_2}$$

$$\frac{1}{T_2} = 0.0033 - \left(1.6 \times 10^{-4}\right)$$

$$\frac{1}{T_2} = 0.00314$$

$$\therefore T_2 = 318.5\ \text{K} = 45.5°\text{C}$$

Answer: At 45.5°C, the rate constant would double for the given reaction.

6. A reaction depicts first-order kinetics. At 298 K, the half-life time of the reaction is 4000 s. At 310 K, the concentration is reduced to half of its initial concentration after 900 s. Calculate: (i) the rate constant for the reaction at 298 K and (ii) the activation energy of the reaction.

Given

i. First-order reaction
ii. $T_1 = 298$ K
iii. $T_2 = 310$ K
iv. $t_{1/2}$ at 298 K = 4000 s
v. $t_{1/2}$ at 310 K = 900 s

To find

i. Rate constant for the reaction at 298 K
ii. Activation energy for the reaction

Solution

For a first-order reaction,

$$t_{1/2} = \frac{0.693}{k}$$

At $T_1 = 298$ K,

$$4000 = \frac{0.693}{k_1}$$

$$\therefore k_1 = \frac{0.693}{4000} = 1.7 \times 10^{-4}\ \text{s}^{-1}$$

At $T_2 = 310$ K,

$$900 = \frac{0.693}{k_2}$$

$$\therefore k_2 = \frac{0.693}{900} = 7.7 \times 10^{-4}\ \text{s}^{-1}$$

$$\ln \frac{k_2}{k_1} = \frac{E_a}{R}\left(\frac{1}{T_1} - \frac{1}{T_2}\right)$$

$$\ln \frac{\left(7.7 \times 10^{-4}\right)}{\left(1.7 \times 10^{-4}\right)} = \frac{E_a}{8.314}\left(\frac{1}{298} - \frac{1}{310}\right)$$

$$1.511 = \left(1.562 \times 10^{-5}\right) E_a$$

$$\therefore E_a = \frac{1.511}{1.562 \times 10^{-5}} = 0.967 \times 10^{5}\ \text{J/mol} = 96.7\ \text{kJ/mol}$$

Answer: (i) Rate constant for the reaction at 298 K = 1.7 × 10⁻⁴s⁻¹
(ii) Activation energy for the reaction = 96.7 kJ/mol

7. Rate constant for a reaction is 2×10^{-4} L/mol s at 25°C and 1.5×10^{-3} L/mol s at 37°C. If the initial reactant concentration is 0.05 M, determine the activation energy for the reaction and the half-life of the reaction at 37°C. Determine the order of the reaction.

Given

i. $T_1 = 25°C = 298$ K
ii. $T_2 = 37°C = 310$ K
iii. $k_1 = 2 \times 10^{-4}$ L/mol s
iv. $k_2 = 1.5 \times 10^{-3}$ L/mol s

Initial reactant concentration = 0.05 M

To find

i. Activation energy for the reaction (E_a)
ii. Half-life of the reaction at 37°C ($t_{1/2}$)
iii. Order of the reaction

Solution

$$\ln\frac{k_2}{k_1} = \frac{E_a}{R}\left(\frac{1}{T_1} - \frac{1}{T_2}\right)$$

$$\ln\frac{(1.5\times10^{-3})}{(2\times10^{-4})} = \frac{E_a}{8.314}\left(\frac{1}{298} - \frac{1}{310}\right)$$

$$2.015 = (1.562\times10^{-5})E_a$$

$$\therefore E_a = \frac{2.015}{1.562\times10^{-5}} = 1.29\times10^5 \text{ J/mol} = 129 \text{ kJ/mol}$$

$$t_{1/2} = \frac{1}{k\times C_{A_0}} = \frac{1}{(1.5\times10^{-3}\times0.05)} = 13333\,\text{s} = 222\,\text{min}$$

From the unit of rate constant, it can be inferred that the reaction is of the second order.

Answer: (i) E_a = 129 kJ/mol
(ii) $t_{1/2}$ at 37°C = 13333 s
(iii) Order of the reaction = 2

8. The rate of this reaction depends only on NO_2: $NO_2 + CO \longrightarrow NO + CO_2$.
The following data were collected:

Time (s)	Concentration (mg/L)
0	0.5
1200	0.444
3000	0.381
4500	0.340
9000	0.250
18000	0.174

Determine

a. Rate law for this reaction
b. Concentration of NO_2 at 2.7×10^4 s after the reaction starts

Given

i. $NO_2 + CO \longrightarrow NO + CO_2$
ii. Rate of the reaction depends only on NO_2
iii. Experimental data

To find

i. Rate law for the reaction
ii. Concentration of NO_2 at 2.7×10^4s after the reaction starts

Solution

Since it is given that the rate of the reaction depends only on one of the reactants, i.e., NO_2, the reaction is of the first order.

For a first order reaction, the rate constant is given by slope of the plot between $\ln\left(C_A/C_{A0}\right)$ versus time.

Time (s)	Concentration (mg/L)	$\left(C_A/C_{A0}\right)$	$\ln\left(C_A/C_{A0}\right)$
0	0.5	1	0
1200	0.444	0.888	–0.119
3000	0.381	0.762	–0.272
4500	0.340	0.680	–0.386
9000	0.250	0.500	–0.693
18000	0.174	0.348	–1.056

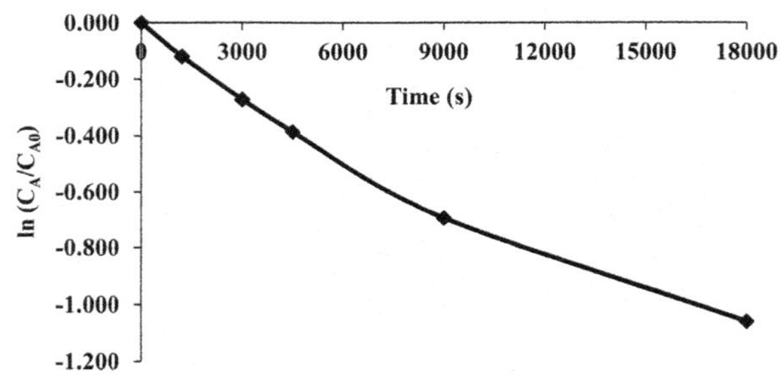

Negative slope of this plot gives $k = 5.781 \times 10^{-5}\,s^{-1}$.

i. Rate law for the reaction

   In general, rate law is given by $r = k[A]^a[B]^b$

   For the given reaction, $r = k[NO_2]^1[CO]^1$

   $r = (5.781 \times 10^{-5})[NO_2][CO]$

ii. Concentration of NO_2 at 2.7×10^4s after the reaction starts

$$\ln\frac{C_A}{C_{A0}} = -kt$$

$$\ln\frac{C_A}{C_{A0}} = -\left(5.781 \times 10^{-5}\right)\left(2.7 \times 10^4\right)$$

$$\ln\frac{C_A}{C_{A0}} = -1.562$$

$$\frac{C_A}{C_{A0}} = 0.210$$

$$\therefore C_A = \left(0.210 \times C_{A0}\right) = 0.210 \times 0.5 = 0.105\,\text{mg/L}$$

Answer: (i) Rate law: r = $(5.781 \times 10^{-5})[NO_2][CO]$

(ii) $[NO_2]$ at 2.7×10^4s after the reaction starts = 0.105 mg/L

9. The following is the data of a kinetics study for a zeroth-order chemical reaction. Determine the rate constant for this reaction.

Time (s)	Concentration (mol)
0	120
60	100
120	90
180	60
240	48
300	26

To find: Rate constant for the reaction

Solution

Plotting the data between concentration and time,

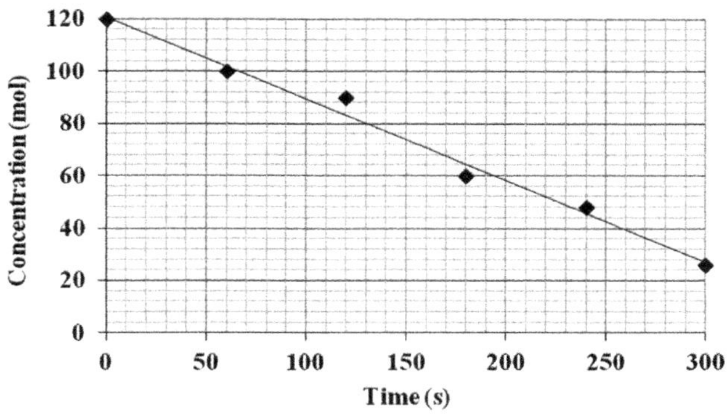

Slope of this plot gives the value for rate constant:

$$\text{Slope} = \frac{dy}{dx} = 0.3124\,\text{mol/s} = k$$

Answer: Rate constant = k = 0.3124 mol/s

10. The following data of reactant concentration versus time were obtained for a first-order reaction. Determine the rate constant and half-life of the reaction.

Time (s)	Concentration (mol/L)
0	320
100	150.2
200	64.6
300	28.4
400	12.4
500	5.0

Given

i. First-order reaction
ii. Experimental data for the concentration of a reactant as a function of time.

To find

i. Rate constant (k)
ii. Half-life of the reaction ($t_{1/2}$)

Solution

Time (s)	Concentration (mol/L)	(C_A/C_{A_0})	$\ln(C_A/C_{A_0})$
0	320	1	0
100	150.2	0.469	−0.757
200	64.6	0.202	−1.599
300	28.4	0.089	−2.419
400	12.4	0.039	−3.244
500	5.0	0.016	−4.135

Negative slope of the plot between $\ln(C_A/C_{A_0})$ and time gives the rate constant for the reaction (k).

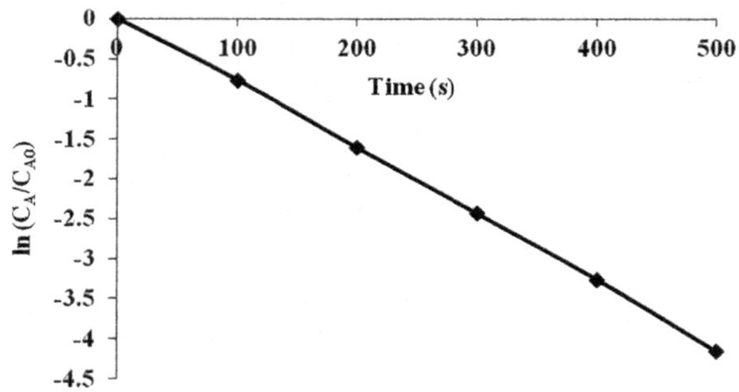

The slope of this plot gives $k = 0.0083\ s^{-1}$.

For a first-order reaction,

$$t_{1/2} = \frac{0.693}{k} = \frac{0.693}{0.0083} = 83.5\ s$$

Answer: (i) $k = 0.0083\,s^{-1}$
(ii) $t_{1/2} = 83.5\,s$

BIBLIOGRAPHY

Ahmed, E. M. 2015. Hydrogel: preparation, characterization, and applications: a review. *Journal of Advanced Research* 6: 105–121.

Ameur, H. 2015. Energy efficiency of different impellers in stirred tank reactors. *Energy Volume* 93: 1980–1988.

Ameur, H., Kamla, Y. and Sahel, D. 2017. Data on the agitation of a viscous Newtonian fluid by radial impellers in a cylindrical tank. *Data in Brief* 15: 752–756.

Burdurlu, H. S., Koca, N. and Karadeniz, F. 2006. Degradation of vitamin C in citrus juice concentrates during storage. *Journal of Food Engineering* 74: 211–216.

Cavalieri, R. P. and Reyes De Corcuera, J. I. 2009. Kinetics of chemical reactions in foods. In *Food Engineering—Vol. I*, ed. G. V. Barbosa-Cánovas, 361–405. Oxford: Eolss Publishers Co. Ltd.

Earle, R. L. and Earle, M. D. 2003. *Fundamentals of Food Reaction Technology*. Palmerston North, New Zealand: The New Zealand Institute of Food Science and Technology (Inc.).

Fennema, O. R. 1996. *Food Chemistry*. New York: Marcel Dekker, Inc.

Food Standards Australia New Zealand 2013. Date marking user guide to standard 1.2.5—Date marking of food. www.foodstandards.gov.au/code/userguide/Documents/Guide%20to%20Standard%201.2.5%20-%20 Date%20Marking%20of%20Food.pdf (accessed June 10, 2018).

IUPAC 1996. *Compendium of Chemical Terminology: The "Gold Book"*, eds. A. D. McNaught and A. Wilkinson. Research Triangle Park, NC: International Union of Pure and Applied Chemistry.

IUPAC 1997. *Compendium of Chemical Terminology: The "Gold Book"*, eds. A. D. McNaught and A. Wilkinson. Research Triangle Park, NC: International Union of Pure and Applied Chemistry.

IUPAC 2014. *Compendium of Chemical Terminology: The Gold Book", Version 2.3.3.*, eds. A. D. McNaught and A. Wilkinson. Research Triangle Park, NC: International Union of Pure and Applied Chemistry. http:goldbook.iupac.org/index.htm (accessed June 10, 2018).

Jirout, T. and Rieger, F. 2011. Impeller design for mixing of suspensions. *Chemical Engineering Research and Design* 89: 1144–1151.

Kumar, K., Yadav, A. N., Vyas, P. and Singh, K. 2016. Chemical changes in food during processing and storage. *5th National Conferences on Chemical Sciences: Emerging Scenario and Global Challenges* (NCCS-2016).

Labuza, T. P. 1984. Application of chemical kinetics to deterioration of foods. *Journal of Chemical Education* 61: 348–358.

Mizrahi, S. 2000. Accelerated shelf life tests. In *The Stability and Shelf Life of Foods*, eds. D. Kilcast and P. Subramaniam, 107–208. Cambridge: Woodhead Publishing.

Polydera, A. C., Stoforos, N. G. and Taoukis, P. S. 2005. Quality degradation kinetics of pasteurised and high pressure processed fresh Navel orange juice: Nutritional parameters and shelf life. *Innovative Food Science and Emerging Technology* 6: 1–9.

Robinson, M. and Cleary, P. W. 2012. Flow and mixing performance in helical ribbon mixers. *Chemical Engineering Science* 84: 382–398.

Sinchaipanit, P., Ahmad, M. and Twichatwitayakul, R. 2015. Kinetics of ascorbic acid degradation and quality changes in guava juice during refrigerated storage. *Journal of Food and Nutrition Research* 3: 550–557.

Stauffer, C. E. 1990. *Functional Additives for Bakery Foods*. New York: Van Nostrand Reinhold.

Zhao, D., Zhu, Q. and Dubbeldam, J. 2015. Terminal sliding mode control for continuous stirred tank reactor. *Chemical Engineering Research and Design* 94: 266–274.

9

Evaporation

In certain food processing operations, raw materials contain more water than is to be attained in the final product. This necessitates the removal of extra water to achieve the required product quality and texture, such as that in a jam-making process. Fresh fruits majorly used for jam preparation such as mango and strawberry, have a high moisture content of 85% (Mercer, 2012) and 89%–91% (Shi et al., 2008), respectively. During the jam-making process, excess water is removed by boiling the fruit pulp with sugar syrup and pectin until a minimum total soluble solid (TSS) content of 65% is reached in the final viscous product. This process of removing excess water is known as *concentration*. In the food industry, the concentration of liquid products is predominantly achieved by *evaporation*.

9.1 The General Principle of Evaporation

Evaporation is a unit operation in which a liquid product is heated to its boiling point for the partial removal of its volatile solvent by vaporization, to increase the concentration of nonvolatile solids. In food products, the volatile solvent is almost always water. In general, removal of water from the product is achieved by indirect heating (Figure 9.1). The product and heating medium (e.g., steam) are separated from each other by a partition, which is usually made of stainless steel (food grade; SS 304/SS 316). The driving force for heat transfer is the difference in temperature between the steam and the product. Steam temperature can be varied by altering the steam pressure. At the atmospheric pressure (1 atm), water boils at 100°C. However, the boiling point is different at other pressures. At its boiling point, the steam condenses in the coils and gives up its latent heat which is transferred to the product via the partition. Subsequently, the product is heated to its boiling point, and a portion of its water content is removed as

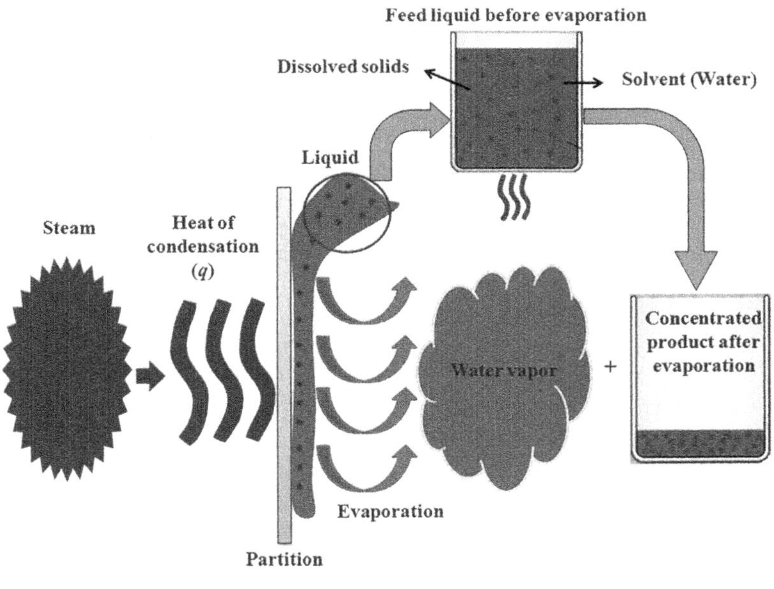

FIGURE 9.1 Schematic of the general principle of evaporation.

the vapor. As a result, the concentration of dissolved solids in the final product is higher than that in the feed liquid before heating (Figure 9.1).

Evaporation is used in food industries for the concentration of sugar, milk, sodium chloride, fruit juices, and many other products. In most cases, the concentrated fraction is the desired product and the water removed is discarded. But, in processes such as desalination, water is the required product. In addition to concentration, evaporation also results in the reduction of water activity and a consequent increase in the microbial stability and shelf life of food products. Apart from achieving the intended quality, stability, safety, and consistency of the final product, evaporation decreases the weight and volume of the product, thus reducing its transportation and storage costs. Evaporation is also used for the pre-concentration of feed liquids before drying, which leads to the economic removal of moisture in the dryers. Although evaporation involves vaporization of a volatile solvent, it is different from distillation, in that, the vapors produced are not fractionated further. Also, it is different from drying as the final product is a flowable liquid instead of a solid.

This chapter deals with the evaporator components and varied evaporator types of relevance to the food industry. Approaches to improve evaporator efficiency are also discussed.

9.2 Evaporator

An evaporator also termed as an *effect* (Figure 9.2) is a device used to remove water from a feed solution or slurry by evaporation. The feed to an evaporator is always in the liquid form, and the concentrate discharged retains its liquid form even after the water is evaporated. Every evaporator requires a source of heat, which is almost always low-pressure steam. The steam required for the operation of the evaporator may be produced in large boilers, gas turbine, vapor compressor, or a combination thereof. Alternatively, an evaporator itself can be used as a low-pressure steam generator or boiler to supply steam to another evaporator. The different components of an evaporator are discussed subsequently.

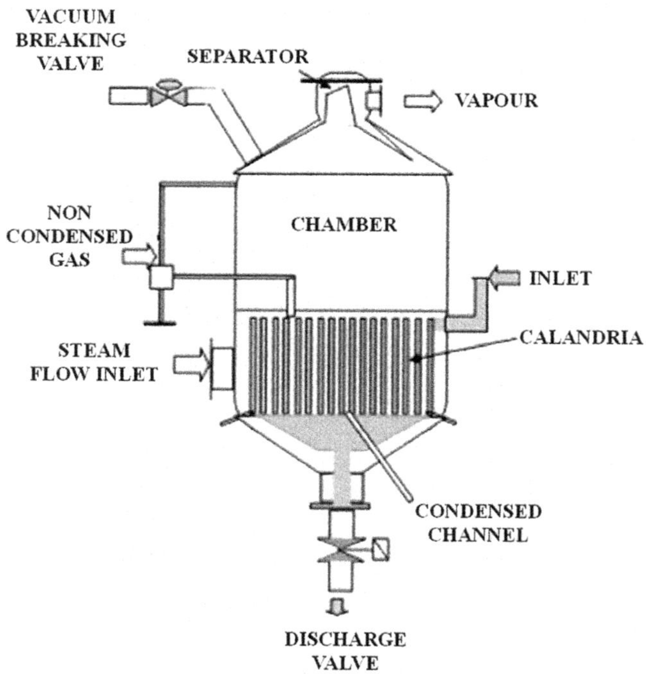

FIGURE 9.2 Schematic diagram of an evaporator effect. (Reproduced with permission from Miranda, V. and Simpson, R. 2005. Modelling and simulation of an industrial multiple effect evaporator: tomato concentrate. *Journal of Food Engineering* 66: 203–210.)

9.2.1 Components of Evaporator

i. **Heat exchanger:** Evaporators require a means to transfer the heat energy from the low-pressure steam to the liquid food product. For this purpose, most evaporators use a shell and tube heat exchanger. Termed as the *calandria*, the heat exchanger is the heart of an evaporator. In the heat exchanger shell, steam condenses on the outside of the tubes after giving up its latent heat (of condensation) to the product. The liquid food product is circulated inside the tubes, which absorbs the heat given up by the steam. The heat, thus, added causes the water in the liquid product to boil. Generally, the product to be evaporated is preheated to its boiling temperature before it enters the calandria. Jacketed vessel (with agitator), straight tube preheaters, or plate heat exchangers are used for this purpose.

ii. **Deaeration/vacuum systems:** As some products are heat-sensitive, the boiling temperature and holding time in the evaporator should be carefully selected to avoid heat damage of the product. To handle heat-sensitive liquids, evaporation is carried out under vacuum, by integrating the evaporator with a vacuum pump. Jet vacuum pumps or liquid ring vacuum pumps are used depending on the size and the operating mode of the evaporation plant. Apart from reducing the working pressure, the vacuum pumps also function to discharge the noncondensable gases which may arise from the dissolved gases introduced into the liquid feed or air leakage.

iii. **Evaporation section:** Here, the liquid product is boiled to vaporization under vacuum to reduce its boiling point, so that it boils at a low temperature to prevent heat damage. The reduced boiling point of the liquid product also increases the temperature gradient between the product and the steam.

iv. **Vapor–liquid separator:** Either the vapor or the concentrated liquid, or both, may be the desired product of evaporation. Therefore, the evaporator should offer a clean separation of vapors from the condensate and feed. To achieve this, an evaporator is equipped with a centrifugal separator or a gravitational separator. The performance of a vapor–liquid separator is judged by its separation efficiency, pressure drop, and the frequency of cleaning required.

v. **Condenser:** A condenser functions to remove the bulk of the volume of vapors separated from the liquid food product. Within a condenser, the vapors are condensed to liquid by removing their latent heat of condensation. Vapor condensers may be of the surface type or barometric type. Surface condensers provide sufficient heat transfer surface, through which the condensing vapors transfer their latent heat to the cooling water circulating in pipes. A barometric condenser is a direct contact condenser in which the vapors and the cooling water are in contact with each other. Condenser duty (q_c) is the amount of heat that should be removed from the vapors to condense them. q_c is determined using the following expression:

$$q_c = V(h_g - h_c) \tag{9.1}$$

where h_g is the enthalpy of vapor in the evaporator, h_c is the enthalpy of the condensate, and V is the quantity of vapor to be condensed.

vi. **Pumps:** In most evaporators, a centrifugal pump is used to circulate the low-viscosity feed liquid through the heat exchanger and separator. The major criteria for pump selection are product properties, suction head conditions, flow rates, and pressure ratios in the evaporation plant. Highly viscous products require the use of positive displacement pumps. On the other hand, for liquids containing solids or crystallized products, propeller pumps are used.

vii. **Cleaning systems:** Scaling and fouling of the internal surfaces of evaporator might occur after a specific time of operation. The extent of scaling and fouling depends on the product's nature. Chemical cleaning is the commonly adopted approach for the removal of deposits. According to the type of deposit, the cleaning agents are chosen which penetrate the incrustation, dissolve or disintegrate it, and thoroughly cleans and sterilizes the evaporator surfaces. The evaporation plant should be equipped with the necessary components, cleaning agent tanks, additional pumps, and fittings. The equipment which ensures the ease of cleaning without disassembly is commonly referred to as *Cleaning-in-Place* or *CIP* system.

viii. **Vapor scrubbers:** A vapor scrubber is installed when the plant is heated with discharge stream such as the exhaust vapors from evaporators or dryers, rather than the live steam. This is significant as the vapors must be cleaned to avoid contamination and fouling before they are transferred into the heating chamber of the evaporation plant.

ix. **Condensate polishing systems:** In certain instances, the condensate quality might not correspond to the required purity. This is possible especially if the product contains volatile constituents. Thus, depending on its end use, the condensate can be further purified by using a rectification column or a membrane filtration system.

9.3 Boiling Point Elevation

The boiling point of a solution increases with the increase in its concentration. Thus, boiling point elevation (BPE) of a solution is defined as the *increase in boiling point of a solution over that of pure water at a given pressure.* Denoted by ΔT_b, BPE can be determined by an empirical expression based on the molality, *m*, of the solution as given in Eq. (9.2):

$$\Delta T_b = 0.51\text{m} \tag{9.2}$$

Another means of determining BPE is by using the Dühring's rule, which explains the linear relationship between the boiling point temperature of the solution and that of water at the same pressure. In other words, the ratio of the temperature is constant at which two solutions exert the same pressure. The Dühring's lines for sodium chloride solution are shown in the Figure 9.3. The methodology for using Dühring's chart to calculate BPE is explained in Example 9.1.

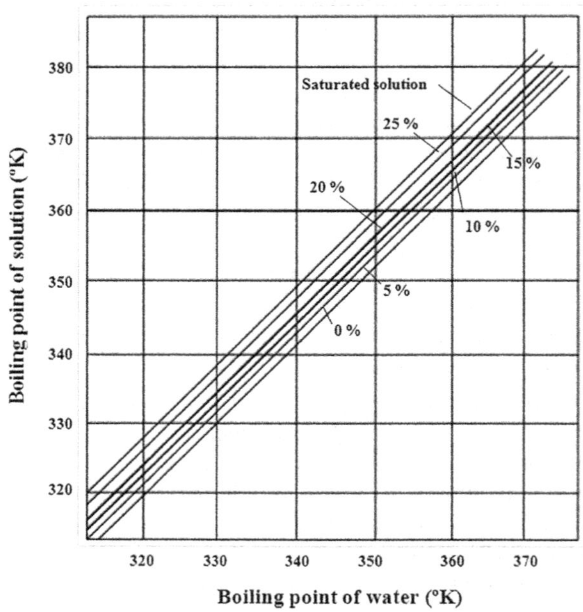

FIGURE 9.3 Schematic of Dühring chart indicating the influence of solute concentration on the BPE of sodium chloride solution.

Example 9.1

Using the Dühring chart, determine the effect on the boiling point of water when 40 g of sodium chloride is added to 120 mL of water.

Solution

Given: Concentration of sodium chloride solution = 40 × 100/(40 + 120) = 25%;

Boiling point of water = 373 K

Thus, from the Dühring chart (Figure 9.3), on constructing a vertical line from the *X*-axis at 373 K to reach the 25% concentration line of sodium chloride and extrapolating it to the *Y*-axis, boiling point of the NaCl solution is 379 K.

Thus, adding 40 g of sodium chloride to 120 mL of water will increase the boiling by (379 − 373) = 6 K.

Answer: Addition of sodium chloride 40 g to 120 mL of water increases its boiling point by 6 K

9.4 Mass and Energy Balance Around the Evaporator

The overall material balance for a single-effect evaporator is given by

$$F = V + P \tag{9.3}$$

where F is the mass flow rate of feed (kg/s), V is the mass flow rate of vapor (kg/s), and P is the mass flow rate of product (kg/s).

Component balance for the solid is expressed as

$$x_F \cdot F = x_P \cdot P \tag{9.4}$$

where x_F is the mass fraction of solids in the feed and x_P is the mass fraction of solids in the product.

Enthalpy (total heat) balance around the evaporator is given by

$$H_F \cdot F + H_{VS} \cdot S = H_V \cdot V + H_P \cdot P + H_{CS} \cdot S \tag{9.5}$$

where H_F is the enthalpy of feed, (kJ/kg); H_{VS} is the enthalpy of saturated vapor at saturation temperature, (kJ/kg); H_V is the enthalpy of saturated vapor at boiling temperature T, (kJ/kg); H_P is the enthalpy of concentrated product, (kJ/kg); H_{CS} is the enthalpy of condensate, (kJ/kg); and S is the mass flow rate of steam (kg/s).

In Eq. (9.5), H_F is determined by the expression

$$H_F = C_{PF}(T_F - 0°\text{C}) \tag{9.6}$$

where C_{PF} is the specific heat content of the feed (kJ/kg °C). It is assumed that saturated steam is used in the evaporator, and therefore, H_{VS} is obtained from the steam table at saturation temperature.

In Eq. (9.5), the term H_V on the right-hand side is obtained from the steam table at temperature T. H_P, the second term on the right-hand side of the Eq. (9.5), is determined by the expression

$$H_P = C_{PP}(T - 0) \tag{9.7}$$

where C_{PP} is the specific heat content of the concentrated product, (kJ/kg °C).

The rate of heat transfer, q(W), in the heat exchanger of an evaporator is given by the expression

$$q = UA(T_S - T) = S \cdot H_{VS} - H_{CS} \cdot S \tag{9.8}$$

where U is the overall heat transfer coefficient (W/[m^2 K]) and A is the area of the heat exchanger (m^2). Overall heat transfer coefficient (U) does not remain constant for the evaporator as it varies with the concentration and boiling point rise of the product. However, a constant value of overall heat transfer coefficient is assumed and used in the energy balance calculations.

9.5 Evaporator Capacity and Steam Economy

Evaporator capacity is defined as the mass of water vaporized per hour. The steam economy is the mass of water vaporized (m_V) per unit mass of the steam fed to the evaporator. It is given by Eq. (9.9), in which m_S is the mass of steam supplied.

$$\therefore \text{Steam economy} = \frac{m_V}{m_S} \tag{9.9}$$

Steam consumption is an important parameter to be considered while designing and evaluating the performance of an evaporator. It is determined from the evaporator capacity and steam economy, using the expression given as follows:

$$\therefore \text{Steam consumption} = \frac{\text{Evaporator capacity}}{\text{Steam economy}}$$

9.6 Types of Evaporators

The working principle and operation of the common types of evaporators (Figure 9.4) used in the food industry are elaborated in this section. Further, case studies on the established application of each evaporator type for the manufacturing of a specific product are also presented in the relevant sections.

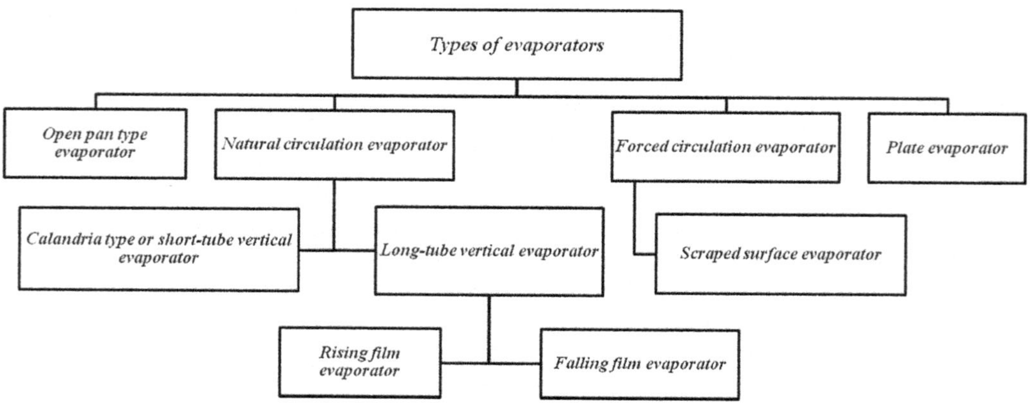

FIGURE 9.4 Classification of evaporators.

9.6.1 Batch Pan Evaporator

i. **Construction**

Batch pan evaporator (Figure 9.5) is one of the oldest and simplest types of evaporators used in the food industry. It consists of a steam jacketed, hemispherical-shaped open pan, in which the liquid food product is boiled to achieve the concentration of soluble solids. Heat is supplied to the pan either through the steam jacket or by using electric heating coils or heaters.

ii. **Working**

In general, open pan evaporation is carried out under atmospheric conditions. Heat-sensitive products are essentially boiled at low temperature under vacuum. Heat transfer occurs through natural convection.

iii. **Advantages and disadvantages**

The heat transfer area per unit volume is less owing to the vessel shape. Consequently, the heat transfer coefficient is low under natural convection conditions and the residence time of the product is longer in the order of several hours. The abovementioned factors limit the evaporation capacity and reduce the heat economy of pan evaporators. Consequently, this evaporator type is expensive with respect to its running cost, despite the lower capital cost. Scrapers or paddles may be fitted to agitate the liquid product within the pan to improve heat transfer.

iv. **Major applications in the food industry**

In the earlier years, batch pan evaporator was predominantly used for the concentration of corn syrups. Amidst the advancements in the evaporator technology, one of the major applications for which batch pan evaporator is still used in the food industry is in the production of jams, jellies, ketchup, and sauces.

9.6.1.1 Application of Pan Evaporator in Jam Manufacturing

This section is intended to explain a vacuum evaporation process for the production of jams containing fruit pieces or whole fruit. A typical jam-processing unit is shown in Figure 9.6. First, the ingredients are weighed into the jacketed premix vessel, the functions of which are to preheat the mix and dissolve the sugar added. After mixing, the premix is drawn under vacuum into the vacuum pan evaporator having zoned dimple panel heating elements. The pan evaporator consists of a combination of scraped

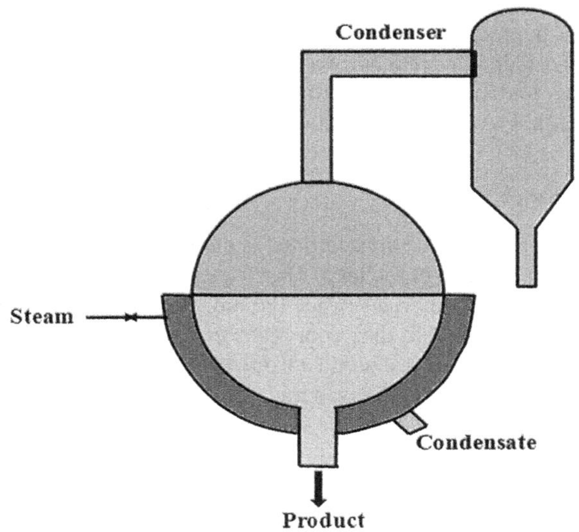

FIGURE 9.5 Schematic diagram of batch pan evaporator.

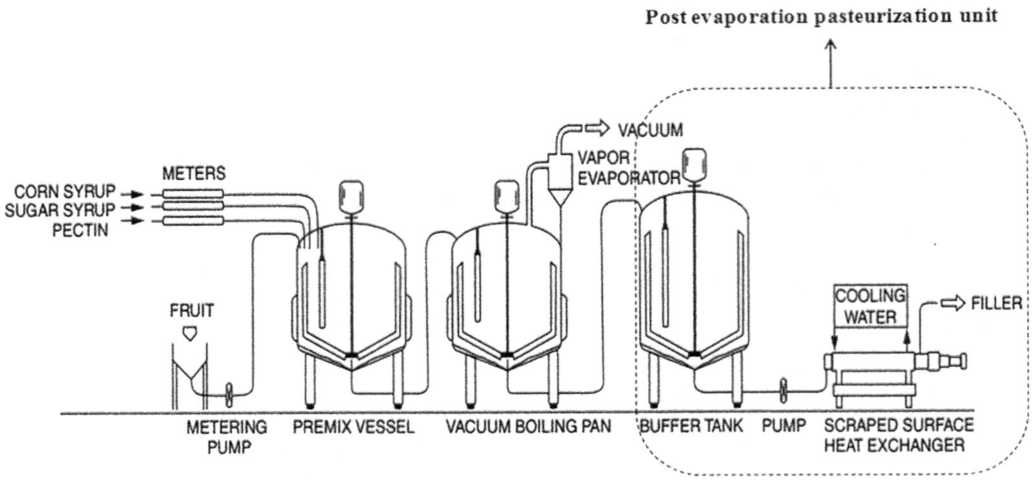

FIGURE 9.6 Schematic diagram of the jam processing unit. (Modified and reproduced with permission from Baker, R. A., Berry, N., Hui, Y. H. and Barrett, D. M. 2005. Fruit preserves and jams. In *Processing Fruits Science and Technology* (Second edition), eds. D. M. Barrett, L. Somogyi and H. S. Ramaswamy, 113–125. Boca Raton, FL: CRC Press.)

surface agitator and baffle which lead to an efficient heat transfer and rapid but gentle blending of all the ingredients. Rapid vacuum evaporation takes place at 60°C–65°C (140°F–149°F). The vacuum-mediated low evaporation temperature reduces thermal degradation and leads to a high-quality product. Finally, the vapor is separated in a cyclone separator. Product carryover if any is returned to the pan evaporator. The vapor condensate may be reused in recipe makeup or CIP. Optionally, a partial condensation volatile recovery system may be integrated with the evaporator to recover the volatile flavors lost during evaporation.

Further, the jam-manufacturing process shown in Figure 9.6 depicts an additional unit operation: "post-evaporation pasteurization," which facilitates the production of jams with a wide range of TSS with or without preservatives. This is because the low-temperature vacuum evaporation may not be effective in inactivating the spoilage microorganisms present in the ingredients. After completion of low-temperature evaporation, the product is transferred using top-filtered air pressure to the buffer tank. From the buffer tank, it is pumped to a scraped surface heat exchanger to achieve flash pasteurization, before filling. The pasteurization is carried out at a temperature of 85°C–95°C (185°F–203°F) for a short holding time followed by rapid cooling to the filling temperature. Pasteurization is carried out under pressure to avoid any volatile loss (Baker et al., 2005).

9.6.2 Tubular Evaporators

In tubular evaporators, the product to be concentrated is circulated inside a bundle of tubes. The saturated steam heats the tubes, horizontal or inclined. The steam condenses on the outer tube surface to heat the product within. As the product vaporizes inside the tube, the vapor pressure at the product surface causes it to proceed as a thin film within the vapor at the center. The tubes are connected to a cyclone separator, which separates the concentrated product from the vapors. While the vapors are directed from the separator towards a condenser or to a next evaporator or *effect*, the product is led out of the equipment or to another effect for further concentration. The circulation velocity of the product is chosen such that it increases the heat transfer coefficient and facilitates the removal of scales deposited along the inner surface of the tube. In the case of viscous products, circulation through the tubes is facilitated by the use of pumps (Palmer, 1993).

The steam-heated tubular evaporators can be classified as natural and forced circulation evaporators, depending on whether the product is introduced into the top or the bottom of the tubes.

9.6.2.1 Natural Circulation Evaporator

As indicated by the name, in this evaporator type, circulation of the liquid product inside the tubes is natural, driven by the difference in specific gravity between the bulk liquid and the heated liquid and vapor generated inside the tubes. As the liquid product (to be concentrated) enters the tubes, it is picked up by a rotating assembly of blades and thrown against the tube wall. The blades are mounted on a shaft placed coaxially with the inner tube. Thus, the product is continuously spread as a thin liquid film on the tube wall by the blades. This facilitates handling of viscous, heat-sensitive, crystallizing, and fouling products. The blade may be of different configurations (rollers, wipers, pitched blades, or fixed clearance blades) and maintain a close clearance of ≤1.5 mm from the inner tube wall. Gravity drives the downward flow of film which becomes concentrated during its transit. The vapor–liquid mixture rises inside the tubes due to the upward movement of vapors. Finally, vapors are separated from the liquid at the top of the heating tubes. The concentrated product moves along the downcomer which is eventually withdrawn from the bottom of the evaporator (Figure 9.7).

Natural circulation evaporators are suitable for dilute liquid food products with no suspended solid particles. The choice of this evaporator type is appropriate for non-Newtonian liquid foods and products which have a high tendency to foul. The operation can be batch, semi-batch, or continuous and do not require pumps for recirculation or intermediate transfer of the liquid product. Larger evaporation capacity is possible with the external heater which is independent of the size or shape of the vessel. However, owing to a longer residence time of the product inside the tube, heat-sensitive foods cannot be concentrated using this evaporator type.

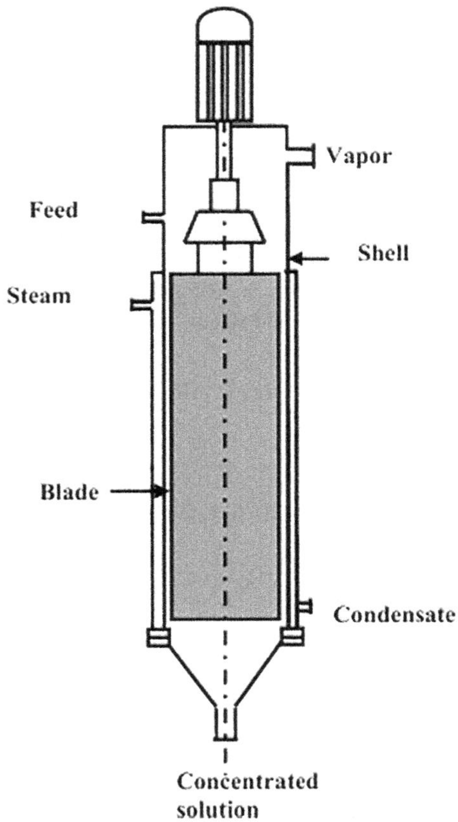

FIGURE 9.7 Natural circulation evaporator. (Reproduced with permission from Pawar, S. B., Mujumdar, A. S. and Thorat, B. N. 2012. CFD analysis of flow pattern in the agitated thin film evaporator. *Chemical Engineering Research and Design* 90: 757–765.)

The natural circulation evaporators can be further classified into calandria-type or short-tube vertical evaporator and long-tube vertical evaporator.

9.6.2.1.1 Short-Tube Vertical or Calandria-Type Evaporators

i. **Construction**

Short-tube vertical evaporators consist of a short tube bundle enclosed within a cylindrical shell, which is collectively called as the *calandria*. Tubes of diameter 1–4 in and length 30 in to 6 ft are commonly used (Perry et al., 1963). The tube bundle extends across the body of the evaporator in the vertical direction and a downspout is located at the center (Figure 9.2). Usually, the cross section of downspout varies from 50% to 100% of the flow area of the tubes, which ensures low frictional resistance to flow through the downspout (Ahmad and Khan, 2013).

ii. **Working**

The tubes are rolled between two tube sheets. Liquid product circulates in the tubes, and the steam is introduced in a tangential direction into the region outside the tubes (Figure 9.2). Typical of a natural circulation evaporator, the circulation of liquid product in the calandria-type evaporator is facilitated by the density difference. The liquid is heated and partially vaporized in the tubes. The maximum heat transfer coefficient is achieved when the level is just halfway up the tubes.

The operating level of the liquid product in the tube sheet has a major influence on the heat transfer. Usually, the liquid level is maintained closer to the top of the tube sheet. The reduction below the optimum level leads to incomplete wetting of the tube wall and a resultant increase in fouling and rapid reduction in evaporation capacity. In this evaporator type, owing to the shorter tube length, to achieve the desired concentration, the liquid product should be circulated many times within the tubes.

iii. **Advantages and disadvantages**

The calandria-type evaporator is suitable for products that have a moderate tendency to scale as the product is on the tube side, which is accessible for cleaning. The heat transfer in this evaporator type is dependent on the viscosity and temperature. Reasonably high heat-transfer coefficients can be obtained for thin liquids with a viscosity in the range of 5–10 cP. However, heat-sensitive and crystalline products cannot be handled in this evaporator type. Other advantages of a short-tube vertical evaporator include easy mechanical descaling and its availability at a relatively inexpensive cost (Ahmad and Khan, 2013).

iv. **Major applications in the food industry**

The principal application of the short-tube vertical evaporator is in the concentration of sugarcane juice.

9.6.2.1.2 Long-Tube Vertical Evaporators

The two types of long-tube vertical evaporators are (i) rising film and (ii) falling film, which are discussed subsequently:

9.6.2.1.2.1 Rising-Film Evaporators

i. **Construction**

Rising-film evaporator is essentially a single-pass vertical shell and tube heat exchanger connected to a vapor–liquid separator at the top for the removal of entrained liquid from the vapor (similar to that shown in Figure 9.8). A deflector is provided just above the tube sheet. In general, the tubes are of an outer diameter (o.d.): 1–2 in (25–50 mm) and length: 12–32 ft (3–10 m) (McCabe and Smith, 1958). The length of the evaporator tube is divided into three main sections (Gupta and Holland, 1966). The first section is termed as the *sensible heating section*, in which heat is supplied to increase the temperature of the cold liquid to its boiling point. The length of the sensible heating section is a function of feed temperature. The second section is known as the *foaming section*, the length of which lies between the point of first

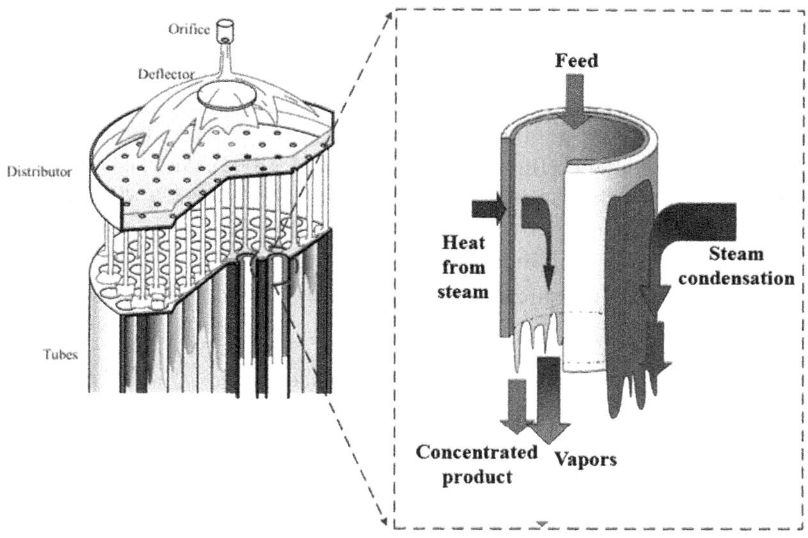

FIGURE 9.8 Falling film evaporator. (Modified and reproduced with permission from Munir, M. T., Zhang, Y., Wilson, D. I., Yu, W. and Young, B. R. 2014. Modelling of a Falling Film Evaporator for Dairy Processes. *Chemeca 2014: Processing Excellence; Powering Our Future*, Perth, Western Australia.)

bubble formation and up to the position where a well-defined film is formed on the tube wall. Thus, the length of the foaming section is a function of the feed flow rate and physical properties of the liquid and is independent of the feed temperature. The last section of the rising-film evaporator is the *evaporating section* which lies above the foaming section. Length of this section is a function of the feed flow rate (Gupta and Holland, 1966).

After the first pass, the concentrated liquid product is recirculated to the evaporator, to achieve a higher heat transfer coefficient and product concentration (Bourgois and Maguer, 1984). The entry temperature of feed is an important parameter as it affects the surface film temperature. The height of the evaporator increases with the decrease in feed temperature. Therefore, it is preferable to preheat the fluid entering the evaporator tube, which also reduces the risk of splashing (Cvengroš et al., 2000).

ii. **Working**

Initially, the dilute liquid food product enters the bottom of the tube sheet and flows up the tubes. The feed is allowed to boil by using steam which remains on the shell side, outside the tubes. When the feed is at the lower point of the tubes, it is heated to its boiling point. As it moves up the tube, bubbles are formed on the tubes, followed by the commencement of boiling. The vapors, thus, generated form a core in the center of the tube and move upward along with a thin film of liquid. As the liquid moves up further, more vapors are formed resulting in a higher central core velocity that force the remaining liquid towards the tube wall. The higher velocities of vapor result in a thinner, more rapidly rising liquid film and a greater rate of heat transfer. Thus, it is evident that the drag exerted by the vapor on the liquid is responsible for the climbing of liquid thin films. Close to the top of the tube, the bubbles grow more rapidly. In this zone, slugs of liquid and bubbles quickly rose through the tubes and discharged at high velocity from the top, at which point they impinge on the vapor–liquid separator to break any foam (McCabe and Smith, 1958). Finally, the concentrated product is collected at the bottom, and the vapor leaves from the top of the evaporator.

iii. **Advantages and disadvantages**

The major advantages of rising film evaporator are the higher heat transfer coefficients and shorter residence time of liquid feed in the evaporator. The former is relevant in reducing the overall heat transfer area requirement and lowering the initial capital cost of the evaporator. The latter facilitates the usage of this evaporator even at higher operating temperatures leading to high-quality final products and permits handling of heat-sensitive liquid feed. Nevertheless, the rising film evaporators are associated with certain limitations. These evaporators require a driving force to move the liquid film against gravity which is provided by a sufficient temperature difference between the heating surfaces (Artal et al., 2002). Another major limitation of rising-film evaporators is the requirement for initial liquid feed with low viscosity and minimal fouling tendencies.

iv. **Major applications in the food industry**

The common food industry application of rising film evaporator is the concentration of cane sugar syrups.

9.6.2.1.2.1.1 Application of Rising Film Evaporator for the Concentration of Cane-Sugar Syrup

Many sugar mills use rising-film evaporators in shell and tube configuration with multiple effects for the concentration of cane-sugar syrup. This section presents a case study on the evaporation of industrial cane-sugar syrup in a pilot scale rising-film evaporator with steam as the heating medium (Ahmad et al., 2018). The sugar syrup had a concentration of 15°Brix–15.5°Brix and a pH of 7–7.5. The syrup was preheated to a temperature of 60°C–80°C in a preheater before entering the evaporator. Usually, the temperature of sugar syrup before entering the evaporator is between 65°C and 70°C. The evaporator comprised of two peristaltic pumps, feed tanks, condenser, condensate vessels, and steam production unit.

The evaporation was conducted at varied flow rates (80–170 mL/min), recirculation ratio (0.2–0.8), and steam pressure (0.2–0.5 bar). Recirculation ratio is defined as the ratio of the volumetric flow rate of the recycle stream to the volumetric flow rate of feed. The steam produced inside steam generator was introduced in the shell side. The overall heat transfer coefficient varied between 1300 and 9300 W/m^2 K, with the higher values occurring at higher feed flow rate and feed temperature and lower recirculation ratio and steam pressure. During the concentration of cane-sugar syrup, the influence of feed flow rate and steam pressure on the heat transfer coefficient was significant than that of the recirculation ratio (Ahmad et al., 2018).

9.6.2.1.2.2 Falling Film Evaporator

i. **Construction**

With respect to its construction, falling film evaporator is an upside-down variant of the rising-film evaporator, such that the tubular heat exchanger is mounted on top of the vapor–liquid separator (Figure 9.8). It comprises a vertical bundle of tubes (diameter 25–80 mm; length: 4–10 m), surrounded by a steam jacket. A critical component in the falling film evaporators is the distributor. It is vital that the product is distributed uniformly at the feed inlet to ensure a sufficient supply of liquid to each tube. The consequence, if the aforesaid requirement is not satisfied, is that the individual heating tubes will not be provided with sufficient liquid, and hence, there would be a formation of deposits. It has been estimated that about 1.3% of total solids in the whole milk feed is lost as deposits on the tube walls (Bouman et al., 1988).

ii. **Working**

In falling film evaporator, the feed enters at the top of the evaporator wherein specially designed distributors (perforated plate placed over the tube sheet) or spray nozzles evenly distribute the feed into every tube in the tube sheet. As the name of the evaporators suggests, the liquid film moves downward in the vertical tubes under gravity. The concentrated product is removed from the bottom of the evaporator after being separated from the vapors within the vapor–liquid separator. The design of vapor–liquid separator depends on the properties of the liquid product that is concentrated and the operating conditions. The ratio of the quantity of feed liquid distilled to the feed rate is a critical parameter. High ratio handled in a single pass through the vertical tube can reduce the flow of liquid to a level which is insufficient to keep the tubes wetted near the bottom. This would potentially lead to fouling of the tubes by the degraded product.

iii. **Advantages and disadvantages**

The thinner and rapidly moving film results in high heat transfer coefficients and short residence time in the heating zone. A falling film evaporator consumes only one-third kilogram of steam consumed per kilogram of water evaporated in a double-effect evaporator. Unlike the rising film evaporator, this evaporator type can handle high viscous liquid foods. The other advantages include relatively low cost and a lesser requirement for floor space. However, the falling film evaporators are not suitable for products that tend to form scales or salt deposits.

iv. **Major applications in the food industry**

Falling film evaporators are widely used for the concentration of dairy products (skim milk, cream, hydrolyzed milk, and whey and milk protein), coffee extract, and sugar solutions.

9.6.2.1.2.2.1 Application of Falling-Film Evaporator for the Concentration of Dairy Products

Falling film evaporators are widely used for the concentration of milk before spray drying. Typically, milk enters the evaporator with about 13% solids content and is evaporated to about 50% solids. The milk flows as a film down the whole length along the inside of tubes (Morison, 2015). For the concentration of milk, the falling film evaporator is operated under reduced pressure (16–32 kPa) to facilitate low temperature evaporation (70°C–80°C), minimize heat damage to the product, and prevent the formation of deposits on the surface of evaporator tubes. Before entering the evaporator section, milk is preheated to temperatures ranging from 72°C to 120°C (Singh, 2007) using plate heat exchangers, shell and tube heat exchangers and/or direct steam injection (Morison, 2015). The aforementioned processing conditions are applicable only if the milk is just to be concentrated. However, the preheating step is avoided if the milk concentrate is to be dried further. The process layout of a milk powder manufacturing plant consists of two evaporation chambers or effects, a thermal compressor for vapor recycling, a condenser to remove heat, and a homogenizer to reduce milk fouling.

Initially, raw liquid milk from the feed tank is fed at a defined flow rate (F_F) to a moderate temperature pasteurizer (70°C–80°C) to deactivate most pathogens. A direct steam injector (DSI) installed after the pasteurizer inactivates the remaining pathogens and preheats the milk to a temperature of 104°C (T_H). On exiting the DSI, the milk is maintained under an increased pressure of 2–3 bar (P_H) to prevent vaporization inside the tubing owing to high temperatures. The temperature of milk in the evaporator chambers is T_{E1}. Immediately after entering the evaporator, a portion of liquid milk forms vapors due to rapid exposure to a low-pressure system, which is termed as *flash evaporation*. The remaining milk flows down along the inner side of long vertical tubes which is heated on the shell side by fresh steam and/or recycled vapor generated by the thermal compressor. The concentrated milk is collected in a holding tank (at a defined level L_1) and then directed to the second effect to recover the discarded heat and to concentrate the product further. The concentrated milk exits the falling film evaporator section at a flow rate of F_P into a holding tank. The concentrated product from the holding tank may be redirected to a spray dryer to produce milk powder (Haasbroek et al., 2013).

9.6.3 Forced Circulation Evaporator

A forced circulation evaporator is one which uses an external driving force such as the mechanical energy from a pump to impart higher circulation velocity to the liquid product to be concentrated. This evaporator type is developed based on the principle that increased velocity of liquid feed in the tubes may potentially reduce fouling and, hence, maintain the evaporation capacity and reduce the downtime. The typical circulation velocities are in the range of 7–20 ft/s (~2–6 m/s).

i. **Construction**

The main components of a forced circulation evaporator are a shell and tube heat exchanger which may be horizontal or vertical, a separator positioned above the heat exchanger, and a circulating pump (Figure 9.9).

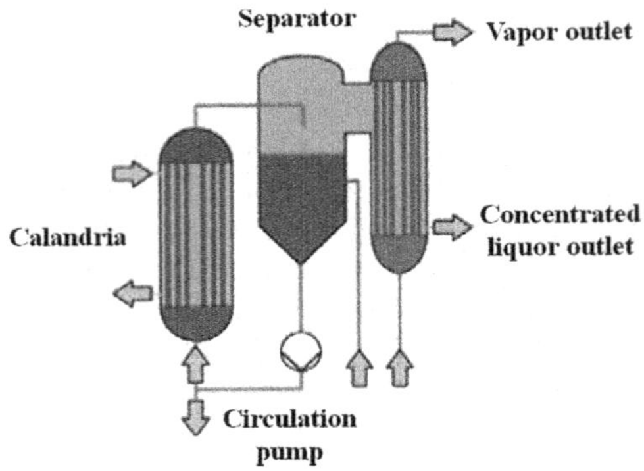

FIGURE 9.9 Vertical-tube forced-circulation evaporator. (Modified and reproduced with permission from Vieira, B. S., Gullberg, D. and Perrot, V. 2016. Recent Experience with Metallic Heaters for Phosphoric Acid Evaporation. *Procedia Engineering* 138: 437–444.)

ii. **Working**

In the forced circulation evaporator, the liquid product is circulated at a high rate through the calandria using a circulation pump (Figure 9.9). Boiling of liquid is prevented within the unit by the hydrostatic head maintained above the top tube plate. Within the calandria, the product is superheated at an elevated pressure, higher than its normal boiling pressure. On entering the separator where the absolute pressure is slightly less than that in the tube bundle, the liquid experiences a considerable pressure drop and flashes to form vapor. The increase in temperature across the tube bundle is maintained at the minimum possible level, often as low as 2°C–3°C. Consequently, a high recirculation ratio is obtained, which ranges between 100 and 150 kg of liquid per kilogram of water evaporated.

iii. **Advantages and disadvantages**

As discussed previously, the high recirculation rates result in high liquid velocities through the tube which aid in minimizing the accumulation of deposits or crystals along the heating surface. Consequently, high heat transfer coefficients are obtained. However, forced circulation evaporators are relatively more expensive than film evaporators because of their requirement for large bore circulating pipe work and large recirculating pumps. Operating costs are also substantially high due to the power consumption for the circulating pumps. A longer residence time of the product within the heating zone is also a major limitation (Perry et al., 1963).

iv. **Major applications in the food industry**

Forced circulation evaporators are mainly used for highly viscous liquids, feed containing solids or susceptible to crystallization, and liquids with a high tendency for fouling.

9.6.3.1 Application of Forced Circulation Evaporator in the Manufacturing of Sweetened Condensed Milk

The production of sweetened condensed milk using a forced circulation evaporator requires certain preprocessing steps. The skim milk or whole milk is preheated to a temperature of 77°C–80°C. Further, sugar syrup is prepared by adding 1 kg of water per kilogram of sugar. This sugar syrup is added to the preheated skim milk or whole milk stored in a tank. Subsequently, the concentration of feed solution containing milk and sugar is carried out in a two-effect forced-circulation evaporator (UNIDO, 1969).

Two typical recipes for the production of sweetened condensed milk are given in Table 9.1.

A typical temperature program for the production of sweetened condensed milk in a forced circulation evaporator is given in Table 9.2.

The forced circulation evaporator is suitable for the production of sweetened condensed milk attributed to the sufficient holding time that leads to the formation of sugar crystals having the optimum size.

TABLE 9.1

Recipes for the Production of Sweetened Condensed Milk Using Forced Circulation Evaporator

Recipe 1	
Skim milk	1000 kg
Sugar	163.5 kg dissolved in some quantity of water
The skim milk + sugar composition prepared in the abovementioned ratio is concentrated until the final product attains 47.5% sugar, 25.2% nonfat substances (such as casein), and 27.3% water.	
Recipe 2	
Whole milk	1000 kg containing 2.85% fat
Sugar	207 kg dissolved in some quantity of water
The whole milk + sugar composition prepared in the aforesaid ratio is concentrated until the final product attains 46.4% sugar, 20.2% nonfat substances, 6.4% fat, and 27.0% water.	

Source: Data from UNIDO (1969).

TABLE 9.2

Temperature Program for the Production of Sweetened Condensed Milk in a Forced Circulation Evaporator

Milk tank	50°C
Evaporator	
• Effect-1	75°C – 50°C
• Effect-2	45°C – 50°C
Finished product	
• After cooling	20°C

Source: Data from UNIDO (1969).

The concentrated product from the evaporator is collected in a jacketed tank fitted with a scraper. The concentrate at a temperature of ~50°C is held in the tank with continuous agitation for about 1 h. Then, the batch is cooled to about 20°C using water at ambient temperature or cold water circulated in the tank jacket. The resultant product is ready to be filled in containers (UNIDO, 1969).

9.6.4 Scraped Surface Evaporator

Many industries use scraped surface evaporators (SSEs) in the final stages of evaporation to achieve a high solid concentration in the end product. Also, SSEs are preferred as they prevent charring and promote mixing of highly viscous products during the concentration process. As the name indicates, scraper blades in the calandria ensure proper mixing and, hence, protect the product from thermal degradation.

i. **Construction**

   The construction of an SSE is similar to the scraped surface heat exchanger (discussed in Chapter 5). It includes a cylinder having an inner tube and an outer tube. The inner tube provides the surface area for heat transfer. The annular space between the inner and outer tubes forms the steam chest which contains the steam (heating medium). Liquid product to be concentrated flows down the inner surface of the evaporator tube. The flow of heating medium and product are countercurrent to each other. A concentric rotor having four scraper blades is present within the inner tube. The speed of this rotor can be changed by a variable frequency drive connected to a motor. In general, the gap between the outer edge of rotor blades and the tube's inner surface is about 1–2 mm. The rotor also consists of a distributing ring which facilitates a uniform distribution of the feed liquid along the inner surface of the evaporator tube.

ii. **Working**

SSEs are usually operated under vacuum conditions. Sub-atmospheric pressure is created outside the evaporator tube by diverting the flow of a part of the steam to a vacuum pump via a pipeline. Consequently, the temperature of steam within the jacket of the evaporator tube is reduced to below 100°C.

The concentric rotor with scraper blades is central to the operation of SSE. As it rotates, the concentric rotor maintains a thin film of liquid product at the tube surface. The turbulent flow under the scraper defines the thickness of the liquid film. However, once the film is formed, it demonstrates laminar behavior with respect to heat transfer. Saturated steam condenses within the jacket at the outer surface of the evaporator tube and supplies the heat required for boiling and concentration of liquid that flows down the evaporator tube. The scraper blades provide mechanical agitation and ensure the contact between the heating wall and the liquid in the form of a thin layer which is renewed continuously. Heat transfer in the liquid film occurs through both conduction and convection. The heated liquid along with the vapors move downward and then enter the vapor–liquid separator. The thickly concentrated liquor is obtained at the bottom of the separator and then collected in tanks. The vapors are condensed in the condenser and collected in a separate tank.

iii. **Advantages and disadvantages**

The SSE is versatile with its ability to handle highly viscous liquids and proteinaceous and heat-sensitive liquids which tend to stick to or foul a heat transfer surface or salt out during evaporation. This is possible because of the high degree of agitation and scraping off of the product from the wall of the heat transfer surface, provided by the scraper blades. The abovementioned conditions do not allow the product to remain in contact with the hot surface and, hence, prevents deposition. Consequently, both residence time and fouling are reduced. High heat transfer coefficients can be obtained with an SSE as the fresh material is continuously replenished by the boundary layer. Also, since the product is in contact with the heating surface for a very short residence time in the order of few seconds, high-temperature differentials can be used without leading to reactions (Rao and Hartel, 2007). The SSEs provide better heat transfer and prevent charring of the product. However, due to the low product-side heat transfer resistance in SSE, condensate and wall heat transfer resistances constitute a substantial part of the total heat transfer resistance than in the other evaporators handling viscous products. Reducing wall thickness and using deflectors that promote condensate shedding are the approaches to mitigate these resistances and provide significantly higher evaporation rates (Schwartzberg, 1989).

With the increase in rotor speed, the liquid is subjected to an increased shear rate which reduces its viscosity. This is because most of the concentrated liquid foods exhibit pseudoplastic behavior (explained in Chapter 4). As a result, SSEs can handle liquid products with viscosities as high as 40000 cP. Residence times are in the range of 0.5–1 min.

iv. **Major applications in the food industry**

SSE is appropriate for heat-sensitive and highly viscous liquid food products such as corn syrup and tomato puree, which cannot be concentrated in any other type of evaporator. The other typical applications of SSE include processing of mashes, pulps, concentrates, and pastes (Lozano, 2006).

9.6.4.1 Application of Scraped Surface Evaporator in the Concentration of Tomato Pulp

This section explains a case study of the concentration of tomato pulp using a thin film SSE. The SSE used for this application was 1.4 m high with a diameter of 0.22 m. The initial concentration of total solids in the tomato pulp was 5.9%. The steam temperature, rotor speed, and feed flow rate were observed to be the parameters with a major influence on the water evaporation rate, overall heat transfer coefficient, and the final concentration of total solids. Accordingly, the abovementioned operational parameters were chosen in the following range:

- Steam temperature (T_S): 65°C–80°C
- Rotor speed (S_R): 355 RPM
- Feed flow rate (Q_f): 40.8–51.0 kg/h.

The overall heat transfer coefficient obtained for the evaporation of tomato pulp varied between 625.6 and 910.9 W/m^2°C. Further, the evaporation rate and final concentration of total solids in the product were in the range of 13.22–33.72 kg/h and 8.02%–19.21%, respectively. Eventually, the optimum values of the aforementioned operational parameters for the concentration of tomato pulp were found to be the following:

- T_S: 73°C
- S_R: 355 RPM
- Q_f: 40.3 kg/h

The optimum process parameters resulted in an overall heat transfer coefficient of 840 W/m^2°C, an evaporation rate of 27 kg/h, and total solids of 18%. At steam temperature and feed flow rate higher than the optimum value (say, at 75°C and 45.86 kg/h, respectively), the ascorbic acid content and color of the tomato pulp were slightly affected (Sangrame et al., 2000).

9.6.5 Plate Evaporator

The advent of plate evaporators dates back to 1957. The plate evaporators were developed and introduced by APV (United Kingdom) as an alternative for tubular systems. In this evaporator type, the heating element is comprised of plates rather than tubes. Rising film or falling film or a combination of both can be used with the plate evaporator.

i. **Construction**

   The construction of a plate evaporator is similar to that of a plate heat exchanger (discussed in Chapter 5). It consists of a plate-and-frame configuration with the alternate product and heating channels. In general, the plates are 1.0–2.2 m (3–7 ft) high, spaced at a distance of 6.0–7.4 mm. However, the space between the plates can be increased to accommodate a much larger volume of vapors when compared to liquid. The embossed plates are welded with openings in all its four corners and supported by carrying bars at their top and bottom ends. Gaskets positioned within specially designed slots around the periphery of plates prevent the mixing of liquid product with the heating medium. Also, gaskets guide the liquid along the flow path through the corner opening of the plates. The gaskets can be inserted and removed without the aid of any special tools.

ii. **Working**

   In a plate evaporator, the feed liquid and heating medium (steam) enter in the alternate spaces. The fluids flow in countercurrent direction through their respective channels. The heat transfer coefficient is greatly enhanced as a result of the highly turbulent flow through the narrow passages caused by the defined plate distances in conjunction with special plate shapes. Addition of heat causes the product to boil on the plate surface in the form of a film. The vapors, thus, formed drive the residual liquid, as a rising film, into the vapor duct of the plate unit. Vapor–liquid separation occurs in the downstream centrifugal separator. The feed liquid may be passed multiple times through the plate-frame configuration until the desired concentration is obtained.

iii. **Advantages and disadvantages**

Similar to the thin-film evaporators, plate evaporators provide high heat transfer coefficient and short residence time of the product. This advantage may be attributed to the complete accessibility to the heat transfer surfaces and boiling of liquid in the form of thin film on the plate surface. In addition, plate evaporators require less energy and less requirement of floor space. Another advantage of plate evaporators is that its handling capacity and evaporation rate can be increased by just adding more plate units. Highly viscous liquids with higher fouling tendencies can be handled conveniently as the inner parts of the plate evaporator are easily accessible for cleaning and maintenance. Overall, the shorter product residence time and a more compact design due to preassembled and portable construction units result in a superior quality product and low installation cost, respectively.

iv. **Major applications in the food industry**

Plate evaporators can handle a wide range of products. Products that are commonly processed in this evaporator include apple juice, coffee, fruit juices (pear, grape, pineapple, orange, lime, and mango), fruit purees, vegetable juices, pectin, whole milk, skim milk, liquid egg, sugars, whey protein, cheese whey, low alcohol beer, chicken broth, and many others.

9.6.5.1 Application of Plate Evaporator in the Dairy Industry

The dairy industry is not just the oldest but also the largest user of plate evaporators. Initially, plate evaporators were used for the concentration of both whole and skim milk up to 50% total solids (Anon, 1959; Holland, 1964). Based on the data from various trials, it has been found that a high-quality milk concentrate was resultant from falling film plate evaporator within a short residence time (0.13 s in the first effect and 0.055 s in the second effect) and with high velocity (approximately 14 and 34 m). Also, the CIP of plate evaporators used for the concentration of milk and milk products is completely effective such that the plant can be operated for several days without dismantling the equipment (Anon, 1975).

Across the years, the product portfolio handled by the plate evaporators has widened to include sweetened milk, buttermilk, whey, cream, and ice-cream mix. The production of sweetened condensed milk is usually operated in a batch mode in which sugar syrup is added to the whole milk or skim milk, and the resultant mixture is concentrated to a total solid content of 74.5% (whole milk) and 73% (skim milk). Not just limited to a batch process, the plate evaporator is also capable of producing the product on a fully continuous mode. The final concentration of product is maintained within a narrow range using a density control system which is generally a U-tube sensing device and modulates a valve in the steam line. The high-temperature operation in plate evaporator results in a sterile product without compromising the product quality. Consequently, after cooling and seeding, the product can be directly filled into the cans.

Plate evaporators are also used as pollution abatement devices by the dairy industries. Whey, a by-product from cheese production which tends to foul, can be easily concentrated using plate evaporators to produce whey protein concentrates of high commercial viability. Thus, the use of plate evaporators in dairy industry alleviates the difficulties associated with the disposal of large quantities of whey without polluting the water stream (Dodeja and Zaidi, 1980).

9.7 Approaches to Improve Evaporator Efficiency

Evaporation is an energy-intensive process. In most of the industrial-scale evaporation processes, the energy content of the spent steam goes off unutilized. The vapors at low pressure and reduced enthalpy generated from single-effect evaporators are merely expelled to the atmosphere or condensed to discard the condensate. The need for reducing the energy cost and increasing the thermal efficiency of evaporators has led to the advent of two key approaches, namely multiple-effect evaporation and vapor recompression.

9.7.1 Multiple-Effect Evaporation

As aforementioned, every single unit of an evaporator is termed as an *effect*. Multiple-effect evaporation can be defined as the use of more than one evaporator unit in series. The arrangement having two evaporators in series is known as the double-effect evaporator. Similarly, extra effects can be added in series to obtain a triple-effect evaporator, quadruple-effect evaporator, and so on. In general, the evaporators added in series have an equal surface area of heat transfer. In multiple-effect evaporators, the vapor or the spent steam obtained from the previous effect acts as the heating medium for the following effect maintained at a lower pressure. In other words, vapors generated in the first effect are condensed to recover their latent heat and the same is used to evaporate the liquid product in the second effect further. A temperature gradient of about 7°C–10°C is maintained between the effects to maximize the efficiency. Also, the partially concentrated liquid product from the preceding effect is passed as feed to the subsequent effect for further concentration. Vacuum is applied to the second and subsequent effects to reduce the boiling point and maintain the temperature gradient since the vapor temperature drops across the following effects. Once the product reaches the desired concentration, the product is pumped out of the evaporator chamber. The vapor produced in the last effect is sent to the condenser and vacuum system. Schematic of a triple-effect evaporator is shown in Figure 9.10.

Multiple-effect evaporators are effective in reducing the steam consumption which is directly proportional to the number of effects (Table 9.3). As the number of effects is increased, the steam economy and the process efficiency also increase as the total mass of vapor produced per kg of steam is more compared to a single-effect evaporator. However, the capital cost is significantly high in multiple-effect evaporators. In general, *n* effects will approximately cost *n* times as high as a single effect. Therefore, multiple-effect evaporation should be considered when the evaporation rate is above 1350 kg/h (SPX Corporation, 2008).

9.7.1.1 Feeding of Multiple-Effect Evaporators

Based on the progress of liquid feed from the first effect to the following effects, multiple-effect evaporators can be classified into different types as shown in Figure 9.11.

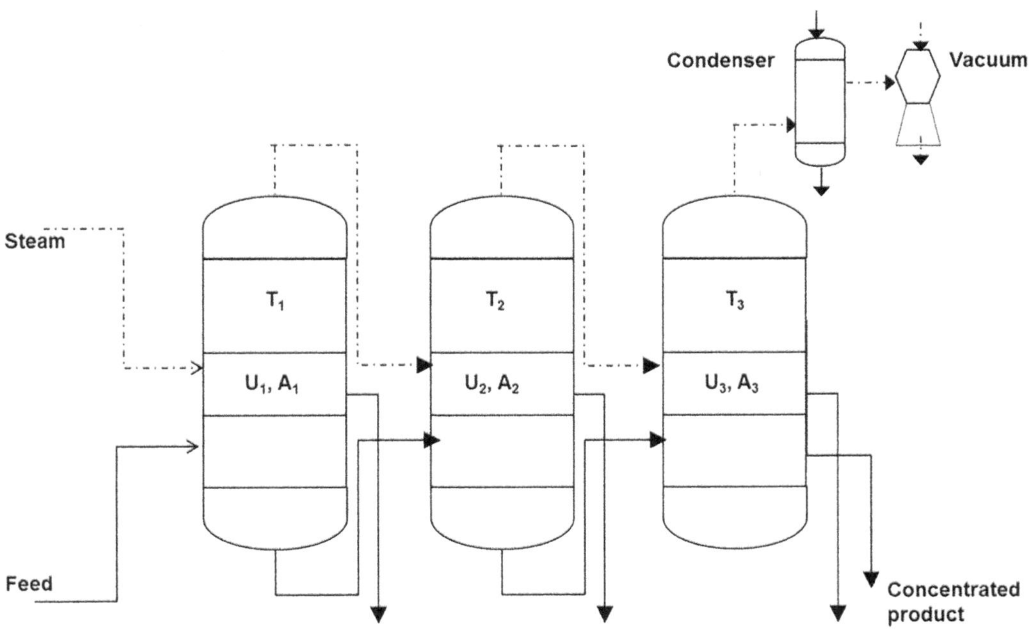

FIGURE 9.10 Schematic of a triple-effect evaporator.

TABLE 9.3

Steam Consumption and Running Costs of Evaporators

Number of Effects	Steam Consumption (kg Steam/kg Water Evaporated)
1	1.1
2	0.57
3	0.4

Source: Data from Grosse and Duffield (1954).

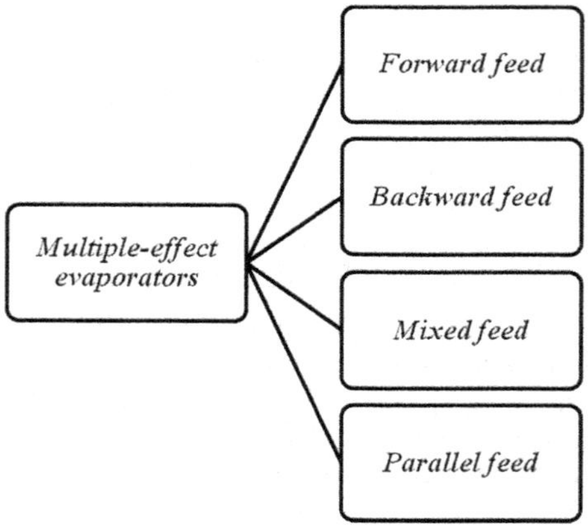

FIGURE 9.11 Classification of multiple-effect evaporators.

9.7.1.1.1 Forward Feed

The progress of liquid feed across different effects is simplest in a forward feed multiple-effect evaporator (Figure 9.12). Concentrated product flows in the direction of first effect → second effect → third effect and so on without pumping. A pump is required just to supply the feed to the first effect and withdraw the concentrated product from the last effect. Steam is also fed to the first effect along with the

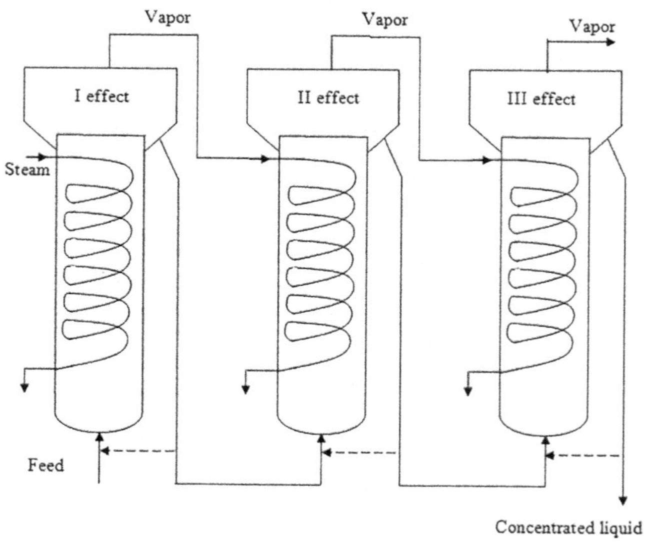

FIGURE 9.12 Forward feed multiple-effect evaporator (NPTEL, 2018).

dilute liquid product. The following conditions of operation facilitate the forward feed flow. Considering the example of a double-effect evaporator, the temperature in the calandria of the first effect is higher than that in the second effect. For the vapor derived from the first effect to function as the heating medium for the second effect, the boiling point of the product in the latter must be lower. Therefore, the second effect is maintained at a lower pressure than the first. With more number of effects, after the first effect, the subsequent effects are maintained under increasingly lower pressure, which is more commonly accomplished by applying vacuum. Thus, after the first effect, the liquid flows under the pressure gradient, thereby eliminating the requirement for a pump to transfer the partially concentrated product from one effect to the other. The highest overall heat transfer coefficient is obtained in the first effect due to a larger temperature gradient. Forward feed arrangement is used when the feed temperature is very close to the vapor temperature of that effect.

9.7.1.1.2 Backward Feed

In the backward feed arrangement (Figure 9.13), the feed moves in reverse direction from the last effect, i.e., n^{th} effect → $(n-1)^{th}$ effect → $(n-2)^{th}$ effect and so on until the first effect from which the concentrated liquid product is obtained. This feeding arrangement requires a pump between each pair of effects to transfer the liquid feed against the pressure drop. Steam is introduced in the first effect. Therefore, the product at the highest concentration will receive the maximum heat which is disadvantageous from the perspective of product quality. However, the aforementioned conditions lead to larger evaporation capacity than the forward feed evaporator, as the concentrated and viscous liquid products are handled at the highest temperatures in the initial effects. Also, in contrast to the forward feed arrangement, the highest overall heat transfer coefficient is obtained in the last effect.

9.7.1.1.3 Mixed Feed

In the mixed feed arrangement, the dilute liquid feed enters at an intermediate effect and then flows forward to the next higher effect until it reaches the last effect of the series. The partially concentrated liquid product is then pumped back to the effect before the one to which the fresh feed was introduced initially (Figure 9.14). Mixed feed system reduces the number of pumps required in the backward configuration, as the feed flows due to pressure difference in the forward direction initially.

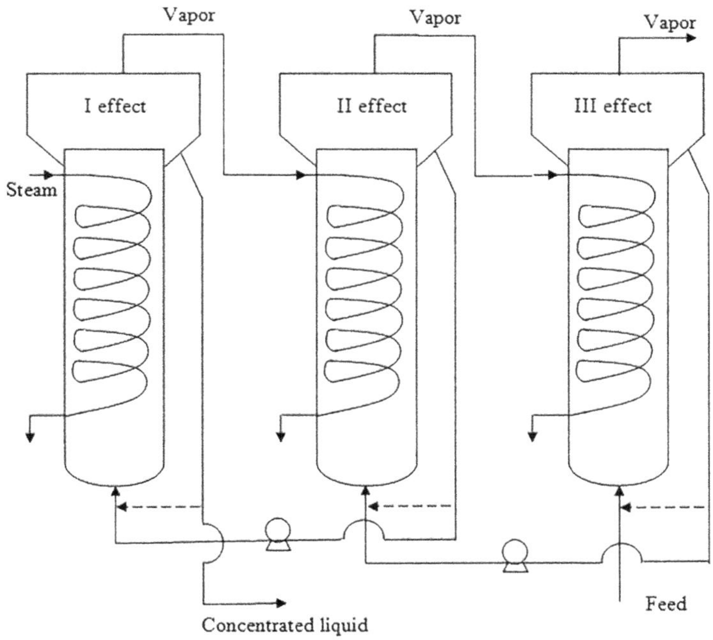

FIGURE 9.13 Backward feed multiple-effect evaporator (NPTEL, 2018).

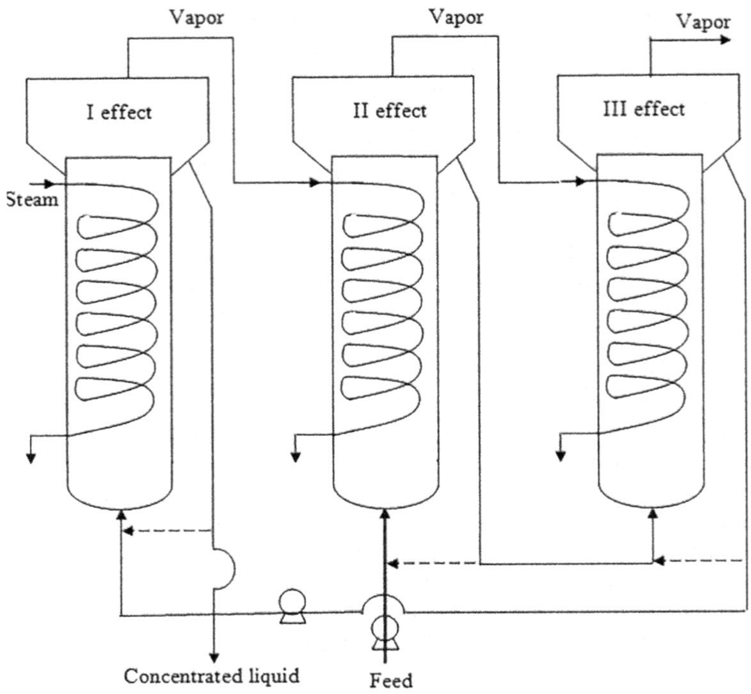

FIGURE 9.14 Mixed feed multiple-effect evaporator (NPTEL, 2018).

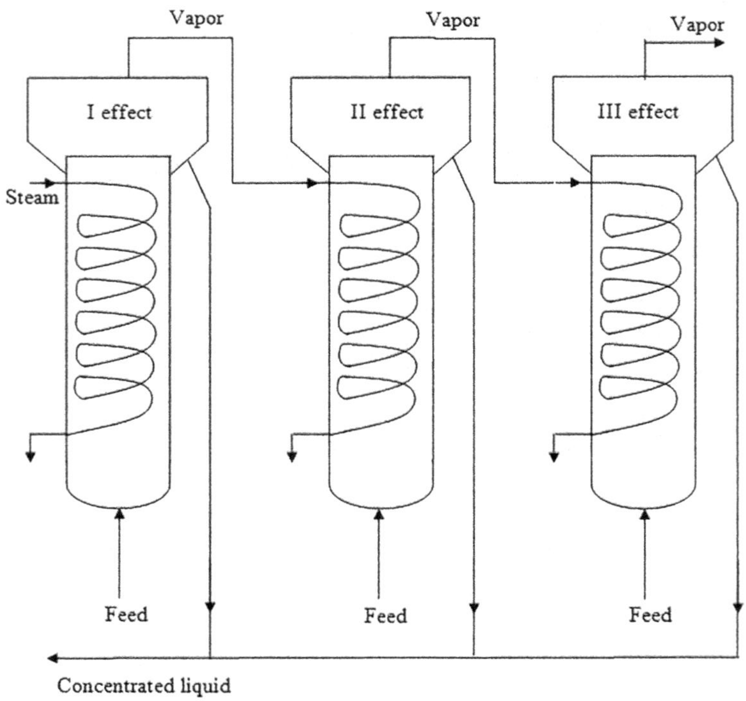

FIGURE 9.15 Parallel feed multiple-effect evaporator (NPTEL, 2018).

9.7.1.1.4 Parallel Feed

In the parallel feed arrangement, the feed is supplied to each effect while removing the concentrated product from every effect. However, steam is supplied in the same manner as in the forward feed arrangement (Figure 9.15). High steam efficiencies are obtained in this system. Parallel feed arrangement is preferred for liquid feeds which are prone to fast crystallization. It is difficult to transport a crystallized concentrated liquid, as the presence of crystals impairs the flowing ability.

9.7.2 Design of Multiple-Effect Evaporators

Designing a multiple-effect evaporator involves a systematic procedure comprising of a logical series of steps. Essentially, the design approach requires estimation of the amount of steam consumed, the surface area of heat transfer, approximate temperatures in the different effects, and the amount of vapor leaving the last effect. The temperatures in different effects can be arrived at by determining the intermediate temperatures. Likewise, the heat transferred in each effect and area of heat transfer can be obtained by solving the material balances for the flow rate of water vapor (Ermis et al., 2017). The steps involved in the procedure for designing a multiple-effect evaporator are listed in Figure 9.16 (McCabe et al., 1993; Singh and Heldman, 2014).

To understand the design procedure, a forward feed triple-effect evaporator as shown in Figure 9.12 is considered.

9.7.2.1 Mass and Energy Balance

The amount of concentrated product and the vapor leaving the evaporator can be calculated by using an overall material balance as given in Eq. (9.10):

$$F = P + V \tag{9.10}$$

where F is the mass flow rate of feed solution (kg/h), P is the mass flow rate of concentrated product (kg/h), and V is the mass flow rate of vapor exiting the evaporator (kg/h). To determine P, a component balance equation on solid content is required, which is given by

$$Fx_F = Px_P + V(0) \tag{9.11}$$

$$Fx_F = Px_P \tag{9.12}$$

where x_F is the weight fraction of solids in the feed solution entering the evaporator and x_P is the weight fraction of solids in the concentrated product.

To calculate the heat load of the evaporator (q), the heat balance around the evaporator should be known, which is given as

$$q + FC_{PF}T_0 = WH_V + PC_{PP}T + q_L \tag{9.13}$$

where C_{PF} and C_{PP} are the specific heat capacities of the feed and the product (kJ/kg °C), respectively, H_V is the enthalpy of vapor (kJ/kg), T is the boiling point of the product (°C), T_0 is the temperature of the feed solution (°C), and q_L is the heat loss.

Since the feed solution is a mixture of concentrated product and water vapor,

$$FC_{PF}T = PC_{PP}T + WC_{Pw}T \tag{9.14}$$

$$\therefore PC_{PP} = FC_{PF} - WC_{Pw} \tag{9.15}$$

where C_{Pw} is the specific heat of liquid water (kJ/kg °C). Substituting Eq. (9.15) in Eq. (9.13),

$$q = FC_{PF}(T - T_0) + W\left(H_V - C_{Pw}T\right) + q_L \tag{9.16}$$

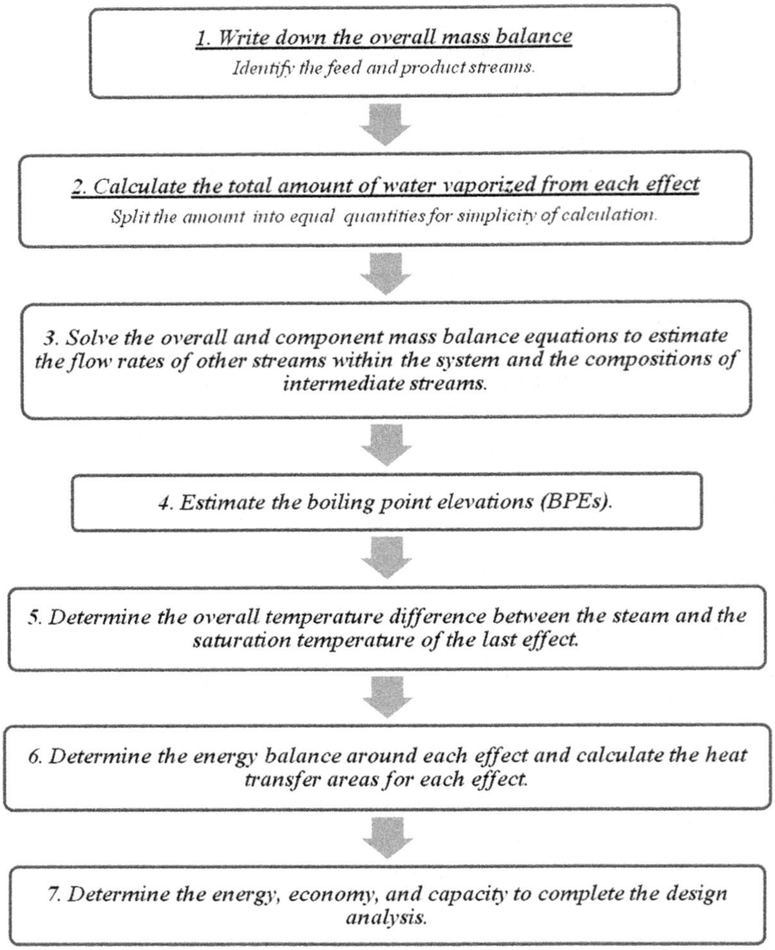

FIGURE 9.16 Procedure for designing a multiple-effect evaporator.

If the heat losses are negligible ($q_L = 0$), then,

$$q = FC_{Pf}(T - T_0) + W\left(H_V - C_{Pw}T\right) \tag{9.17}$$

9.7.2.2 Determination of BPE

To determine the BPE of feed solution, the following experimental procedure should be carried out. The feed solution at a concentration of say, $x\%$ (by weight) should be boiled, and the concentration should be measured at equal intervals of time (ex. 15 min). A plot as shown in Figure 9.17 should be constructed between measured concentration versus the temperature difference between the boiling point of solution and boiling point of water.

9.7.2.3 Calculation of Overall Temperature Difference and Heat Transfer Areas for Each Effect

The driving force for heat transfer during evaporation is the existence of temperature drop at each effect. The temperature of vapors generated in a given effect must be greater than the boiling temperature in the subsequent effect:

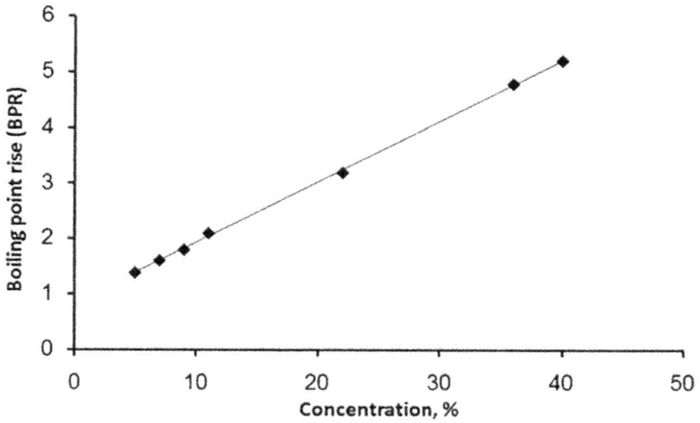

FIGURE 9.17 Determination of BPE. (Reproduced with permission from Sorour, M. A. 2015. Optimization of Multiple Effect Evaporators Designed for Fruit Juice Concentrate. *American Journal of Energy Engineering* 3: 6–11 [Open access].)

$$T_1 > T_2 > \ldots > T_n; \quad \therefore P_1 > P_2 > \ldots \ggg P_n \tag{9.18}$$

where n is the number of effects.

In the design calculations of a multiple-effect evaporator, certain values are known while the others have to be determined, which are listed in Table 9.4.

The unknown values are calculated by solving the equations that represent multiple-effect evaporation which include the equations of material balance, heat balance, and the capacity for each effect given by $q = UA\Delta T$. The equations are formulated using the following scheme of methodology, and the solution is derived from a trial-and-error approach.

1. If the concentration of the final product and pressure in the last effect is known, the boiling point in the last effect can be determined by solving the overall mass balance and component mass balance equations. Subsequently, the flow rate of product (Q_1, Q_2, Q_3, … Q_n) and vapor (V_1, V_2, V_3, … V_n) obtained from each effect (kg/h) can be calculated.
2. The amount of heat transferred per hour in the first effect of a forward feed multiple-effect evaporator (Figure 9.12) is given by

$$q_1 = U_1 A_1 \Delta T_1 \tag{9.19}$$

where ΔT_1 is the difference in temperature between the condensing steam and the boiling point of the liquid in the first effect.

$$\Delta T_1 = T_S - T_1 \tag{9.20}$$

TABLE 9.4

Data for the Design of Multiple-Effect Evaporators

Known Values	Values to Be Determined
1. Area of heating surface in each effect (generally, assumed equal)	1. Steam pressure in the first effect 2. Final pressure in the vapor space of the last effect
2. Amount of steam to be supplied per hour	3. Feed flow rate to the first effect until the nth effect
3. Amount of vapor leaving each effect	4. Final concentration of the product exiting the last effect 5. Enthalpies of the liquid and vapors 6. Overall heat transfer coefficient in each effect

Assuming the absence of BPE and the heat of solution and neglecting the sensible heat necessary to heat the feed to the boiling point, approximately, all the latent heat of condensing steam would be transformed to the latent heat in the vapor. This vapor then condenses in the second effect to give up approximately the same amount of heat. Thus,

$$q_2 = U_2 A_2 \Delta T_2 \tag{9.21}$$

The reasoning mentioned previously holds true until the nth effect and q_n. Therefore,

$$q_1 = q_2 = \cdots = q_n \tag{9.22}$$

$$\therefore U_1 A_1 \Delta T_1 = U_2 A_2 \Delta T_2 = \cdots U_n A_n \Delta T_n \tag{9.23}$$

Usually, in industrial applications, the areas in all effects are equal. Hence,

$$\frac{q}{A} = U_1 \Delta T_1 = U_2 \Delta T_2 = \cdots = U_n \Delta T_n \tag{9.24}$$

3. From the previous step, it is evident that the temperature drops (ΔT) in multiple-effect evaporators are inversely proportional to the overall heat transfer coefficient (U). In the presence of BPE,

$$\Sigma \Delta T = \Delta T_1 + \Delta T_2 + \cdots + \Delta T_n = T_{sf} - T_n - \left[(\text{BPE})_1 + (\text{BPE})_2 + \cdots + (\text{BPE})_n\right] \tag{9.25}$$

where

- T_{sf} = inlet temperature of the steam to the first effect (°C)
- ΔT_1 = temperature drop in the first effect = $(T_{sf} - T_1)$ (°C)
- ΔT_2 = temperature drop in the second effect = $(T_{s1} - T_2)$ (°C)
- ΔT_n = temperature drop in the nth effect = $(T_{s(n-1)} - T_n)$ (°C)
- T_1 = temperature of liquid product in the first effect (°C)
- T_2 = temperature of liquid product in the second effect (°C)
- T_n = temperature of liquid product in the nth effect (°C)
- T_{s1} = inlet temperature of steam to the second effect (°C)
- T_{sn} = inlet temperature of steam to the nth effect (°C)

It is to be noted that,

$$\Delta T\,°\text{C} = \Delta T\,\text{K} \tag{9.26}$$

From Eq. (9.24), since ΔT_l is proportional to $\frac{1}{U_1}$,

$$\Delta T_1 = \Sigma \Delta T \left[\frac{1/U_1}{1/U_1 + 1/U_2 + \cdots + 1/U_n}\right] \tag{9.27}$$

Similarly, the equations for ΔT_2, ΔT_3, ... ΔT_n can be written. Where U_1, U_2, and U_n are the respective heat transfer coefficients for the first, second, and n^{th} effects (kW/m^2 K).

4. Using the heat and material balance equations around the first effect, the amount of water vaporized can be calculated. Subsequently, the amount of steam ($\dot{m}_S$) entering the first evaporator and the heat transferred (q) in each effect can be estimated.

5. From the capacity or rate equation, $q = UA\Delta T$ for each effect, the heat transfer areas for each effect, $A_1, A_2, \ldots A_n$ can be estimated. The average value A_{av} is given by

$$A_{av} = \frac{A_1 + A_2 + \cdots + A_n}{n} \tag{9.28}$$

6. If the heat transfer areas thus estimated at the end of step-5 are reasonably equal, the design calculations are said to be complete, without the need for iteration. Contrarily, if the areas are not nearly equal, a second trial should be conducted by repeating the abovementioned five steps. Then, the new values of $\Delta T_1', \Delta T_2', \ldots \Delta T_n'$ can be obtained from the following equations:

$$\Delta T_1' = \frac{\Delta T_1 + A_1}{A_{av}} \tag{9.29}$$

$$\Delta T_2' = \frac{\Delta T_2 + A_2}{A_{av}} \tag{9.30}$$

$$\Delta T_n' = \frac{\Delta T_n + A_n}{A_{av}} \tag{9.31}$$

The sum $\Delta T_1' + \Delta T_2' + \cdots + \Delta T_n'$ must be equal to the original $\Sigma\Delta T$ (Eq. (9.25)). If this condition is not met, then all the $\Delta T'$ values should be proportionately readjusted followed by the calculation of BPE in each effect. As explained previously, the values of $q_1, q_2, \ldots q_n$ and $A_1, A_2, \ldots A_n$ are obtained using the new values of $\Delta T_1', \Delta T_2', \ldots \Delta T_n'$ (Sorour, 2015).

9.7.2.4 Economic Analysis

To select the optimum number of effects, it is important to understand the economic balance between the fixed cost and the operating cost.

9.7.2.4.1 Annual Fixed Cost

The annual fixed cost (V_{FC}) of a multiple-effect evaporator is approximately proportional to the 0.75th power of the number of effects (Coston and Lindsay, 1956).

$$V_{FC} = \frac{C_E}{A} \times N^{0.75} \tag{9.32}$$

where

C_E = Estimated cost of a single effect
A = Surface area of heat transfer
N = Number of effects

9.7.2.4.2 Annual Operating Cost

The annual operating cost (variable cost) of a multiple-effect evaporator (V_{OC}) is almost inversely proportional to the number of effects. Annual operating costs can be further categorized into two types: annual steam cost and annual evaporator cost.

a. **Annual cost of steam**

The steam consumption for a single effect evaporator is given by

$$\dot{m}_S = \frac{q}{\lambda} \tag{9.33}$$

The steam consumption for a multiple effect evaporator is given by

$$\dot{m}_S \lambda + FC_P T = P_1 C_P T_1 + V_1 H_1 \tag{9.34}$$

where $\dot{m}_S$ is the mass flow rate of steam fed to the evaporator (kg/s), λ is the latent heat of vaporization (kJ/kg), F is the flow rate of the feed solution entering the evaporator (kg/s), P_1 is the flow rate of product from the first effect (kg/s), T_1 is the temperature of the solution in the first effect (°C), C_P is the specific heat of the solution, (kJ/kg °C), V_1 is the amount of vapor leaving the first effect (kg/h), H_1 is the enthalpy of vapor leaving the evaporator (kJ/kg), and q is the heat load of the evaporator (kJ/s).

$$\text{Cost of steam} = \dot{m}_S C_T \tag{9.35}$$

where C_T is the cost per ton of steam.

$$\therefore \text{Annual cost of steam} = C_S = \text{Cost of steam} \times \text{Annual operating days} \times \text{Operating hours} \tag{9.36}$$

b. **Annual cost of an evaporator**
Annual cost of an evaporator may be calculated as follows (Peter and Timmerhaus, 1991):

$$\text{Purchased cost} = 75228.2A^{0.5053} \tag{9.37}$$

$$\text{Annual cost of evaporator} = \text{Purchased cost } (1 + 0.6)(0.15) \tag{9.38}$$

where (1 + 0.6) is the cost of piping and installation, and 0.15 is the depreciation cost. A is the area of heat transfer of the evaporator, calculated according to the equation $q = UA\Delta T$, where, ΔT (°C) is the driving force $= \left(T_{\text{steam}} - T_{\text{boiling}}\right)$, and U is the overall heat transfer coefficient (kW/m^2 K).

c. **Annual total cost**
The total yearly cost is calculated by applying the following equation:

$$\text{Annual total cost} = \text{Annual cost of steam} + \text{Annual cost of the evaporator} \tag{9.39}$$

Example 9.2

A triple-effect evaporator with a forward feed arrangement is used to concentrate a mango fruit juice at the rate of 2000 kg/h flow rate with 15% solids to 30% solids. The temperature of steam in the first-effect evaporator is 120°C, and the boiling point in the last effect is 55°C. The heat transfer coefficient in the first, second, and third effect is 2.5, 2.2, and 2 kW/m^2°C, respectively. Assuming that the specific heat of the mango juice is 4 kJ/kg °C in the entire range of evaporation process temperature with no boiling point rise, calculate the steam consumption, ΔT in each effect, and the heat transfer area of any effect.

Solution

Basis: 1 h operation of mango juice evaporation.

$$T_S = 120\,°\text{C and } T_3 = 55\,°\text{C},$$

Mass balance for the evaporation process is given as

$$F = P + V \text{ (or) } 2000 = P + V$$

Mass balance for the solid content of the mango juice and concentrated

$$x_F \cdot F = x_P \cdot P$$

$$2000 \times 0.15 = 0.3 \times P \quad \therefore P = 1000\,\text{kg}; V = 1000\,\text{kg}$$

Assuming that $V_1 = V_2 = V_3$, $V_1 = V_2 = V_3 = \dfrac{1000}{3} = 333.33\,\text{kg}$.

$$\Delta T = 120 - 55 = 65°\text{C and } \Delta T = \Delta T_1 + \Delta T_2 + \Delta T_3 = 65$$

Therefore, $U_1 \Delta T_1 = U_2 \Delta T_2 = U_3 \Delta T_3$

$$2.5\Delta T_1 = 2.2\Delta T_2 = 2\Delta T_3 \quad \therefore \Delta T_2 = 1.136\Delta T_1 \quad \therefore \Delta T_3 = 1.25\Delta T_1$$

$$\Delta T_1(1 + 1.136 + 1.25) = 65°\text{C} \quad \therefore \Delta T_1 = 19.197°\text{C}$$

Similarly, $\therefore \Delta T_2 = 21.808°\text{C} \quad \therefore \Delta T_3 = 23.996°\text{C}$

$$T_1 = T_S - \Delta T_1 = 120 - 19.197 = 100.803°\text{C}$$

$$T_2 = T_1 - \Delta T_2 = 100.803 - 21.808 = 78.995°\text{C}$$

$$T_3 = T_2 - \Delta T_3 = 78.995 - 23.996 = 54.999°\text{C}$$

$$q = mC_P\Delta T = 2000 \times 4 \times 65 = 5.2 \times 10^5\ \text{kJ/h}.$$

$$q = \frac{5.2 \times 10^5}{3{,}600} = 144.44\,\text{kW}$$

$$q = U_1 A_1 \Delta T_1$$

$$\therefore A_1 = \frac{q}{U_1 \Delta T_1} = \frac{144.44}{2.5 \times 19.197} = 3.01\,\text{m}^2$$

$$\therefore A_1 = \frac{q}{U_2 \Delta T_2} = \frac{144.44}{2.2 \times 21.808} = 3.01\,\text{m}^2$$

$$\therefore A_3 = \frac{q}{U_3 \Delta T_3} = \frac{144.44}{2.0 \times 23.993} = 3.01\,\text{m}^2$$

Total heat transfer coefficient

$$\frac{1}{U} = \frac{1}{U_1} + \frac{1}{U_2} + \frac{1}{U_3} = \frac{1}{2.5} + \frac{1}{2.2} + \frac{1}{2} = 1.355$$

$$\therefore U = 0.738\,\text{kW/m}^2\,°\text{C}$$

$$\therefore A = 3.01\,\text{m}^2$$

Answer: The steam consumption is 1000 kg, and the heat transfer area of each effect is 3.01 m²; the ΔT in the three effects are, respectively, 19.2°C, 21.8°C and 24°C

9.7.3 Vapor Recompression

Vapor recompression is an alternative approach to the multiple-effect evaporation towards improving the evaporator efficiency. During vapor recompression, the pressure and temperature of the vapor generated in a single-effect evaporator are increased to recover its latent heat of condensation and compressed vapor is reused as heating medium in the same effect/evaporator, in which it is allowed to condense.

This is different from the working principle of multiple-effect evaporation wherein heat recovery is accomplished in different effects. The two different systems for achieving vapor recompression are shown in Figure 9.18. The differentiation is based on the methodology used to raise the temperature and pressure of vapor generated.

9.7.3.1 Mechanical Vapor Recompression

During mechanical vapor recompression (MVR), vapors generated after product evaporation are sucked from the evaporator and compressed to higher pressure using a mechanical device. The mechanical device can be a centrifugal compressor (for applications with high compression ratio) or a fan (for lower compression applications). Either of these devices can be driven by an electric motor, steam turbine, internal combustion engine, or gas turbine (SPX Corporation, 2008). The resultant high-pressure steam at a corresponding high temperature is then returned to the heating side of the evaporator and reused as the heating medium for evaporation. An evaporator with MVR system reuses the latent heat of the vapors.

The working principle of MVR is based on the direct relationship between enthalpy and pressure, depicted by the Mollier diagram (Figure 3.6; Chapter 3). While compression elevates the pressure of vapors, the enthalpy of vapor also increases equivalent to that of the original steam supplied. The mechanical energy which drives the compressor is utilized to increase the pressure of vapor. Thus, a compressor is central to the operation of MVR evaporators. The working cycle of an MVR system (Figure 9.19) is explained as follows:

- The vapor is sucked from the evaporator (point *A*) at temperature T_a and pressure P_a.
- Vapor is adiabatically compressed to pressure P_r (point B').
- Due to the heat of compression, the temperature of vapor increases from T_a to T_r.
- Subsequently, the superheated vapor at the exit of the compressor is cooled down to the final saturation point (*B*) at temperature T_f and pressure P_f.
- The compressed vapor is then condensed in the condenser to recover the latent heat.
- The condensate is discharged at a temperature of T_f.

MVR evaporators (Figure 9.20) are typically single-effect evaporators. An evaporator with MVR system does not require additional thermal energy or very less thermal energy except for the steam used for heat treatment before the first effect. Hence, there is almost no residual vapor to be condensed. MVR makes it possible to evaporate 80–100 kg of water with 1 kWh. Using a triple-effect evaporator with MVR can reduce the operating costs by half compared to a conventional seven-effect evaporator equipped with

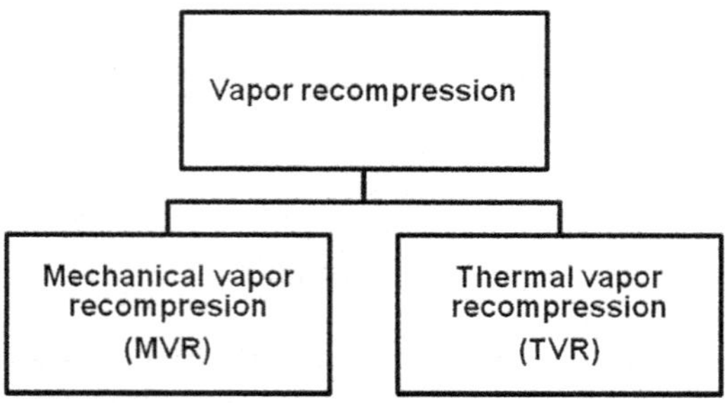

FIGURE 9.18 Classification of vapor recompression systems.

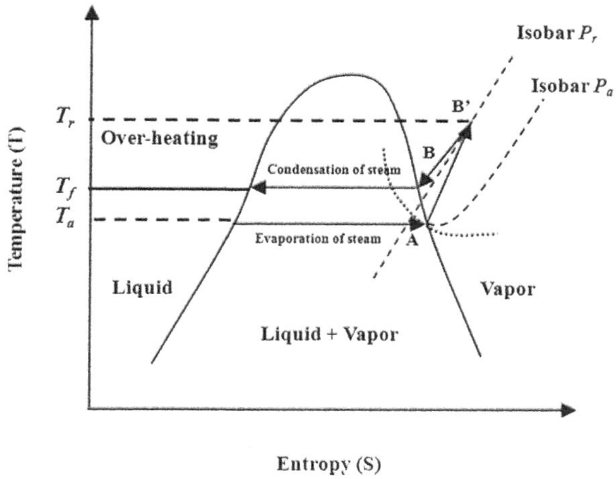

FIGURE 9.19 Working cycle of the mechanical compressor.

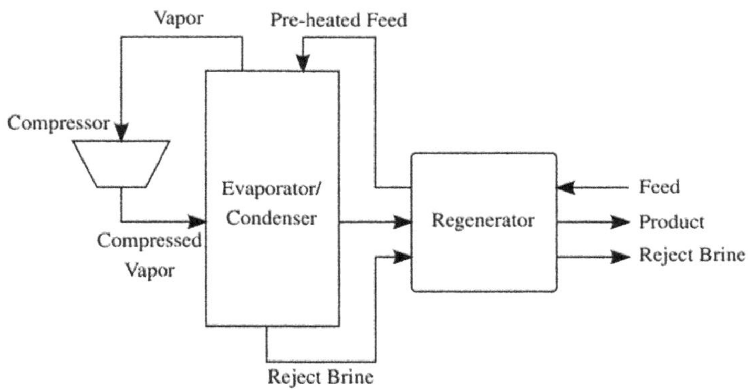

FIGURE 9.20 Single-effect MVR system for the concentration of brine. (Reproduced with permission from Mistry, K. H., McGovern, R. K., Thiel, G. P. Summers, E. K., Zubair, S. M. and Lienhard, J. H. V. 2011. Entropy Generation Analysis of Desalination Technologies. *Entropy* 13: 1829–1864 [Open access].)

thermal vapor recompression (TVR) system (Bylund, 1995). Nevertheless, the major disadvantage of MVR system is the handling of a large volume of vapor and, hence, the requirement for a high capacity compressor. Consequently, MVR evaporators are very noisy during operations. The inclusion of MVR system in an evaporator is justified by striking a balance between the running cost of the compressor and the extent of increase in steam economy. Performance of MVR system is determined based on the ratio of latent heat of saturated steam produced from vapors to work done during compression.

In the food sector, the applications of MVR are mainly encountered in the concentration of skim milk, whey, fruit juices, and vegetable juices (apple, beetroot, tomato) (Grattieri et al., 2001).

9.7.3.2 Thermal Vapor Recompression

TVR system (Figure 9.21) uses high-pressure steam as the motive fluid (source of energy) to compress part of the vapor produced from an effect/evaporator to higher pressure and temperature. For the above-mentioned purpose, a steam jet ejector is used. A steam jet ejector is a device in which high-pressure steam passing through a narrow space creates a reduced pressure zone by increasing the velocity of steam. A part of vapor from the evaporator is allowed to mix with the high-pressure steam near the throat of the nozzle at a right angle. Due to the pressure difference between the evaporator and ejector,

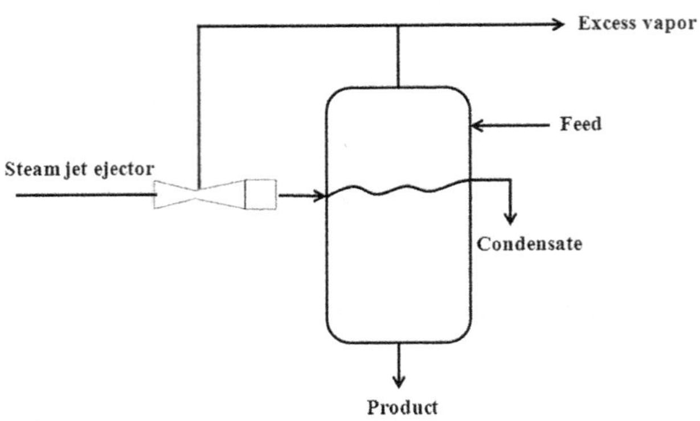

FIGURE 9.21 TVR system.

the vapor is automatically dragged with the high jet steam which is then recirculated to the evaporator and condensed back for recovery. Since only a part of the vapors is compressed by the TVR system, the residual vapors are forwarded to the second effect for the complete recovery of its latent heat. The amount of excess energy in the residual vapor corresponds to the amount of energy supplied towards the operation of the steam jet ejector. This is considered as the additional heat input or work done to recover the large amount of heat of the vapors discarded from single-effect evaporators.

The advantage of TVR system is that it can handle a high volume of vapor without the need for mechanically driven high friction moving parts. A thermal vapor recompressor is as effective as an additional evaporation effect in terms of the steam/energy saving benefit. At the same evaporation rate, an evaporator equipped with a TVR system uses 36% lesser steam and 36% lesser water compared to a straight steam evaporator. Also, the evaporator equipped with thermal vapor recompressor uses less heating surface than the straight steam system. The advantage mentioned previously can be attributed to the observation that with the inclusion of TVR, a majority of the evaporation is completed in the first effect where the product is at its lowest solid concentration. Consequently, the evaporation load is less on the subsequent effects operating at higher product concentration but lower heat transfer coefficient (Croll Reynolds, 2018). However, TVR system requires steam at high pressure, and its usage is limited to single-effect evaporators or the first-effect of multiple-effect evaporators.

In the food and beverages industry, TVR system is utilized in the evaporation of milk, whey, and sugar solutions.

9.8 Problems to Practice

9.8.1 Multiple Choice Questions

1. The evaporator economy depends on
   a. mass transfer rate
   b. energy balance considerations
   c. heat transfer rate

Answer: b

2. Low-temperature ___________ evaporators are used for heat-sensitive foods.
   a. ultrafiltration
   b. vacuum
   c. osmosis
   d. drum

Answer: b

3. Industry A is low on budget and needs an evaporator to be used in the sugar industry. Industry B is ready to invest on an expensive evaporator for the concentration of meat extracts. The appropriate evaporators for Industry A and Industry B are
   a. forced circulation evaporator, forced circulation evaporator
   b. rising film evaporator, forced circulation evaporator
   c. forced circulation evaporator, rising film evaporator
   d. rising film evaporator, rising film evaporator

Answer: b

4. Rate of evaporation is
   a. directly proportional to the temperature of the liquid
   b. directly proportional to the humidity of the surrounding air
   c. independent of the temperature of the liquid
   d. inversely proportional to the temperature of the liquid

Answer: a

5. Multiple-effect evaporation leads to
   a. reduction in operating cost and reduction in capital cost
   b. increase in operating cost and increase in capital cost
   c. increase in operating cost and reduction in capital cost
   d. reduction in operating cost and increase in capital cost

Answer: d

6. Which one of the following is NOT a component of an evaporator?
   a. Heat exchanger
   b. Vacuum separator
   c. Condenser
   d. Cyclone separator

Answer: d

7. In case of calandria-type evaporator, the solution to be evaporated is _________ the tube and steam flows _________ the tubes in steam chest,
   a. outside, outside
   b. inside, inside
   c. outside, inside
   d. inside, outside

Answer: d

8. Vacuum is
   a. any system with pressure at zero atmospheric pressure
   b. any system with pressure at one atmospheric pressure
   c. any system with pressure below that of atmospheric pressure
   d. any system pressure at above the atmospheric pressure

Answer: c

9. Economy of single-effect evaporator is
   a. always more than 1
   b. always less than 1
   c. always 1
   d. always 100

Answer: b

10. In forced circulation evaporator, the velocity of liquid entering the tube is of the order of
    a. 2–6 m/s
    b. 6–10 m/s
    c. 10–14 m/s
    d. 14–18 m/s

Answer: a

11. Economy of multiple-effect evaporator is
    a. always more than 1
    b. always less than 1
    c. always 1
    d. always 100

Answer: a

12. In ________ feed system, vapor and liquid flow in counter-current fashion.
    a. forward
    b. backward
    c. both a and b
    d. neither of a and b

Answer: b

13. In case of evaporation, generally __________ is the valuable product.
    a. concentrated solution
    b. thick liquor
    c. both a and b
    d. neither of a and b

Answer: c

14. In most of the evaporation operation, __________ is discarded.
    a. water
    b. feed liquid
    c. water vapor
    d. concentrated liquid

Answer: c

15. The objective of multiple effect evaporation is
    a. improvement of overall steam economy
    b. increase plant capacity
    c. both a and b
    d. none of the above

Answer: c

16. The weak liquor to be fed to the evaporator is composed of
    a. non-volatile solute and volatile solvent
    b. non-volatile solution and volatile solvent
    c. non-volatile solute and non-volatile solvent
    d. non-volatile solution and non-volatile solvent

Answer: a

17. During boiling
    a. vapor pressure = total pressure = atmospheric pressure
    b. vapor pressure < total pressure < atmospheric pressure
    c. vapor pressure > total pressure > atmospheric pressure
    d. vapor pressure < total pressure > atmospheric pressure

Answer: a

18. If the capacity of an evaporator is 100 kg, it means that
    a. it can vaporize 100 kg of water in 1 h
    b. it can vaporize 100 kg of water in 1 day
    c. it can form 100 kg of steam in 1 h
    d. it can form 100 kg of steam in 1 day

Answer: a

19. Elevation of boiling point of a solution and its relationship with the boiling point of water is explained by
    a. Dühring's rule
    b. Herring's rule
    c. Charles's law
    d. Boyle's law

Answer: a

20. Steam economy is more in
    a. backward feed evaporator
    b. forward feed evaporator
    c. parallel feed evaporator
    d. mixed feed evaporator

Answer: a

21. Superheated steam is not preferred as heating medium in evaporator because
    a. metal tubes cannot withstand high temperature
    b. metal tubes cannot withstand high pressure
    c. superheated steam has very low film coefficient
    d. all of the above reasons

Answer: c

22. The effect of BPE on the multiple-effect evaporation is
    a. reduction of capacity
    b. reduction of steam economy
    c. increase in steam economy
    d. increase in capacity

Answer: a

23. The advantage of backward feed multiple-effect evaporators is
    a. heat-sensitive materials can be handled
    b. there is no additional cost of pumping
    c. most concentrated liquid is at the highest temperature
    d. equal heat transfer coefficient exists in various effects

Answer: c

24. A multiple-effect evaporator as compared to a single-effect evaporator of same capacity has
    a. lower heat transfer area
    b. lower steam economy
    c. higher steam economy
    d. higher solute concentration in the product

Answer: c

25. Evaporators are normally operated under vacuum in order to
    a. reduce the wall thickness of the evaporator body
    b. enable the use of low pressure steam as heating medium
    c. prevent thermal degradation of the solute
    d. increase the steam economy

Answer: c

9.8.2 Numerical Problems

1. Fruit juice (200 kg/h) is to be concentrated from 10% solids to 30% solids in a single-effect circulating-type climbing film evaporator. Pressure in the evaporator is maintained at 0.76 atm absolute. The steam required for evaporation is available at a pressure of 1.97 atm. The overall heat transfer coefficient is 1.7 kJ/m^{2}s °C. The temperature of the feed is 20°C, and the boiling point of the solution at 77 kPa is 90°C. The specific heat of the fruit juice is 3.805 kJ/kg °C. Assume that the latent heat of vaporization of fruit juice is same as that of water (2260 kJ/kg). Calculate the quantity of steam required per hour and the heat transfer area.

Given

i. Temperature of fruit juice (feed) = 20°C
ii. Feed rate of fruit juice = 200 kg/h
iii. Initial solid content of fruit juice = 10%
iv. Final solid content of fruit juice after concentration = 30%
v. Pressure in the evaporator = 0.76 atm (abs)
vi. Steam pressure = 1.97 atm
vii. Overall heat transfer coefficient = 1.7 kJ/m^{2}s °C
viii. Boiling point of the solution at 77 kPa = 90°C
ix. Specific heat of fruit juice = 3.805 kJ/kg °C
x. Latent heat of vaporization of fruit juice = 2260 kJ/kg

To find

i. Quantity of steam required per hour
ii. Area of heat transfer surface

Solution

From the steam table, temperature of steam at 1.97 atm is 134°C and the corresponding latent heat is 2164 kJ/kg. Similarly, the condensing temperature at 0.76 atm (abs) is 91°C.

Mass balance around the evaporator

	Solid Content	Liquid Content	Total
Feed	10	190	200
Product	30	36.7	66.7
Water vapor removed			133.3

Heat balance around the evaporator

Heat per kilogram of steam (to be cooled to 91°C)

$$= \text{latent heat} + \text{sensible heat}$$

$$= 2.164 \times 10^6 + \left\{\left(4.186 \times 10^3\right) \times (134 - 91)\right\}$$

$$= 2.164 \times 10^6 + \left\{1.8 \times 10^5\right\}$$

$$= 2.34 \times 10^6 \text{ J}$$

Heat required by the feed solution (to be heated from 20°C to 90°C)

$$= \text{latent heat} + \text{sensible heat}$$

$$= \left\{2260 \times 10^3 \times 133.3\right\} + \left\{200 \times 3.805 \times 10^3 \times (90 - 20)\right\}$$

$$= \left\{3.012 \times 10^8\right\} + \{53270000\}$$

$$= 3.545 \times 10^8 \text{ J}$$

According to the principle of energy balance,

$$\text{Heat per kilogram of steam} = \text{heat required by the fruit juice}$$

$$\therefore \text{Quantity of steam required per hour} = \frac{3.545 \times 10^8}{2.34 \times 10^6} = 151.5\text{kg}$$

$$\text{Quantity of steam required per hour} = \frac{151.5}{133.3} = 1.14\,\text{kg steam/kg water}$$

Heat transfer area

Temperature of condensing steam = 134°C

Temperature difference across the evaporator = (134 – 91) = 43°C.

The heat transfer (q) is given by

$$q = UA\Delta T$$

$$\frac{3.545 \times 10^8}{3600} = 1700 \times A \times 43$$

$$\therefore A = 1.35\,\text{m}^2$$

Answer: (i) Quantity of steam required per hour = 1.14 kg
(ii) Area of heat transfer surface = 1.35 m^2

2. In problem no. 1, consider that the fruit juice is concentrated in a triple-effect evaporator instead of a single effect evaporator. The temperature of steam to the first effect is 102°C. Boiling point of the solution in the last effect is 50°C. The overall heat transfer coefficient in the first, second and third effect is 2000, 1500 and 1000 W/m^2°C, respectively. Assume that the there is no BPE during the evaporation. Calculate the steam consumption, ΔT in each effect and the heat transfer area.

Given

i. Number of effects = 3
ii. Temperature of steam to the 1st effect (T_S) = 102°C
iii. Boiling point of the solution in the last effect (T_3) = 50°C
iv. $U_1 = 2000\,\text{W/m}^2\,°\text{C}$
v. $U_2 = 1500\,\text{W/m}^2\,°\text{C}$
vi. $U_3 = 1000\,\text{W/m}^2\,°\text{C}$

To find

i. Steam consumption
ii. ΔT in each effect
iii. Heat transfer area

Solution

Basis: 1 h operation of fruit juice evaporation.

Mass balance for the evaporation process

From the mass balance explained in the problem 1, the amount of water evaporated during the evaporation is 133.2 kg. Dividing this quantity equally among the three effects, i.e. $\frac{133.3}{3} = 44.4\,\text{kg}$.

$$\Delta T = 102 - 50 = 52 \text{ and } \Delta T = \Delta T_1 + \Delta T_2 + \Delta T_3 = 52$$

$$\therefore U_1 \Delta T_1 = U_2 \Delta T_2 = U_3 \Delta T_3$$

$$\therefore 2000(\Delta T_1) = 1500(\Delta T_2) = 1000(\Delta T_3)$$

$$\therefore \Delta T_2 = 1.33\ \Delta T_1 \text{ and } \Delta T_3 = 2\Delta T_1$$

$$\Delta T_1 (1 + 1.33 + 2) = 52$$

$$\therefore \Delta T_1 = 12°\text{C}$$

$$\therefore \Delta T_2 = 15.96°\text{C and } \Delta T_3 = 24°\text{C}$$

$$T_1 = T_S - \Delta T_1 = 102 - 12 = 90°\text{C}$$

$$T_2 = T_1 - \Delta T_2 = 90 - 15.96 = 74.04°C$$

$$T_3 = T_2 - \Delta T_3 = 74.04 - 24 = 50.04°C$$

$$q = mC_P\Delta T = 200 \times 3805 \times 52 = 3.9 \times 10^7 \text{ J/h}$$

$$q = \frac{3.9 \times 10^7}{3600} = 10992\,\text{W}$$

$$q = U_1A_1\Delta T_1; \quad \therefore A_1 = \frac{q}{U_1\Delta T_1} = \frac{10992}{2000 \times 12} = 0.458\,\text{m}^2$$

Similarly,

$$A_2 = \frac{q}{U_2\Delta T_2} = \frac{10992}{1500 \times 15.96} = 0.459\,\text{m}^2$$

$$A_3 = \frac{q}{U_3\Delta T_3} = \frac{10992}{1000 \times 24} = 0.458\,\text{m}^2$$

Answer: The steam consumption is 133.2 kg, and the heat transfer area of each effect is 0.458 m^2

3. The evaporation capacity of an evaporator operating under a pressure of 25 kPa is 4000 kg/h. The temperature of condensing water is 20°C and the maximum permissible temperature for the water discharged from the condenser is 30°C. Calculate the amount of water required in a jet condenser to condense the vapors from the abovementioned evaporator.

Given

i. Evaporation capacity = 4000 kg/h
ii. Temperature of condensing water = 20°C
iii. Maximum permissible temperature of the condensate = 30°C
iv. Working pressure of the evaporator = 25 kPa

To find: Amount of water required by the condenser to condense the vapors from the evaporator

Solution

Heat balance around the evaporator

From the steam tables, the condensing temperature of water corresponding to the pressure of 25 kPa is 65°C and the latent heat of vaporization is 2345.52 kJ/kg.

$$\text{Heat removed from condensate per kilogram} = 2345.52 \times 10^3 + 4.186 \times 10^3 \times (65 - 30)$$

$$= 2.49 \times 10^6 \text{ J/kg}$$

$$\therefore \text{Total heat removed from the condensate per hour} = 4000 \times 2.49 \times 10^6 = 9.96 \times 10^9 \text{ J}$$

$$\text{Heat absorbed by cooling water} = 4.186 \times 10^3 \times (30 - 20) = 41860\,\text{J/kg}$$

$$\therefore \text{Quantity of cooling water per hour} = \frac{4000 \times 2.49 \times 10^6}{41860} = 237935.977 = 2.38 \times 10^5 \text{ kg}$$

Answer: Amount of water required by the condenser to condense the vapors from the evaporator = 2.38 × 10^5 kg

4. Orange juice is to be concentrated from 10% solids to 30% solids in a climbing film evaporator of height 2 m and diameter 3 cm. The juice is fed to the evaporator at 55°C. The steam pressure is 150 kPa (abs) and the overall heat transfer coefficient is 5000 J/(m^2s °C). Assuming that the latent heat of vaporization of orange juice is similar to that of water (2256 kJ/kg), estimate the rate of evaporation.

Given

i. Initial concentration of orange juice = 10%
ii. Final concentration of orange juice = 30%
iii. Feed temperature = 55°C
iv. Steam pressure = 150 kPa
v. U = 5000 J/(m^2s °C)
vi. Latent heat of vaporization of orange juice = 2256 kJ/kg
vii. Height of the climbing film evaporator = 2 m
viii. Diameter of the climbing film evaporator = 3 cm

To find: Rate of evaporation

Solution

Mass balance around the evaporator

Basis: 100 kg of feed (orange juice)

	Solid content	Liquid content	Total
Feed	10	90	100
Product	30	3.3	33.3
Water vapor removed			66.7

Heat balance around the evaporator

$$\text{Area of the evaporator tube} = \pi DL = \pi \times \left(3 \times 10^{-2}\right) \times 2 = 0.19\,\text{m}^2$$

From the steam table,

$$\text{Condensing steam temperature at 150 kPa (abs)} = 111.35°\text{C}$$

$$\text{Heat supplied by steam, } q = UA\Delta T = 5000 \times 0.19 \times (111.35 - 55) = 53532.5\,\text{J/s}$$

$$\text{Heat required for the evaporation of unit mass of feed} = \text{Water evaporated} \times \text{latent heat}$$

$$= 66.7 \times 2256 \times 10^3 = 1.5 \times 10^8\,\text{J}$$

$$\therefore \text{Water evaporated } \frac{5.4 \times 10^4}{1.5 \times 10^8} = 0.00036\,\text{kg/s}$$

$$\therefore \text{Rate of evaporation} = 0.00036 \times 3600 = 1.296\,\text{kg/h}$$

Answer: Rate of evaporation = 1.296 kg/h

5. In problem no. 4, consider that the orange juice is concentrated in a short-tube vertical evaporator of area 12.57 m^2. The pressure in the evaporator is 70 kPa absolute, and steam pressure is

100 kPa absolute. If the overall heat transfer coefficient is 450 J/(m²s °C), and each tube is of length 1 m and diameter 4 cm estimate the evaporation rate and number of tubes.

Given

i. Heat transfer area = 12.57 m^2
ii. Length of each tube = 1 m
iii. Diameter of each tube = 4 cm = 0.04 m
iv. Feed temperature = 20°C
v. Steam pressure = 100 kPa (abs)
vi. Pressure in the evaporator = 70 kPa (abs)
vii. $U = 450$ J/(m^2s °C)

To find: Rate of evaporation

Solution

Heat balance around the evaporator

$$\text{Heat transfer area of the evaporator} = 12.57 \text{ m}^2$$

$$\text{Heat transfer area} = n \times (\pi \text{DL}) = n \times (\pi \times 0.04 \times 1) = 12.57$$

$$\therefore n = \frac{12.57}{0.1257} = 100$$

$$\text{No. of tubes} = 100$$

From the steam table,

$$\text{Condensing steam temperature at } 100 \text{ kPa (abs)} = 99.6°\text{C}$$

$$\text{Heat supplied by steam, } q = UA\Delta T = 450 \times 12.57 \times (99.6 - 20) = 450257.4 \text{ J/s}$$

$$\text{Heat required for the evaporation of unit mass of feed} = \text{Water evaporated} \times \text{latent heat}$$

$$= 66.7 \times 2256 \times 10^3 = 1.5 \times 10^8 \text{ J.}$$

$$\therefore \text{Water evaporated } \frac{4.5 \times 10^5}{1.5 \times 10^8} = 0.003 \text{ kg/s}$$

$$\therefore \text{Rate of evaporation} = 0.003 \times 3600 = 10.8 \text{ kg/h}$$

Answer: Rate of evaporation = 10.8 kg/h

BIBLIOGRAPHY

Ahmad, F. and Khan, N. A. 2013. Design of triple effect evaporator developed by a program in C++. *International Journal of Scientific and Engineering Research* 4: 2298–2302.

Ahmad, M. I., Shah, M. U. H., Khan, M. A. Z., Ahmad, M. A. K. A., Irfan, M. and Durrani, A. A. 2018. Concentration of cane-sugar syrup in a pilot scale climbing film evaporator. *Chemical Industry and Chemical Engineering Quarterly* 24: 43–50.

Anon 1959. *Canadian Dairy and lce Cream Journal* 38:19.

Anon 1975. *Food Processing* 36:67.

Artal, J., Serra, L. and Uche, J. 2002. *Comparison of Heat Transfer Coefficient Correlations for Thermal Desalination Units*. Zaragoza, Spain: Centre of Research for Power Plant Efficiency, University of Zaragoza.

Baker, R. A., Berry, N., Hui, Y. H. and Barrett, D. M. 2005. Fruit preserves and jams. In *Processing Fruits Science and Technology* (Second edition), eds. D. M. Barrett, L. Somogyi and H. S. Ramaswamy, 113–125. Boca Raton, FL: CRC Press.

Bouman, S., Brinkman, D. W., de Jong, P. and Waalewijn, R. 1988. Multistage evaporation in the dairy industry: energy saving, product losses and cleaning. In *Preconcentration and Drying of Food Materials*, ed. S. Bruin. Amsterdam: Elsevier Science.

Bourgois, J. and Maguer, M. L. 1984. Modelling of heat transfer in a climbing-film evaporator: III. Application to an industrial evaporator. *Journal of Food Engineering* 3: 39–50.

Bylund, G. 1995. *Dairy Processing Handbook*. Lund, Sweden: Tetra Pak Processing Systems AB.

Coston, M. M. and Lindsay, E. E. 1956. Selection and application of evaporator equipment. *Chemical Engineering Progress* 52: 49.

Croll Reynolds 2018. http://croll.com/library/technical-articles/thermal-recompression/ (accessed July 1, 2018).

Cvengroš, J., Lutišan, J. and Micova, M. 2000. Feed temperature influence on the efficiency of a molecular evaporator. *Chemical Engineering Journal* 78: 61–67.

Dodeja, A. K. and Zaidi, A. H. 1980. Plate evaporator-how it is advantageous for dairy industry. *Indian Dairyman* 33: 243–247.

Ermis, K., Kucukrendeci, I. and Karabektas, M. 2017. Investigation of Multiple Effect Evaporator Design. *Published in 5th International Symposium on Innovative Technologies in Engineering and Science*, 29-30 September 2017 (ISITES2017 Baku - Azerbaijan).

Grattieri, W., Medich, C. and Vanzan, R. 2001. Electro Technologies for Energy and Uses: Application of Mechanical Vapor Recompression to Food Industry. *Published in 16th International Conference and Exhibition on Electricity Distribution Part I*, Amsterdam.

Grosse, J. W. and Duffield, G. M. 1954. *Chemistry and Industry* 1464.

Gupta, A. S. and Holland, F. A. 1966. Heat transfer studies in a climbing film evaporator. Part I. Heat transfer from condensing steam to boiling water. *The Canadian Journal of Chemical Engineering* 44: 77–81.

Haasbroek, A. 2013. Advanced control with semi-empirical and data based modelling for falling film evaporators. M.Sc. Thesis. Submitted to the Stellenbosch University, South Africa.

Holland, R. F. 1964. *American Milk Review*, 26: 56.

Lozano, J. E. 2006. *Fruit Manufacturing. Scientific Basis, Engineering Properties, and Deteriorative Reactions of Technological Importance*. Argentina: Springer.

McCabe, W. L. and Smith, J. C. 1958. *Unit Operations of Chemical Engineering*. New York: McGraw Hill.

McCabe, W. L., Smith, J. C. and Harriott, P. 1993. *Unit Operations of Chemical Engineering* (Fifth edition). New York: McGraw Hill.

Mercer, D. G. 2012. A comparison of the kinetics of mango drying in open-air, solar, and forced-air dryers. *African Journal of Food, Agriculture, Nutrition and Development* 12: 6836–6852.

Miranda, V. and Simpson, R. 2005. Modelling and simulation of an industrial multiple effect evaporator: tomato concentrate. *Journal of Food Engineering* 66: 203–210.

Mistry, K. H., McGovern, R. K., Thiel, G. P., Summers, E. K., Zubair, S. M. and Lienhard, J. H. V. 2011. Entropy generation analysis of desalination technologies. *Entropy* 13: 1829–1864.

Morison, K. R. 2015. Reduction of fouling in falling-film evaporators by design. *Food and Bioproducts Processing* 93: 211–216.

Munir, M. T., Zhang, Y., Wilson, D. I., Yu, W. and Young, B. R. 2014. Modelling of a falling film evaporator for dairy processes. *Chemeca 2014: Processing Excellence; Powering Our Future*, Perth, Western Australia.

NPTEL 2018. Method of feeding: multiple effect evaporators. In *Evaporators* (Lecture 39). https://nptel.ac.in/courses/103103032/module9/lec39/2.html (accessed October 15, 2018).

Palmer, F. J. 1993. Fruit juices and soft drinks. In *Food Industries Manual*, eds. M. D. Ranken, R. C. Kill and C. J. Baker, 211–235. London, UK: Blackie Academic and Professional.

Pawar, S. B., Mujumdar, A. S. and Thorat, B. N. 2012. CFD analysis of flow pattern in the agitated thin film evaporator. *Chemical Engineering Research and Design* 90: 757–765.

Perry, R. H., Chilton, C. H. and Kirkpatrick, S. D. 1963. *Chemical Engineer's Handbook* (Fourth edition). New York: McGraw Hill.

Peter, M. S. and Timmerhaus, K. D. 1991. *Plant Design and Economics for Chemical Engineers* (Fourth edition). New York: McGraw Hill.

Rao, C. S. and Hartel, R. W. 2007. Scraped surface heat exchangers. *Critical Reviews in Food Science and Nutrition* 46: 207–219.

Sangrame, G., Bhagavathi, D., Thakare, H., Ali, S. and Das, H. 2000. Performance evaluation of a thin film scraped surface evaporator for concentration of tomato pulp. *Journal of Food Engineering* 43: 205–211.

Schwartzberg, H. G. 1989. Food property effects in evaporation. In *Food Properties and Computer-Aided Engineering of Food Processing Systems*, eds. R. P. Singh and A. G. Medina, 443–470. the Netherlands: Kluwer Academic Publishers.

Shi, C., Pan, Z., McHugh, T., Wood, D. and Hirschberg, E. 2008. Sequential infrared radiation and freeze-drying method for producing crispy strawberries. *Transactions of the ASABE* 51: 205–216.

Singh, H. 2007. Interactions of milk proteins during the manufacture of milk powders. *Lait* 87: 413–423.

Singh, R. P. and Heldman, D. R. 2014. *Introduction to Food Engineering* (Fifth edition). London, UK: Academic Press.

Sorour, M. A. 2015. Optimization of multiple effect evaporators designed for fruit juice concentrate. *American Journal of Energy Engineering* 3: 6–11.

SPX Corporation 2008. *Evaporator Handbook.* Charlotte, NC: SPX, Global headquarters.

UNIDO (United Nations Industrial Development Organization) 1969. The use of centri-therm expanding-flow and forced-circulation plate evaporators in the food and biochemical industries. *Food Ind. Stud. No. I*, New York: United Nations.

Vieira, B. S., Gullberg, D. and Perrot, V. 2016. Recent Experience with Metallic Heaters for Phosphoric Acid Evaporation. *Procedia Engineering* 138: 437–444.

10

Drying

Drying is the oldest method of food preservation which involves the removal of moisture from a food product. It is an indispensable unit operation in the food and agricultural processing sectors. The term *drying* is often used exchangeably with *evaporation* and *dehydration*. However, these are entirely different phenomena. While the final product in drying is solid, the end product resultant from evaporation is a highly viscous liquid. Also, the final moisture content of a dried food product is more than 2.5%; whereas, dehydrated products have moisture content of less than 2.5% (Vega-Mercado et al., 2001).

The foremost purpose of drying is to achieve food preservation and shelf-life extension by preventing the microbial growth and deteriorative chemical reactions. The preservation effect is achieved by reducing the available water that supports the growth of food spoilage microorganisms and promotes deteriorative reactions such as browning (enzymatic and nonenzymatic) and lipid oxidation. In addition, drying reduces the weight and volume of a food product, thereby decreasing the cost of packaging, transportation, and storage. Dying also leads to structural modifications of macromolecular components in foods (carbohydrates, proteins, and lipids) which impart certain functional characteristics to the final product. Other physical changes brought about by drying include shrinkage, puffing, crystallization, and glass transitions (Mujumdar, 2007). Another key application of drying is the encapsulation of bioactive and sensitive food components (core) within a protective outer layer (wall). Drying-mediated encapsulation is achieved by the difference in drying rate between the wall and the core materials.

Drying is an energy-intensive operation which demands considerable attention towards its optimization, such that the whole process is rendered commercially viable. An ideal drying process for foods is one which provides the optimum heat and mass transfer within the raw material and results in a dried product with defined physical structure. This chapter intends to present a comprehensive discussion of the principle of drying and the classification and design aspects of dryers.

10.1 Theory of Drying

The principle of drying can be well appreciated with a clear understanding of the nature and properties of the solvent to be removed, which is almost always water, in the case of foods. Indeed, the necessity for drying of foods arises since the fresh agricultural commodities always contain some amount of water which limits their storage life. The subsequent sections explain the different phases and forms in which water can be present in a food product.

10.1.1 Phase Diagram of Water

Phase can be defined as a physically unique and chemically uniform part of a system that is identified by a particular chemical composition and structure. Pure water can exist in three different phases: solid, liquid, and gas. The phase or state of water under specific conditions of temperature and pressure can be mapped from a phase diagram (Figure 10.1). The temperature and pressure at which the solid, liquid, and vapor phases of water coexist in equilibrium is called the *triple point* (0.0098°C and 4.58 mm Hg), denoted by point D on the phase diagram.

For almost all food products, drying involves the removal of liquid water from the material via the gaseous phase. Drying techniques can be categorized based on the methods used for heat addition and water vapor removal (Ishwarya et al., 2015). In hot-air (HA) drying, moisture removal is facilitated by direct contact between the product and hot air at atmospheric pressure (Geankoplis, 2006). With respect

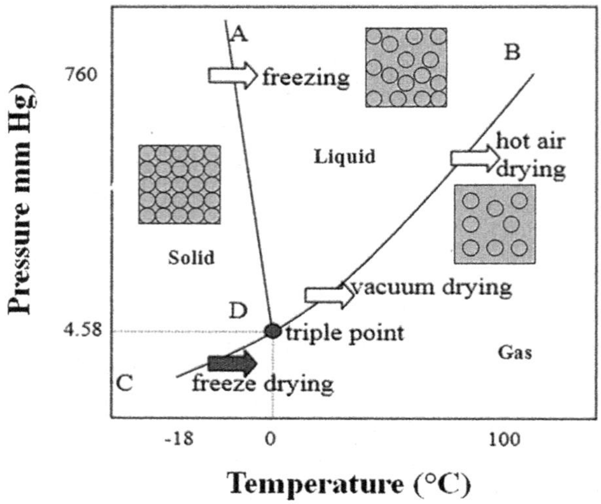

FIGURE 10.1 Phase diagram of water (not to scale).

to the phase diagram, hot air drying occurs in zone AB (Figure 10.1) in which the liquid water is converted to the gas phase. On the other hand, freeze-drying takes place in the zone below the triple point *D* of the phase diagram. In the freeze-drying process, water is removed from the frozen product via sublimation during which the solid water (ice) (zone C) is directly converted to the gaseous phase (zone B) without passing through the liquid phase (zone A). The pressure at the triple point of the phase diagram, 4.58 mm Hg refers to the water vapor pressure and not the total pressure of the system. Thus, from the phase diagram of water, it is evident that sublimation can occur at atmospheric pressure as long as the water vapor pressure is below 4.58 mm Hg. Therefore, the requirement of vacuum is not a prerequisite for the freeze-drying process. In contrast, vacuum drying takes place under reduced pressure involving the gas and liquid phases.

10.1.2 Moisture Content

Moisture content describes the mass of water present in a food product. The moisture content of a food product can be expressed on the wet or dry basis. Moisture content on a wet basis (X_W) is the ratio of the mass of water to the total mass of the sample explained as

$$X_W = \frac{\text{mass of water}}{\text{total mass of sample}} = \frac{\text{mass of water}}{\text{mass of water} + \text{mass of dry solid}} \tag{10.1}$$

Expressing the moisture content on a wet basis is common in the grain industry. Moisture content on a dry basis (X_{db}) is the ratio of moisture content in the sample to the weight of dry solids (bone dry), given by

$$X_{db} = \frac{\text{mass of water}}{\text{mass of dry solid}} = \frac{X_W}{1 - X_W} \tag{10.2}$$

From Eqs. (10.1) and (10.2), the moisture content on a wet and dry basis can be related as given in Eq. (10.3).

$$X_W = \frac{X_{db}}{1 + X_{db}} \tag{10.3}$$

For a given food product, the moisture content on a dry basis is always higher than the wet basis moisture content. With high-moisture foods such as fruits and vegetables, the moisture content on a dry basis may exceed 100% as the weight of moisture may be more than that of the dry solids. For such products, moisture is expressed on the decimal basis as *mass of water per unit mass of dry product*, given by the quantities, X_W or X_{db}. The moisture content may also be expressed as a percentage which is obtained by multiplying the decimal moisture content (X_W or X_{db}) by 100. The disadvantage of expressing the moisture content on a wet basis is that the total mass which is the reference base changes as the moisture is removed. Conversely, the amount of dry matter does not change.

10.1.2.1 Moisture Sorption Isotherm

Relative humidity (RH) is the ratio of partial pressure of water vapor in the gas–vapor mixture to equilibrium vapor pressure at the same temperature. Water activity (a_w) is defined as the ratio of the vapor pressure of water in a food (P) to the vapor pressure of pure water (P_0) at the same temperature, or, in other words, equilibrium relative humidity (ERH) divided by 100. Thus, it can be expressed as

$$a_W = \frac{P}{P_0} = \frac{\text{ERH}}{100} \tag{10.4}$$

Water activity relates to the proportion of available water which facilitates the chemical reactions and microbial growth in foods. A graphical representation of the relationship between moisture content and water activity of a food is known as the *moisture sorption isotherm* (Figure 10.2). The shape of moisture sorption isotherm is characteristic of the chemical composition, physical structure, and nature of the water binding in the food. The moisture sorption isotherms can be plotted from adsorption (wetting or gain of moisture until equilibrium is attained with the surrounding of high RH) or desorption (drying or loss of moisture until equilibrium is attained with the surrounding of low RH). The difference between the adsorption and desorption curves is known as hysteresis (Figure 10.2). Obtaining an understanding of the moisture sorption isotherms is relevant from the perspectives of design and optimization of drying equipment.

The typical shape of an isotherm reflects the nature of binding between water and the other constituents of food, which determines the equilibrium vapor pressure above a food and its drying characteristics. Accordingly, the moisture sorption isotherm can be typically divided into three zones (Figure 10.2).

10.1.2.1.1 Zone A: Bound Moisture Content

Zone A of the moisture sorption isotherm represents the strongly bound moisture content. Bound moisture content is the amount of moisture (Figure 10.2) which is physically or chemically bound to the

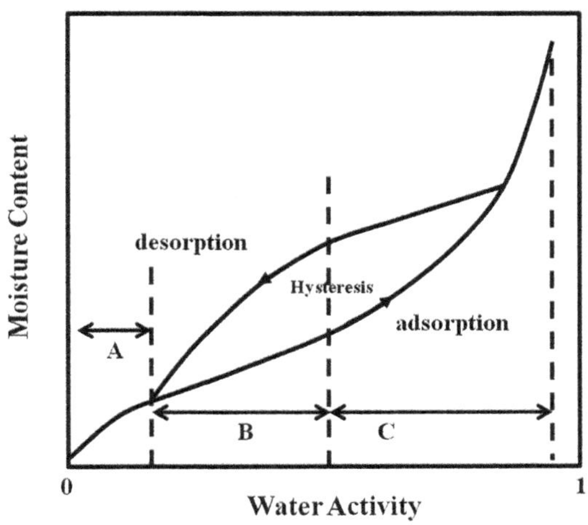

FIGURE 10.2 Moisture sorption isotherm for a typical food product. (Reproduced with permission from Andrade, R. D., Lemus, R. M. and Pérez, C. E. C. 2011. Models of sorption isotherms for food: uses and limitations. *Vitae, Revista De La Facultad De Quõmica Farmaceutica* 325–334.)

microstructure of food matrix through the *adsorption* process. The bound water includes structural water (H-bonded water) and monolayer water, which is sorbed by the hydrophilic and polar groups of food components such as polysaccharides and proteins. It also includes the equilibrium moisture content which is the amount of moisture in equilibrium with its surroundings. At a given temperature and vapor pressure of water in the surrounding air, beyond the equilibrium moisture content, there is no exchange of water between the food and its surroundings. The properties of bound moisture content are different from those of bulk water. It exerts a lower vapor pressure than the pure liquid at the same temperature. Bound moisture is unfreezable by nature and unavailable for chemical reactions or as a plasticizer.

10.1.2.1.2 Zone B: Unbound Moisture Content

The zone B represents the unbound moisture content which is present in excess of the equilibrium moisture content (Figure 10.2). The water molecules of this category are usually present in small capillaries and bind less firmly than those in the zone A. Thus, unbound moisture is a portion of the total moisture content which can be easily removed from the food matrix. At a given temperature, the unbound moisture content exerts vapor pressure equal to that of the pure liquid, thus corresponding to the saturated humidity.

10.1.2.1.3 Zone C: Free Moisture Content

Free moisture content, represented by zone C of the sorption isotherm, is the amount of moisture (Figure 10.2) that is held in voids, large capillaries, and crevices of the food matrix (Fortes and Okos, 1980; Van den Berg and Bruin, 1981; Kinsella and Fox, 1986; Mohsenin, 1986). It is loosely bound to food components. Thus, during drying, only the free moisture content can be easily removed from the food products. The property of free moisture content is nearly similar to that of bulk water, at a particular temperature and humidity.

Example 10.1

A dried food product contains 85 g solid and 15 g moisture. Calculate the moisture content on the dry and wet basis.

Solution

Given
Mass of water = 15 g
Mass of solids = 85 g
Moisture content on wet basis is given by

$$X_W = \frac{\text{mass of water}}{\text{total mass of sample}} = \frac{\text{mass of water}}{\text{mass of water} + \text{mass of dry solid}}$$

$$X_W(\%) = \frac{15}{(85+15)} \times 100 = 15\%$$

Moisture content on dry basis is given by

$$X_{db} = \frac{\text{mass of water}}{\text{mass of dry solid}} = \frac{X_W}{1 - X_W}$$

$$X_{db}(\%) = \frac{15}{85} \times 100 = 17.647\%$$

Conversely, the moisture content of wet basis can also be calculated from the moisture content on dry basis

$$X_W = \frac{X_{db}}{1 + X_{db}}$$

$$X_W(\%) = \frac{0.17647}{(1+0.17647)} \times 100 = 14.999\% \sim 15\%$$

Answer: $X_W = 15\%$; $X_{db} = 17.647\%$

10.1.3 The Concept of Simultaneous Heat and Mass Transfer

Drying is described as a process of simultaneous heat and mass transfer. Heat transfer proceeds from the product surface in contact with the drying medium at a higher temperature (T_h) to the core of the food product and also from one point to another within the product. Consequently, a temperature gradient is created between the product and water surface at some location within the product. This temperature gradient acts as the driving force for the heat transfer phenomenon which occurs by one or a combination of the conduction, convection, and radiation mechanisms (Figure 10.3). The abovementioned mechanisms vary with respect to the type of drying medium, mode of contact between the product and drying medium, the scale of heat transfer within the product, i.e., molecular or bulk transport, and the direction of heat transfer.

The drying medium supplies the required sensible heat to increase the initial product temperature (T_i) to the evaporation temperature (T_v) (Figure 10.4). It also provides the latent heat of vaporization (2200 kJ/kg) to remove the bound moisture from the food product, via the vapor phase. The subsequent diffusion of water vapor from within the product to the surrounding medium constitutes the mass transfer phenomenon. Thus, it is justified that drying is a *process of simultaneous heat and mass transfer.* While the thermal conductivity of the product limits the heat transfer inside the product, molecular diffusion influences the mass transfer during the drying process.

Thus, it is clear that similar to the heat transfer, the mass transfer also occurs both within the product and between the product and drying medium. The former occurs by diffusion governed by Fick's law (explained in Chapter 6: Mass Transfer), through a tortuous pathway (Figure 10.5). The organization of this tortuous pathway is unique for each food product depending upon the structural changes that occur during the drying process.

Capillary forces can also act along with the diffusion mechanism to facilitate the drying of food products (Gorling, 1956, 1958). Here, the surface tension plays a major role in setting the moisture movement. In addition, the porous microstructure of the product is also important in this mechanism as the

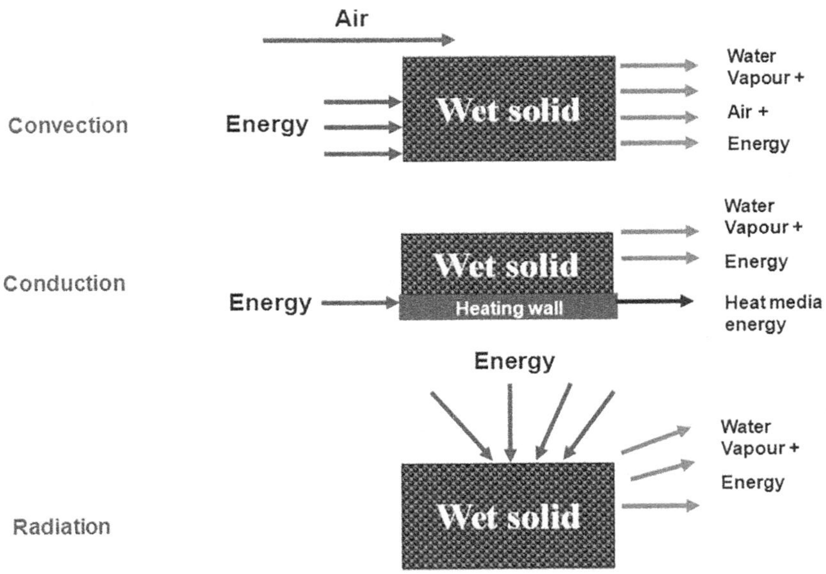

FIGURE 10.3 Modes of heat transfer during drying.

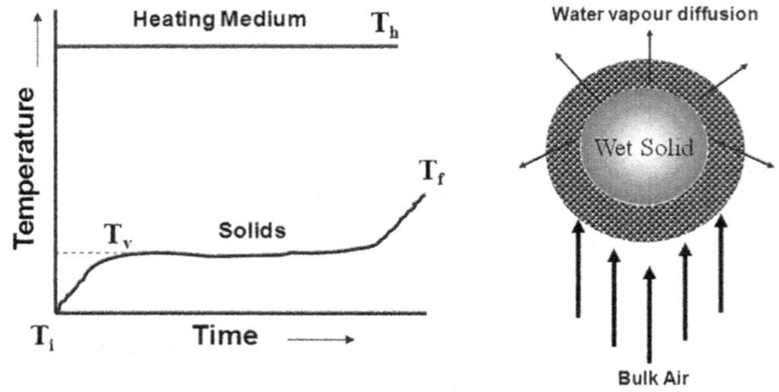

FIGURE 10.4 The concept of simultaneous heat and mass transfer during drying.

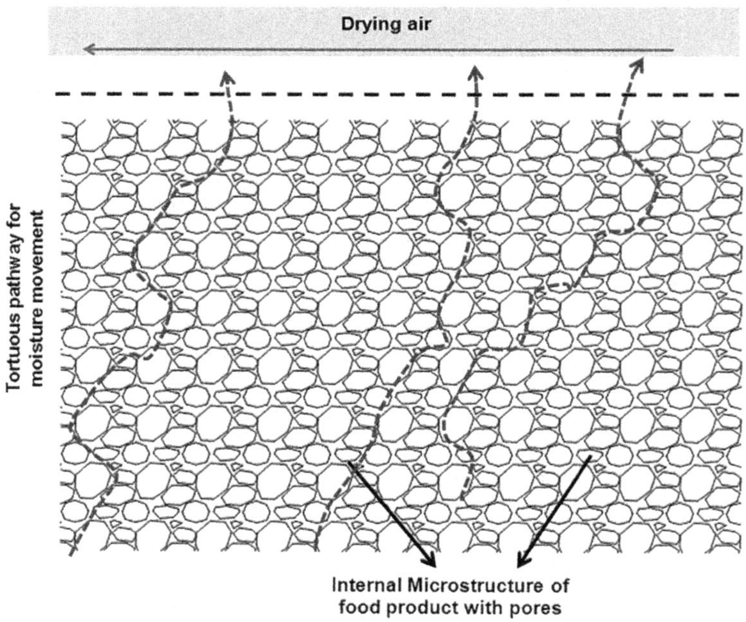

FIGURE 10.5 Tortuous pathway for moisture removal during drying. (Reproduced with permission from Anandharamakrishnan, C. 2017. Introduction to drying. In *Handbook of Drying for Dairy Products*, ed. C. Anandharamakrishnan, 1–14. Chichester, West Sussex, UK: John Wiley and Sons.)

interconnected pores communicate with the external surface by channels through which the moisture is removed. Moisture removal occurs in conjunction with the formation of a meniscus across each pore which sets an interfacial tension at the solid–water boundary. Subsequently, capillary forces are initiated at the interface in a direction perpendicular to the solid surface. These capillary forces act as the driving force for moisture movement from within the porous structure of the product to the surface. Smaller pores lead to greater capillary forces than the larger ones (Geankoplis, 2006). Surfactants that reduce the surface tension of water can be used to enhance the capillary forces (Rahman and Perera, 2007).

On the other hand, mass transfer between the product and drying medium is a function of the free moisture content at the surface. The limiting parameter is the water activity (a_w), which presents the gradient for water movement and an equilibrium condition that marks the end point of the drying process. This justifies the use of high temperature in most of the drying methods, as higher temperatures result in lower equilibrium moisture content and, hence, presents a larger concentration gradient for moisture removal (Singh and Heldman, 2014).

10.2 Drying Rate Curve

From the engineering and economic viewpoints, the rate of drying is important in determining the production capacity of the dryer. Drying rate is defined as the mass of water removed per unit time per unit mass of dry solids. Alternatively, it is defined as the mass of water removed per unit time per unit area. The drying rate curve is a plot between free moisture content and drying rate (Figure 10.6), which is a clear representation of the different stages of drying. The line AB shows the initial warming period of the sample during drying, where the temperature of the water inside the sample increases and the pore opens. This phase is short and often omitted in the calculation of drying time.

10.2.1 Constant Rate Period

During the constant rate period (indicated by the line BC of Figure 10.6), the free moisture available at the product surface is removed at a constant rate. The constant rate drying period is externally mass transfer controlled and continues until a thin film of water exists at the product surface owing to the absence of internal or external mass transfer resistance (Rahman and Perera, 2007). Water within the product's interior moves to the surface through diffusion and capillary mechanisms to replenish the evaporated water. The simultaneous withdrawal of latent heat from the drying medium maintains the product at the wet-bulb temperature (T_{wb}) of air. This phenomenon is known as the *evaporative cooling*. T_{wb} is the temperature of air when it reaches 100% RH, as water evaporates into it due to the supply of latent heat by the drying medium. T_{wb} is the lowest temperature that can be reached under ambient conditions by the evaporation of water. Drying rate during the constant rate period is limited by the rate of heat transfer from the drying medium to a food material.

Case hardening is an important phenomenon to be considered during the constant rate drying period. It can occur if the constant rate drying period occurs at a low RH and high temperature. Under these conditions, moisture is removed from the surface of the material at a much faster rate than it can diffuse from within the material. The result is the formation of a hardened relatively impervious layer or skin on the surface of the material which hinders the movement of moisture from the center of the food. Consequently, the subsequent phases of drying proceed at a much slower rate than it would otherwise be (Wilhelm et al., 2004). Case hardening can be prevented by using cooler drying air at the beginning of the drying process.

10.2.2 Falling Rate Period

Once the product attains its critical moisture content (point C in the drying rate curve shown in Figure 10.6), the drying rate drops sharply which is known as the falling rate period (denoted by line CD in Figure 10.6). During this period, the drying rate shows a linear decrease with the decrease in moisture content. Falling

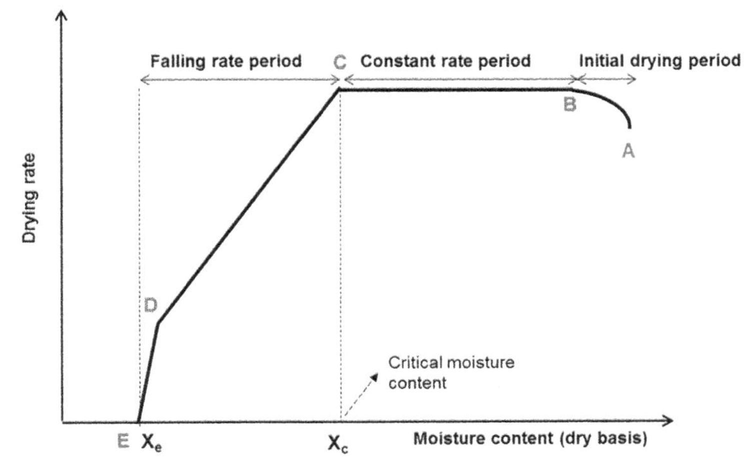

FIGURE 10.6 Schematic of the drying rate curve.

rate period is the longest phase of the drying process, during which the internal diffusion of moisture is the rate-limiting step (similar to the evaporation of surface water during the constant rate drying period). In this period of drying, the surface dries out and the product attains the dry-bulb temperature (T_{db}) of air, which is also the final temperature of the product (T_f). This period is very critical for the product quality as it can cause heat damage to the product due to an increase in temperature.

During the real-time drying of foods, the transition point between the constant and falling rate periods is not distinctly visible. However, it is retained for the purpose of modeling the drying process. While the nonhygroscopic materials have only one falling rate period, hygroscopic materials exhibit two or more falling rate periods. The second falling rate period commences once the food surface is completely dried and the ERH drops below 100%. Moisture held by multimolecular adsorption is removed during this phase. This stage of drying requires more heat input for the vaporization of water as the heat of adsorption is required in addition to the heat of vaporization of pure water. Drying is said to be complete once the equilibrium moisture content (point E in Figure 10.6) is reached.

10.2.3 Factors Influencing the Drying Rate

The factors affecting the drying rate will vary slightly depending upon the type of drying system used. However, in general, the factors to be considered are the temperature, humidity, and velocity of the air. It is proposed that a boundary film is present between the food product and the surrounding air. This boundary film influences the heat and mass transfer during drying. Higher air velocity results in thin boundary film and high drying rate whereas low air velocity causes the thick boundary film and results in low drying rate. Water vapor removed during the drying diffuses to the surrounding air which increases the air humidity. Increase in the humidity of air reduces the water vapor pressure gradient and thus reduces the drying rate. With respect to air velocity (v), the drying rate is approximately proportional to $v^{0.8}$. Yet another factor that controls drying rate is the rate at which moisture can move from within the food product to the surface. Shorter the distance that moisture has to travel, faster the drying rate. This is accomplished by the size reduction of the food products to be dried. Cutting the food product into small pieces prior to drying reduces its size and increases the surface area per unit volume.

10.3 Classification of Dryers

Dryers can be classified into three broad categories based on the product to be dried (Lucideon, 2017):

1. **Static:** In this type of dryers, the product to be dried remains stationary on a tray. The drying medium can be static or circulated and vented. The static dyers can operate on both continuous and batch modes. The common examples of batch static dryers are the tray and superheated steam dryers. The static dryers that can also operate on continuous mode include vacuum and freeze dryers.
2. **Fluidized:** In the fluidized dryers, the products to be dried are converted to fine particulates and suspended in the drying media. Reducing the size of the product maximizes the drying area and hence the heat transfer rate. Fluidized bed and spray dryers are the well-known instances of this type.
3. **Agitated:** Here, the particulate or granulated products are mechanically mixed to enhance their exposure to the hot surface and facilitate uniform drying. Examples include rotary and drum dryers.

Some of the major types of dryers are explained in this section to provide an insight into their working principle, design, and applications.

10.3.1 Cabinet Tray Dryer

The tray dryer is the most useful type of dryer when the production rate is small. It is more usually operated in batch mode. Nevertheless, continuous units with higher capacity are also available for large-scale use.

10.3.1.1 Construction and Working of Cabinet Tray Dryer

A typical cabinet dryer is shown in Figure 10.7. It consists of an insulated heating chamber where several trays holding the food product are stacked. The trays are made up of aluminium or steel, with a solid, perforated, or wire-meshed bottom. Alternatively, the trays can be placed in trucks on wheels, which can be rolled in and out of the heating chamber. Air is heated by heating coils, steam, or gas. Hot air at high velocity generated by the blower passes over or through the food material placed on the trays. Fans are fixed to the wall of the tunnel to blow hot air across the trays. At times, baffles are used to facilitate a uniform distribution of air over the stack of trays. Some moist air is continuously vented through the exhaust duct; make-up fresh air enters through the inlet. As the drying air passes over the trays, it carries away the moisture from the product and loses its heat. The moist air is then discharged through the outlet. The temperature inside the chamber is controlled and maintained at around 50°C–75°C. Tray dryers can also be operated under vacuum with indirect heating for the drying of heat-sensitive food products.

The capacity of a tray dryer can be altered by increasing or decreasing the number of trays. Temperature is the key control parameter (CP) for the tray drying process. When the tray dryer is operated at high temperature, the final product with lowest moisture content can be accomplished within the shortest time duration. Ventilation and tray arrangements are critical in the design of a tray dryer to ensure an even air flow distribution for uniform drying (Lucideon, 2017).

10.3.1.2 Advantages and Limitations of Cabinet Tray Dryer

Cabinet dryers are economical in operation, simple in construction, flexible to use, easy to monitor, and thus, suitable for small-scale drying. However, drying time is longer in the range of a few hours. Another major disadvantage of tray dryer is the non-uniform drying as the temperature distribution within the chamber is not constant. Consequently, local overheating occurs, especially at the edges leading to poor product quality. To overcome the aforementioned limitation, the product is intermittently mixed or shuffled.

Generally, tray dryers are used for the drying of low-cost food products such as fruits and vegetables. Meat and confectionery products are also dried in a tray dryer. This type of dryer is also used for the removal of initial moisture from the product.

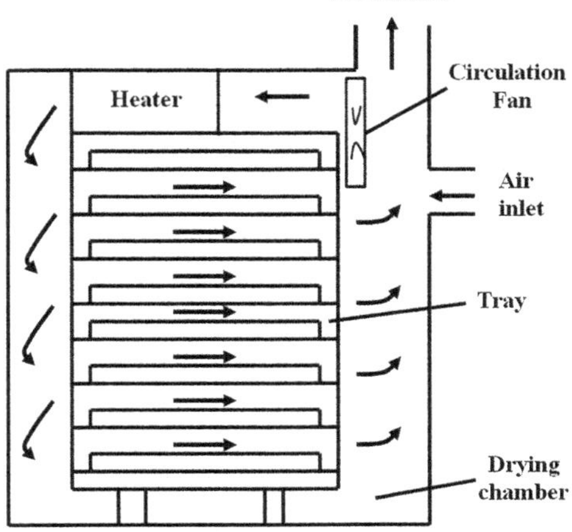

FIGURE 10.7 Schematic diagram of the cabinet or tray dryer. (Reproduced with permission from Yataganbaba, A. and Kurtbaş İ. 2016. A scientific approach with bibliometric analysis related to brick and tile drying: a review. *Renewable and Sustainable Energy Reviews* 59: 206–224.)

10.3.2 Vacuum Dryer

10.3.2.1 Construction and Working of Vacuum Dryer

Vacuum dryer is a variation of the cabinet tray dryer, specially designed for the drying of heat-sensitive and hygroscopic materials. The vacuum dryer operates at reduced pressure to decrease the evaporation temperature of water in food products. For instance, in an applied vacuum of 0.0296–0.059 atm, the boiling point of water is reduced to 25°C–30°C. The water vapor thus formed is removed by the vacuum pump, thus maintaining a high gradient of water vapor pressure between the food product and the surrounding environment in the drying chamber. This facilitates faster drying than that achieved in the dryers operated at normal atmospheric pressure. The vacuum pump is responsible for the vacuum level inside the dryer.

In a vacuum dryer, heat is supplied by passing steam or hot water through hollow shelves (Figure 10.8). Vacuum dryer is an indirect heat dryer. Heat is transferred to the product by conduction, as it contacts the dryer's heated surface. Drying times are usually in the range of 12–48 h. Usually, vacuum dryers are operated in batch mode.

10.3.2.2 Advantages and Limitations of Vacuum Dryer

Vacuum dryer has several advantages including energy conservation, reduced manufacturing costs, faster drying, and shortened processing times. Also, the vacuum dryer retains the original product quality by protecting against heat damage and case-hardening. With respect to personnel safety during industrial scale operation, vacuum dryers outweigh other types of drying equipment. This is because ventilation does not occur with vacuum dryer, thus eliminating vented fumes and particles that can make people sick and demand usage of protective garments. However, the major disadvantage of vacuum dryer is generally related to its mode of heat transfer and the limitation in the rate at which the product temperature can be raised. The abovementioned demerits can be attributed to the restricted surface area available for heat transfer (Parikh, 2015).

10.3.3 Tunnel Dryer

For the continuous industrial-scale operation, tray dryer is extended in the form of a tunnel dryer (Figure 10.9). Food material is loaded on the trays, and the trays are stacked in trolleys. Trolleys are loaded into the tunnel dryer from one end in the direction of hot air. When drying is complete, trolleys are discharged on the other end of the tunnel with the dried material. The movement of trolleys is adjusted to provide sufficient residence time for the complete drying of food material.

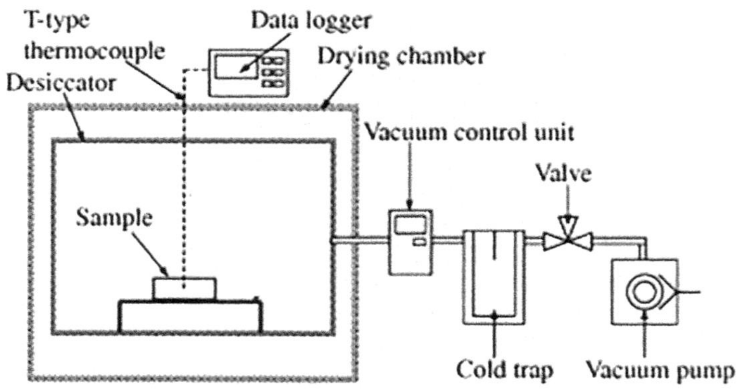

FIGURE 10.8 Schematic diagram of a vacuum dryer. (Slightly modified and reproduced with permission from Orikasa, T., Koide, S., Okamoto, S., Imaizumi, T., Muramatsu, Y., Takeda, J., Shiina, T. and Tagawa, A. 2014. Impacts of hot air and vacuum drying on the quality attributes of kiwifruit slices. *Journal of Food Engineering* 125: 51–58.)

FIGURE 10.9 Schematic diagram of tunnel dryer. (CC BY SA by SolidsWiki; Tunnel dryers. www.solidswiki.com/index.php?title=Tunnel_Dryers (accessed October 17, 2018.))

In the tunnel dryer, hot air enters counter-currently or co-currently according to the heat sensitivity of the food materials. Though the efficiency of a counter-current system is higher, it yields low-quality product compared to a cocurrent system. Hot air recirculation is used to increase the energy efficiency of the tunnel dryer. Onion and garlic flakes are generally dried in tunnel dryers of length 50–100 m depending on the capacity of the processing plant.

10.3.4 Conveyor Belt Dryer

Conveyor belt dryers are the widely used multipurpose continuous dryers for the drying of grains, vegetables, fruits, cereals, and nuts. In conveyor belt dryers, the food material is conveyed through perforated conveyor made up of metal mesh which passes through a tunnel (Figure 10.10). Hot air passes over and through the food product, initially upward and later downward. Metal mesh allows the hot air to pass through it. In general, the conveyors are 1–3 m wide and 20–50 m long. Food material to be dried is spread on the conveyor at a thickness of 50–100 mm. This dryer type may be equipped with a sensor-based temperature control system to manage the drying temperature of the thin bed of product. Conveyor belt dryers require less manpower compared to tunnel dryers, as it does not require loading and unloading of trays.

10.3.5 Fluidized Bed Dryer

10.3.5.1 Theory of Fluidization

> The arrival time of a space probe travelling to Saturn can be predicted more accurately than the behavior of a fluidized bed chemical reactor
>
> *(Geldart, 1986)*

The above mentioned quote implies the challenges in measuring the size, shape, and density of the particles which are fundamental variables involved in the prediction of the dynamic behavior of fluidized beds.

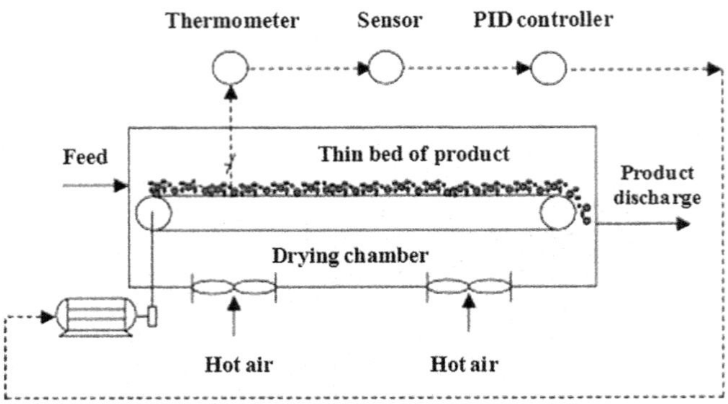

FIGURE 10.10 Schematic diagram of conveyor belt dryer. (Modified and reproduced with permission from Jensen, S., Meleiro, L. A. C. and Zanoelo, E. F. 2011. Soft-sensor model design for control of a virtual conveyor-belt dryer of mate leaves (*Ilex paraguariensis*). *Biosystems Engineering* 108:75–85.)

Fluidization is the underlying concept of fluidized bed drying. Fluidization can be defined as the *completely suspended state of a bed of solid particles in a closed area rested on a gas distributor plate, during which the solid particles behave like a fluid.* This occurs when the solid particles are acted upon by a fluid at a velocity sufficient enough to support the weight of the bed but not that high to elutriate (carry away) the particles out of the bed. The significant parameters in the fluidization process are the minimum fluidization velocity (u_{mf}) and terminal velocity (u_t). u_{mf} is the minimum gas velocity at which the gas stream totally supports the weight of the particulate bed and the bed attains the fluidized state. u_{mf} is given by the following equation, which is applicable for particle diameter <100 μm.

$$u_{mf} = \frac{\left(\rho_P - \rho_g\right)^{0.934} g^{0.934} d_p^{1.8}}{111\mu^{0.87}\ \rho_g^{0.066}} \tag{10.5}$$

where ρ_p and ρ_g are the particle and gas densities, respectively (kg/m^3), d_p is the particle diameter (μm), and μ is the gas viscosity (kg/m s). The value of u_{mf} depends on the physical properties of the fluid and the solid particles. At the u_{mf}, the bed is divided into an emulsion phase and a bubble phase (Figure 10.11). At gas velocity greater than u_{mf}, the bed demonstrates more vigorous bubbling with the diameter of the bubbles approaching the vessel diameter. The behavior of the bed corresponding to the aforementioned stage is known as *slugging*. Further increase in gas velocity leads to more particles being thrown into the *freeboard*, the space above the bed. When the gas superficial velocity exceeds the terminal velocity of the particle (u_t), the solids will be carried over and out of the system, the condition of which is described as *entrainment* (Figure 10.11) (Yates, 1983; Geldart, 1986; Howard, 1989; Kunii and Levenspiel, 1991). U_t is given by

$$u_t = \left(4gd_p\left(\rho_p - \rho_g\right)/3\rho_g C_d\right)^{0.5} \tag{10.6}$$

In Eq. (10.6), C_d is the drag coefficient which is, in turn, a function of Reynolds number (for instance, C_d for $N_{Re} < 0.1$ is given by $24/N_{Re}$; for N_{Re} between 0.1 and 1000, $C_d = 18.5/N_{Re}^{0.6}$; and, for $N_{Re} > 1000$, C_d remains constant at 0.44). The equation for terminal velocity provides the relationship between the fluidization velocity and the size and shape parameters of the solid particles (Anandharamakrishnan and Ishwarya, 2015).

Not all particles can be fluidized satisfactorily. Geldart (1986) proposed a convenient classification of particles in a gas fluidized bed, based on their behavior under atmospheric conditions (Figure 10.12). Accordingly, there are four groups of particles as follows:

- **Group-A:** Particles in the Geldart group A have low density with diameter (d_p) in the range of 30–150 μm. These particles are aeratable and hence fluidize well. For example, milk powder.

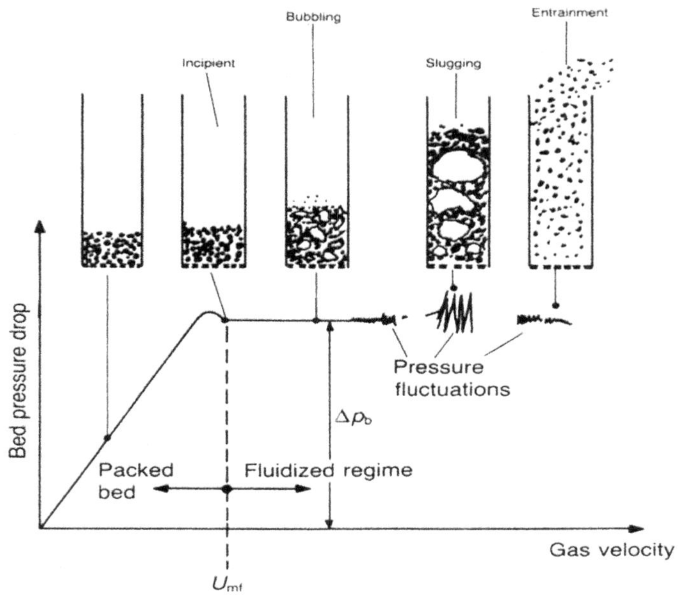

FIGURE 10.11 Fluidization regimes in a gas–solid fluidized bed. (Reproduced with permission from Howard, J. R. 1989. *Fluidized bed technology—Principles and applications*. New York: Adam Hilger.)

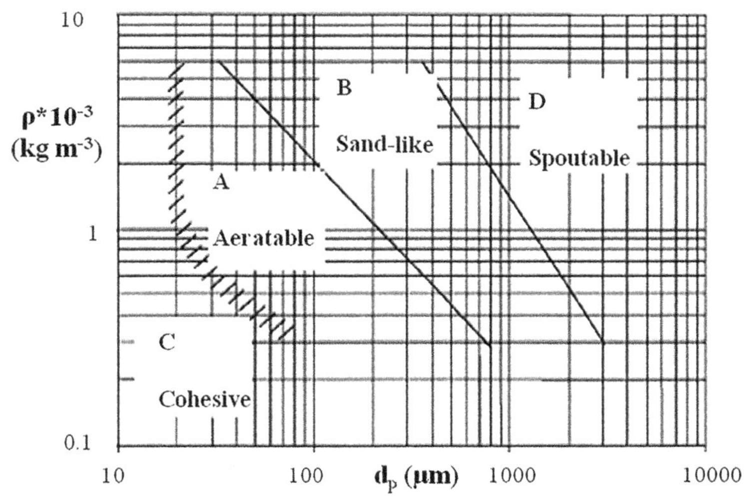

FIGURE 10.12 Geldart classification of particles at atmospheric conditions. (Reproduced with permission from Werther, J. 1987. Gas fluidization technology. In *Berichte der Bunsengesellschaft für physikalische Chemie*, ed. D. Geldart (Volume 91), 678–679. Chichester, New York, Brisbane, Toronto, Singapore: John Wiley & Sons.)

- **Group-B:** These are sand-like particles which are easily fluidizable but do not undergo smooth fluidization; bubbles form at the onset of fluidization. The diameter (d_p) of the group-B particles is in the range of 150–500 µm;
- **Group-C:** The particles of this category are cohesive or very fine powders with a diameter (d_p) < 30 µm; e.g., flour. Fluidization of these particles is difficult due to the interparticle forces. Hence, fluidization of these particles is achieved by the application of an external force, such as mechanical agitation.

- **Group-D:** These particles are of a very large size or very high density. For example, coffee beans, seeds, vegetable pieces, grains, and peas. Fluidization of the particles belonging to this group requires very high fluid energies. During the fluidization of group-D powders, bubble formation is enormous and slugging is observed even in large fluidized beds. Therefore, these powders are processed in spouted beds, which have lower gas requirements than normal fluidized beds (Epstein and Grace, 1997). A spouted bed is a type of fluidized bed in which the gas moves mainly through the center of the bed (Cocco et al., 2014).

10.3.5.1.1 Correlation for the Minimum Fluidization Velocity

The minimum fluidization (u_{mf}) velocity is a primary attribute of a fluidized bed. The velocity of gas must always be maintained above this value during the course of operation. The accurate prediction of u_{mf} is important for the successful design and operation of a fluidized bed. Numerous correlations have been proposed for the prediction of u_{mf} under the atmospheric conditions, of which, Ergun's equation (1952) is the most commonly used. Ergun's equation is given by

$$\frac{\Delta P}{L} = 150\frac{(1-\varepsilon)^2}{\varepsilon^3}\frac{\mu_f u_o}{(\phi d_p)^2} + 1.75\frac{(1-\varepsilon)}{\varepsilon^3}\frac{\rho_f u_o^2}{\phi d_p} \tag{10.7}$$

where ϕ is the sphericity of particles expressed as

$$\phi = \left[\frac{6/d_p}{S_p/V_p}\right] \tag{10.8}$$

where ε is the voidage, S_p is the surface area of a single particle (m^2), V_p is the volume of a single particle (m^3), and u_o is the superficial velocity (m/s).

Also,

$$\frac{\Delta P}{L} = (1-\varepsilon_{mf})(\rho_p - \rho_f)g \tag{10.9}$$

If $\varepsilon = \varepsilon_{mf}$ and $u = u_{mf}$, multiplying each side of Eq. (10.7) by $\dfrac{\rho_f d_p^3}{\mu_f^2(1-\varepsilon_{mf})}$,

$$\frac{\rho_f(\rho_p - \rho_f)g d_p^3}{\mu_f^2} = 150\frac{(1-\varepsilon_{mf})}{\phi^2\varepsilon_{mf}^3}\frac{\rho_f u_{mf} d_p}{\mu_f} + 1.75\frac{1}{\varphi\varepsilon_{mf}^3}\frac{\rho_f^2 u_{mf}^2 d_p^2}{\mu_f^2} \tag{10.10}$$

where $\dfrac{\rho_f(\rho_p - \rho_f)g d_p^3}{\mu_f^2}$ = Archimedes number (Ar), and the Reynolds number in minimum fluidization conditions is $R_{emf} = \dfrac{\rho_f u_{mf} d_p}{\mu_f}$.

Substituting Ar and R_{emf} in Eq. (10.10),

$$Ar = 150\frac{(1-\varepsilon_{mf})}{\varphi^2\varepsilon_{mf}^3}\mathrm{Re}_{mf} + 1.75\frac{1}{\varphi\varepsilon_{mf}^3}\mathrm{Re}_{mf}^2 \tag{10.11}$$

Thus, Eq. (10.11) can be used to calculate the minimum fluidization velocity (u_{mf}), after obtaining the values for voidage (ε_{mf}), mean diameter of the particles (d_p), and sphericity of particles (ϕ).

10.3.5.2 Working of a Fluidized Bed Dryer

Fluidized bed dryers are named so as the product to be dried is evenly distributed over a perforated plate to form a bed which is maintained under fluidized conditions by blowing hot air at high velocity through the perforated plate. Due to the air velocity, there is an increase in pressure drop across the bed. At the incipient velocity, food particles in the bed remain suspended in the hot air. During the operation, particles separate from the bed momentarily and tumble back on it. At the entrainment velocity, particles in the bed are carried away with the air stream. Therefore, fluidized bed dryers are operated between the incipient and entrainment velocities. Higher density particles require higher air velocity to fluidize the bed. Food particles should not be sticky or prone to mechanical damage, and size or the diameter should be in the range of 20–10000 μm to facilitate fluidization.

The schematic diagram of a fluidized bed dryer is shown in Figure 10.13. The gas is distributed across the bed inlet by a perforated distributor plate with orifices of smaller diameter drilled on a square pitch. In addition to the distributor plate, Raschig rings are placed in the wind box to facilitate gas distribution. Raschig rings (named after its inventor Friedrich August Raschig, a German Chemist) are pieces of equal-sized (length and diameter) tube made of ceramic or metal. These rings are used to provide a large surface area for interaction between the liquid and gas vapors, within the volume of fluidization column. The fluidization air is supplied by a blower. The air flow rate is measured using an orifice plate designed according to the standards developed by the American Society of Mechanical Engineers (ASME) and a manometer setup. The pressure drop across the fluidization bed is measured with a differential pressure transducer. The freeboard section as shown in the Figure 10.13 separates any entrained material from the exhaust gas. The freeboard is usually covered to minimize the loss of fines from the dryer (Wormsbecker, 2008).

Vegetables, grains, and peas are the products that are predominantly dried using a fluidized bed dryer. High heat transfer rate due to the close contact of the particles with hot air is the major advantage of fluidized bed dryers. These dryers can be operated in batch or continuous mode. In the latter type, feed enters from one end and the dried product exits over a weir at the other end. The feed should comprise particles of uniform size and moisture content to obtain the best quality product.

A variant of the fluidized bed dryer is the vibro-fluidized bed dryer where the bed is mechanically vibrated. Vibro-fluidized bed dryer can be used for particles that are sticky or agglomerated and those

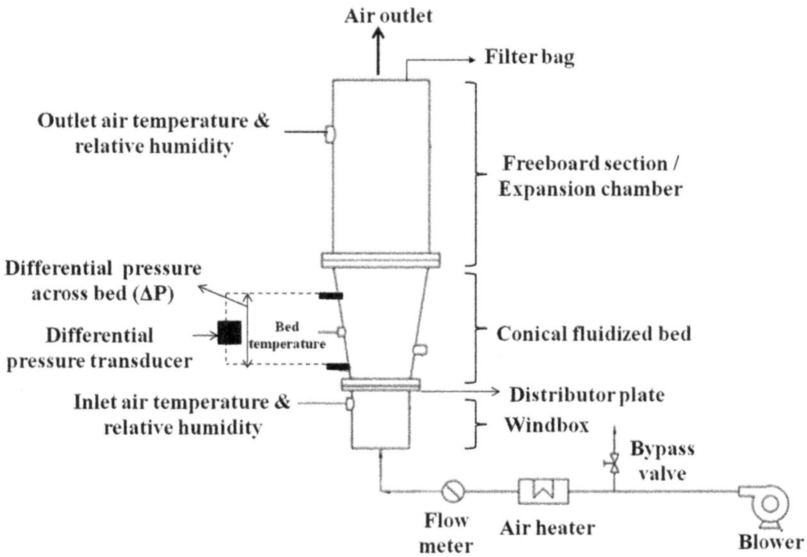

FIGURE 10.13 Fluidized bed dryer. (Modified and reproduced with permission from Ge, R., Ye, J., Wang, H. and Yang, W. 2016. Investigation of gas–solids flow characteristics in a conical fluidized bed dryer by pressure fluctuation and electrical capacitance tomography. *Drying Technology* 34: 1359–1372.)

with wider particle size distribution. Here, the air velocity is low at a level which is sufficient to fluidize the food particles. Consequently, the energy consumption of vibro-fluidized bed dryers is low.

Spouted bed dryer (Figure 10.14) is another form of fluidized bed dryer, in which, high-velocity hot air jet from the bottom of the bed forms a spout at the center. The high rate of evaporation is achieved through the spout. Larger size particles can be handled in this dryer when compared to its stationary counterpart. Granular products that are too coarse to be readily fluidized such as grains or seeds are the potential candidates to be dried using a spouted bed dryer.

10.3.5.3 Mathematical Models for Fluidized Bed Drying

A large number of fluidized bed drying models are available in the literature. Most of the models are based on two-phase (solid–gas) theory of fluidization with different assumptions. Some of the individual models, which are based on a three-phase (emulsion, cloud, and bubble) system and a vacuum fluidized model are discussed in this section.

10.3.5.3.1 Kunii and Levenspiel Model

One of the earliest studies on fluidized bed drying was carried out by Kunii and Levenspiel (1968, 1990). They proposed the three-phase model (bubbling bed) for gas–solid reaction with a porous or a non-porous catalyst with or without mass transfer limitations, along with the following assumptions:

- Fluidized bed consists of three regions: emulsion (solid and gas phase), cloud (solid and gas), and bubble regions.
- The wake region is considered to be part of the cloud region.
- The bubbles exhibit plug flow and are uniform in size.
- The mass interchange occurs in two stages, which is characterized by two mass transfer coefficients i.e., from bubble to the cloud (K_{bc}) and from cloud to emulsion (K_{ce}).

For fine particle beds, the bubble concentration (C_A) at height z is given by

$$\frac{C_A}{C_{A(\text{inlet})}} = \exp\left(-K_f \frac{z}{u_b}\right) \tag{10.12}$$

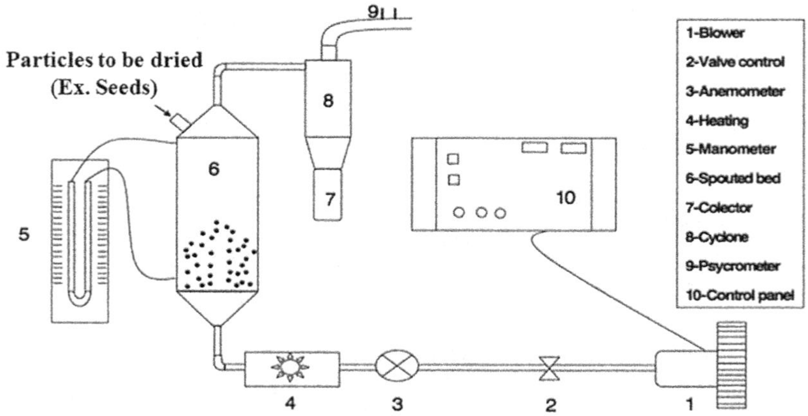

FIGURE 10.14 Spouted bed dryer. (Modified and Reproduced with permission from Chielle, D. P., Bertuol, D. A., Meili, L., Tanabe, E. H. and Dotto, G. L. 2016. Spouted bed drying of papaya seeds for oil production. *LWT—Food Science and Technology* 65: 852–860.)

The outlet bubble concentration (C_{Ao}) at height L is

$$1 - X_A = \frac{C_{Ao(\text{outlet})}}{C_{A(\text{inlet})}} = \exp\left[-K_f \frac{L}{u_b}\right] \tag{10.13}$$

where X_A is the outlet fractional conversion and K_f is the overall rate constant. This can be broken down into individual mass transfer resistances as follows:

$$K_f = \left[\gamma_b K_r + \cfrac{1}{\cfrac{1}{K_{bc}} + \cfrac{1}{\gamma_c K_r + \cfrac{1}{\cfrac{1}{K_{ce}} + \cfrac{1}{\gamma_e K_r}}}}\right] \tag{10.14}$$

where K_r is the reaction rate constant and $\gamma_b, \gamma_c, \gamma_e$ are the dimensionless ratios of the solid dispersed in bubbles, clouds, and emulsion, respectively. The velocity of a bubble rising through a bed is given by

$$u_b = u_o - u_{mf} + u_{br} \tag{10.15}$$

Where u_o is the superficial gas velocity (m/s) and u_{br} is the bubble rise velocity for single particle (Kunii and Levenspiel, 1991), given by

$$u_{br} = 0.711(gd_b)^{0.5} \tag{10.16}$$

The interchange coefficients K_{bc} and K_{ce} are expressed as

$$K_{bc} = 4.5\left(\frac{u_{mf}}{d_b}\right) + 5.85\left(\frac{D_g^{0.5} g^{0.25}}{d_b^{5/4}}\right) \tag{10.17}$$

$$K_{ce} = 6.77\left(\frac{D_g \varepsilon_{mf} u_{br}}{d_b^3}\right)^{0.5} \tag{10.18}$$

For a situation where there is no chemical reaction ($K_r = 0$), Eq. (10.14) reduces to

$$\frac{1}{K_{be}} = \frac{1}{K_{bc}} + \frac{1}{K_{ce}} \tag{10.19}$$

The Kunii and Levenspiel (K–L) model gives the conversion in a fluidized bed as a function of the bed condition, and it is applicable only at $u_o > u_{mf}$. This model is a simple one, based on which many other models have been published later.

10.3.5.3.2 Hoebink and Rietema Model

Hoebink and Rietema (1980a,b) developed a three-phase model based on perfectly mixed dense phase and the bubble phase together with the cloud phase taken to be in plug flow. They assumed that there was no diffusion limitation inside the solids and that the mass transfer was fast in the dense phase. Nevertheless, this model is applicable only for the constant rate drying period.

10.3.5.3.3 Palancz Model

In this continuous fluidized bed (solids in and out) drying model, Palancz (1983) used the K–L-type three-phase model inside the bed. Palancz assumed that the bubble phase was in plug flow (an idealized flow condition in which there is no shearing between the adjacent layers of fluid, and thus, there is no mixing of fluid particles), and the dense phase was perfectly mixed, bubbles were uniform in size, and contained no particles. According to the Palancz model, a higher inlet gas temperature increases the particle temperature as well as the drying rate. However, this model does not account for the heat and mass transfer inside the particles and hence applies only to the constant rate period.

10.3.5.3.4 Thomas and Varma Model

A receding core model for drying kinetics of granular food materials (green pepper, black pepper, and mustard) was developed by Thomas and Varma (1992). These authors found that the interparticle diffusion controls the drying process, and the falling rate might be nonlinear depending on the nature of the material. They concluded that the rate coefficient for the constant drying rate period and critical moisture content depend on the velocity and temperature of the heating medium and the particle size and mass of the solids.

10.3.5.3.5 Kannan et al. Model

A three-phase model for the drying of solids in the fluidized bed was developed by Kannan et al. (1994). This research group considered perfectly mixed phases of solid and interstitial gas and constant bubble size throughout the bed axis. Moreover, they assumed equilibrium between the gas and solid particles and evaluated the effective diffusivity of the solid particles by matching the experimental data with the model.

10.3.5.3.6 Hajidavallo and Hamdullahpur Model

Based on the K–L three-phase model, Hajidavallo and Hamdullahpur (2000) developed a model for the bubbling phase of fluidized bed drying. These authors assumed that the bubbles were of uniform size in the bed, and the voidages of the different phases (bubble and emulsion) were uniform along the bed height. They also included the wall effects. The model exhibited good agreement with their experimental results. They found that the drying rate was influenced by the inlet temperature of the air and the physical properties of the particles, particularly for biological materials. This model was developed based on the behavior of Geldart group D particles (e.g., wheat) including the bubble size, flow rate, and velocity of bubbles which were not constant along the bed height.

10.3.5.3.7 Kozanoglu et al. Model

Kozanoglu et al. model is the first two-phase model developed for vacuum fluidized bed drying. Kozanoglu et al. (2002) assumed that the whole bed was in plug flow and used the Llop et al. (1996) equation (Eq. (10.20)) for the minimum fluidization velocity.

$$\mathrm{Re}_{mf} = \left[\left(\frac{0.909}{Kn_p + 0.0309}\right)^2 + 0.0357\mathrm{Ar}\right]^{1/2} - \left(\frac{0.909}{Kn_p + 0.0309}\right) \tag{10.20}$$

The authors correlated the model predictions with their experimental results in the constant rate period and derived the following equation to estimate the mass transfer coefficient.

$$\frac{kd_p y}{D_g} = 5{,}882\left(\frac{u\rho_g d_p}{\mu_g}\right)^{3.07}\left(\frac{\lambda_m}{d_p}\right)^{2.94} \tag{10.21}$$

where k is the mass transfer coefficient (m/s), d_p is the particle diameter (m), y is the average mole fraction of nondiffusing component, D_g is the diffusivity of gas (m^2/s), u is the superficial gas velocity (m/s), ρ_g is the gas density (kg/m^3), μ_g is the gas viscosity (Pa s), and λ_m is the mean free path of molecule (m), given by $Kn_p = \frac{\lambda_m}{d_p}$, where Kn_p is the Knudsen number of the particle and d_p is the diameter of the particle (m).

This model covers both the constant and falling drying rate periods and is in close agreement with their experimental results. They found that the difference in temperature and moisture concentration of the gas phase between the bottom and top of the bed were 0.25% and 2%, respectively. However, this model was applicable for only the Geldart groups B and A particles, wherein the bubbling regime is encountered (Anandharamakrishnan, 2008).

10.3.6 Drum Drying

Drum dryer is generally used to remove water from slurries and paste. Drum dryer consists of either one or two cylindrical drums (Figure 10.15a and b). Based on the number of drums, it is classified as a single drum dryer and a double drum dryer. In the double drum dryer, two oppositely rotating drums are used with an adjustable narrow gap. The product to be dried should adhere to the surface of the drum. Optionally, additives are used to increase the adherence of feed material. In certain cases, drums are partially immersed in the slurry or paste. The feed can also be sprayed and splashed. Alternatively, applicators are used to spread the feed evenly on the drum surface of single or double drum dryers. The use of applicator rolls is common with viscous food products such as paste and purees, cereal-based breakfast food, weaning food products, and pre-gelatinized starches. Efficient spreading of the feed is achieved with a feed having lesser surface tension and a lower contact angle with the drum surface. A thin film forms on the surface of the drum due to the adherence of feed material, and its thickness is controlled by adjusting the gap.

Double drum dryer is widely used for the drying of low viscous foods such as milk and dairy products. Whereas, a single drum dryer is used for products such as instant mashed potato. With drum dryers, higher drying rate can be obtained by maintaining a high temperature gradient and overall heat transfer coefficient.

10.3.7 Spray Drying

Spray drying is defined as the *transformation of feed from a fluid state into a dried particulate form by spraying the feed into a hot drying medium* (Masters, 1991). Spray dryer works on the principle of convective heat transfer, by which, moisture removal is facilitated by the application of heat to the feed product and by controlling the humidity of drying medium.

10.3.7.1 Components of a Spray Dryer and Their Functions

Spray dryer (Figure 10.16) consists of the following components (Devakate et al., 2009; Ishwarya and Anandharamakrishnan, 2017):

- Atomizer to divide the bulk feed solution into tiny droplets.
- Feed pump to supply the liquid feed to the atomizer. The feed pumps must be selected to handle such concentrated feed material of high viscosity such as pre-concentrated milk. Peristaltic pumps are most commonly used for spray drying.
- Air heater: To raise the temperature of ambient air to the specified inlet air temperature.
- Fan or blower to circulate the air through the system.
- The air disperser is located in the roof of the spray-drying chamber alongside the atomizer. To ensure an equal flow of the drying gas in all directions within the spray chamber, the pressure drop is created using perforated plates or vaned channels.
- The aspirator supplies dry air to the spray-drying chamber by means of a motor operating under pressure conditions. Aspirator flow rate can be altered to regulate the amount of heated drying air entering the spray chamber.

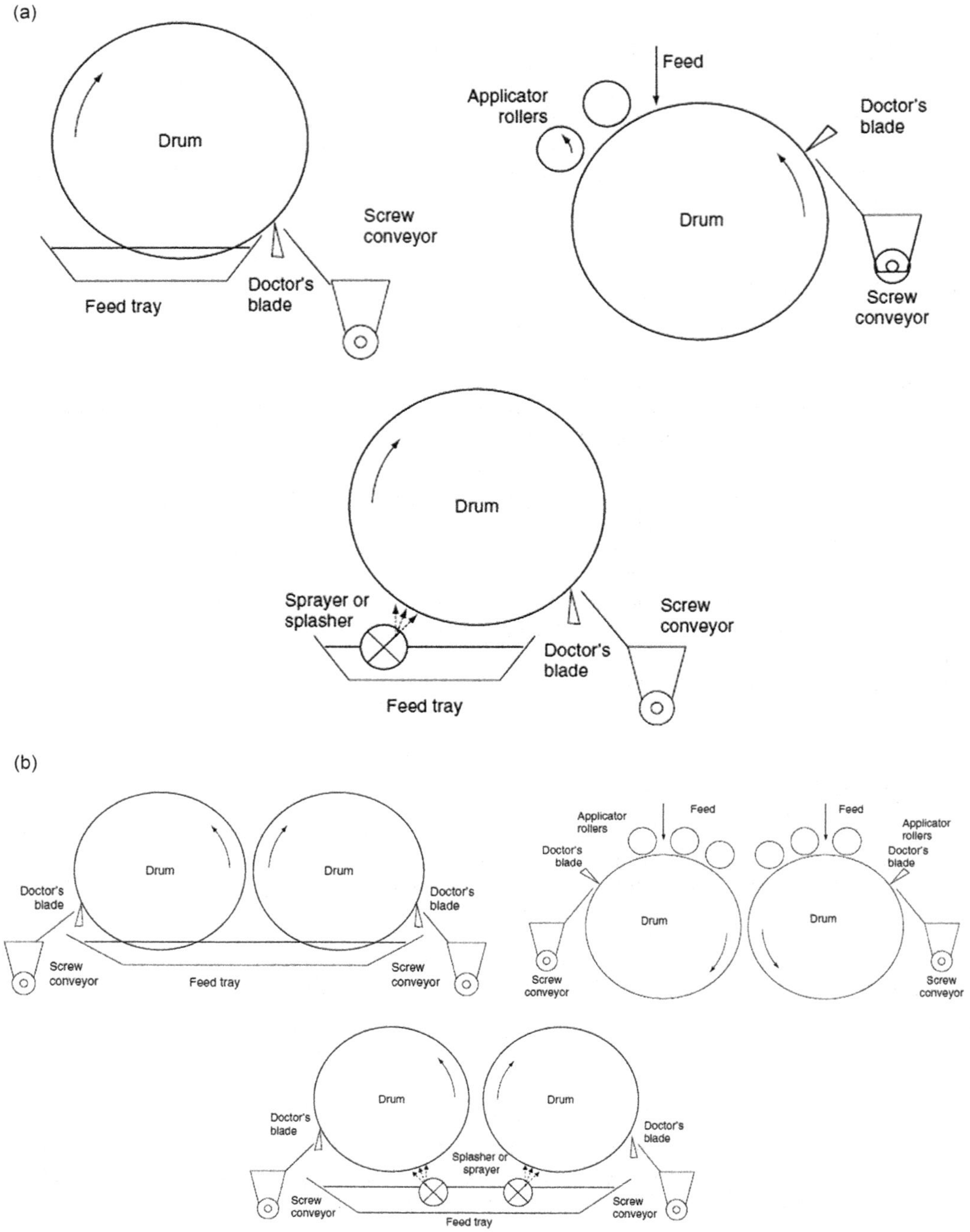

FIGURE 10.15 Schematic diagram of drum dryers: (a) single drum dryer; (b) double drum dryer (Reproduced with permission from Daud, W. R. W. 2007. Drum Dryers. In *Handbook of Industrial Drying*, ed. A. S. Mujumdar, 203–211. Boca Raton, FL: CRC Press, Taylor & Francis Group.)

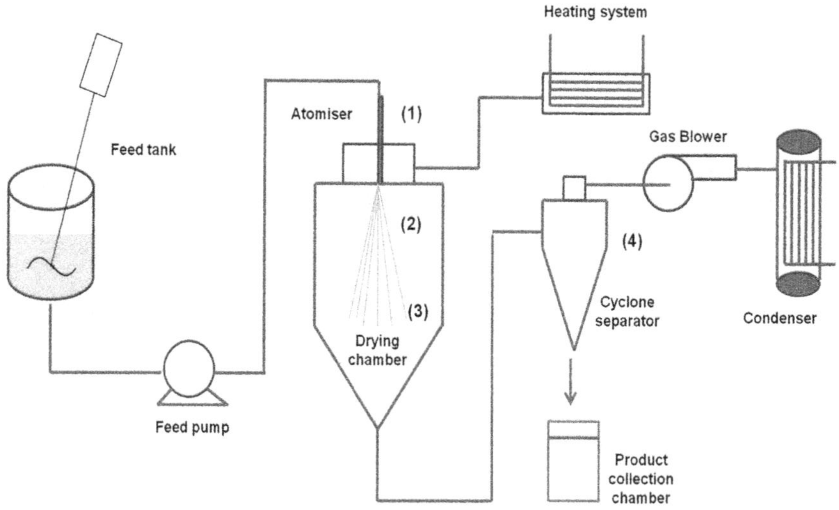

FIGURE 10.16 Schematic representation of spray dryer. (1) Atomization; (2) spray—HA contact; (3) evaporation of moisture; (4) product separation. (Reproduced with permission from Anandharamakrishnan, C. and Ishwarya, S. P. 2015. *Spray Drying Techniques for Food Ingredient Encapsulation.* Hoboken, NJ: John Wiley and Sons, Inc.)

- Drying chamber constituted by a vertical cylindrical section with a conical bottom.
- Cyclone separator to separate the fine particulate product from the air stream.
- Equipment for product discharge, transport, packaging, and removal of air.

10.3.7.2 Stages Involved in Spray Drying

A liquid feed entering the spray dryer goes through the following four stages (Figure 10.16) before being converted to dried powder (Anandharamakrishnan and Ishwarya, 2015):

1. Atomization of the feed solution
2. Contact of spray with the hot gas
3. Evaporation of moisture
4. Particle separation

10.3.7.2.1 Atomization

Atomization can be defined as the disintegration of bulk feed liquid into a large number of finely divided droplets. This reduces the internal resistance to moisture transfer from the droplet to the surrounding medium. Atomization is accomplished using the atomizer which sprays the feed into the hot air within the drying chamber. The size of atomized droplets varies from 10 to 200 μm, depending on the atomizer.

Rotary atomizer (Figure 10.17a), pressure nozzle atomizer (Figure 10.17b), and twin-fluid nozzle (Figure 10.17c) are the commonly used atomizers in the industrial-scale spray-drying operation.

i. **Rotary atomizer**
   - **Principle:** Rotary atomizer is driven by the high-velocity discharge of liquid from the edge of a wheel or disc (Figure 10.17a). Feed liquid is centrifugally accelerated at high velocity to the center of a rotating wheel at a peripheral velocity of 200 m/s. The outward flowing feed with respect to the rotating wheel surface accelerates to the periphery and then disintegrates into a spray of droplets.
   - **Atomization energy:** Centrifugal energy

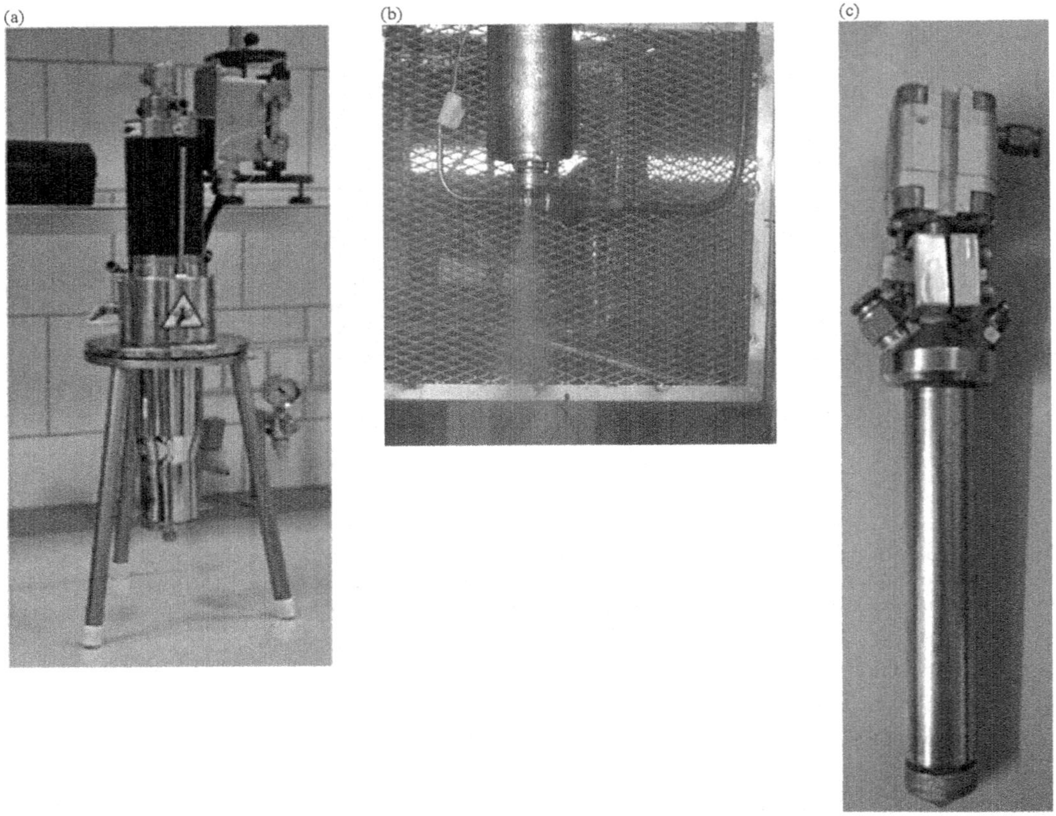

FIGURE 10.17 Different types of atomizer: (a) rotary atomizer (CC BY 3.0 by CBIMO. www.cbimo.zut.edu.pl/ (accessed October 17, 2018)); (b) pressure nozzle atomizer (Reproduced with permission from Ishwarya, S. P., Anandharamakrishnan, C. and Stapley, A. G. F. 2017. Spray freeze drying. In *Handbook of Drying for Dairy Products*, ed. C. Anandharamakrishnan, 123–148. Chichester, West Sussex, UK: John Wiley and Sons.); (c) twin-fluid nozzle atomizer. (Reproduced with permission from Anandharamakrishnan, C. and Ishwarya, S. P. 2015. *Spray Drying Techniques for Food Ingredient Encapsulation*. Hoboken, NJ: John Wiley and Sons, Inc.)

- **Atomization parameters:** Wheel speed in rotation per minute
- **Type of spray:** Fine, coarse, or medium
- **Mean droplet size:** 30–120 μm
- **Relationship between mean droplet size and atomization parameters:** Droplet size is directly proportional to the feed rate and feed viscosity and inversely proportional to the wheel speed and wheel diameter.
- **Physical property of feed:** Rotary atomizers have the ability to handle abrasive feedstock by virtue of atomizer vanes and bushings.
- **Advantages:** Rotary atomizers do not clog and tend to produce more uniform-sized droplets. Since the necessary atomization energy is supplied by the rotating wheel, the feed supply unit can operate at low pressure than that required in the hydraulic and pneumatic nozzle atomizers.
- **Limitations:** Viscous feed is difficult to be handled using the rotary atomizer. The fine particles produced can potentially lead to environmental pollution. Further, it is not possible to accommodate the spray produced by a rotary atomizer in a horizontal spray dryer (Anandharamakrishnan and Ishwarya, 2015).

ii. **Pressure nozzle or hydraulic atomizer**

- **Principle:** The operation of a pressure nozzle atomizer is facilitated by the discharge of liquid under pressure through an orifice (Figure 10.17b). Pressure energy is converted to kinetic energy and feed emerging from the nozzle orifice as a high-speed film readily breaks into a spray of droplets.
- **Atomization energy:** Pressure energy
- **Atomization parameters:** Nozzle pressure
- **Operating pressure range:** 250–10000 psi
- **Type of spray:** Coarse and less homogeneous
- **Mean droplet size:** 120–250 μm
- **Relationship between mean droplet size and atomization parameters:** The droplet diameter is directly proportional to the feed rate and viscosity and inversely related to atomization pressure. The relationship between selected atomization parameters and droplet diameter is given by the following empirical equations (Masters, 1985):

$$\frac{D_2}{D_1} = \left(\frac{P_2}{P_1}\right)^{-0.3} \tag{10.22}$$

where D_1 and D_2 are the initial and final droplet sizes on changing the atomization pressure from P_1 to P_2, respectively. Equation 10.22 implies that higher atomization pressure is required for the efficient transfer of atomization energy to bring about the fission of bulk feed liquid into droplets. The relationship between droplet diameter and feed viscosity is given in Eq. (10.23), which holds good with the feed density as well.

$$\frac{D_2}{D_1} = \left(\frac{\mu_2}{\mu_1}\right)^{0.2} \tag{10.23}$$

where D_1 and D_2 are the initial and final droplet sizes on changing the feed viscosity from μ_1 to μ_2, respectively.
- **Physical property of feed:** Low viscosity feed
- **Advantages:** Pressure nozzles result in particles with less occluded air. As a result, the powdered product is of a higher density with good flow characteristics. Depending on the specifications of the end product, it is also capable of producing particles with relatively greater size using pressure nozzles. Pressure nozzles can be integrated into multiple nozzle arrangements to obtain an increased amount of flow rate and particle size flexibility.
- **Limitations:** At high feed rates, sprays are generally less homogeneous and coarser than rotary atomizers (Anandharamakrishnan and Ishwarya, 2015).

iii. **Twin-fluid nozzle atomizer**

- **Principle:** Twin-fluid atomizers lead to the breakup of liquid on impact with high-velocity air or other gaseous flow. Compressed air creates a shear field, which atomizes the liquid and produces a wide range of droplet sizes (Figure 10.17c).
- **Atomization energy:** Kinetic energy
- **Atomization parameters:** Nozzle pressure
- **Operating pressure range:** 250–10000 psi
- **Type of spray:** Medium coarseness but poor homogeneity
- **Mean droplet size:** 30–150 μm
- **Relationship between mean droplet size and atomization parameters:** The droplet diameter is directly proportional to feed rate and viscosity and inversely related to atomization pressure.

- **Physical property of feed:** Can handle highly viscous feed
- **Advantages:** Twin-fluid nozzles are capable of handling highly viscous feed. These atomizers also produce much finer and homogeneous spray when compared to pressure nozzles. These nozzles exert better control over the droplet size.
- **Limitation:** With the use of a twin-fluid nozzle, the cost of operation of spray dryer is increased due to the requirement of compressed air. Also, this nozzle type results in high occlusion of air content within the particles, thus lowering the product density. The use of these nozzles also introduces extra cold air into the spray chamber in the zone of atomization and hence reduces the temperature gradient between the droplets and the surrounding drying medium. This decreases the effectiveness of heat transfer between the droplet and drying medium. Further, twin-fluid nozzles exhibit a higher tendency to clog especially when the liquid feed is of mucilaginous or fibrous nature (Anandharamakrishnan and Ishwarya, 2015). Another disadvantage of the twin-fluid nozzle is the *downstream turbulence*, which causes the fine particles to be carried away to the atmosphere by the large gas flows used. This phenomenon is known as *overspray* which tends to contaminate the atmosphere which is in close proximity to the nozzle and demands expensive cleanup and tedious maintenance procedures (Sewell, 1987).

10.3.7.2.2 Spray-Air Contact

During the spray-air contact, droplets meet hot air in the drying chamber, either in cocurrent or countercurrent pattern (Figure 10.18). In the cocurrent flow, hot air and feed solution passes through the dryer in the same direction. In this arrangement, the atomized droplets entering the dryer are in contact with the hot air. Due to the evaporative cooling, product temperature is maintained low at the wet-bulb temperature. With the decrease in air temperature and moisture content of the product along the dryer's bottom portion, the particle temperature does not rise substantially. The temperature of the product at the end of the drying operation is low compared to the exhaust air temperature. Cocurrent configuration

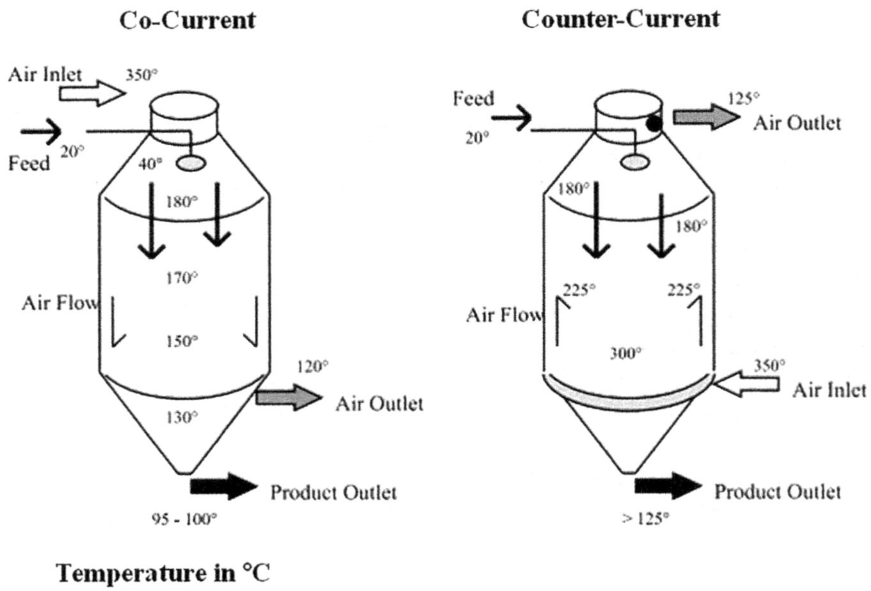

FIGURE 10.18 Schematic representation of cocurrent and countercurrent spray dryers and the temperature profiles. (Modified and reproduced with permission from Vega-Mercado, H., Gongora-Nieto, M. M. and Barbosa-Canovas, G. V. 2001. Advances in dehydration of foods. *Journal of Food Engineering* 49: 271–289 and Masters, K. 1991. *Spray Drying Handbook*. Harlow: Longman Scientific and Technical.)

is most suitable for heat-sensitive food materials. Rapid spray evaporation, shorter evaporation time, and less thermal degradation of the products are the main advantages of spray drying in cocurrent mode.

In the countercurrent configuration, feed solution and hot air enter at the opposite ends of the drying chamber, i.e., feed solution is atomized from the top whereas hot air enters from the bottom of the drying chamber. In this arrangement, the product at the outlet is subjected to high temperature and the feed solution emanating from the atomizer is exposed to low temperature. Countercurrent arrangement is used for non-heat-sensitive products.

The spray-air contact within the spray dryer decides the subsequent drying operation and physical properties of dry powders. Constant rate drying period is observed before the formation of solid crust around the particles. Once the thick crust forms, the falling rate period commences and the drying rate depends on the rate of mass transfer. Excellent rehydration properties are observed in case of spray-dried food particles, due to hollow, low density, and uniformly spherical particles.

Numerous designs of the drying chambers are used in the industry according to the end-product characteristics. One of the commonly used designs used in the food industry is the tall cylindrical tower with a conical base. Tall-type spray dryer is generally used for heat-sensitive liquid products and large droplets. Figure 10.19a shows the tall-type spray dryer with the cocurrent arrangement, where hot air and feed are introduced at the top of the chamber. Some of the tall-type spray dryers are used for the encapsulation of bacteria and enzymes by exposing them to moderate temperature. On the other hand, in a short-type spray dryer (Figure 10.19b), air enters tangentially at the top of the chamber as the feed is sprayed using a centrifugal atomizer. The possibility of heat damage to the product is high in this type of arrangement. Therefore, it is not widely used in the food industries.

10.3.7.2.3 Moisture Evaporation

During this stage, evaporation of moisture takes place in a rapid and uniform manner due to the smaller Biot number of the minute droplets. This step marks the beginning of particle formation during spray drying. The temperature of the heated drying air (T_i) and the temperature of the air stream laden with solid particles at the exit of the drying chamber (T_o) are the process parameters of relevance in this stage

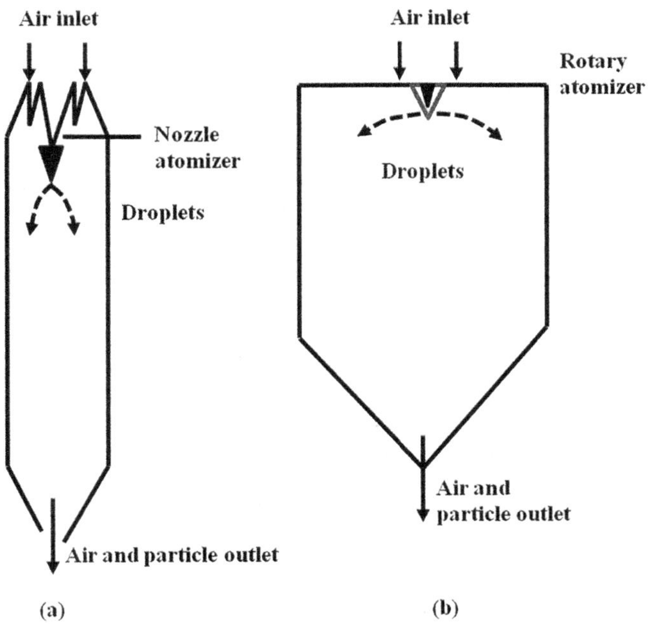

FIGURE 10.19 (a) Tall-type spray dryer fitted with a nozzle atomizer and (b) short-type spray dryer fitted with a rotary atomizer. (Reproduced with permission from Langrish, T. A. G. and Fletcher, D. F. 2001. Spray drying of food ingredients and applications of CFD in spray drying. *Chemical Engineering and Processing: Process Intensification* 40: 345–354.)

of spray drying. But T_o is not an independent parameter, and it is interrelated to T_i and the feed flow rate. An increase in T_i and a decrease in feed flow rate lead to high T_o values.

Once the contact is established between the droplets and the hot air, moisture evaporation occurs in two stages: (i) constant rate period and (ii) falling rate period. During the constant rate period, latent heat of vaporization withdrawn from the drying medium is utilized to evaporate the free water from the droplet surface. The driving force for this simultaneous heat and mass transfer is provided by the gradient in moisture concentration and the difference in vapor pressure and temperature between the interior and exterior of the droplet (Patel and Chen, 2008). Due to the lower internal resistance to mass transfer within the droplet, there is a continuous migration of moisture to the droplet surface. This leads to maintenance of saturated conditions, promotes the constant rate of moisture evaporation from the surface, and prevents the temperature of the droplet from rising above the wet-bulb temperature of drying air. The aforementioned sequence of events is collectively known as the *evaporative cooling*. Thus, the constant rate period can also be termed as the *low-temperature phase* of spray drying. An important phenomenon that occurs during the constant rate drying period is the redistribution of solute components within the drying droplet, which results from the predominantly fluid state of the droplet and the absence of dense skin or membrane around it (Kim et al., 2009).

The beginning of the falling rate period is marked by the reduction in moisture content below a level that limits the maintenance of saturation conditions at the droplet surface. The dissolved solutes from the saturated feed solution form a crust of solids or skin thus initiating the transformation of the droplet to particle phase. The solid crust then continues to build upon and turns as a limiting resistance to the diffusion of water vapor from the droplet interior. This decreases the drying rate and elevates the particle temperature (Charlesworth and Marshall, 1960; Birchal et al., 2006), thus, initiating the transition of spray drying from a low temperature to a high-temperature process. The bubble formation and droplet or core shrinkage completes the moisture evaporation stage and determines the particle morphology. Bubble formation occurs when the partial pressure of water vapor at the center of the droplet increases over the ambient pressure with a concurrent increase in temperature. The droplet inflates to the outer radius and ultimately results in irregular and random shape particles (Etzel et al., 1996). When the droplet with the inflated bubble is subjected to further high temperature, it might lead to a blistered, shriveled, hollow, or inflated puffed particle.

10.3.7.2.4 Particle Separation

Particle separation involves the primary and secondary phases of separation to remove the product from the drying medium. Often, the spray drying chamber has a conical bottom to facilitate the easy collection of the dried powder. During the primary separation, the dry powder is collected at the base of the dryer followed by removal using a screw conveyor or a pneumatic system with a cyclone separator at the time of secondary separation. The gas stream laden with the evaporated moisture is withdrawn from the center of the cone above the conical bottom and discharged through a side outlet. The relatively low efficiency of collection necessitates the use of an additional particle collection system comprising dry collectors followed by wet scrubbers. The dry collectors include the cyclone separator and the bag filter. The choice between a cyclone separator and a bag filter depends on the size of the particles carried by the exhaust gas and the final product specifications.

10.3.7.2.4.1 Cyclone Separator A cyclone separator is a stationary mechanical device that works on the principle of centrifugal separation to remove the solid particles from a carrier gas. It consists of an upper cylindrical part referred to as the *barrel* and a lower conical part referred to as the *cone* (Figure 10.20). The gas stream with the solid particles exiting the spray dryer enters tangentially at the top of the barrel and travels downward into the cone forming an outer vortex. The increasing air velocity in the outer vortex results in a centrifugal force on the particles which separates them from the gas stream. When the gas stream reaches the bottom of the cone, an inner vortex is created thus reversing its direction and leaving at the top as clean gas. The particulates accumulate in the collection chamber which is attached to the bottom of the cyclone.

10.3.7.2.4.2 Bag Filter The bag filter (Figure 10.21) includes a metallic housing designed for continuous operation and automatic cleaning. Under suction or pressure, the particle-loaded air enters the hopper of the bag filter through the center or bottom part of the collector. The air containing the particles travels through the filter bag which retains the product particles on its surface. The clean air passes

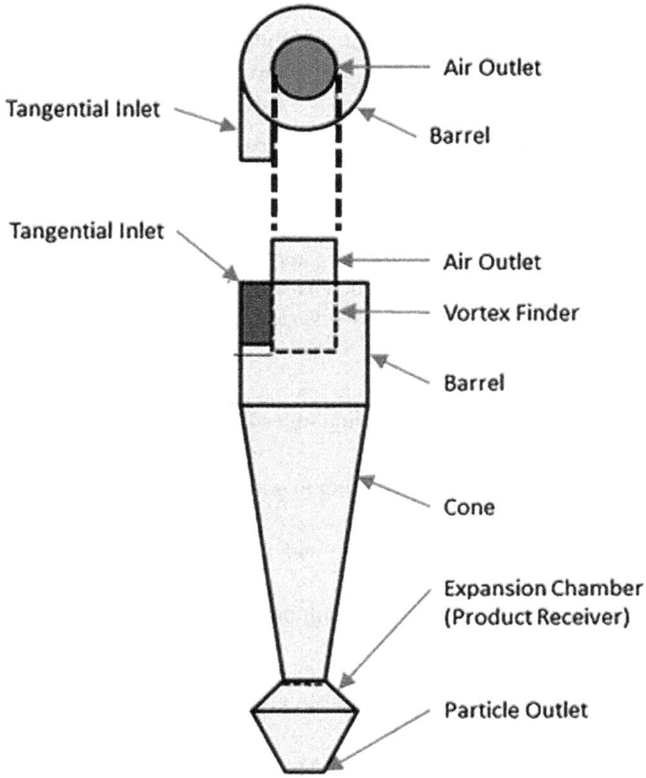

FIGURE 10.20 A typical cyclone separator. (Modified and reproduced with permission from Funk, P. A., Elsayed, K., Yeater, K. M., Holt, G. A. and Whitelock, D. P. 2015. Could cyclone performance improve with reduced inlet velocity? *Powder Technology* 280: 211–218.)

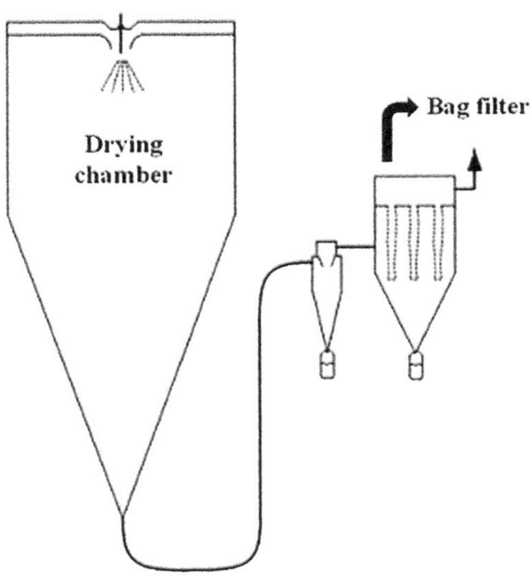

FIGURE 10.21 Schematic of spray dryer with a bag filter. (Modified and reproduced with permission from Lindeløv, J. S. and Wahlberg, M. 2009. Spray drying for processing of nanomaterials. *Journal of Physics Conference Series* 170: 012027.)

through the bags and plenum to the outlet of the bag filter. Above each row of bags, there is a tube with holes that are aligned with the central air passage gap, through which the compressed air is injected to momentarily invert the gas flow and remove the particulate material collected outside the bags.

10.3.7.3 Mass and Energy Balance around the Spray Dryer

10.3.7.3.1 Mass Balance

In Figure 10.22, M_a is the air flow rate (kg dry air/h), M_p is the product flow rate (kg dry solid/h), W is the product moisture content (kg water/kg dry solid), and H is the absolute humidity of air (kg water/kg dry air). In a continuous steady-state operation such as spray drying, the sum of mass flow rate of water in the inlet air and the feed liquid equals the mass flow rate of water in the outlet streams. The individual stream flows (per hour) are as follows:

$$\text{Moisture entering in feed} = M_p W_{\text{in}} \tag{10.24}$$

$$\text{Moisture entering in gas} = M_a H_{\text{in}} \tag{10.25}$$

$$\text{Moisture exiting the dryer along with the dried product} = M_p W_{\text{out}} \tag{10.26}$$

$$\text{Moisture exiting the dryer along with the exhaust drying air} = M_a H_{\text{out}} \tag{10.27}$$

In a continuous steady-state operation such as spray drying, according to the principle of mass balance explained in Chapter 2, the overall mass balance around the spray dryer is given by

$$\text{Mass in} = \text{Mass out} \tag{10.28}$$

The component balance on the system with respect to moisture content is given by

$$M_a H_{\text{in}} + M_p W_{\text{in}} = M_a H_{\text{out}} + M_p W_{\text{out}} \tag{10.29}$$

$$M_a (H_{\text{in}} - H_{\text{out}}) = M_p (W_{\text{out}} - W_{\text{in}}) \tag{10.30}$$

Thus, the mass of the water evaporated from the liquid feed equals that taken up by the air and rearranging Eq. (10.30) in terms of outlet air humidity

$$H_{\text{out}} = H_{\text{in}} - \left(\frac{M_p}{M_a}\right)(W_{\text{out}} - W_{\text{in}}) \tag{10.31}$$

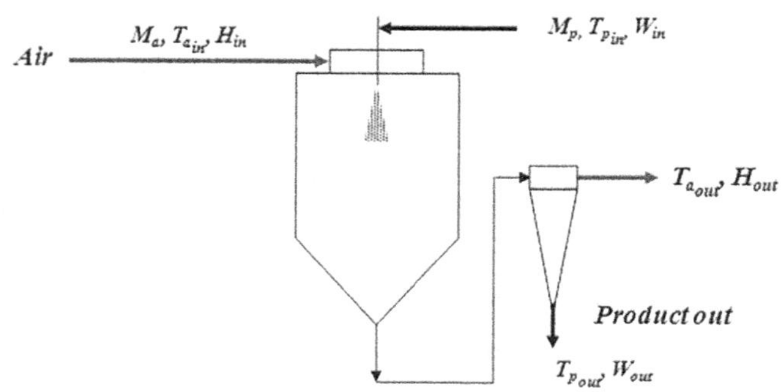

FIGURE 10.22 Schematic of mass and energy balance around spray dryer.

In Eq. (10.31), H_{in} is the absolute inlet gas humidity to the dryer which can be calculated from the combustion reaction of natural gas based on measured inlet air humidity. The remaining terms are measurable which enables the outlet air humidity to be calculated.

10.3.7.3.2 Energy Balance

Spray drying being a continuous steady-state operation, the heat flow in and out of the spray dryer must be balanced. The individual heat flows are as follows:

$$\text{Enthalpy flow of air entering the spray dryer} = M_a Q_{a_{in}} \tag{10.32}$$

$$\text{Enthalpy flow of feed liquid entering the spray dryer} = M_p Q_{p_{in}} \tag{10.33}$$

$$\text{Enthalpy flow of air leaving the dryer} = M_a Q_{a_{out}} \tag{10.34}$$

$$\text{Enthalpy flow of product leaving the dryer} = M_p Q_{p_{out}} \tag{10.35}$$

$$\text{Heat loss from the dryer} = q_L \tag{10.36}$$

Hence, according to the first law of thermodynamics and the principle of energy balance explained in Chapter 3, the overall energy balance around the spray dryer is given by

$$\text{Energy in} = \text{Energy out} \tag{10.37}$$

$$M_a Q_{a_{in}} + M_p Q_{p_{in}} = M_a Q_{a_{out}} + M_p Q_{p_{out}} + q_L \tag{10.38}$$

$$M_a \left(Q_{a_{in}} - Q_{a_{out}}\right) = M_p \left(Q_{p_{out}} - Q_{p_{in}}\right) + q_L \tag{10.39}$$

In Eqs. (10.38) and (10.39),

Q_a: Heat content of air (kJ/kg dry air) $= C_{P_{air}}\left(T_a - T_o\right) + H\lambda$
Q_p: Heat content of product (kJ/kg dry solid) $= C_{P_{solid}}\left(T_p - T_o\right) + WC_{P_{water}}\left(T_p - T_o\right)$
M_a: Mass of air (kg)
M_p: Mass of product (kg)
q_L: Heat loss from the dryer (kJ)

Rearranging Eq. (10.39) in terms of heat loss (q_L),

$$q_L = M_a \left(Q_{a_{in}} - Q_{a_{out}}\right) + M_p \left(Q_{p_{in}} - Q_{p_{out}}\right) \tag{10.40}$$

The heat flow associated with each stream is based on the following equations:

$$Q_{a_{in}} = \left(c_g + c_{wv} H_{in}\right)\left(T_{a_{in}} - T_o\right) + h_{fg} H_{in} \tag{10.41}$$

$$Q_{a_{out}} = \left(c_g + c_{wv} H_{out}\right)\left(T_{a_{out}} - T_o\right) + h_{fg} H_{out} \tag{10.42}$$

$$Q_{p_{in}} = c_p\left(T_{p1} - T_o\right) + W_{s_{in}} c_w \left(T_{a_{in}} - T_o\right) \tag{10.43}$$

$$Q_{p_{out}} = c_p\left(T_{p2} - T_o\right) + W_{s_{out}} c_w \left(T_{a_{out}} - T_o\right) \tag{10.44}$$

$$\text{Specific heat of wet gas, } c_{wg}\left(\text{J/kg K}\right) = c_g + c_{wv} H_{in} \tag{10.45}$$

where

h_{fg}: Latent heat of vaporization for water at T_o = 2502000 J/kg K
H_{in}: Absolute humidity of inlet gas to the dryer (kg water/kg dry air)
H_{in}: Absolute humidity of outlet gas from the dryer (kg water/kg dry air)
c_g: Specific heat of gas = 1012 J/kg dry air K
c_{wv}: Specific heat of water vapor = 1914 J/kg K
c_p: Specific heat of water = 4185 J/kg K
W: Moisture content of the product (kg water/kg dry solid)
$T_{a_{in}}$: Temperature of inlet gas to the dryer (K)
$T_{a_{out}}$: Temperature of outlet gas from the dryer (K)
T_{p1}: Temperature of feed to the dryer (K)
T_{p2}: Temperature of particle exiting the dryer (K)
T_o: Reference temperature (298 K)

10.3.7.4 Advantages and Limitations of Spray Drying

The advantages of the spray-drying process are its continuous operation, versatility to operate under a wide range of operating temperatures, short residence time of the product, ability to produce a powdered product in the stable and dry form, suitability for both heat-sensitive and heat-resistant materials, and rapid drying (Keey, 1972; Katta and Gauvin, 1975; Masters, 1991; Fellows, 1998). However, the quality of spray-dried food products is often susceptible to the loss of volatile aroma components. The extent of volatile loss depends upon the droplet size and boiling point of the flavor oil. In addition, the spray-drying process requires a high capital cost. The two important processing problems associated with spray drying are stickiness and thermal degradation of heat-sensitive products.

Stickiness is a major problem that limits the spray drying of various food products. Glass transition temperature (T_g) contribute to the stickiness problem (Bhandari et al., 1997), where glass transition is the temperature and time-dependent change in physical state from the glassy to rubbery viscous liquid without phase change (Jeremiah, 1995). It is a nonequilibrium metastable state which shows a high degree of hygroscopicity (Bhandari et al., 1997). T_g is the temperature below which the amorphous sugars in foods are stable and exist as solid *glass* and above which exist as *rubber* (Blanshard and Lillford, 1993). Materials containing high sugar content and organic acid-rich materials have a low glass transition temperature (T_g) and therefore may exhibit the sticky behavior. This problem could be avoided by using additives such as glucose syrup or maltodextrin, which increase the glass transition temperature (Truong et al., 2005).

The Gordon and Taylor equation (Eq. (10.46)) can be used to predict T_g for a binary mixture of water and a single solute (Miao and Roos, 2004).

$$T_g = \frac{w_1 T_{g1} + c w_2 T_{g2}}{w_1 + c w_2} \tag{10.46}$$

where w_1 and w_2 are the mass fractions of solute and water, T_{g1} and T_{g2} are the glass transition temperature (K) of solute and water (138 K), respectively, and c is the ratio of specific heat change of solute to water at the glass transition temperature.

Most biological products such as proteins, enzymes, antioxidants, vitamins, and bacteria are heat sensitive. During spray drying, inactivation or denaturation occurs in these products due to high temperature. These effects can be minimized by using small droplets, with low outlet air temperature or by adding encapsulating materials (wall material) to protect these heat-sensitive materials (Etzel et al., 1996; Gouin, 2004).

10.3.7.5 Applications of Spray Drying

Spray drying is widely used in food industry for producing high-quality powders with low moisture content (Charm, 1971; Masters, 1991). It is used for producing a wide range of products such as proteins, vitamins, bacteria, enzymes, yeast, flavors, milk, cream, coffee, tea, juices, eggs, soups, beverages, fruit juices, vegetables, and encapsulated products.

10.3.8 Freeze Dryer

Freeze-drying or *lyophilization* is a drying method in which water in a solution or suspension form is crystallized at low temperatures and sublimated from the solid state directly into the vapor phase (Oetjen, 1999; Kochs et al., 1993). A freeze-drying process provides a product which is stable (dry form), rapidly soluble (large surface area), and elegant (uniformly colored cake). The freeze-drying process consists of four stages: (i) freezing, (ii) primary drying, (iii) secondary drying, and (iv) final treatment.

10.3.8.1 Freezing

In the freezing stage, the water content in the food is converted to ice crystals. The crystallization depends on the cooling velocity, initial concentration, and end temperature of cooling. Most of the food components remain in an amorphous, glassy state (i.e., do not crystallize). The size and structure of the resultant ice crystals vary with the initial solute concentration of feed, the degree of supercooling, freezing temperature, freezing rate, type of freezing, and product composition. The sublimation rate and final product characteristics are dependent on the crystal structure formed during the freezing process. Thus, freezing is a critical stage as it is not possible to alter the crystal structure during the subsequent primary and secondary drying stages.

10.3.8.2 Primary Drying

During the primary drying stage, ice crystals in the frozen food product are removed by sublimation. It is an endothermic process which takes place at the interface between the frozen and dry material and this starts at the ice surface. It essentially occurs at low temperature and low pressure, more specifically, at temperature and pressure conditions below the triple point of water. The enthalpy of sublimation is the sum of the latent heat of fusion and latent heat of vaporization. Sublimation starts at the ice surface and occurs at the interface between the frozen and dry material. Product temperature is maintained constant during the primary drying. During primary drying, water vapor pressure in the drying chamber is lower than the vapor pressure of ice at the corresponding temperature, which causes ice sublimation. This is achieved as the pressure is reduced inside the chamber by applying a vacuum. Sublimation of ice crystals leaves behind pores in the matrix, thus leading to excellent rehydration characteristics of the freeze-dried products.

A major problem during the primary drying stage of freeze-drying is the *collapse* of product. Collapse occurs when the viscosity of the structural material is reduced to a level at which it cannot support its weight against gravity (Bhandari et al., 1997) and leads to structural loss, reduction in pore size, and volumetric shrinkage of the dried food products (Levi and Karel, 1995). The collapse temperature (T_c) decreases with the decreasing molecular weight and increasing moisture content (Oetjen, 1999). Collapse can be prevented by increasing the molecular weight of the product by adding materials such as dextran, fructose, glucose, maltose, or polyethylene glycol. T_c can be related to the glass transition temperature. When the temperature (T) is higher than T_g, the viscosity of amorphous matrix decreases, which is a function of (T–T_g). The abovementioned phenomenon can be explained using the William–Landel–Ferry relationship Eq. (10.47) (Bhandari and Howes, 1999).

$$\log_{10}\frac{\mu}{\mu_g} = -\frac{17.4\left(T - T_g\right)}{51.6 + \left(T - T_g\right)} \tag{10.47}$$

TABLE 10.1
Collapse Temperature of Some Food Products

Material	Collapse Temperature (T_c;°C)
Orange juice	–24.0
Sweetened concord grape juice	–33.5
Grapefruit juice	–30.5
Lemon juice	–36.5
Apple juice	–41.5
Prune extract	–35.0
Pineapple juice	–41.5
Concord grape juice	–46.0
Coffee extract	–20.0

Source: Anandharamakrishnan (2008).

where μ is the viscosity, μ_g is the viscosity at glass transition temperature T_g, and T is the temperature (K). The collapse temperature (T_c) for some of the food products is listed in Table 10.1 (Kudra and Strumillo, 1998).

10.3.8.3 Secondary Drying

Unfrozen water cannot be removed by sublimation and is subsequently removed secondary drying or desorption, at a comparatively higher temperature than that of primary drying. As the ice crystals convert to vapor, product temperature rises and approaches towards the shelf temperature. Secondary drying is the slowest stage consuming about one-third of the processing time of freeze-drying operation and hence regarded as the rate-limiting step. At the end of secondary drying, the residual moisture content of the food product is about 2%–10%.

10.3.8.4 Final Treatment

In the final stage, the drying chamber is filled with an inert gas (nitrogen for foodstuffs, argon for biological products) for preserving the products after drying.

A freeze dryer (Figure 10.23) consists of a cylindrical vacuum chamber. The condenser attached to a vacuum pump creates and maintains the vacuum inside the chamber. The food product is loaded on trays and placed on the shelves inside the chamber. All the three stages of freeze-drying, namely, freezing, primary drying, and secondary drying are carried out synchronously in the product chamber. However, for some tabletop freeze dryers, the prerequisite is that only frozen food products can be loaded.

The pressure inside the chamber is reduced by applying a vacuum. The vacuum pump and condenser are the integral parts of any freeze dryer. A vacuum pump removes the non-condensable gases from the chamber. A condenser maintains the low water vapor pressure inside the chamber by removing the water vapor formed during sublimation. The condenser consisting of coils is located in another chamber and connected by a duct to the drying chamber. The buildup of ice on the condenser coil surface reduces its efficiency; therefore, two condensers are used alternatively in the system. Periodic cleaning is required to prevent loss of efficiency. It is important to keep the refrigerant temperature lower than the corresponding chamber temperature which is in the range of –10°C to –50°C. Similar to the condensers, two vacuum pumps are used to generate the required vacuum in the freeze dryer. During the secondary drying phase, heat is supplied to the shelves (which are hollow), by circulating a thermal fluid through it. The food product is heated by conduction through the slab and trays and by radiation from the top shelf.

The process operates at a low temperature which reduces thermal damage to the products and reduces volatile losses. The absence of shrinkage and porous structure gives the best rehydration characteristics. Consequently, superior product quality in terms of aroma, flavor, color, shape, and structure are the major

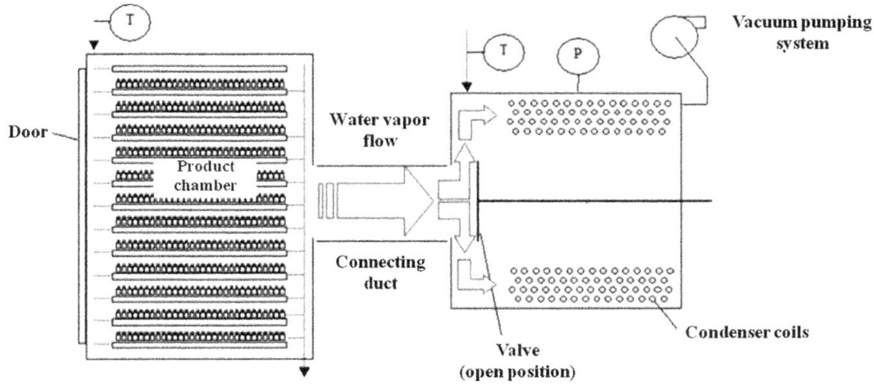

FIGURE 10.23 Freeze dryer. (Slightly modified and reproduced with permission from Alexeenko, A. A., Ganguly, A. and Nail, S. L. 2009. Computational analysis of fluid dynamics in pharmaceutical freeze-drying. *Journal of Pharmaceutical Sciences* 98:3483–3494.)

advantages of the freeze-drying process. Fruit juices and coffee are the well-known examples of freeze-dried foods that have great demand due to their superior product quality. Other freeze-dried products include meat, shrimp, vegetables, and extracts. However, freeze-drying is an expensive dehydration process because of low drying rates and high capital and energy cost due to the refrigeration and vacuum units (Franks, 1998).

10.3.9 Superheated Steam Dryer

10.3.9.1 Working Principle of Superheated Steam Dryer

Superheated steam is the steam which exists at a temperature above the boiling point. As long as its temperature remains higher than the boiling point, the drop in temperature does not lead to condensation of the steam (Karimi, 2010; van Deventer, 2004). Superheated steam drying (SSD) is based on the use of superheated steam to supply the sensible heat required to vaporize the moisture in the product and carry away the evaporated moisture. Therefore, the principle of SSD is based on the recirculation of superheated steam in a closed loop (Figure 10.24). The quantity of steam removed from the system is equal to the quantity of water evaporated during the drying process. Thus, the superheated steam plays the role of the heat source as well as the drying medium to detract the evaporated water. The water vapor removed from the product can be reheated and recirculated to be a part of the incoming superheated steam or can be used elsewhere in the process or plant.

SSD occurs in three distinctive stages as depicted in the drying curve shown in Figure 10.25:

1. **Preheating and condensation period:** During this initial period of drying, the moisture content of the product increases as the heat is transferred through steam condensation. The product is heated by the latent heat of the steam while an intense evaporation begins at 100°C. Subsequently, a layer of moisture envelops the product, and from then, the evaporation proceeds at a constant temperature of the product. In this stage of SSD, evaporation is dominated by the heat transfer coefficient between the surface of the product and superheated steam. Owing to the increase in moisture content, the drying time increases. However, this is relevant only for those food products with low initial moisture content.
2. **Constant drying rate period:** In this stage of SSD, the bulk of the water is removed from the product owing to the absence of diffusive resistance at the boundary layer. The drying rate remains constant due to water vaporization occurring at the product surface. Similar to the preheating stage, during the constant drying rate period as well, evaporation is dominated by

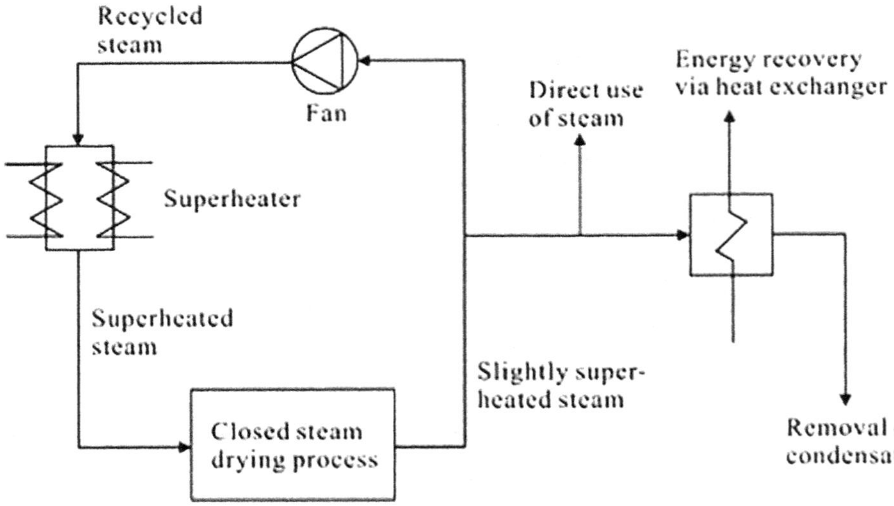

FIGURE 10.24 Line diagram of the superheated steam dryer. (Reproduced with permission from Li, J., Liang, Q- C. and Bennamoun, L. 2016. Superheated steam drying: design aspects, energetic performances, and mathematical modeling. *Renewable and Sustainable Energy Reviews* 60: 1562–1583.)

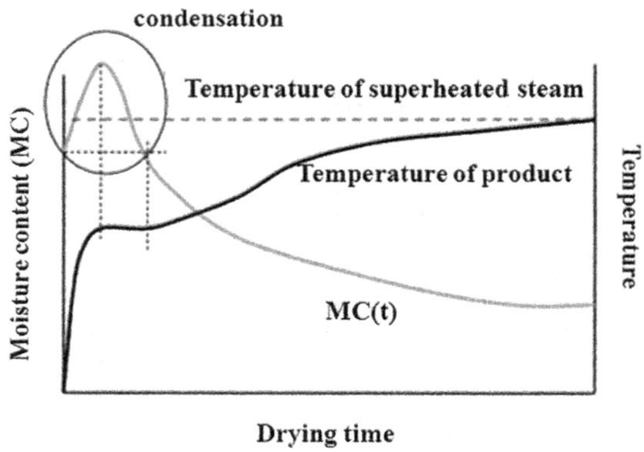

FIGURE 10.25 Drying curve depicting the three stages of SSD.

the heat transfer coefficient between the product surface and superheated steam. The product temperature is constant at the boiling point of water (100°C).

3. **Falling drying rate period:** During this period of SSD, the drying rate reduces due to the formation of a dry layer at the product surface. The product temperature approaches that of the superheated steam (>100°C). Thus, in order to prevent thermal degradation of the food product, SSD is usually performed at reduced pressure. However, lower pressure is accompanied by a lower saturation temperature which reduces the drying rate due to poorer convective heat transfer under reduced pressures. But the quality of products obtained from low-pressure SSD is superior to that obtained from conventional vacuum drying (Devahastin et al., 2004). The evaporation of moisture from the product during this phase of SSD is dominated by the internal diffusion of moisture to the surface of the product.

10.3.9.2 Construction of Superheated Steam Dryer

In general, a superheated steam dryer consists of the following components (Figure 10.24):

- Steam generator
- Fan
- Heater
- Drying chamber
- Heat exchanger
- Condenser

10.3.9.3 Advantages and Limitations of Superheated Steam Dryer

The main advantage of superheated steam dryer is its high thermal efficiency. The dryer exhaust is nothing but steam at lower specific enthalpy, which can be reheated and reused after recovering its latent heat by condensation, mechanical compression, or thermal compression. Therefore, the energy consumption in SSD is lower compared to that in hot air drying. In addition to the heat recovery opportunity, the higher heat transfer coefficients and higher drying rates also contribute to the thermal efficiency of SSD, leading to energy savings of about 50%–80% (Pronyk et al., 2004). For food products, SSD is used in combination with blanching, pasteurization, and sterilization to render it an attractive drying process for food manufacturers (van Deventer and Hejimans, 2001). The other advantages and limitations of SSD are listed in Table 10.2.

TABLE 10.2
Advantages and Limitations of Superheated Steam Dryer

Advantages of SSD	Limitations of SSD
No oxidation or combustion reactions are possible in SSD. This means no fire or explosion hazards and hence assures safe operation.	The system is more complex. No leaks can be allowed. Feeding and discharge of SSDs must not allow infiltration of air. The product itself may bring in non-condensable constituents. Compared to air dryer, start-up and shutdown are more complex operations in an SSD, especially when run under reduced pressure which demands good sealing.
Higher drying rate due to higher thermal conductivity and heat capacity of superheated steam.	As the feed enters at ambient temperature, there is inevitable condensation in the SSD before evaporation begins. This adds about 10%–15% to the residence time in the dryer. In some cases, moisture condensation at the beginning of drying might increase the drying time. To achieve low moisture end levels, the superheat of the drying steam needs to be high.
SSD permits pasteurization, sterilization, and deodorization of food products.	Products that may melt, undergo glass transitions or be otherwise damaged at the saturation temperature of steam as the dryer operating pressure cannot be dried in superheated steam even if they contain only surface moisture.
The quality of dried products from SSD is good owing to the uniform shrinkage and superior rehydration ability.	Products that may require oxidation reactions (e.g., browning of foods) to develop desired quality parameters cannot be dried in superheated steam.
	Cost of the ancillaries such as the feeding systems, product-collection systems, and exhaust steam recovery systems is more than the cost of the superheated steam dryer.
	When used for the drying of heat-sensitive products, SSD should be performed at reduced pressure to reduce the boiling point of water and to lessen the thermal damage.
	The superheated steam dryer requires good insulation as all the parts in contact with the superheated steam must be maintained above the condensation temperature at the given pressure.

10.3.9.4 Major Applications of SSD in the Food Industry

Hitherto, SSD has been used for the drying of the following food products: sugar beet pulp, vegetables, noodles, wheat flour, pork meat, bone meal, fish meal, spent distillers grain, brewers grains, corn fibers, soybeans, rape seed, sunflower seed, soy meal, okra, shrimp, pork, bamboo shoot, paddy/rice, herbs, cacao beans, corn gluten, tobacco, starch, milk, yoghurt, potato products, coffee grounds, and spices.

10.3.10 Supercritical Dryer

10.3.10.1 Working Principle of Supercritical Dryer

Supercritical drying (SCD) is based on the principle of overcoming the surface tension and capillary stress to transform the solvent in a substance into gas by heating or pressurizing the solvent in the product beyond its critical point (temperature/pressure). The following two approaches can be utilized to eliminate the surface tension:

I. **By supercritical transition of the solvent originally present in the product:** As the solvent (liquid) in a food product (almost always water) approaches its critical point (above which it is transformed into a supercritical fluid), its molecules move past each other in a faster manner without adhering to the neighboring molecules. This is possible only when the surface tension and capillary stress exerted by the solvent decreases. Beyond the critical point, there is no distinction between the liquid and vapor phases and the solvent can be removed without introducing a liquid–vapor interface by converting it to a supercritical fluid. When the supercritical transition of the solvent occurs, it loses all its surface tension and can no longer exert capillary stress. At this point, the solvent can be removed from the pores of the product by isothermal depressurization of the fluid, while maintaining it above its critical temperature (if the temperature of the fluid falls below its critical temperature, then the liquid would begin to drain off from the fluid). Subsequently, when the system containing the product is depressurized, the molecules are transferred from the fluid to the surroundings in the form of gas and the fluid becomes less dense. Eventually, a substantial amount of fluid would have been removed from the system such that when it is cooled below its critical point, enough substance would not be available to recondense into a liquid as the fluid is too low in density and instead transforms to the gaseous state. Thus, the solvent which was present in the liquid state in the product has been converted into its gas phase without causing any capillary stress within the structure of the product. Thus, SCD retains the delicate microstructure and pores within the product, intact.

II. **By supercritical transition of CO_2 or other low-critical parameter solvents:** As an alternative to the abovementioned approach, a supercritical fluid such as CO_2 with a low critical temperature (31.1°C) can be used to remove water from the product. This approach is based on a high-pressure spray technique called particles from gas-saturated solutions (PGSS). In the PGSS process, the product to be dried is first melted or suspended in a solvent at a defined temperature in an autoclave. Subsequently, supercritical carbon dioxide (SC-CO_2) is solubilized in the melted or liquid-suspended product to be dried. This is carried out by dosing the melt via a high-pressure pump into a static mixer, where compressed and preheated carbon dioxide is added to it to form a gas-saturated solution or suspension. Optionally, a cosolvent may be added to the SC-CO_2 to increase the ability of SC-CO_2 to dissolve polar compounds. The gas-saturated solution is then depressurized through a nozzle into a spray tower operated at ambient pressure to form fine droplets (Benali and Boumghar, 2015). The solvent in the droplets is evaporated by adjusting the pre-expansion conditions in the spray tower. The solvent is withdrawn with the expanded carbon dioxide to eventually result in the dry powdered product. Thus, SCD is similar to an extraction process in that it uses a supercritical fluid as the solvent to extract water from the food product.

Similar to the other supercritical fluid processes such as extraction, SCD is also carried out at low temperature (30°C–60°C) and in an inert and oxygen-free atmosphere. This renders the process appropriate for sensitive substances (Meterc et al., 2008).

10.3.10.2 Construction of a Supercritical Dryer

During SCD, in order to avoid the damages due to surface tension forces, the product is dried in an autoclave (Figure 10.26) at the supercritical conditions of the liquid phases present in the pores of the product (Kistler et al., 1943; McHugh and Krukonis, 1986). In general, liquid CO_2 is used as the supercritical extraction fluid since it has a much lower critical temperature and thus enables SCD to be performed at near-ambient temperature (Tewari et al., 1985).

The advantages and limitations of supercritical dryer are compiled in Table 10.3.

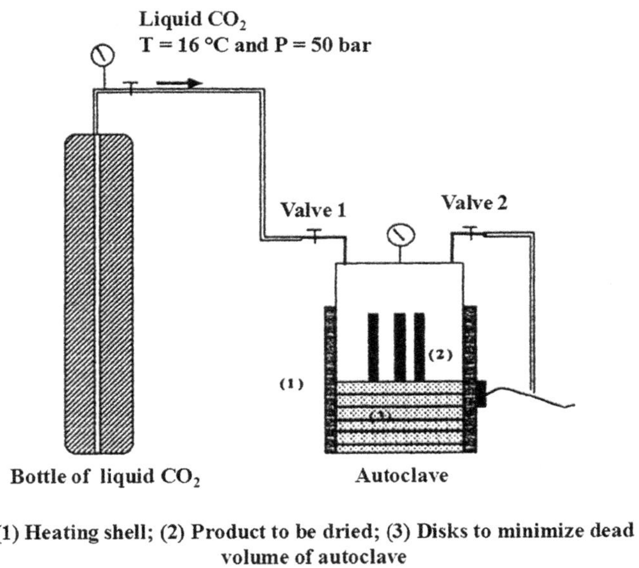

FIGURE 10.26 SCD system. (Reproduced with permission from Begag, R., Pajonk, G. M., Elaloui, E. and Chevalier, B. 1999. Synthesis and properties of some monolithic silica carbogels produced from polyethoxydisiloxanes dissolved in ethylacetoacetate (etac) and acid catalysis. *Materials Chemistry and Physics* 58: 256–263.)

TABLE 10.3
Advantages and Limitations of Supercritical Dryer

Advantages of Supercritical Dryer	Limitations of Supercritical Dryer
• SCD is appropriate for the production of porous products, as it avoids vapor–liquid interfaces by working in a homogeneous phase. • Owing to the high diffusivity of supercritical fluids, the mass transport of water from within the product to the surrounding during SCD is respectable. This renders SCD to outweigh the conventional air drying wherein higher temperatures are required to promote moisture diffusivity, leading to the deterioration of the dried product's structure. • As the critical temperature of supercritical CO_2 is as low as 31.1°C, SCD is highly suitable for heat-sensitive products that are prone to thermal degradation. • The shape and particle size distribution of the dried product can be tailored by controlling the temperature and pressure during SCD. • The enhanced water solubility of supercritical-dried products promotes the powder formulation process and improves the oral bioavailability of nutraceutical powders.	• A major limitation in the use of PGSS process for SCD is its prerequisite that the product to be dried should be in a liquid state. However, most of the products used in the food industry would distort before reaching their melting point.

10.3.10.3 Major Applications of Supercritical Dryer in the Food Industry

The applications of SCD are rather limited in the food sector. Nevertheless, SCD has been successfully used for the drying of green tea extracts containing antioxidants and polyphenols. Dry and free-flowing powders were obtained without degradation of the active ingredients (Meterc et al., 2008).

Carrot pieces were dried at a pressure of 20 MPa and temperature of 40°C–60°C in the supercritical fluid environment using supercritical carbon dioxide containing a co-solvent, ethanol (SC-CO_2-EtOH). The SC-CO_2-EtOH-dried carrot pieces retained their shape much better than the air-dried counterpart which underwent shrinkage. Consequently, the SC-CO_2-EtOH-dried product showed more porous and less dense structures resulting in favorable rehydration and textural properties than the air-dried equivalents. However, a slight color loss was observed in the carrot pieces after the SCD. But, this was largely reversible upon rehydration of the product (Brown et al., 2008; Brown, 2010).

10.3.11 Dielectric Dryers

Dielectric dryers comprise the microwave (MW) and radiofrequency (RF) drying systems. Most foods are dielectric materials and hence are capable of storing electric energy and converting it into heat (Sisquella et al., 2014). This is because the water molecules in food are dipolar in nature with positive and negative charges at their either ends with the latter concentrated more on the oxygen atom of the molecule. The principle of dielectric drying is based on the direct interaction of electromagnetic energy with the dipolar water molecules to cause a rapid increase in the center temperature of the food product.

10.3.11.1 Microwave Dryer

10.3.11.1.1 Working Principle of MW Dryer

MWs are a part of the electromagnetic spectrum in the frequency band of 300 MHz–300 GHz, with their wavelength ranging from 1 m to 1 mm. MWs are used at a frequency of 2450 MHz in domestic applications and at 915 MHz for industrial applications.

The principle of MW drying is based on the conversion of electromagnetic energy into heat energy. The incident MW energy is specifically absorbed by the water in a food product. Consequently, the polarity of water changes at the rate of 2.45 billion cycles per second. The resultant molecular vibration generates heat which selectively heats the water and water-containing components in foods, thus contributing to the heat transfer during MW drying. The direct interaction between MW energy and moisture eliminates the need to transfer heat from the low-moisture surface into the high-moisture interior of the product (Feng et al., 2012).

The heating of food product during MW drying is said to be *volumetric* or *inside-to-out* heating which leads to the accumulation of an internal vapor pressure. The resultant gas pressure gradient acts as the driving force for the mass transfer during the drying process and removes moisture from the product without overheating the surface (Schiffman, 1987).

Thus, MW drying is different from conventional drying as the latter method approaches the food product from the surface and applies heat only to its outer edges leading to case hardening.

10.3.11.1.2 Construction of a MW Dryer

A continuous-type MW dryer is shown in Figure 10.27. It consists of a conveyor-belt assembly, an MW chamber (includes the MW source: a magnetron-type vacuum oscillator, waveguide, and applicator), a fan, and a control panel. The speed of the conveyor belt and the power output of the MW generator can be adjusted as required. The control parameters in continuous MW drying are the MW energy, the belt speed, and the feed input rate, which are precisely controlled by microprocessors/computers. Optimization of the abovementioned parameters ensures proper drying without exposing the product to undesirable high temperature.

The product to be dried is passed through a tunnel on the conveyor belt. The conveyor belt and the interior portion of the tunnel should be made of MW-resistant material. In the tunnel, the product is exposed to a series of MW sources. As a result, the temperature of water within the fresh product is raised to its

FIGURE 10.27 Continuous type MW dryer. 1: Inlet, 2: control panel, 3: microwave chamber, 4: conveyor, 5: thermal camera, 6: data acquisition system. (Reproduced with permission from Manickavasagan, A., Jayas, D. S. and White, N. D. G. 2007. Germination of wheat grains from uneven microwave heating in an industrial microwave dryer. *Canadian Biosystems Engineering* 49: 3.23–3.27.)

boiling point. Subsequently, water is evaporated from the product and the water vapor is removed from the tunnel by air blowers or humidity discharging system.

The applications of MW drying in food processing would be listed in the forthcoming section on hybrid drying techniques involving a combination of MW drying and conventional drying methods. The advantages and limitations of microwave dryer are tabulated in Table 10.4.

10.3.11.2 Radiofrequency Dryer

10.3.11.2.1 Working Principle of RF Dryer

The RF band occupies a broad range of frequency in the electromagnetic spectrum, varying from the kHz range (3 kHz $< f \leq$ 1 MHz) to the MHz range (1 MHz $< f \leq$ 300 MHz). The principle of RF drying is similar to that of MW drying discussed in Section 10.3.11.1.1. However, the free-space wavelength in the RF waves is 20–360 times longer than that of the commonly used MW frequencies. This permits a higher penetration depth of RF energy into the foods to provide uniform heating of the product than the MWs (Wang et al., 2003). For industrial applications, RF waves at the frequencies of 13.56, 27.12, and 40.68 MHz are used.

TABLE 10.4

Advantages and Limitations of MW Dryer

Advantages of MW Dryer	Limitations of MW Dryer
• Significant reduction in drying time (25%–90%) and increase in drying rate (four to eight times), compared to convective drying (Feng et al., 1999a, b; Maskan, 2001; Prabhanjan et al., 1995; Díaz et al., 2003; Soysal, 2004; Brygidyr et al., 1977). • High energy efficiency during the falling rate period due to direct interaction between MW energy and moisture. • Case hardening is avoided due to the volumetric heating effect, surface moisture accumulation, and homogeneous drying without large moisture gradients during MW drying. • Improved product quality in terms of enhanced aroma retention (Feng et al., 1999a), instant rehydration (Giri and Prasad, 2007), improved color retention (Soysal, 2004), and higher porosity (Torringa et al., 1996).	• Non-uniformity due to uneven distribution of MW field in the cavity caused by the superposition of the sinusoid MWs (Zhang et al., 2006). • The nonuniformity in mass transfer rate is a concern which causes puffing (Zhang et al., 2006) and even disintegration of the product. • Relatively high cost and low life span of the magnetron (Feng et al., 2012).

10.3.11.2.2 Construction of RF Dryer

In an RF dryer, an RF generator creates the alternating electric field between a pair of electrodes separated by a distance, *d* (Figure 10.28). The food product to be dried is placed between the two electrodes in a batch system or conveyed between the electrodes in a continuous RF dryer.

The advantages and limitations of radiofrequency dryer are listed in Table 10.5.

10.3.11.2.3 Applications of RF Dryer in Food Processing

The major application of RF drying in food processing is the post-baking drying of cookies, crackers, and snack foods. RF drying prevents *checking*, a major quality defect encountered in biscuits. Checking results from thermal stressing of biscuits on cooling due to the uneven distribution of moisture in the product. RF post-baking drying at 27.12 MHz eliminates the problem of checking as the

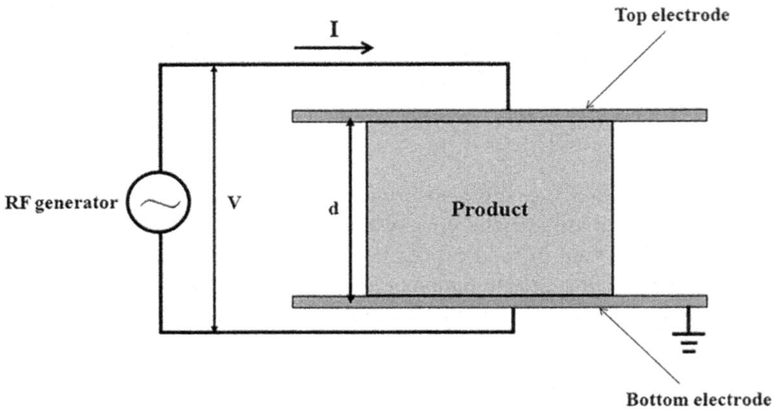

FIGURE 10.28 Schematic diagram of a RF dryer.

TABLE 10.5

Advantages and Limitations of RF Dryer

Advantages of RF Dryer	Limitations of RF Dryer
• The major advantages of RF drying are the consistent drying in a short drying time and the lesser requirement for floor space as heating begins instantaneously throughout the product, due to which the residence time of the product in an RF dryer is significantly lesser than that in a conventional dryer. • Compared to conventional drying, RF drying eliminates surface cracking caused by the stresses of uneven shrinkage in drying. This can be attributed to the uniform heating throughout the product and maintenance of uniform moisture from the center to the surface of the product during the RF drying process. • As RF drying is based on the direct application of heat, there is no wastage of heat. • RF drying is 2–20 times faster than conventional drying methods. • RF drying has the potential to reduce quality degradation during the drying of heat-sensitive food products. • Maintenance cost of an RF dryer is low when compared to other drying techniques. Also, it saves operational cost by saving time, energy, and increased controlled heating • RF drying is an environment-friendly and clean process.	• The major limitation faced by the food industry is the overheating of product in the corners, edges, and centre parts, especially when RF drying is used for foods of intermediate and high moisture content (Alfaifi et al., 2014). • High operating cost and initial high capital investment on the equipment. • Low power density. • For an equivalent power output, RF drying is more expensive than conventional convection, radiation or steam drying systems.

RF waves act specifically on the zones within the biscuit, where the moisture content is more and leads to selective and faster heating of those areas compared to the surrounding. This results in uniform heating of the product from the inside and the removal of a fraction of the residual moisture in biscuits. This leads to even distribution of moisture throughout the biscuit (Mermelstein, 1998). RF post-baking drying facilitates the baker to increase the oven line speed and develop the desired crumb structure, color, and target moisture content of the baked product. It also leads to highest production yields without sacrificing the quality. In addition to post-baking drying, RF drying has also been used for the drying of granular foods with poor thermal characteristics such as coffee beans, cocoa beans, corn, grains, and nuts (Piyasena et al., 2003).

10.3.12 Infrared Dryer

10.3.12.1 Working Principle of Infrared Dryer

Infrared (IR) radiation is a part of the electromagnetic spectrum that can be categorized into three classes based on its wavelength (Mongpraneet et al., 2004):

- Near-infrared (NIR: $\lambda = 0.78–1.4\,\mu m$)
- Middle-infrared (MIR; $\lambda = 1.4–3\,\mu m$)
- Far-infrared (FIR; $\lambda = 3–1000\,\mu m$)

FIR radiations are mostly used for the purpose of drying due to their low temperature compared to NIR and MIR radiations. IR drying in the FIR region has been found more appropriate for the drying of thin layers. However, drying of thicker bodies is considered to be more efficient using the NIR radiation (Riadh et al., 2015).

Drying by IR radiation energy is based on the transfer of heat from the IR heating element to the product surface but without the necessity to heat the surrounding air. On impinging the product, the IR radiation penetrates it after which it is converted to sensible heat (Ginzburg, 1969). As the moisture content is reduced during the drying process, the absorptivity of the dried material is decreased and its reflectivity and transmissivity are increased. Thus, as the drying proceeds, the temperature of the emitter must be changed to enhance the absorption of the radiation during drying (Ratti and Mujumdar, 1995).

IR drying is achieved without direct contact between the heating medium and food product. The following factors are known to influence the IR drying of food products :

- *Wavelength of IR radiation* which has an influence on its penetration depth
- *Surface temperature of the radiator* is directly related to the drying rate (Masamura et al., 1988).
- *IR power:* Holds an inverse relationship with the drying time (Sharma et al., 2005)
- *Thickness of the product layer subjected to drying:* Increase in thickness increases the absorptivity and decreases the transmissivity of radiation (Ratti and Mujumdar, 1995).

10.3.12.2 Construction of Infrared Dryer

Figures 10.29a and b show the schematic diagrams of a batch and a continuous IR dryer, respectively. A batch IR dryer consists of the following nine components (Figure 10.29a):

1. A stainless steel drying chamber
2. A centrifugal fan which is driven by a motor with an adjustable airflow controller
3. Three IR quartz glass tubes with an emission peak wavelength of 5.0 μm, at a power level of 400 W. The tubes are fitted in a row at the top of the drying chamber. The top of each glass lamp is coated with 1 mm gold to prevent entry of unwanted radiation to the top surface of the chamber.

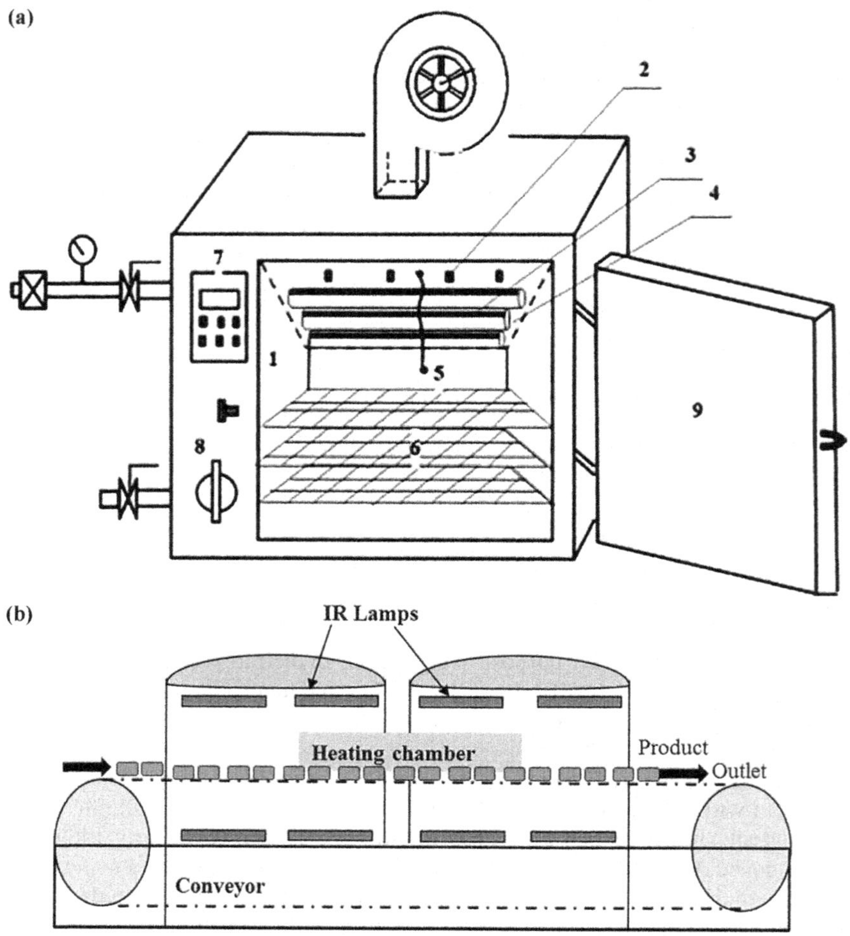

FIGURE 10.29 Schematic diagram of (a) batch IR dryer (Slightly modified and reproduced with permission from Roknul, A. S. M., Zhang, M., Mujumdar, A. S. and Wang, Y. 2014. A comparative study of four drying methods on drying time and quality characteristics of stem lettuce slices (*Lactuca sativa* L.). *Drying Technology* 32: 657–666.); (b) continuous IR dryer.

4. Type K thermocouples to measure the temperature of the air inside the drying chamber
5. Three wire-mesh trays positioned below the IR glass lamps. The distance between the IR heaters and the first tray and the distance between the trays are fixed at a specific value.
6. An air handling unit to control the humidity of the air
7. A hygrometer to measure the RH
8. A temperature controller to maintain a constant temperature in the drying chamber
9. A door that allows the placing and removal of samples

On the other hand, in a continuous IR dryer, the product is conveyed on a conveyor belt and exposed to the IR radiation emitted by the IR lamps (Figure 10.29b)

The advantages and limitations of infrared dryer are compiled in Table 10.6.

In addition to producing shelf-stable dried products, IR drying also finds application in the measurement of moisture content in food products (Hagen and Drawert, 1986; Anonymous, 1995). The key findings obtained from different studies on the applications of infrared dryer in food processing are listed in Table 10.7.

TABLE 10.6

Advantages and Limitations of IR Dryer

Advantages of IR Dryer	Limitations of IR Dryer
• Reduced drying time. • Fast heating rate. • Uniform drying temperature. • High degree of process control. • High thermal efficiency. • IR functions as an alternate source of energy. • Increased energy efficiency. • Maintains uniform drying temperature in food, inactivates microorganism, and inhibits enzymatic reaction to some extent and therefore results in better-quality products. • Drying by IR radiation overcomes the limitations associated with the conventional drying methods, i.e., the low rehydration rate of hot air dried products and the relatively expensive nature of the freeze-dried products. As a result, products dried using an IR dryer demonstrate better quality. compared to that from conventional drying. • IR drying provides high energy density with uniform temperature distribution and rapid increase in surface temperature. • Use of IR enhances the dehydration rate of the food product and thus drastically reduces the drying time. • Cleaner working environment. • The possibility of selective heating.	• Low penetration power. • Prolonged exposure of biological materials may cause fracturing. • Not sensitive to reflective properties of coatings.

TABLE 10.7

Applications of IR Dryer in Food Processing

Product	Findings	Reference
Potato	• FIR radiation was used. • Diffusivity in potato varied with radiation intensity and the thickness of slab at a constant radiation intensity level. • The decrease in the energy of activation increased the drying rate.	Afzal and Abe (1998)
Wine grape pomace	• IR drying resulted in the highest drying rate compared to convective drying. • IR drying reduced the drying time by more than 47.3% than convective drying. • The wine grape pomace dried by IR drying had the highest content of polyphenols and proanthocyanidins, which showed that low drying temperature led to less damage to the abovementioned functional constituents of grape pomace.	Sui et al. (2014)
Wet olive husk	• IR drying was studied in the temperature range from 80°C to 140°C. • Drying rate increased with temperature. • As the drying temperature was increased from 80°C to 140°C, the drying time to reduce the moisture content of the wet olive husk from 91.97% (w/w) to 8.69% (w/w) (dry basis) reduced from 105 to 35 min.	Celma et al. (2008)
Carrot slices	• IR drying was conducted at three levels of IR power (300, 400, and 500 W) and air velocities (1.0, 1.5, and 2.0 m/s). • The drying time at the IR power of 300, 400, and 500 W was 252 and 277 min, 205 and 236 min, and 145 and 155 min at air velocities of 1.0 and 2.0 m/ s, respectively. • The drying rate increased with increasing IR power.	Kocabiyik and Tezer (2009)

(*Continued*)

TABLE 10.7 (*Continued*)
Applications of IR Dryer in Food Processing

Product	Findings	Reference
Red pepper slices	• IR drying of red pepper slices was carried out at three levels of IR power (300, 400, and 500 W) and at three air velocities (1.0, 1.5, and 2.0 m/s). • The drying time at air velocity from 1.0 to 2.0 m/s ranged between 314 and 455 min, 213 and 297 min, and 196 and 230 min at IR power of 300, 400, and 500 W, respectively. • Drying rate increased with increasing IR power and decreasing air velocity.	Nasiroglu and Kocabiyik (2009)
Rough rice	• FIR radiation (λ = 3.3–8 µm) which heats the product to much lower than 250°C was used. • The mean drying rates were 0.34%, 0.57%, and 0.77% per hour for cycle times of 31, 16, and 9 min, respectively. • The electricity to remove 1 kg of moisture from rice was found to be 1.535 kW h minimum.	Bekki (1991)
Blueberries	• IR drying was conducted at four product temperatures (60°C, 70°C, 80°C, and 90°C) and compared with hot air convective drying at 60°C. • IR drying produced firmer-texture products with much-reduced drying time compared to hot air drying. • For fresh blueberries, IR drying conserved drying time by 44% at 60°C.	Shi et al. (2008)
Onion slices	• Single-layer drying of onion slices was carried out at IR power levels of 300, 400, and 500 W; drying air temperatures of 35°C, 40°C, and 45°C; and inlet drying air velocities of 1.0, 1.25, and 1.5 m/s. • The drying time reduced by about 2.25 times on increasing the IR power from 300 to 500 W, air temperature from 35°C to 45°C, and air velocity from 1.0 to 1.5 m/s. Effective moisture diffusivity varied from 0.21×10^{-10} to 1.57×10^{-10} m^2/s and was significantly influenced by IR power and air temperature. • The rehydration ratio of dehydrated onion slices was found to be in the range of 4.5–5.3.	Sharma et al. (2005)
Apple slices	• IR drying was conducted in the NIR region with a peak wavelength of 1200 nm. • The energy efficiency of the IR dryer was between 35% and 45%. • Kinetics of IR drying was dependent on the distance between emitters and the heat-irradiated surface and air velocity. • Comparison between IR drying and convective drying carried out under equivalent conditions showed that drying time can be shortened by up to 50% with IR drying.	Nowak and Lewicki (2004)

10.3.13 Heat Pump Dryer

10.3.13.1 Working Principle of Heat Pump Dryer

The concept of heat pump drying (HPD) is based on the thermodynamic principle of transferring heat from a low-temperature heat source to a high-temperature heat sink, using a device which consumes work. The device which uses work input in the form of mechanical energy (vapor-compression cycle) or heat (absorption cycle) to facilitate heat transfer in the reverse direction is known as the *heat pump.* Mechanical energy input is the most commonly used principle of heat pump operation. Thus, a heat pump can be considered as a closed-loop heat-lifting device or as a heat engine operating in reverse (Lu, 2007), working on the thermodynamic principle of vapor compression cycle.

10.3.13.2 Construction and Working of Heat Pump Dryer

A vapor compression heat pump dryer consists of five major components: the evaporator (dehumidifier), the compressor, a condenser (heater), an expansion valve, and a dryer attachment (see Figure 10.30). Generally, low temperature dehumidified air is used as the source of heat (drying medium) in HPD. These components form a closed cycle in which the drying air is condensed in the condenser to recover its latent heat and convert the same to sensible heat. The heat, thus, recovered is recycled back to the heat pump dryer by using it to heat the dehumidified drying air. The moisture in a food product is evaporated in the evaporator. The water vapor derived from the food product is mechanically compressed after which it is condensed in a heat exchanger to give up the heat for further evaporation of water. Finally, the condensate is withdrawn from the heat pump system.

The advantages and limitations and applications of heat pump dryer are summarized in Tables 10.8 and 10.9, respectively.

10.3.14 Refractance Window Dryer

10.3.14.1 Working Principle of Refractance Window Dryer

The name *Refractance window* provides a prelude to the dryer's principle. Water transmits heat within itself by all the three modes of heat transfer: conduction, convection, and radiation, of which, radiation is the fastest occurring at the speed of light. If the surface of the heated water is covered by a transparent medium such as plastic heat loss by evaporation is prevented and only conduction takes place (Bolland, 2000). The principle of refractance window (RW) drying is based on the finding that the IR transmission is stronger when the plastic interface is in intimate contact with hot water (typically at temperature between 93.3°C and 98.9°C (Bolland, 2000); higher temperatures create air bubbles which interfere with energy transfer through the belt (Clarke, 2004) circulating on one side and a moisture-laden material on the other side. Accordingly, when a product loaded with moisture (e.g., puree) is spread on top

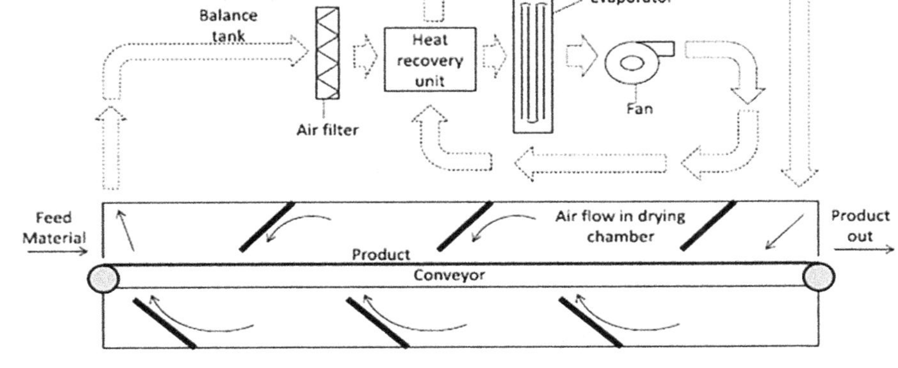

FIGURE 10.30 HPD system. (Adapted and reproduced with permission from Icier, F., Colak, N., Erbay, Z., Kuzgunkaya, E. H. and Hepbasli, A. 2010. A comparative study on exergetic performance assessment for drying of Broccoli Florets in three different drying systems. *Drying Technology* 28: 193–204 and Moses, J. A., Norton, T., Alagusundaram, K. and Tiwari, B. K. 2014. Novel drying techniques for the food industry. *Food Engineering Reviews* 6: 43–55.)

TABLE 10.8

Advantages and Limitations of Heat Pump Dryer

Advantages of Heat Pump Dryer	Limitations of Heat Pump Dryer
• HPD has the ability to make use of the low-grade heat in the humid air for useful operations, which leads to high energy efficiency and up to 60% reduction in energy costs compared to conventional drying methods. • Advantageous for heat-sensitive and high-value food products (e.g., saffron) which require low-humidity drying environments to minimize the loss of volatile organic compounds. • Require minimal regular maintenance and hence have a long life span without loss of efficiency. • Operation of heat pump dryer is clean, quiet, odorless, and environment friendly as it is free from contaminants that cause harm to the environment. • Easy installation and safe operation. • Continuous operation and consistent output of products are possible, thus, leading to a higher production potential than conventional drying techniques. • High product quality and attainment of desired product specifications owing to precise control of HPD conditions such as temperature and air flow rate. • Possibility for a wide range of drying conditions typically from –20°C to 100°C.	• The initial cost of HPD pertaining to its components such as the controllers, compressor, evaporators, and condensers may be higher compared to conventional dryers. • The compressor, condenser, and evaporator of a heat pump dryer need regular maintenance to achieve optimal performance of the dryers. • A steady-state period is required for the system to attain the desired drying conditions.

TABLE 10.9

Major Applications of Heat Pump Dryer in the Food Industry

Product	Findings	Reference
Onion	• When compared to drying by electrical heating without a heat pump, heat pump-assisted drying reduced the drying time by 40.7%. • Consequently, the electrical energy was saved by 40% with heat pump-assisted drying.	Rossi et al. (1992)
Bananas	• HPD is economically feasible. • It is the appropriate drying technique for high moisture foods.	Prasertsan et al. (1997); Prasertsan and Saen-saby (1998)
Banana, guava, and potato	• Intermittent HPD resulted in improved drying kinetics and color change.	Chua et al. (2000)
Lactic acid bacteria	• HPD was found to be more economical in terms of capital and running cost, compared to freeze-drying of lactic acid bacteria.	Cardona et al. (2002)
Mango	• Improved energy efficiency compared to an electrical resistance dryer.	Kohayakawa et al. (2004)
Plum	• The optimum temperature of drying was found to be 70°C–80°C. • The specific moisture extraction rate (SMER) of the heat pump dryer was significantly more than conventional dryers.	Chegini et al. (2007)

(Continued)

TABLE 10.9 ***(Continued)***

Major Applications of Heat Pump Dryer in the Food Industry

Product	Findings	Reference
Sapodilla pulp	• HPD resulted in shorter drying time compared to HA drying. • When conducted at high temperature, HPD reduced the sugar content and rehydration ratio of sapota powder.	Jangam et al. (2008)
Ginger	• Due to the low drying temperatures, ginger dried in HPD was found to retain >26% of gingerol (the principal volatile flavor component responsible for its pungency), compared to the rotary dried product which retained only about 20%.	Carrington (2008)
Fish	• The high quality of the dried fish with respect to porosity, rehydration rates, strength, texture and color, made possible by introducing a temperature controllable program to HPD.	Strommen and Kramer (1994)

of a thin and IR-transparent plastic surface, refraction at the plastic–product interface is minimized, causing the radiant thermal energy to pass through the plastic into the product. The absorptivity of the moisture-laden product is influenced by its thickness and moisture content (Ratti and Mujumdar, 1995). Thus, an IR *window* is created which acts as a passage for the IR energy created at the point of contact between the moist product and the plastic or where the plastic stands on the water. At this point, heat transfer occurs by all the three modes (Figure 10.31). The heat transfer from the product to the ambient air is initially by convection and evaporative cooling of the food product. This evaporation is very rapid and intense and constitutes a major part of energy consumption in RW drying. However, as the product dries, the IR "window" closes to render conduction as the predominant mode of heat transfer to the product. Consequently, the rate of heat transfer to the product slows as the product dries further. Also, since plastic is a poor conductor of heat, the product is safeguarded from overheating. Therefore, RW drying is a self-limiting process with the control elements inherently embedded in the system. Hence, it requires less intervention of the operator in maintaining the quality characteristics of the products to be dried.

Thus, RW drying is a method in which water transmits heat energy into the product and drives out the moisture from the product by evaporation. In other words, RW drying uses water as the tool to evaporate water from the product (Bolland, 2000). Figure 10.31 represents the whole scheme by which heat is transferred from the hot water to the product. As evident from the principle of RW drying, it falls under the category of indirect-contact or film-drying techniques.

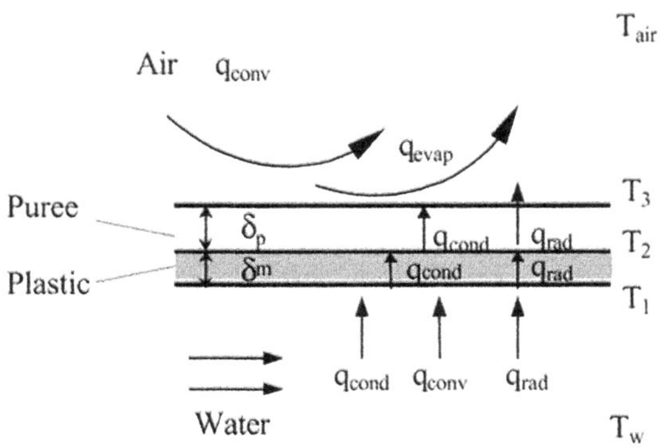

FIGURE 10.31 Schematic representation of heat transfer during RW drying. (Reproduced with permission from Nindo C. I. and Tang, J. 2007. Refractance Window dehydration technology: a novel contact drying method. *Drying Technology* 25: 37–48.)

10.3.14.2 Construction of a Refractance Window Dryer

An industrial RW drying unit is schematically represented in Figure 10.32. A commercial RW drying plant is made of modules, each of which is of length ~6 m and width ~2.4 m. At either end of the dryer, there is a liquid product application section and a dried product removal section. Hot water at atmospheric temperature is used as the source of heat for drying. The water in each module is continually recirculated and maintained at a pre-defined temperature as heat energy is transferred to the product.

A conveyor belt made from a transparent plastic material (e.g., polyester) spans the entire equipment and floats on the surface of hot circulating water. The belt speed can be adjusted up to 3 m/min, depending on the drying regime required. The product to be dried is sprayed on the surface of the conveyor belt, to a thickness of 0.2–1.0 mm. The IR energy of the water passes through the belt into the moist product. Subsequently, water from the product is released into the air above the belt and extracted by the air circulated above the belt. Therefore, as the product moves along the conveyor belt, its moisture level decreases. Consequently, heat transfer from the water by radiation slows down as the RW gradually closes due to the reduction in the water content of the product being dried. In the cooling section at the exit of the RW dryer near the end of the conveyor belt, cooling water is introduced at a temperature below the glass transition temperature (T_g) of the product to reduce the temperature of the dried product, limit its degradation, and facilitate its removal from the belt (Clarke, 2004; Nindo and Tang, 2007). The dry product is then scraped off the conveyor using a doctor blade which extends across the full width of the belt (Nindo and Tang, 2007). The dried product is obtained in the form of flake, sheet, or small particle at the product removal section of the RW dryer. Due to evaporative cooling, the product temperatures remain between 57.2°C and 71.1°C. If required, the flakes are subjected to size reduction before packaging (Bolland, 2000).

The advantages and limitations and applications of refractance window dryer are listed in Tables 10.10 and 10.11, respectively.

10.3.15 Hybrid Drying Techniques

Hybrid drying has been defined as the *successful integration or intelligent combination of two or more conventional drying methods* (Chua and Chou, 2014). It might combine different type of dryers or use more than one mode of heat transfer. Having understood the different methods of drying in the preceding sections, this part of the chapter is to provide an overview of hybrid drying techniques widely used in the food industry.

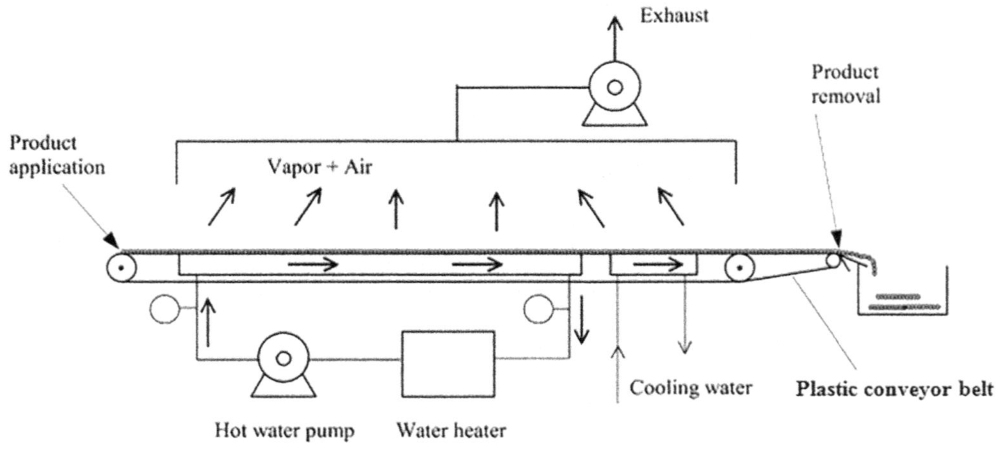

FIGURE 10.32 RW dryer. (Slightly modified and Reproduced with permission from Abonyi, B. I., Feng, H., Tang, J., Edwards, C. G., Chew, B. P., Mattinson, D. S. and Fellman, J. K. 2002. Quality retention in strawberry and carrot purees dried with Refractance Window System. *Journal of Food Science* 67: 1051–1056.)

TABLE 10.10

Advantages and Limitations of RW Dryer

Advantages of RW Dryer	Limitations of RW Dryer
• Compared to convective dryers that require drying time of 12–72 h, RW drying is completed in short drying times of 2–6 min to dry most products. • The product temperature is relatively low at ~70°C. • RW drying is capable of handling liquid food products without the need for wall/carrier materials. • RW drying has a competitive edge over tray drying, tunnel drying, drum drying and spray drying in terms of preserving the sensory quality parameters such as color, flavor, and aroma of heat-sensitive products (e.g., fruits, puree), in addition to the high retention of vitamins and anti-oxidants. • The capital and energy cost of RW drying is comparatively lower than conventional techniques, as it is operated under atmospheric pressure and comparatively lower temperatures. For example, to dry a similar product quantity, the cost of RW dryer is approximately one-third to one-half of that of a freeze dryer. Also, the energy costs to operate RW dryers are less than half of the freeze dryers. • Unlike direct dryers, cross-contamination is not encountered in RW system since the product is not in direct contact with the heat transfer medium.	• RW drying is not suitable for the drying of whole pieces of a food product. • The throughput of RW drying (~20 kg/h) may be lesser than conventional dryers, owing to their low-temperature operation.

TABLE 10.11
Applications of RW Dryer in Food Processing

Product	Conditions of RW Drying	Findings	Reference
Strawberry puree	• Air temperature: 20°C • RH of air: 52% • Air velocity: 0.7 m/s • Water temperature: 95°C • Belt speed: 0.45–0.58 m/min. • Thickness of puree application: ~1 mm. • Residence time of product on the drying bed: 3–5 min	• With RW drying, the retention of vitamin C was 94%, comparable to 93.6% in the freeze-dried product. • The color of RW-dried strawberry purees was similar to that of the freeze-dried product. • The overall flavor of strawberry purees was altered by RW drying, with a decrease in fruity and green notes.	Abonyi et al. (2002)
Carrot puree		• The color of RW-dried carrot product was similar to that of fresh puree. • The carotene losses for RW-dried puree were 8.7% (total carotene), 7.4% (α-carotene), and 9.9% (β-carotene) and were similar to the losses observed in the freeze-dried product.	
Pulp, juice, or sliced food (fruit or vegetable)	• Water temperature: 127°C	• Drying time: 3–10 min. • The rapid and gentle drying minimized the product degradation caused by heat and oxidation. • Maximum retention of color, flavor, and aroma. • The requirement for additives, such as sulfites, was eliminated by RW drying. • Dried product showed high bulk density.	Magoon (1985)
Tomato slices	• RW drying was conducted at two different temperatures: 75°C and 90°C and compared with hot air drying carried out at the same temperatures. • Air velocity: 0.1 m/s • Material of conveyor belt: polyester film	• The drying time in the hot air drier was 140 min at 75°C and 120 min at 90°C. But, the RW drier took 140 min to dry the samples at 75°C. However, the drying time was only 70 min at 90°C. The lower drying time with RW dryer at 90°C can be attributed to the involvement of all the three modes of heat transfer during RW drying. • Substantial changes in the total color difference (ΔE) occurred after hot air drying at 75°C, while the insignificant changes in ΔE were observed after RW drying at 90°C. • Least chemical alteration, especially with respect to phenol content in the product obtained from RW drying at 90°C.	Abbasid et al. (2015)

(Continued)

TABLE 10.11 (*Continued*)

Applications of RW Dryer in Food Processing

Product	Conditions of RW Drying	Findings	Reference
Asparagus	• Circulating water temperature: 95°C–97°C • Residence time of puree on the belt: 4.5 min.	• RW-dried pureed asparagus was closest in greenness to its freeze-dried counterpart. • RW and freeze-drying enhanced the total antioxidant activity of asparagus. • RW drying retained the highest amount of ascorbic acid compared to other drying methods.	Nindo et al. (2003)
Paprika	• Material of conveyor belt: Plastic • Temperature of hot water: 95–97°C • Drying time for each run: 3 min	• RW drying of paprika was compared with freeze-drying, HA oven drying, and natural convective drying. • With respect to reflected color, RW and freeze-drying resulted in products with better characteristics. • RW and freeze-drying resulted in samples with the lowest browning index, indicating the occurrence of less nonenzymatic browning reaction during these processes.	Topuz et al. (2009)
Mango pulp	• Material of conveyor belt: polyester and aluminum. • Thickness of mango pulp layer on belt: 2 mm • Water temperature: 95 ± 1.5°C	• While the product temperature during RW drying varied between 71°C and 75°C when dried on polyester sheet; the product dried on aluminum sheet reached 80°C. • Drying time: 10 min • Higher drying rate was observed for aluminum sheet (0.0029 kg water/m^2s) owing to its high thermal conductivity (200 W/m K) compared to polyester (0.2 W/m K) which resulted in a maximum drying rate of 0.0024 kg water/m^2s.	Kaur et al. (2017)
Kiwifruit	• Drying temperatures: 80°C–100°C • Slice thickness: 0.8–2.4 mm • Thickness of polyester film: 100–300 μm	• Drying time: 30–60 min under different drying temperatures. • Most of the organoleptic properties of RW-dried kiwifruits, including flavor, color, texture, shrinkage and overall acceptance, were judged to be of medium to good quality (scores: 3–4, on a 5-point scale; 1 = very bad, 2 = bad, 3 = mediocre, 4 = good, 5 = very good) by panelists.	Azizi et al. (2016)

10.3.15.1 Spray-Freeze-Drying

Spray-freeze-drying (SFD) is a synergistic drying technique between spray and freeze-drying. SFD is a three-step operation: (i) atomization, (ii) freezing, and (iii) drying (Figure 10.33). Atomization generates fine droplets of core emulsion or suspension formed in a solution of suitable wall material, resulting in a large air–liquid interfacial area which promotes rapid freezing on contact with a cryogenic liquid at ultra-low temperature; freezing concentrates the core compound in the wall matrix by solidifying the solvent (often water) into tiny ice crystals, size of which depends on the freezing rate. This technique emerged based on the relationship that drying times vary approximately with the square of the sample thickness. Reducing the sample dimension by atomization and thus increasing the surface mass transfer coefficient during the ensuing rapid freezing and freeze-drying steps are central to this process. The finely divided droplet size and a large temperature gradient between the droplet and surrounding cold gas drive the rapid cooling and freezing.

The freezing step during the SFD process can be carried out in either of the following three modes:

i. Spray freezing into vapor (SFV): Involves the atomization of a feed solution containing dissolved solids and contacting the spray with cold desiccated gas to freeze the droplets (Figure 10.34).
ii. Spray freezing into vapor over liquid (SFV/L): Feed solution is atomized through a nozzle positioned at a distance above a boiling cryogenic liquid (Figure 10.35). The droplets may begin to solidify while passing through the vapor gap and then freeze completely as contact is made with the liquid (Sauer, 1969; Dunn et al., 1972; Briggs and Maxwell, 1976).
iii. Spray freezing into liquid (SFL): The nozzle is inserted beneath the surface and directly into a cold liquid (Figure 10.36), which can be liquid nitrogen, argon, hydrofluoroether, or pentane. SFL works on the principle of liquid–liquid impingement between the atomized (pressurized) feed solution exiting the nozzle and the cryogenic liquid. The droplets freeze instantaneously after they are formed. An impeller may be used to stir the cryogenic liquid containing the sprayed product, to avoid particle agglomeration.

The subsequent freeze-drying step can be carried out by any of the following methods :

i. Vacuum freeze-drying: Operates on the same principle as that of freeze drying explained in section 10.3.8, except for that the processing time is less due to the reduced sample dimensions.

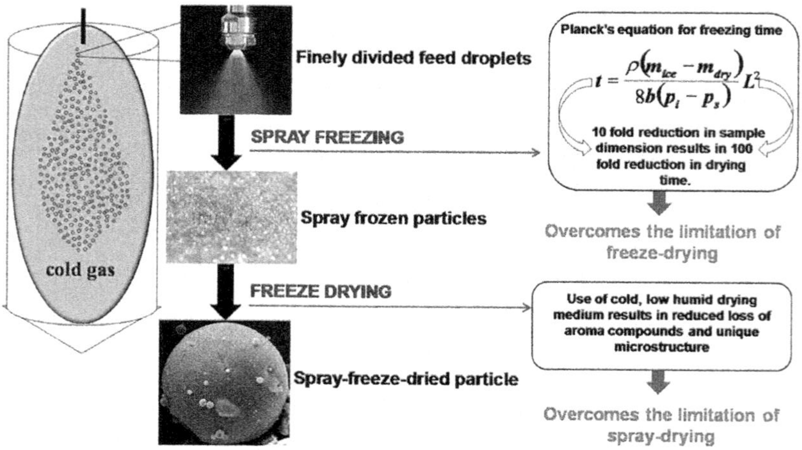

FIGURE 10.33 Principle of SFD process. (Reproduced with permission from Ishwarya, S. P., Anandharamakrishnan, C. and Stapley, A. G. F. 2017. Spray freeze drying. In *Handbook of Drying for Dairy Products*, ed. C. Anandharamakrishnan, 123–148. Chichester, West Sussex, UK: John Wiley and Sons.)

FIGURE 10.34 Photograph of spray-freeze-dryer with SFV arrangement. (Reproduced with permission from Anandharamakrishnan, C. 2008. Experimental and computational fluid dynamics studies on spray-freeze-drying and spray-drying of proteins. *PhD Thesis*. UK: Loughborough University.)

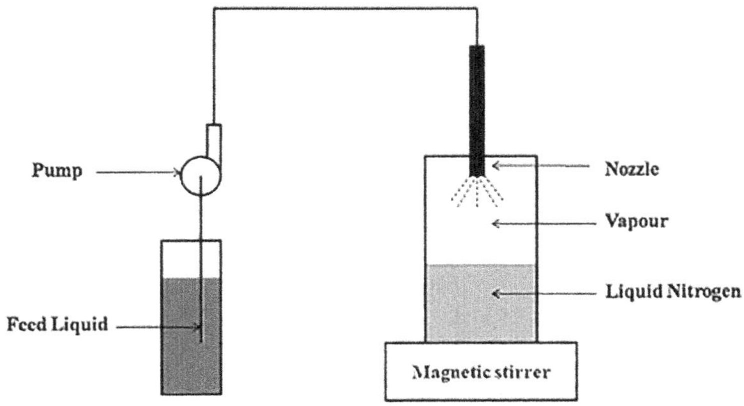

FIGURE 10.35 Schematic diagram of SFV/L. (Reproduced with permission from Anandharamakrishnan, C. 2008. Experimental and computational fluid dynamics studies on spray-freeze-drying and spray-drying of proteins. *PhD Thesis*. UK: Loughborough University.)

ii. Atmospheric fluidized bed freeze-drying: This method is based on the theory that vacuum is not the absolute requirement for freeze-drying to take place and drying rate is a function of only ice temperature and vapor pressure gradient between the site of water vapor formation and the drying media rather than the total pressure in the drying chamber (Meryman, 1959). In an atmospheric SFD apparatus (Figure 10.37), cold gases such as air, nitrogen, or helium can be used to cause sublimation of moisture from a frozen product at or near atmospheric pressure (Mumenthaler and Leuenberger, 1991).

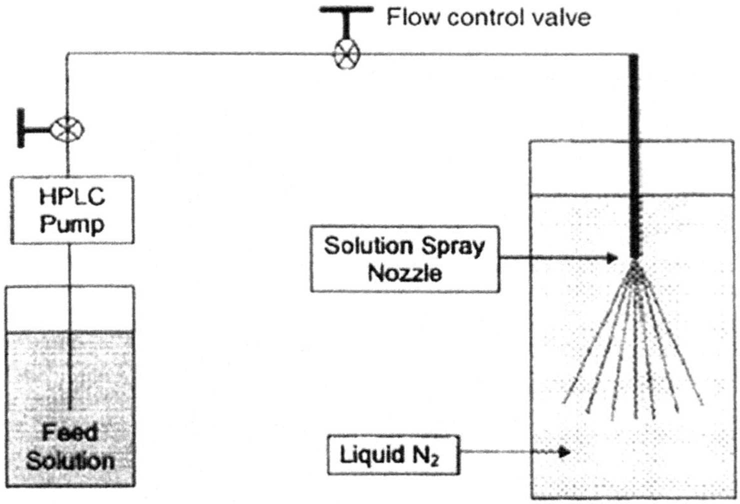

FIGURE 10.36 Schematic diagram of SFL. (Reproduced with permission from Rogers, T. L., Hu, J., Hu, Z., Johnston, K. P., and Williams, R. O. 2002. A novel particle engineering technology: spray-freezing into liquid. *International Journal of Pharmaceutics* 242: 93–100.)

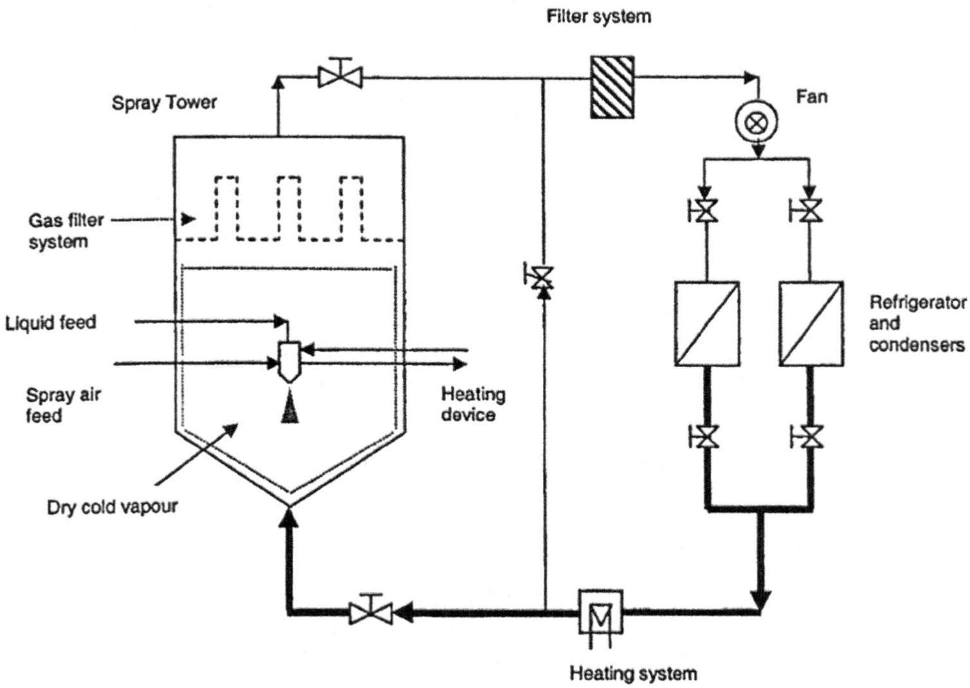

FIGURE 10.37 Schematic diagram of atmospheric SFD apparatus. (Reproduced with permission from Leuenberger, H. 2002. Spray Freeze drying—The process of choice for low water soluble drugs? *Journal of Nanoparticle Research* 4: 111–119.)

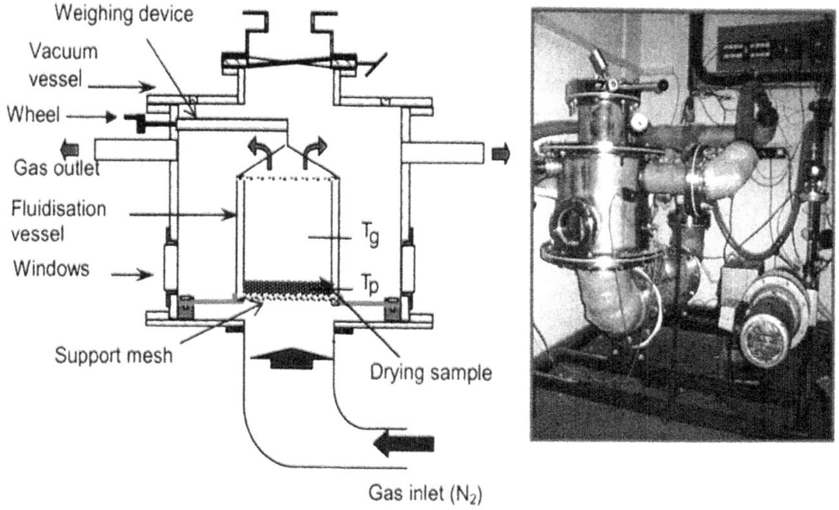

FIGURE 10.38 Subatmospheric fluidized bed freeze-drying apparatus (Slightly modified and Reproduced with permission from Anandharamakrishnan, C. 2008. Experimental and computational fluid dynamics studies on spray-freeze-drying and spray-drying of proteins. *PhD Thesis*. UK: Loughborough University.)

iii. Sub-atmospheric fluidized bed freeze-drying: When the pressure of the system is reduced, the mass of gas required for the fluidized bed freeze-drying process is also reduced by the same factor. This is the concept that governs the sub-atmospheric fluidized bed freeze-drying technique (Figure 10.38). On the other hand, the density of the gas can also be reduced which would decrease the inertial or non-viscous drag forces on the particles and prevent the particles from being carried away from the bed (Anandharamakrishnan et al., 2010). Besides, the sub-atmospheric pressure causes a reduction in drying time due to the higher fluidization velocity of the particles (Leuenberger et al., 2006). Low-pressure operation increases the temperature gradient due to its influence on heat transfer coefficient between the gas and particle bed. Larger differences are also possible between the inlet and outlet gas temperatures (Anandharamakrishnan et al., 2010).

A spray-freeze-dryer combines the aspects of both spray and freeze dryer. Consequently, the resultant product also shares the characteristics of spray- and freeze-dried products, namely, the free-flowing ability and a porous microstructure, respectively. Coffee, milk, maltodextrin, and dairy products such as whey protein are the products that have been dried using spray-freeze-dryer. The advantages and limitations of the spray-freeze-drying process are explained in Table 10.12. Further, the applications of SFD in food processing are summarized in Table 10.13.

10.3.15.2 Spray-Fluidized Bed Drying

Spray-fluidized bed drying (SFBD) is a hybrid between the spray drying and fluidized bed drying techniques. The major application of this hybrid drying technique is the coating of dry solid particles of an active ingredient with a carrier material. The active ingredient is usually termed the *core*, and the carrier material is known as the *wall* (Anandharamakrishnan and Ishwarya, 2015). The process is similar to spray drying in terms of the atomization step. However, SFBD differs from spray drying in that it involves all three states of matter, i.e., solid (core), liquid (wall solution), and gas (fluidization medium), which is not so in the latter. The sequence of events during SFBD is shown in Figure 10.39. Fluidization of the core material and atomization of the solution of wall matrix occur simultaneously such that the wall material spreads on the surface of the core particles, to form a uniform coating after repeated cycles

TABLE 10.12

Advantages and Limitations of the SFD Process

Advantages of SFD	Limitations of SFD
• SFD leads to spherical shape and porous microstructure of the product resulting in superior rehydration properties [rehydration time: 30–50 s (DHA); 10–12 s (coffee)] than freeze-drying [36–48 s (DHA); 21–23 s (coffee)] and spray drying [89–102 s (DHA); 19–21 s (coffee)]. • The percentage of oxidation during drying and the oxidative stability under different storage conditions after drying (of DHA) also proved SFD to be an advantageous technique over spray drying and freeze-drying (33% oxidation). A more than twofold reduction in % oxidations has been observed. • SFD reduced the processing time almost by 2.75 times compared to the conventional freeze-drying process (Karthik and Anandharamakrishnan, 2013; Ishwarya and Anandharamakrishnan, 2015).	• High fixed and operating cost (Wolff and Gibert, 1990) due to the energy-intensive operation demanded by the requirement of vacuum and the batch mode operation (Di Matteo et al., 2003). • In the atmospheric SFD, the cost per kg of the dry product depends on the drying temperature. Hence, the cost increases drastically with feed materials whose frozen solution have a low eutectic (T_e) or glass transition temperature (T_g) and hence require a low drying temperature. • The available prototypes of spray freeze dryers (Mumenthaler and Leuenberger, 1991) are found to be too small for the freeze-drying of foods, owing to the indeterminate T_e and T_g values of most of the food products (Leuenberger, 2002). • The disadvantage with SFV/L, when used for biological products, is the loss of protein stability during atomization. Atomization in ambient gas creates a large gas–liquid interfacial area for protein adsorption and unfolding (Webb et al., 2002).

of wetting and drying. Therefore, SFBD is governed by the particle dynamics of the core substance in the fluidized bed and a combination of heat and mass transfer.

10.3.15.2.1 Atomization

Atomization during SFBD is similar to that in spray drying. However, in addition to fluidized bed coating, atomization is the decider step in determining the droplet impact with the core particle surface and the subsequent wetting and drying that eventually results in a homogeneous coating. The atomization parameters of importance in the context of fluidized bed coating are the angle and penetration length of the atomized jet, the spatial distribution of the resultant droplets and their size, and size distribution. However, the abovementioned factors depend on the atomization energy and the flow rate of the wall material solution.

Higher the atomization energy (pressure in case of pneumatic nozzles and wheel speed in case of rotary atomizers), smaller the spray angle and mean droplet size. A shorter spray angle confines the size of the atomization zone and hence would limit the probability of contact between the atomized wall matrix and the fluidized core particles. Smaller mean droplet diameter results in a higher droplet penetration length but owing to greater surface area undergoes a premature drying by the influence of the hot fluidization air stream prior to reaching the droplet surface. On the contrary, a higher feed flow rate results in the formation of bigger droplets. Although a higher flow rate can be translated as shorter wetting time, it also leads to a longer drying time due to the lower temperature of the fluidized bed. This would ultimately result in a lower production rate. For an efficient encapsulation, the droplet mean diameter should be lesser than that of the particles in the bed. But, too small a droplet diameter would lead to premature droplet evaporation, which results in loss of coating material and lower encapsulation efficiency. While larger droplet diameter reduces the premature droplet evaporation, it renders the drying difficult and eventually results in stickiness.

Depending on the position of the atomizer, SFBD is classified into three types: top SFBD (Figure 10.40a), bottom SFBD (Figure 10.40b), and side SFBD (Figure 10.40c).

A top SFBD (Figure 10.40a) comprises an air handling system with appropriate humidity control to supply the fluidization air to the bed, a product container to hold the core particulates until they are

TABLE 10.13

Applications of SFD in Food Processing

Application	Temperature of Spray Freezing (°C)	Temperature of Freeze-Drying (°C)	Reported Physical Properties	Reference
Development of dried powder of whey protein	−85°C ± 2°C	−10°C	• Moisture content (%): 8.1 ± 2.27 • Particle diameter (μm): 480 ± 53	Anandharamakrishnan et al., (2010)
Development of dried whey powder	−45°C ± 5°C	Self-regulated by sublimative cooling.	• Moisture content (%): 4.11 • Absolute density (g/cm^3): 0.41 • Bulk density (g/cm^3): 0.22 • Insoluble matter (%): ND[b]	Al-Hakim and Stapley, (2004)
Dried coffee powder	−60°C ± 5°C	<−60°C	• Moisture content (%): 15.0	Khwanpruk et al. (2008)
Dried maltodextrin powder	−60°C ± 5°C	<−60°C	• Moisture content (%): 10.5	Khwanpruk et al. (2008)
Production of skimmed and whole milk powders	−8°C to −25°C	-NM-	• Mean particle diameter (μm): 415 • Mean wetting time (s): 2.31	Rogers et al. (2008)
Drying of apple juice	Frozen by liquid nitrogen (−196°C[a])	−34°C	• Size of frozen droplet (μm): 250–600 • Moisture loss (% per hour): 0.6	Malecki et al. (1970)
Drying of egg albumin	Frozen by liquid nitrogen (−196°C[a])	−20°C	• Size of the frozen droplet (μm): 250–600 • Moisture loss (% per hour): 6.5.	Malecki et al. (1970)
Microencapsulation of docosahexaenoic acid with whey protein isolate as carrier material	Frozen by liquid nitrogen (−196°C[a])	−24°C	• Moisture content (%): 3.667 ± 0.014 • Oil content (%): 71	Karthik and Anandharamakrishnan, (2013)
Production of instant coffee powder	Frozen by liquid nitrogen (−196°C[a])	Temperature range of primary drying stage was −25°C to −10°C and secondary drying was carried out at 10°C.	• Moisture content (%): 8.665 ± 0.001 • Solubility time (s): 11 ± 1 • Mean volume diameter (μm): 91.1 • Free bulk density (g/mL): • 0.612 ± 0.007 • Tapped bulk density (g/mL): • 0.679 ± 0.008	Ishwarya and Anandharamakrishnan, (2015)

Source: Adapted from Ishwarya et al. (2015).

[a] Boiling point of liquid nitrogen.

[b] Not detected.

NM: Not mentioned.

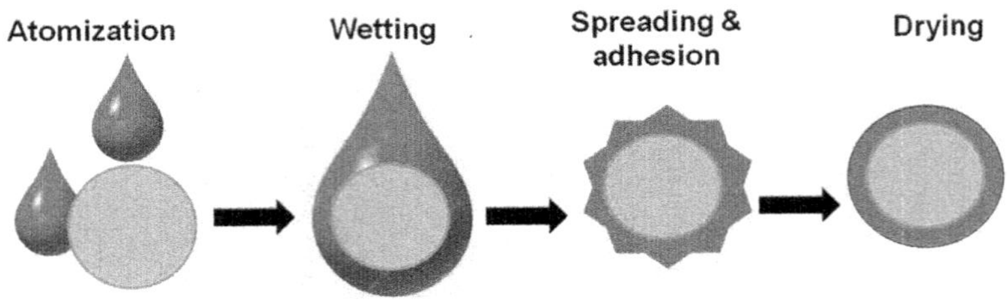

FIGURE 10.39 Sequential events during SFBD. (Reproduced with permission from Anandharamakrishnan, C. and Ishwarya, S. P. 2015. *Spray Drying Techniques for Food Ingredient Encapsulation*. Hoboken, NJ: John Wiley and Sons, Inc.)

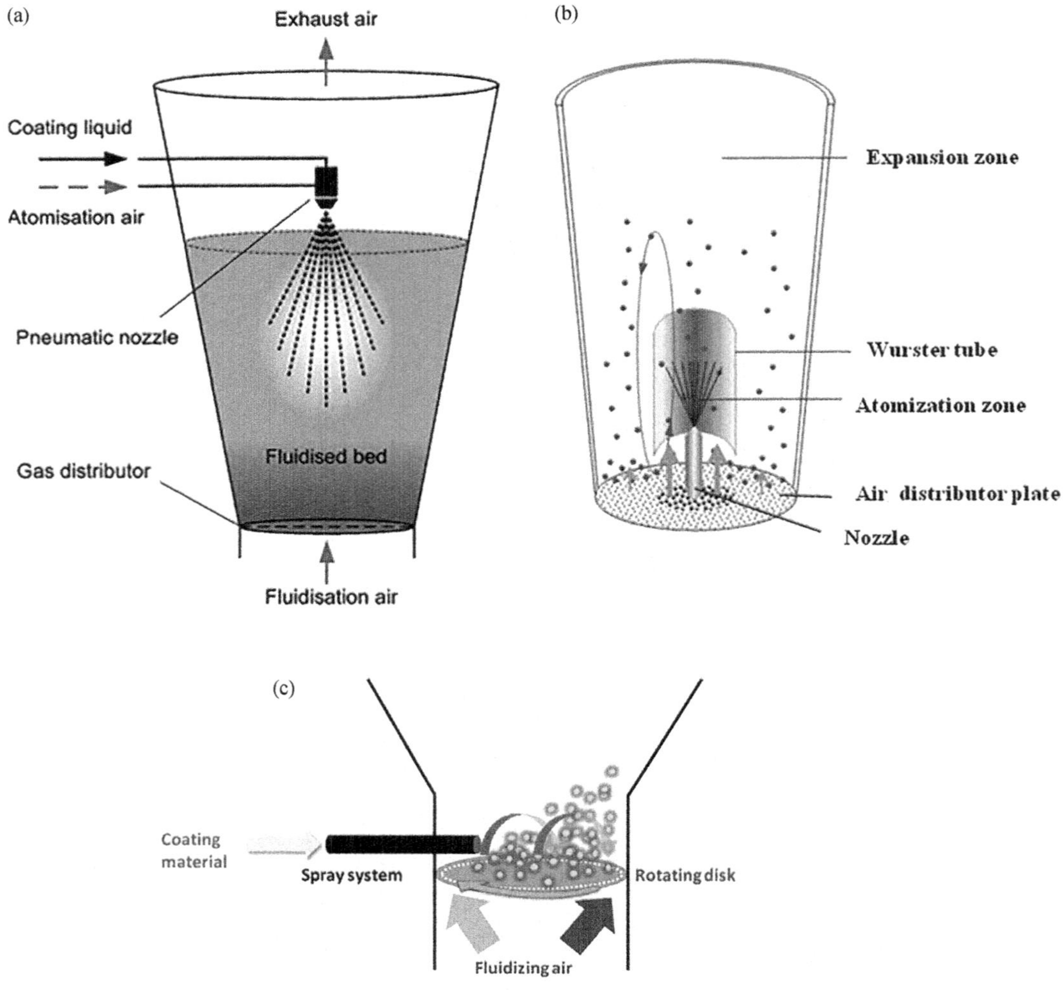

FIGURE 10.40 Schematic diagram of (a) top SFBD (Reproduced with permission from Ronsse, F., Pieters, J. G. and Dewettinck, K. 2008. Modelling side effect spray drying in top-spray fluidised bed coating process. *Journal of Food Engineering* 86: 529–541.); (b) bottom SFBD (Adapted and reproduced with permission from Guignon, B., Duquenoy, A. and Dumoulin, E. D. 2002. Fluid bed encapsulation of particles: principles and practice. *Drying Technology* 20: 419–447.); (c) side SFBD. (Reproduced with permission from Burgain, J., Gaiani, C., Linder, M. and Scher, J. 2011. Encapsulation of probiotic living cells: from laboratory scale to industrial applications. *Journal of Food Engineering* 104: 467–483.)

accelerated for fluidization, collection of coated particles before they are recycled through the coating zone, an expansion chamber, and a filtration system to trap the entrained coating substance accumulated due to their fine droplet size or premature droplet evaporation (Srivastava and Mishra, 2010). This configuration facilitates a countercurrent interaction between the atomized droplets of the coating solution and the fluidized core particles. Thus, it is difficult to control the residence time of the droplets through the fluidized bed. Therefore, premature droplet evaporation and loss of coating material become inevitable in this process. Combating this challenging process environment is possible by a careful monitoring of drop size by selection of suitable atomization parameters. During a top spray-fluidized bed coating process, the size and velocity of the coating droplets are the limiting parameters that determine the coating efficiency. A higher atomization pressure can negatively impact the coating efficiency by premature droplet evaporation due to smaller drop size and decline in fluid-bed temperature, by affecting the initial drop velocity which has an influence on the probability of contact between the drop and the particle to be coated (Dewettinck and Huyghebaert, 1998). Nevertheless, in spite of the aforementioned limitations, top spray fluidization is the predominantly used configuration in the food industries due to their relatively larger batch size and simple operation (Dewettinck, 1997). Coating of substrates as small as 100 μm has been achieved with this configuration (Jones, 1988) and encapsulation of core particles in the size range of 2–5 μm is possible by adsorbing them on a coarser carrier material (Thiel and Nguyen, 1984).

Bottom spray mode exhibits a cocurrent operation, where the nozzle position is at the bottom of the bed close to the distributor plate (Figure 10.40b). Also known as Wurster system, the unique features of this configuration include a draft tube arrangement known as the Wurster tube which serves to organize the fluidization of particles through the atomization zone. The draft tube is an insert, placed coaxially in the bed to guide the particles that are carried in the upward direction by the fluidizing gas stream through it and then fall down around it at the top. This facilitates the circulation of the particles, thus increasing the drying rate and reducing agglomeration, ultimately leading to a smooth and continuous coating enclosing the core particles (Anandharamakrishnan and Ishwarya, 2015).

Another mode is the side spray configuration, where the spray is tangential to the movement of particles (Figure 10.40c). It is similar in principle to the bottom spray process, except to that the particle movement is aided by a motor-driven rotating disc. The rolling pattern of particle movement prevents agglomeration due to intense separation forces (Anandharamakrishnan and Ishwarya, 2015).

10.3.15.2.2 Wetting

Wetting is defined as the impact or collision of drop on the particle surface. The relevant drop characteristics with respect to wetting are its size, viscosity, density, temperature, and surface tension relative to the particle surface. The relationship between wetting and the drop surface tension can be explained by the expression for the wetting energy, W_m (Eq. (10.48)).

$$W_m = -\sigma \operatorname{Cos}\theta \tag{10.48}$$

where σ is the surface tension and θ is the contact angle between the drop and particle surface (Figure 10.41). Therefore, coating is not possible when the angle between the drop and particle

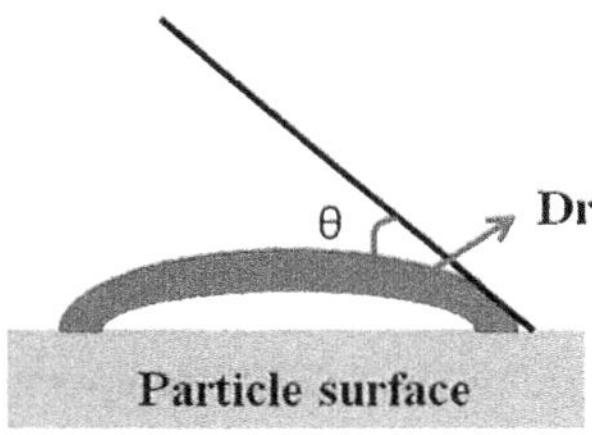

FIGURE 10.41 Contact angle between drop and particle surface on impact. (Reproduced with permission from Anandharamakrishnan, C. and Ishwarya, S. P. 2015. *Spray Drying Techniques for Food Ingredient Encapsulation*. Hoboken, NJ: John Wiley and Sons, Inc.)

surface is 90°. The compatibility between the drop and particle surface increases when the impact is energetically favorable, i.e., a negative wetting energy which is promoted by a combination of an aqueous solution of wall material and a hydrophilic particle surface.

10.3.15.2.3 Spreading and Adhesion

Spreading is governed by Reynolds number $\left(N_{Re} = du\rho/\mu\right)$, Weber number $\left(N_{We} = \rho u^2 d/\sigma\right)$, and Ohnesorge number $\left(N_{oh} = \sqrt{We}/Re\right)$. The influence of different process parameters on droplet-particle interactions at different stages can be identified and quantified by the previously stated dimensionless numbers. Further to elaborate, the extent of spread depends on the Weber number which plays a role during the initial stages of the droplet recoiling or bouncing phenomenon. There arises a capillary force due to the difference between the initial equilibrium state and that after spreading, which then takes over to assist the recoiling process while the flow is continuously resisted by inertia and viscosity. This resisting force can be estimated by Ohnesorge number. When Ohnesorge number is small, the resisting forces are contributed by both inertia and viscosity; whereas, at large Ohnesorge number, viscosity plays the significant role (Kim and Chun, 2001). The dependence of spreading on the abovementioned dimensionless numbers shows that the droplet velocity is the principal factor in deciding the spreading efficiency.

In addition to velocity, the droplet surface tension and viscosity also influences the spreadability. A higher surface tension translates to a strong withdrawal force that interrupts the contact between the droplet and particle surface and enhances the possibility of re-bounce. This can be overcome by the use of surfactants that adsorb at the air-droplet interface thus reducing the droplet surface tension and lowering the contact angle (Werner et al., 2007).

Fluid viscosity affects the spread diameter. A drop with higher viscosity must overcome more dissipation energy during its spreading process. Therefore, the final spread for a given droplet impact velocity decreases inversely with an increase in the liquid viscosity (de Gennes, 1985). In addition, highly viscous solutions present difficulty during atomization.

After impact and spreading, the droplet adheres either to the particle surface or to the coating material previously deposited on the particle surface. The adhesion mechanisms can be classified into four types: pressure, thermal adhesion, diffusion, and chemical reaction (Sobolev et al., 1997). Adhesion under the pressure of impact is due to the close contact between the drop and particle surface, contributed by the van der Waals forces between the nearest neighboring molecules. The roughness of the particle surface due to the presence of pores also contributes to the impact. When pores are present, capillary suction pressures come into existence; the pressure of impact should be able to overcome the capillary pressure so as to obtain a uniform coating. Also, the roughness increases the hydrophobicity of the particle surface, which causes the drop to bounce more intensely than that on the smooth surface (Andrade et al., 2013). The probability of adhesion depends on the kinetic momentum of the drop, characteristics of the surfaces in contact, and physicochemical properties of both the coating solution and the particle. Here, the complexity is induced by the dynamicity in the abovementioned characteristics all along the drying phase (Link and Schlunder, 1997).

10.3.15.2.4 Drying

Hot air is the frequently used drying medium. The driving force for drying is the temperature difference between the three states of matter involved in the fluidized bed coating process. But, under isothermal conditions, the mass transfer phenomenon is driven by the difference in partial water vapor pressure between the droplet that has spread on the particle surface and the drying air. The mass transfer coefficient of the liquid film at the droplet-air interface governs the mass transfer flux. The water vapor pressure of the liquid film is related to its water activity which continues to decrease during drying (Masters, 1988). Thus, the property of this liquid film depends upon the evaporation rate of the wall material used. In turn, the rate of evaporation is governed by the velocity, temperature, and humidity of the fluidization air. Temperature is a critical factor as improper temperature control will result in an incomplete coating of the core material leading to a poor loading efficiency of the microencapsulates.

The volume of the fluidization air also holds importance in terms of controlling the height of the particulate bed in contact with the air stream. This determines the residence time of the fluidized particles in the atomization zone and, thereby, the time for drying or coating. Similar to spray drying, the drying rate also depends on the particle size distribution of the resultant coating material–particle combination, post-droplet–particle interactions. Heat transfer between a single particle and the fluidization air can be given by the conventional equation for convective heat transfer, i.e.,

$$q = h_p A_p \left(T_p - T_g \right) \tag{10.49}$$

where q is the heat transfer rate (W), h_p is the heat transfer coefficient (W/m^2 K), A_p is the surface area of the single particle (m^2), T_p is the temperature of the particle (K), and T_g is the temperature of fluidization air (K).

Further, the optimum circulation of particles to be dried in the bed increases the drying rate which promotes an efficient and uniform heat and mass transfer. The end point of an SFBD process is indicated by a rapid increase in the outlet air temperature and equality between the dew point of outlet and inlet air streams (Hede, 2013).

The advantages and limitations of SFBD are summarized in Table 10.14.

10.3.15.2.5 Applications of SFBD in Food Processing

A number of preservatives, functional, and nutraceutical additives have been encapsulated using the SFBD technique. These include ascorbic acid, acetic acid, vitamins and minerals, acidulants for processed meat, leavening agents used in the bakery industry, and probiotics used in the dairy sector.

The other commonly used hybrid drying techniques with respect to food processing are as follows:

- MW hot air drying
- MW vacuum drying
- RF-assisted heat pump drying
- IR-assisted hot air drying
- IR-assisted HPD

The working principle and advantages of the individual drying techniques involved in the previously listed hybrid drying methods have already been explained in the preceding sections. Therefore, the schematic diagrams of dryers in the hybrid configuration and their food processing applications are presented in Table 10.15 and Table 10.16, respectively.

TABLE 10.14

Advantages and Limitations of SFBD

Advantages of SFBD	Limitations of SFBD

• The choice of carrier materials with SFBD is wide (Gouin, 2004), including:
  • gums (gum arabic, gum tragacanth)
  • hydrocolloids (methyl cellulose, hydroxypropyl methylcellulose, carboxymethylcellulose)
  • starch derivatives (maltodextrin)
  • proteins (casein, whey protein)
  • lipids (waxes, hydrogenated vegetable oil, fatty acids, and alcohols)
  • Synthetic polymers (polymethacrylate, ethylcellulose)—All as aqueous solutions or in molten form.
• The wider choice for carrier materials is considered as an advantage over spray drying, where the availability of wall materials is limited due to short of substances with the unique drying characteristics (Anandharamakrishnan and Ishwarya, 2015).
	• High-temperature operation and direct exposure of the active components to hot air for repeated cycles. This would be deleterious for heat-labile core compounds that are susceptible to oxidative degradation at high temperatures.
	• Submicron-sized particles (with a mean diameter less than 30 μm) cannot be handled by SFBD as they are not fluidizable owing to their cohesive nature due to large interparticle forces that outweigh the drag force exerted by the fluidization gas (Anandharamakrishnan and Ishwarya, 2015).

TABLE 10.15

Schematic Diagrams of Hybrid Dryers[a]

1. MW-HA dryer (Roknul et al., 2014)

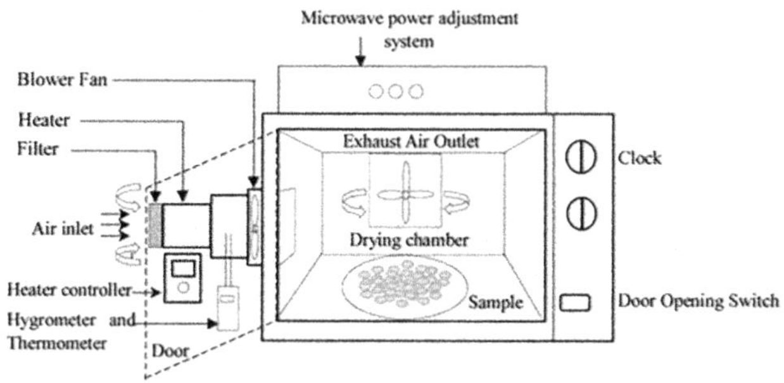

2. MW-vacuum dryer (Monteiro et al., 2015)

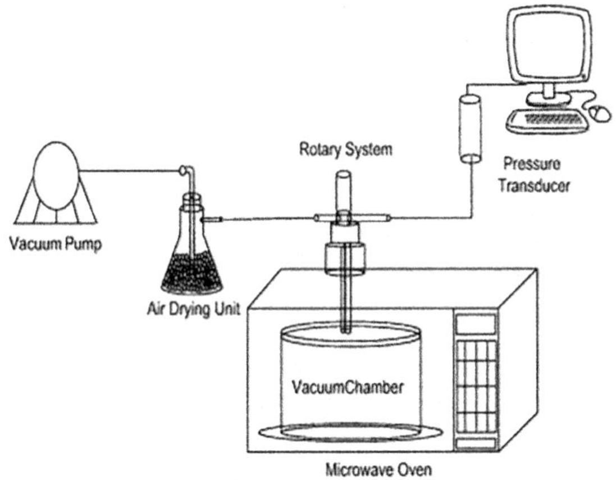

3. RF-assisted heat pump dryer (Marshall and Metaxas, 1998)

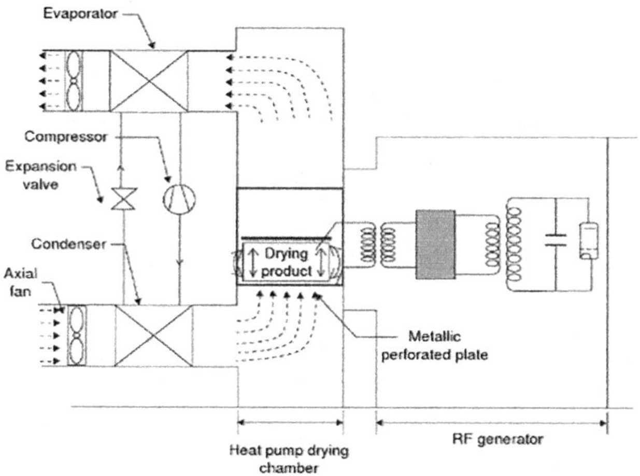

4. Hybrid IR-HA dryer (Vishwanathan et al., 2013)

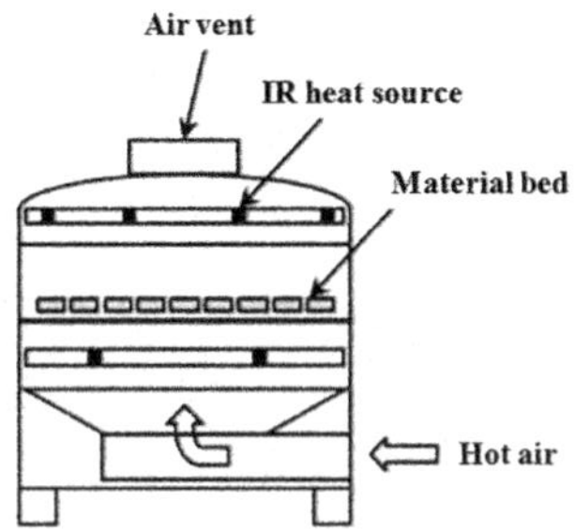

(Continued)

TABLE 10.15 (*Continued*)

Schematic Diagrams of Hybrid Dryers[a]

5. IR-assisted heat pump dryer (Jangam and Mujumdar, 2012)

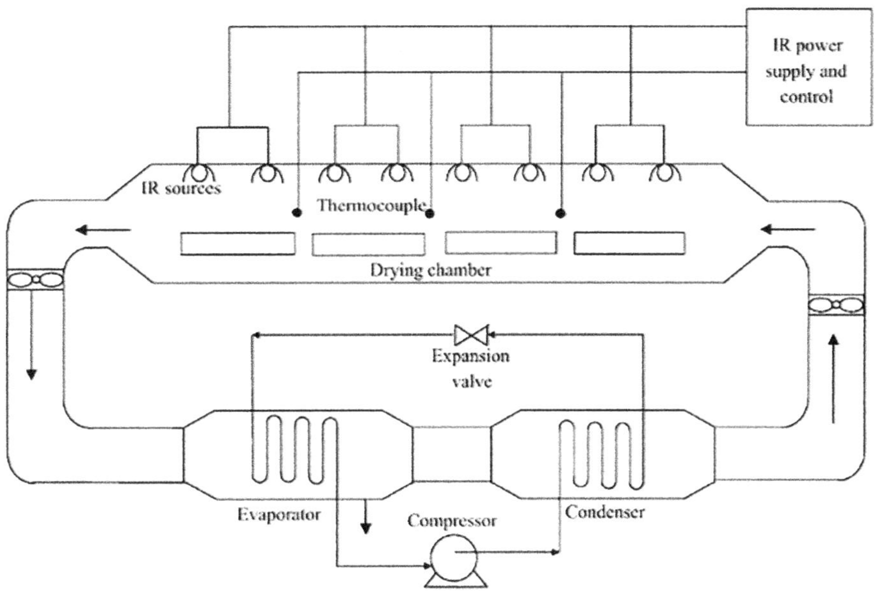

[a] MW–HA dryer (Reproduced with permission from Roknul, A. S. M., Zhang, M., Mujumdar, A. S. and Wang, Y. 2014. A comparative study of four drying methods on drying time and quality characteristics of stem lettuce slices (*Lactuca sativa* L.). *Drying Technology* 32: 657–666.); MW–vacuum dryer (Reproduced with permission from Monteiro, R.L., Carciofi, B. A. M., Marsaioli, A. and Laurindo, J. B. 2015. How to make a MW vacuum dryer with turntable. *Journal of Food Engineering* 166: 276–284.); RF-assisted heat pump dryer (Reproduced with permission from Marshall, M. G. and Metaxas, A. C. 1998. Modeling of the radio frequency electric field strength developed during the RF assisted HPD of particulates. *The Journal of Microwave Power and Electromagnetic Energy* 33: 167–177.); hybrid IR-HA dryer (Reproduced with permission from Vishwanathan, K. H., Giwari, G. K. and Hebbar, H. U. 2013. IR assisted dry-blanching and hybrid drying of carrot. *Food and Bioproducts Processing* 91: 89–94.); IR-assisted heat pump dryer (Reproduced with permission from Jangam, S. V. and Mujumdar, A. S. 2012. Heat pump assisted drying technology—Overview with focus on energy, environment and product quality. In *Modern Drying Technology* (Volume 4), eds. E. Tsotsas and A. S. Mujumdar, 121–162. Weinheim: Wiley-VCH, Verlag GmBH and Co.)

TABLE 10.16

Applications of Hybrid Drying Techniques in Food Processing

Type of Hybrid Drying	Synergistic Advantage	Applications in Food Processing

MW-Hot air (HA) drying	• When MW drying is used in combination with conventional air dryers, the direct dissipation of electromagnetic energy-dried material leads to significant reduction of heat losses.
	• MW-HA drying leads to a substantial increase in drying rate and drying capacity, proportional to each unit of electromagnetic energy added.
	• MW-HA drying is more appropriate for the drying of low-moisture content products, wherein convective heat transfer is less efficient.
	• During MW-HA drying, the air stream carries away the moisture evaporated during MW irradiation. This overcomes the economic limitations when MW drying is used individually.
	• MW is applied to a product after HA drying at a stage when its moisture content is so low such that convective drying is inefficient in removing the residual moisture.
		• Pumpkin slices (Alibas et al., 2007):
		  • MW power: 350 W
		  • Air velocity: 1 m/s
		  • Air temperature: 75°C
		• Apple (Funebo and Ohlsson, 1998):
		  • MW power: 0.5 W/g
		  • Air velocity: 1 m/s
		  • Air temperature: 60°C
		• Mushroom (Funebo and Ohlsson, 1998):
		  • MW power: 0.5 W/g
		  • Air velocity: 1.5 m/s
		  • Air temperature: 80°C
		• Banana slice (Maskan, 2000):
		  • MW power: 350 W
		  • Air velocity: 1.45 m/s
		  • Air temperature: 60°C
		• Kiwi fruits (Maskan, 2001):
		  • MW power: 210 W
		  • Air velocity: 1.29 m/s
		  • Air temperature: 60°C
		• Carrots (Prabhanjan et al., 1995):
		  • MW power: 120 and 240 W
		  • Air velocity: 1.7 m/s
		  • Air temperature: 45°C and 60°C
		• Fish fillets (Duan et al., 2011):
		  • MW power: 400–600 W
		  • Air velocity: 2 m/s
		  • Air temperature: 50°C

MW vacuum drying (MVD)	• Microwave vacuum drying (MVD) results in better sensory properties than conventional vacuum freeze-drying (carrot).
	• MVD leads to simultaneous moisture removal and significant improvement in water absorption capacity (starch).
	• MVD leads to lower energy consumption and shorter drying time than MW freeze-drying (mixed apple with potato chips).
	• During MVD, drying is much faster towards the end of the process (tomatoes).
	• MVD led to a 70%–90% reduction in drying time than HA drying (mushrooms).
	• MVD is better than MW-convective air drying in terms of energy consumption and drying time (cranberries).
	• MVD leads to higher color retention than MW-air drying (mint leaves).
		• Carrot slices (Lin et al., 1998)
		  • MW power: 4 kW
		  • Vacuum range: 100 mm Hg
		• Wild cabbage (Yanyang et al., 2004)
		  • MW power: 1900 W
		  • Vacuum range: 2 kPa
		• Starch (Mollekopf et al., 2011)
		  • MW power: 600–1500 W
		  • Vacuum range: 5 kPa
		• Mixed apple with potato chips (Huang et al., 2011)
		  • MW power: 4 W/g
		  • Vacuum range: 5 kPa
		• Tomatoes (Durance and Wang, 2002)
		  • MW power: 4–20 W
		  • Vacuum range: 6.65 kPa
		• Mushrooms (Giri and Prasad, 2007)
		  • MW power: 115–285 W
		  • Vacuum range: 6.5–23.5 kPa

(Continued)

TABLE 10.16 (*Continued*)

Applications of Hybrid Drying Techniques in Food Processing

Type of Hybrid Drying	Synergistic Advantage	Applications in Food Processing

		• Cranberries (Sunjka et al., 2008)
		  • MW power: 1.25 and 1.5 W/g
		  • Vacuum range: 3.4, 18.6, 33.8 kPa
		• Mint leaves (Therdthai and Zhou, 2009)
		  • MW power: 8–11.2 W/g
		  • Vacuum range: 13.33 kPa

RF-assisted HPD	• The limitations of heat transfer with HA drying alone towards the falling rate period can be overcome by combining RF heating with conventional HPD (Thomas, 1996). • RF heating causes volumetric heating of the wet material by the combined mechanisms of dipole rotation and ionic conduction, which speeds up the drying process. • When operated in combination with RF heating, the heat pump coefficient of performance (COP) of the combined dryer remains unaffected (Marshall and Metaxas, 1998). • The RF-assisted HPD reduces discoloration of dried products. • Cracking caused by stress due to uneven shrinkage during drying can be eliminated by RF-assisted drying (Kudra and Mujumdar, 2002). • RF-assisted HPD improves the SMER and COP of the heat pump system (Marshall and Metaxas, 1999). • RF-assisted HPD increases the product throughput. The throughput of crackers and cookies was enhanced by 30% and 40%, respectively (Clark, 1997; Chua and Chou, 2005).	• Suitable for food materials that are difficult to be dried with convective heating, especially food products that have a film of wax on the surface such as chilies, cherries, and tomatoes.

IR-assisted HA drying	• IR-HA drying leads to higher mass transfer rate, increased energy efficiency, reduction of drying time, and improvement of product quality.
	• Specifically, with respect to apples, IR-HA drying was faster than HA by 39.1%; the rehydration ratio, shrinkage, and color properties of apples dried under IR-HA conditions were better than that obtained from conventional HA drying.
		• Carrots and potatoes (Vishwanathan et al., 2010)
		  • MIR peak wavelength: 2.4–3.0 μm
		  • Air temperature: 40°C
		  • Drying temperature: 80°C
		  • Air velocity: 1.4 m/s
		• Shredded squid and qualities of dried squid products (Wang et al., 2014)
		  • MIR peak wavelength: 2.5–3.0 μm
		  • Air velocity: 0.5 m/s
		  • Air temperature: 50°C
		• Apple (El-Mesery and Mwithiga, 2015)
		  • IR intensity: 2000 W/m^2
		  • Air temperature: 60°C
		  • Air velocity: 0.6 m/s

IR-assisted HPD	• IR-HPD drying is accelerated due to the volumetric heat generation by dipole rotation and ionic conduction.
	• Uniform drying throughout the volume of the product.
		• Banana (Song, 2013)
		  • IR power: 500, 1000, and 2000 W
		  • Air temperature: 40°C
		  • Airflow rate: 0.8 m/s

(*Continued*)

TABLE 10.16 (*Continued*)

Applications of Hybrid Drying Techniques in Food Processing

Type of Hybrid Drying	Synergistic Advantage	Applications in Food Processing
	• IR-HPD drying exerts more precise control of the moisture content and product temperature.	• Squid fillets (Deng et al., 2011) • IR power: 500, 1000 and 2000 W • Air temperature: 50°C • Airflow rate: 0.8 m/s • Pineapple slices (Tan et al., 2001) • IR power: 1000 W • Air temperature: 40°C • Airflow rate: 2.0 m/s • Air RH: 21%

10.4 Selection of Dryer

The selection of dryer for a specific application is mainly based on the thermal (or energy) efficiency of the dryer, drying time, and the dryer economics. Each of the abovementioned elements of dryer selection is elaborated in the following sections:

10.4.1 Thermal Efficiency

Thermal efficiency is central to the economy of the drying process as energy consumption is a substantial component of the drying costs. Thermal efficiency (η) is a simple ratio between the energy utilized for the evaporation of moisture at product temperature (E_{ev}) and total energy supplied to the dryer (E_t).

$$\eta = \frac{E_{ev}}{E_t} \tag{10.50}$$

The sensible heat required to increase the product's temperature to the evaporation temperature (T_{ev}) can also be added to the numerator of Eq. (10.50), as this heat added to the food product often cannot be economically recovered. The complexities in food composition and the interaction between constituents and water demand appropriate energy analysis depending on the type of food product and dryer. In order to simplify the calculation of thermal efficiency, the dryer is considered as adiabatic with no exchange of heat with the surroundings. Thus, Eq. (10.50) can be rewritten as

$$\eta^t = \frac{\Delta H}{W^*(I_0 - I_a)} = \frac{\Delta H(Y_0 - Y_a)}{W^*(I_0 - I_a)} = \frac{\Delta H(Y_0 - Y_a)}{W^*(I_0 - I_a)} \tag{10.51}$$

where

η^t = Thermal efficiency of theoretical (adiabatic) dryer

ΔH = Latent heat of vaporization (kJ/kg)

$W^* = \frac{W_g}{W_{ev}} = \frac{1}{(Y_0 - Y_a)}$ = Specific air consumption (kg air/kg H_2O) is the mass of air required to evaporate 1 kg of water.

I_0 = Initial enthalpy of air (kJ/kg)

I_a = Final enthalpy of air (kJ/kg)

Y_0 = Initial humidity of air (kJ/kg)

Y_a = Final humidity of air (kJ/kg)

Thus, the energy efficiency of a real dryer can be determined from the following equation:

$$\eta = \frac{\Delta H}{W^*(I_0 - I_a) + \Sigma Q_1} \tag{10.52}$$

Expressing Eq. (10.52) in terms of the theoretical dryer,

$$\eta = \frac{1}{\frac{1}{\eta^t} + \frac{\Sigma Q_1}{\Delta H}} \tag{10.53}$$

where ΣQ_1 is the overall heat losses from the dryer. For low-humidity and low-temperature convective drying, the energy efficiency can be approximated by thermal efficiency (η^T) which is based on the inlet air temperature (T_i), outlet air temperature (T_o), and the ambient temperature (T_a), expressed as

$$\eta^T = \frac{(T_i - T_o)}{(T_i - T_a)} \tag{10.54}$$

The temperature differential in the numerator is the major factor which determines the thermal efficiency of a dryer.

Calculation of thermal efficiency is useful in appraising the performance of dryer, improving the dryer design, and choosing the appropriate dryer for a particular drying operation. In order to assess the dryer performance based on the thermal efficiency, the latter should be compared with the maximum thermal efficiency ($\eta_{T,\max}$), which is given by

$$\eta_{T,\max} = \frac{(T_i - T_{WB})}{(T_i - T_a)} = \frac{(T_i - T_{AS})}{(T_i - T_a)} \tag{10.55}$$

Thus, maximum thermal efficiency is attained when the outlet air temperature equals either the wet-bulb temperature (T_{WB}) or the negligibly lower adiabatic saturation temperature (T_{AS}) (Pakowski, 2001; Bond, 2002). From Eq. (10.55), it follows that the maximum energy efficiency exists when the outlet air is saturated with water vapor $\left(Y_o^{sat}\right)$ that can be attained for a given evaporation rate when the air flow is minimal, yet securing both heat and hydrodynamic requirements. Therefore,

$$\eta_{\max} = \frac{\Delta H\left(Y_o^{\text{sat}} - Y_a\right)}{(I_0 - I_a)} \tag{10.56}$$

Nevertheless, complete saturation of exhaust air never occurs in practice because the exhaust gas temperature should be kept below the saturation temperature by 10°C (Baker and McKenzie, 2005) to avoid condensation in exhaust ducts and cyclone or bag filter systems as in the case of the spray dryer. Thermal efficiencies of selected industrial dryers are compiled in Table 10.17.

The thermal efficiency of a dryer can be improved by the following approaches:

- Reducing heat losses from the dryer to the surroundings: This can be accomplished by insulation and installation of heat recovery units within the dryer system (Kemp, 2012).
- Pretreatment of the food product using mechanical pressing, concentration, or osmotic dehydration to reduce the moisture load. The moisture reduction decreases the evaporation load on the dryer as water has a high specific heat of vaporization (2500 kJ/kg at 0°C).

TABLE 10.17

Thermal Efficiency of Selected Industrial Food Dryers

Type of Dryer	Thermal Efficiency (%)
Tray dryer	85
Conveyor belt dryer	40–60
Tunnel dryer	35–40
Spray dryer	20–50
Fluidized bed dryer	40–80
Vibrated fluidized bed dryer	56–80
Drum dryer	35–85
Rotary dryer	75–90
Vacuum rotary dryer	Maximum 70
Tower dryer	20–40
Freeze dryer	~10
Flash dryer	50–75

Source: Data from Ramaswamy and Marcotte (2005), Strumillo et al. (2006), and Marcotte and Grabowski (2008).

- Raising the inlet air temperature (considering the heat sensitivity of food product to be dried) and reducing the outlet air temperature.
- Usage of a heating medium other than air as the source of energy supply to the dryer. This can be done by recycling the exhaust air to indirectly preheat the dryer's supply air or combining the exhaust air stream with inlet air. Nevertheless, the air humidity should be effectively monitored and controlled to avoid saturation that would hinder the drying process (Ramaswamy and Marcotte, 2005).

10.4.2 Drying Time

Drying time depends on the final moisture content of the product, and its prediction is important from the perspective of the quality characteristics of the final product. Drying time during constant rate drying period is predicted by the following expression:

$$t_c = \frac{\lambda_v (X_0 - X_c)}{hA(T_a - T_s)} \tag{10.57}$$

where t_c is the time required for the constant rate drying (s), λ_v is the latent heat of vaporization of water at wet-bulb temperature of hot air (kJ/kg water), X_0 is the initial moisture content of food on dry basis (kg water/kg dry solids); X_c is the critical moisture content of food on dry basis (kg water/kg dry solids), h is the convective heat transfer coefficient (W/m^2 K), A is the surface area of food product exposed to hot air (m^2), T_a is the temperature of hot air (°C), and T_s is the surface temperature of the product (°C).

From Eq. (10.57), it can be noted that constant rate drying time is directly proportional to the difference between initial moisture content and critical moisture content. It is inversely proportional to the temperature difference between the hot air and product surface. The drying time can also be expressed in terms of moisture removal rate as

$$t_c = \frac{(X_0 - X_c)}{r_c} = \frac{0.622RT_A(X_0 - X_c)}{k_m A M_W P(H_s - H_a)} \tag{10.58}$$

where

$$r_c = \frac{k_m A M_W P(H_s - H_a)}{0.622 R T_A} \tag{10.59}$$

where r_c is the moisture removal rate (s^{-1}), k_m is the convective mass transfer coefficient (m/s), M_W is the molecular weight of water, P is the atmospheric pressure (kPa), R is the universal gas constant, T_A is the absolute temperature (K), H_s is the humidity ratio at product surface (kg water/kg dry air), and H_a is the humidity ratio for air (kg water/kg dry air).

During the falling rate period, mass transfer from the internal product structure is the rate-limiting factor. The product temperature also increases beyond the wet-bulb temperature of the air. For a plate or slab drying, the time during the falling rate period is expressed as

$$t_f = \frac{4d_c^2}{\pi^2 D} \ln\left[\frac{8}{\pi^2} \frac{(X_c - X_e)}{(X - X_e)}\right] \tag{10.60}$$

where D is the effective mass diffusivity (m^2/s), d_c is the characteristic dimension, half-thickness of the slab (m), and X_e is the equilibrium moisture content of food on a dry basis (kg water/kg dry solids).

Drying time for the falling rate period can also be calculated by the expression derived by Ranz and Marshall given as,

$$t_f = \frac{\rho_L \lambda_v r_c^2 (X_c - X)}{3k_f \left(\overline{\Delta T}\right)} \tag{10.61}$$

where r_c is the radius of the dried particle and $\overline{\Delta T}$ is the mean temperature difference between the surface of the particle and hot drying air during the falling rate period.

Example 10.2

The drying of a solid food occurs during the constant rate period between the initial moisture content of 0.8 kg water/kg solids and critical moisture content of 0.6 kg water/kg solids. The corresponding temperature of the drying air and the product is 100°C and 30°C. The heat transfer coefficient for the product is 3.15 W/m² K. The product is circular with a radius of 10 cm. Estimate the constant rate drying time.

Solution

Given

Product radius = 10 cm = 0.1 m
Initial moisture content, X_0 = 0.8 kg water/kg solids
Critical moisture content, X_c = 0.6 kg water/kg solids
Heat transfer coefficient, h = 3.15 W/m² K
Temperature of drying air = 100°C
Temperature at the product surface = 30°C
Latent heat of vaporization for water = 2257 kJ/kg
The time of constant rate drying period is given by

$$t_c = \frac{\lambda_v (X_0 - X_c)}{hA(T_a - T_s)}$$

$$t_c = \frac{2257 \times 1000 \times (0.8 - 0.6)}{3.15 \times (\pi \times 0.1 \times 0.1) \times (100 - 30)} = 65163.3\,\text{s} = 18.1\,\text{h}$$

Answer: The duration of constant rate drying period is 18.1 h.

10.4.3 Dryer Economics

Economic analysis is an important component of the dryer design and selection. The dryer economics can be ascertained using three methods: (i) annualized cost; (ii) life-cycle savings, and (iii) payback period (Sreekumar, 2010; Aravindh and Sreekumar, 2014a,b; Prakash et al., 2017).

10.4.3.1 Annualized Cost

The concept of annualized cost compares the drying cost of the product obtained from dryers using different energy sources. A dryer's annualized cost (C_a) is given by the following equation:

$$C_a = C_{ac} + C_{am} + V_{as} + C_{af} + C_{ae} \tag{10.62}$$

where

C_{ac} = Yearly capital cost
C_{am} = Annualized maintenance cost
V_{as} = Annual salvage value
C_{af} = Yearly running fuel cost
C_{ae} = Yearly electricity cost for fans

The expressions for the calculation of yearly capital cost (C_{ac}) and annual salvage value (V_{as}) are as given in the following equations:

$$C_{ac} = C_{cc} \times F_{cr} \tag{10.63}$$

$$V_{as} = V_s \times F_{sr} \tag{10.64}$$

where C_{cc} is the capital cost of the dryer, V_s is the salvage value of the dryer (the estimated resale value of an asset at the end of its useful life), F_{cr} is the capital recovery factor, and F_{sr} is the salvage fund factor, given by

$$F_{cr} = \frac{i(1+i)^n}{(1+i)^n - 1} \tag{10.65}$$

$$F_{sr} = \frac{i}{(1+i)^n - 1} \tag{10.66}$$

In Eqs. (10.65) and (10.66), i is the rate of interest on a long-term investment. The yearly maintenance cost (C_{am}) is taken as a static percentage of year-wise capital cost.

The drying cost per unit weight of the product (C_{dp}) is obtained by dividing the annualized cost by the amount of product dried in a year (Sreekumar, 2010).

$$C_{dp} = \frac{C_a}{M_{dp}} \tag{10.67}$$

where M_{dp} is the mass of dried product removed from the dryer per batch, given by

$$M_{dp} = \frac{M_{pr} N_d}{N_{dd}} \tag{10.68}$$

In Eq. (10.68), M_{pr} is the mass of product dried in the dryer per year (kg), N_d is the number of days of use of domestic dryer per year, and N_{dd} is the number of drying days per batch.

An electric dryer's year-wise running fuel cost (C_{rf}) may thus be quantified by the following equation:

$$C_{rf} = M_{dp}\left[\left(\frac{m_{db}}{100}\right)_{\eta_e} \frac{h_1 C_{ee}}{\eta_{ed} \times 3{,}600}\right] \tag{10.69}$$

where h_l is the latent heat of water (kJ/kg), C_{ee} is the cost per kWh of electrical energy, η_{ed} is the dryer efficiency, and m_{db} is the moisture content on dry basis given by

$$m_{db} = \left(\frac{M_{fp} - M_{dp}}{M_{dp}} \right) \times 100 \tag{10.70}$$

In Eq. (10.70), M_{fp} is the mass of fresh product loaded in the dryer per batch (kg).

On the other hand, for a conventional dryer such as the solar dryer, while the fuel cost is zero, the cost of operation for the blower in the dryer (C_{re}) should be considered which can be calculated using the following equation:

$$C_{re} = N_h \times W \times C_e \tag{10.71}$$

where N_h is the number of operating hours of fans and blowers, W is the rated power of fans and blowers, and C_e is the unit charge of electricity.

The annualized cost is a variable for the electric and fossil fuel-based dryers owing to the fluctuating price of electricity and petroleum products. While the annualized cost analysis strikes a comparative study, it is not precise due to the volatile fossil fuel prices. To have more clarity on the dryer economics, the life-cycle savings method is adopted.

10.4.3.2 Life-Cycle Savings

This approach determines the savings to be made on the dryer in the future years. It also takes into account the present worth of annual savings over the life of the dryer.

10.4.3.2.1 Savings per Day

The cost of fresh product per kg of dried product (C_{fd}) is determined using the following equation:

$$C_{fd} = C_{fp} \times \frac{M_{fp}}{M_{dp}} \tag{10.72}$$

where C_{fp} is the cost per kilogram of fresh product. C_{fp} and the cost of drying per kg of dried product (C_s) constitute the cost of 1 kg of dried product (C_{ds}):

$$C_{ds} = C_{fp} + C_s \tag{10.73}$$

The saving per kilogram of dried product (S_{kg}) occurring in the reference year due to the usage of the dryer is given by

$$S_{kg} = C_b - C_{ds} \tag{10.74}$$

where C_b is the selling price of branded dried product. The batchwise saving (S_b) and the saving on drying day's basis (S_d) in the base year can be obtained by the following equations:

$$S_b = S_{kg} \times M_d \tag{10.75}$$

$$S_d = \frac{S_b^b}{D_b} \tag{10.76}$$

M_d is the mass of dried product removed from the dryer per batch, and b is the number of batches.

10.4.3.2.2 Annual Saving's Present Worth

With respect to the evaluation of the dryer's life, the year-wise savings (S_j) are obtained for drying the product using the following equation:

$$S_j = S_d \times D \times (1+i)^{j-1} \tag{10.77}$$

In Eq. (10.77), D is the net income or price of the substituted conventional energy and j is the number of years. The following equation can then be used for the calculation of the present worth of savings (P_j) in the jth year:

$$P_j = F_{pj} \times S_j \tag{10.78}$$

The present worth factor F_{pj} in the jth year can be computed as

$$F_{pj} = \frac{1}{(1+i)^j} \tag{10.79}$$

The total present worth of the system's life savings, i.e., savings of life cycle is obtained by the summation of year-wise saving's present worth over the time period of the system.

10.4.3.3 Payback Period

The calculation of payback period provides an insight to the investment's return period and thus, determines the acceptability of the drying technology. The following equation can be used to determine the payback period (P_n) (Prakash and Kumar, 2014):

$$P_n = \frac{\ln\left(1 - \frac{C_{cc}}{V_{as}}\left(i_{\text{interest}} - i_{\text{inflation}}\right)\right)}{\ln\left(\frac{1 + i_{\text{interest}}}{1 + i_{\text{inflation}}}\right)} \tag{10.80}$$

A shorter payback period of the dryer is desired while considering for selection among different dryer types.

10.5 Problems to Practice

10.5.1 Multiple Choice Questions

1. Moisture content of an agricultural produce, when expressed on wet basis is 75%. The corresponding moisture content on dry basis will be
   a. 30%
   b. 3%
   c. 300%

Answer: c

2. The major mode of heat transfer in a drum dryer is
   a. conduction
   b. convection
   c. radiation

Answer: a

3. 100 kg of orange juice is dried from 70% to 20% moisture (by weight). The mass of moisture removed is
   a. 50 kg
   b. 37.5 kg
   c. 62.5 kg

Answer: c

4. The prerequisite for the sublimation step of freeze-drying process is
   a. the water vapor pressure should be below the atmospheric pressure
   b. the water vapor pressure should be below the saturation vapor pressure of ice at the temperature of the food being dried
   c. subzero temperature within the freeze dryer

Answer: b

5. The most commonly used dryer for the production of milk powders is
   a. tray dryer
   b. spray dryer
   c. fluidized bed dryer

Answer: b

6. In a spray dryer of co-current configuration, the final product temperature is
   a. lower than the outlet air temperature
   b. equal to the outlet air temperature
   c. equal to the inlet air temperature

Answer: a

7. Drying rate curve is the plot between
   a. moisture content and time
   b. moisture content and drying rate
   c. drying rate and water activity

Answer: b

8. The main purpose of vacuum pump in a freeze dryer is
   a. to remove non-condensable gases
   b. to reduce the pressure inside the dryer
   c. to reduce the temperature inside the dryer

Answer: a

9. When the product dimension is reduced 10-fold, then the drying time
   a. reduces 10-fold
   b. reduces 100-fold
   c. increases 10-fold

Answer: b

10. In a spray dryer of counter-current configuration, the final product temperature is
   a. lower than the outlet air temperature
   b. greater than the outlet air temperature
   c. equal to the outlet air temperature

Answer: b

11. When the gas superficial velocity exceeds the terminal velocity of the particle, the solids will
    a. behave like bubbles
    b. be carried away by the gas stream
    c. reach the minimum fluidization velocity

Answer: b

12. The glass transition temperature is used as the indicator of
    a. feed liquid temperature
    b. collapse
    c. stickiness

Answer: c

13. Which of the following is correct?
    a. Rate = driving force × resistance
    b. Driving force = rate × resistance
    c. Rate = resistance/driving force

Answer: b

14. During the period of constant rate drying,
    a. temperature of material remains constant
    b. drying rate remains constant
    c. both a and b

Answer: c

15. The point where the constant drying period changes to falling rate period is called
    a. equilibrium moisture content
    b. critical moisture content
    c. free moisture content

Answer: b

16. The method of drying which uses water as the tool to evaporate water from the product is
    a. SFD
    b. RW drying
    c. SCD

Answer: b

17. The resistances to be overcome during SCD are
    a. viscosity and sample thickness
    b. surface tension and capillary stress of the solvent
    c. thermal conductivity of the food product

Answer: b

18. The method of drying which works on the thermodynamic principle of vapor compression cycle is
    a. spray dryer
    b. freeze dryer
    c. heat pump dryer

Answer: c

19. The longest phase of freeze-drying is
    a. secondary drying
    b. primary drying
    c. freezing

Answer: a

20. The initial drying period is followed by a much slower rate of drying, during which the moisture content of the product
    a. remains constant
    b. increases
    c. is removed at a constant rate

Answer: c

10.5.2 Numerical Problems

1. Fresh strawberries are dried from an initial moisture content of 90% to a final moisture content of 30% (wet basis; w.b). Express the moisture content of raw material and dried product on dry basis (d.b) and calculate the amount of water that is removed from 100 kg of strawberries.

Given

i. Initial moisture content of strawberries before drying (M_i) = 90% (w.b)
ii. Final moisture content of strawberries after drying (M_f) = 30% (w.b)

To find

i. Moisture content of raw material and dried product on dry basis
ii. Amount of water removed from 100 kg of strawberries

Solution

Moisture content of raw material on dry basis

$$\text{MC}_{db} = \frac{X_w}{1 - X_w} \times 100$$

$$\text{MC}_{db} = \frac{0.9}{1 - 0.9} \times 100 = 900\%$$

Moisture content of dried product on dry basis

$$\text{MC}_{db} = \frac{0.3}{1 - 0.3} \times 100 = 42.9\%$$

Amount of water removed from 100 kg of strawberries

Basis: 100 kg of strawberries

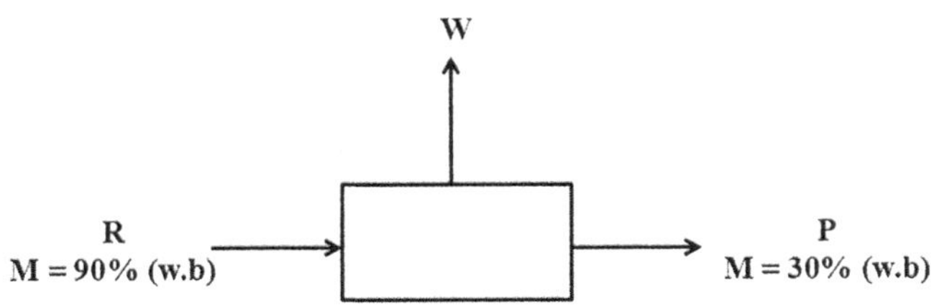

In the block diagram,

R = Quantity of fresh strawberries (kg)
P = Quantity of dried strawberries (kg)
W = Amount of water removed during drying (kg)
M = Moisture content (%)

Overall mass balance equation:

$$100 = W + P$$

Solid balance equation

$$100(1-0.9) = W(0) + P(1-0.3)$$

$$10 = 0.7P$$

$$\therefore P = \frac{10}{0.7} = 14.3\,\text{kg}$$

From the overall mass balance equation,

$$W = 100 - P = 100 - 14.3 = 85.7\,\text{kg}$$

Answer: (i) Moisture content of raw material on dry basis = 900%
(ii) Moisture content of dried product on dry basis
(iii) Amount of water removed = 85.7 kg

2. Liquid milk with 88% water is dried at 100°C to reduce its moisture content to 4%. If the initial temperature of the food product is 20°C, calculate the quantity of heat energy required for drying per unit weight of the raw material. The latent heat of vaporization of water at 100°C and at standard atmospheric pressure is 2257 kJ/kg. The specific heat capacity of milk is 3.9 kJ/kg K and that of water is 4.2 kJ/kg K.

Given

i. Initial water content of liquid milk = 88%
ii. Drying temperature = 100°C
iii. Final moisture content of milk powder = 4%
iv. Initial temperature of the food product = 20°C
v. Latent heat of vaporization of water = 2257 kJ/kg
vi. Specific heat capacity of milk = 3.9 kJ/kg K
vii. Specific heat capacity of water = 4.2 kJ/kg K

To find

i. Quantity of heat energy required for drying per unit weight of the raw material.

Solution

Basis: 1 kg of milk

Given that the initial moisture content of milk is 88%, it implies that 880 g of moisture is associated with 120 g of dry matter. Similarly the final moisture content of 4% in the milk powder implies that 40 g of moisture is associated with 960 g of dry matter.

$$\therefore \frac{40 \times 120}{960} = 5\,\text{g of moisture is associated with 120 g of dry matter.}$$

Hence, 1 kg of liquid milk should loose (880 – 5) g of moisture, i.e., 875 g or 0.875 kg of moisture.

$\therefore$ Heat energy required for drying per kg of liquid milk

= heat energy to raise the temperature to 100°C

+ latent heat to remove the water in vapor phase

$$= \{(100-20)\times 3.9\} + \{0.875\times 2257\} = 312 + 1974.9 = 2286.9 \approx 2287\,\text{kJ}$$

Answer: 2287 kJ of heat energy per unit weight of liquid milk is required for drying.

3. A food product of weight 25 kg is dried from an initial moisture content of 300% (d.b) to a final moisture content of 15% (w.b). Calculate the amount of moisture removed from the product per kg of dry solids.

Given

i. Quantity of food product = 25 kg
ii. Initial moisture content of the food product = 300% (d.b)
iii. Final moisture content of the food product = 15% (w.b)

To find: Amount of moisture removed from the product per kg of dry solids.

Solution

$$\text{Initial moisture content on dry basis} = \frac{\text{mass of water}}{\text{mass of dry solids}} \times 100 = 300$$

$$\frac{\text{mass of water}}{\text{mass of dry solids}} = 3\,\text{kg water/kg dry solids}$$

$$\therefore \text{MC}_{wb} = \frac{3}{1+3} = 0.75\,\text{kg water/kg product}$$

Final moisture content

$$= 0.15\,\text{kg water/kg product} \equiv 0.85\,\text{kg solid/kg product} = \frac{1-0.85}{0.85} = 0.18\,\text{kg water/kg dry solids}$$

Hence, amount of moisture removed from the product per kg of dry solids

$$= 3 - 0.18 = 2.82\,\text{kg water/kg dry solids}$$

Answer: 2.82 kg of water is removed from the product per kg of dry solids.

4. The initial moisture content and critical moisture content of a food product are 80% and 25% (w.b), respectively. The constant drying rate is 0.15 kg water/(m^2s). Calculate the time for constant rate drying period, if the product is cube-shaped with sides of 4 cm and has an initial product density of 900 kg/m^3.

Given

i. Initial moisture content of the food product = 80% (w.b)
ii. Critical moisture content of the food product = 25% (w.b)
iii. Constant drying rate = 0.15 kg water/m^2s
iv. Side of cube-shaped product = 4 cm = 4×10^{-2} m = 0.04 m
v. Initial product density = 900 kg/m^3

To find: Time for constant rate drying period

Solution

i. Basis: 1 kg of product

ii. Initial moisture content

$$= 0.80\,\text{kg water/kg product} \equiv 0.20\,\text{kg solids/kg product} \equiv 4\,\text{kg water/kg solids}$$

iii. Critical moisture content

$$= 0.25\,\text{kg water/kg product} \equiv 0.75\,\text{kg solids/kg product} \equiv 0.33\ \text{kg water/kg solids}$$

iv. Amount of moisture to be removed from the product during

$$\text{constant rate drying period} = 4 - 0.33 = 3.67\,\text{kg water/kg solids}$$

v. Surface area of the product $= 6a^2 = 6 \times (0.04)^2 = 9.6 \times 10^{-3} = 0.0096\,\text{m}^2$

vi. $\therefore$ Drying rate $= 0.15 \times 0.0096 = 1.44 \times 10^{-3}\,\dfrac{\text{kg water}}{\text{s}}$

vii. From the initial product density, the initial product mass

$$= 900\,\text{kg/m}^3 \times (0.04)^3\,\text{m}^3 = 0.0576\,\text{kg}$$

viii. $0.0576\,\text{kg product} \times 0.20\,\text{kg solids/kg product} = 0.01152\,\text{kg solids}$

ix. Total amount of water to be removed

$$= 3.67\,\text{kg water/kg solids} \times 0.01152\,\text{kg solids} = 0.0423\,\text{kg water}$$

x. Using the drying rate, the time for constant rate drying period

$$= 0.0423/1.44 \times 10^{-3}\left(\text{kg water}/\left(\text{kg water/s}\right)\right) = 29.36\,\text{s}$$

Answer: Time for constant rate drying period = 29.36 s

5. A food product contains 70% moisture content on wet basis. A tray dryer takes 3 h to at 80°C air temperature to remove 80% of the initial moisture content of the product. Calculate the final moisture content of the product on wet basis.

Given

i. Initial moisture content of food product = 70% (w.b)
ii. Drying time = 3 h
iii. Temperature of drying air = 80°C
iv. Moisture to be removed = 80% of initial moisture content of the food product.

To find: Final moisture content of the product on wet basis

Solution

Basis: 1 kg of food product

Given that the initial moisture content of the food product is 70%, it implies that 1 kg of food product contains 0.7 kg water/kg product.

$$80\% \text{ of initial moisture content of the food product} = 0.8 \times 0.7 = 0.56\,\text{kg}$$

$$\text{Final moisture content of the food product} = 0.7 - 0.56 = 0.14\,\text{kg water/kg product}$$

Answer: Final moisture content of the product on wet basis is 0.14 kg water/kg product.

6. 80 kg of wheat containing 10 kg of moisture has been dried to a moisture content of 8% (w.b) in 3 h under constant rate period of drying. Calculate the drying rate in kg/h.

Given

i. Quantity of wheat = 80 kg
ii. Initial moisture content of 80 kg of wheat = 10 kg
iii. Final moisture content of wheat = 8% (w.b)
iv. Drying time = 3 h

To find: Drying rate in kg/h

Solution

$$\text{Initial moisture content of wheat}(\text{w.b}) = \frac{10\,\text{kg of moisture}}{80\,\text{kg of wheat}(\text{product})}$$

$$= 0.125\frac{\text{kg water}}{\text{kg product}} \equiv 0.875\frac{\text{kg solid}}{\text{kg product}}$$

$$\text{Initial moisture content of wheat}(\text{d.b}) = \frac{1-0.875}{0.875} = 0.143\frac{\text{kg water}}{\text{kg solids}}$$

$$\text{Final moisture content of wheat}(\text{d.b}) = \frac{1-0.92}{0.92} = 0.087\frac{\text{kg water}}{\text{kg solids}}$$

$$\therefore \text{Amount of moisture to be removed} = 0.143 - 0.087 = 0.056\frac{\text{kg water}}{\text{kg solids}}$$

$$\text{Initial solid content of wheat} = 80\,\text{kg} \times 0.875\frac{\text{kg solid}}{\text{kg product}} = 70\,\text{kg solids}$$

$$\text{Total amount of water to be removed} = 0.056\frac{\text{kg water}}{\text{kg solids}} \times 70\,\text{kg} = 3.92\,\text{kg water}$$

$$\text{Drying rate}\left(\frac{\text{kg}}{\text{h}}\right) = \frac{\text{Total amount of water to be removed}}{\text{Drying time}} = \frac{3.92}{3} = 1.307$$

Answer: Drying rate = 1.307 kg/h

7. Calculate the fold change in freeze-drying time when the characteristic sample dimension is reduced from 0.5 cm to 50 μm.

Given

i. Size of sample-1$\left(d_c^1\right) = 0.5\,\text{cm} = 0.5 \times 10^{-2}\,\text{m} = 0.005\,\text{m}$

ii. Size of sample-2$\left(d_c^2\right) = 50\,\mu\text{m} = 50 \times 10^{-6}\,\text{m} = 0.00005\,\text{m}$

To find: Number of folds reduction in the freeze-drying time on reducing the size of the sample.

Solution

$$\text{Number of folds reduction in the sample dimension} = \frac{D_1}{D_2} = \frac{0.005}{0.00005} = 100$$

Thus, the size reduction is 100-fold. From the equation for drying time, it is evident that $t_f \propto d_c^2$.

$$t_f = \frac{4d_c^2}{\pi^2 D} \ln\left[\frac{8}{\pi^2}\frac{(X_c - X_e)}{(X - X_e)}\right]$$

Therefore, the reduction in drying time would be $(100)^2 = 10^4$ fold.

Answer: The freeze-drying time would be reduced by 10^4-fold or 10000-fold when the sample size is reduced from 0.5 cm to 50 µm

8. Calculate the weight reduction during the drying of curry leaves in a tray dryer from an initial moisture content of 63% to final moisture content of 4%.

Given

i. Initial moisture content of curry leaves (before tray drying) = 63%
ii. Final moisture content of curry leaves (after tray drying) = 4%

To find: Weight reduction in the curry leaves during the drying process

Solution

Basis: 100 kg of curry leaves

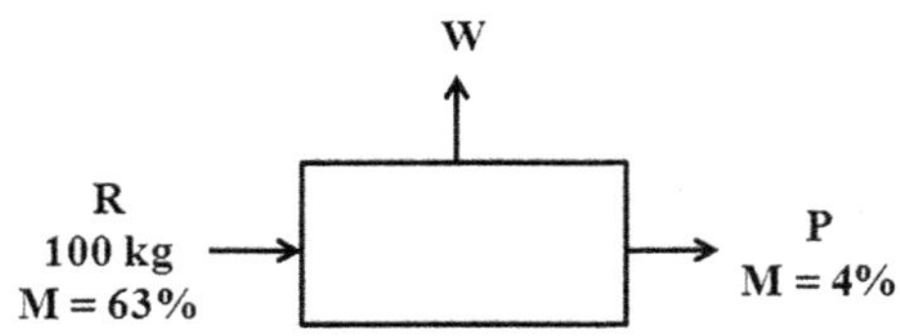

In the block diagram,

R = Quantity of fresh curry leaves (kg)

P = Quantity of dried curry leaves (kg)

W = Amount of moisture removed during drying (kg)

M = Moisture content (%)

Overall mass balance around the dryer

$$R = W + P$$

$$100 = W + P$$

Solid component balance

$$100(1 - 0.63) = W(0) + P(1 - 0.04)$$

$$37 = 0.96P$$

$$\therefore P = 37/0.96 = 38.542\,\text{kg}$$

$$\text{Weight reduction during drying} = R - P = 100 - 38.542 = 61.5\,\text{kg}$$

Answer: Weight reduction in curry leaves during drying = 61.5 kg

9. 100000 kg of paddy was harvested at 21% moisture content (w.b.) and should be conditioned to a moisture content of 15% (w.b.). Calculate the weight of paddy after drying.

Given

i. Quantity of paddy = 100000 kg
ii. Initial moisture content of paddy (M_i) = 21% (w.b)
iii. Required final moisture content (M_f) = 15% (w.b)

To find: Weight of paddy after drying

Solution

Basis: 100000 kg of paddy

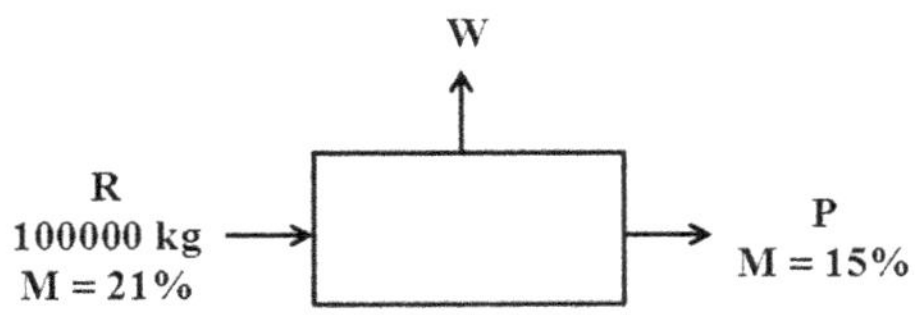

In the block diagram,

R = Amount of freshly harvested paddy (kg)
P = Amount of dried paddy (kg)
W = Amount of water removed during drying (kg)
M = Moisture content (%)

Overall mass balance

$$R = W + P$$

$$100000 = W + P$$

Solid component balance

$$100000(1 - 0.21) = W(0) + P(1 - 0.15)$$

$$79000 = 0.85P$$

$$\therefore P = \frac{79000}{0.85} = 92941.2\,\text{kg}$$

Answer: Weight of paddy after drying = 92941.2 kg

10. A solid food product undergoes drying in the constant rate period between the initial moisture content of 42.8% (w.b) and critical moisture content of 33.3% (w.b). The temperatures of the drying air and the product are 100°C and 20°C, respectively. The heat transfer coefficient for the product is 5 W/m² K. The product is of circular shape with radius 5 cm. Calculate the constant rate drying time. (Note: Latent heat of vaporization for water is 2257 kJ/kg.)

Given

i. Initial moisture content of solid food product (M_i) = 42.8% (w.b)
ii. Critical moisture content (M_c) = 33.3% (w.b)
iii. Temperature of the drying air (T_a) = 100°C
iv. Temperature of the product (T_s) = 20°C
v. Heat transfer coefficient for the product (h) = 5 W/m² K
vi. Radius of the circular-shaped product (r) = 5 cm = 5 × 10^{-2} m = 0.05 m
vii. Latent heat of vaporization for water (λ_v) = 2257 kJ/kg

To find: Constant rate drying time

Solution

$$\text{Initial moisture content (d.b)}, X_0 = \frac{0.428}{1-0.428} = \frac{0.428}{0.572} = 0.75 \frac{\text{kg water}}{\text{kg dry solids}}$$

$$\text{Critical moisture content (d.b)}, X_c = \frac{0.333}{1-0.333} = \frac{0.333}{0.667} = 0.5 \frac{\text{kg water}}{\text{kg dry solids}}$$

$$t_c = \frac{\lambda_v (X_0 - X_c)}{hA(T_a - T_s)}$$

$$\text{Area, } A = \pi r^2 = \pi \times (0.05)^2 = 0.0079\,\text{m}^2$$

$$t_c = \frac{2257 \times 1000 \times (0.75 - 0.5)}{(5 \times 0.0079 \times (100 - 20))} = \frac{564250}{3.16} = 178560.1\,\text{s} = 49.6\,\text{h}$$

Answer: Constant rate drying time = 49.6 h

BIBLIOGRAPHY

Abbasid, A., Niakousari, M., Ardekani, S. A. Y. 2015. The advantages of the refractance window method of dehydrating fresh tomato slices and the relevant characteristics thereof. *Journal of Applied Environmental and Biological Sciences* 4: 6–13.

Abonyi, B. I., Feng, H., Tang, J., Edwards, C. G., Chew, B. P., Mattinson, D. S. and Fellman, J. K. 2002. Quality retention in strawberry and carrot purees dried with refractance window system. *Journal of Food Science* 67: 1051–1056.

Afzal, T. M. and Abe, T. 1998. Diffusion in potato during far infrared radiation drying. *Journal of Food Engineering* 37: 353–365.

Alexeenko, A. A. Ganguly, A. and Nail, S. L. 2009. Computational analysis of fluid dynamics in pharmaceutical freeze-drying. *Journal of Pharmaceutical Sciences* 98: 3483–3494.

Alfaifi, B., Tang, J., Jiao, Y., Wang, S. J., Rasco, B., Jiao, S. and Sablani, S. 2014. Radio frequency disinfestation treatments for dried fruit: model development and validation. *Journal of Food Engineering* 120: 268–276.

Al-Hakim, K. and Stapley, A. G. F. 2004. Morphology of spray-dried and spray-freeze-dried whey powders. Drying—*Proceedings of the 14th International Drying Symposium (IDS 2004)*, Vol. C, 1720–1726.

Alibas, O. I., Akbudak, B. and Akbudak, N. 2007. Microwave drying characteristics of spinach. *Journal of Food Engineering* 78: 577–583.

Anandharamakrishnan, C. 2008. Experimental and computational fluid dynamics studies on spray-freeze-drying and spray-drying of proteins. *PhD Thesis*. UK: Loughborough University.

Anandharamakrishnan, C. 2017. Introduction to drying. In *Handbook of Drying for Dairy Products*, ed. C. Anandharamakrishnan, 1–14. Chichester, West Sussex, UK: John Wiley and Sons.

Anandharamakrishnan, C. and Ishwarya, S. P. 2015. *Spray Drying Techniques for Food Ingredient Encapsulation*. Hoboken, NJ: John Wiley and Sons, Inc.

Anandharamakrishnan, C., Rielly, C. D. and Stapley, A. G. F. 2010. Spray-freeze-drying of whey proteins at sub-atmospheric pressures. *Dairy Science and Technology* 90: 321–334.

Andrade, R., Skurtys, O. and Osorio, F. 2013. Drop impact behavior on food using spray coating: fundamentals and applications. *Food Research International* 54: 397–405.

Andrade, R. D., Lemus, R. M. and Pérez, C. E. C. 2011. Models of sorption isotherms for food: uses and limitations. *Vitae, Revista De La Facultad De Quõmica Farmaceutica* 18: 325–334.

Anonymous. 1995. Determination of moisture content in Finnish honey using an infrared dryer. *Food Marketing & Technology* 9: 40–41.

Aravindh, M. A. and Sreekumar, A. 2014a. An energy efficient solar drier. *Spice India* 27: 10–12.

Aravindh, M. A. and Sreekumar, A. 2014b. Experimental and economic analysis of a solar matrix collector for drying application. *Current Science* 107: 350–355.

Azizi, D., Jafari, S. M., Mirzaei, H. and Dehnad, D. 2016. The influence of refractance window drying on qualitative properties of kiwifruit slices. *International Journal of Food Engineering* 13: 20160201.

Baker, C. G. J. and McKenzie, K. A. 2005. Energy consumption of industrial spray dryers. *Drying Technology* 23: 365–386.

Begag, R., Pajonk, G. M., Elaloui, E. and Chevalier, B. 1999. Synthesis and properties of some monolithic silica carbogels produced from polyethoxydisiloxanes dissolved in ethylacetoacetate (etac) and acid catalysis. *Materials Chemistry and Physics* 58: 256–263.

Bekki, E. 1991. Rough rice drying with a far-infrared panel heater. *Journal of the Japanese Society of Agricultural Machinery* 53: 55–63.

Benali, M. and Boumghar, Y. 2015. Supercritical fluid-assisted drying. In *Handbook of Industrial Drying* (Fourth edition), ed. A. S. Mujumdar, 1261–1270. Boca Raton, FL: CRC Press.

Bhandari, B. R., Datta, N. and Howes, T. 1997. Problems associated with spray drying of sugar-rich foods. *Drying Technology* 15: 671–684.

Bhandari, B. R. and Howes, T. 1999. Implication of glass transition for the drying and stability of dried foods. *Journal of Food Engineering* 40: 71–79.

Birchal, V. S., Huang, L., Mujumdar, A. S. and Passos, M. L. 2006. Spray dryers: modeling and simulation. *Drying Technology* 24: 359–371.

Blanshard, J. M. V. and Lillford, P. J. 1993. *The Glassy State in Foods*. Loughborough: Nottingham University Press.

Bolland, K. M. 2000. Refractance Window food drying system delivers quality product efficiently. MCD Technologies, Inc., Food Online. www.foodonline.com/doc/refractance-window-food-drying-system-deliver-0001 (accessed June 15, 2018).

Bond, J. F. 2002. *Visual Metrix*. Verdun, QC: Drying Doctor, Inc.

Briggs, A. R. and Maxwell, T. J. 1976. Method of preparation of lyophilized biological products. US Patent 3932943.

Brown, Z. K. 2010. The drying of foods using supercritical carbon dioxide. *PhD Thesis*. Birmingham: University of Birmingham.

Brown, Z. K., Fryer, P. J., Norton, I. T., Bakalis, S. and Bridson, R. H. 2008. Drying of foods using supercritical carbon dioxide—Investigations with carrot. *Innovative Food Science and Emerging Technologies* 9: 280–289.

Brygidyr, A. M., Rzepecka, M. A. and McConnell, M. B. 1977. Characterization and drying of tomato paste foam by hot air and microwave energy. *Canadian Institute of Food Science and Technology Journal* 10: 313–319.

Burgain, J., Gaiani, C., Linder, M. and Scher, J. 2011. Encapsulation of probiotic living cells: from laboratory scale to industrial applications. *Journal of Food Engineering* 104: 467–483.

Cardona, T. D., Driscoll, R. H., Paterson, J. L., Srzednicki, G. S. and Kim, W. S. 2002. Optimizing conditions for heat pump dehydration of lactic acid bacteria. *Drying Technology* 20: 1611–1632.

Carrington, C. G. 2008. Heat pump and dehumidification drying. In *Food Drying Science and Technology: Microbiology, Chemistry, Applications*, eds. Y. H. Hui, C. Clary, M. M. Faird, O. O. Fasina, A. Noomhorm and J. Welti-Chanes, 249–274. Lancaster, PA: DEStech Publications, Inc.

CBIMO. www.cbimo.zut.edu.pl/ (accessed October 17, 2018).

Celma, A. R., Rojas, S. and Lopez-Rodriguez, F. 2008. Mathematical modelling of thin-layer infrared drying of wet olive husk. *Chemical Engineering and Processing* 47: 1810–1818.

Charlesworth, D. H. and Marshall, W. R. 1960. Evaporation from drops containing dissolved solids. *AIChE Journal* 6: 9–23.

Charm, S. E. 1971. *The Fundamentals of Food Engineering*. Westport, CT: The AVI Publishing Company.

Chegini, G., Khayaei, J., Rostami, H. A. and Sanjari, A. R. 2007. Designing of a heat pump dryer for drying of plum. *Journal of Research and Applications in Agricultural Engineering* 52: 63–65.

Chielle, D. P., Bertuol, D. A., Meili, L., Tanabe, E. H. and Dotto, G. L. 2016. Spouted bed drying of papaya seeds for oil production. *LWT—Food Science and Technology* 65: 852–860.

Chua, K. J. and Chou, S. K. 2005. A comparative study between intermittent microwave and infrared drying of bioproducts. *International Journal of Food Science and Technology* 40: 23–39.

Chua, K. J. and Chou, S. K. 2014. Recent advances in hybrid drying technologies. In *Emerging Technologies for Food Processing*, ed. D. W. Sun, 447–160. Salt Lake City, UT: Academic Press.

Chua, K. J., Mujumdar, A. S., Chou, S. K., Hawlader, M. N. A. and Ho, J. C. 2000. Convective drying of banana, guava and potato pieces: effect of cyclical variations of air temperature on convective drying kinetics and colour change. *Drying Technology* 18: 907–936.

Clark, T. D. 1997. The current status of radio frequency post-baking drying technology. *72nd Annual Technical Conference of the Biscuit and Cracker Manufacturers' Association*. Forth Worth, TX.

Clarke P. T. 2004. Refractance Window™ "down under". *Proceedings of the 14th International Drying Symposium (IDS 2004)*, eds. M. A. Silva and S. C. S. Rocha. Sao Paulo: UNICAMP.

Cocco, R., Karri, S. B. R. and Knowlton, T. 2014. Introduction to fluidization. American Institute of Chemical Engineers (AIChE), November 2014. www.aiche.org/cep (accessed October 17, 2018).

Daud, W. R. W. 2007. Drum Dryers. In *Handbook of Industrial Drying*, ed. A. S. Mujumdar, 203–211. Boca Raton, FL: CRC Press, Taylor & Francis Group.

de Gennes, P. G. 1985. Wetting: statics and dynamics. *Reviews of Modern Physics* 57: 827–863.

Deng, Y., Liu, Y., Qian, B., Su, S., Wu, J., Song, J. and Yang, H. 2011. *European Food Research and Technology* 232: 761–768.

Devahastin, S., Suvarnakuta, P., Soponronnarit, S. and Mujumdar, A. S. 2004. A comparative study of low-pressure superheated steam and vacuum drying of a heat-sensitive material. *Drying Technology* 22: 1845–1867.

Devakate, R. V., Patil, V. V., Waje, S. S. and Thorat, B. N. 2009. Purification and drying of bromelain. *Separation and Purification Technology* 64: 259–264.

Dewettinck, K. 1997. Fluidized bed coating in food technology: process and product quality. *PhD Dissertation*. Ghent, Belgium: Submitted to Ghent University.

Dewettinck, K. and Huyghebaert, A. 1998. Top-spray fluidized bed coating: effect of process variables on coating efficiency. *Lebensmittel-Wissenschaft und Technologie* 31: 568–575.

Di Matteo, P., Donsi, G. and Ferrari, G. 2003. The role of heat and mass transfer phenomena in atmospheric freeze-drying of foods in a fluidized bed. *Journal of Food Engineering* 59: 267–275.

Díaz, R. G., Martínez-Monzo, J. and Chiralt, P. F. A. 2003. Modelling of dehydration-rehydration of orange slices in combined microwave/air drying. *Innovative Food Science and Emerging Technologies* 4: 203–209.

Duan, Z., Jiang, L., Wang, J., Yu, X. and Wang, T. 2011. Drying and quality characteristics of tilapia fish fillets dried with hot air-microwave heating. *Food and Bioproducts Processing* 89: 472–476.

Dunn, D. B., Masavage, G. J. and Sauer, H. A. 1972. Method of freezing solution droplets and the like using immiscible refrigerants of differing densities. US Patent 3653222.

Durance, T. D. and Wang, J. H. 2002. Energy consumption, density and rehydration rate of vacuum microwave and hot-air convection dehydrated tomatoes. *Journal of Food Science* 67: 2212–2216.

El-Mesery, H. S. and Mwithiga, G. 2015. Performance of a convective, infrared and combined infrared-convective heated conveyor-belt dryer. *Journal of Food Science and Technology* 52: 2721–2730.

Epstein, N. and Grace, J. R. 1997. Spouting of particulate solids. In *Handbook of Powder Science and Technology*, eds. M. E. Fayed and L. Otten, 532–567. New York: Springer.

Ergun, S. 1952. Fluid flow through packed columns. *Chemical Engineering Progress* 48: 89–94.

Etzel, M. R., Suen, S. Y., Halverson, S. L. and Budijono, S. 1996. Enzyme inactivation in a droplet forming a bubble during drying. *Journal of Food Engineering* 27: 17–34.

Fellows, P. J. 1998. *Food Processing Technology—Principles and Practice*. Cambridge, UK: Woodhead Publishing Limited.

Feng, H., Tang, J. and Cavalieri, R. 1999b. Combined microwave and spouted bed drying of diced apples: effect of drying conditions on drying kinetics and product temperature. Invited contribution to the special Hall Issue. *Drying Technology* 17: 1981–1998.

Feng, H., Tang, J., Mattinson, D. S. and Fellman, J. K. 1999a. Microwave and spouted bed drying of blueberries: the effect of drying and pretreatment methods on physical properties and retention of flavor volatiles. *Journal of Food Processing and Preservation* 23: 463–479.

Feng, H., Yin, Y. and Tang, J. 2012. Microwave drying of food and agricultural materials: basics and heat and mass transfer modeling. *Food Engineering Reviews* 4: 89–106.

Fortes, M. and Okos, M. R. 1980. Drying theories: their bases and limitations as applied to foods and grains. In *Advances in Drying* (Volume 1), ed. A. S. Mujumdar, 301. Washington, DC: Hemisphere Publishing Corporation.

Franks, F. 1998. Freeze-drying of bioproducts: putting principles into practice. *European Journal of Pharmaceutics and Biopharmaceutics* 45: 221–229.

Funebo, T. and Ohlsson, T. 1998. Microwave-assisted air dehydration of apple and mushroom. *Journal of Food Engineering* 38: 353–367.

Funk, P. A., Elsayed, K., Yeater, K. M., Holt, G. A., Whitelock, D. P. 2015. Could cyclone performance improve with reduced inlet velocity? *Powder Technology* 280: 211–218.

Ge, R., Ye, J., Wang, H. and Yang, W. 2016. Investigation of gas–solids flow characteristics in a conical fluidized bed dryer by pressure fluctuation and electrical capacitance tomography. *Drying Technology* 34: 1359–1372.

Geankoplis, C. J. 2006. *Transport Processes and Separation Process Principles (Includes Unit Operations)* (Fourth edition). New Delhi: Prentice Hall of India Private Limited.

Geldart, D. 1986. Single particles, fixed and quiescent beds. In *Gas Fluidization Technology*, ed. D. Geldart, 11–32. Chichester: John Wiley & Sons.

Ginzburg, A. S. 1969. *Application of Infrared Radiation in Food Processing, Chemical and Process Engineering Series*. London, UK: Leonard Hill.

Giri, S. and Prasad, S. 2007. Drying kinetics and rehydration characteristics of microwave-vacuum and convective hot-air dried mushrooms. *Journal of Food Engineering* 78: 512–521.

Gorling, P. 1956. Drying behavior of vegetable substances. *V.D.I. Forsch. Gebeite Ingenieurw* 22: 5.

Gorling, P. 1958. *Fundamental Aspects of Dehydrated Foods*. London, UK: Macmillan.

Gouin, S. 2004. Microencapsulation: industrial appraisal of existing technologies and trends. *Trends in Food Science and Technology* 15: 330–347.

Guignon, B., Duquenoy, A., Dumoulin, E. D. 2002. Fluid bed encapsulation of particles: principles and practice. *Drying Technology* 20: 419–447.

Hagen, W. and Drawert, F. 1986. Determination of water content by infrared. *Monatsschrift Brauwissenschaft* 40: 240–246.

Hajidavalloo, E. and Hamdullahpur, F. 2000. Thermal analysis of a fluidized bed drying process for crops. Part II: experimental results and model verification. *International Journal of Energy Research* 24: 809–820.

Hede, P. D. 2013. *Fluid Bed Particle Processing* (First edition). London, UK: bookboon.com Ltd.

Hoebink, J. H. B. J. and Rietema, K. 1980a. Drying of granular solids in a fluidised bed I. *Chemical Engineering Science* 35: 2135–2139.

Hoebink, J. H. B. J. and Rietema, K. 1980b. Drying of granular solids in a fluidised bed II. *Chemical Engineering Science* 35: 2257–2265.

Howard, J. R. 1989. *Fluidized Bed Technology—Principles and Applications*. New York: Adam Hilger.

Huang, L., Zhang, M., Mujumdar, A. S. and Lim, R. 2011. Comparison of four drying methods for re-structured mixed potato with apple chips. *Journal of Food Engineering* 103: 279–284.

Icier, F., Colak, N., Erbay, Z., Kuzgunkaya, E. H. and Hepbasli, A. 2010. A comparative study on exergetic performance assessment for drying of Broccoli Florets in three different drying systems. *Drying Technology* 28: 193–204.

Ishwarya, S. P. and Anandharamakrishnan, C. 2015. Spray-freeze-drying approach for soluble coffee processing and its effect on quality characteristics. *Journal of Food Engineering* 149: 171–180.

Ishwarya, S. P. and Anandharamakrishnan, C. 2017. Spray drying. In *Handbook of Drying for Dairy Products*, ed. C. Anandharamakrishnan, 57–94. Chichester, West Sussex, UK: John Wiley and Sons.

Ishwarya, S. P., Anandharamakrishnan, C. and Stapley, A. G. F. 2015. Spray-freeze-drying: a novel process for the drying of foods and bioproducts. *Trends in Food Science and Technology* 41: 161–181.

Ishwarya, S. P., Anandharamakrishnan, C. and Stapley, A. G. F. 2017. Spray freeze drying. In *Handbook of Drying for Dairy Products*, ed. C. Anandharamakrishnan, 123–148. Chichester, West Sussex, UK: John Wiley and Sons.

Jangam, S. V., Joshi, V. S., Mujumdar, A. S. and Thorat, B. N. 2008. Studies on dehydration of sapota (*Achras zapota*). *Drying Technology* 26: 369–377.

Jangam, S. V. and Mujumdar, A. S. 2012. Heat pump assisted drying technology—overview with focus on energy, environment and product quality. In *Modern Drying Technology* (Volume 4), eds. E. Tsotsas and A. S. Mujumdar, 121–162. Weinheim: Wiley-VCH, Verlag GmBH and Co.

Jensen, S., Meleiro, L. A. C. and Zanoelo, E. F. 2011. Soft-sensor model design for control of a virtual conveyor-belt dryer of mate leaves (*Ilex paraguariensis*). *Biosystems Engineering* 108: 75–85.

Jeremiah, L. E. 1995. *Freezing Effects on Food Quality*. New York: Marcel Dekker, Inc.

Jones, D. J. 1988. Controlling particle size and release properties: secondary processing techniques. In *Flavour Encapsulation*, eds. S. J. Risch and G. A. Reineccius. Washington, DC: Am. Chem. Soc.

Kannan, C. S., Rao, S. S. and Varma, Y. B. G. 1994. A kinetic model for drying of solids in batch fluidized beds. *Industrial & Engineering Chemistry Research* 33: 363–370.

Karimi, F. 2010. Applications of superheated steam for the drying of food products. *International Agrophysics* 24: 195–204.

Karthik, P. and Anandharamakrishnan, C. 2013. Microencapsulation of docosahexaenoic acid by spray-freeze-drying method and comparison of its stability with spray-drying and freeze-drying methods. *Food and Bioprocess Technology* 6: 2780–2790.

Katta, S. and Gauvin, W. H. 1975. Some fundamental aspects of spray drying, *AIChE Journal* 27: 143–153.

Kaur, G., Saha, S., Kumari, K. and Datta, A. K. 2017. Mango pulp drying by refractance window method. *AgricEngInt: CIGR Journal* 19: 145–151. www.cigrjournal.org.

Keey, R. B. 1972. *Drying Principles and Practice*. Oxford, UK: Pergamon Press.

Kemp, I. C. 2012. Fundamentals of energy analysis of dryers. In *Modern Drying Technology, Volume 4: Energy Savings* (First edition), eds. E. Tsotsas and A. S. Mujumdar, 1–45. Weinheim: Wiley-VCH Verlag and Co. KGaA.

Khwanpruk, K., Anandharamakrishnan, C., Rielly, C. D. and Stapley, A. G. F. 2008. Volatiles retention during the sub-atmospheric spray freeze drying of coffee and maltodextrin. *Proceedings of the 16th International Drying Symposium (IDS2008)* November 9–12, 2008 (Hyderabad, India), 1066–1072.

Kim, E. H. J., Chen, X. D. and Pearce, D. 2009. Surface composition of industrial spray-dried milk powders. 2. Effects of spray drying conditions on the surface composition. *Journal of Food Engineering* 94: 169–181.

Kim, H. Y. and Chun, J. H. 2001. The recoiling of liquid droplets upon collision with solid surfaces. *Physics of Fluids* 13: 643–659.

Kinsella, J. E. and Fox, P. F. 1986. Water sorption by proteins: milk and whey proteins. *Critical Reviews in Food Science and Nutrition* 24: 91–139.

Kistler, S. S., Fischer, E. A., Freeman, I. R. 1943. Sorption and surface area in silica aerogel. *Journal of the American Chemical Society* 65: 1909–1919.

Kocabiyik, H. and Tezer, D. 2009. Drying of carrot slices using infrared radiation. *International Journal of Food Science and Technology* 44: 953–959.

Kochs, M., Korber, C. H., Heschel, I. and Nunner, B. 1993. The influence of the freezing process on vapour transport during sublimation in vacuum freeze drying of macroscopic samples. *International Journal of Heat Mass Transfer* 36: 1727–1738.

Kohayakawa, M. N., Silveria-Junior, V. and Telis-Romero, J. 2004. Drying of mango slices using heat pump dryer. *Paper presented at the Proceedings of the 14th International Drying Symposium* August 22–25, 2004 (Sao Paulo, Brazil).

Kozanoglu, B., Chanes, J. W., Cuautle, D. G. and Jean, J. S. 2002. Hydrodynamics of large particle fluidization in reduced pressure operations: an experimental study. *Powder Technology* 125: 55–60.

Kudra, T. and Mujumdar, A. S. 2002. *Advanced Drying Technologies*. Basel: Marcel Dekker, Inc.

Kudra, T. and Strumillo, C. 1998. *Thermal Processing of Bio-Materials*. Amsterdam, Netherlands: Gordon and Breach Science Publishers, Inc.

Kunii, D. and Levenspiel, O. 1968. Bubbling bed model. Model for flow of gas through a fluidized bed. *Industrial & Engineering Chemistry Process Design and Development* 7: 446–452.

Kunii, D. and Levenspiel, O. 1990. Entrainment of solids from fluidized beds II operation of fast fluidized bed. *Powder Technology* 61: 193–206.

Kunii, D. and Levenspiel, O. 1991. *Fluidization Engineering*. Boston, MA: Butterworth-Heinemann.

Langrish, T. A. G. and Fletcher, D. F. 2001. Spray drying of food ingredients and applications of CFD in spray drying. *Chemical Engineering and Processing: Process Intensification* 40: 345–354.

Leuenberger, H. 2002. Spray freeze drying—The process of choice for low water soluble drugs? *Journal of Nanoparticle Research* 4: 111–119.

Leuenberger, H., Plitzko, M. and Puchkov, M. 2006. Spray freeze drying in a fluidized bed at normal and low pressure. *Drying Technology* 24: 711–719.

Levi, G. and Karel, M. 1995. Volumetric shrinkage (collapse) in freeze-dried carbohydrates above their glass transition temperature. *Food Research International* 28: 145–151.

Li, J., Liang, Q. C. and Bennamoun, L. 2016. Superheated steam drying: design aspects, energetic performances, and mathematical modeling. *Renewable and Sustainable Energy Reviews* 60: 1562–1583.

Lin, T. M., Durance, T. D. and Scaman, C. H. 1998. Characterization of vacuum microwave air and freeze dried carrot slices. *Food Research International* 4: 111–117.

Lindeløv, J. S. and Wahlberg, M. 2009. Spray drying for processing of nanomaterials. *Journal of Physics Conference Series* 170: 012027.

Link, K. C. and Schlunder, E. U. 1997. Fluidized bed spray granulation. Investigation of the coating process on a single sphere. *Chemical Engineering and Processing* 36: 443–457.

Llop, M. F., Madrid, F., Arnaldos, J. and Casal, J. 1996. Fluidization at vacuum conditions. A generalized equation for the prediction of minimum fluidization velocity. *Chemical Engineering Science* 51: 5149–5157.

Lu, A. 2007. Heat pumps. In *Encyclopedia of Energy Engineering and Technology—3 Volume Set (Print Version)*, eds. S. Anwar and B. L. Capehart, 814. Boca Raton, FL: CRC Press.

Lucideon. 2017. The importance of drying—choosing the right equipment and methods. www.azom.com/article.aspx?ArticleID=14179 (accessed May 1, 2018).

Magoon, R. E. 1985. Method and apparatus for drying fruit pulp and the like. US patent 4631837A.

Malecki, G. J., Shinde, P., Morgan, A. I. and Farkas, D. F. 1970. Atmospheric fluidized bed freeze drying. *Food Technology* 24: 601–603.

Manickavasagan, A., Jayas, D. S. and White, N. D. G. 2007. Germination of wheat grains from uneven microwave heating in an industrial microwave dryer. *Canadian Biosystems Engineering* 49: 3.23–3.27.

Marcotte, M. and Grabowski, S. 2008. Minimising energy consumption associated with drying, baking and evaporation. In *Handbook of Water and Energy Management in Food Processing*, eds. J. Klemes, R. Smith and J. K. Kim, 481–522. Cambridge, UK: Woodhead Publishing.

Marshall, M. G. and Metaxas, A. C. 1998. Modeling of the radio frequency electric field strength developed during the Rf assisted heat pump drying of particulates. *The Journal of Microwave Power and Electromagnetic Energy* 33: 167–177.

Marshall, M. G. and Metaxas, A. C. 1999. Radio frequency assisted heat pump drying of crushed brick. *Applied Thermal Engineering* 19: 375–388.

Masamura, A., Sado, H., Nabetani, H. and Nakajima, M. 1988. Drying of potato by far-infrared radiation. *Nippon Shokuhin Kogyo Gakkaishi* 35: 309–314.

Maskan, M. 2000. Microwave/air and microwave finish drying of banana. *Journal of Food Engineering* 44: 71–78.

Maskan, M. 2001. Drying, shrinkage and rehydration characteristics of kiwifruits during hot air and microwave drying. *Journal of Food Engineering* 48: 177–182.

Masters, K. 1985. *Spray Drying Handbook*. New York: Longman Scientific and Technical.

Masters, K. 1988. *Spray Drying Handbook*. New York: Longman/Wiley.

Masters, K. 1991. *Spray Drying Handbook*. Harlow: Longman Scientific and Technical.

McHugh, M. A. and Krukonis, V. J. 1986. *Supercritical Fluid Extraction Principles and Practice*. Boston, MA: Butterworths.

Mermelstein, N. 1998. Microwave and radio frequency drying. *Food Technology* 52: 84–86.

Meryman, H. T. 1959. Sublimation freeze-drying without vacuum. *Science* 130: 628–629.

Meterc, D., Petermann, M. and Weidner, E. 2008. Drying of aqueous green tea extracts using a supercritical fluid spray process. *Journal of Supercritical Fluids* 45: 253–259.

Miao, S. and Roos, Y. H. 2004. Nonenymatic browning kinetics of carbohydrate-based low moisture food system at temperatures applicable to spray drying. *Journal of Agricultural and Food Chemistry* 52: 5250–5257.

Mohsenin, N. N. 1986. *Physical Properties of Plant and Animal Materials* (Second edition). New York: Gordon and Breach Publishing.

Mollekopf, N., Treppe, K., Fiala, P. and Dixit, O. 2011. Vacuum microwave treatment of potato starch and the resultant modification of properties. *Chemie Ingenieur Technik* 83: 262–272.

Mongpraneet, S., Abe, T. and Tsurusaki, T. 2004. Kinematic model for a far infrared vacuum dryer. *Drying Technology* 22: 1675–1693.

Monteiro, R. L., Carciofi, B. A. M., Marsaioli, A. and Laurindo, J. B. 2015. How to make a microwave vacuum dryer with turntable. *Journal of Food Engineering* 166: 276–284.

Moses, J. A., Norton, T., Alagusundaram, K. and Tiwari, B. K. 2014. Novel drying techniques for the food industry. *Food Engineering Reviews* 6: 43–55.

Mujumdar, A. S. 2007. Principles, classification and selection of dryers. In *Handbook of Industrial Drying* (Third edition), ed. A. S. Mujumdar, 4–31. Boca Raton, FL: Taylor and Francis Group.

Mumenthaler, M. and Leuenberger, H. 1991. Atmospheric spray freeze-drying: a suitable alternative in freeze drying technology. *International Journal of Pharmaceutics* 72: 97–110.

Nasiroglu, S. and Kocabiyik, H. 2009. Thin-layer infrared radiation drying of red pepper slices. *Journal of Food Process Engineering* 32: 1–16.

Nindo, C. I., Sun, T., Wang, S. W., Tang, J. and Powers, J. R. 2003. Evaluation of drying technologies for retention of physical quality and antioxidants in asparagus (*Asparagus officinalis*, L.). *LWT—Food Science and Technology* 36: 507–516.

Nindo, C. I. and Tang, J. 2007. Refractance window dehydration technology: a novel contact drying method. *Drying Technology* 25: 37–48.

Nowak, D. and Lewicki, P. P. 2004. Infrared drying of apple slices. *Innovative Food Science and Emerging Technologies* 5: 353–360.

Oetjen, G. W. 1999. *Freeze-Drying*. New York: Wiley-VCH.

Orikasa, T., Koide, S., Okamoto, S., Imaizumi, T., Muramatsu, Y., Takeda, J., Shiina, T. and Tagawa, A. 2014. Impacts of hot air and vacuum drying on the quality attributes of kiwifruit slices. *Journal of Food Engineering* 125: 51–58.

Pakowski, Z. 2001. *DryPak 2000LE: Psychrometric and Drying Computations*. Lodz: OMNIKON Ltd.

Palancz, B. A. 1983. A mathematical model for continuous fluidised bed drying. *Chemical Engineering Science* 38: 1045–1049.

Parikh, D. 2015. Vacuum drying: basics and application. *Feature Report, Chemical Engineering* April 1: 48–54.

Patel, K. C. and Chen, X. D. 2008. Surface-center temperature differences within milk droplets during convective drying and drying-based Biot number analysis. *AIChE Journal* 54: 3273–3290.

Piyasena, P., Dussault, C., Koutchma, T., Ramaswamy, H. S. and Awuah, G. B. 2003. Radio frequency heating of foods: principles, applications and related properties—A review. *Critical Reviews in Food Science and Nutrition* 43: 587–606.

Prabhanjan, D. G., Ramaswamy, H. S. and Raghavan, G. S. V. 1995. Microwave-assisted convective air drying of thin layer carrots. *Journal of Food Engineering* 25: 283–293.

Prakash, O. and Kumar, A. 2014. Solar greenhouse drying: a review. *Renewable and Sustainable Energy Review* 29: 905–910.

Prakash, O., Ranjan, S., Kumar, A. and Gupta, R. 2017. Economic analysis of various developed solar dryers. In *Solar Drying Technology: Concept, Design, Testing, Modeling, Economics, and Environment*, eds. O. Prakash and A. Kumar, 495–514. Singapore: Springer Nature Singapore Pte. Ltd.

Prasertsan, S. and Saen-saby, P. 1998. Heat pump drying of agricultural materials. *Drying Technology* 16: 235–250.

Prasertsan, S., Sean-saby, P., Prateepchaikul, G. 1997. Heat pump dryer. Part 3: experiment verification of the simulation. *International Journal of Energy Research* 21: 1–20.

Pronyk, C., Cenkowski, S. and Muir, W. E. 2004. Drying foodstuffs with superheated steam. *Drying Technology* 22: 899–916.

Rahman, M. S. and Perera, C. O. 2007. Drying and food preservation. In *Handbook of Food Preservation*, ed. M. S. Rahman, 404–427. Boca Raton, FL: CRC Press, Taylor and Francis Group.

Ramaswamy, H. and Marcotte, M. 2005. *Food Processing Principles and Applications*. Boca Raton, FL: CRC Press.

Ratti, C. and Mujumdar, A. S. 1995. Infrared drying. In *Handbook of Industrial Drying*, ed. A. S. Mujumdar, 567–588. New York: Marcel Dekker.

Riadh, M. H., Ahmad, S. A. B., Marhaban, M. H. and Soh, A. C. 2015. Infrared heating in food drying: an overview. *Drying Technology* 33: 322–335.

Rogers, S., Wu, W. D., Saunders, J. and Chen, X. D. 2008. Characteristics of milk powders produced by spray freeze drying. *Drying Technology* 26: 404–412.

Rogers, T. L., Hu, J., Hu, Z., Johnston, K. P. and Williams, R. O. 2002. A novel particle engineering technology: spray-freezing into liquid. *International Journal of Pharmaceutics* 242: 93–100.

Roknul, A. S. M., Zhang, M., Mujumdar, A. S. and Wang, Y. 2014. A comparative study of four drying methods on drying time and quality characteristics of stem lettuce slices (*Lactuca sativa* L.). *Drying Technology* 32: 657–666.

Ronsse, F., Pieters, J. G. and Dewettinck, K. 2008. Modelling side effect spray drying in top-spray fluidised bed coating process. *Journal of Food Engineering* 86: 529–541.

Rossi, S. J., Neues, C., Kicokbusch, T. G. 1992. Thermodynamics and energetic evaluation of a heat pump applied to drying of vegetables. In *Drying '92*, ed. A. S. Mujumdar. Amsterdam: Elsevier Science.

Sauer, H. A. 1969. Method and apparatus for freeze-drying. US Patent 3484946.

Schiffman, R. F. 1987. Microwave and dielectric drying. In *Handbook of Industrial Drying*, ed. A. S. Mujumdar, 327–356. New York and Basel: Marcel Dekker, Inc.

Sewell, P. C. 1987. Low pressure atomization nozzle. EP 0244204 A1.

Sharma, G. P., Verma, R. C. and Pathare, P. B. 2005. Thin-layer infrared radiation drying of onion slices. *Journal of Food Engineering* 67: 361–366.

Shi, J. L., Pan, Z. L., McHugh, T. H., Wood, D., Hirschberg, E. and Olson, D. 2008. Drying and quality characteristics of fresh and sugar-infused blueberries dried with infrared radiation heating. *LWT—Food Science and Technology* 41: 1962–1972.

Singh, R. P. and Heldman, D. R. 2014. *Introduction to Food Engineering* (Fifth edition). London, UK: Elsevier, Inc.

Sisquella, M., Vinas, I., Picouet, P., Torres, R. and Usall, J. 2014. Effect of host and Monilinia spp. variables on the efficacy of radio frequency treatment on peaches. *Postharvest Biology and Technology* 87: 6–12.

Sobolev, V. V., Guilemany, J. M., Nutting, J. and Miquel, J. R. 1997. Development of substrate-coating adhesion in thermal spraying. *International Materials Reviews* 42: 117–136.

SolidsWiki. Tunnel dryers. www.solidswiki.com/index.php?title=Tunnel_Dryers (accessed October 17, 2018).

Song, X. 2013. Banana chip drying using far infrared-assisted heat pump. *Philippine Agricultural Scientist* 96: 275–281.

Soysal, Y. 2004. Microwave drying characteristics of parsley. *Biosystems Engineering* 89: 167–173.

Sreekumar, A. 2010. Techno-economic analysis of a roof-integrated solar air heating system for drying fruit and vegetables. *Energy Conversion and Management* 51: 2230–2238.

Srivastava, S. and Mishra, G. 2010. Fluid bed technology: overview and parameters for process selection. *International Journal of Pharmaceutical Sciences and Drug Research* 2: 236–246.

Strommen, I. and Kramer, K. 1994. New applications of heat pumps in drying process. *Drying Technology* 12: 889–901.

Strumillo, C., Jones, P. L. and Zylla, R. 2006. Energy aspects in drying. In *Handbook of Industrial Drying* (Third edition), ed. A. S. Mujumdar, 1075–1101. Boca Raton, FL: CRC Press.

Sui, Y., Yang, J., Ye, Q., Li, H. and Wang, H. 2014. Infrared, convective, and sequential infrared and convective drying of wine grape pomace. *Drying Technology* 32: 686–694.

Sunjka, P. S., Orsat, V. and Raghavan, G. S. V. 2008. Microwave/vacuum drying of cranberries (*Vacccinium macrocarpon*). *American Journal of Food Technology* 3: 100–108.

Tan, M., Chua, K. J. Mujumdar, A. S. and Chou, S. K. 2001. Effect of osmotic pre-treatment and infrared radiation on drying rate and color changes during drying of potato and pineapple. *Drying Technology* 19: 2193–2207.

Tewari, P. H., Hunt, A. J. and Lofflus, K. D. 1985. Ambient-temperature supercritical drying of transparent silica aerogels. *Materials Letters* 3: 363–367.

Therdthai, N. and Zhou, W. 2009. Characterization of microwave vacuum drying and hot air drying of mint leaves (*Mentha cordifolia* Opiz ex Fresen). *Journal of Food Engineering* 91: 482–489.

Thiel, W. J. and Nguyen, L. T. 1984. Fluidized bed film coating of an ordered powder mixture to produce microencapsulated ordered units. *Journal of Pharmacy and Pharmacology* 36: 145–152.

Thomas, P. P. and Varma, Y. B. G. 1992. Fluidized bed drying of granular food materials. *Powder Technology* 69: 213–222.

Thomas, W. J. 1996. RF drying provides process savings: new systems optimize radio frequency drying for the ceramic and glass fiber industries. *Ceramic Industry Magazine* 30–34.

Topuz, A., Feng, H. and Kushad, M. 2009. The effect of drying method and storage on color characteristics of paprika. *LWT—Food Science and Technology* 42: 1667–1673.

Torringa, H. M., Dijk, E. J. V. and Bartels, P. V. 1996. Microwave puffing of vegetables: modeling and measurements. *Proceedings of 31st Microwave Power Symposium*, 16–19.

Truong, V., Bhandari, B. R. and Howes, T. 2005. Optimization of concurrent spray drying process for sugar-rich foods. Part II—optimization of spray drying process based on glass transition concept. *Journal of Food Engineering* 71: 66–72.

Van den Berg, C. and Bruin, S. 1981. Water activity and its estimation in food systems: theoretical aspects. In *Water Activity: Influences on Food Quality*, eds. L. B. Rockland and G. F. Stewart, 236. New York: Academic Press.

van Deventer, H. C. 2004. *Industrial Superheated Steam Drying*. Apeldroon: TNO Environment, Energy and Process Innovation, TNO Report, R 2004/239.

van Deventer, H. C. and Heijmans, R. M. H. 2001. Drying with superheated steam. *Drying Technology* 19: 2033–2045.

Vega-Mercado, H., Gongora-Nieto, M. M. and Barbosa-Canovas, G. V. 2001. Advances in dehydration of foods. *Journal of Food Engineering* 49: 271–289.

Vishwanathan, K. H., Hebbar, H. U. and Raghavarao, K. S. M. S. 2010. Hot air assisted infrared drying of vegetables and its quality. *Food Science and Technology Research* 16: 381–388.

Vishwanathan, K. H., Giwari, G. K. and Hebbar, H. U. 2013. Infrared assisted dry-blanching and hybrid drying of carrot. *Food and Bioproducts Processing* 91: 89–94.

Wang, Y., Wig, T. D., Tang, J. and Hallberg, L. M. 2003. Sterilization of foodstuff using radiofrequency heating. *Journal of Food Science* 68: 539–544.

Wang, Y., Zhang, M., Mujumdar, A. S. and Chen, H. 2014. Drying and quality characteristics of shredded squid in an infrared-assisted convective dryer. *Drying Technology* 32: 1828–1839.

Webb, S. D., Golledge, S. L., Cleland, J. L., Carpenter, J. F. and Randolph, T. W. 2002. Surface adsorption of recombinant human interferon-c in lyophilized and spray-lyophilized formulations. *Journal of Pharmaceutical Sciences* 91: 1474–1487.

Werner, S. R., Jones, J. R., Paterson, A. H., Archer, R. H. and Pearce, D. L. 2007. Droplet impact and spreading: droplet formulation effects. *Chemical Engineering Science* 62: 2336–2345.

Werther, J. 1987. Gas fluidization technology. In *Berichte der Bunsengesellschaft für physikalische Chemie*, ed. D. Geldart (Volume 91), 678–679. Chichester, New York, Brisbane, Toronto, Singapore: John Wiley & Sons.

Wilhelm, L. R., Suter, D. A. and Brusewitz, G. H. 2004. *Food and Process Engineering Technology*. St. Joseph, MI: ASAE

Wolff, E. and Gibert, H. 1990. Atmospheric freeze-drying Part 2: modelling drying kinetics using adsorption isotherms. Drying Technology, 8: 405–428.

Wormsbecker, M. 2008. Study of hydrodynamic behaviour in a conical fluidized bed dryer using pressure fluctuation analysis and X-ray densitometry. *PhD Thesis*. Saskatoon: University of Saskatchewan.

Yanyang, X., Zhang, M., Mujumdar, A. S., Le-qun, Z. and Jin-cai, S. 2004. Studies on hot air and microwave vacuum drying of wild cabbage. *Drying Technology* 22: 2201–2209.

Yataganbaba, A. and Kurtbaş, Í. 2016. A scientific approach with bibliometric analysis related to brick and tile drying: a review. *Renewable and Sustainable Energy Reviews* 59: 206–224.

Yates, J. G. 1983. *Fundamentals of Fluidized Bed Chemical Processes*. London, UK: Butterworth.

Zhang, M., Tang, J., Mujumdar, A. S. and Wang, S. 2006. Trends in microwave-related drying of fruits and vegetables. *Trends in Food Science and Technology* 17: 524–534.

11

Refrigeration and Freezing of Foods

Using low temperature to preserve foods dates back to the prehistoric times when man stored meat of hunted game in snow and blocks of ice. Refrigeration and freezing are unit operations that extend the shelf life of perishable food products based on the well-known principle in physical chemistry which states that *molecular mobility is depressed and consequently chemical reactions and biological processes are slowed down at low temperature.* The reaction rate rule governs the deceleration of deteriorative changes in food products by refrigeration and freezing according to which *decreasing temperature of the product by 10°C reduces the reaction rate by half.*

The fundamental difference between refrigeration and freezing of foods is the range of low temperature applied. Refrigeration involves bringing the food product to a moderately low temperature (15°C–20°C) to a temperature not significantly above its freezing point (0°C to –2°C). Nevertheless, conventional freezing reduces the temperature of the food product to below its freezing point, down to the range of –12°C to –40°C. On the other hand, with cryogenic freezing, the food product is brought down to much lower temperatures than conventional freezing by establishing direct contact with cryogens such as CO_2 (commonly known as dry ice) and liquid nitrogen. While dry ice exhibits a sublimation temperature of –78.5°C, liquid nitrogen has a boiling point of –196°C. During freezing, apart from the reduction in the temperature of a food product to below its freezing point, a part of liquid water undergoes a change in its physical state to ice crystals. Thus, in addition to the reduction in temperature, the principle of food preservation by freezing is also based on the removal of liquid water and rendering it unavailable for microbial growth, physicochemical, and biochemical reactions, all of which lead to deterioration of product quality. Refrigeration temperature slows down the chemical reactions, enzyme activity, and microbial growth. The sub-zero temperatures of freezing prevent the growth of most food-borne microorganisms and stops or extremely slows down the enzymatic reactions and microbial growth. Consequently, the shelf life of foods is extended in the order of few days and several months by refrigeration and freezing, respectively. Apart from the preservation effect, refrigeration and freezing also bring about textural changes in food products.

Earlier, the terms *refrigeration* and *freezing* were merely associated with domestic refrigerators at the households and freezers installed at the supermarkets that sold frozen desserts and confectioneries. But, today, refrigeration and freezing, together known as the *cold storage facility* occupy a key position in the food production chain. Although many industries utilize cold storage, the food industry is the first on the list of sectors that require this infrastructure. This is because of the increasing demand for food supply and increased production of perishable goods such as fruit and vegetables. Cold storage helps to prolong the shelf life of fresh goods and protect them from spoilage. It is also vital in reducing food waste and in prolonging the time frame for marketing the food products. Furthermore, the frozen food market is one of the largest sectors of the food industry. Substantial quantities of frozen foods are being consumed all over the world (Barbosa-Cǻnovas et al., 2005). Some of the popular frozen food categories include frozen vegetables, frozen desserts and bakery, frozen meat, and frozen seafood. With the current advancements, the cold storage infrastructure is evolving continuously. Thus, it is essential to obtain knowledge about the principle, operation, equipment, and recent advancements pertaining to refrigeration and freezing.

11.1 Glossary of Food Refrigeration and Freezing

- **Average product temperature:** It is the temperature of the product calculated as the average of temperatures at the local freezing sites (Barbosa-Cǻnovas et al., 2005).

- **Refrigerant:** It is a low boiling point liquid, which changes in state from liquid to vapor by absorbing the heat from the product.
- **Freezing capacity (tonnes/h):** Ratio between the quantity of the product that can be loaded into the freezer and the holding time of the product in that particular freezer (Barbosa-Cánovas et al., 2005).
- **Pre-freezing stage:** The duration of time between the starting of freezing process and the appearance of the first ice crystal (Barbosa-Cánovas et al., 2005).
- **Freezing period:** The period after pre-freezing during which the phase change occurs with the transformation of water to ice. This period is considered the second stage of the freezing process (Barbosa-Cánovas et al., 2005).
- **Freezing point depression:** The decrease in freezing point of a solvent caused by the presence of a solute (Barbosa-Cánovas et al., 2005).
- **Freezing rate (°C/h):** The ratio of the difference between initial and final temperatures of the product to freezing time (Barbosa-Cánovas et al., 2005).
- **Quick freezing:** Freezing carried out within the rapid time duration to retain the flavor and nutritional value of foods (Barbosa-Cánovas et al., 2005).

11.2 Refrigeration of Foods

With refrigeration, the range of cooling temperature is significant as it determines the cost of refrigeration; a lower cooling temperature increases the refrigeration cost. Industrial-scale food refrigeration can be categorized under two major ranges of cooling temperature: moderate cooling (15°C–20°C) and chilling (0°C to –2°C). Moderate cooling is achieved by a simpler and economical approach using cold water or evaporative cooling. Chilling is accomplished by mechanical refrigeration which involves evaporating and compressing a refrigerant in a continuous cycle. Before proceeding to understand the principle of mechanical refrigeration, it is important to gain knowledge about the refrigerants and properties governing the selection of refrigerants for specific applications.

11.2.1 Refrigerants

Refrigerants absorb heat from the food product to be cooled and transfer it to the surroundings. Since the 19th century, a variety of refrigerants have been introduced for different applications (Table 11.1). Usually, refrigerants are designated as *R* followed by a two- or three-digit number and, in some cases, one or two letters. The designation *Rxyz* is determined by the criteria described in Table 11.2. Earlier, ammonia was considered to be more advantageous among the several refrigerants due to its lower energy cost, higher heat transfer coefficients, greater detectability in the event of a leak, and no effect on the

TABLE 11.1

Chronology of Developments in Refrigerant Usage

Period	Refrigerants
1800–1900	Ethyl ether, ethyl alcohol, methyl amine, ethyl amine, methyl chloride, ethyl chloride, sulfur dioxide, carbon dioxide, ammonia
1900–1930	Ethyl bromide, carbon tetrachloride, water, propane, isobutene, gasoline, methylene chloride
1931–1990	CFCs, hydrochlorofluorocarbons, ammonia, water
1990–2010	Hydrofluorocarbons, ammonia, isobutene, propane, carbon dioxide, water
Present/immediate future	Hydrofluoroolefins, hydrofluorocarbons, hydrocarbons, carbon dioxide, water

Source: Venkatarathnam and Murthy (2012).

TABLE 11.2

Designation of Refrigerants

Category	*x*	*y*	*z*	Examples
Methane	Refers to the number of carbon atoms minus one; (*x*) = 0; but for these compounds, the '0' is ignored in the designation.	Refers to the number of hydrogen atoms plus one.	Refers to the number of fluorine atoms.	R12 (dichlorodifluoromethane); R22 (chlorodifluoromethane)
Ethane	Refers to the number of carbon atoms minus one; (*x*) = 1	Refers to the number of hydrogen atoms plus one.	Refers to the number of fluorine atoms.	R114 (1,2-Dichlorotetrafluoroethane); R124 (1-Chloro-1,2,2,2-tetrafluoroethane); R134a (1,1,1,2-Tetrafluoroethane)
Propane	Refers to the number of carbon atoms minus one; (*x*) = 2	Refers to the number of hydrogen atoms plus one.	Refers to the number of fluorine atoms.	R290 (propane)
Zeotropic (components in a mixture having different boiling points)	(*x*) = 4	(*y*) and (*z*) are ordinal numbers		R407A (combination of R32/R125/R134a in the ratio of 20:40:40) R407C (combination of R32/R125/R134a in the ratio of 23:25:52)
Azeotropic mixtures (homogeneous substances with one specific boiling point)	(*x*) = 5	(*y*) and (*z*) are ordinal numbers		R502 (combination of R22 and R115 in the ratio of 48.8:51.2) R507 (combination of R125 and R143a in the ratio of 50:50)
High organic compounds (Comprises hydrocarbons, oxygen compounds, sulphuric compounds and nitrogen compounds)	(*x*) = 6	The subgroups are assigned different number series within the main group; (*y*) and (*z*) describe the subgroup and order within the subgroup.		R600 (butane); R600a (isobutene)
Inorganic compounds	(*x*) = 7	(*y*) and (*z*) denote the molar mass.		R717 (ammonia); R718 (water); R744 (carbon dioxide)
Unsaturated ethane compounds	(*x*) = 11	Refers to the number of hydrogen atoms plus one.	Refers to the number of fluorine atoms.	R1150 (ethylene)
Unsaturated propane compounds	(*x*) = 12	Refers to the number of hydrogen atoms plus one.	Refers to the number of fluorine atoms.	R1270 (propylene)

- The lower case letters at the end describe the structure of the molecule.
  - For example: R600 (butane) and R600a (isobutane) are two compounds that have the same chemical formula but different spatial arrangements.
- Capital letters describe specific mixing proportions of different components.
  - For example: R407 A-E are mixtures of the refrigerants R32, R125, and R134a. R407A has the following mixing proportions: 20% R32, 40% R125, and 40% R134a. R407C consists of 23% R32, 25% R125, and 52% R134a.

ozone layer. Ammonia has been predominantly used for the cooling of fresh fruits, vegetables, meat, and fish and refrigeration of beverages such as beer and wine and dairy products such as milk and cheese. However, the major drawback of ammonia is its toxicity, which renders it inappropriate for domestic use. Similar to ammonia, owing to toxicity, many of the refrigerants used earlier are not in use today. In the 1920s, instances of accidental leaks and fatal consequences necessitated the development of safe refrigerants for the domestic and commercial use. As a result, "dichlorofluoromethane," the first member of the chlorofluorocarbon (CFC) family of refrigerants was introduced. Later, among the various CFCs developed, "dichlorodifluoromethane" was ascertained the most suitable refrigerant for commercial use. This marked the beginning of identifying the CFC family of refrigerants with the trade name, "Freon." Accordingly, dichlorodifluoromethane is popularly known under the brand name, "Freon-12." Soon after their introduction, CFCs were appreciated for their versatility and low cost. A mixture of R115 (chloropentafluoroethane) and R22 (chlorodifluoromethane) known as R502 is the dominant refrigerant used in commercial refrigeration system installed in supermarkets. Nevertheless, in the 1970s, it was established that CFCs deplete the ozone layer which protects the earth from harmful ultraviolet radiation and thereby enhance the greenhouse effect and prevent infrared radiation from escaping the earth. The greenhouse effect of CFCs turned out to be a major scientific and public concern. Hence, various national and international legislative actions imposed a ban on the use of some CFCs. Consequently, the research focus was shifted towards refrigerants with zero ozone depletion potential (ODP) and low global warming potential (GWP). An example of such refrigerant is the chlorine-free R134a (1, 1, 1, 2-tetrafluoroethane) which was developed to replace R12 (dichlorodifluoromethane) (Papadopoulos and Koroneos, 2004).

A food engineer should be aware of the criteria based on which a refrigerant must be chosen. Apart from safety, a refrigerant should also satisfy other critical requirements from the perspectives of sustainable environment, design, and operation. The suitability of a refrigerant based on the abovementioned criteria is measured by specific indices and properties which are compiled in Table 11.3.

TABLE 11.3

Selection Criteria for Refrigerants (Based on Venkatarathnam and Murthy, 2012)

Criteria	Index/Desired Property
Environmental sustainability	• Low GWP • Zero ODP
Safety	• Non-toxic • Non-flammable • Non-corrosive • Chemically stable • Easy detection of the leak (distinct color/odor) • Must be easily disposable
Design and operation	• The high latent heat of vaporization for a large refrigeration effect or a small mass flow rate for a given cooling load to minimize the amount of refrigerant used. • High thermal conductivity in both liquid and vapor states to improve heat transfer. • Moderate temperature rise during compression to reduce the risk of compressor overheating and to avoid a chemical reaction between refrigerant oil and other materials. • Small specific volume for large mass flow rate per unit volume of compression. • Suitable freezing and boiling points with low viscosity at the temperature and pressure of operation to reduce pressure loss. • Critical temperature higher than the condensing temperature to have a broad range of isothermal energy transfer. • Low condensing pressure to allow the use of lightweight materials for heat exchangers, compressors, and piping. • Evaporating pressure must be slightly higher than atmospheric pressure to prevent any air leakage into the refrigeration system. As a specific case, a refrigerant should have a saturation pressure of 1 atm (0.101325 MPa) or higher, at 20°C. R717, R12, and R134a are substances with such properties. • Higher suction pressure than atmospheric pressure for ease of leak detection and to prevent ingression of air and moisture into the system. • Lower freezing temperature than the evaporator temperature. • Low compression ratio to give high volumetric efficiency and low power consumption. • Compatibility with the materials used in refrigeration systems (metals, polymers, lubricating oils). • Availability at a low cost per unit mass (to reduce the operational cost of the refrigeration process).

11.2.2 Theory of Mechanical Refrigeration System

Mechanical refrigeration works on the principle of removing and transferring the heat from a low-temperature heat source to a high-temperature heat sink. However, this challenges the Clausius statement of the second law of thermodynamics which states that *it is impossible to construct a device operating in a cyclic fashion whose sole purpose is to transfer heat from a low-temperature reservoir to a higher temperature reservoir.* Thus, a device is required which consumes work and transfers the heat from a zone of low temperature to high temperature. Such a device that drives the refrigeration operation against the second law of thermodynamics is termed as the *heat engine* or *refrigerator.* Reversed Carnot cycle is the underlying principle that governs the operation of a mechanical refrigeration system. It involves the following four stages:

1. **Isothermal expansion:** Heat is rejected by the system.
2. **Adiabatic compression:** Work is done on the system.
3. **Isothermal compression:** Heat is absorbed by the system.
4. **Adiabatic expansion:** Work is produced by the system.

In accordance with the aforesaid four steps, the mechanical refrigeration system (Figure 11.1) is constituted of four major components:

1. Evaporator
2. Compressor
3. Condenser
4. Expansion valve

A refrigerant circulates through these four components to change in state from liquid to vapor and back to liquid state. The product is placed inside a cooling chamber, from where the refrigerant absorbs the heat.

11.2.2.1 Components of a Mechanical Refrigeration System

11.2.2.1.1 Evaporator

An evaporator is a heat exchanger by its operation. Inside the evaporator, the product temperature is lowered as the liquid refrigerant vaporizes to the gaseous state by absorbing the latent heat of vaporization

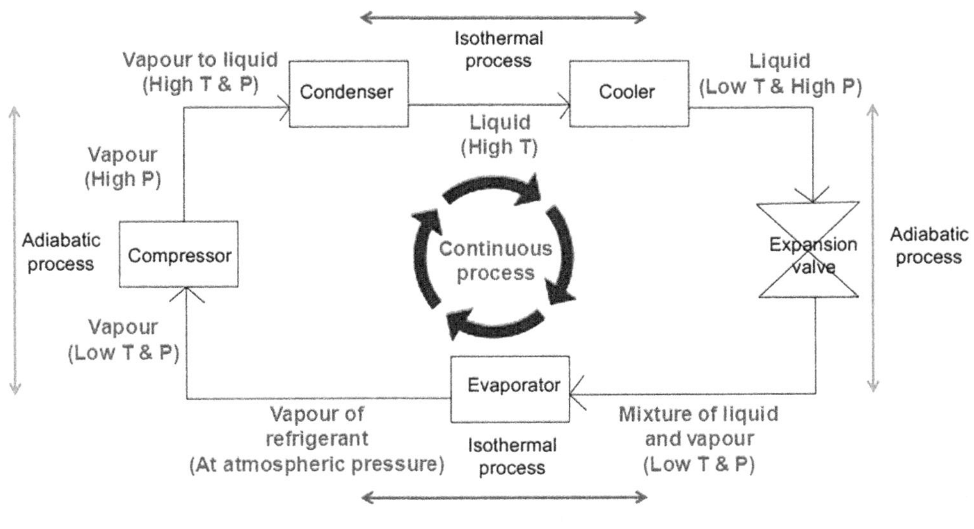

FIGURE 11.1 Mechanical refrigeration system and its components.

from the food product. Thus, in the evaporator, heat is rejected by the food product, and the vapors of refrigerant undergo isothermal expansion. The aforementioned events constitute the first step of the reversed Carnot cycle of mechanical refrigeration.

Evaporators are generally of finned tube or plate type. In the finned type evaporator (Figure 11.2(a)), air is forced to flow through the fins. As air flows over the fins, it gains heat energy from the food product and delivers the heat to the evaporator. The function of fins is to decrease the resistance to heat transfer and increase the contact area of the evaporator. In the plate type evaporator (Figure 11.2(b)), cooling occurs by indirect contact between the refrigerant and the product to be cooled.

11.2.2.1.2 Compressor

The vapors of refrigerant that exit the evaporator at low temperature and pressure enter the compressor. Compressor (Figure 11.3) draws low pressure on the cooling side and compresses the gas into the high-pressure side of the mechanical refrigeration system (Barbosa-Cánovas et al., 2005). It increases the enthalpy of refrigerant vapor by elevating its pressure and temperature. Thus, inside the compressor, work is done on the refrigerant vapors without the transfer of heat or matter, which constitutes the second step of the mechanical refrigeration cycle, i.e., the adiabatic compression.

Subsequently, the refrigerant discharges the heat in the condenser. The temperature difference between the heated refrigerant emerging from the compressor and the ambient temperature around the condenser is the driving force for the heat transfer between the refrigerant and the surrounding medium.

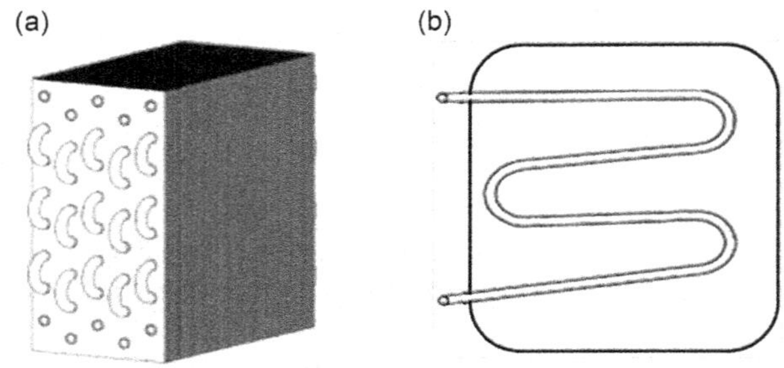

FIGURE 11.2 Types of evaporator: (a) finned tube evaporator (Reproduced with permission from Starace, G., Fiorentino, M., Meleleo, B. and Risolo, C. 2018. The hybrid method applied to the plate-finned tube evaporator geometry. *International Journal of Refrigeration* 88: 67–77.); (b) plate evaporator.

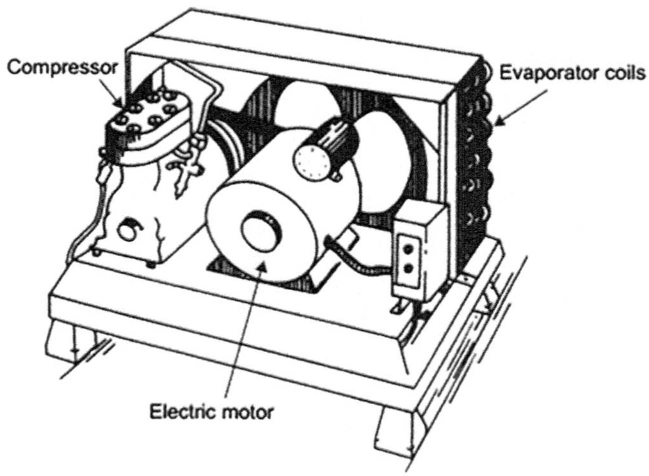

FIGURE 11.3 A typical compressor of a mechanical refrigeration system with a two-cylinder, air-cooled condenser driven by an electric motor. (Reproduced with permission from Singh, R. P. and Heldman, D. 2014. *Introduction to Food Engineering* (Enhanced Fifth edition). London: Academic Press.)

11.2.2.1.3 Condenser

A condenser is also a heat exchanger by its function, in which the heat from the refrigerant is transferred to a medium such as air and water. By rejecting heat, the gaseous refrigerant condenses to form liquid inside the condenser. The water or air cooled environment in a condenser reduces the temperature of the superheated refrigerant vapors exiting the compressor. Consequently, the vapors cool and condense to form the liquid, while sustaining their high pressure acquired within the compressor. The abovementioned events constitute the "isothermal compression," the third stage of the mechanical refrigeration cycle.

The major types of condensers used in the mechanical refrigeration system are (i) water-cooled, (ii) air-cooled, and (iii) evaporative condensers which use both air and water. Three common types of water-cooled condensers are (i) double pipe (Figure 11.4 (b)), (ii) shell and tube (Figure 11.4 (a)), and (iii) shell and coil.

11.2.2.1.4 Expansion Valve

An expansion valve is a temperature-sensitive valve located between the condenser and the evaporator. It controls the flow of refrigerant to the evaporator by maintaining the pressure gradient (Barbosa-Cánovas et al., 2005). It reduces the pressure of the liquefied refrigerant, after which the refrigerant exists as a typical mixture of liquid (75%) and vapors (25%) and enters the evaporator to expand adiabatically and complete one cycle of the mechanical refrigeration. The expansion valve separates the high-pressure zone of the mechanical refrigeration system from the low-pressure zone. It can be operated manually or automated using sensors for pressure or temperature (thermostatic expansion valve).

11.2.3 Pressure–Enthalpy Charts

During the refrigeration cycle, a refrigerant passes through a series of isothermal and adiabatic processes during which the pressure and enthalpy of the refrigerant change continuously. Inside the evaporator and condenser, while the enthalpy of refrigerant changes, its pressure remains constant. Within the compressor, the refrigerant is compressed to increase both its enthalpy and pressure. Enthalpy of the refrigerant remains constant when it passes through the expansion valve. Consequently, the refrigerant absorbs and releases the heat energy during a cycle. The pressure–enthalpy chart is one which explains the changes in refrigerant properties during the refrigeration cycle in a straightforward manner. Familiarizing with the pressure–enthalpy diagram would facilitate a food engineer in designing a refrigeration system.

Figure 11.5 depicts the pressure–enthalpy chart under saturated conditions, in which y-axis (ordinate) represents the pressure of the refrigerant and x-axis (abscissa) denotes the enthalpy of the refrigerant. Points ABCDE indicate various stages of the mechanical refrigeration process. In the compressor

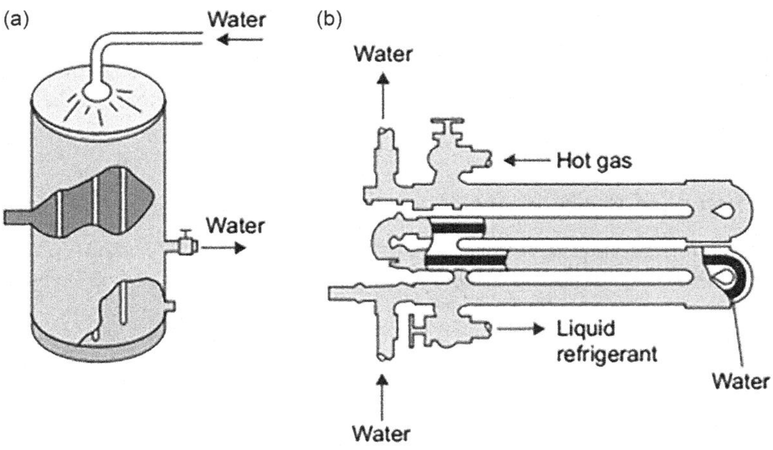

FIGURE 11.4 Types of water-cooled condenser: (a) shell and tube and (b) double-pipe (Reproduced with permission from Singh, R. P. and Heldman, D. 2014. *Introduction to Food Engineering* (Enhanced Fifth edition). London: Academic Press.)

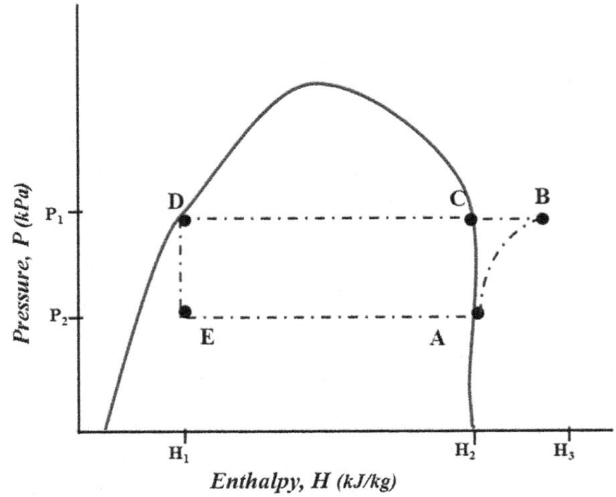

FIGURE 11.5 Pressure–enthalpy chart for mechanical refrigeration.

(from point A to B), the refrigerant undergoes compression at constant entropy. The pressure of the refrigerant is increased from P_2 to P_1, and the enthalpy of the refrigerant is increased from H_2 to H_3. In the condenser (from point B to C), enthalpy is reduced from H_3 to H_2 at constant pressure. Further, in the condenser (from point C to D), latent heat is removed to condense the refrigerant from vapor to liquid state at a constant pressure while reducing its enthalpy from H_2 to H_1. Now the saturated liquid enters the expansion valve at point D. Pressure decreases from P_1 to P_2 in the expansion valve at constant entropy while the enthalpy (H_1) remains constant. At point E, the refrigerant exists in the form of a vapor–liquid mixture, which enters the evaporator and then converted to the vapor form by increasing the enthalpy from H_1 to H_2, at constant pressure, P_2.

11.2.4 Mathematical Expressions for Calculations in Mechanical Refrigeration

11.2.4.1 Cooling Load

Cooling load is defined as the heat removal rate from a given space or product for lowering the temperature to the desired level. The typical unit for the cooling load is the ton of refrigeration. One ton capacity of a refrigeration unit is 3.5168 kW, which is approximately equivalent to the latent heat of fusion of 1 ton of ice in 24 h. In the earlier days before the introduction of mechanical refrigeration, ice was the most widely used cooling medium. Hence, the cooling capacity was often related to the melting of ice. Thus, a mechanical refrigeration system that can absorb heat from the refrigerated space at the rate of 3.5168 kW is rated at 1 ton of refrigeration.

11.2.4.2 Coefficient of Performance

The efficiency of the refrigeration system is given by its coefficient of performance (CoP). It is the ratio of heat absorbed by the refrigerant in the evaporator to the heat equivalent work done by the compressor. The following expression gives the CoP,

$$\text{CoP} = \frac{(H_2 - H_1)}{(H_3 - H_2)} \tag{11.1}$$

where H_1 is the enthalpy of the refrigerant at the exit of the condenser, H_2 is the enthalpy of the refrigerant at the entry of compressor, H_3 is the enthalpy of the refrigerant at the exit of the compressor. It is always important to have a higher CoP for a refrigeration cycle. Alternatively, the CoP can also be expressed in terms of temperature as

$$\text{CoP} = \frac{T_1}{(T_2 - T_1)} \tag{11.2}$$

where T_1 and T_2 are, respectively, the lower (temperature of refrigerant in the evaporator) and higher (temperature of refrigerant in the condenser) values of the refrigerator's working range of temperature.

11.2.4.3 Refrigerant Flow Rate

Refrigerant flow rate $(\dot{m})$ depends on the total cooling load (Q). Thus, it can be expressed in terms of Q as

$$\dot{m} = \frac{Q}{H_2 - H_1} \tag{11.3}$$

In Eq. (11.3), the total cooling load is computed from the heat to be removed from the refrigerated space.

11.2.4.4 Work Done by the Compressor

Work done by the compressor (W_c) on the refrigerant during constant entropy compression can be determined based on the enthalpy rise of the refrigerant and mass flow rate of the refrigerant.

$$W_C = \dot{m}(H_3 - H_2) \tag{11.4}$$

11.2.4.5 Heat Exchanged in the Condenser and Evaporator

In the condenser and evaporator, heat exchange occurs at constant pressure. The amount of heat gained or rejected by the refrigerant in the evaporator and condenser can be expressed as given in Eqs. (11.5) and (11.6), respectively.

$$Q_E = \dot{m}(H_2 - H_1) \tag{11.5}$$

$$Q_C = \dot{m}(H_3 - H_1) \tag{11.6}$$

11.2.4.6 Refrigeration Effect

Refrigeration effect is the difference in enthalpy of the refrigerant at the inlet and outlet of an evaporator. Refrigeration effect is given by ($H_2 - H_1$). Thus, CoP (Eq. (11.1)) can also be defined as the refrigerant effect divided by the work done externally.

11.2.4.7 Theoretical Power to Drive the Compressor

Theoretical power (w) required for driving the compressor per ton of cooling load can be calculated by applying the following equation:

$$w = \dot{m}(H_2 - H_1) \tag{11.7}$$

11.2.4.8 Power per Unit Ton of Refrigeration Capacity

Power per unit ton of refrigeration capacity is given by the inverse of CoP which is equal to $\frac{(H_3 - H_2)}{(H_2 - H_1)}$.

Example 11.1

A refrigeration system is maintained at 20°C using a vapor-compression refrigeration system that uses R717. The evaporator and condenser temperatures for the R717 refrigerant are −34.4°C and 37.8°C, respectively. The enthalpies of the refrigerant at the exit of the condenser, entry of the compressor, and exit of the compressor are 36, 139, and 182.3 J/kg, respectively. The refrigeration load is 10 tons. Calculate the mass flow rate of refrigerant and the CoP.

Solution

Mass flow rate of refrigerant is given by $m = \dfrac{Q}{H_2 - H_1}$

$$m = \frac{10 \times 303852}{(24 \times 3600) \times (139 - 36)} = 0.341\,\text{kg/s}$$

CoP is given by

$$\text{CoP} = \frac{H_2 - H_1}{H_3 - H_2} = \frac{(139 - 36)}{(182.3 - 139)} = 2.379$$

Answer: Mass flow rate: 0.341 kg/s; CoP = 2.379

11.2.5 Calculation of Cooling Time

11.2.5.1 For Liquid Food Products

For a liquid food which is well stirred, the temperature (T) is uniform throughout the product. Thus, considering Newton's law of cooling,

$$mC_P \frac{dT}{dt} = hA(T - T_a) \tag{11.8}$$

where m is the mass of liquid food product (kg), C_P is the specific heat of liquid product (J/kg K), and t is the cooling time. Based on Eq. (11.8), it is possible to predict the change in temperature of the liquid product with time under constant conditions.

$$T = T_a + (T_i - T_a)\exp\left(-\frac{hA}{mc_P}t\right) \tag{11.9}$$

where T_i is the initial temperature of the product (°C), T_a is the temperature of freezing medium (°C), and h is the heat transfer coefficient (W/m^2 K). The rate of cooling (or, heat release), Q (W) is given by

$$Q = hA(T - T_a) \tag{11.10}$$

Substituting Eq. (11.9) in Eq. (11.10),

$$Q = hA(T_i - T_a)\exp\left(-\frac{hA}{mc_p}t\right) \tag{11.11}$$

11.2.5.2 For Solid Food Products

Unlike liquid foods, the temperature within a solid food product is not uniform during the cooling process. This is due to the internal resistance to heat transfer as the heat from the center is conducted through layers of solid before reaching the surface and transferred to the surroundings either directly or through layers of packaging. As discussed in Chapter 5, Biot number is the measure of the ratio of internal to external resistance, given by

$$N_{Bi} = \frac{hd}{k} \tag{11.12}$$

where d and k are the smallest dimension and thermal conductivity of the product, respectively. If $N_{Bi} \ll 1$, under the conditions of high thermal conductivity, small thickness, and low heat transfer coefficient, the internal resistance can be neglected, thus, limiting the case to chilling of liquid products. Conversely, if $N_{Bi} > 1$ owing to the high internal resistance, the cooling rate can be improved by reducing the product dimension rather than reducing the layers of packaging or increasing the air velocity. Also, the cooling time is influenced by the shape of the solid product. The mathematical expressions for cooling time pertaining to different solid shapes and the explanation of different parameters in the equation are listed in Table 11.4. For solid food products with irregular shape, the solution is obtained by approximating to the nearest regular shape. The mathematical expressions listed in Table 11.4 and the plots shown in Figures 11.6–11.8 can then be used to predict the cooling time (Pham, 2015).

TABLE 11.4

Cooling Time of Slabs, Infinite Cylinders, and Spheres

Mathematical Expression	Explanation of Parameters
$Y_c \equiv \frac{T_c - T_a}{T_i - T_a} = j_c \exp\left(-2.303\frac{t}{f}\right)$ $Y_m \equiv \frac{T_m - T_a}{T_i - T_a} = j_m \exp\left(-2.303\frac{t}{f}\right)$ (Reference: Pham, 2015)	T_c = Center temperature T_m = Mean temperature j_c, j_m = Lag factors can be calculated from the plots shown in Figures 11.6–11.8. f = Time for the residual temperature difference and heat load to decrease by a factor of 1/10 Y_c, Y_m = Dimensionless residual temperatures

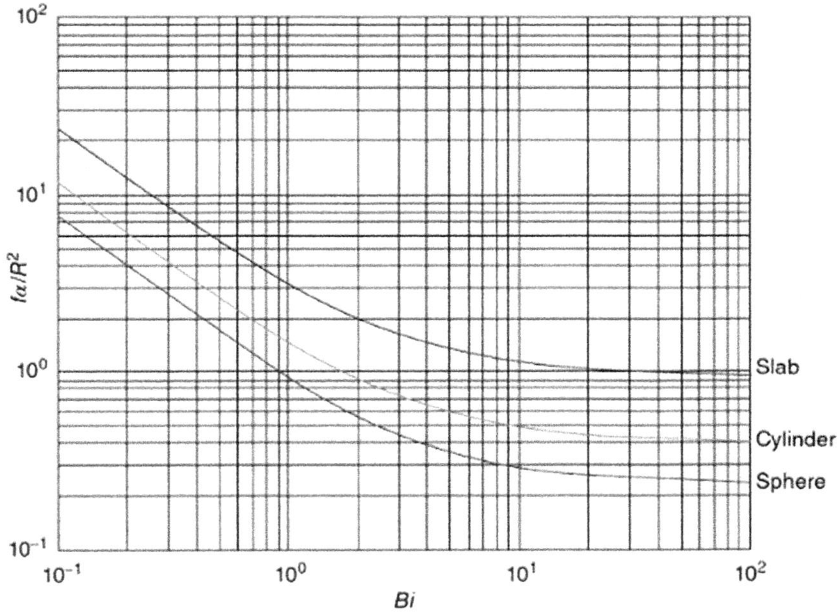

FIGURE 11.6 Plot between $f\alpha/R^2$ and Biot number. (α: thermal diffusivity (m^2/s); $\alpha = k/\rho c_p$; ρ =density (kg/m^3). (Reproduced with permission from Pham, Q. T. 2015. Refrigeration in food preservation and processing. In *Conventional and Advanced Food Processing Technologies*, ed. S. Bhattacharya, 357–386. Chichester, West Sussex, UK: John Wiley and Sons, Ltd.)

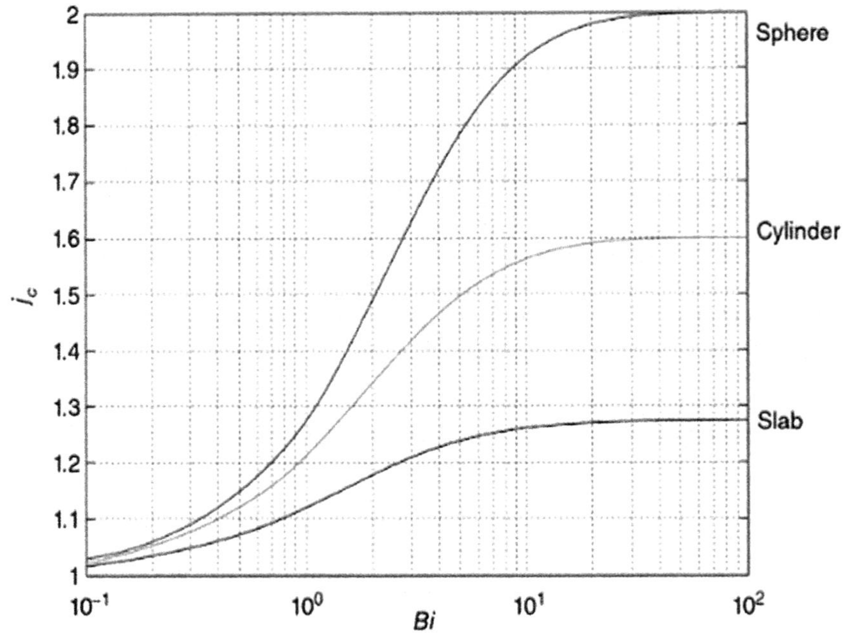

FIGURE 11.7 The plot between j_c and Biot number. (Reproduced with permission from Pham, Q. T. 2015. Refrigeration in food preservation and processing. In *Conventional and Advanced Food Processing Technologies*, ed. S. Bhattacharya, 357–386. Chichester, West Sussex, UK: John Wiley and Sons, Ltd.)

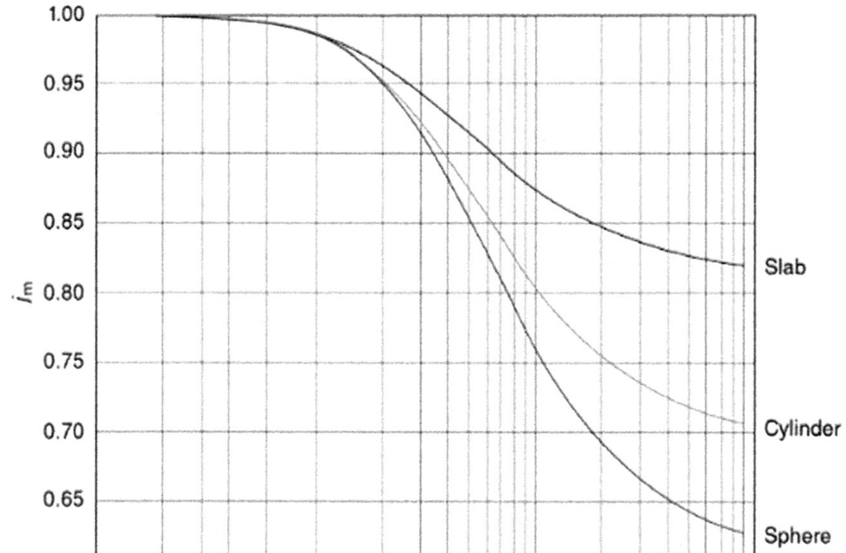

FIGURE 11.8 The plot between j_m and Biot number. (Reproduced with permission from Pham, Q. T. 2015. Refrigeration in food preservation and processing. In *Conventional and Advanced Food Processing Technologies*, ed. S. Bhattacharya, 357–386. Chichester, West Sussex, UK: John Wiley and Sons, Ltd.)

11.3 Freezing of Foods

11.3.1 Theory of Freezing

Freezing is a complicated process than refrigeration, as it involves the phase change from liquid water to solid ice. The freezing process begins with the removal of sensible heat to lower the temperature of food product to below its freezing point. After the supercooling of the product down to its freezing point, a substantial amount of energy is required for the ice crystal formation. The specific heat and latent heat of water are reasonably high in the order of 4200 J/kg K and 335 kJ/kg, respectively. A freezing curve explains the temperature history at the thermal center of a food product during the freezing process. The freezing curve is a plot of product temperature versus time, which is unique for each food product. Figure 11.9 depicts the freezing curve for pure water.

- Line AB shows the temperature history of the product when it is precooled from its initial temperature to the freezing temperature (T_f), in conjunction with the removal of sensible heat. Initially, at the freezing temperature, ice nucleation does not occur. But, the product is further cooled which is known as *supercooling*. During the supercooling of the food product, water remains in the liquid state. Once the temperature reaches the point B (freezing point), the nucleation starts.
- Line BC shows the stage where ice crystallization begins, and the latent heat is released. During this period, the product temperature increases due to the release of latent heat.
- Line CD indicates the conversion of a significant portion of water inside the food to ice crystals. This occurs at a nearly constant temperature of the product as phase change involves only latent heat and does not entail sensible heat. The rate of ice formation depends on the removal of latent heat.
- Line DE shows that once all the water freezes, the temperature of the frozen product falls to the temperature of the freezing chamber.

Freezing point is defined as the temperature at which the first ice crystal appears, and the liquid at that temperature is in equilibrium with the solid (Barbosa-Cánovas et al., 2005). The freezing point of pure water is 0°C (273 K). During the freezing of pure water, a sharp shift is observed between the different stages of freezing (Figure 11.9). But, this is not the case with food products which include both free and bound water. The bound water is one which does not freeze even at very low temperatures. The interaction of this unfreezable water with its constituents such as salts, carbohydrates, proteins, fibers, and ash causes a decrease in the freezing point of water. The higher the solute content, the lower the freezing

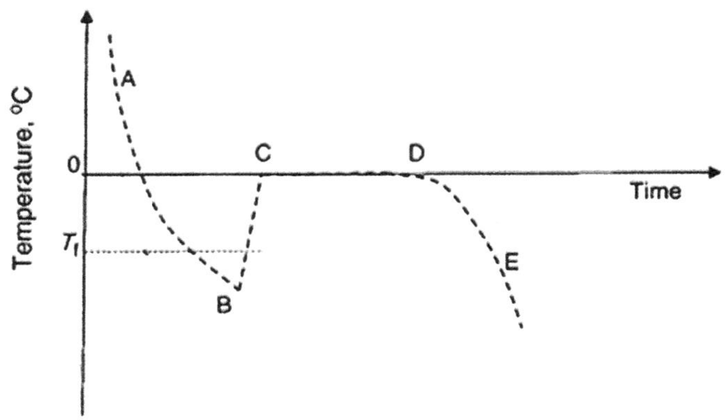

FIGURE 11.9 The freezing curve for pure water. (Modified and reproduced with permission from Kozlowski, T. 2009. Some factors affecting supercooling and the equilibrium freezing point in soil–water systems. *Cold Regions Science and Technology* 59: 25–33.)

point. This phenomenon is known as *freezing point depression.* As a result, the freezing point of food products is always less than 0°C. Due to ice crystallization, the solute concentration increases and further reduces the freezing point. Therefore, freezing does not take place at a constant temperature but occurs over a temperature range. However, the temperature at which the first ice crystal appears is commonly regarded as the initial freezing temperature (Barbosa-Cánovas et al., 2005). The thermal properties of a food product also vary during the freezing process. In the case of food products, during the pre-freezing stage, the temperature decreases below the freezing temperature (Point A in Figure 11.10). With the formation of the first ice crystal, the temperature again increases to the freezing temperature (Point B in Figure 11.10). Subsequently, during the freezing period, the food product's temperature decreases slightly due to the increasing concentration of solutes in the unfrozen water. This is in contrast to the constant temperature encountered during the freezing of pure water.

Ice formation ceases once the temperature of food product reaches the glass transition temperature (temperature below which the product exists in the glassy solid or rigid state). Nevertheless, the freezing rate influences the ice crystal formation and eventually the product quality. While rapid freezing leads to the formation of smaller ice crystals, slow freezing forms larger ice crystals. While small ice crystals do not affect the product structure and texture, larger ice crystals have a deteriorative effect on the product. During slow freezing, the large ice crystals grow between the cells of the food structure and hence rupture the cell wall. The detrimental effect of ice crystals is attributed to their lower water vapor pressure when compared to that within the cells. Due to this gradient, water moves from the cells to ice crystals. With the progress in the freezing process, the solute concentration increases and the cells become dehydrated and ultimately lead to structural collapse. On the other hand, during rapid freezing, fine ice crystals form in both the intracellular and intercellular spaces. With fast freezing, due to the absence of a gradient in water vapor pressure, the dehydration effect is reduced and the structural and textural damages are prevented (Fellows, 2000).

11.3.2 Freezing Time

A variable temperature distribution exists within the food product during the freezing process. Freezing time varies depending on the location within the product at which the temperature is monitored. *Thermal center* is defined as the location in the food product which cools most slowly. It is generally used as the reference location for defining the freezing time. Accordingly, the definition

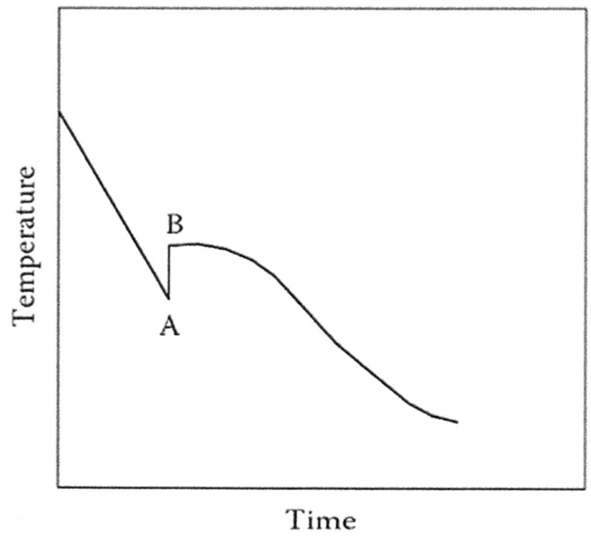

FIGURE 11.10 Typical freezing curve for food products. (Reproduced with permission from Wang, L. and Weller, C. J. 2012. Thermophysical properties of frozen foods. In *Handbook of Frozen Food Processing and Packaging*, ed. D. W. Sun, 101–128. Boca Raton, FL: CRC Press, Taylor & Francis Group.)

provided by the International Institute of Refrigeration (1972) states that *effective freezing time is the total time required to lower the temperature of a food material at its thermal center to a desired temperature below the initial freezing point.*

Freezing time is influenced by multiple factors including (Persson and Löndahl, 1993)

- Dimensions (mainly thickness) and shape of the food product
- Initial and final temperatures of the food product
- The surface heat transfer coefficient of the food product
- The thermal conductivity of the food product
- Change in enthalpy
- The temperature gradient between the freezing medium and the food product
- Barriers to heat transfer such as the packaging material

11.3.2.1 Calculation of Freezing Time

Prediction of freezing time is important from the perspectives of assessing the food product quality, processing requirements, and economic aspects of food freezing (Mittal et al., 1993). For uniformly shaped food products, the freezing time can be calculated reasonably accurate, provided all the required information is available. However, in general, calculation of freezing time is complex owing to the change in thermophysical properties (density, thermal conductivity, specific heat, and thermal diffusivity) of the food with the reduction in temperature and change in phase (Rolfe, 1967). This is because freezing does not occur at a single temperature as the foods undergoing freezing release latent heat over a range of temperature. Therefore, the commonly used approach to calculate freezing time is to seek empirical relationships, rather than exact analytical equations (Mittal et al., 1993). The subsequent sections discuss the various empirical equations for the calculation of freezing time.

11.3.2.1.1 Plank's Equation

An approximate solution for the freezing time was derived by Plank (1941) which is based on the assumptions that, initially, the unfrozen food is at the freezing temperature and the thermal conductivity of the frozen part is constant. Further, the Plank's equation considers that all the material freezes at the freezing point with constant latent heat. In the frozen layer, the rate of conductive heat transfer is slow such that the conditions can be approximated to pseudo-steady state.

Consider a slab of thickness a (m) which is cooled from both sides by convection. At a given time t (s), a frozen layer of thickness x (m) forms on both sides. The surrounding temperature is constant at T_l (°C), and the freezing temperature is constant at T_f (°C). An unfrozen layer is present in the center at temperature T_f (Figure 11.11).

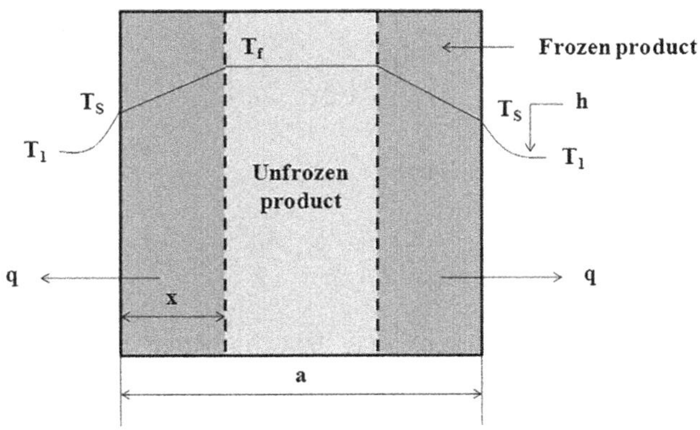

FIGURE 11.11 Temperature profile during freezing.

The heat removed from the slab at time t is q W. Under the pseudo-steady-state conditions, at time t, the heat removed from the slab to the surrounding by convection is given by

$$q = hA(T_s - T_1) \tag{11.13}$$

where A is the surface area, h and T_s are the convective heat transfer coefficient and temperature at the surface of the material, respectively. Also, the heat being conducted through the frozen layer of x thickness at steady state is

$$q = \frac{kA}{x}(T_f - T_s) \tag{11.14}$$

where k is the thermal conductivity of the frozen material (W/m °C). Within a given time dt, a layer of thickness dx of the material freezes. The mass (kg) of frozen product is obtained by multiplying A times dx times ρ. Multiplying this quantity by the latent heat of food product λ (J/kg) and dividing by dt,

$$q = \frac{Adx\rho\lambda}{dt} = A\rho\lambda\frac{dx}{dt} \tag{11.15}$$

where ρ is the density of the unfrozen material (kg/m³). To eliminate T_s from Eqs. (11.13) and (11.14), Eq. (11.13) is solved for T_s and substituted into Eq. (11.14), which gives

$$q = \frac{(T_f - T_1)A}{x/k + 1/h} \tag{11.16}$$

Equating Eqs. (11.15) and (11.16),

$$\frac{(T_f - T_1)A}{x/k + 1/h} = A\rho\lambda\frac{dx}{dt} \tag{11.17}$$

Rearranging and integrating Eq. (11.17) between $t = 0$ and $x = 0$, to $t = t$ and $x = a/2$,

$$(T_f - T_1)\int_0^t dt = \lambda\rho\int_0^{a/2}\left(\frac{x}{k} + \frac{1}{h}\right)dx \tag{11.18}$$

Solving the Eq. (11.18) for t,

$$t = \frac{\lambda\rho}{T_f - T_1}\left(\frac{a}{2h} + \frac{a^2}{8k}\right) \tag{11.19}$$

To obtain a generalized equation for other shapes,

$$t = \frac{\lambda\rho}{T_f - T_1}\left(\frac{Pa}{h} + \frac{Ra^2}{k}\right) \tag{11.20}$$

where a can be the thickness of an infinite slab (as shown in Figure 11.11), the diameter of a sphere, diameter of a long cylinder, or smallest dimension of a rectangular block or brick (in the unit of meter, m). The constants P and R account for the product shape as mentioned in Table 11.5.

From Eq. (11.20), it is evident that freezing time is directly proportional to the density, latent heat, and the thickness of the product. It is inversely proportional to the thermal conductivity, temperature

TABLE 11.5

Shape Constants of the Plank's Equation for Freezing Time

Constant	Infinite Slab	Infinite Cylinder	Sphere
P	1/2	1/4	1/6
R	1/8	1/16	1/24

gradient, and convective heat transfer coefficient. Therefore, to reduce the freezing time, product thickness should be minimal, and the temperature gradient should be high. However, the limitation of Plank's equation is that it considers only the period of phase change during the freezing process.

11.3.2.1.2 Pham's Equation

Pham proposed a method for the prediction of freezing and thawing time for foods (Pham, 1986). In this method, the mean freezing temperature (T_{fm}) of a product is determined by the following empirical expression, which finds application in the calculation of freezing time (t).

$$T_{fm} = 1.8 + 0.263T_C + 0.105T_a \tag{11.21}$$

where T_c is the final center temperature of the product and T_a is the temperature of the freezing medium (cold air). Further, Pham equation for calculating freezing time is given by

$$t = \frac{d_c}{E_f h}\left[\frac{\Delta H_1}{\Delta T_1} + \frac{\Delta H_2}{\Delta T_2}\right]\left(1 + \frac{N_{Bi}}{2}\right) \tag{11.22}$$

where d_c is the characteristic dimension of product, E_f is the shape factor, an equivalent heat transfer dimension similar to P and R of Plank's equation (E_f is 1 for infinite slab, 2 for an infinite cylinder, and 3 for sphere), h is the convective heat transfer coefficient (W/m^2 K), N_{Bi} is the Biot number, ΔT_1 and ΔT_2 are the temperature gradients for the initial precooling and post-cooling periods, respectively, ΔH_1 is the change in volumetric enthalpy (J/m^3) for the precooling period, and ΔH_2 is the change in volumetric enthalpy (J/m^3) for the phase change and post-cooling period. In Eq. (11.22), ΔH_1, ΔH_2, ΔT_1, and ΔT_2 are determined based on the following expressions:

$$\Delta H_1 = \rho_l C_{Pl}(T_i - T_{fm}) \tag{11.23}$$

$$\Delta H_2 = \rho_f\left[L_f + C_{Pf}\left(T_{fm} - T_c\right)\right] \tag{11.24}$$

$$\Delta T_1 = \left(\frac{T_i + T_{fm}}{2}\right) - T_a \tag{11.25}$$

$$\Delta T_2 = T_{fm} - T_a \tag{11.26}$$

where ρ_l is the density of liquid (kg/m^3), ρ_f is the density of frozen material (kg/m^3), C_{Pl} is the specific heat of liquid material (J/kg K), C_{Pf} is the specific heat of frozen material (J/kg K), T_i is the initial temperature of the material (K), and L_f is the latent heat of fusion of food material (J/kg).

Various other mathematical expressions for the calculation of freezing time have been proposed by modification of the Plank and Pham equations, which are compiled in Table 11.6.

TABLE 11.6

Mathematical Expressions for the Calculation of Freezing Time

Mathematical Expression	Equation	Explanation of Parameters
Nagaoka et al. equation (1956) (modified Plank model, for regular shaped foods)	$t_F = \frac{\Delta H' \rho}{T_F - T_\infty}\left[\frac{Pa}{h_c} + \frac{Ra^2}{k_1}\right]$ $\Delta H' = (1 + 0.008T_i)\left[C_{Pu}(T_i - T_F) + L_f + C_{PI}(T_F - T)\right]$	T_i: Initial temperature of the food product, T: Final temperature of the frozen food product, C_{Pu}: Specific heat of unfrozen food, C_{PI}: Specific heat of the frozen food
Levy equation (1958) (modification of Nagaoka et al. equation; for regular shaped foods)	$\Delta H' = (1 + 0.008(T_i - T_F))\left[C_{Pu}(T_i - T_F) + L_f + C_{PI}(T_F - T)\right]$	
Cleland and Earle equation (1984) (modified Plank model; applicable to any shape) valid within Biot number = 0.5 to 4.5, Plank number = 0 to 0.55, and Stephan number = 0.155 to 0.345	$t_f = \frac{\rho \Delta H_{ref}}{E(T_f - T_\infty)}\left[\frac{Pa}{h_c} + \frac{Ra^2}{k_1}\right]\left(1 - \frac{1.65N_{Ste}}{k_1}\ln\left[\frac{(T - T_\infty)}{(T_{ref} - T_\infty)}\right]\right)$ $N_{Ste} = \text{Stefan number} = \frac{C_{PI}(T_F - T_\infty)}{\Delta H_{ref}}$ $N_{PK} = \text{Plank's number} = \frac{C_{Pu}(T_i - T_F)}{\Delta H_{ref}}$ $P = 0.5\left[1.026 + 0.5808N_{PK} + N_{Ste}(0.2296N_{PK} + 0.105)\right]$ $R = 0.125\left[1.202 + N_{Ste}(3.410N_{PK} + 0.7336)\right]$	T_{ref}: Reference temperature (considered as −10°C) E: 1 for an infinite slab, 2 for an infinite cylinder, and 3 for a sphere. ΔH_{ref}: enthalpy change from T_F to T_{ref}
Cleland et al. equation (1987)	$t_f = \frac{1.3179\rho C_{PI} a^2}{k_1 E}\left[\frac{0.5}{N_{Bi}N_{Ste}} + \frac{0.125}{N_{Ste}}\right]^{0.9576}$ $N_{Ste}^{0.0550}10^{0.0017N_{Bi}+0.1727N_{PK}}\left[1 - \frac{1.65N_{Ste}}{k_1}\ln\frac{T - T_\infty}{T_{ref} - T_\infty}\right]$ $N_{Bi} = \text{Biot number} = \frac{ha}{k_1}$	T_{ref}: Reference temperature (considered as −10°C)

Example 11.2

A spherical food product is frozen in the air-blast freezer. The initial temperature of the product is 20°C, and the temperature of the cold air is −50°C. The product has a diameter of 5 cm, with a density of 1000 kg/m³. The initial freezing temperature is −2.2°C. The thermal conductivity of frozen product is 1.2 W/m K; latent heat of fusion is 250 kJ/kg, and the heat transfer coefficient is 50 W/m² K. Calculate the freezing time.

Solution

Given

$T_i = 20°C = 293.15$ K
$T_1 = -50°C = 223.15$ K
$T_f = -2.2°C = 270.95$ K
$D_p = 5$ cm $= 5 \times 10^{-2}$ m $= 0.05$ m
$\rho_f = 1000$ kg/m³
$k = 1.2$ W/m K
$\lambda = 250$ kJ/kg $= 250000$ J/kg
$h = 50$ W/m² K
Since the product is spherical, the shape factors are: $P = 1/6$; $R = 1/24$

Therefore, according to the Plank's equation, $t_f = \frac{\lambda\rho}{T_f - T_1}\left(\frac{Pa}{h} + \frac{Ra^2}{k}\right)$

$$t_f = \frac{250000 \times 1000}{270.95 - 223.15}\left(\frac{\frac{1}{6} \times 0.05}{50} + \frac{\frac{1}{24} \times 0.05^2}{1.2}\right)$$

Thus, the freezing time is 1326 s or 22.1 min.

Answer: Freezing time is 22.1 min

11.3.3 Types of Freezing Equipment

The criteria for the classification of freezing equipment include the following:

i. *Type of contact established between the product and freezing medium:* The contact between freezing medium and product can be direct or indirect. In direct contact freezers, the rate of heat transfer is high due to the effective contact between the product and freezing medium. This method of rapid freezing in direct contact systems is also known as *individual quick freezing*. In indirect contact freezers, a non-permeable barrier exists between the food product and the freezing medium.
ii. *Condition of the freezing medium:* The freezing medium can be still or stationary or it can be circulated at a high velocity to establish contact with the food product to be frozen.
iii. *Rate of ice front movement:* The rate of movement of ice from the product surface to the thermal center of the food depends on the properties of food product and the efficiency of heat transfer between the freezing medium and the food product, quantified by the heat transfer coefficient. According to the rate of ice front movement, freezers are classified into four types:
   - Slow and sharp freezer (rate of ice front movement: 0.2–1 cm/h)
   - Quick freezers (rate of ice front formation: 0.5–3 cm/h)
   - Rapid freezers (rate of ice front formation: 5–10 cm/h)
   - Ultrarapid freezers (rate of ice front formation: 10–100 cm/h)

Based on these criteria, various types of freezing equipment are available as listed in Table 11.7.

11.3.3.1 Plate Freezers

In the plate freezer (Figure 11.12), series of plates are arranged in parallel either horizontally or vertically, through which the coolant is circulated. Opening (for loading and unloading of the food product) and closing (to establish adequate contact between the plate and food product during freezing) of space between the plates is accomplished using a hydraulic (by applying pressure) system. The product in contact with the plates freezes rapidly by conduction. A uniform and effective heat transfer between the plate and product is facilitated by the high packing density or low void space of the product within the package and a minimum headspace. Spacers between the plates and a pressure relief valve in the hydraulic system prevent the product from being crushed between the plates during the closure. While the horizontal plate freezers are suitable for regular-shaped products and products packed into rectangular cartons, vertical plate freezers are used for the freezing of unpackaged and deformable products such as fish and meat.

The plate freezers can be operated in batch or continuous mode. In the batch operation, the residence time depends on the time to reduce the product temperature from its initial temperature to the freezing point, which in turn depends on the product thickness. At the end of the freezing process, ice seal between the plate and the food product is broken by circulating hot fluid through the plates. Cleaning

TABLE 11.7

Classification of Freezing Equipment

Type of Freezer	Type of Contact	Condition of Coolant	Heat Transfer Coefficient (*h*) (W/m²K)	Rate of Ice Front Movement (cm/h)	Characteristic Freezing Time (min) to reach –18°C and Corresponding Examples of Food Products
Plate freezer	Indirect	Freezing medium in contact with the plate surface	600	0.5–3	120 (blocks of fish/minced meat) 25 (vegetables, seafood, ice cream, hamburgers packed in cartons)
Air-blast freezer	Direct or Indirect	Air velocity: 2.5 m/s Air velocity: 5 m/s	17 10–50	0.5–3	15–20 (unpacked vegetables)
Fluidized bed freezer	Direct	Air in *fluidized* state Air velocity: ~30 m/s	110–160	5–10	3–4 (peas, beans, sweet corn)
Scraped surface freezer	Indirect	Freezing medium in contact with the outer wall of the heat exchanger barrel	900	5–10	0.3–0.5 (ice cream of 1 mm thick layer)
Spiral belt freezer	Indirect	Air velocity: 3–8 m/s	25	0.5–3	12–19 (fish fingers)
Immersion freezer	Direct	Freezing medium in the liquid state	500 (Freon) 100–950 (aqueous solutions)	5–10	10–15 (orange juice packed in cartons) 0.5 (peas)
Cryogenic freezer	Direct	Gas zone (pre-cooling) Spray zone (freezing)	1500 (liquid nitrogen)	10–100	2–5 (seafoods) 0.5–6 (fruits and vegetables)
Cold storage or still air freezer	Direct	Still air	6–9	0.2–1	180–4000 (meat carcass)

Source: Data from Heldman and Singh (1981), Fellows (2000), Mascheroni (2001), Anon (2007), Arce and Sweat (1980), and Dempsey and Bansal (2010).

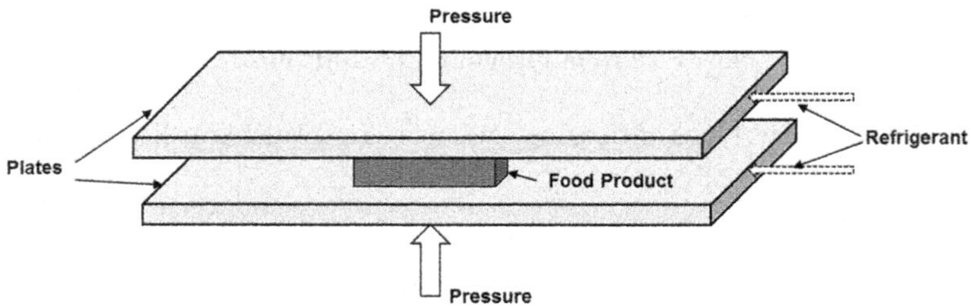

FIGURE 11.12 Schematic diagram of a plate freezer.

may be required before loading the next batch of product. During the continuous operation, automated systems enable the opening of individual plates to facilitate simultaneous loading and unloading of the product from feed conveyors. A roller plate moves, the speed of which is set according to the freezing time of food product to control its residence time within the freezer.

Plate freezers are advantageous concerning providing a high freezing rate for packaged products. The consistent size and shape of the frozen product enables it to be bulk stacked with high packing density and renders it stable during the transit. Further, the plate freezers are very compact with a less frequent defrosting requirement. Nevertheless, high capital cost limits the commercial application of fully automated plate freezers while the manual operation of batch plate freezers is labor intensive. Moreover, the thickness of the product loaded between each plate should be the same, thus confining the product types that can be frozen using a plate freezer.

11.3.3.2 Air-Blast Freezers

An air-blast freezer (Figure 11.13) is the most commonly used food freezer owing to its simpler design. The product to be frozen is placed on trays in racks or suspended so that air flow is possible around each product item. Using the refrigerant, air is cooled down to a temperature of −30°C and −40°C, by the refrigerant in a heat exchanger. The air-blast freezer uses high power fans to blow the ultra-low temperature air at a velocity of 1.5 m/s to 6 m/s around the product for the defined freezing time. High air velocity ensures high heat transfer rate. However, this is possible only when a stage is reached where heat transfer through the food is the rate-limiting factor. Thus, in the air-blast freezers, low temperature and high velocity of air lead to a synergistic effect of the high rate of convective heat transfer from the product. The typical freezing time in air blast freezers varies between 12 and 48 h.

Air-blast freezers can be operated in batch or continuous mode. Unlike plate freezers, air-blast freezers are used irrespective of the product shape. The predominant types of foods frozen in air-blast freezers are medium to large products in which the size limits the freezing rate. The primary advantage of the air-blast freezer is that it can be used to freeze food product of any shape. However, due to the lower temperature gradient between the food and cold air, freezing time is longer with the air-blast freezer. The air-blast freezers can be further classified into room or tunnel and belt freezers.

11.3.3.2.1 Tunnel Air-Blast Freezers

In batch tunnel freezers, the products are manually loaded in racks and positioned inside a tunnel. In continuous tunnel freezers, a mechanical system moves the racks through the tunnel in a cyclic manner, with an automated device for loading and unloading of the product. In this type of freezer, air is circulated over the product at a temperature of −30°C and −40°C and velocity between 1.5 and 6.0 m/s. The air flow is limited to the cross section where the product is located. The product should be spaced evenly through the length of the tunnel to achieve uniform air distribution and high air velocity for a low total air flow rate and fan power.

11.3.3.2.2 Belt Air-Blast Freezers

In belt freezers, the product is transported on a perforated conveyor belt. The cold air of high velocity is directed vertically up through the perforations of the belt and perpendicular to the product layer. The

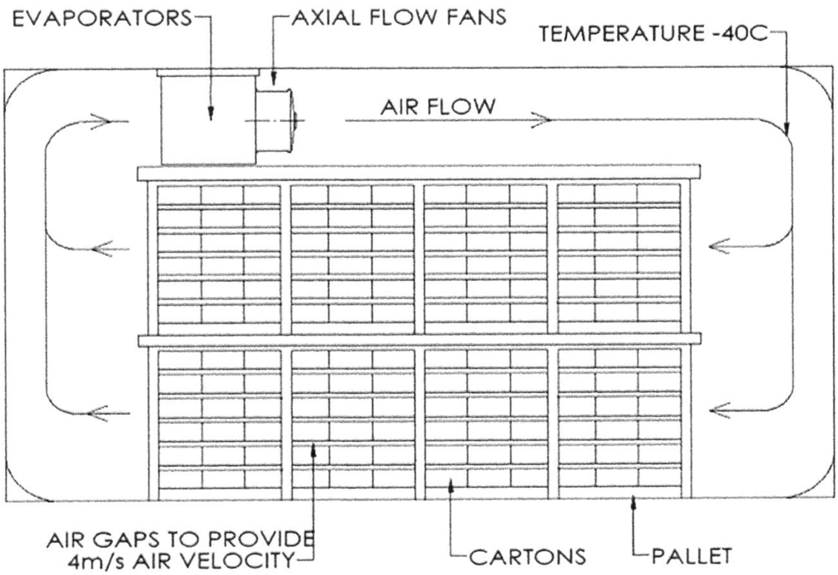

FIGURE 11.13 Schematic of a typical batch air blast freezer. (Reproduced with permission from Dempsey, P. and Bansal, P. 2010. Air blast freezers and their significance to food freezing: A Review. *Proceedings of ENCIT 2010, 13th Brazilian Congress of Thermal Sciences and Engineering* December 05–10, 2010, Uberlandia, MG, Brazil.)

air flow may also be parallel to the flow of product on the conveyor belt. The air velocities are typically in the range of 1–6 m/s. This facilitates the high rate of heat transfer between the air and product. The speed of the conveyor belt is adjusted and set according to the freezing time, and multiple belt passes may be used. The product should be evenly distributed across the belt and spread at a uniform thickness to achieve consistent air distribution and freezing rate.

To increase the productivity of belt freezers and maximize the belt surface area for given floor space, the conveyor belts may be positioned one above the other which run in the same or opposite directions or be arranged spirally. The former is termed as *multi-tier freezer*, and the latter is known as *spiral belt freezer*. Spiral belt freezers (Figure 11.14) comprise of a flexible mesh conveyor which is laterally bent around a rotating drum and stacked one above the other up to 30 spiral tiers. The cold air which moves down through the tiers of the belt contacts the product lifted by the belt in a countercurrent configuration. The temperature and velocity of cold air are similar to that used in tunnel freezers. The advantages of this freezer type are the high throughput of up to 3000 kg/h, and its flexibility to handle both packed and unpacked food products with wide range of freezing times (10 min–3 h) (ASHRAE Handbook, 1994).

11.3.3.3 Fluidized Bed Freezer

Fluidized bed freezer is a modified form of the air-blast freezer. This type of freezer comprises a bed with a perforated bottom through which cold air is blown vertically upward at a high air velocity (Figure 11.15). At the fluidization velocity of cold air, the solid food particles behave like a fluid and remain suspended in the stream of the freezing medium, without adhering to each other. In the fluidized bed freezer, effective contact between the food product and freezing medium leads to high heat transfer rate and rapid freezing.

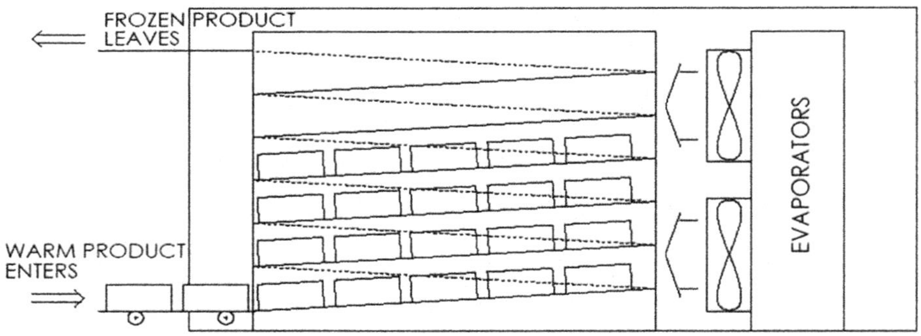

FIGURE 11.14 Schematic of a typical spiral belt freezer. (Reproduced with permission from Dempsey, P. and Bansal, P. 2010. Air blast freezers and their significance to food freezing: A Review. *Proceedings of ENCIT 2010, 13th Brazilian Congress of Thermal Sciences and Engineering* December 05–10, 2010, Uberlandia, MG, Brazil.)

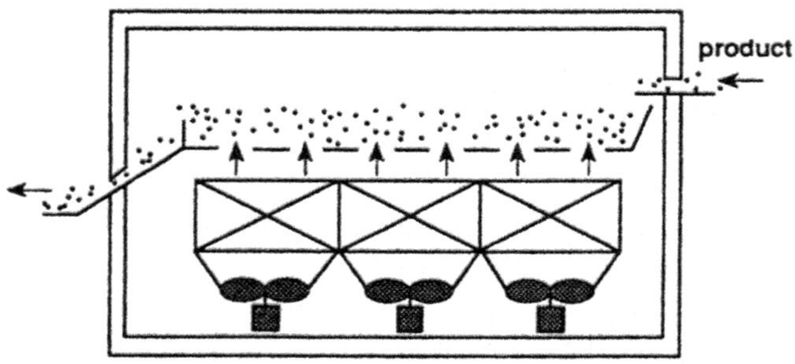

FIGURE 11.15 Fluidized bed freezer. (Reproduced with permission from Cleland, D. J. and Valentas, K. J. 1997. Prediction of freezing time and design of food freezers. In *Handbook of Food Engineering Practice*, eds. K. J. Valentas, E. Rotstein and R. P. Singh, 71–124. Boca Raton, FL: CRC Press, Taylor & Francis Group.)

The bed thickness and air velocity are adjusted according to the size and shape of the food product to be frozen. The average residence time of the individual pieces of food in a fluidized bed freezer depends on the flow pattern in the bed, the feed rate, and the volume of the bed which is controlled by the height of overflow weir. The foods that are commonly frozen in fluidized bed freezers include peas, cut corn, diced carrots, strawberries, diced meat, sliced fruits and vegetables, and small pieces of seafood and shrimps.

11.3.3.4 Scraped Surface Freezers

Scraped surface freezers are used for those food products in which a consistently smooth texture is an essential indicator of quality. The product types that are predominantly handled in this freezer are ice cream and margarine.

The scraped surface freezer is a tubular heat exchanger, constituted of two concentric cylinders (Figure 11.16). A refrigerant such as ammonia or Freon flows in the annular space between the two concentric cylinders such that the refrigerant temperature is maintained along the length of the freezer. The refrigerant is maintained at a high pressure to maintain its liquid state, reduce the air phase volume, and increase the heat transfer. The pressure is controlled such that the refrigerant remains at a temperature of about −30°C. When the high-pressure refrigerant is pumped into the freezer at low pressure, it expands and vaporizes to provide the refrigeration effect. The product temperature in contact with the refrigerated metal wall of the barrel reduces rapidly to below the freezing point after losing its heat to the refrigerant. Then, nucleation occurs, and the freezing cycle of the product proceeds further. The scraper blades held by a dasher rotating at 150–300 RPM within the barrel of the freezer scrapes the frozen food and newly formed ice crystals from the refrigerated wall of freezer barrel. The abovementioned operation leads to controlled crystal growth. The scraper blades also function as beaters to whip the product and incorporate air into it to impart the smooth and light texture. The occluded air content is known as *overrun*, which is the percentage increase in the volume of the product (e.g., ice cream) over and above the amount of mix used to produce that product.

The rate of freezing is high in continuous scraped surface freezers with 50% of water frozen (Jaspersen, 1989), within say, 10–30 s after the product is pumped through the freezer. Rapid freezing leads to the formation of fine ice crystals which gives a smooth and creamy consistency to the product. However, the freezing time in batch freezers is in the range of 10–15 min.

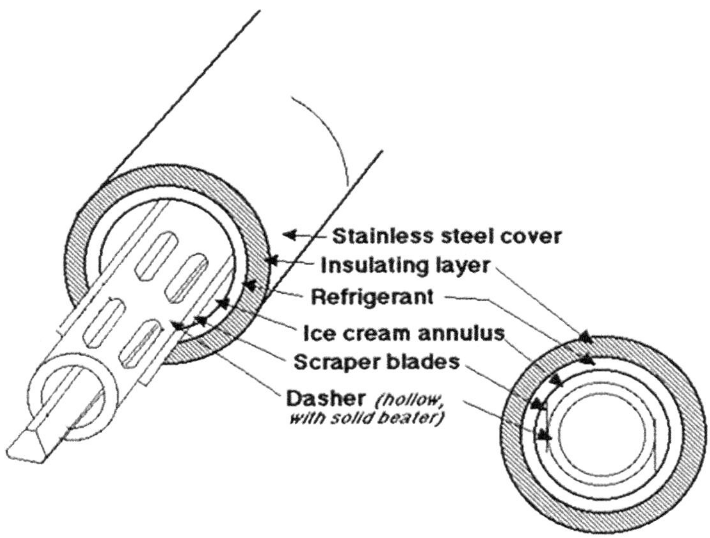

FIGURE 11.16 Schematic of a scraped surface freezer. (Reproduced with permission from Goff, H. D. 2018. Dairy science and technology education series. University of Guelph, Canada. www.uoguelph.ca/foodscience/book-page/freezingwhipping-ice-cream)

11.3.3.5 Immersion Freezers

In the immersion freezers, packed or unpacked food products are passed on a conveyor and directly immersed in the low-temperature liquid freezing medium contained in a tank (Figure 11.17). The low temperature of the liquid freezing medium is achieved by circulating through a heat exchanger or cooling coils or a jacket built into the tank. Alternatively, the freezing medium can be sprayed on the product while being conveyed through the freezer. Rapid freezing ($h > 100$ W/m^{2}K) occurs due to the direct immersion of food product in the freezing medium. Freezing time of the product can be controlled by regulating the speed of the conveyor. Immersion freezers are widely used for surface freezing followed by air blast freezing to achieve complete freezing of the product. Products with irregular shapes can be easily handled in an immersion freezer. Although high rates of freezing can be achieved, these types of freezers are now seldom used except for some fish, meat, and poultry products.

The freezing medium used in immersion food freezers must be of food grade. It should neither influence the flavor of the food product nor leave any residue on the product after freezing. Brine and glycol are the commonly used freezing media for the immersion freezing systems.

11.3.3.6 Cryogenic Freezers

Immersion freezers can also be operated with cryogens such as liquid nitrogen (LN_2; boiling point: −196°C at atmospheric pressure) and liquid carbon dioxide (LCO_2; boiling point: −79°C at atmospheric pressure) as freezing medium. The product is either sprayed with or immersed in the cryogen at atmospheric pressure. The design of cryogenic freezers is such that the cryogen is circulated in a direction countercurrent to that of the movement of food product to result in efficient heat transfer. Cryogenic freezers are generally used for small- to medium-sized products such as fish fillets, fruits, and berries because in larger products the rate of freezing is limited by heat transfer internal to the product. Cryogenic freezing can also be used in conjunction with tunnel and fluidized bed freezers.

The major advantage of cryogenic freezers is the larger temperature gradient, which leads to higher heat transfer rate. Cryogens are colorless, odorless, chemically inert, and non-toxic at normal concentrations. Therefore, they are safe for direct contact with food. An initially rapid decrease in surface temperature minimizes the product moisture and leads to a higher quality product. In addition, a low refrigerant-to-product surface heat transfer resistance results in lower weight loss. Ease of operation, compact size, low equipment cost, low operation cost, low maintenance cost, and rapid installation and start-up are among the other advantages of cryogenic freezers. However, the main disadvantage of cryogenic freezers is the high cost of cryogens and their storage system. Effective insulation and refrigeration of the storage tank are necessary to prevent the entry of heat and the consequent loss of cryogen.

11.3.3.7 Still Air Freezers

This type of freezer is used for both freezing and cold storage of frozen food products. Cold air at a temperature of −10°C to −29°C is obtained by natural convective cooling on contact with the

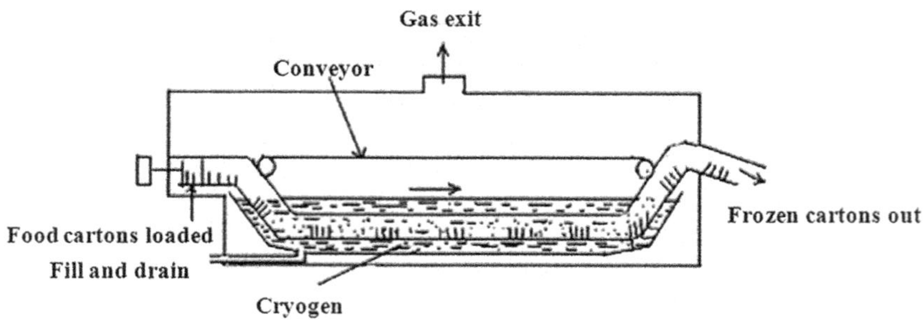

FIGURE 11.17 Immersion freezing system. (Modified and reproduced with permission from Khadatkar, R. M., Kumar, S. and Pattanayak, S. C. 2004. Cryofreezing and cryofreezer. *Cryogenics* 44: 661–678.)

evaporator. The products are stacked on the shelves of a refrigerated room fitted with refrigerated coils at one side. The products are exposed to cold air which flows past the product at minimum velocity. Consequently, the convective heat transfer coefficients are very low leading to slow freezing, longer freezing time, higher weight loss, and poor quality of the product due to the formation of large ice crystals. Further, in still air freezers, the stored products may be subjected to dehydration and undesirable temperature fluctuations owing to the removal of heat from the adjacent freezing product.

11.4 Refrigerated Transportation of Foods

Refrigerated transportation or reefer freight can be defined as the method for shipping temperature-sensitive, perishable food products in an enclosed refrigerated space which can be a truck, container, van, or any carrier vehicle. The vehicle is built in with a refrigeration unit which provides a controlled temperature in the range of −25°C to 15°C to maintain the product at the desired temperature throughout the transit. The essential components of a refrigerated transportation system include the vehicle, refrigeration unit, and air delivery systems.

11.4.1 Vehicle

By design, the vehicle for refrigerated transportation is a semitrailer (an unpowered vehicle pulled by a powered vehicle but without a front axle) insulated rigid box. The typical dimensions of a refrigerated vehicle are as follows (Marshall et al., 2006):

- External dimensions
  - Length: 13.56 m
  - Width: 2.6 m
  - Height: 2.75 m
- Internal dimensions
  - Length: 13.35 m
  - Width: 2.46 m
  - Height: 2.5 m

Insulating the vehicle is important to prevent the flow of heat through its walls. Expanded polyurethane foam is the commonly used material for the insulation of refrigerated vehicles. It is a lightweight, waterproof, and noncorrosive material with thermal conductivity in the range of 0.022 W/m K. Extruded polystyrene is another commonly used insulation material. The insulation is further sandwiched between two external layers of plywood of few millimeters thickness covered with a glass reinforced polyester, steel, or aluminum skin. These layers reflect the heat rays from the sun or road surfaces and improve the quality of insulation. Apart from the insulation, the door seals must be intact and properly fitted to avoid heat loss.

The vehicles for refrigerated transport can comprise single or multiple compartments. In multi-compartment vehicles, the refrigerated space is subdivided into many sections (Figure 11.18). Each compartment can be set at independently controlled temperature set points to provide flexibility in logistics and cater to many business operations. The temperature in each chamber is achieved using distributed evaporator coils fed from a single condensing unit. Typically, the coldest compartment is located at the front, and the warmest is at the rear. Nevertheless, the system is adaptable to any temperature configuration. The influential factors of importance in the design of multi-compartment refrigeration systems are the heat transfer between the compartments, product loading patterns, door openings, and the method of temperature control (LeBlanc and Hui, 2005). Supermarket chains are the primary users of multi-compartment vehicles to deliver multiple products with varied temperature requirements (Tassou et al., 2009).

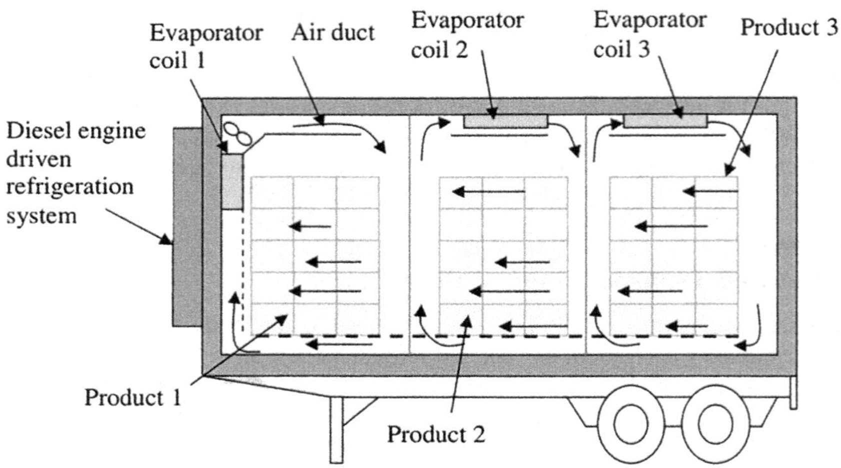

FIGURE 11.18 Schematic of a multi-compartment vehicle for refrigerated transport. (Reproduced with permission from Tassou, S. A., De-Lille, G. and Ge, Y. T. 2009. Food transport refrigeration—approaches to reduce energy consumption and environmental impacts of road transport. *Applied Thermal Engineering* 29: 1467–1477.)

11.4.2 Refrigeration Unit

Mechanical refrigeration with vapor compression system (Figure 11.19) is the most commonly used refrigeration system for the refrigerated food transport applications. The drive systems for refrigerated transport vapor compression systems are as follows (Marshall et al., 2006; Department for Transport—Transfrigoroute, 2016):

- Vehicle alternator unit
  - The crankshaft of vehicle engine drives a single upgraded alternator and a 70 Ah battery.
  - The alternator charges the vehicle battery which is sufficient to feed a small refrigeration system with a 12 V DC supply.

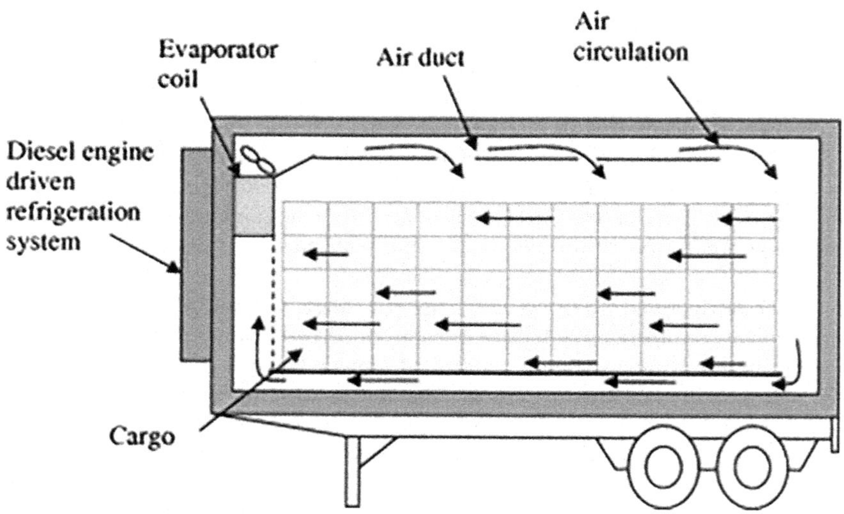

FIGURE 11.19 Refrigeration system and air circulation in a semitrailer. (Reproduced with permission from Tassou, S. A., De-Lille, G. and Ge, Y. T. 2009. Food transport refrigeration—approaches to reduce energy consumption and environmental impacts of road transport. *Applied Thermal Engineering* 29: 1467–1477.)

  - During standby, the system can also be driven with a 230 V mains electric supply.
  - Generally used in small delivery vans.
- Direct belt drive
  - The compressor of the refrigeration unit is directly driven from the vehicle engine through a belt.
  - Finds application in van size vehicles.
- Auxiliary alternator unit
  - In this type, a dedicated large alternator is driven by a belt from the main traction engine which generates power to drive an electric motor in the refrigeration unit.
  - Fan motors for the heat exchangers and control system are also fed from the alternator output.
- Auxiliary diesel unit
  - A diesel engine built into the refrigeration unit.
  - This drive type is used in vast majority of medium to large vehicles.

The choice among the aforementioned drive systems is based on the duty required, weight, noise limits, maintenance requirements, installation cost, environmental considerations, and fuel taxation. The performance and power requirements of the refrigeration unit are typically assessed under full load. In reality, however, transport refrigeration systems operate over a wide range of loads. In general, the CoP for a transport refrigeration unit ranges between 0.5 and 1.5 (Tassou et al., 2009).

11.4.3 Air Delivery Systems

Top air delivery is the principally adopted approach in refrigerated transportation. Fans in the refrigeration unit facilitate the circulation of temperature-controlled air around the inside of the vehicle, roof, walls, floor, and doors. Consequently, the heat is removed from the products, which is conducted and infiltrated from the outside and returned to the cooling coil through the floor or space under pallets. In chilled cargoes carrying fruit and vegetables, horizontal channels are provided between two rows of cartons to permit good return airflow through the load. This is important in commodities in which heat of respiration forms a significant part of the heat load (Tassou et al., 2009).

11.5 Problems to Practice

11.5.1 Multiple Choice Questions

1. One ton of refrigeration is equal to the refrigeration effect corresponding to melting of 1000 kg of ice in
   a. 1 h
   b. 1 min
   c. 24 h
   d. 12 h

Answer: c

2. The vapor compression refrigerator uses the following cycle:
   a. Rankine
   b. Carnot
   c. Reversed Rankine
   d. Reversed Carnot

Answer: d

3. The value of CoP in vapor compression cycle is
   a. always less than unity
   b. always more than unity
   c. equal to unity
   d. any of the above

Answer: b

4. In vapor compression refrigeration system, refrigerant occurs as liquid between
   a. condenser and expansion valve
   b. compressor and evaporator
   c. expansion valve and evaporator
   d. compressor and condenser

Answer: a

5. The condition of refrigerant after passing through the condenser in a vapor compression system is
   a. saturated liquid
   b. wet vapor
   c. dry saturated vapor
   d. superheated vapor

Answer: a

6. In air blast freezers
   a. cold air at high velocity is used
   b. liquid refrigerant is used
   c. plate heat exchanger is used
   d. still air is used

Answer: a

7. Cryogenic freezing results in frozen products with
   a. large ice crystals
   b. small ice crystals
   c. combination of both large and small ice crystals
   d. large or small ice crystals

Answer: b

8. At a fixed evaporator temperature, when the condenser temperature increases,
   a. the CoP decreases
   b. the CoP increases
   c. the CoP remains constant
   d. the CoP may increase or decrease

Answer: a

9. According to Plank's equation of freezing time, freezing time is
   a. directly proportional to the square of product thickness
   b. inversely proportional to the square of product thickness
   c. inversely proportional to density of the product
   d. directly proportional to thermal conductivity of the product

Answer: a

10. The portion of water in a food which does not freeze is
    a. unbound water
    b. bound water
    c. free water
    d. none of the above

Answer: b

11. Freon-12 is a compound consisting of
    a. carbon, hydrogen, and fluorine
    b. carbon, hydrogen, and oxygen
    c. carbon, hydrogen, fluorine, and chlorine
    d. carbon, fluorine, and chlorine

Answer: d

12. In a mechanical refrigeration system the highest temperature of refrigerant occurs
    a. between condenser and evaporator
    b. in evaporator
    c. between compressor and condenser
    d. before expansion valve

Answer: c

13. The flow of the refrigerant in a refrigeration cycle is controlled by
    a. expansion valve
    b. compressor
    c. condenser
    d. evaporator

Answer: a

14. The condition of refrigerant as it leaves the compressor in a vapor compression refrigeration cycle is
    a. saturated liquid
    b. wet vapor
    c. dry saturated vapor
    d. superheated vapor

Answer: d

15. The vapor pressure of refrigerant should be
    a. lesser than atmospheric pressure
    b. greater than atmospheric pressure
    c. equal to atmospheric pressure
    d. May be lesser or greater than atmospheric pressure

Answer: b

16. The refrigerant with the lowest freezing point is
    a. R12
    b. R22
    c. R11
    d. R21

Answer: b

17. The lowest temperature during the vapor compression cycle occurs after
    a. compression
    b. expansion
    c. condensation
    d. evaporation

Answer: d

18. Ammonia is preferred as a refrigerant in large commercial installations because of
    a. lower energy cost
    b. higher heat transfer coefficients
    c. greater detectability in the event of a leak
    d. all of the above

Answer: d

19. The evaporator changes the low pressure liquid refrigerant from the expansion valve into
    a. high pressure liquid refrigerant
    b. low pressure liquid and vapor refrigerant
    c. low pressure vapor refrigerant
    d. high pressure vapor refrigerant

Answer: c

20. A condenser is used in the ________ pressure side of the refrigeration system.
    a. low
    b. high
    c. both a and b
    d. none of the above

Answer: b

11.5.2 Numerical Problems

1. Calculate the following for a refrigeration cycle with Freon-12 as the refrigerant flowing at a rate of 1.5 kg/min:
    a. CoP of the cycle
    b. Mass flow rate of the refrigerant
    c. Work done by the compressor

   The evaporator and condenser temperatures for Freon-12 are −10°C and 40°C, respectively. The enthalpies of refrigerant at the exit of condenser, at the beginning of compressor, and at the end of compression are 26.85, 183.1, and 220 kJ/kg, respectively. The refrigeration load is 10 tons.

   Given

   i. Evaporator temperature (T) = −10°C
   ii. Condenser temperature = 40°C
   iii. $H_1 = 26.85$ kJ/kg
   iv. $H_2 = 183.1$kJ/kg
   v. $H_3 = 220$ kJ/kg
   vi. Refrigeration load = 10 tons

To find

i. CoP of the cycle
ii. Mass flow rate of the refrigerant
iii. Work done by the compressor

Solution

$$\text{CoP} = \frac{(H_2 - H_1)}{(H_3 - H_2)} = \frac{(183.1 - 26.85)}{(220 - 183.1)} = 4.2$$

$$\dot{m} = \frac{Q}{H_2 - H_1} = \frac{(10\,\text{tons}) \times (303852\,\text{kJ})}{(183.1 - 26.85) \times \left(\frac{24\,\text{h} \times 3600\,\text{s}}{h}\right)} = 0.225$$

(Since, 1 ton of refrigeration = 303852 kJ/24 h)

$$W_C = \dot{m}(H_3 - H_2) = 0.225(220 - 183.1) = 8.302$$

Answer: (i) CoP = 4.2
(ii) Mass flow rate of the refrigerant = 0.225 kg/s
(iii) Work done by the compressor = 8.302 kW

2. For a 20 ton capacity refrigeration system, the pressure of refrigerant in the evaporator and condenser are 200 and 700 kPa, respectively. If ammonia (R-717) is used under saturated conditions, calculate the theoretical power required to operate the compressor.

Given

i. Cooling load (Q) = 20 ton
ii. Pressure of refrigerant in the evaporator = 200 kPa
iii. Pressure of refrigerant in the condenser = 700 kPa

To find: Theoretical power required to operate the compressor (w)

Solution

$$w = \dot{m}(H_2 - H_1)$$

where $\dot{m} = \dfrac{Q}{H_2 - H_1}$

$$\therefore w = Q$$

$$w = Q = 20\,\text{ton}$$

$$1\,\text{ton} = 3.5168\,\text{kW}$$

$$\therefore 20\,\text{ton} = 70.336\,\text{kW}$$

$$1\,\text{HP} = 0.746\,\text{kW}$$

$$\therefore 70.336\,\text{kW} = 94.284\,\text{HP}$$

Answer: Theoretical power required to operate the compressor = 70.336 kW or 94.284 HP

3. 200 kg of bananas are supplied to a cold storage at 20°C. The cold storage is maintained at −4°C and the fruits are cooled to the storage temperature in 5 h. The latent heat of freezing for bananas is 255 kJ/kg and the specific heat of fruit is 3.35 kJ/kg K. Use the given information to determine the refrigeration capacity of the plant.

Given:

i. Quantity of bananas (m) = 200 kg
ii. $T_2 = 20°C = 20+273 = 293$ K
iii. $T_1 = -4°C = -4+273 = 269$ K
iv. Latent heat of freezing of banana ($h_{fg(banana)}$) = 255 kJ/kg
v. Specific heat of banana (C_P) = 3.35 kJ/kg K
vi. Cooling time = 5 h

To find: Refrigeration capacity of the plant

Solution

The sensible heat removed from the bananas in 5 h is

$$q_S = mC_P\Delta T = 200 \times 3.35 \times (293 - 269) = 16080\,\text{kJ}$$

The total latent heat of freezing is

$$q_L = m.h_{fg(\text{banana})} = 200 \times 255 = 51000\,\text{kJ}$$

Therefore, the total heat removed from the bananas in 5 h is

$$q = q_S + q_L = 16080 + 51000 = 67080\,\text{kJ}$$

Thus, the total heat removed in 1 min is $\dfrac{67080}{5 \times 60} = 223.6\,\text{kJ/min}$

$$\text{Refrigeration capacity of the plant} = 223.6/211 = 1.059\,\text{TR}$$

(Since, $1\,\text{TR} = 3.5168\,\text{kJ/s} = 3.5168 \times 60 = 211\,\text{kJ/min}$)

Answer: Refrigeration capacity of the plant = 1.059 TR

4. Find the CoP of a refrigeration system if the work input is 50 kJ/kg and the heat extraction capacity is 150 kJ/kg.

Given

i. Work input or work done by the compressor (W_c) = 50 kJ/kg
ii. Heat extraction capacity (Q_E) = 150 kJ/kg (which is nothing but the heat extracted from the product by the refrigerant or the heat gained by the refrigerant in the evaporator)

To find: CoP

Solution

$$\text{CoP} = \frac{(H_2 - H_1)}{(H_3 - H_2)}$$

Therefore, for unit mass flow of refrigerant,

$$\text{CoP} = \frac{Q_E}{W_C} = \frac{150}{50} = 3$$

Answer: CoP of the refrigeration system = 3

5. For a refrigerator with working temperature between −5°C and 20°C, the cooling load is 100 TR. If the cycle operates on reversed Carnot cycle and the latent heat of fusion of ice is 335 kJ/kg, calculate the mass of ice produced per day from water at 20°C and the CoP of the refrigeration system.

Given

i. Cooling load (Q) = 100 TR
ii. $T_1 = -5°C = 268$ K
iii. $T_2 = 20°C = 293$ K
iv. $T_w = 20°C$
v. $\lambda = 335$ kJ/kg

To find

i. Mass of ice produced per day
ii. CoP

Solution

i. Mass of ice produced per day

Given that the cooling load of the refrigerator is 100 TR and 1 TR = 3.5168 kJ/s,

$$100\,\text{TR} = 100 \times 3.5168 = 351.68\,\text{kJ/s}$$

And, heat removed from 1 kg of water at 20° C to form ice at 0°C

$$= (\text{Mass} \times \text{specific heat} \times \text{rise in temperature}) + (h_{fg(\text{ice})})$$

$$= 4.187 \times (20 - 0) + 335 = 418.74\,\text{kJ/kg}$$

Therefore, mass of ice produced per day $= \frac{351.68}{418.74} = 0.83985\,\text{kg/s}$, which is equal to $\frac{0.83985}{\left(\frac{1}{86400}\right)} =$ 72563.04 kg (or) 72.563 tons of ice (since 1 day = 86400 s).

ii. CoP of the refrigeration system

CoP for a reversed Carnot cycle is given by $\frac{T_1}{T_2 - T_1} = \frac{268}{293 - 268} = 10.72$

Answer: (i) Mass of ice produced per day = 72.563 tons
(ii) CoP = 10.7

6. A refrigerated room is maintained at the desired temperature using an evaporator temperature of 0°C and condenser pressure of 800 kPa, with an R-134a vapor-compression refrigeration system. Calculate the CoP and power per unit ton of refrigeration capacity of this refrigeration unit.

Given

i. $T_l = 0°C = 273$ K
ii. Refrigerant: R134a (1, 1, 1, 2-tetrafluoroethane)
iii. Condenser pressure = 800 kPa

To find

i. CoP
ii. Power per unit ton of refrigeration capacity

Solution

i. ***CoP***

CoP for a reversed Carnot cycle is given by $\dfrac{T_1}{T_2 - T_1}$

The temperature of R134a corresponding to condenser pressure of 800 kPa is 35.8°C or 308.95 K.

$$\therefore \text{CoP} = \frac{273}{309 - 273} = 7.6$$

Power per unit ton of refrigeration capacity is the inverse of CoP which is given by 1/7.6 = 0.132.

7. A spherical product of diameter 50 mm is frozen in an air-blast tunnel freezer. The initial product temperature is 10°C and the temperature of cold air circulated in the tunnel is −12°C. The product density is 900 kg/m³ and convective heat transfer coefficient is 45 W/m²K. The thermal conductivity of the product is 1 W/m K and the initial freezing temperature is −5°C. The latent heat of fusion is 335 kJ/kg. Compute the freezing time using Plank's Equation.

Given:

i. $T_i = 10°C$
ii. $T_l = -12°C$
iii. $T_f = -5°C$
iv. $D_p = 50 \text{ mm} = 50 \times 10^{-3} \text{ m} = 0.05 \text{ m}$
v. $\rho_f = 900 \text{ kg/m}^3$
vi. $k = 1$ W/m K
vii. $\lambda = 335$ kJ/kg
viii. $h = 45 \text{ W/m}^2 \text{ K}$
ix. Since the product is spherical, the shape factors are: P = 1/6; R = 1/24

To find: Freezing time (*t*)

Solution

Therefore, according to Plank's equation,

$$t = \frac{\lambda\rho}{T_f - T_1}\left(\frac{Pa}{h} + \frac{Ra^2}{k}\right) = \frac{(335\times 900)}{-5-(-12)}\left(\left(\frac{(1/6)\times(0.05)}{(45\times 10^{-3})}\right) + \frac{((1/24)\times(0.05)^2)}{(1\times 10^{-3})}\right) = 12447.6\text{s}$$

$= 12447.6/3600 = 3.5\text{h}$

Answer: Freezing time = 3.5 h

8. If the freezing time is 3 h in problem no. 7, calculate the diameter of the spherical product being frozen in the air blast tunnel freezer.

Given

Freezing time (t) = 3 h = 3 × 3600 = 10800 s

To find: Diameter of the product

Solution

Using Plank's equation,

$$t = \frac{\lambda \rho}{T_f - T_1}\left(\frac{Pa}{h} + \frac{Ra^2}{k}\right)$$

$$10800 = \frac{(335 \times 900)}{-5-(-12)}\left(\left(\frac{(1/6)\times(a)}{\left(45\times 10^{-3}\right)}\right) + \frac{\left((1/24)\times(a)^2\right)}{\left(1\times 10^{-3}\right)}\right)$$

$$10800 = 43071.4\left(3.704\ a + 41.667\ a^2\right)$$

$$0.251 = \left(3.704\ a + 41.667\ a^2\right)$$

$$\therefore a = 0.045\,\text{m}$$

Answer: Diameter of the product = 0.045 m or 45 mm

9. Orange juice contained in a can of diameter 5 cm and height 15 cm is frozen in an air-blast freezer. The surface heat transfer coefficient is 15 W/m²K, initial product temperature is 4°C, and temperature of the freezing medium is −10°C. Using Plank's equation, calculate the time required to freeze the product to −2°C. Assume infinite cylinder geometry and λ = 200 kJ/kg.

Given

i. $T_i = 4°C$
ii. $T_l = -10°C$
iii. $T_f = -2°C$
iv. D_p = 5 cm = 5 × 10^{-2} m = 0.05 m
v. From the literature, density of orange juice (ρ_f) = 1038 kg/m³ and k = 0.435 W/m K
vi. λ = 200 kJ/kg
vii. h = 15 W/m² K
viii. Since the can is cylindrical, the shape factors are: P = 1/4; R = 1/8

To find: Freezing time (t)

According to Plank's equation,

$$t = \frac{\lambda \rho}{T_f - T_1}\left(\frac{Pa}{h} + \frac{Ra^2}{k}\right)$$

$$= \frac{(200\times 1038)}{-2-(-10)}\left(\left(\frac{(1/4)\times\left(5\times 10^{-2}\right)}{\left(15\times 10^{-3}\right)}\right) + \frac{\left((1/16)\times\left(5\times 10^{-2}\right)^2\right)}{\left(0.435\times 10^{-3}\right)}\right) = 30937.5\,\text{s}$$

Answer: Freezing time = 30937.5 s or 8.6 h

10. Determine the diameter of a spherical product which is frozen in an air-blast freezer. The initial temperature of the product is 30°C and the temperature of cold air in the tunnel is –30°C. Density and thermal conductivity of the product are 1200 kg/m³ and 1.5 W/m K, respectively. The convective heat transfer coefficient for the air blast is 60 W/m² K. The initial freezing temperature is –1.5°C, the latent heat of fusion is 250 kJ/kg, and the freezing time is 2 h.

Given

i. $T_i = 30°C$
ii. $T_1 = -30°C$
iii. $T_f = -1.5°C$
iv. $\rho_f = 1200\,kg/m^3$
v. $k = 1.5$ W/m K
vi. $\lambda = 250$ kJ/kg
vii. $h = 60\,W/m^2$ K
viii. $t = 2$ h = 7200 s
ix. Since the product is spherical, the shape factors are: P = 1/6; R = 1/24

To find: Diameter of the product

Solution

According to Plank's equation, $t = \frac{\lambda\rho}{T_f - T_1}\left(\frac{Pa}{h} + \frac{Ra^2}{k}\right)$

$$7200 = \frac{(250 \times 1200)}{-1.5-(-30)}\left(\frac{(1/6)a}{(60 \times 10^{-3})} + \frac{(1/24)a^2}{(1.5 \times 10^{-3})}\right)$$

$$7200 = 10526.3\left(2.778a + 27.778a^2\right)$$

$$0.684 = \left(2.778a + 27.778a^2\right)$$

$$\therefore a = 0.115 \text{ m}$$

Answer: Diameter of the product = 0.115 m or 11.5 cm

BIBLIOGRAPHY

Anon 2007. Freezing of standard fish blocks in horizontal plate freezers, company information from Beck Pack Systems A/S. www.beck-liner.com/media/Fishblockmanual.pdf (accessed June 1, 2018).

Arce, J. and Sweat, V. E. 1980. Survey of published heat transfer coefficients encountered in food refrigeration processes. *ASHRAE Transactions* 86: 235–260.

ASHRAE Handbook 1994. *Refrigeration Systems and applications*. Atlanta, GA: American Society of Heating, Refrigerating, and Air-conditioning Engineers.

Barbosa-Cánovas, G. V., Altunakar, B. and Mejía-Lorío, D. J. 2005. *Freezing of Fruits and Vegetables: An Agribusiness Alternative for Rural and Semi-Rural Areas*. FAO Agricultural Services Bulletin 158. Rome, Italy: Food and Agriculture Organization of the United Nations (ISBN 92–5–105295–6).

Cleland, A. C. and Earle, R. L. 1984. Freezing time predictions for differential final product temperatures. *Journal of Food Science* 49: 1230–1232.

Cleland, D. J., Cleland, A. C., Earle, R. L. and Byrne, S. J. 1987. Prediction of freezing and thawing times for multi-dimensional shapes by numerical methods. *International Journal of Refrigeration* 10: 32.

Cleland, D. J. and Valentas, K. J. 1997. Prediction of freezing time and design of food freezers. In *Handbook of Food Engineering Practice*, eds. K. J. Valentas, E. Rotstein and R. P. Singh, 71–124. Boca Raton, FL: CRC Press, Taylor & Francis Group.

Dempsey, P. and Bansal, P. 2010. Air blast freezers and their significance to food freezing: a review. *Proceedings of ENCIT 2010, 13th Brazilian Congress of Thermal Sciences and Engineering* December 05–10, 2010, Uberlandia, MG, Brazil.
Department for Transport—Transfrigoroute 2016. Buyer's guide to refrigerated transport equipment. www.freightbestpractice.org.uk/imagebank/TE248.pdf (accessed June 1, 2018).
Fellows, P. 2000. *Food Processing Technology* (Third edition). Duxford: Woodhead Publishing.
Goff, H. D. 2018. Dairy science and technology education series. University of Guelph, Canada. www.uoguelph.ca/foodscience/book-page/freezingwhipping-ice-cream (accessed July 10, 2018).
Heldman, D. R. and Singh, R. P. 1981. *Food Process Engineering*. Westport, CT: AVI Publishing Co.
International Institute of Refrigeration 1972. *Recommendations for the Processing* and *Handling* of *Frozen Foods* (Second edition). Paris, France: International Institute of Refrigeration.
Jaspersen, W. S. 1989. Speciality ice cream extrusion technology. In *Food Technology International Europe*, ed. A. Turner, 85–88. London, UK: Sterling Publications International.
Khadatkar, R. M., Kumar, S. and Pattanayak, S. C. 2004. Cryofreezing and cryofreezer. *Cryogenics* 44: 661–678.
Kozlowski, T. 2009. Some factors affecting supercooling and the equilibrium freezing point in soil–water systems. *Cold Regions Science and Technology* 59: 25–33.
LeBlanc, D. I. and Hui, K. P. C. 2005. Land transportation of fresh fruits and vegetables: an update. http://epe.lacbac.gc.ca/100/201/300/stewart_postharvest/2005/v01n01/June_2005/Le Blanc4.pdf (accessed June 2, 2018).
Levy, F. L. 1958. Calculating freezing time of fish in air blast freezers. *Journal of Refrigeration* 1: 55–58.
Marshall, R., Lawton, R. and Lawson, I. 2006. *Energy Usage during Refrigerated Transport*. Technical document prepared for the School of Engineering and Design. Uxbridge, Middlesex: Brunel University London.
Mascheroni, R. H. 2001. Plate and air-blast cooling/freezing. In *Advances in Food Refrigeration*, ed. D-W. Sun, 193–219. Leatherhead: Leatherhead Publishing, LFRA.
Mittal, G. S., Hanenian, R. and Mallikarjunan, P. 1993. Evaluation of freezing time prediction models for meat patties. *Canadian Agricultural Engineering* 35: 75–81.
Nagaoka, J. S., Takagi, S. and Hotani, S. 1956. Experiments on the freezing of fish by air blast freezer. *Journal Tokyo University of Fisheries* 42: 65–73.
Papadopoulos, A. M. and Koroneos, C. J. 2004. Refrigeration and cryogenic systems. In *Air Conditioning—Energy Consumption and Environmental Quality, Encyclopedia of Life Support Systems (EOLSS)*, ed. M. Santamouris. Oxford, UK: Developed under the Auspices of the UNESCO, EOLSS Publishers. www.eolss.net (accessed June 5, 2018).
Persson, P. O. and Löndahl, G. 1993. Freezing technology. In *Frozen Food Technology*, ed. C. P. Mallett, 20–58. London, UK: Blackie Academic and Professional.
Pham, Q. T. 1986. Simplified equation for predicting the freezing time of foodstuffs. *Journal of Food Technology* 21: 209219.
Pham, Q. T. 2015. Refrigeration in food preservation and processing. In *Conventional and Advanced Food Processing Technologies*, ed. S. Bhattacharya, 357–386. Chichester, West Sussex, UK: John Wiley and Sons, Ltd.
Plank, R. 1941. Beitrage zur Berechrung und Bewertung der Gefrigeschwindikeit von Lebensmittelm zeitschrift fur die gesamte kalte Industrie. *Beih Rcihe* 3: 1–16.
Rolfe, E. J. 1967. The chilling and freezing of foodstuffs. In *Biochemical and Biological Engineering Science*, ed. N. Blakebrough, 137–208. London, UK: Academic Press London.
Singh, R. P. and Heldman, D. 2014. *Introduction to Food Engineering* (Enhanced Fifth edition). London, UK: Academic Press.
Starace, G., Fiorentino, M., Meleleo, B. and Risolo, C. 2018. The hybrid method applied to the plate-finned tube evaporator geometry. *International Journal of Refrigeration* 88: 67–77.
Tassou, S. A., De-Lille, G. and Ge, Y. T. 2009. Food transport refrigeration—approaches to reduce energy consumption and environmental impacts of road transport. *Applied Thermal Engineering* 29: 1467–1477.
Venkatarathnam, G. and Murthy, S.S. 2012. Refrigerants for vapor compression refrigeration systems. *Resonance* 17: 139–162.
Wang, L. and Weller, C. J. 2012. Thermophysical properties of frozen foods. In *Handbook of Frozen Food Processing and Packaging*, ed. D. W. Sun, 101–128. Boca Raton, FL: CRC Press, Taylor & Francis Group.

12

Mixing and Separation Processes

12.1 Mixing

Even a great recipe turns out to be a delicious food only when the right amounts of ingredients and flavors are well mixed. *Mixing* is a unit operation that involves combining two or more components by dispersing one throughout the other. The component in larger proportion is termed as the *continuous* phase, whereas that in smaller proportion is termed as the *dispersed* phase. The end product obtained after mixing is a uniform mixture. Thus, mixing can be formally defined as *any operation that changes a non-uniform system into a uniform one*. In other words, mixing is the *random distribution into and through one another of two or more initially separated phases*.

The importance of proper mixing has been realized at all levels: right from the small-scale mixing done manually at home kitchens to the industrial-scale blending of multiple ingredients in high volumes. With respect to food products, while consumers expect a high level of batch-to-batch consistency, the food industry strives to achieve the facility-to-facility consistency. In this regard, mixing is an inevitable unit operation to reduce nonuniformities in properties such as concentration, color, texture, or taste between different parts of a product (Uhl and Gray, 1986). Many food products have gone through at least one stage of mixing before attaining their final form. The well-known examples of food mixtures include infant foods, spice mixes, cereal mixes, topical seasoning for snacks, soup mixes, and bakery mixes that combine a diverse range of food ingredients such as sugar, salt, fruits, vegetables, milk powders, flour, and dried cereals.

Initially, food mixing was merely understood as a process that leads to product homogeneity. Nevertheless, it is beyond the scope of homogeneity. The purpose of food mixing is manifold as represented in Figure 12.1.

The purview of food mixing ranges from the mere blending of dry powders, preparation of suspensions containing large particles, high-viscosity pastes, batters, and dough with or without the inclusion of gas, to the production of nanoemulsions comprising droplets of dimension in the range of 50–1000 nm. A thorough understanding of the mixing operation is significant as the failed product resulting from the poor mixing jeopardizes the process economics. Poor mixing often results due to the lack of understanding of component characteristics, inadequate clarity on the mixing objectives, improper selection of mixers, wrong scale-up approach, and limited knowledge on the design and parameters of the mixing equipment.

12.1.1 Classification of Food Mixing

Food mixing is classified according to the state of the continuous and dispersed phases: solid, liquid, or gas, and the miscibility of one substance into the other. The different classes of food mixing are schematically represented in Figure 12.2. Each type of mixing is characterized by a mechanism that determines the rate and degree of mixing. The theory underlying each of the category and the corresponding equipment will be discussed in the forthcoming sections.

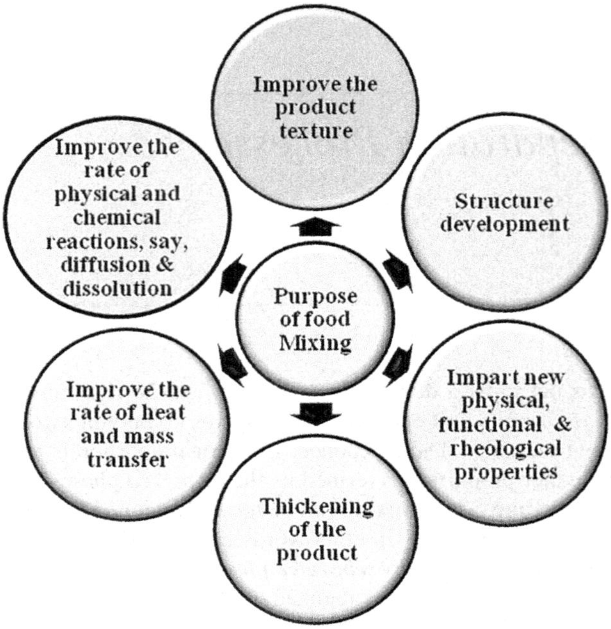

FIGURE 12.1 Purpose of food mixing.

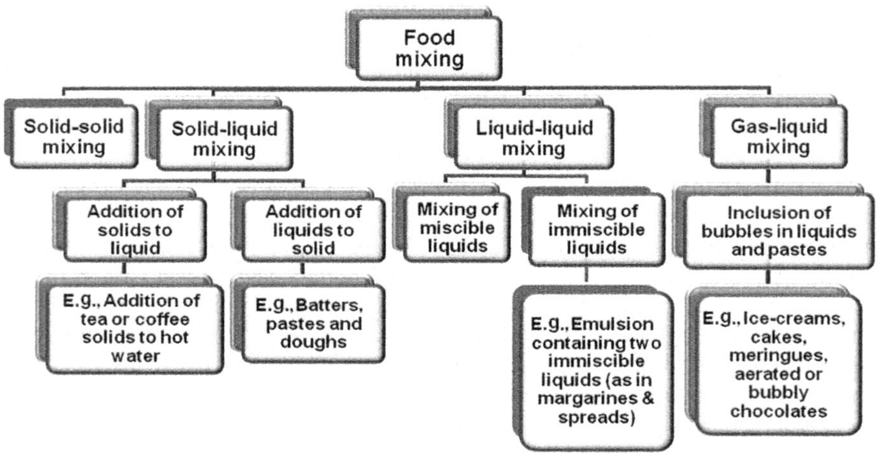

FIGURE 12.2 Classification of food mixing.

12.1.2 Theory of Solid Mixing

The mixing of free-flowing or particulate solids is a common processing operation in the food industry. The rationale of solid mixing is to achieve a completely homogeneous mixture of dry powders. Solid mixing occurs by three different mechanisms, which are as follows:

i. Convective or macro-mixing
ii. Diffusive or micro-mixing
iii. Shear mixing

12.1.2.1 Convective Mixing

Convective mixing is the most commonly used principle of mixing in the food industry. It is achieved by mechanical agitation, in which the rotation of mixing elements such as blades, paddles, or screws leads to the bulk movement of lumps of powder around the powder mass (Figure 12.3a). Blades and paddles achieve mixing by the inversion of the powder bed. Whereas, screw elements achieve it by means of a revolving action. Considerable shear may occur during the convective mixing.

12.1.2.2 Diffusive Mixing

Mixing is said to be diffusive when it occurs by the random motion of particles driven by gravity or vibration but without mechanical agitation. The bed of particles is tilted, and hence, the particles undergo a relative change in their position (Figure 12.3b). The gravitational forces cause the upper layers to slip, and diffusion of individual particles takes place over newly developed surfaces. There is no shear in the diffusive mixing of solids. This is the preferred mechanism for microscopic homogenization and mixing of fragile particulates or agglomerates that require gentle handling. However, diffusive mixing is not suitable for cohesive particles.

12.1.2.3 Shear Mixing

Shear mixing occurs in two stages: initially, the apparition of a shearing surface occurs followed by the reorganization of particles by the convection and diffusion mechanisms (Mayer-Laigle et al., 2015) (Figure 12.3c). In this mechanism of mixing, the forces of attraction between the particles are broken down using the shear stresses created by the arms of an agitator or a blast of air, such that each particle is free to move on its own, both between the zones of different components and parallel to their surface. This type of solid mixing reduces the degree of separation by thinning of the dissimilar layers of particles.

All the three mechanisms may exist in a single mixer, but often, one or two might be predominant.

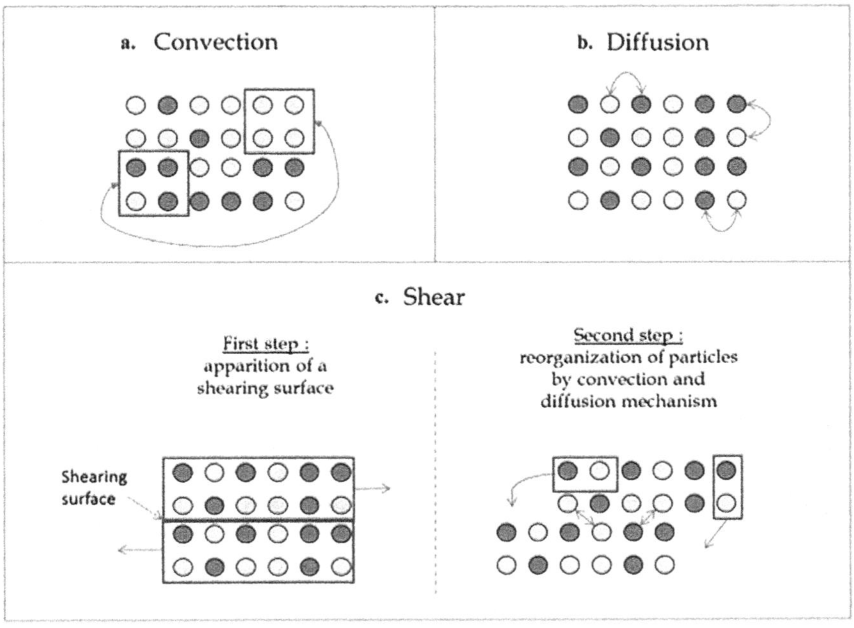

FIGURE 12.3 Mechanisms of solid mixing. (Reproduced with permission from Mayer-Laigle, C., Gatumel, C. and Berthiaux, H. 2015. Mixing dynamics for easy flowing powders in a lab scale Turbula® mixer. *Chemical Engineering Research and Design* 95: 248–261.)

12.1.2.4 Assessing Mixedness during Solid Mixing

The degree of solid mixing depends on the properties of each component, which include particle size and shape, density, moisture content, surface characteristics (roughness, surface charge, polarity), and flow characteristics [cohesive (e.g., maltodextrin powder) and non-cohesive (e.g., grains)]. In addition, the tendency of the particles to aggregate and the mixer efficiency also determine the extent of mixing. Deviations in the aforementioned properties lead to *de-mixing* or *segregation* of components in the mixture. To achieve an optimal solid mixing, it is important to avoid the phenomenon of *segregation* (Figure 12.4). Poor flow characteristics of the particles (cohesiveness or non-free-flowing); differences in particle size, shape, and density (it is easy to mix particles of similar size, shape, and density compared to the dissimilar ones); and particle attraction due to electrostatic charges can lead to the segregation. A common example for the segregation in food mixtures is the pack of breakfast cereals, in which the larger particles find their way to the top of the packet and the finer ones accumulate at the bottom. Thus, solid mixing is a reversible process, and the resultant level of product homogeneity is the net effect of mixing and de-mixing phenomena. In other words, the degree of homogeneity depends on the equilibrium achieved between mixing and segregation, which is again determined by the type of mixer, operating conditions, and the nature of food component. Compared to a mixture of liquids or gases, a solid mixture will be less homogeneous. This is because the particles tend to segregate and fluid molecules tend to mix.

At the beginning of mixing, the components remain as segregated groups. Towards the end of the process, the components would be randomly distributed in the mixture (Figure 12.5).

With respect to the scheme shown in Figure 12.5, the degree of mixing can be judged by a *mixing index* (*M*). Mixing index is calculated based on the expected variances (s^2) of the sample compositions, which is in turn estimated from the mean sample composition. Mixing index can be calculated by considering a two-component mixture consisting of a fraction x of component X and a fraction y of

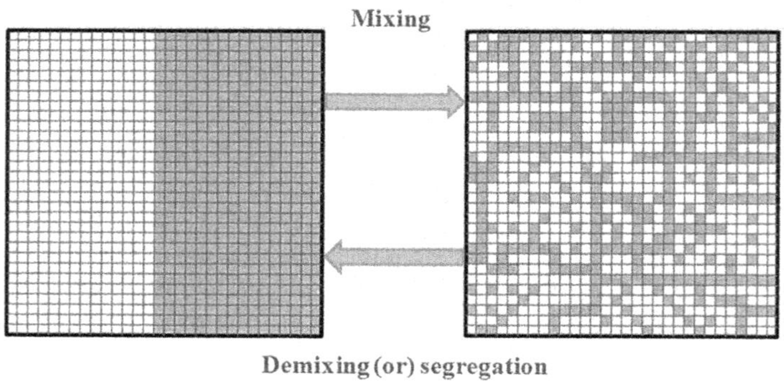

FIGURE 12.4 Mixing and segregation.

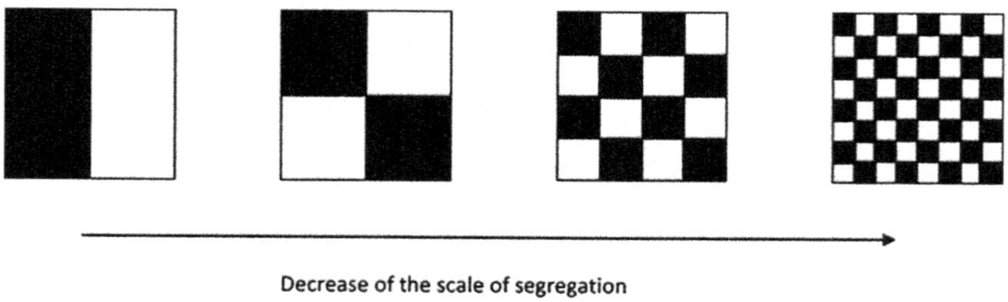

FIGURE 12.5 Scheme of mixing process. (Reproduced with permission from Mayer-Laigle, C., Gatumel, C. and Berthiaux, H. 2015. Mixing dynamics for easy flowing powders in a lab scale Turbula® mixer. *Chemical Engineering Research and Design* 95: 248–261.)

component *Y*. Initially, in the unmixed state, when a large number of samples are drawn, a proportion *x* of the samples will contain pure component *X* and a proportion *y* of the samples will contain pure component *Y*. Accordingly, for the proportion *x* of samples containing pure component *X*, the deviation from the mean composition would be (1−*x*), as it has a fractional composition 1 for component *X*. Similarly, in the proportion *y* of the samples containing pure *Y*, the fractional composition of component *X* is 0, and thus, the deviation from the mean composition is (0−*x*). Collating these in terms of the fractional composition of component *X*, for *n* samples, the initial variance (in the unmixed state) can be written as

$$s_o^2 = \frac{1}{n}\left[xn(1-x)^2 + (1-x)n(0-x)^2\right] = x(1-x), \left(\because, x + y = 1\right) \tag{12.1}$$

Subsequently, after the complete dispersion of the mixture, it can be assumed that the components are distributed throughout the volume according to their overall proportions. The probability that any particle picked at random is component *Y* will be *y*. Therefore, the probability that any particle picked at random is not *y* is (1−*y*). Extrapolating this probability to samples containing *N* particles and applying the theory of probability, the expected variance at the end of mixing can be written as

$$s_r^2 = \frac{x(1-x)}{N} = \frac{s_o^2}{N} \tag{12.2}$$

The subscripts *o* and *r* denote the initial and the random values of s^2. Equation 12.2 is based on the following assumptions:

1. All the particles in the mixture are of the same size.
2. Each particle is either pure *X* or pure *Y*.

From Eqs. (12.1) and (12.2), it is apparent that during the mixing process, the value of s^2 decreases from $x(1-x)$ to 1/Nth of the value of $x(1-x)$. Therefore, the intermediate values between s_o^2 and s_r^2 can be used to explain the progress of mixing. Based on the aforementioned discussion, the equation for mixing index is given by

$$M = \frac{\left(s_o^2 - s^2\right)}{\left(s_o^2 - s_r^2\right)} \tag{12.3}$$

where

$$s^2 = \frac{1}{n}\left[(x_1 - \bar{x})^2 + (x_2 - \bar{x})^2 + (x_n - \bar{x})^2\right] \tag{12.4}$$

In Eq. (12.4), $\bar{x}$ is the mean mass by weight of the particles. During the course of the mixing process, the value of *M* varies between 0 and 1. The calculation of mixing index is important in controlling the product uniformity and to attain a homogeneous distribution of components in the mixture. The usage of *M* as a measure of mixedness is valid for mixtures of particles and heavy pastes. The plot between mixing factor (*M*, on a logarithmic scale) and mixing time (*t*, on a linear scale) is known as the *characteristic curve of mixing* (Figure 12.6). In the case of batch mixing, mixing time is measured from the time point at which the mixing begins. However, in a continuous mixing operation, it corresponds to the mean residence time, given by the powder volume in a mixer per unit volumetric flow rate. The characteristic curve of mixing finds its application in evaluating the performance of mixers.

From Figure 12.6, it is apparent that the convection is the predominant mechanism of mixing during the initial stages (Stage I). Subsequently, mixing proceeds in a steady fashion by both convective and shear mechanisms, the time period of which is termed as the "intermediate stage" (Stage II). The final stage of mixing (Stage III) is marked by the attainment of equilibrium between mixing and segregation. This stage is dominated by the diffusion mechanism. The degree of mixedness at the last stage is termed

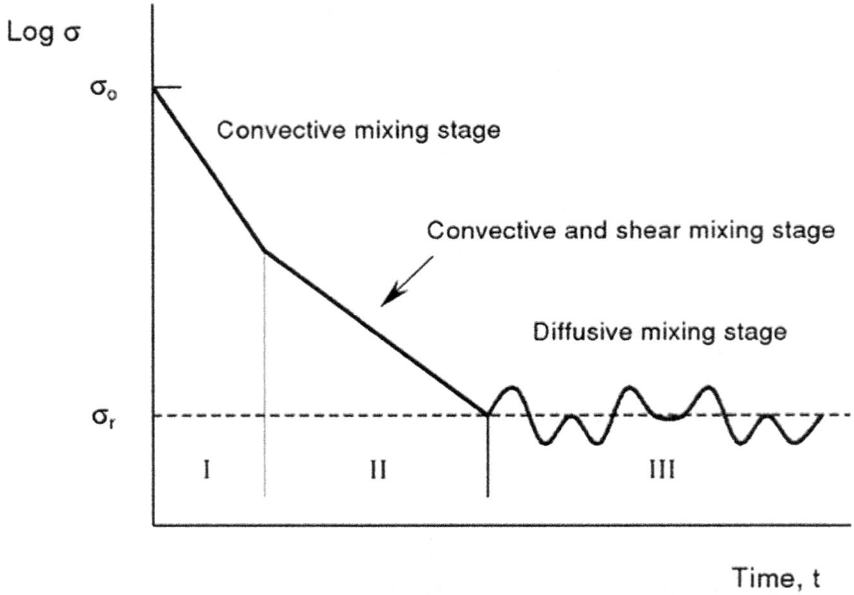

FIGURE 12.6 Characteristic curve of mixing. (Reproduced with permission from Miyanami, K. 2006. Mixing. In *Powder Technology Handbook* (Third edition), eds. H. Masuda, K. Higashitani and H. Yoshida, 577–590. Boca Raton, FL: CRC Press, Taylor & Francis Group.)

as the "final degree of mixedness" (M_∞). The relationship between the mixing index and the mixing time is given by

$$dM/dt = K(1-M) \tag{12.5}$$

where K is the mixing rate constant, which varies with the mixer type and the nature of components in the mixture. Integrating Eq. (12.5) between the limits $t=0$ to $t=t_m$ and $M=0$ to $M=M$ and subsequently eliminating the natural logarithm,

$$[1-M] = e^{-Kt_m} \tag{12.6}$$

where t_m is the mixing time (s).

Example 12.1

For a baking trial, the dough was mixed in 95 kg batch, followed by the blending of 5 kg of yeast at a later stage. The composition of yeast, by percentage, in 50 g of dough samples is given in the following table:

After 4 min (%)	2.0	18.6	2.6	4.6	12.1	16.6	0.5	1.8	4.0	2.7
After 8 min (%)	3.0	8.6	8.5	4.1	4.2	5.5	6.9	4.4	3.5	12.5

Calculate the mixing index after the 4th and 8th min of mixing.

Solution

Given

$$\text{Fraction of yeast}: x = \frac{5}{100} = 0.05$$

$$\therefore \text{Dough fraction}: y = 1 - x = 1 - 0.05 = 0.95$$

From Eq. (12.1), $s_o^2 = x(1-x) = 0.05 \times 0.95 = 0.0475$

If the number of yeast particles in the sample is very large, then $s_r^2 \approx 0$.

After 4 min:

$$\bar{x} = \frac{0.02 + 0.186 + 0.026 + 0.046 + 0.121 + 0.166 + 0.005 + 0.018 + 0.004 + 0.027}{10} = 0.0655$$

After 8 min:

$$\bar{x} = \frac{0.03 + 0.086 + 0.085 + 0.041 + 0.042 + 0.055 + 0.069 + 0.044 + 0.035 + 0.125}{10} = 0.0612$$

$$s_R^2 = \frac{\Sigma(x - \bar{x})^2}{n-1}$$

$$\therefore s_4^2 = 4.4 \times 10^{-3}$$

$$\therefore s_8^2 = 9 \times 10^{-4}$$

From Eq. (12.3),

$$M_4 = \frac{\left(0.0475 - \left[4.4 \times 10^{-3}\right]\right)}{(0.0475 - 0)} = 0.907$$

$$M_8 = \frac{\left(0.0475 - \left[9 \times 10^{-4}\right]\right)}{(0.0475 - 0)} = 0.981$$

Answer: (i) Mixing index after the 4th min of mixing = 0.907
(ii) Mixing index after the 8th min of mixing = 0.981

12.1.3 Theory of Liquid Mixing

The principle of liquid mixing can be better understood under two categories: low-viscosity liquids and high-viscosity liquids, pastes, or dough.

12.1.3.1 Low-Viscosity Liquids

A mixer induces three different component velocities in low-viscosity liquids (Figure 12.7), which are as follows:

1. Longitudinal velocity, parallel to the mixer shaft
2. Rotational velocity, tangential to the mixer shaft
3. Radial velocity, perpendicular to the mixer shaft

Often, the radial and longitudinal velocities in the liquid are increased by the use of baffles, angled blades, and angled mixer shafts. In addition, turbulence must be introduced in the bulk liquid to draw its slow-moving parts into the faster-moving ones. Nevertheless, a vortex should be avoided as it hinders mixing. When a vortex is formed, the adjacent layers of circulating liquid move at a similar speed and merely rotate around the mixer, without being mixed.

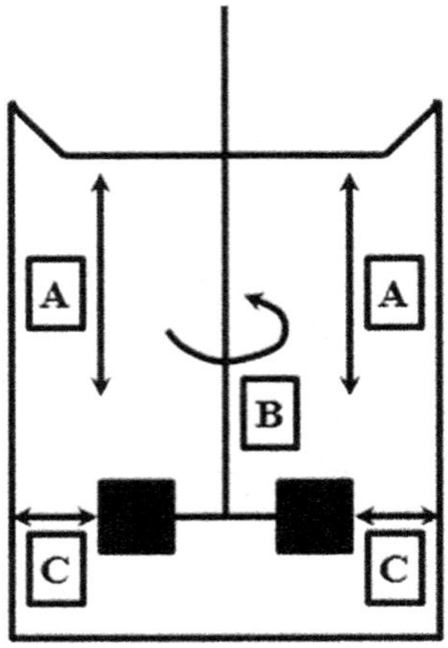

FIGURE 12.7 Component velocities in fluid mixing A: longitudinal; B: rotational; C: radial.

12.1.3.2 High-Viscosity Liquids, Pastes, and Dough

The approach of mixing in high-viscosity liquids is entirely different from that of their low-viscosity counterparts. Here, the mixing occurs by the following procedures:

1. *Kneading*: A food material is pressed against the wall of the mixing bowl or into another material, for example, mixing of bread dough (Figure 12.8a).
2. *Folding*: The unmixed food is folded into the mixed portion (Figure 12.8b).
3. *Stretching*: The material is subjected to shearing (Figure 12.8b).

As in solid mixing, the rate of liquid mixing is explained by a mixing index. In this case, the mixing rate constant (K) in Eq. (12.5) depends upon the liquids being mixed and the mixer. It is given by

$$K \propto \frac{D^3 N}{D_t^2 z} \tag{12.7}$$

where D is the diameter of the agitator (m), N is the agitator speed (rev/s), D_t is the diameter of the mixing vessel (m), and z is the height of the liquid (m).

The liquid flow during mixing is defined by the dimensionless numbers, namely, Reynolds number (N_{Re}) (Eq. 12.8), Froude number (N_{Fr}) (Eq. 12.9), and the Power number (N_{Po}) (Eq. 12.10).

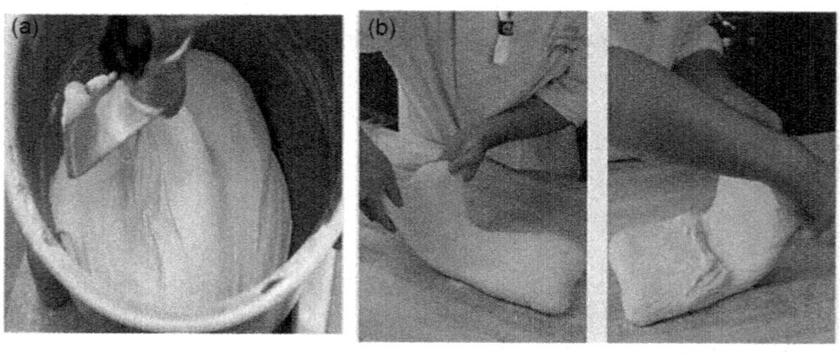

FIGURE 12.8 (a) Kneading of bread dough and (b) mixing the dough in a stretch-and-fold sequence (Reproduced with permission from Hitz, C. 2008. *Baking Artisan Bread*. Beverly, MA: Quarry Books.)

$$N_{Re} = \frac{D^2 N \rho_m}{\mu_m} \tag{12.8}$$

From Eq. (12.8), the calculation of Reynolds number involves the product *DN* instead of velocity (v). The value of *DN* differs by a factor of p from the actual velocity at the tip of the propeller.

$$N_{Fr} = \frac{DN^2}{g} \tag{12.9}$$

$$N_{Po} = \frac{P}{\rho_m N^3 D^5} \tag{12.10}$$

where P is the power transmitted via the agitator (W), ρ_m is the density of the mixture (kg/m^3), and μ_m is the viscosity of the mixture (N s m^{-2}). From Eqs. (12.9) and (12.10), Froude number is the relationship between inertial and gravitational forces and Power number relates drag forces to inertial forces: These numbers provide an understanding of all the important forces involved in mixing. Further, they also relate the dimensions, type, and operating conditions of the mixer. The relationship between these dimensionless groupings provides the expression for the power consumed in liquid mixing, which is given by,

$$N_{Po} = K\left(N_{Re}\right)^n \left(N_{Fr}\right)^m \tag{12.11}$$

where K, n, and m are the factors related to the geometry of the agitator, determined from the experiments. The Froude number is significant only at $N_{Re} > 300$ and if the vessel is unbaffled due to which vortex occurs. Thus, except for the aforementioned conditions, Eq. (12.11) becomes

$$N_{Po} = K\left(N_{Re}\right)^n \tag{12.12}$$

The density of a mixture is additive in nature and thus can be obtained by adding together the component densities of the continuous and dispersed phases, which is given by

$$\rho_m = V_1 \rho_1 + V_2 \rho_2 \tag{12.13}$$

where V is the volume fraction of the components. The subscripts 1 and 2 denote the continuous phase and dispersed phase, respectively.

The viscosity of a mixture depends on whether the mixer is baffled or unbaffled, and thus, two different equations are available for the calculation of viscosity as given in Eqs. (12.14) and (12.15) (Jackson and Lamb, 1981):

$$\mu_m\left(\text{unbaffled}\right) = \mu_1^{V_1} \mu_2^{V_2} \tag{12.14}$$

$$\mu_m\left(\text{baffled}\right) = \frac{\mu_1}{V_1}\left(\frac{1 + 1.5\mu_2 V_2}{\mu_1 + \mu_2}\right) \tag{12.15}$$

However, a general formula is not available yet. Therefore, the results are limited to the particular propeller configurations used in the experiments. But if the experimental curves are available, the values for n and K in Eq. (12.12) can be calculated, after which the equation can be used to predict the power consumption of mixers. For instance, the relationship between the Power number and Reynolds number for different turbine impellers as a function of the ratio between size of the impeller and diameter of the vessel (*w*/*D*) is shown in Figure 12.9.

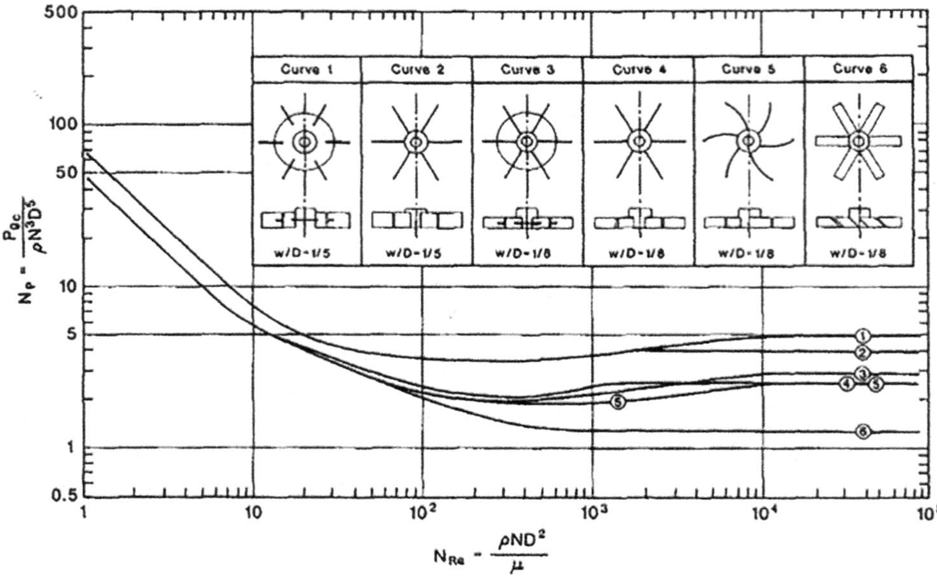

FIGURE 12.9 Power number versus Reynolds number for turbine impellers as a function of *w/D*. (Reproduced with permission from Bates, R. L., Fondy, P. L. and Corpstein, R. R. 1963. Examination of some geometric parameters of impeller power. *Industrial & Engineering Chemistry Process Design and Development* 2: 310–314.)

12.1.4 Theory of Gas–Liquid Mixing

Bubbles are arguably the world's most important ingredient...

(Campbell, 2009).

The inclusion of bubbles in foods permits creation of very novel structures...

(Sahu and Niranjan, 2009).

Gas–liquid mixing is of importance in the food industry, mainly with respect to aerated products in which air bubbles are the major ingredient. The most popular aerated food products include bread, cakes, ice cream, bubbly chocolates, breakfast cereals, cappuccino (a coffee-based foamed beverage), beer, and champagne. The two principal mechanisms of gas–liquid (G-L) mixing are as follows:

1. *Mechanical agitation under positive pressure*: This is the most commonly encountered mechanism of G-L mixing. A common example is an electric whisker used in home kitchens. In this process, air is effectively beaten into the liquid, in the beginning, to incorporate larger bubbles from the headspace; the bubble size reduces as the agitation proceeds (Prins, 1988). Here, the degree of mixing depends on the headspace pressure, whisk speed, gas solubility, time, and area of contact (Sahu and Niranjan, 2009).

   Examples

   i. Occlusion of gas bubbles in bread dough
   ii. Carbonation of soft drinks in a carbonator that facilitates a close contact between the gas (carbon dioxide) and the liquid

2. *Steam-induced mixing*: In this mechanism, the G-L mixing is achieved by a direct injection of steam into the product or by exposing the product to high temperatures, which generates steam within the product.

Examples

i. Preparation of cappuccino by mixing a shot of espresso and frothed milk, in which frothed milk is obtained by injecting steam into the milk and entraining air in the headspace of the coffee cup. Espresso is the highly concentrated black coffee drink with a thick layer of foam at the top of the cup. It is obtained using a brewing machine, which pressurizes hot water against finely ground coffee.
ii. Cereals, puffed rice, and popcorn: The cereal grains, rice, or corn is exposed to a maximum temperature of up to 300°C for 90 s, in a toasting oven. The extreme temperature leads to the flash evaporation of the trace amounts of moisture in the porous structure to form steam. Subsequently, the steam is forced out of the product, which creates the expanded product (Sahu and Niranjan, 2009).
iii. Application of dry heat to bread dough or cake batter inside a baking oven majorly contributes to the volume of the product.

An important aspect of gas–liquid mixing is the estimation of the amount of gas incorporated in the continuous phase after mixing, which is termed as *gas holdup*. Gas holdup (Ø, %) is a measure of the level of bubbles in liquid foods, given by

$$\varnothing = \left(1 - \frac{m_f}{m_i}\right) \times 100 \qquad (12.16)$$

where m_i is the mass of the continuous phase (kg) and m_f is the mass of the foam (kg). Eq. (12.16) is applicable to liquids and medium-to-high-viscosity pastes such as ice creams and whipping creams. For solid foams such as aerated chocolate, the flotation method is used for the calculation of gas holdup (Haedelt et al., 2005). For the low-viscosity liquid foods such as beer and cappuccino, the gas holdup can be determined if the heights of the foam and liquid layers are known. The value of gas holdup ranges between 15% and 20% in milkshakes, and it is more than 90% in extruded products such as popcorn and rice cakes (Sahu and Niranjan, 2009).

12.1.4.1 Structure Development by Gas–Liquid Mixing: A Case Study on Dough Mixing during Breadmaking Process

A unique application of gas–liquid mixing in the food industry is the structural development in aerated products. The example of dough mixing during the breadmaking process is considered here.

From a food engineer's perspective, bread can be defined as solid foam since it constitutes a final gas volume of more than 70%. This is justified as materials with a high gas fraction in the order of more than 50% are usually designated as foams (Scanlon and Zghal, 2001). Thus, it is also reasonable to consider breadmaking as an aeration process. Mixing marks the commencement of aeration and foam formation during the breadmaking process. It is the stage where bakers have the utmost control over bread aeration. The final gas cells in bread are the result of the bubbles occluded in the dough initially during mixing. The source of bubbles in the dough is the gas pockets entrapped from the mixer headspace atmosphere during mixing. As the dough is mixed, it folds over itself, trapping pockets of air bubbles between its layers (Trinh et al., 2013). The bubbles thus included into the dough matrix are further broken down due to the mixing action, which reduces the mean bubble size and increases the number of bubbles. The bubble size distribution in dough depends on the amount of energy input during mixing (Hanselmann and Windhab, 1998). The typical size of gas bubbles in the mixed dough was found to be in the range of 21.9–26.2 μm (Figure 12.10). The shape of the gas bubbles was almost spherical with roundness value in the range of 0.75 ± 0.022 (Ishwarya et al., 2017).

The required input of heat and mechanical energy into the dough is achieved through the stretching and pulling actions of the mixing operation. The shear field experienced by the bubbles in a higher viscosity system such as the bread dough tends to produce smaller bubbles for a given mixing speed (Mills et al., 2003).

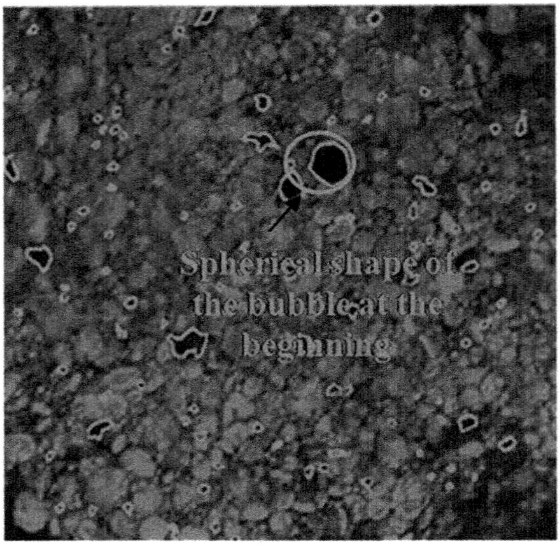

FIGURE 12.10 Gas bubbles in mixed bread dough visualized by confocal laser scanning microscopy. (Reproduced with permission from Ishwarya, S. P., Desai, K. M., Naladala, S. and Anandharamakrishnan, C. 2017. Bran-induced effects on the evolution of bubbles and rheological properties in bread dough. *Journal of Texture Studies* 48: 415–426.)

Mixing may also bring both soluble and insoluble components to the bubble surface, the movement of which might be otherwise hindered by the structure and viscosity of the gluten–starch matrix of dough (Ornebro et al., 2000). In addition to the entrapment of air bubbles, mixing has two important functions: first to evenly distribute the ingredients and second to develop the gluten (Trinh, 2013), which is important for providing the structure of bread. Bubbles in bread also have an influence on the texture, that is, sponginess or softness and mouth-feel properties of the product (Haedelt et al., 2007). Specifically, the aerated structure created in mixer has been considered to have a direct effect on the final baked loaf texture (Chamberlain and Collins, 1979; Cauvain et al., 1999; Campbell, 2003). Because of its vital role, Cauvain (2000) rightly stated that 90% of the final bread quality depends upon mixing. Mixing is also central to the development of dough, referred to as "ripening" or "maturing" (Cauvain and Young, 2006). This is of importance because, when the dough is developed, it transforms into a viscoelastic material, which is capable of holding the expanding gas bubbles (Ishwarya, 2017).

12.1.5 Mixing Equipment

Similar to the classification of mixing, mixing equipment is also categorized based on the state of continuous and dispersed phases being mixed. Accordingly, four classes of mixers are available, which are as follows:

i. Mixers for solid–solid mixing (for mixing dry powders or particulate solids)
ii. Mixers for liquid–liquid mixing (for mixing low-viscosity, medium-viscosity, and high-viscosity liquids)
iii. Mixers for dispersal of powders in liquids
iv. Mixers for semisolids (pastes and dough)

12.1.5.1 Mixers for Solid–Solid Mixing (or) Powder Mixers

The working principle of solid–solid mixers is based on displacing parts of the mixture from one location to another within the mixing vessel. To set the solid particles in motion, an external force such as mechanical agitation is required (Miyanami, 2006). Based on the type of external force or action that causes mixing, two types of design are available for solid mixers: (i) tumbling action of rotating vessels (e.g., drum mixers, double-cone mixers, Y-cone mixers, and V-cone mixers) and (ii) positive movement of materials in screw-type mixers (e.g., ribbon blenders).

The tumbling mixers are also known as the segregating mixers that operate by diffusive mechanism and are usually of the non-impeller type. These mixers are most appropriate for free-flowing powders containing particles of an identical size and density. Small amounts of liquids may also be added. In general, the tumbling mixers are enclosed containers that rotate about a horizontal axis. These mixers are usually operated in the batch mode, with 50%–60% of the vessel volume filled with the solids. The vessels rotate at variable speeds with the value ranging up to 100 rev/min. Usually, the operational speed of rotation for the tumbling mixers is set at 50%–80% of the critical rotational speed (N_{cr}). N_{cr} is the speed at which all the solids would be thrown upward towards the walls of the mixer, given by

$$N_{cr} = \frac{0.498}{\sqrt{R_{\max}}} \tag{12.17}$$

where $R_{\max}$ is the maximum radius of rotation of the mixer (m). The inner walls of the mixers may be fitted with baffles or ploughs that assist in inducing convection by lifting the solids. Although the rate of mixing is quite low, a good final degree of mixedness can be obtained with the tumbling mixers.

On the other hand, the screw-type mixers are of the nonsegregating type, governed by the convective mechanism. These mixers include screws, blades, and ploughs as the mixing elements.

The solid mixers are commonly used for blending dry ingredients during the preparation of instant potato mixes, soup mixes, cake mixes, blending of flour and grains, flavorings, and spices. The different solid mixers used in the food industry are explained in the forthcoming sections.

12.1.5.1.1 Drum Mixers

Drum mixers are the simplest among the industrial powder mixers (Figure 12.11). The mechanism of mixing in a drum mixer is based on the random motion of powder particles as they tumble down the moving surface of the powder. In the drum mixers, while parts of the mixer move, the mixing container remains in a fixed position with respect to the mixer geometry and the driving mechanism that activates the mixer. The dry ingredients or powders to be mixed are added to the drum. The drum is then set into a rolling motion. Subsequently, the powder rises up and falls down along the wall of the drum. The volume

FIGURE 12.11 Drum mixer. (Reproduced with permission from Metal Powder Report (2008). Rotary or double cone? Take your pick for quality. *Metal Powder Report* 63: 8–10.)

of the drum that is filled with powder, expressed in percentage, is referred to as the *powder charge*. This measure is also expressed as a fraction of the internal volume of the powder mixer. For effective mixing, the powder charge must not be too large, as it leads to insufficient freedom of motion within the mixer for the powders to mix. Similarly, when the speed of rotation of the drum is too high, the particles will be centrifuged towards the walls and would remain in that position without undergoing mixing. Generally, drum mixers are used in the production of mineral and protein premixes.

12.1.5.1.2 Double-Cone Mixers

The double-cone mixers include two cone-shaped sections sloped at 45° (Figure 12.12). The cones are arranged with their open ends welded together to a center band. The cones rotate about an axis that passes through their common base. The two trunnions between which the double-cone mixer is mounted enable the unit to tumble end-over-end. Inlet to the mixer can be in one end of the cone, and the outlet at the other end. Alternatively, an opening in one of the ends of the cones can also serve as inlet and outlet. Cleaning of these mixers is carried out through the outlet. Mixing occurs as the double-cone mixer tumbles and the product in the vessel spreads out.

The zone of transition at the band between the cones guides the product to fold over itself without merely sliding along the inner wall. The double-cone mixer is suitable for dry powders that require gentle mixing. The particles may be subjected to a very light shear during the mixing. Examples of food products mixed using a double-cone blender include starch, semolina flour, coffee beans and ground coffee, cocoa, chocolate flakes, powdered milk, infant food, and powders used in the preparation of dehydrated soups and creams.

12.1.5.1.3 V-Cone and Y-Cone Mixers

V-cone mixers are similar to the double-cone mixers but, in addition, include two larger diameter pipe sections cut at an angle of 45° and welded together to form a V (Figure 12.13(a)). The Y-cone mixer has a third section that extends the volume of the mixer in a bisectional direction relative to the other pipe sections (Figure 12.13(b)). Whereas the inlets are located at both the ends of V or Y sections, the outlets are positioned at the V point or at the bottom of the Y. A batch of mixed product is discharged from the mixer outlet into drums or containers (for transportation), through an optional retractable sleeve. Similar to the double-cone mixers, the end-to-end tumbling action of V- and Y-cone mixers is facilitated by the trunnions on which the unit is mounted. These mixers can be integrated with a spray line for the addition of liquids and an agitator for the de-lumping of solids. Compared to a double-cone blender, the V- and Y-cone mixers result in less gentle mixing owing to the free-falling action and the increased frictional contact between the material and the sides of the long vessel. A major limitation of V- and Y-cone mixers is the segregation during discharge due to the formation of funnel flow pattern. To prevent the aforementioned limitation, the hopper surfaces must be steep enough and sufficiently low in friction to provide

FIGURE 12.12 Double-cone mixer. (Reproduced with permission from Metal Powder Report (2008). Rotary or double cone? Take your pick for quality. *Metal Powder Report* 63: 8–10.)

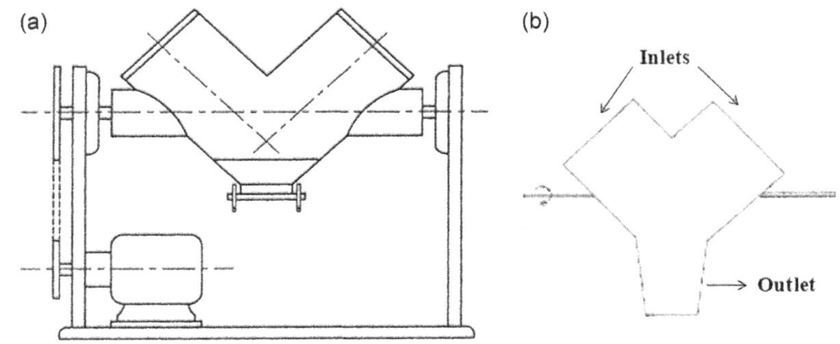

FIGURE 12.13 (a) V-cone mixer (Reproduced with permission from Harnby, N. 1992. The selection of powder mixers. In *Mixing in the Process Industries*, eds. N. Harnby, M. F. Edwards and A. W. Nienow, 42–61. Oxford: Butterworth-Heinemann.) and (b) Y-cone mixer (Modified and Reproduced with permission from Smith P. 2011. *Introduction to Food Process Engineering*. Boston, MA: Springer. Food Science Text Series.)

mass-flow discharge or the unit should be connected to a mass-flow container (Carson et al., 1996), which also maximizes the vessel capacity.

In general, Y-cone mixers are used for mixing of dry powders, and V-cone mixers are used for the dry blending of free-flowing solids such as milk powder, coffee, dry flavors, spice blends, and baby foods.

12.1.5.1.4 Ribbon Blender

A ribbon blender works on the principle of convective mixing. It consists of a static trough or a cylinder with a ribbon or shaft attached to two open helical screws: one right-handed and the other left-handed (Figure 12.14). The helical ribbons are pitched to move the material in opposite directions, both axially and radially. The helical ribbons rotate inside the trough, at a speed of 15–60 rev/min. Consequently, the particles move in opposite directions and thereby undergo a vigorous displacement relative to each other.

The ribbon blender is a cost-effective mixer for the production of dry food mixes and beverage blends. Cake and muffin mixes, bread improvers, flour, cereals, snack bars, spices and herbs, tea leaves, iced tea powders, and whole or ground coffee beans are some of the dry food mixes produced using a ribbon blender. The beverage blends handled in ribbon blenders comprise coffee premixes, whey protein shakes, energy drinks, chocolate drinks, and juice mixes.

Ribbon blenders may also be used for the addition of small amounts of liquid flavorings, colorings, and oils to coat or adsorb on to the surface of dry solids. In such applications, the liquid ingredients are added through a charge port on the cover of the ribbon blender. However, the liquid addition can be accomplished more precisely by the use of spray nozzles fitted to a spray bar placed just above the ribbon agitator. The parameters such as liquid flow rate and blender speed are adjusted such that the overflow and formation of wet clumps of powder are prevented. In addition, the ribbon blenders can also be used

FIGURE 12.14 Ribbon blender (Reproduced with permission from Muzzio, F. J., Llusa, M., Goodridge, C. L., Duong, N.-H. and Shen, E. 2008. Evaluating the mixing performance of a ribbon blender. *Powder Technology* 186: 247–254.)

for food extrusion in the preparation of products such as pasta, ready-to-eat cereals, snack chips, and pet food. Here, the function of the ribbon blender is to achieve a homogeneous mixture of two or more grains, flours, oil, sugar, emulsifiers, extrusion aid, and other powders. After blending, water is added to the batch to raise the moisture content to a level that is appropriate for extrusion.

12.1.5.2 Mixers for Liquid–Liquid Mixing and Dispersal of Solids in Liquids

12.1.5.2.1 High-Speed Propeller Mixers

Propeller mixer is the most commonly used equipment for the mixing of miscible, low-viscosity liquids to result in smooth and homogeneous emulsions. It includes a mixing vessel and angled blades or impellers attached to shafts. The shaft is connected to a motor through an inferior coupling held by set screws (Figure 12.15). The propeller mixers work on the principle of shearing action imparted by the edges and tips of the blades, on the low-viscosity liquid foods. The blades lead to the circulation of fluid in both axial and radial directions. More often, the propeller mixers operate at a high speed of rotation up to 8000 rpm. This gives a satisfactory axial and radial flow pattern to the liquids. Typically, the ratio between the diameters of propeller and vessel ranges from 1:10 to 1:20. For a satisfactory performance of the propeller mixer, the liquid must be sufficiently mobile to enable the fast flow of fluid towards the propeller. However, this is hindered in the case of viscous liquids.

As mixing proceeds, the propeller blades impart a centrifugal force to the liquid, which causes it to swirl around the sides of the vessel and create a depression at the shaft to form a vortex. As the speed of rotation increases, pockets of air may be entrapped in the liquid by the vortex formation. This can lead to frothing and possible oxidation of the liquid product. Vortex formation can be suppressed by the off-centric mounting of the propeller, by mounting the propeller at an angle (Figure 12.15a), or by fitting vertical baffles into the vessel (Figure 12.15b). The baffles divert the rotating fluid from its circular path to the center of the vessel, which is the spot of vortex formation otherwise.

The applications of propeller mixers have been realized in the dilution of concentrated solutions, preparation of syrups or brines, and dissolving ingredients.

12.1.5.2.2 High-Pressure Homogenizers

High-pressure homogenizers comprise a high-pressure pump operating at 1×10^4–7×10^4 kPa (P_h), fitted with a homogenization valve at the outlet. The liquid is pumped through a narrow adjustable gap of <300 μm between the valve and the valve seat (Figure 12.16). The resultant high pressure imparts a high velocity to the liquid. Subsequently, when this liquid emerges from the valve, it experiences a sudden drop in velocity and pressure (reaches the atmospheric pressure, P_{atm}) leading to very high turbulence that creates intense shearing forces. The shear force disrupts the droplets of the dispersed phase and reduces its mean size. Further reduction in droplet size is brought about by the *cavitation* phenomenon (collapse of air bubbles) and the impact forces created in some valves by placing an impact ring (also termed as the "breaker ring," a hard surface) in the path of the liquid.

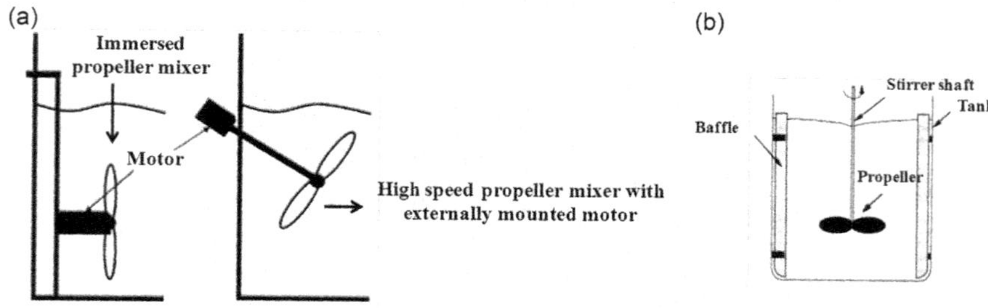

FIGURE 12.15 (a) High-speed propeller mixer (Modified and Reproduced with permission from Wiedemann, L., Conti, F., Janus, T., Sonnleitner, M., Zörner, W. and Goldbrunner, M. 2017. Mixing in biogas digesters and development of an artificial substrate for laboratory-scale mixing optimization. *Chemical Engineering and Technology* 40: 238–247.); (b) Propeller mixer with baffle.

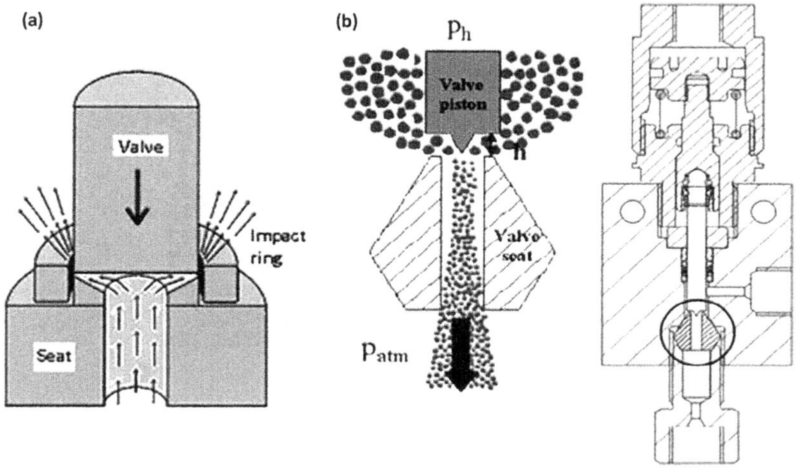

FIGURE 12.16 High-pressure homogenizer: (a) Components of the homogenizer; (b) Cross-section of the valve. (Reproduced with permission from Jacobo, A. S., Saldo, J. and Gervilla, R. 2014. Influence of high-pressure and ultra-high-pressure homogenization on antioxidants in fruit juice. In *Processing and Impact on Antioxidants in Beverages*, ed. V. Preedy, 185–193. Oxford, UK: Elsevier Inc.)

Pressure homogenizers are mainly used in the dairy industry for the homogenization of milk before the pasteurization and ultra-high-temperature sterilization processes.

12.1.5.2.3 Microfluidizer

Microfluidizers are mainly used for the production of nanoemulsions and nanodispersions. In a microfluidizer, the liquid product enters the system through the inlet reservoir. It is then forced at a pressure of 170–2750 bar by a high-pressure pump and at a high velocity of up to 500 m/s into the Y-type or Z-type interaction chamber (Figure 12.17a). On entering the Y-type interaction chamber, the fluid stream is split into two, each of which flows through separate microchannels of defined geometry (Figure 12.17b). The fluid streams continue to flow until they meet to undergo a head-on impact in a microliter volume element in which very rapid mixing occurs. On the other hand, in a Z-type interaction chamber, the pressurized fluid stream is forced through one or many zigzag microchannels that create intense shear forces to reduce the droplet size (Figure 12.17c). Fluid stream emanating from the microchannels impacts the walls of the microliter volume element. A stable emulsion with finer droplets of the dispersed phase is achieved by the high shearing forces (maximum shear rate of $10^7 s^{-1}$) resulting from the velocity gradient between the adjacent layers of liquid leads, cavitation, and ultra-high-frequency vibration. The emulsion thus obtained is then cooled aided by the cooling jacket and collected in the output reservoir.

12.1.5.2.4 Ultrasonic Homogenizers

In an ultrasonic homogenizer, a high-frequency sound wave in the range of 18–30 kHz causes alternate cycles of compression and tension in low-viscosity liquids (Figure 12.18). The aforementioned phenomenon along with cavitation results in the formation of emulsions with droplet sizes in the range of 1–2 μm.

Ultrasonic homogenizers are used for the production of ice creams, baby foods, and essential oil emulsions. It is also used for dispersing powders in liquids.

12.1.5.2.5 Colloid Mills

A colloid mill is more appropriate for high-viscosity liquids, as it creates high shear than pressure homogenizers (Figure 12.19). It is a vertical disk mill with a narrow gap (0.05–1.3 mm) between the stationary and rotating disks that rotate at 3000–15000 rpm. Different designs of the disk are available for varied applications: flat, corrugated, and conical-shaped and those made up of carborundum (silicon carbide). The colloid mills require cooling due to the greater friction created during the size reduction of high-viscosity liquid foods. Cooling is achieved by the water that circulates in the jacketed layer of the mill.

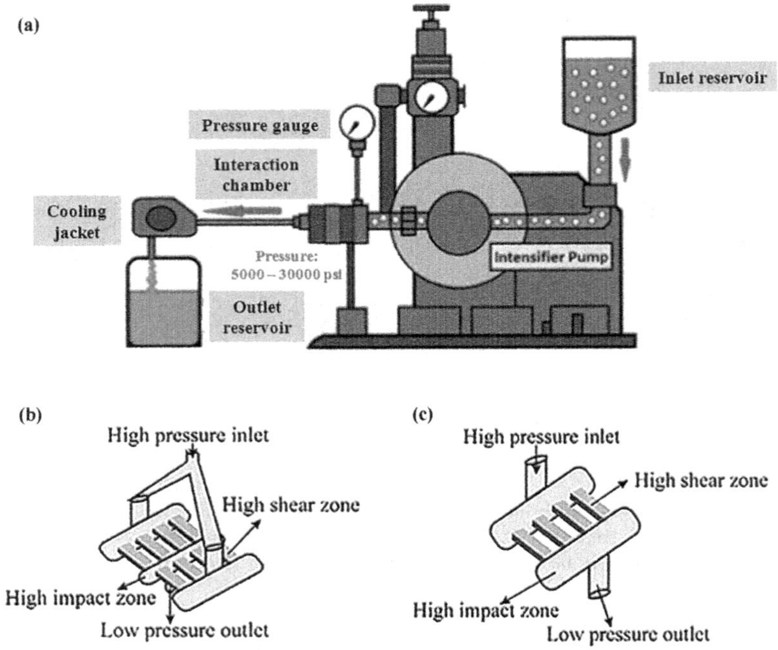

FIGURE 12.17 (a) Microfluidizer (Modified and Reproduced with permission from Jafari, S. M., He, Y. and Bhandari, B. 2007. Optimization of nano-emulsions production by microfluidization. *European Food Research and Technology* 225: 733–741; Kaialy, W. and Shafiee, M. A. 2016. Recent advances in the engineering of nanosized active pharmaceutical ingredients: promises and challenges. *Advances in Colloid and Interface Science* 228: 71–91.); (b) Y-type interaction chamber; (c) Z-type interaction chamber. (Reproduced with permission from Bai, L., Huan, S., Gu, J., McClements, D.J. 2016. Fabrication of oil-in-water nanoemulsions by dual-channel microfluidization using natural emulsifiers: saponins, phospholipids, proteins, and polysaccharides. *Food Hydrocolloids* 61: 703–711.)

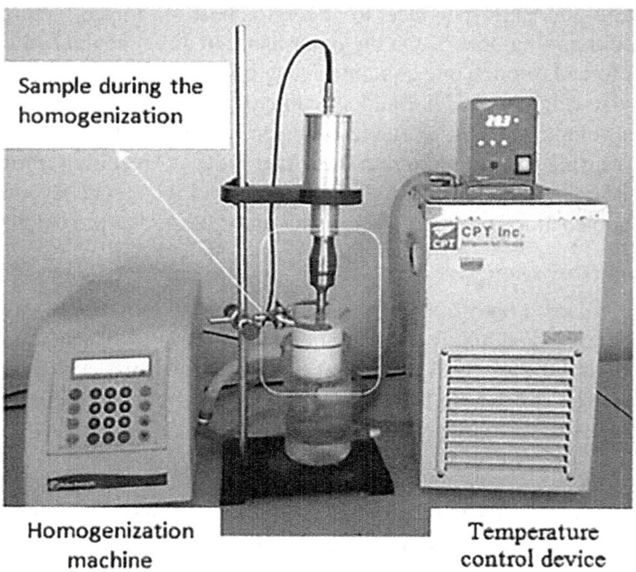

FIGURE 12.18 Ultrasonic homogenizer. (Reproduced with permission from Mahbubul, I. M., Elcioglu, E. B., Saidur, R. and Amalina, M. A. 2017. Optimization of ultrasonication period for better dispersion and stability of TiO_2–water nanofluid. *Ultrasonics Sonochemistry* 37: 360–367.)

FIGURE 12.19 Colloid mill. (Adapted and Reproduced with permission from Hoffmeister, C. R. D., Fandaruff, C., da Costa, M. A., Cabral, L. M., Pitta, L. R., Bilatto, S. E. R., Prado, L. D., Corrêa, D. S., Tasso, L., Silva, M. A. S. and Rocha, H. V. A. 2017. Efavirenz dissolution enhancement III: colloid milling, pharmacokinetics and electronic tongue evaluation. *European Journal of Pharmaceutical Sciences* 99: 310–317.)

12.1.5.3 Mixers for Semisolids (Dough and Paste Mixers)

12.1.5.3.1 Sigma Mixers

Sigma mixer is the commonly used one for high-viscosity and sticky food products. It includes a saddle-shaped, stationary bowl, usually made of stainless steel (SS 304 or SS 316). The bowl is jacketed for temperature control when heating or cooling of the product is required. The mixing elements are two steel-casted sigma-shaped blades (Figure 12.20) that intermesh and rotate inward towards each other. The blades rotate at either similar or differential speed with respect to each other. The sigma blades are fitted at close proximity to the container, in order to provide a uniform mixing. The complete unit as described previously is connected to a fixed speed drive (which permits the sigma blades to rotate at different speeds) and mounted on a steel stand of suitable strength to endure the vibration and thereby ensure a noise-free performance. The components to be mixed are loaded in the bowl from the top at 40%–65% of the mixer volume. The mixed product is unloaded by tilting the entire bowl by means of a rack-and-pinion drive. Alternatively, the sigma mixer may be equipped with discharge arrangements such as a bottom discharge valve, extruder, or screw located in the lower section between the two compartments of the stationary bowl.

In this mixer type, mixing is achieved by the intermeshing of sigma-shaped blades that pull, shear, compress, knead, and fold the material both between the rotating blades and against the walls of the

FIGURE 12.20 Sigma mixer. (Reproduced with permission from Spitz, L. 2016. Bar soap finishing. In *Soap Manufacturing Technology* (Second edition), ed. L. Spitz, 167–202. London: AOCS Press.)

stationary bowl. Thus, in sigma mixers, the mixing action is a combination of bulk movement, shearing, stretching, folding, dividing, and recombining of the material.

In a sigma mixer, the effectiveness of mixing depends upon the design of the kneader blade. The shape of the blades is significant from the perspective of facilitating a steady flow of material from the side walls of the bowl to the middle of the bowl. Different types of design are available based on the curvature and concavity of the kneading blades. For certain applications, the blades are made very rugged and serrated to give higher shredding action. The orientation of the blades also plays an important role in mixing. Further, the ratio of the rotation speed of the blades and the helical angle of the blade also influence the shearing action.

Single-arm sigma mixers, also known as the *sweet dough mixers*, are predominantly used for mixing during the production of cookies, crackers, biscuits, pet foods, and tortillas. Sigma mixers are also used in the preparation of candy and confectionery.

12.1.5.3.2 Planetary Mixers

Planetary mixers are widely used for bakery application, at both the domestic and the commercial scales (Figure 12.21). It comprises three major components: bowl for placing the ingredients, mixing element or agitator, and the motor that facilitates the movement of the bowl and the mixing element. The bowl can

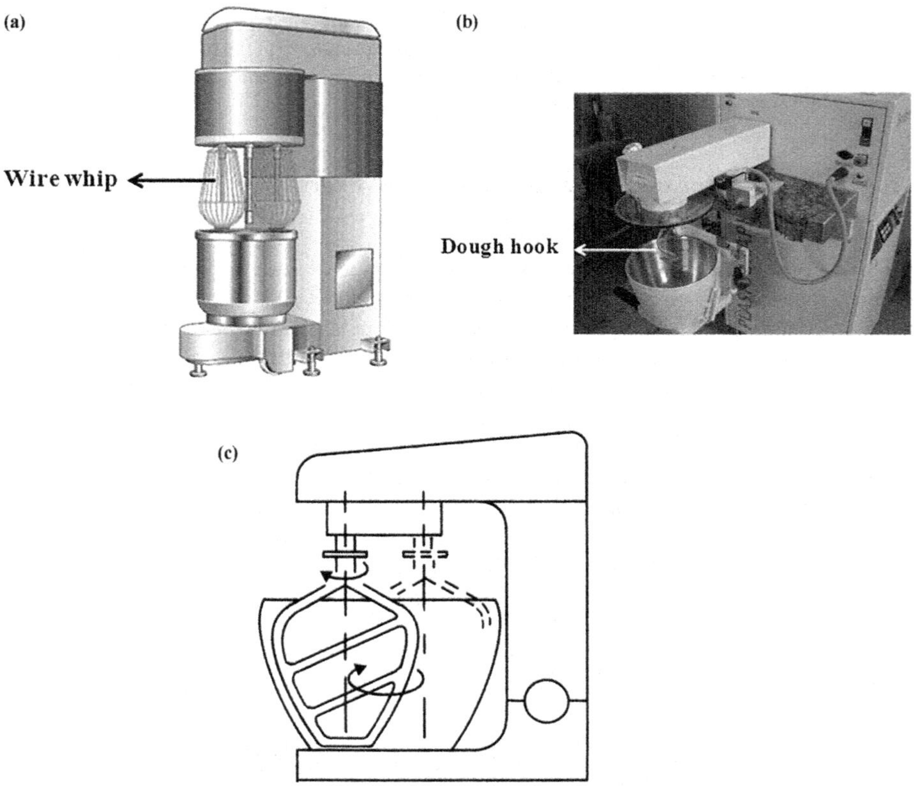

FIGURE 12.21 Planetary mixer fitted with (a) wire whip (Slightly modified and Reproduced with permission from Rabone, P. 2007. *Level 2 Certificate in Professional Cookery*. Oxford: City and Guilds and Harcourt Education Ltd.); (b) dough hook (Slightly modified and Reproduced with permission from Auger, F., Delaplace, G., Bouvier, L., Redl, A., André, C. and Morel, M-H. 2013. Hydrodynamics of a planetary mixer used for dough process: influence of impeller speeds ratio on the power dissipated for Newtonian fluids. *Journal of Food Engineering* 118: 350–357.); (c) gate beater. (Reproduced with permission from Prins, M. and Boeke, W. 2015. Equipment. In *Practical Pharmaceutics. An International Guideline for the Preparation, Care and Use of Medicinal Products*, eds. Y. Bouwman-Boer, V. Fenton-May and P. L. Brun, 609–649. Basel, Switzerland: KNMP and Springer International Publishing.)

be lifted or lowered manually or by an auxiliary motor. The name *planetary* is derived from the pattern of rotation followed by the agitator, which revolves around its own vertical axis at a relatively high speed. The axis also moves in a circular pattern and rotates around the bowl at a relatively lower speed in the opposite direction. Thus, all the parts of the mixer are involved in the mixing action. The agitators are available in different designs such as whisk or wire whip (Figure 12.21a), hook (Figure 12.21b), flat or gate beater, or scrapper (Figure 12.21c). The dough hook is the one that is used for yeast-raised dough, which assists in the gluten development, by its stretching and kneading action. It is a single-curved arm made of aluminum or steel. The wire whip contains a set of wires that is wide at the top and pointed at the bottom. It is suitable for the preparation of cake batters, in which air inclusion and diminution of bubbles are important. The batter beaters or gate beaters are either two-winged or four-winged, to fit the bowl side wall. These mixing elements are used for cake batters and general mixing applications.

Planetary mixers can be used for any bakery product such as sweet dough, pie dough, bread dough, and cake batters.

12.2 Separation Processes

Separation can be defined as the process of isolating specific ingredients from a mixture without the occurrence of a chemical reaction (Ortega-Rivas and Perez-Vega, 2011). Separation processes play multiple roles in the food industry: recovery of residual solids after an extraction process (e.g., spent tea and coffee grounds); concentration, clarification, purification, and refining of liquid foods (e.g., fruit juice, vegetable oil, and drinking water); physical removal of microorganisms; recovery of valuable product from a process (e.g., recovery of sugar crystals); and in effluent treatment. Separation techniques are classified on the basis of the phases involved (Figure 12.22). The following sections provide a detailed discussion on each of the types of separation processes.

12.2.1 Filtration

Filtration is commonly used to separate solid particles from a liquid food (Figure 12.23). It can also be used for the separation of particulate matter from gases. During filtration, a fluid with suspended solid particles is passed through the filter media having a porous structure, to yield a clear liquid termed as the *filtrate*, while entrapping the suspended solid particles in the porous structure. As filtration proceeds, the solid particles accumulate in the form of a porous cake, which adds to the filtration effect. With an increase in its buildup, the cake resists the flow of feed and filtrate.

12.2.1.1 Filtration Rate

Filtration rate is defined as the amount or quantity of filtrate obtained per unit time. Filtration rate can be derived from Darcy's law for fluid flow through porous media, which is given by the following equation:

$$Q = \frac{dV}{dt} = \frac{A\Delta P}{\mu R} \tag{12.18}$$

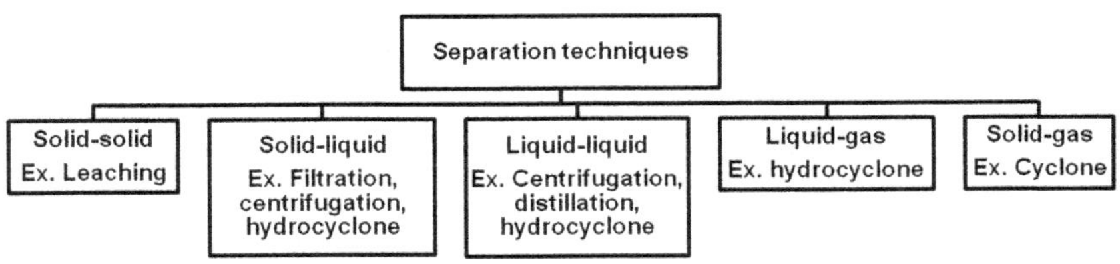

FIGURE 12.22 Classification of separation techniques.

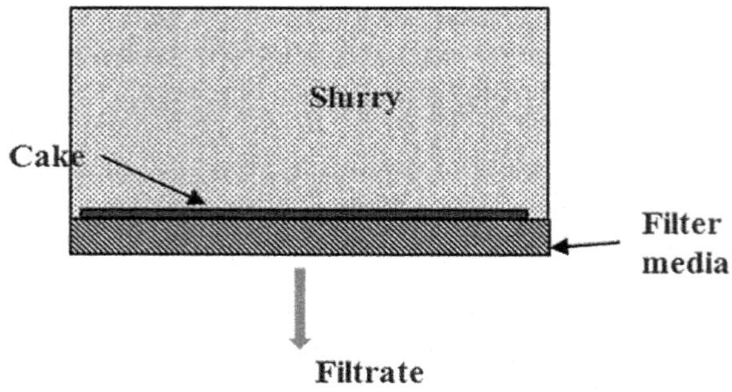

FIGURE 12.23 Schematic diagram of the filtration process.

where Q is the flow rate of the filtrate (m^3/s), V is the filtrate volume (m^3), t is the time (s), A is the surface area of filter (m^2), μ is the viscosity of fluid (Pa s), R is the resistance offered by the bed to flow (1/m), and ΔP is the pressure drop (N/m^2).

The resistance to flow, R, can be expressed as

$$R = \alpha(L + L_f) \tag{12.19}$$

where α is the specific cake resistance ($1/m^2$), L is the cake thickness (m), and L_f is the factitious cake thickness (m); thus, transforming Eq. (12.18) as

$$Q = \frac{A\Delta P}{\mu\alpha(L + L_f)} \tag{12.20}$$

L_f, in turn, is given by

$$L_f = \frac{SV}{A} \tag{12.21}$$

where S is the solid concentration of the fluid and V is the volume of cake formed per unit volume of the filtrate obtained. Therefore, the final form of Darcy's equation is given as follows:

$$\frac{dV}{dt} = \frac{A\Delta P}{\alpha\mu\left(\frac{SV}{A} + L\right)} \tag{12.22}$$

As aforementioned, cake formation over the filter media has a major influence on the filtration rate. With an increase in cake thickness, pressure on the filter media increases and reduces the flow of filtrate. In addition to cake thickness, filtration rate is also influenced by the pressure drop across the filter medium, the surface area of the filter, and the viscosity of the filtrate.

Filtration is usually carried out under two conditions: constant filtration rate and constant pressure drop. As the names suggest, in constant rate filtration, the volume of filtrate per unit time (Q) remains constant. However, in constant pressure drop filtration, the pressure drop (ΔP) is constant.

12.2.1.2 Constant Rate Filtration

From Eq. (12.22), the flow rate of filtrate during the constant rate filtration is given by

$$\frac{V}{t} = \frac{A\Delta P}{\alpha\mu\left(\frac{SV}{A} + L\right)} \tag{12.23}$$

Therefore,

$$\Delta P = \alpha\mu\left(\frac{SV}{A} + L\right)\frac{V}{tA} = \alpha\mu\left(\frac{SV + LA}{A}\right)\frac{V}{tA} \tag{12.24}$$

$$\Delta P = \frac{\alpha\mu}{tA^2}\left(SV^2 + LAV\right) \tag{12.25}$$

If cake thickness is negligible ($L \equiv 0$),

$$\Delta P = \frac{\alpha\mu}{tA^2}\left(SV^2\right) \tag{12.26}$$

12.2.1.3 Constant Pressure Drop Filtration

From Eq. (12.22), the flow rate of filtrate during the constant pressure drop filtration is given by

$$\mu\alpha\left(\frac{SV}{A} + L\right)dV = A\Delta P dt \tag{12.27}$$

$$\frac{\mu\alpha S}{A}VdV + \mu\alpha LdV = A\Delta P dt \tag{12.28}$$

Integrating on both sides

$$\frac{\mu\alpha S}{A}\int_0^V V\,dV + \mu\alpha L\int_0^V dV = A\Delta P\int_0^t dt \tag{12.29}$$

$$\frac{\mu\alpha S}{A}\frac{V^2}{2} + \mu\alpha LV = A\Delta Pt \tag{12.30}$$

Rearranging Eq. (12.30) gives

$$t = \frac{\mu\alpha SV^2}{2A^2\Delta P} + \frac{\mu\alpha LV}{A\Delta P} \tag{12.31}$$

From Eq. (12.31), it can be observed that the relation between t and V is not linear.

If cake thickness is negligible ($L \equiv 0$),

$$t = \frac{\mu\alpha SV^2}{2A^2\Delta P} \tag{12.32}$$

Dividing Eq. (12.31) by V gives

$$\frac{t}{V} = \frac{\mu\alpha SV}{2A^2\Delta P} + \frac{\mu\alpha L}{A\Delta P} \tag{12.33}$$

Or

$$\frac{t}{V} = K_pV + \beta \tag{12.34}$$

where $K_p = \frac{\mu \alpha S}{2A^2 \Delta P}$, $\beta = \frac{\mu \alpha L}{A \Delta P}$

$$\therefore \alpha = \frac{K_p 2A^2 \Delta P}{\mu S} \tag{12.35}$$

A straight line can be obtained by plotting t/V against V, the slope or gradient of which yields K_p, and the intercept provides the value of β.

Further, the specific cake resistance can also be determined from the gradient of the line as shown in Figure 12.24. In the case of constant pressure filtration, the viscosity of the filter medium (μ) is measurable and the values of the surface area of filter press (A) and pressure drop (ΔP) are known. Therefore, cake thickness (L) can be calculated from the value of intercept after substituting the values for μ, α, A, and ΔP. The calculation of the aforementioned parameters facilitates the development of an optimized filtration system.

Example 12.2

Fruit pulp is treated with pectinase enzyme and filtered in a plate-and-frame filter press to make a clear juice. The slurry contains 25 g of solids per liter, and the density of solids is 900 kg/m^3. The total filtration area is 1 m^2. The following data on volume versus time are obtained at 25°C, keeping a pressure drop of 0.1 MN/m^2 (gauge). Calculate the specific cake resistance and the equivalent cake thickness.

Time (min)	2	10	20	30	50	70
Volume (L)	50	100	150	200	250	300

Solution

Given

$$\rho_s = 900\,\text{kg/m}^3,\ S = 25\,\text{kg/m}^3,\ A = 1\,\text{m}^2,\ \Delta P = 0.1\,\text{MN/m}^2 = 1\times10^5\,\text{Pa}$$

The viscosity of water at 25°C is from the literature, $\mu = 1\times10^{-3}$ Pa s

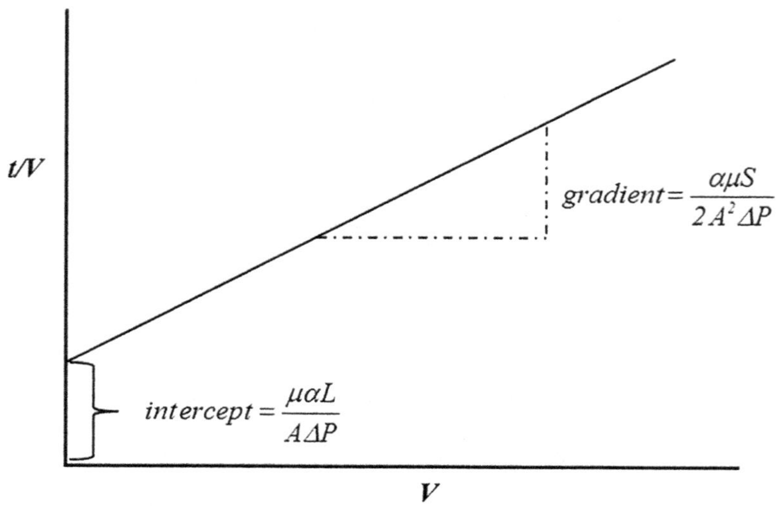

FIGURE 12.24 Constant pressure filtration.

Time (min)	Time (s)	Volume (L)	Volume (m^3)	t/V (s/m^3)
2	120	50	0.05	2400
10	600	100	0.10	6000
20	1200	150	0.15	8000
30	1800	200	0.20	9000
50	3000	250	0.25	12000
70	4200	300	0.30	14000

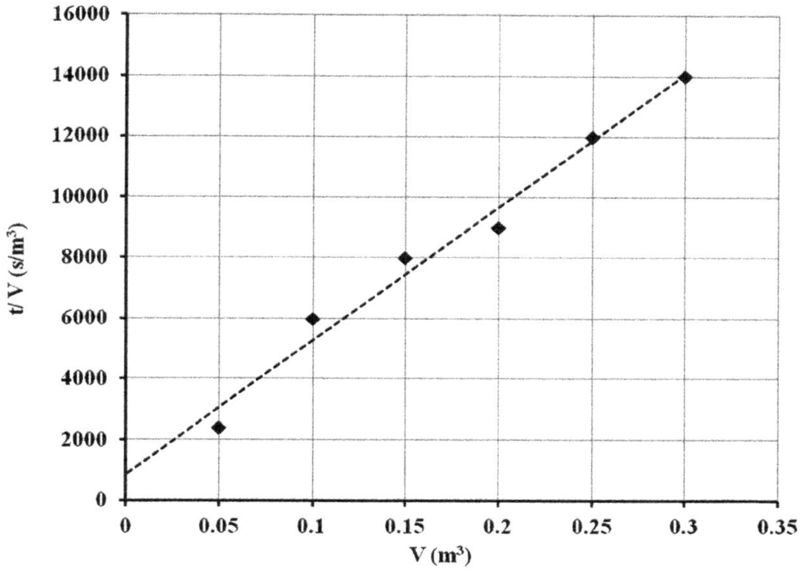

$$\text{Slope} = \frac{12000 - 2400}{0.25 - 0.05} = 48000$$

$$\text{Intercept} = 800$$

$$K_p = \frac{\mu \alpha S}{2A^2 \Delta P} = 48000$$

Therefore, $\alpha = 3.84 \times 10^{11}$

$$\beta = \frac{\mu \alpha L}{A \Delta P} = 800$$

Therefore, $L = 0.208$

The specific cake resistance (α) is 3.84×10^{11} m/kg, and the equivalent cake thickness (L) is 0.208 m.

12.2.1.4 Filter Media

The primary function of filter media is to facilitate the removal of solids from the liquid feed to produce a clear filtrate. It helps in the formation of porous cake and supports the cake once formed. Filter media should be sufficiently strong to withstand the pressure exerted by the feed material and cake. It may be rigid or flexible. Generally, flexible filter media are made up of woven cloth, woven glass fibers, or synthetic polymers that are non-toxic to food. Apart from the aforementioned materials, sand, charcoal, and gravel can also be used. Rigid filter media are made in the form of disks or cartridges using diatomaceous earth or foamed plastics such as polyethylene (PE) or polyvinyl chloride. These are available with different pore sizes and are tailor-made for the defined requirement. The pores of filter media should not be plugged so as to prevent the slowdown of filtration rate. Eventually, the filter medium should permit a clean and complete discharge of the cake formed.

12.2.1.5 Filter Aid

The filter aid consists of the large-sized, rigid, and chemically inert particles, which facilitates the filtration. The filter aid is precoated on the filter medium or can be mixed with the feed liquid, which has small particles due to which it is unable to form a porous cake on the filter bed. Diatomaceous earth, perlite, cellulose powders, activated carbon, and bentonites are the commonly used filter aid. However, filter aid can be used only when the filtrate is the desired product, rather than the cake.

12.2.1.6 Filtration Equipment

Depending on the driving force used, the filtration process may be classified as gravity filtration, pressure filtration, vacuum filtration, centrifugal filtration, and membrane filtration. Accordingly, the major filtration equipment types used in food industries are as follows:

1. Plate-and-frame filter press
2. Continuous rotary vacuum filter
3. Centrifugal filter
4. Membrane filters

12.2.1.6.1 Plate-and-Frame Filter Press

The plate-and-frame filter press is one of the widely used pressure filters in the food processing industry. It comprises a set of plates, stacked together horizontally. The plates are covered with filter cloth on both the sides. Hollow frames are used to hold the plates (Figure 12.25).

Stacking of more plates provides a higher surface area for filtration. Plates and frames are stacked on a stand where the plates are separated by the frames. These plates are pressed together mechanically using a screw or jack. The slurry is fed through the feed line using a positive displacement pump. The pump is used to create the high-pressure gradient required for the filtration process. The filtrate is collected through the discharge line. As the time progresses, cake thickness increases and clogs the filter cloth, which reduces the filtration rate. Then, the plates and frames are dismantled to remove the cake.

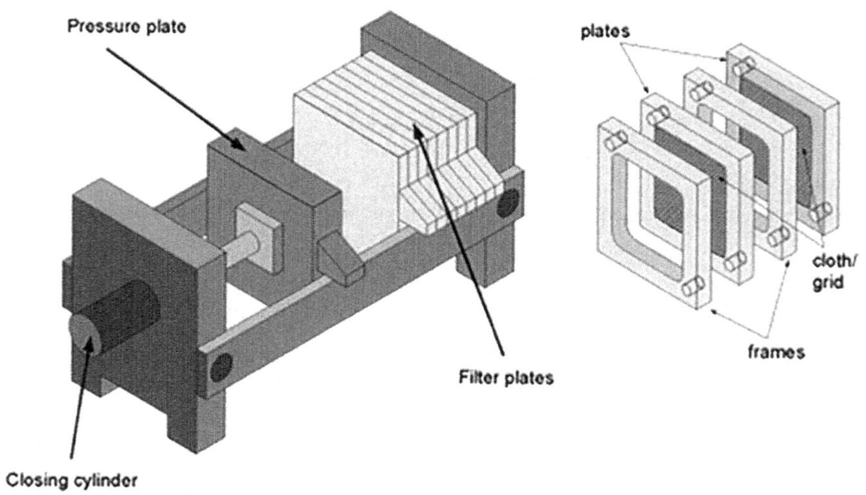

FIGURE 12.25 Schematic of plate-and-frame filter press. (Reproduced with permission from Sparks, T. 2012. Solid-liquid filtration: understanding filter presses and belt filters. *Filtration + Separation* 49: 20–24.)

The plate-and-frame filter press systems have various advantages including low floor space requirement, easy operation and maintenance, increased capacity with an increase in the number of plates, and the versatility that permits its use for a wide variety of slurries. However, it suffers from certain disadvantages such as frequent clogging of filter media, the shorter life of filter media due to wear and tear, leakage of filtrate through filter media, and high manpower requirement.

12.2.1.6.2 Continuous Rotary Drum Vacuum Filter

Continuous rotary drum vacuum filter consists of a hollow rotating drum. In this filter type, vacuum is applied instead of pressure. In this filter, the rolling drum is submerged in the slurry. Drum surface is covered with the porous filter medium. The drum continuously moves through the slurry and sucks the filtrate. Filtrate sucked through the filter medium is further discharged through the pipes connected to the shaft of the drum. Solids adhere on to the surface of the filter medium, thus forming a cake that is further scrapped by a blade, usually known as the *doctor's blade* (Figure 12.26). Rotary vacuum filter needs low labor cost but demands a high initial capital cost.

12.2.1.6.3 Centrifugal Filter

In this type, the flow of filtrate through the filter medium and the cake is facilitated by the centrifugal force. Centrifugal filters consist of a perforated cylindrical basket open at one end (Figure 12.27). The basket wall is perforated so that any liquid in the solid material drips. The basket is mounted on a drive shaft connected to a motor. The slurry is fed to the perforated basket. Under the action of centrifugal force generated due to the rotation of the basket, the slurry is thrown to the perforated wall. Filtrate passes through the perforated wall, and the aggregated solid forms the cake. An adjustable unloader knife is used to remove the filter cake. Alternatively, wash liquid can be sprayed on the cake in order to loosen it prior to removal. The removal of cake is done either through the open end or via openings at the base below the detachable valve plate. The filtrate is then discharged through the liquid outlets.

12.2.1.6.4 Membrane Filtration

The commencement of the application of membrane technology in the food industry dates back to the 1960s. In the last decade, membrane processes have replaced the energy-intensive thermal processes in the beverage and dairy industry, mainly for the fractionation and concentration purposes. The non-thermal operation of membrane processes renders them suitable for thermally sensitive materials and leads to improved retention of product characteristics such as flavor, color, and nutritional components. This is because membrane filtration removes water from a liquid product, without the involvement of phase change, thus eliminating the application of thermal energy. Membrane filtration also results in higher process efficiency with the lowest waste generation.

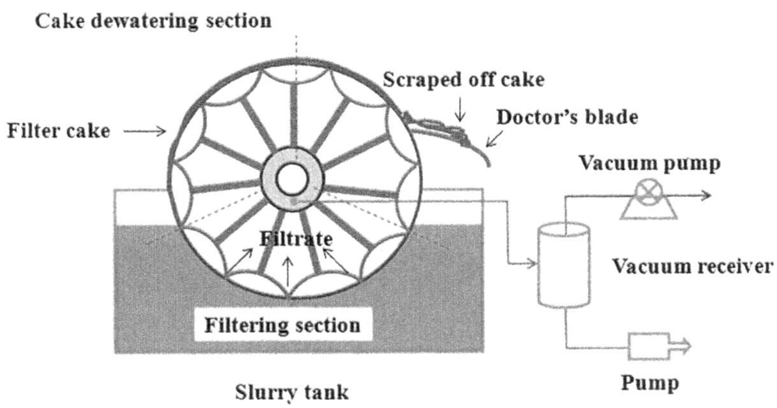

FIGURE 12.26 Schematic of continuous rotary drum vacuum filter. (Modified and Reproduced with permission from Shao, P., Darcovich, K., McCracken, T., Ordorica-Garcia, G., Reith, M. and O'Leary, S. 2015. Algae-dewatering using rotary drum vacuum filters: process modeling, simulation and techno-economics. *Chemical Engineering Journal* 268: 67–75.)

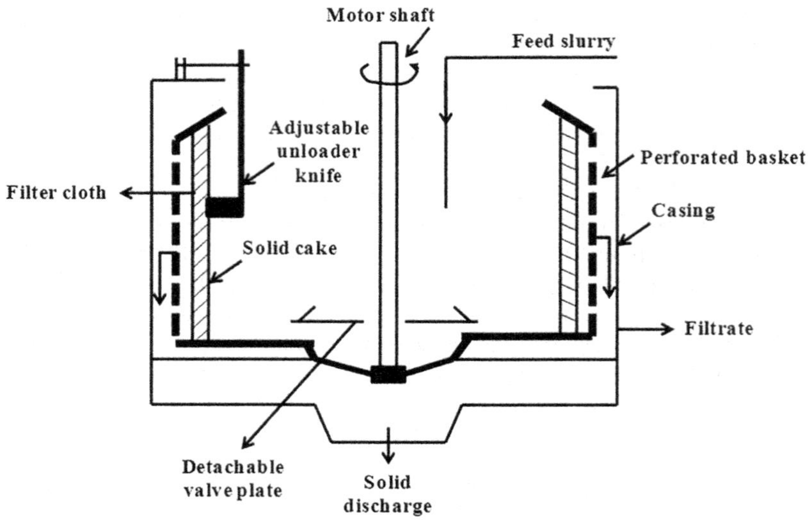

FIGURE 12.27 Centrifugal filter.

The membrane is defined as a thin film material with selective permeability. The pressure difference across the membrane is the major driving force for the membrane separation processes. The major membrane separation processes include electrodialysis (ED), reverse osmosis (RO), nanofiltration (NF), microfiltration (MF), and ultrafiltration (UF). Figure 12.28 provides an overview of the different membrane processes available for particles of the different size range.

12.2.1.6.4.1 Material of Construction for Membranes In membrane filtration, the membrane is the most important part of the system. Earlier, the membranes were made up of cellulose acetate that exhibited thermal instability at a temperature above 30°C. Thus, they were replaced by thermally stable and chemically inert membranes made up of ceramics and polyamides.

Currently, the membranes are mainly made up of organic polymers such as cellulose derivatives, polyamides, polyolefins, and halogen-substituted hydrocarbons. Inorganic membranes are ceramic based, made up of oxides of titanium, silicium, and aluminum. The characteristics expected of an ideal membrane are as follows:

- Good mechanical strength, chemical stability, and inert nature
- Thermal stability
- Resistance against the microbial action and sanitizing agents
- Smooth and fouling-resistant surface
- High permeate flux
- Cost-effective with long service life

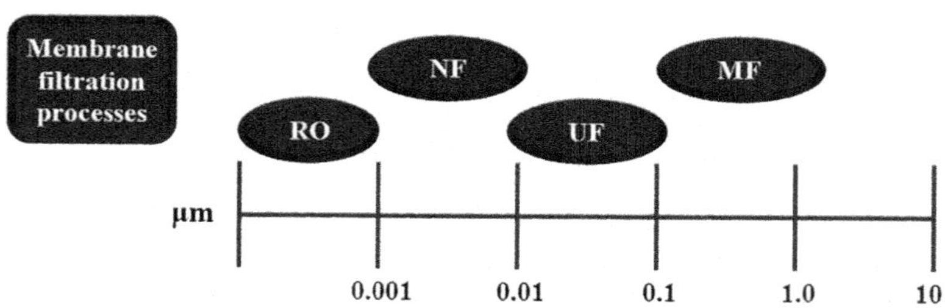

FIGURE 12.28 Particle size chart for membrane filtration processes.

Thermal stability of membrane is the most important factor as the filtration rate is enhanced at increased temperature leading to reduced microbial activity. Also, cleaning-in-place is easier at an increased temperature. Therefore, thermally stable membranes are preferred. Chemical inertness of the membrane is also important as it should not react with the feed material to result in undesirable products. Further, chemical reactions may also lower the life of the membrane. Though ceramic membranes are expensive compared to organic polymer membranes, they are preferred due to their longer operational life.

12.2.1.6.4.1.1 Membrane Modules Membranes are thin and therefore require a porous support to withstand high pressure. The geometrical configuration including the membrane and its porous support is known as the *module*, the orientation of which varies with respect to the flow of feed and permeate. The important characteristics of membrane modules are compactness, less friction and low pressure drop, uniform velocity distribution throughout the membrane, high turbulence, easy cleaning, and maintenance with low cost per unit area of the membrane. The membrane modules commonly used in the food industry are explained in Sections 12.2.1.6.4.1.1.1–12.2.1.6.4.1.1.4 that follow.

12.2.1.6.4.1.1.1 Plate-and-Frame Module Plate-and-frame module configuration bears a resemblance to the plate-and-frame filter press, except for that in the former case, a membrane replaces the filter media (Figure 12.29). This module has a low surface area-to-volume ratio. The membranes may be rectangular or circular in shape and may be stacked vertically or horizontally. Plate-and-frame module is generally used for MF and UF. However, it cannot withstand high operational pressure.

12.2.1.6.4.1.1.2 Tubular Module Tubular module is similar to a shell and tube heat exchanger (Figure 12.30). The inner wall of this module is made up of a porous structure, on which the membrane is fixed.

Tubes are installed inside the shell and connected to end plates with their diameter in the range of 10–25 mm. Due to the larger diameter of the tubes, the surface area-to-volume ratio of this module is low. The permeate flows in the shell side, whereas the retentate flows inside the tubes. The flow direction for the operation is from inside to outside, and the membrane can be easily cleaned by reversing the flow direction, that is, from outside to inside. This configuration allows high tangential velocity and thus suitable for feed with particulate suspended solids.

12.2.1.6.4.1.1.3 Spiral Wound Module Spiral wound module provides a high surface area-to-volume ratio at a low membrane cost (Figure 12.31). Therefore, commercially, it is more preferable than other membrane modules. In the spiral wound module, two membranes are heat-sealed on three sides to form a bag and porous mesh. The permeate flows through space that is sandwiched between the two membranes.

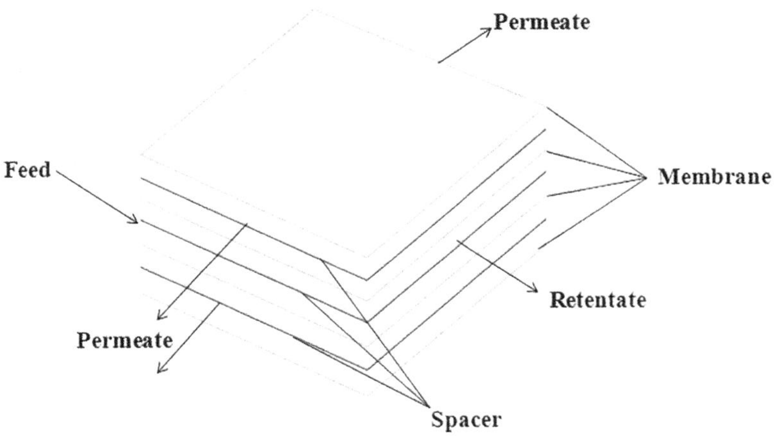

FIGURE 12.29 Plate-and-frame module.

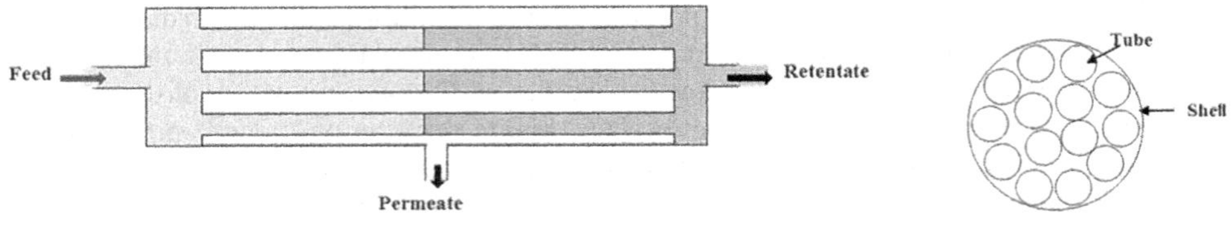

FIGURE 12.30 Tubular module.

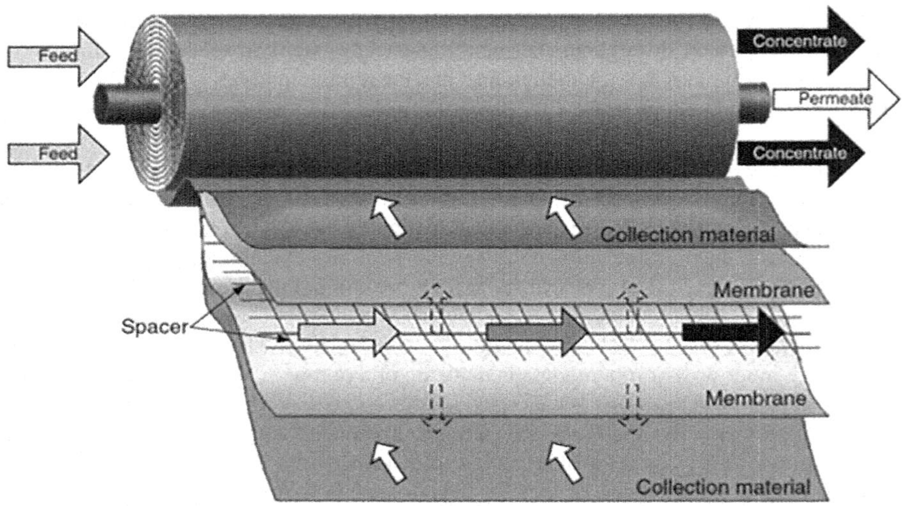

FIGURE 12.31 Spiral wound module. (Reproduced with permission from Park, J. and Lee, K. S. 2017. A two-dimensional model for the spiral wound reverse osmosis membrane module. *Desalination* 416: 157–165.)

This sandwiched assembly is wound spirally around a perforated tube. Feed is pumped from one side of the spiral wound module to enter inside the mesh. Generally, a spiral wound tube has a diameter of 12 cm and a length of 1 m.

12.2.1.6.4.1.1.4 Hollow Fiber Module Hollow fiber module is similar to the tubular module, but in this case, the tubes have a very small diameter in the range of 40–80 µm (Figure 12.32). The small diameter of the tube provides good mechanical strength to the module. A large number of tubes are bundled together and inserted in a vessel while connected by a perforated plate. Compactness and high surface area-to-volume ratio are the major advantages of hollow fiber module. However, they are prone to clogging and fouling, and hence can be used only for low-viscous fluids. Hollow fiber module is mainly used in the RO process for the purification of water. Also, these modules are expensive, and if one or several fibers are impaired, it demands a change of the whole unit.

12.2.1.6.4.2 Types of Membrane Processes

12.2.1.6.4.2.1 Electrodialysis ED is an electrochemical membrane process, where the driving force is an electrical field instead of the pressure gradient. It operates under the application of electrical field where the ions (cations and anions) are carried out selectively through the membrane. Membranes are selective for anions or cations depending on the fixed charge attached to them. If the membrane has a fixed positive charge, it will allow only anions and repel any cation on its encounter. Thus, these membranes are resistant to water permeability and allow only selective ions to permeate through.

Schematic of the ED system is shown in Figure 12.33. It contains cation-exchange and anion-exchange membranes, arranged alternately. A compartment containing the solution is sandwiched between the

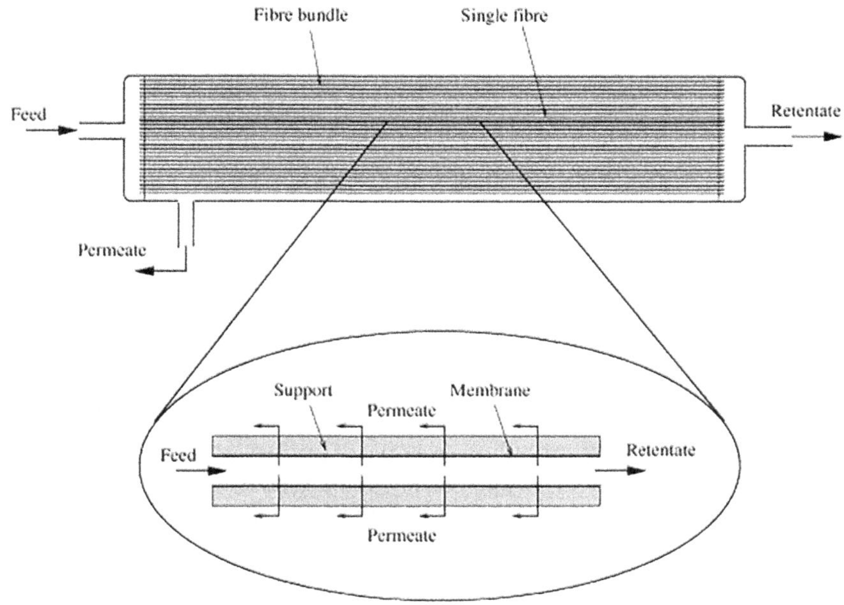

FIGURE 12.32 Hollow fiber module. (Reproduced with permission from Marriott, J. and Sorensen, E. 2003. A general approach to modelling membrane modules. *Chemical Engineering Science* 58: 4975–4990.)

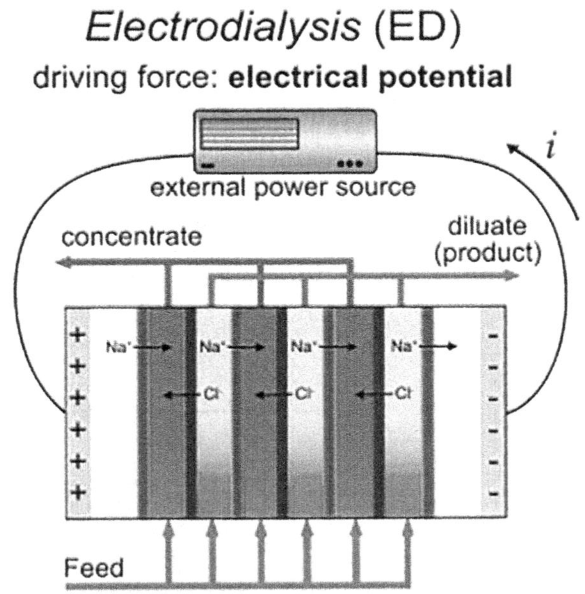

FIGURE 12.33 Schematic of ED membrane system. (Modified and Reproduced with permission from Tedesco, M., Hamelers, H. V. M. and Biesheuvel, P. M. 2016. Nernst-Planck transport theory for (reverse) electrodialysis: I. Effect of co-ion transport through the membranes. *Journal of Membrane Science* 510: 370–381.)

cation-exchange membrane and the anion-exchange membrane. The electric charge on the membrane is applied using the electrode, which is placed at the assembly end.

The working of ED can be further explained by the example of Na^+ and Cl^- ions. On applying electric charge using the electrode, anions (Cl^-) are attracted towards the cathode as they move through the anion-exchange membrane. Moreover, the anions will be repelled by the cation-exchange membrane and thus remain in the solution. Finally, solutions that are ion-concentrated and ions depleted are obtained in the alternative streams.

The energy required for the ED process can be calculated based on the following equation:

$$E = (nR_c)\left(\frac{zFm\Delta c}{U}\right)^2 \tag{12.36}$$

where E is the energy required (J), n is the number of cells in stack, R_c is the cell resistance (Ω), z is the electrochemical valence, m is the flow rate of the feed solution (L/s), F is Faraday's constant (96500 A/s equivalent), U is the current utilization factor, and Δc is the concentration gradient between product and feed.

ED process is mainly used for the desalination of water, for obtaining table salt from seawater and removal of salts and ions from whey and orange juice. Energy consumption of the process is mainly influenced by total dissolved solids (TDS). ED process is economically feasible for TDS in the range of 1000–5000 mg/L. If the TDS is higher than this range, the process turns economically unfeasible due to higher energy consumption. ED process is limited only to the removal of ions, and it fails to eliminate microorganisms, colloidal particles, and any other nonionized material.

12.2.1.6.4.2.2 Microfiltration MF separates larger molecular weight suspended or colloidal compounds from the dissolved solids by applying low pressure (Figure 12.34). The difference in size between the molecules is the underlying principle of MF. Particles in the range of 1–10 μm are separated using MF.

MF is widely used in the food industry to clarify cloudy fluids and for the removal of microorganisms from liquids, as MF proves to be economical compared to centrifugation and filtration processes. Standard materials for the MF membrane are polyethersulfonate and ceramics. Ceramic membranes are useful in the food industry due to their mechanical and pH stability. A membrane with a diameter less than 0.5 μm is capable of processing liquids against any microbial contamination. Therefore, MF is widely used for the purification of drinking water. In addition, it is used in the clarification of vinegar, sugar syrups, and wine. MF is generally used as a pretreatment method before UF and RO to remove the suspended solids. This prevents fouling or clogging in UF and RO processes. Furthermore, MF is also used for waste stream treatment to remove the oil droplets or fat from the waste. MF has been extensively used in the skim milk processing and cheese manufacturing process. It also shows wide applications in downstream processing for the recovery of extracellular protein from biomass. Flux reduction due to fouling is the major problem associated with MF, which restricts its use in the food processing sector.

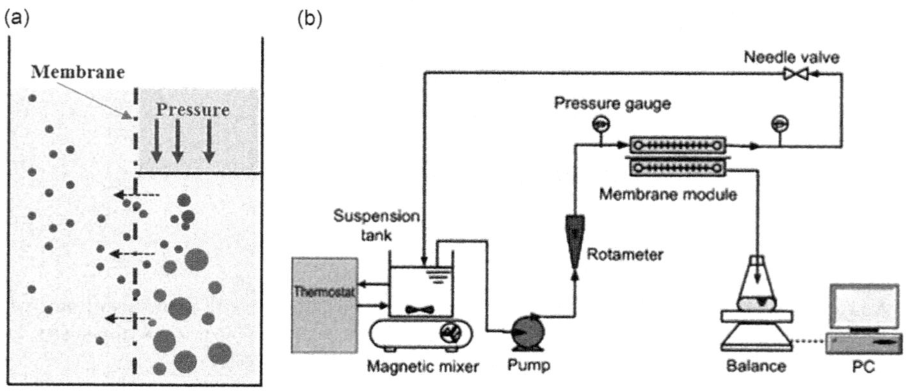

FIGURE 12.34 (a) Working principle and (b) schematic diagram of a MF membrane system. (Reproduced with permission from Hwang, K- J. and Chiang, Y- C. 2014. Comparisons of membrane fouling and separation efficiency in protein/polysaccharide cross-flow microfiltration using membranes with different morphologies. *Separation and Purification Technology* 125: 74–82.)

12.2.1.6.4.2.3 RO and NF Membrane Systems RO is defined as a process in which an external pressure is applied that exceeds the osmotic pressure to cause a reversal in the direction of the movement of water against its chemical potential, through a semipermeable membrane (Figure 12.35). Thus, it is the converse of the osmosis process and therefore known as RO. In the osmosis process, due to the presence of a concentration gradient, water passes from the region of higher concentration to the zone of lower concentration. In RO, water is referred to as *permeate*; the leftover concentrated solution is termed as *retentate*.

The energy required for the RO depends on the concentration of salt in the water. Solutions with higher salt concentration need more energy for the RO process when compared to those with lower salt concentration. RO uses cross-flow filtration to avoid any film buildup and maintain the membrane surface clean. Using van't Hoff equation, osmotic pressure of the system can be obtained as

$$\Pi = \frac{cRT}{M} \tag{12.37}$$

where Π is the osmotic pressure (Pa), c is the concentration of solute (kg/m^3), M is the molecular weight, T is the absolute temperature (K), and R is the universal gas constant (J/mol K).

In RO process, transmembrane flux is given by the following equation:

$$J = \frac{k_W}{x}(P - \Delta\Pi) \tag{12.38}$$

where J is the transmembrane flux (kg/m^{2}s), x is the membrane thickness (m), k_W is the mass transfer coefficient (s), and P is the transmembrane pressure (Pa).

Percentage salt rejection gives the efficiency of the RO membrane system. A well-designed RO system can reject 98%–99% of the contaminants in the feed solution. % salt rejection is calculated as follows:

$$\%\ \text{Salt rejection} = \frac{\text{Conductivity of feed water} - \text{Conductivity of permeate water}}{\text{Conductivity of feed water}} \times 100 \tag{12.39}$$

Higher the salt rejection, higher the efficiency of the RO system. Moreover, low salt rejection indicates fouling of the membrane system and indicates a requirement for the cleaning of RO module. Fouling of membrane can occur due to particulate or colloid matter and biofilms produced by the microorganism. Fouling in RO system can be prevented by pretreatments such as MF, multimedia filtration, and use of softened water.

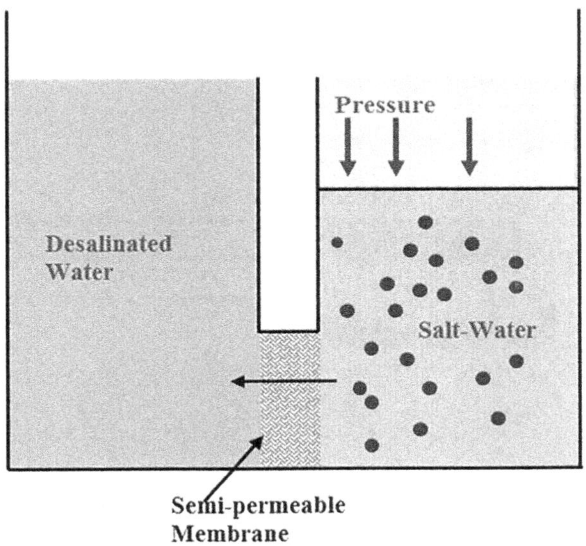

FIGURE 12.35 Schematic diagram of a RO membrane system.

Cellulose acetate membrane, made up of highly organized polymer, with a tight structure and thin membrane thickness is used for the RO process due to its water-permeating ability. Membranes made from aromatic polyamide are also used for the RO process. Membranes used for the RO process have very small pores with a diameter in the range of 0.5–1.5 nm. These membranes allow water, uncharged solutes, and smallest organic molecule to pass through them.

An advantage of RO is that it requires low power for the recovery of salt and chemicals from the effluent. The energy requirement is limited to the operation of the pressure pump to generate external pressure for the process. It has a compact process unit that requires low floor space. Retention of volatile compounds and aromas during RO processing is higher compared to the conventional process. Installation of RO unit is less costly compared to the units used for conventional methods of concentration. However, a major disadvantage of RO is the susceptibility of the membrane to fouling. Another limitation of the RO system is a phenomenon known as the *concentration polarization*. This refers to the concentration gradient of salts on the high-pressure side of the RO membrane surface, caused by the re-dilution of residual salts as water permeates through the membrane. The salt concentration in this boundary layer surpasses the concentration of the bulk water. Concentration polarization affects the membrane performance of an RO system by increasing the osmotic pressure at the membrane's surface leading to reduced flux, increased salt leakage, and greater probability of scale formation. Concentration polarization can be alleviated by increasing the velocity or turbulence of the feed stream. Yet another disadvantage is with respect to the low temperature (18°C–30°C) limited operation of RO filtration, owing to the thermolabile nature of the cellulose acetate membrane. Moreover, the membranes used in RO process are not chemically inert and therefore susceptible to damage caused by strong acidic or basic solutions.

RO is mainly used at the commercial level for water desalination. In the food industry, RO is mainly applied to the concentration of liquid foods. RO water is mainly used in the food and beverage industries. Furthermore, it also finds an application in water softening and wastewater treatment. It is also used for the pre-concentration of very dilute juice or solutions such as maple juice, before the final concentration by the evaporator. Skim milk, whole milk, and whey are concentrated using the RO. It is also used for the pre-concentration of cane juice and sugar beet juice.

In NF, particles are separated based on their particle size in the nanometer range. NF is similar to RO, except for its partial rejection of monovalent ions compared to the total rejection by RO (Figure 12.36). Polyvalent ions are rejected by the NF membrane. NF membranes are less dense compared to the RO

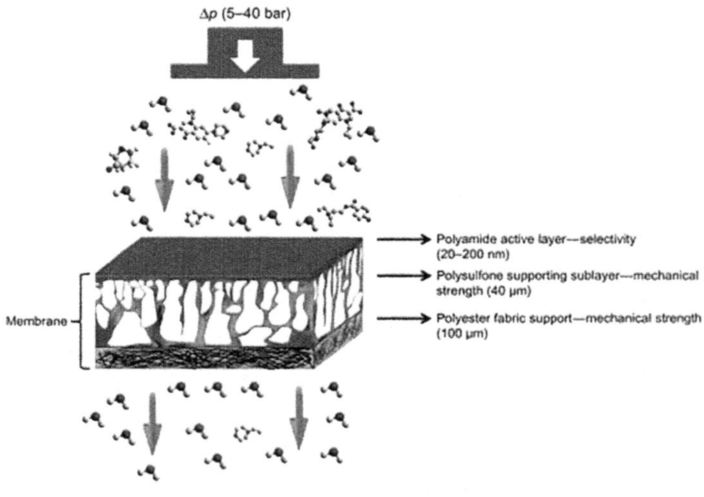

FIGURE 12.36 Schematic of NF membrane system. (Reproduced with permission from Dolar, D. and Košutić, K. 2013. Removal of pharmaceuticals by ultrafiltration (UF), nanofiltration (NF), and reverse osmosis (RO). In *Comprehensive Analytical Chemistry* (Volume 62), eds. M. Petrovic, D. Barcelo and S. Pérez, 319–344. Amsterdam, The Netherlands: Elsevier.)

membrane. Porous membranes are used for NF, whereas non-porous membranes are used for RO. NF is mostly used for the deacidification and demineralization of whey.

12.2.1.6.4.2.4 Ultrafiltration UF membrane system (Figure 12.37) separates molecules based on the molecular weight (in the range of 5–500 kDa) and structure. It allows passage of the molecules with lower molecular weight and rejects the passage of the molecules with higher molecular weight. The membranes used in UF have a pore diameter of 10–100 µm. It removes inorganic and organic polymers, colloidal materials, and viruses. The permeability of UF membrane is influenced by pore size, solute, electrostatic charge on the membrane surface, applied pressure, process temperature, nature and concentration of feed material, and flow rate.

UF membranes are made up of synthetic polymers (Table 12.1). These membranes are chemically inert and thermally stable. The selection of the UF membrane depends on the molecular cutoff, hydrophilicity, and the thermal and chemical stability. Flat sheet, tubular, and hollow fiber modules are commercially used for the UF process.

The UF membrane flux rate (N) is determined by the following equation:

$$N = KA\Delta P \tag{12.40}$$

where K is the membrane permeability constant (kg /m^{2}kPa s), A is the area (m^2), and ΔP is the pressure gradient (Pa).

UF has wide applications in the dairy industry for the concentration of whole milk before cheese manufacturing and for the concentration of whey protein. Concentration and fractionation of protein using electro- UF and ultrasonic UF are also carried out commercially. Soy protein is extracted and concentrated from the defatted soy flour using UF.

12.2.1.6.4.3 Applications of Filtration in the Food Processing Industry The major applications of filtration (Brennan and Grandison, 2011; Ortega-Rivas and Perez-Vega, 2011) in the food processing industry are as follows:

- Plate-and-frame filters, rotary drum vacuum filters, and shell-and-leaf filters are used for the filtration of wine at various stages of the wine-making process. These filters are also used for the filtration of matured beer, to separate the yeast deposits that form at the bottom of the maturation tank.

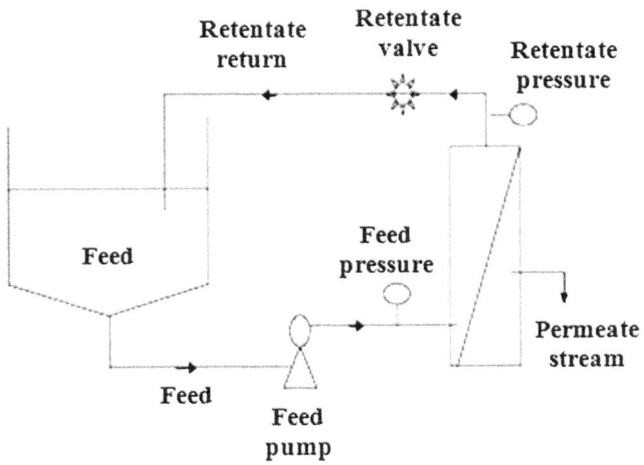

FIGURE 12.37 Schematic of UF membrane system. (Reproduced with permission from Adjalle, K. D., Brar, S. K., Verma, M., Tyagi, R. D., Valero, J. R. and Surampalli, R. Y. 2007. Ultrafiltration recovery of entomotoxicity from supernatant of *Bacillus thuringiensis* fermented wastewater and wastewater sludge. *Process Biochemistry* 42: 1302–1311.)

TABLE 12.1

Comparison between the Different Membrane Processes

	RO	NF	UF	MF
Membrane	Asymmetrical	Asymmetrical	Asymmetrical	Symmetrical
Thickness (μm)	150	150	150–250	10–150
Thin film (μm)	1	1	1	
Pore size (μm)	<0.002	<0.002	0.2 – 0.02	4 – 0.02
Components rejected	Sodium chloride, glucose, amino acids	Mono-, di-, and oligosaccharides, polyvalent negative ions	Macromolecules, proteins, polysaccharides, viruses	Particles, clay, bacteria
Membrane material	Cellulose acetate, thin-film polyamide	Cellulose acetate, thin-film polyamide	Ceramic., polysulfone (PSO), polyvinylidenedifluoride (PVDF), cellulose acetate	Ceramic, polypropylene (PP), PSO, PVDF
Membrane module	Tubular spiral wound, plate and frame	Tubular spiral wound, plate and frame	Tubular hollow fiber, spiral wound, plate and frame	Tubular hollow fiber
Operating pressure (bar)	15–150	5–35	1–10	<2
Applications	Lactose, concentration of whey permeate, delactosed, deproteinized whey	Bioactive milk and whey proteins, hydrolysates	WPC, whey protein isolate, milk protein concentrate, β-lactoglobulin, α-lactalbumin	Micellar casein, native whey proteins

Source: Data from Pouliot (2008).

- Vegetable oils are filtered to remove impurities such as seed coat, nuts, and other residues. The plate-and-frame filters are used on a small scale, whereas the rotary filters are used for the large-scale operation.
- In the sugar manufacturing industry, filtration is used extensively at different stages of processing, from the filtration of raw juice to mother liquor. The plate-and-frame filters and rotary drum vacuum filters are commonly used for the aforementioned purpose. After crystallization, the sugar crystals and juice are separated using the centrifugal filters.

12.2.1.6.4.4 Case Study: Application of Filtration in the Bottled Water Industry Bottled water industry is the fastest growing liquid beverage market. The increasing consumption of bottled water and technological advancements in the water processing equipment are the major reasons behind the booming market. Filtration is a key step in the production of bottled water (Figure 12.38). The method of filtration to be used in bottled water production depends on the source of water, which determines the physical, chemical, and microbiological composition of water. The water is usually obtained from widely varied sources, namely, a natural well, spring, and groundwater source, or from the municipal water supply. Accordingly, the filtration process may involve a series of filters, with or without the requirement for chemical pretreatment (Gordon, 2017). Specifically, the non-municipal water demands a chemical pretreatment prior to filtration.

Figure 12.38 shows the general scheme for water filtration when the pretreated municipal water is sourced for the production of purified bottled water. The groundwater derived from well, spring, or artesian is pumped to the raw water-holding tanks, after the chemical pretreatment (if required). The water is then passed through a series of sand and activated carbon filters. The sand filter removes the larger suspended particles and large-molecule impurities. Whereas, the carbon filter eliminates the organic contaminants, residual chemicals, if any, from the pretreatment, and the undesirable odors and colors from the raw water. The water thus emanating from the sand and carbon filters passes through a sedimentation filter, also known as the *polisher.* The polisher removes the small suspended particles in the

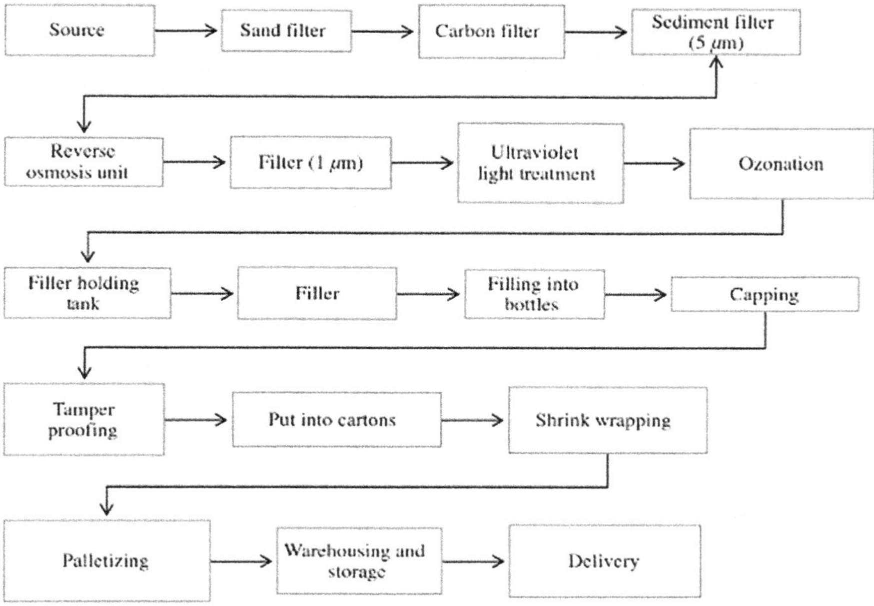

FIGURE 12.38 Process flowchart for the production of bottled water. (Reproduced with permission from Gordon, A. 2017. Case study: food safety and quality systems implementation in small beverage operations—Mountain Top Springs Limited. In *Food Safety and Quality Systems in Developing Countries* (Volume 2: Case studies of effective implementation), ed. A. Gordon, 81–116. London: Elsevier Inc.)

water, which would have escaped the previous filtration steps, fine rust, or any other contaminants that can potentially damage the membrane during the RO filtration. Subsequently, the water is transferred to the holding tanks, which supplies the process water for usage in the plant or directly passed on to the filler-holding tank. Finally, the RO unit removes any remaining organic substances, colloidal material, and some bacteria from the water. Destruction of the remaining bacteria is accomplished by the use of additional filtration steps by using a double-stage filter (passing water through a 5 μm filter, followed by a 1 μm filter). The filtered water is thus ready for further processing including ultraviolet (UV) light treatment and ozonation (Figure 12.38) (Gordon, 2017).

12.3 Centrifugation

In the centrifugation process, heterogeneous food mixture is separated by the application of centrifugal force. The principle of centrifugal separation is based on the density difference between components. Centrifugation is achieved by the angular rotation of the feed material. During centrifugation, the particles move radially away from the center of the axis of rotation. Centrifugal force acting on the particle is dependent on the distance of particles from the center, mass, and rotation speed. Heavier particles tend to move away from the center and experience the highest centrifugal force. Centrifugation is used for the separation of solid–liquid, liquid–liquid, and gas–solid mixtures. Apart from separation, the rationale of centrifugation can be diverse as depicted in Figure 12.39.

12.3.1 Derivation of Expression for the Centrifugal Force Acting on a Particle

Centrifugal force acting on a particle (F_C) is determined by the following equation:

$$F_C = mr\omega^2 \tag{12.41}$$

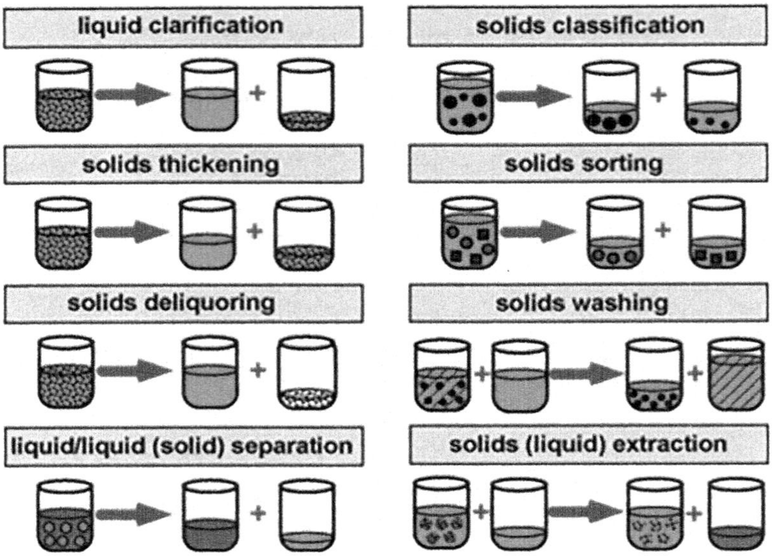

FIGURE 12.39 Purposes of centrifugal separation. (Reproduced with permission from Anlauf, H. 2007. Recent developments in centrifuge technology. *Separation and Purification Technology* 58: 242–246.)

where F_C is the centrifugal force acting on the particle, m is the mass of the particle, r is the radius of the path traveled by the particle, and ω is the angular velocity of the particle. Angular velocity is expressed in terms of tangential velocity (v), which is given by

$$\omega = \frac{v}{r} \tag{12.42}$$

Therefore, Eq. (12.41) can be written as

$$F_C = \frac{mv^2}{r} \tag{12.43}$$

The speed of centrifuge is expressed in terms of revolution per minute (N), and therefore, the angular speed can be expressed as

$$\omega = \frac{2\pi N}{60} \tag{12.44}$$

Equation 12.41 can be written as

$$F_C = mr\left(\frac{2\pi N}{60}\right)^2 = 0.011\,mrN^2 \tag{12.45}$$

From Eq. (12.45), it can be observed that the centrifugal force acting on the particle is dependent on the mass of the particle, the rotation rate, and the radius of revolution. Thus, dense particles experience higher centrifugal force and hence displaced to a greater extent than the lighter particles that tend to be closer to the center of rotation. Hence, dense particles are thrown outward by centrifugal force and separated easily from a heterogeneous mixture. Figure 12.40 shows the pattern of separation between the dense and lighter liquids during the centrifugation process.

Consider an elemental concentric cylinder of fluid thickness dr and mass dm having a radius r and height h. Centrifugal force acting on the element is given by

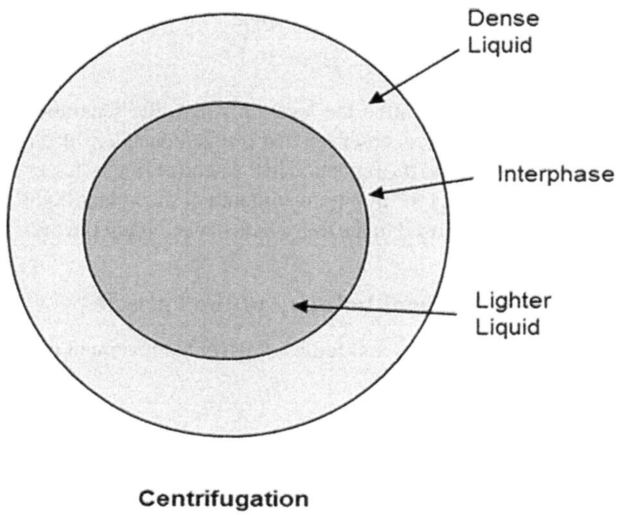

FIGURE 12.40 Schematic of centrifugation.

$$dF_C = (dm) r\omega^2 \tag{12.46}$$

where $dm = \rho \cdot 2\pi hr(dr)$

Therefore,

$$dF_C = 2\pi\omega^2 r^2 h\rho(dr) \tag{12.47}$$

As, $dF_C = dP \cdot A = dP \cdot 2\pi rh$

where dP is the pressure acting on the fluid element.

Therefore,

$$dP \cdot 2\pi rh = 2\pi\omega^2 r^2 h\rho(dr) \tag{12.48}$$

$$dP = \rho\omega^2 r(dr) \tag{12.49}$$

Integrating on both sides,

$$\int_{P_1}^{P_2} dP = \rho\omega^2 \int_{r_1}^{r_2} r(dr) \tag{12.50}$$

$$P_2 - P_1 = \frac{\rho\omega^2\left(r_2^2 - r_1^2\right)}{2} \tag{12.51}$$

If hydrostatic pressure is equal, that is, $P_2 = P_1$,

$$\rho_A\left(r_n^2 - r_1^2\right) = \rho_B\left(r_n^2 - r_2^2\right) \tag{12.52}$$

where r_n is the radius of neutral zone, ρ_A is the density of heavier liquid, and ρ_B is the density of lighter liquid. Rearranging Eq. (12.52) gives

$$r_n^2\left(\rho_A - \rho_B\right) = \rho_A r_1^2 - \rho_B r_2^2 \tag{12.53}$$

$$r_n^2 = \frac{\rho_A r_1^2 - \rho_B r_2^2}{(\rho_A - \rho_B)} \tag{12.54}$$

Eq. (12.54) provides an expression to determine the location where the separation takes place. Separation plates can be fixed based on the location. It is observed that if r_1 is reduced, r_n also reduces, and thus, neutral zone is reached near the central axis. Due to the reduction in the neutral zone, a lesser centrifugal force will act on a lighter liquid. This will help to extract the maximum amount of the lighter liquid. This expression for the calculation of centrifugal force is mainly used in the dairy industry to obtain the maximum amount of cream.

12.3.2 Derivation for the Centrifugal Velocity of the Particle

Considering a centrifuge with bowl radius, r_2 and radius of particle movement path, r_B, the terminal velocity for centrifugation (u_t) is given by

$$u_t = \frac{D_p^2 \cdot a(\rho_p - \rho_l)}{18\mu} \tag{12.55}$$

where D_p is the particle diameter, a is the acceleration due to centrifugal force, ρ_p is the density of the particle, ρ_l is the density of the liquid, and μ is the viscosity of the liquid.

However, $a = r\omega^2$; therefore, Eq. (12.55) becomes

$$u_t = \frac{D_p^2 \cdot r\omega^2(\rho_p - \rho_l)}{18\mu} \tag{12.56}$$

As

$$u_t = \frac{dr}{dt} \tag{12.57}$$

Equation 12.56 changes to

$$\frac{dr}{dt} = \frac{D_p^2 \cdot r\omega^2(\rho_p - \rho_l)}{18\mu} \tag{12.58}$$

And

$$dt = \frac{18\mu}{D_p^2 \cdot \omega^2(\rho_p - \rho_l)} \frac{dr}{r} \tag{12.59}$$

Integrating Eq. (12.59) between the limits $t = 0$, $t = t_T$ and $r = r_A$, $r = r_B$

$$\int_0^{t_T} dt = \frac{18\mu}{D_p^2 \cdot \omega^2(\rho_p - \rho_l)} \int_{r_A}^{r_B} \frac{dr}{r} \tag{12.60}$$

$$t_T = \frac{18\mu}{D_p^2 \cdot \omega^2(\rho_p - \rho_l)} \ln\left[\frac{r_B}{r_A}\right] \tag{12.61}$$

where t_T is the residence time, which can be described as

$$t_T = Residence\ time = \frac{\text{Volume of liquid}}{\text{Volumetric flow rate}} = \frac{V}{Q} \tag{12.62}$$

Therefore,

$$Q = \frac{D_p^2 \cdot \omega^2 (\rho_p - \rho_l) V}{18\mu \left[\ln\left(\frac{r_B}{r_A} \right) \right]} \tag{12.63}$$

However,

$$V = \pi r^2 h = \pi h \left(r_2^2 - r_1^2 \right) \tag{12.64}$$

Therefore,

$$Q = \frac{D_p^2 \cdot \omega^2 (\rho_p - \rho_l) \pi h \left(r_2^2 - r_1^2 \right)}{18\mu \left[\ln\left(\frac{r_B}{r_A} \right) \right]} \tag{12.65}$$

To determine r_A and r_B, an assumption is made that the particle moves half the way from liquid flow and the particle settles at the radius of r_2.

Therefore, $r_A = \frac{r_1 + r_2}{2}$ and $r_B = r_2$

Hence,

$$Q = \frac{D_p^2 \cdot \omega^2 (\rho_p - \rho_l) \pi h \left(r_2^2 - r_1^2 \right)}{18\mu \left[\ln\left(\frac{r_2}{(r_1 + r_2/2)} \right) \right]} \tag{12.66}$$

Multiplying and dividing the Eq. (12.66) by $2g$ and rearranging give

$$Q = \frac{2gD_p^2 (\rho_p - \rho_l)}{18\mu} \frac{\omega^2 \pi h \left(r_2^2 - r_1^2 \right)}{2g \left[\ln\left(\frac{r_2}{(r_1 + r_2/2)} \right) \right]} \tag{12.67}$$

$$Q = 2u_T \Sigma \tag{12.68}$$

where $u_T = \frac{gD_p^2 (\rho_p - \rho_l)}{18\mu}$ and $\Sigma = \frac{\omega^2 \pi h \left(r_2^2 - r_1^2 \right)}{2g \left[\ln\left(\frac{r_2}{(r_1 + r_2/2)} \right) \right]}$

where Σ is a property, which changes with the type of centrifuge.

For the tubular centrifuge, Σ is given by

$$\Sigma = \frac{\pi h \omega^2}{4g} \left(3r_2^2 + r_1^2 \right) \tag{12.69}$$

For the disk centrifuge, Σ is given by

$$\Sigma = \frac{2\pi (N-1) \left(r_x^3 - r_y^3 \right) \omega^2}{3g \tan\theta} \tag{12.70}$$

where N is the number of disks in the stack, r_x and r_y are, respectively, the inner and outer radius of the disk, and θ is the angle of the disk.

Example 12.3

An oil-in-water emulsion is separated using a bowl centrifuge. Determine the radius of the neutral zone so that the feed pipe can be positioned correctly. (Assume that the density of continuous phase is 1000 kg/m^3 and the density of oil is 900 kg/m^3. The outlet radii from the centrifuge are 5 and 6.5 cm.)

Solution

Given

$$\rho_A = 1000\,\text{kg/m}^3,\ \rho_B = 900\,\text{kg/m}^3,\ r_1 = 6.5\,\text{cm} = 0.065\,\text{m},\ r_2 = 5\,\text{cm} = 0.05\,\text{m}$$

The radius of the neutral zone is determined by the equation:

$$r_n^2 = \frac{\rho_A r_1^2 - \rho_B r_2^2}{(\rho_A - \rho_B)}$$

$$r_n = \sqrt{\frac{\rho_A r_1^2 - \rho_B r_2^2}{(\rho_A - \rho_B)}}$$

Therefore, $r_n = \sqrt{\left[\dfrac{1000(0.065)^2 - 900(0.05)^2}{1000 - 900}\right]}$

$$r_n = \sqrt{\left[\frac{4.225 - 2.25}{100}\right]}$$

$$r_n = 0.14\,\text{m} = 14\,\text{cm}$$

Hence, the radius of the neutral zone is 14 cm.

Example 12.4

An oil emulsion has a density of 800 kg/ m^3, and the average droplet diameter is 70 µm. A centrifuge is used for the separation of oil-in-water emulsion, which rotates at 3000 rpm, and the effective radius at which separation takes place is 35 mm. The flow rate of the feed is 150 liters per minute. Calculate the velocity of oil globules in water and the Σ factor of the centrifuge.

Solution

Given

$$D_p = 70\,\mu\text{m} = 7\times10^{-5}\,\text{m},\ \rho_p = 800\,\text{kg/m}^3,\ r = 35\,\text{mm} = 0.035\,\text{m},\ \rho_l = 1000\,\text{kg/m}^3,\ \mu = 1\times10^{-3}\,\text{Pa s}$$

$$N = 3000\,\text{rpm} = \left(\frac{3000}{60}\right) = 50\,\text{rps}$$

$$u_t = \frac{D_p^2 \cdot r\omega^2(\rho_p - \rho_l)}{18\mu} = \frac{D_p^2 \cdot r(2\pi N)^2(\rho_p - \rho_l)}{18\mu}$$

$$u_t = \frac{(7\times10^{-5})^2 \times 0.035 \times (2\pi\times50)^2 \times (800 - 1{,}000)}{18\times10^{-3}}$$

$$u_t = -0.188\,\text{m/s}$$

The negative sign indicates that the oil moves towards the center.

Example 12.5

Beer having 1.5% solids has a specific gravity of 1.042, a viscosity of 1.40×10^{-3} Ns/m^2, and a density of 1160 kg/m^3 is clarified at the rate of 240 L/h. The clarification is done in a bowl centrifuge having an operating volume of 0.09 m^3 and rotating at a speed of 5000 rpm. The diameter of the bowl fitted with an outlet of 6 cm is 7.5 cm. Determine the effect of feed rate if the bowl speed is increased to 10000 rpm and the minimum particle size that can be removed at higher speed. All conditions remain the same except the bowl speed.

Solution

Given

$$\rho_p = 1160\,\text{kg/m}^3,\ \rho_l = 1.042\times10^3 = 1042\,\text{kg/m}^3,\ \mu = 1.4\times10^{-3}\,\text{Pa s},$$

$$N_1 = 5000\,\text{rpm},\ N_2 = 10000\,\text{rpm}$$

$$\text{Initial flow rate} = Q_1 = \frac{D_p^2(2\pi N_1/60)^2(\rho_s-\rho_l)V}{18\mu[\ln r_B/r_A]}$$

$$\text{New flow rate} = Q_2 = \frac{D_p^2(2\pi N_2/60)^2(\rho_s-\rho_l)V}{18\mu[\ln(r_B/r_A)]}$$

As all conditions remain the same except the bowl speed,

$$\frac{Q_2}{Q_1} = \frac{(2\pi N_2/60)^2}{(2\pi N_1/60)^2}$$

$$Q_1 = 240\,\text{L/h} = 0.24\,\text{m}^3/\text{h} = 0.24/3{,}600\,\text{m}^3/\text{s}$$

$$Q_1 = 6.67\times10^{-5}\,\text{m}^3/\text{s}$$

$$\frac{Q_2}{6.67\times10^{-5}} = \frac{(2\times3.142\times10{,}000/60)^2}{(2\times3.142\times5{,}000/60)^2}$$

$$Q_2 = 2.67\times10^{-4}\,\text{m}^3/\text{s}$$

To find the minimum particle size,

$$D_p^2 = \frac{Q_2[18\mu\ln(r_B/r_A)]}{\omega^2(\rho_p-\rho_l)V}$$

$$D_p^2 = \frac{Q_2[18\mu\ln(r_B/r_A)]}{(2\pi N_2/60)^2(\rho_p-\rho_l)V}$$

$$D_p^2 = \frac{2.67\times10^{-4}(18\times1.40\times10^{-3}\times\ln(0.0375/0.03))}{(2\times3.142\times10000/60)^2\times(1160-1042)\times0.09}$$

Therefore, $D_p^2 = 1.289\times10^{-13}\text{m}^2$

Therefore, $D_p = 3.59\times10^{-7}\text{m} = 0.359\,\mu\text{m}$

Hence, the diameter of the particle that can be removed at a higher speed is 0.359 μm.

12.3.3 Equipment for Centrifugation

12.3.3.1 Disk Bowl Centrifuge

Disk bowl centrifuge is used for the solid–liquid or liquid–liquid separation (Figure 12.41). This centrifuge type is ideal for a fluid mixture having low solid concentration. Disk bowl centrifuge comprises a stationary casing within which there are a shallow bowl with a cylindrical body and a conical top with the central spindle. Truncated metal disks are stacked in the bowl, which rotates within the bowl. These disks have holes that form a channel on stacking. The gap between the stacked disks acts like a mini-centrifuge. Feed is introduced from the top and falls on the rotating centrifuge disks. Feed reaches to spaces between the rotating disks. Light phase is pushed towards the center of the bowl, whereas the dense phase travels towards the bowl wall and is separated out from the respective outlet. The solids are accumulated in the disk bowl, which is removed manually at the end of the operation. For continuous operation, nozzle centrifuge is used in which the solid particles are discharged through the nozzle. Disk bowl centrifuge is suitable for suspensions containing fine solid particles, such as milk and coconut milk.

12.3.3.2 Tubular Centrifuge

Tubular centrifuge is a simple mechanical centrifuge consisting of a vertical tubular rotor, feed inlet, and feed outlet for the separated liquids, housed in a stationary casing. Tubular centrifuge rotates at high speed (15000–50000 rpm) and therefore can be applied on a liquid feed with the low difference in density between its components. The length-to-diameter ratio of the tubular bowl centrifuge is 4–8 with a bowl diameter of 10–15 cm. Feed enters from the bottom of the centrifuge and is subjected to centrifugation. Due to centrifugal action, fluid rises in the bowl while being separated based on the density. Lighter fluid moves towards the central axis, whereas the dense fluid moves towards the bowl wall. The tubular centrifuge is mainly used for and limited to the two-phase system (liquid–liquid slurry with 1% solid concentration) and is not used for the three-phase (liquid–liquid–solid) system.

12.3.3.3 Decanter Centrifuge

Decanter centrifuges are mainly used as clarifiers. These are suitable for slurry having a solid concentration of 40%–60%. Decanter consists of a horizontal bowl with a cylindrical and conical zone with a screw

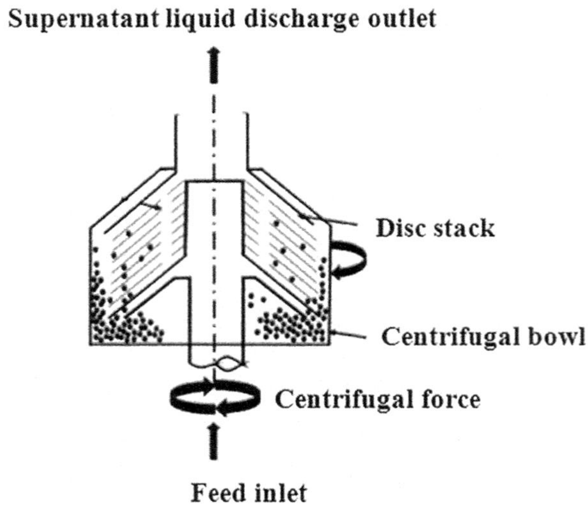

FIGURE 12.41 Schematic of disk bowl centrifuge. (Modified and Reproduced with permission from Shekhawat, L. K., Sarkar, J., Gupta, R., Hadpe, S. and Rathore, A. S. 2018. Application of CFD in Bioprocessing: separation of mammalian cells using disc stack centrifuge during production of biotherapeutics. *Journal of Biotechnology* 267: 1–11.)

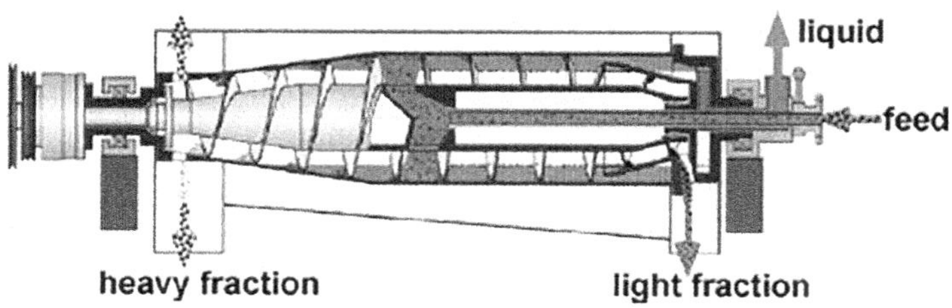

FIGURE 12.42 Decanter centrifuge. (Reproduced with permission from Anlauf, H. 2007. Recent developments in centrifuge technology. *Separation and Purification Technology* 58: 242–246.)

conveyor (Figure 12.42). Solid particles of the feed move towards the wall, whereas the clear liquid moves to the center. Solid particles decanted on the wall are scraped and passed to the solid outlet. Decanters are widely used in mango pulp processing to remove the black and fibrous particles from the pulp.

12.3.3.4 Hydrocyclone

Hydrocyclone or liquid cyclone is a device for the separation of suspended solid particles from a liquid. In a hydrocyclone, the particles in suspension are subjected to centrifugal forces, which separate them from the surrounding fluid based on the difference in density. Apart from being used as solid–liquid separators, hydrocyclones are also used in the liquid–solid, liquid–liquid, and liquid–gas separations, and as clarifiers and classifiers.

As shown in Figure 12.43, hydrocyclone is a cylindroconical assembly with the cylindrical section at the top and a conical base. The liquid suspension is fed tangentially to the cylindrical section. The tangential feeding subjects the fluid to a spiral motion, which is termed as the "development of vortex." The vortex leads to the centrifugal acceleration, which causes the coarse particles or liquid of larger density to be thrown against the wall of the hydrocyclone, which is then discharged through the underflow orifice.

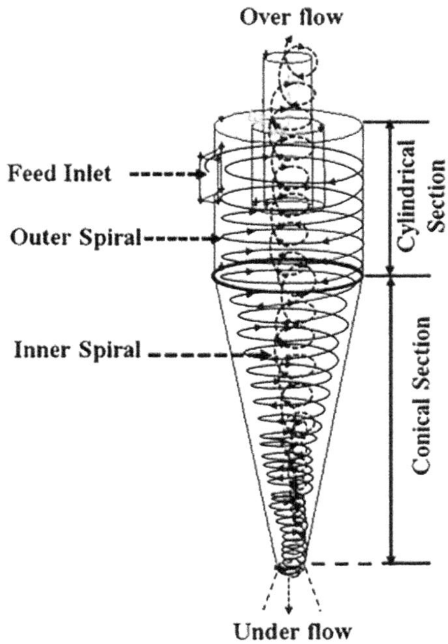

FIGURE 12.43 Schematic of hydrocyclone. (Modified and Reproduced with permission from Murthy, Y. R. and Bhaskar, K. U. 2012. Parametric CFD studies on hydrocyclone. *Powder Technology* 230: 36–47.)

On the other hand, the fine particles and the material of lower density remain around the central axis of the cyclone and form a "free outer vortex" (outer spiral; Figure 12.43) and a "forced inner vortex" (inner spiral; Figure 12.43), which is subsequently carried away in the overflow stream (Vieira et al., 2005).

The major applications of hydrocyclones have been established in the refining of corn and potato starch, processing of cottonseed oil, extraction of soluble coffee, and clarification of fruit juices (Ortega-Rivas and Perez-Vega, 2011). Other applications of hydrocyclones in the food industry include the following (FLSmidth A/S, 2017):

- *Ketchup*: For the removal of contaminants before milling and cooking stages
- *Wine processing*: For the removal of seeds, dirt, and solids from the crushed grape juice
- *Jams, sauces, and purees*: For the removal of dirt and metal particles ahead of the milling step so as to protect the milling equipment
- *Potato starch recovery*: For the removal of starch in cutting circuits that facilitates the reuse of water and the recovered starch as additives in animal feed
- *Cooking oil*: For the removal of crumbs and grit to extend the life of the cooking oil.

12.3.3.5 Cyclone

Cyclone is primarily a solid–gas separator that removes solid particles (e.g., food powders or dust) from air or any other gas stream, aided by the centrifugal force. The construction of a cyclone separator is similar to that of a hydrocyclone with a vertical cylindrical section attached to a conical base (Figure 12.44). The solid–gas mixture enters the cylindrical section of the cyclone in a tangential direction. Due to the rotational movement, the pressurized mixture forms a vortex inside the cyclone, which leads to the development of centrifugal force. Driven by the centrifugal force, the heavier solid particles move towards the wall of the cyclone and settle down at the conical section. However, the solid-free gas stream along with a minor proportion of finer particles forms a small vortex near the center and moves upward. A concentric vertical cylinder directs the particulate-free gas (air) from the center of the cyclone to the air exhaust (Figure 12.44). The development of vortex within the cyclone provides sufficient separation time.

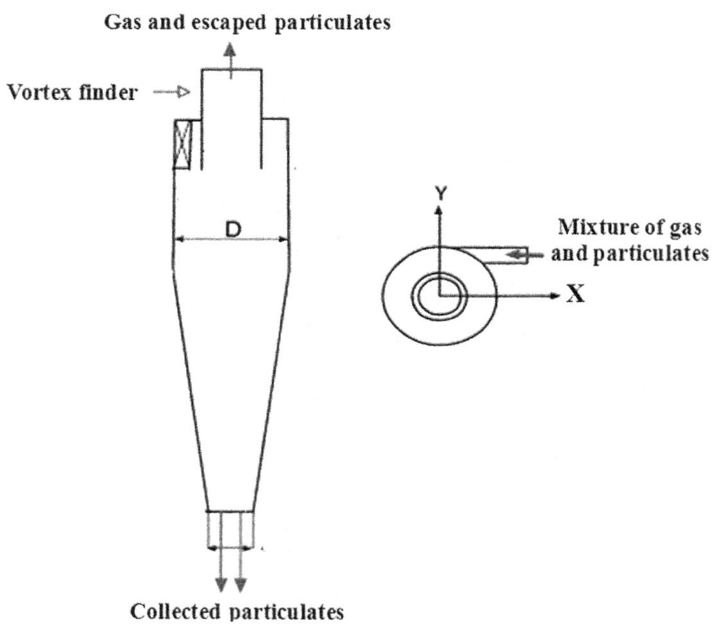

FIGURE 12.44 Cyclone separator. (Reproduced with permission from Elsayed, K. 2015. Optimization of the cyclone separator geometry for minimum pressure drop using Co-Kriging. *Powder Technology* 269: 409–424.)

Cyclones are often used to remove particles of diameter ~10 μm or more, from the airstream. The separation efficiency of cyclones is influenced by the particle size, diameter and length of the cyclone, and air velocity. Studies have shown that the particle collection efficiency of a cyclone separator increases with increasing mean diameter and density of the particles, increasing gas tangential velocity, decreasing cyclone diameter, and increasing cyclone length.

According to Stokes' law, the terminal velocity of the particles is proportional to the force acting on them. Thus, in a cyclone separator, as given in Eq. (12.43),

$$F_c = \frac{mv^2}{r}$$

where F_c is the centrifugal force acting on the particle, m is the mass of the particle, v is the tangential velocity of the particle, and r is the radius of the cyclone. Thus, from the equation of Stokes' law, when the force on the particle increases, the radius decreases at a constant velocity. Therefore, cyclones having the smallest diameter are the most efficient for removing small particles. However, a balance should be achieved between the capital cost of small-diameter cyclones and its throughput and pressure drop. As already discussed in Chapter 10, cyclone separator finds its application in the spray-drying operation for the removal of dried food powders from the stream of hot air.

12.3.4 Case Study: Applications of Centrifugation in the Dairy Industry

Centrifugation is widely used in the dairy industry. The two major applications of centrifugation in the dairy industry will be discussed in the subsequent sections:

1. Skimming (separation of cream from whole milk)
2. Clarification (removal of solid impurities from milk)

12.3.4.1 Skimming

Skimming of milk by centrifugation is governed by the following two principles:

i. *High centrifugal acceleration of fat globules*: This causes the fat globules to move at a faster rate than natural creaming. When subjected to centrifugal force, the fat globules (or cream) in whole milk move in through the separation channels towards the axis of rotation. This is because the cream is less dense than the skim milk. Consequently, the skim milk moves outward and exits the centrifuge through a separate outlet.
ii. *Dividing the zone of cream separation inside the centrifuge*: This limits the distance over which the fat globules should move.

The milk enters the centrifugal separator along the central axis and flows into the bowl rotating at a high speed of 5000–6000 rpm, through the openings in the bowl. The inflow of milk is controlled using a valve installed at the inlet to the separator. The milk then flows through a stack of conical disks with openings of similar size, followed by bifurcation of the flow over the multiple slits between the disks. As the disk stack revolves, the centrifugal force (3000–6000 times greater than the gravitational force) acting on the milk drives the fat globules along the upper side of each disk towards the center of the bowl (in the inward direction) and forces the skim milk against the bottom side of each disk and then towards the periphery of the bowl (in the outward direction). Subsequently, the fat globules are channelized towards the central core and discharged in the form of cream through the cream outlet. The skim milk fraction with few small fat globules that have escaped the separation is discharged through the skim milk outlet. This centrifugal separation is achieved by the higher density of skim milk (1036 kg/m^3 or 1.036 kg/L) compared to the cream (900 kg/m^3 or 0.9 kg/L).

The centrifugal separation of fat globules from the whole milk can be explained by the Stokes' law, which gives the force of viscosity on a small sphere moving through a viscous fluid. The following equation results when the Stokes' law is applied to the centrifugal separation process:

$$v = \frac{2r^2(\rho_s - \rho_f)(4\pi^2 RN^2)}{9\eta} \tag{12.71}$$

where v is the velocity of movement of the fat globule, r is the radius of the fat globule (μm), ρ_s is the density of skim milk (kg/m^3), ρ_f is the density of fat (kg/m^3), R is the distance of fat globule from the axis of rotation (m), N is the bowl speed (rpm), and η is the viscosity of skim milk (kg m^{-1}s^{-1}). As evident from Eq. (12.71), the velocity of creaming is inversely related to the viscosity of skim milk. Therefore, to increase the creaming velocity, the milk is preheated to 37°C–50°C, as warm milk has a lower viscosity than the cold milk.

The efficiency of the centrifugal separation between skim milk and cream is evaluated by the skimming efficiency (E_s), given by

$$E_S(\%) = \frac{f_c}{f_m} \times 100 \tag{12.72}$$

where f_c is the total fat content of cream, f_m is the total fat content of milk, and E_s is the measure of the amount of fat transferred from the milk to the cream. The modern centrifugal separators are capable of achieving a skim milk product with a fat content of 0.04%–0.07%. The factors influencing the skimming efficiency are listed and explained in Figure 12.45.

The other important parameters to judge the efficiency of centrifugal separation of whole milk are the yields of cream (Y_c) and skim milk (Y_s), given by

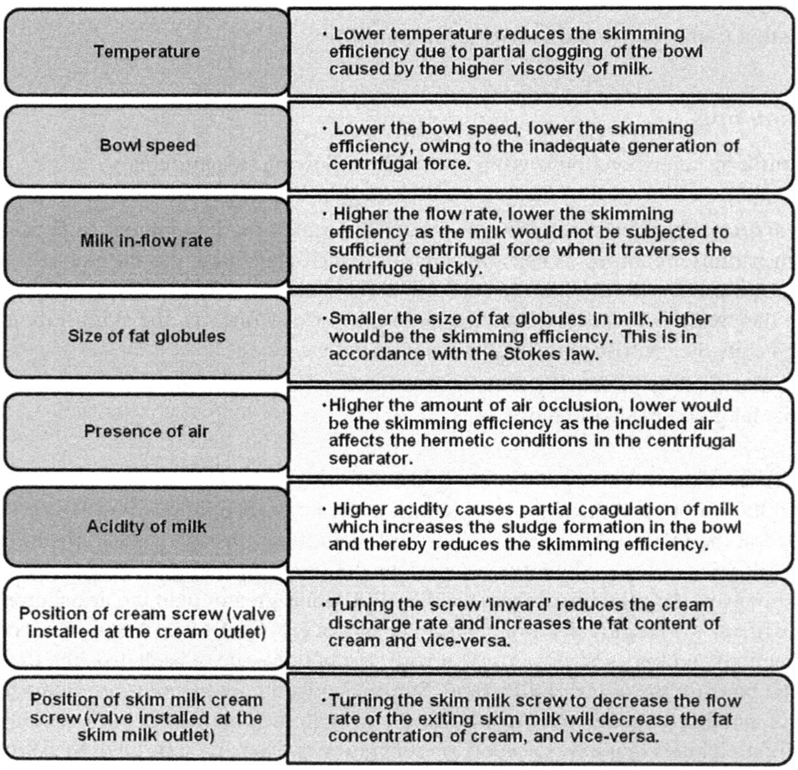

FIGURE 12.45 Factors influencing skimming efficiency.

$$Y_c = \frac{f_m - f_c}{f_c - f_s} \times 100 \quad (12.73)$$

$$Y_s = \frac{f_c - f_m}{f_c - f_s} \times 100 \quad (12.74)$$

where f_m, f_s, and f_c are the fat contents of whole milk, skim milk, and cream, respectively, in percentage. All the factors that affect the skimming efficiency have an influence on the aforementioned yields as well (Figure 12.43).

12.3.4.2 Clarification

During clarification, particles that are denser than the continuous milk phase are thrown back to the periphery of the centrifuge. The solids that accumulate in the centrifuge are composed of dirt, epithelial cells, leukocytes, corpuscles, bacteria sediment, and sludge. A centrifuge used for the purpose of clarification is referred to as a *clarifier.*

Bactofugation is a specific application of centrifugation for clarification in the dairy industry. It is the process of removing microorganisms (both active and inactive), mainly the spore formers such as *Bacillus cereus* and *Clostridium tyrobutyricum* and the spores formed by them from the low-pasteurized milk, using centrifugal force. The centrifuge used for this purpose is known as a *bactofuge*, which is hermetic by its design. A bactofuge is a high-speed clarifier that can operate up to a speed of 20000 rpm. The microorganisms and spores thus separated from the milk, known as the *bactofugate*, are concentrated on the periphery of the bowl and then thrown outside through the discharge nozzles fitted to the wall of the bowl. Eventually, the bactofugate is collected in the sludge discharge pipe. The bactofugate constitutes approximately 3% of the feed volume. It also contains some amount of milk proteins.

As centrifugation is based on the density difference, the smaller size of bacteria (0.5–5 μm) and spores (1–1.5 μm) does not limit their separation from the milk. The density difference between the bacteria (1.07–1.13 g/cm^3) and the skim milk (1.036 g/cm^3) facilitates a substantial removal of spores (90%–95%) at the optimized temperature of separation. The separation temperature should be between 55°C and 62°C as the milk viscosity is relatively low in this temperature range. This is in accordance with the Stokes' law (Eq. (12.71)), by which the sedimentation velocity of the bacteria would be greater at higher temperatures owing to the low viscosity of milk. However, too high temperatures would lead to protein damage.

Further, by using two bactofuges in series at a separation temperature of 75°C, the spore count in milk can be reduced by more than 99.9%. Nevertheless, bactofugation is never equivalent to sterilization as the product emerging from the bactofuge still contains heat-resistant bacteria and a small number of spores. The major advantage of this process is that it improves the hygienic quality and bacteriological purity of milk and thereby prolongs the shelf life of pasteurized milk under refrigeration. It also enables the milk to be sterilized at a lower temperature and within a shorter processing time. Consequently, bactofugation is of high relevance in the cheese-making industry wherein the removal of anaerobic (Clostridial) spores from milk is useful in avoiding the *late blowing defect* (butyric acid fermentation) in hard and semihard cheeses, which leads to serious quality losses with respect to flavor and texture.

Apart from the dairy industry, centrifugation finds its major applications in the fruit juice industry. It helps to reduce the pulp content and yield clear fruit juice. A centrifuge is also used as a decanter to remove fibrous particles from the fruit pulp. In the edible oil industry, centrifugation is applied to remove traces of water from the extracted edible oil. Further, centrifugation is also used in the extraction of essential oil from citrus fruit, recovery of yeast from biomass, and removing traces of oil from the snack food products.

12.4 Leaching

Leaching is a mass transfer operation involving solid–liquid extraction. During leaching, a soluble component from the solid is extracted using a liquid solvent. Concentration gradient that exists between the feed material and solvent is the major driving force for the leaching process. The solution phase (solid–liquid stream) that exits each stage of a leaching process is known as the *underflow* or *raffinate*. The term *overflow* is applied to the liquid stream. The leaching process can be represented by an equilibrium diagram. The equilibrium curve is a plot between N and y_A, where $N = 0$ on the axis represents the overflow which is devoid of solids (Figure 12.46). The upper curve N versus y_A represents the underflow. The tie lines are vertical on an x-y diagram, and the equilibrium line is given by $y_A = x_A$, which is always straight (angle = 45°) as the concentration of solute in an overflow stream is equal to that in the underflow stream.

Leaching is carried out in a single stage or multiple stages. Multiple stages are applied to increase the recovery of the desired component. In each stage of the process, feed material and solvent are mixed together to ensure a proper contact for the extraction of the desired component. Shape, size, and structure of the material affect the leaching process. The leaching process occurs in three steps:

I. Contact of solvent with the feed material and dissolution of the solute in the solvent
II. The occurrence of mass transfer based on the concentration gradient, during which the solute will diffuse from the feed material to the surface
III. The emergence of solute in the bulk solution (solvent + solute)

The major factors that influence the leaching process or the rate of solid–liquid extraction are as follows:

- *Interface area of solid–liquid*: Solid–liquid interface area vastly influences the mass transfer. This is because greater the surface area, higher the mass transfer. Hence, small particles show a higher mass transfer due to their larger surface area. In addition, solute has to cross less distance to diffuse out from the surface.
- *Solute concentration gradient*: Rate of extraction is influenced by the concentration gradient of solute on the particle surface and in the bulk solution. Extraction rate decreases as the leaching process attains an equilibrium condition. Therefore, the use of multistage countercurrent systems can result in the complete extraction with higher extraction rate.
- *Mass transfer coefficient*: Extraction rate is also influenced by the mass transfer coefficient. Mass transfer coefficient increases with an increase in solvent temperature. Therefore, the solvent may be preheated to increase the extraction or leaching rate.

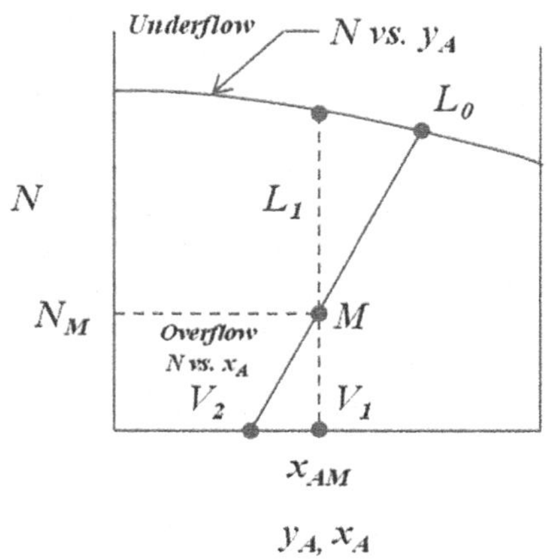

FIGURE 12.46 Equilibrium diagram for leaching. (Redrawn from Geankoplis, C. J. 2003. *Transport Processes and Separation Process Principles (Includes Unit Operations)* (Fourth edition). New Delhi: Prentice Hall of India Private Limited.)

12.4.1 Single-Stage Leaching

A single-stage leaching process is shown in Figure 12.47. Three components are involved in the leaching process: the solute (A), leached solid in the slurry (B), and the leaching solvent (C). In the slurry, the concentration of B is defined by

$$N = \frac{\text{kg } B}{\text{kg } A + \text{kg } C} = \frac{\text{kg solid}}{\text{kg solution}} \tag{12.75}$$

It is to be noted that for the overflow stream, $N = 0$, whereas for the underflow stream, N depends on the composition. The overflow solution has a mass flow rate of V kg/h with composition x_A (Eq. (12.76)), and the liquid in the slurry B is denoted by L in kg/h, with composition y_A (Eq. (12.77)).

$$x_A = \frac{\text{kg } A}{\text{kg } A + \text{kg } C} = \frac{\text{kg solute}}{\text{kg solution}} \tag{12.76}$$

$$y_A = \frac{\text{kg } A}{\text{kg } A + \text{kg } C} = \frac{\text{kg solute}}{\text{kg solution}} \tag{12.77}$$

For the solid feed, $y_A = 1.0$, and for pure solvent that enters the system, $N = 0$ and $x_A = 0$.

The overall material balance on the single-stage leaching system is given by

$$L_0 + V_2 = L_1 + V_1 = M \tag{12.78}$$

where M is the total flow rate in kg A. The component balance is given by

$$L_0 y_{A0} + V_2 x_{A2} = L_1 y_{A1} + V_1 x_{A1} = M x_{AM} \tag{12.79}$$

$$B = N_0 L_0 = N_1 L_1 = N_M M \tag{12.80}$$

where x_{AM} and N_M are the coordinates of the point M on the equilibrium diagram for leaching (Figure 12.46) (Geankoplis, 2003).

12.4.2 Countercurrent Multiple-Stage Leaching

Figure 12.48 shows the countercurrent leaching process. In the flow diagram of the leaching process, upper flow stream is denoted by V, underflow stream is denoted by L, and their weight fractions are represented by y and x, respectively.

The overall mass balance is given by

$$L_0 + V_{n+1} = L_n + V_1 \tag{12.81}$$

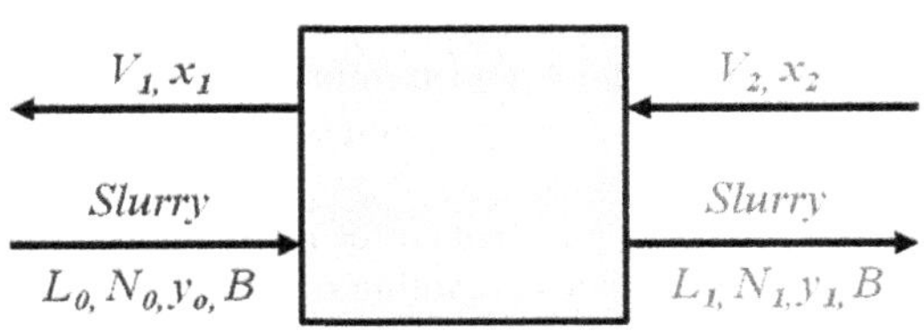

FIGURE 12.47 Process flow in single-stage leaching. (Redrawn from Geankoplis, C. J. 2003. *Transport Processes and Separation Process Principles (Includes Unit Operations)* (Fourth edition). New Delhi: Prentice Hall of India Private Limited.)

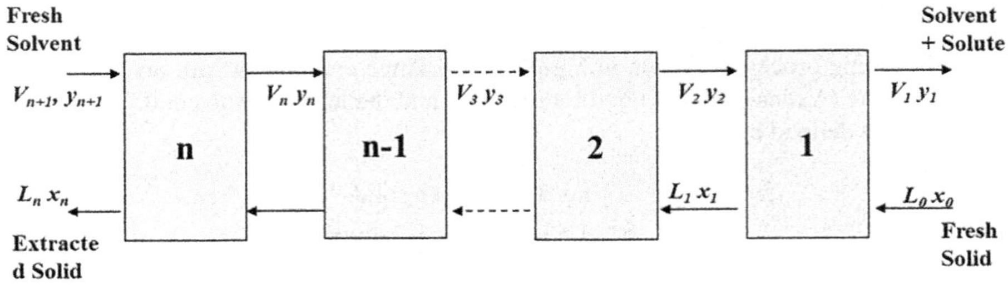

FIGURE 12.48 Process flow diagram of countercurrent leaching process.

Mass balance on the solute is given by

$$L_0x_0 + V_{n+1}y_{n+1} = L_nx_n + V_1y_1 \tag{12.82}$$

$$y_{n+1} = \frac{L_nx_n}{V_{n+1}} + \frac{V_1y_1 - L_0x_0}{V_{n+1}} \tag{12.83}$$

Eq. (12.83) is also called the operating line equation.

And,

$$V_{n+1} = L_n - L_0 + V_1 \tag{12.84}$$

$$y_{n+1} = \frac{L_nx_n}{L_n - L_0 + V_1} + \frac{V_1y_1 - L_0x_0}{L_n - L_0 + V_1} \tag{12.85}$$

Throughout the leaching process, equal overflows and underflows are maintained.

Therefore,

$$V_1 = V_2 \cdots = V_n = V_{n+1} = V \text{ and } L_1 = L_2 \cdots = L_n = L \tag{12.86}$$

Operating line equation for the constant overflows and constant underflows is given by

$$y_{n+1} = \left(\frac{L}{V}\right)x_n + y_1 - \left(\frac{L}{V}\right)x_0 \tag{12.87}$$

At the stage when equilibrium is reached, that is, when $y_n = x_n$, the line is said to be an equilibrium line.

12.4.3 Leaching Equipment

12.4.3.1 Fixed Bed Extractor

Fixed bed extractors are simple with a single stage and are operated in batch mode. In the fixed bed extractor, a solid bed of the feed material is kept stationary and the fresh solvent is introduced by spraying. The extraction operation is carried out for a predetermined time, after which the solids are removed and the liquid is drained off. The process is explained with the example of coffee extraction in the paragraph that follows.

For the extraction of soluble fraction during instant coffee manufacturing, a fixed bed extractor is used as shown in Figure 12.49. There are five tanks, containing roasted coffee bean. Hot water is introduced in the tank 1. Coffee extract from the tank 1 is passed to the tank 2. Subsequently, coffee extract is collected at the end of tank 5. Tanks may be jacketed and equipped with the heater to maintain the solvent temperature. Pumps are used for the circulation of the solution, and the spent solids of coffee beans are manually removed.

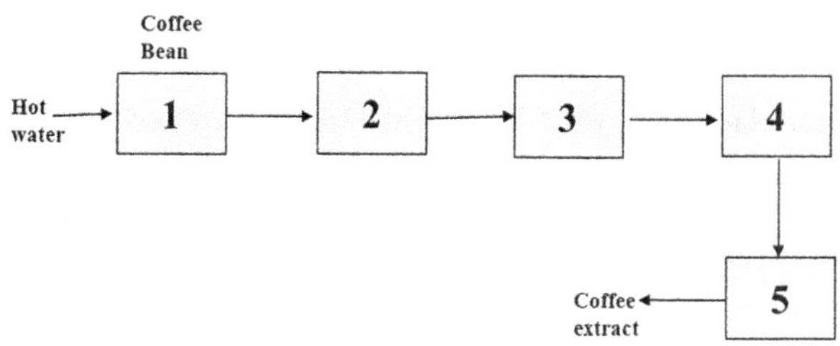

FIGURE 12.49 Fixed bed coffee extraction process.

12.4.3.2 Bollman Extractor

This is a moving bed extractor wherein the bed moves in contrary to being static as in the fixed bed counterpart. Here, the device is an enclosed bucket elevator (see Figure 12.50). Dry flakes are added at the top of a perforated bucket. As the buckets descend, they come into contact with a dilute solution of the solute in a solvent called as *half miscella*. The leaching liquid percolates downward through the moving buckets and is eventually collected at the bottom as the strong solution, known as *full miscella*. The buckets in transit in the upward direction are leached by the fresh solvent moving countercurrently, which is sprayed onto the top bucket. The wet flakes are removed continuously. This type of extractor is commonly used for the extraction of vegetable oil from the flaked oilseeds of peanuts and soybeans (Geankoplis, 2003).

12.4.4 Applications of Leaching in Food Processing

- Leaching process is useful for the extraction of oil using food-grade hexane, provided that the oil content is less than 20%. Mainly rice bran oil and soybean oil are extracted by the solvent extraction method.
- Fish oil and corn oil are also extracted using the solvent extraction method.
- Natural food pigments such as that from annatto seeds (yellow dye), paprika (red color), and turmeric (yellow color) are extracted using the leaching process.

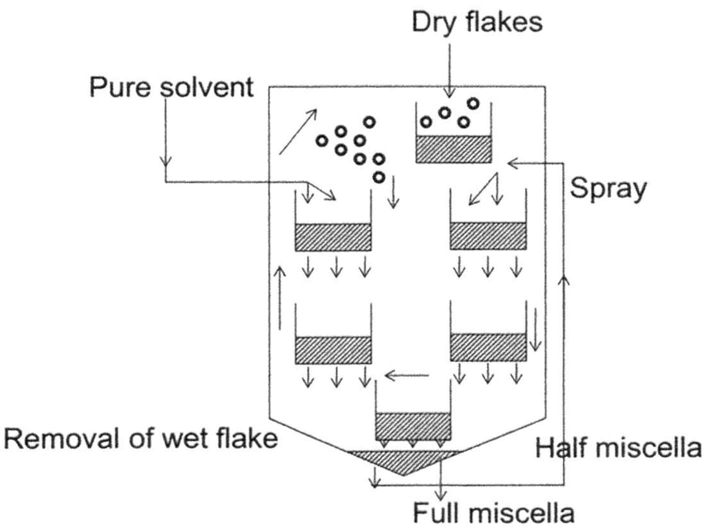

FIGURE 12.50 Schematic of a Bollman extractor.

- Oleoresins from spices such as pepper, turmeric, and ginger are extracted using hexane as a solvent.
- Instant coffee is manufactured by the extraction of coffee from roast and ground coffee with hot water.
- For the manufacturing of instant tea, tea leaves are subjected to hot water extraction. On completion of the extraction process, tea solution contains 2%–2.5% of solid content, which is further concentrated.

12.4.5 Case Study: Application of Leaching in Instant Coffee Manufacturing

Extraction of soluble solids from the roast and ground coffee is the heart of instant coffee manufacturing process. In an industrial-scale instant coffee manufacturing process, the extraction is usually carried out in two or more stages, to attain improved yield. Hot water is the commonly used extraction liquid. Heeb and Mandralis (2000) described a two-stage continuous, countercurrent process for the extraction of water-soluble solids from roast and ground coffee.

The coffee extraction process was carried out in a series of two tubular reactors. During the first stage of the extraction process, a readily soluble fraction was extracted from the fresh roast and ground coffee particles of size 1.3–1.4 mm. In general, the small particle size of the coffee grounds promotes an efficient and fast extraction. A slurry comprising these coffee granules to be extracted and a weak coffee extract were introduced immediately above the solid–liquid (S-L) separator of the first reactor at a temperature of 90°C. The weak extract is nothing but a portion of the coffee extract obtained from the second extraction reactor combined with the filtrate obtained from the first extraction reactor. Thus, this weak extract acts as the extraction liquid in the first reactor. The temperature of the extraction liquid was about 130° C. The solids formed an upward moving packed bed of height 0.5 m above the S-L separator. Whereas, the extract drained through the solid–liquid separator.

Subsequently, the partially extracted coffee grounds derived from the first extraction reactor were fed into a solubilization reactor and subjected to thermal solubilization at 198°C for 5 min. The solubilized grounds were slurried with the remaining portion of the coffee extract obtained from the second extraction reactor and fed into the second extraction reactor at a temperature of 96°C. The yield of soluble coffee solids in the coffee extract derived from the first extraction reactor was found to be 49% on the basis of fresh coffee grounds entering the system. The coffee extract thus obtained is then concentrated by evaporation, followed by spray drying or freeze drying to obtain the soluble coffee powder.

12.5 Aqueous Two-Phase Extraction (ATPE)

Aqueous two-phase extraction (ATPE) is an important liquid–liquid separation method for the purification of biomolecules such as proteins and microbial cells. An ATPE system (ATPS) is composed of an aqueous (water-based) solution of two structurally different and immiscible polymers that are water soluble at a specific concentration, or of one water-soluble polymer and a few salts. -The former is known as the polymer–polymer system (e.g., gelatin–agar; gelatin–soluble starch; PE glycol (PEG)–maltodextrin; PEG–cashew nut gum; PEG–guar gum) and the latter is termed as the polymer–salt system (e.g., PEG–sodium sulfate, PEG–magnesium sulfate, and PEG–ammonium sulfate). The polymers and salts are referred to as "phase-forming components."

12.5.1 Principle of Separation in ATPE

In an ATPS, two immiscible aqueous phases are formed when the limiting concentrations of the phase-forming components are exceeded (Albertsson, 1986). Here, the principle of separation is based on the differential partitioning of the biomolecule of interest and impurities between the two immiscible liquid phases formed (Figure 12.51).

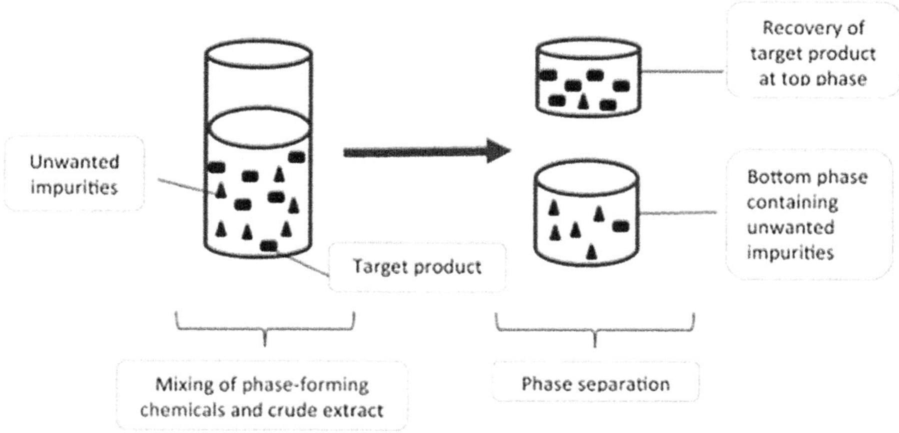

FIGURE 12.51 Schematic of the principle of separation in an ATPS. (Reproduced with permission from Phong, W. N., Show, P. L., Chow, Y. H. and Ling, T. C. 2018 Recovery of biotechnological products using aqueous two phase systems. *Journal of Bioscience and Bioengineering* 126: 273–281.)

The phase diagram shown in Figure 12.52 explains the separation in an ATPS under defined conditions (temperature and pH) (Iqbal et al., 2016). It provides information on the concentration of components required for the formation of two immiscible phases and their concentration in the top and bottom

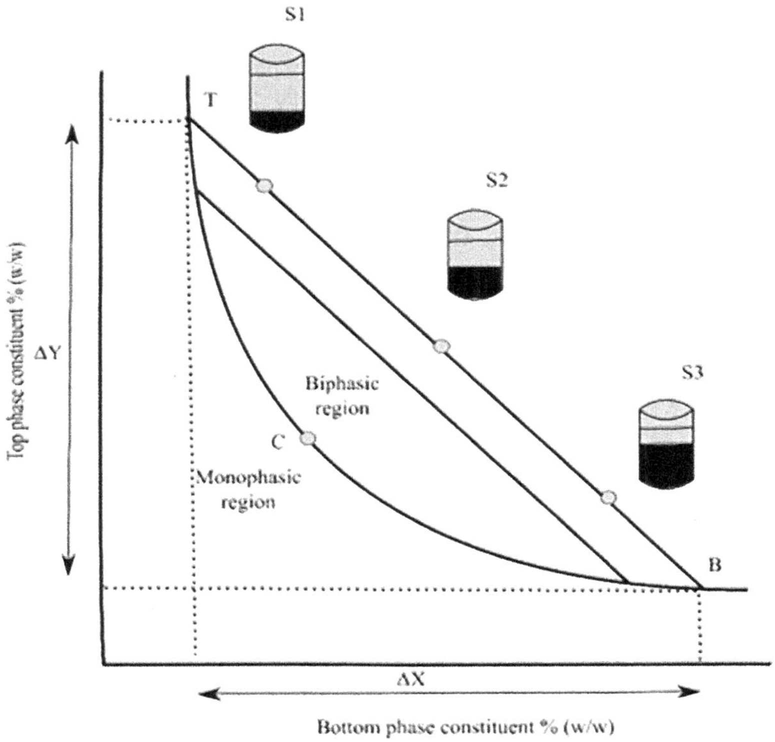

FIGURE 12.52 Schematic representation of phase diagram for ATPE. (Reproduced with permission from Iqbal, M., Tao, Y., Xie, S., Zhu, Y., Chen, D., Wang, X., Huang, L., Peng, D., Sattar, A., Abu Bakr Shabbir, M., Iftikhar Hussain, H., Ahmed, S. and Yuan, Z. 2016. Aqueous two-phase system (ATPS): an overview and advances in its applications. *Biological Procedures Online* 18: 18.)

phases (Hatti-Kaul, 2000; Raja et al., 2012). The phase diagram shows a binodal curve, which divides the region of component concentration into two zones (Figure 12.52). The zone above the binodal curve gives the concentrations of components that form the two immiscible aqueous phases. The zone below the binodal curve gives the concentration of components that constitute a single phase. The line (TB) in the phase diagram is known as the "tie line," which connects the two nodes on the binodal curve (Figure 12.52). The three systems S1, S2, and S3 shown in Figure 12.52 are said to have the same top-phase and bottom-phase equilibrium composition as they all are positioned on the same tie line. Point C on binodal is said to be the critical point, above which the volume of both the phases is theoretically equal. At point C, the value of tie line length (TLL) is equal to zero. The TLL and component concentration have the same units. The TLL can be estimated by the weight ratio, which is given by

$$\frac{V_t \rho_t}{V_b \rho_b} = \frac{SB}{ST} \tag{12.88}$$

where V is the volume of top (t) and bottom (b) phases, ρ is the density of top (t) and bottom (b) phases, and SB and ST are the lengths or segments as shown in Figure 12.50.

An alternative precise method involves the analysis of the top and bottom phases, given by

$$\text{TLL} = \sqrt{\Delta X^2 + \Delta Y^2} \tag{12.89}$$

In general, the tie lines are straight and the slope of tie line (STL) can be calculated using the following equation:

$$\text{STL} = \frac{\Delta Y}{\Delta X} \tag{12.90}$$

The partition of biomolecules between the top and the bottom phases of the ATPS is determined by the equilibrium relationship, which is given by the partition coefficient (K) defined as

$$K = \frac{C_{AT}}{C_{AB}} \tag{12.91}$$

where C_{AT} and C_{AB} are the equilibrium concentrations of target component A in the top and bottom phases, respectively (Raja et al., 2012). Partition coefficient deviates from unity when the interaction of solute with the two immiscible phases is unequal.

Recovery of target component in the top phase is calculated as percentage yield (Y_T %) on weight basis using the following equations (Anandharamakrishnan et al., 2005):

$$Y_T(\%) = \frac{(C_{AT} V_{AT}) \times 100}{(C_0 V_0)} \tag{12.92}$$

$$Y_B(\%) = \frac{(C_{AB} V_{AB}) \times 100}{(C_0 V_0)} \tag{12.93}$$

where V_{AT} and V_{AB} are the volumes of top and bottom phases, respectively, C_0 is the concentration of target component in the phase system, and V_0 is the volume of total phase system.

In a polymer–polymer system, the polymers begin to separate between two different phases as a result of steric exclusion. Steric exclusion occurs when a solute molecule in aqueous phase exhibits a relatively larger hydrodynamic radius than water, thus leading to a deficit of the solute molecule surrounded by a second solute molecule which is of interest (Arakawa et al., 1990). The balance between the gravitational, flotation, and frictional forces acting on the solute determines its movement during the phase separation. The gravitational force depends on the density of phases and the flotation and frictional forces depend on the flow properties of phases (Asenjo and Andrews, 2012; Salamanca et al., 1998). For example, an ATPS comprising dextran and PEG turned turbid above certain polymer concentrations, and the two phases remained in equilibrium (Silva and Franco, 2000). The lighter phase was enriched in PEG and the heavier phase was rich in dextran. Thus, even when both the polymers are completely water soluble, as their limiting concentrations are exceeded, they turn incompatible and separate into two phases (Albertsson, 1986).

In a polymer–salt ATPS, the salt preferentially absorbs large amount of water to bring about the steric exclusion and phase separation (Asenjo and Andrews, 2011; Harris and Angal, 1989). The aforementioned fact was found to hold true when a polymer–salt system comprising PEG and sodium sulfate in water was used for the extraction of caffeine. Partitioning of caffeine was temperature dependent, and under all the conditions studied, the caffeine was found to preferentially accumulate in the PEG-rich top phase (Figure 12.53; Sampaio et al., 2016). This was attributed to the major influence of the *salting-out* effect, by which the solubility of caffeine in the water was decreased by the addition of sodium sulfate (salt).

12.5.2 Factors Influencing Separation in ATPE

The limiting concentrations and the partitioning of biomolecules between the two phases are influenced by the following factors:

- Affinity of the biomolecule towards the phase-forming polymer
- Components of the ATPS and their surface properties (Asenjo and Andrews, 2011). Interfacial tension (γ) is an important surface property that influences the partitioning behavior of particles in ATPS (Albertsson, 1986). The interfacial adsorption of particles at the liquid–liquid interface reduces the interfacial area between the phases and decreases the interfacial tension (Wu et al., 1996; Bamberger et al., 1984) to lower the surface free energy.
- *Concentration and molecular weight/size of the polymer*: Higher the molecular weight, lower the concentration of polymer required for the phase formation (Iqbal et al., 2016). In a polymer–polymer ATPS, the partition of biomolecule decreases towards the phase having a polymer of high molecular weight. The aforesaid fact was ascribed to the steric exclusion of the biomolecule from that phase or due to changes in the hydrophobicity of phases (Grilo et al., 2016; Asenjo and Andrews, 2011; Walter and Johansson, 1994). When the molecular weight of the polymer increases, the hydrophobicity is increased by reducing the hydrophilic groups/hydrophobic area

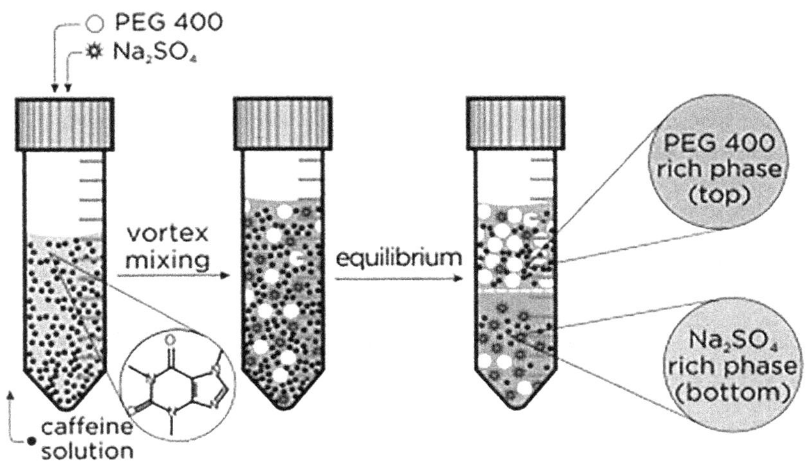

FIGURE 12.53 Separation of caffeine in a polymer–salt ATPS. (Reproduced with permission from Sampaio, D. D. A., Mafra, L. I., Yamamoto, C. I., Andrade, E. F. D., de Souza, M. O., Mafra, M. R. and Castilhos, F. D. 2016. Aqueous two-phase (polyethylene glycol + sodium sulfate) system for caffeine extraction: equilibrium diagrams and partitioning study. *The Journal of Chemical Thermodynamics* 98: 86–94.)

(Asenjo and Andrews, 2011). On the other hand, in a polymer–salt ATPS, the movement of biomolecule towards the polymer-rich phase reduces as the concentration of polymer increases.

- Concentration and composition of salt (Zafarani-Moattar et al., 2011; Mazzola et al., 2008).
- pH and ionic strength.
- *Temperature of the aqueous solution*: Temperature exerts a profound influence on the composition of two phases in an ATPS and consequently affects the changes in the phase diagram. The changes in temperature affect the partition of biomolecules via the variations in viscosity and density. Generally, phase separation occurs at lower temperature in a polymer–polymer ATPS with lower concentrations of polymer, and vice versa has been observed in a polymer–salt system (Walter and Johansson, 1994).

12.5.3 Case Study: Application of ATPE for Recovery of Proteins from Whey

In general, when a polymer–salt ATPS is used for protein purification, the protein will usually accumulate in the top phase that is hydrophobic (less soluble in water, for instance, in the PEG phase). Eventually, the proteins can be separated by changing the molecular weight of the polymer, ion type, or ionic strength in salt phase with the help of an additional salt (e.g., sodium sulfate and ammonium sulfate). The aforementioned principle was found to hold true when a polymer–salt ATPS was used for the recovery of proteins (devoid of fat) from cheese whey. The whey proteins partitioned to the PEG-rich phase; whereas, the fat partitioned to the salt-rich phase (Table 12.2). In the polymer-salt ATPS, three different types of polymer–salt combinations were used: PEG 6000/ammonium sulfate, PEG 6000/potassium dihydrogen phosphate, and PEG 6000/(potassium dihydrogen phosphate+dipotassium hydrogen phosphate) (Anandharamakrishnan et al., 2005).

In the aforementioned polymer–salt ATPS, depending on the concentration or solubility limit, the proteins exist in soluble or particulate form. In turn, the solubility limit depends on the pH and concentration of phase component. Conversely, fat exists as dispersed droplets of varied size depending on the ionic strength of the surrounding continuous phase. The presence of salts reduces the solubility and dispersibility of fat. Consequently, fat droplets of bigger size occur in the salt-rich phase compared to that in cheese whey.

The differential partitioning of fat and proteins in the polymer–salt ATPS depends upon the following mechanism (Figure 12.54):

- The fat droplets migrate from the high energy level (PEG-rich phase where the fat droplets have high surface free energy) to the interface between dispersed phase and continuous phase.
- The fat droplets adsorb at the phase interface.
- The fat droplets migrate from the interface to the low energy level (salt-rich phase where the fat droplets have low surface free energy). This step will not take place if the droplets attain lower energy at the interface.

TABLE 12.2
Differential Partitioning of Fat and Proteins in Polymer–Salt ATPS

Phase System and Composition (% w/w)	pH	Volume Ratio (mL/mL)	Concentration of Protein in the Phase System (mg/100 g of System)	Partition Coefficient of Protein (K)	Recovery of Protein in the Top Phase (%)
PEG 6000/$(NH_4)_2SO_4$ (17.88/14.26% w/w)	4.5	38/50	28.5	1.9	24.6
PEG 6000/KH_2PO_4 (12/16% w/w)	3.8	50/38	30.2	0.5	24.5
PEG 6000/KH_2PO_4 + K_2HPO_4 (11.7/10% w/w)	7.0	72/12	32.8	0.9	83.4

Source: Adapted from Anandharamakrishnan et al. (2005).

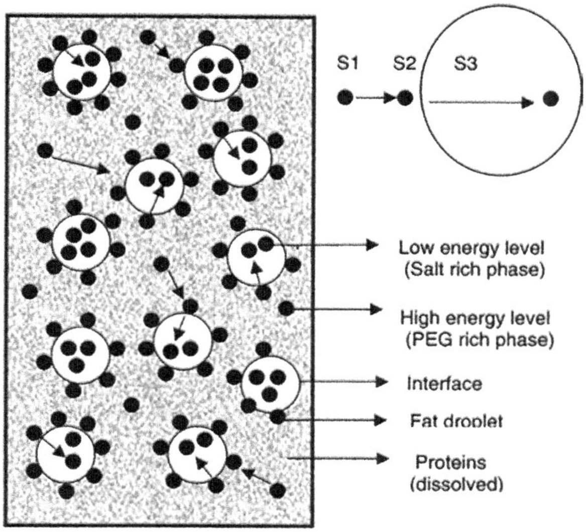

FIGURE 12.54 Mechanism for differential partitioning of fat and proteins in ATPS. (Reproduced with permission from Anandharamakrishnan, C., Raghavendra, S. N., Barhate, R. S., Hanumesh, U. and Raghavarao, K. S. M. S. 2005. Aqueous two-phase extraction for recovery of proteins from cheese whey. *Food and Bioproducts Processing* 83: 191–197.)

The polymer-rich and salt-rich phases wet the fat droplets depending on the surface free energy of droplets in these phases. The wetting phenomenon involves the interaction between the surface of fat droplet and the contacting phase leading to reduced surface free energy of fat droplets. The interactions could be electrostatic (or) hydrophobic (or) attractive (or) repulsive. In order to obtain one-sided partitioning of fat droplets to the salt-rich phase, it is necessary that the salt-rich phase is capable of preferentially wetting the fat droplets over the polymer-rich or protein-extracting phase (PEG). The wetting criteria for fat droplets can be determined using the following expression as proposed for solid surfaces (Schultz and Nardin, 1992):

$$\left|\gamma_{P1} - \gamma_{P2}\right| < \gamma_{TB} \tag{12.94}$$

where γ_{P1} is the interfacial tension between fat droplets and top phase (N/m), γ_{P2} is the interfacial tension between fat droplets and bottom phase (N/m), and γ_{TB} is the interfacial tension of phase system (N/m). The value of γ_{P1} and γ_{P2} depends on the interaction between fat droplets and the contacting phases (polymer and salt). The polarity indices of phases and the surface charge on fat droplets decide the magnitude of these interactions. Preferential wetting of fat droplets by one of the phases is determined by the polarity difference between the phase-forming components (i.e., between polymer and polymer or polymer and salt). The polymer–salt ATPSs have higher probability of obtaining one-sided partitioning of fat owing to their higher polarity difference than the polymer–polymer ATPSs. The polar surface-active lipids present in fat impart a charge to the surface of droplets. The high concentration of counterions present in the salts neutralizes the charged surface of fat droplets. Thus, the interaction between the ions in the salt phase and fat droplets plays a key role in achieving the lower surface free energy and the resultant wetting of fat droplets. The aforementioned interactions between the ions and the fat droplets are responsible for the preferential portioning of fat to the salt-rich phase rather than to the PEG-rich phase.

Further, whey proteins belong to the category of acidic proteins that have their isoelectric point in the range of 4.2–4.5 for α-lactalbumin and β-lactoglobulin (Marshall, 1982). The lower solubility of proteins in PEG/$(NH_4)_2SO_4$ and PEG/KH_2PO_4 is due to the pH of the system, which is close to the isoelectric pH of whey proteins. The higher solubility of proteins in the PEG/potassium phosphate system is due to the higher pH of the system (pH 7). Aggregation of protein plays an important role in partitioning.

The higher aggregation of protein reduces the soluble amount of protein available for partitioning, which in turn reduces the recovery of proteins. For this reason, it is necessary to operate the extraction at suitable phase compositions such that aggregation is low, which would lead to high solubility of proteins (Anandharamakrishnan et al., 2005).

12.6 Distillation

Distillation is yet another mass transfer operation in which the components of a liquid–liquid mixture are separated by boiling, based on their relative volatility. Relative volatility (α) of a compound is a measure of its vapor pressure in comparison with that of water. Consider a binary solution of two immiscible liquids: "A" and "B" with different boiling points. When heated at constant pressure, "A" separates out as vapor first if it is more volatile than "B." On collecting and condensing this vapor, liquid A can be obtained. Distillation is mostly used for the separation of alcohol from the fermented liquids. Apart from that, it is also used for the recovery and concentration of volatile aromas.

A continuous distillation or fractionation column is shown in Figure 12.55. It consists of a series of plates that are designed to bring in contact between the liquid phase and the vapor phase at equilibrium. The plates are perforated disks placed horizontally at equidistance, throughout the distillation column. Liquid flows through the perforated disk downward, and the vapor bubbles rise upward. As the vapor rises upward through the column, it is enriched with more volatile components. Reboiler/heat exchanger is placed at the bottom, which is heated by steam. The liquid in the reboiler is at the bubble point, and the residue is withdrawn at a specified concentration (x_W) from the bottom of the column. The section above the feed plate in the column is known as "rectifying section," and the one below is termed as the "stripping section." The condenser condenses the overhead vapor to liquid at its bubble point. Some of the liquid at concentration x_D is removed as product, and the remaining liquid re-enters as reflux. The ratio of reflux to the product withdrawn is known as *reflux ratio*. The reflux ratio is important to strike a balance between the capital cost of the distillation column and the operating cost of the distillation process. Lower reflux ratio indicates the requirement of more number of stages (high capital cost), whereas higher reflux ratio corresponds to less number of stages with a wider diameter column to handle larger vapor flow (high operating cost for reboiler).

12.6.1 McCabe–Thiele Method

The Mccabe–Thiele method is a graphical representation of the material balance equations, in the form of operating lines on a vapor–liquid equilibrium (VLE) diagram. VLE diagram is the plot between mole fraction of the component to be separated (A) in the liquid fraction (abscissa) against that in the vapor fraction (ordinate) (Figure 12.56). The equilibrium data from the VLE diagram provide the relationship

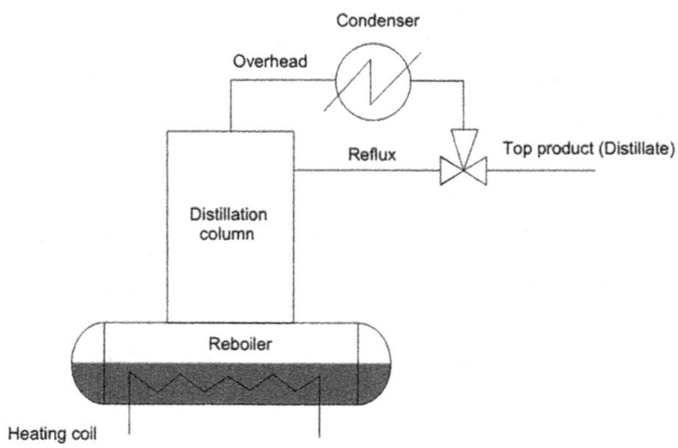

FIGURE 12.55 Schematic diagram of a continuous distillation column.

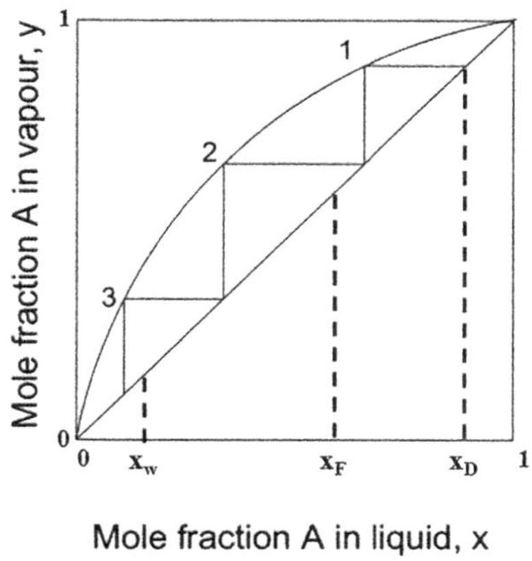

FIGURE 12.56 The McCabe–Thiele method for calculation of the number of stages.

between the composition of liquid leaving a plate and the composition of vapor beneath the plate. This facilitates the graphical means of calculating the number of equilibrium stages in the distillation process (McCabe et al., 1993). The procedure of using the McCabe–Thiele method for the calculation of the number of stages will be described in the sections that follow.

Consider a continuous distillation column where *D*, *F*, and *W* are the molar flow rates of distillate, feed, and residue streams, respectively, and their corresponding mole fractions are x_D, x_F, and x_w. *L* and *V* are the molar flow rates of liquid and vapor, respectively (Figure 12.57). The overall material balance and component balance are given by

$$F = D + W \tag{12.95}$$

$$x_F F = x_D D + x_W W \tag{12.96}$$

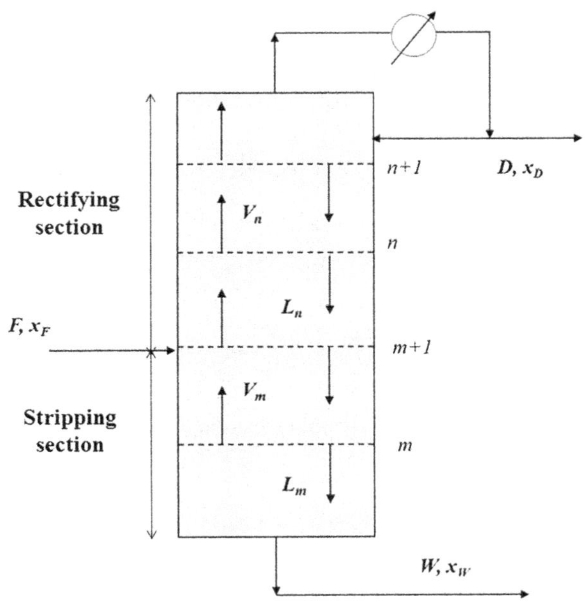

FIGURE 12.57 Material balance across continuous distillation column.

12.6.1.1 Rectifying Section

Overall material balance and component balance for the plates n and $(n + 1)$ by considering a constant molal overflow are given by

$$V_n = L_n + D \tag{12.97}$$

$$y_n V_n = x_{n+1} L_n + x_D D \tag{12.98}$$

$$y_n = \frac{L_n}{V_n} x_{n+1} + \frac{D}{V_n} x_D \tag{12.99}$$

Reflux ratio (R) is given by

$$R = \frac{L_n}{D} = \frac{V_n - D}{D} \tag{12.100}$$

$$y_n = \frac{R}{R+1} x_{n+1} + \frac{x_D}{R+1} \tag{12.101}$$

Considering $x_{n+1} = x_D$

$$y_n = \frac{R}{R+1} x_D + \frac{x_D}{R+1} = x_D \tag{12.102}$$

Considering $x_{n+1} = 0$

$$y_n = \frac{x_D}{R+1} \tag{12.103}$$

The coordinates of the operating line for the rectifying section or top operating line are (x_D, x_D) and $\left(0, \frac{x_D}{R+1}\right)$.

12.6.1.2 Stripping Section

For the stripping section, overall material balance and component balance for the plates n and $(n + 1)$ by considering constant molal overflow are given by

$$V_m = L_m - W \tag{12.104}$$

$$y_m V_m = x_{m+1} L_m + x_W W \tag{12.105}$$

Consider $x_{m+1} = x_W$

$$y_m = \frac{L_m}{V_m} x_W - \frac{W}{V_m} x_W = x_W \tag{12.106}$$

The coordinates of the operating line for the stripping section are (x_W, x_W), and the gradient is $\frac{L_m}{V_m}$.

At the intersection of operating line,

$$y_q V_n = x_q L_n + x_D D \tag{12.107}$$

$$y_q V_m = x_q L_m - x_W W \tag{12.108}$$

Subtracting Eq. (12.107) from Eq. (12.108) gives

$$y_q(V_m - V_n) = x_q(L_m - L_n) - (x_D D + x_W W) \tag{12.109}$$

Material balance across the feed plate is given by

$$F + L_n + V_m = L_m + V_n \tag{12.110}$$

Considering the enthalpy per kmol of feed to be h', at which the boiling point is h_f, and considering the molar latent heat of vaporization to be h_{fg}, Eq. (12.110) can be written as

$$\frac{F(h_f - h')}{h_{fg}} = V_m - V_n \tag{12.111}$$

$$L_m - L_n = \frac{F(h_f - h')}{h_{fg}} + F \tag{12.112}$$

q is the quantity of heat required to vaporize 1 kmol of the actual feed divided by the molar latent heat of feed, which is expressed as

$$q = \frac{h_{fg} + (h_f - h')}{h_{fg}} \tag{12.113}$$

As $L_m - L_n = qF$ and $V_m - V_n = qF - F$

$$y_q(qF - F) = x_q qF - x_F F \tag{12.114}$$

Rearranging Eq. (12.114) gives

$$y_q = \frac{q}{q-1} x_q - \frac{x_F}{q-1} \tag{12.115}$$

The equation is for the *q-line* and has the coordinates (x_F, x_F) and gradient $\left(\frac{q}{q-1}\right)$.

12.6.1.3 Steps Involved in Constructing the McCabe–Thiele Diagram and Calculation of the Number of Stages

1. VLE curve for the binary mixture is constructed (Figure 12.58).
2. The 45° line is drawn.
3. The feed line is then drawn, which for a saturated liquid, is a vertical line running from x_F through the equilibrium curve. However, for saturated vapor, the feed line is a vertical line from x_F to the 45° line. Then, a horizontal line is drawn from the 45° line through the equilibrium curve.
4. The residue composition, x_W, and the desired distillate composition, x_D, are fixed on the X-axis.
5. The rectifying line is drawn corresponding to the calculated reflux ratio, R. This rectifying line begins at the point x_D on the 45° line and has a slope of $R/(R + 1)$.
6. Stripping line is constructed by connecting the intersection of the feed line and rectifying line and the point x_W on the 45° line. The slope of the stripping line is given by $(V_B + 1)/V_B$, where V_B is the boil-up ratio. *Boil-up ratio* is defined as the fraction of the amount of liquid that is boiled back into the column to the amount of liquid that exits.
7. Starting from the point x_D on the rectifying line, a horizontal line is drawn to the equilibrium curve, and then, a vertical line is drawn to the operating (rectifying or stripping) line.

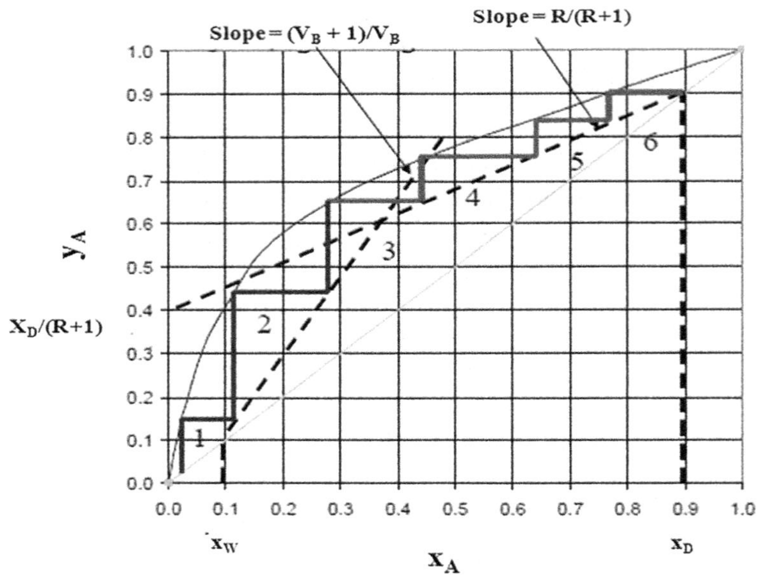

FIGURE 12.58 Calculation of number of stages using the McCabe–Thiele procedure.

8. A staircase is formed by repeating the step 7 until the point x_W on the 45° line is reached.
9. Each point where the staircase intersects the equilibrium curve is considered as one stage of the column.
10. The number of stages drawn is counted.
11. Finally, the number of theoretical trays is obtained by subtracting one for the reboiler section.

Example 12.6

A binary mixture of volatile liquids containing 55% A and 45% B is to be separated by continuous fractionation into a distillate with a composition of 96% A and a residue containing no more than 5% A. All compositions are on a molar basis. The vapor–liquid equilibrium data for this system are as follows:

x	0.10	0.20	0.30	0.40	0.50	0.60	0.70	0.80	0.90
y	0.330	0.550	0.680	0.765	0.835	0.885	0.925	0.955	0.98

How many ideal stages are required if the feed liquid is at its bubble point as it enters the distillation column and the column is operated at a reflux ratio of 1.4?

The steps in the graphical solution are as follows:

1. Plot the data given in the table (Figure 12.58).
2. Draw the line $y = x$.
3. As the feed enters at its bubble point in the distillation column, $q = 1$. Draw a *q-line* from the coordinates (0.55, 0.55) vertically upward.
4. The composition of distillate is 0.96, and the gradient is (0.96/2.4) = 0.4. Therefore, operating line for the rectifying section has coordinates (0, 0.40) and (0.96, 0.96).
5. The residue composition is 0.05, and therefore, the operating line for stripping section passes through the coordinates (0.05, 0.05) and intersects with the top operating line on *q-line*.
6. Beginning from the 0.96 drawn in the ideal stages as shown in Figure 12.56.
7. Totally, six ideal stages are needed to give a residue composition of 0.05 (Figure 12.58).

Answer: Six ideal stages are required.

12.6.2 Equipment for Distillation

12.6.2.1 Batch Distillation

Batch distillation is the simplest method of distillation used in the food industry. In this method, a fixed amount of liquid to be distilled is added to the still, which is a column placed on the top of the reboiler or *still pot* (Figure 12.59). The product (vapor) is withdrawn from the top of the reboiler continuously. As a result, the liquid level in the still pot decreases constantly. The distillate is collected after condensing the vapors using a condenser. During the initial stages of batch distillation, the composition of boiling liquid is pure. But, later with the progress in the distillation process, the vapor also carries other volatiles along with the desired volatile component at which stage the distillation is stopped. A run of batch distillation may last from a few hours (8–16 h) to several days.

The major application of batch distillation is the separation of solvent from the liquid, after the extraction process.

12.6.2.2 Fractional Distillation

In fractional distillation, distillation occurs in a column (Figure 12.60). It contains multiple contact stages through which vapor and liquid move in a countercurrent manner. The name "fractional distillation" implies that at any position of the column, the distillate fraction has a different composition. Feed is introduced at a certain point of the column. The portion above the feed inlet is known as the rectification zone, whereas that below the feed inlet is known as the stripping zone (as shown in Figure 12.57). During the process, vapor moves in the upward direction and liquid moves in the downward direction. Heat exchange occurs between the vapor and the liquid, and thus, the less-volatile vapor component condenses into a liquid and high-volatile component vaporizes. Thus, the vapor is rich in the high-volatile component. The column is connected to a condenser to achieve a continuous heat exchange.

12.6.2.3 Steam Distillation

The steam distillation apparatus is supplied with water or steam to lower the boiling points of the compounds and thereby facilitate their evaporation at temperatures below the point at which their thermal degradation occurs (Figure 12.61). The working principle of steam distillation can be put forth as follows. It is known that a liquid boils when its vapor pressure is equal to the atmospheric pressure. During the steam distillation, water and organic compounds vaporize together and their vapor pressures sum up to equal the atmospheric pressure. As the vapor pressure of the organic compound should be lower than

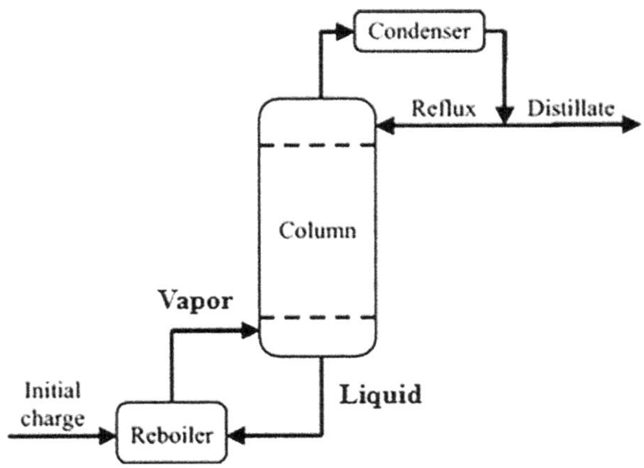

FIGURE 12.59 Schematic diagram of a batch distillation apparatus. (Reproduced with permission from Lopes, M. M., and Song, T. W. 2010. Batch distillation: better at constant or variable reflux? *Chemical Engineering and Processing: Process Intensification* 49: 1298–1304.)

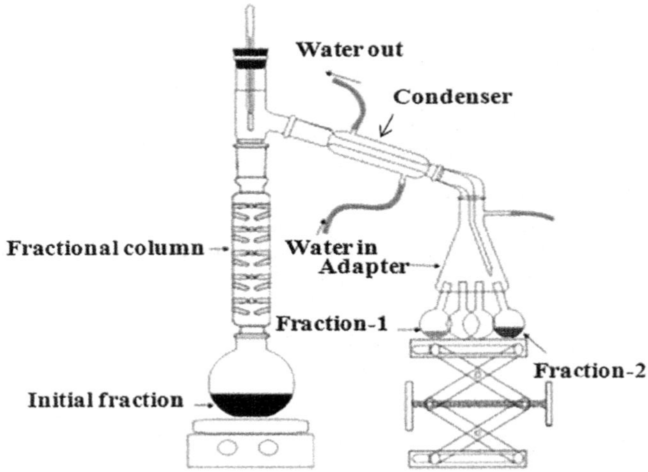

FIGURE 12.60 Schematic diagram of a fractional distillation apparatus. (Modified and Reproduced with permission from Nam, H., Choi, J. and Capareda, S. C. 2016. Comparative study of vacuum and fractional distillation using pyrolytic microalgae (Nannochloropsis oculata) bio-oil. *Algal Research* 17: 87–96.)

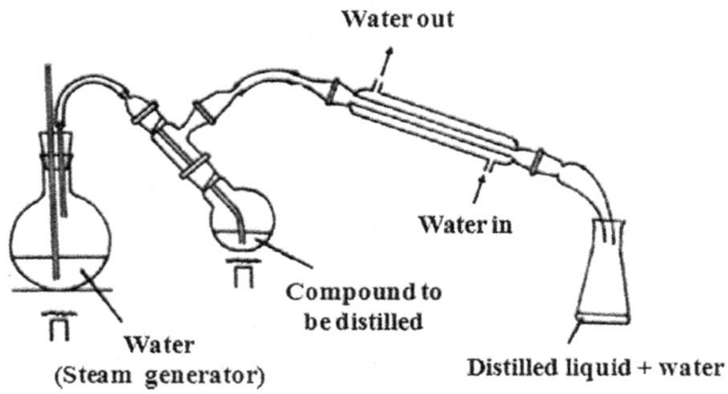

FIGURE 12.61 Schematic diagram of a conventional steam distillation apparatus. (Modified and Reproduced with permission from Kusuma, H. S., Putra, A. F. P. and Mahfud, M. 2016. Comparison of two isolation methods for essential oils from orange peel (*Citrus aurantium* L) as a growth promoter for fish: microwave steam distillation and conventional steam distillation. *Journal of Aquaculture Research and Development* 7: 409.)

the atmospheric pressure, it vaporizes at lower temperatures. The following equation relates the amount of organic compound that distills over with steam and the vapor pressures and molecular masses.

$$\frac{W_1}{W_2} = \frac{P_1 M_1}{P_2 M_2} \tag{12.116}$$

where W is the mass, P is the vapor pressure, and M is the molecular mass. The subscripts 1 and 2 denote water and organic compound, respectively. The separation of components is achieved by condensing the mixture of steam and the volatile organic compounds that distills out. These volatile materials are not soluble in water. Therefore, on condensing the vapor, two immiscible layers of liquid are formed containing water and volatile oil, respectively. Water is heavier than the essential oil and thus settles at the bottom. The essential oil is recovered from the top layer.

Steam distillation is mainly used for the separation of heat-sensitive compounds with a boiling point lower than 100°C, such as the essential oils from spices. It is also used for the deodorization and removal of off-flavors from the food products. Due to the high boiling temperature of essential oils, steam distillation is an economical process.

12.6.3 Applications of Distillation in Food Processing

- Distillation finds wide applications in the alcoholic beverage industry. Ethyl alcohol is separated from the fermented broth using distillation. Dilute alcohol obtained earlier is further concentrated to produce whisky.
- Solvent recovery is essential during the extraction process to minimize the processing cost. Solvent separated by distillation can be reused for the extraction process.
- Aromatic essential oils obtained from spices have culinary and medicinal applications. Production of these essential oils from ginger, clove, cinnamon, and so on is carried out using steam distillation.

12.6.4 Case Study: Application of Distillation in the Production of Alcoholic Beverages

The applications of distillation in the production of alcoholic beverages can be appreciated by the following discussion on Scotch whisky:

According to the Scotch Whisky Act (1988), "Scotch whisky" refers to the whisky *that has been produced at a distillery in Scotland from water and malted barley (to which only whole grains of other cereals may be added) all of which have been processed at that distillery into a mash, converted to a fermentable substrate only by endogenous enzyme systems fermented only by the addition of yeast.* Scotch whisky is classified into two types depending on the substrate used for fermentation and the method of distillation: malt whisky and grain whisky. Scotch malt whisky is prepared from malted barley and the alcohol resulting from the fermentation is concentrated by the batch distillation process in copper pot stills.

The malted barley is ground to a roughly cut grist. It is mixed with hot water in a vessel known as a *mash tun*. This process converts the starch in barley into a sugary liquid known as *wort*. Subsequently, the wort is separated from the mash and transferred to a fermenting vat, or *washback*. At this stage, yeast is added, which converts the sugary wort into an alcoholic fraction known as the *'wash'*. The wash is distilled in the copper pot stills, which are heated by steam to a temperature just below the boiling point of alcohol to selectively volatilize the alcohol and other components from the wash.

Usually, the Scotch malt whisky is distilled twice: the first distillation takes place in a larger wash still, and the second one is carried out in a slightly smaller spirit still. During the first distillation, the temperature of the fermented liquid within the wash still is increased gradually until the alcohol in the wash vaporizes. These vapors rise upward through the neck and pass above the head of the still, after which they are guided through the water-cooled condensers. The liquid distillate obtained from the wash still or first fermentation is known as the *low wines*, which contain about 20% (v/v) alcohol. The alcohol is then separated from the water, yeast, and residue, which is called *pot ale*, and collected in a receiver before being passed through the second spirit still to repeat the process. More control is exerted on the process during the second distillation, from which only the heart or *middle-cut* of the spirit flow (containing 68% alcohol; v/v) will be collected for maturation. However, even the middle-cut is collected by the distiller only after ensuring that it has reached the desired quality. *Foreshots*, the first to distill off containing more volatile compounds, and *feints*, the final part in which more oily compounds are vaporized, are returned for redistillation with the next batch of low wines.

On the other hand, Scotch grain whisky is fermented from a *mash*, which is a combination of non-malted cooked cereals such as barley, wheat, or maize and a smaller amount of malted barley. The alcohol derived from this fermentation process is concentrated in a continuous distillation unit, known as the *Coffey or patent still*. The patent still consists of an *analyzer* and a *rectifier*. The wash is pumped to the top of the rectifier in which as it flows down the coil, it is warmed by the heat of vapors rising upward through the still. As the wash is heated, it enters the top of the analyzer and then descends the column through the perforated plates. Simultaneously, the low-pressure steam moves up the analyzer that extracts the alcohol vapors and forwards it to the bottom of the rectifier. Later, the alcohol vapors begin to rise after which they are steadily condensed by the cool wash coils, and then, the Scotch grain whisky is extracted from the column.

According to the Scotch Whisky Act (1988), Scotch whisky is distilled at an alcoholic strength of less than 98.4% (v/v) such that the distillate possesses the characteristic aroma and taste of the raw materials used and the method of its production. Both the Scotch malt whisky and the Scotch grain whisky are matured in oak casks of a capacity, not more than 700 liters, for a minimum of 3 years. Before filling into casks for maturation, the distilled Scotch whisky is diluted to an alcoholic strength of 63.5% (v/v) of alcohol. According to the law, the minimum bottling strength should be 40% alcohol (v/v) (The Scotch Whisky Association, 2017).

12.7 Problems to Practice

12.7.1 Multiple Choice Questions

1. Substances that are mixed together but are not chemically combined is termed as
   a. mixture
   b. solution
   c. solute
   d. solvent

Answer: a

2. The power required by the impeller during liquid mixing is the function of
   a. viscosity of the mixture
   b. mass of the mixture
   c. speed of agitator
   d. conductivity of the mixture

Answer: c

3. Mixing of two fluids is
   a. a reversible process
   b. an irreversible process
   c. an isothermal process
   d. none of the above

Answer: a

4. Ribbon mixer is generally used for the blending of
   a. solid–solid
   b. liquids
   c. solid–liquid
   d. dough

Answer: a

5. The value of mixing index for a complete mixing is
   a. 6
   b. 1
   c. less than 1
   d. greater than 1

Answer: b

6. During the course of the mixing process, the value of mixing index
   a. remains constant at a value between 0 and 1
   b. is 1
   c. is 0
   d. varies between 0 and 1

Answer: d

7. The screw-type mixers are governed by the mechanism of
   a. convection
   b. diffusion
   c. shear
   d. convection and diffusion

Answer: a

8. The mixers that are used when one of the components is in very less quantity and to mix items in savory or a snack item are, respectively,
   a. planetary mixer and ribbon mixer
   b. Y-cone mixer and double-cone mixer
   c. double-cone mixer and double-cone mixer
   d. planetary mixer and sigma mixer

Answer: c

9. Batch distillation is also termed as
   a. equilibrium distillation
   b. flash distillation
   c. differential distillation
   d. none of the above

Answer: c

10. In general, filtration involves the separation of particles of size
    a. $d_p > 5\mu m$
    b. $d_p > 10\,\mu m$
    c. $d_p > 15\mu m$
    d. $d_p > 20\mu m$

Answer: b

11. MF is the separation of suspended material using a membrane of pore size
    a. 0.02–10 µm
    b. 1–10 A°
    c. 20–30 µm
    d. 10–200 A°

Answer: a

12. An example of continuous-type filter is
    a. plate and frame
    b. spiral wound membrane
    c. rotary drum vacuum
    d. tubular membrane

Answer: c

13. Filter aid is used to
    a. increase the filtering efficiency
    b. decrease the filtering efficiency
    c. to give body to the filtrate
    d. to increase the mass of cake

Answer: a

14. The centrifuge that is commonly used in the dairy industry is
    a. solid bowl centrifuge
    b. disk bowl centrifuge
    c. tubular bowl centrifuge
    d. cylinder bowl centrifuge

Answer: b

15. Rate of filtration is
    a. maximum at early stage
    b. maximum at middle stage
    c. maximum at final stage
    d. uniform throughout the filtration process

Answer: a

16. Distillation is used for the separation of
    a. a soluble solid from a solution
    b. a liquid from a solution
    c. a solid from a solid
    d. an insoluble solid from a solution

Answer: b

17. In fractional distillation, a mixture of liquids is separated based on their
    a. boiling point
    b. solubility
    c. density
    d. chemical composition

Answer: a

18. A fruit juice of viscosity μ and density ρ is agitated using an impeller of diameter D at a speed of N revolutions per minute. The terms $X = \frac{P}{\rho_m N^3 D^5}$, $Y = \frac{D^2 N\rho}{\mu}$, and $Z = \frac{DN^2}{g}$ represent three dimensionless numbers, where P is the power imparted by impeller and g is the acceleration due to gravity. Which of the following is the correct representation of these numbers?
    a. X = Power, Y = Froude, and Z = Reynolds
    b. X = Power, Y = Reynolds, and Z = Froude
    c. X = Froude, Y = Reynolds, and Z = Power
    d. X = Reynolds, Y = Power, and Z = Froude

Answer: b

19. The method of membrane filtration that uses the highest pressure than the other methods is
    a. NF
    b. RO
    c. UF
    d. MF

Answer: b

20. The terminal velocity of centrifugation is directly proportional to
    a. viscosity of the liquid
    b. temperature of the liquid
    c. square of the particle diameter
    d. particle diameter

Answer: c

12.7.2 Numerical Problems

1. A mixed infant food sample was analyzed for its protein content. Six samples each weighing 10 g were taken for the analysis. The weights of protein in the samples were

   2.4 g, 0.64 g, 1.34 g, 1.66 g, 2.74 g, and 0.46 g.

   Calculate the standard deviation of the sample compositions from the mean composition.

Given

i. Number of samples = 6
ii. Weight of each sample = 10 g
iii. Weights of protein in the samples: 2.4 g, 0.64 g, 1.34 g, 1.66 g, 2.74 g, and 0.46 g

To find: Standard deviation of the sample compositions from the mean composition.

Solution

Fractional compositions of protein in the samples (the weight of protein divided by the sample weight) are, respectively,

0.24, 0.064, 0.134, 0.166, 0.274, and 0.046 (x)

Mean composition of protein in the sample = 0.154 ($\overline{x}$)

Deviation of samples from the mean is given by

i. (0.24 − 0.154) = 0.086
ii. (0.064 − 0.154) = −0.09

iii. (0.134 − 0.154) = −0.02
iv. (0.166 − 0.154) = 0.012
v. (0.274 − 0.154) = 0.12
vi. (0.046 − 0.154) = −0.108

$$\therefore s^2 = \frac{1}{n}\left[(x_1 - \bar{x})^2 + (x_2 - \bar{x})^2 + (x_n - \bar{x})^2\right]$$

$$s^2 = \frac{1}{6}\left[(0.086)^2 + (-0.09)^2 + (-0.02)^2 + (0.012)^2 + (0.12)^2 + (-0.108)^2\right] = 0.007$$

$$\therefore s = 0.084\ g$$

Answer: Standard deviation of the sample composition = 0.084 g

2. In a ribbon blender used for the preparation of a mixture of dry sugar and creamer, the initial proportions of nondairy creamer and dry sugar were 51% and 49%, respectively. The variance of the sample compositions measured in terms of fractional compositions of creamer was found to be 0.08 after 5 min of mixing. Calculate the mixing rate constant. Assume that the creamer and sugar particles are approximately of the same size and that a sample contains 25 particles.

Given

i. Initial proportion of nondairy creamer = 51%
ii. Initial proportion of dry sugar = 49%
iii. Variance of the sample composition after 5 min of mixing = 0.08
iv. Creamer and sugar particles are approximately of the same size and that a sample contains 25 particles.

To find: Mixing rate constant

Solution

Let the fractional content of sugar be $p = 0.49$,

$$(1 - p) = (1 - 0.49) = 0.51$$

$$s_o^2 = 0.49 \times 0.51 = 0.25$$

$$s_r^2 = \frac{x(1-x)}{N} = \frac{s_o^2}{N} = \frac{0.25}{25} = 0.01$$

For $s^2 = 0.08$, substituting the values of s_o^2 and s_r^2 in the equation for mixing index

$$M = \frac{\left(s_o^2 - s^2\right)}{\left(s_o^2 - s_r^2\right)}$$

$$M = \frac{(0.25 - 0.08)}{(0.25 - 0.01)} = 0.708$$

Substituting the value of M in Eq. (12.6): $[1 - M] = e^{-Kt_m}$

$$[1 - 0.708] = e^{-300K}$$

$$0.292 = e^{-300K}$$

$$\therefore K = 4 \times 10^{-3}$$

Answer: The mixing rate constant is 4 × 10^{-3}

3. For the data given in problem no. 2, determine the extra time period for which the mixing should continue to reach the specified maximum sample composition variance of 0.03.

Given

i. Maximum sample composition variance = 0.03

To find: Extra time period for which the mixing should continue to reach the specified maximum sample composition variance

Solution

For $s^2 = 0.03$,

$$M = \frac{(0.25 - 0.03)}{(0.25 - 0.01)} = 0.917$$

$$[1 - 0.917] = e^{-0.004 t_m}$$

$$\therefore t_m = 622\,\text{s}$$

Thus, the additional mixing time required is 622 s – 300 s = 322 s.

Answer: 322 s

4. In a liquid mixer, a propeller stirrer of diameter 0.4 m is rotating in a 64% aqueous solution of whey protein concentrate (WPC), at a speed of 50 rev/min. If the diameter of the tank is 1.5 m, determine the power required to operate the stirrer. At 20°C, the viscosity and density of the 64% WPC solution are 6.6 N s m^{-2} and 1203 kg/m^3, respectively.

Given

i. Diameter of propeller's stirrer = 0.4 m
ii. Concentration of WPC solution = 64%
iii. Rotation speed = 50 rev/min
iv. Diameter of the tank = 1.5 m
v. Viscosity of WPC solution at 20°C = 6.6 N s m^{-2}
vi. Density of WPC solution at 20°C = 1203 kg/m^3

To find: Power required to operate the stirrer.

Solution

For the impeller of diameter 0.4 m,

$$N_{Re} = \frac{D^2 N \rho_m}{\mu_m} = \frac{0.4^2 \times 0.83 \times 1203}{6.6} = 24.2$$

$$\left(50\,\text{rpm} = \frac{50}{60} = 0.83 \text{ revolution per second}\right)$$

Since $N_{Re} < 300$, $N_{Po} = K(N_{Re})^n$, where $N_{Po} = \dfrac{P}{\rho_m N^3 D^5}$

Assuming that pitch is equal to the diameter, Rushton chart gives $n = -1$ and $K = 41$.

$$\frac{P}{\rho_m N^3 D^5} = K(N_{Re})^n$$

$$\therefore P = 41(24.2)^{-1}\left(1203 \times 0.83^3 \times 0.4^5\right) = 11.9 \approx 12\,\text{J/s}$$

1 horsepower = 746 W

$$\therefore \text{Required power} = \frac{12}{746} = 0.016\,\text{HP}$$

Answer: 12 J/s or 0.016 HP

5. If the same condition specified in problem no. 4 is used for stirring a 70% WPC solution, estimate the power required.

Given

i. Concentration of WPC solution = 70%

To find: Power required to operate the stirrer.

Solution

For a 70% solution of WPC, the viscosity and density are 15.1 N s m^{-2} and 1222 kg/m^3, respectively.

$$N_{Re} = \frac{D^2 N \rho_m}{\mu_m} = \frac{0.4^2 \times 0.83 \times 1222}{15.1} = 10.7$$

Since $N_{Re} < 300$, $N_{Po} = K(N_{Re})^n$, where $N_{Po} = \dfrac{P}{\rho_m N^3 D^5}$

Assuming that pitch is equal to the diameter, Rushton chart gives $n = -1$ and $K = 41$.

$$\frac{P}{\rho_m N^3 D^5} = K(N_{Re})^n$$

$$\therefore P = 41(10.7)^{-1}\left(1222 \times 0.83^3 \times 0.4^5\right) = 27.4\,\text{J/s}$$

1 horsepower = 746 W

$$\therefore \text{Required power} = \frac{27.4}{746} = 0.037\,\text{HP}$$

Answer: 27.4 J/s or 0.037 HP

6. The mixing index for a mixing operation carried out for 5 min is 0.8. Calculate the mixing index after 10 min of mixing.

Given

i. Mixing index for a mixing operation carried out for 5 min = 0.8

To find: Mixing index after 10 min of mixing

Solution

$$[1 - M] = e^{-Kt_m}$$

$$5\,\text{min} = 300\,\text{s},$$

$$[1 - 0.8] = e^{-300K}$$

$$\therefore K = 0.005$$

After 10 min of mixing,

$$10\,\text{min} = 600\,\text{s}$$

$$[1 - M] = e^{-0.005 \times 600}$$

$$\therefore M = 0.95$$

Answer: Mixing index after 10 min is 0.95.

7. Filtration of $CaCO_3$ slurry in water was carried out at a constant pressure of 7 psia, at 30 °C, using a plate-and-frame filter press. The area of the plate-and-frame press is 0.0485 m^2, and the viscosity of feed slurry is 1200 cP. The concentration of the feed slurry is 24.58 kg solid/m^3. Calculate α and β for the data given in the following table, where V is the volume of filtrate collected and t is the time.

V ($\times 10^3$ m^3)	t (s)
0.5	19.5
1.0	43.5
1.5	74.0
2.0	110.5
2.5	154.0
3.0	204.0

Given

i. Filter used: plate-and-frame filter press
ii. Area of the plate-and-frame filter press, $A = 0.0485$ m^2
iii. Concentration of feed slurry, S: 24.58 kg solid/m^3
iv. Viscosity of feed slurry, $\mu = 1200$ cP $= 1.2$ kg/m s
v. Pressure = 7 psia
vi. Temperature = 30°C
vii. Experimental data: V versus t

To find: α and β

Solution

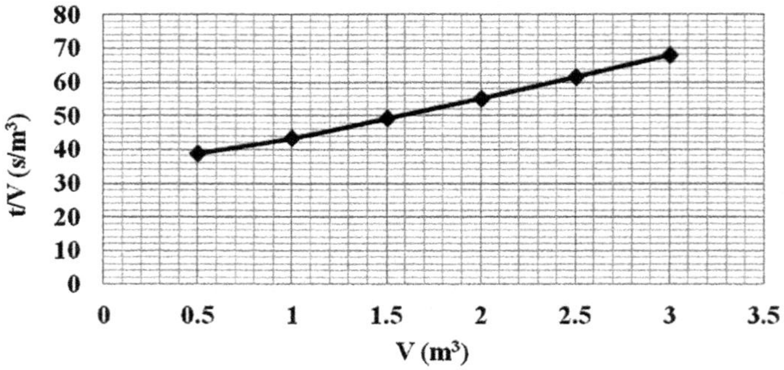

Slope of plot between t/V and $V = 11.729$, which is equal to K_p

Intercept of plot between t/V and $V = 32.25$, which is equal to β

Therefore, $\alpha = \dfrac{K_p 2A^2 \Delta P}{\mu S} = \dfrac{11.729 \times 2 \times 0.0485^2 \times 48263.3}{1.2 \times 24.58} = 90.288\,\text{m/kg}$

(7 psia = 48263.3 Pa)

Answer: α = 90.288 m/kg; β = 32.25 s/m³

8. For the data given in problem no. 7, if the plate-and-frame filter press used has an area of 0.0775 m², calculate the time required to obtain 2.4 m³ of filtrate.

Given

i. Area = 0.0775 m²
ii. Volume of the filtrate = 2.4 m³

To find: Time required for obtaining 2.4 m³ of filtrate.

Solution

$$\Delta P = \frac{\alpha\mu}{tA^2}\left(SV^2\right)$$

$$t = \frac{\alpha\mu SV^2}{\Delta PA^2} = \frac{90.288 \times 1.2 \times 24.58 \times 2.4^2}{48263.3\ (0.0775)^2} = 52.9$$

Answer: Time required to obtain 2.4 m³ of filtrate = 52.9 s

9. A centrifuge has an outlet discharge radius of 52.6 mm and an outlet radius of 78.5 mm. The density of product centrifuged is 1032 kg/m³, and that of the separated product is 865 kg/m³. Calculate the radius of neutral zone.

Given

i. Outlet radius (r_1) = 78.5 mm
ii. Outlet discharge radius (r_2) = 52.6 mm
iii. Density of product centrifuged (ρ_A) = 1032 kg/m³
iv. Density of product separated (ρ_B) = 865 kg/m³

To find: Radius of neutral zone

Solution

$$r_n^2 = \frac{\rho_A r_1^2 - \rho_B r_2^2}{(\rho_A - \rho_B)} = \frac{1032(78.5\times10^{-3})^2 - 865(52.6\times10^{-3})^2}{(1032-865)} = 0.024$$

$$\therefore r_n = 0.154\,\text{m}$$

Answer: Radius of the neutral zone = 0.154 m

10. Calculate the number of "g" that can be obtained in a centrifuge that can centrifuge a liquid at 1500 rpm at a maximum radius of 10 cm.

Given

i. $N = 1500\,\text{rpm}$

ii. $r = 10\,\text{cm}$

To find: Number of "g" that can be obtained in the centrifuge.

Solution

$$F_C = mr\left(\frac{2\pi N}{60}\right)^2 = 0.011mrN^2$$

$$F_g = mg$$

$$\frac{F_C}{F_g} = \frac{0.011\,rN^2}{g} = \frac{0.011\times(10\times10^{-2})\times 1500^2}{9.8} = 252$$

Answer: 252 g

BIBLIOGRAPHY

Adjalle, K. D., Brar, S. K., Verma, M., Tyagi, R. D., Valero, J. R. and Surampalli, R. Y. 2007. Ultrafiltration recovery of entomotoxicity from supernatant of *Bacillus thuringiensis* fermented wastewater and wastewater sludge. *Process Biochemistry* 42: 1302–1311.

Albertsson, P. Å. 1986. *Partition of Cell Particles and Macromolecules* (Third edition). New York: Wiley Interscience.

Anandharamakrishnan, C., Raghavendra, S. N., Barhate, R. S., Hanumesh, U. and Raghavarao, K. S. M. S. 2005. Aqueous two-phase extraction for recovery of proteins from cheese whey. *Food and Bioproducts Processing* 83: 191–197.

Anlauf, H. 2007. Recent developments in centrifuge technology. *Separation and Purification Technology* 58: 242–246.

Arakawa, T., Carpenter, J. F., Kita, Y. A. and Crowe, J. H. 1990. The basis for toxicity of certain cryoprotectants: a hypothesis. *Cryobiology* 27: 401–415.

Asenjo, J. A. and Andrews, B. A. 2011. Aqueous two-phase systems for protein separation: a perspective. *Journal of Chromatography A* 1218: 8826–8835.

Asenjo, J. A. and Andrews, B. A. 2012. Aqueous two-phase systems for protein separation: phase separation and applications. *Journal of Chromatography A* 1238: 1–10.

Auger, F., Delaplace, G., Bouvier, L., Redl, A., André, C. and Morel, M. H. 2013. Hydrodynamics of a planetary mixer used for dough process: influence of impeller speeds ratio on the power dissipated for Newtonian fluids. *Journal of Food Engineering* 118: 350–357.

Bai, L., Huan, S., Gu, J. and McClements, D. J. 2016. Fabrication of oil-in-water nanoemulsions by dual-channel microfluidization using natural emulsifiers: saponins, phospholipids, proteins, and polysaccharides. *Food Hydrocolloids* 61: 703–711.

Bamberger, S., Seaman, G. V. F., Sharp, K. A. and Brooks, D. E. 1984. The effects of salts of the interfacial tension of aqueous dextran poly(ethylene glycol) phase systems. *Journal of Colloid and Interface Science* 99: 194–200.

Barbosa-Canovas, G. V., Ortega-Rivas, E., Juliano, P. and Yan, H. 2005. *Food Powders: Physical Properties, Processing, and Functionality*. New York: Kluwer Academic/Plenum Publishers.

Bates, R. L., Fondy, P. L. and Corpstein, R. R. 1963. Examination of some geometric parameters of impeller power. *Industrial & Engineering Chemistry Process Design and Development* 2: 310–314.

Brennan, J. G. and Grandison, A. S. 2011. Separations in food processing part 1. In *Food Processing Handbook* (Second edition), eds. J. G. Brennan and A. S. Grandison. Weinheim: Wiley-VCH Verlag GmbH and Co.

Campbell, G. M. 2003. Bread aeration. In *Breadmaking: Improving Quality*, ed. S. P. Cauvain, 352–374. Cambridge, UK: Woodhead Publishing Ltd.

Campbell, G. M. 2009. A history of aerated foods. *Cereal Foods World* 54: 8–14. Adapted from a chapter in Bubbles in Food 2: Novelty, Health and Luxury published by AACC Intl., St. Paul, MN.

Carson, J. W., Royal, T. A. and Troxel, T. G. 1996. Mix dry bulk solids properly and maintain blend integrity. *Chemical Engineering Progress* 92: 72–80.

Cauvain, S. 2000. Breadmaking: an overview. In *Bread making Improving Quality*, ed. S. P. Cauvain, 8–28. Boca Raton, FL; Cambridge, UK: CRC Press; Woodhead Publishing Limited.

Cauvain, S. P., Whitworth, M. B. and Alava, J. M. 1999. The evolution of bubble structure in bread doughs and its effect on bread structure. In *Bubbles in Food*, eds. G. M. Campbell, C. Webb, S. S. Pandiella and K. Niranjan, 85–88. St. Paul, MN: Eagan Press.

Cauvain, S. P. and Young, L. S. 2006. *Baked Products: Science, Technology and Practice*. Oxford, UK: Blackwell Publishing Limited.

Chamberlain, N. and Collins, T. H. 1979. The chorleywood bread process: the role of oxygen and nitrogen. *Bakers Digest* 53: 18–24.

Dolar, D. and Košutić, K. 2013. Removal of pharmaceuticals by ultrafiltration (UF), nanofiltration (NF), and reverse osmosis (RO). In *Comprehensive Analytical Chemistry* (Volume 62), eds. M. Petrovic, D. Barcelo and S. Pérez, 319–344.Amsterdam, The Netherlands: Elsevier.

Elsayed, K. 2015. Optimization of the cyclone separator geometry for minimum pressure drop using Co-Kriging. *Powder Technology* 269: 409–424.

FLSmidth A/S. 2017. Krebs® cyclones for food & beverage processing industries. www.flsmidth.com/~/media/PDF%20Files/Krebs/KrebsHydrocyclonesforFoodandBeverageApplicationJAN62017.ashx (accessed September 8, 2018).

Geankoplis, C. J. 2003. *Transport Processes and Separation Process Principles (Includes Unit Operations)* (Fourth edition). New Delhi: Prentice Hall of India Private Limited.

Gordon, A. 2017. Case study: food safety and quality systems implementation in small beverage operations—Mountain Top Springs Limited. In *Food Safety and Quality Systems in Developing Countries* (Volume 2: Case studies of effective implementation), ed. A. Gordon, 81–116. London, UK: Elsevier Inc.

Grilo, A. L., Aires-Barros, R. M. and Azevedo, A. M. 2016. Partitioning in aqueous two-phase systems: fundamentals, applications and trends. *Separation & Purification Reviews* 45: 68–80.

Haedelt, J., Beckett, S. T. and Niranjan, K. 2007. Bubble-included chocolate: relating structure with sensory response. *Journal of Food Science* 72: B138–B142.

Haedelt, J., Leo Pyle, D., Beckett, S. and Niranjan, K. 2005. Vacuum-induced bubble formation in liquid-tempered chocolate. *Journal of Food Science* 70: E159–E164.

Hanselmann, W. and Windhab, E. J. 1998. Flow characteristics and modelling of foam generation in a continuous rotor/stator mixer. *Journal of Food Engineering* 38: 393–405.

Harnby, N. 1992. The selection of powder mixers. In *Mixing in the Process Industries*, eds. N. Harnby, M. F. Edwards and A. W. Nienow, 42–61. Oxford, UK: Butterworth-Heinemann.

Harris, E. L. and Angal, S. 1989. *Protein Purification Methods*. Oxford, UK: IRL Press at Oxford University Press.

Hatti-Kaul, R. 2000. *Aqueous Two-Phase Systems: Methods and Protocols*. Berlin: Springer Science & Business Media.

Heeb, T. G. and Mandralis, Z. I. 2000. Extraction product and process. US Patent 6165536.

Hitz, C. 2008. *Baking Artisan Bread.* Beverly, MAQuarry Books.

Hoffmeister, C. R. D., Fandaruff, C., da Costa, M. A., Cabral, L. M., Pitta, L. R., Bilatto, S. E. R., Prado, L. D., Corrêa, D. S., Tasso, L., Silva, M. A. S. and Rocha, H. V. A. 2017. Efavirenz dissolution enhancement III: colloid milling, pharmacokinetics and electronic tongue evaluation. *European Journal of Pharmaceutical Sciences* 99: 310–317.

Hwang, K. J. and Chiang, Y. C. 2014. Comparisons of membrane fouling and separation efficiency in protein/polysaccharide cross-flow microfiltration using membranes with different morphologies. *Separation and Purification Technology* 125: 74–82.

Iqbal, M., Tao, Y., Xie, S., Zhu, Y., Chen, D., Wang, X., Huang, L., Peng, D., Sattar, A., Abu Bakr Shabbir, M., Iftikhar Hussain, H., Ahmed, S. and Yuan, Z. 2016. Aqueous two-phase system (ATPS): an overview and advances in its applications. *Biological Procedures Online* 18: 18.

Ishwarya, S. P. 2017. Development of a combined experimental and computational modeling approach to investigate the influence of bran addition on the volume and structural development in bread. *PhD Thesis.* AcSIR, India.

Ishwarya, S. P., Desai, K. M., Naladala, S. and Anandharamakrishnan, C. 2017. Bran-induced effects on the evolution of bubbles and rheological properties in bread dough. *Journal of Texture Studies* 48: 415–426.

Jackson, A. T. and Lamb, J. 1981. *Calculations in Food Chemical Engineering*. London, UK: Macmillan.

Jacobo, A. S., Saldo, J. and Gervilla, R. 2014. Influence of high-pressure and ultra-high-pressure homogenization on antioxidants in fruit juice. In *Processing and Impact on Antioxidants in Beverages*, ed. V. Preedy, 185–193. Oxford, UK: Elsevier Inc.

Jafari, S. M., He, Y. and Bhandari, B. 2007. Optimization of nano-emulsions production by microfluidization. *European Food Research and Technology* 225: 733–741.

Kaialy, W. and Shafiee, M. A. 2016. Recent advances in the engineering of nanosized active pharmaceutical ingredients: promises and challenges. *Advances in Colloid and Interface Science* 228: 71–91.

Kusuma, H. S., Putra, A. F. P. and Mahfud, M. 2016. Comparison of two isolation methods for essential oils from orange peel (*Citrus auranticum* L) as a growth promoter for fish: microwave steam distillation and conventional steam distillation. *Journal of Aquaculture Research and Development* 7: 409.

Lopes, M. M. and Song, T. W. 2010. Batch distillation: better at constant or variable reflux? *Chemical Engineering and Processing: Process Intensification* 49: 1298–1304.

Mahbubul, I. M., Elcioglu, E. B., Saidur, R. and Amalina, M. A. 2017. Optimization of ultrasonication period for better dispersion and stability of TiO_2–water nanofluid. *Ultrasonics Sonochemistry* 37: 360–367.

Marriott, J. and Sorensen, E. 2003. A general approach to modelling membrane modules. *Chemical Engineering Science* 58: 4975–4990.

Marshall, K. R. 1982. Industrial isolation of milk proteins: whey proteins. In *Developments in Dairy Chemistry* (Volume 1), ed. P. F. Fox, 339–373. New York: Elsevier Applied Science Publishers Ltd.

Mayer-Laigle, C., Gatumel, C. and Berthiaux, H. 2015. Mixing dynamics for easy flowing powders in a lab scale Turbula® mixer. *Chemical Engineering Research and Design* 95: 248–261.

Mazzola, P. G., Lopes, A. M., Hasmann, F. A., Jozala, A. F., Penna, T. C., Magalhaes P. O., Rangel-Yagui, C. O. and Pessoa Jr, A. 2008. Liquid–liquid extraction of biomolecules: an overview and update of the main techniques. *Journal of Chemical Technology & Biotechnology* 83: 143–157.

McCabe, W. L., Smith, J. C. and Harriot, P. 1993. *Unit Operations of Chemical Engineering* (Fifth edition). New York: McGraw-Hill.

Metal Powder Report 2008. Rotary or double cone? Take your pick for quality. *Metal Powder Report* 63: 8–10.

Mills, E. N. C., Wilde, P. J., Salt, L. J. and Skeggs, P. 2003. Bubble formation and stabilization in bread dough. *Food and Bioproducts Processing* 81: 189–193.

Miyanami, K. 2006. Mixing. In *Powder Technology Handbook* (Third edition), eds. H. Masuda, K. Higashitani and H. Yoshida, 577–590. Boca Raton, FL: CRC Press, Taylor & Francis Group.

Murthy, Y. R. and Bhaskar, K. U. 2012. Parametric CFD studies on hydrocyclone. *Powder Technology* 230: 36–47.

Muzzio, F. J., Llusa, M., Goodridge, C. L., Duong, N.-H. and Shen, E. 2008. Evaluating the mixing performance of a ribbon blender. *Powder Technology* 186: 247–254.

Nam, H., Choi, J. and Capareda, S. C. 2016. Comparative study of vacuum and fractional distillation using pyrolytic microalgae (Nannochloropsis oculata) bio-oil. *Algal Research* 17: 87–96.

Ornebro, J., Nylander, T. and Eliasson, A. C. 2000. Interfacial behaviour of wheat proteins. *Journal of Cereal Science* 31: 195–221.

Ortega-Rivas, E. and Perez-Vega, S. B. 2011. Solid-liquid separations in the food industry: operating aspects and relevant applications. *Journal of Food and Nutrition Research* 50: 86–105.

Park, J. and Lee, K. S. 2017. A two-dimensional model for the spiral wound reverse osmosis membrane module. *Desalination* 416: 157–165.

Phong, W. N., Show, P. L., Chow, Y. H. and Ling, T. C. 2018 Recovery of biotechnological products using aqueous two phase systems. *Journal of Bioscience and Bioengineering* 126: 273–281.

Pouliot, Y. 2008. Membrane processes in dairy technology—From a simple idea to worldwide panacea. *International Dairy Journal* 18: 735–740.

Prins, A., 1988. Principles of foam stability. In *Advances in Food Emulsions and Foams*, eds. E. Dickinson and G. Stainsby, 91–122. London, UK: Elsevier Applied Science.

Prins, M. and Boeke, W. 2015. Equipment. In *Practical Pharmaceutics. An International Guideline for the Preparation, Care and Use of Medicinal Products*, eds. Y. Bouwman-Boer, V. Fenton-May and P. L. Brun, 609–649. Basel, Switzerland: KNMP and Springer International Publishing.

Rabone, P. 2007. *Level 2 Certificate in Professional Cookery*. Oxford, UK: City and Guilds and Harcourt Education Ltd.

Raja, S., Murty, V. R., Thivaharan, V., Rajasekar, V. and Ramesh, V. 2012. Aqueous two phase systems for the recovery of biomolecules-a review. *Science and Technology* 1: 7–16.

Sahu, J. K. and Niranjan, K. 2009. Gas-liquid mixing. In *Food Mixing: Principles and Applications*, ed. P. J. Cullen, 230–252. Chichester, West Sussex, UK: John Wiley and Sons Ltd.

Salamanca, M., Merchuk, J., Andrews, B. and Asenjo, J. 1998. On the kinetics of phase separation in aqueous two-phase systems. *Journal of Chromatography B* 711: 319–329.

Sampaio, D. D. A., Mafra, L. I., Yamamoto, C. I., Andrade, E. F. D., de Souza, M. O., Mafra, M. R. and Castilhos, F. D. 2016. Aqueous two-phase (polyethylene glycol + sodium sulfate) system for caffeine extraction: equilibrium diagrams and partitioning study. *The Journal of Chemical Thermodynamics* 98: 86–94.

Scanlon, M. G. and Zghal, M. C. 2001. Bread properties and crumb structure. *Food Research International* 10: 841–864.

Schultz, J. and Nardin, M. 1992. Determination of the surface free energy of solids by the two-liquid-phase method. In *Modern Approaches to Wettability: Theory and Applications*, eds. M. E. Schrader and L. George, 76–77. New York: Plenum Press.

The Scotch Whisky Association. 2017. Questions and Answers. Edinburgh: The Scotch Whisky Association. www.scotch-whisky.org.uk/ (accessed June 10, 2018).

Shao, P., Darcovich, K., McCracken, T., Ordorica-Garcia, G., Reith, M. and O'Leary, S. 2015. Algae-dewatering using rotary drum vacuum filters: process modeling, simulation and techno-economics. *Chemical Engineering Journal* 268: 67–75.

Shekhawat, L. K., Sarkar, J., Gupta, R., Hadpe, S. and Rathore, A. S. 2018. Application of CFD in bioprocessing: separation of mammalian cells using disc stack centrifuge during production of biotherapeutics. *Journal of Biotechnology* 267: 1–11.

Silva, M. E. and Franco, T. T. 2000. Liquid–liquid extraction of biomolecules in downstream processing—A review paper. *Brazilian Journal of Chemical Engineering* 17: 1–17.

Smith, P. 2011. *Introduction to Food Process Engineering*. Boston, MA: Springer. Food Science Text Series.

Sparks, T. 2012. Solid-liquid filtration: understanding filter presses and belt filters. *Filtration + Separation* 49: 20–24.

Spitz, L. 2016. Bar soap finishing. In *Soap Manufacturing Technology* (Second edition), ed. L. Spitz, 167–202. London, UK: AOCS Press.

Tedesco, M., Hamelers, H. V. M. and Biesheuvel, P. M. 2016. Nernst–Planck transport theory for (reverse) electrodialysis: I. Effect of co-ion transport through the membranes. *Journal of Membrane Science* 510: 370–381.

Trinh, L. 2013. Gas cells in bread dough. *PhD Thesis*. The University of Manchester.

Trinh, L., Lowe, T., Campbell, G. M., Withers, P. J. and Martin, P. J. 2013. Bread dough aeration dynamics during pressure step-change mixing: studied by X-ray tomography, dough density and population balance modelling. *Chemical Engineering Science* 101: 470–477.

Uhl, V. W. and Gray, J. B. 1986. *Mixing* (Volumes 1–3). Orlando, FL: Academic Press.

Vieira, L. G. M., Barbosa, E. A., Damasceno, J. J. R. and Barrozo, M. A. S. 2005. Performance analysis and design of filtering hydrocyclones. *Brazilian Journal of Chemical Engineering* 22: 143–152.

Walter, H. and Johansson, G. 1994. *Aqueous Two-Phase Systems*. Berlin: Elsevier.

Wiedemann, L., Conti, F., Janus, T., Sonnleitner, M., Zörner, W. and Goldbrunner, M. 2017. Mixing in biogas digesters and development of an artificial substrate for laboratory-scale mixing optimization. *Chemical Engineering and Technology* 40: 238–247.

Wu, Y. T., Zhu, Z. Q. and Mei, L. H. 1996. Interfacial tension of poly(ethylene glycol) salt + water systems. *Journal of Chemical & Engineering Data* 41: 1032–1035.

Zafarani-Moattar, M. T., Hamzehzadeh, S. and Nasiri, S. 2011. A new aqueous biphasic system containing polypropylene glycol and a water-miscible ionic liquid. *Biotechnology Progress* 28: 146–156.

13

Thermal Processing of Foods

Thermal processing is a method of food preservation that involves the controlled use of heat (heating at a specified temperature for a defined time interval) to inhibit or inactivate the vegetative microbes and degradative enzymes and destroy the dormant microbial spores. Heating of solid foods is achieved by conduction, and the distribution of heat within liquid foods is brought about by convection. Apart from extending the shelf life, thermal processing also improves the palatability of food products, destroys antinutritional components, and improves the availability of some nutrients. This chapter will explain the different types of thermal processing techniques, their principle, and intended purpose. Further, the different methods of thermal processing calculations will be elaborated in the latter parts of the chapter.

13.1 Classification of Thermal Processing Techniques

Based on the amount and severity of heat applied to the food product, thermal processing techniques can be classified into two types: mild heat treatment and severe heat treatment (Figure 13.1). Pasteurization and blanching are the two important types of mild heat treatment, and sterilization is the example of severe heat treatment. Depending on the intended objective of the preservation process, a food processor can choose between the mild and severe heat treatments. The mild heat treatment methods are used in conjunction with other preserving factors or processes such as refrigeration and freezing. Whereas, the severe heat treatment processes are used as the sole preservation technique (Stoforos, 1995). Mild heat treatment causes a minimal damage to the flavor, texture, and nutritional quality of foods. On the contrary, severe heat treatment leads to major changes in the organoleptic and nutritional quality of the food product.

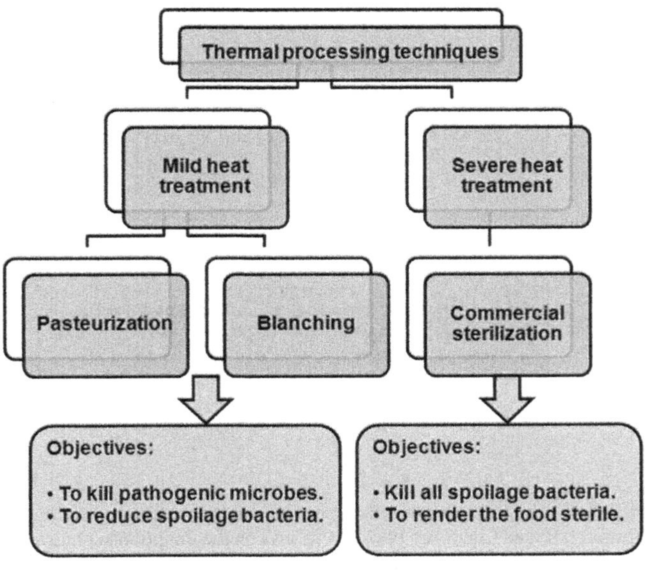

FIGURE 13.1 Classification of thermal processing techniques.

The choice between mild and severe heat treatments and subsequent implementation of the selected thermal processing technique depend on the following factors (Lund, 1975; Fellows, 1988; Awuah et al., 2007):

i. **Type of food product**

According to the Good Manufacturing Practice (GMP) regulations provided by the Food and Drug Administration (FDA), food products can be classified into four types: pH controlled foods, *F*-value and hermetically sealed controlled foods, pH and water activity (a_w) controlled foods and a_w controlled foods (Figure 13.2) (Johnston and Lin, 1987). The criteria for categorization include pH and a_w, with their critical limits at 4.6 and 0.85, respectively. Low-acid foods have pH > 4.6, whereas high-acid foods have pH < 4.6. Similarly, foods with $a_w > 0.85$ are classified as high-moisture foods, and those with $a_w < 0.85$ are designated as low-moisture foods. The purpose and extent of heat treatment during thermal pasteurization vary with the pH of the food product. In low-acid foods, the heat treatment is intended to inactivate the pathogenic (disease-causing) bacteria; whereas, in high-acid foods, pasteurization aims at the destruction of spoilage microorganisms and inactivation of enzymes. Low-acid foods with high moisture content packaged in hermetically sealed containers must be subjected to severe heat treatment (retorting) or a combination of mild heat treatment (pasteurization) and acidification (Rahman, 2015).

ii. **Heat resistance of the target microorganism and enzyme**

Pasteurization temperature is determined by the heat resistance of the most heat-resistant microorganism or enzyme in the food product. The heat resistance of a microorganism is given by its *D* value or decimal reduction value. *D* value is defined as the time in minutes at a given temperature required to destroy 1 log or 90% of the population of the target microorganism. Accordingly, the inactivation of *Mycobacterium tuberculosis*, a pathogenic organism having the highest thermal resistance, is considered as the marker to evaluate the effectiveness of milk pasteurization. Thus, the temperature for milk pasteurization is decided based on the *D* value of this bacillus, which is 0.2–0.3 minute at 65.6°C, denoted as $D_{65.6°C}$ (Stumbo, 1973). The type and composition of food influence the heat resistance of microorganisms. Based on the optimal (T_{opt}) and maximum (T_{max}) growth temperatures, microorganisms can be classified into psychrophilic (T_{opt}: 12°C–15°C; T_{max}: 20°C), psychrotrophic (T_{opt}: 20°C–30°C; T_{max}: 0°C), mesophilic (T_{opt}: 30°C–42°C; T_{max}: 15°C–47°C), and thermophilic (T_{opt}: 55°C–65°C; T_{max}: 40°C–90°C). A more severe process may be required in the case of food products in which

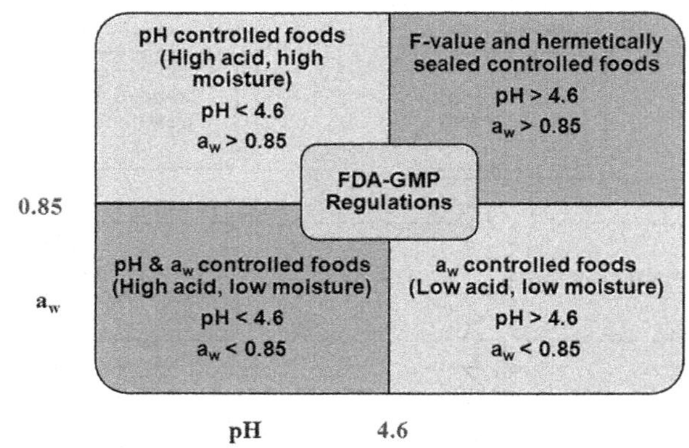

FIGURE 13.2 FDA-GMP regulations governing the selection of thermal processing treatment. (Modified and Reproduced with permission from Johnston, I. R. and Lin, R. C. 1987. FDA views on the importance of a_w in good manufacturing practice. In *Water Activity: Theory and Applications to Food*, eds. L. B. Rockland and L. R. Beuchat, 287. New York: Dekker.)

thermophilic spoilage could be a concern because of very high heat resistance of the thermophilic microorganism or spores.

Alkaline phosphatase is a naturally occurring enzyme in the raw milk, which has a *D* value similar to that of most heat-resistant pathogens. Hence, phosphatase activity is commonly used as the marker to determine the adequacy of milk pasteurization. Similar to alkaline phosphatase in milk, certain enzymes in fruits and vegetables also indicate the efficiency of the pasteurization process. The examples include peroxidase (POD), pectin methylesterase, polygalacturonase (PG), and polyphenol oxidase (PPO). Understanding the time–temperature relationships of the marker enzymes in food products is important for the optimization of thermal pasteurization processes for commercial applications.

iii. **Initial microbial load in the raw material**

If the initial microbial load in the food product is high, the heating time required for the thermal processing of the product should be increased accordingly to reduce the microbial population down to an acceptable level.

iv. **Mode of heat transfer and heat transfer properties of the food**

By increasing the rate of heat transfer, the time required for thermal processing can be considerably reduced. This can be achieved by agitating the containers to generate circulating eddies in the food product inside (Potter and Hotchkiss, 1998). If the physical state of a food formulation is altered such that the mechanism of heat transfer changes from convection (as in liquid foods) to conduction (as in solid foods), the processing time should also be altered to account for the change in the mode of heat transfer. Else, the food would be under-processed and lead to a potential health hazard with respect to botulism.

v. **Storage conditions of the food product after thermal processing**

Control of storage temperature and ensuring a hermetic seal of the container are important to prevent recontamination of the treated food after thermal processing and during storage.

13.2 Pasteurization

13.2.1 Definition

Pasteurization acquired its name from Louis Pasteur, a French scientist who invented the process of heating liquids (wine and beer) at a relatively mild temperature (about 55°C) for a short time to prevent spoilage (Silva et al., 2014; Wilbey, 2014). Pasteurization is defined by the International Dairy Federation (IDF, 1986) as

> A process applied to a product with the objective of minimizing possible health hazards arising from pathogenic microorganisms associated with the product which is consistent with minimal chemical, physical, and organoleptic changes in the product.

An updated definition of pasteurization was provided by the National Advisory Committee on Microbiological Criteria for Foods (NACMCF, 2006) of the United States as

> Any process, treatment, or combination thereof that is applied to food to reduce the most resistant microorganism(s) of public health significance to a level that is not likely to present a public health risk under normal conditions of distribution and storage.

13.2.2 Classification of the Pasteurization Process

Pasteurization is classified based on two criteria: the mode of operation and the temperature–time combination. Based on the mode of operation, pasteurization can be classified into the batch and continuous processes.

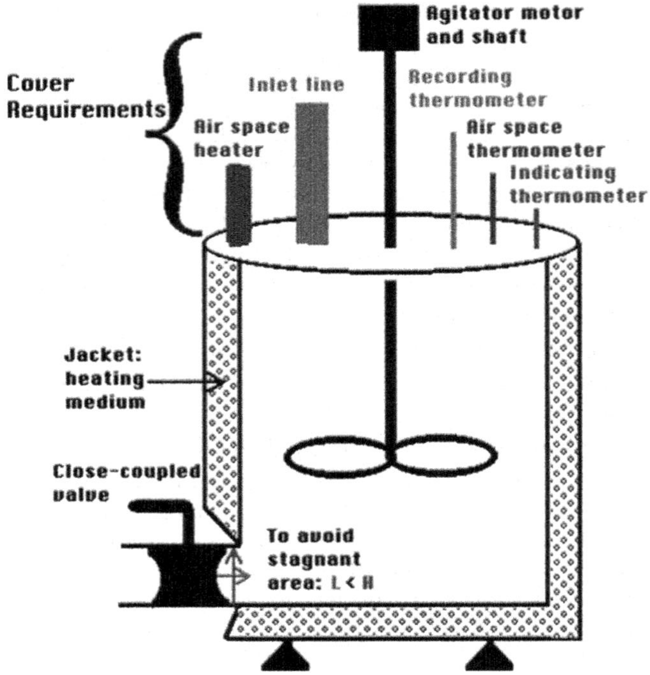

FIGURE 13.3 Schematic of a batch pasteurizer. (Reproduced with permission from Goff, H. D. 2018. *Dairy Science and Technology Education Series*. University of Guelph, Canada. www.uoguelph.ca/foodscience/book-page/batch-method.)

13.2.2.1 Batch Pasteurization

In the batch pasteurization process, a large quantity of the liquid product is taken in a jacketed stainless steel vessel, termed as a *vat*. The vat comprises an inner vessel, surrounded by an insulated outer casting that forms a jacket through which the heating or cooling medium is circulated (Figure 13.3). Thus, during the batch pasteurization process, the product is heated and cooled in the same vat. Hot water or steam accomplishes heating of product to the specific temperature. Alternatively, heating coils of water or steam are also used. An agitator is mounted onto the vessel to facilitate uniform heat transfer and avoid localized overheating of the product. While being agitated, the product is heated for a defined holding period. The liquid product is cooled in the vat by the refrigerant or chilled water circulated through the outer jacket. Instead, the hot product can be removed from the vat and rapidly cooled in a plate or surface chiller. Batch or vat pasteurization is used extensively in the ice cream industry.

The limitations of a batch pasteurizer system are the requirements for a higher amount of heat energy and longer cooling time. Also, heat regeneration is not possible in a batch pasteurizer due to the considerable loss of latent heat through the heated product. Consequently, the thermal efficiency of a batch pasteurization process is low.

13.2.2.2 Continuous-Flow Pasteurization

In the continuous-flow pasteurization process, the liquid food is typically treated in a plate heat exchanger (PHE) using hot water or steam as the heating medium. Continuous-flow pasteurization is appropriate for the bulk heat treatment of pumpable products such as liquids, puree, semiliquids, and suspensions with small particles. A typical unit for continuous-flow pasteurization includes the following components (Figure 13.4):

1. Balance tank
2. PHE

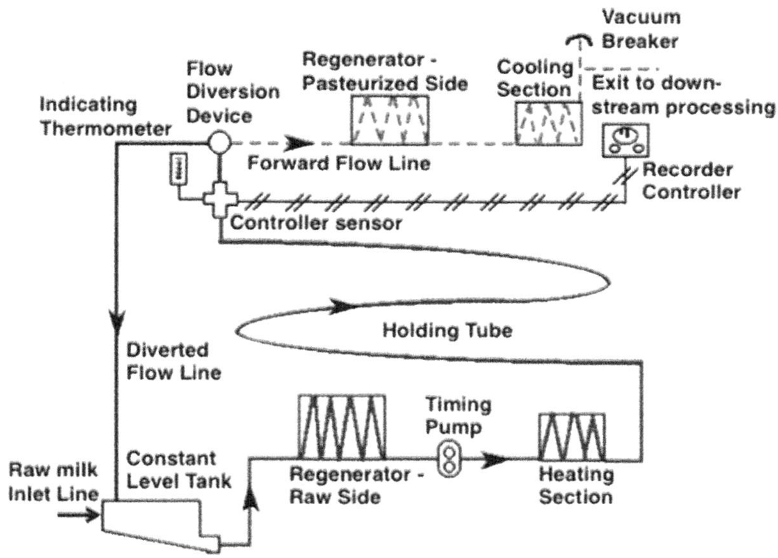

FIGURE 13.4 Schematic of a continuous-flow pasteurization system. (Reproduced with permission from Goff, H. D. 2018. *Dairy Science and Technology Education Series*. University of Guelph, Canada. www.uoguelph.ca/foodscience/book-page/htst-milk-flow-overview.)

3. Holding tube
4. Timing pump
5. Indicating thermometer
6. Flow diversion valve (FDV)
7. Vacuum breaker

1. **Balance tank**

   The balance tank provides a constant supply of the liquid product (Figure 13.5). It is equipped with a float valve assembly that maintains a constant liquid level. In order to maintain a higher pressure on the pasteurized product side of the heat exchanger, the level of overflow must always be below the level of lowest product passage in the regenerator. The balance tank prevents the air from entering the pasteurizer by the placement of the top of the outlet pipe at a position lower than the lowest point in the tank. It facilitates the recirculation of diverted or pasteurized product.

2. **Plate heat exchanger**

   Heating of food product to the defined pasteurization temperature is accomplished within the PHE, the construction of which was explained in Chapter 5. In a PHE, pasteurization temperatures are maintained at relatively higher values for viscous products, so as to overcome the difficulties in heat transfer through the viscous mass.

3. **Holding tube**

   The holding tube is a key component of the continuous-flow pasteurizer system. The function of a holding tube is to provide the required residence time to the product at the defined high temperature of pasteurization. According to the FDA, the lethality accumulated in the holding section is considered important (Dignan et al., 1989). The flow of product in the holding tube is controlled such that the residence time is just above the heating time required to achieve lethality. The required product flow rate in the holding section is maintained by a timing pump or metering pump, located upstream. The timing pump is a positive displacement pump or a centrifugal pump, which drives the product through the raw material side of the regenerator

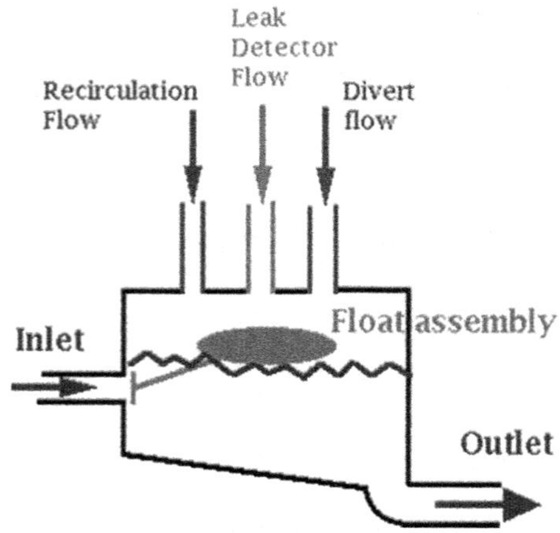

FIGURE 13.5 Balance tank of a continuous-flow pasteurizer. (Reproduced with permission from Goff, H. D. 2018. *Dairy Science and Technology Education Series*. University of Guelph, Canada. www.uoguelph.ca/foodscience/book-page/balance-tank.)

and passes it to the pasteurized product side of the regeneration section under pressure. While flowing under pressure within the holding tube, the product is maintained at the specific high temperature for the required time. Two metering pumps are used in the continuous-flow pasteurizer system to ensure pressure difference and thereby avoid the contamination between the product streams that are pasteurized and unpasteurized. To ensure an infallible operation, the timing pump operates only when the FDV is fully forward or fully diverted, wherein FDV is a device that is installed downstream of the holding tube. FDV functions to divert the unpasteurized (inadequately heated) product back to the heating section. It is a three-way valve, which, at temperatures greater than the set temperature, opens to forward flow. At temperatures less than the set temperature, the valve recloses to the normal position and redirects the product back to the balance tank. The aforementioned mechanism of FDV is activated by a temperature-actuated sensor located at the exit of the holding tube, which works based on the decision tree shown in Figure 13.6. Thus, FDV operates on the measured temperature, but not the time, at the end of the holding period. It ensures that the product which has not received the intended temperature–time treatment does not reach the packaging section.

There are two types of FDV:

i. *Single-stem FDV*: This is an older valve system with the disadvantage that it cannot be cleaned in place (CIP) (Figure 13.7a).
ii. *Dual-stem FDV*: This type of FDV is a foolproof system that consists of two valves in series (Figure 13.7b). Dual-stem FDV is more appropriate for automation and can be cleaned-in-place.

4. **Safety thermal limit recorder**

   The safety thermal limit recorder (STLR) monitors, controls, and records the position of the FDV and supplies power to the FDV during the forward flow. STLR controllers can be of either pneumatic or electronic type.

5. **Indicating thermometer**

   The indicating thermometer provides the most accurate temperature reading to which the STLR is adjusted. The probe of the thermometer should be located close to the probe of STLR at a distance not greater than 18-inch upstream of the FDV.

6. **Vacuum breaker**

   A vacuum breaker that breaks to the atmospheric pressure is installed at the discharge end of the pasteurized product. It must be located at a point greater than 12 inch above the highest

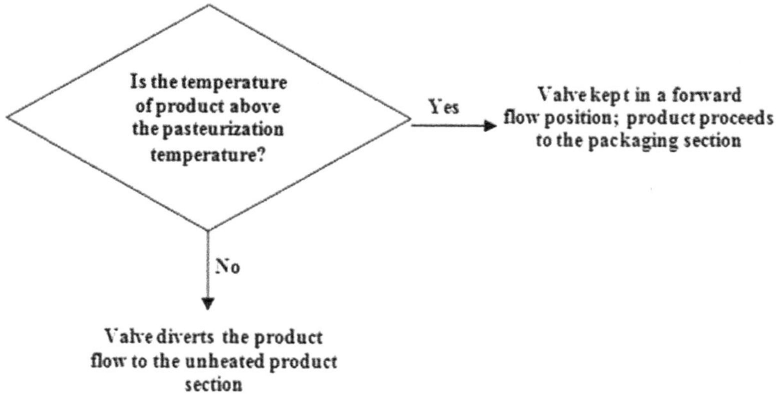

FIGURE 13.6 The operational algorithm of a FDV.

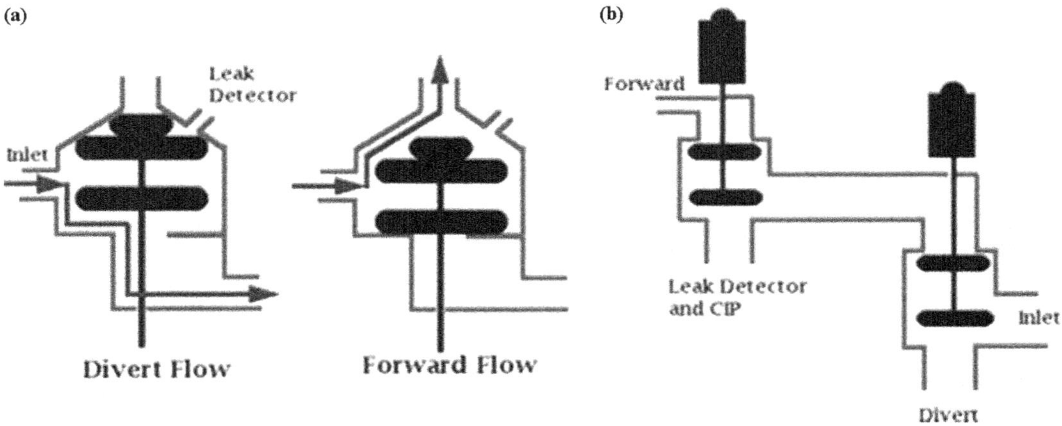

FIGURE 13.7 (a) Single-stem FDV and (b) dual-stem FDV. (Reproduced with permission from Goff, H. D. 2018. *Dairy Science and Technology Education Series*. University of Guelph, Canada. www.uoguelph.ca/foodscience/book-page/flow-diversion-device-fdd.)

point of raw product in the system. It ensures that suction is not created downstream of the pasteurized product side.

The pasteurized food product is cooled followed by hermetic sealing and packing. The product is then stored under refrigeration until transportation before retailing. Instead, the pasteurized food products can also be hot-filled and sealed, followed by cooling with water at ambient temperature. Hot filling is adopted for viscous products that behave as free-flowing liquids at higher temperatures.

Based on the time–temperature combination, pasteurization can be classified into many types, namely, low-temperature–long-time (LTLT) pasteurization, high-temperature–short-time (HTST) pasteurization, higher heat–shorter time (HHST) pasteurization, ultra-pasteurization (UP), and ultrahigh-temperature (UHT) pasteurization. Table 13.1 summarizes the details on the different types of pasteurization treatments based on the time–temperature combination.

TABLE 13.1

Classification of Pasteurization Based on Temperature–Time Combination

Pasteurization Type	Temperature	Time	Food Product	Description
LTLT or vat pasteurization	63°C	30 min	Milk	• Best conducted in a batch mode. • Kills all pathogenic bacteria and reduces the load of spoilage bacteria in the food products. • Preserves the physicochemical properties of the product. • Undesirable quality changes occur due to the prolonged heat treatment.
	69°C	30 min	Viscous products or products with more than 10% fat or added sweetener; eggnog (a sweetened, dairy-based beverage prepared from milk, cream, sugar, whipped egg whites, and egg yolks)	
HTST pasteurization or flash pasteurization	72°C	15 s	Liquid milk	• Conducted in a continuous mode. • Kills vegetative pathogenic and spoilage bacteria. • HTST treatment also causes the denaturation of protein in milk. • Treated products do not need refrigeration during the storage period. • Preferred for products with heat-labile nutrients. • Accordingly, milk is pasteurized by the HTST method for the retention of heat-sensitive vitamins and to prevent the development of cooked flavor.
	80°C	15 s	Ice cream mix	
	88°C	15 s	Fruit juices	
	80°C 83°C	25 s 15 s	Viscous products or products with more than 10% fat or added sweetener; eggnog	
Higher heat shorter time (HHST) pasteurization	89°C 90°C 94°C 96°C 100°C	1 s 0.5 s 0.1 s 0.05 s 0.01 s	Milk	• Kills vegetative pathogens.
Ultra pasteurization (UP)	138°C	2 s	Milk, cream, and eggnog	• After heating, the product is aseptically placed in sterile conventional containers. • Ultra-pasteurized milk and cream when refrigerated will last at least 45 days. • This process does a minimal damage to the flavor.
Ultra-high temperature (UHT) pasteurization	138°C	2 s	Milk, cream	• Similar to ultra-pasteurization. • But the product (milk) then goes into a sterile five-layer laminated "box" or a metal can. • The product can be stored without refrigeration and has a shelf life of 6 months to a year.

13.2.2.2.1 Tunnel Pasteurization

Tunnel pasteurization is an alternative to flash pasteurization. However, it operates at a lower processing temperature (60°C) for a longer time than flash pasteurization. The arrangement of this process enables *in-package pasteurization*; that is, the pasteurization is carried out after the product is filled in bottles and sealed. Consequently, tunnel pasteurization is the primarily used technique for beer packed in cans or bottles. During the process, beer bottles pass through a tunnel that is divided into various temperature

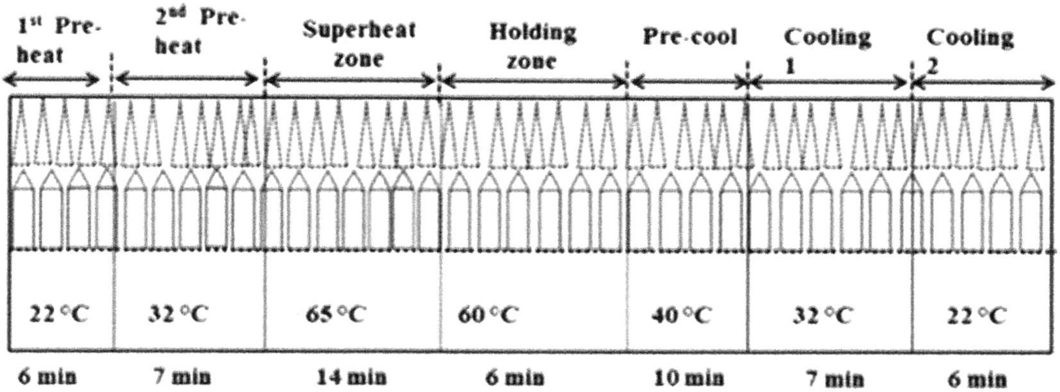

FIGURE 13.8 Schematic diagram of tunnel pasteurizer. (Reproduced with permission from Bhuvaneswari, E. and Anandharamakrishnan, C. 2014. Heat transfer analysis of pasteurization of bottled beer in a tunnel pasteurizer using computational fluid dynamics. *Innovative Food Science and Emerging Technologies* 23: 156–163.)

zones. The temperature within each zone is controlled by the temperature of water that is sprayed down onto the packages. The bottles remain in each temperature zone for the definite residence time.

Figure 13.8 depicts the schematic representation of a tunnel pasteurizer consisting of seven zones. The first two zones are the preheating zones where the temperature of the beer is gradually raised from 2°C to around 20°C. This avoids the risk of bottle breakage due to pressure buildup. Zone 3 is the superheating zone, where the water spray temperature is maintained at 65°C for 14 min and the temperature of the beer rises to 60°C. This zone is the most critical section as any lapse in temperature control here will lead to over-pasteurization and flavor deterioration in the beer. Zone 4 is the holding section, in which the temperature of both the water spray and the beer is maintained at 60°C for 6 min. Pasteurization of beer takes place in this holding section. The next three zones are the cooling sections where the temperature of the beer is reduced to 28°C. The bottles or cans move on a conveyor belt through the tunnel pasteurizer at a controlled speed. The tunnel pasteurization process is measured in "pasteurization unit" (PU), in which 1 PU denotes a treatment time of 1 min at 60°C. Thermal processing of beer is carried out in levels from 5 to 15 PU (Dilay et al., 2006; Buzrul, 2007).

13.3 Blanching

13.3.1 Definition, Principle, and Applications of Blanching

During freezing and dehydration, the maximum processing temperature is not high enough to inactivate enzymes. Similarly, with canning, the time taken to reach sterilizing temperatures may permit enzyme activity to occur, particularly with large cans. To overcome the aforementioned limitations, *blanching*, a mild thermal processing technique, is used for pretreating the vegetables or fruits before freezing, canning, and drying. Thus, the primary objectives of a blanching process are to inactivate specific deteriorative enzymes in vegetables or fruits that cause quality loss (development of off-flavors and off-colors) during frozen storage and to reduce the microbial load on the product surface. A few examples of the target enzymes for blanching include lipoxygenase (LOX), peroxidase (POD), polyphenol oxidase (PPO), polygalacturonase (PG), and chlorophyllase. Of the aforementioned enzymes, POD and LOX are the most heat-stable enzymes present in vegetables and thus used as markers to indicate the adequacy of heat treatment during blanching. It is established that if POD is destroyed, then it is quite unlikely that other enzymes survive such a heat treatment (Halpin and Lee, 1987).

The principle and application of blanching with respect to freezing, canning, and drying are presented as follows (Corcuera et al., 2004):

Freezing: The major purpose of blanching before freezing is to remove trapped air and metabolic gases within the vegetable cells and replace them with water. Consequently, a semicontinuous water phase is formed which favors the growth of more uniform ice crystals during the freezing process that follows. However, fruits are usually not blanched, or blanched at low temperature, before freezing because blanching produces undesirable changes in texture.

Canning: Gas removal is the major objective of blanching done before canning. Removal of air from the intracellular spaces permits easier filling of the cans by increasing the density of the product, lessens the strain on the can during heating, and reduces can corrosion. Enzyme inactivation can also take place during the blanching treatment (Downing, 1996).

Drying: Fruits and vegetables are blanched before drying to inactivate enzyme, which is not accomplished by drying when used individually. Blanching treatment also leads to an enhanced drying rate. The reasons for the blanching-mediated improvement in drying rate include the complex interaction between the following factors:

- Improved water permeability of the surface due to decreased hardening of the sample surface (Orikasa et al., 2008)
- Reduction of the internal resistance to moisture diffusion by changes in the microstructure caused by the physical damage to the product (Watanabe et al., 2014)
- Improved internal water permeability due to the heat stress-induced damage to the cell membranes (Ando et al., 2016)

Other applications of the blanching process are:

- *Peeling*: Blanching softens the tissues of fruits and vegetables that assist in the peeling of skin.
- *Brightening the color of green vegetables and fruits*: Blanching inactivates the chlorophyllase enzyme in green vegetables, which degrades the chlorophyll pigment and leads to discoloration. Blanching removes the dust on the surface of fruits and vegetables and thereby changes the wavelength of reflected light (Chamorro and Vidaurreta, 2012).
- *Reduction of surface contamination and microbial load*: Blanching is different from pasteurization in that it is applied to solid products such as vegetable and fruits. Similar to pasteurization, in the blanching process, food is subjected to thermal treatment at a preset temperature in the range of 70°C–100°C and held at this temperature for a specified duration ranging from seconds to minutes. Unsteady-state conduction and convection govern the heat transfer during blanching. Blanching is followed by a rapid cooling, which is accomplished by spraying cold water or blowing cold dry air, or by conveying them to a flume of cold water that transports the solid foods to the next process (drying, freezing, or canning).

13.3.2 Blanching Equipment

The design of equipment for blanching depends on the product to be treated, the process following it, and the intended use of the product (Corcuera et al., 2004). The blanching equipment can be classified into two major types: steam blanchers and hot water blanchers.

13.3.2.1 Steam Blancher

A steam blancher consists of a mesh conveyor belt to carry the product through a tunnel, which is 15 m long and 1–1.5 m wide where "food-grade" steam at approximately 100°C is directly injected. The holding time of the product at the steam temperature is controlled by the speed of the conveyor belt and length of the tunnel. The conveyor belt may be replaced with rotary valves (Fellows, 2000; Scott et al., 1981). The energy recovery from steam is accomplished using water sprays for condensation. Steam can also be reused by passing through venturi valves.

Steam blanching is advantageous in terms of minimal nutrient losses in the blanched product, plausibly due to less leaching effects. The lessened leaching of soluble solids also leads to increased retention of natural sugars, flavor, and color. However, uneven blanching may occur due to high piling up of product on the conveyor belt and high-temperature gradients between the surface and the center of the product. With wider temperature gradients, larger pieces of the product can be *over-blanched* near the surface and *under-blanched* at the center (Corcuera et al., 2004). Tedious cleaning procedures and complex equipment design are the other limitations of steam blanchers. Novel technologies that have emerged to improve the thermal efficiency of steam blanching are as follows:

- *Use of steam seals*: Steam is confined in the chamber using rotary locks or forced water to prevent evaporation to the atmosphere.
- *Insulation*: The steam chamber is completely insulated to maximize heat retention.
- *Forced convection*: A forced convection blancher uses positive and negative pressure areas within the steam chamber to force steam through the product depth, thus reducing the blanching time (Key Technology's Turbo-Flow™). These blanchers comprise nested chambers that allow recirculating steam with a fan that interconnects both chambers. The fan forces the flow of steam through a packed bed of product conveyed by a mesh belt. This technology allows higher product bed depths and greater product throughput. The steam that is not absorbed during the first pass is recirculated back to the product. Therefore, the forced convection blanchers reduce the amount of steam, save energy, improve product quality, and reduce waste generation by 80%. All of the above is accomplished through the recycling of steam (Parreño and Torres, 2012). These blanchers are appropriate for potatoes, other vegetable products and fruit, meat, seafood, poultry, and most canned/frozen fruit and vegetable products (Office of Industrial Technologies and Key Technology, Inc., 2002).
- *Heat-and-hold process*: In this process, only an adequate amount of steam is transferred to heat the surface of the product. The heat is then allowed to spread evenly throughout the product in a separate compartment. Consequently, the central core temperature is increased to the desired level of blanching without the requirement for additional steam. This technology shortens the blanching time and improves the retention of color, nutrient, and flavor (Anonymous, 2010). Decreased steam quantity reduces the energy cost and effluent volume (Parreño and Torres, 2012).
- *Individual quick blanching*: A single layer of product is conveyed through the steam chamber, and each *individual* piece of the product immediately enters in contact with the steam (Corcuera et al., 2004).

13.3.2.2 Hot Water Blancher

Hot water blanching is performed with water at temperatures in the range of 70°C–100°C. Hot water blanching can be carried out in a batch or continuous mode of operation. In the batch mode, a defined quantity of the solid food product is dipped in hot water at 90°C–99°C for the necessary time duration. A continuous hot water blancher (Figure 13.9) comprises the preheating, blanching, and cooling sections. The food remains on the conveyor belt through each stage. In the first section, that is, heating zone, the product is subjected to preheating with water that is recirculated through a heat exchanger, followed by blanching at the specified temperature. The hot water is obtained by indirect heating with steam in a heat exchanger. Therefore, unlike steam blanching, the steam quality need not be "food grade" for water blanching. Subsequently, in the cooling zone, a recirculation system cools the food. The same heat exchanger simultaneously heats and cools the preheating water and cooling water, respectively, leading to 70% of heat recovery. Other continuous hot water blanchers are of screw type, drum type, or pipe type. Although hot water blanchers are better than steam blanchers with respect to their lower capital cost, and uniform heating, their use is limited by the loss of water-soluble nutrients due to leaching of solids.

Apart from the hot water and steam blanchers, gas blanchers and microwave blanchers are also available. Hot gas blanching uses a combustion of flue gases with the addition of steam. This type of blanching reduces waste production but often results in weight loss of the product, which limits its industrial

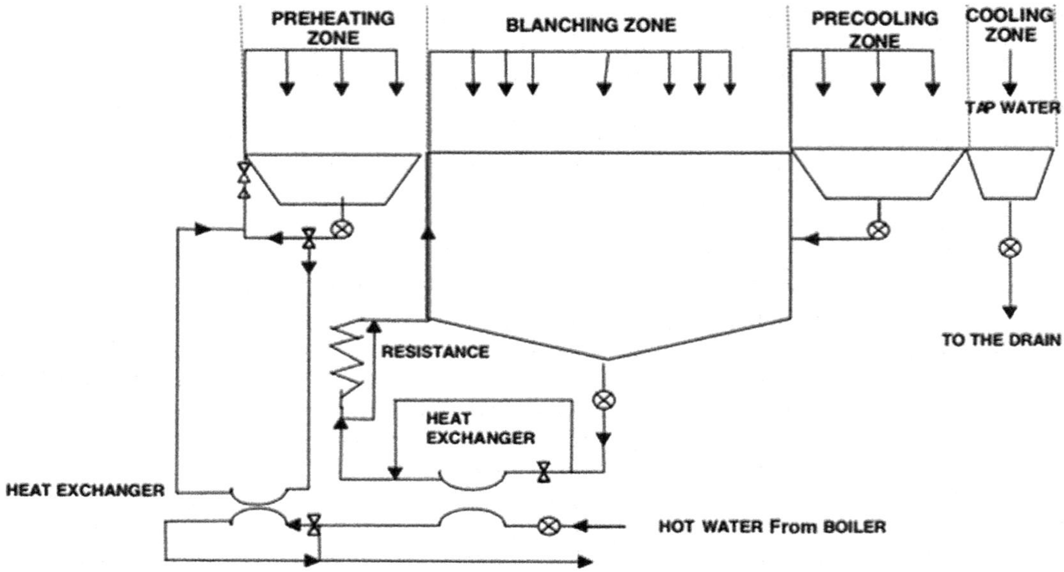

FIGURE 13.9 Hot water blancher. (Modified and Reproduced with permission from Arroqui, C., López, A., Esnoz, A. and Vírseda, P. 2003. Mathematic model of an integrated blancher/cooler. *Journal of Food Engineering* 59: 297–307.)

applications and demands further research (Downing, 1996). Microwave blanching is considered advantageous with respect to higher nutrient retention, improved product quality, and shorter processing times. However, its commercial application is not well established as the use of microwave ovens in the food industry is limited (Corcuera et al., 2004).

13.4 Commercial Sterilization

Commercial sterilization can be defined as an "intense thermal process in which a product is exposed to very high temperature (>100°C, the boiling point of water), at which all the pathogenic and toxin-forming organisms are destroyed, which if present can grow in the product and cause spoilage under normal handling and storage conditions" (Vaclavik and Christian, 2003). In general, the reference temperature of a commercial sterilization process is 121.1°C. *Clostridium botulinum* is the major target microorganism in the commercial sterilization process. This organism forms spores that are heat resistant and also produces an exotoxin named "botulin," which is lethal even at a low quantity of about 0.001 μg/kg body weight (parts per billion) in humans. Even at the end of severe heat treatment, some heat-resistant bacterial spores may be present in the sterilized product; however, it cannot grow under normal conditions. The shelf life of a commercially sterilized product is around 6 months to 2 years, depending on the product characteristics.

Commercial sterilization process can be classified into three categories: batch (still retort), continuous (crateless retort), and aseptic processing (Figure 13.10). The food product is subjected to *in-package* sterilization in the batch and continuous sterilization processes. In aseptic processing, the thermal treatment of product and container is carried out independently followed by the packaging of the product into containers in an aseptic environment.

13.4.1 Batch Retort System

The batch system for commercial sterilization process is generally referred to as a still retort (Figure 13.11). The retort consists of a cylindrical pressure vessel with a steam inlet, water drain, thermometer, pressure

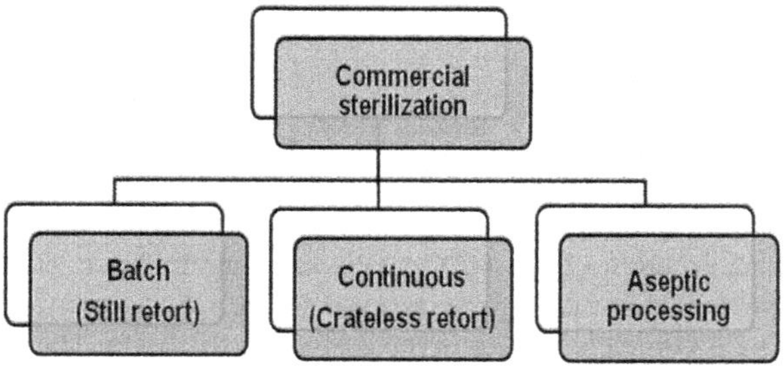

FIGURE 13.10 Classification of the commercial sterilization process.

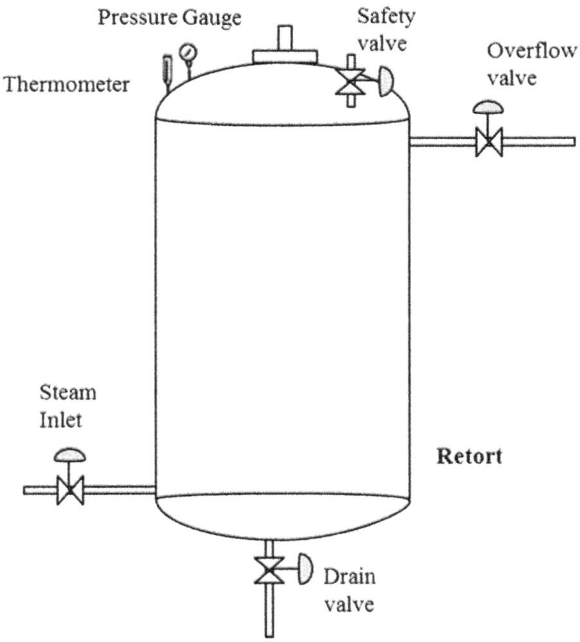

FIGURE 13.11 Batch retort.

gauge safety, and pressure relief valves. Initially, the retort is preheated to around 100°C, using steam. The containers or cans filled with the product are loaded into the crates and placed inside the retort. Steam enters the retort at a temperature (121°C), which is higher than the boiling point of water. The retort is then closed, and the pressure is allowed to build inside. Temperature and pressure inside the retort system are controlled by altering the amount of the inlet steam, with the aid of inlet valve. Monitoring of temperature and pressure inside the retort is done using the thermometer and pressure gauge, respectively. A safety valve is placed at the top of the retort to avoid buildup of excess pressure inside the retort vessel. The retort is maintained at the specified temperature for the desired time, the combination of which depends on the heat transfer properties of the container (can or glass bottles) and the product. On attaining the required process temperature and pressure, the steam valve is closed, and a mixture of cold water and air is introduced inside the retort. While air helps to maintain the pressure inside the retort and avoids the deformation of container shape, water leads to rapid cooling of the product to 40°C after which it is removed from the retort.

13.4.2 Continuous Retort System

13.4.2.1 Hydrostatic Retort System

Hydrostatic retort consists of the three columns (Figure 13.12): the preheating, sterilization, and cooling columns through which the cans or containers are passed to achieve the commercial sterility. Product container enters the retort through the preheating column containing hot water where the product temperature begins to increase gradually from 80°C at the entrance to reach 117°C–120°C at the exit of this column. Then, the product-filled cans are conveyed to the steam environment in the sterilization chamber, where the heating is completed to attain the desired sterilization. The height of water in the two columns maintains the required temperature and pressure of the steam. In the final chamber, the heated product is cooled by the cold water. The residence time of the container is controlled by the speed of the conveyor or chain.

13.4.2.2 Aseptic Processing

Aseptic processing is a HTST thermal treatment, which is defined as the processing and packaging of a commercially sterile product into sterilized containers followed by hermetic sealing with a sterilized closure in a manner that prevents viable microbiological recontamination of the sterile product (Betta et al., 2011). The containers are sterilized using hydrogen peroxide, ethylene oxide, superheated steam, ultraviolet (UV) radiations, and their combination thereof. Product sterilization is achieved by pumping it through a heat exchanger in which steam is used as the heating medium to increase the temperature up to 100°C. The product is then pumped through a holding tube to maintain it at the sterilization temperature for a defined residence time required to achieve the commercial sterility. Finally, the product is cooled in a heat exchanger using cold water as the cooling medium. Aseptic conditions of product filling are achieved by superheated steam and UV radiation. In general, plastic-coated paper cartons, plastic pouches, cans, and drums are used for the filling of aseptically processed products. The HTST processing facilitates the retention of product quality.

Liquid food products that are aseptically processed include fruit juices, milk, coffee creamers, purees, puddings, soups, baby foods, and cheese sauces (Graves, 1996). The applications of aseptic processing have also been tailored for high-viscosity foods and liquid products containing solid particles (Singh and Heldman, 2014). Factors of significance in the aseptic filling process are the type of product, type of package, attaining and maintaining a sterile environment for filling, and the sealing process (Mauer, 2003).

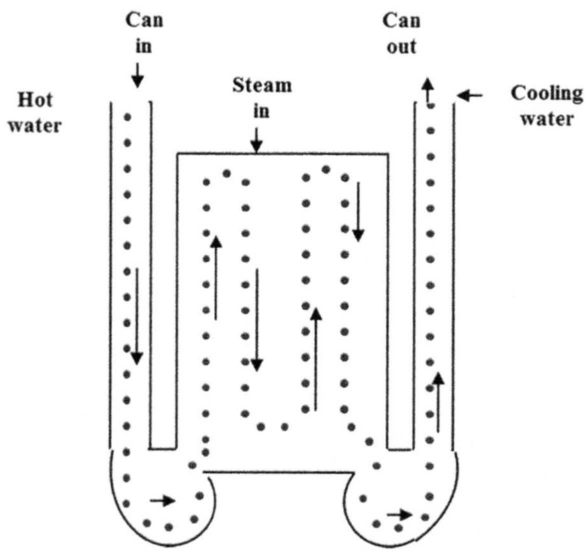

FIGURE 13.12 Hydrostatic retort system.

The advantages of aseptic processing encompass longer product shelf life of 1–2 years at ambient temperature, thus eliminating the need for refrigeration, lower energy consumption, wider packaging sizes, variety of choice for container materials, ease of automation, and improved nutritional and sensory properties. However, the limitations of aseptic processing are slower filler speeds, higher overall initial cost, need for better quality control of raw ingredients, the requirement for trained personnel, better control of process variables and equipment, and stringent validation procedures (Kumar and Sandeep, 2014).

13.5 Thermal Process Calculations

Process calculation is an essential element in the design of a safe thermal process. Understanding the process calculation method and the governing conditions would facilitate the selection of the appropriate thermal processing technique for food products. The building blocks of thermal process calculations are depicted in Figure 13.13.

13.5.1 The Microbial Survivor Curve and *D* value

During thermal processing, heat is the external agent that is used to reduce the microbial population in foods. The modes of action exerted by the applied heat energy against microbes include the denaturation of proteins and destruction of enzyme activity. In response to the thermal treatment, the pattern in which the population of vegetative pathogens (*Escherichia coli, Salmonella*) decreases with time is known as the *survivor curve*. The survivor curve is obtained by plotting the logarithm of the number of microorganisms (log N) against time. The shape of the survivor curve may be linear, convex, concave, or sigmoidal (Figure 13.14), depending on the nature of microorganism and the treatment applied.

Inactivation of microorganisms by radiation results in a convex survivor curve, which features an extended lag at lower doses, followed by a nearly exponential decay phase (curve *A* in Figure 13.14). A convex survivor curve has also been depicted by the heating and chemical methods of microbial inactivation. A sigmoid-shaped survivor curve is characterized by a relatively pronounced tail occurring after initial shoulder and exponential decay phases (curve *B* in Figure 13.14). Concave survivor curves (curves *C* and *D* in Figure 13.14) show a continuously decreasing death rate with an increase in the treatment time or size. This type of curve is demonstrated by many physical and chemical lethal agents (van Dam et al., 1988).

In general, the survivor curve is described by the following equation (Eq. 13.1):

$$-\frac{dN}{dt} = kN^n \tag{13.1}$$

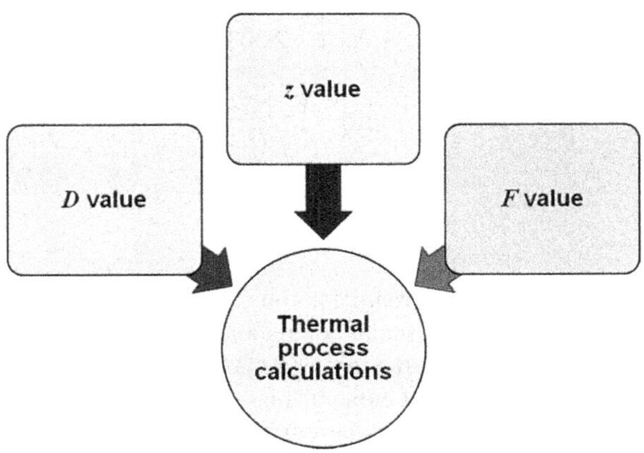

FIGURE 13.13 Building blocks of thermal process calculations.

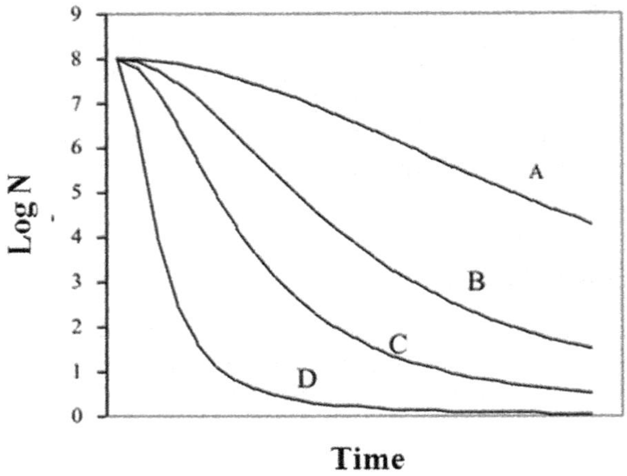

FIGURE 13.14 Survivor curve of microorganisms (A: convex curve; B: sigmoidal curve; C and D: concave and continuously decreasing curve). (Modified and Reproduced with permission from van Dam, K., Mulder, M. M., de Mattos, J. T. And Westerhoff, H. V. A. 1988. Thermodynamic View of Bacterial Growth. In *Physiological Models in Microbiology* (Volume I), eds. M. J. Bazin and J. I. Prosser, 24–47. Boca Raton, FL: CRC Press.)

where N is the number of microorganisms, k is the rate constant and n is the order of the model. With a first-order model, where $n = 1$, Eq. (13.1) reduces to

$$-\frac{dN}{dt} = kN \tag{13.2}$$

Under the initial conditions, when $t = 0$ and $N = N_0$, integration of Eq. (13.2) gives

$$\ln\left(\frac{N}{N_0}\right) = -kt \tag{13.3}$$

Further, Eq. (13.3) can be written in the logarithmic form as

$$2.303\log\left(\frac{N}{N_0}\right) = -kt \tag{13.4}$$

$$\log\left(\frac{N}{N_0}\right) = \frac{-kt}{2.303} \tag{13.5}$$

$$\log\left(\frac{N}{N_0}\right) = \frac{-t}{D} \tag{13.6}$$

$$\left(\frac{N}{N_0}\right) = 10^{\frac{-t}{D}} \tag{13.7}$$

where $D = (2.303/k)$, called as the "decimal reduction time." D value is defined as the "time required at a specified temperature T, to reduce the microbial population by 90% or by one logarithmic value." The concept of D value arises from the fact that the microbial population is reduced at a "logarithmic death rate," which is proportional only to the initial numbers present in the food, irrespective of the size of the surviving population. In other words, at a constant temperature, within a given time interval, the same percentage of a bacterial population will be destroyed, as depicted by the survivor curve data presented on the semilogarithmic coordinate (Figure 13.15). Therefore, the D value can be calculated using a

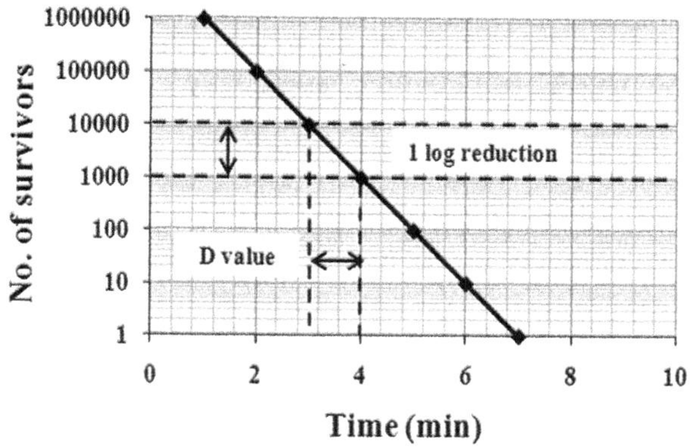

FIGURE 13.15 Microbial survivor curve on semilogarithmic plot and calculation of D value.

graphical approach as shown in Figure 13.15. In the absence of experimental data, D value at 121.1°C ($D_{121.1°C}$) is usually assumed as 1 min.

13.5.2 Thermal Resistance Constant (z Value)

D value is mainly influenced by the change in temperature. According to the reaction rate kinetics, the rate of a reaction increases with an increase in temperature. Similarly, the rate of thermal destruction of microorganisms also accelerates with the increase in temperature. Consequently, as the temperature increases, D value decreases. Reduction in the D value from D_1 to D_2 when the temperature is increased from T_1 to T_2 is explained by Eq. (13.8),

$$\log \frac{D_1}{D_2} = \frac{T_2 - T_1}{z} \tag{13.8}$$

where z, the thermal resistance constant or "z value," is defined as the *increase in temperature to achieve a ten-fold or one log reduction in D value*. For most of the spore-forming bacteria, the z value is in the range of 8°C–12°C. A linear relationship is observed in the plot between temperature and log D (Figure 13.16). *z value* can be calculated graphically from the slope of T versus log D plot. Generally, z value is assumed to be 10°C in the absence of required experimental data.

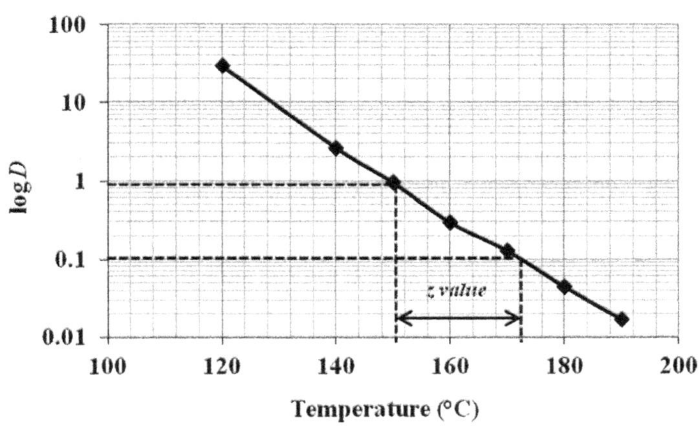

FIGURE 13.16 Determination of z value.

The mean D value and z value for some important microorganisms targeted by the thermal processes, such as *C. botulinum*, *Bacillus subtilis*, *C. sporogenes*, and *C. histolyticum*, are 0.2 min/10°C, 0.5 min/10°C, 0.8–1.4 min/13°C, and 0.01 min/10°C, respectively.

13.5.3 Thermal Death Time (*F* Value)

Thermal death time or "F value" is defined as the *total time required to achieve a defined reduction in the population of vegetative cells or spores.* It is expressed in multiples of D value, with the unit of time, as

$$F = D\log\left(\frac{N_0}{N}\right) \tag{13.9}$$

The aim of a commercial sterilization process for low-acid foods is the achievement of a 12 log reduction ($12D$ or 10^{12}) in the initial microbial population. If a food product packed in 1 kg container or can has an initial spore load of 10^2 spores of *C. botulinum* per gram of product prior to thermal sterilization, a $12D$ thermal sterilization process will result in 10^{-10} spores per gram of product or a corresponding probability of one in 10 million (10^7) cans having one viable spore. As a conventional practice, F value is denoted as $F_{T_0}^{z}$, with the subscript denoting the process temperature and the superscript indicating the z value for the target microorganism under consideration. Then, the F value is given by the following expression:

$$F_{T_0}^{z} = \int_0^t 10^{\left(\frac{T-T_0}{z}\right)}dt \tag{13.10}$$

A reference F value is often designated at temperature, $T_0 = 121°C$ and z value $= 10°C$. This reference value is denoted as F_0, which is the time taken for a given reduction in the population of a microbial spore with a z value of 10°C at a temperature of 121°C and given by the following expression:

$$F_{121}^{10} = F_0 = \int_0^t 10^{\left(\frac{T-121}{10}\right)}dt \tag{13.11}$$

F value of the target microorganism decides the amount of heat and the length of time used in pasteurization. The minimum combination should target the most heat-resistant pathogen (e.g., *Coxiella burnetii*). When deriving the F value for a microorganism, the temperature is maintained at a constant value and the holding time is varied to kill the specified number of cells.

13.5.3.1 Methods to Determine the Process Lethality (F_0 Value)

13.5.3.1.1 The Improved General Method

A graphical method for the determination of process lethality was developed in the 1920s by Bigelow and others and later improved by Schulz and Olson in 1940. The improved general method is based on the relationship between the survival curve and F value. Here, the temperature history of the slowest heating zone within the container or can is recorded by a thermocouple placed inside the container. The heat penetration curve is obtained by plotting the product temperature against the processing time (Figure 13.17). The heat penetration curve signifies that the temperature and time exert a combined lethal effect on the target microbial population. Therefore, the integration of time and temperature histories will provide the process lethality.

Now, process lethality is determined upon the base temperature, 121.1°C, considering that the lethality is 1 min at 121.1°C. Base temperature is fixed with reference to the thermal processing values of *C. botulinum*. Lethality at a particular temperature is determined using Eq. (13.12).

$$L = 10^{-\left(\frac{121.1-T}{z}\right)} \tag{13.12}$$

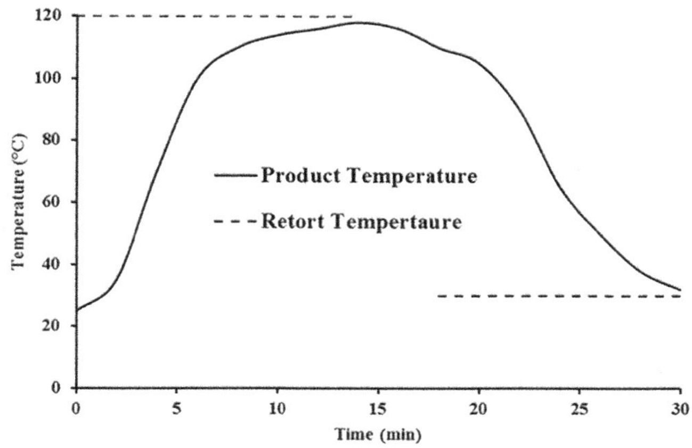

FIGURE 13.17 Heat penetration curve.

According to Eq. (13.12), the lethality at 101.1°C is 0.01 min, that at 111.1°C is 0.1 min, and that at 131.1°C is 10 min. From the lethality value, the process lethality or F_0 value is calculated by the principle of integration (Eq. 13.13).

$$F_0 = \int_0^t L\,dt \tag{13.13}$$

This is done by plotting the lethality rate (L) value against time (t) (Figure 13.18). Since lethality values are additive, area under the curve can be used to determine the process lethality. Here, it should be noted that the lethality of the process also occurs during the cooling period. Therefore, the rate of cooling is also important to avoid the overprocessing of the food product.

The general method is the widely used method due to its simplicity. However, it just gives an approximation of process lethality. Therefore, it may underpredict the process lethality and hence lead to overprocessing. The following example will help in understanding the use of the general method to calculate the process lethality.

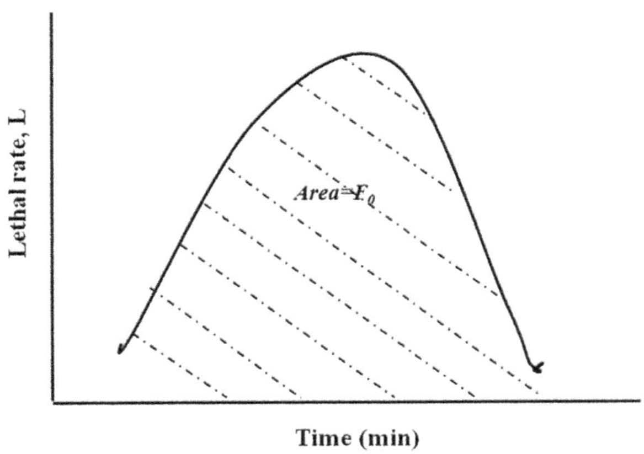

FIGURE 13.18 Lethality of a thermal process as a function of time.

Example 13.1

A time–temperature history for the fruit juice sterilization is given in the following table. Determine the F_0 value for the process.

Time (min)	Temperature (°C)	Time (min)	Temperature (°C)
0	25	16	118
2	35	18	116
4	70	20	114
6	100	22	112
8	110	24	110
10	114	26	107
12	116	28	105
14	118	30	100

Solution

Lethality rate is determined using Eq. (13.11) and presented in the following table:

Time (min)	Temperature (°C)	Lethal Rate	Time (min)	Temperature (°C)	Lethal Rate
0	25	2.512×10^{-10}	16	118	5.012×10^{-1}
2	35	2.512×10^{-9}	18	116	3.162×10^{-1}
4	70	7.943×10^{-6}	20	114	1.995×10^{-1}
6	100	7.943×10^{-3}	22	112	1.259×10^{-1}
8	110	7.943×10^{-2}	24	110	7.943×10^{-2}
10	114	1.995×10^{-1}	26	107	3.981×10^{-2}
12	116	3.162×10^{-1}	28	105	2.512×10^{-2}
14	118	5.012×10^{-1}	30	100	7.943×10^{-3}
					$\Sigma = 2.399$

The sum of the lethal rate is 2.399, and as the width of the time interval is 2 min (Figure 13.19), F_0 is equal to $2 \times 2.399 = 4.799$ or 4.8 min.

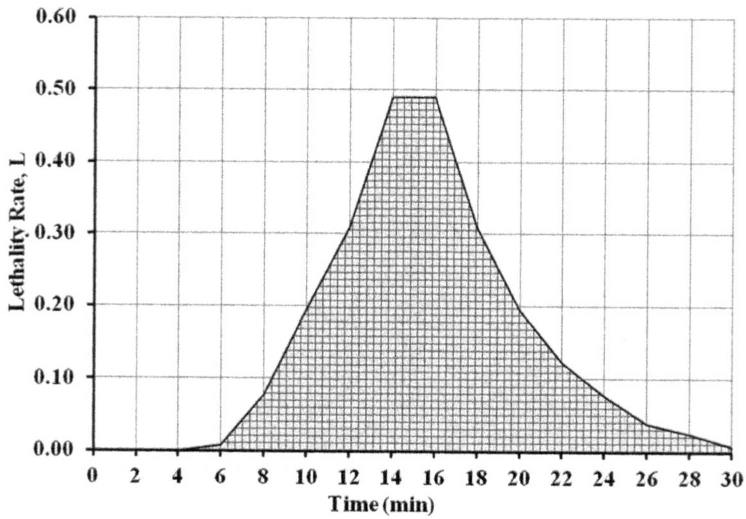

FIGURE 13.19 Lethality curve for fruit juice sterilization.

13.5.3.1.2 Formula Method

To predict the exact thermal processing time, the mathematical or formula method was developed. This method is useful in determining the effect of container size on the thermal process. Formula method is based on the plot of the inverse logarithm of temperature deficit against the heating time (Figure 13.20). Temperature deficit is the difference between container's temperature and retort temperature. It can be seen that the temperature deficit decreases with time. A linear plot is obtained except during the initial come-up time for the retort, where steam enters the retort and the retort pressure rises. However, it was observed that 42% of this come-up time contributes to the process lethality. Hence, $t = 0$ is set at the 58% of the come-up time.

The linear part of the plot (Figure 13.20) is described by the following equation:

$$B = f_h \log\left(\frac{j_h I}{g}\right) \tag{13.14}$$

where B is the process time from the adjusted 0th (min), f_h is the penetration factor (min) or slope of the heating line, j_h is the thermal lag factor, I is the initial temperature deficit, and g is the temperature deficit at the end of the thermal process. Here, the penetration factor is defined as the time required for one log cycle of temperature deficit, and it decreases as the heat penetration increases.

Thermal lag factor (j_h) is determined from the following equation:

$$j_h = \frac{(T_R - T_{pih})}{(T_R - T_{ih})} \tag{13.15}$$

where T_R is the retort temperature, T_{pih} is the pseudo-initial temperature at the start of heating, and T_{ih} is the initial temperature of the product.

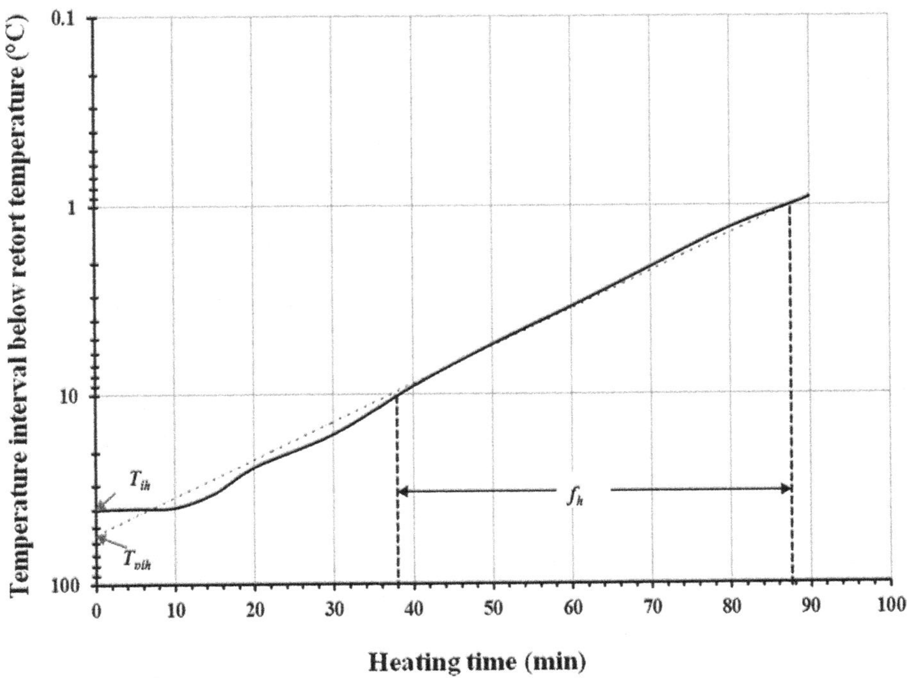

FIGURE 13.20 Heat penetration curve on a semilog scale.

Similarly, the cooling lag factor is calculated as

$$j_c = \frac{(T_c - T_{pic})}{(T_c - T_{ic})} \tag{13.16}$$

where T_c is the retort temperature, and T_{pic} and T_{ic} are the pseudo-initial can temperature and the initial temperature of the product, respectively, at the beginning of cooling. If the data is unavailable for Eq. (13.16), it can be considered that $j_c = j_h$.

F value is determined by Eq. (13.17), and U, which is equivalent to F_0 value, is given by the following equation.

$$U = F \cdot 10^{\left(\frac{121.1 - T_R}{z}\right)} \tag{13.17}$$

Thermal process time (B) can be calculated, provided f_h, j_h, the temperature difference between retort temperature (T_R) and initial product temperature, U, and g are known. The former four variables are obtained from the process parameters, and g is determined as a function of $\frac{f_h}{U}$ and j_c.

Example 13.2

Calculate the processing time for retort processing of fruit juice at 115°C using a process based on an F_0 value of 7 min. The retort took 5 min to reach the process temperature. The heat penetration data gives the following information:

$T_{ih} = 78°C, f_h = 20\,min, f_c = 20\,min, j_c = 1.8, T_{pih} = 41°C, F_1 = 4.084$, and $g = 0.241°C$.

Solution

$$I = 115 - 78 = 37°C$$

$$j_h = \frac{(115-41)}{(115-78)} = 2$$

Thus,

$$B = f_h \log\left(\frac{j_h I}{g}\right) = 20 \times \log\left(\frac{2 \times 37}{0.241}\right) = 49.75\,min$$

Therefore, processing time = $49.7 - (0.4 \times 5) = 47.75$ min = 47 min and 45 s.

Answer: Processing time is 47 min and 45 s

13.6 Problems to Practice

13.6.1 Multiple Choice Questions

1. Pasteurization is a process that involves
   a. subjecting the food product to HTST treatment to destroy harmful microorganisms
   b. heating the food product to inactivate deteriorative enzymes
   c. fortification of foods with vitamins
   d. irradiation of foods to destroy pathogenic microorganisms

Answer: a

2. Pasteurization is the process of heating a food product
   a. above 150°C
   b. above 121.1°C
   c. above boiling point
   d. below boiling point

Answer: d

3. Which one of the following processes for milk involves heating to 72°C for 15 s and rapid cooling?
   a. HTST pasteurization
   b. Aseptic processing
   c. LTLT pasteurization
   d. Ultra-pasteurization

Answer: a

4. The marker enzyme that is indicative of an efficient milk pasteurization is
   a. PPO
   b. POD
   c. alkaline phosphatase
   d. LOX

Answer: c

5. According to the FDA, the lethality accumulated in the ____________________ of pasteurizer is significant.
   a. heat exchanger
   b. holding tube
   c. regenerator
   d. balance tank

Answer: b

6. Statement 1: Blanching is a process in which steam is used to inactivate degradative enzymes in the fruits and vegetables.

   Statement 2: Aseptically processed products should be stored under refrigeration and pasteurized products can be stored at room temperature.

   Which of the following holds true for statement 1 and statement 2, respectively?
   a. True, true
   b. False, false
   c. True, false
   d. False, true

Answer: c

7. The D value is the time required at a specific temperature to
   a. increase the microbial population by 10%
   b. decrease the microbial population by 90%
   c. decrease the z value by 90%
   d. decrease the F value by 90%

Answer: b

8. The operation of FDV in a continuous-flow pasteurizer is based on
   a. temperature
   b. time
   c. product flow rate
   d. pressure

Answer: a

9. If the D value for destruction of vitamin C at 121°C is 245 min, then the corresponding reaction velocity constant k is
   a. 0.94 min^{-1}
   b. 0.245 min^{-1}
   c. 0.0094 min^{-1}
   d. 0.121 min^{-1}

Answer: c

10. If $D_{121.1°C}$ and z value for a thermophilic spore in milk are 30 s and 10°C, respectively, the D value at 150°C is
    a. 0.00644 min
    b. 0.000644 min
    c. 0.000854 min
    d. 0.644 min

Answer: b

11. In a thermal processing operation, the D value for microorganisms was found to be 2.78 minutes and 9.98 minutes at 121.1°C and 115.5°C, respectively. The z value of the microorganism is
    a. 10.09°C
    b. 9.1°C
    c. 1.99°C
    d. 0.11°C

Answer: a

12. A medium-acid food product is sterilized at 100°C in a can to reduce the number of heat-resistant organisms ($D_{120°C}$ = 0.2 min; z = 10°C) from an initial count of 10000 per can to a probability of survival of 1 in million. $D_{100°C}$ value of this organism is
    a. 0.4 min
    b. 20 min
    c. 4 min
    d. 10 min

Answer: b

13. Generally, the reference temperature for a commercial sterilization process is
    a. 150°C
    b. 100°C
    c. 273.15°C
    d. 121.1°C

Answer: d

14. The thermal processing method that eliminates the need for refrigeration of the treated food product is
    a. pasteurization
    b. aseptic processing
    c. blanching
    d. both a and b

Answer: b

15. Temperature used in UHT pasteurization is
    a. 90–100°C
    b. 100–120°C
    c. 130–140°C
    d. 120–125°C

Answer: c

16. The purpose of blanching is
    a. to soften products to obtain better filling
    b. to denature enzymes that affect the color and texture
    c. to reduce microbial population
    d. all of the above

Answer: d

17. Survivor curve is the plot between
    a. D value and temperature
    b. logarithm of the number of microorganisms and time
    c. temperature and time
    d. number of microorganisms and temperature

Answer: b

18. z value is defined as
    a. an increase in temperature to achieve a tenfold reduction in the microbial population
    b. the total time required to achieve a defined reduction in the population of vegetative cells or spores
    c. the time required to reduce the microbial population by 90%
    d. the increase in temperature to achieve a ten-fold reduction in D value

Answer: d

19. Commercial sterilization uses temperature that is
    a. less than the boiling point of water
    b. more than the boiling point of water
    c. equal to the boiling point of water
    d. lesser or greater than the boiling point of water

Answer: b

20. The major function of blanching when done before canning is
    a. inactivation of enzymes
    b. inactivation of microorganisms
    c. removal of air from the intracellular spaces to facilitate easy filling of the cans
    d. prevent discoloration of the product

Answer: c

13.6.2 Numerical Problems

1. Determine the *D* value and the first-order rate constant (*k*) for the microbial population with the following survivor curve data:

Time (min)	No. of Survivors
1	1000000
2	100000
3	10000
4	1000
5	100
6	10
7	1
8	0.1
9	0.01

Given

Survivor curve data

To find

i. *D* value
ii. First-order rate constant (*k*)

Solution

Using the graphical approach, a semilogarithmic plot between number of survivors and time is constructed as shown in the Figure.

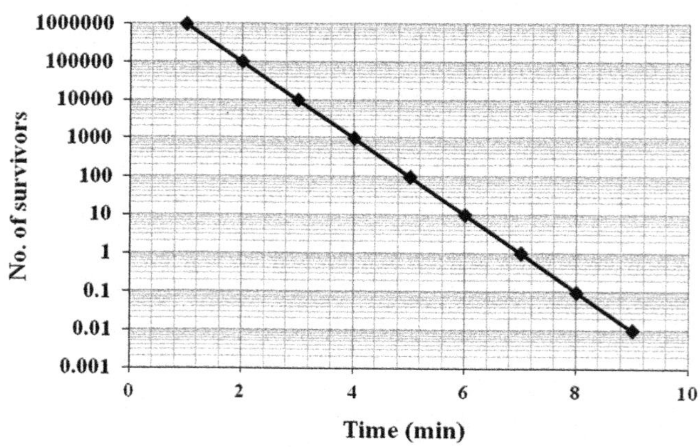

From the graphical approach, *D* value = 1 min. The first-order rate constant (*k*) is given by

$$D = \frac{2.303}{k}$$

$$\therefore k = \frac{2.303}{D} = \frac{2.303}{1} = 2.303\,\text{min}^{-1}$$

Answer: (i) *D* value = 1 min; (ii) *k* = 2.303 min⁻¹

2. Determine the *z* value for a microorganism that exhibits the *D* values listed in the following table with an increase in temperature from 120°C to 190°C.

Temperature (°C)	*D* Value (min)
120	30
140	2.7
150	0.967
160	0.3
170	0.13
180	0.045
190	0.017

Given

i. $T_1 = 120°C$
ii. $T_2 = 190°C$
iii. Temperature versus *D* value data

To find: *z* value

Solution

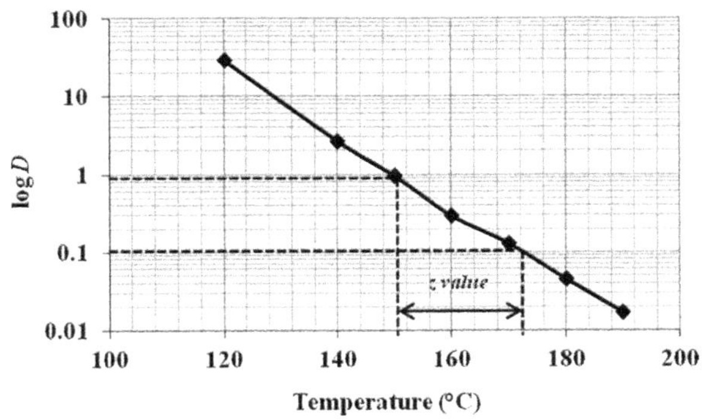

From the graphical method, *z* value is 22°C.

Answer: (i) *z* value = 22°C

3. The *D* value for an organism at 121°C (D_{121}) is 0.35 min, and the *z* value is 10°C. Calculate the *D* value at 100°C (D_{100}).

Given

i. $T_1 = 121°C$
ii. $T_2 = 100°C$
iii. $D_{121} = D_1 = 0.35\,\text{min}$
iv. z value = 10°C

To find: D value at 100°C (D_{100}) (D_1)

Solution

$$D_{T_1} = D_{T_2} 10^{\frac{T_2 - T_1}{z}}$$

$$D_{121} = D_{100} 10^{\frac{100 - 121}{10}}$$

$$0.35 = D_{100} 10^{\frac{100 - 121}{10}}$$

$$0.35 = D_{100} 10^{\frac{100 - 121}{10}}$$

$$0.35 = D_{100}(0.0079)$$

$$D_{T_2} = \frac{0.35}{0.0079} = 44.3\,\text{min}$$

Answer: *D* value at 100°C is 44.3 min

4. Determine the process time for a F_0 value of 8 min at a temperature of 118°C.

Given

i. $F_0 = 8\,\text{min}$
ii. $T = 118°\text{C}$

To find: Process time

Solution

$$F_0 = \int_0^t 10^{\left(\frac{T-121}{10}\right)} dt = 8\,\text{min}$$

$$F_0 = \int_0^t 10^{\left(\frac{118-121}{10}\right)} dt = 8\,\text{min}$$

$$\int_0^t 10^{(-0.3)} dt = 8$$

$$10^{(-0.3)}(t) = 8$$

$$t = \frac{8}{10^{-0.3}} = \frac{8}{0.5} = 16\,\text{min}$$

Answer: Process time = 16 min

5. The reaction rate constant (k) for the degradation of vitamin C at 121°C is 0.0015 min^{-1} as determined experimentally. Determine the D value for the degradation reaction.

Given

i. $k = 0.0015\,\text{min}^{-1}$
ii. $T = 121°\text{C}$

To find: D value

Solution

$$D = \frac{2.303}{k} = \frac{2.303}{(0.0015)} = 1535.3 \text{ s} = 25.6\,\text{min}$$

Answer: D value = 25.6 min

6. For an initial spore load equal to 50 spores per can, calculate the spoilage probability if an F equivalent to $F_{121} = 11.7$ min was applied. The D_{121} for the concerned spoilage organism was determined as 2.5 min.

Given

i. Initial spore load = 50 spores per can
ii. $F_{121} = 11.7$ min
iii. $D_{121} = 2.5$ min

To find: Spoilage probability

Solution

$$F_T = D_T \log\frac{N_0}{N}$$

$$F_{121} = D_{121} \log\frac{50}{N}$$

$$11.7 = 2.5 \log\frac{50}{N}$$

$$4.68 = \log\frac{50}{N}$$

$$47863 = \frac{50}{N}$$

$$\therefore N = \frac{50}{47863} = 0.001 = 10^{-3}$$

Answer: Spoilage probability = 1 can in 1000 cans

7. Determine the required heating time for the process described in Problem no. 3 (F_{121} value), to obtain a 12 log cycle reduction in the microbial population.

Given

i. $D_{121} = 0.35$ min

To find: Lethality at 121°C for a 12 log cycle reduction in the population.

Solution

$$F_T = D_T \log\frac{N_0}{N}$$

$$\because \log\frac{N_0}{N} = 12,\ F_{121} = D_{121} \times 12 = 0.35 \times 12 = 4.2\,\text{min}$$

Answer: The required heating time for the process is F = 4.2 min

8. Calculate the F_0 value for the sterilization of a canned product with the time–temperature history mentioned in the table that follows.

Time (min)	Temperature (°C)
0	20
5	50
10	60
15	75
20	88
25	90
30	90
35	79
40	60
45	50
50	30

Given

i. Time–temperature history for the sterilization of a canned product:

To find: F_0 value

Solution

Lethality rate is determined using the following equation and presented in the tabular form.

$$F_{121}^{10} = F_0 = \int_0^t 10^{\left(\frac{T-121}{10}\right)} dt$$

Time (min)	Temperature (°C)	Lethal Rate
0	20	7.943×10^{-11}
5	50	7.943×10^{-8}
10	60	7.943×10^{-7}
15	75	2.512×10^{-5}
20	88	0.0005012
25	90	0.0007943
30	90	0.0007943
35	79	6.31×10^{-5}
40	60	7.943×10^{-7}
45	50	7.943×10^{-8}
50	30	7.943×10^{-10}
		$\Sigma = 0.002$

The sum of the lethal rate is 0.002, and as the width of the time interval is 5 min, F_0 is equal to $5 \times 0.002 = 0.01$ min.

Answer: F_0 = 0.01 min

BIBLIOGRAPHY

Ando, Y., Maeda, Y., Mizutani, K., Wakatsuki, N., Hagiwara, S. and Nabetani, H. 2016. Impact of blanching and freeze-thaw pretreatment on drying rate of carrot roots in relation to changes in cell membrane function and cell wall structure. *LWT - Food Science and Technology* 71: 40–46.

Anonymous. 2010. Breakthrough Blanching Technology Combines Benefits of Steam and Rotary Drum Design. LYCO Manufacturing, Inc. at, 115 Commercial, Drive, P.O. Box 31, Columbus Wisconsin 53925. www.lycomfg.com (accessed June 15, 2018).

Arroqui, C., López, A., Esnoz, A. and Vírseda, P. 2003. Mathematic model of an integrated blancher/cooler. *Journal of Food Engineering* 59: 297–307.

Awuah, G. B., Ramaswamy, H. S. and Economides, A. 2007. Thermal processing and quality: Principles and overview. *Chemical Engineering and Processing* 46: 584–602.

Betta, G., Barbanti, D. and Massini, R. 2011. Food Hygiene in aseptic processing and packaging system: A survey in the Italian food industry. *Trends in Food Science and Technology* 22: 327–334.

Bhuvaneswari, E. and Anandharamakrishnan, C. 2014. Heat transfer analysis of pasteurization of bottled beer in a tunnel pasteurizer using computational fluid dynamics. *Innovative Food Science and Emerging Technologies* 23: 156–163.

Bigelow, W. D., Bohart, G. S., Richardson, A. C. and Ball, C. O. 1920. *Heat Penetration in Processing Canned Foods. Bull. Nr 16-L Res.* Washington, DC: Lab. Natl. Canners Assn.

Buzrul, S. 2007. A suitable model of microbial survival curves for beer pasteurization. *LWT-Food Science and Technology* 40: 1330–1336.

Chamorro, P. V. and Vidaurreta, C. A. 2012. Blanching of fruits and vegetable products. In *Operations in Food Refrigeration*, ed. R. H. Mascheroni, 93–112, Boca Raton, FL: CRC Press, Taylor and Francis Group.

Corcuera, J. I. De., Cavalieri, R. P. and Powers, J. R. 2004. Blanching of foods. In *Encyclopedia of Agricultural, Food, and Biological Engineering*, ed. D. Heldman, 1–5. New York: Marcel Dekker.

Dignan, D. M., Barry, M. R., Pflug, I. J. and Gardine, T. D. 1989. Safety considerations in establishing aseptic processes for low-acid foods containing particulates. *Food Technology* 43: 118–121.

Dilay, E., Vargas, J. V. C., Amico, S. C. and Ordonez, J. C. 2006. Modeling, simulation and optimization of a beer pasteurization tunnel. *Journal of Food Engineering* 77: 500–513.

Downing, D. L. 1996. *A Complete Course in Canning and Related Processes: Book 1. Fundamental Information on Canning*. Baltimore, MD: CTI Publications, Inc.

Fellows, P. 1988. *Food Processing Technology* (First edition). Chichester: Elis Horwood, Ltd.

Fellows, P. 2000. *Food Processing Technology* (Third edition). Chichester: Elis Horwood, Ltd.

Frazier, W. C. and Westhoff, D. C. 1988. *Food Microbiology* (Fourth edition). New York: McGraw-Hill Inc.

Goff, H. D. 2018. *Dairy Science and Technology Education Series*. University of Guelph, Canada. www.uoguelph.ca/foodscience/book-page/batch-method; www.uoguelph.ca/foodscience/book-page/htst-milk-flow-overview; www.uoguelph.ca/foodscience/book-page/balance-tank; www.uoguelph.ca/foodscience/book-page/flow-diversion-device-fdd (accessed July 10, 2018).

Graves, R. 1996. United States history and evolution. In *Aseptic Processing and Packaging of Food: A Food Industry Perspective*, eds. J. R. D. David, R. H. Graves and V. R. Carlson, 21–30. Boca Raton, FL: CRC Press.

Hallstrom, B., Skjolderbrand, C. and Tragardh, C. 1988. *Heat Transfer and Food Products*. London: Elsevier Applied Science.

Halpin, B. E. and Lee, C. Y. 1987. Effect of blanching on enzyme activity and quality changes in green peas. *Journal of Food Science* 52: 1002–1005.

IDF 1986. *Monograph on Pasteurised Milk, Document No. 20*. Brussels: International Dairy Federation.

Johnston, I. R. and Lin, R. C. 1987. FDA views on the importance of a_w in good manufacturing practice. In *Water Activity: Theory and Applications to Food*, eds. L. B. Rockland and L. R. Beuchat, 287. New York: Dekker.

Kumar, P. and Sandeep, K. P. 2014. Thermal principles and kinetics. In *Food Processing: Principles and Applications* (Second edition), eds. S. Clark, S. Jung and B. Lamsal, 17–31. Chichester: John Wiley and Sons, Ltd.

Lund, D. B. 1975. Effects of blanching, pasteurization, and sterilization on nutrients. In *Nutritional Evaluation of Food Processing*, eds. R. S. Harris and E. Karmas, 205. New York: AVI Publishing Co.

Mauer, L. 2003. PACKAGING | Aseptic Filling. In *Encyclopedia of Food Sciences and Nutrition* (Second edition), ed. B. Caballero, 4316–4322. Oxford: Academic Press.

National Advisory Committee on Microbiological Criteria for Foods (NACMCF) 2006. Requisite scientific parameters for establishing the equivalence of alternative methods of pasteurization. *Journal of Food Protection* 69: 1190–1216.

Office of Industrial Technologies Energy Efficiency and Renewable Energy, U.S. Department of Energy 2002. *Energy-Efficient Food-Blanching System.* http://attfile.konetic.or.kr/konetic/xml/use/31B2A0205535.pdf (accessed June 15, 2018).

Orikasa, T., Wu, L., Shina, T. and Tagawa, A. 2008. Drying characteristics of kiwifruit during hot air drying. *Journal of Food Engineering* 85: 303–308.

Parreño, W. C. and Torres, M. D. A. 2012. Quality and safety of frozen vegetables. In *Handbook of Frozen Food Processing and Packaging* (Second edition), ed. Da-Wen Sun, 387–434. Boca Raton, FL: CRC Press, Taylor and Francis Group.

Potter, N. N. and Hotchkiss, J. H. 1998. *Food Science.* Berlin: Springer.

Rahman, M. S. 2015. Hurdle technology in food preservation. In *Minimally Processed Foods, Food Engineering Series*, eds. M. W. Siddiqui and M. S. Rahman, 17–33. Switzerland: Springer International Publishing.

Schultz, O. T. and Olson, F. C. 1940. Thermal processing of canned foods in tin containers. III. Recent improvements in the General Method of thermal process calculation. A special coordinate paper and methods of converting initial retort temperature. *Food Research* 5: 399.

Scott, E. P., Carroad, P. A., Rumsey, T. R., Horn, J., Buhlert, J. and Rose, W. W. 1981. Energy consumption in steam blanchers. *Journal of Food Process Engineering* 5: 77–88.

Silva, F. V. M. and Gibbs, P. A. 2010. Non-proteolytic Clostridium Botulinum spores in low-acid cold-distributed foods and design of pasteurization processes. *Trends in Food Science and Technology* 21: 95–105.

Silva, F. V. M., Gibbs, P. A., Nunez, H., Almonacid, S. and Simpson, R. 2014. Pasteurization. In *Encyclopedia of Food Microbiology* (Second edition), eds. C. A. Batt and M. L. Tortorello, 577–595. San Diego: Academic Press.

Singh, R. P. and Heldman, D. R. 2014. *Introduction to Food Engineering* (Fifth edition). San Diego, CA: Academic Press.

Stoforos, N. G. 1995. Thermal process design. *Food Control* 6: 81–94.

Stumbo, C. R. 1973. *Thermobacteriology in Food Processing* (Second edition). New York: Academic Press.

Turbo-flow™. *Blancher/Cooker Heat Treatment Principles, Product Documentation.* Key Technology. USA: Key Technology, 150 Avery Street, Walla Walla, WA 99362. www.key.net (accessed June 15, 2018).

Vaclavik, V. A. and Christian, E. W. 2003. *Essentials of Food Science* (Second edition). New York: Springer Science+Business Media.

van Dam, K., Mulder, M. M., de Mattos, J. T. and Westerhoff, H. V. A. 1988. Thermodynamic view of bacterial growth. In *Physiological Models in Microbiology* (Volume I), eds. M. J. Bazin and J. I. Prosser, 24–47. Boca Raton, FL: CRC Press.

Watanabe, T., Orikasa, T., Sasaki, K., Koide, S., Shiina, T. and Tagawa, A. 2014. Influence of blanching on water transpiration rate and quality changes during far-infrared drying of cut cabbage. *Journal of the Japanese Society of Agricultural Machinery and Food Engineers* 76: 387–394.

Wilbey, R. A. 2014. Principles of pasteurization. In *Encyclopedia of Food Microbiology* (Second edition), eds. C. A. Batt and M.-L. Tortorello, 169–174. San Diego: Academic Press.

14

Nonthermal and Alternative Food Processing Technologies

Conventional methods of food processing depend on heat-mediated inactivation of foodborne pathogens such as bacteria, viruses, and parasites to render food safe for consumption. Nevertheless, certain foods are susceptible to cause risk of bacterial or viral foodborne diseases for which heat is undesirable or unusable. In recent years, demand is on the rise for minimally processed foods. This scenario is obvious as innovations in the arena of food processing are always targeted at improving the product quality through preservation of nutritional and nutraceutical components. Consequently, the demand and research on nonthermal food processing methods (methods that do not rely upon heat to achieve food preservation) are on the rise.

Food processing sector is a field of continuous developments and technological advancements. This century has witnessed an extensive transformation in the scale of food processing from *macro to micro*, and more recently, there is a boom in the concept of *nanofood technology*. Consequently, nanotechnology-based food processing techniques are evolving perpetually.

Many of the advanced food processing methods have been successfully scaled up and commercialized from the laboratory to industrial scale, whereas some others show potential scope for commercial applications. This chapter provides an overview of such recent and emerging technologies in the food processing sector.

14.1 Nonthermal Techniques for Food Processing

14.1.1 High-Pressure Processing (HPP)

The fact that high pressure (HP) can kill microorganisms was discovered in the late nineteenth century. However, the application of high-pressure processing (HPP) as a nonthermal technique for food preservation was appreciated only in the 1980s, during which the consumers' demand for minimally processed foods was high. The HP-processed jam was the first commercial product to be launched in the 1980s in the Japanese market. Since then, the HPP technology has continuously evolved to show a multitude of applications in the preservation and processing of liquid food products such as fruit juices and high-moisture foods including ready-to-eat meals, meat products, and seafood products (Anandharamakrishnan, 2013).

HPP can be defined as "an athermal decontamination process which comprises of subjecting the packaged food product to a high level of water pressure, in the range of 100–900 MPa." Strikingly, the pressure range mentioned in the definition is about 1000–10000 times higher than the atmospheric pressure. In spite of the high pressure applied, owing to its nonthermal operation, HPP maintains the food product's freshness intact, over and above preserving nutrients such as vitamins.

14.1.1.1 Principle of HPP

HPP is governed by the following principles:

- **Le Chatelier's principle:** The Le Chatelier's principle (Le Chatelier, 1884) states that "a chemical system under equilibrium condition would experience a reaction change, accompanied by a decrease in volume when enhanced by pressure and vice versa" (Yordanov and Angelov, 2010; Jaeger et al., 2012). According to this principle, when the equilibrium of a system is disturbed, the system responds to minimize the disturbance by accelerating the reactions associated with volume reduction and suppressing the reactions leading to an increase in volume.

- **Principle of microscopic ordering:** Increasing the pressure at a constant temperature mutually increases the degree of ordering of molecules of a given substance. Consequently, the pressure and temperature exert opposing forces on the molecular structure (Benet, 2005).
- **Isostatic rule:** The isostatic rule states that "the pressure applied on a product is uniformly and instantaneously transmitted throughout the product, irrespective of the size of the product." As a result, there is no pressure gradient in the sample. This renders HPP to have a competitive edge over the conventional thermal processing techniques, wherein the presence of a thermal gradient in the product affects the process lethality.

During HPP, microbial inactivation occurs as a result of the permeabilization of the cell membrane. At the lower pressure range of HPP, the cytoplasmic organelles are deformed leading to the leakage of the intracellular material. At higher pressure, there is a complete loss of intracellular material, and the function of the nucleus is impaired. However, the sensory properties of the product are retained since the low-molecular-weight components containing covalent bonds which are responsible for the nutritional and sensory characteristics of the food remain unaffected during the HP processing. But, the non-covalent bonds (hydrogen bonds and ionic bonds) and high-molecular-weight components such as protein and fat are altered during the process.

14.1.1.2 Methodology of HPP

In the HP processing, a processing chamber which is capable of sustaining high pressure is used. The food product may be packed or unpacked, depending on its nature and then placed inside the processing chamber. A pressure-transmitting fluid is introduced, which facilitates the uniform distribution of pressure within the chamber. Water is the most commonly used pressure-transmitting fluid. Nevertheless, glycol, castor oil, silicon oil, and ethanol can also be used for this purpose. The pressure-transmitting fluids are chosen based on their properties such as viscosity under the operational conditions of HPP and their corrosion-resistant nature.

During HPP of food, water is pressurized until the required chamber pressure is achieved. The pressure in the chamber is maintained for the specified time duration without further energy expenditure. After subjecting the product to the intended high pressure for a defined time interval, the system is depressurized, and the product is removed from the chamber. Total processing time includes the duration for pressurization, holding, and depressurization of the food product. Temperature rise during HPP is attributed to the adiabatic heating, which in turn is influenced by the product composition. High-fat foods show a higher temperature rise than the low-fat counterparts.

14.1.1.3 Equipment for HPP

The equipment for HPP can be classified into direct and indirect pressurization systems (Figure 14.1). In the direct pressurization systems, a piston is used to compress the pressure-transmitting fluid inside the chamber. On the other hand, in indirect pressurization systems, a pump is used for the compression of fluid. Further, based on the mode of operation, the HPP can be categorized as batch, continuous, and semi-continuous systems.

14.1.1.3.1 Batch HPP System

The batch HPP equipment comprises a cylindrical pressure vessel, end closure(s), threads or yoke to fasten the closures, and an intensifier pump which uses a low-pressure pump to supply the pressurizing fluid and essential system controls (Rao et al., 2014). Potable water is used as the pressure-transmitting medium which is filled in the pressure vessel. The food product to be HP-treated is packed in flexible plastic pouches and bottles, or vacuum-packed in containers to prevent mixing. The capacity of a batch HPP unit is generally in the range of 35–700L, with its operating pressure around 600MPa.

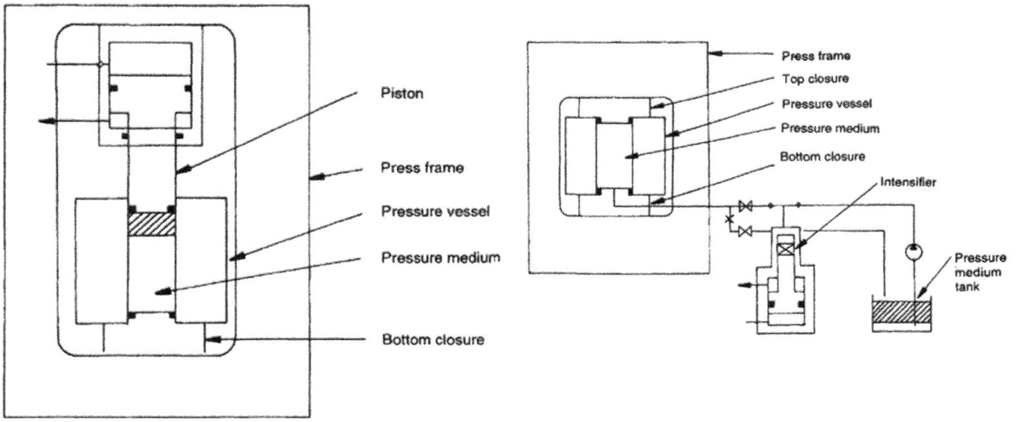

FIGURE 14.1 Schematic diagram of HP food processing systems (left: direct; right: indirect). (Reproduced with permission from Mertens B. 1995. Hydrostatic pressure treatment of food: equipment and processing. In *New Methods of Food Preservation*, ed. G. W. Gould, 135–158. Boston, MA: Springer.)

During the operation of batch HPP system, the pressure vessel is loaded with the containers and closed. Then, the pressure-transmitting fluid is pumped isostatically into the vessel. Once the required pressure is attained, the pump is stopped by closing the inlet valves. On applying pressure, the package is compressed along with its contents (food product). Four to fifteen percentage compression (reduction ratio in volume) is achieved in this system at an applied pressure in the range of 100–680 MPa. Usually, products are held at the defined HP for 3–5 min, and five to six such cycles are sufficient for loading, compression, holding, decompression, and unloading the product (Elamin et al., 2015). After holding the product at the defined HP for the required time, the vessel is depressurized by releasing the pressure-transmitting fluid back to its initial tank reservoir (Farkas and Hoover, 2000). On depressurization, the package regains its shape after which it is discharged from the pressure vessel and taken forward for storage using the conventional methods (Yordanov and Angelova, 2010).

Batch system is preferred for food products packed in packages of different shapes and styles. It is important that the package is devoid of gas inclusion, headspace, and moisture content. In addition, the package should have sufficient strength to withstand the HP encountered during the operation. An important parameter of batch HP processing is the *fill ratio.* Fill ratio is defined as the proportion of product volume per batch divided by the capacity of the HPP equipment. Better fill ratio is achieved with a large volume system compared to the small vessels. Fruit juice, shellfish, ready meals, and cooked meat are generally processed in batch systems.

14.1.1.3.2 Continuous HPP System

The continuous HP processing system (Figure 14.2) is specifically used for the processing of liquid foods (Balasubramaniam et al., 2015), such as fruit juice. The product is processed in its unpacked form. In a continuous HPP system, the products flow through an open-end tube system. The pressure applied is in the order of 100 MPa. Pressure greater than 100 MPa can also be used by installing HP intensifiers. Subsequently, the depressurization is carried out in such a manner that avoids extreme shear and heating of the product. The depressurized product is then forwarded to a sterile tank for filling under hygienic conditions (van den Berg et al., 2001). HP-processed fruit juice can be packed in the suitable packages such as bottles or aseptic pouches. This system is also suitable for vegetable juices, soups, fruit purees, and other pumpable liquid food products.

The microbial inactivation during the continuous HPP is attributed to the heat generated when the fluid is subjected to shearing, cavitations, and/or frictional effects during its transit through the pressure releasing component. The high shear conditions also cause modifications in the functional properties of the product. To overcome the limitations arising from the high shear forces and frictional heating, the semicontinuous operation can be adopted, as explained next.

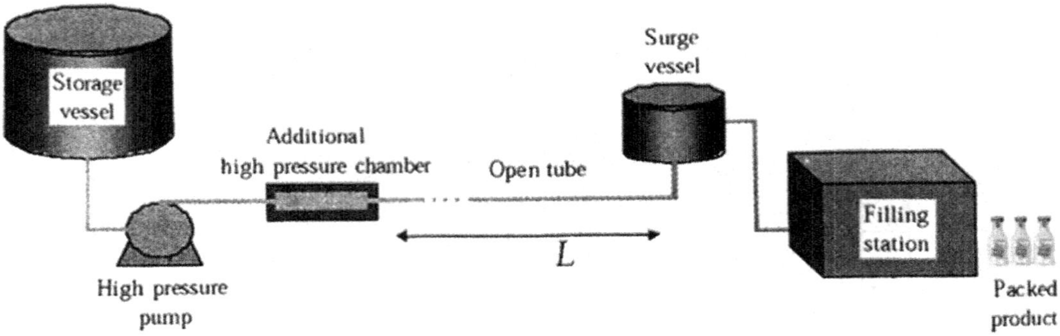

FIGURE 14.2 Continuous HPP system. (Reproduced with permission from van den Berg, R. W., Hoogland, H., Lelieveld, H. and Van Schepdael, L. 2001. High pressure equipment designs for food processing applications. In *Ultra High Pressure Treatment of Foods*, eds. M.E.G. Hendrickx and D. Knorr, 297–314. New York: Springer.)

14.1.1.3.3 Semicontinuous HPP System

A typical semicontinuous HPP system includes two or more pressure vessels, low-pressure pump to fill the vessels, HP transfer pump, holding tank, sterilized tanks, and control valves (Cavender, 2011). Inside the pressure vessel, a free-to-move divider piston separates the product from the pressurizing fluid. The function of control valves is to prevent cross-contamination between the outgoing stream of treated product and the incoming stream of untreated product (Balci and Wilbey, 1999).

Initially, the pressure vessel is filled with the liquid food product using the low-pressure pump. During filling, the free moving piston remains in a displaced position. Once the vessel is filled with the food product, the inlet valve is locked, and the pressure-transmitting fluid is compressed into the vessel, thereby moving the piston to compress the liquid food. The percentage compression obtained is similar to that in a batch HPP system. After the appropriate residence time, the vessel is depressurized. On depressurization, the food product is decompressed, and the free moving piston returns to its initial position. A low-pressure pump is used to move the piston towards the direction of the discharge port. The treated liquid product is then discharged into a sterile tank via a sterilized outlet, followed by aseptic packing in sterilized containers (Hogan et al., 2014).

14.1.1.4 Significant Operational Factors of HPP System

Working pressure of the HPP equipment is the most important factor in determining the product quality and capital cost of the equipment. It has been reported that a low working pressure increases the working life of the instrument and reduces the chances of operational failure. In general, the product temperature increases by 3°C for every 100 MPa increase in the chamber pressure.

Pressure vessel should withstand at least 8 h of operation/day in continuous or batch mode. It should preferably be made of thick steel wall to hold the pressure within. When exposed to HP in a continuous mode for a longer time, the steel wall may expand and stresses may be imposed on the steel structure. This may result in a leak eventually leading to safety hazards. HPP systems require a standard hydraulic pump to drive the intensifier pump which houses a large piston. The piston moves back and forth aided by the hydraulic oil within a low-pressure pump cylinder. In the HP cylinder, two small pistons are connected to the large piston of the low-pressure cylinder. This arrangement gives the high pressure output in HP cylinders. State-of-the-art control systems are used to ensure the safety during the HPP operation.

14.1.1.5 Advantages of HPP

The major advantage of HPP is that the characteristics of the processed product are retained as the sensorial and nutritional properties are maintained intact during processing. This is due to the uniform distribution of pressure inside the food leading to consistent treatment of the product. HPP enhances the product shelf life owing to the inactivation of pathogens such as *Listeria*, *Salmonella*, *Vibrio*, and *Norovirus*, and deteriorative enzymes. In addition, HPP eliminates the need for food preservatives, because of which HP-treated foods are free of chemical additives and hence considered as clean label foods (Hicks et al., 2009; Heinz and Buckow, 2010; Patterson, 2005). Heat-sensitive food products that cannot be treated thermally can be handled by HPP as it can be carried out at ambient temperature or even lower temperatures. Apart from the preservation effect, HPP at the pressure range of 500–700 MPa also imparts functional effects such as gelatinization of starch, gelation of proteins and polysaccharides, and tenderization of meat and fish. Thus, it is possible to prepare modified textured foods using the HPP technology. An added advantage is that the HPP technique is independent of the size and geometry of the samples.

HPP operates at a low operational cost owing to its lower energy requirement and reduced processing time. Due to this reason, HPP is an environment-friendly and energy-efficient technology. However, the major disadvantages of the HPP are the high equipment or capital cost and lack of clear guidelines about the HP-processed foods. The food processing applications of HPP technique are summarized in Table 14.1.

14.1.2 Cold Plasma

Plasma is referred to as the fourth state of matter after the solid, liquid, and gaseous states. Plasma formation occurs when the temperature of a gas is increased to a level where ionization occurs. Ionization of gas to the plasma state is achieved by using energy from a variety of sources such as electromagnetic radiation [microwave, radiofrequency (RF)], laser light, heat, and electricity, or by rapid compression. Plasma is composed of free electrons, photons, and ions. Thus, plasma has a net neutral charge. It possesses the highest energy level among all states. The increase in energy is in the order of solid < liquid < gas < plasma. Plasma is classified into thermal plasma and nonthermal or cold plasma, based on the conditions under which the plasma is generated. Spark plug and welding arc are the common examples of thermal plasma. Cold plasmas are found in plasma displays and neon signs. The temperature of thermal plasma is very high in the range of 10^7 K, which renders it inappropriate for food processing applications. On the other hand, the temperature of cold plasma is close to ambient temperature since the gas temperature remains at temperature levels below 40°C, even if the electron temperature reaches several ten thousands of kelvin (Mastwijk and Groot, 2010). Thus, cold plasma is suitable for food processing applications.

14.1.2.1 The Concept of Cold Plasma Technology

For food processing applications, cold plasma generated at atmospheric pressure is of great interest due to its microbial destruction ability. The principal mechanism of microbial inactivation by cold plasma is the chemical interaction between the cell membrane and the free radicals and the production of reactive species and charged particles during the plasma generation. Cold plasma inactivates the microorganisms by the following three mechanisms:

a. **Destruction of genetic material of microorganisms by UV irradiation:** UV radiation induces the formation of thymine dimers in DNA and hence inhibits the replication ability of bacteria (Laroussi et al., 2000).
b. **Erosion of microorganisms due to intrinsic photodesorption:** UV photons break the chemical bonds in the cellular material of microorganisms and lead to the formation of volatile compounds (e.g., CO and CH_x) from atoms intrinsic to the microorganism.
c. **Erosion of microorganisms through etching:** Etching results from the adsorption of reactive species on the microorganism and the subsequent formation of volatile compounds (e.g., CO_2, H_2O). The reactive species can be atoms (e.g., O), molecular radicals (O_3), or excited molecules in a metastable state (e.g., $^1O_2$ singlet state). The etching mechanism may be enhanced by UV photons (known as UV-induced etching) where the photons act synergistically with the reactive species and accelerate the inactivation rate of microorganisms (Philip et al., 2002).

TABLE 14.1

Applications of HPP in Food Industry

Product	Conditions of HPP	Salient Results	References
Potato cubes	Pressure: 400 MPa Holding time: 15 min Temperature: 5°C–50°C	• Reduction in the microbial count. • Complete inactivation of polyphenol oxidase at 20°C with the use of a dilute citric acid solution (0.5% or 1.0%) as the immersion medium. • Leaching of potassium was reduced by 20% after the HP treatment on vacuum-packaged samples. • Retention of ascorbic acid in HP-treated samples was temperature-dependent ranging from 90% at 5°C to at 50°C.	Eshtiaghi and Knorr (1993)
Apricot nectar, distilled water	Pressure: 600–900 MPa Holding time: 1–20 min Temperature: 20°C	• Complete inactivation of *Talaromyces flavus* ascospores at 900 MPa for 20 min at 20°C. • 2 $\log_{10}$ cycle reduction in *Neosartorya fischeri*. • No effect on *Byssochlamys fulva* and *Byssochlamys nivea* populations. • Complete inactivation of *Talaromyces flavus*, *Neosartorya fischeri*, *Byssochlamys fulva*, *Byssochlamys nivea* when the apricot nectar was preheated at 50°C followed by HPP at 800 MPa for 1–4 min.	Maggi et al. (1994)
Guava puree	Pressure: 400, 600 MPa Holding time: 15 min Temperature: 25°C	• After HPP at 600 MPa and 25°C for 15 min, the microorganisms in guava puree were inactivated to less than 10 CFU/mL. • No change in color, pectin, cloud, and ascorbic acid content. • The microbial count was reduced to 200 CFU/mL. • The pectin content, cloud, ascorbic acid content, and color gradually decreased in the puree treated at 400 MPa. • Retention of good quality in the product treated at 600 MPa and 25°C for 15 min after storage at 4°C for 40 days.	Yen and Lin (1996)
Milk	Pressure: 200–250 MPa Holding time: DNM[a] Temperature: DNM[a]	• In combination with bacteriocin (lacticin), the HP treatment resulted in significant reduction of microbial flora in milk. • No significant changes in the cheese-making properties.	Morgan et al. (2000)
Milk	Pressure: 100–600 MPa Holding time: 0–30 min Temperature: 20°C	• The rennet coagulation time of heated milk decreased with increasing pressure and time. • The strength of the HP-treated coagulum from heated milk was significantly greater. • The yield of cheese curd was 15% higher than that obtained from the unheated/non-HP-treated milk. • The protein content of whey resultant from HP-treated milk was 30% lower than the untreated counterpart.	Huppertz et al. (2002)
Cheese milk	Pressure: 200–400 MPa Holding time: DNM[a] Temperature: DNM[a]	• Increase in the yield of low-fat cheese with improved protein and moisture retention. • Improvement in coagulation properties. • Rapid degradation of protein and faster development of texture and flavor in the product. • The product was found to be less hard and cohesive and thus obtained high sensory scores.	Molina et al. (2000)
Milk	Pressure: 200–500 MPa Holding time: 60 min Temperature: 20°C	• Effective destruction of pathogens such as *Listeria monocytogenes*, *Escherichia coli*, and *Salmonella enteritidis*.	Vachon et al. (2002)
Cheese milk	Pressure: 300–600 MPa Holding time: DNM[a] Temperature: DNM[a]	• Increase in cheese yield due to denaturation of whey proteins and increased moisture retention. • Higher moisture content in HP-treated milk as the casein molecules and fat globules may not aggregate closely and thus might have allowed the moisture to be trapped in the cheese.	Drake et al. (1997)

(*Continued*)

TABLE 14.1 (*Continued*)

Applications of HPP in Food Industry

Product	Conditions of HPP	Salient Results	References
Milk	Pressure: 200–400 MPa Holding time: DNM[a] Temperature: DNM[a]	• The rennet coagulation time was reduced.	Needs et al. (2000)
Milk	Pressure: 250–500 MPa Holding time: 5 min Temperature: 20°C	• In combination with bacteriocin (nisin), HP treatment resulted in greater inactivation of Gram-positive and Gram-negative bacteria.	Black et al. (2005)
Cheese	Pressure: 345 and 483 MPa Holding time: 3 and 7 min	• Improved shredability of Cheddar cheese. • High visual acceptability and improved tactile handling were observed in the shreds from unripe milled curd Cheddar cheese.	Serrano et al. (2005)
Milk	Pressure: 400 MPa Holding time: 3 min	• No significant variations were observed in the contents of vitamin B1 and vitamin B6.	Sierra et al. (2000)
Pasteurized milk	Pressure: 400, 500 MPa Holding time: 15, 3 min Temperature: DNM[a]	• Increase in shelf life by 10 days.	Rademacher and Kessler (1997)
Cheddar cheese	Pressure: 50 MPa Holding time: 3 days Temperature: 25°C	• Accelerated ripening of commercial Cheddar cheese due to the degradation of αS1-Casein.	O'Reilly et al. (2000)
Fruits and vegetables	Pressure: 680 MPa Holding time: 10 min Temperature: Room temperature	• 5–6 log cycle reduction in microorganisms. • In combination with high-temperature treatment (67°C –71°C), HPP resulted in increased shelf life.	Hite et al. (1914)
Raw minced meat	Pressure: 700 MPa Holding time: 1 min Temperature: 15°C	• Reduction in *E. coli* O157:H7 from an initial count of 5.9 $\log_{10}$ CFU/g to 5 $\log_{10}$ CFU/g, after treatment.	Gola et al. (2000)
Pork slurry	Pressure: 400 MPa Holding time: 25 min Temperature: 10°C	• Reduction in *Campylobacter jejuni* from an initial count of 6–7 $\log_{10}$ CFU/g to 6 $\log_{10}$ CFU/g, after treatment.	Shigehisa et al. (1991)
Freshly ground raw chicken meat	Pressure: 408, 616, 818 MPa Holding time: 10 min Temperature: <28°C	• Applications of 408, 616, and 888 MPa resulted in estimated microbial spoilage times of 27, 70, and > 98 days, respectively.	O'Brien and Marshall (1996)
Cooked ham	Pressure: 600 MPa Holding time: 6 min Temperature: 31°C	• Reduction in *Satmonetia* spp. from an initial count of 3.8 $\log_{10}$ CFU/g to total inactivation (<1 CFU/g) after 120 days, at 4°C.	Garriga et al. (2002)
Bovine meat	Pressure: 520 MPa Holding time: 4.2 min Temperature: 10°C	• Reduction in total viable count from 4.0 to 2.5 $\log_{10}$ CFU/g, after treatment.	Jung et al. (2000)
Raw milk	Pressure: 100–400 MPa Holding time: 10–60 min Temperature: 20°C	• Improved microbiological quality of milk without modifying the activity of native milk enzymes such as lactoperoxidase or plasmin. • β-Lactoglobulin was denatured by pressures >100 MPa. • α-Lactalbumin and bovine serum albumin were resistant to pressures ≤400 MPa for 60 min. • Cheese yield increased with pressurization at 300 and 400 MPa, due to additional β-lactoglobulin and greater moisture retention. • The coagulation time decreased at pressure ≤200 MPa and then increased again until 400 MPa, reaching values comparable with those of the raw milk.	Lopez-Fandiño et al. (1996)

(*Continued*)

TABLE 14.1 (*Continued*)

Applications of HPP in Food Industry

Product	Conditions of HPP	Salient Results	References
Skim milk	Pressure: 310 MPa Holding time: 0.3 s Temperature: 25°C	• Pressurized milk (9% and 18% solids) produced stable foams. • The emulsion stability index was lower for the pressurized concentrated milk samples. • HP treatment produced more stable emulsions in unconcentrated milk. • No significant effect on interfacial tension. • Values for surface tension and viscosity were higher for HP-treated milk samples. • HP-treated milk had lower L* (less white), a* (greener), and b* (bluer) values than the unpressurized milk. • In concentrated milk, HP treatment imparted greener color only.	Adapa et al. (1997)
Marinated beef	Pressure: 600 MPa Holding time: 6 min Temperature: 31°C	• Reduction in lactic acid bacteria from 4.94 to 3.99 $\log_{10}$ CFU/g, after treatment.	Hugas et al. (2002)
Dry-cured ham	Pressure: 600 MPa Holding time: 9 min Temperature: DNM[a]	• Reduction in *L. monocytogenes* from 4.65 $\log_{10}$ CFU/g to total inactivation (<1 CFU/g), after treatment.	Tanzi et al. (2004)
Minced beef muscle	Pressure: 300 MPa Holding time: 10 min Temperature: 20°C	• Reduction in *C. freundii* from 7 to 5 $\log_{10}$ CFU/g, after treatment.	Carlez et al. (1993)

[a] DNM, Details not mentioned.

14.1.2.2 Cold Plasma Equipment

Cold plasma equipment can be broadly classified into three types, based on the placement of the food product.

i. **Food is placed significantly away from the plasma generation point:** Plasma is generated and then brought closer to the surface to be treated, by magnetic field or gas. In this system, the plasma generation section and treatment chamber are separated and thus, this system offers more flexibility with respect to the shape and size of the product. However, the inactivation activity is imparted only by the long-living chemical species. Hence, the extent of microbial destruction is low since the species with higher inactivation activity are short-lived and thus do not reach the surface intended to be treated.
ii. **Food is placed close to the plasma generation point:** Food material to be treated is kept in direct contact with the active plasma. These systems are operated in pulsed mode (frequency range: 100–1000/s). However, this system is not suitable for high water activity foods. Foods with high water activity conduct electricity and hence are susceptible to charring which affects the sensory quality and leads to coagulation of protein in the product. This may also alter the texture, aroma, and appearance of the product. Hence, the direct contact plasma systems should be designed in such a manner to avoid point source concentrated electrical conduction through the product.
iii. **Food is placed within the plasma generation field:** In this case, the food material to be treated is placed between the electrodes, within the plasma generation field (see Figure 14.3). Here, the food product is exposed to the highest plasma intensity. The spacing between the electrodes leads to some limitations with respect to the shape and size of the food material that can be treated in this system. This system is appropriate for the treatment of nuts, seeds, eggs, berries, or flat-shaped products.

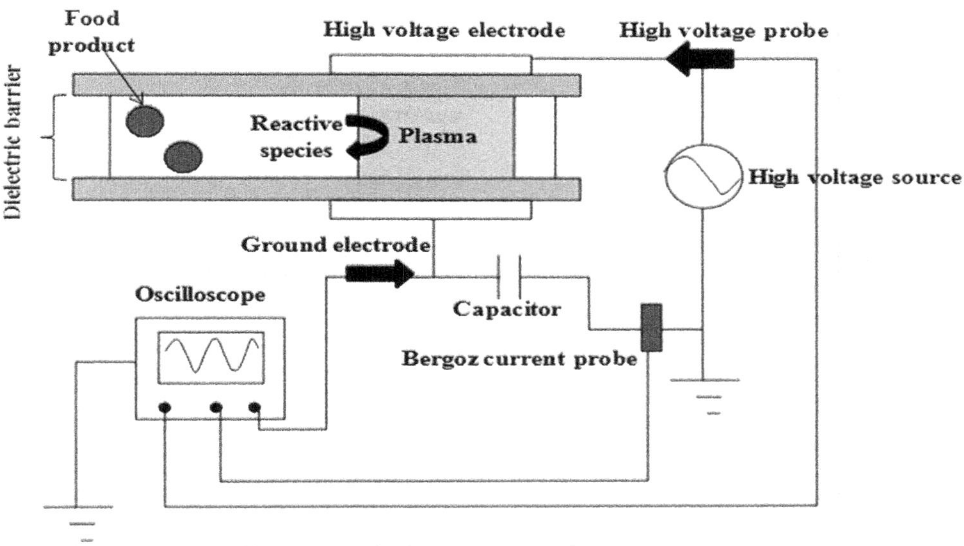

FIGURE 14.3 Atmospheric cold plasma equipment for surface decontamination of strawberry. (Redrawn from Misra, N., Patil, S., Moiseev, T., Bourke, P., Mosnier, J., Keener, K. and Cullen, P. J. 2014. In-package atmospheric pressure cold plasma treatment of strawberries. *Journal of Food Engineering* 125: 131–138.)

14.1.2.3 Method of Cold Plasma Generation

14.1.2.3.1 Atmospheric Pressure Plasma Jet

Atmospheric pressure plasma jet consists of two electrodes (Figure 14.4a). One is a needle electrode, and the other is a ring or grounded electrode. The distance between the two electrodes is maintained at several millimeters (mm). Nobel gases such as argon or neon are used for the plasma generation. Voltage is applied to generate the plasma. Plasma flame is generated in the RF range. Plasma jet thus generated has a smaller dimension and greater penetration in narrow gaps.

14.1.2.3.2 Dielectric Barrier Discharge

Dielectric barrier discharge consists of two electrodes and a dielectric material placed between these electrodes (Figure 14.4b). Here, the plasma is generated between the two electrodes. The presence of a dielectric barrier between the electrodes eliminates the phenomena of electrode etching and corrosion. This system uses a specific type of alternating current (AC) discharge that provides cold plasma at atmospheric pressure. The presence of capacitive coupling requires time-varying voltages to drive the dielectric barrier discharge. An AC voltage with an amplitude of 1–100 kV and a frequency up to several megahertz (MHz) are applied to this configuration (Becker et al., 2006; Chirokov et al., 2005). In this system, the major process parameters include the gas used for plasma generation, the distance between the electrodes, and the voltage supplied.

14.1.2.3.3 Corona Discharge

This system consists of an asymmetric electrode pair, i.e., a pointed or wire electrode with its geometry similar to a nozzle and another electrode of plate-type geometry (Figure 14.4c). The pointed electrode is powered by high voltage and may carry a positive or negative charge, and the plate electrode is grounded. This leads to the generation of a nonuniform electric field in the electrode gap. Plasma is created when the electric field near the point electrode exceeds the breakdown strength of gas. This phenomenon of the local breakdown of gases is termed as *corona discharge*. Corona discharge for plasma generation can be operated in direct current (DC) or pulse mode.

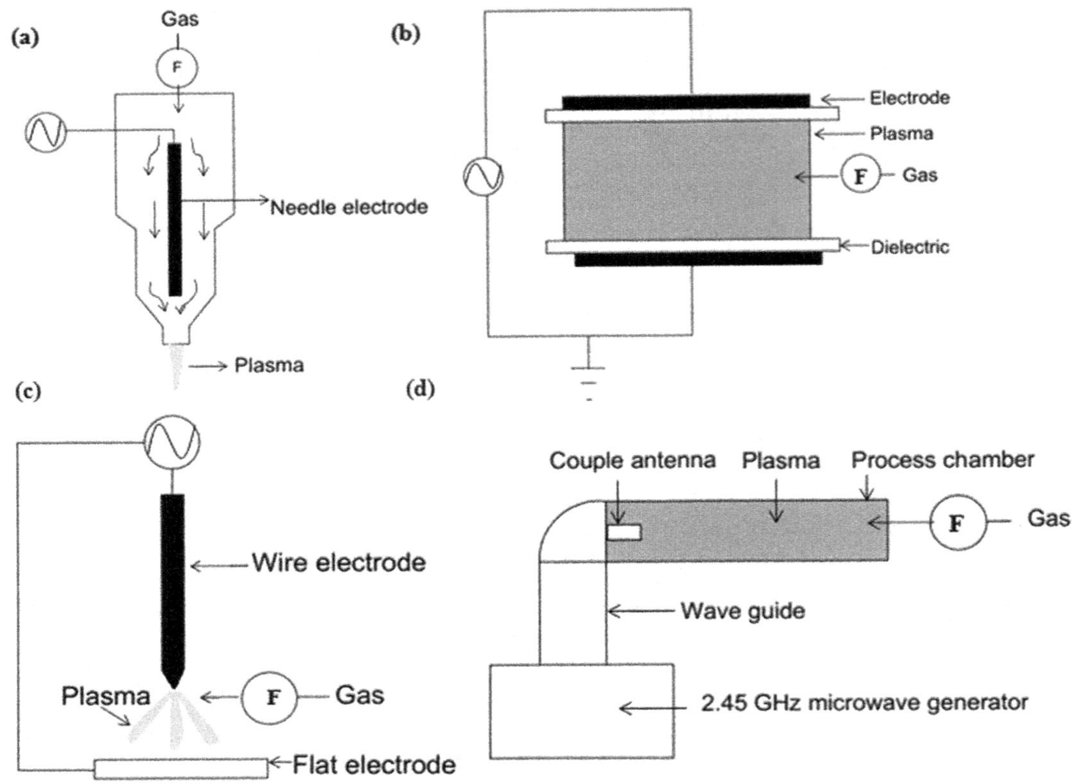

FIGURE 14.4 Different schematic setups of atmospheric pressure plasma sources: (a) Atmospheric pressure plasma jet; (b) dielectric barrier discharge; (c) streamer corona discharge; (d) autonomous through the chamber moving microwave discharge. (Reproduced with permission from Surowsky, B., Schlüter, O. and Knorr, D. 2015. Interactions of Non-Thermal Atmospheric Pressure Plasma with Solid and Liquid Food Systems: A Review. *Food Engineering Reviews* 7: 82–108.)

14.1.2.3.4 Microwave-Driven Discharge

Microwave-driven discharge systems operate without the requirement of electrodes for plasma generation. Microwaves generated in magnetron are passed to a gas chamber (Figure 14.4d). The plasma is generated when the gas ionizes by absorbing the microwave energy. The principle of microwave plasma jets is based on the surface wave sustained discharges in the dielectric tubes (Moisan et al., 1974, 1984). In this type of discharge, the surface wave (microwave) propagates along the interface between the generated plasma and the dielectric tube which encloses the plasma. While traveling along the plasma column surface, the wave continuously transfers a fraction of its power to the plasma it maintains. When the wave power is too low to sustain the plasma, the column ends on ensuring the appropriate gas flow rate. Then, the plasma exits out of the tube and forms plasma jet (Czylkowski et al., 2012).

14.1.2.4 Advantages and Limitations of Cold Plasma Technology

The cold plasma technology is a rapid nonthermal sterilization technique. This process is carried out at the ambient temperature and atmospheric pressure. Surface decontamination is achieved without altering any nutritional characteristics of the food product. Environment-friendly gases are used in the cold plasma process. Application of this process for surface treatment of food and packaging materials eliminates the use of chemical agents.

The cold plasma process is at its early stage, and the current research is focused on the antimicrobial efficiency of the plasma process, optimization of the process parameters for higher antimicrobial activity, feed-gas composition, and different design considerations for varied products. At present, cold plasma is mainly applied for the surface decontamination of food material. Currently, the operational cost of the cold plasma process is higher due to the use of feed gases (helium, oxygen, nitrogen) and use of high voltage (low-pressure cold plasma requires low electrical energy for generating plasma). Use of a mixture of feed gases or air as the feed gas may lower the operational cost of the cold plasma process. Also, the feed-gas composition directly influences the energy requirement for plasma generation. Generation of plasma from helium gas requires low energy, but the cost of helium gas limits its use as a sole feed gas for the process. Further, the fabrication and installation cost of the atmospheric cold plasma processing unit is high.

The food processing applications of cold plasma technique are summarized in Table 14.2.

TABLE 14.2

Applications of Cold Plasma Technology in Food Preservation

Product	Conditions of Cold Plasma Processing (CPP)	Salient Results	Reference
Orange juice	• Dielectric barrier discharge • Air/modified atmosphere 65 (65% O_2, 30% CO_2, 5% N_2) • 90 kV • 30–120 s	• No significant change in Brix or pH. • Vitamin C is reduced by 22% in air. • Pectin methylesterase (PME) activity reduced by 74% in air and 82% in MA65. • The maximum total color difference is less than 1.2. • Up to 5 $\log_{10}$ reduction of *Salmonella enterica*.	Xu et al. (2017)
Prebiotic orange juice	• Dielectric barrier discharge • 70 kV (50 Hz) • 15–60 s	• Degradation of oligosaccharides in the juice. • Decrease in pH. • Increase in L^* value and a slight reduction in chroma and hue angle. • Decrease in total phenolic content and antioxidant capacity in some cases.	Almeida et al. (2015)
White grape juice	• Dielectric barrier discharge • Source: air • 60 Hz; 80 kV • Treatment time: 1–4 min	• No significant change in pH, acidity, and electrical conductivity of the juice. • An increase in nonenzymatic browning with a minimal total color difference. • Decrease in total phenolics, total flavonoids, 2,2-diphenyl-1-picrylhydrazyl (DPPH) free radicals scavenging, and antioxidant capacity. • An increase in total flavonols content. • 7.4 $\log_{10}$ CFU/mL reduction in *Saccharomyces cerevisiae* at 80 kV for 4 min.	Pankaj et al. (2017)
Pomegranate juice	• Plasma jet • Source: Ar • 25 kHz • 0.75–1.25 dm^3/min • Treatment time: 3–7 min	• Increase in total anthocyanin content. • No visual differences in color.	Kovačević et al. (2016)
Radish sprouts	• Microwave plasma • Source: N_2 • Flow rate: 1 L/min • 2.45 GHz, 900 W, 669 Pa • Treatment time: 1–20 min	• No change in color, water activity, ascorbic acid concentration, and antioxidant activity. • Lower moisture content during storage. • 2.6 $\log_{10}$ reduction in *Salmonella typhimurium*. • 0.8 $\log_{10}$ reduction in total mesophilic aerobes	Oh et al. (2017)

(Continued)

TABLE 14.2 (*Continued*)

Applications of Cold Plasma Technology in Food Preservation

Product	Conditions of Cold Plasma Processing (CPP)	Salient Results	Reference
Blueberry	• Dielectric barrier discharge • Source: air • 50 Hz, 60–80 kV • Treatment time: 0–5 min	• Decrease in firmness, total phenol, flavonoid, and anthocyanin on extended cold plasma treatment at the higher voltage level. • A significant increase in total soluble solid. • No significant change in acidity and color (except fruit darkening at 80 kV for 5 min).	Sarangapani et al. (2017)
Blueberry	• Plasma jet • Source: air • 47 kHz, 549 W • 4–7 cubic feet/min • 7.5 cm • Treatment time: 0–120 s	• Significant reductions in firmness, color, and anthocyanins at higher treatment times. • Up to 2 $\log_{10}$ reduction in total aerobic plate count.	Lacombe et al. (2015)
Strawberry	• Dielectric barrier discharge • 60 kV, 50 Hz • 65% O_2 + 16% N_2 + 19% CO_2 and 90% N_2 + 10% O_2. • Indirect exposure • Treatment time: 5 min	• Strawberries in high oxygen mixture showed higher firmness with similar respiration rates. • Some changes in L* and a* values were observed. • ~3.0 $\log_{10}$ reduction in microbes in both gas mixtures.	Misra et al. (2014)
Mandarins	• Microwave plasma • 2.45 GHz, 900 W, 1 L/min • 0.7 kPa, N_2, He, N_2 + O_2 (4:1) • Treatment time: 10 min	• Increased total phenolic content and antioxidant activity. • No significant change in CO_2 generation, weight loss, soluble solids, acidity, pH, ascorbic acid, and color. • Significant inhibition of *Penicillium italicum* (84% reduction in disease incidence).	Won et al. (2017)
Kiwifruit	• Dielectric barrier discharge • 15 kV • Treatment time: 10–20 min	• Improved color retention and reduced darkened area formation during storage. • No significant changes in color, hardness, vitamin C, and antioxidant activity. • Longer treatment increased the soluble solid content. • 15% decrease in *chlorophyll a* on day 0 with no difference on day 4.	Ramazzina et al. (2015)
Apple (Pink Lady apples)	• Dielectric barrier discharge • Source: air • 12.7 kHz, 150 W • Treatment time: 30, 120 min	• Up to 10% reduction of antioxidant content and antioxidant capacity. • No significant difference in total phenolic content but significant decrease in the total phenolic index was observed.	Ramazzina et al. (2016)
Melon	• Dielectric barrier discharge • Source: air • 15 kV, 12.5 kHz • Treatment time: 30, 60 min	• No change in acidity, soluble solid content, dry matter, color, and texture. • 17% and 7% reduction in peroxidase and PME activities, respectively. • 3.4 and 2 $\log_{10}$ reductions in mesophilic and lactic acid bacteria, respectively.	Tappi et al. (2016)
Cherry tomatoes	• Dielectric barrier discharge • Source: air • 100 kV • Treatment time: 150 s	• No significant difference in color, firmness, pH, or total soluble solids. • >5 and 3.5 $\log_{10}$ CFU/sample reduction in *E. coli* and *Listeria innocua*. • Up to 3.5 $\log_{10}$ CFU/sample reduction of spoilage microflora (mesophiles, yeast, and mold).	Ziuzina et al. (2016)

(*Continued*)

TABLE 14.2 (*Continued*)

Applications of Cold Plasma Technology in Food Preservation

Product	Conditions of Cold Plasma Processing (CPP)	Salient Results	Reference
Fresh fruit and vegetable slices (pears, cucumbers, and carrots)	• Plasma micro-jet • 30 mA; 500 V • Treatment time: 1–8 min	• Less than 5% moisture loss in all three samples after 8-min treatment. • Minimal change in total color difference. • 3.6%, 3.2%, and 2.8% reduction of vitamin C in cucumber, carrot, and pear slice, respectively. • 90%, 60%, and 40% *Salmonella* inactivation in carrot, cucumber, and pear slice, respectively	Wang et al. (2012)
Red chicory	• Dielectric barrier discharge • 19.15 V, 3.15 A • Treatment time: 15 min • Deionized water	• No detrimental effects on color, freshness, and texture. • Odor and overall acceptability slightly decreased during storage. • >4 $\log_{10}$ CFU/cm^2 reduction of *L. monocytogenes* and >5 $\log_{10}$ reduction of *E. coli*.	Trevisani et al. (2017)
Red chicory (radicchio)	• Dielectric barrier discharge • Source: air • 15 kV, 12.5 kHz • 1.5 m/s • Treatment time: 15–30 min	• No significant effects on antioxidant activity and external appearance. • 1.35 $\log_{10}$ MPN/cm^2 reduction of *E. coli* O158:H7. • 2.2 $\log_{10}$ CFU/cm^2 reduction of *L. monocytogenes*.	Pasquali et al. (2016)
Romaine lettuce	• Dielectric barrier discharge • Source: air • 42.6 kV, 1.5 A • Treatment time: 10 min	• No significant change in the surface morphology, color, respiration rate, and weight loss. • 0.4–0.8 $\log_{10}$ CFU/g reduction of *E. coli* O157:H7 in the leaf samples in the 1-, 3-, and 5-layer configurations. • 1.1 $\log_{10}$ CFU/g reduction in bulk stacking with seven layers.	Min et al. (2017)
Fresh produce (romaine lettuce, baby carrots, and cocktail tomatoes)	• Atmospheric pressure cold plasma • Source: argon • 3.95–12.83 kV • 60 Hz • Treatment time: 0.5–10 min	• No significant changes in color in any samples. • 0.5, 1.7, and 1.5 $\log_{10}$ reduction of *E. coli* in carrot, tomato, and lettuce, respectively.	Bermúdez-Aguirre et al. (2013)
Grains: wheat, bean, chickpea, soybean, barley, oat, rye, lentil, and corn	• Low-pressure cold plasma • Source: air and SF_6 • 1 kHz, 20 kV, 500 mTorr; 300 W • Treatment time: 5–20 min	• A slight change in moisture content of legume and wheat. • No difference in water soaking, yield, and cooking time of legumes. • No change in wet gluten content, gluten index, and sedimentation in wheat. • 3 $\log_{10}$ reduction of *Aspergillus* spp. and *Penicillium* spp. after 15 min treatment in SF_6.	Selcuk et al. (2008)
Wheat flour (soft and hard)	• Dielectric barrier discharge • Source: air • 60–70 kV • Treatment time: 5–10 min	• An increase in the peak time, peak integral, elastic modulus, viscous modulus, dough strength, and optimum mixing time.	Misra et al. (2015)
Onion powder	• Microwave plasma • Source: He • 170 and 250 mW/m^2, 2.45 GHz, 400–900 W, 0.7 kPa; 1 L/min • Treatment time: 10–40 min	• No effect on color, antioxidant activity and quercetin concentration. • 2.1 $\log_{10}$ spores/cm^2, 1.6 $\log_{10}$ spores/cm^2 and 1.9 CFU/cm^2 reduction of *Bacillus cereus*, *A. brasiliensis* spore, and *E. coli* O157:H7, respectively.	Kim et al. (2017)

(*Continued*)

TABLE 14.2 (*Continued*)

Applications of Cold Plasma Technology in Food Preservation

Product	Conditions of Cold Plasma Processing (CPP)	Salient Results	Reference
Fresh and frozen pork	• Plasma jet • Source: air • 20 kV; 58 kHz; 1.5 amp • Treatment time: 0–120 s	• No significant changes in volatile basic nitrogen, peroxide value, and thiobarbituric acid reactive substances (TBARS). • No significant impact on the sensory characteristics of frozen pork. • Significant changes in color for both fresh and frozen pork. • 1.5 $\log_{10}$ reduction of *E. coli* O157:H7 and >1.0 $\log_{10}$ reduction of *L. monocytogenes*.	Choi et al. (2016)
Raw pork	• Low-pressure plasma • Source: He • 20 kPa • Treatment time: 0–10 min	• Significant changes in total color difference, hue angle, and chroma. • Decreased ferric reducing ability after 14 days of storage. • 3% increase in polyunsaturated fatty acids during storage. • No oxidative processes were observed.	Ulbin-Figlewicz et al. (2016)

14.1.3 Pulsed Electric Field Processing

14.1.3.1 The Concept of Pulsed Electric Field Processing

In the pulsed electric field (PEF) processing, a food sample is placed or passed between two electrodes, and pulses of high-intensity electric field are applied to inactivate the microorganisms in the food product. Treatment time in case of PEF processing is very less (1 s), and electric field in the range of 20–80 kV is applied in the form of 2-μs pulse duration.

In PEF, inactivation of microorganisms occurs due to the electroporation and electrical breakdown of the cell. In electroporation, the membrane of cell is perforated due to the high electric field strength. Swelling followed by cell rupture results in the death of microorganisms.

PEF system comprises a capacitor, high voltage supply, high voltage switch of high power (to discharge capacitor's energy), two electrodes, and a treatment chamber (Figure 14.5a). Sometimes, a cooling mechanism is also required to dissipate the heat generated inside the food. High voltage is continuously supplied and stored in the capacitor. The quantity of energy stored in the capacitor is given by

$$Q = \frac{CE^2}{2} \tag{14.1}$$

where Q is the energy stored in the capacitor (J), E is the charging voltage (V), and C is the effective capacitance of the capacitor (F).

Based on the dielectric constant of the food, the effective capacitance can be determined by the following equation:

$$C = \frac{\tau \sigma A}{d} = \frac{\varepsilon_0 \varepsilon_r A}{d} \tag{14.2}$$

where τ is the pulse duration (s), σ is the conductivity of food (S/m), A is the surface area of the electrode (m^2), d is the distance between two electrodes (m), ε_0 is the permittivity of free space, and ε_r is the relative permittivity.

The stored energy in the capacitor is discharged instantaneously in pulse form through the electrodes into the food. Electrodes are in direct contact with the food and arranged in parallel or as concentric cylinders, through which the fluid food is passed.

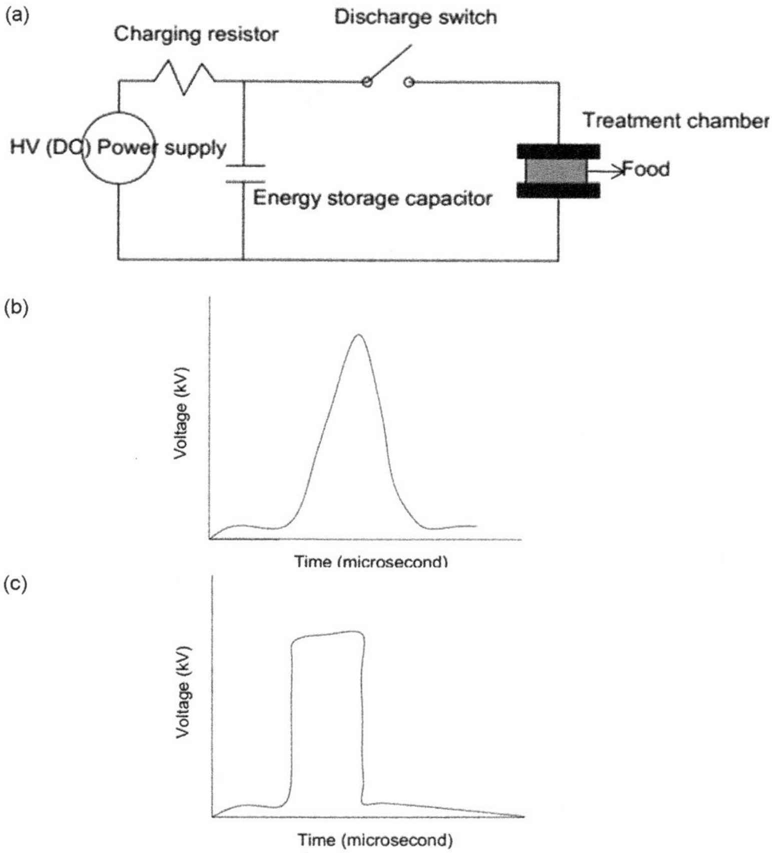

FIGURE 14.5 (a) Schematic of PEF system; (b) exponential decay form; (c) square wave form.

Two different forms of pulse are observed on the instantaneous discharge of voltage from the capacitor: exponential decay and square wave form. Figure 14.5a shows the arrangement of a PEF system consisting of one capacitor and the resultant exponential wave form (Figure 14.5b). In this form, the voltage rises quickly to its maximum value and gradually slows down to zero. On arranging two or more capacitors in parallel, a pulse of the square wave form is generated. In the square wave form (Figure 14.5c), voltage is increased from zero to maximum and held for the specific time duration, followed by a rapid fall to zero.

Pulse frequency (v) is given by the following equation:

$$v = \frac{nF}{V} \tag{14.3}$$

where n is the number of pulses, V is the volume of the chamber, and F is the volumetric flow rate.

14.1.3.2 Factors Affecting Microbial Inactivation by PEF Treatment

Increase in electric field strength facilitates microbial inactivation as it surpasses a critical level to cause irreversible damage to the cell membrane. However, at very high field strength, arcing may result. PEF treatment time is given by the product of pulse width and the number of pulses applied. Treatment time should exceed the critical treatment time in order to inactivate the microbial cells. Longer treatment time may result in the nutrient loss as well as heating of the food product. Nevertheless, treatment time contributes less towards the microbial inactivation, when compared to the electric field strength. In addition to the electric field strength and treatment time, the shape of the pulse also affects the lethality of the process. The square wave form is more lethal than the exponential decay form for the inactivation of microorganisms. Using pulse of short time duration (1–3 μs) minimizes the absorption of energy that restricts the heating of food and reduces the energy loss. Further, a bipolar pulse is more lethal than the monopolar wave. Bipolar pulse is generated by the reversal of polar orientation of the electric field. In addition, bipolar pulse reduces energy consumption. However, the presence of bivalent ions such as Ca^{2+} in foods shows a protective effect on the microbial cell membrane. But, at low pH, PEF leads to an enhanced microbial inactivation rate. Microbial cells in the logarithmic phase are more susceptible to PEF treatment. Gram-negative bacteria are more prone to inactivation by PEF than the Gram-positive counterparts.

14.1.3.3 Advantages and Limitations of PEF Processing

Nutritional quality of the processed food product is enhanced with PEF methodology. PEF requires less energy compared to the thermal treatment methods. PEF is rapid and instantaneous, and requires very low processing time (<1 s). There is no requirement for temperature gradient inside the food to be treated by PEF. However, only liquid foods can be processed using PEF. Liquid food containing particles of dimensions larger than the gap between the electrodes cannot be treated with PEF. The presence of bubbles in the food may lead to arcing of electrodes and ultimately results in a non-uniform treatment. Though PEF requires very less processing time and low operational cost, it needs a high capital investment. This restricts the wide commercial use of PEF. Applications of PEF processing in the food sector are summarized in Table 14.3.

TABLE 14.3

Applications of PEF Technology in Food Processing

Product	PEF Conditions	Salient Results	Reference
Orange juice	• Electric field strength, *E* (kV/cm): 6.7 • Total time, *t* (μs): 20 • No. of pulses (*n*): 5 • Temperature of the product after PEF treatment, *T* (°C): 45–50	Microflora inactivated: *Saccharomyces cerevisiae* Log cycles D: 4.9 D	Grahl and Märkl (1996)
Pasteurized Milk	• *E* (kV/cm): 43 • *t* (μs): 100 • *n*: 23 • *T* (°C): 42.8	Microflora inactivated: *E. coli* Log cycles D: 3 D	Dunn and Pearlman (1987)
Milk	• *E* (kV/cm): 36.7 • *t* (μs): 100 • *n*: 40 • *T* (°C): 63	Microflora inactivated: *Salmonella dublin* Log cycles D: 3 D	Dunn and Pearlman (1987)
Skim milk	• *E* (kV/cm): 35 • *t* (ns): 500 • *n*: 64 • *T* (°C): <50	Microflora inactivated: *Salmonella enteritidis* Log cycles D: 1.0 D	Floury et al. (2006)
Whole milk, skim milk, dairy cream	• *E* (kV/cm): 17–46 • *t* (μs): 1.51 • *n*: 545 • *T* (°C): 21.5–42	Microflora inactivated: *L. innocua* Log cycles D: 6 D	Picart et al. (2002)
Skim milk	• *E* (kV/cm): 35 • *t* (μs): 8 • *n*: 150 • *T* (°C): 25	Microflora inactivated: *Staphylococcus aureus* Log cycles D: 4.5 D	Sobrino-Lopez et al. (2006)

(*Continued*)

TABLE 14.3 (*Continued*)

Applications of PEF Technology in Food Processing

Product	PEF Conditions	Salient Results	Reference
Sterile dairy cream (33% fat)	• E (kV/cm): 34 • t (μs): 7 • n: 64 • T (°C): <30	Microflora inactivated: *E. coli* Log cycles D: 5 D	Manas et al. (2001)
Liquid whole egg with a portion of the yolk removed	• E (kV/cm): 36 • n: 25 • Exponential, static chamber, parallel electrodes, d = 2 cm	• The pasteurized sample treated with PEF+Heat (60°C)+Additives (potassium sorbate and citric acid) resulted in the maximum shelf life of 28 days at 4°C and 10 days at 10°C.	Dunn et al. (1989)
Ovalbumin solution 2% and dialyzed fresh egg white	• E (kV/cm): 27–33 • t (μs): 0.3 (20 nF) and 0.9 (80 nF) • n: 50–400 pulses in series of 20 or 50, at 1.1 Hz • Temperature during treatment (°C): <29°C • Discontinuous and exponential pulse. • 0.7–2.3 J/pulse*mL; 100 Hz, V = 5.7 mL; Ø = 0.38 cm	• No significant differences in the sensory score of the product. • No changes in polarity and conformation of the amino acids: tyrosine and tryptophan. • A slight reduction was observed in the gelling properties.	Fernández-Díaz et al. (2000)
Beef muscle	• E (kV/cm): 1.1–2.8 • Energy density (kJ/kg): 12.7–226 • Frequency (Hz): 5–200 • n: 152–300	• PEF treatment that induced a ΔT of 22°C significantly affected the weight loss of samples after the treatment, indicating slight changes in the cell membrane and more water loss. • Particle size analysis of the extracted myofibrils showed significant changes in the myofibrils. • No significant changes in the instrumental texture.	O'Dowd et al. (2013)
Apple juice	• E (kV/cm): 50, 58, 66 • n: 2, 4, 8, 16	• Reduced aerobic plate count, yeasts and molds, and aciduric bacteria. • No significant changes in the quality attributes such as pH, acidity, and soluble solids. • Fading of juice color was observed.	Ortega-Rivas et al. (1998)
β-lactoglobulin solution and fresh egg white	• E (kV/cm): 12.5 • n: 1–20 • Electroporation system	• High denaturation was observed in both β-lactoglobulin solution and fresh egg white. • Low aggregation in β-lg. • Reduced gelling time.	Qin et al. (1995)
Blend of orange juice and carrot juice (4:1 v/v)	• E (kV/cm): 25 • Total treatment time (t, μs): 340 • Maximum outlet temperature (T_{max}, °C): 63 • Continuous treatment with 2.5-μs rectangular pulses applied in bipolar mode	• 81.4% inactivation of the enzyme, PME was observed.	Rodrigo et al. (2003)

(*Continued*)

TABLE 14.3 (*Continued*)

Applications of PEF Technology in Food Processing

Product	PEF Conditions	Salient Results	Reference
Tomato juice	• *E* (kV/cm): 35 • *t* (μs): 1000 • Pulse width (μs): 1–7 • Frequency (Hz): 50–250 • Monopolar or bipolar mode	• An increase in frequency or pulse width resulted in a decrease of both residual lipoxygenase and hydroperoxide lyase activities. • The lowest residual activity of lipoxygenase (81%) was observed when tomato juice was treated at 250 Hz for 7 μs in the bipolar mode. • The aforesaid conditions reduced the residual activity of hydroperoxide lyase down to 10%.	Aguiló-Aguayo et al. (2009)
Sponge cake made with PEF-treated egg	• *E* (kV/cm): 48 • *n*: 20 • *t* (μs): 2	• No significant differences in density and whiteness of the cake. • The strength of the cake made with PEF-treated egg was greater than that prepared with untreated egg.	Ma et al. (1997)
Potato	• *E* (kV/cm): 0.35–3.0 • *n*: 1–70 • *t* (μs): 2	• Treatment under the optimum PEF processing conditions of $E = 0.9–2.0$ kV/cm and $n = 15–30$) resulted in improved mass transfer within the product. • Reduction in the drying time during fluidized bed drying of potato cubes, by 20%–25%. • No significant changes in the rehydration properties and quality characteristics of the products after the PEF treatment.	Angersbach and Knorr (1997)
Coconut	• *E* (kV/cm): 0.1–2.5 • *n*: 0–200 • Pulse width (μs): 575 • Frequency: 1 Hz	• Treatment with PEF under optimized conditions ($E = 2.5$ kV/cm; $n = 20$ pulses; pulse width = 575 μs and frequency = 1 Hz) resulted in 20% increase in milk yield compared to untreated samples. • The fat contents of the milk extracted from PEF-treated and PEF-untreated samples were 58.0% and 61.2%, respectively. • The protein contents of the aqueous medium in the milk extracted from PEF-treated and PEF-untreated samples, after separation by centrifugation were 50.0% and 51.6%. • The synergistic effect of PEF and centrifugation lead to a reduction in drying time by ~22% relative to the untreated samples; this is considered advantageous in terms of copra production.	Ade-Omowaye et al. (2000)
Clementine juice	• $E = 25$ kV/cm • Total treatment time: 330 μs • Maximum outlet temperature (T_{max}): ~72°C • Continuous treatment with 2-μs rectangular pulses applied in the bipolar mode	• Inactivation of PME achieved: 88.3%	Sentandreu et al. (2006)

(*Continued*)

TABLE 14.3 (*Continued*)

Applications of PEF Technology in Food Processing

Product	PEF Conditions	Salient Results	Reference
Grapefruit juice	• $E = 20$ kV/cm; Total treatment time: 25 µs; Maximum outlet temperature (T_{max}): ~25°C • $E = 30$ kV/cm; Total treatment time: 100 µs; Maximum outlet temperature (T_{max}): 60°C • Continuous treatment in a parallel plate system with 1 µs rectangular pulses in monopolar mode	• Inactivation of PME achieved: 59% • Inactivation of PME achieved: 90%	Riener et al. (2009)
Skim milk, whole milk, and yogurt	• Field strength (kV/cm): 15–28 • Treatment time (µs): 0–2000 • Specific energy (kJ/kg): 0–3490	• A maximum inactivation of *L. monocytogenes*: 4.77 $\log_{10}$ cycles was observed after the PEF treatment at 28 kV/cm, 2000 µs and 3490 kJ/kg	Álvarez et al. (2003)
Cranberry juice	• E (kV/cm): 20, 40 • t (µs): 50,150	• Higher field strength and longer treatment time reduced more viable microbial cells • No significant effect on the overall volatile profile of the juice after the PEF treatment • No significant difference in color was observed in the PEF-treated samples • Under the conditions of 40 kV/cm and 150 µs, no growth of molds and yeasts was observed during storage at 22°C and 4°C, and no growth of aerobic bacteria was noted during storage at 4°C	Jin and Zhang (1999)
Orange juice	• $E = 40$ kV/cm • Total treatment time: 150 µs • Maximum outlet temperature (T_{max}): 56°C • pH of juice: 3.7	• A 5.5 $\log_{10}$ reduction in the population of *Staphylococcus aureus* in the juice was achieved after the PEF treatment	Walkling-Ribeiro et al. (2009)

14.2 Electromagnetic Radiation-Based Food Processing Techniques

14.2.1 Infrared Heating

14.2.1.1 The Principle of Infrared (IR) Heating

Infrared or IR heating works on the principle of radiative heat transfer (as discussed in Chapter 5). IR radiation is a part of the electromagnetic spectrum (Figure 14.6), with its wavelength in the range of 0.74–1000 µm and positioned in between the ultraviolet and microwave radiations. The underlying mechanism of IR heating is the absorption of radiation by the product, which in turn solely depends on the wavelength of radiation. Based on wavelength, the IR radiation can be classified into three regions, given as follows (Sakai and Hanzawa, 1994):

1. Near-infrared (NIR) → Wavelength: 0.75–1.4 µm
2. Mid-infrared (MIR) → Wavelength: 1.4–3 µm
3. Far-infrared (FIR) → Wavelength: 3–1000 µm

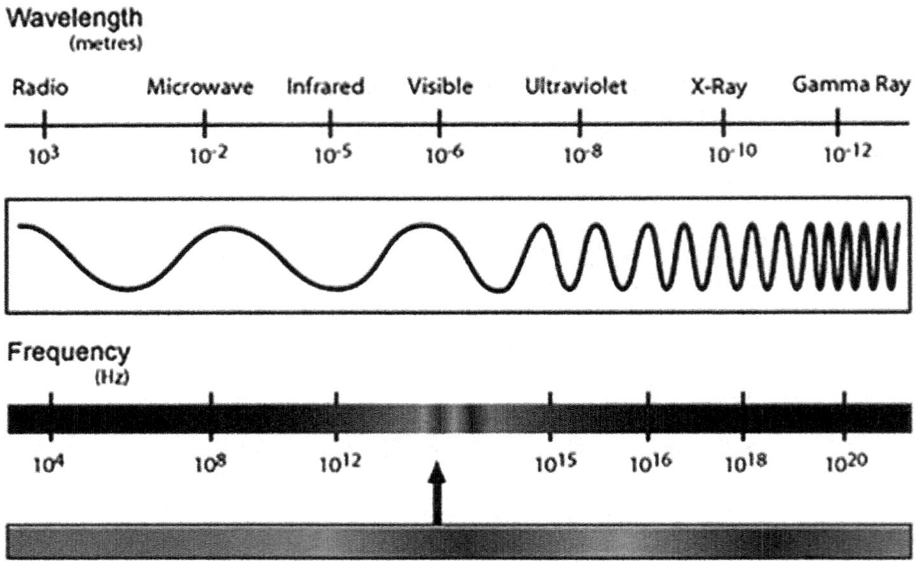

FIGURE 14.6 The electromagnetic wave spectrum. (Reproduced with permission from Motasemi, F. and Ani, F. N. 2012. A review on microwave-assisted production of biodiesel. *Renewable and Sustainable Energy Reviews* 16: 4719–4733.)

Penetration ability of the shorter wavelength IR radiation is ten times as that of its longer wavelength counterparts. The wavelength of IR radiation is of relevance in heating applications as it is inversely proportional to temperature, according to the Wien's displacement law (discussed in Chapter 5: Heat Transfer). With the increase in temperature of the source, maximum emitting power is shifted to the shorter wavelength. Therefore, shorter wavelength results in higher temperature and vice versa. In accordance with the above, NIR is commonly used for industrial applications. Nevertheless, FIR is preferred for food processing applications as most of the food components absorb radiative energy in the wavelength region of FIR (Sandu, 1986; Krishnamurthy et al., 2008).

As the FIR radiation penetrates the food product, the specific food molecules absorb the associated radiant energy at their corresponding range of wavelength (Table 14.4). This leads to an increase in the vibrational activity of molecules. When the molecules return to their normal state, the absorbed energy is released as heat. In food products, the molecules absorbing IR radiation are almost always that of water. When the hydroxyl (O–H) bonds in water absorb the energy of radiation, they begin to rotate in a frequency same as that of the incident radiation. The conversion of radiation energy to rotational energy results in the evaporation of water. Other food molecules that absorb efficiently in the FIR region include proteins, lipids, and carbohydrates that contain polar groups such as –OH, –NH, –CO, and C=C (Staack, 2008) (Table 14.4).

With this fundamental insight on the principle of IR heating, it is important to understand the basic laws governing IR radiation (Table 14.5). The laws listed in Table 14.5 collectively indicate the following (Krishnamurthy et al., 2008):

- The amount of IR radiation incident on any surface has spectral dependence, as energy from an emitter contains different wavelengths.
- The fraction of radiation in each band relies on the temperature and emissivity of the emitter. Accordingly, the rate of IR heating increases with the increase in the surface temperature of the radiator (Masamura et al., 1988).
- The wavelength at which the maximum radiation occurs is dependent on the temperature of IR heating elements. A blackbody at a temperature of 1000 K will have a peak emission at a wavelength of 2.9 μm, whereas a tungsten lamp at a temperature of 2900 K will have a peak emission at a wavelength of 1 μm.

TABLE 14.4

IR Absorption Wavelength Region of Chemical Groups and the Corresponding Food Components

Food Component	Chemical Group	Absorption Wavelength (μm)
Water, sugars	Hydroxyl group (–OH)	2.7–3.3
Lipids, sugars, and proteins	Aliphatic carbon–hydrogen bond	3.25–3.7
Lipids	Carbonyl group (C=O) (ester)	5.71–5.76
Proteins	Carbonyl group (C=O) (amide)	5.92
Proteins	Nitrogen–hydrogen group (–NH–)	2.83–3.33
Unsaturated lipids	Carbon–carbon double bond (C=C)	4.44–4.76

Source: Adapted from Rosenthal (1992).

TABLE 14.5

Governing Laws of IR Radiation

Governing Law	Equation	Physical Meaning and Applications
Planck's law	$E_{b\lambda}(T,\lambda) = \dfrac{2\pi hc_0^2}{n^2\lambda^5\left[e^{\frac{hc_0}{n\lambda kT}} - 1\right]}$ where k is the Boltzmann's constant, 1.381×10^{-3} J/K, n is the refractive index of the medium, λ is the wavelength (μm), T is the source temperature (K), c_o is the speed of light (km/s), and h is the Planck's constant (6.626×10^{-34} J s).	Provides distribution of the emissive power for a spectral blackbody, given by $E_{b\lambda}$ (T,λ).
Wien's displacement law	$\lambda_{max} = \dfrac{2898}{T}$ where λ_{max} is the peak wavelength (μm), and T is the source temperature (K).	Gives the peak wavelength at which the radiation spectral distribution emitted by a blackbody achieves maximum emissive power. Shows that wavelength and temperature are inversely related.
Stefan–Boltzmann law	$E_b(T) = \int_0^\infty E_{b\lambda}(T,\lambda)d\lambda = n^2T^4 \dfrac{2\pi hc_0^2 d(n\lambda T)}{n^2\lambda^5\left[e^{\frac{hc_0}{n\lambda kT}} - 1\right]} = n^2\sigma T^4$ where σ is Stefan–Boltzmann's constant (5.670×10^{-8} W/m² K⁴).	Gives the total power radiated by an IR source at a specific temperature.
Beer's law	$H_\lambda = H_{\lambda 0}\exp(-\sigma_\lambda^* u)$ where σ^*_λ is the spectral extinction coefficient (m²/kg), $H_{\lambda o}$ is the incident spectral irradiance (W/m²·μm), H_λ is the transmitted spectral irradiance (W/m²·μm), and u is the mass of absorbing medium per unit area (kg/ m²).	Gives the transmitted spectral irradiance in a non-homogenous system. It states that the amount of light absorbed by the solution varies exponentially with the length of the light path in the solution and concentration of the absorbing species in the solution.

In addition to the above, the other factors affecting IR penetration into the food product and absorption of IR by the food products are the optical characteristics including absorptivity, transmissivity, and reflectivity, and the size, shape, and chemical composition of the product.

14.2.1.2 Advantages of IR Heating

The major advantages of IR heating over conventional heating methods are the high rate of heat transfer, uniform heating, reduced processing time, and improved product quality. All of these merits of IR heating can be attributed to the fundamental differences in the pattern of heat and mass transfer between the conventional and IR heating. When the product is exposed to IR heating, intense heating occurs from inside the product to the surface, leading to a consequent reduction in temperature and moisture gradient within the product in a short time. This prevents the phenomenon of *case-hardening* which occurs during conventional heating since the heat transferred by convection acts mainly on the product surface and then passed on to the center of the product by conduction (Hebbar and Rastogi, 2001). Case-hardening hinders mass transfer and leads to non-uniform heating, incomplete moisture removal, and reduced product quality. The other merits of IR heating include the following:

- High thermal efficiency and heat density
- Shorter start-up time
- A medium is not required for energy transfer (as discussed in Chapter 5)
- Efficient and uniform heating of food product.
- Easy process and control of parameters
- Less processing time and less energy cost
- Can be used for a wide range of food products
- Environment-friendly technique compared to conventional heating methods

14.2.1.3 Limitations of IR Heating

Irrespective of the abovementioned advantages, IR heating suffers from certain limitations, listed as follows (Dostie et al., 1989; Jones, 1992; Sakai and Hanzawa, 1994; Mongpraneet et al., 2002).

- The penetration power of IR radiation is limited to only a few millimeters below the surface of the product, because of which the heat transfer to within the product occurs by conduction.
- With the increase in sample volume, heating is conduction-limited at positions within the depth of the food material which leads to a wider distribution of moisture.
- Swelling and fracturing of biological materials on prolonged exposure to IR heating.
- Insensitiveness to the reflective properties of the coatings.

14.2.1.4 Food Processing Applications of IR Heating

IR radiation heating is a well-known technology in the food industry for the last few decades. Recently, its use in different food processing operations has increased due to its rapid surface heating capability. The major food processing applications of IR heating include drying (discussed in Chapter 10: Drying), baking, and microbial inactivation. The subsequent sections provide an insight into each of these applications.

14.2.1.4.1 Baking

Near-IR radiation is mainly used for IR baking. IR baking occurs in three stages. During the first stage, the surface temperature increases to 100°C, and there is only a slight loss of weight of the product. The mass transfer begins during the second stage, during which the evaporation zone formed moves towards the central portion of the product. The product core reaches the temperature of 98°C–99°C during the final stage of IR baking (Skjoldebrand, 2001). The main difference in IR baking compared to conventional baking is that in the former, the formation of crust and baking of crumb occur simultaneously (Skjoldebrand and Andersson, 1989). The effective heat transfer during IR heating leads to reduced baking time compared to conduction or convection heating. A hybrid heating oven (IR+electrical) fabricated using two IR lamps and electric heating coils reduced the baking time (18 min) by 7 min compared to the conventional baking oven (25 min) (at the same baking temperature of 220°C). A 7-min reduction in the baking time was reported to be equivalent to a 28% reduction in energy demand with the IR-cum-electrical baking process compared to the conventional baking by electrical heating (Chhanwal et al., 2015). Uniform moisture distribution and high volume rise are the major advantages of bakery products baked in IR ovens.

IR heating finds applications in the baking of bread, biscuit, cookies, pizza, and other bakery products using an IR oven. Apart from baking, IR ovens are also used for the removal of residual moisture from the baked biscuits and preheating of the baking pans.

14.2.1.4.2 Microbial Inactivation

IR heating also finds application as a pretreatment method to inactivate microbes from the food surface. IR heating, more specifically FIR radiation is used for the inactivation of bacteria, spores, yeast, and mold present in both liquid and solid foods. The mode of microbial inactivation by IR heating is through DNA damage and thermal effects (Hamanaka et al., 2000). Microbial inactivation by IR heating follows the first-order reaction kinetics. The advantage is that the IR heating shows a higher death rate constant when compared to thermal conductive heating, at any given temperature. Microbial inactivation by IR heating is influenced by the following factors:

- Temperature of food sample
- IR power level
- Sample depth
- The peak wavelength of IR heating
- Moisture content and composition of the food product
- The type and growth phase of the microorganism

14.2.2 Dielectric Heating

Dielectric heating includes the microwave and radiofrequency (RF) heating techniques, which are frequency specific. The penetration depth of microwaves and RF waves is directly dependent on their wave frequency. Dielectric heating of foods is governed by the dipolar nature of water and its ability to absorb the electromagnetic energy and convert it into heat. This occurs by two mechanisms: *dipole rotation* and *ionic conduction* (Figure 14.7a and b). The permanently polarized dipolar water molecules respond to the oscillating electric field incident on them by realigning in the direction of the electric field. Due to the high frequency of the applied electric field, this realignment of water molecules occurs at the rate of million times per second. This phenomenon is termed the *dipole rotation* which leads to internal friction between the molecules. Molecular friction resulting from the rotation of dipolar water molecules generates heat and results in the *volumetric heating* of the food product. The volumetric heating from within

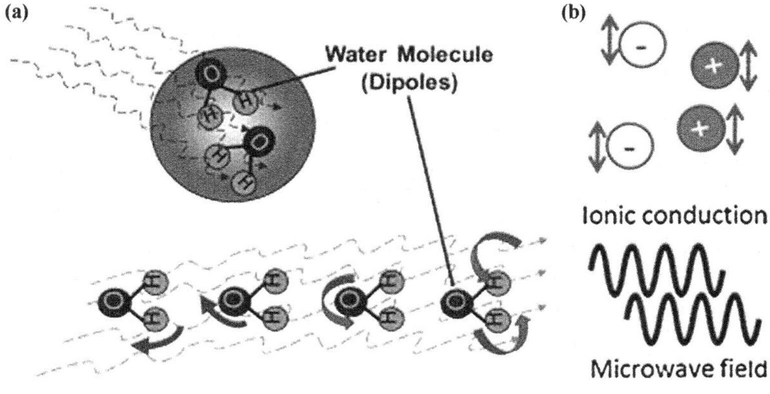

FIGURE 14.7 Principle of dielectric heating. (a) Dipole rotation. (Adapted and reproduced with permission from Tyagi, V. K. and Lo, S- L. 2013. Microwave irradiation: a sustainable way for sludge treatment and resource recovery. *Renewable and Sustainable Energy Reviews* 18: 288–305.); (b) ionic conduction. (Adapted and reproduced with permission from Gude, V. G., Patil, P., Martinez-Guerra, E., Deng, S. and Nirmalakhandan, N. 2013. Microwave energy potential for biodiesel production. *Sustainable Chemical Processes* 1: 5.)

the food product leads to internal vapor generation. This results in the accumulation of an internal vapor pressure which drives the moisture out of the food product.

In addition, when an electric field is applied, the positive and negative ions of dissolved salts in the food product move at an accelerated rate by virtue of their charge, which results in collision between the ions. The collision converts the kinetic energy of moving ions into thermal energy and increases the temperature of food. This phenomenon of the oscillatory migration of ions in foods, which generates heat in the presence of a high-frequency oscillating electric field, is known as the *ionic conduction*. Thus, the presence of water is central to dielectric heating of foods. Water, a major constituent of many food products, is the primary component that interacts with the microwaves and RF waves, due to its strong dipole rotation.

The increase in temperature of a material due to dielectric heating can be calculated using the following equation (Nelson, 1996):

$$\Delta T = \left(5.563 \times 10^{-11} fE^2 \varepsilon'' \Delta t\right)\left(\rho C_P\right)^{-1} \tag{14.4}$$

where ΔT is the increase in temperature of the product (°C), f is the frequency of the applied electromagnetic field (Hz), E is the electric field intensity (V/m), ε'' is the dielectric loss factor, Δt is the time duration (s), ρ is the product density material (kg/m^3), and C_P is the specific heat of the material (J/kg °C). From Eq. (14.4), it is evident that the increase in temperature is proportional to the electric field intensity, frequency, and treatment time in addition to the product's dielectric loss factor.

14.2.2.1 Microwave Heating

Microwaves are part of the electromagnetic spectrum with their frequency ranging between 300 MHz and 300 GHz, corresponding to a wavelength from 1 mm to 1 m. The domestic microwave appliances operate at a frequency of 2450 MHz. Whereas, the industrial microwave systems operate at 915 MHz (Datta and Anantheswaran, 2000).

14.2.2.1.1 Mechanism of Microwave Heating

In the presence of the applied microwave field, the polarity of water molecules changes at the rate of 2.45 billion cycles per second (corresponding to the frequency of microwaves at 2450 MHz). The dipole rotation at such high rates leads to friction with the surrounding medium. The heat thus generated is transferred throughout the product by the ionic conduction mechanism. The rapid and volumetric heating during the microwave heating can be attributed to the complete interaction of microwaves with the polar water molecules and charged ions in the food.

The differences between microwave heating and conventional heating are depicted in Figure 14.8. In conventional methods, the product heating is comparatively slow and less-energy-efficient, as the transfer of heat energy is *outside-in*; i.e., initially, the heat energy is transferred from the heating medium to the product, and then from the hot surface to within the product. Consequently, the hot surface can lead to local overheating which is disadvantageous with respect to heat-sensitive food products.

On the contrary, in the microwave heating, the site of heat generation is the food product itself. Microwave heating is energy efficient attributed to its *inside-out* heating resultant from a direct interaction between the microwave energy and the dipoles and/or ions in the product, which is almost always water in the case of food products. In other words, there is no initial heating of the product surface during microwave heating. The microwaves penetrate through the product and interact specifically with the dipoles and ions at a molecular level. The direct conversion of microwave energy into heat energy results in extremely fast heating rates than the conventional heating.

The two important factors which influence microwave heating are the dielectric properties and penetration depth. Dielectric property (ε^*) can be defined as the ability of a material to convert the energy from the incident microwave radiation to heat energy, given by

$$\varepsilon^* = \varepsilon' - j\varepsilon'' \tag{14.5}$$

In Eq. (14.5), the real part of dielectric property ε' is known as the dielectric constant which signifies the ability to store electric energy. The imaginary part of dielectric property ε'' is termed as the dielectric

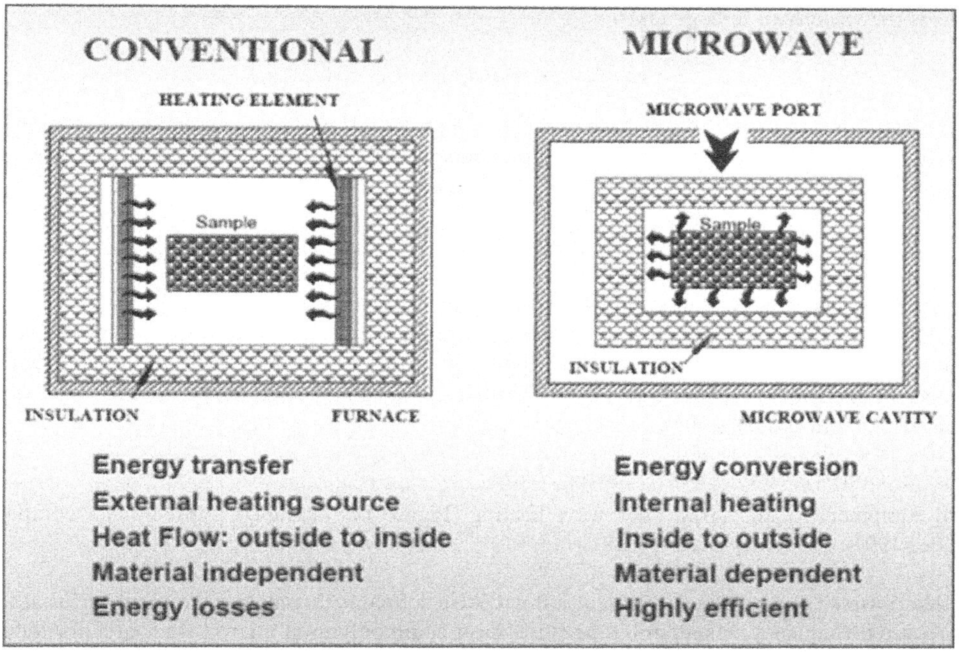

FIGURE 14.8 Differences between microwave heating and conventional heating. (Reproduced with permission from Hines, J. and Nickels, L. 2011. Hot topic—The growth of high temperature microwave technology. *Metal Powder Report* 66: 7–9.)

loss, which signifies the ability of a material to convert electric energy into heat. Further, in Eq. (14.5), $j = \sqrt{-1}$. The ratio between dielectric loss and dielectric constant is known as the *loss tangent*, which is expressed as

$$\tan\delta = \frac{\kappa''}{\kappa'} = \frac{\varepsilon''}{\varepsilon'} \tag{14.6}$$

where κ' and κ'' are the relative dielectric constant and relative dielectric loss, respectively, which are given by

$$\kappa' = \frac{\varepsilon'}{\varepsilon_0}$$

$$\kappa'' = \frac{\varepsilon''}{\varepsilon_0}$$

In these equations, ε_0 is the permittivity of free space ($\varepsilon_0 = 8.854 \times 10^{-12}$F/m). The dielectric properties are majorly influenced by the operating temperature and the frequency of microwave radiation and water content in the food product as water is a good absorber of microwaves.

The penetration depth (D_p) is the distance at which the power density drops to a value of 1/e from its value at the surface, given by (Metaxas and Meredith, 1983)

$$D_p = \frac{c}{2\sqrt{2}\pi f\sqrt{\varepsilon'}\left[\sqrt{1+\tan^2\delta}-1\right]^{\frac{1}{2}}} \tag{14.7}$$

where c is the velocity of light given by

$$c = \left(\mu_0 \varepsilon_0\right)^{-\frac{1}{2}} \tag{14.8}$$

In Eq. (14.8), μ_0 is the permeability of free space ($\mu_0 = 4\pi \times 10^{-7}$ H/m). The equation for penetration depth (Eq. (14.7)) is applicable for non-magnetic food products ($\mu_r = 1$). The power which varies with the square of the electric field is given by

$$q = \frac{1}{2} \omega \varepsilon_0 \kappa'' E^2 \tag{14.9}$$

where E is the electric field intensity and ω is the angular frequency.

In addition to the dielectric properties and penetration depth, other factors which influence the microwave food processing are the size and geometry of the microwave oven, microwave frequency, placement of food material inside the oven, density, composition, load, shape, and the size of food materials (Icier and Baysal, 2004).

14.2.2.1.2 Construction of a Microwave Heater

Typical equipment for industrial microwave heating (Figure 14.9) comprises three major components (Metaxas, 1991):

1. ***Microwave source***: The function of a microwave source is to generate the microwaves at the required frequency. Magnetron tube is the most commonly used microwave source for industrial and domestic applications (Schubert and Regier, 2005). The magnetron includes a vacuum

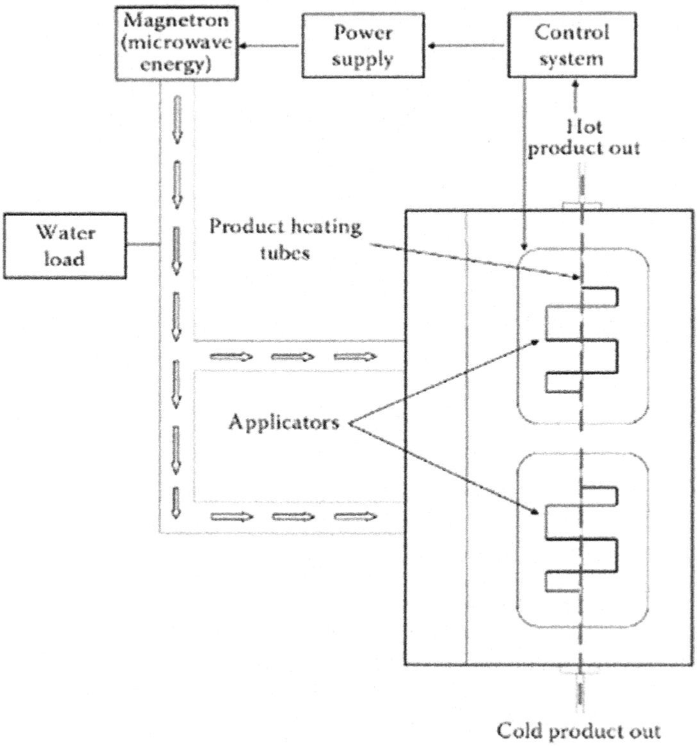

FIGURE 14.9 Typical construction of a microwave heater used in the food industry. (Reproduced with permission from Benlloch-Tinoco, M., Salvador, A., Rodrigo, A. and Martínez-Navarrete, N. 2016. Microwave heating technology. In *Handbook of Food Processing: Food Preservation*, eds. T. Varzakas and C. Tzia, 297–318. Boca Raton, FL: CRC Press, Taylor and Francis Group.)

tube with a central electron-emitting cathode of highly negative potential. The cathode is surrounded by a structured anode which forms cavities. The cavities are integrated to the surrounding fields and have the intended microwave resonant frequency (Schubert and Regier, 2006).

2. ***Waveguide*:** Waveguide is an element composed of hollow conductors (rectangular or circular) with a constant cross section. The main role of a waveguide is to direct the microwaves from the microwave source (magnetron) to the microwave applicator. A rectangle-shaped waveguide is more commonly used in industrial and domestic microwave heating. Within the waveguide, the microwaves travel in certain modes which determine the distribution of electromagnetic field inside it.
3. ***Applicator*:** An applicator contains the food product to be heated and subjects it to strong microwave fields by distributing the microwave energy around the product. Any auxiliary process equipment such as a pump can be connected to the applicator. Based on the field configuration, the applicators can be classified into three types—(i) near field, (ii) single mode, and (iii) multimode—of which the multimode applicators are generally used for industrial and domestic applications (Schubert and Regier, 2005).

In addition to these components, a control circuit integrated into the system optimizes and regulates the overall performance of the microwave heater. The advantages and limitations of microwave heating are compiled in Table 14.6.

14.2.2.1.3 Applications of Microwave Heating in Food Processing

The major applications of microwave heating in the food industry have been realized in the processes including blanching, baking, drying (discussed in Chapter 10: Drying), thawing and tempering, and pasteurization and sterilization. The conditions used for different applications of microwave heating and the results obtained are summarized in Table 14.7.

14.2.2.2 Radiofrequency Heating

RF waves fall in the frequency range of 1–300 MHz with their wavelength ranging between several meters. To avoid the interference with telecommunications, frequencies used in radio frequency heating

TABLE 14.6

Advantages and Limitations of Microwave Heating

Advantages of Microwave Heating	Limitations of Microwave Heating
• The main advantage of microwave heating is the reduced energy consumption, resultant from its volumetric heating. • Rapid and volumetric heating of foods reduces the heating time and thereby increases the production capacity and flexibility. • Due to reduced processing time and volumetric heating, microwave heating leads to improved product quality, in terms of superior aroma retention, instantaneous rehydration, enhanced color retention, and higher porosity. • Case-hardening is prevented with microwave heating due to the accumulation of surface moisture.	• The major limitation of microwave heating is the ***thermal runaway*** *or* ***runaway heating***. It refers to a process wherein the product to be heated absorbs an increased amount of the microwave energy, leading to a rapid increase in its temperature. For example, thermal runaway occurs when microwave heating is used for the thawing of frozen foods. In frozen foods, water molecules on the surface are relatively free to move than those within the product. Thus, when microwave energy is applied to frozen foods, heating commences at the surface which causes the neighboring ice crystals to melt and the surface temperature to rise. But, the inside temperature remains unchanged. Since more liquid water is available at the surface, rapid microwave heating occurs leading to excessive heating at the surface while the inner portion of the product is still frozen. To prevent this, microwave energy should be transferred to the product at a slow rate, to prolong the time for the heat to be conducted from the surface to within the food product. Thus, the use of microwave heating for food applications requires prior knowledge of the process and adequate experience to prevent runaway heating. • Microwave heating requires electrical energy and thus proves to be expensive.

TABLE 14.7

Applications of Microwave Heating in Food Processing

Application of Microwave Heating	Food Product	Conditions of Microwave Heating	Salient Results		Reference
Blanching	Broccoli	2450 MHz at 950 W for 3 min	• Higher retention of protein by 3.9% when compared to the conventional process (92°C for 0.5–4 min). • Higher retention of ash content by 53.7% when compared to the conventional process. • Higher retention of vitamin C by 18.7% when compared to the conventional process. • Higher retention of iron by 33.9% when compared to the conventional process. • Higher retention of phosphorus by 42.9% when compared to the conventional process.		Patricia et al. (2011)
	Green beans	Domestic; 650 W; 2450 MHz, for 60 s	88.5% retention of vitamin C, compared to 84.7% retained after conventional blanching with boiling water (90 ± 2°C for 3.5 min).		Muftugil (1986)
Baking	Pound cake	Double-cycle microwave-baked (2450 MHz, with 1000 W) and 10 power levels).	• Volume (cm^3): 98.3 ± 6.2 • Weight loss (g/100 g): 19.3 ± 1.0 • Luminosity (L): 84.5 ± 0.4 • Weight (g): 36.3 ± 0.42 • Moisture (g/100 g): 21.3 ± 0.67 • Water activity: 0.87 ± 0.004 • Density (g/cm^3): 0.37 ± 0.02		Zia-ur-Rehman et al. (2003)
		Conventional oven-baked (commercial electric oven at 180°C for 35 min).	• Volume (cm^3): 88.4 ± 5.5 • Weight loss (g/100 g): 9.8 ± 2.0 • Luminosity (L): 82.8 ± 0.4 • Weight (g): 40.6 ± 0.95 • Moisture (g/100 g): 36.2 ± 0.3 • Water activity: 0.93 ± 0.005 • Density (g/cm^3): 0.46 ± 0.02		
	Bread	Microwave-baked: 50% power for 2 min and 100% power for 1 min.	• Weight loss (%): 10.80 • Specific volume (ml/g): 2.04 • Firmness (N): 2.88 • Color change (ΔE): 3.0		Daomukda et al. (2011)
		Conventional-baked (preheated to the set) Temperatures 175°C, 200°C, and 225°C for 12, 13, and 14 min)	• Weight loss (%): 4.06 • Specific volume (ml/g): 1.60 • Firmness (N): 0.67 • Color change (ΔE): 47.7		
Pasteurization	Apple juice	Conventional heating at 50°C–70°C	*Saccharomyces cerevisiae*	$D_{50°C} = 58$ s $D_{55°C} = 25$ s $D_{60°C} = 10$ s	Tajchakavit et al. (1998)
		Microwave heating at 700 W, 2450 MHz		$D_{52.5°C} = 4.8$ s $D_{55°C} = 2.1$ s $D_{67.5°C} = 1.1$ s	
		Conventional heating at 50°C–80°C	*Lactobacillus plantarum*	$D_{55°C} = 52$ s $D_{60°C} = 22$ s $D_{70°C} = 8.4$ s	
		Microwave heating at 700 W, 2450 MHz		$D_{57.5°C} = 14$ s $D_{60°C} = 3.8$ s $D_{62.5°C} = 0.79$ s	

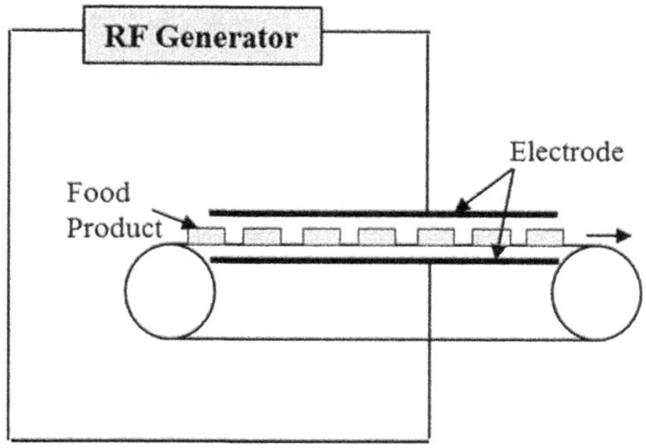

FIGURE 14.10 Line diagram of RF heating system (through-field applicator).

are specified by the international regulation. Accordingly, the RF frequencies that are permitted to be used for industrial, scientific, and medical (ISM) applications are 13.56, 27.12, and 40.68 MHz, respectively. The wavelengths corresponding to these frequencies are 22, 11, and 7.4 m.

In RF heating, the RF generator (Figure 14.10) creates an alternating electric field between the two electrodes. Food is placed between these electrodes. Polar molecules present in the food material reorient themselves with the alternating field. This reorientation causes dissipation of electrical energy to the food product. The resistance offered by the food material to the flow of high-frequency current through them results in heat generation across the food. Here, food material is not in direct contact with the electrodes. RF heating of the food material is directly proportional to the electrical conductivity of the material and square of the electric field strength. Electric field strength is defined as the voltage divided by the distance of separation between the electrodes.

Amount of RF energy required for any process can be calculated based upon the following equation:

$$E = \frac{m(T_f - T_i)C_P}{863} \tag{14.10}$$

where E is the energy supplied (kW), m is the mass flow rate of product (kg/h), and T_i and T_f are the initial and final temperatures of the product (°C), respectively.

Compared to microwave heating, RF heating offers greater penetration depth inside the food product due to the longer wavelength of electromagnetic waves in the RF spectrum. The thickness of the food material is restricted only to the distance between the two electrodes.

Three types of RF applicators are widely used:

1. *Through-field applicator*: Through-field applicator contains two electrodes which form a parallel plate capacitor (Figure 14.10). High voltage for the RF application is supplied by the generator. Generally, a through-field applicator is used for the heating of thick food products, where the product can be either placed between or conveyed through flat electrodes.
2. *Fringe-field applicator*: Fringe-field applicator also called as the stray-field electrode is shown in Figure 14.11. In this system, a thin layer of food material passes over several pairs of electrodes, placed parallel to its plane. Each electrode pair is maintained at opposite polarity on connection with the RF generator. Product and electrode are in contact which ensures a constant electric field. Fringe-field applicator is mainly used for thin-layered products.
3. *Staggered through-field applicator*: In this type, electrodes are arranged above and below the food product in a staggered manner (Figure 14.12). It is used for the products of intermediate thickness.

The advantages and disadvantages of RF heating are tabulated in Table 14.8.

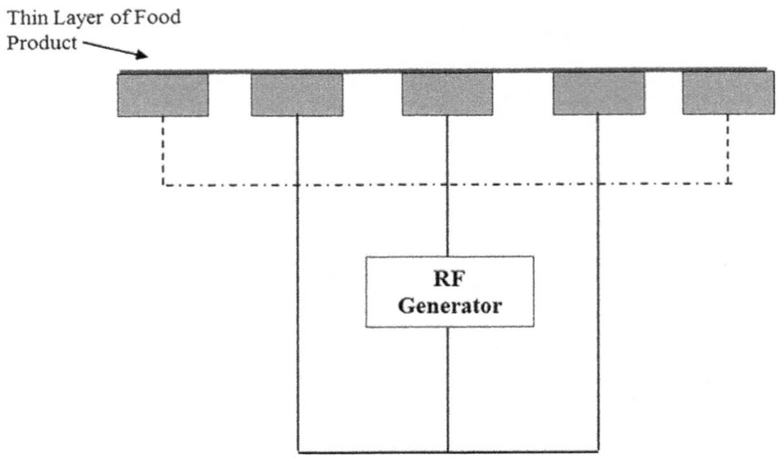

FIGURE 14.11 Fringe-field RF heating system.

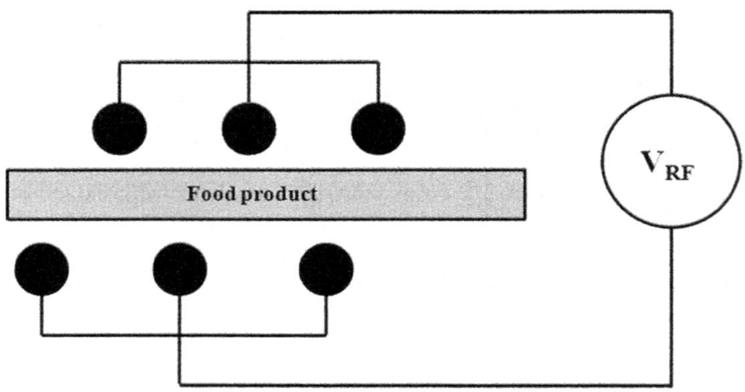

FIGURE 14.12 Staggered through-field applicator system.

TABLE 14.8

Advantages and Disadvantages of RF Heating

Advantages	Disadvantages
Increased heating and processing rate, leading to two-fold to tenfold faster dehydration than conventional heating.	A large requirement for floor space.
Uniform heating results in improved product quality.	The possibility of arcing hazard.
Instant ON/OFF and temperature control.	Not suitable for food products containing high moisture content.
Energy-efficient process.	Limited to low-temperature processing.
Uniform moisture distribution throughout the product.	Band frequency is narrow.
Low maintenance is sufficient for the RF heating equipment.	

14.2.2.2.1 Applications of RF Heating in Food Processing

RF heating has been used for the *preheating of yeasted dough*. Proving of dough during the bread-making process requires a very long proving oven to bring about a gradual increase in the temperature of the dough. This is significant as overheating leads to inactivation of yeast. With conventional heating, which initiates from the surface, dough acts as an insulator and hence prolongs the time for the heat energy to reach the center of the product. This is eliminated by the volumetric and rapid preheating with RF. RF preheating facilitates a reduction in the length of proofing tunnel by up to 60% (Monga, 2014).

Apart from the above, RF heating has wide applications in the areas of drying (discussed in Chapter 10: Drying), baking, blanching, sterilization, and preheating of food products (summarized in Table 14.9).

TABLE 14.9

Applications of RF Heating in Food Processing

Process	Frequency (MHz)	Food Items	Salient Results	References
Thawing of frozen foods	14–17	Eggs, fruits, vegetables	• RF thawing times were shorter (in minutes) when compared to hours in conventional thawing. • The resultant product quality was better due to minimum discoloration and loss of flavor.	Cathcart et al. (1947)
	36–40	Fish	• The duration of RF thawing was 12.5 min, compared to 16 h using air and 3 h using water. • Less drip loss, better odor, and flavor.	Jason and Sanders (1962)
	36–40	Meat	• Thawing was achieved within the time duration of 10–50 min, depending on the uniformity of the blocks, size, and dielectric properties. • Low drip losses.	Sanders (1966)
Pasteurization	9	Boned ham	• 2.7 kg of meat could be heated to 80°C in about 10 min. • An energy conversion efficiency of 56.6% was achieved.	Pircon et al. (1953)
	60	Cured ham	• Reduced juice losses and improved quality compared to traditional processing in hot water.	Bengtsson and Green (1970)
	27	Sausage emulsion	• RF heat treatments showed a lethal effect on the test organism, comparable to the conventional heat treatment at the same pasteurization values. • Final products were well coagulated with good appearance and taste.	Houben et al. (1991, 1994)
Roasting	60	Cocoa beans	• Reduction in the moisture content of cocoa beans from 6% to 1%.	Electrotechnology application center, PA, Cresko and Anantheswaran (1998)
Heating	14–17	Loaves of wrapped, sliced white bread and wrapped Boston bread	• No mold growth was observed in the samples heated to 60°C.	Cathcart et al. (1947)
Blanching	15	Vegetables	• Catalase was inactivated in the vegetables heated to 77°C and after few days of storage at −23°C. • Samples blanched at 88°C were found to possess the highest ascorbic acid content. • However, RF blanching resulted in poor flavor and color when compared to those blanched by traditional method using water and steam.	Moyer and Stotz (1947)

14.3 Ohmic Heating

14.3.1 Principle of Ohmic Heating

Ohmic heating or Joule heating is a process in which heat is generated inside the food by the passage of alternating current (AC) through it. Earlier in the twentieth century, ohmic heating was used for the pasteurization of milk. However, due to the high electricity cost, it was discontinued. Later in the 1980s, the technique was revived for the sterilization of liquids containing large particles. The underlying principle of ohmic heating is based on the resistance offered by the foods to the applied electrical current. Therefore, it is also known as resistive heating or electroconductive heating. Ohmic heating is based on the Joule effect, according to which the electrical energy supplied by the electrical conductor to the food material dissipates in the form of heat. Heat generated depends on the electrical conductivity of food material and square of the electrical field applied as given in Eq. (14.11).

$$P = \frac{V^2 \sigma A}{L} \tag{14.11}$$

where P is the electric power (W), V is the voltage applied, σ is the electrical conductivity (S/m), A is the area (m^2), and L is the distance between the electrodes (m).

In the ohmic heating process, the electrical conductivity of the food material is the most influential parameter. Electrical conductivity (Eq. (14.12)) is defined as the quantity of electricity transferred through a unit area, per unit potential gradient per unit time.

$$\sigma = \frac{L}{AR} \tag{14.12}$$

where L is the length of the sample (m), A is the cross-sectional area of the sample (m^2), and R is the resistance offered by the sample (Ω).

The electrical conductivity of a food sample is influenced by its moisture content, temperature, ionic constituents, and the field strength. It increases with an increase in the moisture content. At higher temperature, the electrical conductivity of the sample increases due to increase in ionic mobility.

Internal heat generation (E) in the product is given by

$$E = |\nabla V|^2 \sigma \tag{14.13}$$

During ohmic heating, the temperature distribution within the food product can be predicted by the following equation which accounts for the internal heat generation along with conductive heat transfer:

$$\nabla(k\nabla T) + E = \rho C_P \tag{14.14}$$

where k is the thermal conductivity of the product (W/m K), ρ is the density of the product (kg/m^3), C_P is the specific heat of the product (J/kg K), T is the temperature (K), and t is the time (s).

Due to the volumetric heating effect similar to that in RF heating, the food product is uniformly and rapidly heated during the ohmic heating. This also ensures the intactness of product quality characteristics such as color, flavor, and texture. Ohmic heating leads to coagulation of protein and starch gelatinization in foods, which is similar to that exerted by the conventional heating methods.

14.3.2 Ohmic Heating System

A typical ohmic heating system as shown in Figure 14.13 consists of a pipe as a treatment chamber, power source, and electrodes. Maintenance of turbulent flow at a uniform velocity is important for ohmic heating. Therefore, the pipe diameter varies with the product characteristics. Accordingly, the pipe diameter for liquid foods is 2.5 cm, whereas that for the particulate–liquid mixture is 7.5 cm. Two electrodes are built into the pipe wall, placed on the opposite sides. Electrodes are made up of titanium-coated carbon rods and insulated by lining the pipe wall with Teflon; 5–12 kV voltage with 60 Hz AC is applied to

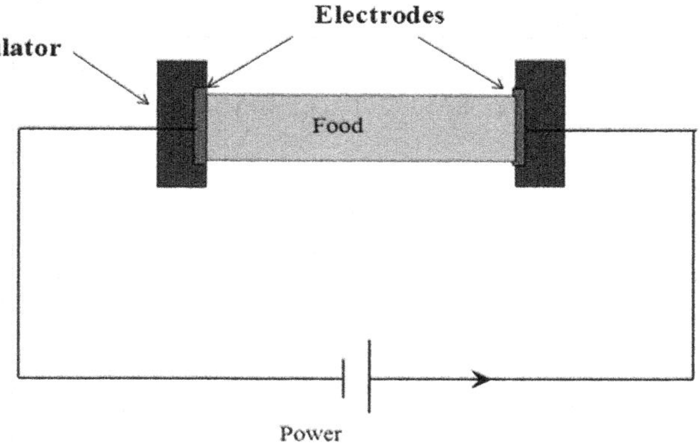

FIGURE 14.13 Schematic of ohmic heating system.

achieve ohmic heating. With AC at lower frequencies (<50 Hz), the electrode may contaminate the product by releasing metal ions in the food due to electrolysis. Therefore, AC at higher frequency is preferred. At higher frequencies in the range of 20–100 kHz, stainless steel electrode is used.

The advantages and disadvantages of ohmic heating (derived from Skudder, 1988; Kim et al., 1996) are tabulated in Table 14.10.

14.3.3 Applications of Ohmic Heating in Food Processing

14.3.3.1 Sterilization and Pasteurization

Ohmic heating is used for the sterilization and pasteurization of both Newtonian and non-Newtonian food materials. It is more appropriate for viscous liquid foods wherein the conventional processes fail to heat the product uniformly. The established applications of ohmic heating include commercial-scale sterilization and pasteurization of milk, tomato sauce, liquid egg products, processing of strawberries, fruit purees jam, juices, and ready meals. Soups containing particulate pieces of vegetables and seafood are also sterilized using ohmic heating.

TABLE 14.10

Advantages and Limitations of Ohmic Heating

Advantages of Ohmic Heating	Limitations of Ohmic Heating
Due to the absence of temperature gradient and volumetric heating, uniform heating of the liquid–particulate mixture is possible.	Cannot be used for foods having low electric conductivity such as those with high-fat content.
Ultrahigh temperature in particulates can be achieved rapidly.	The possibility of contamination by metal ions at low frequencies (<50 Hz).
No burning or charring of the food product leading to better nutrient retention.	Arcing may occur if air bubbles are present in the product.
High-energy-efficient process as 90% of electrical energy is converted to heat.	Complex coupling between temperature and voltage distributions.
No moving parts and low maintenance cost.	Sensitivity to particle shape and orientation.
Continuous and quiet in operation.	Ohmic heating is a complex process as the heat generation rate is influenced by a number of factors including electrical heterogeneity of the particle, heat channeling, complex coupling between temperature and electrical field distributions, and particle shape and orientation.
Instant start and shutdown, easy to control, no residual current.	
Process temperature can be maintained accurately and controlled easily.	
Low capital cost.	

14.3.3.2 Extraction of Bioactive Compounds

This application is based on the capability of ohmic heating to accelerate moisture loss and increase the diffusion coefficient of the bioactive compounds to be extracted.

14.3.3.3 Microbial Inactivation

The mode of microbial inactivation by ohmic heating has been attributed to the combination of thermal effects and a mild electroporation mechanism. The latter mechanism occurs with low-frequency ohmic heating at 50–60 Hz, during which there is a buildup of electrical charges and a consequent formation of pores across the cell walls of microbes (Ruan et al., 2001).

14.3.3.4 Thawing

This application of ohmic heating is based on the concept that the increase in electrical conductivity with temperature is lower for frozen foods by two orders of magnitude compared to that for thawed foods. This allows the thawed portion to be cooked, while the remaining portion is still frozen (Goullieux and Pain, 2014).

14.3.3.5 Detection of Starch Gelatinization

The application of ohmic heating for the detection of starch gelatinization is based on the observation that the plot between electrical conductivity and temperature of starch solutions shows negative peaks, similar to the gelatinization peaks observed in the thermographs obtained from differential scanning calorimetry (Wang, 1997; Wang and Sastry, 1997).

For this application, a starch gelatinization measuring system (Figure 14.14) was developed by Wang (1997), which includes the following components:

i. An ohmic heater mounted on a shaker which holds and heats the starch-containing food product.

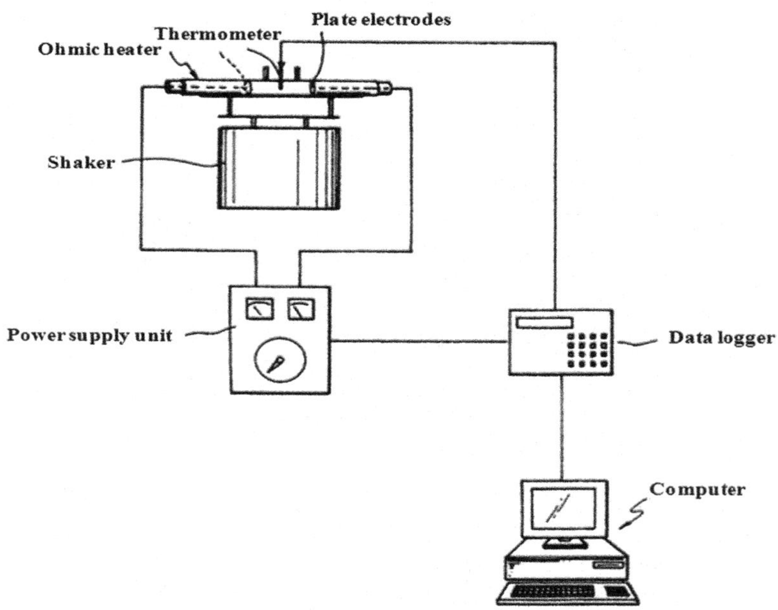

FIGURE 14.14 Starch gelatinization measuring system. (Adapted from Wang, W.-C. 1997. Starch gelatinization measuring system. US Patent 5604302.)

ii. Two plate electrodes for heating.
iii. A power supply unit which is electrically connected to the ohmic heater. This unit controls the supply of electric current and electric voltage to the plate electrodes.
iv. A thermometer to detect the temperature of the starch sample being heated.
v. A data logger to record the temperature, electric current, and electric voltage, during heating.
vi. A computer to analyze the data provided by the data logger (Figure 14.14).

The methodology for the calculation of the degree of starch gelatinization based on the conductivity measurements is explained in the forthcoming section with the example of a conductivity curve obtained from potato starch during its gelatinization.

14.3.3.5.1 Calculation of Conductivity

$$\sigma = \frac{L}{AR} \tag{14.15}$$

where σ is the conductivity, L is the distance between the plate electrodes of the ohmic heater, A is the cross section of the ohmic heater, and R is the resistance.

14.3.3.5.2 Calculation of the Degree of Starch Gelatinization

The degree of starch gelatinization can be calculated using the reference value of starch gelatinization (PA_0) and the starch gelatinization supplement value (PA_1).

Step 1: Determination of PA_0

For the estimation of PA_0, the non-gelatinized starch sample is mixed with distilled water to form a suspension. Then, the conductivity of the suspension is measured in order to obtain a conductivity curve as shown in Figure 14.15. From the conductivity curve, it can be observed that the conductivity of the starch sample dropped during the occurrence of starch gelatinization. This can be attributed to the swelling of starch granules. After gelatinization, the conductivity of the starch sample increased as electrical conductivity is proportional to temperature. The peak curve is integrated to acquire the total area of the peaks, which is then used as the value of PA_0.

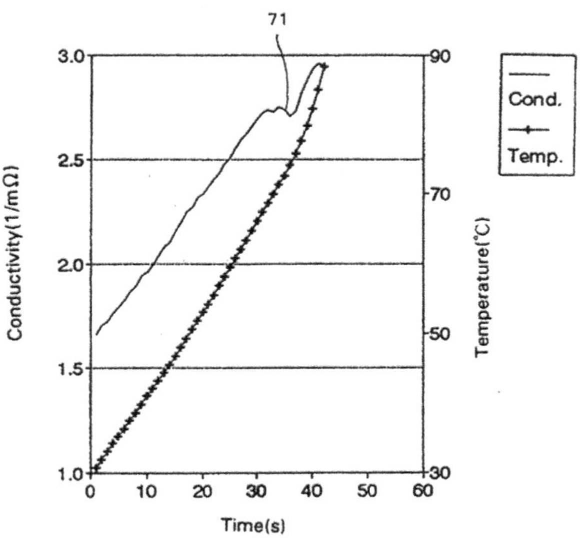

FIGURE 14.15 Conductivity curve of potato starch during its gelatinization. (Wang, W.-C. 1997. Starch gelatinization measuring system. US Patent 5604302.)

Step 2: Determination of PA_1

Similar to that done in the calculation of PA_0, the conductivity of the gelatinized starch sample was measured to obtain its conductivity curve. Subsequently, the peak curve can be integrated to obtain the total area of the peaks, which is nothing but the value of PA_1.

Step 3: Equation for the calculation of the degree of starch gelatinization

The degree of gelatinization of the starch sample under test can be obtained by comparing the value of PA_1 against PA_0. The formula for determination of the degree of starch gelatinization (%SG) is given by

$$\%SG = \left(1 - \frac{PA_1}{PA_0}\right) \times 100 \tag{14.16}$$

The starch gelatinization system described previously can be used for measuring the degree of starch gelatinization by means of sampling measurement or online measurement. When used for online measurement, the ohmic heater, temperature detector, and the electrodes could be installed on the production line. The starch gelatinization data thus obtained can be instantaneously fed back to the main control unit of the production line for the automatic adjustment of control parameters to maintain the product quality at the desired level. For instance, when the starch of a liquid food product is gelatinized, it turns thick. Consequently, its flowing speed would be relatively slowed down which may lead to overheating of the product. The measuring system when installed online can immediately detect the starch gelatinization, send information to the main control unit, and thereby accelerate the delivering speed of the liquid food (Wang, 1997).

A summary of the applications of ohmic heating in food processing is presented in Table 14.11.

TABLE 14.11

Applications of Ohmic Heating in Food Processing and Preservation

Applications	Advantages	Food Items	References
Sterilization, heating liquid foods containing large particulates and heat-sensitive liquids, aseptic processing	• Attractive appearance. • Firmness properties. • Pasteurization of milk without protein denaturation.	Cauliflower florets, soups, stews, fruit slices in syrups and sauces, ready to cook meals containing particulates, milk, juices, fruit purees.	Eliot-Godéreaux et al. (2001), Pataro et al. (2011), Icier and Ilicali (2005) and Içi'er et al. (2008)
Cooking of solid foods	• The cooking time could be reduced significantly. • The center temperature rises much faster than in conventional heating, improving the final sterility of the product, less power consumption, and safer product.	Hamburger patties, meat patties, minced beef, vegetable pieces, chicken, pork cuts	Shirsat et al. (2004), Piette et al. (2004), de Halleux et al. (2005), Wills et al. (2006), Liu et al. (2007), Zell et al. (2009, 2010) and Bozkurt and Icier (2009)
Space food and military ration	• Food reheating and waste sterilization. • Less energy consumption for heating food to serving temperature, products in reusable pouches with long shelf life. • Additive-free foods with good keeping quality of 3 years.	Stew-type foods	Jun et al. (2007), Jun and Sastry (2005) and Yang et al. (1997)
Thawing	• Thawing without an increase in moisture content of the product.	Shrimp blocks	Roberts et al. (1996)

(Continued)

TABLE 14.11 (*Continued*)

Applications of Ohmic Heating in Food Processing and Preservation

Applications	Advantages	Food Items	References
Inactivation of spores and enzymes	• To improve food safety and enhance shelf life. • Increased stability and energy efficiency. • Reduced time for inactivation of lipoxygenase and polyphenol oxidase. • Inactivation of enzymes without affecting the flavor of the product.	Processed fish cake, orange juice, juices	Loypimai et al. (2009)
Blanching and extraction	• Enhanced moisture loss and increase in juice yield	Potato slices, vegetable purees, extraction of sucrose from sugar beets, extraction of soy milk from soybeans	Katrokha et al. (1984), Kim and Pyun (1995) and Wang and Sastry (2000)

Source: Varghese et al. (2014).

14.4 Nanotechnology-Based Food Processing Techniques

14.4.1 Nanospray Drying

The advent of nanospray drying was driven by the limitations of the conventional spray-drying technique concerning its limited collection efficiency of particles with size <2 μm. This is of relevance as it negatively impacts the productivity and jeopardizes the commercial viability of the process. For the production of nanoparticles, the ability of atomizer to generate micron and sub-micron sized droplets and the air flow pattern inside the spray chamber should be such that it prevents the dried nanoparticles from sticking on to the chamber wall. Thus, a nanospray dryer is different from the conventional spray dryer with respect to its atomizer, spray chamber, and product collection system. The principle of operation and modifications in each of the aforementioned components of nanospray dryer would be explained in the following sections.

14.4.1.1 Working Principle of Nanospray Dryer

Similar to the conventional spray dryer, the operation of nanospray dryer (Figure 14.16) comprises the following four stages (Anandharamakrishnan and Ishwarya, 2015):

1. The dry air is heated in a porous metal foam-based heater, which serves the twin purpose of providing a laminar air flow in the spray chamber in addition to hot air generation. This arrangement is different from the electrical heating and turbulent air flow in conventional spray dryers.
2. Atomization of the feed solution using a piezoelectric-driven vibrating mesh atomizer to generate micron and sub-micron sized feed droplets. This is in contrast to the use of centrifugal/pressure/twin-fluid atomizer in conventional spray dryers.
3. The contact between the finely divided droplets and heated air and the subsequent drying occurs in a spray chamber of vertical configuration. Unlike the cylindroconical configuration of spray chamber in conventional spray dryers, the vertical configuration of spray chamber in nanospray dryer directs the submicron particles towards the particle collector and minimizes the sticking of particles to the sidewalls of the drying chamber. This increases the particle collection efficiency.

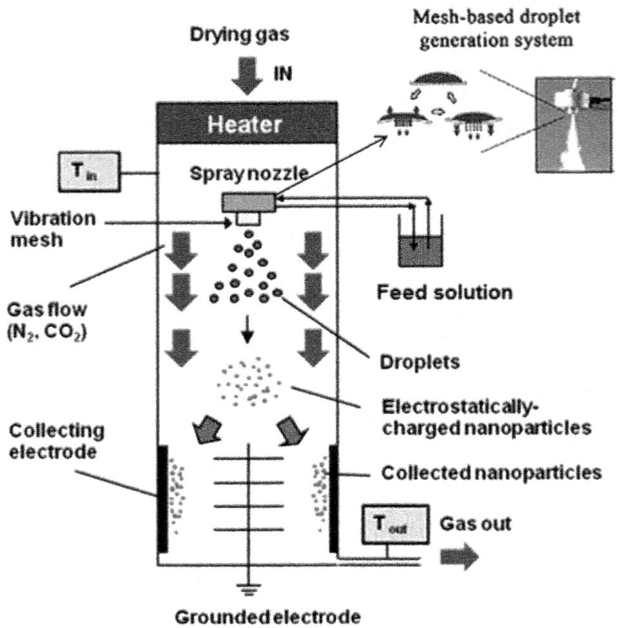

FIGURE 14.16 Schematic diagram of nanospray dryer. (Adapted and reproduced with permission from Baba, K. and Nishida, K. 2013. Steroid nanocrystals prepared using the nano spray dryer B-90. *Pharmaceutics* 5: 107–114; Li, X., Anton, N., Arpagaus, C., Belleteix, F. and Vandamme, T. F. 2010. Nanoparticles by spray drying using innovative new technology: the Büchi Nano Spray Dryer B-90. *Journal of Controlled Release* 147: 304–310.)

4. Separation of dried particles by an electrostatic precipitator (ESP), which is different from the cyclone separator or bag filter in conventional spray dryers. Unlike the cyclone separators in conventional spray dryers which depend on particle mass, the ESP achieves particle separation by virtue of the difference in particle charge.

14.4.1.2 Vibrating Mesh Atomization

The atomizer of a nanospray dryer includes a piezoelectric-driven actuator, a vibrating mesh, and a spray cap. This assembly is connected to the feed reservoir as shown in Figure 14.17. Here, droplet generation is based on the vibration induced by the piezoelectric element around the mesh with which it bears a direct contact. The mesh consists of a thin perforated stainless steel membrane, housed inside a spray cap and comprising of an array of micron-sized holes. This membrane is also termed as the spray mesh.

The piezoelectric actuator is driven at an ultrasonic frequency (~60 kHz). Consequently, the spray mesh that is in contact with the actuator vibrates upward and downward and forces the feed solution through it to inject millions of precisely sized droplets with a very narrow size distribution (Lee et al., 2011). The droplet size decreases with a decrease in pore size of the spray mesh. This is similar to the influence of wheel speed and diameter in rotary atomizer and nozzle pressure in hydraulic and twin-fluid nozzles, on the droplet size. In general, commercially available spray meshes come in three different pore sizes: 4.0, 5.5, and 7.0 µm. The influence of spray mesh pore size on the particle size of spray-dried bioactive compounds is shown in Figure 14.18.

14.4.1.3 Heating Mode, Hot Air Flow Pattern, and Configuration of the Spray Chamber

The heating system of a nanospray dryer includes open-pore metal foam, arranged between the air inlet and outlet; the outlet directly opens to the spray-drying chamber. Metal foam encompasses a large

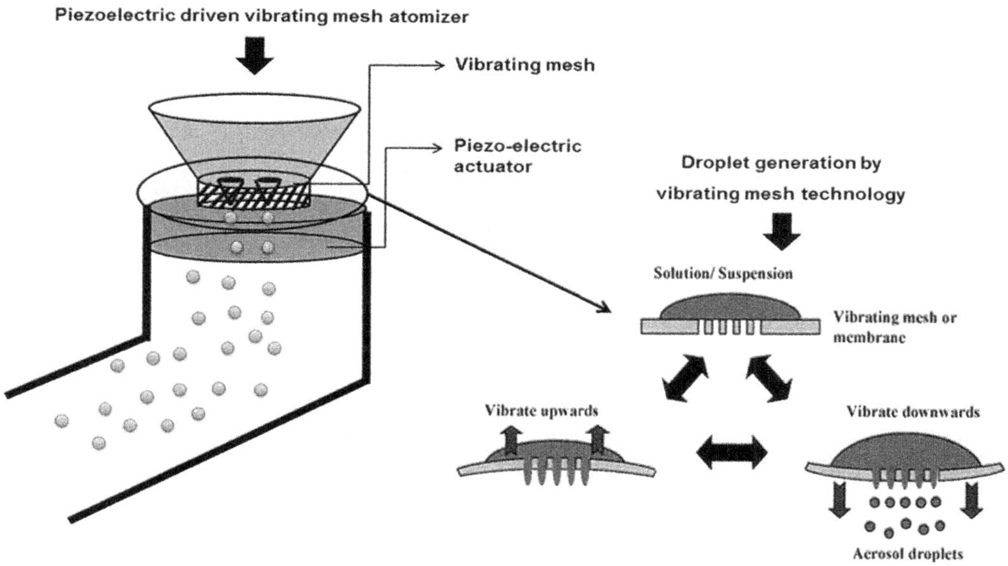

FIGURE 14.17 The operational principle of the piezoelectrically driven atomizer. (Adapted and redrawn from Vecillio, L. 2006. The mesh nebuliser: a recent technical innovation for aerosol delivery. *Breathe* 2: 253–260; Lee, S. H., Heng, D., Ng, W. K., Chan, H. K. and Tan, R. B. 2011. Nano spray drying: a novel method for preparing protein nanoparticles for protein therapy. *International Journal of Pharmaceutics* 403: 192–200.)

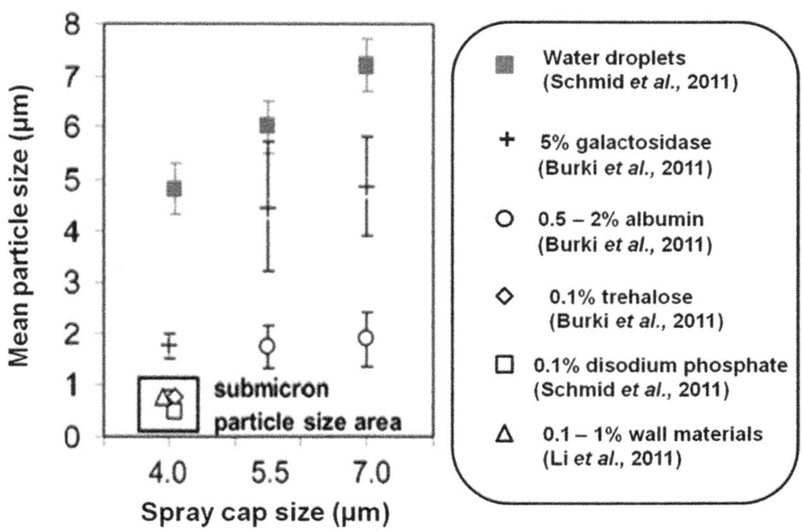

FIGURE 14.18 Correlation between spray cap size and particle size; the submicron particle size is typically reached when using the 4.0-μm spray cap and diluted solid concentrations of about 0.1 w%. (Reproduced with permission from Arpagus, C. 2012. A novel laboratory-scale spray dryer to produce nanoparticles. *Drying Technology* 30: 1113–1121.)

volume fraction of gas-filled pores in a solid metal to form a cellular structure. The advantage of metallic foam is its ability to uniformly heat the entire volume of air flowing through it (Schön and Baumgartner, 2009). Pores in the metal foam structure can be sealed to result in closed-cell foam, or they can be interconnected to form an open-cell foam. The open-celled metal foams enable an effective heat transfer from the metal foam to the drying medium, owing to its large surface area.

In the heating system of nanospray dryers, one of the surfaces of the metal foam forms the fluid inlet and/or the fluid outlet. The arrangement of metal foam within the air heating unit is such that the flow pattern of hot gas, which flows through the fluid inlet and emerges out of the fluid outlet, is laminar rather than turbulent as in the conventional spray dryer. Further, the metal foam heater requires substantially less working space than a conventional electric heater (Anandharamakrishnan and Ishwarya, 2015).

14.4.1.4 Product Separation by the Electrostatic Precipitator (ESP)

ESP uses electrostatic force to separate the solid particles from the gas stream (Mizuno, 2000). This is accomplished by imparting a charge on the individual particles in the gas stream and driving the charged particulates towards the oppositely charged collection electrode. Then, the particles are separated from the surface of the collector electrode, usually with a rubber spatula or by rappers which vibrate the collection surface to assist in particle recovery.

An ESP system consists of a grounded star electrode (cathode) and a cylindrical particle collecting electrode (anode) (Figure 14.19). The working of an ESP comprises the following steps:

1. ***Establishment of an electrical field***: The electric field is generated by the application of high voltage between the cathode and anode of the ESP. The electric field initiates the generation of charging ions, which collide with the particles to ionize them and create the necessary driving force to migrate them away from the surrounding gas stream towards the collection electrode. The strength of the electric field is directly related to the performance of ESP.
2. ***Corona generation***: Corona is the electrically active region of the gas stream, formed as a result of stripping of electrons from the gas molecules by the electric field initiated in the previous step. The corona is created when the magnitude of the applied electric field is high enough to accelerate the free electrons.
3. ***Gas stream ionization***: The corona discharge as explained previously leads to the generation of positively charged gas molecules and a stream of free electrons. These free electrons are accelerated to cause additional ionization, and this chain of events is popularly known as the '*electron avalanche*' (Helfritch, 1993).

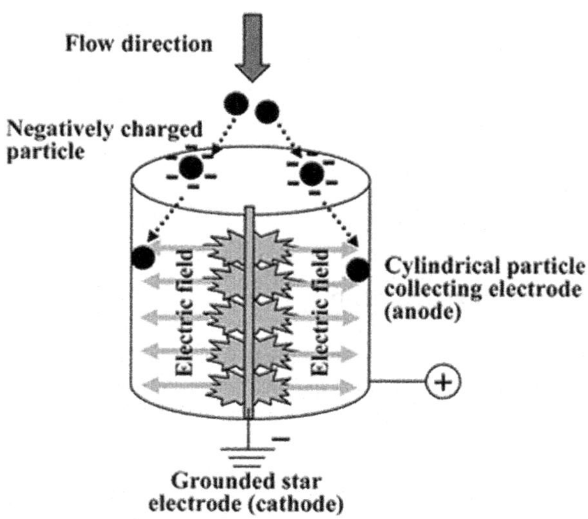

FIGURE 14.19 Schematic of the working principle of ESP. (Reproduced with permission from Lee, S. H., Heng, D., Ng, W. K., Chan, H. K. and Tan, R. B. 2011. Nano spray drying: a novel method for preparing protein nanoparticles for protein therapy. *International Journal of Pharmaceutics* 403: 192–200.)

4. ***Charging and migration of the particles to the collection electrode*:** Particle charging and collection occurs at an interface region between the corona glow and collection electrode. In this region, the gas particles are subjected to the generation of negative ions from the corona generation step. Also, the uncharged particulates in the gas stream are exposed to free electrons produced from the gas ionization. This leads to the generation of dipoles within the uncharged particulates during which the particle itself remains neutral, while positive and negative charges within the particle accumulate in separate zones. The positive charges inside the particle concentrate within it in an area which is in close proximity to the approaching negative ion. Here, it retains some electrical charge from the ion. This imparts a net negative charge on the previously neutral particulate, which is the prerequisite for the electrostatic forces to act on the particle in order to divert and separate them from the gas stream. The final step in particle collection by ESP is the movement of the charged particles towards an oppositely charged electrode, which is governed by the action of the electric field (Anandharamakrishnan and Ishwarya, 2015). The advantages and limitations of nanospray dryer are summarized in Table 14.12.

14.4.1.5 Applications of Nanospray Drying in Food Processing

Folic acid, an essential water-soluble vitamin of nutraceutical importance, was encapsulated using nanospray drying within two types of wall materials: whey protein concentrates (WPCs) and resistant starch (RS). A 0.7-μm membrane cap was used for the atomization. The solutions were introduced into the nanospray dryer through a silicone wire, which was connected to the spraying head of the equipment. The air flow was 140 L/h with an inlet and outlet temperatures of 90°C and 45°C, respectively. Spherical nano-, submicron-, and micron-sized folic acid capsules were obtained from the nanospray dryer (Figure 14.20). Nanospray drying resulted in stable folic acid encapsulates with compact structure and improved moisture resistance (Pérez-Masiá et al., 2015).

The other food applications of nanospray dryer include the production of:

- **Trehalose particles for end use as protein stabilizer (Schmid et al., 2009):** The trehalose particles depicted the size of 600 nm and 1.2 μm at feed concentrations of 0.1% and 1%, respectively.
- **Sodium alginate to be used as an emulsifier (Blasi et al., 2010):** The process yield was >90%, resulting in spherical and smooth surfaced particles. The spray cap size had a prominent influence on the final particle size which reduced from 5.5 μm to 761 nm with the decrease in mesh size from 7 to 4 μm.

TABLE 14.12

Advantages and Limitations of the Nanospray Dryer

Advantages of Nanospray Dryer	Limitations of Nanospray Dryer
• The major advantage of nanospray dryer is its ability to process small volumes of feed (in mg or mL) to result in high yield of small particles. • Narrow size distribution of droplets from the vibrating mesh atomizer. • The particle size of dry powder obtained from nanospray dryer ranges between 300 nm and 5 μm. • Suitable for heat-labile compounds with minimal loss of activity. • Convenient simple setup and cleaning procedures.	• The major limitation of the vibrating mesh atomizer is the difficulty in handling highly viscous feed streams which could not be sprayed even with the use of 7.0 μm mesh. • Mesh nozzle is prone to clogging when the feed viscosity is greater than 10 cP. • Expensive. • Nanospray dryer is limited to laboratory scale operation, which has a low throughput and long process time. • Particle size is limited by the commercially available nozzle mesh sizes (4.0, 5.5, and 7.0 μm). • The collected powder is in the form of a cake which is not free-flowing and requires scrapping off using a rubber spatula.

Source: Adapted from Heng et al. (2011).

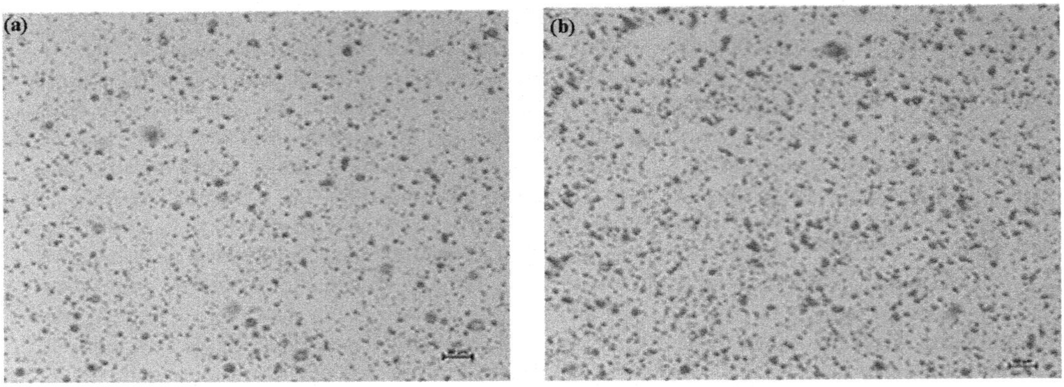

FIGURE 14.20 Optical micrographs of folic acid capsules with (a) WPC and (b) RS as wall materials. (Reproduced with permission from Pérez-Masiá, R., López-Nicolás, R., Jesús Periago, M., Ros, G., Lagaron, J. M. and López-Rubio, A. 2015. Encapsulation of folic acid in food hydrocolloids through nanospray drying and electrospraying for nutraceutical applications. *Food Chemistry* 168: 124–133.)

- **Vitamin E acetate (Li et al., 2010):** A closed loop spray drying was performed with an organic solvent as suspension medium and nitrogen as drying gas (oxygen content <4%) to prevent oxidation.
- **L-leucine amino acid for use as a dispersing agent and food additive (Feng et al., 2011):** The crystalline, low-density, and well-dispersing capsules of L-leucine particles obtained from the nanospray dryer showed an aerodynamic particle diameter of 2.1–5.4 µm with >25% leucine.

14.4.2 Electrohydrodynamic Techniques: Electrospraying and Electrospinning

The electrohydrodynamic processes such as electrospinning and electrospraying utilize electrically charged jet of polymer solution for the production of fibers and particles, respectively, of micron, submicron, and nanoscale dimensions. A typical electrospinning and electrospraying system consists of four major components: (i) a high voltage source (1–30kV), operated in the DC or AC mode (Kessick et al., 2004); (ii) a blunt-ended stainless steel needle or capillary; (iii) a syringe pump; and (iv) a grounded collector which is a rotating drum in an electrospinning unit (Figure 14.21) and a flat plate in an electrospray system (Figure 14.22).

In electrospinning, the polymer solution within a capillary is induced with free charges by a high voltage potential. At the tip of the capillary, due to electrostatic repulsion of like charges and the Coulombic force of the external electric field (known as the electrostatic forces), the hemispherical surface of the droplet is distorted into a conical shape known as the *Taylor cone*. Once the electrostatic force overcomes the surface tension, a charged polymer jet is ejected from the tip of the Taylor cone. As the jet is driven towards the collector, the unevenly distributed charges cause whipping or bending motion of the jet. As a result, the jet extends followed by the rapid evaporation of the solvent to result in the deposition of solid, thin fibers on the grounded collector (Kriegel et al., 2008; Li and Xia, 2004; Reneker and Yarin, 2008).

In contrast, electrospraying (Figures 14.22 and 14.23) is a liquid atomization process, governed by the electrical forces. The difference between electrospinning and electrospraying techniques is based on the concentration of the polymer solution (Bhushani and Anandharamakrishnan, 2014). At high concentration of the solution, the jet is destabilized due to varicose instability, leading to the formation of fine droplets. The highly charged droplets are self-dispersing in space. Therefore, droplet agglomeration and coagulation are prevented (Brandenberger et al., 1999; Jaworek and Sobczyk, 2008). Subsequent evaporation of the solvent leads to contraction and solidification of droplets and deposition of solid polymeric particles on the grounded collector (Bock et al., 2012).

The droplet size or dimensions of the fiber is decided by the intensity of electric field across the electrodes, flow rate, and surface tension of the polymer solution, according to the following equation (Jaworek, 2007):

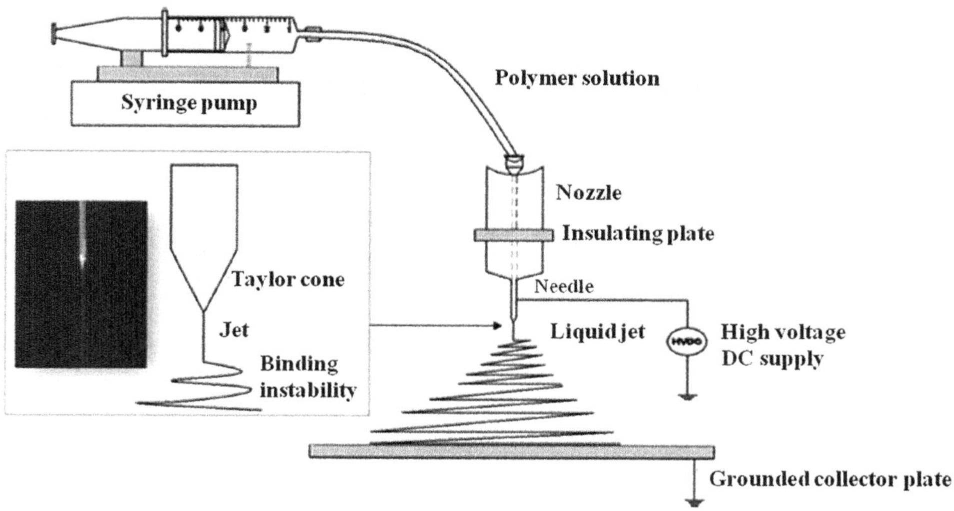

FIGURE 14.21 A typical laboratory-scale electrospinning setup. (Reproduced with permission from Bhushani, J. A. and Anandharamakrishnan, C. 2014. Electrospinning and electrospraying techniques: potential food based applications. *Trends in Food Science and Technology* 38: 21–33.)

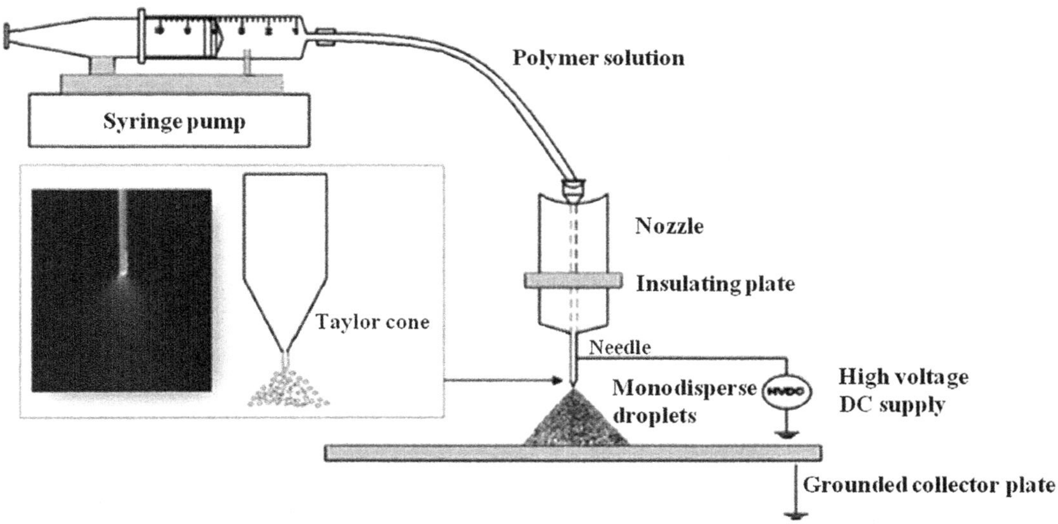

FIGURE 14.22 A typical electrospraying setup. (Reproduced with permission from Bhushani, J. A. and Anandharamakrishnan, C. 2014. Electrospinning and electrospraying techniques: potential food based applications. *Trends in Food Science and Technology* 38: 21–33.)

$$d_D = \alpha \left(\frac{Q^3 \varepsilon_0 \rho}{\pi^4 \sigma \gamma} \right)^{1/6} \tag{14.17}$$

where d_D is the droplet size, Q is the flow rate, ε_0 is the permittivity of vacuum, ρ, σ, and γ are the density, conductivity, and surface tension of the feed liquid, respectively, and α is a constant which is generally equal to 2.9.

These techniques eliminate the use of organic solvents for dissolving the polymers and thus compatible for aqueous formulations by altering the process parameters in Eq. (14.17) or by means of changing the solution

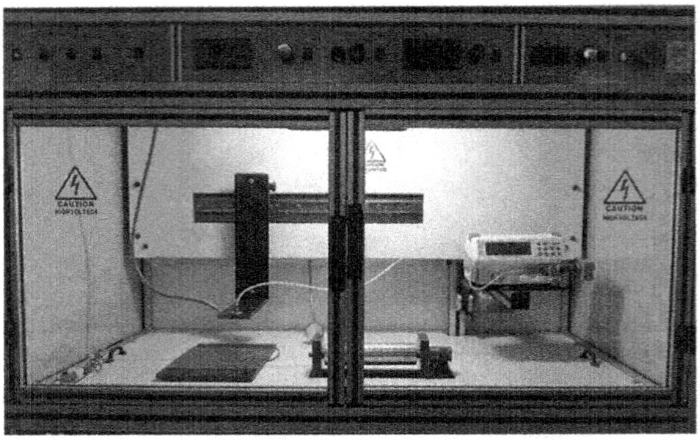

FIGURE 14.23 Electrospraying-cum-electrospinning equipment. (Reproduced with permission from Anandharamakrishnan, C. and Ishwarya, S. P. 2015. *Spray Drying Techniques for Food Ingredient Encapsulation*. Hoboken, NJ, USA: John Wiley and Sons, Ltd.)

properties using suitable additives (Lopez-Rubio et al., 2012). However, the extremely low feed flow rate used in these electrohydrodynamic techniques limits their scale-up (Anandharamakrishnan and Ishwarya, 2015).

The applications of electrohydrodynamic techniques have been mainly realized in the disciplines of encapsulation of bioactive compounds, food packaging, food coating, and production of filter aid (Table 14.13).

TABLE 14.13

Food Processing Applications of Electrospinning and Electrospraying Techniques

Application	Examples	References
Encapsulation of food bioactive compounds	1. Electrospraying was used for the nanoencapsulation of green tea catechins in zein, a biocompatible, biodegradable, and hydrophobic protein extracted from corn. Among the various concentrations of zein studied (1% to 40% w/w), 5% w/w zein solution yielded spherical, monodisperse nanoparticles with a mean diameter of 157 ± 36 nm.	Bhushani et al. (2017)
	2. Electrospraying was used for the production of micron- (1724 ± 524 nm), submicron- (336.6 ± 218.8 nm), and nano- (83.1 ± 11.5 nm) capsules of WPC.	Lopez-Rubio and Lagaron (2012)
	3. Electrospinning was used for the production of nanofibers from hydrolyzed chitosan. The fibers showed a mean diameter of 140 nm.	Homayoni et al. (2009)
	4. Electrospraying: polysaccharides were electrosprayed for the production of polyelectrolyte complex (polycation: sodium alginate solution, polyanion:chitosan, or calcium chloride aqueous solution) to result in micro- (80–230 μm) and nanocapsules (80 nm) appropriate for bioactive encapsulation.	Fukui et al. (2010), Ghayempour and Mortazavi (2013)
	5. Electrospinning: the soluble fiber fraction obtained by alkaline treatment of cereal wastes (corncob and wheat straw) has been electrospun along with 6% (w/v) polyvinyl alcohol (PVA) to obtain ultrafine fibers of dimension 200–800 nm.	Kuan et al. (2011)
	6. Electrospinning was used for the encapsulation of oxygen-sensitive and light-sensitive β-carotene using zein as an encapsulating material. The fibers depicted minimum and maximum cross sections of 540 and 3580 nm, respectively.	Fernandez et al. (2009)
	7. Electrospun curcumin-in-zein nanofiber had a mean diameter of 310 nm.	Brahatheeswaran et al. (2012)
	8. The stability of epigallocatechin gallate was enhanced by electrospinning in zein nanofibers (mean diameter: 472 ± 46 nm).	Li et al. (2009)

(*Continued*)

TABLE 14.13 (*Continued*)

Food Processing Applications of Electrospinning and Electrospraying Techniques

Application	Examples	References
	9. Electrospinning technique has been used for the encapsulation of living probiotic cells and bacteriocins. The electrospun nanofibers of Bifidobacterium strains (specifically *Bifidobacterium animalis* spp. *lactis* Bb12) within PVA had an average diameter of 150 nm.	Lopez-Rubio et al. (2009)
	10. Bacteriocin-producing cells of *Lactobacillus plantarum* was electrospun using polyethylene oxide (PEO) to obtain nanofibers of diameter 250–300 nm.	Heunis et al. (2010)
	11. Zein was used as the polymer to encapsulate docosahexaenoic acid and curcumin to obtain submicron (490 ± 200 nm) and nanoparticles (175–250 nm).	Torres-Giner et al. (2010)
Food packaging	1. A natural antimicrobial compound, allyl isothiocyanate, was electrospun using fiber-forming solutions of soy protein isolate (SPI) or poly(lactic acid), by using β-cyclodextrin inclusion complex. The electrospun fibers of diameters ranging from 200 nm to 2 μm exhibited humidity triggered release of the active compound, which can be potentially used for active packaging applications.	Vega-Lugo and Lim (2009)
	2. Antimicrobial activity of chitosan was used to produce zein/chitosan electrospun insoluble fiber mats with biocide properties.	Torres-Giner et al. (2009)
	3. The PVA electrospun fibers containing nano-sized tea powder of Pu-erh tea, a variety of fermented dark tea, showed antibacterial activity against *E. coli*.	Su et al. (2012)
	4. Red raspberry extract, a rich source of anthocyanin, was electrospun into SPI or denatured SPI nanocomposite fiber mats, and gallic acid was incorporated into zein fibers. The fiber mats obtained from both the studies revealed its potential for food preservation by acting as functionalized nanomaterials in active packaging.	Wang et al. (2013), Neo et al. (2013a)
	5. Electrospun gallic acid-loaded zein fibers had high chemical and thermal stability as assessed by storage at 21.5°C for 60 days at 58% relative humidity. The fibers displayed low water activity along with antibacterial and reasonable antifungal properties. These fibers possessing fast release of the active compound can be used as safe food contact materials even for use on dry foods as edible coatings.	Neo et al. (2013b)
Food coating	1. The electrospraying system was used to produce starch films (40 mm diameter) that serve as edible coatings. 2. Coating of chocolate by electrohydrodynamic spraying: It was found that good quality electrospraying can be achieved using samples with high-fat content and lower than 1.5% lecithin (surfactant) concentration.	Pareta and Edirisinghe (2006) and Gorty and Barringer (2011)
In the production of filtration aid	1. Electrospun polyvinylidene fluoride nanofibrous membrane was similar to that of a normal microfiltration membrane such that more than 90% of the polystyrene microparticles (1, 5, and 10 μm) were rejected from filtration.	Gopal et al. (2006)
	2. Poly(ethylene terephthalate) (PET) nanofibrous electrospun membranes with an average fiber diameter of 420 nm and mean thickness of 0.20 mm were produced for usage in apple juice clarification process. The new process of using electrospun PET membrane was found to be a faster, simple, economical, and efficient filtration media than the traditional process of using filtering aids.	Veleirinho and Lopes-da-Silva (2009)

14.5 3D Food Printing

Three-dimensional (3D) food printing is an "additive" manufacturing (AM) process which involves successive addition or precise layering of tiny semi-liquefied food particles on top of each other to create novel processed foods (Lam et al., 2012), without human intervention. In this AM process, a 3D solid object is created from its 3D digital model obtained from computer-aided design (CAD) software or based on the information acquired from a 3D scanner (Figure 14.24a and b). Subsequently, specialized software slices the model into hundreds or thousands of horizontal cross-sectional layers. The sliced model in the form of a computer file is fed to the 3D printer (the AM machine) through a USB drive, SD memory, card or Wi-Fi. Eventually, the 3D printer which is programmed with the material and process parameters creates the object by forming each layer via the selective placement (or forming) of the material (Campbell et al., 2011) (Figure 14.24). Hence, the functioning of a 3D food printer is similar to

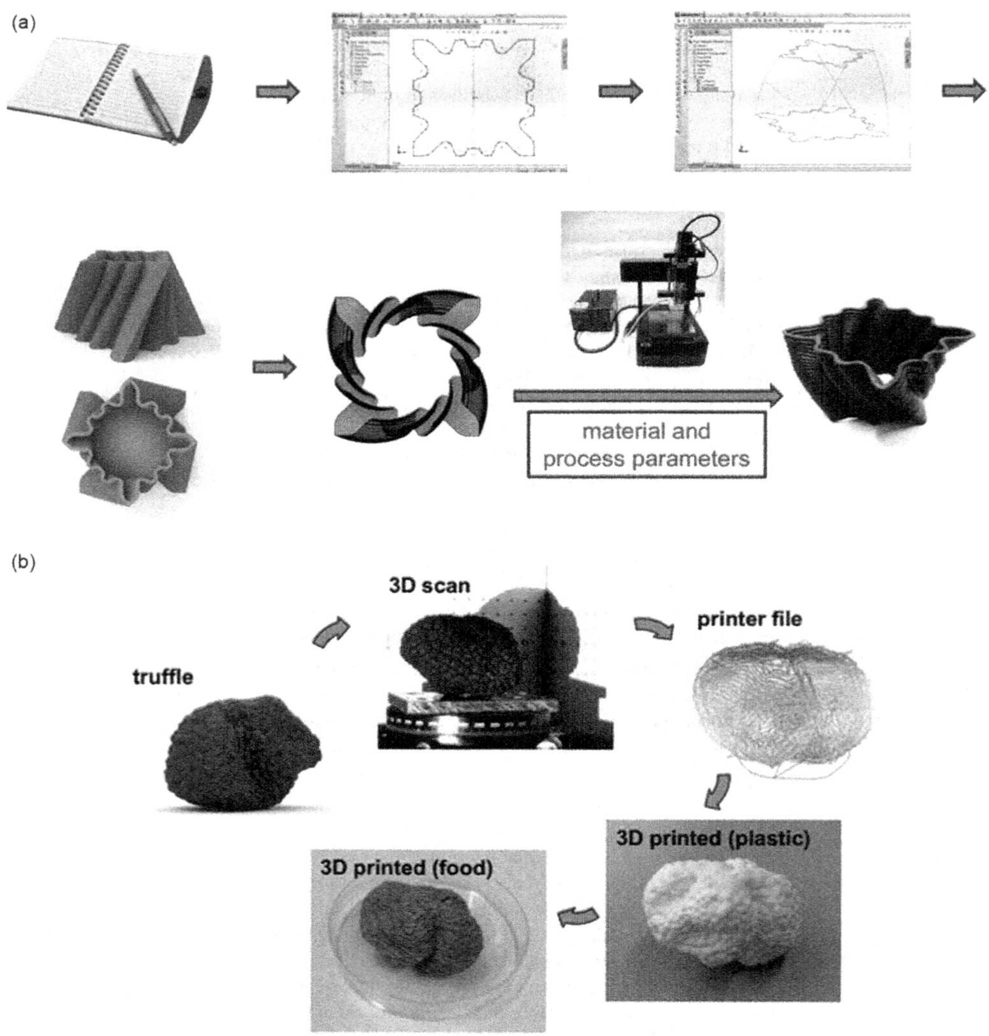

FIGURE 14.24 Steps in 3D food printing. (a) From CAD software; (b) based on data from 3D scanner. (Reproduced with permission from Linden, V. D. 2015. 3D food printing creating shapes and textures. www.tno.nl/media/5517/3d_food_printing_march_2015.pdf)

printing a word processor text using an inkjet printer which moves repeatedly over a page in a row-wise manner and prints each line of the document until it is complete.

14.5.1 Rationale and Advantages of 3D Food Printing Technology

Three-dimensional food printing is perceived as an innovative food structuring technique which facilitates the customization of composition (proportion of ingredients), structure, texture, and taste of foods. Accordingly, the rationales of 3D food printing are discussed subsequently.

14.5.1.1 Creation of Personalized Food Products

Personalization refers to the choice of foods based on self-knowledge (German and Watzke, 2004). For any food product, the following parameters can be personalized: composition (proportion of food ingredients), total number of calories, added macronutrients (protein, fat, minerals, vitamins, and polyunsaturated fatty acids), flavor, size, structure, and hardness. Three-dimensional food printing can also be used to customize food products based on the health, lifestyle, like, and dislike of the consumers. It is possible to feed the information on the health of a consumer to the 3D printers and print personalized foods accordingly. Further, 3D printing can customize nutrition by preparing the same food for different people, based on their personal biometrics (Lipson, 2014) and calorie requirement. Three-dimensional food printing was used to develop a personalized fruit-based snack which is nutritionally designed to meet the energy requirements of children in the age group of 3–10 years. The formula was rich in essential vitamins such as vitamin D, calcium, and iron which are generally absent in commercial fruit-based products. The quantities of these nutrients in the 3D-printed fruit snack were customized so as to adequately fulfill the children's daily intake (Derossi et al., 2018). Another example is the 3D-printed powder mix developed for elderly people with dysphagia and chewing disability. This product solidifies after printing and dissolves when placed in the mouth (Kodama et al., 2017).

14.5.1.2 Enhancement in the Efficiency and Flexibility of Food Production

Three-dimensional food printing technology has the potential to create novel products with unique shape, structure, texture, and flavors, which is generally not possible or involves relatively longer time and high cost with other methods. Three-dimensional printing offers an economic benefit to a food industry with respect to its ability to replace multiple steps in a food production process. Therefore, the cost of mass customization and human errors in food production are reduced while the production efficiency is increased (Sun et al., 2015a, b). Also, with the 3D printing technology, products can be readily made after the receipt of orders and payment. This enhances the flexibility of the production process, improves the capital working management, prevents the accumulation of stocked products, and reduces energy consumption and food waste (Berman et al., 2012; Kietzmann et al., 2015; Rayna and Striukova, 2016). PepsiCo has been using the 3D printing technology for the production of ridge-shaped potato chips, which has cut down the time for product designing from 12 to 18 months (using a hand slicer) to 3 months (using a 3D printer) (Richard Dunham, PepsiCo, in an Interview to TheStreet; Sozzi, 2015).

14.5.1.3 Designing Novel Foods Using Alternative Ingredients

Three-dimensional food printing facilitates the use of alternative sources of nutrients such as the algal, fungal, and insect proteins and lupine seeds for the production of tasty food products with novel structure. The above ingredients are advantageous with respect to both health and environment. For instance, 3D printing has been used to produce meat analogues from soy-based or gluten-based materials, which taste very similar to meat (Sun et al., 2015a, b).

14.5.2 Principle and Classification of 3D Food Printers

Three-dimensional printer (Figure 14.25a) is a device for manufacturing a 3D object that stacks materials based on 3D data (Kodama et al., 2017). In general, a 3D food printer comprises a food printing

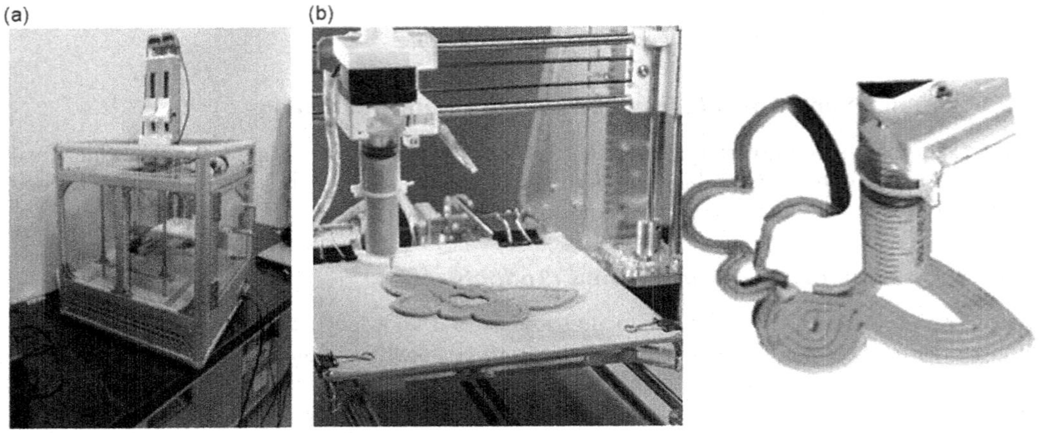

FIGURE 14.25 (a) A 3D food printer; (Reproduced with permission from Kodama, M., Takita, Y., Tamate, H., Saito, A., Gong, J., Makino, M, Khosla A., Kawakami M., and Furukawa H. 2017. Novel soft meals developed by 3D printing. In *Future Foods*, ed. H. Mikkola. IntechOpen. www.intechopen.com/books/future-foods/novel-soft-meals-developed-by-3d-printing); (b) food printing platform and print head. (Reproduced with permission from Sun, J., Peng, Z., Zhou, W., Fuh, J. Y. H., Hong, G. S. and Chiu, A. 2015b. A review on 3D printing for customized food fabrication. *Procedia Manufacturing* 1: 308–319.)

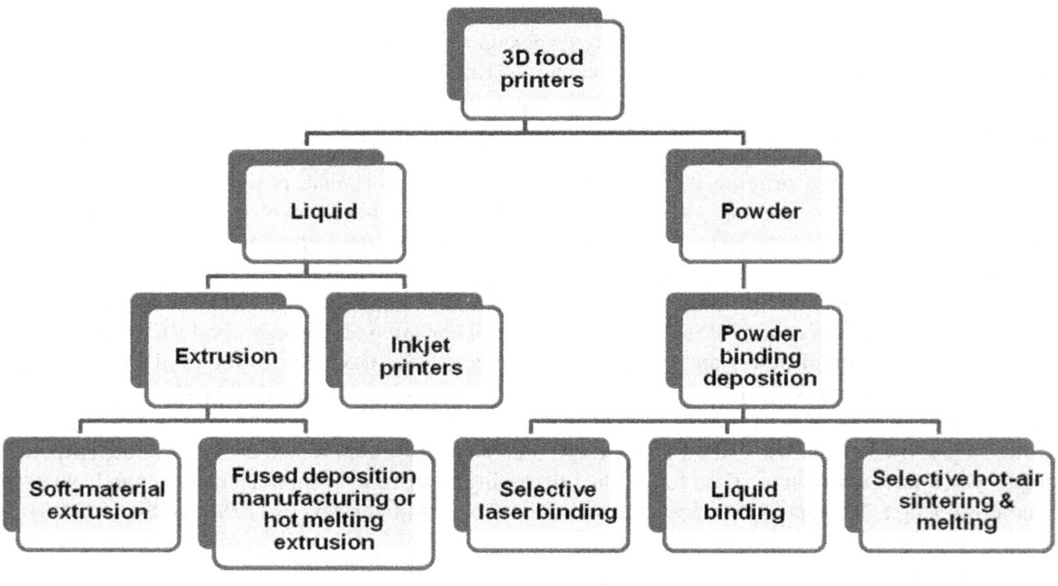

FIGURE 14.26 Classification of 3D food printers.

platform and a print head (Figure 14.25b). Based on the nature of food substrates used for printing and the method of stacking, 3D printers can be classified into the types shown in Figure 14.26.

14.5.2.1 Extrusion

In general, the extrusion-based 3D food printers are appropriate for the preparation of liquid or soft foods from materials such as melted chocolate, pasta dough, fondant, fruit, and vegetable purées. In these printers, the mass is first filled into a cartridge and then squeezed under pressure through an extrusion

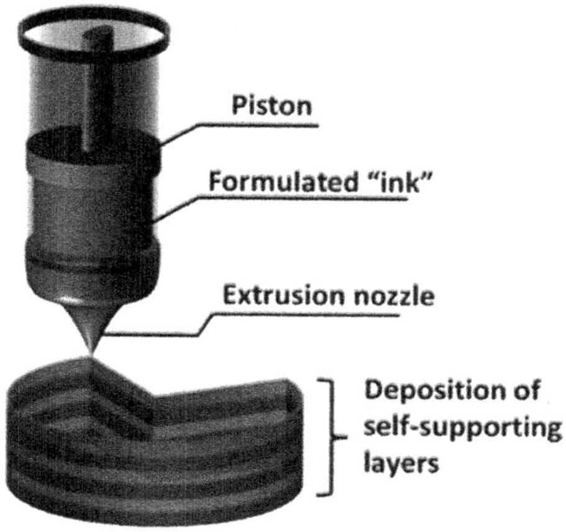

FIGURE 14.27 Schematic diagram of 3D food printing extrusion processes. (Reproduced with permission from Godoi, F. C., Prakash, S. and Bhandari, B. R. 2016. 3D printing technologies applied for food design: status and prospects. *Journal of Food Engineering* 179: 44–54.)

nozzle (Figure 14.27). The printed objects maintain their shape owing to their rheological properties and the cooling phase after extrusion. Optionally, hydrogels may be used to retain the structure after printing. Here, the process parameters that influence the form and stability of printed foods include output pressure, output quantity, movement coordinates, and movement speed of the print head (DLG Expert Report 4, 2017). The different types of extrusion-based 3D food printers are discussed next.

14.5.2.1.1 Soft-Materials Extrusion

With the soft-materials extrusion, 3D structures are created by mixing and depositing self-supporting layers of soft materials such as dough, meat paste, and processed cheese. Here, the key processing parameter is the material's viscosity which should be sufficiently low to permit extrusion of the soft material through a fine nozzle and high enough to support the structure after the deposition. Food-grade additives may be used to desirably modify the rheological properties of the material to be printed (Godoi et al., 2016).

14.5.2.1.2 Fused Deposition Manufacturing or Hot Melting Extrusion

Fused deposition manufacturing (FDM) also termed as hot-melt extrusion involves extrusion of a molten semisolid material through a nozzle which may be heated (Figure 14.28). The extruded material is then deposited onto a substrate in a layer-by-layer manner. In the FDM printers, the material is heated slightly above its melting point such that it solidifies instantaneously after extrusion and combines with the preceding layer (Sun et al., 2015a, b). The working temperature of an FDM printer ranges between 28°C and 40°C (Hao et al., 2010a, b; Schaal, 2009). This printer type has been extensively used for 3D chocolate printing. FDM printers can be operated with multiple extruders to facilitate simultaneous printing of several components. In addition, a dual-feed extruder may be used to obtain a third color by extruding two materials of different colors in exact ratios (Yang et al., 2017). The factors influencing the geometrical precision of food material deposition include (i) nozzle aperture diameter, (ii) optimum nozzle height from the forming bed, and (iii) the extrusion-axis movement (Hao et al., 2010a, b). FDM-based 3D food printers are compact in size and involve low expenditure on their maintenance.

14.5.2.2 Inkjet Printers

Inkjet 3D food printers are similar to the conventional printers used for printing books and labels, with the difference that an edible ink (in the former) replaces the printing ink (in the latter). An inkjet 3D food printer uses a syringe-type, drop-on-demand print head with a single channel that dispenses a mixture

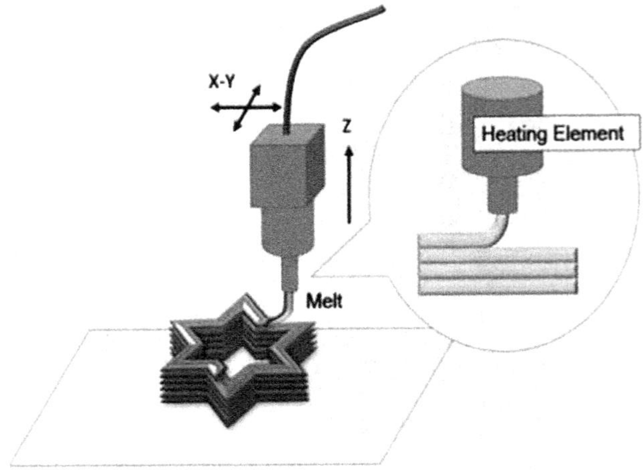

FIGURE 14.28 FDM or hot-melt extrusion. (Reproduced with permission from Sun, J., Peng, Z., Zhou, W., Fuh, J. Y. H., Hong, G. S. and Chiu, A. 2015b. A review on 3D printing for customized food fabrication. *Procedia Manufacturing* 1: 308–319.)

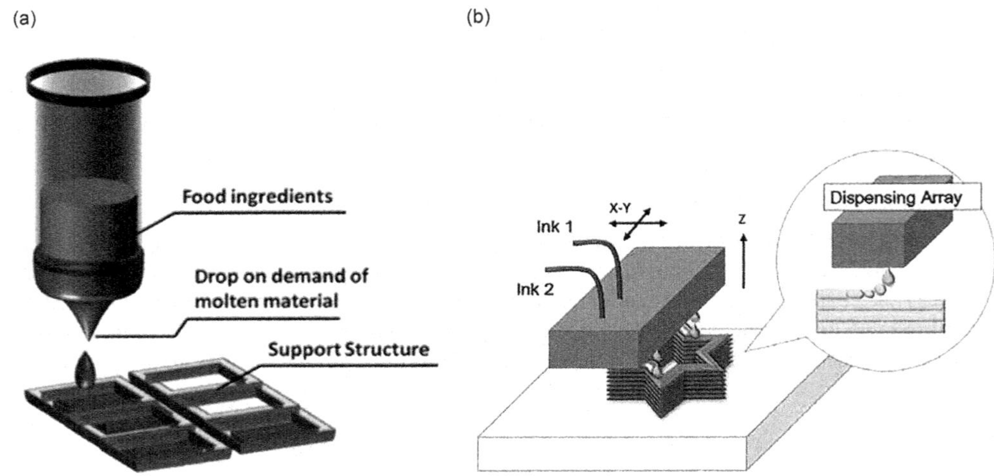

FIGURE 14.29 Schematic diagram of inkjet printer. (a) Single channel print head (Reproduced with permission from Godoi, F. C., Prakash, S. and Bhandari, B. R. 2016. 3D printing technologies applied for food design: status and prospects. *Journal of Food Engineering* 179: 44–54.); (b) multichannel print head (Reproduced with permission from Sun, J., Peng, Z., Zhou, W., Fuh, J. Y. H., Hong, G. S. and Chiu, A. 2015b. A review on 3D printing for customized food fabrication. *Procedia Manufacturing* 1: 308–319.)

of multiple materials (Figure 14.29a) or multiple channels that dispense streams/droplets of two or more materials (Figure 14.29b). The stream/droplets thus dispensed fall under gravity, contact the substrate, and dry by solvent evaporation. The drops form a 2.5D digital image which serves as decoration or surface fill. Cookies, cakes, or pastries are the commonly printed food products using inkjet food printers (Sun et al., 2015a, b; Yang et al., 2017).

14.5.2.3 Powder Binding Deposition

The second most popular method of 3D food printing after extrusion is powder binding deposition which involves the powder deposition in bed (Godoi et al., 2016). This method can be further classified into the following three types.

14.5.2.3.1 Selective Hot-Air and Laser Sintering

In selective hot-air sintering technology, particles of the powder are melted and fused together layer-by-layer through the application of hot air or IR laser (Figure 14.30a and b). The heat source moves along the *X* and *Y* axes of the bed on which a thin and uniform layer of powder is applied. Then, the base is lowered by one layer thickness, and another layer of powder is applied and sintered in a similar manner as described previously. The steps of powder application and sintering are repeated until the 3D structure is formed (Deckard and Beaman, 1988). Finally, the unfixed powder is removed from the final product. Sugar and fat-based materials with relatively low melting points are effectively processed by the selective sintering technology. This technology is capable of fabricating more complex food structures in a short time without the need for post-processing (Izdebska, 2016).

14.5.2.3.2 Liquid Binding

The liquid binding technology is similar to the sintering method described previously, except that a liquid binder spray is overprinted onto layers of powder to combine them (Figure 14.30c). Usually, water spray is used to stabilize the powder material and minimize the disturbance caused by the dispensing of binder (Sun et al., 2015a, b). Here, the binding between the layers is based on the adhesive forces or chemical reactions between the powder and the binder. But, there is no phase change during layer solidification (Wegrzyn et al., 2012). Nevertheless, unlike the sintering technology, products created by liquid binding method may require post-processing which includes additional curing and improvement of the fused layers (Izdebska, 2016). A wide range of confectionary recipes including sugar, fondant, and sweet and sour candy in a variety of flavors have been produced using liquid binding (Hasseln et al., 2014).

14.5.3 Types of Materials for 3D Food Printing

Available food materials are pre-processed to render them suitable for 3D printing and increase their thermal stability during post-processing. The raw materials filled in the cartridge of 3D food printers usually have longer shelf life than the final food products. Consequently, the filled cartridges containing different materials can be transported to distant countries and even for space missions. With 3D food printing, a high energy source is not mandatory to completely remove the liquid content from the food product. However, the fabricated layer should be adequately rigid and strong to support its own weight and the weight of subsequent layers, without significant deformation. Based on printability, the food materials used in 3D food printing can be classified into two categories as follows (Sun et al., 2015a, b).

14.5.3.1 Natively Printable Materials

Natively printable materials are those that can be smoothly extruded from a syringe (Cohen et al., 2009). Printability of a material is evaluated by its viscosity, consistency, and solidifying properties. Based on the abovementioned criteria, cake frosting, cheese, pasta dough, and chocolate are considered natively printable. Natively printable materials are substantially stable to retain their shape after deposition and do not require post-processing. Also, food products prepared from natively printable materials can be completely personalized for taste, nutritional value, and texture (Sun et al., 2015a, b).

14.5.3.2 Non-Printable Materials

Non-printable materials are those which are not inherently printable but rendered printable by the addition of food-grade hydrocolloids. Staple foods such as rice, meat, fruit, and vegetables are non-printable by nature. As an approach to convert a wide range of non-printable traditional food materials to printable ones, a small group of ingredients with extensive degrees of freedom of texture and flavor was created. By modifying the concentration of hydrocolloids, a broad range of mouthfeel was achieved with products such as cooked spaghetti and cake icing. Similarly, the flavors can be altered by fine-tuning the amount of flavor additives used. Most of the aforementioned traditional foods require post-processing

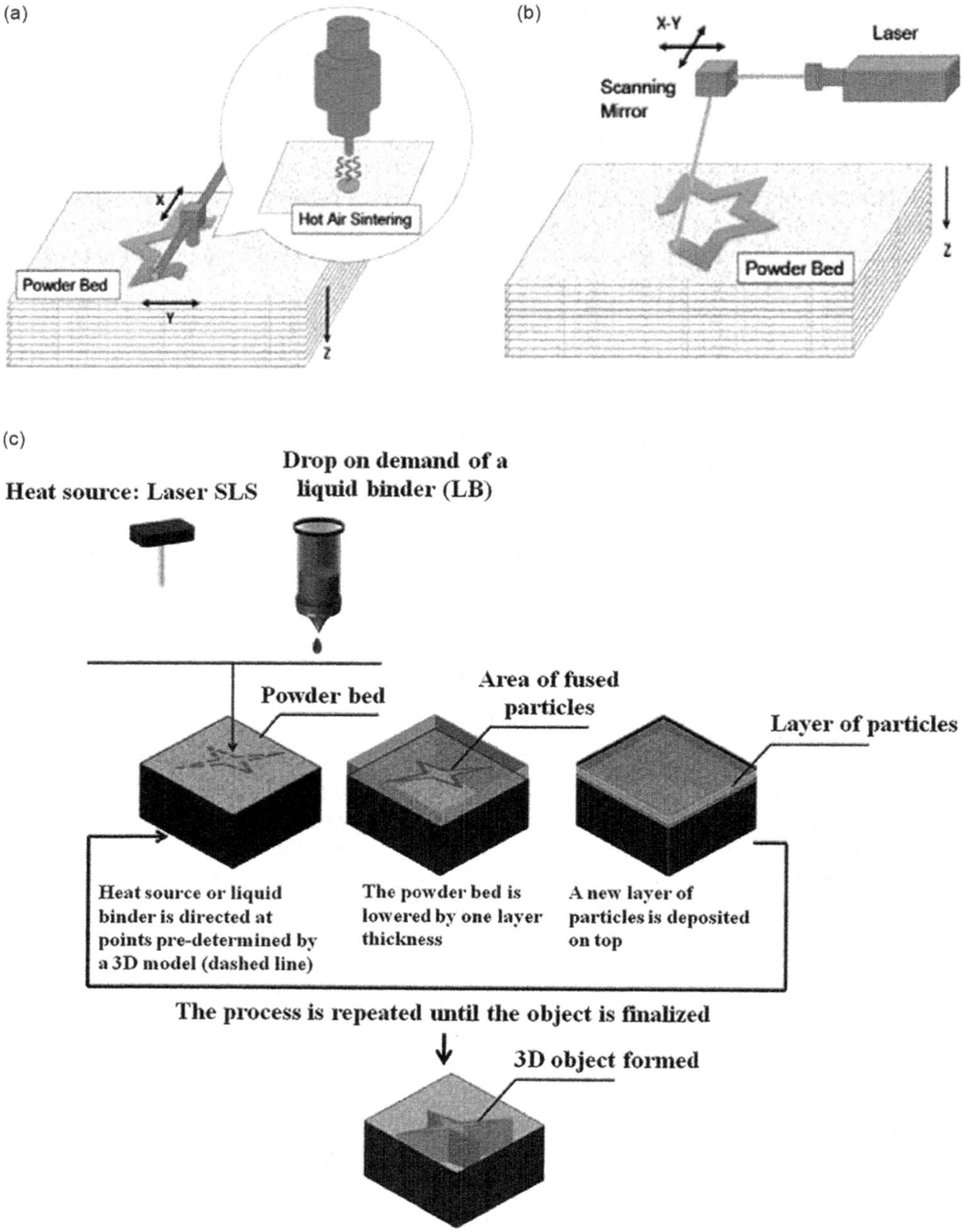

FIGURE 14.30 (a) Selective hot-air sintering; (b) selective laser sintering; (Reproduced with permission from Sun, J., Peng, Z., Zhou, W., Fuh, J. Y. H., Hong, G. S. and Chiu, A. 2015b. A review on 3D printing for customized food fabrication. *Procedia Manufacturing* 1: 308–319.); (c) liquid binding. (Reproduced with permission from Godoi, F. C., Prakash, S. and Bhandari, B. R. 2016. 3D printing technologies applied for food design: status and prospects. *Journal of Food Engineering* 179: 44–54.)

such as baking, steaming, or frying after the 3D structures are created. This demands the application of heat in different levels. Thus, the ingredients for 3D printing should be appropriately modified to withstand the heat treatment to avoid non-homogenous texture (Sun et al., 2015a, b).

The applications of 3D food printing are tabulated in Table 14.14.

TABLE 14.14

Applications of 3D Food Printers

Technique	Application	Reference
Raw material: liquid		
Soft-materials extrusion	Frosting, processed cheese, dough (e.g., pizza), meat puree	Chang et al. (2014), Kuo et al. (2014); Lipton et al. (2010), Periard et al. (2007)
Melting extrusion	Chocolate, confection	Hao et al. (2010b), Schaal (2009)
Inkjet printing	Chocolate, liquid dough, sugar icing, meat paste, cheese, jams, gels	Grood and Grood (2011)
Raw material: powder		
Liquid binding	Chocolate	3D Systems 2015, Diaz et al. (2015a, b) and Hasseln et al. (2014)
Selective laser sintering	Sugar, Nesquik	Diaz et al. (2014a, b)
Hot-air sintering and melting	Sugar	CandyFab (2006)

Source: Adapted from Godoi et al. (2016).

14.5.4 Challenges in 3D Food Printing

With all its advantages aside, 3D food printing is still associated with a number of challenges. These include long fabrication time, low throughput, delamination caused by temperature fluctuation, limitations with respect to texture (rough surface finish) and taste of the final 3D structured foods, and high machine cost. Also, the number of stable printable food materials available is limited. In addition, safety standards need to be established for the ingredients used in 3D food printing technology (Jin, 2017).

BIBLIOGRAPHY

Adapa, S., Schmidt, K. A. and Toledo, R. 1997. Functional properties of skim milk processed with continuous high pressure throttling. *Journal of Dairy Science* 80: 1941–1948.

Ade-Omowaye, B. I. O., Eshtiaghi, N. and Knorr, D. 2000. Impact of high intensity electric field pulses on cell permeabilisation and as pre-processing step in coconut processing. *Innovative Food Science and Emerging Technologies* 1: 203–209.

Aguiló-Aguayo, I., Soliva-Fortunity, R. and Martin-Belloso, O. 2009. Effects of high-intensity pulsed electric fields on lipoxygenase and hydroperoxide lyase activities in tomato juice. *Journal of Food Science* 74: 595–601.

Almeida, F. D. L., Cavalcante, R. S., Cullen, P. J., Frias, J. M., Bourke, P., Fernandes, F. A. N. and Rodrigues, S. 2015. Effects of atmospheric cold plasma and ozone on prebiotic orange juice. *Innovative Food Science and Emerging Technologies* 32: 127–135.

Álvarez, I., Pagán, R., Condón, S. and Raso, J. 2003. The influence of process parameters for the inactivation of *Listeria monocytogenes* by pulsed electric fields. *International Journal of Food Microbiology* 87: 87–95.

Anandharamakrishnan, C. 2013. *Techniques for Nanoencapsulation of Food Ingredients*. New York: Springer.

Anandharamakrishnan, C. and Ishwarya, S. P. 2015. *Spray Drying Techniques for Food Ingredient Encapsulation*. Hoboken, NJ: John Wiley & Sons, Ltd.

Angersbach, A. and Knorr, D., 1997. High intensity electric field pulses as pretreatment for affecting dehydration characteristics and rehydration properties of potato cubes. *Nahrung* 55: 143–146.

Arpagus, C. 2012. A novel laboratory-scale spray dryer to produce nanoparticles. *Drying Technology* 30: 1113–1121.

Baba, K. and Nishida, K. 2013. Steroid nanocrystals prepared using the nano spray dryer B-90. *Pharmaceutics* 5: 107–114.

Balasubramaniam, V. M., Martínez-Monteagudo, S. I. and Gupta, R. 2015. Principles and application of high pressure–based technologies in the food industry. *Annual Review of Food Science and Technology* 6: 435–462.

Balci, A. T. and Wilbey, R. A. 1999. High pressure processing of milk—the first 100 years in the development of a new technology. *International Journal of Dairy Technology* 52: 149–155.

Becker, K. H., Schoenbach, K. H. and Eden, J. G. 2006. Microplasmas and applications. *Journal of Physics D: Applied Physics* 39: R55–R70.

Benet, G. U. 2005. High-pressure low-temperature processing of foods: impact of metastable phases on process and quality parameters. *PhD Dissertation*. Berlin: University of Technology.

Bengtsson, N. E. and Green, W. 1970. Radio-frequency pasteurisation of cured hams. *Journal of Food Science* 35: 681–687.

Benlloch-Tinoco, M., Salvador, A., Rodrigo, A. and Martínez-Navarrete, N. 2016. Microwave heating technology. In *Handbook of Food Processing: Food Preservation*, eds. T. Varzakas and C. Tzia, 297–318. Boca Raton, FL: CRC Press, Taylor and Francis Group.

Berman, B., Zarb, F. G. and Hall, W. 2012. 3D printing: the new industrial revolution. *Business Horizons* 55: 155–162.

Bermúdez-Aguirre, D., Wemlinger, E., Pedrow, P., Barbosa-Cánovas, G. and Garcia-Perez, M. 2013. Effect of atmospheric pressure cold plasma (APCP) on the inactivation of *Escherichia coli* in fresh produce. *Food Control* 34: 149–157.

Bhushani, J. A. and Anandharamakrishnan, C. 2014. Electrospinning and electrospraying techniques: potential food based applications. *Trends in Food Science and Technology* 38: 21–33.

Bhushani, J. A., Kurrey, N. K. and Anandharamakrishnan, C. 2017. Nanoencapsulation of green tea catechins by electrospraying technique and its effect on controlled release and in-vitro permeability. *Journal of Food Engineering* 199: 82–92.

Black, P. E., Kelly, L. A. and Fitzgerald, F. G. 2005. The combined effect of high pressure and nisin on inactivation of microorganisms in milk. *Innovative Food Science and Emerging Technologies* 6: 286–292.

Blasi, P., Schoubben, A., Giovagnoli, S., Rossi, C. and Ricci, M. 2010. Alginate micro- and nanoparticle production by spray drying. *Proceedings of Meeting on Lactose as a Carrier for Inhalation Products*, Parma, Italy, September 26th–28th, 2010, 137–138.

Bock, N., Dargaville, T. R. and Woodruff, M. A. 2012. Electrospraying of polymers with therapeutic molecules: state of the art. *Progress in Polymer Science* 37: 1510–1551.

Bozkurt, H. and Icier, F. 2009. Electrical conductivity changes of minced beef-fat blends during ohmic cooking. *Journal of Food Engineering* 96: 86–92.

Brahatheeswaran, D., Mathew, A., Aswathy, R. G., Nagaoka, Y., Venugopal, K., Yoshida, Y., Maekawa, T. and Sakthikumar, D. 2012. Hybrid fluorescent curcumin loaded zein electrospun nanofibrous scaffold for biomedical applications. *Biomedical Materials* 7: 045001.

Brandenberger, H., Nussli, D., Piech, V. and Widmer, F. 1999. Monodisperse particle production: a method to prevent drop coalescence using electrostatic forces. *Journal of Electrostatics* 45: 227–238.

Campbell, T., Williams, C., Ivanova, O. and Garrett, B. 2011. Could 3D printing change the world? Technologies, potential, and implications of additive manufacturing. Atlantic Council Strategic Foresight Report.

CandyFab. 2006. The CandyFab project. https://candyfab.org/ (accessed October 2, 2018).

Carlez, A., Chefel, J. C., Rosec, J. P. and Richard, N. 1993. High pressure inactivation of *Citrobacter freundii*, *Pseudomonas fluorescens* and *Listeria innocua* in inoculated minced beef muscle. *LWT—Food Science and Technology* 26: 357–363.

Cathcart, W. H., Parker, J. J. and Beattie, H. G. 1947. The treatment of packaged bread with high frequency heat. *Food Technology* 1: 174–177.

Cavender, G. A. 2011. Continuous high pressure processing of liquid foods: an analysis of physical, structural and microbial effects. *PhD Thesis*. Georgia: University of Georgia.

Chang, C., Chen, S., Delgado, V., Hsu, T., Huang, S., Kuo, C., Mao, C., Olive, X., Rodriguez, L. and Sepulveda, E. 2014. *Additive Manufacturing Printer System for Printing e.g. Food Product, has Processor that Provides Controller with Position Coordinates for Movement of Tool, Instructions for Exchange of Capsule Holders, and Adjustment of Heating Device*. Barcelona, Spain: Natural Machines LLC (NATU-Nonstandard).

Chhanwal, N., Ezhilarasi, P. N., Indrani, D. and Anandharamakrishnan, C. 2015. Influence of electrical and hybrid heating on bread quality during baking. *Journal of Food Science and Technology* 52: 4467–4474.

Chirokov, A., Gutsol, A. and Fridman, A. 2005. Atmospheric pressure plasma of dielectric barrier discharges. *Pure and Applied Chemistry* 77: 487–495.

Choi, S., Puligundla, P. and Mok, C. 2016. Corona discharge plasma jet for inactivation of *Escherichia coli* O157:H7 and *Listeria monocytogenes* on inoculated pork and its impact on meat quality attributes. *Annals of Microbiology* 66: 685–694.

Cohen, D. L., Jeffrey, I. L., Cutler, M., Coulter, D., Vesco, A. and Lipson, H. 2009. Hydrocolloid printing: a novel platform for customized food production. *Proceedings of Solid Freeform Fabrication Symposium (SFF'09)*, Austin, TX, USA 3–5.

Cresko, J. W. and Anantheswaran, R. C. 1998. Dielectric drying and roasting for food manufacturing. *Proceedings of 33rd Microwave Power Symposium*, Chicago, IL. 95–98.

Czylkowski, D., Jasinski, M. and Mizeraczyk, J. 2012. Novel low power microwave plasma sources at atmospheric pressure. *Przegląd Elektrotechniczny (Electrical Review)* 88(8): 39–42.

Daomukda, N., Moongngarm, A., Payakapol, L. and Noisuwan, A. 2011. Effect of cooking methods on physicochemical properties of brown rice. *2nd International Conference on Environmental Science and Technology IPCBEE 6:V1-1-V1-4*. Singapore: IACSIT Press.

Datta, A. K. and Anantheswaran, R. C. 2000. *Handbook of Microwave Technology for Food Applications*. New York: Marcel Dekker Inc.

de Halleux, G., Piett, M., Buteau, M. L. and Dostie, A. 2005. Ohmic cooking of processed meats: energy evaluation and food safety considerations. *Canadian Biosystems Engineering* 47: 3.41–3.47.

Deckard, C. and Beaman, J. 1988. Process and control issues in selective laser sintering. *American Society of Mechanical Engineers, Production Engineering Division (Publication) PED* 33: 191–197.

Derossi, A., Caporizzi, R., Azzollini, D. and Severini, C. 2018. Application of 3D printing for customized food. A case on the development of a fruit-based snack for children. *Journal of Food Engineering* 220: 65–75.

Diaz, J. V., Noort, M. W. and Van Bommel, K. J. C. 2015a. *Producing Edible Object Used in Food Product, Comprises Subjecting Edible Powder Composition Comprising Water Soluble Protein, Hydrocolloid and Plasticizer to Powder Bed Printing by Depositing Edible Liquid onto Powder in Layer-wise Manner*. The Hague, Netherlands: Nederlandse Org Toegepast Natuurwetensch (Nede-C).

Diaz, J. V., Noort, M. W. J. and Van Bommel, K. J. C. 2015b. Method for the production of an edible object by powder bed (3d) printing and food products obtainable therewith. Patent WO2015115897A1.

Diaz, J. V., Van Bommel, K. J. C., Noort, M. W. J., Henket, J. and Brier, P. 2014a. Method for the production of edible objects using Sls and food products. US Patent US20160100621A1.

Diaz, J. V., Van Bommel, K. J. C., Noort, M. W., Henket, J. and Brier, P., 2014b. *Preparing Edible Product, Preferably Food Product Including Bakery Product, and Confectionary Product, Involves Providing Edible Powder Composition, and Subjecting Composition to Selective Laser Sintering*. The Hague, Netherlands: Nederlandse Org Toegepast Natuurwetensch (Nede-C).

DLG Expert Report 4. 2017. 3D food printing: what options the new technology offers. DLG e.V. Competence Center Food, Germany. www.DLG.org.

Dostie, M., Seguin, J. N., Maure, D., Tonthat, Q. A. and Chatingy, R. 1989. Preliminary measurements on the drying of thick porous materials by combinations of intermittent infrared and continuous convection heating. In *Drying'89*, eds. A. S. Mujumdar and M. A. Roques, 92: 513–520. New York: Hemisphere Press.

Drake, M. A., Harrison, S. L., Asplund, M., Barbosa, C. G. and Swanson, B. G. 1997. High-pressure treatment of milk and effects on microbiological and sensory quality of Cheddar cheese. *Journal of Food Science* 62: 843–845.

Dunn, J. E. and Pearlman, J. S. 1987. Methods and apparatus for extending the shelf life of fluid food products. US Patent 4695472.

Dunn, J. E., Pearlman, J. S. and La Costa, R. 1989. Methods and apparatus for extending the shelf life of fluid food products. US Patent 4838154.

Elamin, W. M., Endan, J. B., Yosuf, Y. A., Shamsudin, R. and Ahmedov, A. 2015. High pressure processing technology and equipment evolution: a review. *Journal of Engineering Science and Technology Review* 8: 75–83.

Eliot-Godéreaux, S. C., Zuber, F. and Goullieux, A. 2001. Processing and stabilisation of cauliflower by ohmic heating technology. *Innovative Food Science and Emerging Technologies* 2: 279–287.

Eshtiaghi, M. N. and Knorr, D. 1993. Potato cubes response to water blanching and high hydrostatics pressure. *Journal of Food Science* 58: 1371–1374.

Farkas, D. F. and Hoover, D. G. 2000. High pressure processing. *Journal of Food Science* 65: 47–64.

Feng, A. L., Boraey, M. A., Gwin, M. A., Finlay, P. R., Kuehl, P. J. and Vehring, R. 2011. Mechanistic models facilitate efficient development of leucine containing microparticles for pulmonary drug delivery. *International Journal of Pharmaceutics* 409: 156–163.

Fernandez, A., Torres-Giner, S. and Lagaron, J. M. 2009. Novel route to stabilization of bioactive antioxidants by encapsulation in electrospun fibers of zein prolamine. *Food Hydrocolloids* 23: 1427–1432.

Fernández-Díaz, M. D., Barsotti, L., Dumay, E. and Cheftel, J. C. 2000. Effects of pulsed electric fields on ovalbumin and dialyzed egg white. *Journal of Agricultural and Food Chemistry* 48: 2332–2339.

Floury, J., Grosset, N., Leconte, N., Pasco, M., Madec, M. and Jeantet, R. 2006. Continuous raw skim milk processing by pulsed electric field at non-lethal temperature: effect on microbial inactivation and functional properties. *Le Lait* 86: 43–57.

Fukui, Y., Maruyama, T., Iwamatsu, Y., Fujii, A., Tanaka, T., Ohmukai, Y.and Matsuyama, H. 2010. Preparation of monodispersed polyelectrolyte microcapsules with high encapsulation efficiency by an electrospray technique. *Colloids and Surfaces A: Physicochemical and Engineering Aspects* 370: 28–34.

Garriga, M., Aymerich, M. T. and Hugas, M. 2002. Effect of high pressure processing on the microbiology of skin-vacuum packaged sliced meat products: cooked pork ham, dry cured pork ham and marinated beef loin. Profit Final Project Report FIT060000200066.

German, J. B. and Watzke, H. J. 2004. Personalizing foods for health and delight. *Comprehensive Reviews in Food Science and Food Safety* 3: 145–151.

Ghayempour, S. and Mortazavi, S. M. 2013. Fabrication of micro-nanocapsules by a new electrospraying method using coaxial jets and examination of effective parameters on their production. *Journal of Electrostatics* 71: 717–727.

Godoi, F. C., Prakash, S. and Bhandari, B. R. 2016. 3D printing technologies applied for food design: status and prospects. *Journal of Food Engineering* 179: 44–54.

Gola, S., Mutti, P., Manganelli, E., Squarcina, N. and Rovere, P. 2000. Behaviour of E. coli O157:H7 strains in model system and in raw meat by HPP: microbial and technological aspects. *High Pressure Research* 19: 481–487.

Gopal, R., Kaur, S., Ma, Z., Chan, C., Ramakrishna, S. and Matsuura, T. 2006. Electrospun nanofibrous filtration membrane. *Journal of Membrane Science* 281: 581–586.

Gorty, A. V. and Barringer, S. A. 2011. Electrohydrodynamic spraying of chocolate. *Journal of Food Processing and Preservation* 35: 542–549.

Goullieux, A. and Pain, J. P. 2014. Ohmic heating. In *Emerging Technologies for Food Processing* (Second edition), ed. D. W. Sun, 399–426. London, UK: Academic Press.

Grahl, T. and Märkl, H. 1996. Killing of microorganisms by pulsed electric fields. *Applied Microbiology and Biotechnology* 45: 148–157.

Grood, J. P. W. and Grood, P. J. 2011. Method and device for dispensing a liquid. US Patent US20110121016A1.

Gude, V. G., Patil, P., Martinez-Guerra, E., Deng, S. and Nirmalakhandan, N. 2013. Microwave energy potential for biodiesel production. *Sustainable Chemical Processes* 1: 5.

Hamanaka, D., Dokan, S., Yasunaga, E., Kuroki, S., Uchino, T. and Akimoto, K. 2000. The sterilization effect on infrared ray of the agricultural products spoilage microorganisms (Part 1). *ASAE Annual Meeting International*, Milwaukee, 9th–12th July 2000, 1–9.

Hao, L., Mellor, S., Seaman, O., Henderson, J., Sewell, N. and Sloan, M. 2010a. Material characterisation and process development for chocolate additive layer manufacturing. *Virtual and Physical Prototyping* 5: 57–64.

Hao, L., Seaman, O., Mellor, S., Henderson, J., Sewell, N. and Sloan, M., 2010b. Extrusion behavior of chocolate for additive layer manufacturing, innovative developments in design and manufacturing. *Advanced Research in Virtual and Rapid Prototyping* 245–250.

Hasseln, V. K. W., Hasseln, V. E. M., Williams, D. X. and Gale, R. R. 2014. *Making an Edible Component, Comprises Depositing Successive Layers of Food Material According to Digital Data that Describes the Edible Component, and Applying Edible Binders to Regions of the Successive Layers of the Food Material*. Rock Hill, South Carolina, CA: 3d Systems Inc (Thde-C).

Hebbar, H. U. and Rastogi, N. K. 2001. Mass transfer during infrared drying of cashew kernel. *Journal of Food Engineering* 47: 1–5.

Heinz, V. and Buckow, R. 2010. Food preservation by high pressure. *Journal of Consumer Protection and Food Safety* 5: 73–81.

Helfritch, D. J. 1993. Pulsed corona discharge for hydrogen sulfide decomposition. *Non-Thermal Plasma Techniques for Pollution Control* 34: 211–221.

Heng, D., Lee, S. H., Ng, W. K. and Tan, R. B. H. 2011. The nano spray dryer B-90. *Expert Opinion on Drug Delivery* 8: 965–972.

Heunis, T. D. J., Botes, M. and Dicks, L. M. T. 2010. Encapsulation of *Lactobacillus plantarum* 423 and its bacteriocin in nanofibers. *Probiotics and Antimicrobial Proteins* 2: 46–51.

Hicks, D. T., Pivarnik, L. F., McDermott, R., Richard, N., Hoover, D. G. and Kniel, K. E. 2009. Consumer awareness and willingness to pay for high-pressure processing of ready-to-eat food. *Journal of Food Science* 8: 32–38.

Hines, J. and Nickels, L. 2011. Hot topic—the growth of high temperature microwave technology. *Metal Powder Report* 66: 7–9.

Hite, B. H., Giddings, N. J. and Weakley, C. E. 1914. Effect of pressure on certain micro-organisms encountered in the preservation of fruits and vegetables. *West Virginia University Agricultural Experiment Station* 146: 2–67.

Hogan, E., Kelly, A. L. and Sun, D. W. 2014. High pressure processing of foods: an overview, In *Emerging Technologies for Food Processing*, ed. D. W. Sun, 3–24. London: Elsevier Academic.

Homayoni, H., Ravandi, S. A. H. and Valizadeh, M. 2009. Electrospinning of chitosan nanofibers: processing optimization. *Carbohydrate Polymers* 77: 656–661.

Houben, J., Schoenmakers, L., van Putten, E., van Roon, P. and Krol, B. 1991. Radio-frequency pasteurization of sausage emulsions as a continuous process. *Journal of Microwave Power and Electromagnetic Energy* 26: 202–205.

Houben, J., van Roon, P. and Krol, B. 1994. Continuous radio-frequency pasteurisation in sausage manufacturing lines. *Fleischewirtsch. International* 2: 35–36.

Hugas, M., Garriga, M. and Monfort, J. M. 2002. New mild technologies in meat processing: high pressure as a model technology. *Meat Science* 62: 359–371.

Huppertz, T., Alan, L. and Kelly Fox, P. F. 2002. Effect of high pressure on constituents and properties of milk. *International Dairy Journal* 12: 561–572.

Icier, F. and Baysal, T. 2004. Dielectrical properties of food materials—1: factors affecting and industrial uses. *Critical Reviews in Food Science and Nutrition* 44: 465–471.

Icier, F. and Ilicali, C. 2005. Temperature dependent electrical conductivities of fruit purees during ohmic heating. *Food Research International* 38: 1135–1142.

Icier, F., Yildiz, H. and Baysal, T. 2008. Polyphenoloxidase deactivation kinetics during ohmic heating of grape juice. *Journal of Food Engineering* 85: 410–417.

Izdebska, J. 2016. 3D food printing—facts and future. *Agro FOOD Industry Hi Tech* 27: 33–37.

Jaeger, H., Reineke, K., Schoessler, K. and Knorr, D. 2012. Effects of emerging processing technologies on food material properties. In *Food Materials Science and Engineering*, ed. B. Bhandari and Y. H. Roos, 222–262. Oxford, UK: Wiley-Blackwell.

Jason, A. C. and Sanders, H. R. 1962. Dielectric thawing of fish. experiments with frozen herrings. Experiments with frozen white fish. *Food Technology* 16: 101–112.

Jaworek, A. 2007. Micro- and nanoparticle production by electrospraying. *Powder Technology* 176: 18–35.

Jaworek, A. T. S. A. and Sobczyk, A. T. 2008. Electrospraying route to nanotechnology: an overview. *Journal of Electrostatics* 66: 197–219.

Jin, J. 2017. The potential of 3D printing in the food industry. How 3D printing is redefining industries. PreScouter, Inc. www.prescouter.com.

Jin, Z. and Zhang, Q. H. 1999. Pulsed electric field inactivation of microorganisms and preservation of quality of cranberry juice. *Journal of Food Processing and Preservation* 23: 481–497.

Jones, W. 1992. *A Place in the Line for Micronizer*, Special Report. Woodbridge, UK: Micronizing Co.

Jun, S. and Sastry, S. 2005. Modeling and optimization of ohmic heating of foods inside a flexible package. *Journal of Food Process Engineering* 28: 417–436.

Jun, S., Sastry, S. and Samaranayake, C. 2007. Migration of electrode components during ohmic heating of foods in retort pouches. *Innovative Food Science and Emerging Technologies* 8: 237–243.

Jung, S., Anton, M. L., Taylor, R. G. and Ghoul, M. 2000. High-pressure effects on lysosome integrity and lysosomal enzyme activity in bovine muscle. *Journal of Agricultural and Food Chemistry* 48: 2467–2471.

Katrokha, I., Matvienko, A., Vorona, I., Kupchik, M. and Zaets, V. 1984. Intensification of sugar extraction from sweet sugar beet cossettes in an electric field. *Sakharnaya Promyshlennost* 7: 28–31.

Kessick, R., Fenn, J. and Tepper, G. 2004. The use of AC potentials in electrospraying and electrospinning processes. *Polymer* 45: 2981–2984.

Kietzmann, J., Pitt, L. and Berthon, P. 2015. Disruptions, decisions, and destinations: enter the age of 3D printing and additive manufacturing. *Business Horizons* 58: 209–215.

Kim, H. J., Choi, Y. M., Yang, T. C. S., Taub, I. A., Tempest, P., Skudder, P. J., Tucker, G. and Parrott, D. L. 1996. Validation of ohmic heating for quality enhancement of food products. *Food Technology* 50: 253–261.

Kim, J. and Pyun, Y. 1995. Extraction of soy milk using ohmic heating. *Abstract, 9th Congress of Food Science and Technology*, Budapest, Hungary.

Kim, J. E., Oh, Y. J., Won, M. Y., Lee, K. S. and Min, S. C. 2017. Microbial decontamination of onion powder using microwave-powered cold plasma treatments. *Food Microbiology* 62: 112–123.

Kodama, M., Takita, Y., Tamate, H., Saito, A., Gong, J., Makino, M., Khosla, A., Kawakami, M. and Furukawa, H. 2017. Novel soft meals developed by 3D printing. In *Future Foods*, ed. H. Mikkola, 161–181. London, United Kingdom: IntechOpen. www.intechopen.com/books/future-foods/novel-soft-meals-developed-by-3d-printing (accessed October 2, 2018).

Kovačević, D. B., Putnik, P., Dragović-Uzelac, V., Pedisić, S., Jambrak, A. R. and Herceg, Z. 2016. Effects of cold atmospheric gas phase plasma on anthocyanins and color in pomegranate juice. *Food Chemistry* 190: 317–323.

Kriegel, C., Arrechi, A., Kit, K., McClements, D. J. and Weiss, J. 2008. Fabrication, functionalization, and application of electrospun biopolymer nanofibers. *Critical Reviews in Food Science and Nutrition* 48: 775–797.

Krishnamurthy, K., Khurana, H. K., Jun, H., Irudayaraj, J. and Demirci, A. 2008. Infrared heating in food processing: an overview. *Comprehensive Reviews in Food Science and Food Safety* 7: 1–13.

Kuan, C. Y., Yuen, K. H., Bhat, R. and Liong, M. T. 2011. Physicochemical characterization of alkali treated fractions from corncob and wheat straw and the production of nanofibers. *Food Research International* 44: 2822–2829.

Kuo, C. J., Huang, S. H., Hsu, T. H., Rodriguez, L., Olive, X., Mao, C. Y., Chang, C. T., Chen, S. C. and Sepulveda, E. 2014. *Manufacturing Food Using 3d Printing Technology*. Barcelona: Natural Machines LLC.

Lacombe, A., Niemira, B. A., Gurtler, J. B., Fan, X., Sites, J., Boyd, G. and Chen, H. 2015. Atmospheric cold plasma inactivation of aerobic microorganisms on blueberries and effects on quality attributes. *Food Microbiology* 46: 479–484.

Lam, C. X. F., Mo, X. M., Teoh, S. H. and Hutmacher, D. W. 2012. Scaffold development using 3D printing with a starch-based polymer. *Materials Science and Engineering* 20: 49–56.

Laroussi, M., Alexeff, I., Kang, W. L. 2000. Biological decontamination by nonthermal plasmas. *IEEE Transactions on Plasma Science* 28: 184–188.

Le Chatelier, H. L. 1884. A general statement of the laws of chemical equilibrium. *Comptes Rendus* 99: 786–789.

Lee, S. H., Heng, D., Ng, W. K., Chan, H. K. and Tan, R. B. 2011. Nano spray drying: a novel method for preparing protein nanoparticles for protein therapy. *International Journal of Pharmaceutics* 403: 192–200.

Li, D. and Xia, Y. 2004. Electrospinning of nanofibers: reinventing the wheel? *Advanced Materials* 16: 1151–1170.

Li, X., Anton, N., Arpagaus, C., Belleteix, F. and Vandamme, T. F. 2010. Nanoparticles by spray drying using innovative new technology: the Büchi nano spray dryer B-90. *Journal of Controlled Release* 147: 304–310.

Li, Y., Lim, L. T. and Kakuda, Y. 2009. Electrospun zein fibers as carriers to stabilize (−)-epigallocatechin gallate. *Journal of Food Science* 74: C233–C240.

Linden, V. D. 2015. 3D food printing creating shapes and textures. www.tno.nl/media/5517/3d_food_printing_march_2015.pdf (accessed October 2, 2018).

Lipson, H. 2014. Personal communication. (Information from an interview with Dr. Hod Lipson at Cornell University on March 15, 2014).

Lipton, J., Arnold, D., Nigl, F., Lopez, N., Cohen, D., Noren, N. and Lipson, H. 2010. Mutli-material food printing with complex internal structure suitable for conventional post-processing. *21st Annual International Solid Freeform Fabrication Symposium - an Additive Manufacturing Conference, SFF 2010*, 809–815.

Liu, Z., Jayasinghe, S., Gao, W. and Farid, M. 2007. Corrosion mechanism of electrodes in ohmic cooking. *Asia-Pacific Journal of Chemical Engineering* 2: 487–492.

Lopez-Fandiño, F. R., Carrascosa, A. V. and Olano, A. 1996. The effect of high pressure whey protein denaturation and cheese making properties of raw milk. *Journal of Dairy Science* 79: 929–936.

Lopez-Rubio, A. and Lagaron, J. M. 2012. Whey protein capsules obtained through electrospraying for the encapsulation of bioactives. *Innovative Food Science and Emerging Technologies* 13: 200–206.

Lopez-Rubio, A., Sanchez, E., Sanz, Y. and Lagaron, J. M. 2009. Encapsulation of living bifidobacteria in ultrathin PVOH electrospun fibers. *Biomacromolecules* 10: 2823–2829.

Lopez-Rubio, A., Sanchez, E., Wilkanowicz, S., Sanz, Y. and Lagaron, J. M. 2012. Electrospinning as a useful technique for the encapsulation of living bifidobacteria in food hydrocolloids. *Food Hydrocolloids* 28: 159–167.

Loypimai, P., Moonggarm, A. and Chottano, P. 2009. Effects of ohmic heating on lipase activity, bioactive compounds and antioxidant activity of rice bran. *Australian Journal of Basic and Applied Sciences* 3: 3642–3652.

Ma, L., Chang, F. J. and Barbosa-Cánovas, G. V. 1997. Inactivation of E. coli in liquid whole egg using pulsed electric fields technology. *Proceedings of the Fifth Conference of Food Engineering*, New Frontiers in Food Engineering, 216-221. New York, USA.

Maggi, A., Gola, S., Spotti, E., Rovere, P. and Mutti, P. 1994. High pressure treatment of ascospores of heat resistant moulds and patulin in apricot nectar and water. *Industria Conserve* 69: 26–29.

Manas, P., Barsotti, L. and Cheftel, J. C. 2001. Microbial inactivation by pulsed electric fields in a batch treatment chamber: effects some electric parameters and food constituents. *Innovative Food Science and Emerging Technologies* 2: 239–249.

Mertens, B. 1995. Hydrostatic pressure treatment of food: equipment and processing. In *New Methods of Food Preservation*, ed. G. W. Gould, 135–158. Boston, MA: Springer.

Masamura, A., Sado, H., Nabetani, H. and Nakajima, M. 1988. Drying of potato by far-infrared radiation. *Nippon Shokuhin Kogyo Gakkaishi* 35: 309–314.

Mastwijk, H. C. and Groot, M. N. N. 2010. Cold plasmas used for food processing. In *Encyclopedia of Biotechnology in Agriculture and Food*, eds. D. R. Heldman, M. B. Wheeler and D. G. Hoover, 174–177. Boca Raton, FL: CRC Press.

Metaxas, A. C. 1991. Microwave heating. *Power Engineering Journal* 5: 237–247.

Metaxas, A. C. and Meredith, R. J. 1983. *Industrial Microwave Heating*. London: Polu Pelegrinus Ltd.

Min, S. C., Roh, S. H., Niemira, B. A., Boyd, G., Sites, J. E., Uknalis, J. and Fan, X. 2017. In-package inhibition of *E. Coli* O157:H7 on bulk romaine lettuce using cold plasma. *Food Microbiology* 65: 1–6.

Misra, N., Kaur, S., Tiwari, B. K., Kaur, A., Singh, N. and Cullen, P. 2015. Atmospheric pressure cold plasma (ACP) treatment of wheat flour. *Food Hydrocolloids* 44: 115–121.

Misra, N., Patil, S., Moiseev, T., Bourke, P., Mosnier, J., Keener, K. and Cullen, P. J. 2014. In-package atmospheric pressure cold plasma treatment of strawberries. *Journal of Food Engineering* 125: 131–138.

Mizuno, A. 2000. Electrostatic precipitation. *IEEE Transactions on Dielectrics and Electrical Insulation* 7: 615–624.

Moisan, M., Beaudry, C. and Leprince, P. 1974. A new HF device for the production of long plasma columns at a high electron density. *Physics Letters* 50A: 125–126.

Moisan, M., Zakrzewski, Z., Pantel, R. and Leprince, P. 1984. Waveguide-based launcher to sustain long plasma columns through the propagation of an electromagnetic surface wave. *IEEE Transactions on Plasma Science* 12: 203–213.

Molina, E., Alvarez, M. D., Ramos, M., Olano, A. and Lopez, F. R. 2000. Use of high pressure treated milk for the production of reduced fat cheese. *International Dairy Journal* 10: 467–475.

Monga, A. 2014. R.F. applications in bakeries. United Kingdom: Innovators in Dielectric Heating Technology. www.rf-drying.com/documents/2014/12/rf-applications-in-bakeries.pdf (accessed June 20, 2018).

Mongpraneet, S., Abe, T. and Tsurusaki, T. 2002. Accelerated drying of welsh onion by far infrared radiation under vacuum conditions. *Journal of Food Engineering* 55: 147–156.

Morgan, S. M., Ross, R. P., Beresford, T. and Hill, C. 2000. Combination of high hydrostatic pressure and lacticin 3147 causes increased killing of Staphylococcus and Listeria. *Journal of Applied Microbiology* 88: 414–420.

Motasemi, F. and Ani, F. N. 2012. A review on microwave-assisted production of biodiesel. *Renewable and Sustainable Energy Reviews* 16: 4719–4733.

Moyer, J. C. and Stotz, E. 1947. The blanching of vegetables by electronics. *Food Technology* 1: 252–257.

Muftugil, N. 1986. Effect of different types of blanching on the color and the ascorbic acid and chlorophyll contents of green beans. *Journal of Food Processing and Preservation* 10: 69–76.

Needs, E. C., Capells, M., Bland, P., Manoj, P., MacDougal, D. B. and Gopal, P. 2000. Comparison of heat and pressure treatment of skimmed milk, fortified with whey protein concentrate for set yoghurt preparation: effects on milk proteins and gel structure. *Journal of Dairy Research* 67: 329–348.

Nelson, S. O. 1996. Review and assessment of radio-frequency and microwave energy for stored-grain insect control. *Trans ASAE* 39: 1475–1484.

Neo, Y. P., Ray, S., Jin, J., Gizdavic-Nikolaidis, M., Nieuwoudt, M. K., Liu, D. and Quek, S. Y. 2013a. Encapsulation of food grade antioxidant in natural biopolymer by electrospinning technique: a physicochemical study based on zein-gallic acid system. *Food Chemistry* 136: 1013–1021.

Neo, Y. P., Swift, S., Ray, S., Gizdavic-Nikolaidis, M., Jin, J. and Perera, C. O. 2013b. Evaluation of gallic acid loaded zein submicron electrospun fiber mats as novel active packaging materials. *Food Chemistry* 141: 3192–3200.

O'Brien, J. K. and Marshall, R. T. 1996. Microbiological quality of raw ground chicken processed at high isostatic pressure. *Journal of Food Protection* 59: 146–150.

O'Dowd, L. P., Arimi, J. M., Noci, F., Cronin, D. A. and Lyng, J. G. 2013. An assessment of the effect of pulsed electrical fields on tenderness and selected quality attributes of post rigor beef muscle. *Meat Science* 93: 303–309.

O'Reilly, C. E., O'Connor, P. M., Murphy, P. M., Kelly, A. L. and Beresford, T. P. 2000. The effect of exposure to pressure of 50 MPa on Cheddar cheese ripening. *Innovative Food Science and Emerging Technologies* 1: 109–117.

Oh, Y. J., Song, A. Y. and Min, S. C. 2017. Inhibition of *Salmonella typhimurium* on radish sprouts using nitrogen-cold plasma. *International Journal of Food Microbiology* 249: 66–71.

Ohlsson, T. 1987. Sterilization of foods by microwaves. *International Seminar on New Trends in Aseptic Processing and Packaging of Food stuffs*; 22nd October 1987, Munich. SLK Report No. 564. Goteborg, Sweden: The Swedish Institute for Food and Biotechnology.

Ortega-Rivas, E., Zárate-Rodríguez, E. and Barbosa-Cánovas, G. V. 1998. Apple juice pasteurization using ultrafiltration and pulsed electric fields. *Food and Bioproducts Processing* 76C: 193–198.

Pankaj, S. K., Wan, Z., Colonna, W. and Keener, K. M. 2017. Effect of high voltage atmospheric cold plasma on white grape juice quality. *Journal of the Science of Food and Agriculture* 97: 4016–4021.

Pareta, R. and Edirisinghe, M. J. 2006. A novel method for the preparation of starch films and coatings. *Carbohydrate Polymers* 63: 425–431.

Pasquali, F., Stratakos, A. C., Koidis, A., Berardinelli, A., Cevoli, C., Ragni, L., Mancusi, R., Manfreda, G. and Trevisani, M. 2016. Atmospheric cold plasma process for vegetable leaf decontamination: a feasibility study on radicchio (red chicory, *Cichorium intybus* L.). *Food Control* 60: 552–559.

Pataro, G., Donsì, G. and Ferrari, G. 2011. Aseptic processing of apricots in syrup by means of a continuous pilot scale ohmic unit. *LWT–Food Science and Technology* 44: 1546–1554.

Patricia, C. M., Bibiana, D. Y. and Jose, P. M. 2011. Evaluation of microwave technology in blanching of broccoli (*Brassica oleracea* L. var. botrytis) as a substitute for conventional blanching. *Procedia Food Science* 1: 426–432.

Patterson, M. F. 2005. A review: microbiology of pressure-treated foods. *Journal of Applied Microbiology* 98: 1400–1409.

Pérez-Masiá, R., López-Nicolás, R., Jesús Periago, M., Ros, G., Lagaron, J. M. and López-Rubio, A. 2015. Encapsulation of folic acid in food hydrocolloids through nanospray drying and electrospraying for nutraceutical applications. *Food Chemistry* 168: 124–133.

Periard, D., Schaal, N., Schaal, M., Malone, E. and Lipson, H. 2007. Printing food. *18th Solid Freeform Fabrication Symposium*, SFF 2007, 564–574.

Philip, N., Saoudi, B., Crevier, M. C., Moisan, M., Barbeau, J. and Pelletier, J. 2002. The respective roles of UV photons and oxygen atoms in plasma sterilization at reduced gas pressure: the case of N_2–O_2 mixtures. *IEEE Transactions on Plasma Science* 30: 1429–1436.

Picart, L., Dumay, E. and Cheftel, J. C. 2002. Inactivation of *Listeria innocua* in dairy fluids by pulsed electric fields: influence of electric parameters and food composition. *Innovative Food Science and Emerging Technologies* 3: 357–369.

Piette, G., Buteau, M. E., Halleux, D., Chiu, L., Raymond, Y. and Ramaswamy, H. S. 2004. Ohmic cooking of processed meats and its effects on product quality. *Journal of Food Science* 69: 71–78.

Pircon, L. J., Loquercio, P. and Doty, D. M. 1953. High-frequency heating as a unit operation in meat processing. *Journal of Agricultural and Food Chemistry* 1: 844–847.

Qin, B. L., Pothakamury, U. R., Vega-Mercado, H., MartínBelloso, O. M., Barbosa-Cánovas, G. V. and Swanson, B. G. 1995. Food pasteurization using high-intensity pulsed electric fields. *Food Technology* 12: 55–60.

Rademacher, B. and Kessler, H. G. 1997. High-pressure inactivation of microorganisms and enzymes in milk and milk products. *High Pressure Research in Biosciences and Biotechnology: Proceedings of the XXXIVth Meeting of the European High Pressure Research Group*, Leuven, Belgium, September 1–5, 1996, ed. K. Heremans, 291–293. Belgium: Leuven University Press.

Ramazzina, I., Berardinelli, A., Rizzi, F., Tappi, S., Ragni, L., Sacchetti, G. and Rocculi, P. 2015. Effect of cold plasma treatment on physico-chemical parameters and antioxidant activity of minimally processed kiwifruit. *Postharvest Biology and Technology* 107: 55–65.

Ramazzina, I., Tappi, S., Rocculi, P., Sacchetti, G., Berardinelli, A., Marseglia, A. and Rizzi, F. 2016. Effect of cold plasma treatment on the functional properties of fresh-cut apples. *Journal of Agricultural and Food Chemistry* 64: 8010–8018.

Rao, P. S., Chakraborty, S., Kaushik, N., Paul Kaur, B. and Swami Hulle, N. R. 2014. High hydrostatic pressure processing of food materials. In *Introduction to Advanced Food Process Engineering*, ed. J. K. Sahu, 151–186. Boca Raton, FL: CRC Press.

Rayna, T. and Striukova, L. 2016. From rapid prototyping to home fabrication: how 3D printing is changing business model innovation. *Technological Forecasting and Social Change* 102: 214–224.

Reneker, D. H. and Yarin, A. L. 2008. Electrospinning jets and polymer nanofibers. *Polymer* 49: 2387–2425.

Riener, J., Noci, F., Cronin, D. A., Morgan, D. J. and Glyng, J. G. 2009. Combined effect of temperature and pulsed electric fields on pectin methyl esterase inactivation in red grapefruit juice (*Citrus paradisi*). *European Food Research and Technology* 228: 373–379.

Roberts, J., Balaben, M. and Luziriaga, D. 1996. Automated ohmic thawing of shrimp blocks. *Seafood Science and Technology Society of the America* FLSGP-W-95-001, SGR 115: 72–81.

Rodrigo, D., Barbosa-Canovas, G. V., Martinez, A. and Rodrigo, M. 2003. Pectin methylesterase and natural microflora of fresh mixed orange and carrot juice treated with pulsed electric fields. *Journal of Food Protection* 66: 2336–2342.

Rosenthal, I. 1992. *Electromagnetic Radiations in Food Science*. Berlin, Heidelberg: SpringerVerlag.

Ruan, R., Chen, Y. P., Doona, C. J. and Taub, I. 2001. Ohmic heating. In *Thermal Technologies in Food Processing*, ed. P. Richardson, 241–265. Cambridge, England; Boca Raton, FL: Woodhead Publishing Limited; CRC Press.

Sakai, N. and Hanzawa, T. 1994. Applications and advances in far-infrared heating in Japan. *Trends in Food Science and Technology* 5: 357–362.

Sanders, H. R. 1966. Dielectric thawing of meat and meat products. *Journal of Food Technology* 1: 183–192.

Sandu, C. 1986. Infrared radiative drying in food engineering: a process analysis. *Biotechnology Progress* 2: 109–119.

Sarangapani, C., O'Toole, G., Cullen, P. and Bourke, P. 2017. Atmospheric cold plasma dissipation efficiency of agrochemicals on blueberries. *Innovative Food Science and Emerging Technologies* 44: 235–241.

Schaal, N. 2009. Preliminary experiments with chocolate [online]. http://fabathome.org/wiki/index.php?title=Materials:Chocolate (accessed October 2, 2018).

Schmid, K., Arpagaus, C. and Friess, W. 2009. Evaluation of a vibrating mesh spray dryer for preparation of submicron particles. *Respiratory Drug Delivery Europe* 2: 323–326.

Schön, M. and Baumgartner, R. 2009. Heating, method for heating and laminating, electrostatic separator, spray drier, separating device and method for separating particles. European Patent, EP 2056037.

Schubert, H. and Regier, M. 2005. *The Microwave Processing of Foods*. Cambridge, UK: Woodhead.

Schubert, H. and Regier, M. 2006. Novel and traditional microwave applications in the food industry. In *Advances in Microwave and Radio Frequency Processing: Report from the Eighth Conference on Microwave and High-Frequency Heating*, ed. M. Willert-Porada, 259–270. Berlin: Springer.

Selcuk, M., Oksuz, L. and Basaran, P. 2008. Decontamination of grains and legumes infected with *Aspergillus* spp. and *Penicillium* spp. by cold plasma treatment. *Bioresource Technology* 99: 5104–5109.

Sentandreu, E., Carbonell, L., Rodrigo, D. and Carbonell, J. V. 2006. Pulsed electric fields versus thermal treatment: equivalent processes to obtain equally acceptable citrus juices. *Journal of Food Protection* 69: 2016–2018.

Serrano, J., Velazquez, G., Lopetcharat, K., Ramirez, J. A. and Torres, J. A. 2005. Moderately high hydrostatic pressure processing to reduce production costs of shredded cheese: microstructure, texture, and sensory properties of shredded milled curd cheddar. *Journal of Food Science* 70: 286–293.

Shigehisa, T., Ohmori, T., Saito, A., Taji, S. and Hayashi, R. 1991. Effects of high hydrostatics pressure on characteristics of pork slurries and inactivation of microorganisms associated with meat products. *International Journal of Food Microbiology* 12: 207–215.

Shirsat, N., Lyng, J. G., Brunton, N. P. and McKenna, B. 2004. Ohmic processing: electrical conductivities of pork cuts. *Meat Science* 67: 507–514.

Sierra, I., Vidal, V. C. and Lopez, F. R. 2000. Effect of high pressure on the vitamin B1 and B6 content of milk. *Milchwissenschaft* 55: 365–367.

Skjoldebrand, C. 2001. Infrared heating. In *Thermal Technologies in Food Processing*, ed. P. Richardson, 208–228. Cambridge, UK; Boca Raton, FL: Woodhead Publishing Limited; CRC Press.

Skjoldebrand, C. and Andersson, C. 1989. A comparison of infrared bread baking and conventional baking. *Journal of Microwave Power and Electromagnetic Energy* 24: 91–101.

Skudder, P. J. 1988. Ohmic heating: new alternative for aseptic processing of viscous foods. *Food Engineering* 60: 99–101.

Sobrino-Lopez, A., Raybaudi-Massilia, R. and Martin-Belloso, O. 2006. High-Intensity pulsed electric field variables affecting *Staphylococcus aureus* inoculated in milk. *Journal of Dairy Science* 89: 3739–3748.

Sozzi, B. 2015. Here is pepsico using 3D printing to make ruffles potato chips. *The Street*, September 22nd 2015.

Staack, N. 2008. Potential of infrared heating as a method for decontaminating food powder, Process development and impact on product quality. *PhD thesis*. Faculty III—School of Process Sciences and Engineering, Institute of Food Technology and Food Chemistry, Department of Food Biotechnology and Food Process Engineering, Berlin.

Su, Y., Zhang, C., Wang, Y. and Li, P. 2012. Antibacterial property and mechanism of a novel Pu-erh tea nanofibrous membrane. *Applied Microbiology and Biotechnology* 93: 1663–1671.

Sun, J., Peng, Z., Yan, L. K., Fuh, J. Y. H. and Hong, G. S. 2015a. 3D food printing—An innovative way of mass customization in food fabrication. *International Journal of Bioprinting* 1: 27–38.

Sun, J., Peng, Z., Zhou, W., Fuh, J. Y. H., Hong, G. S. and Chiu, A. 2015b. A review on 3D printing for customized food fabrication. *Procedia Manufacturing* 1: 308–319.

Surowsky, B., Schlüter, O. and Knorr, D. 2015. Interactions of Non-Thermal Atmospheric Pressure Plasma with Solid and Liquid Food Systems: A Review. *Food Engineering Reviews* 7: 82–108.

Tajchakavit, S., Ramaswamy, H. S. and Fustier, P. 1998. Enhanced destruction of spoilage microorganisms in apple juice during continuous flow microwave heating. *Food Research International* 31: 713–722.

Tanzi, E., Saccani, G., Barbuti, S., Grisenti, M. S., Lori, D., Bolzoni, S. and Paroalri, G. 2004. High-pressure treatment of raw ham. Sanitation and impact to quality. *Industria Conserve* 79: 37–50.

Tappi, S., Gozzi, G., Vannini, L., Berardinelli, A., Romani, S., Ragni, L. and Rocculi, P. 2016. Cold plasma treatment for fresh-cut melon stabilization. *Innovative Food Science and Emerging Technologies* 33: 225–233.

Torres-Giner, S., Martinez-Abad, A., Ocio, M. J. and Lagaron, J. M. 2010. Stabilization of a nutraceutical omega-3 fatty acid by encapsulation in ultrathin electrosprayed zein prolamine. *Journal of Food Science* 75: N69–N79.

Torres-Giner, S., Ocio, M. J. and Lagaron, J. M. 2009. Novel antimicrobial ultrathin structures of zein/chitosan blends obtained by electrospinning. *Carbohydrate Polymers* 77: 261–266.

Trevisani, M., Berardinelli, A., Cevoli, C., Cecchini, M., Ragni, L. and Pasquali, F. 2017. Effects of sanitizing treatments with atmospheric cold plasma, sds and lactic acid on verotoxin-producing *Escherichia coli* and *Listeria monocytogenes* in red chicory (radicchio). *Food Control* 78: 138–143.

Tyagi, V. K. and Lo, S. L. 2013. Microwave irradiation: a sustainable way for sludge treatment and resource recovery. *Renewable and Sustainable Energy Reviews* 18: 288–305.

Ulbin-Figlewicz, N. and Jarmoluk, A. 2016. Effect of low-pressure plasma treatment on the color and oxidative stability of raw pork during refrigerated storage. *Food Science and Technology International* 22: 313–324.

Vachon, J. F., Kheadr, E. E., Giasson, J., Paquin, P. and Fliss, I. 2002. Inactivation of foodborne pathogens in milk using dynamic high pressure. *Journal of Food Protection* 65: 345–352.

van den Berg, R. W., Hoogland, H., Lelieveld, H. and Van Schepdael, L. 2001. High pressure equipment designs for food processing applications. In *Ultra High Pressure Treatment of Foods*, eds. M. E. G. Hendrickx and D. Knorr, 297–314. New York: Springer.

Varghese, K. S., Pandey, M. C., Radhakrishna, K. and Bawa, A. S. 2014. Technology, applications and modelling of ohmic heating: a review. *Journal of Food Science and Technology* 51: 2304–2317.

Vecillio, L. 2006. The mesh nebuliser: a recent technical innovation for aerosol delivery. *Breathe* 2: 253–260.

Vega-Lugo, A. C. and Lim, L. T. 2009. Controlled release of allyl isothiocyanate using soy protein and poly (lactic acid) electrospun fibers. *Food Research International* 42: 933–940.

Veleirinho, B. and Lopes-da-Silva, J. A. 2009. Application of electrospun poly (ethylene terephthalate) nanofiber mat to apple juice clarification. *Process Biochemistry* 44: 353–356.

Walkling-Ribeiro, M., Noci, F., Riener, J., Cronin, D. A., Lyng, J. G. and Morgan, D. J. 2009. The impact of thermosonication and pulsed electric fields on *Staphylococcus aureus* inactivation and selected quality parameters in orange juice. *Food and Bioprocess Technology* 2: 422–430.

Wang, R., Nian, W., Wu, H., Feng, H., Zhang, K., Zhang, J., Zhu, W., Becker, K. and Fang, J. 2012. Atmospheric-pressure cold plasma treatment of contaminated fresh fruit and vegetable slices: inactivation and physiochemical properties evaluation. *The European Physical Journal D* 66: 1–7.

Wang, S., Marcone, M. F., Barbut, S. and Lim, L. T. 2013. Electrospun soy protein isolate-based fibre fortified with anthocyanin-rich red raspberry (*Rubus strigosus*) extracts. *Food Research International* 52: 467–472.

Wang, W. and Sastry, S. 1997. Changes in electrical conductivity of selected vegetables during multiple thermal treatments. *Journal of Food Process Engineering* 20: 449–516.

Wang, W. and Sastry, S. 2000. Effects of thermal and electrothermal pretreatments on hot air drying rate of vegetable tissue. *Journal of Food Process Engineering* 23: 299–319.

Wang, W. C. 1997. Starch gelatinization measuring system. US Patent 5604302.

Wegrzyn, T. F., Golding, M. and Archer, R. H. 2012. Food layered manufacture: a new process for constructing solid foods. *Trends in Food Science and Technology* 27: 66–72.

Wills, T. M., de Witt, C. A. M., Sigfusson, H. and Bellmer, D. 2006. Effect of cooking method and ethanolic tocopherol on oxidative stability and quality of beef patties during refrigerated storage (oxidative stability of cooked patties). *Journal of Food Science* 71: 109–114.

Won, M. Y., Lee, S. J. and Min, S. C. 2017. Mandarin preservation by microwave-powered cold plasma treatment. *Innovative Food Science and Emerging Technologies* 39: 25–32.

Xu, L., Garner, A. L., Tao, B. and Keener, K. M. 2017. Microbial inactivation and quality changes in orange juice treated by high voltage atmospheric cold plasma. *Food and Bioprocess Technology* 10: 1–14.

Yang, F., Zhang, M. and Bhandari, B., 2017. Recent development in 3D food printing. *Critical Reviews in Food Science and Nutrition* 57: 3145–3153.

Yang, T. C. S., Cohen, J. S., Kluter, R. A., Tempest, P., Manvell, C., Blackmore, S. J. and Adams, S. 1997. Microbiological and sensory evaluation of six ohmically heated stew type products. *Journal of Food Quality* 20: 303–313.

Yen, G. C. and Lin, H. T. 1996. Comparison of high pressure treatment and thermal pasteurization effects on the quality and shelf life of guava puree. *International Journal of Food Science and Technology* 31: 205–213.

Yordanov, D. and Angelova, G. 2010. High pressure processing for foods preserving. *Biotechnology and Biotechnological Equipment* 24: 1940–1945.

Zell, M., Lyng, J. G., Cronin, D. A. and Morgan, D. J. 2009. Ohmic heating of meats: electrical conductivities of whole meats and processed meat ingredients. *Meat Science* 83: 563–570.

Zell, M., Lyng, J. G., Cronin, D. A. and Morgan, D. J. 2010. Ohmic cooking of whole beef muscle—evaluation of the impact of a novel rapid ohmic cooking method on product quality. *Meat Science* 86: 258–264.

Zia-ur-Rehman, Z., Islam, M. and Shah, W. H. 2003. Effect of microwave and conventional cooking on insoluble dietary fiber components of vegetables. *Food Chemistry* 80: 237–240.

Ziuzina, D., Misra, N., Cullen, P., Keener, K., Mosnier, J., Vilaró, I., Gaston, E. and Bourke, P. 2016. Demonstrating the potential of industrial scale in-package atmospheric cold plasma for decontamination of cherry tomatoes. *Plasma Medicine* 6: 397–412.

15

Food Packaging

To a consumer, packaging is the first source of information about a food product. Packaging holds a product, keeps it safe, and acts as an interface between the manufacturers and consumers. Thus, food packaging has been precisely defined as *a coordinated system of preparing food for transport, distribution, storage, retailing, and end-use to satisfy the ultimate consumer with optimal cost* (Coles et al., 2003). Food industry relies on packaging engineering and technology to contain, protect, and preserve its products during the transit from factory until it reaches the consumer. Attractive package design is considered an effective marketing strategy to influence consumer choice. Food packaging is a field of continuous developments driven by the consumer demand for convenient and high-quality food products, the advent of novel polymers and newer technologies.

This chapter is ordered in five major headings. Understanding the role of food packaging is vital not just for the development of quality packaging but also to avoid overpackaging and reduce product waste. Thus, the first part of the chapter elaborates the primary and secondary functions of food packaging. The second and third parts deal with the different types of food packaging materials and techniques, respectively. Mass transfer is an important phenomenon which determines the performance of food packaging materials. The fourth section of this chapter discusses the different modes of mass transfer in food packaging and their relevance. The final section details the standard test methods to evaluate the quality and performance of packaging materials.

15.1 Functions of Food Packaging

The primary functions of food packaging are given as follows (Coles, 2003; Marsh and Bugusu, 2007):

- ***Containment*:** to contain or hold the food. This is the most fundamental role of packaging that avoids contamination of food products and thereby facilitates their transportation, storage and distribution.
- ***Protection*:** to protect the food products against the external influences; this encompasses the protection of food against physical (mechanical damage), chemical (exposure to oxygen, moisture, or light), and biological (spoilage and pathogenic microorganisms, insects, rodents, and other animals) deterioration.
- ***Information*:** Package labeling is important towards complying with legal requirements on providing information such as product identification, nutritional value, ingredient declaration, net weight, and manufacturer information. In addition, the package provides information such as product shelf life, best-before and use-by dates, storage conditions, cooking instructions, brand identification, pricing, and the reuse or recycling features of the package. Certain manufacturers use larger labels to even accommodate recipes.

The secondary functions of food packaging include the following (Marsh and Bugusu, 2007):

- ***Traceability*:** Food manufacturers include unique codes in the form of printed barcodes or electronic radio frequency identification (RFID) on the packaging labels of their products. These codes can be read manually or by machines which enable the producers to track their products throughout the supply chain.

- ***Convenience*:** Convenience attribute of packaging refers to minimizing the efforts required to prepare and serve foods. It includes ease of access and handling (e.g., spouted pouches for liquid food products), disposal, product visibility (see-through or transparent packs), resealability, and microwavability.
- ***Tamper-evidence*:** Tamper-evident features of food packaging material include shrink bands, tamper-evident seals and tapes, special membranes, and breakaway closures. These also include holograms that cannot be easily duplicated and specially printed graphics or text on bottle liners or composite cans, which undergo an irreversible change when opened.

15.2 Types of Food Packaging Materials

Depending on the extent of contact with the food product and their intended function, packaging materials can be classified into three types: primary, secondary, and tertiary. A primary container is a direct-food-contact surface which holds the food product. It is also referred to as *sales, consumer, or retail packaging* which includes the other packaging components such as cap and label to complete the sales unit. Glass bottles, sachets, overwraps, plastic bottles, and metal cans are examples of primary containers. The primary packaging materials should be mandatorily tested for any possible migration of its constituents into the food. Hence, these materials are subject to approval by the Food and Drug Administration (FDA) which conducts the migration tests.

A secondary container also called *grouped packaging* holds several primary containers together. It does not have direct contact with the food product and serves the purposes of displaying and handling multiple unit products for sales and logistics. Corrugated fiberboard box and shrink film are typical examples of secondary packaging material. Further, a tertiary container or *transport packaging* holds several secondary containers together for distribution or palletization. It offers additional protection to the unit product from the impact of dropping, denting, or crushing of cartons during storage and distribution. Examples of tertiary packaging include sacks, shrink-wrapped containers/pallets, corrugated fiber board cartons, drums, barrels, crates, foil bags, and wooden pallet.

The most important criterion in the selection of materials for food packaging is the safety of the polymeric film or membrane. This is relevant as the packaging material comes in direct contact with the food and interacts with it during processing, storage, and transportation. However, the choice of packaging material is not just limited to food safety but also includes other key factors such as barrier properties, energy, and material costs. Food packaging must also be carefully designed by considering the packaging and labeling regulations of the region and the methods of disposal that cause a minimal environmental impact. In accordance with the aforementioned criteria, several food packaging materials are currently in use, which can be broadly classified into four categories: glass, paper, metal, and plastic (Figure 15.1). Table 15.1 presents an overview of the different food packaging materials (Figure 15.2), their properties, and applications for defined categories of food products.

15.2.1 Biopolymers

A shift in trend towards the biodegradable food packaging materials has been observed in recent times. An increased awareness of the non-biodegradable nature of packaging materials based on petrochemical products is the underlying driving force for the identification of novel biopolymers. Biopolymers are natural polymers derived from the biomass of plant or animal source, biologically derived monomers and microorganisms (Figure 15.3).

- **Biopolymers from biomass:** Biopolymers extracted from biomass are hydrophilic and moderately crystalline. Therefore, these polymers are not suitable for the packaging of moist food products. Nevertheless, these polymers show excellent gas barrier properties which render them suitable for food packaging (Avérous and Pollet, 2012).

FIGURE 15.1 Different types of food packaging materials. (Slightly modified and reproduced with permission from Sanyang, M. L., Ilyas, R. A., Sapuan, S. M. and Jumaidin, R. 2018. Sugar palm starch-based composites for packaging applications. In *Bionanocomposites for Packaging* Applications, eds. M. Jawaid and S. Swain, 125–147. Cham: Springer.)

- **Biopolymers from bio-derived monomers:** This includes polymeric materials synthesized by polymerization reaction from renewable bio-based monomers. A typical example is polylactic acid (PLA), synthesized from the fermentation of wheat starch, and cornstarch to form dextrose which is subsequently converted into lactic acid. PLA is the first commercial biopolymer to be formed into injection-molded objects, blown films, and coatings (Rasal et al., 2010). PLA has effectively replaced high-density polyethylene (HDPE), low-density polyethylene (LDPE), polyethylene terephthalate (PET), and polystyrene (PS) in packaging applications.
- **Biopolymers from microorganisms:** Polymers derived from microorganisms or genetically modified bacteria constitute this category of biopolymers. The microbial biopolymers have unique rheological properties, high purity, and regular structure. Further, microbial biopolymers remain active at extreme temperature, pH, and salinity (Jindal and Khattar, 2018). Microbial polymers including pullulan, kefiran, bacterial cellulose (BC), gellan and levan fall under the category of exopolysaccharides (EPS). Notably, EPS demonstrate film formation property. Apart from EPS, a number of bacterial genera synthesize carbon energy reserves in their cells. These are known as polyhydroxyalkanoates (PHAs), and microbial polyesters, which can be extruded into films with excellent moisture and oxygen barrier properties. Lactic acid bacteria produce blow moldable biopolymers such as PHA and polylactic acid (PLA) through fermentation. The aforesaid products of microbial fermentation have been projected as potential biodegradable polymers for packaging applications (Vijayendra and Shyamala, 2014).

Hitherto, various biopolymers have been developed for the preparation of packaging films which act as a barrier against moisture, oxygen, aroma, flavor, and oil (Rhim and Ng, 2007). The food industry applications of major biopolymer types are presented in Table 15.2. Biopolymers are abundant, renewable, inexpensive, biodegradable, and eco-friendly, and thereby outweigh the conventional packaging materials. In addition, natural polymers are excellent carriers of food additives such as antioxidants, antifungal agents, antimicrobials, colors, and other nutrients (Rhim and Ng, 2007). Consequently, biopolymers contribute to the overall improvement of food quality and extension of product shelf life. Nonetheless, the commercial application of biopolymers in food packaging is limited by their low heat distortion temperature and

TABLE 15.1
Types of Packaging Materials and Their Properties

Packaging Material	Description/Physical Properties	Mechanical, Chemical, and Other Properties	Barrier Properties	Food Product Category
Jute sacks	• Density: 1482 kg/m^3 • Temperature range for transportation: 18°C–22°C	• Good stack stability • Poor mechanical strength	• Poor barrier to moisture	• Food grains • Fruits • Vegetables • Dried fish
Glass	• Transparent and hence permits the consumer to see the product • Can be colored for light-sensitive products	• Odorless • Chemically inert • Impermeable to gases and vapors • Rigid • Provides good insulation • Reusable and recyclable • Brittle and susceptible to breakage from internal pressure, impact, or thermal shock	• Excellent barrier to oxygen and moisture	• Beverages • Confectionery • Coffee • Milk and milk products
Metals				
Tinplate containers	• Produced from low-carbon steel by coating both sides with thin layers of tin	• Low weight and high mechanical strength • Can be heat-treated and sealed hermetically and hence suitable for sterile products • Good ductility and formability	• Excellent barrier properties to gases, water vapor, light, and odors	• Bakery products • Breakfast cereals • Confectionery • Canned fish • Canned meat • Coffee
Tin-free steel (TFS)	• Also known as electrolytic chromium or chrome oxide coated steel, TFS is created by electrolytically coating black plate with a thin layer of chrome/chrome oxide	• Strong • Withstands heat processing • Requires a coating of organic material to provide complete corrosion resistance	• Excellent barrier properties to gases, water vapor, moisture and microbes	• Food cans, can ends, trays, bottle caps, and closures can be made from TFS • Can be used to make large containers (such as drums) for bulk sale and bulk storage of ingredients or finished goods
Aluminum	• Metal derived from bauxite ore, where it exists in combination with oxygen as alumina • Lightweight • Good portability	• Resistant to corrosion • Withstands heat processing • Limited structural strength • Unbreakable	• Highly effective barrier to air, temperature, moisture, odors, light, microorganisms, and chemical attack	• Light packaging of primarily soft drink cans, pet food, seafood, and pre-threaded closures

(Continued)

TABLE 15.1 (*Continued*)
Types of Packaging Materials and Their Properties

Packaging Material	Description/Physical Properties	Mechanical, Chemical, and Other Properties	Barrier Properties	Food Product Category
Paper				
• Paper (kraft paper, sulfite paper, grease-proof paper, glassine, parchment paper)	• Very good strength-to-weight characteristics • White board: made from several thin layers of bleached chemical pulp, the white board is typically used as the inner layer of a carton	• Kraft paper: strongest of all papers • Sulfite paper: lighter and weaker than kraft paper; can be glazed to improve its appearance wet strength and oil resistance	• Poor barrier to gases, moisture, and light • Moisture sensitive	• Flour, sugar (kraft paper) • Small bags or wrappers for packaging biscuits and confectionary (sulfite paper) • As liner for biscuits, cooking fats, fast foods, and baked goods (glassine)
• Paperboard (white board, fiberboard)	• Fiberboard can be solid or corrugated. • Solid fiberboard has an inner white board layer and an outer kraft layer	• Generally, fiberboards are resistant to impact, abrasion, and crushing damage • White board may be coated with wax or laminated with polyethylene for heat sealability	• Loses strength with increasing humidity • Tears easily	• Snack foods, cookies, candy bars, and other oily food (greaseproof paper) • Butter and lard (parchment paper)
	• Corrugated fiberboard is made of two layers of kraft paper (liners) with a central corrugating (or fluting) material (Figure 15.2b)	• White board is the only form of paperboard recommended for direct food contact		• Fiberboards are used for shipping bulk food and case packing of retail food products. Solid fiberboard is used for packing coffee and milk powder
• Paper laminates	• Paper laminates are coated or uncoated papers based on kraft and sulfite pulp. • Can be laminated with plastic or aluminum to improve various properties	• Paper laminated with polyethylene makes it heat sealable	• Paper laminated with polyethylene exhibits improved gas and moisture barrier properties	• Soups • Herbs • Spices
Plastic				
LDPE	• Density: 910-925 kg/m^3 • Transparency: Poor-fair • Low crystallinity • Temperature range: –50°C to 80°C	• Tough • Flexible • Resistant to grease and chemicals • Good sealing properties	• High moisture barrier • Very low gas barrier	• Bread • Frozen food bags • Flexible lids • Squeezable food bottles
Linear low-density polyethylene (LLDPE)	• Density: 910–940 kg/m^3 • Transparency: Poor-fair • High crystallinity • Temperature range: –30°C to 100°C	• Tough • Extensible • Good resistance to grease • Good sealing properties	• High moisture barrier • Very low gas barrier	• Stretch/cling wrap • Heat sealant coating

(*Continued*)

TABLE 15.1 (*Continued*)
Types of Packaging Materials and Their Properties

Packaging Material	Description/Physical Properties	Mechanical, Chemical, and Other Properties	Barrier Properties	Food Product Category
HDPE	• Density: 945–967 kg/m^3 • Transparency: Poor • High crystallinity • Temperature range: –40°C to 120°C	• Tough • Stiff and strong • Resistant to grease and chemicals • Good sealing properties • Easy to process and form	• High moisture barrier • Low gas barrier	• Bottles of milk, juice and water • Cereal box liners • Margarine tubs • Hot filled and microwavable packaging • Trash and retail bags
Polypropylene (PP)	• Density: 900–915 kg/m^3 • Transparency: Fair • Low crystallinity • Temperature range: –40°C to 120°C • Melting point: 160°C	• Moderately stiff • Strong • Good resistance to grease and chemicals	• High moisture barrier • Low gas barrier	• Bottles of milk, juice and water • Cereal box liners • Margarine tubs • Hot filled and microwavable packaging • Trash and retail bags
Polyesters	• Density: 900–915 kg/m^3 • Transparency: high • Temperature range: –10°C to 220°C • Melting point: >200°C	• High impact resistance • Low scratch resistance • Tough and strong • Resistant to grease and oil	• High moisture and gas barrier	• Refillable water bottles • Sterilizable baby bottles
PET (Figure 15.2a)	• Density: 1380–1410 kg/m^3 • Transparency: high • Low crystallinity • Temperature range: –60°C to 200°C	• Stiff • Strong • Good resistance to grease and chemicals	• Good barrier to gases and moisture	• As containers (bottles, jars, and tubs) • Semirigid sheets (trays and blisters) • Thin-oriented films (bags and snack food wrappers)
Polyvinylidene chloride	• Density: 1600–1700 kg/m^3 • Transparency: Good • Temperature range: –20°C to 130°C	• Strong • Stiff • Ductile • Resistant to chemicals • Stable electrical properties • Very good grease and oil resistance	• Excellent oxygen and moisture barrier properties	• Poultry • Cured meats • Cheese • Tea • Coffee • Snack foods • Confectionery • Can be used in hot-filling, low-temperature storage, and modified atmosphere storage conditions

(*Continued*)

TABLE 15.1 (*Continued*)

Types of Packaging Materials and Their Properties

Packaging Material	Description/Physical Properties	Mechanical, Chemical, and Other Properties	Barrier Properties	Food Product Category
PS	• Density: 1030–1100 kg/m^3 • Transparency: Very good • Temperature range: –20°C to 90°C	• Hard and brittle with low melting point • Fair to good resistance to oil and grease	• Low moisture and air barrier	• Used as protective packaging for eggs, disposable plastic ware, cups, plates, bottles and trays. • Expanded form may be used as cushion material.
Ethylene vinyl alcohol (EVOH)	• Density: 1140–1210 kg/m^3 • Transparency: Good • Temperature range: –20°C to 150°C	• Stiff • Strong • Very strong oil and grease resistance	• An excellent moisture barrier, high air barrier	• Used in co-extruded films to avoid its contact with water
Polyamide	• Density: 1130–1160 kg/m^3 • Transparency: Good • Temperature range: –2°C to 120°C	• Stiff • Strong • Good resistance to grease and chemicals	• High air and moisture barrier	• Used for boil-in-bag packaging
Polyethylene naphthalate (PEN)	• Density: 1.36 g/cm^3 • Transparency: Good • Applicable at both high and low temperatures	• Stiff • Chemical and hydrolytic resistance • Thermal and thermo-oxidative resistance • Suitable for hot refills • Re-washable and recyclable	• Good gas and moisture barrier, UV light barrier	• Suitable for beer and wine bottles to preserve the flavor

Source: Adapted from Marsh and Bugusu (2007), Lee et al. (2008), and Fellows and Axtell (2002).

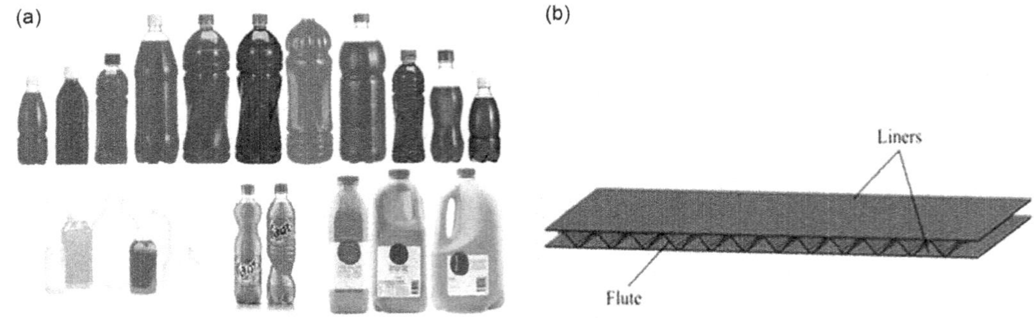

FIGURE 15.2 (a) PET bottles for packaging of juices. (Reproduced with permission from Naz, R. 2018. Storage in polyethylene terephthalate bottles: changes and shelf life. In *Fruit Juices Extraction, Composition, Quality and Analysis*, eds. G. Rajauria and B. K. Tiwari, 621–635. London: Academic Press.); (b) Structure of corrugated board. (Reproduced with permission from Gospodinov, D., Stefanov, S. and Hadjiiski, V. 2011. Use of the finite element method in studying the influence of different layers on mechanical characteristics of corrugated paperboard. *Tehnicki Vjesnik/Technical Gazette* 18: 357–361.)

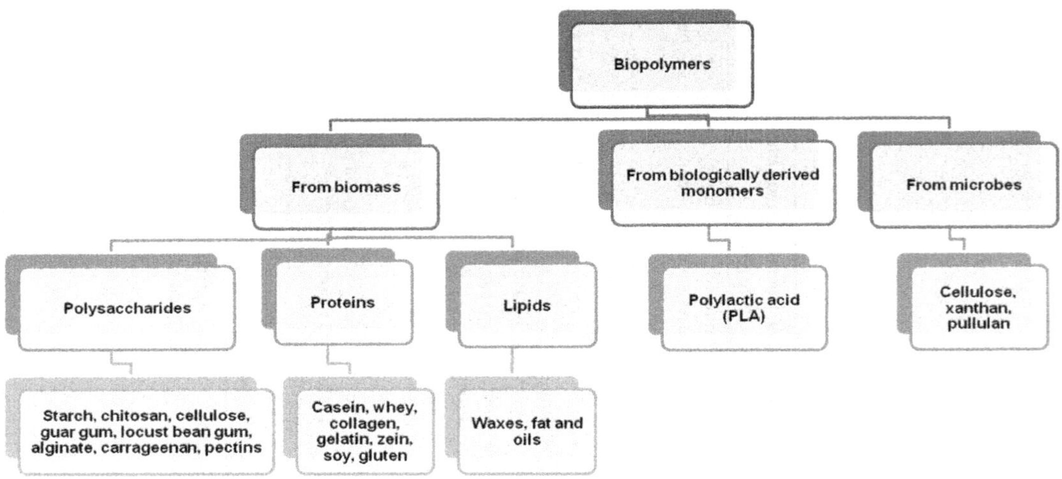

FIGURE 15.3 Classification of biopolymers.

poor resistance to processing operations (Anandharamakrishnan and Kolli, 2015). The advantages and disadvantages of some well-studied biopolymers are compiled in Table 15.3.

15.2.2 Bionanocomposites

The advent of nanotechnology in the field of food packaging has been viewed as a promising sign to overcome the environmental concerns of conventional petroleum-based packaging materials and poor mechanical and barrier properties of biopolymers. Bionanocomposite is formed when a defined weight percentage of nano-reinforcements are randomly and homogeneously dispersed in a biopolymer matrix at a molecular level. Nano-reinforcements are either inorganic or organic fillers with certain geometries such as fibers, flakes, spheres, and particulates, which have at least one dimension in the nanometric range (10^{-9} m). The incorporation of nano-reinforcements improves the mechanical and physical properties of the biopolymer. Bionanocomposites are readily biodegradable without accumulation in the landfills (Anandharamakrishnan and Kolli, 2015).

Montmorillonite (MMT) or nanoclay is one of the widely exploited nano-reinforcements for the preparation of bionanocomposites (Almasi et al., 2010). MMT consists of an aluminum layer sandwiched

TABLE 15.2

Food Industry Applications of Biopolymers

Packaging Application	Biopolymer	Company
PLA		
Coffee and tea	Cardboard cups coated with PLA	KLM
Beverages	PLA cups	MOS Burger (Japan)
Fresh salads	PLA bowls	McDonald's
Fresh cut fruits and vegetables	Rigid PLA trays and packs	ASDA (retailer)
Potato chips	PLA bags	Pepsico Frito Lay
Yoghurt	PLA jars	Stonyfield (Danone)
Bread	Paper bags with PLA window	Delhaize (retailer)
Starch-based biopolymers		
Milk chocolates	Cornstarch trays	Cadbury Schweppes
Organic tomatoes	Corn-based packaging	Iper supermarkets (Italy) Coop Italia
Cellulose-based biopolymers		
Kiwi	Bio-based trays wrapped with a cellulose film	Wal-Mart
Potato chips	Metalized cellulose film	Boulder Canyon
Sweets	Metalized cellulose film	Quality Street, Thornton

Source: Jabeen et al. (2015).

TABLE 15.3

Advantages and Disadvantages of Biopolymers

Raw Material	Origin	Advantages	Disadvantages
Zein	Maize protein	Good film-forming properties	More brittle
		Good tensile and moisture properties	
		Antimicrobial and antifungal activity	
		Good mechanical properties	
		Low oxygen and carbon dioxide permeability	
Chitosan	Chitin derivate	Antimicrobial and antifungal activity	High water sensitivity
		Good mechanical properties	
		Low oxygen and carbon dioxide permeability	
Whey protein isolate	Cheese derivative	Good oxygen and aroma barrier	Moderate moisture barrier
Gluten	Wheat derivative	Low cost	High sensitivity to moisture
		Good oxygen barrier	Brittle
		Good film-forming property	

Source: Modified from Jabeen et al. (2015).

between two tetrahedral layers of silica, in the ratio of 2:1. The silica layers are of thickness 1 nm and diameter 100–500 nm (Figure 15.4). These clay platelets with a high surface area of 750 m^2/g and aspect ratios in the order of 100–500 are separated by thin layers of water known as a gallery (Arora and Padua, 2010). The aforesaid features render MMT an excellent filler material that adds up strength and barrier properties to the biopolymer.

The total clay structure of MMT is formed by hundreds of layered platelets stacked into particles or tactoids of diameter 8–10 μm. The imbalance of the surface negative charges is compensated by the exchangeable cations of sodium (Na^+) and calcium (Ca^{2+}), resulting in the formation of two different types of MMTs: sodium MMT (Na^+MMT) and calcium MMT (Ca^{2+}MMT), respectively (de Azeredo, 2009). In the presence of water and organic cations, MMT undergoes intercalation and swelling. Depending on the interaction between MMT and biopolymer matrix, there are three possible types of polymer–clay formations (Carrado, 2003; Ray and Okamoto, 2003; Okada and Usuki, 2006) (Figure 15.5):

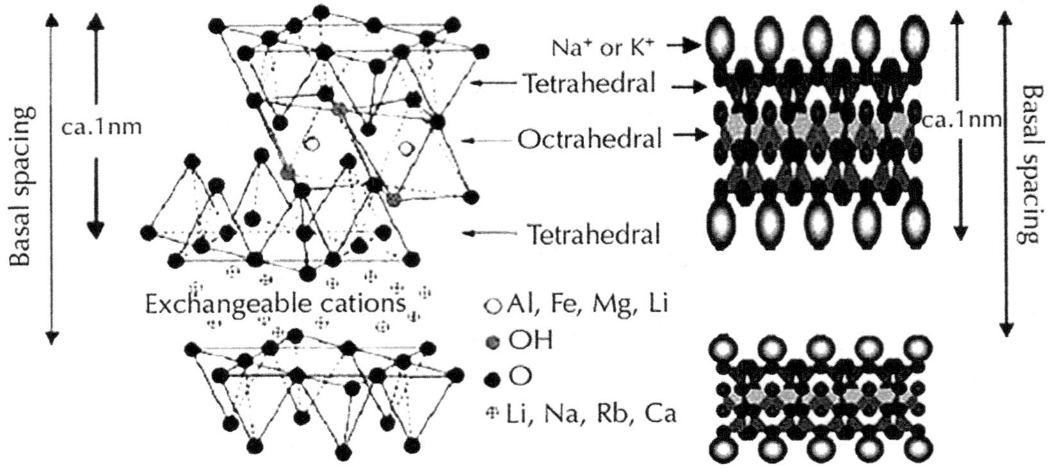

FIGURE 15.4 Structure of MMT. (Reproduced with permission from Sorrentino, A., Gorrasi, G. and Vittoria, V. 2007. Potential perspectives of bio-nanocomposites for food packaging applications. *Trends in Food Science & Technology* 18: 84–95.)

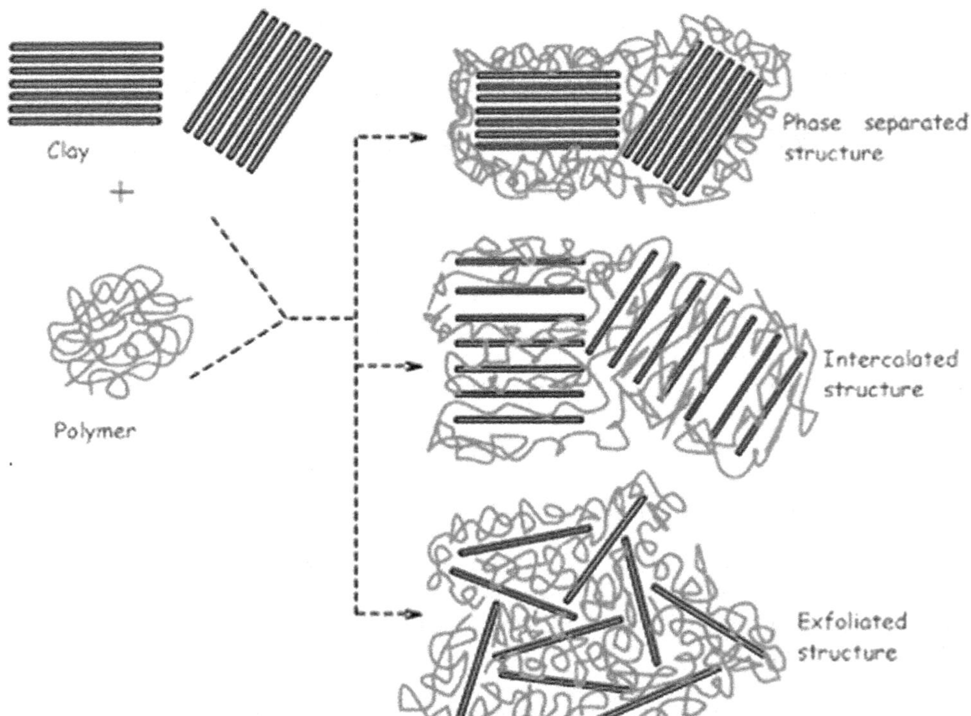

FIGURE 15.5 Types of composite derived from the interaction between clays and polymers. (Reproduced with permission from Valapa, R. B., Loganathan, S., Pugazhenthi, G., Thomas, S. and Varghese, T. O. 2017. An overview of polymer–clay nanocomposites. In *Clay-Polymer Nanocomposites*, eds. K. Jlassi, M. M. Chehimi and S. Thomas, 29–81. Amsterdam, Netherlands: Elsevier.)

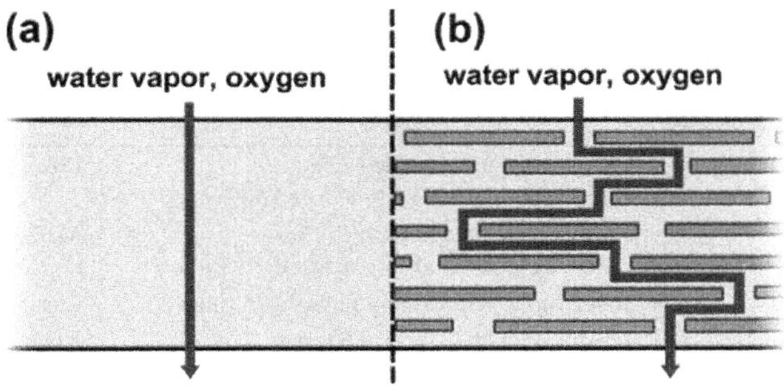

FIGURE 15.6 Penetration of gas molecules in (a) pure polymer matrix and (b) tortuous pathway created by the incorporation of the nanocomposite in a polymer matrix. (Slightly modified and reproduced with permission from Duncan, T. V. 2011. Applications of nanotechnology in food packaging and food safety: barrier materials, antimicrobials, and sensors. *Journal of Colloid and Interface Science* 363: 1–24.)

a. ***Phase-separated microcomposite or tactoid***: Due to the poor affinity of filler material to the polymer, the interlayer spaces of the clay gallery remain unexpanded to result in a tactoid structure. However, this formation is not desirable in nanocomposites (Alexandre and Dubois, 2000).
b. ***Intercalated***: Intercalated structures are formed when there is a moderate expansion of the clay interlayer due to the fair affinity between the polymer and clay. Polymer chains penetrate the basal spacing of clay resulting in the insight expansion of spaces between the layers. However, the shape of the layered stack remains unchanged.
c. ***Exfoliated***: An exfoliated structure results from a high affinity between the polymer and the clay. In this formation, the clay clusters lose their layered identity. Subsequently, the clay clusters separate into single sheets and remain uniformly distributed in the continuous polymer phase. The exfoliated nanocomposite is the most desirable formation as it completely utilizes the high surface area of nanoclays (Arora and Padua, 2010).

The nanomaterial reinforced in the matrix of bionanocomposite directs the gas molecules to pass through a tortuous path (Figure 15.6b) rather than a straight path (Figure 15.6a) as observed in pure virgin polymer films. This increase in effective path length for gas diffusion improves the barrier properties of the bionanocomposite (Rhim and Ng, 2007). Thus, with bionanocomposites, a manufacturer can attain larger effective film thickness by using only a smaller amount of polymer (Duncan, 2011).

The effects of different loading levels of MMT on the strength and barrier properties of various biopolymers are tabulated in Table 15.4.

15.3 Innovative Food Packaging Technologies

15.3.1 Active Packaging

Active packaging can be defined as "a system in which the product, the package, and the environment interact in a positive way to extend shelf life or to achieve some characteristics that cannot be obtained otherwise" (Miltz et al., 1995). This definition justifies the placement of active packaging above the protection function in Figure 15.7. Active packaging protects the food product against oxygen, moisture, and microorganisms. The rationale of active packaging is to extend the shelf life and enhance the margin of food safety by altering the condition of the food (de Kruijf et al., 2002). The intended protective function of active packaging is achieved by deliberately including subsidiary constituents either in the packaging material or in the package headspace (Robertson, 2006a, b). Depending on the constituent included,

TABLE 15.4
Different Loading Levels of MMT and Its Effects

Biopolymer	% Loading	Effects	Reference
Agar	10	• Tensile strength increased 22% • Water vapor permeability reduced 50%	Rhim (2011)
Starch	6	• Tensile strength increased 1.9 times • Water vapor permeability reduced 2.1 times	Maksimov et al. (2009)
Pectin	10	• Water vapor permeability reduced 35 times	Castello et al. (2010)
Soy protein	10	• Tensile strength increased 6 times • Water vapor permeability reduced 43%	Kumar et al. (2010)
Wheat protein	5	• Tensile strength increased 6 times	Angellier-Coussy et al. (2008)
Bovine gelatin	5	• Water vapor permeability reduced 3 times	Martucci and Ruseckaite (2010)
Fish gelatin	5	• Tensile strength increased 33% • Water vapor permeability reduced 71%	Bae et al. (2009)
Whey protein	5	• Water vapor permeability reduced 28% • Antibacterial properties	Sothornvit et al. (2009)

Source: Modified from Anandharamakrishnan and Kolli (2015).

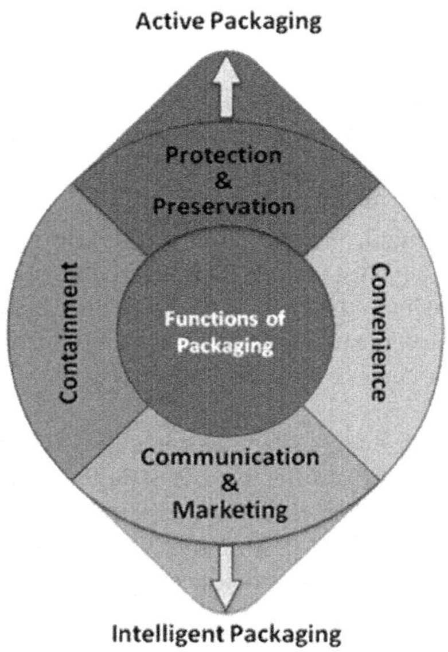

FIGURE 15.7 Extended packaging functions in advanced packaging systems: active and intelligent packaging. (Reproduced with permission from Mihindukulasuriya, S. D. F. and Lim, L. -T. 2014. Nanotechnology development in food packaging: a review. *Trends in Food Science & Technology* 40: 149–167.)

different types of active packaging systems are available (Figure 15.8), which are briefly described in Sections 15.3.1.1–15.3.1.5.

15.3.1.1 Oxygen Scavengers

The presence of oxygen inside a package is detrimental to the food product as it supports the growth of aerobic microbes and molds and accelerates oxidative reactions leading to off-odors, off-flavors, adverse color changes, and reduced nutritional quality. Oxygen scavengers are compounds that react with oxygen (which permeates from the external environment into the package through the packaging material or residual oxygen that remains inside the package) to reduce its concentration and thereby retard the

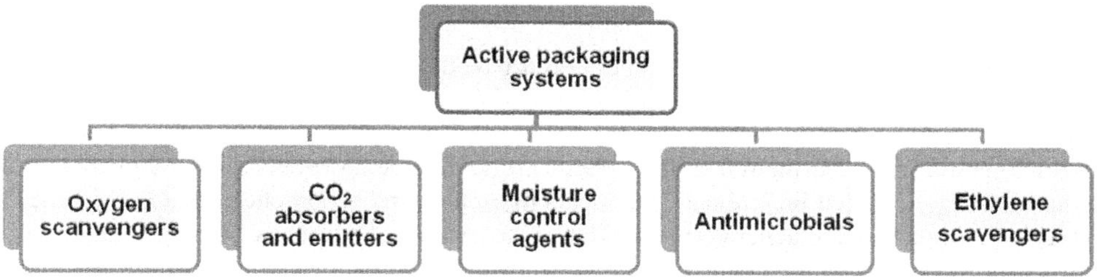

FIGURE 15.8 Active packaging systems.

TABLE 15.5
Classification and Mechanism of Oxygen Scavenging Agents

Classification	Oxygen Scavenging Agents	Oxidation Mechanism
Metallic	Iron powder, activated iron, ferrous oxide, iron salt, Cobalt (II), Zinc	Oxidation of iron with the supply of moisture and action of an optional catalyst
Organic	Ascorbic acid, ascorbic acid salts, isoascorbic acid, tocopherol, hydroquinone, catechol, rongalit, sorbose, lignin, gallic acid, polyunsaturated fatty acids	Oxidation of organic substrate with a metallic catalyst or an alkaline substance
Inorganic	Sulfate, thiosulfate, dithionite, hydrogen sulfate, titanium dioxide	Oxidation of inorganic substrate by UV light
Polymer-based	Oxidation-reduction resin, polymer metallic complex	Oxidation of polymer components with metallic catalyst (mostly cobalt)
Enzyme-based	Glucose oxidase, laccase, ethanol oxidase	Immobilization as coatings on polymers such as PS, polyethylene and polypropylene.

Source: Gaikwad et al. (2018).

oxidative reactions. The classification and oxidation mechanism of different oxygen scavenging agents are tabulated in Table 15.5.

Oxygen scavengers may be used in the form of sachets containing various iron-based powders, films labels, card, closure liners, or concentrates (Suppakul et al., 2003) (Figure 15.9). The oxygen scavenging sachets used in food packages must be easy to handle and compact in size. More importantly, the sachets should not produce toxic substances or unpleasant odors/gases (Harima, 1990). Oxygen scavengers may also be incorporated directly into the packaging film and bottle crowns.

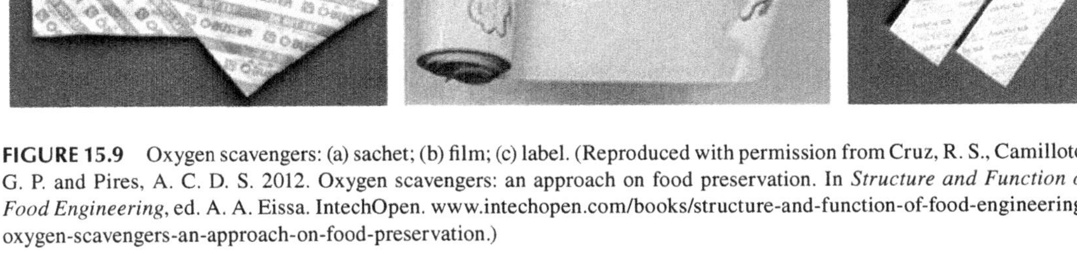

FIGURE 15.9 Oxygen scavengers: (a) sachet; (b) film; (c) label. (Reproduced with permission from Cruz, R. S., Camilloto, G. P. and Pires, A. C. D. S. 2012. Oxygen scavengers: an approach on food preservation. In *Structure and Function of Food Engineering*, ed. A. A. Eissa. IntechOpen. www.intechopen.com/books/structure-and-function-of-food-engineering/oxygen-scavengers-an-approach-on-food-preservation.)

15.3.1.2 Carbon Dioxide Absorbers and Emitters

In products such as fresh meat, poultry, cheese, and baked goods, the addition of carbon dioxide is favorable towards the suppression of microbial growth, for reducing the respiration rate of fresh produce and to overcome package collapse or partial vacuum caused by oxygen scavengers (Labuza 1996; Lopez-Rubio et al., 2004; Vermeiren et al., 1999). Carbon dioxide absorbers are used in various forms including moisture-activated bicarbonate chemicals in sachets and absorbent pads. On the other hand, high levels of carbon dioxide generated by oxidative reactions can adversely affect the quality of food products. The excess carbon dioxide can be removed by using highly permeable plastics whose permeability increases with a rise in temperature (Brody et al., 2008).

15.3.1.3 Moisture Control Agents

Excess moisture in packages can lead to caking of powdered products (e.g., soluble coffee), softening of crispy products (e.g., crackers), and moistening of hygroscopic products (e.g., sweets and candy). Conversely, excessive moisture loss from the food product leads to product desiccation. Moisture control agents are chemical components that are capable of the following actions:

- Control water activity and thereby retard microbial growth.
- Remove melting water from frozen products and blood or fluids from meat products.
- Prevent condensation from fresh produce.
- Control the rate of lipid oxidation (Vermeiren et al., 1999).

The commonly used moisture control agents in dry foods are silica gels, natural clays, and calcium oxide. These are used in the form of internal porous sachets or perforated water vapor barrier plastic cartridges containing desiccants. They can also be incorporated into packaging material. For high-moisture foods such as meat, poultry, fruits, and vegetables, internal humidity controllers are used which maintain optimum in-package relative humidity (RH) (~85% for cut fruits and vegetables), reduce moisture loss, and prevent excess moisture in headspace and interstices where microorganisms can grow (Brody et al., 2008).

15.3.1.4 Antimicrobials

The function of antimicrobials in food packaging is to reduce the surface contamination of processed foods and thereby improve their quality and safety (Brody et al., 2001; Cooksey, 2005). The mechanism of action by which the antimicrobials reduce the growth rate of spoilage and pathogenic microorganisms is based on extending the lag phase or inactivating the microbes (Quintavalla and Vicini, 2002). Some of the antimicrobial agents on which research is underway include (Wilson, 2007) the following:

- Silver salts
- Ethyl alcohol
- Chlorine dioxide
- Nisin
- Organic acids (e.g., acetic, benzoic, lactic, tartaric, and propionic acids)
- Allyl isothiocyanate
- Spice-based essential oils (e.g., oregano oil, mustard oil, clove, carvacrol, thymol, and cinnamon)
- Metal oxides (magnesium oxide and zinc oxide)

Antimicrobial agents are either incorporated directly into the packaging materials such that they release slowly to the food surface or used in vapor form (Brody et al., 2008).

15.3.1.5 Ethylene Scavengers

Ethylene is a natural plant hormone released during the ripening of fruits, which accelerates the respiration and leads to maturity and senescence. Removal of ethylene from a package environment extends the shelf life of fresh produce. Potassium permanganate is a commonly used ethylene scavenger which oxidizes ethylene to acetate and ethanol (Lopez-Rubio et al., 2004). Ethylene may also be removed by physical adsorption onto active surfaces such as activated carbon or zeolite. While potassium permanganate is used in the form of sachets, other adsorbent or absorbent chemicals may be incorporated into the packaging materials (Brody et al., 2008).

15.3.2 Intelligent Packaging

Intelligent packaging has been defined as "a packaging system that is capable of carrying out intelligent functions (such as detecting, sensing, recording, tracing, communicating, and applying scientific logic) to facilitate decision-making to extend shelf life, enhance safety, improve quality, provide information, and warn about possible problems." An intelligent package monitors the quality or safety condition of a food product and provides early warning to the consumer or food manufacturer. Thus, the intelligent packaging is placed beside the communication function in Figure 15.7. The difference between intelligent and active packaging is that the former is responsible for sensing the environment and processing information, and the latter is responsible for executing a specific action (such as releasing an antimicrobial) to protect the food product (Yam et al., 2005).

The sequential flow of information in an intelligent packaging system occurs between the following four components, before an action is initiated:

1. Smart package devices
2. Data layers
3. Data processing, and
4. Information highway (wire or wireless communication networks) in the food supply chain

The smart package devices enable a package to acquire, store, and transfer data. On the other hand, the data layers, data processing, and information highway constitute the decision support system. In the context of intelligent packaging, data refers to the information indicative of food quality and safety such as the time–temperature history, microbial count, pH, and water activity. The data processing is based on food models and business models that facilitate effective decision-making. Food models refer to the scientific principles; e.g., heat transfer principle and the business models include the rules for processing information to maximize profits (Yam et al., 2005). Depending on the functions carried out, there are two different types of smart packaging devices, as shown in Figure 15.10.

15.3.2.1 Barcodes

Barcodes are the most widely used data carriers. It is an optical machine-readable symbol relating to the object to which it is attached (Fang et al., 2017). A meaningful interpretation of a barcode is possible only when its communication is enabled with scanner and printer. "Symbology" is an accepted language of communication between the barcodes, scanners, and printers (Pearce and Bushnell, 1997). The Universal Product Code barcode is one of the initially introduced linear symbologies which consist of a pattern of bars and spaces to represent 12 digits of data (Figure 15.11a). However, its storage capacity was very less to hold merely limited information such as manufacturer identification number and item number. In order to enable the barcodes to store more data in a smaller space, several new barcode symbologies have been introduced (Uniform Code Council, 2004) (Figure 15.11a).

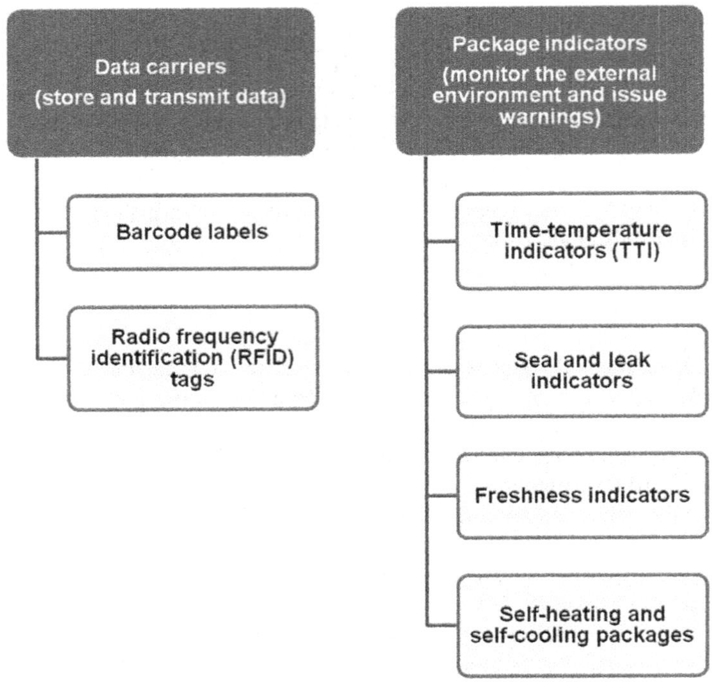

FIGURE 15.10 Different types of smart package devices.

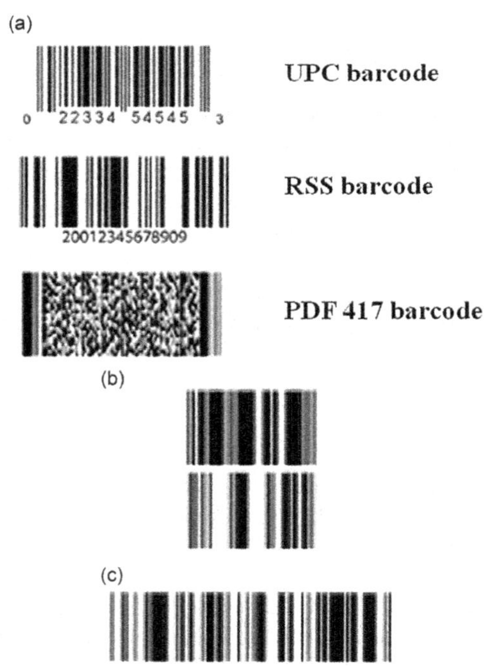

FIGURE 15.11 (a) Barcode symbologies (Reproduced with permission from Fang, Z., Zhao, Y., Warner, R. D., and Johnson, S. K. 2017. Active and intelligent packaging in meat industry. *Trends in Food Science & Technology* 61: 60–71.); (b) RSS 14-stacked omnidirectional barcode; (c) RSS Expanded Barcode. (Reproduced with permission from Yam, K. L., Takhistov, P. T. and Miltz, J. 2005. Intelligent packaging: concepts and applications. *Journal of Food Science* 70: R1–R10.)

i. ***Reduced Space Symbology*****:** Reduced space symbology (RSS) is appropriate for the identification of product at the point-of-sale and for product traceability in the grocery industry. The RSS-14 (Figure 15.11b), one of the members of the RSS family, is a stacked omnidirectional barcode which encodes the complete 14-digit Global Trade Item Number. RSS-14 is applicable for loose produce items such as apples or oranges where space limitation requires a narrow symbol. On the other hand, the RSS Expanded Barcode (Figure 15.11c) encodes up to 74 alphanumeric characters. This may be used for variable measure products such as meat and seafood that are sold by weight, wherein larger data capacity is required to encode more information such as date of packing, batch or lot number, and weight of the package (Yam et al., 2005).

ii. ***Portable data file (PDF) 417*****:** PDF 417 is a two-dimensional symbology having a data space of 1.1 kilobytes (Anonymous, n.d.). It allows the encoding of additional information such as nutritional information, cooking instructions, website address of the food manufacturer, and graphics. The portable data can be accessed immediately without the need for an external database (Yam et al., 2005).

15.3.2.2 Radio Frequency Identification

RFID is an advanced form of radio wave-based wireless data information carrier, which can effectively identify and trace a product (Figure 15.12a). It comprises the following components:

- ***Reader*****:** A reader, also known as the receiver, emits a radio signal to capture data from an RFID tag which is a minuscule microchip connected to a tiny antenna (Figure 15.12b). This facilitates the RFID tags to be read for a range of 15–100 ft (Yam et al., 2005).

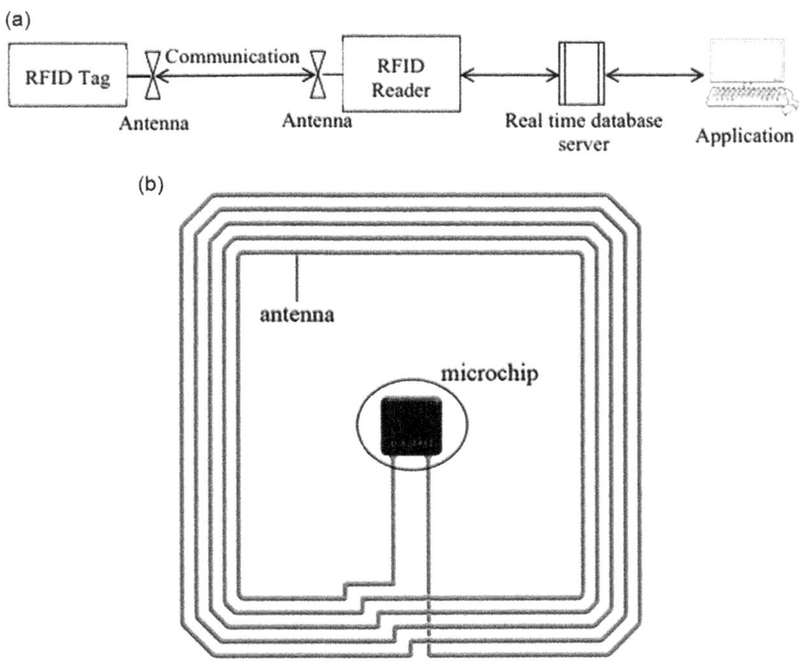

FIGURE 15.12 (a) Schematic of an RFID system. (Reproduced with permission from Fang, Z., Zhao, Y., Warner, R. D. and Johnson, S. K. 2017. Active and intelligent packaging in meat industry. *Trends in Food Science & Technology* 61: 60–71.); (b) microchip of an RFID tag. (Reproduced with permission from Vanderroost, M., Ragaert, P., Devlieghere, F. and Meulenaer, B. D. 2014. Intelligent food packaging: the next generation. *Trends in Food Science & Technology* 39: 47–62.)

- ***Labels or radar***: This functions as the data carrier to a computer for analysis.
- ***Computer***: It may be connected to a local network or to the Internet for analysis and decision-making (Want, 2004).

RFID being a wireless technique has a competitive edge over the conventional barcodes as they need not be in a direct line of sight to be recognized by a scanner. Thus, RFID can be used in a grocery store to simplify the stock checking activities. These can also store information such as temperature and RH data, nutritional information, and cooking instructions. Further, RFID could be integrated with a time-temperature indicator (TTI) or a biosensor to pass on time–temperature information or microbiological data (Yam et al., 2005). Food industry perceives RFID as a potential traceability tool to improve food safety and effectiveness of the supply chain (Brody et al., 2008).

15.3.2.3 Time–Temperature Indicators

The simple and more commonly used TTIs are the visual indicators which change color in response to cumulative exposure to temperature. Visual indicators work on the principles of enzymatic reactions, polymerization, or chemical diffusion. Applications of visual TTIs are established in monitoring the exposure of foods to any temperature abuse during transport and storage. TTIs ensure that the product reaches the consumer in optimal conditions (Welt et al., 2003).

An interesting application of TTI is in the cooking appliances such as an intelligent microwave oven system shown in Figure 15.13 (Yam, 2000). Microwave ovens are available in different sizes and power outputs. Hence, it is difficult to generalize or customize the heating instructions printed on microwave-able food packages for the different ovens sold in the market. This concern may be overcome by scanning the barcode to facilitate the decision support system to match between the microwave oven and the food package. As shown in Figure 15.13, a PDF 417 barcode in the package carries data about the food product. The data processing system generates the appropriate heating instructions for controlling the magnetron and turntable of the microwave oven, by using algorithms that are based on heat transfer principles. In addition, temperature and moisture sensors may be placed inside the microwave oven to provide feedback for the data processing system (Yam et al., 2005).

15.3.2.4 Seal and Leak Indicators

The composition of the gas in the package headspace fluctuates due to the activity of food product, leaks, nature of the package, or environmental conditions. Usually, O_2 and CO_2 are used as seal and leak indicators to verify the effectiveness of an oxygen absorber. These indicators change color in response to chemical or enzymatic reactions and signify the increase in oxygen concentration beyond the defined limit in a sealed food package (Hu et al., 2009).

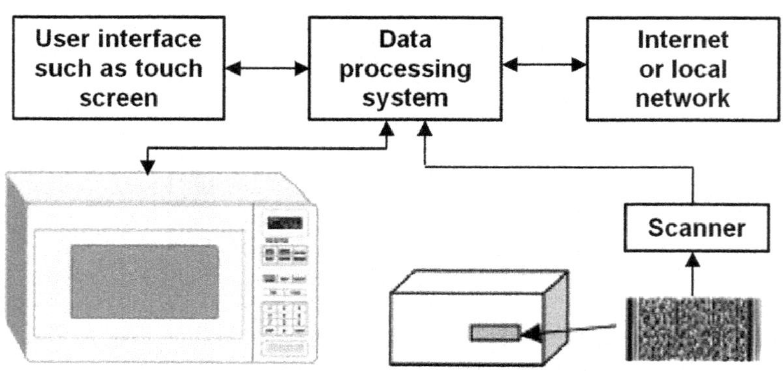

FIGURE 15.13 Intelligent microwave oven system. (Reproduced with permission from Yam, K. L., Takhistov, P. T. and Miltz, J. 2005. Intelligent packaging: concepts and applications. *Journal of Food Science* 70: R1–R10.)

15.3.2.5 Freshness Indicators

Freshness indicators signal the loss in the freshness of the packaged goods by detecting the changes in the concentration of chemicals produced during the storage of meat (see Figure 15.14), fish, fruits, and vegetables. These chemical markers include (Hong and Park, 2000; Smolander et al., 2002; Rokka et al., 2004; Kerry et al., 2006; Nopwinyuwong et al., 2010; Wanihsuksombat et al., 2010) the following:

- Volatile metabolites (diacetyl, amines, carbon dioxide)
- Organic acids (*n*-butyrate, L-lactic, D-lactate, and acetic acid),
- Microbial metabolites (glucose, carbon dioxide, hydrogen sulfide, sulfur dioxide, ammonia)
- Biogenic amines (putrescine, cadaverine, histamine, and others) formed by the degradation of protein-containing food to amino acids.

The indication is usually a color change in response to the variation in pH due to the interaction between chemical markers and pH-sensitive indicators which are usually dyes. For example, bromothymol blue pH dyes are used to monitor the formation of carbon dioxide released as a result of microbial growth in meat (Holte, 1993). Under increased carbon dioxide levels, the pH dye reacts with the generated gas and changes in color. Other pH dyes used in freshness indicators are xylenol blue, bromocresol purple, bromocresol green, cresol red, phenol red, methyl red, and alizarin (Horan, 2000).

15.3.2.6 Temperature Control: Self-Heating and Self-Cooling Packaging

In self-heating packaging, heating occurs due to the exothermic reaction generated by calcium or magnesium oxide and water. The applications of self-heating packages have been established in plastic coffee cans, military rations, and on-the-go meal platters. However, the limitation of self-heating packages is that the heating device occupies almost half of the volume within the package.

On the other hand, self-cooling packaging functions on the principle of "evaporative cooling" by evaporating the external components such as water, which removes heat from the package contents and adsorbs onto the surface (Brody et al., 2008). This is relevant as the evaporation of 10 mL of water can theoretically cool 330 mL of water by 18°C (Steeman, 2012). Self-cooling technology is mainly used in beverage cans to enhance the flavor experience of the consumers.

A summary of the principles of operation and applications of the previously discussed smart packaging devices is provided in Table 15.6.

15.3.3 Modified Atmosphere Packaging

Modified atmosphere packaging (MAP) aims at extending the shelf life and improving the quality of fresh produce by altering the atmosphere inside the package to obtain an optimum gaseous environment.

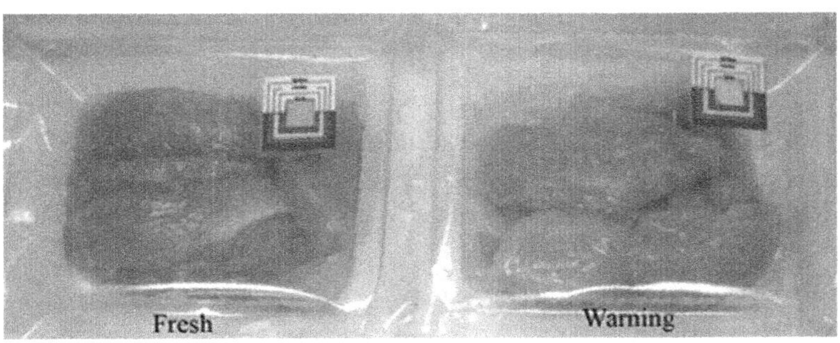

FIGURE 15.14 Freshness indicator for monitoring skinless chicken breast. (Reproduced with permission from Rukchon, C., Nopwinyuwong, A., Trevanich, S., Jinkarn, T. and Suppakul, P. 2014. Development of a food spoilage indicator for monitoring freshness of skinless chicken breast. *Talanta* 130: 547–554.)

TABLE 15.6

Principle of Operation of Different Smart Packaging Devices

Smart Devices	Principle/Reagents	Information	Application
Barcodes	Symbology	Product and manufacturer information	Product identification, facilitating inventory control, stock reordering, and checkout
RFID identification tags	Radio waves	Product and manufacturer information	Product identification, supply chain management, asset tracking, security control
TTIs	Mechanical, chemical, enzymatic, microbiological	Storage conditions	Foods stored under chilled and frozen conditions
Freshness indicators	pH dyes; Dyes reacting with (non-) volatile metabolites	Microbial quality of food (i.e. spoilage)	Perishable foods such as meat, fish and poultry

Source: Hurme et al. (2002).

The MAP can be achieved through active or passive methods. In the active MAP, the optimal gaseous composition is attained by displacing the air inside the package with a flush of a controlled mixture of gases. Thus, the active MAP is generally referred to as *gas flushing*. Contrastingly, the passive approach achieves the optimum gaseous milieu inside the package by a combination of the respiration and metabolism of the contained food. Under the conditions of the MAP, the degradative reactions in foods including moisture loss, oxidation, enzyme activity, microbial growth, and postharvest metabolic activities are delayed.

The typical composition of air (by volume) is 78.08% nitrogen, 20.95% oxygen, 0.93% argon, 0.03% carbon dioxide, and traces of nine more gases. Active MAP involves the individual or combined use of O_2, CO_2, and N_2. The proportion of these gases and temperature applied in the MAP are variable with the commodity in the package (Table 15.7). Packaging materials for MAP should be chosen carefully by considering the respiration rate of the fresh produce and the package permeability to gases.

15.3.4 Controlled Atmosphere Packaging

According to FDA, controlled atmosphere packaging (CAP) is an "active system which continuously maintains the desired atmosphere within a package throughout the shelf-life of a product by the use of agents to bind or scavenge oxygen or a sachet containing compounds to emit a gas." Thus, CAP involves packaging a product in a modified atmosphere and subsequently maintaining that atmosphere to preserve the quality of foods. It is an effective alternative to pesticides and preservatives. CAP and MAP are extremely useful in standardizing the variable storage temperatures and conditions of distribution for the fresh and processed foods (Vaclavik and Christian, 2003).

As the packaged products respire and allow microbial growth in the presence of oxygen, the following reaction occurs:

$$C_6H_{12}O_6 \xrightarrow{O_2} CO_2 + H_2O + \text{Heat}$$

The CAP containers control the exchange between oxygen and carbon dioxide by reducing the available O_2, elevating CO_2, and controlling the water vapor and ethylene concentration. Thus, senescence is lessened, and the nutritional value is retained.

15.3.5 Aseptic Packaging

Aseptic packaging involves independent sterilization of both the foods and packaging material, followed by hermetic sealing of the food inside the package under sterile environmental conditions. This technique extends the shelf life of low-acid foods by inactivating the spores of *Clostridium botulinum*.

TABLE 15.7

Recommended MAP Conditions for Various Products

Commodity	Temperature Range (°C)	O_2 (%)	CO_2 (%)	N_2 (%)
Bread	23	0	60–70	30–40
Pizza	5	0	80	20
Hard cheeses	1–4	0	60	40
Soft cheeses	1–4	0	30	70
Lean fish/white fish	0–2	30	40	30
Oily and smoked fish	0–2	0	60	40
Apple (whole)	0–5	2–3	1–5	—
Avocado	5–13	2–5	3–10	—
Banana	12–15	2–5	2–5	—
Kiwi fruit	0–5	2	5	—
Mango	10–15	5	5	—
Pineapple	10–15	5	10	—
Strawberry	0–5	10	15–20	—
Beef	−1 to 2	60–80	20–40	—
Pork	−1 to 2	40–80	20	0
Poultry	−1 to 2	0	20–100	—
Asparagus	0–5	20	5–10	—
Broccoli	0–5	1–2	5–10	—
Lettuce	0–5	2–3	5–6	—
Mushrooms	0–5	21	10–15	—
Tomato	7–12	4	4	—

Source: Adapted from Kader et al. (1989), Parry (1993), Floros and Matsos (2005), FDA (2001), and Brody (2000).

In aseptic packaging, the packaging material consists of layers of polyethylene, paperboard, and foil (Figure 15.15a). The package is sterilized by heat (superheated steam or dry hot air) or a combination of heat and hydrogen peroxide. It may also be sterilized by nonchemical techniques by employing ionizing and non-ionizing radiation (e.g., low-energy electron beam ~200 keV) to avoid chemical sterilant residues. The sterilized package is then roll-fed through the packer to create the typical brick/block shape (Figures 15.15b and 15.16). Subsequently, the sterile container is filled with the sterile or commercially sterile liquid food product and sealed inside a closed and sterile chamber. Creamers, milk, and juices are common examples of aseptically packed products. These products need not be refrigerated. Nowadays, the aseptic cartons include attached straws and easy-to-open, easy-to-pour plastic devices that are injection-molded and attached to the top of the package (Vaclavik and Christian, 2003).

15.4 Mass Transfer in Food Packaging

Permeation, diffusion, and absorption are the predominant mass transfer phenomena occurring through a food packaging material. Each of these phenomena is elaborated in the subsequent sections.

15.4.1 Permeation

Permeation, permeance, or migration is the ability of a component to penetrate and pass through a packaging material in response to a difference in partial pressure (ASTM, 2009; Segall and Scanlon, 1996). The permeating component is referred to as a penetrant molecule or "permeant." The permeant can be a gas which obeys Henry's law or a solute which does not abide by Henry's law. Thus, it can be inferred that the surface concentration of a gas component is related to the partial pressure in the atmosphere which remains in contact with the packaging material. But, the permeability of a solute is directly related

(a)

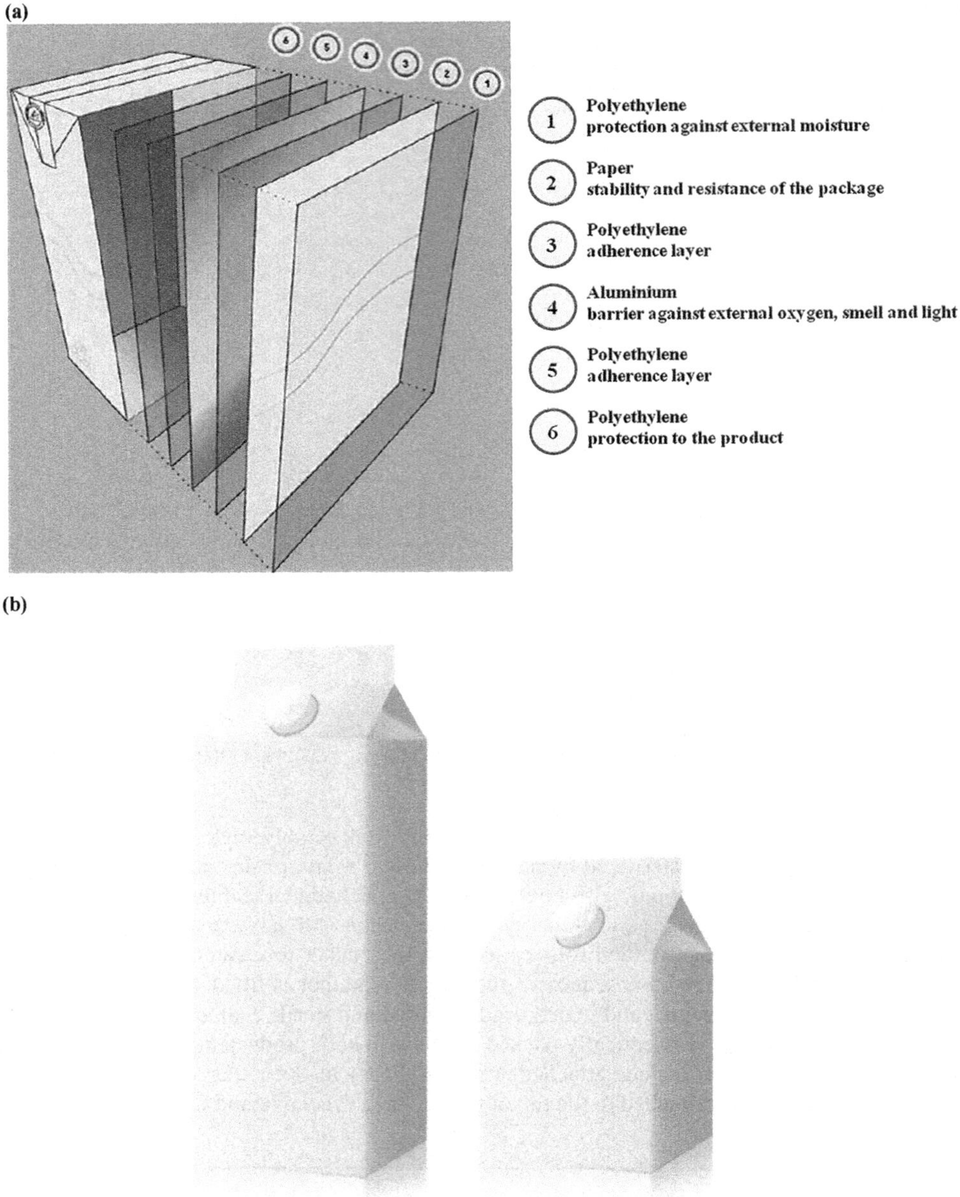

(b)

FIGURE 15.15 (a) Structure (Reproduced with permission from Ebadi, M., Farsi, M., Narchin, P. and Madhoushi, M. 2015. The effect of beverage storage packets (Tetra Pak™) waste on mechanical properties of wood–plastic composites. *Journal of Thermoplastic Composite Materials* 29: 1–10.) and (b) shape of carton used in aseptic packaging. (Reproduced with permission from Matche, R. S. 2018. Packaging technologies for fruit juices. In *Fruit Juices Extraction, Composition, Quality and Analysis*, eds. G. Rajauria and B. K. Tiwari, 637–666. London: Academic Press.)

to the overall mass transfer coefficient, which includes both the surface mass transfer coefficient (k_m) and the diffusional resistance (L/D). k_m is determined by the partition of solute between the solvent and packaging film, governed by the partition coefficient (K) which is the ratio between the concentrations of solute at the surface and in the solution (Han and Scanlon, 2014).

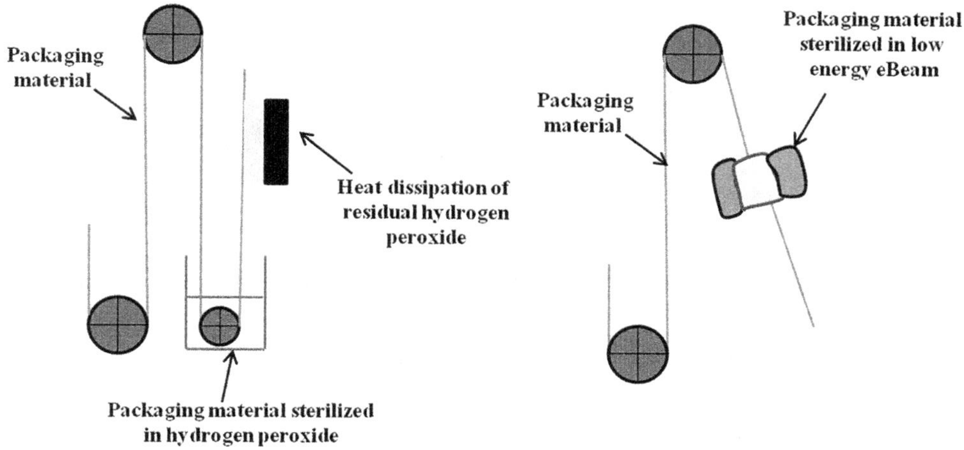

FIGURE 15.16 Schematic representation of the sterilization systems for aseptic packaging materials. (Reproduced with permission from Pillai, S. D. and Shayanfar, S. 2015. Aseptic packaging of foods and its combination with electron beam processing. In *Electron Beam Pasteurization and Complementary Food Processing Technologies*. Woodhead Publishing Series in Food Science, Technology and Nutrition, eds. S. D. Pillai and S. Shayanfar, 83–93. Cambridge, UK: Woodhead Publishing.)

15.4.2 Diffusion

Diffusion is the movement of a component in a medium caused by a concentration difference acting as the driving force. The diffusing component is termed as the diffusant. As already discussed in Chapter 6, diffusion obeys Fick's law which is given by

$$J = -D\frac{\partial C}{\partial x} \tag{15.1}$$

where J is the flux per unit cross-sectional area, D is the diffusivity, C is the concentration of the diffusant, and x is the distance traveled by the diffusant. Diffusivity is defined as the transfer rate of an amount of the diffusant across a known distance in the material. While diffusivity is measured in units of m^2/s, flux is expressed in mol/m^{2}s or kg/m^{2}s (Han and Scanlon, 2014). Integrating Eq. (15.1) under the conditions of constant concentrations C_1 and C_0 and diffusivity (D) gives the steady-state flux of the diffusant Eq. (15.2):

$$J = \frac{Q}{A \times t} = D\frac{(C_1 - C_0)}{L} \tag{15.2}$$

where Q is the amount of diffusant (mol/kg), A is the cross-sectional area of diffusion (m^2), and L is the thickness of the packaging film or membrane (m).

$$D = \frac{J \times L}{\Delta C} = \frac{Q \times L}{A \times t \times \Delta C} \tag{15.3}$$

15.4.3 Absorption

Absorption is the measure of the affinity of a given substance for two media with which it comes into contact. It is the surface sorption of the molecules from the surroundings to the material. Solubility (S) is the mass transfer coefficient of the absorption phenomenon. Flavor scalping is a good example of absorption in packaging. *Flavor scalping* is a terminology used in the packaging industry to describe the loss of quality of a packaged product due to the following reasons (Sajilata et al., 2007):

- The package absorbing the volatile flavors from the food product (or)
- The food absorbing undesirable flavors from the packaging material

Flavor losses from a food product occur due to its interaction with the constituent polymers of the packaging material by permeation through the package or from sorption by the container (Strandburg et al., 1991). Flavor scalping is a significant concern in the food industry as foods packaged in plastic materials are susceptible to changes in both the intensity and characteristics of the food flavors owing to their absorption by the packaging material (Roland and Hotchkiss, 1991).

Permeation begins as the permeant molecule collides with the polymer followed by its sorption into the polymer. Then, the penetrant diffuses through the polymer matrix and desorbs from the polymer to complete the process (Sajilata et al., 2007) (Figure 15.17).

From this description, it is evident that Fick's law of diffusion and Henry's law can be effectively applied to predict the permeability of gases through a packaging material or film (Figure 15.17). Accordingly, gas permeability through a polymeric film can be expressed as

$$\frac{\dot{m}}{A} = J = -D\frac{dc}{dx} = D\frac{(c_1 - c_2)}{x} \tag{15.4}$$

where $\frac{\dot{m}}{A}$ or J is the molar flux which is the amount of component diffused per unit area per unit time (kmol/m^{2}s), and x is the film thickness. In turn, concentration depends on the partial pressure and solubility of the component, which is governed by Henry's law. Therefore, Eq. (15.4) can be written as follows:

$$J = D \times H\frac{(p_1 - p_2)}{x} \tag{15.5}$$

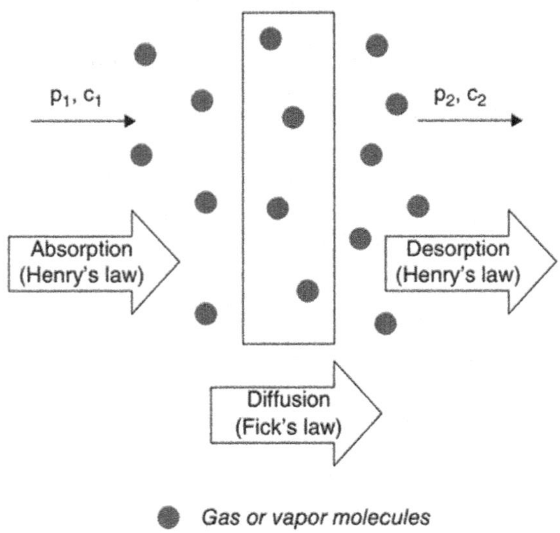

FIGURE 15.17 Mechanism of gas permeation through the packaging material. (Reproduced with permission from Shin, J. and Selke, S. E. M. 2014. Food packaging. In *Food Processing: Principles and Applications*, eds. S. Clark, S. Jung and B. Lamsal, 249–273. Chichester, West Sussex, UK: John Wiley and Sons.)

TABLE 15.8

Gas Permeability of Food Packaging Materials

Material	Permeability (*P*) (×10[11] mL(STP) cm/cm s (cm Hg))			
	O_2	CO_2	N_2	H_2O at 90% RH
LDPE	30–69	130–280	1.9–3.1	800
HDPE	6–11	45	3.3	180
PP	9–15	92	4.4	680
Polyethylene terephthalate	0.3–0.75	1.6–3.0	0.04–0.06	1300
PS	15–27	105	7.8	12000–18000

Source: Adapted from Robertson (2006a, b) and Mangaraj and Goswami (2009).

where $(p_1 - p_2)$ is the difference in partial pressure of the gas between the product inside the packaging material and the storage environment outside the container. The product $(D \times H)$ is referred to as the permeability coefficient or permeability constant (*P*).

$$\therefore J = P\frac{(p_1 - p_2)}{x} \tag{15.6}$$

From Eq. (15.6), permeability coefficient (*P*) can be defined as the amount of gas, by volume, which penetrates unit thickness and unit area of specimen per unit time, under constant temperature and unit pressure difference when the permeation is stable. Thus, the permeability coefficient is considered as the mass transfer coefficient for permeation, and it is expressed in cm^3 cm/cm^2s Pa or cm^3 cm/cm^2s atm. Thus, Fick's law in combination with Henry's law provides an expression which relates permeation rate and the area and thickness of the membrane.

Equation 15.6 is useful in calculating the amount of gas diffusion through the film. Also, it can be applied to design the thickness and area of the polymeric film constituting the packaging material to reduce the flux of gas transmission or the flavor scalping phenomenon. This is of great relevance as the major function of packaging material is to act as a barrier against the permeation of gases such as oxygen, carbon dioxide, nitrogen, and water vapor. This is because the interaction of a food product with these gases influences its susceptibility to rancidity, ripening, hydration, or dehydration. Therefore, these gases determine the quality and period of shelf life of the food products. The gas permeability of majorly used food packaging materials is listed in Table 15.8. The gas and water vapor transmission rates (WVTR) of packaging materials give a good indication of permeation. Based on the mass transfer phenomenon, oxygen transmission rate (OTR) and WVTR are the parameters used in the selection and evaluation of packaging materials (Han and Scanlon, 2014).

15.4.4 Oxygen Transmission Rate

OTR can be defined as the rate at which oxygen gas permeates through a film under steady-state conditions. It is expressed as the volume of oxygen that penetrates a given area in a one-day period. Thus, the unit of OTR is cc/m^2/24 h.

15.4.5 Water Vapor Transmission Rate

Controlling moisture migration is central to maintaining the taste, texture, microbial stability, and overall quality of packaged food products. WVTR, also referred to as the moisture vapor transfer rate, indicates the ease of permeation of moisture through a packaging film. The higher the value of WVTR for a packaging material, the greater would be its permeability and hence lower will be its ability to retain the dryness and moisture of dry (e.g., coffee and grains) and moist foods (dairy, meat, seafood), respectively. The value of WVTR is expressed in g/m^2.

The standard methods for the measurement of OTR and WVTR would be elaborated in Section 15.5.

Further, the Eq. (15.6) can also be applied to determine the shelf life of a food product where gas permeability is the critical factor. The shelf life can be calculated using the following equation:

$$t_{SL} = \frac{Q \times x}{PA(p_1 - p_2)} \tag{15.7}$$

where t_{SL} is the shelf life of the food product (in days), Q is the maximum quantity of gas that is permissible through the package (in moles or g or mL), x is the film thickness (μm), and P is the permeability of gas (mol μm/m²day atm).

Example 15.1

A food product is packaged in a plastic package covered with polyvinyl chloride (PVC) film. Calculate the required thickness of the film so that 2×10^{-5} mol of oxygen enters the package through the film in 48 h. The permeability of PVC to oxygen is 456×10^{-11} cm³/(s cm² atm/cm), the surface area of the film is 400 cm², and the partial pressure of oxygen inside and outside the package is 0 and 0.24 atm, respectively.

Given

$$m = 2 \times 10^{-5} \text{ mol of oxygen}$$

$$t = 48\,\text{h} = 172800\,\text{s}$$

$$P = 456 \times 10^{-11}\,\text{cm}^3/(\text{s}\,\text{cm}^2\,\text{atm/cm}) = 456 \times 10^{-15}\,\text{m}^3/(\text{s}\,\text{m}^2\,\text{atm/m})$$

$$A = 400\,\text{cm}^2 = 400 \times 10^{-4}\,\text{m}^2$$

$$p_1 = 0.24\,\text{atm}$$

$$p_2 = 0\,\text{atm}$$

Solution

$$J = {}^{m}\!/_{A \times t} = \frac{2 \times 10^{-5}}{400 \times 10^{-4} \times 172{,}800} = 2.894 \times 10^{-9}$$

From Eq. (15.6),

$$x = P\frac{(p_1 - p_2)}{J}$$

$$x = \frac{P \times (p_1 - p_2)}{J} = \frac{456 \times 10^{-15} \times (0.24 - 0)}{2.894 \times 10^{-9}} = 37.816 \times 10^{-6}\,\text{m} = 37.816\,\mu\text{m}$$

Answer: The thickness of the film should be 37.816 μm which would permit 2×10^{-5} mol of oxygen to enter the package through the film in 48 h.

15.5 Quality Evaluation of Packaging Materials

The performance of a packaging material for industrial and commercial applications is evaluated by different test methods for various quality parameters. In the food industry, the quality evaluation of packaging materials is important from the following perspectives:

- Selection of an established packaging material for a new product formulation
- Comparison between packaging materials supplied by different vendors
- Evaluation of the suitability of a newly developed packaging material for an existing food product
- Ensuring the consistency of quality for the packaging materials supplied by an approved vendor
- Regular quality checks during the production of packaging materials and before using the material for processed food products

Testing of packaging can be classified into two categories:

- Testing of the materials used, e.g., paper, film, foil, laminate.
- Testing of the formed package, e.g., pouch, bottle, carton

The different tests for the quality evaluation of packaging materials and their intended purposes are listed in Table 15.9. Usually, the quality testing of packaging materials is carried out according to the standard and approved test procedures published by the American Society of Testing and Materials Standards (ASTM), Technical Association for the Pulp and Paper Industry (TAPPI), International Organization for Standardization (ISO), Federation of Corrugated Box Manufacturers (FCBM), or British Standards (BS). Alternatively, the methods approved by the national standards authority of the country are also followed.

15.5.1 Conditioning of Samples

The quality evaluation of packaging materials should be carried out under standard atmospheric conditions. The standard atmospheric conditions have been established country-wise. For countries including Argentina, Australia, Belgium, France, Germany, the Netherlands, New Zealand, and United Kingdom, the standard conditions are 20°C and 65% RH; in the United States, Canada, Burma, Mexico, and South Africa, the conditions are 23°C and 50% RH; in India, the standard conditions are 27°C and 65% RH (ISO 187, 1990). Thus, quality testing laboratories are designed and constructed to maintain the standard atmospheric conditions.

Before testing, the samples should be allowed to reach equilibrium. The time for sample equilibration is usually 24 h. Sample conditioning is vital as the properties of many packaging materials depend on the climatic conditions of the surroundings to which they are exposed. For instance, the physical properties of paper are affected by its moisture content which varies according to the RH and temperature of the surrounding atmosphere. The quality tests of packaging material are mostly based on the evaluation of their physical, mechanical, and barrier properties, which are discussed in the following sections.

TABLE 15.9

Tests for Quality Evaluation of Packaging Materials

Test Method	Purpose
Tensile strength and elongation	To evaluate the basic strength of the packaging material.
Heat/cold seal strength	To evaluate the strength of closing while the package is formed.
Interlayer bond strength	To evaluate the strength of holding different layers together.
Tear resistance	To evaluate the strength required to propagate a tear through the laminate.
Impact resistance	To evaluate the strength of a package to withstand an impact load.
Puncture resistance	To evaluate the strength required to puncture a pack using sharp protrusions.

15.5.2 Basic Quality Tests for Packaging Materials

15.5.2.1 Grammage

Grammage is one of the basic quality tests for packaging materials. It is the weight in grams of one square meter (1 m^2) of the sample and hence referred to as grams per square meter (GSM). Generally, papers, paperboards, foils, and films are sold and purchased on the basis of grammage. Hence, grammage is of significance to both the manufacturer and the consumer in defining price. Other properties such as bursting strength and thickness are specified with respect to the grammage of the packaging material.

At least 20 test pieces are cut from a minimum of five specimens. Each test piece shall have an area of not less than 50000 mm^2 (preferably 200 mm × 250 mm) and not more than 100000 mm^2. The area is determined for each test piece by calculation from measurements recorded to the nearest 0.5 mm. Each test piece is weighed on the balance, and the mass is expressed to three significant figures. The grammage (g) in g/m^2 to three significant figures for each test specimen is calculated using the following equation (ISO 536, 2012):

$$g = \frac{m}{A} \times 10^6 \tag{15.8}$$

where m is the mass of test piece (g), and A is the area of the test piece (mm^2).

15.5.2.2 Thickness

Thickness can be defined as the perpendicular distance between the two principal surfaces of the sample substrate. Thickness is an important property for paper, paperboard, foil, and laminates. Variations in thickness of the these materials can lead to a difference in their other properties such as stiffness, permeability, tensile strength, sealability, seal strength, moisture, gas, and light barrier properties.

For single-sheet thickness, maximum two test pieces of minimum dimensions 60 mm × 60 mm are cut from each specimen. At least 20 test pieces should be prepared. The thickness of the sample is measured by placing it between two points of a dial type or digital thickness gauge or micrometer. (ISO 534, 2011). A minimum of ten readings are recorded, and the mean thickness is expressed in micron, inches, or mil (0.001 in = 25 micron = 100 gauge = 1 mil).

15.5.2.3 Water Penetration

Testing the water penetration is important to evaluate the suitability of packaging materials especially paper and paperboards for applications which may involve exposure to water. The test for water penetration is known as the "Cobb test." During the test, the amount of water absorbed by the sample in a specified time under standard conditions is measured. Thus, water absorptiveness or Cobb value is defined as the mass of water absorbed in a specific time by 1 m^2 of the sample under 1 cm of water (TAPPI T-441, 2013).

The Cobb test (or water absorption) apparatus (Figure 15.18a) is a device which permits one side of the sample to be wetted uniformly at the moment the soaking period begins. It facilitates controlled rapid removal of the water from the sample after the test period. The specimen holder comprises a metal ring with a machined lower face of cross-sectional area 100 cm^2, height 2.5 cm (1 in), and thickness 0.6 cm, clamped to a flat base of the same or bigger size. The specimen is clamped to a rubber mat of size larger than the outside dimensions of the ring placed on the baseplate.

During the test, a weighed sample of defined dimension (e.g., 125 mm × 125 mm) is placed in the sample-holder device (Figure 15.18a), into which distilled water is poured to a height of 1 cm (TAPPI T-441, 2013). The area exposed to water is 100 cm^2. After the specified duration, the sample is removed, blotted, and reweighed. Usually, for most well-sized papers, the standard test time is 120 s on a single-sheet thickness when 100 mL of water is used to wet an area of 1 cm deep. For combined board, the

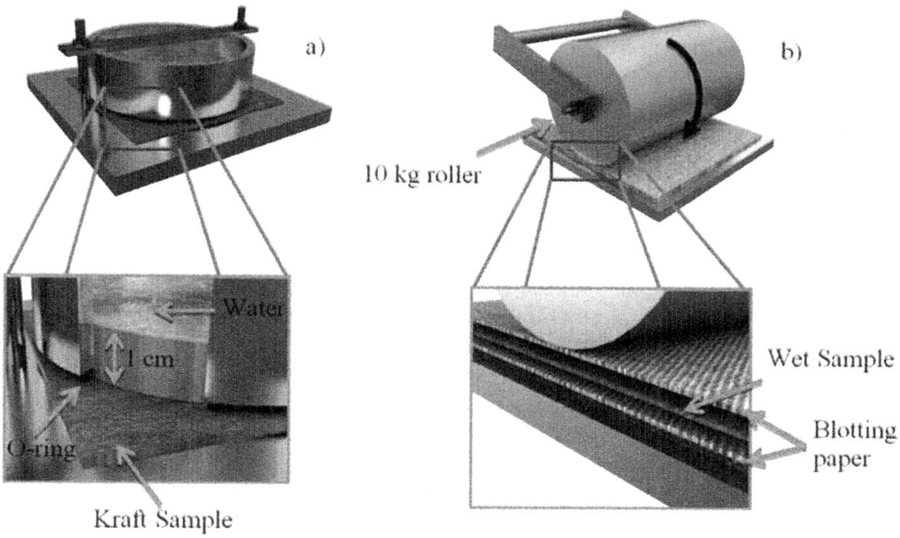

FIGURE 15.18 (a) Cobb test for water penetration; (b) drying of sample. (Reproduced with permission from Jimenez-Francisco, M., Caamal-Canche, J. A., Carrillo, J. G. and Cruz-Estrada, R. H. 2018. Performance assessment of a composite material based on kraft paper and a resin formulated with expanded PS waste: a case study from Mexico. *Journal of Polymers and the Environment* 26: 1573–1580.)

standard test time is 1800 s. The test duration may be varied according to the water absorptiveness and nature of the paper or paperboard and depending on the end-users' requirement.

Ten seconds before the end of the test, the water is removed carefully, and the sample is placed with the unexposed side on a flat, rigid surface. Subsequently, a sheet of absorbent or blotting paper is placed on the humid side so as to eliminate excess water on the surface. The time between the removal of excess water and blotting shall be 10 ± 2 s. Then, a metal roller is passed over the sample once back-and-forth without applying additional pressure on the roller (Figure 15.18b). The metal roller is made of stainless steel or other corrosion-resistant material, having a smooth face of width 20 cm and weight 10.0±0.5 kg. Finally, the sample is weighed on an analytical scale to the nearest 0.01 g. The difference in weight {[Final weight (g) − Conditioned weight (g)] × 100} gives the amount of water absorbed by the sample. The results are expressed in g/m^2.

15.5.3 Tensile Strength

Tensile strength is a mechanical property of packaging materials such as paper, board, and plastics. The test for tensile strength is carried out to determine the maximum load that can be applied to a material before it ruptures or tears. The tensile test is carried out at 23°C±1°C and 50% RH (ASTM D882-18, 2018). Knowledge of tensile properties is useful from the perspectives of research and development, engineering design, quality control, and specification. During the tensile test, the material demonstrates an elastic behavior up to a certain point and then ruptures. The tensile test can be modified to suit different requirements such as measuring a material's strength or elongation, tearing resistance and seal strength. The different methods under tensile strength testing are discussed in Sections 15.5.3.1–15.5.3.4.

15.5.3.1 Seal Strength

Seal strength of flexible laminates and formed pouches is a quantitative parameter to judge the ability of packaging processes to result in consistent seals. Thus, it is a measure of process validation, process control, and capability. Seal strength is also useful in evaluating the opening force and package integrity.

A calibrated tensile tester is used for the measurement. Samples (width: 1 in perpendicular to the seal) are cut from a formed pouch. Alternatively, laminates can be sealed under controlled temperature, pressure, and residence time to create sample seals. Subsequently, the test specimen is gripped in a tensile testing machine, with the seal area approximately equidistant between the grips. The grips are then separated at the rate of 8–12 in/min, and the maximum force required to separate the seal is reported as seal strength (ASTM F88/F88M-15, 2015).

15.5.3.2 Peel Test or Interlayer Bond Strength

The test for interlayer bond strength is applicable to ply adhesion of similar laminates made from flexible materials such as cellulose, paper, plastic film, and foil (ASTM F904, 2016). Improper bonding between the layers affects the performance of the packaging. Therefore, the performance of laminates is judged by their ability to function as a single element.

During the test, ply separation of the test sample is initiated mechanically by the application of heat or by using a solvent. The sample of separated plies having a width of 1-inch is then mounted between the rubber padded grips of a tensile testing machine. To prevent any slippage, the pressure on the grips is controlled pneumatically. Then, the grips are pulled apart at a constant rate (cm/min). The bond strength is measured as the force required to further separate the plies (ASTM F904, 2016). The program of the testing machine is set to record a defined number of data points per second, and the maximum load and the average load to break are reported.

15.5.3.3 Tear Strength

Tear strength test is highly relevant for evaluating the strength of papers and thin films (including plastic films and biofilms). Initiation of tear and tear propagation resistance are the two mechanical properties associated with tear resistance of plastic films. The force to initiate tearing of thin plastic films at very low rates of loading is related to its corresponding tensile strength (Briassoulis, 2006; Scarascia-Mugnozza et al., 2006). On the other hand, the tear propagation resistance measures the tearing behavior of a film after an initial localized failure due to accidental initial tearing, puncturing, and rupture due to penetration or impact. Therefore, tear propagation resistance should be adequate to protect the packaging against accidental damages (Briassoulis and Giannoulis, 2018). The required tear strength may be high or low depending on the end use of the packaging material.

The principle of this test is based on measuring the energy absorbed by the test sample in spreading a tear that has already been initiated by making a small cut in the test piece. Tear strength test is performed using the Elmendorf tearing tester which has two grips set side by side with only a small separation. One of the grips is stationary and mounted upright on the instrument base; whereas, the other grip is movable and mounted on a pendulum. In turn, the pendulum is mounted on a frictionless bearing and swings on a shaft. The sample of dimensions 50×62 mm is clamped in the two grips, and a cut is made using a sharp knife fixed on the tester. As the pendulum is released, it swings down on the precut sample. This indicates the residual energy lost in tearing. The tear strength is expressed in mN (millinewton) (ASTM D689-79, Part 20, 2003; the active standard is ASTM D689-17, 2017).

15.5.3.4 Puncture or Penetration Resistance

Loss of package integrity may result in gases and odors that contaminate the headspace of a food package (Briassoulis and Giannoulis, 2018). Measurement of penetration resistance is done using a puncture test fixture with pneumatic sample clamping (Figure 15.19). A conical probe is used according to the ASTM method (ASTM F1306-90 e1, 2008) and a cylindrical one is used according to the European Standard (EN) method (EN 14477, 2004). The penetration resistance of a package is measured at a constant speed of 25 mm/min (ASTM F1306-90 e1, 2008) and within the speed range of 1–100 mm/min (EN 14477, 2004).

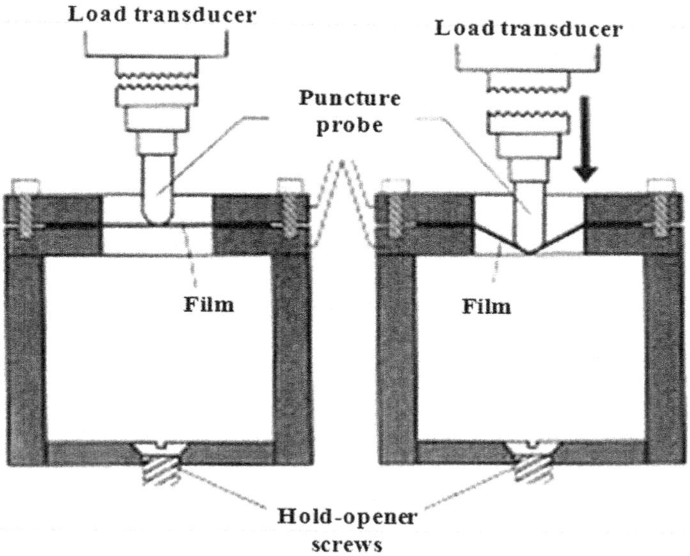

FIGURE 15.19 Schematic of the working of puncture strength testing equipment. (Reproduced with permission from Radebaugh, G. W., Murtha, J. L., Julian, T. N. and Bondi, J. N. 1988. Methods for evaluating the puncture and shear properties of pharmaceutical polymeric films. *International Journal of Pharmaceutics* 45: 39–46.)

15.5.4 Leak Test

Leak may be defined as an opening in a flexible package that either allows the contents of the package to escape or permits external substances to enter. Leak test is of relevance to judge the package integrity, especially of nonporous packages such as pouches, blister packs, cans, and bottles. Also, the test is useful to understand the performance of a package when subjected to low atmospheric pressures. A typical example could be the encountering of puffed pouches of dairy creamer and milk blister packs served with tea in aircraft; if the pouches are not strong enough, they would burst open under the low-pressure conditions inside the aircraft. There are two types of leak tests: the dry test and wet test.

15.5.4.1 Dry Test

In the dry test, the package is filled with the actual liquid product or more preferably colored water as the viscosity of water is usually lower than the actual product. The package is placed on a white tissue paper and kept inside a vacuum desiccator. The desiccator is closed and gradually increasing vacuum is applied to the tank to a predefined level. Then, the vacuum is maintained at that level for a specified time period and released after that. In case of leakage in the pouch, the product (or colored water) would come out and stain the white tissue paper underneath it. In case of bottles, they are placed inside the desiccator in an inverted position and tapped slightly from the cap. The cap is then removed to check if any product is visible on the rim of the bottle or the threads of the cap.

15.5.4.2 Wet Test

Wet test is used for packages containing a solid product such as a powder or which have a substantial amount of air in the headspace. Alternatively, the package can be filled with prior to and then checked for a leak during the test. Bottles are usually tested in an empty condition. The tank of the vacuum desiccator is filled with colored or normal water into which the package is dipped and restrained throughout the test. The tank is then closed, and steadily increasing vacuum is applied to the tank to a preset level.

The vacuum is then maintained at that level for a set time period during which leaks can be identified by visual observation. Detection of a constant stream of air bubbles indicates a confirmed leakage. After the preset time period, the vacuum is released. With semirigid or rigid packs, sometimes the air bubbles are random and less distinct. In this case, the packages are opened after the completion of the test and the product is checked for wet or stained spots. On releasing the vacuum, water enters the packs from the leak spots, which confirms leakage (ASTM F2338-09, 2013).

It is to be noted that the leak tests explained above are limited to detect leaks due to larger pinholes or gaps which can permit the fluid or gas to pass out or water to enter the package. Nevertheless, this test is not appropriate for detecting micro-leakages, for packages containing high viscosity fluids or vacuum-sealed packs.

15.5.5 Bursting Strength

Assessment of the bursting strength of paper, paperboard, and corrugated boxes holds significance with respect to evaluating the strength and toughness of the material. Bursting strength can be defined as the ability of the sample to absorb energy.

The test sample is fixed between the clamps of a bursting strength tester. The sample is subjected to gradually increasing hydrostatic pressure exerted perpendicular to the face of the sample through a rubber diaphragm till the sample cannot bear the pressure and bursts. The maximum or peak pressure required to rupture the sample is recorded by a pressure gauge as the bursting strength (TAPPI T 810 om-17, 2017).

15.5.6 Compression Strength

The primary purpose of a corrugated shipper or a carton is to protect the unit packages placed inside from excessive stress. Further, the carton is expected to withstand the stress of stack load in warehousing and during transportation. If the carton is stronger than the stack load, the products would remain intact. On the contrary, if the carton is weak, the carton will collapse and pass on the load to the unit packs placed inside. The carton compression specification is given by the product of the stacking load and safety factor. Usually, the safety factor ranges between 3 and 7. It depends on multiple factors including product fragility, product cost, distribution cycle, and the period and methods for transportation. Thus, the test for compression strength is considered as a performance test for cartons. Good compression strength signifies a good carton.

The test sample (carton) is placed between two parallel platens of a compression tester and pressed. The maximum load it can bear before the carton collapses is recorded as the compression strength (TAPPI T 804 om-12, 2012).

15.5.7 Drop Test

An important performance parameter for packaging is its ability to withstand accidental drops due to incorrect handling practices. Unit packs may be accidentally dropped during various stages of production, packing, loading, and even while being arranged on the retail store shelves or being picked up from the shelf by the consumer. Similarly, the secondary or transport packs may encounter accidental drops in the warehouse or during transportation. In all the cases, the package is expected to remain intact and hold the product. The drop test is used

- To evaluate the capability of a container to withstand the sudden shock resulting from a free fall.
- To evaluate the capability of a container and its inner packing to protect its contents during the sudden shock resulting from a free fall.
- To test the durability of internal packages.
- To test the effectiveness of partitioning and cushioning arrangements.

- To evaluate the product fragility.
- To determine the drop failure height; this can be used to specify the maximum storage height of the packed items in warehouses.

Unit packs or secondary packages are dropped from different heights and at different orientations to evaluate the durability. Generally, the test height is the height from which the product is expected to be dropped in practical situations or slightly higher than it. Since heavier packages are handled at lower heights, these are dropped from lower heights during the test. Similarly, lighter weight products are placed higher in shelves or stacked higher and hence dropped from higher heights (ASTM D5276-98, 2017).

15.5.8 Grease Resistance

Grease resistance test is important for packaging materials that contain fat-rich food products such as butter, ghee, and oil. This test is carried out by adding 5 g of sand on the specimen through a hollow cylinder metallic piece and then topping the sand with 1.1 mL of colored turpentine dye. These contents are then placed on a sheet of white paper. At specified intervals, the white paper is examined for the first spot after which the experiment is discontinued. The time between the application of turpentine dye and appearance of the first spot is recorded as transudation time in seconds (ASTM D722-93, 2007).

15.5.9 Permeation Tests

The previous section on mass transfer explained the significance of the permeability of plastic films and laminates to gases, water vapor, low-molecular-weight compounds such as aromas, and flavor and solutes like additives present in the food product.

The ASTM desiccant method (ASTM E96-80) is generally used for the measurement of WVTR of a film. Anhydrous calcium chloride is placed in the bottom of the test dish, and molten wax is used to seal the preconditioned film samples to the dish. Moisture permeation is calculated by weighing the dishes to the nearest 0.0001 g. The WVTR is calculated using the following equation (ASTM E96-80, 1990):

$$\text{WVTR} = \frac{\Delta m \times \Delta T}{A} \tag{15.9}$$

where Δm is the weight gain of the cell with calcium chloride (g), A is the film area (m^2), and ΔT is the test time (days). In turn, the water vapor permeability coefficient is calculated as

$$\text{WVPC} = \frac{\text{WVTR} \times l}{\Delta P} \tag{15.10}$$

where l is the thickness of the film (m), and ΔP is the difference in the partial pressure of water vapor across the film (Pa).

On the other hand, the OTR is generally analyzed using the ASTM D3985 standard method (ASTM, 1995). OTR is the measurement of the ease with which the oxygen passes through a film when subjected to a partial pressure of oxygen across the film. It is expressed as

$$\text{OTR} = \frac{Q}{At\Delta P} \tag{15.11}$$

where Q is the quantity of oxygen molecules passing through the film, A is the surface area of exposure of the film, and t is the time of exposure of the film at steady state under a partial pressure difference (ΔP) in oxygen on the either sides of the film (Massey, 2003).

Specimen of defined measurement area (e.g., 50 cm^2) is mounted onto the diffusion cells inside which 100% oxygen (test gas) is routed through one side of the film sample. A mixture of 98% N_2 and 2% H_2

(carrier gas) is passed on the other side. As the oxygen permeates through the film sample, it is picked up by the carrier gas. Subsequently, the carrier gas is passed through a coulometric oxygen sensor to determine the amount of oxygen contained in it at equilibrium (Mokwena et al., 2009).

15.6 Problems to Practice

15.6.1 Multiple Choice Questions

1. What are the primary functions of packaging?
    - i. containment
    - ii. communication
    - iii. protection
    - iv. convenience
        - a. i, ii, and iii
        - b. ii, iii, and iv
        - c. i, ii, and iv
        - d. i, ii, iii, and iv

Answer: a

2. What are the secondary functions of packaging?
    - i. convenience
    - ii. communication
    - iii. traceability
    - iv. tamper-evidence
        - a. i, ii, and iii
        - b. i, iii, and iv
        - c. i, ii, and iv
        - d. i, ii, iii, and iv

Answer: b

3. The exact definition of packaging is
    - a. reflecting the dynamic nature of the product it contains
    - b. the procedure for creating a package
    - c. a coordinated system of preparing goods for transport, distribution, storage, and retailing
    - d. a physical form that is intended to contain, protect, transport, and inform

Answer: c

4. The statement which precisely defines the primary packaging is
    - i. the first wrap or containment
    - ii. directly holds product for sale
    - iii. grouping of product for bulk handling
    - iv. No decorations
        - a. i and ii
        - b. i and iii
        - c. i and iv
        - d. ii and iii

Answer: a

5. The correct order in ascending form for the following layers of packaging material is
   i. instant coffee in LDPE pouch
   ii. corrugated brown carton
   iii. paperboard carton
      a. iii, i, and ii
      b. i, iii, and ii
      c. ii, i, and iii
      d. i, ii, and iii

Answer: b

6. The packaging system used to package ultra-heat-treated milk is
   a. MAP
   b. intelligent packaging
   c. aseptic packaging
   d. active packaging

Answer: c

7. An example of tertiary packaging among the following options is
   a. a crate of beer bottles
   b. a packet of dried apricot
   c. a box of chocolates
   d. a shrink-wrapped pallet of pineapple juice

Answer: d

8. The chemically inert packaging material among the following is
   a. glass
   b. metal
   c. paper
   d. plastics

Answer: a

9. The function of a TTI is
   a. tampering-evidence
   b. containment
   c. communication
   d. protection

Answer: c

10. The most appropriate type of active packaging for fruits is
   a. oxygen absorbers
   b. moisture absorbers
   c. ethylene absorber
   d. carbon dioxide absorbers

Answer: c

11. The combination of gases used in MAP is
    a. oxygen, argon, and nitrogen
    b. helium, carbon dioxide, and argon
    c. argon, helium, and nitrogen
    d. oxygen, carbon dioxide, and nitrogen

Answer: d

12. WVTR and OTR, respectively, stand for
    a. water-to-vapor transient rate and odor transfer rate
    b. water vapor transmission rate and oxygen transmission rate
    c. water vapor transfer rate and oxygen testing result
    d. water-to-vapor total ratio and odor testing result

Answer: b

13. The unit of WVTR is
    a. $g/cm^2/24\,h$
    b. g/cm^2
    c. $g/24\,h$
    d. g/m^2

Answer: a

14. The acronym ASTM stands for
    a. Asian Society for Testing and Materials Standards
    b. American Society for Trade and Marketing Standards
    c. African Society for Traceability and Manufacturing Standards
    d. American Society for Testing and Materials Standards

Answer: d

15. CAP involves
    a. monitoring and controlling the gas composition inside the package
    b. introducing a controlled gas mixture composition into the package
    c. introducing air into the package
    d. both b and c

Answer: a

16. The correct definition for aseptic packaging is
    a. the filling of a commercially sterile product into sterile containers under aseptic conditions and hermetically sealing the containers
    b. the filling of a processed product into containers under aseptic conditions and hermetically sealing the containers
    c. the filling of a commercially sterile product into sterile containers under aseptic conditions and sealing the containers
    d. the filling of a commercially sterile product into containers and sealing the containers

Answer: a

17. The most desirable polymer–clay formation of MMT for bionanocomposite application is
    a. phase-separated microcomposite
    b. tactoid
    c. exfoliated
    d. intercalated

Answer: c

18. A secondary package is one which
    a. directly holds the sales unit of the product
    b. contains the primary package
    c. protects the product during transportation
    d. protects the product during distribution

Answer: b

19. The chemical used in aseptic packaging is
    a. hydrogen peroxide
    b. sodium sulfate
    c. sulfur dioxide
    d. potassium permanganate

Answer: a

20. The test for water penetration of packaging materials in
    a. Elmendorf test
    b. Cobb test
    c. static test
    d. peel test

Answer: b

15.6.2 Numerical Problems

1. At a uniform temperature of 20°C, oxygen gas at 4 bar and 1 bar is separated by a polyethylene film of thickness 0.5 mm. The diffusion coefficient of oxygen in polyethylene is 4.5×10^{-11} m²/s. The solubility of oxygen in the film is 1.9709×10^{-7} mol/m³ Pa. Calculate the molar concentration of oxygen on either side of the polyethylene film and the molar flux of oxygen.

   Given
   i. Temperature = 20°C
   ii. $P_{\text{inner}} = 4$ bar; $P_{\text{outer}} = 1$ bar
   iii. Film thickness (x) = 0.5 mm = 0.5×10^{-3} m = 0.0005 m
   iv. $D_{O_2} = 4.5 \times 10^{-11}$ m²/s
   v. Solubility of oxygen in polyethylene = 1.9709×10^{-7} mol/m³ Pa = 1.9709×10^{-2} mol/m³ bar (since, 1 Pa = 10^{-5} bar)

   To find
   i. The molar concentration of oxygen on either side of the film
   ii. The molar flux of oxygen

Solution

Molar concentration of oxygen on inner side$\left(c_{O_{2inner}}\right)$ = Solubility × Inner pressure

$$= 1.9709 \times 10^{-2} \times 4 = 0.0788\,\text{mol/m}^3$$

Molar concentration of oxygen on outer side$\left(c_{O_{2outer}}\right)$ = Solubility × Outer pressure

$$= 1.9709 \times 10^{-2} \times 1 = 0.0197\,\text{mol/m}^3$$

$$\text{Molar flux of oxygen}\left(J_{O_2}\right) = \frac{D_{O_2}}{x}\left(c_{O_{2inner}} - c_{O_{2outer}}\right)$$

$$= \frac{4.5 \times 10^{-11}}{0.5 \times 10^{-3}}(0.0788 - 0.0197)$$

$$= 0.53 \times 10^{-8}\,\text{mol}\,\text{s}^{-1}\,\text{m}^{-2}$$

Answer: (i) molar concentration of oxygen = 0.0788 mol/m^3 & 0.0197 mol/m^3; (ii) molar flux of oxygen = 0.53 × 10^{-8} mol/s m^2

2. Instant coffee powder is packed in a LDPE pouch. Calculate the aroma release rate if the mean diameter of instant coffee powder is 300 µm, the film thickness is 30 µm, the concentration of the aroma compound is 0.2 and 0.02 g/cm^3 on the inside and outside surface of the LDPE pouch, respectively, and the diffusion coefficient of the aroma compound through the packaging material is 0.5 × 10^{-10} m^2/s.

   Given

   i. Mean diameter of instant coffee powder particles = 300 µm = 300 × 10^{-6} m
   ii. Film thickness = 30 µm = 30 × 10^{-6} m
   iii. Concentration of aroma compound on the inner surface of the LDPE pouch = 0.2 g/cm^3 = 200 kg/m^3
   iv. Concentration of aroma compound on the inner surface of the LDPE pouch = 0.02 g/cm^3 = 20 kg/m^3
   v. Diffusion coefficient of the aroma compound through LDPE = 0.5 × 10^{-10} m^2/s

   To find: Aroma release rate

   Solution

$$\text{Area of spherical instant coffee particles}(A) = 4\pi r^2 = 4 \times \pi \times \left(\frac{300 \times 10^{-6}}{2}\right)^2 = 2.827 \times 10^{-7}\,\text{m}^2$$

$$\text{Aroma release rate} = \text{Molar flux} \times \text{Area} = J \times A = \frac{A \times D}{x}(c_i - c_o)$$

$$= \frac{2.827 \times 10^{-7} \times 0.5 \times 10^{-10}}{30 \times 10^{-6}}(200 - 20) = 8.481 \times 10^{-11}\,\text{kg/s}$$

Answer: Aroma release rate = 8.481 × 10^{-5} mg/s

3. Frozen meat is packaged in 0.1 mm-thick polypropylene film. The permeability of polypropylene to O_2 and water vapor is 0.5 × 10^{-8} cm^3/(s cm^2 atm/cm) and 4 × 10^{-15} g/(s cm^2 Pa/cm), respectively. Inside the package, the partial pressure of O_2 is 0 atm and that of water vapor is 0.1 Pa, whereas outside the package, the partial pressure of O_2 is 0.24 atm and that of water

vapor is 0.05 Pa. Calculate the amount of oxygen and water vapor that will permeate through the polypropylene film of 500 cm^2 surface area in 5 days.

Given

i. Thickness of polypropylene film = 0.1 mm = 0.1×10^{-3} m = 0.0001 m
ii. t = 5 days = $5 \times 24 \times 3600$ s = 432000 s
iii. Permeability of polypropylene to O_2 (P_{O_2}) = 0.5×10^{-8} cm^3/(s cm^2 atm/cm) = 0.5×10^{-12} m^3/(s m^2 atm/m)
iv. Permeability of polypropylene to water vapor (P_{wv}) = 4×10^{-15} g/(s cm^2 Pa/cm) = 0.405×10^{-10} kg/(s m^2 atm/m)
v. Surface area of the film (A) = 500 cm^2 = 500×10^{-4} m^2 = 0.05 m^2
vi. Partial pressure of oxygen outside the package = 0.24 atm
vii. Partial pressure of oxygen inside the package = 0 atm
viii. Partial pressure of water vapor outside the package = 0.05 Pa = 4.935×10^{-7} atm
ix. Partial pressure of water vapor inside the package = 0.1 Pa = 9.8692×10^{-7} atm

To find: Amount of oxygen (m_{O2}) and water vapor (m_{wv}) that will permeate through the polypropylene film.

Solution

i. Amount of oxygen (m_{O2}) that will permeate through the polypropylene film:

$$J_{O_2} = \frac{P \times (p_1 - p_2)}{x} = \frac{0.5 \times 10^{-12} \times (0.24 - 0)}{0.0001} = 1200 \times 10^{-12}\ \text{mol/m}^2\,\text{s}$$

$$m_{O_2} = J_{O_2} \times A \times t = 1200 \times 10^{-12} \times 0.05 \times 432000\,\text{s} = 2.592 \times 10^{-5}\ \text{mol} = 25.92\,\mu\text{mol}$$

ii. Amount of water vapor (m_{wv}) that will permeate through the polypropylene film:

$$J_{wv} = \frac{P \times (p_1 - p_2)}{x} = \frac{0.405 \times 10^{-10} \times \left(\{9.692 \times 10^{-7}\} - \{4.935 \times 10^{-7}\}\right)}{0.0001} = 1.927 \times 10^{-13}\ \text{kg/m}^2\,\text{s}$$

$$m_{wv} = J_{wv} \times A \times t = 1.927 \times 10^{-13} \times 0.05 \times 432000\,\text{s} = 4.162 \times 10^{-9}\ \text{kg} = 4.162\,\mu\text{g}$$

Answer: Amount of oxygen (m_{O2}) and water vapor (m_{wv}) that will permeate through the polypropylene film is 25.92 μmol and 4.162 μg, respectively.

4. Meat is packaged in a plastic container covered with PVC film. Calculate the required film thickness so that 0.5×10^{-3} mol of oxygen enters the package through the film in 36 h. The permeability of PVC to oxygen is 456×10^{-11} cm^3/(s cm^2 atm/cm), the surface area of the film is 250 cm^2, and the partial pressure of oxygen inside and outside the package is 0 and 0.1 atm, respectively.

Given

i. Amount of oxygen (m) that enters through the film = 0.5×10^{-3} mol
ii. t = 36 h = 129600 s
iii. Oxygen permeability through PVC film (P) = 456×10^{-11} cm^3/(s cm^2 atm/cm) = 456×10^{-15} m^3/(s m^2 atm/m)
iv. Surface area of the film (A) = 250 cm^2 = 250×10^{-4} m^2 = 0.025 m^2
v. Partial pressure of oxygen outside the package (p_1) = 0.1 atm
vi. Partial pressure of oxygen inside the package (p_2) = 0 atm

To find: Film thickness

Solution

$$J = \frac{m}{A \times t} = \frac{0.5 \times 10^{-3}}{0.025 \times 129600} = 1.543 \times 10^{-7} \text{ mol/m}^2\text{ s}$$

$$x = \frac{P \times (p_1 - p_2)}{J} = \frac{456 \times 10^{-15} \times (0.1 - 0)}{1.543 \times 10^{-7}} = 29.55 \times 10^{-8} \text{ m} = 0.295 \mu\text{m}$$

Answer: Thickness of the film should be 0.295 μm

5. Potato chips are packed in a plastic bag flushed with nitrogen. Calculate the amount of N_2 that will diffuse out of the plastic bag in 1 month if the thickness of the film is 500 μm, the absolute pressure of N_2 in the bag is 1.1 atm, the surface area of the bag is 0.1 m^2, and the permeability of the plastic film to N_2 is 1.5×10^{-10} cm^3/s cm^2 atm/cm.

Given

i. Thickness of plastic film = 500 μm = 500×10^{-6} m
ii. t = 1 month = 30 days = 30 × 24 × 3600 = 2592000 s
iii. Permeability of plastic film to nitrogen (P_{N_2}) = 1.5×10^{-10} cm^3/s cm^2 atm/cm = 1.5×10^{-14} m^3/(s m^2 atm/m)
iv. Surface area of the film (A) = 0.1 m^2
v. Absolute pressure of N_2 in the bag = 1.1 atm

To find: Amount of N_2 that will diffuse out of the plastic bag

Solution

Assuming outside pressure to be 1 atm

$$J_{N_2} = \frac{P_{N_2} \times (p_1 - p_2)}{x} = \frac{1.5 \times 10^{-14} \times (1.1 - 1)}{500 \times 10^{-6}} = 3 \times 10^{-12} \text{ m/s}$$

$$m_{N_2} = J_{N_2} \times A \times t = 3 \times 10^{-12} \times 0.1 \times 2592000 = 7.776 \times 10^{-7} \text{ m}^3$$

$$= 7.776 \times 10^{-7} \times 1.165 \text{ kg/m}^3 = 9.059 \times 10^{-7} \text{ kg}$$

$$\left(\text{Since, } \rho_{N_2} = 1.165 \text{ kg/m}^3\right)$$

$$1 \text{ kg of } N_2 = 35.697 \text{ moles}$$

$$\therefore 9.059 \times 10^{-7} \text{ kg of } N_2 = 3.233 \times 10^{-5} \text{ moles of } N_2$$

Answer: Amount of N_2 that will diffuse out of the plastic bag is 3.233×10^{-5} mol

BIBLIOGRAPHY

Alexandre, M. and Dubois, P. 2000. Polymer-layered silicate nanocomposites preparation, properties and uses of a new class of materials. *Materials Science and Engineering* 28: 1–63.

Almasi, H., Ghanbarzadeh, B. and Entezami, A. A. 2010. Physicochemical properties of starch–CMC–nanoclay biodegradable films. *International Journal of Biological Macromolecules* 46: 1–5.

Anandharamakrishnan, C. and Kolli, U. K. 2015. Bionanocomposites and their potential applications in food packaging. In *Polymers for Packaging Applications*, eds. S. Alavi, S. Thomas, K. P. Sandeep, N. Kalarikkal, J. Varghese and S. Yaragalla, 229–262. Boca Raton, FL: CRC Press, Taylor & Francis Group & Oakville, ON: Apple Academic Press, Inc.

Angellier-Coussy, H., Torres-Giner, S., Morel, M., Nathalie Gontard, N. and Gastaldi, E. 2008. Functional properties of thermoformed wheat gluten/montmorillonite materials with respect to formulation and processing conditions. *Journal of Applied Polymer Science* 107: 487–496.

Anonymous. n.d. PDF417: The new symbol of data management [white paper]. Holtsville, NY: Symbol Technologies Inc.

Arora, A. and Padua, G. W. 2010. Review: nanocomposites in food packaging. *Journal of Food Science* 75: R43–R49.

ASTM 1995. Standard test method for oxygen gas transmission rate through plastic film and sheeting using a coulometric sensor. In *Annual Book of ASTM Standards*, 532–537. Philadelphia, PA: American Society for Testing Materials (ASTM).

ASTM 2009. Designation: d1434-82 (reapproved 2009): standard test method for determining gas permeability characteristics of plastic film and sheeting. In *Annual Book of ASTM Standards* (Volume 15.10). West Conshohocken, PA: ASTM International.

ASTM D5276-98 2017. *Standard Test Method for Drop Test of Loaded Containers by Free Fall*. West Conshohocken, PA: ASTM International.

ASTM D689-17 2017. *Standard Test Method for Internal Tearing Resistance of Paper*. West Conshohocken, PA: ASTM International.

ASTM D689-79 2003. *Standard Test Method for Internal Tearing Resistance of Paper (Withdrawn 2009)*. West Conshohocken, PA: ASTM International.

ASTM D722-93 2007. *Standard Test Method for Grease Resistance of Paper (Withdrawn 2010)*. West Conshohocken, PA: ASTM International.

ASTM D882-18 2018. *Standard Test Method for Tensile Properties of Thin Plastic Sheeting*. West Conshohocken, PA: ASTM International.

ASTM E96-80 1990. Standard test method for water vapour transmission of materials. In *Annual Book of ASTM Standards*. Philadelphia, PA: American Society for Testing and Materials.

ASTM F1306-90 e1 2008. *Standard Test Method for Slow Rate Penetration Resistance of Flexible Barrier Films and Laminates*. West Conshohocken, PA: ASTM International.

ASTM F2338-09 2013. *Standard Test Method for Nondestructive Detection of Leaks in Packages by Vacuum Decay Method*. West Conshohocken, PA: ASTM International.

ASTM F88/F88M-15 2015. *Standard Test Method for Seal Strength of Flexible Barrier Materials*. West Conshohocken, PA: ASTM International.

ASTM F904 2016. *Standard Test Method for Comparison of Bond Strength or Ply Adhesion of Similar Laminates Made from Flexible Materials*. West Conshohocken, PA: ASTM International.

Avérous, L. and Pollet, E. 2012. Biodegradable polymers. In *Environmental Silicate Nano-Biocomposites, Green Energy and Technology*, eds. L. Avérous and E. Pollet, 13–39. London: Springer-Verlag.

Bae, H. J., Park, H. J., Hong, S. I., Byun, Y. J., Darby, D. O., Robert, M. and Whiteside, W. S. 2009. Effect of clay content, homogenization RPM, pH, and ultrasonication on mechanical and barrier properties of fish gelatin/montmorillonite nanocomposite films. *LWT—Food Science and Technology* 42: 1179–1186.

Bibi, F., Guillaume, C., Gontard, N. and Sorli, B. 2017. A review: RFID technology having sensing aptitudes for food industry and their contribution to tracking and monitoring of food products. *Trends in Food Science & Technology* 62: 91–103.

Briassoulis, D. 2006. Mechanical behaviour of biodegradable agricultural films under real field conditions. *Polymer Degradation and Stability* 91: 1256–1272.

Briassoulis, D. and Giannoulis, A. 2018. Evaluation of the functionality of bio-based food packaging films. *Polymer Testing* 69: 39–51.

Brody, A. L. 2000. The when and why of aseptic packaging. *Food Technology* 54: 101–102.

Brody, A. L., Bugusu, B., Han, J. H., Sand, C. K. and McHugh, T. H. 2008. Innovative food packaging solutions. *Journal of Food Science* 73: R107–R116.

Brody, A. L., Strupinsky, E. R. and Kline, L. R. 2001. *Active Packaging for Food Applications*. Lancaster: Technomic Publishing.

Carrado, K. A. 2003. Synthetic organo- and polymer-clays: preparation, characterization, and materials applications. *Applied Clay Science* 17: 1–23.

Castello, R., Ferreira, A. R., Costa, N., Fonseca, I. M., Alves, V. D. and Coelhoso, I. M. 2010. Nanocomposite films obtained from carrageenan/pectin biodegradable polymers. *Presented at International Conference on Food Innovation. Food Innova.* 25–29.

Coles, R. 2003. Introduction. In *Food packaging technology*, eds. R. Coles, D. McDowell and M. J. Kirwan, 1–31. London, UK: Blackwell Publishing, CRC Press.

Cooksey, K. 2005. Effectiveness of antimicrobial food packaging materials. *Food Additives & Contaminants* 22: 980–987.

Cruz, R. S., Camilloto, G. P. and Pires, A. C. D. S. 2012. Oxygen scavengers: an approach on food preservation. In *Structure and Function of Food Engineering*, ed. A. A. Eissa. London, UK: IntechOpen. www.intechopen.com/books/structure-and-function-of-food-engineering/oxygen-scavengers-an-approach-on-food-preservation (accessed October 17, 2018).

de Azeredo, H. M. C. 2009. Nanocomposites for food packaging applications. *Food Research International* 42: 1240–1253.

De Kruijf, N., Van Beest, M., Rijk, R., Sipiläinen-Malm, T., Losada, P. P. and De Meulenaer, B. 2002. Active and intelligent packaging: applications and regulatory aspects. *Food Additives & Contaminants* 19: 144–162.

Duncan, T. V. 2011. Applications of nanotechnology in food packaging and food safety: barrier materials, antimicrobials, and sensors. *Journal of Colloid and Interface Science* 363: 1–24.

Ebadi, M., Farsi, M., Narchin, P. and Madhoushi, M. 2015. The effect of beverage storage packets (Tetra Pak™) waste on mechanical properties of wood–plastic composites. *Journal of Thermoplastic Composite Materials* 29: 1–10.

EN 14477 2004. Packaging—Flexible tubes—Determination of puncture resistance-test method. www.cen.eu/Pages/default.aspx/ (accessed October 17, 2018).

Fang, Z., Zhao, Y., Warner, R. D., Johnson, S. K. 2017. Active and intelligent packaging in meat industry. *Trends in Food Science & Technology* 61: 60–71.

FDA (Food and Drug Administration) 2001. *Fish & Fisheries Products Hazards & Control Guidance*. Rockville, MD: Centre for Safety and Applied Nutrition, Department of Health and Human Services.

Fellows, P. and Axtell, B. 2002. *Appropriate Food Packaging: Materials and Methods for Small Businesses*. Essex, UK: ITDG Publishing.

Floros, J. D. and Matsos, H. I. 2005. Introduction to modified atmosphere packaging. In *Innovations in Food Packaging*, ed. J. H. Han, 159–171. London: Elsevier Academic Press.

Gaikwad, K. K., Singh, S. and Lee, Y. S. 2018. Oxygen scavenging films in food packaging. *Environmental Chemistry Letters* 16: 523–528.

Gospodinov, D., Stefanov, S. and Hadjiiski, V. 2011. Use of the finite element method in studying the influence of different layers on mechanical characteristics of corrugated paperboard. *Tehnicki Vjesnik/Technical Gazette* 18: 357–361.

Han, J. H. and Scanlon, M. G. 2014. Mass transfer of gas solute through packaging materials. In *Innovations in Food Packaging*, ed. J. H. Han, 37–50. London, UK: Academic Press.

Harima, Y. 1990. *Food Packaging*, 229–252. London, UK: Academic Press.

Holte, B. 1993. An apparatus for indicating the presence of CO_2 and a method of measuring and indicating bacterial activity within a container or bag. PCT International Patent Application WO 93/15402.

Hong, S. and Park, W. 2000. Use of color indicators as an active packaging system for evaluating kimchi fermentation. *Journal of Food Engineering* 46: 67–72.

Horan, T. J. 2000. Method for determining deleterious bacterial growth in packaged food utilizing hydrophilic polymers. US Patent 6149952.

Hu, C. T., Liu, C. K., Huang, M. W., Syue, S. H., Wu, J. M., Chang, Y. S., Yeh, J. W. and Shih, H. C. 2009. Plasma-enhanced chemical vapor deposition carbon nanotubes for ethanol gas sensors. *Diamond and Related Materials* 18: 472–477.

Hurme, E., Sipilainen-Malm, T., Ahvenainen, R. and Nielsen, T. 2002. Active and intelligent packaging. In *Minimal Processing Technologies in the Food Industry*, eds. T. Ohlsson and N. Bengtsson, 87–123. Cambridge, UK: Woodhead Publishing Limited.

ISO 187 1990. Paper, board and pulps—Standard atmosphere for conditioning and testing and procedure for monitoring the atmosphere and conditioning of samples (Last reviewed and confirmed in 2013).

ISO 534 2011. Paper and board—Determination of thickness, density and specific volume (reviewed and confirmed in 2017).

ISO 536 2012. Paper and board—Determination of grammage (Last reviewed and confirmed in 2018).

Jabeen, N., Majid, M. and Nayik, G. A. 2015. Bioplastics and food packaging: a review. *Cogent Food & Agriculture* 1: 1117749.

Jimenez-Francisco, M., Caamal-Canche, J. A., Carrillo, J. G. and Cruz-Estrada, R. H. 2018. Performance assessment of a composite material based on kraft paper and a resin formulated with expanded polystyrene waste: a case study from Mexico. *Journal of Polymers and the Environment* 26: 1573–1580.

Jindal, N. and Khattar, J. S. 2018. Microbial polysaccharides in food industry. In *Handbook of Food Bioengineering, Biopolymers for Food Design*, eds. A. M. Grumezescu and A. M. Holban, 95–123. London, UK: Academic Press.

Kader, A. A., Zagory, D. and Kerbel, E. L. 1989. Modified atmosphere packaging of fruits and vegetables. *Critical Reviews in Food Science and Nutrition* 28: 1–30.

Kerry, J. P., O'Grady, M. N. and Hogan, S. A. 2006. Past, current and potential utilisation of active and intelligent packaging systems for meat and muscle-based products: a review. *Meat Science* 74: 113–130.

Kumar, P., Sandeep, K. P., Alavi, S., Truong, V. D. and Gorga, R. E. 2010. Preparation and characterization of bio-nanocomposite films based on soy protein isolate and montmorillonite using melt extrusion. *Journal of Food Engineering* 100: 480–489.

Labuza, T. P. 1996. An introduction to active packaging for foods. *Food Technology* 50: 68–71.

Lee, D. S., Yam, K. L. and Piergiovanni, L. 2008. *Food Packaging Science and Technology*. Boca Raton, FL: CRC Press.

Lopez-Rubio, A., Almenar, E., Hernandez-Munoz, P., Lagaron, J. M., Catala, R. and Gavara, R. 2004. Overview of active polymer based packaging technologies for food applications. *Food Reviews International* 20: 357–387.

Maksimov, R. D., Lagzdins, A., Lilichenko, N. and Plume, E. 2009. Mechanical properties and water vapor permeability of starch/montmorillonite nanocomposites. *Polymer Engineering and Science* 49: 2421–2429.

Mangaraj, S. and Goswami, T. K. 2009. Modified atmosphere packaging—An ideal food preservation technique. *Journal of Food Science and Technology* 46: 399–410.

Marsh, K. and Bugusu, B. 2007. Food packaging—Roles, materials, and environmental issues. *Journal of Food Science* 72: R39–R55.

Martucci, J. F. and Ruseckaite, R. A. 2010. Biodegradable bovine gelatin/Na+-montmorillonite nanocomposite films. Structure, barrier and dynamic mechanical properties. *Polymer-Plastics Technology and Engineering* 49: 581–588.

Massey, L. L. 2003. *Permeability Properties of Plastics and Elastomers: A Guide to Packaging and Barrier Materials*. PDL Handbook Series. Norwich, New York: Plastic Design Library/William Andrew Publishing.

Matche, R. S. 2018. Packaging technologies for fruit juices. In *Fruit Juices Extraction, Composition, Quality and Analysis*, eds. G. Rajauria and B. K. Tiwari, 637–666. London: Academic Press.

Mihindukulasuriya, S. D. F. and Lim, L. T. 2014. Nanotechnology development in food packaging: a review. *Trends in Food Science & Technology* 40: 149–167.

Miltz, J., Passy, N. and Mannheim, C. H. 1995. Trends and applications of active packaging systems. In *Food and Packaging Materials—Chemical Interaction* (Volume 162), eds. P. Ackerman, M. Jagerstad and M. Ohlsson, 201–210. London, England: The Royal Society of Chemistry.

Mokwena, K. K., Tang, J., Dunne, C. P., Yang, T. C. S. and Chow, E. 2009. Oxygen transmission of multilayer EVOH films after microwave sterilization. *Journal of Food Engineering* 92: 291–296.

Naz, R. 2018. Storage in polyethylene terephthalate bottles: changes and shelf life. In *Fruit Juices Extraction, Composition, Quality and Analysis*, eds. G. Rajauria and B. K. Tiwari, 621–635. London: Academic Press.

Nopwinyuwong, A., Trevanich, S. and Suppakul, P. 2010. Development of a novel colorimetric indicator label for monitoring freshness of intermediate-moisture dessert spoilage. *Talanta* 81: 1126–1132.

Okada, A. and Usuki, A. 2006. Twenty years of polymer-clay nanocomposites. *Macromolecular Materials and Engineering* 291: 1449–1476.

Parry, R. T. 1993. Introduction. In *Principles and Applications of Modified Atmosphere Packaging of Foods*, ed. R. T. Parry, 1–17. London: Chapman & Hall.

Pearce, S. and Bushnell, R. D. 1997. *The Bar Code Implementation Guide: Using Bar Codes in Distribution*. Surf City, NJ: Quad II Inc.

Pillai, S. D. and Shayanfar, S. 2015. Aseptic packaging of foods and its combination with electron beam processing. In *Electron Beam Pasteurization and Complementary Food Processing Technologies*. Woodhead Publishing Series in Food Science, Technology and Nutrition, eds. S. D. Pillai and S. Shayanfar, 83–93. Cambridge, UK: Woodhead Publishing.

Quintavalla, S. and Vicini, L. 2002. Antimicrobial food packaging in meat industry. *Meat Science* 62: 373–380.

Radebaugh, G. W., Murtha, J. L., Julian, T. N. and Bondi, J. N. 1988. Methods for evaluating the puncture and shear properties of pharmaceutical polymeric films. *International Journal of Pharmaceutics* 45: 39–46.

Rasal, R. M., Janorkar, A. V. and Hirt, D. E. 2010. Poly(lactic acid) modifications. *Progress in Polymer Science* 35: 338–356.

Ray, S. S. and Okamoto, M. 2003. Polymer/layered silicate nanocomposites: a review from preparation to processing. *Progress in Polymer Science* 28: 1539–1641.

Rhim, J. W. 2011. Effect of clay contents on mechanical and water vapor barrier properties of agar based nanocomposite films. *Carbohydrate Polymers* 86: 691–699.

Rhim, J. W. and Ng, P. K. W. 2007. Natural biopolymer-based nanocomposite films for packaging applications. *Critical Reviews in Food Science and Nutrition* 47: 411–433.

Robertson, G. 2006a. *Food Packaging Principles and Practices* (Second edition). Boca Raton, FL: Taylor & Francis.

Robertson, G. L. 2006b. Food packaging. In *Food Science and Technology* (Volume 1), ed. G. Campbell-Platt, 279–298. Chichester, West Sussex, UK: John Wiley & Sons, Ltd.

Rokka, M., Eerola, S., Smolander, M., Alakomi, H. and Ahvenainen, R. 2004. Monitoring of the quality of modified atmosphere packaged broiler chicken cuts stored in different temperature conditions B. Biogenic amines as quality-indicating metabolites. *Food Control* 15: 601–607.

Roland, A. M. and Hotchkiss, J. H. 1991. Determination of flavor-polymer interactions by vacuum microgravimetric method. In *Food and Packaging Interactions II*, eds. S. J. Risch and J. H. Hotchkiss, 149–160. Washington D.C.: American Chemical Society.

Rukchon, C., Nopwinyuwong, A., Trevanich, S., Jinkarn, T. and Suppakul, P. 2014. Development of a food spoilage indicator for monitoring freshness of skinless chicken breast. *Talanta* 130: 547–554.

Sajilata, M. G., Savitha, K., Singhal, R. S. and Kanetkar, V. R. 2007. Scalping of flavors in packaged foods. *Comprehensive Reviews in Food Science and Food Safety* 6: 17–35.

Sanyang, M. L., Ilyas, R. A., Sapuan, S. M. and Jumaidin, R. 2018. Sugar palm starch-based composites for packaging applications. In *Bionanocomposites for Packaging Applications*, eds. M. Jawaid and S. Swain, 125–147. Cham: Springer.

Scarascia-Mugnozza, G., Schettini, E., Vox, G., Malinconico, M., Immirzi, B. and Pagliara, S. 2006. Mechanical properties decay and morphological behaviour of biodegradable films for agricultural mulching in real scale experiment. *Polymer Degradation and Stability* 91: 2801–2808.

Segall, K. I. and Scanlon, M. G. 1996. Design and analysis of a modified-atmosphere package for minimally processed romaine lettuce. *Journal of the American Society for Horticultural Science* 121: 722–729.

Shin, J. and Selke, S. E. M. 2014. Food packaging. In *Food Processing: Principles and Applications*, eds. S. Clark, S. Jung and B. Lamsal, 249–273. Chichester, West Sussex, UK: John Wiley and Sons.

Smolander, M., Hurme, E., Latva-Kala, K., Louma, T., Alakomi, H. and Ahvenainen, R. 2002. Myoglobin-based indicators for the evaluation of freshness of unmarinated broiler cuts. *Innovative Food Science and Emerging Technologies* 3: 279–288.

Sorrentino, A., Gorrasi, G. and Vittoria, V. 2007. Potential perspectives of bio-nanocomposites for food packaging applications. *Trends in Food Science & Technology* 18: 84–95.

Sothornvit, R., Rhim, J. and Hong, S. 2009. Effect of nanoclay type on the physical and antimicrobial properties of whey protein isolate/clay composite films. *Journal of Food Engineering* 91: 468–473.

Steeman, A. 2012. The self-cooling technology and the future—Part 3. *Best in Packing*, 17 December 2012. https://bestinpackaging.com/2012/12/17/the-self-cooling-technology-and-the-future-part-3/ (accessed October 17, 2018).

Strandburg, G., DeLassus, P. T. and Howell, B. A. 1991. Thermodynamics of permeation of flavours in polymers: prediction of solubility coefficients. In *Food and Packaging Interactions II*, eds. S. J. Risch and J. H. Hotchkiss, 133–148. Washington, DC: American Chemical Society.

Suppakul, P., Miltz, J., Sonneveld, K. and Bigger, S. W. 2003. Active packaging technologies with an emphasis on antimicrobial packaging and its applications. *Journal of Food Science* 68: 408–420.

TAPPI T-441 2013. Water absorptiveness of sized (non-bibulous) paper, paperboard and corrugated fiberboard (Cobb test).

TAPPI T 804 om-12 2012. Compression test of fiberboard shipping containers.

TAPPI T 810 om-17 2017. Bursting strength—Corrugated and solid fiberboard.

Uniform Code Council 2004. *Small Symbologies (RSS and Composite)*. Lawrenceville, NJ: Uniform Code Council. www.uc-council.org/ean_ucc_system/stnds_and_tech/small_sym.html (accessed October 17, 2018).

Vaclavik, V. A. and Christian, E. W. 2003. *Essentials of Food Science* (Second edition). New York: Springer Science+Business Media.

Valapa, R. B., Loganathan, S., Pugazhenthi, G., Thomas, S. and Varghese, T. O. 2017. An overview of polymer–clay nanocomposites. In *Clay-Polymer Nanocomposites*, eds. K. Jlassi, M. M. Chehimi and S. Thomas, 29–81. Amsterdam, the Netherlands: Elsevier.

Vanderroost, M., Ragaert, P., Devlieghere, F. and Meulenaer, B. D. 2014. Intelligent food packaging: the next generation. *Trends in Food Science & Technology* 39: 47–62.

Vermeiren, L., Devlieghere, F., VanBeest, M., de Kruijf, N. and Debevere, J. 1999. Developments in the active packaging of foods. *Trends in Food Science & Technology* 10: 77–86.

Vijayendra, S. V. N. and Shamala, T. R. 2014. Film forming microbial biopolymers for commercial applications—A review. *Critical Reviews in Biotechnology* 34: 338–357.

Wanihsuksombat, C., Hongtrakul, V. and Suppakul, P. 2010. Development and characterization of a prototype of a lactic acid-based time temperature indicator for monitoring food product quality. *Journal of Food Engineering* 100: 427–434.

Want, R. 2004. RFID: a key to automating everything. *Scientific American* 290: 56–65.

Welt, B. A., Sage, D. S. and Berger, K. L. 2003. Performance specification of time-temperature integrators designed to protect against botulism in refrigerated fresh foods. *Journal of Food Science* 68: 2–9.

Wilson, C. 2007. *Frontiers of Intelligent and Active Packaging for Fruits and Vegetables*. Boca Raton, FL: CRC Press.

Yam, K. L. 2000. Intelligent packaging and the future smart kitchen. *Packaging Technology and Science* 13: 83–85.

Yam, K. L., Takhistov, P. T. and Miltz, J. 2005. Intelligent packaging: concepts and applications. *Journal of Food Science* 70: R1–R10.

16

Fundamentals of Computational Fluid Dynamics Modeling and Its Applications in Food Processing

Food processing is often intricate with the complexities originating from different sources of its varied steps. For instance, during food processing, fluid flow occurs in conjunction with the following:

- Heat transfer in more than one form and involving all the three modes, i.e., conduction, convection, and radiation (e.g., baking)
- Phase change (e.g., melting, freezing)
- Chemical reactions (e.g., fermentation);
- Mechanical movement (e.g., movement of blades during mixing, movement of the piston during pumping, and fans during blowing operations)
- Deformation of solids (e.g., elastic deformation of bread dough during proofing and baking)

To add to the complexity, food products exhibit variable consistency, irregular shape, and their physical properties change with respect to time, temperature, shear, and composition during processing. Besides, biochemical and microbiological changes also take place during food processing. Moreover, in unsteady-state operations, prior estimation of processing time is imperative to avoid overprocessing and under-processing. While overprocessing has a negative influence on food quality, under-processing leads to spoilage of foods. The consequences of the complexities mentioned previously are the high degree of difficulty and expenditure involved in the measurement of related physical properties such as temperature, pressure, specific heat, thermal conductivity, and so forth.

Computational fluid dynamics (CFD) modeling is an economical, faster, and adaptable tool, when compared to the experimental approach. CFD can be defined as, *a simulation tool that uses powerful computers in combination with applied mathematics to model fluid flow and aid in the optimal design of industrial processes* (Anandharamakrishnan, 2003, 2013). CFD is aimed at solving the governing differential equations pertaining to the unsteady-state heat, mass, and momentum transfer. It offers an efficient and powerful tool for simulating and understanding the transport processes in the food industry.

CFD is considered as the third dimension of fluid dynamics, the other two dimensions being the approaches of *pure experiment* and *pure theory* (Figure 16.1). Nevertheless, it outweighs the experimental approach in terms of considering the actual flow domain under all possible conditions of practical operation, while experimental approach limits itself to a laboratory-scale prototype within a limited range of operating conditions. The other differences which render CFD advantageous when compared to the experiment are listed in Table 16.1. Thus, CFD modeling synergistically complements the approaches of *pure theory* and *pure experiment*, but it is never a replacement for either of these approaches (Anderson, 2013). Therefore, the experimental validation of the CFD model is necessary.

Applications of CFD in food processing operations commenced only towards the end of the 20th century. Since then, CFD modeling has been applied to various disciplines of food engineering such as design optimization, scale-up, and troubleshooting, in addition to product development. This chapter intends to provide an insight into the basics of CFD modeling and its applications in food processing.

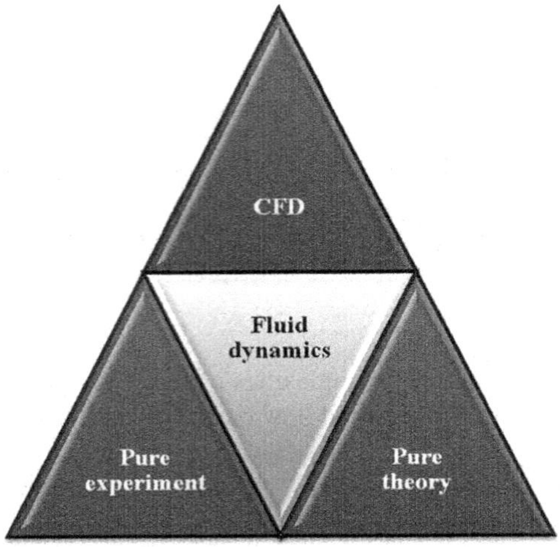

FIGURE 16.1 Three dimensions of fluid dynamics.

TABLE 16.1

Comparison between CFD Modeling and Experiment

	CFD Modeling	Experiment
Flow phenomena	Predicted using CFD software	Described by measurements
Measurement	Determines all the required quantities	Measures one quantity at a time
Scale	For any actual flow domain	Limited to laboratory-scale prototype
Scope	For any problem and realistic operating conditions	For a limited range of problems and operating conditions
Time	A short period of simulation (faster and parallel)	Requires long time (short and sequential)
Repeatability	Repeatable	May or may not be repeatable
Safety	Safe to use	Some experiments are hazardous and hence require appropriate precautionary measures
Ease of handling	CFD software is portable, easy to use, and modify	Equipment and personnel are difficult to transport
Cost	Cheaper	Expensive

16.1 Theory of CFD Modeling

The principal objective of CFD modeling is to predict the fluid properties such as velocity, pressure, temperature, density and viscosity, without actual measurement. This can be accomplished if the initial values of a minimum number of quantities are known. Using the known quantities, the values at some other points within a system can be obtained by using fundamental relationships. But these relationships are empirical in nature which are valid only under a defined set of conditions and hence cannot be used for different conditions. These empirical relationships are referred to as the "basic laws." In the context of fluid dynamics, there are three applicable basic laws:

- Conservation of mass (continuity equation)
- Conservation of momentum (Newton's second law of motion)
- Conservation of energy (First law of thermodynamics).

The aforementioned laws are generally expressed in the form of partial differential equations (PDEs), which are collectively termed *Navier–Stokes equations* and considered as the governing equations of CFD. The Navier–Stokes equations are completed by adding other algebraic equations from thermodynamics such as the equation of state for density and a constitutive equation to describe the rheology (Fletcher, 2000; Anandharamakrishnan, 2013). With CFD modeling, the fluid properties are predicted by seeking a solution for the Navier–Stokes equations. It is important that these equations are clearly understood before proceeding to understand the applications of CFD in a specific food processing operation.

16.1.1 Navier–Stokes Equations

16.1.1.1 Conservation of Mass Equation

The conservation of mass equation, also known as the continuity equation, can be derived by considering a control volume of a defined system (Figure 16.2). Thus, it is necessary to understand the concepts of *system* and *control volume.* In fluid dynamics, *system* is defined as a portion of the fluid particles with constant identity during the course of flow. Constant identity means that the portion is composed of the same fluid particles as it flows. The physical meaning of this definition is nothing but the statement of the law of conservation of mass, which states that the mass of a system is constant irrespective of whether any process (chemical reaction or heating) takes place within it. Mathematically, it is expressed as

$$\left.\frac{dm}{dt}\right|_{\text{system}} = 0 \tag{16.1}$$

From Eq. (16.1), it can be inferred that the rate of change of mass of a system is zero. The same can also be written as

$$\text{Net mass flux into the control volume} = \text{Rate of accumulation of mass within the control volume} \tag{16.2}$$

A *control volume* (V) is a volume or region in space whose identity is not the same as fluid and can enter and leave through the control surface (S) which encloses this volume (Figure 16.2). The shape and size of the control volume may be constant or variable depending upon the coordinate system selected to analyze the fluid flow problem.

Initially, the conservation of mass equation can be derived for a control volume of linear dimensions dx, dy, and dz as shown in Figure 16.3.

Net inward flux of mass in the y-direction is given by

$$= \rho v dx dz - \left(\rho v dx dz + \frac{\partial}{\partial y}(\rho v dx dz) dy \right) \tag{16.3}$$

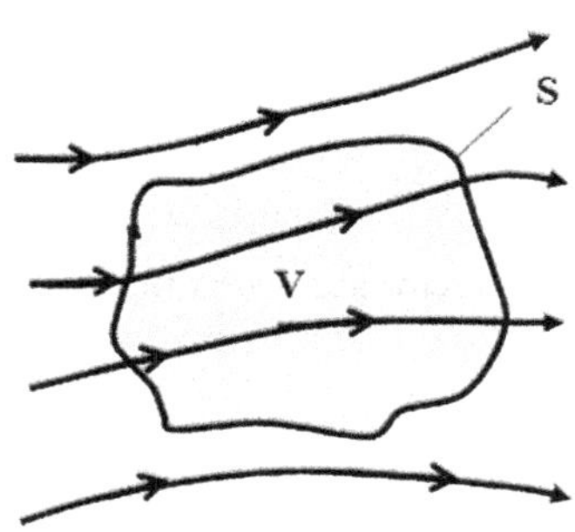

FIGURE 16.2 Schematic of control volume and control surface.

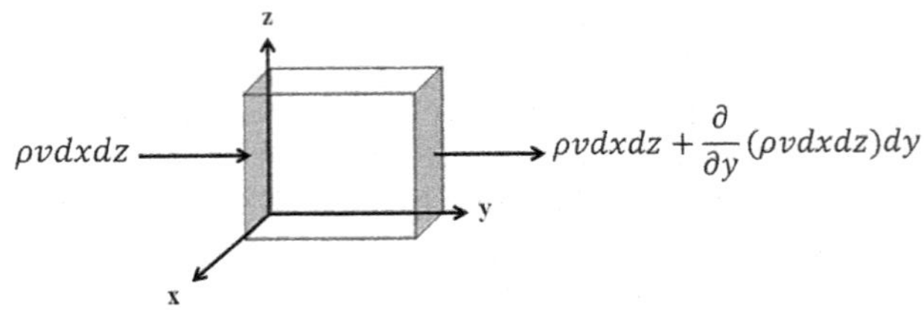

FIGURE 16.3 Control volume of linear dimensions.

$$= -\frac{\partial}{\partial y}(\rho v dxdz)dy \tag{16.4}$$

Since dx and dz are constants, Eq. (16.4) can be written as,

$$-\frac{\partial(\rho V)}{\partial y}(dxdzdy) \tag{16.5}$$

Similarly, the net influx in the x- and z-directions can be obtained. Then, the net influx for the entire control volume can be expressed as,

$$-\left(\frac{\partial(\rho u)}{\partial x} + \frac{\partial(\rho v)}{\partial y} + \frac{\partial(\rho w)}{\partial z}\right)dydzdx \tag{16.6}$$

Thus, the rate of accumulation of mass within the control volume is given by

$$\frac{\partial}{\partial t}(\rho dydxdz) \equiv \frac{\partial \rho}{\partial t}(dydxdz) \tag{16.7}$$

As mentioned previously, according to the law of conservation of mass, the rate of accumulation of mass within the control volume is equal to the net influx of mass. Therefore,

$$-\left(\frac{\partial(\rho u)}{\partial x} + \frac{\partial(\rho v)}{\partial y} + \frac{\partial(\rho w)}{\partial z}\right)dydzdx = \frac{\partial \rho}{\partial t}(dydxdz) \tag{16.8}$$

Since the volume is finite, the law of conservation of mass can be written as

$$\frac{\partial \rho}{\partial t} + \frac{\partial \rho u}{\partial x} + \frac{\partial \rho v}{\partial y} + \frac{\partial \rho w}{\partial z} = 0 \tag{16.9}$$

Eq. (16.9) can also be written in the vector notation as given in the following equation.

$$\frac{\partial \rho}{\partial t} + \nabla \cdot \rho \bar{u} = 0 \tag{16.10}$$

where $\rho \bar{u}$ is known as the mass velocity vector and ∇ has the dimension of reciprocal length, given by, $\nabla = \frac{\partial}{\partial x}u + \frac{\partial}{\partial y}v + \frac{\partial}{\partial z}w$.

Using standard vector identities in Eq. (16.10),

$$\frac{\partial \rho}{\partial t} + \rho \nabla \cdot \bar{u} + \bar{u} \cdot \nabla \rho = 0 \tag{16.11}$$

$$\frac{d\rho}{dt} + \rho\nabla \cdot \bar{u} = 0 \tag{16.12}$$

For an incompressible or constant density fluid, Eq. (16.12) reduces to

$$\nabla \cdot \bar{u} = 0 \tag{16.13}$$

Equation 16.13 is the unsteady, three-dimensional (3D), mass conservation, or continuity equation for the simplified case of an incompressible or constant density fluid (NPTEL, 2018a).

16.1.1.2 Conservation of Momentum Equation

The conservation of momentum is associated with a force balance. Force is a vector or directional quantity. Stress is given by force per unit area. Since both force and area are vector quantities, stress is a tensor and always expressed along with its sign. Generally, stress (τ_{ij}) is considered to act on the ith plane and in the jth direction. Accordingly,

- Force resultant from the stress in j^{th} direction and positive i^{th} plane acts in the positive j^{th} direction.
- Force resultant from the stress in j^{th} direction and negative i^{th} plane acts in the negative j^{th} direction.
- $\tau_{ij}dA_i$ acts in the positive j^{th} direction for positive and in the negative j^{th} direction for negative dA_i.

In a small control volume (in Cartesian coordinates) having sides of dimension, dx, dy, and dz, the stress tensor in the positive direction is $\tau_{zy}dxdy$. For $z = 0$, plane is towards the left, and for $z = dz$, plane is towards the right (Figure 16.4a).

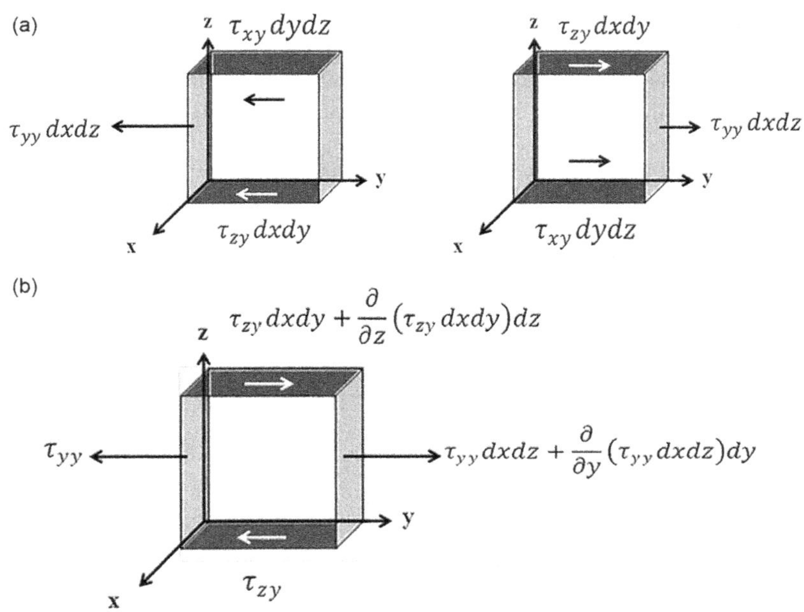

FIGURE 16.4 (a and b) Schematic of the control volume for the derivation of conservation of momentum equation.

The law of conservation of momentum can be written as

$$\text{Rate of accumulation of momentum} = \text{Net rate of influx of momentum due to velocity} + \text{Net rate of influx due to forces} \quad (16.14)$$

First, the forces due to the stresses on the faces of the control volume should be determined. This can be accomplished by considering the y direction and writing the expressions for the force due to stress on the faces, $y = 0$ and $z = 0$. Then, for the faces $y = dy$ and $z = dz$, the Taylor series expansion with the leading term is established. In order to simplify the derivation, the terms on the faces $x = 0$ and $x = dx$ are neglected (Figure 16.4b).

Force F_y in the y-direction is given by

$$\left[\tau_{yy} + \frac{\partial}{\partial y}\tau_{yy}dy - \tau_{yy}\right]dxdz + \left[\tau_{zy} + \frac{\partial}{\partial z}\tau_{zy}dz - \tau_{zy}\right]dxdy + \left[\tau_{xy} + \frac{\partial}{\partial x}\tau_{xy}dx - \tau_{xy}\right]dydz \quad (16.15)$$

Simplifying and rewriting Eq. (16.15),

$$F_y = \frac{\partial}{\partial y}\tau_{yy}dxdydz + \frac{\partial}{\partial z}\tau_{zy}dxdydz + \frac{\partial}{\partial x}\tau_{xy}dxdydz \quad (16.16)$$

Force balance in the x- and z-directions can now be performed, in an identical manner, to get the net force in these directions.

The forces F_z and F_x due to the stresses in the x- and z-directions are as follows:

$$F_z = \frac{\partial}{\partial y}\tau_{yz}dxdydz + \frac{\partial}{\partial z}\tau_{zz}dxdydz + \frac{\partial}{\partial x}\tau_{xz}dxdydz \quad (16.17)$$

$$F_x = \frac{\partial}{\partial y}\tau_{yx}dxdydz + \frac{\partial}{\partial z}\tau_{zx}dxdydz + \frac{\partial}{\partial x}\tau_{xz}dxdydz \quad (16.18)$$

In the next step, the change in momentum due to the mass entering and leaving the control volume is evaluated. To begin with, the influx of momentum due to mass entering the control volume is considered. Let the velocity be u, v, and w in the x-, y-, and z-directions. Accordingly,

- Momentum entering the control volume in the y-direction due to the mass entering the $y = 0$, $z = 0$, and $x = 0$ faces is $(\rho v dxdz)v$, $(\rho w dydxv)$, and $(\rho u dydzv)$, respectively.
- Momentum leaving the control volume due to mass exiting from the faces, $y = dy$, $z = dz$, and $x = dx$ is obtained from the Taylor series expansion by retaining only the leading term.
- Momentum terms in the y-direction due to mass entering or leaving the control volume are shown in Figure 16.5. The term on $x = 0$ face is neglected for the purpose of simplification.

Net influx of momentum in the y-direction due to mass influx is given by

$$\rho vvdxdz + \rho wvdxdy + \rho uvdydz - \left(\rho vv + \frac{\partial}{\partial y}\rho vv + \rho wv + \frac{\partial}{\partial z}\rho wv + \rho uv + \frac{\partial}{\partial x}\rho uv\right)dxdydz$$

$$= -\left(\frac{\partial}{\partial y}\rho vv + \frac{\partial}{\partial z}\rho wv + \frac{\partial}{\partial x}\rho uv\right)dxdydz \quad (16.19)$$

Besides the surface forces due to stresses, the body forces are also assumed to exist. Body force vector per unit mass is represented by

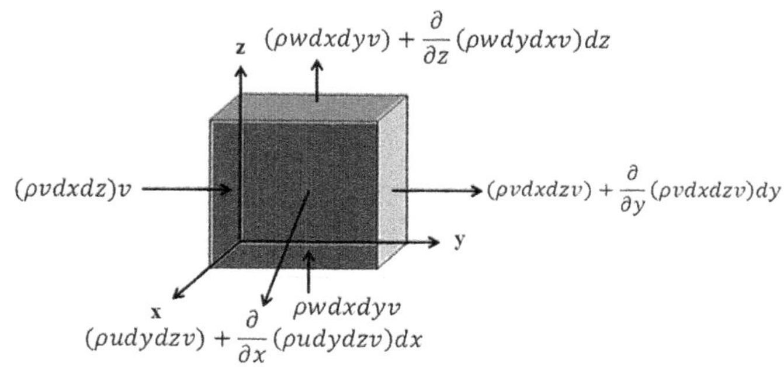

FIGURE 16.5 Momentum terms in the y-direction.

$$\left(X_x\hat{i} + X_y\hat{j} + X_z\hat{k}\right) \tag{16.20}$$

Net influx of momentum into the control volume is due to the mass input Eq. (16.19), force on the faces of control volume Eq. (16.16) and the body force Eq. (16.20). The net rate of accumulation rate is $\frac{\partial}{\partial t}\rho v dxdydz$. Hence, the overall momentum balance equation is given by,

$$-\left(\frac{\partial}{\partial y}\rho v^2 + \frac{\partial}{\partial z}\rho wv + \frac{\partial}{\partial x}\rho uv\right)dxdydz + \left(\frac{\partial}{\partial y}\tau_{yy} + \frac{\partial}{\partial z}\tau_{zy} + \frac{\partial}{\partial x}\tau_{xy}\right)dxdydz$$

$$-\frac{\partial}{\partial t}\rho v dxdydz + \rho X_y dxdydz = 0 \tag{16.21}$$

As measurement and quantification of stress is difficult, the stress term should be converted to a measurable term. This is done by adopting a constitutive relationship. Here, the discussion is limited to the Newtonian fluids. As already discussed in Chapter 4, the relationship for shear stress proposed by Newton's law is $\tau = \mu\frac{du}{dy}$ (for one-dimension). Therefore, by applying the Newton's law, the relationship between velocities and stress can be established using the following equation:

$$\tau_{ij} = \mu\left(\frac{\partial u_i}{\partial x_j} + \frac{\partial u_j}{\partial x_i}\right) \text{ for } i \neq j \tag{16.22}$$

Subsequently, the Stokes constitutive relationship as given in the following equation is considered.

$$\tau_{ij} = \left(-P - \frac{2}{3}\mu\nabla\cdot\bar{u}\right)\delta_{ij} + \mu\left(\frac{\partial u_i}{\partial x_j} + \frac{\partial u_j}{\partial x_i}\right) \tag{16.23}$$

$$\delta_{ij} = 0 \text{ for } i \neq j \tag{16.24}$$

$$\delta_{ij} = 1 \text{ for } i = j \tag{16.25}$$

$$\bar{u} = u_1\hat{i} + u_2\hat{j} + u_3\hat{k} \tag{16.26}$$

By considering the stress terms in the momentum equation,

$$\frac{\partial}{\partial y}\tau_{yy} + \frac{\partial}{\partial x}\tau_{xy} + \frac{\partial}{\partial z}\tau_{zy} = \left(\text{From the momentum equation}\right) \tag{16.27}$$

And substituting the Stokes relationship as given in Eq. (16.28),

$$-\frac{\partial P}{\partial y}-\frac{\partial}{\partial y}\left(\frac{2}{3}\mu\nabla\cdot\bar{u}\right)+\frac{\partial}{\partial y}\left(2\mu\frac{\partial v}{\partial y}\right)$$
$$+\frac{\partial}{\partial x}\left(\mu\left(\frac{\partial u}{\partial y}+\frac{\partial v}{\partial x}\right)\right)+\frac{\partial}{\partial z}\mu\left(\frac{\partial v}{\partial z}+\frac{\partial w}{\partial y}\right) \tag{16.28}$$

Further, if μ is assumed constant, then Eq. (16.28) becomes,

$$\frac{\partial}{\partial y}\tau_{yy}+\frac{\partial}{\partial x}\tau_{xy}+\frac{\partial}{\partial z}\tau_{zy}=-\frac{\partial P}{\partial y}-\frac{2}{3}\mu\frac{\partial}{\partial y}(\nabla\cdot\bar{u})+2\mu\frac{\partial}{\partial y}\left(\frac{\partial v}{\partial y}\right)+\mu\frac{\partial}{\partial x}\left(\frac{\partial u}{\partial y}+\frac{\partial v}{\partial x}\right)+\mu\frac{\partial}{\partial z}\left(\frac{\partial v}{\partial z}+\frac{\partial w}{\partial y}\right) \tag{16.29}$$

As mentioned in the preceding section on continuity equation, for an incompressible fluid, $\nabla.\bar{u}=0$. Since velocity is a continuous function, cross differentiation is acceptable. Therefore,

$$\frac{\partial}{\partial y}\frac{\partial u}{\partial x}=\frac{\partial}{\partial x}\frac{\partial u}{\partial y} \tag{16.30}$$

Substituting the continuity equation and Eq. (16.30) in Eq. (16.28),

$$\frac{\partial}{\partial y}\tau_{yy}+\frac{\partial}{\partial x}\tau_{xy}+\frac{\partial}{\partial z}\tau_{zy}\equiv-\frac{\partial P}{\partial y}-\cancel{\frac{2}{3}\frac{\partial}{\partial y}(\mu\nabla.\bar{u})}(=0)+\mu\left(\frac{\partial^2 v}{\partial x^2}+\frac{\partial^2 v}{\partial y^2}+\right.$$
$$\left.\frac{\partial^2 v}{\partial z^2}\right)+\mu\frac{\partial}{\partial y}\cancel{\left(\frac{\partial u}{\partial x}+\frac{\partial v}{\partial y}+\frac{\partial w}{\partial z}\right)}(=0) \tag{16.31}$$

Substituting Eq. (16.31) in Eq. (16.21),

$$\frac{\partial}{\partial t}(\rho v)+\frac{\partial}{\partial x}\rho uv+\frac{\partial}{\partial y}\rho v^2+\frac{\partial}{\partial z}\rho wv=\rho X_y-\frac{\partial P}{\partial y}+\mu\left(\frac{\partial^2 v}{\partial x^2}+\frac{\partial^2 v}{\partial y^2}+\frac{\partial^2 v}{\partial z^2}\right) \tag{16.32}$$

Equation 16.32 is known as the conservative form of the momentum equation. The name is justified as it is derived in the original form from the conservation equations without applying any simplifications.

Expanding the left-hand side (LHS) of Eq. (16.32),

$$\frac{\partial}{\partial t}(\rho v)+\frac{\partial}{\partial x}\rho uv+\frac{\partial}{\partial y}\rho v^2+\frac{\partial}{\partial z}\rho wv=\rho\left[\frac{\partial v}{\partial t}+u\frac{\partial v}{\partial x}+v\frac{\partial v}{\partial y}+w\frac{\partial v}{\partial z}\right]+v\left[\frac{\partial \rho}{\partial t}+\frac{\partial}{\partial x}\rho u+\frac{\partial}{\partial y}\rho v+\frac{\partial}{\partial z}\rho w\right] \tag{16.33}$$

From the continuity equation, the second term in Eq. (16.33) is zero $\left(v\left[\frac{\partial \rho}{\partial t}+\frac{\partial}{\partial x}\rho u+\frac{\partial}{\partial y}\rho v+\frac{\partial}{\partial z}\rho w\right]=0\right)$.

Therefore, the y-component of the momentum equation becomes $\left(\vartheta=\frac{\mu}{\rho}\right)$:

$$\left[\frac{\partial v}{\partial t}+u\frac{\partial v}{\partial x}+v\frac{\partial v}{\partial y}+w\frac{\partial v}{\partial z}\right]=X_y+\vartheta\left[\frac{\partial^2 v}{\partial x^2}+\frac{\partial^2 v}{\partial y^2}+\frac{\partial^2 v}{\partial z^2}\right]-\frac{1}{\rho}\frac{\partial p}{\partial y} \tag{16.34}$$

A more precise form of the above Y-momentum equation is given by

$$\frac{dv}{dt} = X_y + \vartheta\nabla^2 v - \frac{1}{\rho}\frac{\partial p}{\partial y} \tag{16.35}$$

$$\nabla^2 v \equiv \frac{\partial^2 v}{\partial x^2} + \frac{\partial^2 v}{\partial y^2} + \frac{\partial^2 v}{\partial z^2} \tag{16.36}$$

$$\frac{dv}{dt} \equiv \frac{\partial v}{\partial t} + u\frac{\partial v}{\partial x} + v\frac{\partial v}{\partial y} + w\frac{\partial v}{\partial z} \tag{16.37}$$

Similarly, the X- and Z-momentum equations can be derived. The final momentum equations are

$$\frac{dv}{dt} = X_y + \vartheta\nabla^2 v - \frac{1}{\rho}\frac{\partial p}{\partial y} \tag{16.38}$$

$$\frac{du}{dt} = X_x + \vartheta\nabla^2 u - \frac{1}{\rho}\frac{\partial p}{\partial x} \tag{16.39}$$

$$\frac{dw}{dt} = X_z + \vartheta\nabla^2 w - \frac{1}{\rho}\frac{\partial p}{\partial z} \tag{16.40}$$

Eqs. (16.38) – (16.40) are applicable for laminar, incompressible, constant viscosity, and Newtonian fluids (NPTEL, 2018b).

16.1.1.3 Conservation of Energy Equation

According to the law of conservation of energy (see Chapter 3), the net rate of energy input into the control volume is equal to the rate of accumulation of energy within it. Three modes of energy transfer are possible, i.e., through conduction, convection, or work done on the fluid in the control volume. The energy per unit mass of fluid is composed of three components: the kinetic, potential, and thermal energies Eq. (16.41).

$$\text{Energy} = e_{\text{thermal}} + e_{\text{kinetic}} + e_{\text{potential}} = \hat{u} + \frac{V^2}{2} + gh \tag{16.41}$$

A control volume of size dx, dy, and dz is considered for the purpose of deriving the conservation of energy equation. In Figure 16.6, the energy transfer due to the mass entering in y-direction is shown. The mass entering in the x- and z-directions would also have similar terms but are not shown in the figure.

The rate of energy transferred into the control volume through the convection mode by means of the mass entering in y-direction is shown on the $y = 0$ plane (Figure 16.6). On the $y = dy$ plane, the Taylor series expansion is applied and only the leading terms are retained for the conduction and convection terms. The net rate of heat input through conduction is given by

$$\left[\frac{\partial}{\partial y}\left(k\frac{\partial T}{\partial y}\right) + \frac{\partial}{\partial x}\left(k\frac{\partial T}{\partial x}\right) + \frac{\partial}{\partial z}\left(k\frac{\partial T}{\partial z}\right)\right]dxdydz \tag{16.42}$$

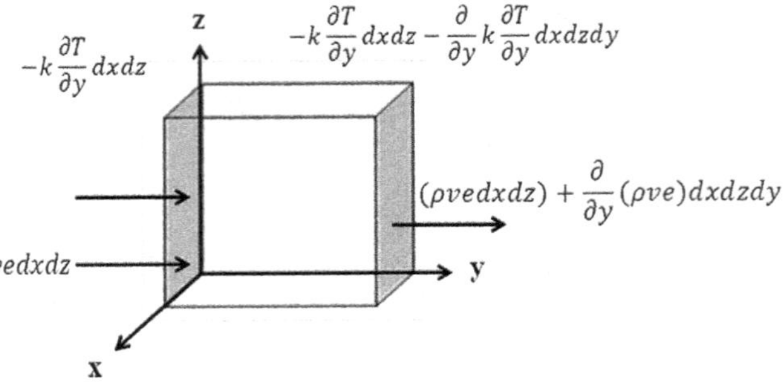

FIGURE 16.6 Energy transfer into the control volume due to the mass entering in y-direction.

The net rate of heat input through convection is

$$-\left[\frac{\partial(\rho ue)}{\partial x}+\frac{\partial(\rho ve)}{\partial y}+\frac{\partial(\rho we)}{\partial z}\right]dxdydz \tag{16.43}$$

The rate of energy accumulation is as given in the following equation:

$$\frac{\partial}{\partial t}(\rho e)dxdydz \tag{16.44}$$

Subtracting Eq. (16.43) from Eq. (16.44) and grouping the appropriate terms, the following expression is obtained:

$$\rho\left[\frac{\partial e}{\partial t}+u\frac{\partial e}{\partial x}+v\frac{\partial e}{\partial y}+w\frac{\partial e}{\partial z}\right]+e\left[\frac{\partial \rho}{\partial t}+\frac{\partial}{\partial x}\rho u+\frac{\partial}{\partial y}\rho v+\frac{\partial}{\partial z}\rho w\right] \tag{16.45}$$

From the continuity equation,

$$\frac{\partial}{\partial y}\rho v = 0 \tag{16.46}$$

The rate of energy transfer due to work done is calculated as the dot product of force and velocity Eq. (16.47)

$$W = \bar{F}\cdot\bar{v} \tag{16.47}$$

When the force and velocity vectors are in the same direction, the rate of work done can be assumed positive. The derivation of the energy equation is continued further by considering the positive and negative work terms on the different faces of the control volume (Figure 16.7).

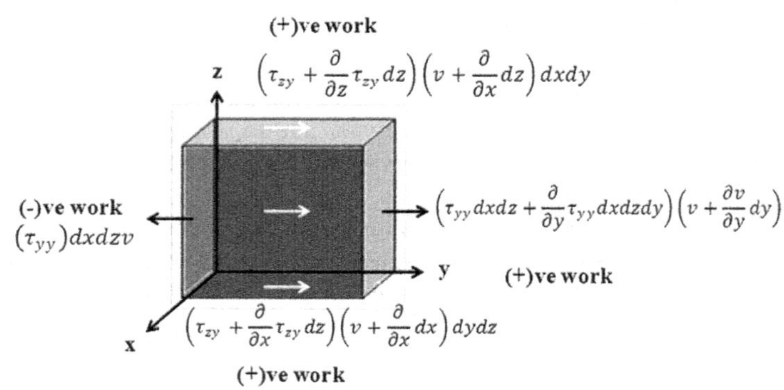

FIGURE 16.7 Schematic of a control volume showing positive and negative work terms on its different faces.

Consider the top, front, and right faces of the control volume in Figure 16.7. The force due to stress acts in the left-to-right direction. The positive v direction is also from the left-to-right. Consequently, the dot product is a positive quantity. However, the left face is acted upon by the force due to stress from the right-to-left direction and v in the left-to-right direction. Thus, the work becomes negative. The bottom and backside faces also have negative values which are not shown in Figure 16.7.

Thus, the net rate of work done on the $y = 0$ and $y = dy$ faces due to the force in the y-direction is given by

$$\begin{aligned}
&\left(\tau_{yy}dxdz + \frac{\partial}{\partial y}\tau_{yy}dxdzdy\right)\left(v + \frac{\partial v}{\partial y}dy\right) - \left[\tau_{yy}dxdz\right]v \\
&= \left[\frac{\partial}{\partial y}\tau_{yy}dxdzdy\right]v + \left[\tau_{yy}dxdz\right]\left[\frac{\partial v}{\partial y}dy\right] + \left[\frac{\partial}{\partial y}\tau_{yy}dxdzdy\right]\left[\frac{\partial v}{\partial y}dy\right] \\
&= \frac{\partial}{\partial y}\left(v\tau_{yy}dxdzdy\right) + \left[\frac{\partial}{\partial y}\tau_{yy}dxdzdy\right]\left[\frac{\partial v}{\partial y}dy\right]
\end{aligned} \tag{16.48}$$

Therefore, the rate of work done per unit volume is given by

$$\frac{W}{dxdydz} = \frac{\partial}{\partial y}\left(v\tau_{yy}\right) + \left[\frac{\partial}{\partial y}\tau_{yy}\right]\left[\frac{\partial v}{\partial y}dy\right] \tag{16.49}$$

Let the volume $dxdydz = 0$. Then Eq. (16.49) becomes

$$\frac{W}{dxdydz} = \frac{\partial}{\partial y}\left(v\tau_{yy}\right) + \left[\frac{\partial}{\partial y}\tau_{yy}\right]\left[\frac{\partial v}{\partial y}dy\right] \tag{16.50}$$

The last term in Eq. (16.50) is zero as it is multiplied by $dy \to 0$ (tending to zero).

$$\left[\frac{\partial}{\partial y}\tau_{yy}\right]\left[\frac{\partial v}{\partial y}dy\right] = 0 \tag{16.51}$$

$$\therefore \frac{W}{dxdydz} = \frac{\partial}{\partial y}\left(v\tau_{yy}\right) \tag{16.52}$$

The other two terms due to forces acting in the y-direction on $x = 0$, $x = dx$ and $z = 0$, $z = dz$ planes are given by

$$\frac{\partial}{\partial x}\left(v\tau_{xy}\right), \frac{\partial}{\partial z}\left(v\tau_{zy}\right) \tag{16.53}$$

Thus, the total rate of work done due to forces in the y-direction is

$$\frac{W}{dxdydz} = \frac{\partial}{\partial x}\tau_{xy}v + \frac{\partial}{\partial y}\tau_{yy}v + \frac{\partial}{\partial z}\tau_{zy}v \tag{16.54}$$

Since work is a scalar quantity, it is expressed as an algebraic sum. Likewise, there would be three terms each for the work due to forces in the x- and z-directions. All these forces can be added together to the total work on the control volume. The resultant total rate of work is given by

$$\frac{W}{dxdydz} = \frac{\partial}{\partial x}(\tau_{xx}u) + \frac{\partial}{\partial y}(\tau_{yx}u) + \frac{\partial}{\partial z}(\tau_{zx}u)$$
$$+ \frac{\partial}{\partial x}(\tau_{xy}v) + \frac{\partial}{\partial y}(\tau_{yy}v) + \frac{\partial}{\partial z}(\tau_{zy}v) \tag{16.55}$$
$$+ \frac{\partial}{\partial x}(\tau_{xz}w) + \frac{\partial}{\partial y}(\tau_{yz}w) + \frac{\partial}{\partial z}(\tau_{zz}w)$$

Each of the product terms in Eq. (16.55) can be divided into two terms. Therefore, a total of 18 terms would be on the right-hand side (RHS) of the equation. Out of the 18 terms in Eq. (16.55), nine terms are also observed in the momentum equation [see Eq. (16.21) with $X_z = -g$]. Therefore, these nine terms can be simplified by using the same approximations applied initially to derive the overall momentum equation (Eq. 16.21). The set of nine terms which can be simplified using the momentum equation is as given in Eq. (16.56):

$$u\left[\frac{\partial}{\partial x}(\tau_{xx}) + \frac{\partial}{\partial y}(\tau_{yx}) + \frac{\partial}{\partial z}(\tau_{zx})\right] = \rho u\left[\frac{\partial u}{\partial t} + u\frac{\partial u}{\partial x} + v\frac{\partial u}{\partial y} + w\frac{\partial u}{\partial z}\right] + v\left[\frac{\partial}{\partial x}(\tau_{xy}) + \frac{\partial}{\partial y}(\tau_{yy}) + \frac{\partial}{\partial z}(\tau_{zy})\right]$$
$$+ \rho v\left[\frac{\partial v}{\partial t} + u\frac{\partial v}{\partial x} + v\frac{\partial v}{\partial y} + w\frac{\partial v}{\partial z}\right] + w\left[\frac{\partial}{\partial x}(\tau_{xz}) + \frac{\partial}{\partial y}(\tau_{yz}) + \frac{\partial}{\partial z}(\tau_{zz})\right]$$
$$+ \rho w\left[\frac{\partial w}{\partial t} + u\frac{\partial w}{\partial x} + v\frac{\partial w}{\partial y} + w\frac{\partial w}{\partial z}\right] + \rho g w \tag{16.56}$$

The following equation is obtained by simplifying the RHS of Eq. (16.56).

$$\rho u\left[\frac{\partial u}{\partial t} + u\frac{\partial u}{\partial x} + v\frac{\partial u}{\partial y} + w\frac{\partial u}{\partial z}\right] \equiv \rho\left[\frac{1}{2}\frac{\partial u^2}{\partial t} + \frac{1}{2}u\frac{\partial u^2}{\partial x} + \frac{1}{2}v\frac{\partial u^2}{\partial y} + \frac{1}{2}w\frac{\partial u^2}{\partial z}\right]$$
$$+\rho v\left[\frac{\partial v}{\partial t} + u\frac{\partial v}{\partial x} + v\frac{\partial v}{\partial y} + w\frac{\partial v}{\partial z}\right] + \rho\left[\frac{1}{2}\frac{\partial v^2}{\partial t} + \frac{1}{2}u\frac{\partial v^2}{\partial x} + \frac{1}{2}v\frac{\partial v^2}{\partial y} + \frac{1}{2}w\frac{\partial v^2}{\partial z}\right]$$
$$+\rho w\left[\frac{\partial w}{\partial t} + u\frac{\partial w}{\partial x} + v\frac{\partial w}{\partial y} + w\frac{\partial w}{\partial z}\right] + \rho g w$$
$$+\rho\left[\frac{1}{2}\frac{\partial w^2}{\partial t} + \frac{1}{2}u\frac{\partial w^2}{\partial x} + \frac{1}{2}v\frac{\partial w^2}{\partial y} + \frac{1}{2}w\frac{\partial w^2}{\partial z}\right] + \rho g w$$
$$\equiv \rho\left[\frac{1}{2}\frac{\partial V^2}{\partial t} + \frac{1}{2}u\frac{\partial V^2}{\partial x} + \frac{1}{2}v\frac{\partial V^2}{\partial y} + \frac{1}{2}w\frac{\partial V^2}{\partial z}\right] + \rho g w \tag{16.57}$$

where $V^2 = u^2 + v^2 + w^2$.

Then, 9 of the 18 stress work terms would become

$$\rho\left[\frac{1}{2}\frac{\partial V^2}{\partial t} + \frac{1}{2}u\frac{\partial V^2}{\partial x} + \frac{1}{2}v\frac{\partial V^2}{\partial y} + \frac{1}{2}w\frac{\partial V^2}{\partial z}\right] + \rho g w = \rho\frac{d\frac{V^2}{2}}{dt} + \rho g w \tag{16.58}$$

The other nine terms are as follows:

$$\tau_{xx}\frac{\partial u}{\partial x}+\tau_{yx}\frac{\partial u}{\partial y}+\tau_{zx}\frac{\partial u}{\partial z}$$
$$+\tau_{xy}\frac{\partial v}{\partial x}+\tau_{yy}\frac{\partial v}{\partial y}+\tau_{zy}\frac{\partial v}{\partial z} \tag{16.59}$$
$$+\tau_{xz}\frac{\partial w}{\partial x}+\tau_{yz}\frac{\partial w}{\partial y}+\tau_{zz}\frac{\partial w}{\partial z}$$

Substituting the Stoke's constitutive equation Eq. (16.23) in Eq. (16.59) gives

$$\left[\left(-\boldsymbol{P}-\frac{2}{3}\mu\nabla\cdot\bar{\boldsymbol{u}}\right)+2\mu\frac{\partial \boldsymbol{u}}{\partial \boldsymbol{x}}\right]\frac{\partial \boldsymbol{u}}{\partial \boldsymbol{x}}+\mu\left[\frac{\partial u}{\partial y}+\frac{\partial v}{\partial x}\right]\frac{\partial u}{\partial y}+\mu\left[\frac{\partial w}{\partial x}+\frac{\partial u}{\partial z}\right]\frac{\partial u}{\partial z}$$
$$+\left[\left(-\boldsymbol{P}-\frac{2}{3}\mu\nabla\cdot\bar{\boldsymbol{u}}\right)+2\mu\frac{\partial \boldsymbol{v}}{\partial \boldsymbol{y}}\right]\frac{\partial \boldsymbol{v}}{\partial \boldsymbol{y}}+\mu\left[\frac{\partial u}{\partial y}+\frac{\partial v}{\partial x}\right]\frac{\partial v}{\partial x}+\mu\left[\frac{\partial v}{\partial z}+\frac{\partial w}{\partial y}\right]\frac{\partial v}{\partial z} \tag{16.60}$$
$$+\left[\left(-\boldsymbol{P}-\frac{2}{3}\mu\nabla\cdot\bar{\boldsymbol{u}}\right)+2\mu\frac{\partial \boldsymbol{w}}{\partial \boldsymbol{z}}\right]\frac{\partial \boldsymbol{w}}{\partial \boldsymbol{z}}+\mu\left[\frac{\partial u}{\partial z}+\frac{\partial w}{\partial x}\right]\frac{\partial w}{\partial x}+\mu\left[\frac{\partial v}{\partial z}+\frac{\partial w}{\partial y}\right]\frac{\partial w}{\partial y}$$

Subsequently, the bold and unbold terms in Eq. (16.60) would be dealt separately for the simplicity of algebraic manipulations. First, the bold terms are taken and simplified, followed by the simplification of the unbold terms. The simplified equations are added together at the end.

Now, considering the bold terms in Eq. (16.60), the terms on the LHS are combined together algebraically to result in the terms on the RHS as given in Eq. (16.61)

$$-P\frac{\partial u}{\partial x}-\frac{2}{3}\mu\left[\frac{\partial u}{\partial x}+\frac{\partial v}{\partial y}+\frac{\partial w}{\partial z}\right]\frac{\partial u}{\partial x}+2\mu\left(\frac{\partial u}{\partial x}\right)^2 \equiv -P\frac{\partial u}{\partial x}-\frac{2}{3}\mu\left[2\left(\frac{\partial u}{\partial x}\right)^2-\frac{\partial v\partial u}{\partial y\partial x}-\frac{\partial u\partial w}{\partial x\,\partial z}\right]$$
$$-P\frac{\partial v}{\partial y}-\frac{2}{3}\mu\left[\frac{\partial u}{\partial x}+\frac{\partial v}{\partial y}+\frac{\partial w}{\partial z}\right]\frac{\partial v}{\partial y}+2\mu\left(\frac{\partial v}{\partial y}\right)^2 \quad -P\frac{\partial v}{\partial y}-\frac{2}{3}\mu\left[2\left(\frac{\partial v}{\partial y}\right)^2-\frac{\partial v\partial u}{\partial y\partial x}-\frac{\partial v\partial w}{\partial y\,\partial z}\right] \tag{16.61}$$
$$-P\frac{\partial w}{\partial z}-\frac{2}{3}\mu\left[\frac{\partial u}{\partial x}+\frac{\partial v}{\partial y}+\frac{\partial w}{\partial z}\right]\frac{\partial w}{\partial z}+2\mu\left(\frac{\partial w}{\partial z}\right)^2 \quad -P\frac{\partial w}{\partial z}-\frac{2}{3}\mu\left[2\left(\frac{\partial w}{\partial z}\right)^2-\frac{\partial v\partial w}{\partial y\partial z}-\frac{\partial u\partial w}{\partial x\,\partial z}\right]$$

In the Eq. (16.61), the identical terms on the RHS are grouped together to obtain the final precise form as given in the following equation:

$$-P\frac{\partial u}{\partial x}-\frac{2}{3}\mu\left[\frac{\partial u}{\partial x}+\frac{\partial v}{\partial y}+\frac{\partial w}{\partial z}\right]\frac{\partial u}{\partial x}+2\mu\left(\frac{\partial u}{\partial x}\right)^2 \equiv \frac{2}{3}\mu\left[\left(\frac{\partial u}{\partial x}\right)^2+\left(\frac{\partial u}{\partial x}\right)^2-\frac{\partial v}{\partial y}\frac{\partial u}{\partial x}-\frac{\partial u}{\partial x}\frac{\partial w}{\partial z}\right]-P\frac{\partial u}{\partial x}$$
$$-P\frac{\partial v}{\partial y}-\frac{2}{3}\mu\left[\frac{\partial u}{\partial x}+\frac{\partial v}{\partial y}+\frac{\partial w}{\partial z}\right]\frac{\partial v}{\partial y}+2\mu\left(\frac{\partial v}{\partial y}\right)^2 \quad \frac{2}{3}\mu\left[\left(\frac{\partial v}{\partial y}\right)^2+\left(\frac{\partial v}{\partial y}\right)^2-\frac{\partial v}{\partial y}\frac{\partial u}{\partial x}-\frac{\partial v}{\partial y}\frac{\partial w}{\partial z}\right]-P\frac{\partial v}{\partial y}$$
$$-P\frac{\partial w}{\partial z}-\frac{2}{3}\mu\left[\frac{\partial u}{\partial x}+\frac{\partial v}{\partial y}+\frac{\partial w}{\partial z}\right]\frac{\partial w}{\partial z}+2\mu\left(\frac{\partial w}{\partial z}\right)^2 \quad \frac{2}{3}\mu\left[\left(\frac{\partial w}{\partial z}\right)^2+\left(\frac{\partial w}{\partial z}\right)^2-\frac{\partial v}{\partial y}\frac{\partial w}{\partial z}-\frac{\partial u}{\partial x}\frac{\partial w}{\partial z}\right]-P\frac{\partial w}{\partial z}$$

(16.62)

Adding together,

$$-P\nabla\cdot\bar{u}+\frac{2}{3}\mu\left[\left(\frac{\partial u}{\partial x}-\frac{\partial v}{\partial y}\right)^2+\left(\frac{\partial u}{\partial x}-\frac{\partial w}{\partial z}\right)^2+\left(\frac{\partial w}{\partial z}-\frac{\partial v}{\partial y}\right)^2\right] \quad (16.63)$$

Next, the unbold terms in Eq. (16.60) are considered. Identical terms are combined together to obtain

$$\mu\left[\frac{\partial u}{\partial y}+\frac{\partial v}{\partial x}\right]\frac{\partial u}{\partial y}+\mu\left[\frac{\partial w}{\partial x}+\frac{\partial u}{\partial z}\right]\frac{\partial u}{\partial z}+\mu\left[\frac{\partial u}{\partial y}+\frac{\partial v}{\partial x}\right]\frac{\partial v}{\partial x}+\mu\left[\frac{\partial v}{\partial z}+\frac{\partial w}{\partial y}\right]\frac{\partial v}{\partial z}$$

$$+\mu\left[\frac{\partial u}{\partial z}+\frac{\partial w}{\partial x}\right]\frac{\partial w}{\partial x}+\mu\left[\frac{\partial v}{\partial z}+\frac{\partial w}{\partial y}\right]\frac{\partial w}{\partial y}\mu\left[\left[\frac{\partial u}{\partial y}+\frac{\partial v}{\partial x}\right]^2+\left[\frac{\partial w}{\partial x}+\frac{\partial u}{\partial z}\right]^2+\left[\frac{\partial v}{\partial z}+\frac{\partial w}{\partial y}\right]^2\right] \quad (16.64)$$

Thus, the total input from the Eq. (16.60) is

$$\frac{2}{3}\mu\left[\left(\frac{\partial \boldsymbol{u}}{\partial \boldsymbol{x}}-\frac{\partial \boldsymbol{v}}{\partial \boldsymbol{y}}\right)^2+\left(\frac{\partial \boldsymbol{u}}{\partial \boldsymbol{x}}-\frac{\partial \boldsymbol{w}}{\partial \boldsymbol{z}}\right)^2+\left(\frac{\partial \boldsymbol{w}}{\partial \boldsymbol{z}}-\frac{\partial \boldsymbol{v}}{\partial \boldsymbol{y}}\right)^2\right]+\mu\left[\left(\frac{\partial \boldsymbol{u}}{\partial \boldsymbol{y}}+\frac{\partial \boldsymbol{v}}{\partial \boldsymbol{x}}\right)^2+\left(\frac{\partial \boldsymbol{w}}{\partial \boldsymbol{x}}+\frac{\partial \boldsymbol{u}}{\partial \boldsymbol{z}}\right)^2+\left(\frac{\partial \boldsymbol{v}}{\partial \boldsymbol{z}}+\frac{\partial \boldsymbol{w}}{\partial \boldsymbol{y}}\right)^2\right]-P\nabla\cdot\bar{u}$$

(16.65)

The portion marked bold in Eq. (16.65) is termed as the viscous dissipation (would be denoted by Q) and it would always remain positive. Therefore, the energy equation would be

$$\rho\frac{De}{Dt}=\left[\frac{\partial}{\partial x}\left(k\frac{\partial T}{\partial x}\right)+\frac{\partial}{\partial y}\left(k\frac{\partial T}{\partial y}\right)+\frac{\partial}{\partial z}\left(k\frac{\partial T}{\partial z}\right)\right]+\rho\frac{D}{Dt}\frac{V^2}{2}+\rho gw-P\nabla\cdot\bar{u}+Q \quad (16.66)$$

Applying Eq. (16.41) $\left(e=\hat{u}+\frac{V^2}{2}+gz\right)$ and simplifying Eq. (16.66) results in

$$\rho\frac{D}{Dt}(\hat{u})=\nabla\cdot k\nabla T-P\nabla\cdot\bar{u}+Q \quad (16.67)$$

Since temperature is the desired variable, the principle of thermodynamics is used to appropriately modify the Eq. (16.67). Accordingly, it is considered to use enthalpy (h), which is a function of pressure and temperature.

$$h=h(T,P) \quad (16.68)$$

By maintaining one variable as constant at a time and differentiating Eq. (16.68) gives

$$dh=\left.\frac{\partial h}{\partial T}\right)_P dT+\left.\frac{\partial h}{\partial P}\right)_T dP \quad (16.69)$$

$$=C_P dT+\left.\frac{\partial h}{\partial P}\right)_T dP \quad (16.70)$$

The second term in Eq. (16.70) is relatively complex and hence, requires manipulation. Thus, entropy (s) which is also a function of T and P is introduced.

$$s = s(T, P) \tag{16.71}$$

$$ds = \left.\frac{\partial s}{\partial T}\right)_P dT + \left.\frac{\partial s}{\partial P}\right)_T dP \tag{16.72}$$

For an isothermal process,

$$dh = \left.\frac{\partial h}{\partial p}\right)_T dP \tag{16.73}$$

$$ds = \left.\frac{\partial s}{\partial p}\right)_T dP \tag{16.74}$$

According to the principles of thermodynamics,

$$dh = Tds + vdp \tag{16.75}$$

Considering an isothermal process and substituting Eq. (16.74) in Eq. (16.75),

$$\left.\frac{\partial h}{\partial P}\right)_T dP = T\left.\frac{\partial s}{\partial P}\right)_T dP + vdP \tag{16.76}$$

Again, the second term in the Eq. (16.76) is not easy to determine. Thus, the equation for Gibbs free energy from thermodynamics is applied, which is given by

$$dg = vdP - sdT \tag{16.77}$$

g is a function of P and T as well. Thus,

$$g = g(T, P) \tag{16.78}$$

$$dg = \left.\frac{\partial g}{\partial P}\right)_T dP + \left.\frac{\partial g}{\partial T}\right)_P dT \tag{16.79}$$

On comparing Eq. (16.77) and (16.79),

$$\left.\frac{\partial g}{\partial P}\right)_T = \frac{1}{\rho}, \tag{16.80}$$

$$\left.\frac{\partial g}{\partial T}\right)_P = -s \tag{16.81}$$

Differentiating Eq. (16.80) with respect to T,

$$\frac{\partial^2 g}{\partial T\,\partial P} = \frac{\partial}{\partial T}\left(\frac{1}{\rho}\right)_P \tag{16.82}$$

It is to be accounted that the derivative of g with respect to P can be a function of T, although T remains constant in the process.

Differentiating Eq. (16.81) with respect to P,

$$\frac{\partial^2 g}{\partial P \partial T} = -\left.\frac{\partial s}{\partial P}\right)_T \tag{16.83}$$

Equating Eq. (16.82) to (16.83):

$$\left.\frac{\partial s}{\partial P}\right)_T = -\frac{\partial}{\partial T}\left(\frac{1}{\rho}\right)_P \tag{16.84}$$

Substituting Eq. (16.70) in Eq. (16.73),

$$\left.\frac{\partial h}{\partial P}\right)_T = T\left.\frac{\partial s}{\partial P}\right)_T + \vartheta \tag{16.85}$$

$$= -T\frac{\partial}{\partial T}\left(\frac{1}{\rho}\right) + \frac{1}{\rho} \tag{16.86}$$

$$= -T\frac{1}{\rho^2}\left.\frac{\partial \rho}{\partial T}\right)_P + \frac{1}{\rho} \tag{16.87}$$

By using the definition of coefficient of thermal expansion (β),

$$\beta = -\frac{1}{\rho}\left.\frac{\partial \rho}{\partial T}\right)_P \tag{16.88}$$

Using Eq. (16.88) in Eq. (16.87),

$$\left.\frac{\partial h}{\partial P}\right)_T = -\frac{\beta T}{\rho} + \frac{1}{\rho} = \frac{1}{\rho}(1 - \beta T) \tag{16.89}$$

From the Eq. (16.70),

$$dh = C_P dT + \frac{1}{\rho}(1 - \beta T)dP \tag{16.90}$$

It is known that, for an ideal gas and an incompressible fluid, $\beta = 1/T$. Therefore,

$$dh = C_P dT \tag{16.91}$$

For an incompressible fluid, $h = h(T)$. Substituting $\beta = 0$ in Eq. (16.90) is insignificant. From thermodynamic definition of enthalpy,

$$h = \hat{u} + P\vartheta,\ \vartheta \equiv \frac{1}{\rho} \tag{16.92}$$

Differentiating Eq. (16.92)

$$\frac{d\hat{u}}{dt} = \frac{dh}{dt} - P\frac{dv}{dt} - v\frac{dP}{dt} \tag{16.93}$$

$$\Rightarrow \rho \frac{d\hat{u}}{dt} = \rho \frac{dh}{dt} + P \frac{1}{\rho} \frac{d\rho}{dt} - \frac{dP}{dt} \tag{16.94}$$

Using the continuity equation in Eq. (16.94),

$$\rho \frac{d\hat{u}}{dt} = \rho \frac{dh}{dt} - P\nabla \cdot \bar{u} - \frac{dP}{dt} \tag{16.95}$$

$$= \rho C_P \frac{dT}{dt} + \frac{\rho}{\rho}(1 - \beta T)\frac{dP}{dt} - P\nabla \cdot \bar{u} - \frac{dP}{dt} \tag{16.96}$$

Therefore, the final relationship between internal energy and temperature is given by

$$\rho \frac{d\hat{u}}{dt} = \rho C_P \frac{dT}{dt} - \beta T \frac{dP}{dt} - P\nabla \cdot \bar{u} \tag{16.97}$$

Substituting in the energy equation

$$\rho C_P \frac{dT}{dt} - \beta T \frac{dP}{dt} - P\nabla \cdot \bar{u} = \nabla \cdot (k\nabla T) - P\nabla \cdot \bar{u} + Q \tag{16.98}$$

(or)

$$\rho C_P \frac{dT}{dt} = \nabla \cdot (k\nabla T) + \beta T \frac{dP}{dt} + Q \tag{16.99}$$

Equation 16.99 is the final form of the energy equation (NPTEL, 2018c, d).

CFD modeling aims at solving the Navier–Stokes equations inside a defined *computational domain* or *geometry*. The geometry defines the shape of the problem to be analyzed, it can be two-dimensional (2D) or 3D. It is divided into many sub-domains termed as *cells* which meet at *nodes* or *grid points*. The various elements of a computational domain are shown in Figure 16.8.

CFD, as a simulation tool, involves two major components as shown in Figure 16.9. The accuracy of the solution sought by a CFD model is governed by the algebraic equations obtained from the discretization of the Navier–Stokes equations. On the other hand, the efficiency of the solution is dictated by the algorithm for solving the same (Figure 16.9).

Discretization is the process of transforming the Navier–Stokes equations into discrete counterparts, such that they are appropriate for numerical evaluation and ready to be solved digitally by computers

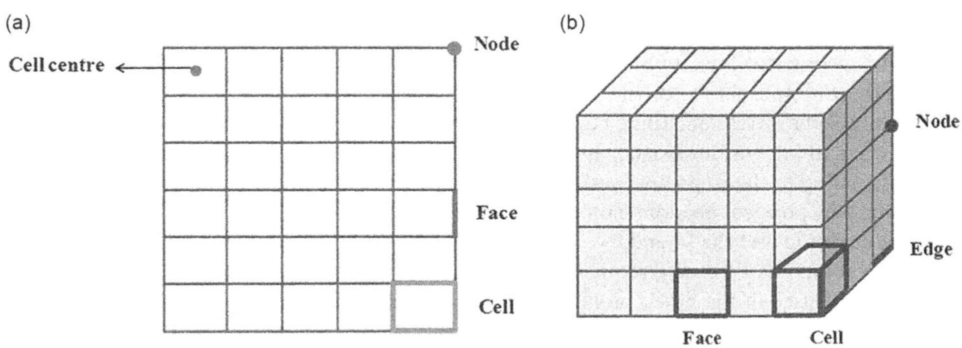

FIGURE 16.8 Geometry of a CFD model: (a) 2D computational grid; (b) 3D computational grid. Edge: the boundary of a face; Face: the boundary of a cell.

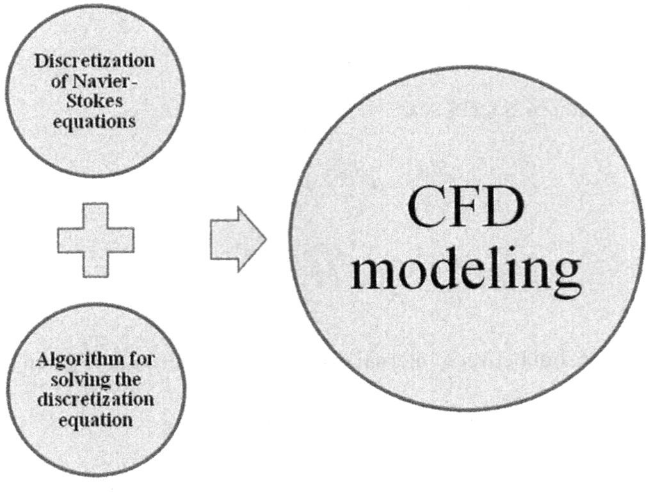

FIGURE 16.9 Components of CFD modeling tool.

using the CFD tool. The well-established computational approaches available for the discretization of Navier–Stokes equations are as follows:

- Finite element method (FEM)
- Finite volume method (FVM)
- Finite difference method (FDM)

These are collectively known as the *numerical methods in CFD modeling*, elaborated in the forthcoming section.

16.2 Numerical Methods in CFD Modeling

Numerical methods are considered as an alternative to solving the PDEs analytically. These methods approximate the PDEs to the numerical model equations, which can then be solved using numerical methods. In turn, the solution thus obtained to the numerical model equations is an approximation of the real solution to the PDEs.

16.2.1 Finite Element Method

The name *finite element* is derived from the concept of FEM, which involves subdividing a computational domain (geometry) into infinitesimal but finite-sized elements of geometrically simple shapes. Collectively, these simple shapes constitute the so-called *finite element mesh*. Also, the response of each element is expressed in terms of a finite number of degrees of freedom (DOF). The DOF is characterized as the value of an unknown function(s), at a set of nodal points.

Before proceeding to the steps involved in FEM, it is important to understand the meaning of boundary conditions. Any physical phenomenon in food engineering involving PDEs can be solved by defining its governing equation and the boundary conditions (Figure 16.10). Boundary conditions (B.C.) are the limits prescribed for solving a differential equation. It involves specifying the fluid behavior and properties at the boundaries of the given problem. For example, in the case of conductive heat transfer, the governing equation is provided by the Fourier's law equation.

$$q_{\text{conduction}} = -kA\frac{dT}{dx} \tag{16.100}$$

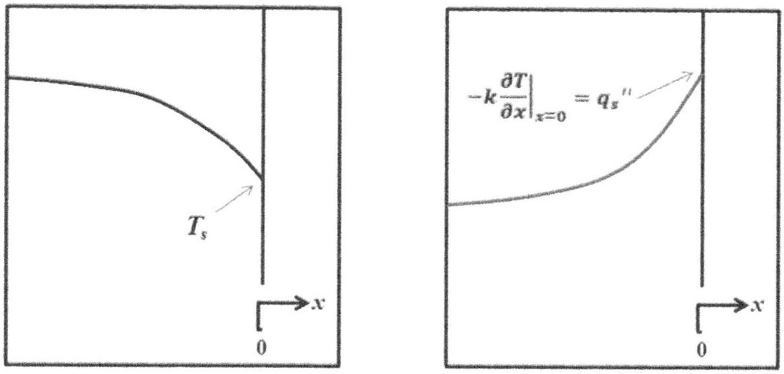

FIGURE 16.10 Boundary conditions for heat transfer.

In general, the boundary conditions for this heat transfer problem could be defined by specifying the following surface conditions of the body (Figure 16.10):

$$\text{Surface temperature}: T\Big|_{x=0} = T_S;\ \text{Ex}: T\Big|_{x=0} = 100 \tag{16.101}$$

$$\text{Surface heat flux}: -k\frac{dT}{dx}\Big|_{x=0} = q_S'';\ \text{Ex}: -k\frac{dT}{dx}\Big|_{x=0} = 3000 \tag{16.102}$$

The initial conditions are the conditions at the time, $t=0$. The boundary and initial conditions are required to solve the governing equation for a particular physical situation. In general, boundary conditions can be classified into four types: no-slip, axisymmetric, inlet–outlet, and periodic. Consider the example of a pipe in which the fluid is flowing from left to right (Figure 16.11). While the left part represents the input, the right part shows the output of the system. As the fluid enters the pipe from the inlet located at the left, the velocity can be set manually. Nevertheless, the outlet (on the RHS) boundary condition may be used to keep all the properties constant, implying that all the gradients are zero. In the no-slip boundary condition, the velocity at the wall of the pipe is set at zero. In other words, the no-slip boundary condition implies that the velocity of a fluid which is in contact with the wall of the pipe is equal to the velocity of the wall. But at the center of the pipe, the axisymmetric boundary condition is used as the geometry and the pattern of flow solution, having mirror symmetry (Figure 16.11). The asymmetric boundary condition can be defined by a zero flux of all quantities across the symmetric boundary. Since there is no convective flux across a symmetry plane, the normal velocity component at the symmetry plane is zero. Also, due to the absence of diffusion flux across a symmetry plane, the normal gradients of all flow variables are zero at the symmetry plane.

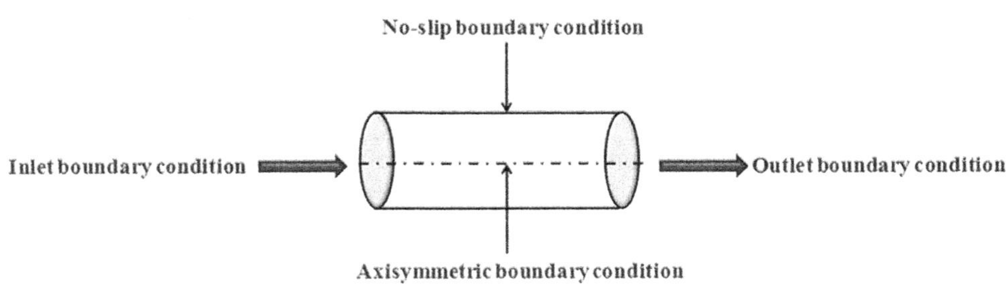

FIGURE 16.11 Boundary conditions of pipe flow.

On the other hand, periodic boundary conditions are applied when the geometry under study and the expected pattern of the flow/thermal solution are of periodically repeating nature. Ultrasound and pulsed electric field processing are the typical examples wherein periodic boundary conditions are used.

The scheme of FEM is shown in Figure 16.12. After discretization of the geometry and defining the boundary conditions for the specified physical phenomenon, a suitable interpolation function (a linear or quadratic polynomial, with a finite number of DOFs) is selected to approximate the unknown solution within an element. Then, the governing equation in the form of PDE is approximated to algebraic equation known as the *finite element equation*, which can be linear or nonlinear. The algebraic equations are obtained for each element, and then the element equations are collated, which is relatively simple than obtaining the algebraic equations for the entire domain. The resultant large matrix equation system can be solved by any of the available matrix solvers, to obtain the unknown variables at the nodes.

The major advantages of FEM include its ability to handle very complex geometry and a broad range of engineering problems such as heat transfer, electromagnetic heating, and all types of structural analysis. Also, FEM is more appropriate for problems that involve more than one physical phenomenon, i.e., multiphysics analysis. FEM enables mixed formulations which involves combining different kinds of functions that approximate the solution within each element. An example of multiphysics analysis could

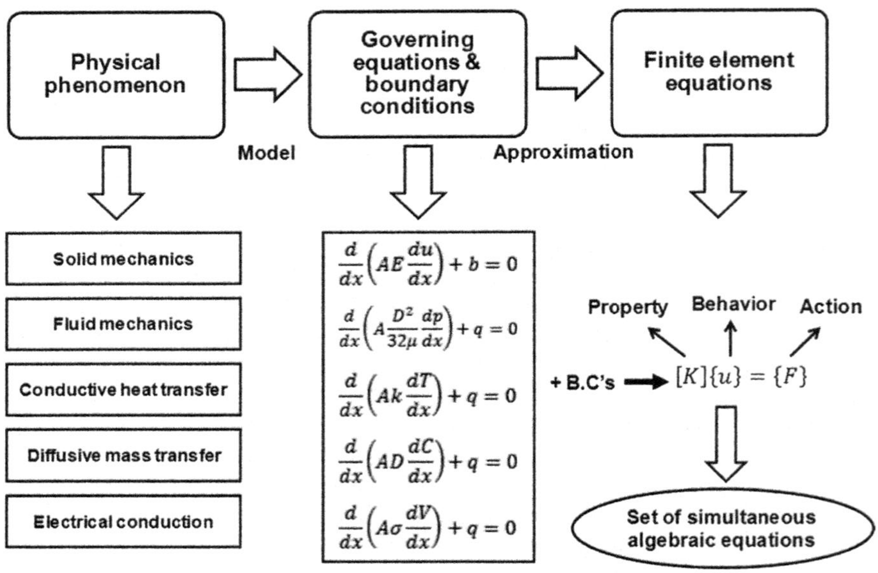

Unknown

	Property [K]	Behavior {u}	Action {F}
Elastic	Stiffness	Displacement	Force
Thermal	Conductivity	Temperature	Heat source
Fluid	Viscosity	Velocity	Body force
Electrostatic	Dielectric permittivity	Electric potential	Charge

FIGURE 16.12 Procedure of FEM.

be that of electromagnetic heating (microwave/radiofrequency heating), which requires one type of function for the electromagnetic field and a different kind of function for heat transfer. FEM facilitates the coupling of both the functions to obtain an accurate solution to the problem.

16.2.2 Finite Volume Method

FVM is the classical approach used in both the commercial and research applications of CFD. Similar to FEM, in this method, the flow geometry is first divided into minute but finite-sized and nonoverlapping *control volumes* or *cells*, where the variable of interest is located at the centroid of the control volume. The governing equations are discretized or approximated over the control volumes in terms of the nodal values. Finally, the discretized equations are solved numerically. The applications of FVM are well established in heat transfer-related food engineering problems.

16.2.3 Finite Difference Method

In the FDM, a point in space is considered where the continuum representation of the equations is considered and replaced with a set of discrete equations, called *finite-difference equations*. It is the most direct approach of approximating the PDEs. The prerequisite for the FDM approach is that it requires structured meshes and is typically defined on a regular grid. Therefore, this method is often adopted for rectangular or block-shaped models but cannot be used for irregular geometries. Both FDM and FVM are well established for handling problems related to convective heat transfer and fluid flow.

16.3 Reference Frames in CFD Modeling

The prerequisite for the modeling of the interaction between two phases (gas-droplet) involves defining the modeling frames, also known as the *reference frames*. Reference frames are important in determining how the particles or droplets interact with the surrounding gas medium in the computational domain. For instance, the reference frames for modeling a spray-drying operation deals with two phases, namely, the dispersed and continuous phases constituted by the droplets and gas medium, respectively. As already known, the designation of phases depends on the volume of the component, the one in larger volume is the continuous phase, and that in smaller volume is the dispersed phase. Reference frames differ in the manner how they consider the interaction between the abovementioned two phases (Anandharamakrishnan and Ishwarya, 2015). Accordingly, there are three different types of reference frames, which are discussed subsequently.

16.3.1 Volume of Fluid

The volume of fluid (VOF) reference frame (Figure 16.13a) is designed for two or more immiscible fluids. VOF solves a single set of momentum equations and tracks the volume fraction of each fluid all over the domain. As the fluids do not mix, each computational cell is filled with either one of the pure fluids or the

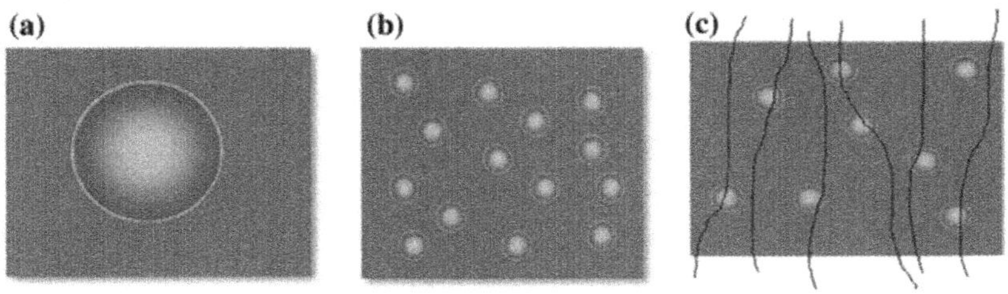

FIGURE 16.13 Reference frames: (a) volume of fluid; (b) Eulerian–Eulerian; (c) Eulerian–Lagrangian. (Reproduced with permission from Anandharamakrishnan, C. 2013. *Computational Fluid Dynamics Applications in Food Processing (Springer Briefs in Food, Health, and Nutrition)*. New York: Springer.)

interface between two (or more) fluids. Typical applications of VOF modeling include the prediction of a jet breakup, the motion of large bubbles in a liquid, and the steady or transient tracking of any liquid–gas interface (ANSYS Fluent User Guide, 2006).

16.3.2 Eulerian–Eulerian

The most commonly used two-phase modeling frames are the Eulerian–Eulerian (E–E) and the Eulerian–Lagrangian (E–L) methods. The E–E approach (Figure 16.13b) considers that the dispersed and continuous phases completely interpenetrate each other. This signifies that each of the two phases is treated as a continuum (Huang and Mujumdar, 2006). There are two different cases under the E–E approach. In the first case, each computational cell contains specific fractions of each of the two phases, i.e., droplets and gas. The dynamics in this computational domain is evaluated by means of averaged transport equations of motion. This approach solves the Navier–Stokes equation individually for each phase, and the subsequent coupling is done by the applied pressure and interphase exchange systems. The transport equations are written such that their volume fractions sum up to one (unity).

In the second case of the E–E approach, the computational cell consists of a single phase, and consequently, the transport equations for the two phases revert to the conventional single-phase system. The E–E approach is considered cheaper in terms of its computational requirements as it obtains the solution with one set of transport equations. However, the latter is true only when a single particle diameter can describe the dispersed phase. If the particle size distribution is multimodal, then the solution can be achieved only with separate sets of transport equations for different particle sizes. Thus, the E–E approach is appropriate for flows with a narrow range of particle sizes or when the droplet and gas phases are present at similar fractions. Also, the E–E approach is more suitable when the focus is not on the particle properties (Mostafa and Mongia, 1987; Jakobsen et al., 1997).

16.3.3 Eulerian–Lagrangian

The E–L approach (Figure 16.13c) falls under the category of discrete phase modeling in which the droplets and surrounding gas medium are treated as two separate entities, unlike the E–E approach. For example, this approach has been applied to the spray-drying process in which the gas phase is modeled using the standard Eulerian approach as described earlier, and the spray is characterized by a number of discrete computational *particles*. The E–L frame provides a particle tracking approach (Figure 16.13c). Individual particles are tracked through the geometry from their injection point until they escape the domain in a Lagrangian framework (Nijdam et al., 2006) (Figure 16.14).

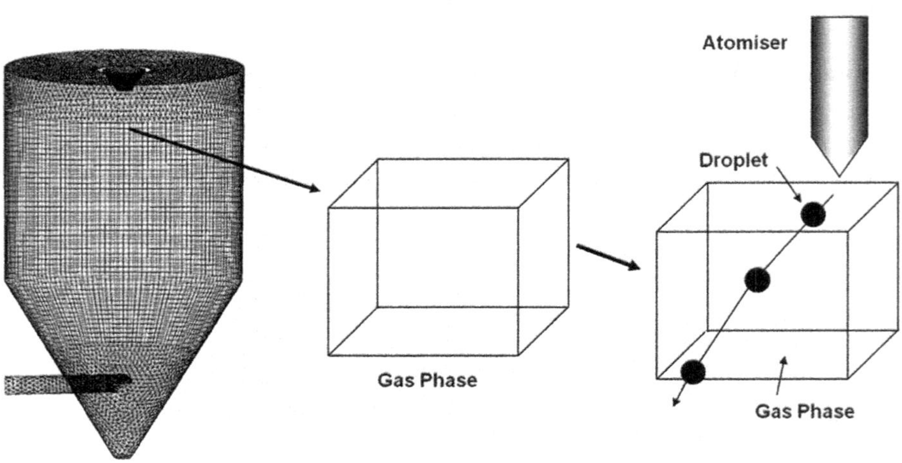

FIGURE 16.14 Schematic representation of the E–L approach applied to the spray-drying process. (Reproduced with permission from Anandharamakrishnan, C. and Ishwarya, S. P. 2015. *Spray Drying Techniques for Food Ingredient Encapsulation*. Hoboken, NJ, USA: John Wiley and Sons Ltd.)

The E–L model overcomes the disadvantage associated with the E–E approach concerning the handling droplets of broader particle size distribution. It is computationally cheaper than the E–E method for a broad range of particle sizes. However, the approach can be expensive if a large number of particles have to be tracked. The E–L approach is best when the dispersed phase does not exceed 10% by volume of the mixture in any region (Marshall and Bakker, 2002). Further, it can provide added details on the behavior and residence times (RTs) of individual particles and can potentially approximate mass and heat transfer more accurately.

16.4 Steps in CFD Modeling

16.4.1 Preprocessing

During preprocessing, a user defines the following:

- The model
- The geometry of the problem
- The element type(s) to be used
- Material properties of the elements
- Geometric properties of the elements (e.g., length, area)

The geometry thus defined is then divided into discrete cells. This is known as *meshing*, which can be done using standard computer-aided design (CAD) program. Meshes can be classified depending on their dimension and type of element that comprise them. The meshes generated could be 2D, that lie in a single plane (generally, in *XY* plane) or 3D, that are not limited to lie in a single plane. The commonly used elements in 2D and 3D meshes are shown in Figure 16.15.

The mesh can be finer or coarser depending on the problem. In general, coarser the mesh, faster the computation and finer the mesh (i.e., more elements), more accurate is the discretization. Mesh

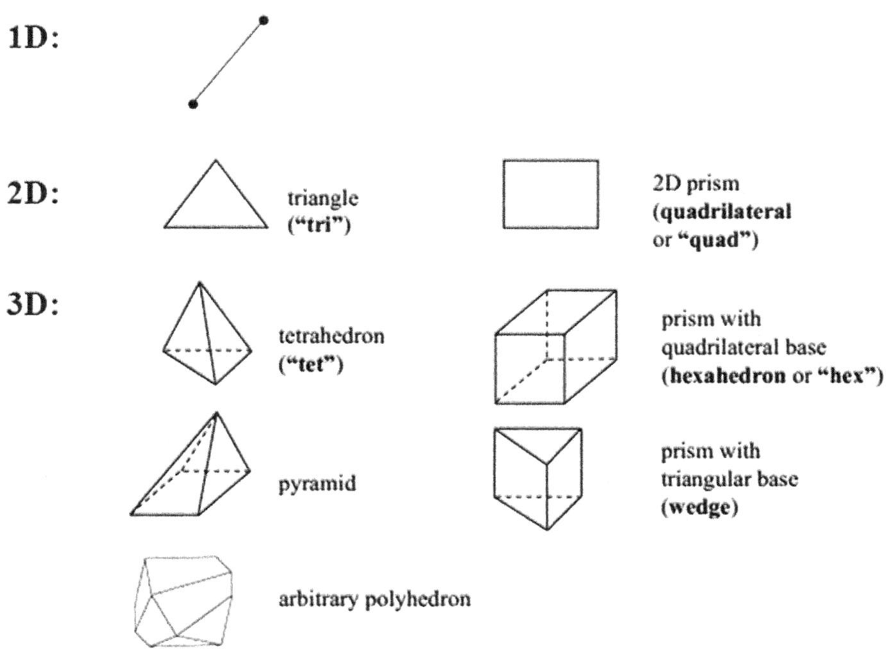

FIGURE 16.15 Types of mesh elements.

refinement is often done on a trial-and-error basis. For instance, the mesh size is decreased by 50%, and the computational simulation is redone. The resultant solutions are compared, and if found similar, the initial mesh configuration is considered appropriate for the defined geometry. However, if there are considerable differences between the two, a more refined mesh needs to be adapted, and the analysis should be repeated until *convergence* is established. *Convergence* refers to arriving at a solution which is close to the exact solution within predefined error tolerance or any other specific criterion. The solution is said to be *mesh convergent* if there is no significant change in the results with the mesh refinement. To obtain an effective model, it is important to arrive at a mesh which fulfils the mesh convergence but is not finer than that which is necessary. This is relevant as mesh refining is done at the expense of longer computation times. Similarly, the hit-and-trial approach can be done with varied element types as well.

16.4.2 Solving

Solving involves defining the analysis type (steady state or transient) and specifying the boundary conditions. For unsteady-state problems, the initial conditions are also specified. At this stage, the model is ready to be solved using the commercially available CFD software packages.

16.4.3 Post-Processing

Post-processing involves time-specific analysis of results and visualization of the resultant solution using contour plots, line graphs, vector displays, and deformed shapes. Alternatively, the results can also be compiled in the tabular format (Madenci and Guven, 2006).

16.4.4 Steps in CFD Modeling: A Case Study on Heat and Mass Transfer during Potato Drying

Step 1: Preprocessing

i. **Description of the problem**

   Among the various techniques available for the drying of fruits and vegetables, convective drying is the most widely used owing to its simplicity and cost-effectiveness. This case study to explain the different steps involved in the CFD modeling of simultaneous heat and mass transfer that occurs during the convective drying of potato slices (Figure 16.16) is excerpted from the study done by Aprajeeta et al. (2015).

ii. **Definition of physics**

   The model deals with two physical phenomena: heat transfer (from the surrounding hot air to the potato slices) and mass transfer (diffusion of moisture from the potato slices to the surrounding air). Heat transfer is assumed to occur solely by conduction, governed by the temperature gradient. Similarly, mass transfer is assumed to occur only by diffusion,

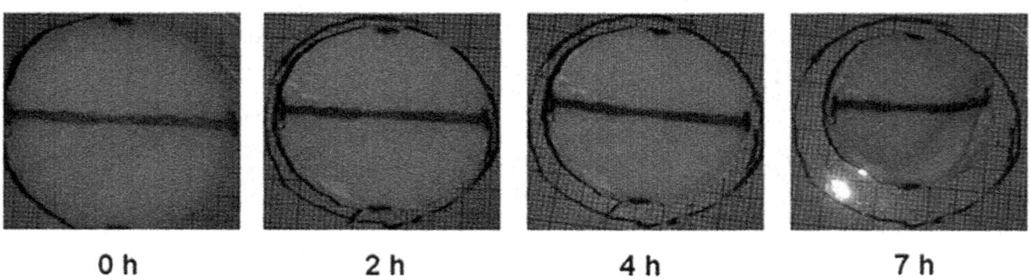

FIGURE 16.16 Images of potato slices during the experimental convective drying process. (Reproduced with permission from Aprajeeta, J., Gopirajah, R. and Anandharamakrishnan, C. 2015. Shrinkage and porosity effects on heat and mass transfer during potato drying. *Journal of Food Engineering* 144: 119–128.)

controlled by the moisture concentration gradient. The governing equations for the above physical phenomena are given as follows:

Heat transfer

The temperature profile resultant from the heat transfer by conduction through the porous matrix of a food product can be described by the Fourier equation as follows:

$$\left(\rho C_P\right)_{eq}\frac{\partial T}{\partial t}=\nabla\cdot\left(k_{eq}\nabla T\right)+Q \tag{16.103}$$

$$k_{eq}=\theta_P k_P+(1-\theta_P)k_L \tag{16.104}$$

$$\left(\rho C_P\right)_{eq}=\theta_P C_P\rho_P+(1-\theta_P)C_{PL}\rho_L \tag{16.105}$$

where ρ is the density of potato sample (kg/m^3); C_P is the specific heat (J/kg K); T is the temperature (K); k is the thermal conductivity (W/m K); θ is the volume fraction; the subscripts *eq*, *P*, and *L* stand for equivalence, particle (sample), and fluid (drying air), respectively. Under steady-state conditions, Q is considered as zero, owing to the absence of heat generation within the sample.

Mass transfer

Under the assumption that mass transport occurs solely by diffusion, Fick's law was applied in this model to define the mass transport equation (Eq. (16.106)).

$$(\varepsilon)\frac{\partial C}{\partial t}+\left(C-\rho_P C_P\right)\frac{\partial\varepsilon}{\partial t}=-\nabla\cdot\left(D_{\text{eff}}\nabla C\right) \tag{16.106}$$

$$D_{\text{eff}}=\varepsilon\tau D_f \tag{16.107}$$

$$\rho_P=\frac{\rho_b}{(1-\varepsilon)} \tag{16.108}$$

where ε is the porosity, C is the moisture concentration (mol/m^3), D_{eff} is the diffusion coefficient of potato (m^2/s), τ is the tortuosity factor, D_f is the free diffusion coefficient, and ρ_b is the density of bulk material.

A 2D finite element model was applied to predict the simultaneous heat and mass transfer occurring in the potato sample during the drying process. The computational software COMSOL Multiphysics (version 4.4) was used to solve the abovementioned PDEs. The modules, heat transfer in porous media and species transport in porous media were used for predicting heat and mass transfer phenomena, respectively. An HP workstation with dual quad-core Xeon processor of speed 2.5 GHz and RAM capacity 12 GB was used to study this unsteady-state (transient) problem by utilizing a direct linear solver with a time step of 30 s for a duration of 25200 s (7 h) (Aprajeeta et al., 2015).

iii. **Defining the model parameters**

The thermophysical properties of the matrix subjected to drying (potato slices) were calculated from the expressions listed in Table 16.2.

iv. **Creation of geometry**

The geometry for this model is a circle of diameter 4.5 cm (Figure 16.17), which represents the initial dimension of the potato slice subjected to drying during the experimental study.

TABLE 16.2
List of Model Parameters

Parameter	Expression	Reference
Thermal conductivity of porous matrix (k_p)	$k_p = 0.148 + 0.493M_{wb}$	Chen et al. (1998)
Thermal conductivity of water (k_w)	$k_w = 0.57109 + 0.0017625T - 0.000067036T^2$	Singh and Heldman (2009)
Bulk density (ρ_b)	$\rho_b = \frac{(1+M_{db})}{\left(\frac{1}{\rho_{b0} + \beta\frac{M_{db}}{\rho_w}}\right)}$, where $\beta = 1$ and $\rho_{b0} = 1500\,\text{kg/m}^3$	Korokida and Maroulis (1997)
Particle density (ρ_P)	$\rho_P = \frac{(1+M_{db})}{\left(\frac{1}{\rho_s} + \frac{M_{db}}{\rho_w}\right)}$, Where $\rho_s = 1780 - 1.05T$	Tasami and Katsioti (2000)
Fluid density (ρ_w)	$\rho_w = 997.18 + 0.0031439T - 0.0037574T^2$	Singh and Heldman (2009) Boukouvalas et al. (2006)
Porosity (ε)	$\varepsilon = 1 - \rho_b/\rho_P$	Korokida and Maroulis (1997)
Tortuosity factor (τ)	$\tau = (1-\theta_P)^{1/3}$, where $\theta_P = 0.2497$	Comsol user guide 4.4
Specific heat capacity (C_{Pp})	$C_{P_p} = 837 + (33.5M_{wb})$	Siebel (1892)
Specific heat capacity fluid (C_{Pw})	$C_{P_w} = 4.1762 + 9.0864\ e^{-5}\ T - 5.4731\ e^{-6}\ T^2$	Singh and Heldman (2009)
Heat transfer coefficient (h)	$h = (N_{Nu} \times k_a)/d$, where $N_{Nu} = 0.664 \times N_{Re}^{0.5} \times N_{Pr}^{0.33}$. The heat transfer coefficient was estimated to be $7.315\,\text{W/m}^2\,\text{K}$	Wang and Brennan (1995); Yang et al. (2001); Perry and Green (1984); Singh and Heldman (2009)
Diffusion coefficient (D_f)	$3.5 \times 10^{-9}\,\text{m}^2/\text{s}$	Hassini et al. (2007)

Source: Aprajeeta et al. (2015).

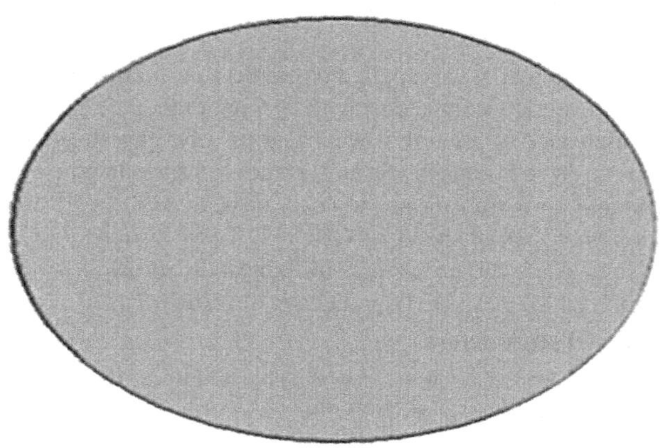

FIGURE 16.17 Geometry of potato slice. (Reproduced with permission from Aprajeeta, J., Gopirajah, R. and Anandharamakrishnan, C. 2015. Shrinkage and porosity effects on heat and mass transfer during potato drying. *Journal of Food Engineering* 144: 119–128.)

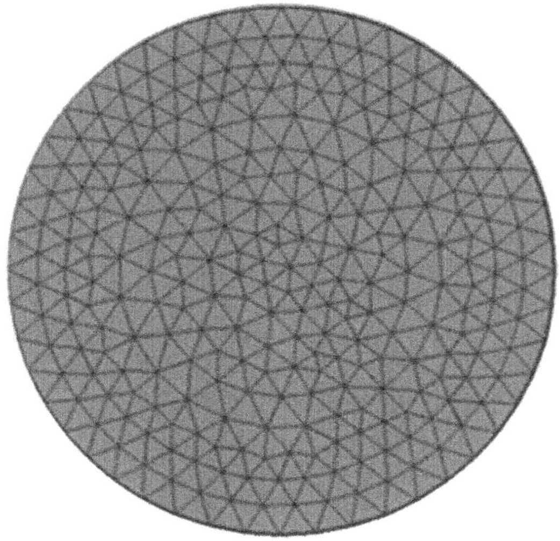

FIGURE 16.18 Meshed geometry of potato slice. (Reproduced with permission from Aprajeeta, J., Gopirajah, R. and Anandharamakrishnan, C. 2015. Shrinkage and porosity effects on heat and mass transfer during potato drying. *Journal of Food Engineering* 144: 119–128.)

v. **Meshing**

The mesh comprised of triangular finite elements, as shown in Figure 16.18.

Step 2: Solving

i. **Defining the type of analysis**

This is an unsteady-state analysis, and it models the time-dependent change in moisture of the potato slices during drying.

ii. **Defining the initial conditions and boundary conditions**

– **Initial conditions for heat transfer and mass transfer**

$$\text{Heat transfer}: T = T_i;\ t = 0, 0 \le r \le R \tag{16.109}$$

$$\text{Mass transfer}: C = C_{\text{int}};\ t = 0, 0 \le r \le R \tag{16.110}$$

– **Boundary conditions for heat transfer and mass transfer**

i. **Heat transfer**

There is no temperature change at the geometric center of the sample during the final stage of drying process. Hence, the following condition was applied:

$$\frac{\partial T}{\partial r};\ t > 0, r = 0 \tag{16.111}$$

In the boundary condition, the flux in the absence of any temperature surface layer is given by

$$k\frac{\partial T}{\partial r} = h(T_o - T_i);\ t > 0,\ \ r = R(t) \tag{16.112}$$

ii. **Mass transfer**

There is no moisture change at the geometric center of the sample at the final stage of drying process, and hence, the following conditions can be applied:

$$\frac{\partial C}{\partial r};\ t > 0, r = 0 \tag{16.113}$$

$$C = C_o;\ t > 0;\ r = R \tag{16.114}$$

Step 3: Post-processing

The results from the simulation yielded the temperature profile (Figure 16.19) and moisture content profile (Figure 16.20) at different time intervals. The surface contours of the temperature profile showed that the major heat penetration within the potato slice occurs during the early initial stages of drying. Subsequently, a slight increase in temperature was observed inside the sample. From Figure 16.19, it can also be observed that the temperature distribution within the sample is nonuniform throughout the process. The prediction of spatial temperature profile is useful in tracking the glass transition temperature and various other chemical reactions to optimize the product and process parameters.

Similar to the temperature profile, the surface contours shown in Figure 16.20 represents the moisture distribution in every spatial position at each time step. This is of relevance in finding the slowest heating zone (SHZ), which is key to assessing the effectiveness of any heat transfer process (Padmavati and Anandharamakrishnan, 2013). From Figure 16.20, it is evident that

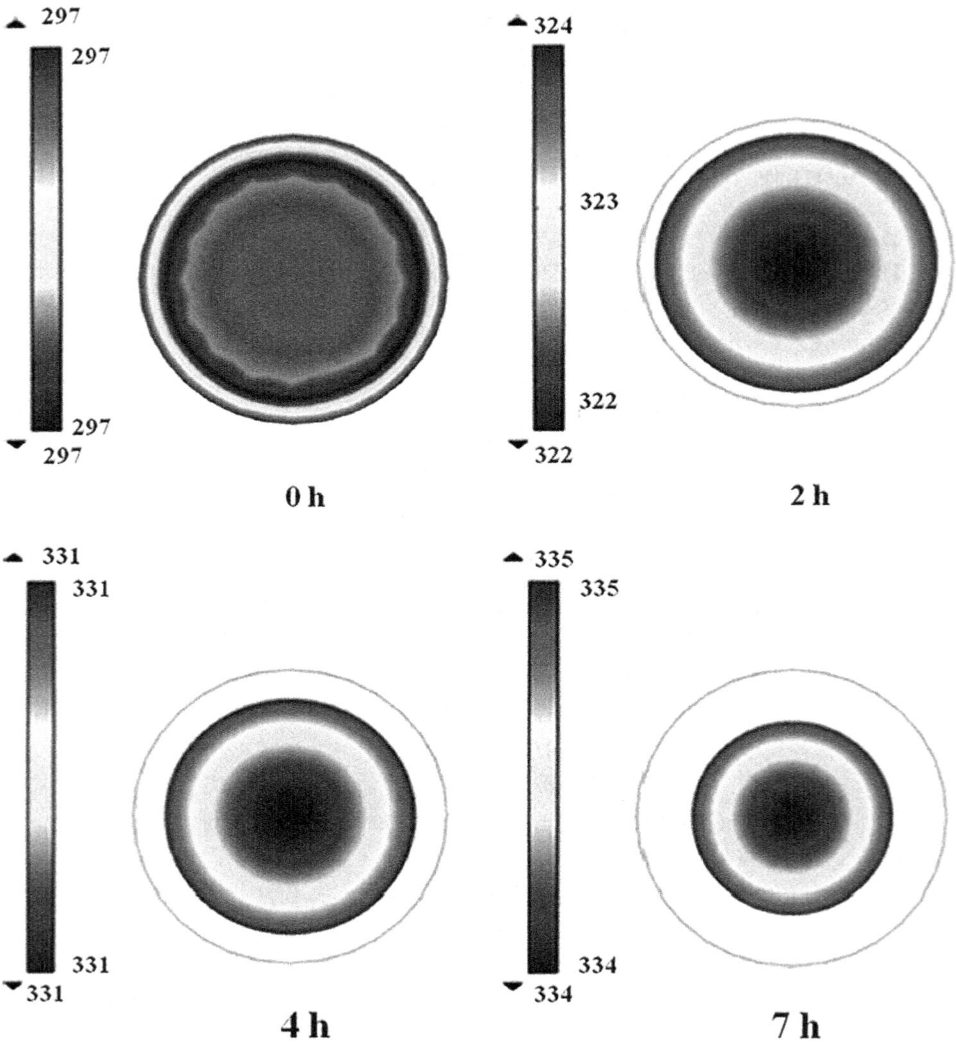

FIGURE 16.19 Predicted temperature profile at different time intervals. (Reproduced with permission from Aprajeeta, J., Gopirajah, R. and Anandharamakrishnan, C. 2015. Shrinkage and porosity effects on heat and mass transfer during potato drying. *Journal of Food Engineering* 144: 119–128.)

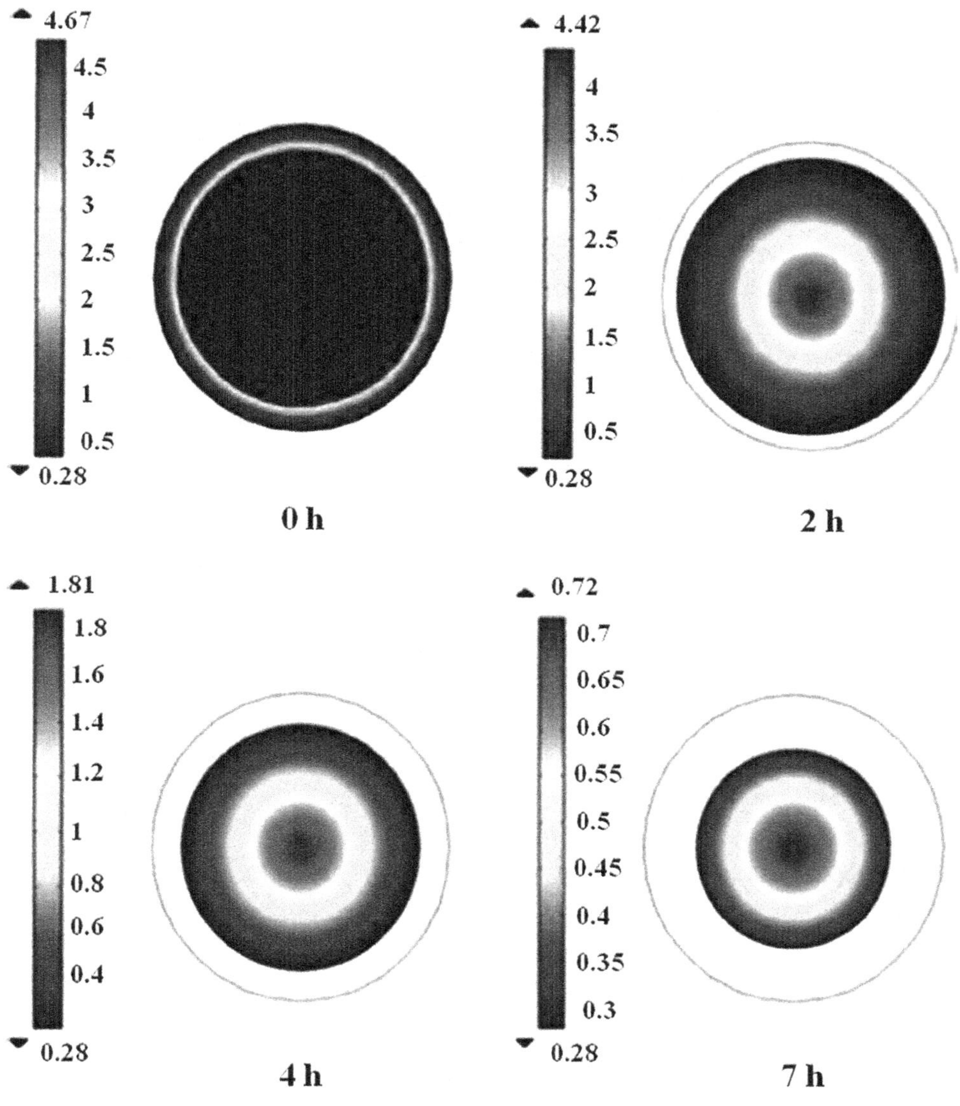

FIGURE 16.20 Predicted moisture content (on dry basis) profile at different time intervals. (Reproduced with permission from Aprajeeta, J., Gopirajah, R. and Anandharamakrishnan, C. 2015. Shrinkage and porosity effects on heat and mass transfer during potato drying. *Journal of Food Engineering* 144: 119–128.)

the SHZ is located at the geometric center of the sample. Thus, the practical application of this model is to estimate the moisture content at a particular location within the food sample at any instance of time during its drying (Aprajeeta et al., 2015).

16.5 An Overview of CFD Applications in Food Processing

The applications of CFD commenced in 1960 and gained momentum as the computational capabilities evolved and improved across the years. Although the aerospace, automobile, and nuclear industries were the pioneers in adopting the CFD approach, application in the food sector began in the late 20th century. The major applications of CFD in food industries are represented in Figure 16.21. The forthcoming

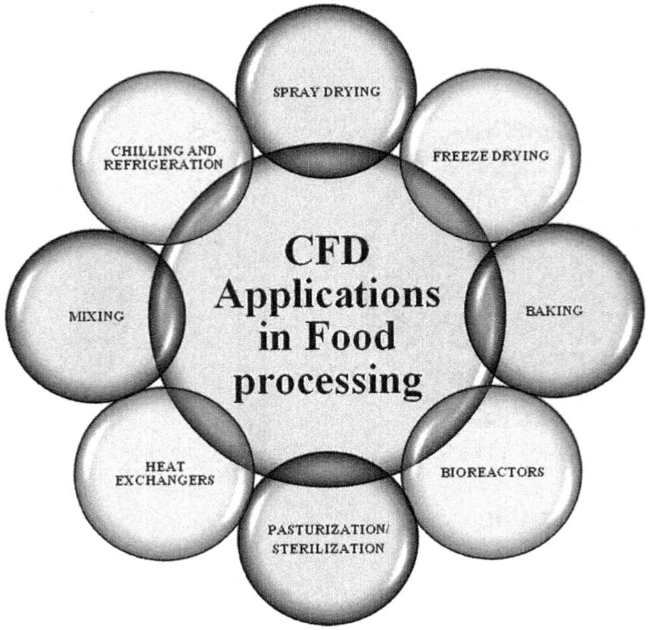

FIGURE 16.21 Applications of CFD in food processing.

discussions would provide an insight to the applications of CFD in pasteurization of canned milk, spray drying, and bread baking.

16.5.1 CFD Modeling of Canned Milk Pasteurization

The objective of an industrial pasteurization process is to extend the shelf life without compromising the product quality. An adequate degree of the heat treatment is necessary to achieve the abovementioned objective. Thermal pasteurization includes the following mechanisms of heat transfer (Chen and Ramaswamy, 2007):

- Conduction (in solid foods)
- Natural convection (in low viscosity liquid foods)
- Combination of convection and conduction (in liquid foods containing solid particles or viscosity modifiers such as starch).

During pasteurization, the analysis of the temperature profile inside the product is a transient problem, as it depends on both the position and time. To conduct an experimental investigation of the same, thermocouples need to be placed at different positions to record the temperature in a container during heating. But, the presence of thermocouple wires may disturb the flow patterns and restrict the free movement of the liquid, thus leading to erroneous temperature measurements (Stoforos and Merson, 1990). Therefore, studying the thermal behavior of foods in real time is laborious owing to its complexities such as variation in initial temperature, nonlinear and non-isotropic thermal properties, irregular-shaped geometries, and transient boundary conditions (Puri and Anantheswaran, 1993). Also, the complex nature of heat transfer during natural convection heating leads to difficulties in locating the SHZ. It is defined as the location within the food product which receives the minimum heat (Ghani et al., 2003). Also, it is intricate to measure the temperature at the SHZ due to the continuous change in the position of this region during the heating progress (Ghani et al., 2002). In the aforementioned perspective, CFD modeling facilitates the prediction of temperature profiles by finding numerical solutions for the PDEs that govern the heat transfer process.

The time–temperature combination for the thermal pasteurization of canned milk should be selected after considering the heat sensitivity of various nutrients such as thiamine, α-lactoalbumin, and β-lactoglobulin present in milk (Walstra et al., 2006). It is vital to attaining a uniform and effective heat distribution inside the canned milk. Thus, the processing time is determined based on the time required to achieve a homogeneous temperature distribution within the can. Nevertheless, arriving at the temperature profile within the canned milk is challenging during an industrial scale continuous operation. To address this limitation, a CFD model was developed for the pasteurization process of canned milk. This model aimed to study the effect of thermal processing time on the temperature distribution inside the can and validate the same with experimental measurements (Paul et al., 2011). In addition, this model also simulated the effect of can rotation on the temperature distribution, processing time, and pasteurization unit (PU). The degree of heat treatment received by the product was expressed in terms of the PU given by the following equation (Shapton et al., 1971; Holdsworth and Simpson, 2007), which is similar to the *F*-value equation:

$$PU_{T_{\text{ref}}} = \int_0^t 10^{(T-T_{\text{ref}})/z} dt \tag{16.115}$$

where T_{ref} is the reference processing temperature, and T is the temperature of the milk. Eq. (16.115) can be interpreted as follows: a PU value of 1 is equivalent to a process of 1 min at a specified processing temperature. PU value is applied to determine the influence of time and temperature on the inactivation of microorganisms.

The CFD model utilized a 3D cylindrical structure of the milk can (Figure 16.22a). A hexahedral mesh of size 0.001 m with 73584 grid cells (Figure 16.22b) was used to minimize the discretization error. The FVM was used to solve the partial differential forms of the continuity equations using the SIMPLE (Semi-Implicit Pressure-Linked Equations) approach for pressure–velocity coupling and a second-order upwind scheme to interpolate the variables on the surface of the control volume due to its stability with respect to time step size. The thermophysical properties of milk listed in Table 16.3 were used as the model parameters. CFD model predictions were validated with experimental measurements of

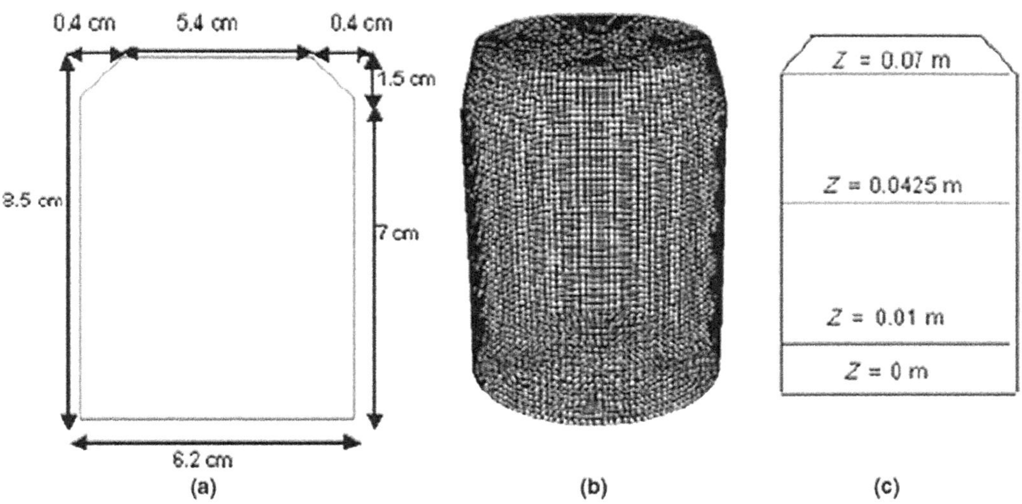

FIGURE 16.22 (a) Can geometry; (b) meshed can; (c) radial planes. (Reproduced with permission from Paul, D. A., Anishaparvin, A. and Anandharamakrishnan, C. 2011. Computational fluid dynamics studies on pasteurisation of canned milk. *International Journal of Dairy Technology* 64: 305–313.)

TABLE 16.3

Thermophysical Properties of Milk

Properties	Value/Equation	Reference
Density (kg/m^3)	1029 (Boussinesq approximation)	Walstra et al. (2006)
Thermal conductivity (W/m K)	$k = \left(326.8 + 1.0412T - 0.00337 \times T^2\right)$ $\left(0.44 + 0.54X_m^{\text{Water}}\right)1.73 \times 10^{-3}$	Riedel (1949)
Thermal expansion coefficient (K^{-1})	0.0002	Short (1955)

Source: Paul et al. (2011).
T: temperature (K); X_m: mass fraction.

temperature recorded at three different positions along the height of can in the radial direction (bottom: $z = 0.01$ m, middle: $z = 0.0425$ m, and top: $z = 0.07$ m) (Figure 16.22c).

The transport equations for conservation of mass, momentum, and energy were the governing equations for the model. The study considered a combination of conduction (heat transfer to milk near the walls of the can) and natural convection (heat transfer within the milk) heat transfer mechanisms. A transient 3D CFD simulation was carried out with three-side heating without slip conditions. A 0.2 mm wall thickness was considered with bottom and side wall set at two different temperatures (T_{wall}): 74°C and 85°C, approximated to be equivalent to the hot water temperature, as the thermal resistance of can material was negligible. The temperature of the top wall was set at 32°C (ambient temperature at the location where the experiment was conducted).

In the stationary mode, the model predictions were in good agreement with the experimental results at the positions, point-1 and point-2, for both the wall temperatures considered (Figure 16.23a and b for $T_{\text{wall}} = 74$°C and Figure 16.24a and b for $T_{\text{wall}} = 85$°C). At 74°C, the temperature increased linearly during the first 180 s of the heating following which it attained a steady-state temperature of 69.5°C at point-1 and 69.2°C at point-2 by 480 s (Figure 16.23a and b). Similar observations were noted at 85°C. However, the steady-state temperature was attained at 540 s, which was 79.6°C at point-1 and 79.2°C at point-2 (Figure 16.24a and b). The rapid increase in temperature was ascribed to the low viscosity of milk. The variation in milk properties led to a slight difference between the experimental and simulation results.

The model showed that during the initial stage of heating (60 s), the SHZ was formed at the bottom portion of the can due to change in viscosity, density, and recirculating pattern of the milk (Figure 16.25a and b). With time, the zone of minimal heating moved towards the core region, indicated by a lower temperature of milk at the central plane compared to the surrounding regions (Figure 16.25d).

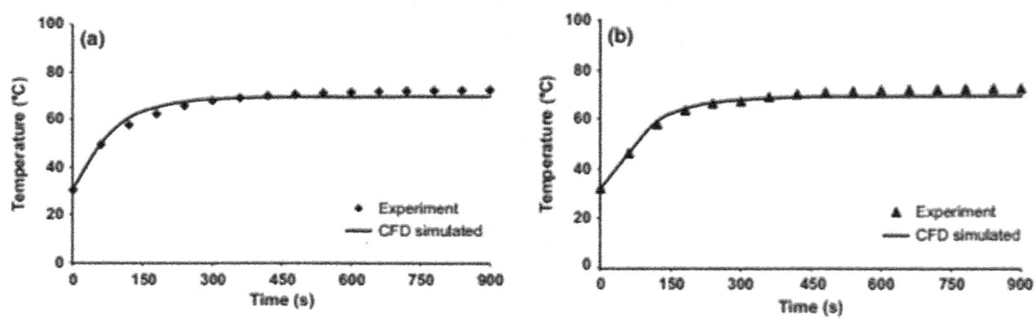

FIGURE 16.23 Comparison of experimental and CFD simulation results at 74°C (a) point-1 (from can top to 0.05 down and 0.01 m from the center); (b) point-2 (from can top to 0.065 and 0.02 m from the center). (Reproduced with permission from Paul, D. A., Anishaparvin, A. and Anandharamakrishnan, C. 2011. Computational fluid dynamics studies on pasteurisation of canned milk. *International Journal of Dairy Technology* 64: 305–313.)

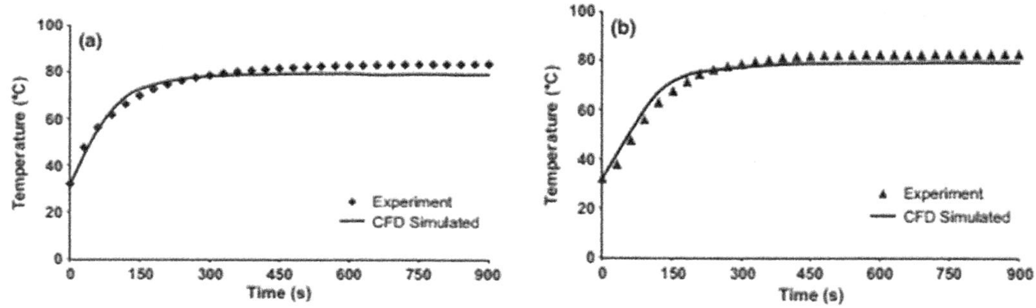

FIGURE 16.24 Comparison of experimental and CFD simulation results at 85°C (a) point-1 (from can top to 0.05 down and 0.01 m from the center); (b) point-2 (from can top to 0.065 and 0.02 m from the center). (Reproduced with permission from Paul, D. A., Anishaparvin, A. and Anandharamakrishnan, C. 2011. Computational fluid dynamics studies on pasteurisation of canned milk. *International Journal of Dairy Technology* 64: 305–313.)

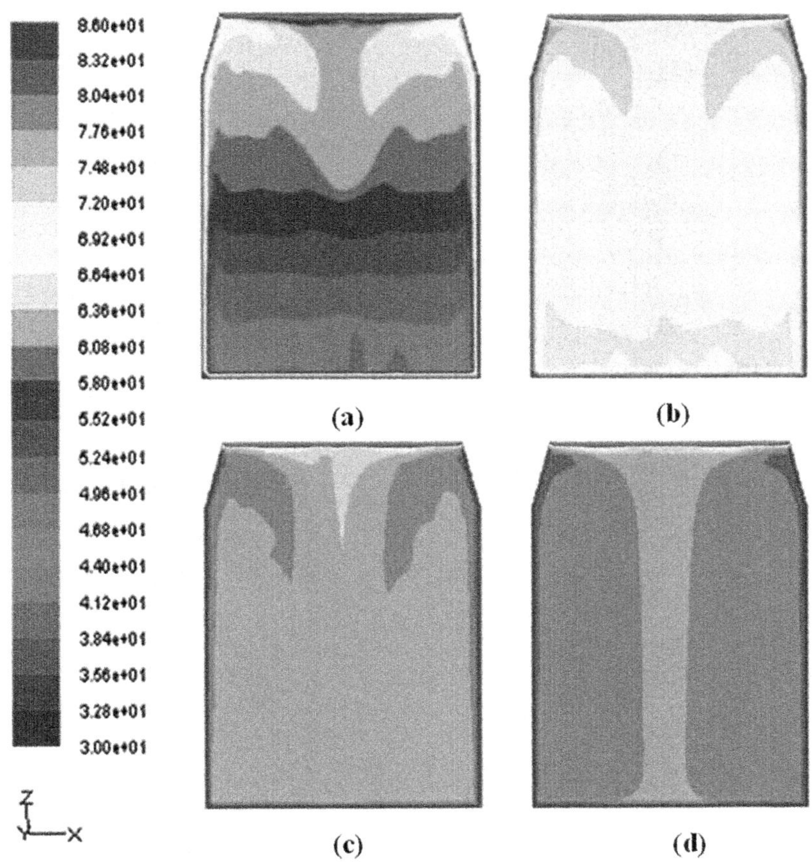

FIGURE 16.25 Temperature (°C) profiles of pasteurization (85°C) process of milk in the stationary position of the can at (a) 60 s, (b) 120 s, (c) 240 s, and (d) 360 s. (Reproduced with permission from Paul, D. A., Anishaparvin, A. and Anandharamakrishnan, C. 2011. Computational fluid dynamics studies on pasteurisation of canned milk. *International Journal of Dairy Technology* 64: 305–313.)

In the stationary can, the velocity profiles demonstrated a gradual increase in velocity at the top central region of the can (Figure 16.26a) due to the recirculatory flow of milk as a result of convective heating. The temperature gradient between the wall and interior regions of the can created density difference in canned milk. As a result, the buoyancy force was induced in the fluid leading to natural convective flow pattern. The velocity of milk at the core region was slightly low when heated for 60 s (Figure 16.26a) compared to that observed when heated for 360 s (Figure 16.26b). Conversely, when the can was rotated at 5 rpm, the velocity distribution was uniform compared to that in a stationary position (Figure 16.26c, d). Due to radial mixing, the temperature distribution was almost homogeneous, and the SHZ disappeared to a greater extent. Hence, the important inference from this CFD study was that, with the rotation of can, uniform heating of milk was achieved at a reduced processing time, due to the absence of SHZ. This is relevant as reduced processing time results in lower cost of processing and reduced price of the final product.

The study also showed that rotation of a can at a higher rpm for a shorter time (100 rpm for 360 s) or lower rpm for a longer time (5 rpm for 480 s) was ideal for the thermal pasteurization of milk. This is because the rotation of can leads to maximum PU value and thereby higher inactivation of the

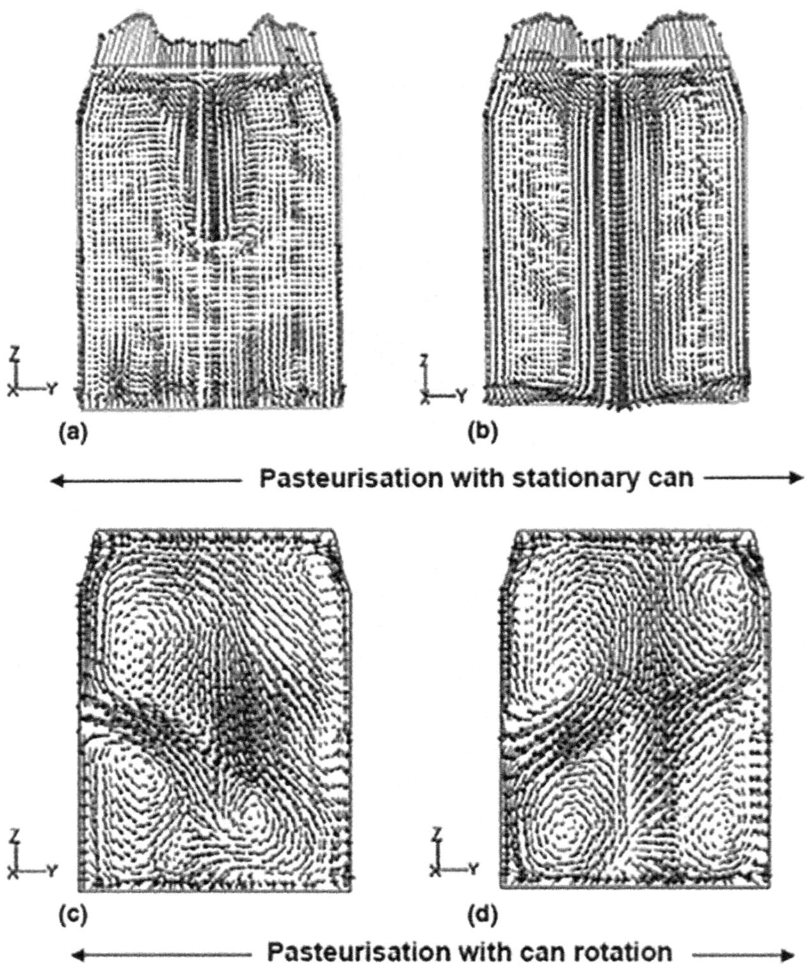

FIGURE 16.26 Velocity profile of milk inside the can during pasteurization (85°C) of milk inside the stationary can at time intervals [(a) 60 s and (b) 360 s] and the can rotated at 5 rpm at time intervals [(c) 60 s and (d) 360 s]. (Reproduced with permission from Paul, D. A., Anishaparvin, A. and Anandharamakrishnan, C. 2011. Computational fluid dynamics studies on pasteurisation of canned milk. *International Journal of Dairy Technology* 64: 305–313.)

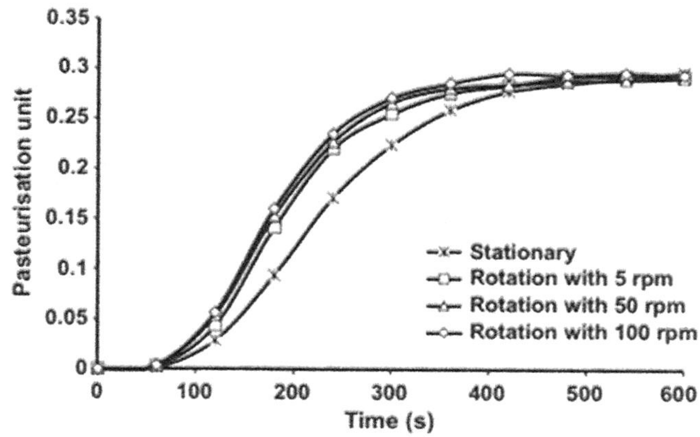

FIGURE 16.27 Effect of can rotation speed on PU during pasteurization of milk at 85°C in stationary, rotation of 5, 50, and 100rpm. (Reproduced with permission from Paul, D. A., Anishaparvin, A. and Anandharamakrishnan, C. 2011. Computational fluid dynamics studies on pasteurisation of canned milk. *International Journal of Dairy Technology* 64: 305–313.)

microorganism when compared to that attained with the stationary can (Figure 16.27). Also, the PU value increased when the can was rotated at a higher speed (100rpm). Maximum PU (1 unit) can be achieved only when the product temperature is equal to the process temperature. This confirms the positive effect of the pasteurization conducted in rotation mode on the microbial inactivation.

16.5.2 CFD Modeling of the Thermal Pasteurization of Egg

Thermal pasteurization of egg is carried out to inactivate *Salmonella enteritidis*, a pathogenic bacterium which causes food-borne illnesses (Ishwarya and Anandharamakrishnan, 2018). In the earlier years, CFD models based on conductive and convective heat transfer and first-order kinetics were developed for predicting the surface heat transfer coefficient and inactivation of *S. enteritidis* with respect to the position of egg yolk during thermal processing (Denys et al., 2003, 2004, 2005). The preceding works considered egg in stationary position. Later, a CFD model was developed which considered conductive heat transfer to compare the stationary and rotational modes of egg pasteurization. The model was used to investigate the influence of rotation on the pasteurization time during the thermal pasteurization of whole egg (with yolk) at 55.6°C. Pasteurization time is the time required by the egg to reach the pasteurization temperature of 55.6°C. In the rotation mode of egg pasteurization, two different speeds of rotation were chosen: 2.5 and 5 rpm, based on the fact that minimum rotation is required to avoid damage to yolk and mixing of yolk and albumin (Ramachandran et al., 2011).

16.5.2.1 Geometry Creation and Meshing

A 3D geometry of egg was used in this study, considering the yolk to be at the center (Figure 16.28). First, the ellipsoid shape of the egg was created in two dimension. Subsequently, the final 3D structure was obtained by revolving the contour 360° about its axis of symmetry. To account for the conductive heat transfer through the eggshell, the geometry included the thickness of the shell wall. The geometry shown in Figure (16.28) was meshed using 50000 tetrahedral/hybrid elements (including that for both albumin and yolk). Finally, the surface and the volume mesh were exported to CFD code for solving.

16.5.2.2 Solving

In this study, a finite volume approach was adopted to carry out an unsteady-state CFD simulation. The conservative equations of mass, momentum, and energy were solved using the SIMPLE method. The model considered the egg white and yolk as non-Newtonian fluids. The shell walls were coupled for

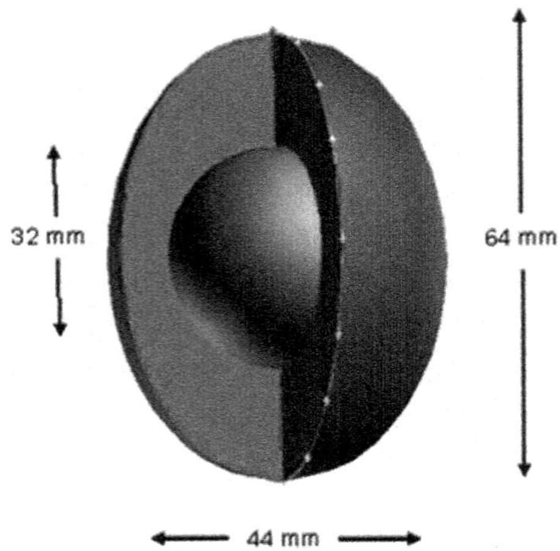

FIGURE 16.28 Three-dimensional geometry of egg. (Reproduced with permission from Ramachandran, R., Malhotra, D., Anishaparvin, A. and Anandharamakrishnan, C. 2011. Computational fluid dynamics simulation studies on pasteurization of egg in stationary and rotation modes. *Innovative Food Science and Emerging Technologies* 12: 38–44.)

simulation studies. The thermal and physical properties of egg white and yolk were expressed as a piecewise linear function of temperature (Table 16.4).

By considering that the eggshell offers a very low thermal resistance, the wall temperature (T_{wall}) was set at 56.5°C, approximately equal to the temperature of hot water. Also, the temperature of hot water was assumed to remain constant along the heating wall boundaries. A no-slip condition with all-side heating was used. During the CFD simulation of the pasteurization of rotating egg, the momentum equation was applied to include the effects of Coriolis acceleration and the centripetal acceleration.

The inactivation kinetics of *S. enteritidis* was studied using the *F*-value equation (Eq. (16.116)).

$$F = \int_0^t 10^{(T-T_{ref})/z} dt \tag{16.116}$$

TABLE 16.4

Thermal Properties of Egg White, Yolk, and Egg Shell

	Value	Reference
Egg white		
Thermal conductivity (W/m K)	$0.43 + 5.5 \times 10^{-4} \times T$	Romanoff and Romanoff (1949)
Specific heat capacity (J/kg K)	3560	Romanoff and Romanoff (1949)
Viscosity (kg/m s)	$3.12 - 8.9 \times 10^{-3} \times T$	Romanoff and Romanoff (1949)
Density (kg/m^3)	$-0.0074 \times T^2 - 0.0039 \times T + 1038.9$	Atlgan and Unluturk (2008)
Yolk		
Thermal conductivity (W/m K)	0.337	Romanoff and Romanoff (1949)
Specific heat capacity (J/kg K)	3560	Romanoff and Romanoff (1949)
Density (kg/m^3)	1035	Romanoff and Romanoff (1949)
Viscosity (kg/m s)	$1.6 - 4.8 \times 10^{-3} \times T$	Romanoff and Romanoff (1949)
Egg shell		
Thermal conductivity (W/m K)	2.25	Denys et al. (2003)
Specific heat capacity (J/kg K)	888	Sabliov et al. (2002)
Density (kg/m^3)	2300	Romanoff and Romanoff (1949)

Source: Ramachandran et al. (2011).

Egg white and yolk have dissimilar effects on the thermal inactivation of *S. enteritidis*. Therefore, two different z-values of 4.37°C and 3.29°C were used for yolk and white, respectively (Denys et al., 2005). The value of T_{wall} was used as T_{ref} in Eq. (16.116), which was then solved with respect to time and temperature of the process to calculate the F-value (Ramachandran et al., 2011).

16.5.2.3 Post-Processing

During post-processing, the simulated results were extracted for three points in the egg:

- Point-1: at 0.025 m, from center to top
- Point-2: at the center
- Point-3: at 0.0195 m, from center to right

The plot of temperature versus time (Figure 16.29) obtained from CFD simulation showed that in the stationary mode, thermal pasteurization of a whole egg requires longer time to reach a uniform temperature. When compared to the rest of the egg, the egg yolk heated up very slowly due to the difference in thermophysical properties (thermal conductivity and density) between the egg yolk and egg white. The temperature of egg white (at point-1 and point-3) increased exponentially with time (Figure 16.29) as it was located close to the eggshell. In contrast, the temperature of egg yolk (point-2) showed a linear increase with time (Figure 16.29), as it remained at room temperature until heat reached the surface of the yolk. The pasteurization time for the center of the yolk was found to be 30 min.

Further, the CFD simulation showed that the SHZ was located at a position slightly below the geometric center of egg in its yolk portion. This was further confirmed by the temperature contours obtained from the CFD simulation at various time intervals during the process: 30, 150, and 300 s (Figure 16.30a). The stationary egg had a SHZ which remained at 37°C even at the end of the processing time, i.e., 300 s (Figure 16.30c).

Compared to stationary heating, the rotational mode resulted in more uniform heating and thereby improved the efficiency of the heating process. Further, in the rotation mode, the egg white attained the defined pasteurization temperature at a faster rate due to improved heat penetration throughout the egg. However, the egg yolk took more time to heat up compared to egg white (Figure 16.30d–f) but at a shorter time than the stationary egg (Figure 16.30c). Unlike the stationary egg (Figure 16.30b), the SHZ

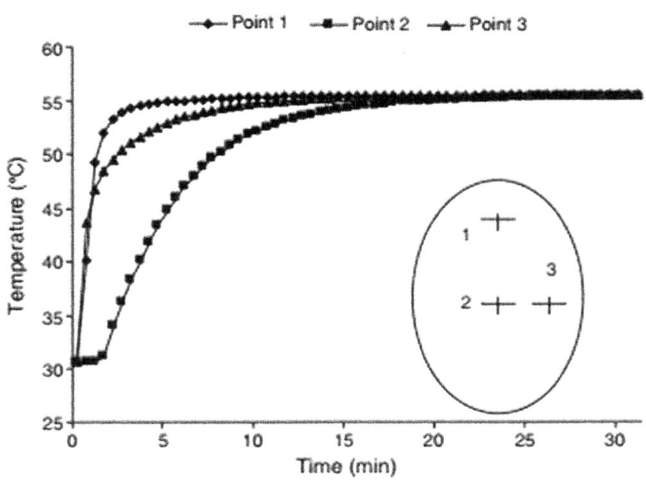

FIGURE 16.29 CFD simulated temperature profiles at three points for whole egg heating at 55.6°C (stationary mode). (Reproduced with permission from Ramachandran, R., Malhotra, D., Anishaparvin, A. and Anandharamakrishnan, C. 2011. Computational fluid dynamics simulation studies on pasteurization of egg in stationary and rotation modes. *Innovative Food Science and Emerging Technologies* 12: 38–44.)

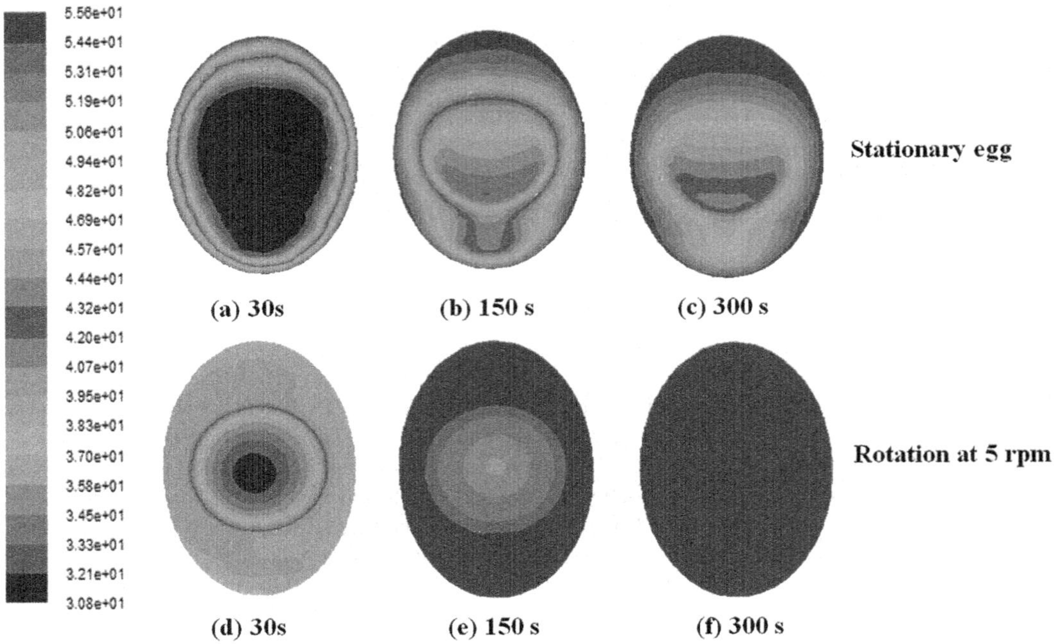

FIGURE 16.30 Comparison of CFD simulated temperature contours of stationary (a–c) and egg rotating at 5 rpm (d–f) heated at 55.6°C. (Reproduced with permission from Ramachandran, R., Malhotra, D., Anishaparvin, A. and Anandharamakrishnan, C. 2011. Computational fluid dynamics simulation studies on pasteurization of egg in stationary and rotation modes. *Innovative Food Science and Emerging Technologies* 12: 38–44.)

of rotating egg was concentrated at the center of the yolk (Figure 16.30d–e). The rotating egg achieved a uniform temperature in approximately 300 s (Figure 16.30f).

Also, the CFD simulation showed that with an increase in the rotational speed from 2.5 to 5 rpm, the rate of heating improved. At 2.5 rpm, the maximum temperature was attained in 7 min. But, at 5 rpm, the egg white (point-1; Figure 16.31a) reached the maximum temperature in 6.5 min. Nevertheless, the pasteurization times attained with rotation at 2.5 and 5 rpm were significantly lesser when compared to the stationary egg, which was 30 min. The specified point inside the yolk (Figure 16.31b) took a longer time to attain the pasteurization temperature than the point within the egg white (Figure 16.31a). With respect to the effect of rotation, the egg yolk exhibited a similar trend in heating rate as the egg white. The egg yolk rotated at 2.5 and 5 rpm took about 13 and 9 min, respectively, to reach the maximum pasteurization temperature (Figure 16.31b).

The effect of rotation on the inactivation kinetics of *S. enteritidis* during pasteurization was similar to that observed with the temperature profile of egg. In both stationary and rotating modes, the *F*-value was calculated for every 1-min interval of the thermal process. In the stationary position, egg attained the maximum *F*-value (higher inactivation of *Salmonella*) only after 30 min of heating (Figure 16.32a). But, the rotated egg reached the same within 7 and 13 min of heating at 5 and 2.5 rpm, respectively. Nevertheless, the time for inactivation of *S. enteritidis* in the yolk (Figure 16.32b) was longer than that in the egg white.

16.5.3 Spray Drying

Spray drying is a complex operation owing to the highly unsteady flow patterns in the spray chamber (Chen, 2004) and the difficulty and expenditure involved in the measurement of the related parameters such as airflow, temperature, humidity, particle size, and residence time (RT) within the drying chamber. To add further to the complexity, a large number of variables intervene during the scale-up of a

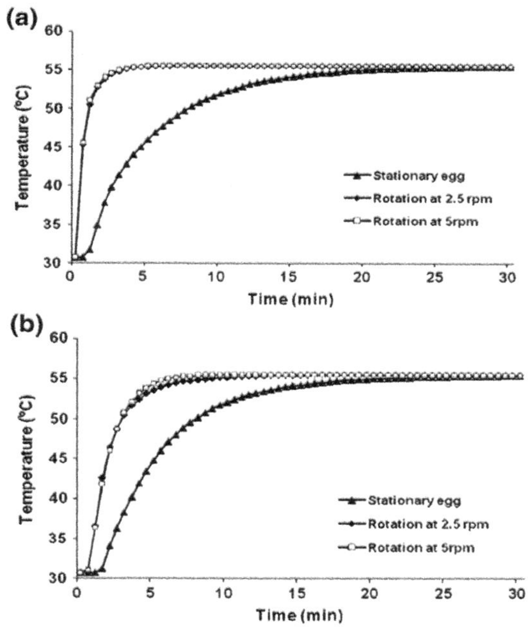

FIGURE 16.31 Comparison of CFD simulated temperature profiles of stationary and rotated egg conductively heated at 55.6°C: (a) point-1 (bottom); (b) point-2 (center). (Reproduced with permission from Ramachandran, R., Malhotra, D., Anishaparvin, A. and Anandharamakrishnan, C. 2011. Computational fluid dynamics simulation studies on pasteurization of egg in stationary and rotation modes. *Innovative Food Science and Emerging Technologies* 12: 38–44.)

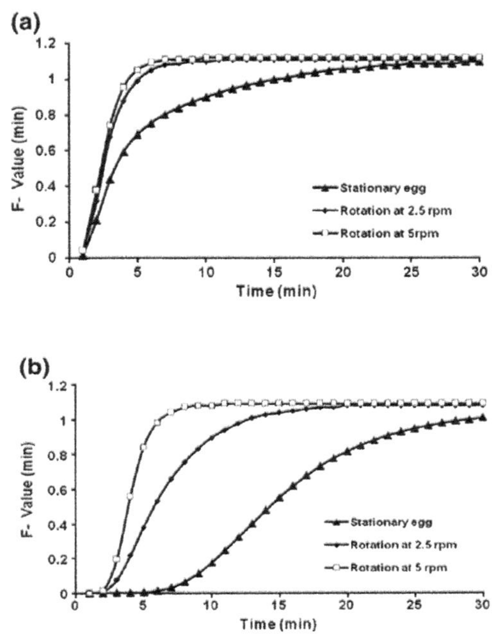

FIGURE 16.32 *F*-value predictions for stationary and rotated egg conductively heated at 55.6°C: (a) top point (0.025 m from center) and (b) center point. (Reproduced with permission from Ramachandran, R., Malhotra, D., Anishaparvin, A. and Anandharamakrishnan, C. 2011. Computational fluid dynamics simulation studies on pasteurization of egg in stationary and rotation modes. *Innovative Food Science and Emerging Technologies* 12: 38–44.)

spray-drying process. These include the diameter of spray chamber and feed droplets, dimensions of the atomizer, and velocity of air (Anandharamakrishnan and Ishwarya, 2015).

Inside a spray dryer, the particle encounters a series of phenomena such as entrapment within the wall material, agglomeration, and breakage (Saleh and Guigon, 2007). The RT of droplets in the spray chamber and drying rate hold high relevance especially when spray drying is used for encapsulation, wherein an effective film formation by the wall material around the core is expected. Further, the relationship between the rate of variation in the number of particles within the domain (in a given time interval) and the change in particle-related properties (i.e., moisture content, temperature, and film thickness of wall material around the droplet) also holds significance in terms of spray drying (Anandharamakrishnan and Ishwarya, 2015).

Understanding the events occurring during spray drying is vital from the perspectives of avoiding highly unsteady flows inside the spray chamber that result in significant product loss, and for an effective industrial scale-up of the spray drying process. However, studying the abovementioned phenomena in real time is often complex, and this is where CFD is of potential use owing to its versatile capabilities (Figure 16.33). A few studies are presented subsequently to obtain insight into the usefulness of CFD in predicting the particle histories such as trajectory, temperature, and RT during the spray-drying process.

16.5.3.1 Droplet/Particle Trajectory

With respect to a rotary atomizer, droplet trajectory can be defined as the distance traveled by the droplet from the edge of the disc against the height of the disc (Teunou and Poncelet, 2005). In the case of nozzle atomizers, droplet trajectory is measured from the tip of the nozzle. Droplet trajectory is influenced by many factors including the spray angle and gas flow pattern, which are difficult to be measured simultaneously. The E–L model is most the commonly used approach for the prediction of droplet trajectory by solving the force balance equation (Anandharamakrishnan and Ishwarya, 2015). The particle trajectories with respect to its diameter and RT are shown in the Figure 16.34a and b.

16.5.3.2 Droplet Temperature

The droplet temperature and RT have a significant influence on the retention of volatile and thermally sensitive core compounds during spray-drying. Generally, droplets of smaller size have higher temperatures

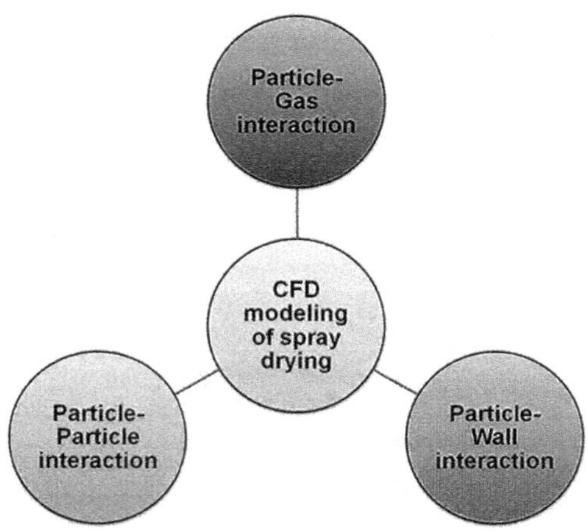

FIGURE 16.33 Capabilities of CFD modeling of spray drying. (Reproduced with permission from Anandharamakrishnan, C. and Ishwarya, S. P. 2015. *Spray Drying Techniques for Food Ingredient Encapsulation*. Hoboken, NJ, USA: John Wiley and Sons Ltd.)

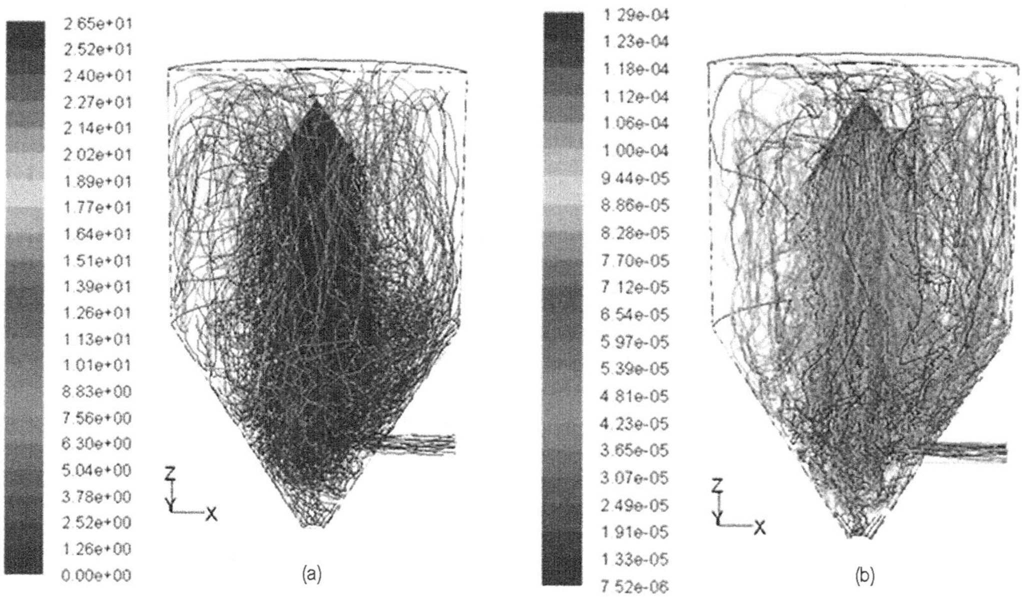

FIGURE 16.34 Particle trajectories colored by (a) RT and (b) diameter. (Reproduced with permission from Anandharamakrishnan, C., Gimbun, J., Stapley, A. G. F. and Rielly, C. D. 2010. A study of particle histories during spray drying using computational fluid dynamic simulations. *Drying Technology* 28: 566–576.)

than the larger ones because of their high surface area-to-volume ratio, consequent higher evaporation rate (Crowe et al., 1977) and a substantial level of evaporative cooling. Using a CFD model, the droplet temperature during spray-drying of whey protein solution was simulated. The simulation results showed that in both short- and tall-form spray dryer, the droplet temperature was high and was almost equal to the surrounding gas temperature, irrespective of the droplet diameter (Figure 16.35). Also, it was observed that the droplet nature was predominantly affected by the outlet air temperature than the inlet air temperature (Anandharamakrishnan et al., 2010).

16.5.3.3 Droplet Residence Time

Calculation of droplet RT comprises of two parts: the estimation of primary RT and secondary RT. The primary RT is computed from the time taken by the droplets leaving the nozzle to impact on the wall or exit at the outlet. A secondary RT is valid for the particles that hit the wall, which is calculated from the time taken by the particle to slide along the wall from the impact position to the outlet of spray-drying chamber. However, this estimation was based on the assumption that particles move with constant velocity along the wall from the impact position (Kieviet, 1997; Anandharamakrishnan et al., 2010).

The trajectory of the particles according to their RT distribution is shown in Figure 16.34a, in which the lighter color of the trajectories indicates a longer particle primary RT. Thus, the simulation results showed that the dried particles tended to recirculate by the up flow of gas at the walls. This recirculated hot gas laden with the dried particles, then mix with the cold gas containing dried particles resulting in the exposure of dried particles to the high inlet gas temperatures. This phenomenon is significant as the repeated exposure of particles comprising a heat-sensitive food ingredient to the high temperature may lead to its deterioration (Anandharamakrishnan and Ishwarya, 2015).

The simulated RT distribution as a function of particle size is depicted in Figure 16.36. From the model results, it was inferred that the larger diameter particles have longer RTs than smaller particles (Anandharamakrishnan, 2008). Smaller particles are more likely to follow the gas flow and thus leave the chamber in less time (Jin and Chen, 2009).

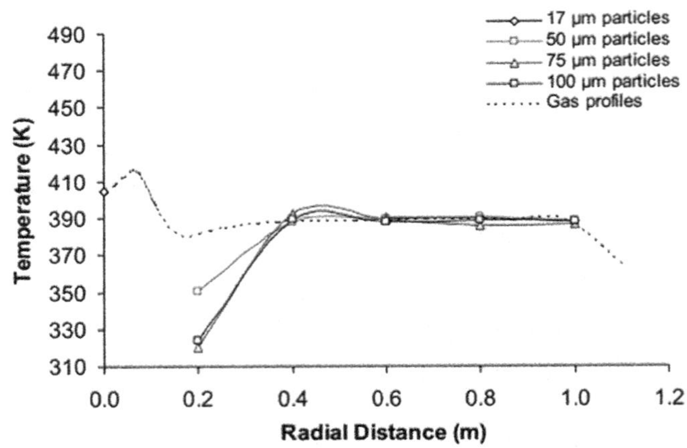

FIGURE 16.35 Simulated radial temperature profile of droplets at a distance of 0.6 m from the chamber top. (Reproduced with permission from Anandharamakrishnan, C., Gimbun, J., Stapley, A. G. F. and Rielly, C. D. 2010. A study of particle histories during spray drying using computational fluid dynamic simulations. *Drying Technology* 28: 566–576.)

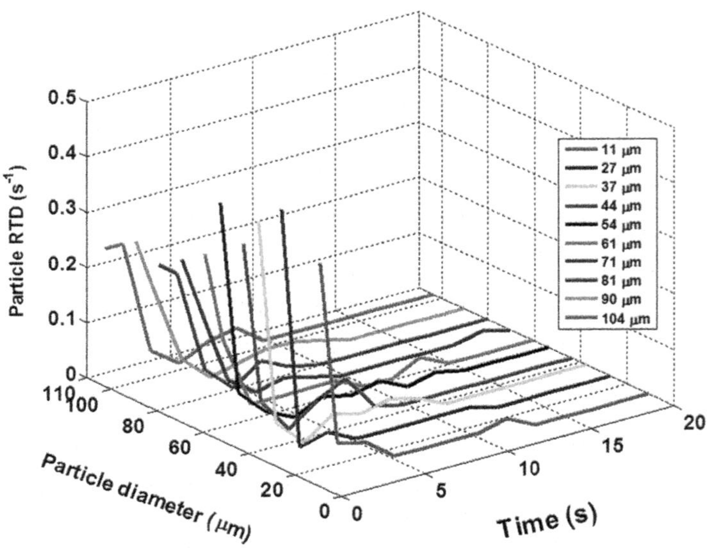

FIGURE 16.36 Simulated RT distributions for different particle diameters. (Reproduced with permission from Anandharamakrishnan, C. 2008. Experimental and computational fluid dynamics studies on spray-freeze-drying and spray-drying of proteins. *PhD Thesis*. UK: Loughborough University.)

16.5.4 CFD Modeling of Bread Baking Process

Bread baking is a complex process involving simultaneous heat and mass transfer, with the involvement of all three modes of heat transfer, thus rendering it an interesting candidate for CFD modeling. The application of CFD modeling is well established in the design and development of baking ovens and optimization of the baking process (Wong et al., 2006; Zhou and Therdthai 2007).

A CFD model was developed to determine the effect of hot air distribution and placement of bread on the temperature profile and starch gelatinization index of bread. Figure 16.37a shows that the 3D geometry of the oven and the meshed oven containing the bread. The mesh size of oven chamber was

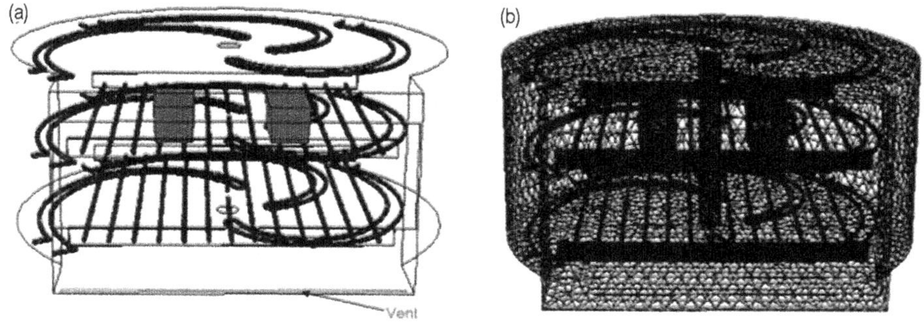

FIGURE 16.37 Oven with bread: (a) geometry; (b) meshed oven. (Reproduced with permission from Anishaparvin, A., Chhanwal, N., Indrani, D., Raghavarao, K. S. M. S. and Anandharamakrishnan, C. 2010. An investigation of bread-baking process in a pilot-scale electrical heating oven using computational fluid dynamics. *Journal of Food Science* 75: E605–E611.)

optimized for higher accuracy to contain 1.23 million cells. The mesh size was set dense near the trays and coils (Figure 16.37b).

Figure 16.38 depicts the change in air flow due to placement of bread. In the upper part of the oven, the air movement is caused by the heating-mediated natural convection. Air velocity near the heat source and around the bread was high due to the diversion of flow pattern and slight recirculation in those regions. Figure 16.39 shows the effect of placement of bread on the temperature profile within the oven chamber after 60 s of baking. Nine heating sources in three layers were placed inside the oven to achieve uniform heating. The simulation results showed that nonuniform air flow pattern (Figure 16.38) inside the oven cavity leads to uneven temperature distribution (Figure 16.39). With respect to placement of

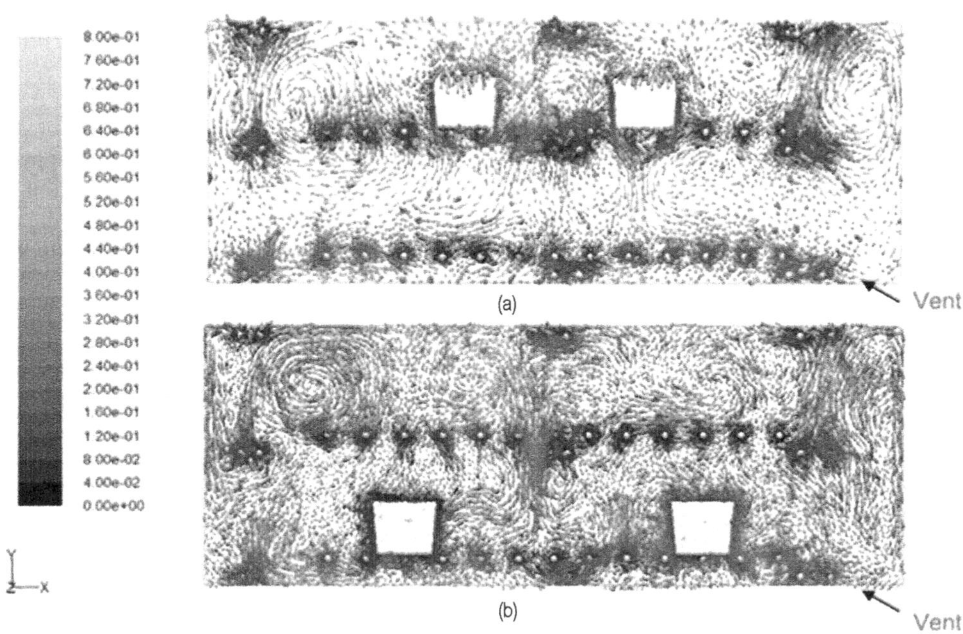

FIGURE 16.38 Effect of placement of bread on velocity profiles (m/s) inside the baking oven at 600 s: (a) top tray (bread-1 and bread-2) and (b) bottom tray (bread-3 and bread-4). (Reproduced with permission from Anishaparvin, A., Chhanwal, N., Indrani, D., Raghavarao, K. S. M. S. and Anandharamakrishnan, C. 2010. An investigation of bread-baking process in a pilot-scale electrical heating oven using computational fluid dynamics. *Journal of Food Science* 75: E605–E611.)

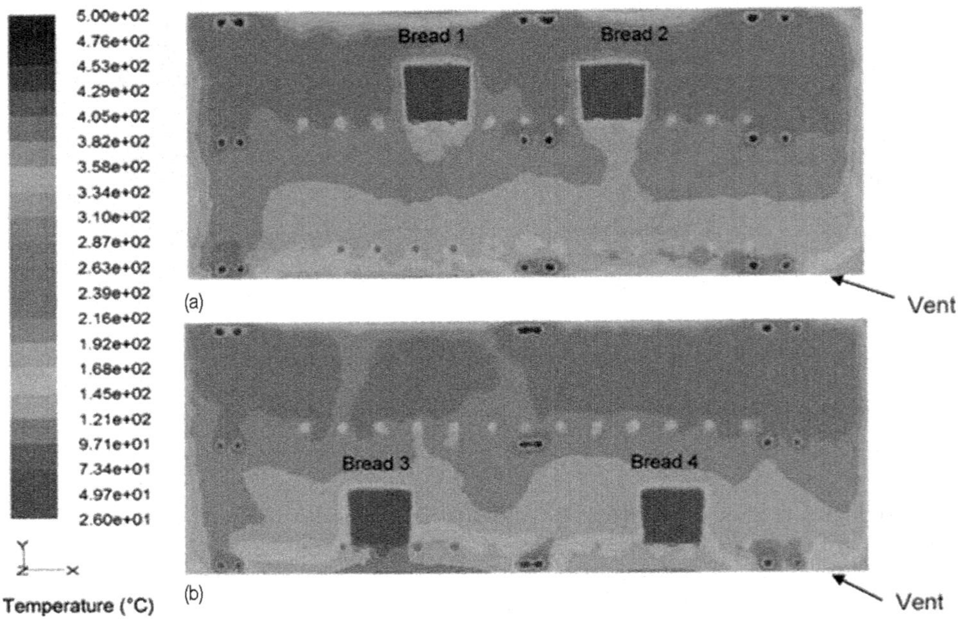

FIGURE 16.39 Effect of placement of bread on temperature profiles (°C) at 60 s of baking inside the baking oven: (a) top tray (bread-1 and bread-2) and (b) bottom tray (bread-3 and bread-4). (Reproduced with permission from Anishaparvin, A., Chhanwal, N., Indrani, D., Raghavarao, K. S. M. S. and Anandharamakrishnan, C. 2010. An investigation of bread-baking process in a pilot-scale electrical heating oven using computational fluid dynamics. *Journal of Food Science* 75: E605–E611.)

bread, baking of bread in upper trays received more heat and hence required shorter baking time and gelatinization index compared to those in the bottom tray (Figure 16.39).

The bread center in the upper tray reached 100°C in 1200 s (Figure 16.40). This was the result of higher air temperature in the upper tray (Figure 16.39). The starch gelatinization profile obtained from the CFD model showed that the gelatinization progressed from the crust to the crumb as the temperature increased in the crumb region. At the center of the bread, gelatinization was completed within 900 s. Degree of starch gelatinization at the top and bottom of bread was slower when compared to that at the left and right sides (Figure 16.41). This observation was attributed to the temperature of air circulation around the bread (Anishaparvin et al., 2010).

The quality of bread depends on the physiochemical and biological changes that occur during the bread baking process. The two major factors which influence the bread quality are the moisture content and temperature (Chhanwal et al., 2012). The change in the temperature profile of bread dough was studied during the entire baking process with preheating of the domestic electrical heating oven (Chhanwal et al., 2010). However, differences were observed between the experimental and CFD simulated temperatures of bread center (Wong et al., 2007; Anishaparvin et al., 2010; Chhanwal et al., 2010). The predictions on bread surface temperature were closer to experimental results when compared to those at the center temperature. Also, the center temperature was found to exceed 100°C. But, practically, the crumb temperature never increases beyond 98–99°C due to the evaporation–condensation mechanism, which was stated as the reason for why the earlier CFD models were unable to predict temperature profile over the entire bread.

Later, a CFD model was developed with the integration of evaporation–condensation mechanism to replicate the actual bread baking process (Chhanwal et al., 2011). The evaporation–condensation mechanism was incorporated in the model by defining specific heat of bread as a function of temperature Eq. (16.117), including the enthalpy jump at the phase change and thermal conductivity of bread as a function of temperature [$k(T)$] Eq. (16.118), wherein the thermal conductivity increases until the center

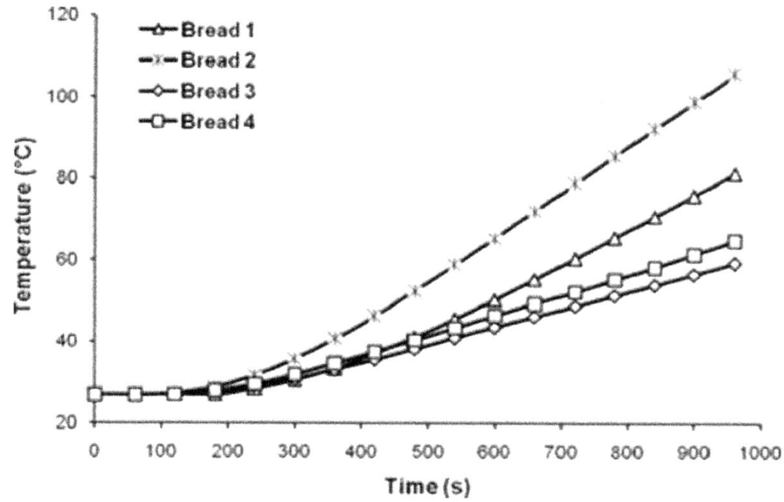

FIGURE 16.40 Effect of placement of bread on bread center temperatures. (Reproduced with permission from Anishaparvin, A., Chhanwal, N., Indrani, D., Raghavarao, K. S. M. S. and Anandharamakrishnan, C. 2010. An investigation of bread-baking process in a pilot-scale electrical heating oven using computational fluid dynamics. *Journal of Food Science* 75: E605–E611.)

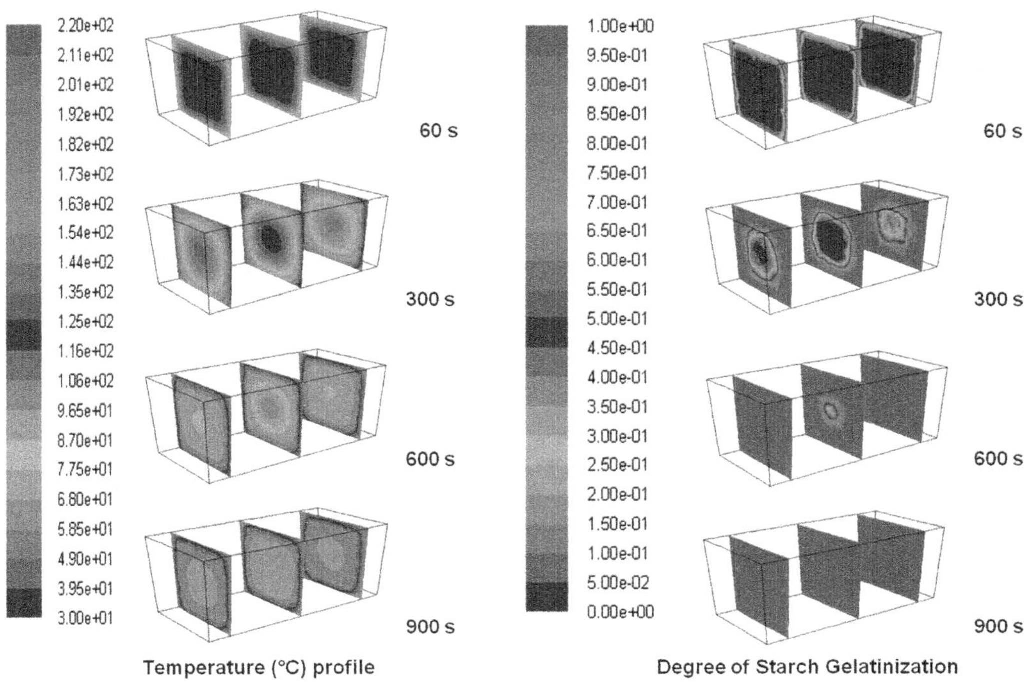

FIGURE 16.41 Comparison of temperature profile and starch gelatinization profile in crumb during bread-baking process (bread-1) at 4 cm from center of the bread on both sides. (Adapted and Reproduced with permission from Anishaparvin, A., Chhanwal, N., Indrani, D., Raghavarao, K. S. M. S. and Anandharamakrishnan, C. 2010. An investigation of bread-baking process in a pilot-scale electrical heating oven using computational fluid dynamics. *Journal of Food Science* 75: E605–E611.)

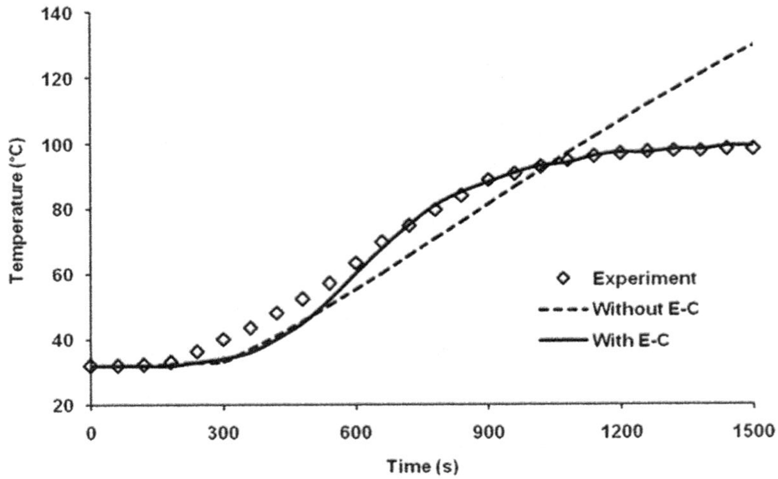

FIGURE 16.42 Comparison between the bread center temperature profile during bread baking process obtained from experiment and CFD modeling with and without the inclusion of evaporation–condensation (E–C) mechanism. (Reproduced with permission from Chhanwal, N., Indrani, D., Raghavarao, K. S. M. S., and Anandharamakrishnan, C. 2011. Computational fluid dynamics modeling of bread baking process. *Food Research International* 44: 978–983.)

temperature reaches 100°C. The thermal conductivity was fixed constant at 0.2 W/m K (Eq. (16.119)), once the condensation begins (Purlis and Salvadori, 2009).

$$C_P = \left(0.013T^2 - 0.217T + 4179\right)X_W + (5T + 1390)(1 - X_W) + \lambda\left(T - T_f, \nabla T\right) \quad (16.117)$$

where X_W is the water content (kg/kg), T is the temperature (°C), T_f is the temperature of phase change (°C), and λ is the latent heat of phase change (J/m³).

$$k(T) = \frac{0.9}{1 + \exp\left(-0.1(T - 353.16)\right)} + 0.2, \quad \text{if } T \leq 100°\text{C} \quad (16.118)$$

$$k(T) = 0.2, \quad \text{if } T \geq 100°\text{C} \quad (16.119)$$

Thus, with the incorporation of the evaporation–condensation mechanism, the temperature predictions for crumb obtained from the model were closer to the experimental values (Figure 16.42).

This chapter provided an essence of the essentials of computational food dynamics modeling and its resourcefulness in predicting certain complex phenomena during food processing operations. The understanding obtained in this chapter is envisaged to aid the readers in further exploring the advanced concepts in CFD modeling, which are evolving continuously since its inception in the 20th century until now.

BIBLIOGRAPHY

Anandharamakrishnan, C. 2003. Computational fluid dynamics (CFD)—Applications for the food industry. *Indian Food Industry* 22: 62–68.

Anandharamakrishnan, C. 2008. Experimental and computational fluid dynamics studies on spray-freeze-drying and spray-drying of proteins. *PhD Thesis*. UK: Loughborough University.

Anandharamakrishnan, C. 2013. *Computational Fluid Dynamics Applications in Food Processing (Springer Briefs in Food, Health, and Nutrition)*. New York: Springer.

Anandharamakrishnan, C., Gimbun, J., Stapley, A. G. F. and Rielly, C. D. 2010. A study of particle histories during spray drying using computational fluid dynamic simulations. *Drying Technology* 28: 566–576.

Anandharamakrishnan, C. and Ishwarya, S. P. 2015. *Spray Drying Techniques for Food Ingredient Encapsulation*. Hoboken, NJ: John Wiley and Sons Ltd.

Anderson, J. D. 2013. *Computational Fluid Dynamics—The Basics With Applications*. New York: McGraw-Hill Inc.

Anishaparvin, A., Chhanwal, N., Indrani, D., Raghavarao, K. S. M. S. and Anandharamakrishnan, C. 2010. An investigation of bread-baking process in a pilot-scale electrical heating oven using computational fluid dynamics. *Journal of Food Science* 75: E605–E611.

ANSYS Fluent User Guide 2006. www.fluent.com (accessed July 10, 2018).

Aprajeeta, J., Gopirajah, R. and Anandharamakrishnan, C. 2015. Shrinkage and porosity effects on heat and mass transfer during potato drying. *Journal of Food Engineering* 144: 119–128.

Atlgan, M. R. and Unluturk, S. 2008. Rheological properties of liquid egg products (LEPS). *International Journal of Food Properties* 11: 296–309.

Boukouvalas, C. J., Krokida, M. K., Maroulis, Z. B. and Marinos-Kouris, D. 2006. Effect of material moisture content and temperature on the true density of foods. *International Journal of Food Properties* 9: 109–125.

Chen, C. R. and Ramaswamy, H. S. 2007. Visual basic computer simulation package for thermal process calculations. *Chemical Engineering and Processing: Process Intensification* 46: 603–613.

Chen, X. D. 2004. Heat-mass transfer and structure formation during drying of single food droplets. *Drying Technology* 22: 179–190.

Chen, X. D., Xie, G. Z. and Rahman, M. S. 1998. Application of the distribution factor concept in correlating thermal conductivity data for fruits and vegetables. *International Journal of Food Properties* 1: 35–44.

Chhanwal, N., Anishaparvin, A., Indrani, D., Raghavarao, K. S. M. S. and Anandharamakrishnan, C. 2010. Computational fluid dynamics (CFD) modeling of an electrical heating oven for bread-baking process. *Journal of Food Engineering* 100: 452–460.

Chhanwal, N., Indrani, D., Raghavarao, K. S. M. S. and Anandharamakrishnan, C. 2011. Computational fluid dynamics modeling of bread baking process. *Food Research International* 44: 978–983.

Chhanwal, N., Tank, A., Raghavarao, K. S. M. S. and Anandharamakrishnan, C. 2012. Computational fluid dynamics (CFD) modeling for bread baking process—A review. *Food and Bioprocess Technology* 5: 1157–1172.

COMSOL. 2018. Convection cooking of chicken patties. COMSOL Multiphysics® Application ID: 448. www.comsol.com/model/convection-cooking-of-chicken-patties-448 (accessed July 10, 2018).

Crowe, C. T., Sharam, M. P. and Stock, D. E. 1977. The particle source in cell (PSI-Cell) model for gas-droplet flows. *Journal of Fluid Engineering* 9: 325–332.

Denys, S., Pieters, J. G. and Dewettinck, K. 2003. Combined CFD and experimental approach for determination of the surface heat transfer coefficient during thermal processing of eggs. *Journal of Food Science* 68: 943–951.

Denys, S., Pieters, J. G. and Dewettinck, K. 2004. Computational fluid dynamics analysis of combined conductive and convective heat transfer in model eggs. *Journal of Food Engineering* 63: 281–290.

Denys, S., Pieters, J. G. and Dewettinck, K. 2005. CFD analysis for process impact assessment during thermal pasteurization of intact eggs. *Journal of Food Protection* 68: 366–374.

Fletcher, A. J. 2000. *Computational Techniques for Fluid Dynamics* (Second edition). New York: Springer-Verlag.

Ghani, A. A. G., Farid, M. M. and Chen, X. D. 2002. Theoretical and experimental investigation of the thermal inactivation of Bacillus stearothermophilus in food pouches. *Journal of Food Engineering* 51: 221–228.

Ghani, A. A. G., Farid, M. M. and Chen, X. D. 2003. A computational and experimental study of heating and cooling cycles during thermal sterilization of liquid foods in pouches using CFD. Proceedings of the Institution of Mechanical Engineers, Part E. *Journal of Process Mechanical Engineering* 217: 1–9.

Hassini, L., Azzouz, S., Peczalski, R. and Belghith, A., 2007. Estimation of potato moisture diffusivity from convective drying kinetics with correction for shrinkage. *Journal of Food Engineering* 79: 47–56.

Holdsworth, S. D. and Simpson, R. 2007. Sterilization, pasteurization and cooking criteria. In *Thermal Processing of Packaged Foods* (Third edition), eds. S. D. Holdsworth and R. Simpson, 123–138. New York: SpringerLink.

Huang, L. X. and Mujumdar, A. S. 2006. Numerical study of two-stage horizontal spray dryers using computational fluid dynamics. *Drying Technology* 24: 727–733.

Ishwarya, P. and Anandharamakrishnan, C. 2018. CFD Analysis of Food Pasteurization Processes. In *Computational Fluid Dynamics in Food Processing* (Second edition), ed. D-W. Sun, 259–285. Boca Raton, FL: CRC Press, Taylor & Francis Group.

Jakobsen, H. A., Sannaes, B. H., Grevskott, S. and Svendsen, H. F. 1997. Modelling of vertical bubble-driven flows. *Industrial Engineering Chemistry Research* 36: 4052–4074.

Jin, Y. and Chen, X. D. 2009. Numerical study of the drying process of different sized particles in an industrial scale spray dryer. *Drying Technology* 27: 371–381.

Kieviet, F. G. 1997. Modeling quality in spray drying. *PhD Thesis*. Netherlands: Endinhoven University of Technology.

Korokida, M. K. and Maroulis, Z. B. 1997. Effect of drying method on shrinkage and porosity. *Drying Technology* 15: 2441–2458.

Madenci, E. and Guven, I. 2006. *The Finite Element Method and Applications in Engineering Using ANSYS®. LLC*. New York: Springer Science+Business Media.

Marshall, E. M. and Bakker, A. 2002. *Computational Fluid Mixing*. Lebanon: Fluent Inc.

Mostafa, A. A. and Mongia, H. C. 1987. On the modeling of turbulent evaporating sprays: Eulerian versus Lagrangian approach. *International Journal of Heat and Mass Transfer* 30: 2583–2593.

Nijdam, J. J., Guo, B., Fletcher, D. F. and Langrish, T. A. G. 2006. Lagrangian and Eulerian models for simulating turbulent dispersion and coalescence of droplets within a spray. *Applied Mathematical Modelling* 30: 1196–1211.

NPTEL 2018a. Derivation of conservation of mass and momentum equations (Lecture 10). In Convection *(Module 2)*. http://nptel.ac.in/courses/112101001/downloads/lec10.pdf (accessed October 15, 2018).

NPTEL 2018b. Derivation of conservation of momentum (Lecture 11). In Convection *(Module 2)*. https://nptel.ac.in/courses/112101001/downloads/lec11.pdf (accessed October 15, 2018).

NPTEL 2018c. Derivation of conservation of energy (Lecture 12). In Convection *(Module 2)*. http://nptel.ac.in/courses/112101001/downloads/lec12.pdf (accessed October 15, 2018).

NPTEL 2018d. Derivation of conservation of energy (contd.) (Lecture 13). In Convection *(Module 2)*. https://nptel.ac.in/courses/112101001/downloads/lec13.pdf (accessed October 15, 2018).

Padmavati, R. and Anandharamakrishnan, C. 2013. Computational fluid dynamics modeling of the thermal processing of canned pineapple slices and tidbits. *Food and Bioprocess Technology* 6: 882–895.

Paul, D. A., Anishaparvin, A. and Anandharamakrishnan, C. 2011. Computational fluid dynamics studies on pasteurisation of canned milk. *International Journal of Dairy Technology* 64: 305–313.

Perry, R. H., Green, D. 1984. *Perry's Chemical Engineers' Handbook*. New York: McGraw-Hill.

Puri, V. M. and Anantheswaran, R. C. 1993. The finite-element method in food processing: a review. *Journal of Food Engineering* 19: 247–274.

Purlis, E. and Salvadori, V. O. 2009. Bread baking as a moving boundary problem. Part 2: model validation and numerical simulation. *Journal of Food Engineering* 91: 434–442.

Ramachandran, R., Malhotra, D., Anishaparvin, A. and Anandharamakrishnan, C. 2011. Computational fluid dynamics simulation studies on pasteurization of egg in stationary and rotation modes. *Innovative Food Science and Emerging Technologies* 12: 38–44.

Riedel, L. 1949. Thermal conductivity measurements of sugar solutions, fruit juice and milk. *Chemie-Ingenieur-Technik* 21: 340–341.

Romanoff, A. L. and Romanoff, A. J. 1949. *The Avian Egg*. New York: John Wiley and sons.

Sabliov, C. M., Farkas, B. E., Keener, K. M. and Curtis, P. A. 2002. Cooling of shell eggs with cryogenic carbon dioxide: A finite element analysis of heat transfer. *LWT Food Science and Technology* 35: 568–574.

Saleh, K. and Guigon, P. 2007. Coating and encapsulation processes in powder technology. In *Handbook of Powder Technology* (Volume 11), eds. A. D. Salman, M. J. Hounslow and J. P. K. Seville, 323–375. Amsterdam, the Netherlands: Elsevier.

Shapton, D. A., Lovelock, D. W. and Laurita-Longo, R. 1971. The evaluation of sterilization and pasteurization processes for temperature measurements in degrees Celsius (°C). *Journal of Applied Bacteriology* 34: 491–500.

Short, A. L. 1955. The temperature coefficient of expansion of raw milk. *Journal of Dairy Research* 22: 69–73.

Siebel, J. E. 1892. Specific heat of various products. *Ice and Refrigeration* 2: 256–257.

Singh, R. P. and Heldmen, D. R. 2009. *Introduction to Food Engineering* (Fourth edition). San Deigo, CA: Elsevier.

Stoforos, N. G. and Merson, R. L. 1990. Estimating heat transfer coefficients in liquid/particulate canned food using only liquid temperature data. *Journal of Food Science* 55: 478–483.

Tasami, E. and Katsioti, M. 2000. Drying kinetics for some fruits: predicting porosity and colour during dehydration. *Drying Technology* 18: 1559–1581.

Teunou, E. and Poncelet, D. 2005. Rotary disc atomisation for microencapsulation applications—prediction of the particle trajectories. *Journal of Food Engineering* 71: 345–353.

Walstra, P., Wouters, J. and Geurts, T. 2006. *Dairy Science and Technology* (Second edition). New York: CRC Press.

Wang, N. and Brennan, J. G. 1995. A mathematical model of simultaneous heat and moisture transfer during drying of potato. *Journal of Food Engineering* 24: 47–60.

Wong, S. Y., Zhou, W. and Hua, J. 2006. Robustness analysis of CFD model to the uncertainties in its physical properties for a bread baking process. *Journal of Food Engineering* 77: 784–791.

Wong, S. Y., Zhou, W. and Hua, J. 2007. CFD modeling of an industrial continuous bread-baking process involving U-movement. *Journal of Food Engineering* 78: 888–896.

Yang, H., Sakai, N. and Watanabe, M. 2001. Drying model with non-isotropic shrinkage deformation undergoing simultaneous heat and mass transfer. *Drying Technology* 19: 1441–1460.

Zhou, W. and Therdthai, N. 2007. Three-dimensional modeling of a continuous industrial baking process. In *Computational Fluid Dynamics in Food Processing*, ed. D. W. Sun, 287–312. Boca Raton: CRC Press.

Appendix I: Saturated Water and Steam (Temperature) Tables

Temperature (in °C) (T)	Absolute Pressure (in bar) (p)	Specific Volume (in m³/kg) Water (v_f)	Steam (v_g)	Specific Enthalpy (in kJ/kg) Water (h_f)	Evaporation (h_{fg})	Steam (h_g)	Specific Entropy (in kJ/kg K) Water (s_f)	Evaporation (s_{fg})	Steam (s_g)
0	0.00611	0.001	206.31	0	2501.6	2501.6	0	9.158	9.158
1	0.00657	0.001	192.61	4.2	2499.2	2503.4	0.015	9.116	9.131
2	0.00706	0.001	179.92	8.4	2496.8	2505.2	0.031	9.074	9.105
3	0.00758	0.001	168.17	12.6	2494.5	2507.1	0.046	9.033	9.079
4	0.00813	0.001	157.27	16.8	2492.1	2508.9	0.061	8.992	9.053
5	0.00872	0.001	147.16	21	2489.7	2510.7	0.076	8.951	9.027
6	0.00935	0.001	137.78	25.2	2487.4	2512.6	0.091	8.911	9.002
7	0.01001	0.001	129.06	29.4	2485	2514.4	0.106	8.87	8.976
8	0.01072	0.001	120.97	33.6	2482.6	2516.2	0.121	8.83	8.951
9	0.01147	0.001	113.44	37.8	2480.3	2518.1	0.136	8.791	8.927
10	0.01227	0.001	106.43	42.0	2477.9	2519.9	0.151	8.751	8.902
11	0.01312	0.001	99.909	46.2	2475.5	2521.7	0.166	8.712	8.878
12	0.01401	0.001	93.835	50.4	2473.2	2523.6	0.181	8.673	8.854
13	0.01497	0.001001	88.176	54.6	2470.8	2525.4	0.195	8.635	8.83
14	0.01597	0.001001	82.9	58.7	2468.5	2527.2	0.21	8.597	8.806
15	0.01704	0.001001	77.978	62.9	2466.1	2529.1	0.224	8.559	8.783
16	0.01817	0.001001	73.384	67.1	2463.8	2530.9	0.239	8.52	8.759
17	0.01936	0.001001	69.095	71.3	2461.4	2532.7	0.253	8.483	8.736
18	0.02062	0.001001	65.087	75.5	2459.0	2534.5	0.268	8.446	8.714
19	0.02196	0.001002	61.341	79.7	2456.7	2536.4	0.282	8.409	8.691
20	0.02337	0.001002	57.838	83.9	2454.3	2538.2	0.296	8.372	8.668
21	0.02485	0.001002	54.561	88.0	2452.0	2540.0	0.31	8.336	8.646
22	0.02642	0.001002	51.492	92.2	2449.6	2541.8	0.325	8.299	8.624
23	0.02808	0.001002	48.619	96.4	2447.2	2543.6	0.339	8.263	8.602
24	0.02982	0.001002	45.926	100.6	2444.9	2545.5	0.353	8.228	8.581
25	0.03166	0.001003	43.402	104.8	2442.5	2547.3	0.367	8.192	8.559
26	0.0336	0.001003	41.034	108.9	2440.2	2549.1	0.381	8.157	8.538
27	0.03564	0.001003	38.813	113.1	2437.8	2550.9	0.395	8.122	8.517
28	0.03778	0.001004	36.728	117.3	2435.4	2552.7	0.409	8.087	8.496
29	0.04004	0.001004	34.769	121.5	2433.1	2554.5	0.423	8.052	8.475
30	0.04242	0.001004	32.929	125.7	2430.7	2556.4	0.437	8.018	8.455
31	0.04491	0.001005	31.199	129.8	2428.3	2558.2	0.45	7.984	8.434
32	0.04753	0.001005	29.572	134.0	2425.9	2560.0	0.464	7.95	8.414
33	0.05029	0.001005	28.042	138.2	2423.6	2561.8	0.478	7.916	8.394
34	0.05318	0.001006	26.601	142.4	2421.2	2563.6	0.491	7.883	8.374
35	0.05622	0.001006	25.245	146.6	2418.8	2565.4	0.505	7.849	8.354
36	0.0594	0.001006	23.967	150.7	2416.4	2567.2	0.518	7.817	8.335
37	0.06274	0.001007	22.763	154.9	2414.1	2569.0	0.532	7.783	8.315

(Continued)

Temperature (in °C) (T)	Absolute Pressure (in bar) (p)	Specific Volume (in m³/kg) Water (v_f)	Steam (v_g)	Specific Enthalpy (in kJ/kg) Water (h_f)	Evaporation (h_{fg})	Steam (h_g)	Specific Entropy (in kJ/kg K) Water (s_f)	Evaporation (s_{fg})	Steam (s_g)
38	0.06624	0.001007	21.627	159.1	2411.7	2570.8	0.545	7.751	8.296
39	0.06991	0.001007	20.557	163.3	2409.3	2572.6	0.559	7.718	8.277
40	0.07375	0.001008	19.546	167.5	2406.9	2574.4	0.572	7.686	8.258
41	0.07777	0.001008	18.592	171.6	2404.5	2576.2	0.585	7.654	8.239
42	0.08199	0.001009	17.692	175.8	2402.1	2577.9	0.599	7.622	8.221
43	0.08639	0.001009	16.841	180.0	2399.7	2579.7	0.612	7.591	8.203
44	0.091	0.001009	16.036	184.2	2397.3	2581.5	0.625	7.559	8.184
45	0.09582	0.00101	15.276	188.4	2394.9	2583.3	0.638	7.528	8.166
46	0.1008	0.00101	14.557	192.5	2392.5	2585.1	0.651	7.497	8.148
47	0.10612	0.001011	13.877	196.7	2390.1	2586.9	0.664	7.466	8.13
48	0.11162	0.001011	13.233	200.9	2387.7	2588.6	0.678	7.435	8.113
49	0.11736	0.001012	12.623	205.1	2385.3	2590.4	0.691	7.404	8.095
50	0.12335	0.001012	12.046	209.3	2382.9	2592.2	0.704	7.374	8.078
51	0.12961	0.001013	11.499	213.4	2380.5	2593.9	0.716	7.344	8.06
52	0.13611	0.001013	10.98	217.6	2378.1	2595.7	0.729	7.314	8.043
53	0.14293	0.001014	10.488	221.8	2375.7	2597.5	0.742	7.284	8.026
54	0.15002	0.001014	10.022	226.0	2373.2	2599.2	0.755	7.254	8.009
55	0.15741	0.001015	9.5789	230.2	2370.8	2601	0.768	7.225	7.993
56	0.16511	0.001015	9.1587	234.3	2368.4	2602.7	0.78	7.196	7.976
57	0.17313	0.001016	8.7598	238.5	2366.0	2604.5	0.793	7.166	7.959
58	0.18147	0.001016	8.3808	242.7	2363.5	2606.2	0.806	7.137	7.943
59	0.19016	0.001017	8.0208	246.9	2361.1	2608	0.818	7.109	7.927
60	0.1992	0.001017	7.6785	251.1	2358.6	2609.7	0.831	7.08	7.911
61	0.20861	0.001018	7.3532	255.3	2356.1	2611.4	0.844	7.051	7.895
62	0.21838	0.001018	7.0437	259.5	2353.7	2613.2	0.856	7.023	7.879
63	0.22855	0.001019	6.7493	263.6	2351.3	2614.9	0.868	6.995	7.863
64	0.23912	0.001019	6.4690	267.8	2348.3	2616.6	0.881	6.967	7.848
65	0.25009	0.00102	6.2023	272.0	2346.4	2618.4	0.893	6.939	7.832
66	0.2615	0.00102	5.9482	276.2	2343.9	2620.1	0.906	6.911	7.817
67	0.27334	0.001021	5.7062	280.4	2341.4	2621.8	0.918	6.884	7.802
68	0.28563	0.001022	5.4756	284.6	2338.9	2623.5	0.93	6.856	7.786
69	0.29838	0.001022	5.2558	288.8	2336.4	2625.2	0.943	6.828	7.771
70	0.31162	0.001023	5.0463	293.0	2333.9	2626.9	0.955	6.802	7.757
71	0.32535	0.001024	4.8464	297.2	2331.4	2628.6	0.967	6.775	7.742
72	0.33958	0.001024	4.6557	301.3	2329.0	2630.3	0.979	6.748	7.727
73	0.35434	0.001025	4.4737	305.5	2326.5	2632.0	0.991	6.721	7.712
74	0.36964	0.001025	4.3000	309.7	2324	2633.7	1.003	6.695	7.698
75	0.38549	0.001026	4.1341	313.9	2321.5	2635.4	1.015	6.668	7.683
76	0.40191	0.001027	3.9757	318.1	2318.9	2637.0	1.027	6.642	7.669
77	0.41891	0.001027	3.8243	322.3	2316.4	2638.7	1.039	6.616	7.655
78	0.43652	0.001028	3.6796	326.5	2313.9	2640.4	1.051	6.59	7.641
79	0.45474	0.001029	3.5413	330.7	2311.4	2642.1	1.063	6.564	7.627
80	0.4736	0.001029	3.4091	334.9	2308.9	2643.8	1.075	6.538	7.613
81	0.49311	0.001030	3.2826	339.1	2306.3	2645.4	1.087	6.512	7.599
82	0.51329	0.001031	3.1616	343.3	2303.8	2647.1	1.099	6.487	7.586
83	0.53416	0.001031	3.0458	347.5	2301.2	2648.7	1.111	6.461	7.572

(Continued)

Temperature (in °C) (T)	Absolute Pressure (in bar) (p)	Specific Volume (in m³/kg)		Specific Enthalpy (in kJ/kg)			Specific Entropy (in kJ/kg K)		
		Water (v_f)	Steam (v_g)	Water (h_f)	Evaporation (h_{fg})	Steam (h_g)	Water (s_f)	Evaporation (s_{fg})	Steam (s_g)
84	0.55573	0.001032	2.9350	351.7	2298.7	2650.4	1.123	6.436	7.559
85	0.57803	0.001033	2.8288	355.9	2296.1	2652.0	1.134	6.411	7.545
86	0.60108	0.001033	2.7272	360.1	2293.5	2653.6	1.146	6.386	7.532
87	0.62489	0.001034	2.6298	364.3	2291.0	2655.3	1.158	6.361	7.519
88	0.64948	0.001035	2.5365	368.5	2288.4	2656.9	1.169	6.337	7.506
89	0.67487	0.001035	2.4470	372.7	2285.8	2658.5	1.181	6.312	7.493
90	0.70109	0.001036	2.3613	376.9	2283.2	2660.1	1.193	6.287	7.48
91	0.72815	0.001037	2.2791	381.1	2280.6	2661.7	1.204	6.263	7.467
92	0.75606	0.001038	2.2002	385.4	2278.0	2663.4	1.216	6.238	7.454
93	0.78489	0.001038	2.1245	389.6	2275.4	2665	1.227	6.215	7.442
94	0.81461	0.001039	2.0519	393.8	2272.8	2666.6	1.239	6.19	7.429
95	0.84526	0.001040	1.9822	398.0	2270.1	2668.1	1.25	6.167	7.417
96	0.87686	0.001041	1.9153	402.2	2267.5	2669.7	1.261	6.143	7.404
97	0.90944	0.001041	1.8510	406.4	2264.9	2671.3	1.273	6.119	7.392
98	0.94301	0.001042	1.7893	410.6	2262.3	2672.9	1.284	6.096	7.38
99	0.97761	0.001043	1.7300	414.8	2259.6	2674.4	1.296	6.072	7.368
100	1.0133	0.001044	1.673	419.1	2256.9	2676.0	1.307	6.048	7.355
102	1.0876	0.001045	1.5655	427.5	2251.6	2679.1	1.329	6.002	7.331
104	1.1668	0.001047	1.4662	435.9	2246.3	2682.2	1.352	5.956	7.308
106	1.2504	0.001048	1.3742	444.4	2240.9	2685.3	1.374	5.91	7.284
108	1.339	0.001050	1.2889	452.9	2235.4	2688.3	1.396	5.865	7.261
110	1.4327	0.001052	1.2099	461.3	2230	2691.3	1.418	5.821	7.239
112	1.5316	0.001054	1.1366	469.8	2224.5	2694.3	1.44	5.776	7.216
114	1.6362	0.001055	1.0685	478.3	2218.9	2697.2	1.462	5.732	7.194
116	1.7465	0.001057	1.0052	486.7	2213.5	2700.2	1.484	5.688	7.172
118	1.8628	0.001059	0.94634	495.2	2207.9	2703.1	1.506	5.645	7.151
120	1.9854	0.001061	0.89152	503.7	2202.3	2706	1.528	5.601	7.129
122	2.1145	0.001063	0.84045	512.2	2196.6	2708.8	1.549	5.559	7.108
124	2.2504	0.001064	0.79283	520.7	2190.9	2711.6	1.57	5.517	7.087
126	2.3933	0.001066	0.74840	529.2	2185.2	2714.4	1.592	5.475	7.067
128	2.5435	0.001068	0.70691	537.8	2179.4	2717.2	1.613	5.433	7.046
130	2.7013	0.001070	0.66814	546.3	2173.6	2719.9	1.634	5.392	7.026
132	2.867	0.001072	0.63188	554.8	2167.8	2722.6	1.655	5.351	7.006
134	3.0407	0.001074	0.5975	563.4	2161.9	2725.3	1.676	5.31	6.986
136	3.2229	0.001076	0.56618	572.0	2155.9	2727.9	1.697	5.27	6.967
138	3.4138	0.001078	0.53641	580.5	2150.0	2730.5	1.718	5.229	6.947
140	3.6139	0.00108	0.50849	589.1	2144.0	2733.1	1.739	5.189	6.928
142	3.8231	0.001082	0.4823	597.7	2137.9	2735.6	1.76	5.15	6.91
144	4.0420	0.001084	0.45771	606.3	2131.8	2738.1	1.78	5.111	6.891
146	4.2709	0.001086	0.4346	614.9	2125.7	2740.6	1.801	5.071	6.872
148	4.5101	0.001089	0.41288	623.5	2119.5	2743	1.821	5.033	6.854
150	4.7600	0.001091	0.39245	632.2	2113.2	2745.4	1.842	4.994	6.836
155	5.4333	0.001096	0.34644	653.8	2097.4	2751.2	1.892	4.899	6.791
160	6.1806	0.001102	0.30676	675.5	2081.2	2756.7	1.943	4.805	6.748
165	7.0077	0.001108	0.2724	697.2	2064.8	2762	1.992	4.713	6.705
170	7.9202	0.001114	0.24255	719.1	2048.0	2767.1	2.042	4.621	6.663

(Continued)

Temperature (in °C) (T)	Absolute Pressure (in bar) (p)	Specific Volume (in m³/kg)		Specific Enthalpy (in kJ/kg)			Specific Entropy (in kJ/kg K)		
		Water (v_f)	Steam (v_g)	Water (h_f)	Evaporation (h_{fg})	Steam (h_g)	Water (s_f)	Evaporation (s_{fg})	Steam (s_g)
175	8.9244	0.001121	0.21654	741.1	2030.7	2771.8	2.091	4.531	6.622
180	10.027	0.001128	0.1938	763.1	2013.2	2776.3	2.139	4.443	6.582
185	11.233	0.001135	0.17386	785.3	1995.1	2780.4	2.187	4.355	6.542
190	12.551	0.001142	0.15632	807.5	1976.8	2784.3	2.236	4.268	6.504
195	13.987	0.001149	0.14084	829.9	1957.9	2787.8	2.283	4.182	6.465
200	15.549	0.001156	0.12716	852.4	1938.5	2790.9	2.331	4.097	6.428
205	17.243	0.001164	0.11503	875.0	1918.8	2793.8	2.378	4.013	6.391
210	19.077	0.001172	0.10424	897.7	1898.5	2796.2	2.425	3.929	6.354
215	21.06	0.001181	0.094625	920.6	1877.7	2798.3	2.471	3.846	6.317
220	23.198	0.00119	0.086038	943.7	1856.2	2799.9	2.518	3.764	6.282
225	25.501	0.001199	0.078349	966.9	1834.3	2801.2	2.564	3.682	6.246
230	27.976	0.001209	0.07145	990.3	1811.7	2802	2.61	3.601	6.211
235	30.632	0.001219	0.065245	1013.8	1788.5	2802.3	2.656	3.519	6.175
240	33.478	0.001229	0.059654	1037.6	1764.6	2802.2	2.702	3.439	6.141
245	36.523	0.00124	0.054606	1061.6	1740	2801.6	2.748	3.358	6.106
250	39.776	0.001251	0.050037	1085.8	1714.6	2800.4	2.794	3.277	6.071
255	43.246	0.001263	0.045896	1110.2	1688.5	2798.7	2.839	3.197	6.036
260	46.943	0.001276	0.042134	1134.9	1661.5	2796.4	2.885	3.116	6.001
265	50.877	0.001289	0.03871	1159.9	1633.6	2793.5	2.931	3.035	5.966
270	55.058	0.001303	0.035588	1185.2	1604.7	2789.9	2.976	2.954	5.93
275	59.496	0.001317	0.032736	1210.8	1574.7	2785.5	3.022	2.873	5.895
280	64.202	0.001332	0.030126	1236.8	1543.6	2780.4	3.068	2.79	5.858
285	69.186	0.001349	0.027733	1263.2	1511.3	2774.5	3.115	2.707	5.822
290	74.461	0.001366	0.025535	1290.0	1477.6	2767.6	3.161	2.624	5.785
295	80.037	0.001384	0.023513	1317.3	1442.5	2759.8	3.208	2.539	5.747
300	85.927	0.001404	0.021649	1345	1406	2751	3.255	2.453	5.708
305	92.144	0.001425	0.019927	1373.4	1367.7	2741.1	3.303	2.366	5.669
310	98.7	0.001448	0.018334	1402.4	1327.6	2730	3.351	2.277	5.628
315	105.61	0.001473	0.016856	1432.1	1285.5	2717.6	3.4	2.186	5.586
320	112.89	0.0015	0.01548	1462.6	1241.1	2703.7	3.45	2.092	5.542
325	120.56	0.001529	0.014195	1494	1194	2688	3.501	1.996	5.497
330	128.63	0.001562	0.012989	1526.5	1143.7	2670.2	3.553	1.896	5.449
335	137.12	0.001598	0.011854	1560.2	1089.5	2649.7	3.606	1.792	5.398
340	146.05	0.001639	0.01078	1595.5	1030.7	2626.2	3.662	1.681	5.343
345	155.45	0.001686	0.097631	1632.5	966.4	2598.9	3.719	1.564	5.283
350	165.35	0.001741	0.087991	1671.9	895.8	2567.7	3.78	1.438	5.218
355	175.77	0.001809	0.078592	1716.6	813.8	2530.4	3.849	1.295	5.144
360	186.75	0.001896	0.069398	1764.2	721.2	2485.4	3.921	1.139	5.06
365	198.33	0.002016	0.060116	1818	610.0	2428.0	4.002	0.956	4.958
370	210.54	0.002214	0.049728	1890.2	452.6	2342.8	4.111	0.703	4.814
374.15	221.20	0.00317	0.003170	2107.4	0.0	2107.4	4.443	0.000	4.443

Appendix II: Saturated Water and Steam (Pressure) Tables

Absolute Pressure (in bar) (p)	Temperature (in °C) (T)	Specific Volume (in m³/kg) Water (v_f)	Steam (v_g)	Specific Enthalpy (in kJ/kg) Water (h_f)	Evaporation (h_{fg})	Steam (h_g)	Specific Entropy (in kJ/kg K) Water (s_f)	Evaporation (s_{fg})	Steam (s_g)
0.0061	0	0.001000	206.31	0	2501.6	2501.6	0	9.158	9.158
0.01	6.983	0.001000	129.21	29.3	2485.1	2514.4	0.106	8.871	8.977
0.015	13.04	0.001001	87.982	54.7	2470.8	2525.5	0.196	8.634	8.83
0.02	17.51	0.001001	67.006	73.5	2460.1	2533.6	0.261	8.464	8.725
0.025	21.1	0.001002	54.256	88.4	2451.8	2540.2	0.312	8.333	8.645
0.03	24.1	0.001003	45.667	101	2444.6	2545.6	0.354	8.224	8.578
0.035	26.69	0.001003	39.479	111.8	2438.6	2550.4	0.391	8.132	8.523
0.04	28.98	0.001004	34.802	121.4	2433.1	2554.5	0.423	8.053	8.476
0.045	31.03	0.001005	31.141	130	2428.2	2558.2	0.451	7.983	8.434
0.05	32.9	0.001005	28.194	137.8	2423.8	2561.6	0.476	7.92	8.396
0.06	36.18	0.001006	23.741	151.5	2416	2567.5	0.521	7.81	8.331
0.07	39.03	0.001007	20.531	163.4	2409.2	2572.6	0.559	7.718	8.277
0.08	41.53	0.001008	18.105	173.9	2403.2	2577.1	0.593	7.617	8.23
0.09	43.79	0.001009	16.204	183.3	2397.8	2581.1	0.622	7.566	8.188
0.1	45.83	0.00101	14.675	191.8	2392.9	2584.7	0.649	7.502	8.151
0.11	47.71	0.001011	13.416	199.7	2388.4	2588.1	0.674	7.444	8.118
0.12	49.45	0.001012	12.362	206.9	2384.4	2591.2	0.696	7.391	8.087
0.13	51.06	0.001013	11.466	213.7	2380.3	2594	0.717	7.342	8.059
0.14	52.57	0.001013	10.694	220	2376.7	2596.7	0.737	7.296	8.033
0.15	54	0.001014	10.023	226	2373.2	2599.2	0.755	7.254	8.009
0.16	55.34	0.001015	9.4331	231.6	2370	2601.6	0.772	7.215	7.987
0.17	56.62	0.001015	8.9111	236.9	2366.9	2603.8	0.788	7.178	7.966
0.18	57.83	0.001016	8.4452	242	2363.9	2605.9	0.804	7.142	7.946
0.19	58.98	0.001017	8.0272	246.8	2361.1	2607.9	0.818	7.109	7.927
0.2	60.09	0.001017	7.6498	251.5	2358.4	2609.9	0.832	7.077	7.909
0.21	61.15	0.001018	7.3073	255.9	2355.8	2611.7	0.845	7.047	7.892
0.22	62.16	0.001018	6.9951	260.1	2353.4	2613.5	0.858	7.018	7.876
0.23	63.14	0.001019	6.7093	264.2	2351	2615.2	0.870	6.991	7.861
0.24	64.08	0.001019	6.4467	268.2	2348.6	2616.8	0.882	6.964	7.846
0.25	64.99	0.00102	6.2045	272	2346.3	2618.3	0.893	6.939	7.832
0.26	65.87	0.00102	5.9803	275.7	2344.2	2619.9	0.904	6.915	7.819
0.27	66.72	0.001021	5.7724	279.2	2342.1	2621.3	0.915	6.891	7.806
0.28	67.55	0.001021	5.5778	282.7	2340	2622.7	0.925	6.868	7.793
0.29	68.35	0.001022	5.3982	286	2338.1	2624.1	0.935	6.847	7.781
0.3	69.12	0.001022	5.2293	289.3	2336.1	2625.4	0.944	6.825	7.769
0.32	70.62	0.001023	4.922	295.6	2332.4	2628	0.962	6.785	7.747
0.34	72.03	0.001024	4.6504	301.5	2328.9	2630.4	0.98	6.747	7.727
0.36	73.37	0.001025	4.4076	307.1	2325.5	2632.6	0.996	6.711	7.707
0.38	74.66	0.001026	4.19	312.5	2322.3	2634.8	1.011	6.677	7.688
0.4	75.89	0.001027	3.9934	317.7	2319.2	2636.9	1.026	6.645	7.671

(Continued)

Absolute Pressure (in bar) (p)	Temperature (in °C) (T)	Specific Volume (in m³/kg) Water (v_f)	Steam (v_g)	Specific Enthalpy (in kJ/kg) Water (h_f)	Evaporation (h_{fg})	Steam (h_g)	Specific Entropy (in kJ/kg K) Water (s_f)	Evaporation (s_{fg})	Steam (s_g)
0.42	77.06	0.001027	3.8148	322.6	2316.3	2638.9	1.04	6.614	7.654
0.44	78.19	0.001028	3.6522	327.3	2313.4	2640.7	1.054	6.584	7.638
0.46	79.28	0.001029	3.5032	331.9	2310.7	2642.6	1.067	6.556	7.623
0.48	80.33	0.001029	3.3663	336.3	2308	2644.3	1.079	6.53	7.609
0.5	81.35	0.00103	3.2401	340.6	2305.4	2646	1.091	6.504	7.595
0.52	82.33	0.001031	3.1233	344.7	2302.9	2647.6	1.103	6.478	7.581
0.54	83.28	0.001031	3.0148	348.7	2300.5	2649.2	1.114	6.455	7.569
0.56	84.19	0.001032	2.9139	352.5	2298.2	2650.7	1.125	6.431	7.556
0.58	85.09	0.001033	2.8197	356.3	2295.8	2652.1	1.135	6.409	7.544
0.6	85.95	0.001033	2.7317	359.9	2293.7	2653.6	1.145	6.388	7.533
0.62	86.8	0.001034	2.6491	363.5	2291.4	2654.9	1.155	6.367	7.522
0.64	87.62	0.001034	2.5715	366.9	2289.4	2656.3	1.165	6.346	7.511
0.66	88.42	0.001035	2.4985	370.3	2287.3	2657.6	1.174	6.326	7.5
0.68	89.2	0.001036	2.4297	373.6	2285.2	2658.8	1.183	6.307	7.49
0.7	89.96	0.001036	2.3647	376.8	2283.3	2660.1	1.192	6.288	7.48
0.72	90.7	0.001037	2.3031	379.9	2281.4	2661.3	1.201	6.27	7.471
0.74	91.43	0.001037	2.2448	382.9	2279.5	2662.4	1.209	6.253	7.462
0.76	92.14	0.001038	2.1895	385.9	2277.7	2663.6	1.217	6.235	7.452
0.78	92.83	0.001038	2.1369	388.9	2275.8	2664.7	1.225	6.219	7.444
0.8	93.51	0.001039	2.0869	391.7	2274.1	2665.8	1.233	6.202	7.435
0.85	95.15	0.001040	1.9721	398.6	2269.8	2668.4	1.252	6.163	7.415
0.9	96.71	0.001041	1.8691	405.2	2265.7	2670.9	1.27	6.125	7.395
0.95	98.2	0.001042	1.7771	411.5	2261.7	2673.2	1.287	6.091	7.378
1	99.63	0.001043	1.6938	417.5	2257.9	2675.4	1.303	6.057	7.36
1.01325	100.00	0.001044	1.673	419.1	2256.9	2676	1.307	6.048	7.355
1.05	101.0	0.001045	1618.1	423.3	2254.3	2677.6	1.318	6.025	7.343
1.1	102.3	0.001046	1.5492	428.8	2250.8	2679.6	1.333	5.995	7.328
1.15	103.6	0.001047	1.4861	434.2	2247.4	2681.6	1.347	5.966	7.313
1.2	104.8	0.001048	1.4281	439.3	2244.1	2683.4	1.361	5.937	7.298
1.25	106.0	0.001049	1.3746	444.4	2240.8	2685.2	1.374	5.911	7.285
1.3	107.1	0.001050	1.3250	449.2	2237.8	2687	1.387	5.885	7.272
1.35	108.2	0.001050	1.2791	453.4	2234.8	2688.7	1.399	5.86	7.259
1.4	109.3	0.001051	1.2363	458.4	2231.9	2690.3	1.411	5.836	7.247
1.45	110.4	0.001052	1.1963	462.8	2229	2691.8	1.423	5.812	7.235
1.5	111.4	0.001053	1.1590	467.1	2226.3	2693.4	1.433	5.79	7.223
1.6	113.3	0.001055	1.0911	475.4	2220.8	2696.2	1.455	5.747	7.202
1.7	115.2	0.001056	1.0309	483.2	2215.8	2699	1.475	5.706	7.181
1.8	116.9	0.001058	0.97718	490.7	2210.8	2701.5	1.494	5.668	7.162
1.9	118.6	0.001059	0.92895	497.9	2206.1	2704	1.513	5.631	7.144
2.0	120.2	0.001061	0.88540	504.7	2201.6	2706.3	1.530	5.597	7.127
2.1	121.8	0.001062	0.84586	511.3	2197.2	2708.5	1.547	5.564	7.111
2.2	123.3	0.001064	0.80980	517.6	2193	2710.6	1.563	5.532	7.095
2.3	124.7	0.001065	0.77677	523.7	2188.9	2712.6	1.578	5.502	7.080
2.4	126.1	0.001066	0.74641	529.6	2184.9	2714.5	1.593	5.473	7.066
2.5	127.4	0.001068	0.71840	535.3	2181.1	2716.4	1.607	5.445	7.052
2.6	128.7	0.001069	0.69247	540.9	2177.3	2718.2	1.621	5.418	7.039
2.7	130.0	0.001070	0.66840	546.2	2173.7	2719.9	1.634	5.392	7.026

(Continued)

Absolute Pressure (in bar) (p)	Temperature (in °C) (T)	Specific Volume (in m³/kg) Water (v_f)	Steam (v_g)	Specific Enthalpy (in kJ/kg) Water (h_f)	Evaporation (h_{fg})	Steam (h_g)	Specific Entropy (in kJ/kg K) Water (s_f)	Evaporation (s_{fg})	Steam (s_g)
2.8	131.2	0.001071	0.64600	551.4	2170.1	2721.5	1.647	5.367	7.014
2.9	132.4	0.001072	0.62509	556.5	2166.6	2723.1	1.660	5.342	7.002
3.0	133.5	0.001074	0.60553	561.5	2163.2	2724.7	1.672	5.319	6.991
3.1	134.7	0.001075	0.58718	566.2	2159.9	2726.1	1.683	5.297	6.980
3.2	135.8	0.001076	0.56995	570.9	2156.7	2727.6	1.695	5.274	6.969
3.3	136.8	0.001077	0.55373	575.5	2153.5	2729.0	1.706	5.253	6.959
3.4	137.9	0.001078	0.53843	579.9	2150.4	2730.3	1.717	5.232	6.949
3.5	138.9	0.001079	0.52397	584.3	2147.3	2731.6	1.727	5.212	6.939
3.6	139.9	0.001080	0.51029	588.5	2144.4	2732.9	1.738	5.192	6.930
3.7	140.8	0.001081	0.49733	592.7	2141.4	2734.1	1.748	5.173	6.921
3.8	141.8	0.001082	0.48502	596.7	2138.6	2735.3	1.758	5.154	6.912
3.9	142.7	0.001083	0.47333	600.8	2135.7	2736.5	1.767	5.136	6.903
4.0	143.6	0.001084	0.46220	604.7	2132.9	2737.6	1.776	5.118	6.894
4.1	144.5	0.001085	0.45159	608.5	2130.2	2738.7	1.786	5.100	6.886
4.2	145.4	0.001086	0.44147	612.3	2127.5	2739.8	1.795	5.083	6.878
4.3	146.3	0.001087	0.43181	616.0	2124.9	2740.9	1.803	5.067	6.870
4.4	147.1	0.001088	0.42257	619.6	2122.3	2741.9	1.812	5.050	6.862
4.5	147.9	0.001089	0.41373	623.2	2119.7	2742.9	1.820	5.035	6.855
4.6	148.7	0.001090	0.40526	626.7	2117.2	2743.9	1.829	5.018	6.847
4.7	149.5	0.001090	0.39714	630.1	2114.7	2744.8	1.837	5.003	6.840
4.8	150.3	0.001091	0.38934	633.5	2112.2	2745.7	1.845	4.988	6.833
4.9	151.1	0.001092	0.38186	636.8	2109.8	2746.6	1.853	4.973	6.826
5.0	151.8	0.001093	0.37466	640.1	2107.4	2747.5	1.860	4.959	6.819
5.2	153.3	0.001095	0.36106	646.5	2102.7	2749.2	1.875	4.931	6.806
5.4	154.8	0.001096	0.34844	652.8	2098.1	2750.9	1.890	4.903	6.793
5.6	156.2	0.001098	0.33669	658.8	2093.7	2752.5	1.904	4.877	6.781
5.8	157.5	0.001099	0.32572	664.7	2089.3	2754.0	1.918	4.851	6.769
6.0	158.8	0.001101	0.31546	670.4	2085.1	2755.5	1.931	4.827	6.758
6.2	160.1	0.001102	0.30584	676.1	2080.8	2756.9	1.944	4.803	6.747
6.4	161.4	0.001104	0.29680	681.5	2076.7	2758.2	1.956	4.780	6.736
6.6	162.6	0.001105	0.28829	686.8	2072.7	2759.5	1.968	4.757	6.725
6.8	163.8	0.001107	0.28026	692.0	2068.8	2760.8	1.980	4.735	6.715
7.0	165.0	0.001108	0.27268	697.1	2064.9	2762.0	1.992	4.713	6.705
7.2	166.1	0.001110	0.26550	702.0	2061.2	2763.2	2.003	4.693	6.696
7.4	167.2	0.001111	0.25870	706.9	2057.4	2764.3	2.014	4.672	6.686
7.6	168.3	0.001112	0.25224	711.7	2053.7	2765.4	2.025	4.652	6.677
7.8	169.4	0.001114	0.24610	716.3	2050.1	2766.4	2.035	4.633	6.668
8.0	170.4	0.001115	0.24026	720.9	2046.5	2767.4	2.046	4.614	6.660
8.2	171.4	0.001116	0.23469	725.4	2043.0	2768.4	2.056	4.595	6.651
8.4	172.4	0.001118	0.22938	729.9	2039.6	2769.4	2.066	4.577	6.643
8.6	173.4	0.001119	0.22431	734.2	2036.2	2770.4	2.075	4.560	6.635
8.8	174.4	0.001120	0.21946	738.5	2032.8	2771.3	2.085	4.542	6.627
9.0	175.4	0.001121	0.21482	742.6	2029.5	2772.1	2.094	4.525	6.619
9.2	176.3	0.001123	0.21037	746.8	2026.2	2773.00	2.103	4.509	6.612
9.4	177.2	0.001124	0.20610	750.8	2023.0	2773.8	2.112	4.492	6.604
9.6	178.1	0.001125	0.20201	754.8	2019.8	2774.6	2.121	4.476	6.597
9.8	179.0	0.001126	0.19808	758.7	2016.7	2775.4	2.130	4.460	6.590

(Continued)

Absolute Pressure (in bar) (p)	Temperature (in °C) (T)	Specific Volume (in m³/kg)		Specific Enthalpy (in kJ/kg)			Specific Entropy (in kJ/kg K)		
		Water (v_f)	Steam (v_g)	Water (h_f)	Evaporation (h_{fg})	Steam (h_g)	Water (s_f)	Evaporation (s_{fg})	Steam (s_g)
10.0	179.9	0.001127	0.19430	762.6	2013.6	2776.2	2.138	4.445	6.583
10.5	182.0	0.001130	0.18548	772.0	2006.0	2778.0	2.159	4.407	6.566
11.0	184.1	0.001133	0.17739	781.1	1998.6	2779.7	2.179	4.371	6.550
11.5	186.0	0.001136	0.17002	789.9	1991.4	2781.3	2.198	4.336	6.534
12.0	188.0	0.001139	0.16321	798.4	1984.3	2782.7	2.216	4.303	6.519
12.5	189.8	0.001141	0.15696	806.7	1977.5	2784.2	2.234	4.271	6.505
13.0	191.6	0.001144	0.15114	814.7	1970.7	2785.4	2.251	4.240	6.491
13.5	193.3	0.001146	0.14576	822.5	1964.2	2786.7	2.267	4.211	6.478
14.0	195.0	0.001149	0.14073	830.1	1957.7	2787.8	2.284	4.181	6.465
14.5	196.7	0.001151	0.13606	837.5	1951.4	2788.9	2.299	4.154	6.453
15.0	198.3	0.001154	0.13167	844.6	1945.3	2789.9	2.314	4.127	6.441
15.5	199.8	0.001156	0.12756	851.6	1939.2	2790.8	2.329	4.100	6.429
16.0	201.4	0.001159	0.12370	858.5	1933.2	2791.7	2.344	4.074	6.418
16.5	202.9	0.001161	0.12006	865.3	1927.3	2792.6	2.358	4.049	6.407
17.0	204.3	0.001163	0.11664	871.8	1921.6	2793.4	2.371	4.025	6.396
17.5	205.7	0.001166	0.11340	878.2	1915.9	2794.1	2.384	4.001	6.385
18.0	207.1	0.001168	0.11033	884.5	1910.3	2794.8	2.398	3.977	6.375
18.5	208.5	0.001170	0.10742	890.7	1904.8	2795.5	2.410	3.955	6.365
19.0	209.8	0.001172	0.10467	896.8	1899.3	2796.1	2.423	3.933	6.356
19.5	211.1	0.001174	0.10204	902.7	1894.0	2796.7	2.435	3.911	6.346
20.0	212.4	0.001177	0.09955	908.5	1888.7	2797.2	2.447	3.890	6.337
21.0	214.8	0.001181	0.094902	919.9	1878.3	2798.2	2.470	3.849	6.319
22.0	217.2	0.001185	0.090663	930.9	1868.1	2799.1	2.492	3.809	6.301
23.0	219.6	0.001189	0.086780	941.6	1858.2	2799.8	2.514	3.771	6.285
24.0	221.8	0.001193	0.083209	951.9	1848.5	2800.4	2.534	3.735	6.269
25.0	223.9	0.001197	0.079915	961.9	1839.1	2801.0	2.554	3.699	6.253
26.0	226.0	0.001201	0.076865	971.7	1829.7	2801.4	2.574	3.665	6.239
27.0	228.1	0.001205	0.074033	981.2	1820.5	2801.7	2.592	3.632	6.224
28.0	230.0	0.001209	0.071396	990.5	1811.5	2802.0	2.611	3.600	6.211
29.0	232.0	0.001213	0.068935	999.5	1802.7	2802.2	2.628	3.569	6.197
30.0	233.8	0.001216	0.066632	1008.3	1794.0	2802.3	2.646	3.538	6.184
31.0	235.7	0.001220	0.064473	1017.1	1785.4	2802.3	2.662	3.509	6.171
32.0	237.4	0.001224	0.062443	1025.4	1776.9	2802.3	2.679	3.480	6.159
33.0	239.2	0.001227	0.060533	1033.7	1768.6	2802.3	2.694	3.452	6.146
34.0	240.9	0.001231	0.058731	1041.8	1760.3	2802.1	2.710	3.424	6.134
35.0	242.5	0.001235	0.057028	1049.7	1752.3	2802.0	2.725	3.398	6.123
36.0	244.2	0.001238	0.055417	1057.5	1744.2	2801.7	2.740	3.371	6.111
37.0	245.8	0.001242	0.053889	1065.2	1736.2	2801.4	2.755	3.345	6.100
38.0	247.3	0.001245	0.052439	1072.7	1728.4	2801.1	2.769	3.321	6.090
39.0	248.8	0.001249	0.051061	1080.1	1720.7	2800.8	2.783	3.296	6.079
40.0	250.3	0.001252	0.049749	1087.4	1712.9	2800.3	2.797	3.272	6.069
42.0	253.2	0.001259	0.047306	1101.6	1697.8	2799.4	2.823	3.225	6.048
44.0	256.1	0.001266	0.045078	1115.4	1682.9	2798.3	2.849	3.180	6.029
46.0	258.8	0.001273	0.043036	1128.8	1668.2	2797.0	2.874	3.136	6.010
48.0	261.4	0.001279	0.041158	1141.8	1653.9	2795.7	2.897	3.094	5.991
50.0	263.9	0.001286	0.039425	1154.5	1639.7	2794.2	2.921	3.053	5.974
52.0	266.4	0.001293	0.037820	1166.9	1625.7	2792.6	2.943	3.013	5.956

(*Continued*)

Absolute Pressure (in bar) (p)	Temperature (in °C) (T)	Specific Volume (in m³/kg) Water (v_f)	Steam (v_g)	Specific Enthalpy (in kJ/kg) Water (h_f)	Evaporation (h_{fg})	Steam (h_g)	Specific Entropy (in kJ/kg K) Water (s_f)	Evaporation (s_{fg})	Steam (s_g)
54.0	268.8	0.001299	0.036330	1179.0	1611.8	2790.8	2.965	2.974	5.939
56.0	271.1	0.001306	0.034942	1190.8	1598.2	2789.0	2.986	2.937	5.923
58.0	273.4	0.001312	0.033646	1202.4	1584.6	2787.0	3.007	2.899	5.906
60.0	275.6	0.001319	0.032433	1213.7	1571.3	2785.0	3.027	2.863	5.890
62.0	277.7	0.001325	0.031295	1224.9	1558.0	2782.9	3.047	2.828	5.875
64.0	279.8	0.001332	0.030225	1235.8	1544.8	2780.6	3.066	2.794	5.860
66.0	281.9	0.001338	0.029218	1246.5	1531.8	2778.3	3.085	2.760	5.845
68.0	283.9	0.001345	0.028267	1257.1	1518.8	2775.9	3.104	2.727	5.831
70.0	285.8	0.001351	0.027368	1267.4	1506.0	2773.4	3.122	2.694	5.816
72.0	287.7	0.001358	0.026517	1277.7	1493.2	2770.9	3.140	2.662	5.802
74.0	289.6	0.001365	0.025711	1287.8	1480.4	2768.2	3.157	2.631	5.788
76.0	291.4	0.001371	0.024944	1297.7	1467.8	2765.5	3.174	2.600	5.774
78.0	293.2	0.001378	0.024215	1307.5	1455.4	2762.7	3.191	2.569	5.760
80.0	295.0	0.001384	0.023521	1317.2	1442.7	2759.9	3.208	2.539	5.747
82.0	296.7	0.001391	0.022860	1326.7	1430.3	2757.0	3.224	2.510	5.734
84.0	298.4	0.001398	0.022228	1336.2	1417.8	2754.0	3.240	2.481	5.721
86.0	300.1	0.001404	0.021624	1345.4	1405.5	2750.9	3.256	2.452	5.708
88.0	301.7	0.001411	0.021046	1354.7	1393.1	2747.8	3.271	2.424	5.695
90.0	303.3	0.001418	0.020493	1363.8	1380.8	2744.6	3.287	2.395	5.682
92	304.9	0.001425	0.019962	1372.8	1368.5	2741.3	3.302	2.367	5.669
94	306.5	0.001432	0.019453	1381.7	1356.3	2738.0	3.317	2.340	5.657
96	308.0	0.001439	0.018964	1390.6	1344.1	2734.7	3.332	2.313	5.644
98	309.5	0.001446	0.018493	1399.4	1331.9	2731.2	3.346	2.286	5.632
100	311.0	0.001453	0.018041	1408.0	1319.7	2727.7	3.361	2.259	5.620
105	314.6	0.001470	0.016981	1429.5	1289.2	2718.7	3.396	2.194	5.590
110	318.0	0.001489	0.016007	1450.5	1258.8	2709.3	3.430	2.129	5.560
115	321.4	0.001508	0.015114	1471.3	1228.2	2699.5	3.464	2.066	5.530
120	324.6	0.001527	0.014285	1491.7	1197.5	2698.2	3.497	2.003	5.500
125	327.8	0.001547	0.013518	1511.9	1166.5	2678.4	3.530	1.941	5.471
130	330.8	0.001567	0.012800	1531.9	1135.1	2667.0	3.561	1.880	5.441
135	333.8	0.001588	0.012130	1551.8	1103.3	2655.1	3.593	1.818	5.411
140	336.6	0.001611	0.011498	1571.5	1070.9	2642.4	3.624	1.756	5.380
145	339.4	0.001634	0.010905	1591.3	1037.9	2629.2	3.655	1.694	5.349
150	342.1	0.001658	0.010343	1610.9	1004.2	2615.1	3.686	1.632	5.318
155	344.8	0.001683	0.009813	1630.7	969.7	2600.4	3.716	1.570	5.286
160	347.3	0.001710	0.009310	1650.4	934.5	2584.9	3.747	1.506	5.253
165	349.7	0.001739	0.008833	1670.4	898.5	2568.9	3.778	1.442	5.220
170	352.3	0.001770	0.008372	1691.6	860.0	2551.6	3.811	1.375	5.186
175	354.6	0.001803	0.007927	1713.3	820.0	2533.3	3.844	1.306	5.150
180	357.0	0.001840	0.007497	1734.8	779.1	2513.9	3.877	1.236	5.113
185	359.2	0.001881	0.007082	1756.5	736.5	2493.0	3.910	1.164	5.074
190	361.4	0.001926	0.006676	1778.7	691.8	2470.5	3.943	1.090	5.033
195	363.6	0.001978	0.006276	1801.9	643.9	2445.8	3.978	1.011	4.989
200	365.7	0.002037	0.005875	1826.6	591.6	2418.2	4.015	0.926	4.941
205	367.8	0.002110	0.005462	1854.2	532.0	2386.2	4.056	0.830	4.886
210	369.8	0.002202	0.005023	1886.3	461.2	2347.5	4.105	0.717	4.822
215	371.8	0.002342	0.004509	1929.4	365.2	2294.6	4.170	0.566	4.736

(*Continued*)

Absolute Pressure (in bar) (p)	Temperature (in °C) (T)	Specific Volume (in m³/kg)		Specific Enthalpy (in kJ/kg)			Specific Entropy (in kJ/kg K)		
		Water (v_f)	Steam (v_g)	Water (h_f)	Evaporation (h_{fg})	Steam (h_g)	Water (s_f)	Evaporation (s_{fg})	Steam (s_g)
220	373.7	0.002668	0.003735	2010.3	186.3	2196.6	4.293	0.288	4.581
221.2	374.15	0.003170	0.003170	2107.4	0.000	2107.4	4.443	0.000	4.443

Appendix III: Specific Volume of Superheated Steam

Absolute Pressure (in bar) (p)	Saturation Temperature (in °C) (t_S)	Specific Volume at Different Temperatures (in °C) (in m³/kg) (v)										
		100	150	200	250	300	350	400	500	600	700	800
0.02	17.5	86.08	97.63	109.2	120.7	132.2	143.8	155.3	178.4	201.5	224.6	247.6
0.04	29.0	43.03	48.81	54.58	60.35	66.12	71.89	77.66	89.20	100.7	112.3	123.8
0.06	36.2	28.68	32.53	36.38	40.23	44.08	47.93	51.77	59.47	67.16	74.85	82.54
0.08	41.5	21.50	24.40	27.28	30.17	33.06	35.94	38.83	44.60	50.37	56.14	61.91
0.10	45.8	17.20	19.51	21.83	24.14	26.45	28.75	31.06	35.68	40.30	44.91	49.53
0.15	54.0	11.51	13.06	14.61	16.16	17.71	19.25	20.80	23.89	26.98	30.07	33.16
0.20	60.1	8.585	9.748	10.91	12.07	13.22	14.37	15.53	17.84	20.15	22.45	24.76
0.25	65.0	6.874	7.808	8.737	9.665	10.59	11.52	12.44	14.29	16.14	17.99	19.84
0.30	69.1	5.714	6.493	7.268	8.040	8.811	9.581	10.35	11.89	13.43	14.70	16.51
0.35	72.7	4.898	5.568	6.233	6.896	7.557	8.218	8.879	10.20	11.52	12.84	14.16
0.40	75.9	4.279	4.866	5.448	6.028	6.607	7.185	7.763	8.918	10.07	11.23	12.38
0.45	78.7	3.803	4.325	4.844	5.360	5.875	6.389	6.903	7.930	8.957	9.984	10.99
0.50	81.3	3.418	3.889	4.356	4.821	5.284	5.747	6.209	7.134	8.057	8.981	9.904
0.60	86.0	2.844	3.238	3.628	4.016	4.402	4.788	5.174	5.944	6.714	7.484	8.254
0.70	90.0	2.434	2.773	3.108	3.441	3.772	4.103	4.434	5.095	5.755	6.415	7.074
0.80	93.5	2.126	2.425	2.718	3.010	3.300	3.590	3.879	4.457	5.035	5.613	6.190
0.90	96.7	1.887	2.153	2.415	2.674	2.933	3.190	3.448	3.962	4.475	4.989	5.502
1.00	99.6	1.696	1.936	2.172	2.406	2.639	2.871	3.103	3.565	4.028	4.490	4.952
1.50	111.4	—	1.285	1.444	1.601	1.757	1.912	2.067	2.376	2.685	2.993	3.301
2.00	120.2	—	0.9595	1.080	1.199	1.316	1.433	1.549	1.781	2.013	2.244	2.475
2.5	127.4	—	0.7641	0.8620	0.9574	1.052	1.145	1.239	1.424	1.610	1.795	1.980
3.0	133.5	—	0.6337	0.7164	0.7964	0.8753	0.9535	1.031	1.187	1.341	1.496	1.650
3.5	138.9	—	0.5406	0.6123	0.6814	0.7493	0.8166	0.8835	1.017	1.149	1.282	1.414
4.0	143.6	—	0.4707	0.5343	0.5952	0.6549	0.7139	0.7725	0.8892	1.005	1.121	1.237
4.5	147.9	—	0.4165	0.4738	0.5284	0.5817	0.6343	0.6865	0.7905	0.8939	0.9971	1.100
5.0	151.8	—	—	0.4250	0.4744	0.5226	0.5701	0.6172	0.7108	0.8040	0.8969	0.9896
6.0	158.8	—	—	0.3520	0.3939	0.4344	0.4742	0.5136	0.5918	0.6696	0.7471	0.8245
7.0	165.0	—	—	0.2999	0.3364	0.3714	0.4057	0.4396	0.5069	0.5737	0.6402	0.7066
8.0	170.4	—	—	0.2608	0.2932	0.3241	0.3543	0.3842	0.4432	0.5017	0.5600	0.6181
9.0	175.4	—	—	0.2303	0.2596	0.2874	0.3144	0.3410	0.3936	0.4458	0.4976	0.5493
10.0	179.9	—	—	0.2059	0.2328	0.2580	0.2824	0.3065	0.3540	0.4010	0.4477	0.4943
11.0	184.1	—	—	0.1859	0.2108	0.2339	0.2563	0.2782	0.3215	0.3644	0.4069	0.4492
12.0	188.0	—	—	0.1692	0.1924	0.2139	0.2345	0.2547	0.2945	0.3338	0.3729	0.4118
13.0	191.6	—	—	0.1551	0.1769	0.1969	0.2161	0.2348	0.2716	0.3080	0.3441	0.3800
14.0	195.0	—	—	0.1429	0.1636	0.1823	0.2002	0.2177	0.2520	0.2859	0.3194	0.3528
15.0	198.3	—	—	0.1324	0.1520	0.1697	0.1865	0.2029	0.2350	0.2667	0.2980	0.3292
16.0	201.4	—	—	—	0.1419	0.1587	0.1745	0.1900	0.2202	0.2499	0.2793	0.3086
17.0	204.3	—	—	—	0.1329	0.1489	0.1640	0.1786	0.2070	0.2351	0.2628	0.2904
18.0	207.1	—	—	—	0.1250	0.1402	0.1546	0.1684	0.1954	0.2219	0.2481	0.2742
19.0	209.8	—	—	—	0.1179	0.1325	0.1461	0.1593	0.1849	0.2101	0.2350	0.2597
20.0	212.4	—	—	—	0.1115	0.1255	0.1386	0.1511	0.1756	0.1995	0.2232	0.2467
22.0	217.2	—	—	—	0.1004	0.1134	0.1255	0.1370	0.1593	0.1812	0.2028	0.2242
24.0	221.8	—	—	—	0.09108	0.1034	0.1146	0.1252	0.1458	0.1659	0.1858	0.2054
26.0	226.0	—	—	—	0.08321	0.09483	0.1053	0.1153	0.1344	0.1530	0.1714	0.1895

(Continued)

Absolute Pressure (in bar) (p)	Saturation Temperature (in °C) (t_S)	Specific Volume at Different Temperatures (in °C) (in m³/kg) (v)										
		100	150	200	250	300	350	400	500	600	700	800
28.0	230.0	—	—	—	0.07644	0.08751	0.09740	0.1067	0.1246	0.1419	0.1590	0.1759
30.0	233.8	—	—	—	0.07055	0.08116	0.09053	0.09931	0.1161	0.1323	0.1483	0.1641
32.0	237.4	—	—	—	0.06538	0.07559	0.08451	0.09283	0.1087	0.1239	0.1390	0.1538
34.0	240.9	—	—	—	0.06080	0.07068	0.07920	0.08711	0.1021	0.1165	0.1307	0.1447
36.0	244.2	—	—	—	0.05670	0.06630	0.07448	0.08202	0.09626	0.1100	0.1234	0.1366
38.0	247.3	—	—	—	0.05302	0.06237	0.07025	0.07747	0.09104	0.1041	0.1168	0.1294
40.0	250.3	—	—	—	—	0.05883	0.06645	0.07338	0.08634	0.09876	0.1109	0.1229
42.0	253.2	—	—	—	—	0.05563	0.06300	0.06967	0.08209	0.09397	0.1056	0.1170
44.0	256.0	—	—	—	—	0.05270	0.05986	0.06630	0.07823	0.08961	0.1007	0.1116
46.0	258.8	—	—	—	—	0.05003	0.05699	0.06322	0.07470	0.08562	0.09626	0.1067
48.0	261.4	—	—	—	—	0.04757	0.05436	0.06039	0.07147	0.08197	0.09219	0.1022
50.0	263.9	—	—	—	—	0.04530	0.05194	0.05779	0.06849	0.07862	0.08845	0.09809
55.0	269.9	—	—	—	—	0.04343	0.04666	0.05213	0.06202	0.07131	0.08031	0.08912
60.0	275.6	—	—	—	—	0.03615	0.04222	0.04738	0.05659	0.06518	0.07348	0.08159
70.0	285.8	—	—	—	—	0.02946	0.03523	0.03992	0.04809	0.05559	0.06279	0.06980
75.0	290.5	—	—	—	—	0.02672	0.03243	0.03694	0.04469	0.05176	0.05852	0.06509
80.0	295.0	—	—	—	—	0.02426	0.02995	0.03431	0.04170	0.04839	0.05477	0.06096
85.0	299.2	—	—	—	—	0.02191	0.02776	0.03200	0.03908	0.04544	0.05148	0.05732
90.0	303.3	—	—	—	—		0.02579	0.02993	0.03674	0.04280	0.04853	0.05408
95.0	307.2	—	—	—	—		0.02403	0.02808	0.03465	0.04045	0.04591	0.05119
100.0	311.0	—	—	—	—	—	0.02242	0.02641	0.03276	0.03832	0.04355	0.04858
110.0	318.0	—	—	—	—	—	0.01961	0.02351	0.02950	0.03466	0.03947	0.04408
120.0	324.6	—	—	—	—	—	0.01721	0.02108	0.02679	0.03160	0.03607	0.04033
130.0	330.8	—	—	—	—	—	0.01510	0.01902	0.02449	0.02902	0.03319	0.03716
140.0	336.6	—	—	—	—	—	0.01321	0.01723	0.02251	0.02680	0.03072	0.03444
150.0	342.1	—	—	—	—	—	0.01146	0.01566	0.02080	0.02488	0.02859	0.03209
160.0	347.3	—	—	—	—	—	0.00976	0.01428	0.01929	0.02320	0.02672	0.03003
170.0	352.3	—	—	—	—	—	—	0.01303	0.01797	0.02172	0.02507	0.02821
180.0	357.0	—	—	—	—	—	—	0.01191	0.01679	0.02040	0.02300	0.02659
190.0	361.4	—	—	—	—	—	—	0.01089	0.01573	0.01922	0.02229	0.02515
200.0	365.7	—	—	—	—	—	—	0.00995	0.01477	0.01816	0.02111	0.02385
210.0	369.8	—	—	—	—	—	—	0.00907	0.01391	0.01720	0.02004	0.02267
220.0	373.7	—	—	—	—	—	—	0.00825	0.01312	0.01633	0.01907	0.02160
221.2	374.15	—	—	—	—	—	—	0.00816	0.01303	0.01622	0.01895	0.02135

Appendix IV: Specific Enthalpy of Superheated Steam

Absolute Pressure (in bar) (p)	Saturation Temperature (in °C) (t_s)	Specific Enthalpy at Different Temperatures (in °C) (in kJ/kg) (h)										
		100	**150**	**200**	**250**	**300**	**350**	**400**	**500**	**600**	**700**	**800**
0.02	17.5	2688.5	2783.7	2880.0	2977.7	3076.8	3177.5	3279.7	3489.2	3705.6	3928.8	4158.7
0.04	29.0	2688.3	2783.5	2879.9	2977.6	3076.8	3177.4	3279.7	3489.2	3705.6	3928.8	4158.7
0.06	36.2	2688.0	2783.4	2879.8	2977.6	3076.7	3177.4	3279.6	3489.2	3705.6	3928.8	4158.7
0.08	41.5	2687.8	2783.2	2879.7	2977.5	3076.7	3177.3	3279.6	3489.1	3705.5	3928.8	4158.7
0.10	45.8	2687.5	2783.1	2879.6	2977.4	3076.6	3177.3	3279.6	3489.1	3705.5	3928.8	4158.7
0.15	54.0	2686.9	2782.4	2879.5	2977.3	3076.5	3177.7	3279.5	3489.1	3705.5	3928.7	4158.7
0.20	60.1	2686.3	2782.3	2879.2	2977.1	3076.4	3177.1	3279.4	3489.0	3705.4	3928.7	4158.7
0.25	65.0	2685.7	2782.0	2879.0	2977.0	3076.3	3177.0	3279.3	3489.0	3705.4	3928.7	4158.6
0.30	69.1	2685.1	2781.6	2878.7	2976.8	3076.1	3176.9	3279.3	3488.9	3705.4	3928.7	4158.6
0.35	72.7	2684.5	2781.2	2878.5	2976.7	3076.0	3176.8	3279.2	3488.9	3705.3	3928.7	4158.6
0.40	75.9	2683.8	2780.9	2878.2	2976.5	3075.9	3176.8	3279.1	3488.8	3705.3	3928.6	4158.6
0.45	78.7	2683.2	2780.5	2878.0	2976.3	3075.8	3176.7	3279.1	3488.8	3705.2	3928.6	4158.5
0.50	81.3	2682.6	2780.1	2877.7	2976.1	3075.7	3176.6	3279.0	3488.7	3705.2	3928.6	4158.5
0.60	86.0	2681.3	2779.4	2877.3	2975.8	3075.4	3176.4	3278.8	3488.6	3705.1	3928.5	4158.5
0.70	90.0	2680.0	2778.6	2876.8	2975.5	3075.2	3176.2	3278.7	3488.5	3705.0	3928.4	4158.4
0.80	93.5	2678.8	2777.8	2876.3	2975.2	3075.0	3176.0	3278.5	3488.4	3705.0	3928.4	4158.4
0.90	96.7	2677.5	2777.1	2875.8	2974.8	3074.7	3175.8	3278.4	3488.3	3704.9	3928.3	4158.3
1.00	99.6	2676.2	2776.3	2875.4	2974.5	3074.5	3175.6	3278.2	3488.1	3704.8	3928.2	4158.3
1.50	111.4	—	2772.5	2872.9	2972.9	3073.3	3174.7	3277.5	3487.6	3704.4	3927.9	4158.0
2.00	120.2	—	2768.5	2870.5	2971.2	3072.1	3173.8	3276.7	3487.0	3704.0	3927.6	4157.8
2.5	127.4	—	2764.5	2868.0	2969.6	3070.9	3172.8	3275.9	3486.5	3703.6	3927.3	4157.6
3.0	133.5	—	2760.4	2865.5	2967.9	3069.7	3171.9	3275.2	3486.0	3703.2	3927.0	4157.3
3.5	138.9	—	2756.3	2863.0	2966.2	3068.4	3170.9	3274.4	3485.4	3702.7	3926.7	4157.1
4.0	143.6	—	2752.0	2860.4	2964.5	3067.2	3170.0	3273.6	3484.9	3702.3	3926.4	4156.9
4.5	147.9	—	2746.7	2857.8	2962.8	3066.0	3169.1	3272.9	3484.3	3701.9	3926.1	4156.7
5.0	151.8	—	—	2855.1	2961.1	3064.8	3168.1	3272.1	3483.8	3701.5	3925.8	4156.4
6.0	158.8	—	—	2849.7	2957.6	3062.3	3166.2	3270.6	3482.7	3700.7	3925.1	4155.9
7.0	165.0	—	—	2844.2	2954.0	3059.8	3164.3	3269.0	3481.6	3699.9	3924.5	4155.5
8.0	170.4	—	—	2838.6	2950.4	3057.3	3162.4	3267.5	3480.5	3699.1	3923.9	4155.0
9.0	175.4	—	—	2832.7	2946.8	3054.7	3160.5	3266.0	3479.4	3698.2	3923.3	4154.5
10.0	179.9	—	—	2826.8	2943.0	3052.1	3158.5	3264.4	3478.3	3697.4	3922.7	4154.1
11.0	184.1	—	—	2820.7	2939.3	3049.6	3156.6	3262.9	3477.2	3696.6	3922.0	4153.6
12.0	188.0	—	—	2814.4	2935.4	3046.9	3154.6	3261.3	3476.1	3695.8	3921.4	4153.1
13.0	191.6	—	—	2808.0	2931.5	3044.3	3152.7	3259.7	3475.0	3695.0	3920.8	4152.7
14.0	195.0	—	—	2801.4	2927.6	3041.6	3150.7	3258.2	3473.9	3694.1	3920.2	4152.2
15.0	198.3	—	—	2794.7	2923.5	3038.9	3148.7	3256.6	3472.8	3693.3	3919.6	4151.7
16.0	201.4	—	—	—	2919.4	3036.2	3146.7	3255.0	3471.7	3692.5	3918.9	4151.3
17.0	204.3	—	—	—	2915.3	3033.5	3144.7	3253.5	3470.6	3691.7	3918.3	4150.8
18.0	207.1	—	—	—	2911.0	3030.7	3142.7	3251.9	3469.5	3690.9	3917.7	4150.3
19.0	209.8	—	—	—	2906.7	3027.9	3140.7	3250.3	3468.4	3690.0	3917.1	4149.8

(Continued)

Absolute Pressure (in bar) (p)	Saturation Temperature (in °C) (t_s)	Specific Enthalpy at Different Temperatures (in °C) (in kJ/kg) (h)										
		100	**150**	**200**	**250**	**300**	**350**	**400**	**500**	**600**	**700**	**800**
20.0	212.4	—	—	—	2902.4	3025.0	3138.6	3248.7	3467.3	3689.2	3916.5	4149.4
22.0	217.2	—	—	—	2893.4	3019.3	3134.5	3245.5	3465.1	3687.6	3915.2	4148.4
24.0	221.8	—	—	—	2884.2	3013.4	3130.4	3242.3	3462.9	3685.9	3914.0	4147.5
26.0	226.0	—	—	—	2874.7	3007.4	3126.1	3239.0	3460.6	3684.3	3912.7	4146.6
28.0	230.0	—	—	—	2864.9	3001.3	3121.9	3235.8	3458.4	3682.6	3911.5	4145.6
30.0	233.8	—	—	—	2854.8	2995.1	3117.5	3232.5	3456.2	3681.0	3910.3	4144.7
32.0	237.4	—	—	—	2844.4	2988.7	3113.2	3229.2	3454.0	3679.3	3909.0	4143.8
34.0	240.9	—	—	—	2833.6	2982.2	3108.7	3225.9	3451.7	3677.7	3907.8	4142.8
36.0	244.2	—	—	—	2822.5	2975.6	3104.2	3222.5	3449.5	3676.1	3906.5	4141.9
38.0	247.3	—	—	—	2811.0	2968.9	3099.7	3219.1	3447.2	3674.4	3905.3	4141.0
40.0	250.3	—	—	—	—	2962.0	3095.1	3215.7	3445.0	3672.8	3904.1	4140.0
42.0	253.2	—	—	—	—	2955.0	3090.4	3212.3	3442.7	3671.1	3902.8	4139.1
44.0	256.0	—	—	—	—	2947.8	3085.7	3208.8	3440.5	3669.5	3901.6	4138.2
46.0	258.8	—	—	—	—	2940.5	3080.9	3205.3	3438.2	3667.8	3900.3	4137.2
48.0	261.4	—	—	—	—	2933.1	3076.1	3201.8	3435.9	3666.2	3899.1	4136.3
50.0	263.9	—	—	—	—	2925.5	3071.2	3198.3	3433.7	3664.5	3897.9	4135.3
55.0	269.9	—	—	—	—	2905.8	3058.7	3189.3	3427.9	3660.4	3894.8	4133.0
60.0	275.6	—	—	—	—	2885.0	3045.8	3180.1	3422.2	3656.2	3891.7	4130.7
65.0	280.8	—	—	—	—	2863.0	3032.4	3170.8	3416.4	3652.1	3888.6	4128.8
70.0	285.8	—	—	—	—	2839.0	3018.7	3161.2	3410.6	3647.9	3885.4	4126.0
75.0	290.5	—	—	—	—	2814.1	3004.5	3151.6	3404.7	3643.7	3882.4	4123.7
80.0	295.0	—	—	—	—	2786.6	2989.9	3141.6	3398.8	3639.5	3879.2	4121.3
85.0	299.2	—	—	—	—	2757.1	2974.7	3131.5	3392.8	3635.4	3876.1	4119.0
90.0	303.3	—	—	—	—	—	2959.0	3121.2	3386.8	3631.1	3873.0	4116.7
95.0	307.2	—	—	—	—	—	2942.7	3110.7	3380.7	3627.0	3869.9	4114.4
100.0	311.0	—	—	—	—	—	2925.8	3099.9	3374.6	3622.7	3866.8	4112.0
110.0	318.0	—	—	—	—	—	2889.6	3077.8	3362.2	3614.2	3860.5	4107.3
120.0	324.6	—	—	—	—	—	2849.7	3054.8	3349.6	3605.7	3854.3	4102.7
130.0	330.8	—	—	—	—	—	2805.0	3030.7	3336.8	3597.1	3848.0	4098.0
140.0	336.6	—	—	—	—	—	2754.2	3005.6	3323.8	3588.5	3841.7	4093.3
150.0	342.1	—	—	—	—	—	2694.8	2979.1	3310.6	3579.8	3835.4	4088.6
160.0	347.3	—	—	—	—	—	2620.8	2951.3	3297.1	3571.0	3829.1	4084.0
170.0	352.3	—	—	—	—	—	—	2921.7	3283.5	3562.2	3822.8	4079.3
180.0	357.0	—	—	—	—	—	—	2890.3	2269.6	3553.4	3816.5	4074.6
190.0	361.4	—	—	—	—	—	—	2856.7	3255.4	3544.5	3810.2	4070.0
200.0	365.7	—	—	—	—	—	—	2820.5	3241.1	3535.5	3803.8	4065.3
210.0	369.8	—	—	—	—	—	—	2781.3	3226.5	3526.5	3797.5	4060.6
220.0	373.7	—	—	—	—	—	—	2738.8	3211.7	3517.4	3791.1	4055.9
221.2	374.15	—	—	—	—	—	—	2734.5	3210.7	3516.4	3789.1	4054.7

Appendix V: Specific Entropy of Superheated Steam

Absolute Pressure (in bar) (p)	Saturation Temperature (in °C) (t_s)	Specific Entropy at Different Temperatures (in °C) (in kJ/kg K) (s)										
		100	150	200	250	300	350	400	500	600	700	800
0.02	17.5	9.193	9.433	9.648	9.844	10.025	10.193	10.351	10.641	10.904	11.146	11.371
0.04	29.0	8.873	9.113	9.328	9.524	9.705	9.874	10.031	10.321	10.585	10.827	11.051
0.06	36.2	8.685	8.925	9.141	9.337	9.518	9.686	9.844	10.134	10.397	10.639	10.864
0.08	41.5	8.552	8.792	9.008	9.204	9.385	9.554	9.711	10.001	10.265	10.507	10.731
0.10	45.8	8.449	8.689	8.905	9.101	9.282	9.450	9.608	9.898	10.162	10.404	10.628
0.15	54.0	8.261	8.502	8.718	8.915	9.096	9.264	9.422	9.712	9.975	10.217	10.442
0.20	60.1	8.126	8.368	8.584	8.781	8.962	9.130	9.288	9.578	9.842	10.084	10.309
0.25	65.0	8.022	8.264	8.481	8.678	8.859	9.028	9.186	9.476	9.739	9.981	10.206
0.30	69.1	7.936	8.179	8.396	8.593	8.774	8.943	9.101	9.391	9.654	9.897	10.121
0.35	72.7	7.864	8.107	8.325	8.522	8.703	8.872	9.030	9.320	9.583	9.826	10.050
0.40	75.9	7.801	8.045	8.263	8.460	8.641	8.810	8.968	9.258	9.522	9.764	9.989
0.45	78.7	7.745	7.990	8.208	8.405	8.587	8.755	8.914	9.204	9.467	9.709	9.934
0.50	81.3	7.695	7.941	8.159	8.356	8.538	8.707	8.865	9.155	9.419	9.661	9.886
0.60	86.0	7.609	7.855	8.074	8.272	8.454	8.622	8.781	9.071	9.334	9.576	9.801
0.70	90.0	7.535	7.783	8.002	8.200	8.382	8.551	8.709	9.000	9.263	9.505	9.730
0.80	93.5	7.470	7.720	7.940	8.138	8.320	8.489	8.648	8.938	9.214	9.444	9.669
0.90	96.7	7.413	7.664	7.884	8.083	8.266	8.435	8.593	8.884	9.147	9.389	9.614
1.00	99.6	7.362	7.614	7.835	8.034	8.217	8.386	8.544	8.835	9.098	9.341	9.565
1.50	111.4	—	7.419	7.644	7.845	8.028	8.198	8.356	8.647	8.911	9.153	9.378
2.00	120.2	—	7.279	7.507	7.710	7.894	8.064	8.223	8.514	8.778	9.020	9.245
2.5	127.4	—	7.169	7.400	7.604	7.789	7.960	8.119	8.410	8.674	8.917	9.142
3.0	133.5	—	7.077	7.312	7.518	7.703	7.874	8.034	8.326	8.590	8.833	9.058
3.5	138.5	—	6.998	7.237	7.444	7.631	7.802	7.962	8.254	8.518	8.761	8.986
4.0	143.6	—	6.929	7.171	7.380	7.568	7.740	7.899	8.192	8.456	8.699	8.925
4.5	147.9	—	6.866	7.112	7.323	7.512	7.684	7.844	8.137	8.402	8.645	8.870
5.0	151.8	—	—	7.059	7.272	7.461	7.634	7.795	8.088	8.353	8.596	8.821
6.0	158.8	—	—	6.966	7.183	7.374	7.548	7.709	8.003	8.268	8.511	8.737
7.0	165.0	—	—	6.886	7.107	7.300	7.475	7.636	7.931	8.196	8.440	8.665
8.0	170.4	—	—	6.815	7.040	7.235	7.411	7.573	7.868	8.134	8.377	8.603
9.0	175.4	—	—	6.751	6.980	7.177	7.354	7.517	7.812	8.079	8.323	8.549
10.0	179.9	—	—	6.692	6.926	7.125	7.303	7.467	7.763	8.029	8.273	8.500
11.0	184.1	—	—	6.638	6.876	7.078	7.257	7.421	7.718	7.985	8.229	8.455
12.0	188.0	—	—	6.587	6.831	7.034	7.214	7.379	7.677	7.944	8.188	8.415
13.0	191.6	—	—	6.539	6.788	6.994	7.175	7.340	7.639	7.906	8.151	8.378
14.0	195.0	—	—	6.494	6.748	6.956	7.139	7.305	7.603	7.871	8.116	8.343
15.0	198.3	—	—	6.451	6.710	6.921	7.104	7.271	7.570	7.839	8.084	8.311
16.0	201.4	—	—	—	6.674	6.887	7.072	7.239	7.540	7.808	8.054	8.281
17.0	204.3	—	—	—	6.640	6.856	7.042	7.210	7.511	7.779	8.025	8.252
18.0	207.1	—	—	—	6.607	6.826	7.013	7.182	7.483	7.752	7.998	8.226
19.0	209.8	—	—	—	6.576	6.797	7.986	7.155	7.457	7.727	7.973	8.200
20.0	212.4	—	—	—	6.545	6.770	6.960	7.130	7.432	7.702	7.949	8.176
22.0	217.2	—	—	—	6.488	6.718	6.911	7.082	7.386	7.657	7.904	8.132

(Continued)

Absolute Pressure (in bar) (p)	Saturation Temperature (in °C) (t_s)	Specific Entropy at Different Temperatures (in °C) (in kJ/kg K) (s)										
		100	150	200	250	300	350	400	500	600	700	800
24.0	221.8	—	—	—	6.434	6.670	6.866	7.038	7.344	7.615	7.862	8.091
26.0	226.0	—	—	—	6.382	6.625	6.824	6.998	7.305	7.577	7.825	8.053
28.0	230.0	—	—	—	6.333	6.582	6.784	6.960	7.269	7.541	7.789	8.018
30.0	233.8	—	—	—	6.286	6.542	6.747	6.925	7.235	7.508	7.756	7.986
32.0	237.4	—	—	—	6.240	6.504	6.712	6.891	7.203	7.477	7.726	7.955
34.0	240.9	—	—	—	6.195	6.467	6.679	6.860	7.172	7.447	7.697	7.927
36.0	244.2	—	—	—	6.151	6.432	6.647	6.829	7.144	7.420	7.669	7.900
38.0	247.3	—	—	—	6.109	6.397	6.616	6.801	7.117	7.393	7.643	7.874
40.0	250.3	—	—	—	—	6.364	6.587	6.773	7.091	7.368	7.619	7.850
42.0	253.2	—	—	—	—	6.332	6.559	6.747	7.066	7.344	7.595	7.826
44.0	256.0	—	—	—	—	6.301	6.532	6.722	7.043	7.321	7.573	7.804
46.0	258.8	—	—	—	—	6.270	6.505	6.697	7.020	7.299	7.551	7.783
48.0	261.4	—	—	—	—	6.240	6.479	6.674	6.998	7.278	7.531	7.763
50.0	263.9	—	—	—	—	6.211	6.455	6.651	6.977	7.258	7.511	7.743
55.0	269.9	—	—	—	—	6.139	6.395	6.597	6.928	7.210	7.464	7.697
60.0	275.6	—	—	—	—	6.069	6.339	6.546	6.882	7.166	7.422	7.655
65.0	280.8	—	—	—	—	6.001	6.285	6.499	6.839	7.126	7.382	7.617
70.0	285.8	—	—	—	—	5.933	6.233	6.454	6.799	7.088	7.346	7.581
75.0	290.5	—	—	—	—	5.864	6.184	6.411	6.762	7.053	7.311	7.547
80.0	295.0	—	—	—	—	5.794	6.135	6.369	6.726	7.019	7.279	7.516
85.0	299.2	—	—	—	—	5.744	6.088	6.330	6.692	6.987	7.249	7.486
90.0	303.3	—	—	—	—	—	6.041	6.292	6.660	6.957	7.220	7.458
95.0	307.2	—	—	—	—	—	5.995	6.254	6.629	6.929	7.192	7.431
100.0	311.0	—	—	—	—	—	5.949	6.218	6.599	6.901	7.166	7.406
110.0	318.0	—	—	—	—	—	5.857	6.148	6.543	6.850	7.117	7.358
120.0	324.6	—	—	—	—	—	5.764	6.081	6.491	6.802	7.072	7.315
130.0	330.8	—	—	—	—	—	5.666	6.016	6.441	6.758	7.030	7.274
140.0	336.6	—	—	—	—	—	5.562	5.951	6.394	6.716	6.991	7.237
150.0	342.1	—	—	—	—	—	5.447	5.888	6.349	6.676	6.954	7.201
160.0	347.3	—	—	—	—	—	5.311	5.824	6.305	6.639	6.919	7.168
170.0	352.3	—	—	—	—	—	—	5.760	6.264	6.603	6.886	7.137
180.0	357.0	—	—	—	—	—	—	5.695	6.223	6.569	6.854	7.107
190.0	361.4	—	—	—	—	—	—	5.628	6.184	6.536	6.824	7.078
200.0	365.7	—	—	—	—	—	—	5.559	6.146	6.504	6.795	7.051
210.0	369.8	—	—	—	—	—	—	5.486	6.108	6.474	6.768	7.025
221.2	374.15	—	—	—	—	—	—	5.399	6.068	6.441	6.738	6.994

Index

C

D

F

G

H

I

Q

R

S

T

U

V

W

Y

Z